COLD SPRING HARBOR SYMPOSIA
ON QUANTITATIVE BIOLOGY

VOLUME XXXVII

COLD SPRING HARBOR SYMPOSIA
ON QUANTITATIVE BIOLOGY

Founded in 1933

by

REGINALD G. HARRIS

Director of the Biological Laboratory
1924 to 1936

LIST OF PREVIOUS VOLUMES

COLD SPRING HARBOR SYMPOSIA ON QUANTITATIVE BIOLOGY

VOLUME XXXVII

The Mechanism of Muscle Contraction

COLD SPRING HARBOR LABORATORY

1973

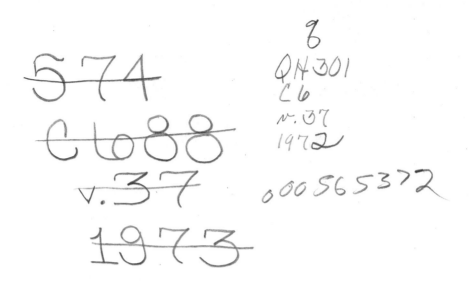

COLD SPRING HARBOR SYMPOSIA
ON QUANTITATIVE BIOLOGY
VOLUME XXXVII

International Standard Book Number
0-87969-036-4 (clothbound)

Library of Congress Catalog Card Number
34-8174

Printed in the United States of America

The Symposium Volumes are published by The Cold
Spring Harbor Laboratory, Cold Spring Harbor, New
York, 11724, and may be purchased directly or through
booksellers.
Price of Volume 37—$30.00 (inc. postage). Price subject
to change without notice.

Symposium Participants

ABBOTT, ROGER H., Department of Zoology, ARC Unit of Muscle Mechanisms, Oxford, England.

ADELSTEIN, ROBERT S., National Heart and Lung Institute, NIH, Bethesda, Md.

APRIL, ERNEST W., Department of Anatomy, College of Physicians and Surgeons, Columbia University, New York.

ARNDT, ISABELLE O., Department of Anatomy, University of Pennsylvania, Philadelphia.

AVIS, VANCE R., Department of Physiology, University of California, Los Angeles.

BARANY, GEORGE, Department of Biochemistry, The Rockefeller University, New York.

BÁRÁNY, KATE, Department of Physical Chemistry, Institute for Muscle Disease, Inc., New York.

BÁRÁNY, MICHAEL, Department of Physical Chemistry, Institute for Muscle Disease, Inc., New York.

BARTON, JANICE S., Department of Biology, Johns Hopkins University, Baltimore.

BASKIN, RONALD J., Department of Zoology, University of California, Davis.

BASS, BERL G., Department of Physiology, Einstein College of Medicine, Bronx, New York.

BENSON, ELLIS S., Department of Laboratory Medicine, University of Minnesota, Minneapolis.

BERNARDINI, ISA, Laboratory of Structural and Molecular Biology, Children's Cancer Research Foundation, Boston.

BHATNAGAR, GHOPAL, Department of Muscle Research, Boston Biomedical Research Institute, Boston.

BIRÓ, N. A., Biochemistry Group, Eötvös Lóránd University, Budapest, Hungary.

BOLAND, RICARDO, Department of Biochemistry, St. Louis University School of Medicine, St. Louis.

BONNER, ROBERT F., Department of Biophysics, Johns Hopkins University, Baltimore.

BRANDT, PHILIP W., Department of Anatomy, Columbia University, New York.

BRAY, DENNIS, MRC Laboratory of Molecular Biology, Cambridge, England.

BREMEL, ROBERT D., Department of Biochemistry, St. Louis University School of Medicine, St. Louis.

BULLARD, BELINDA, Department of Zoology, Oxford University, Oxford, England.

BURKE, MORRIS, Department of Biology, Johns Hopkins University, Baltimore.

BUTLER, THOMAS M., Department of Animal Biology, University of Pennsylvania School of Veterinary Medicine, Philadelphia.

BUZASH, ELIZABETH, Department of Biology, University of Connecticut, Storrs.

CARLSON, FRANCIS D., Department of Biophysics, Johns Hopkins University, Baltimore.

CARNEY, HOWARD, JR., Department of Biophysics, Johns Hopkins University, Baltimore.

CASPAR, DONALD L. D., Children's Cancer Research Foundation, Boston.

CHEUNG, HERBERT C., Department of Engineering Biophysics, University of Alabama Medical Center, Birmingham.

CHOWRASHI, PROKASH K., Department of Anatomy, University of Pennsylvania Medical School, Philadelphia.

COHEN, CAROLYN, Children's Cancer Research Foundation, Boston.

COLLINS, JOHN H., Department of Muscle Research, Boston Biomedical Research Institute, Boston.

COOKE, ROGER, Departments of Biochemistry and Biophysics, University of California, San Francisco.

COSTANTIN, LEROY L., Department of Physiology and Biophysics, Washington University School of Medicine, St. Louis.

CUNNINGHAM, BRUCE A., Department of Biochemistry, The Rockefeller University, New York.

CURTIN, NANCY, Department of Animal Biology, University of Pennsylvania School of Veterinary Medicine, Philadelphia.

DAVIES, ROBERT E., Department of Animal Biology, University of Pennsylvania School of Veterinary Medicine, Philadelphia.

DAWSON, JOAN, Department of Pharmacology, University of Pennsylvania, Philadelphia.

D'HAESE, JOCHEN, Department of Cytology and Micromorphology, University of Bonn, Bonn, Germany.

DOUGLAS, WILLIAM B., JR., Department of Physiology, Michigan State University, East Lansing.

DOW, JOCELYN W., MRC Laboratory of Molecular Biology, Cambridge, England.

DRABIKOWSKI, WITOLD, Department of Biochemistry of Nervous Systems and Muscles, Nencki Institute of Experimental Biology, Warsaw, Poland.

DREIZEN, PAUL, State University of New York, Downstate Medical Center, Brooklyn.

EATON, BARBRA, Biochemistry Graduate Group, University of Pennsylvania, Philadelphia.

*EBASHI, SETSURO, Department of Pharmacology, Faculty of Medicine, University of Tokyo, Tokyo, Japan.

ECKSTEIN, FRITZ, Department of Chemistry, Max-Planck-Institut, Gottingen, Germany.

EDWARDS, CHARLES, Department of Biological Sciences, State University of New York, Albany.

EISENBERG, BRENDA, Department of Physiology, University of California School of Medicine, Los Angeles.

EISENBERG, EVAN, National Heart and Lung Institute, NIH, Bethesda, Md.

EISENBERG, ROBERT, Department of Physiology, University of California School of Medicine, Los Angeles.

ELFVIN, MYRA, Department of Anatomy, Medical College of Pennsylvania, Philadelphia.

ELLIOTT, GERALD F., Department of Physics, The Open University, Bletchley, England.

ELZINGA, MARSHALL, Boston Biomedical Research Institute, Boston.

ENDO, MAKOTO, Department of Biological Sciences, Purdue University, Lafayette, Ind.

EPPENBERGER, HANS M., Laboratory of Developmental Biology, Swiss Federal Institute of Technology, Zurich, Switzerland.

ETLINGER, JOSEPH, Department of Biophysics, University of Chicago, Chicago.

FAY, FREDRIC S., Department of Physiology, University of Massachusetts Medical School, Worcester.

FEIT, THEODORE S., Department of Physiology, Einstein College of Medicine, Bronx, New York.

*Denotes Chairman

FISCHMAN, DONALD A., Department of Biology, University of Chicago, Chicago.

FRASER, ALLAN, Department of Biophysics, Johns Hopkins University, Baltimore.

FRIEDMAN, ZVIEL, Department of Biochemistry, St. Louis University School of Medicine, St. Louis.

GAETJENS, ERIC, Department of Contractile Proteins, Institute for Muscle Disease, Inc., New York.

GAUTHIER, GERALDINE, Laboratory of Electron Microscopy, Wellesley College, Wellesley, Mass.

GERGELY, JOHN, Department of Muscle Research, Boston Biomedical Research Institute, Boston.

GIAMBALVO, ANTHONY, Department of Biophysics, State University of New York, Brooklyn.

GIBBONS, IAN R., Pacific Biomedical Research Center, University of Hawaii, Honolulu.

GOLDMAN, ROBERT, Department of Biology, Case Western Reserve University, Cleveland, Ohio.

GOLL, DARREL, Muscle Biology Group, Iowa State University, Ames.

GRANT, ROBERT J., Department of Biological Sciences, Hunter College, New York.

GREASER, MARION, Muscle Biology Laboratory, University of Wisconsin, Madison.

*HANSON, JEAN, MRC Biophysics Unit, King's College, London, England.

HARRINGTON, WILLIAM F., Department of Biology, Johns Hopkins University, Baltimore.

HARTSHORNE, DAVID J., Department of Chemistry and Biological Sciences, Carnegie Mellon University, Pittsburgh.

HASELGROVE, JOHN, MRC Laboratory of Molecular Biology, Cambridge, England.

HERMAN, LAWRENCE, Department of Pathology, State University of New York, Downstate Medical Center, Brooklyn.

HEUMANN, HANS-GÜNTHER, Department of Zoology, Lehrstuhl für Zellenlehre, Heidelberg, Germany.

HEYWOOD, STUART, Department of Biology, University of Connecticut, Storrs.

HITCHCOCK, SARAH E., Department of Biology, Brandeis University, Waltham, Mass.

*HOLMES, KENNETH C., Department of Biophysics, Max-Planck-Institut, Heidelberg, Germany.

HOLT, JOHN C., Department of Structural Biology, Children's Cancer Research Foundation, Boston.

HOLTZER, HOWARD, Department of Anatomy, University of Pennsylvania School of Medicine, Philadelphia.

HOMSHER, EARL, Department of Physiology, University of California Medical Center, Los Angeles.

*HUXLEY, ANDREW F., Department of Physiology, University College, London, England.

*HUXLEY, HUGH E., MRC Laboratory of Molecular Biology, Cambridge, England.

IKEMOTO, NORIAKI, Department of Muscle Research, Boston Biomedical Research Institute, Boston.

JACOBSON, DAVID, McArdle Laboratory, University of Wisconsin, Madison.

JOHNSON, JAN P., Laboratory of Structural Molecular Biology, Children's Cancer Research Foundation, Boston.

JORGENSON, ANNELISA, Department of Biology, University of Connecticut, Storrs.

JULIAN, FRED, Boston Biomedical Research Institute, Boston.

KALDOR, GEORGE, Department of Physiology, Medical College of Pennsylvania, Philadelphia.

KAMINER, BENJAMIN, Department of Physiology, Boston University School of Medicine, Boston.

KASSLER, HELENE, Department of Biology, Clark University, Worcester, Mass.

KATZEFF, HARVEY, Department of Biophysics, State University of New York, Brooklyn.

KAY, CYRIL M., Department of Biochemistry, University of Alberta, Edmonton, Canada.

KENDRICK-JONES, JOHN, MRC Laboratory of Molecular Biology, Cambridge, England.

KING, MURRAY V., Laboratory of Physical Biochemistry, Massachusetts General Hospital, Boston.

KOMINZ, DAVID R., Laboratory of Biophysical Chemistry, NIAMD, NIH, Bethesda, Md.

KUSHMERICK, MARTIN J., Department of Physiology, Harvard Medical School, Boston.

LAMVIK, MICHAEL K., Enrico Fermi Institute and Department of Biophysics, University of Chicago, Chicago.

LANSMAN, STEVEN, Department of Medicine and Program in Biophysics, State University of New York, Brooklyn.

LEHMAN, WILLIAM, Department of Biology, Brandeis University, Waltham, Mass.

LEHRER, SHERWIN, Department of Muscle Research, Boston Biomedical Research Institute, Boston.

LEVINE, RHEA J., Department of Anatomy, Medical College of Pennsylvania, Philadelphia.

LEVY, RONNIE, Department of Physiology, Harvard Medical School, Boston.

LIGHT, SUSAN, Department of Biology, Clark University, Worcester, Mass.

LOEWY, ARIEL G., Department of Biology, Haverford College, Haverford, Penn.

LOWEY, SUSAN, Children's Cancer Research Foundation, Boston.

*LOWY, JACK, Biophysics Institute, Aarhus University, Aarhus, Denmark.

LYMN, RICHARD W., MRC Laboratory of Molecular Biology, Cambridge, England.

MACLENNAN, DAVID H., Banting and Best Department of Medical Research, University of Toronto, Toronto, Canada.

MAKINOSE, MADOKA, Department of Physiology, Max-Planck-Institut, Heidelberg, Germany.

MALIK, MAZHAR N., Department of Biochemistry, State University of New York, Downstate Medical Center, Brooklyn.

MANNHERZ, HANS G., Department of Biophysics, Max-Planck-Institut, Heidelberg, Germany.

MARCHAND, JOHN F., Department of Clinical Investigations, Institute for Muscle Disease, Inc., New York.

MARGOSSIAN, SARKIS S., Department of Structural Biology, Children's Cancer Research Foundation, Boston.

MARTONOSI, ANTHONY, Department of Biochemistry, St. Louis University School of Medicine, St. Louis.

MATACIC, SLAVICA SMIT, Department of Biology, Haverford College, Haverford, Penn.

MENDELSON, ROBERT, Cardiovascular Research Institute, University of California, San Francisco.

*MILLER, ANDREW, Department of Zoology, Oxford University, Oxford, England.

MILLER, RICHARD, Department of Biophysics and Medicine, State University of New York, Brooklyn.

MOMMAERTS, WILFRIED F. M. M., Department of Physiology, University of California Medical Center, Los Angeles.

MOOS, CARL, Department of Biochemistry, State University of New York, Stony Brook.

MORALES, Manuel F., Cardiovascular Research Institute, University of California, San Francisco.

MORRIS, GLENN, Department of Biology, University of Connecticut, Storrs.

MOWERY, PATRICK, Cardiovascular Research Institute, University of California Medical School, San Francisco.

MURRAY, JOHN, Department of Biochemistry, St. Louis University School of Medicine, St. Louis.

NACHMIAS, V. T., Haverford College, Haverford, Penn.

NAGY, BELA, Department of Muscle Research, Boston Biomedical Research Institute, Boston.

NAUSS, KATHLEEN M., Laboratory of Structural and Molecular Biology, Children's Cancer Research Foundation, Boston.

NOLAN, A. C., Laboratory of Physical Biology, NIH, Bethesda, Md.

OFFER, G. W., Department of Biophysics, King's College, London, England.

O'HARA, DONALD, Department of Physiology, Massachusetts General Hospital, Boston.

*OOSAWA, FUMIO, Institute of Molecular Biology, Nagoya University, Nagoya, Japan.

PAUL, RICHARD J., Department of Physiology, Harvard Medical School, Boston.

PEACHEY, LEE D., Department of Biology, University of Pennsylvania, Philadelphia.

PEMRICK, SUZANNE M., Department of Biological Sciences, State University of New York, Albany.

PEPE, FRANK A., Department of Anatomy, University of Pennsylvania School of Medicine, Philadelphia.

*PERRY, S. V., Department of Biochemistry, University of Birmingham, Birmingham, England.

PODOLSKY, RICHARD J., National Institute of Arthritis and Metabolic Diseases, NIH, Bethesda, Md.

POLLARD, THOMAS D., National Heart and Lung Institute, NIH, Bethesda, Md.

POTTER, JAMES D., Department of Muscle Research, Boston Biomedical Research Institute, Boston.

PULLMAN, JAMES M., Department of Biophysics, University of Chicago, Chicago.

REEDY, MICHAEL K., Department of Anatomy, Duke University, Durham, N.C.

REGENSTEIN, JOE M., Department of Biophysics, Brandeis University, Waltham, Mass.

REISLER, EMIL, Department of Biology, Johns Hopkins University, Baltimore.

RICE, ROBERT V., Department of Biological Sciences, Carnegie Mellon University, Pittsburgh.

RICHARDS, DENNIS, Department of Medicine and Program in Biophysics, State University of New York, Brooklyn.

RICHARDS, E. GLEN, Department of General Medical Research, Veterans Administration Hospital, Dallas.

ROME, ELIZABETH, MRC Muscle Biophysics Unit, King's College, London, England.

SABBADINI, ROGER A., Department of Zoology, University of California, Davis.

SAIDE, JUDITH, Department of Physics and Biochemistry, Massachusetts General Hospital, Boston.

SANDOW, ALEXANDER, Department of Physiology, Institute for Muscle Disease, Inc., New York.

SANGER, JOSEPH W., Department of Anatomy, University of Pennsylvania School of Medicine, Philadelphia.

SANTERRE, ROBERT, Department of Biology, Massachusetts Institute of Technology, Cambridge.

SARKAR, SATYAPRIYA, Department of Muscle Research, Boston Biomedical Research Institute, Boston.

SCHAEFFER, SUSAN F., Laboratory of Electron Microscopy, Wellesley College, Wellesley, Mass.

SCHAUB, MARCUS C., Institute of Pharmacology, University of Switzerland, Zurich.

SCHEIN, STANLEY J., Department of Molecular Biology, Einstein College of Medicine, Bronx, New York.

SCHLISELFELD, LOUIS H., Department of Contractile Proteins, Institute for Muscle Disease, Inc., New York.

SCHOENBERG, MARK, Laboratory of Physical Biology, NIAMD, NIH, Bethesda, Md.

SCORDILIS, STYLIANOS P., Department of Biological Sciences, State University of New York, Albany.

SEIDEL, JOHN, Department of Muscle Research, Boston Biomedical Research Institute, Boston.

SHAHN, EZRA, Department of Biological Science, Hunter College, New York.

SIEMANKOWSKI, RAYMOND, Department of Medicine and Program in Biophysics, State University of New York, Brooklyn.

SIMS, JOHN M., Department of Physiology and Cell Biology, University of Kansas, Lawrence.

SMALL, J. V., Biophysics Institute, Aarhus University, Aarhus, Denmark.

SMILLIE, L. B., Department of Biochemistry, University of Birmingham, Birmingham, England.

SORENSON, MARTHA M., Department of Neurology, College of Physicians and Surgeons, New York.

SPUDICH, JAMES A., Department of Biochemistry and Biophysics, University of California, San Francisco.

SPURWAY, N. C., Department of Biology, University of Pennsylvania, Philadelphia.

STEIGER, G. J., Department of Cell Physiology, Ruhr University, Bochum, Germany.

STOCK, GREGORY B., Department of Biophysics, Johns Hopkins University, Baltimore.

STONE, DEBORAH B., Cardiovascular Research Institute, University of California, San Francisco.

STRACHER, ALFRED, Department of Biochemistry, State University of New York, Downstate Medical Center, Brooklyn.

STULL, JAMES T., JR., Department of Biological Chemistry, University of California School of Medicine, Davis.

SZENT-GYÖRGYI, ALBERT, Marine Biological Laboratory, Woods Hole, Mass.

SZENT-GYÖRGYI, ANDREW, Department of Biology, Brandeis University, Waltham, Mass.

SZENT-GYÖRGYI, EVE, Department of Biology, Brandeis University, Waltham, Mass.

TADA, MICHIHIKO, Department of Cardiology, Mt. Sinai School of Medicine, New York.

TAM, DOMINIC, Department of Biophysics and Medicine. State University of New York, Brooklyn.

TAYLOR, DOUGLASS L., Department of Biology, State University of New York, Albany.

TAYLOR, EDWIN W., Department of Biophysics, University of Chicago, Chicago.

THOMPSON, WILLIAM C., Department of Genetics and Cell Biology, University of Connecticut, Storrs.

TONOMURA, YUJI, Department of Biology. Faculty of Science, Osaka University, Osaka, Japan.

*TREGEAR, RICHARD T., Department of Zoology, University of Oxford, Oxford, England.

TRENTHAM, D. R., Department of Biochemistry, University of Bristol School of Medicine, Bristol, England.

TWAROG, BETTY M., Department of Biology, Tufts University, Medford, Mass.

VIBERT, PETER J., Biophysics Institute, Aarhus University, Aarhus, Denmark.

WACHSBERGER, PHYLLIS, Department of Anatomy, University of Pennsylvania School of Medicine, Philadelphia.

*WEBER, ANNEMARIE, Department of Biochemistry, St. Louis University School of Medicine, St. Louis.

WEEDS, A. G., MRC Laboratory of Molecular Biology, Cambridge, England.

WEISEL, JOHN W., Department of Biophysics, Brandeis University, Waltham, Mass.

WHITE, DAVID C. S., Department of Biology, University of York, York, England.

WIENER, STANLEY L., Department of Medical Research, Long Island Jewish Medical Center, New Hyde Park, New York.

*WILKIE, D. R., Department of Physiology, University College, London, England.

WOLEDGE, ROGER C., Department of Physiology, University College, London, England.

YAFFE, DAVID, Department of Biology, Massachusetts Institute of Technology, Cambridge.

YOUNG, MICHAEL, Department of Medicine, Massachusetts General Hospital, Boston.

YOUNT, RALPH G., Department of Chemistry, Washington State University, Pullman.

Albert Szent-Györgyi
Opening Remarks

First row: S. V. Perry - - H. E. Huxley, S. Ebashi - - E. Eisenberg.
Second row: D. R. Trentham - - F. Oosawa, N. A. Biró, S. Ebashi - - L. H. Schliselfeld.
Third row: R. S. Adelstein - - R. J. Podolsky - - M., K., and G. Bárány.

First row: N. A. Biró - - R. W. Lymn, J. Hanson - - W. F. Harrington.
Second row: L. D. Peachey - - M. Elzinga - - R. V. Rice, D. Bray.
Third row: E. Shahn - - J. Haselgrove - - H. Carney.

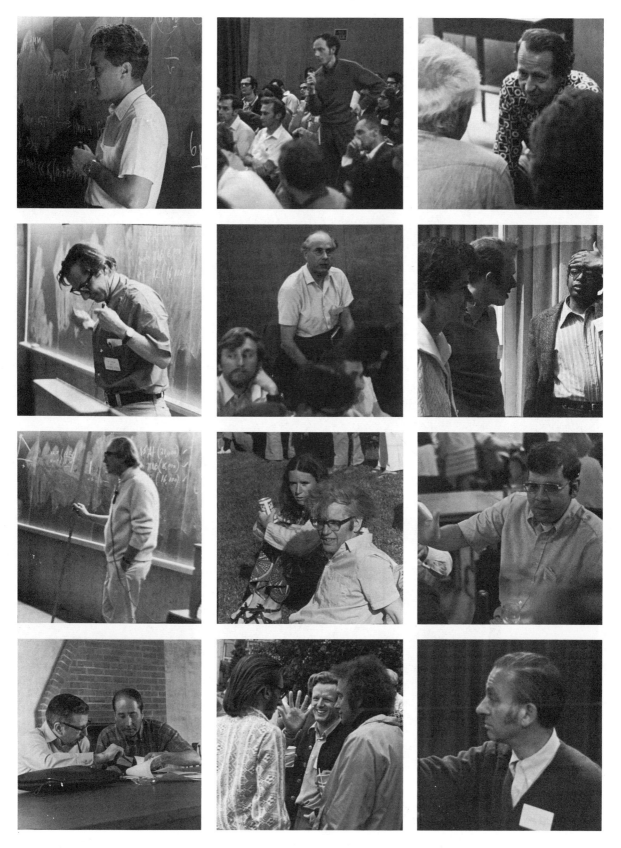

First row: C. Moos - - A. Miller - - W. F. M. M. Mommaerts.
Second row: P. Dreizen - - T. Pollard, A. F. Huxley - - S. Lowey, A. Miller, S. Sarkar.
Third row: M. Young - - S. E. Hitchcock, R. T. Tregear - - R. G. Yount.
Fourth row: F. A. Pepe, E. Gaetjens - - J. V. Small, S. Heywood, G. Morris - - J. Gergely.

Foreword

Our symposium each year gives us the opportunity to seek an exploding phase of biology and to bring together most of the key practitioners.

With the emergence of the sliding filament model the field of muscle contraction has become in itself a major intellectual discipline, with its own well-defined objectives. On no occasion, however, since the sliding filament model was first presented has it been possible to bring together all the many people whose research directly bears on the molecular events underlying the contraction process. So it seemed most appropriate to devote this year's symposium to this theme.

In preparing this program, I greatly profited from the advice of Carolyn Cohen, John Gergely, Andrew Huxley, Hugh Huxley, and Andrew Szent-Györgyi. They suggested many more speakers than we could accommodate even in a very full week's program. So, regretfully, we were not able to include on the final program everyone who had something important to say. But, hopefully, the high level of animated discussion that occurred throughout the week allowed us to hear about all new facts of importance. It is our belief that this symposium volume will be of great value, not only in telling us about the current state of muscle research, but also in pointing out what problems remain to be solved.

Fortunately we were able to entice Albert Szent-Györgyi to open the program, thereby allowing us again to appreciate his vast contributions to the way we think about contraction. During the succeeding seven days, the formal program sessions were generally held in the mornings and evenings, with the afternoons reserved for more informal gatherings.

The meeting ran from the evening of June 6th to noon on June 13th. Approximately 210 people were in attendance. As in the past, the program was supported by the National Institutes of Health, the National Science Foundation, and the United States Atomic Energy Commission. In charge of putting together the resulting volume has been Judy Gordon, who served as managing editor, ably assisted by Mary Lewis and Kate Alston.

James D. Watson
Director

Contents

MUSCLE REGULATORY SYSTEMS

MUSCLE STRUCTURE

SARCOPLASMIC RETICULUM

MYOGENESIS

CONTRACTILE PROTEINS IN NON-MUSCLE TISSUE

ENERGETICS AND MECHANICAL PROPERTIES

Summary

The Amino Acid Sequence of Rabbit Skeletal Muscle Actin

Marshall Elzinga* and John H. Collins

Department of Muscle Research, Boston Biomedical Research Institute, Boston, Mass. 02114
and
Dept. of Neurology, Harvard Medical School, Cambridge, Mass. 02138

Actin participates in a variety of molecular interactions which appear to be functionally important for muscle contraction; these include the binding of a divalent cation (Ca^{++} or Mg^{++}) and a nucleotide (ATP or ADP), polymerization and interaction of the polymer with tropomyosin and troponin, and interaction with myosin to form the interfilament cross-bridges that appear to be essential for the generation of force.

In attempting to explain the basic mechanism of muscle contraction in terms of events that occur in the cross-bridge, it is likely that an understanding of the topography of the actin-myosin interface would be helpful. One step toward this goal is to describe the structures of actin and myosin, and over the past few years we have been studying these proteins. Studies on the amino acid sequence of actin are almost complete, and in this paper we report the sequences of 17 peptides that account for 374 residues in rabbit skeletal muscle actin. These peptides have been aligned (Kuehl et al., 1972) and can be written as two segments: the N-terminal 119 residues and the C-terminal 255 residues.

Methods and Results

Preparation of protein for sequence analysis.
The actin used in this study was prepared from an acetone powder of the back and leg muscles of New Zealand white rabbits. The powder was extracted at 2° and the actin was purified by two polymerization cycles. The actin used for the nitration studies was further purified by Sephadex G-200 gel filtration. The sulfhydryl groups were alkylated either by reaction with iodoacetamide or by amino-ethylation. Gel electrophoresis of alkylated actin in buffers that contain sodium dodecyl sulfate gives a single band of mol wt 42–44,000.

Digestion of protein and separation of peptides.
The protein was cleaved at its methionine residues with the use of cyanogen bromide in 70% formic acid, and the peptides were separated by gel filtration and ion exchange chromatography. We

have previously described the isolation of the peptides designated CB-1 through CB-13 (Elzinga, 1970); the isolation of these peptides, as well as four additional peptides, was reported concurrently by Adelstein and Kuehl (1970). We were unable to obtain these four peptides by the methods we had been using, and it became apparent to us that they were insoluble at pH's near neutrality.

We devised a procedure by which to separate CB-15, 16, and 17 from CB-10, 11, 12, and 13, and the scheme is shown in Fig. 1. The digest (4–5 μmoles) is first passed over a column of Sephadex G-50, and a broad cut is taken as indicated, pooled, and dried. This fraction is then dissolved in 5 ml of 70% formic acid and applied to a 1.9×200 cm column of Sephadex G-10, which has been equilibrated with a pH 6.0 pyridine (10 mM)-acetate buffer. All of these peptides are 30–44 residues long and would thus be excluded from the Sephadex G-10. As shown in Fig. 1b, a sharp peak is obtained at the breakthrough of the column, and rechromatography of this peak (Fig. 1d) and/or ion exchange chromatography on SP-Sephadex (not shown) indicates that it contains only CB-10, 11, 12, and 13. A second peak is obtained later (Fig. 1b), and its position coincides with the volume at which the formic acid appears. Rechromatography of this material on Sephadex G-50 (Fig. 1c) yields CB-16 as a single peak, as well as two other peaks that are mixtures of CB-15 and CB-17. The region between the peaks (Fig. 1b) also contains CB-15, 16, and 17 and appears to be slightly enriched in CB-16; the overall yields of CB-16 and the mixture of CB-15 and 17 are 60–80%. The amino acid compositions of CB-15, CB-16, and CB-17 obtained from the sequences correspond well with those reported by Adelstein and Kuehl (1970).

The separation of peptides CB-10 through CB-17 into two major groups on Sephadex G-10 is clearly not due to gel filtration, since all of them are of a size that would be excluded from the gel. We interpret the separation as being due to differences in solubility at pH 6.0; CB-10, 11, 12, and 13 readily pass from the formic acid in which the sample is applied into the pH 6.0 buffer and are

---

* Present address: Dept. of Biophysics, Max-Planck-Institute for Medical Research, Heidelberg, Germany.

1

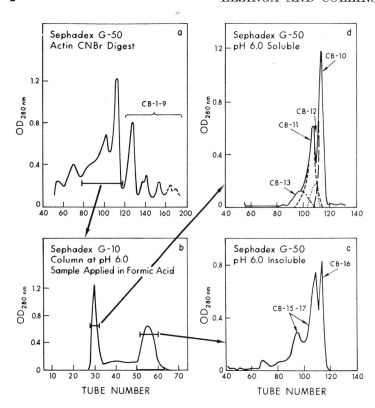

Figure 1. Chromatography of the cyanogen bromide peptides from actin. *a*) Gel filtration of digest on G-50 (25% acetic acid, 1.9 × 400 cm). CB-1–9 are purified by ion exchange chromatography (not shown). The region enclosed by the bar is pooled, dried, dissolved in 5 ml of 70% formic acid, and applied to a 1.9 × 200 cm column of Sephadex G-10. *b*) Rechromatography of the G-10 peaks on G-50 (same conditions as in *a*) is shown in *c* and *d*.

eluted at the void volume, whereas CB-15, 16, and 17 precipitate at the leading edge of the formic acid band and are essentially "washed out" by the formic acid. The presence of small amounts of CB-15, 16, and 17 between the two main peaks means that these peptides are slightly soluble at pH 6.0.

Sequence determinations. The sequences reported here were determined by quantitative subtractive Edman degradations. Amides were determined by enzymic digestion or high voltage paper electrophoresis of appropriate peptides. Details of the procedures have been described (Elzinga, 1970).

Table 1. Amino Acid Compositions of Cyanogen Bromide Peptides

| | CB-9 | | CB-10 | | | CB-15 | CB-16 | | CB-17 | |
	Rabbit	Bovine Heart	Rabbit	Bovine Heart	Ameba	Rabbit	Rabbit	Ameba	Rabbit	Ameba
Lys	2	2.01	3	3.05	2.46	0	3	1.77	3	3.06
Lys-(Me)$_2$	0	0	0	0	0	0	0	0.42	0	0
His	1	0.99	0	0	0.16	2	0	0.08	0	0
His-(^{r}Me)	0	0	1	0.84	0.75	0	0	0	0	0
Arg	1	1.03	1	1.04	0.94	1	1	1.16	3	3.35
Cys	1	(1)	0	0	0	0	0	0	1	(1)
Asp	1	1.12	5	4.85	4.61	3	0	1.58	3	2.54
Thr	1	0.19	2	1.94	2.01	3	1	1.59	3	4.58
Ser	1	1.79	2	1.87	2.34	3	4	3.05	1	1.28
Glu	3	2.92	4	3.72	4.36	2	3	4.18	6	8.83
Pro	1	0.96	1	0.93	1.16	2	2	1.98	0	0.29
Gly	1	1.18	4	4.04	3.85	5	2	2.72	1	1.70
Ala	1	1.08	1	1.30	1.66	5	2	2.64	2	2.61
Val	1	0.82	1	0.95	1.39	5	1	1.13	3	1.62
Ile	2	1.95	4	3.06	1.97	4	5	2.98	3	2.21
Leu	0	0.11	2	2.06	2.12	4	2	2.53	3	3.40
Tyr	1	1.06	2	1.96	1.67	4	1	1.19	2	2.15
Phe	1	1.15	0	0	0.40	0	1	1.21	2	2.30
Trp	1	(1)	1	1	2.32	0	1	1.21	0	0
Hse	0	0	1	1	1	1	1	1	1	1

The numbers given for aspartate and glutamate actually represent the sums of the amide and free acid forms of these amino acids. The compositions given for the rabbit peptides are calculated from the sequences; the data for ameba actin peptides are from Weihing and Korn (1972).

The sequences of CB-1, 2, 3, 4, 5, 6, 7, 8, 9, 10, 11, and 13 have been published (Elzinga, 1969, 1970, 1971a; Collins et al., 1971).

CB-12 and CB-16 were purified and sequenced, but the sequences of CB-15 and CB-17 were determined without actually separating the two peptides. This was accomplished by preparing mixtures of these peptides and digesting them with trypsin, chymotrypsin, or pepsin. The peptides in each of these digests were then purified and sequenced, and they were aligned by inspection of peptides generated from complementary digests. In this way unambiguous sequences for 15 and 17 were established, and their compositions, as well as that of CB-16, are listed in Table 1.

Alignment of the cyanogen bromide peptides. The alignment of the peptides has been established by W. M. Kuehl and R. S. Adelstein at the NIH (personal communication; see also Kuehl et al., 1972); based upon their results, the 16 peptides (plus the tetrapeptide Thr-Gln-Ile-Hse,

see below) may be written as two large blocks of sequence. One of these represents the amino terminal 119 residues, and the other contains 255 residues and extends to the carboxyl terminus. These are shown in Fig. 2. The sequences of the individual cyanogen bromide peptides may be obtained from Fig. 2 by noting the residue numbers as summarized in Table 2.

In addition to the 16 peptides in these segments, we have isolated from three separate digests a tetrapeptide that has the sequence Thr-Gln-Ile-Hse. It was isolated in low yields but appears to arise from a unique cyanogen bromide fragment. Kuehl et al. (1972) have found an overlap peptide that represents this tetrapeptide plus all of CB-4 and the N-terminal segment of CB-15. Thus the tetrapeptide is adjacent to the amino terminus of CB-4. The overlap of the residues designated 119 and (120) has not yet been established; there may be an additional short segment of sequence between these two residues. Thus, while the numbering of residues 1–119 may be considered definitive, the

CB-13→ 10
Ac-Asp-Glu-Thr-Glu-Asp-Thr-Ala-Leu-Val-Cys-
20
Asp-Asp-Gly-Ser-Gly-Leu-Val-Lys-Ala-Gly-
30
Phe-Ala-Gly-Asp-Asp-Ala-Pro-Arg-Ala-Val-
40
Phe-Pro-Ser-Ile-Val-Gly-Arg-Pro-Arg-His-
CB-1→   CB-10→   50
Gln-Gly-Val-Met-Val-Ser-Met-Gly-Gln-Lys-
60
Asp-Ser-Tyr-Val-Gly-Asp-Gly-Ala-Gln-Ser-
70
Lys-Arg-Gly-Ile-Leu-Thr-Leu-Lys-Tyr-Pro-
80
Ile-Glu-His(7Me)-Trp-Gly-Ile-Ile-Thr-Asn-Asp-
CB-11→   90
Asp-Met-Glu-Lys-Ile-Trp-His-His-Thr-Phe-
100
Tyr-Asn-Glu-Leu-Arg-Val-Ala-Pro-Glu-Glu-
110
His-Pro-Thr-Leu-Leu-Thr-Glu-Ala-Pro-Leu-
(120)
Asn-Pro-Lys-Ala-Asn-Arg-Glu-Lys-Met/Thr-
CB-4
→   (130)
Gln-Ile-Met-Phe-Glu-Thr-Phe-Asn-Val-Pro-
CB-15
→   (140)
Ala-Met-Tyr-Val-Ala-Ile-Gln-Ala-Val-Leu-
(150)
Ser-Leu-Tyr-Ala-Ser-Gly-Arg-Thr-Thr-Gly-
(160)
Ile-Val-Leu-Asp-Ser-Gly-Asp-Gly-Val-Thr-
(170)
His-Asn-Val-Pro-Ile-Tyr-Glu-Gly-Tyr-Ala-
CB-7
→   (180)
Leu-Pro-His-Ala-Ile-Met-Arg-Leu-Asp-Leu-
(190)
Ala-Gly-Arg-Asp-Leu-Thr-Asp-Tyr-Leu-Met-
CB-17
→   (200)
Lys-Ile-Lys-Thr-Glu-Arg-Gly-Tyr-Ser-Phe-

(210)
Val-Thr-Thr-Ala-Glu-Arg-Glu-Ile-Val-Arg-
(220)
Asp-Ile-Lys-Gln-Lys-Leu-Cys-Tyr-Val-Ala-
CB-12
→   (230)
Leu-Asp-Phe-Glu-Asn-Glu-Met-Ala-Thr-Ala-
(240)
Ala-Ser-Ser-Ser-Leu-Glu-Lys-Ser-Tyr-Glu-
(250)
Leu-(Pro, Asx, Glx, Gly, Ile, Val)-Thr-Ile-Gly-
(260)
Asn-Glu-Arg-Phe-Arg-Cys-Pro-Glu-Thr-(Phe,
CB-6 (270)
Leu, Phe, Gln, Pro, Ser, Ile, Gly)-Met-Glu-Ser-
(280)
Ala-Gly-Ile-His-Glu-Thr-Thr-Tyr-Asn-Ser-
CB-8
→   (290)
Ile-Met-Lys-Cys-Asp-Ile-Asp-Ile-Arg-Lys-
CB-2
→   (300)
Asp-Leu-Tyr-Ala-Asn-Asn-Val-Met-Ser-Gly-
CB-3
→   (310)
Gly-Thr-Thr-Met-Tyr-Pro-Gly-Ile-Ala-Asp-
CB-5
→   (320)
Arg-Met-Gln-Lys-Glu-Ile-Thr-Ala-Leu-Ala-
CB-16
→   (330)
Pro-Ser-Thr-Met-Lys-Ile-Lys-Ile-Ile-Ala-
(340)
Pro-Pro-Glu-Arg-Lys-Tyr-Ser-Val-Trp-Ile-
(350)
Gly-Gly-Ser-Ile-Leu-Ala-Ser-Leu-Ser-Thr-
CB-9   (360)
Phe-Gln-Gln-Met-Trp-Ile-Thr-Lys-Gln-Glu-
(370)
Tyr-Asp-Glu-Ala-Gly-Pro-Ser-Ile-Val-His-
(374)
Arg-Lys-Cys-Phe.

Figure 2. The tentative amino acid sequence of rabbit skeletal muscle actin. The (/) between residues 119 and (120) indicates that this overlap has not yet been established. Because there may be an additional segment here, all residue numberings after 119 are tentative and are enclosed by parentheses. This sequence is based upon our work on the amino acid sequences of the cyanogen bromide peptides and the alignments established by Kuehl et al. (1972).

Table 2. Residue Numbers for Actin Cyanogen Bromide Peptides

Cyanogen Bromide Peptide	Residues
CB-13	1–44
CB-1	45–47
CB-10	48–82
CB-11	83–119
Tetrapeptide	(120)–(123)
CB-4	(124)–(132)
CB-15	(133)–(176)
CB-7	(177)–(190)
CB-17	(191)–(227)
CB-12	(228)–(268)
CB-6	(269)–(282)
CB-8	(283)–(298)
CB-2	(299)–(304)
CB-3	(305)–(312)
CB-5	(313)–(324)
CB-16	(325)–(354)
CB-9	(355)–(374)

Alignment of the peptides was established by Kuehl et al., 1972.

numbering of the rest of the molecule is tentative; they are numbered here only for the sake of convenience in discussing the sequence and are always enclosed by parentheses.

Discussion

The results reported here represent substantial completion of the first step in establishing the structure of rabbit skeletal muscle actin—that of determining its amino acid sequence. This sequence can be used as a reference with which to compare sequences of peptides from other actins and thus to estimate the relationship among actins from various sources. Also, the ability to fragment the molecule at predictable bonds and isolate peptides that contain selected residues permits one to study the specificity of a wide variety of chemical modifications. Interesting aspects of the sequence, as well as the comparative structure and chemical modification of actin, are discussed below.

Analysis of sequence. Inspection of the sequence reveals some interesting features, although assignment of functional roles to most of the residues in performing the characteristic functions of this protein must obviously await an analysis of its tertiary structure.

The N-terminus of the chain carries a strong negative charge—the α-amino group is blocked, and four of the first five residues have acidic side chains. The single residue of N^τ-methyl histidine (previously called 3-methyl histidine) occurs at position 73, and because of the unusual nature of this amino acid, it is of interest to examine this sequence. The methyl group is transferred to histidine enzymically (from S-adenosyl methionine) and the methyl group appears to be added after

the protein is folded; thus the appropriate histidine (position 73) must be accessible to the methylating enzyme, and it is probably at or near the surface of the actin sphere. Tyrosine-69, which is only four residues away from N^τ-methyl histidine, is the first tyrosine to be nitrated in G-actin, and nitration is accompanied by inhibition of polymerizability (Elzinga, 1971b; Mühlrad et al., 1969; see also discussion below). One of the four (or five) tryptophans is also found in this region, at position 74. Indirect evidence suggests that Trp-74 is one of the principal sources of the intrinsic fluorescence of G-actin (Lehrer and Elzinga, 1972). Taken together, the various bits of evidence suggest that a substantial part of this region, including Tyr-69 and His($^\tau$Me)-73, lies at or near the surface of actin.

Residues 86–91 represent an usual sequence in that five of six residues have heterocyclic or aromatic side chains.

Another side chain that appears to be near the surface is Lys-113; its amino group is available for reaction with FDNB, and its reactivity is far greater in F- than in G-actin (Collins et al., 1971).

There are three long segments that consist primarily of nonpolar residues; the first of these is (120)–(153). Of the 34 residues in this stretch, only two are charged: Glu-(125) and Arg-(147). The second of these uncharged segments is (292)–(309), and the third is the 22-residue segment extending from residue (336) through (357). It is likely that these three segments lie in the interior of the molecule.

The three SH groups that are reactive in G-actin (Bridgen, 1972) occur at positions 10, (284), and (373), while cysteines-(217) and (256) are not available for reaction in the native molecule.

Comparative structure of actin. A substantial body of information exists regarding the degree of species variation of some proteins, most notably cytochrome c and hemoglobin. In general the surfaces of proteins tend to be more variable than the interior, and this presumably is because the variations in size or charge of the inside are more likely to disturb the overall structural stability of a protein. In the case of cytochrome c, however, both the inside of the molecule and the surface features are highly conserved throughout nature (Dickerson et al., 1971). This constancy has been interpreted by Dickerson (1971) as owing to the fact that the surface of the cytochrome c molecule may be considered to be a "carefully tailored substrate for the two enzyme complexes: reductase and oxidase." Thus because the surface must interact with other proteins just as the interior residues must all interact with complimentary residues, the same selection pressure is exerted upon the surface of cytochrome *c* as upon the interiors of other proteins. In its functional role in the muscle, actin, like cytochrome *c*, must participate in a variety of protein-protein interactions, and thus one would predict that it too is a "carefully tailored substrate." These interactions include three or four points of contact with adjacent actin molecules in F-actin, as well as regions that interact with tropomyosin (probably nonspecifically) and a highly specific interaction with myosin. There may also be a point of interaction with troponin, and there is a binding site for a nucleotide and a metal. Thus if Dickerson's argument is correct, one would expect the surface (as well as the interior) of actin to be highly conserved. There is a substantial amount of information suggesting that this is in fact the case. Actins from many sources can bind muscle heavy meromyosin (Ishikawa et al., 1969). Actins from widely divergent sources apparently copolymerize (Totsuka and Hatano 1970). The amino acid compositions and peptide maps of a wide spectrum of actins are essentially indistinguishable (Carsten and Katz, 1964).

The critical test for relatedness among proteins is a comparison of their amino acid sequences, and most of the available information on the comparative structure of actin is presented in Fig. 3 and Table 1. In Fig. 3 the sequence data on trout actin is taken from Bridgen (1971).

An inspection of the trout and rabbit sequences reveals two cases in which there is clearly an amino acid substitution. One is in the region of (258)–(259), where there may be two replacements. The sequences should be rechecked but it is clear that there must be at least a Glu → Ser substitution. At position (368) there is a Leu → Ile substitution. The peptides that contain Cys-10 and Cys-(284)

are identical in composition but differ somewhat in reported sequence. These should also be rechecked in order to be sure whether or not there are actual differences in primary structure among these peptides.

We have compared the compositions of some of the cyanogen bromide fragments of bovine cardiac actin with corresponding peptides from rabbit; these results are summarized in Table 1. CB-9 from bovine heart has 2 Ser and no Thr; this probably means that there is a Thr → Ser substitution at position (357), although we have not proven this by sequence analysis. The composition of CB-10 from bovine heart actin is indistinguishable from that of the rabbit skeletal muscle actin peptide (Table 1). We have also isolated CB-1, 2, 3, 4, and 8, and the compositions of all of them are indistinguishable from those of the corresponding peptides from rabbit. Thus a comparison of the compositions of peptides that represent 97 residues, or about ¼ of the molecule, reveals only one conservative amino acid substitution. It is interesting to note that in cytochrome *c*, a protein that exhibits comparatively little variability throughout the phylogenetic scale, the rabbit sequence differs from the bovine protein at 4 out of 104 positions. Thus actin appears to be even less variable than cytochrome *c*.

Weihing and Korn (1972) have isolated and analyzed three cyanogen bromide peptides from ameba actin, and their compositions are compared with the corresponding rabbit peptides in Table 1. Although the ameba peptides are not completely pure (Weihing and Korn, 1972), the similarities in composition are impressive, and one must conclude that the sequences of rabbit and ameba actin are similar; the precise degree of similarity will have to await further sequence analysis of ameba actin.

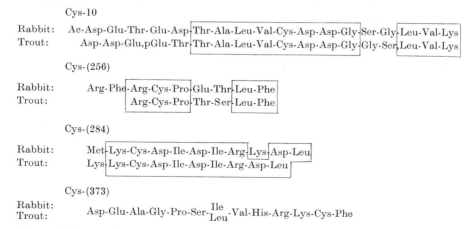

Figure 3. A comparison of the amino acid sequences around four of the five cysteine residues in rabbit and trout actin. Trout data is from Bridgen (1971). Boxes enclose sequences that are identical in the two proteins.

Myosin: Leu-Leu-Gly-Ser-Ile-Asp-Val-Asp-His($^\tau$Me)-Gln-Thr-Tyr-Lys
Actin: Leu-Thr-Leu-Lys-Tyr-Pro-Ile-Glu-His($^\tau$Me)-Trp-Gly-Ile-Ile

Figure 4. A comparison of the amino acid sequences around N^τ-methyl histidine (His($^\tau$Me)) in rabbit skeletal muscle actin and myosin. From Huszar and Elzinga (1971).

Homology with other proteins. There are several instances in nature in which it is clear that proteins which serve unlike functions have evolved from a common precursor; examples are trypsin and chymotrypsin (Walsh and Neurath, 1964) as well as lysozyme and α-lactalbumin (Brew et al., 1967). In the case of muscle proteins it was shown by Asatoor and Armstrong (1967) and Johnson et al. (1967) that both actin and skeletal muscle myosin contain a single residue of the very rare amino acid N^τ-methyl histidine. This suggested the possibility that these two proteins contained similar sequences that included this unusual amino acid. The sequence around N^τ-methyl histidine in rabbit skeletal muscle myosin has been established (Huszar and Elzinga, 1971), and it is compared with the corresponding sequence in actin in Fig. 4. The sequences are clearly different, and there is no evidence for homology of actin and myosin in this region.

Implications for understanding structure and function of actin. As we attempt to define some of the basic aspects of force generation in muscle in terms of events that occur at or near the interface between actin and myosin, it is essential to define the structures of these proteins; determination of the sequence of actin must be considered to be complimentary to eventual studies on its tertiary structure, and only when both are complete can we expect a clear picture of the structure of actin to emerge. X-ray diffraction studies on actin have as yet been impossible because actin has not been crystallized. The probable reason for the lack of suitable crystals of actin is the tendency of the protein to polymerize in the presence of salt, and the polymer is unlikely to be suitable for crystallization. It is not clear how to circumvent this problem, and one possibility is to prepare a derivative in which the ability to polymerize is inhibited. Two derivatives that seem promising have been reported. Martonosi (1968) reported that photooxidation selectively inhibits polymerization, and Mühlrad et al. (1968) have reported that limited nitration of G-actin also selectively inhibits polymerization. It seems reasonable that it would be best to begin with a derivative that is chemically homogeneous, has been subjected to only limited modification, and retains its native conformation. The technology developed during a sequence study permits one to determine the degree of chemical homogeneity by selectively isolating peptides that contain the modified residue(s).

In the case of photooxidation of histidine this is difficult because the oxidized histidine(s) is not conveniently analyzed. Modification by tetranitromethane is amenable to analysis because the product, 3-nitrotyrosine, is stable and may be detected on the amino acid analyzer. We have studied actin in which about one mole of 3-nitrotyrosine is introduced into the globular form of the protein, and have found that, while polymerization is inhibited, the protein seems to retain its native conformation (Lehrer and Elzinga, 1972). The primary site of modification is tyrosine-69; about 70% of the NO_2-tyrosine may be found at this position (Elzinga, 1971b). This is remarkable specificity when one considers that actin has a total of 16 (or 17) tyrosine residues. It is interesting to note that tyrosine-69 is near in the sequence to His($^\tau$Me)-73, and it is tempting (but premature) to suggest that this region is involved in polymerization of the protein. In any case the photooxidized and nitrated derivatives may be suitable for crystallization, and if not, the techniques developed in this study will facilitate chemical studies on other potentially crystallizable derivatives of actin.

Acknowledgments

This work was facilitated by the excellent technical assistance of N. Jackman and G. Bautista and was supported by grants from the USPHS (AM-14728) and the American Heart Association. M. E. is an Established Investigator of the American Heart Association.

References

ADELSTEIN, R. S. and W. M. KUEHL. 1970. Structural studies on rabbit skeletal actin. I. Isolation and characterization of the peptides produced by cyanogen bromide cleavage. *Biochemistry* **9**: 1355.

ASATOOR, A. M. and M. D. ARMSTRONG. 1967. 3-methyl histidine, a component of actin. *Biochem. Biophys. Res. Comm.* **26**: 168.

BREW, K., J. C. VANAMAN, and R. L. HILL. 1967. Comparison of the amino acid sequence of bovine α-lactalbumin and hen egg white lysozyme. *J. Biol. Chem.* **242**: 3747.

BRIDGEN, J. 1971. The amino acid sequence around four cysteine residues in trout actin. *Biochem. J.* **123**: 591.

————. 1972. The reactivity and function of thiol groups in trout actin. *Biochem. J.* **126**: 21.

CARSTEN, M. E. and A. M. KATZ. 1964. Actin: A comparative study. *Biochim. Biophys. Acta* **90**: 534.

COLLINS, J. H., E. MORKIN, and M. ELZINGA. 1971. Structure-function studies on rabbit skeletal muscle actin. *Fed. Proc.* **30**: 558.

DICKERSON, R. E. 1971. The structure of cytochrome *c* and the rates of molecular evolution. *J. Mol. Evolution* **1**:26.

DICKERSON, R. E., T. TAKANO, D. EISENBERG, O. B. KALLAI, L. SANSON, A. COOPER, and E. MARGOLIASH. 1971. Ferricytochrome *c*. I. General features of the horse and bonito proteins at 2.9 Å resolution. *J. Biol. Chem.* **246**: 1511.

ELZINGA, M. 1969. Comparative sequence studies on actin: The C-terminus and the region containing 3-methyl histidine. *Abstr. 3rd Int. Biophys. Congr.*, Cambridge, Mass. IID.8.

————. 1970. Amino acid sequence studies on rabbit skeletal muscle actin. Cyanogen bromide cleavage of the protein and determination of the sequences of seven of the resulting peptides. *Biochemistry* **9**: 1365.

————. 1971a. Amino acid sequence around 3-methyl histidine in rabbit skeletal muscle actin. *Biochemistry* **10**: 224.

————. 1971b. Selective nitration of tyrosine in actin. *Abstr. Fall 1971 Meet. Amer. Chem. Soc.*, Wash., D.C., p. 61.

HUSZAR, G. and M. ELZINGA. 1971. Amino acid sequence around the single 3-methyl histidine residue in rabbit skeletal muscle myosin. *Biochemistry* **10**: 229.

ISHIKAWA, H., R. BISCHOFF, and H. HOLTZER. 1969. Formation of arrowhead complexes with heavy meromyosin in a variety of cell types. *J. Cell Biol.* **43**: 312.

JOHNSON, P., C. I. HARRIS, and S. V. PERRY. 1967. 3-methyl histidine in actin and other muscle proteins. *Biochem. J.* **105**: 361.

KUEHL, W. M., M. A. CONTI, and R. S. ADELSTEIN. 1972. The structure of rabbit skeletal actin: Alignment of the peptide fragments produced by cyanogen bromide. *Abstr. 4th Int. Biophys. Congr.*, Moscow. In press.

LEHRER, S. S. and M. ELZINGA. 1972. Fluorescence studies on NO₂-actin. *Fed. Proc.* **31**: 1627.

MARTONOSI, A. 1968. The sulfhydryl groups of actin. *Arch. Biochem. Biophys.* **123**: 29.

MÜHLRAD, A., A. CORSI, and A. L. GRENATA. 1968. Studies on the properties of chemically modified actin. I. Photo-oxidation, succinylation, nitration. *Biochim. Biophys. Acta* **162**: 435.

TOTSUKA, T. and S. HATANO. 1970. ATPase activity of plasmodium actin polymer formed in the presence of Mg⁺⁺. *Biochim. Biophys. Acta* **223**: 189.

WALSH, K. A. and H. NEURATH. 1964. Trypsinogen and chymotrypsinogen as homologous proteins. *Proc. Nat. Acad. Sci.* **52**: 884.

WEIHING, R. R. and E. D. KORN. 1972. *Acanthamoeba* actin. Composition of a peptide that contains 3-methyl histidine and a peptide that contains Nε-methyl lysine. *Biochemistry* **11**: 1538.

Structural Studies on the Light Chains of Myosin

A. G. WEEDS AND G. FRANK*

MRC Laboratory of Molecular Biology, Cambridge CB2 2QH, England

Rabbit skeletal muscle myosin contains two large polypeptide chains of molecular weight about 200,000 daltons (the heavy chains) (Gershman et al., 1969; Gazith et al., 1970) and several small polypeptide chains of about 20,000 daltons (the light chains). These light chains are nonconvalently bound to the heavy chains and can be dissociated under a variety of conditions, all of which result in loss of ATPase activity: with alkali (Kominz et al., 1959), urea (Tsao, 1953), guanidine hydrochloride (Dreizen et al., 1967), or by acylation of the ε-amino groups of lysine with acetic anhydride (Locker and Hagyard, 1967a). They appear to be required for the enzymic activity of myosin, since they can be removed under much milder conditions with 4 M LiCl, and recombination of the separated light and heavy chains results in a significant recovery of ATPase activity (Stracher, 1969; Dreizen and Gershman, 1970).

There are two distinct classes of light chain in skeletal muscle myosin. Reaction of the myosin thiol groups with 5,5'-dithiobis-(2-nitrobenzoic acid) (DTNB) liberates a single class of light chains of 18,000 daltons without significant loss of ATPase activity (Gazith et al., 1970; Weeds and Lowey, 1971). This light chain is termed the DTNB light chain. The other light chains cannot be removed without loss of activity and these we call the Alkali light chains. The DTNB and alkali light chains can be distinguished chemically, since the former is characterized by two thiol sequences whereas the latter contains only a single thiol group (Weeds, 1969).

The presence of light chains with no ascribed function raises the question whether the DTNB light chain is an impurity in the myosin preparations. Lowey and Steiner (1972) have shown by antibody staining that the DTNB light chain is present in the A-band of muscle. Furthermore the stoichiometry of the light chains, determined by radioisotope dilution using the thiol peptides as markers, indicates that myosin contains two moles of DTNB light chains and two moles of alkali light chains per mole (Weeds and Lowey, 1971). Although this evidence is not unequivocal, it seems unlikely that the DTNB light chains are adventitious contaminants absorbed by the myosin during extraction and purification. Thus we believe

that the myosin molecule is a hexamer comprised of two heavy chains and two pairs of light chains.

The heterogeneity of the light chains on gel electrophoresis first observed by Locker and Hagyard (1967a) raises further complications. A minimum of three electrophoretic components is obtained on polyacrylamide gels, two of which are chemically related and classified as alkali light chains on the basis of their single thiol sequence, yet differ in molecular weight as shown by electrophoresis in the presence of SDS (Weeds and Lowey, 1971). Values of 25,000 for alkali 1 and 16,000 for alkali 2 were obtained from SDS gels. Sedimentation equilibrium measurements have given molecular weights of 21,000 and 17,000 (Holt and Lowey, 1972) for these light chains, which are the same as the minimum molecular weights calculated on the basis of the amino acid compositions (Weeds and Lowey, 1971).

The sequence homology between the two alkali light chains apparent from the common thiol sequence (Weeds, 1967) appears to be very extensive on the basis of peptide mapping, since all the peptides stained for histidine, tyrosine, and arginine have similar mobilities in both proteins. Indeed it could not be ruled out that the smaller light chain might be a degradation product of the larger one, although we have been unable to demonstrate reduced yields of alkali 2 when myosin was prepared in the presence of proteolytic inhibitors such as phenyl methane sulfonyl fluoride or di-isopropyl fluorophosphate. Thus we have undertaken detailed sequence analysis of the alkali light chains to examine the extent of homology between them and to investigate the nature of the additional polypeptide sequence in the alkali 1 protein.

Cyanogen bromide cleavage of both alkali 1 and alkali 2 yielded five fragments, whose total compositions summed up to those of the whole proteins (Table 1). The amino acid compositions of four of these five fragments were identical in both light chains, suggesting that these regions of the molecules have similar sequences. The largest CNBr fragments however were very different in composition (Table 1), that from alkali 1 (CNBr1-A1) being 41 residues larger than the corresponding fragment from alkali 2 (CNBr1-A2). This 41-residue sequence is at the N-terminal end of CNBr1-A1 which is itself at the N-terminus of the light chain. With the exception of a region of 17 residues

* Present address: Eïdg. Technische Hochschule Zürich-Hönggerberg, Institut für Molekularbiologie und Biophysik, CH-8029 Zürich, Switzerland.

Table 1. Amino Acid Compositions of Alkali Light Chains and CNBr1 Fragments

	Alkali 1*	Alkali 2*	Diff.*	CNBr1-A1	CNBr1-A2	Diff.
Lys	21.0	11.6	9	13.0	4.3	9
His	2.0	1.9	—	—	—	—
Arg	4.2	4.3	—	2.1	2.0	—
CMCys	0.96	0.93	—	—	—	—
Asp	19.8	18.3	1–2	9.1	7.6	1–2
Thr	8.0	7.1	1	3.8	3.7	—
Ser	8.8	8.2	0–1	4.8	4.6	—
Glu	29.2	24.4	5	12.2†	7.8†	4–5
Pro	12.1	3.7	8–9	11.1	1.9	9
Gly	12.0	12.0	—	4.3	4.2	—
Ala	22.8	12.5	10	15.7	5.0	11
Val	10.6	10.2	—	4.9	3.8	1
Met	6.0	5.6	—	—	—	—
Ile	9.0	7.0	2	4.2	2.1	2
Leu	13.8	12.8	1	7.3	6.2	1
Tyr	3.0	3.0	—	1.0	1.0	—
Phe	8.1	8.5	—	3.3	3.1	—
Total	192	152		96	58	

* Weeds and Lowey, 1971.
† Homoserine and glutamic acid chromatograph in identical positions under these conditions.

containing only proline and alanine, the sequence of these 41 residues is given in Table 2. Thus the primary difference between the two light chains is a deletion of 4100 daltons at the N-terminal end of alkali 2. However, the composition of these 41 residues, although closely similar, is not identical to the composition difference between the two CNBr fragments. Comparison of the tryptic peptides shows that the remaining sequence of CNBr1-A1 is identical to CNBr1-A2 except for a single peptide at the N-terminus of alkali 2. The sequence of this 11-residue peptide is given in Table 2, showing that five of the first eight residues of alkali 2 differ from the equivalent region of alkali 1.

A further feature of the two sequences which should be noted is that both proteins appear to be blocked at the N-terminal end. The evidence to support this conclusion comes from our inability to identify the N-terminal residue both in the proteins and the N-terminal tryptic peptides; also from the mobilities of the N-terminal tryptic peptides on paper electrophoresis and their poor staining properties with ninhydrin. We have not identified the blocking group to date, although

the experiments of Offer (1965) would lead one to predict an acetyl group. Indeed, Offer reported the presence of N-acetylserine which could have arisen from the N-terminus of the alkali 2 light chain or alternatively as a further pronase digestion product of the pentapeptide N-acetyl-Ser-Ser-Asp-Ala-Asp (Offer, 1965). Offer's method of isolating the acetylated derivatives relied on a primary screening of acidic peptides on sulfonic acid resins, and for this reason the N-terminal pronase peptide of alkali 1 would not be detected.

The results of the sequence studies may be summarized as follows: (1) The sequence of the major part of both proteins appears to be identical, though we cannot rule out other amino acid replacements in the four smaller CNBr fragments as long as they do not alter the compositions of these fragments. (2) The molecular weights of the two proteins are 21,000 and 17,000, and the difference is accounted for by a 41-residue peptide at the N-terminal end of alkali 1. (3) Amino acid replacements occur in five of the first eight residues of alkali 2 when this sequence is compared with the equivalent region of alkali 1. Three of these

Table 2. Amino Acid Sequence of CNBr Fragment 1

A   X-Pro-Pro-Lys-Lys-Asn-Val-Lys-Ala-Ala-Ala-Lys-Lys-(Pro₈, Ala₉)-Lys-Glu-Glu-Lys-Ile-Asp-Leu-Ser-Ala-Ile-Lys-Ile-

B   Glu-Phe-Ser-Lys-Glu-Gln-Gln-Asp-Glu-Phe-Lys-

C   X-SER-PHE-SER-ALA-ASX-GLX-ILE-ALA-GLX-PHE-LYS-

D   Glu-Ala-Phe-Leu-Leu-Tyr-Asp-Arg-Thr-Gly-Asp-Ser-Lys-Ile-Thr-Leu-Ser-Gln-Val-Gly-Asp-Val-Leu-Arg-Ala-Leu-Gly-Thr-Asn-Pro-Thr-Asx-Ala-Glx-Val-Lys-Lys-Val-Leu-Gly-Asn-Pro-Ser-Asp-Glu-Gln-Met

A—Residues 1–41 of alkali 1. B—Residues 42–52 of alkali 1, the difference peptide. C—N-terminal 11 residues of alkali 2. D—Residues 53–99 of alkali 1; sequence common to both proteins.

replacements (Glu to Ser, Lys to Ala, and Gln to Ile) cannot be ascribed to single base changes, nor can the sequences be explained on the basis of a single frame shift in translating the nucleotide sequence. However, two of these three substitutions occur at an average or above average frequency in homologous proteins and only one (Gln to Ile) is very much below average (McLachlan, 1972). It is clear from these observations that the two proteins must be coded by separate genes, though the extent of sequence homology suggests a common ancestral gene.

The reason for the existence of two alkali light chains remains obscure. We have been unable to separate subfragment-1 "heads" with different light chains and it is impossible to distinguish between "heads" in a mixed population. Thus if the light chains are essential for myosin ATPase activity, the significance of the proline-rich region of alkali 1 in this respect is questionable; yet one might expect some specific function from such a remarkably unusual structure. The two light chains may reflect the asymmetry of the two "heads" of myosin, or they may arise from different isoenzymes of myosin present in the muscles. If the former hypothesis were true, then we should find an equal amount of the two light chains in myosin, and further, that other myosins might also contain related light chains of different molecular weights. Lowey and Risby (1971) estimated the stoichiometry of the light chains in rabbit myosin by densitometry of polyacrylamide gels. Their overall results were in good agreement with the isotope dilution studies mentioned earlier, but they further showed that of the total light chain material, 17% was alkali 2 and 32% alkali 1 (as measured in densitometer areas). Correcting these areas for the molecular weights of the proteins calculated from the sequence data gives a ratio of alkali 1: alkali 2 of 1.5/1, which indicates that the two light chains do not occur in equal yield. Thus it seems unlikely that there is a single population of myosin molecules which contain one mole each of the two alkali light chains. It is more probable that there are at least two populations of myosin present in the adult rabbit. Myosin from fetal rabbits contains very little of the alkali 2 protein (Dow and Stracher, 1971a; Perrie and Perry, 1970), from which it might be concluded that the alkali 2 protein is an "adult type" and the alkali 1 a "fetal type" light chain. However, the specific ATPase activities of the adult and fetal myosins are very similar (Dow and Stracher, 1971b), and the alkali 1 protein remains the predominant species in the adult rabbit.

The occurrence of two alkali light chains appears to be a general property of fast muscle myosin.

Myosins from the fast muscles of chicken (Lowey and Risby, 1971) and of sheep and cat (Weeds and Pope, 1971) contain alkali light chains of molecular weight similar to those from rabbit and also characterized by the same single thiol sequence. The presence of two alkali light chains cannot be ascribed to the mixed muscles in these preparations, since myosin prepared from a single fast wing muscle of chicken, the posterior latissimus dorsi, has light chains similar to those of chicken breast muscle myosin (Lowey and Risby, 1971). The cat myosin was also prepared from a single muscle, the flexor digitorum longus. However, it remains to be demonstrated whether the heterogeneity arises from mixed fiber types in the same muscle or occurs within a single muscle cell.

Subfragment-1 prepared by papain digestion retains the alkali light chains, but the DTNB light chains are partially lost (Weeds and Lowey, 1971). Attempts to isolate subfragment-1 species with a single class of alkali light chain have been unsuccessful, and until this is achieved it is unlikely that we will be able to understand the functional significance of these two proteins.

Cardiac Light Chains

Light chains from fast and slow muscle myosins vary both in net charge (Locker and Hagyard, 1967b) and apparent molecular weight. Using polyacrylamide gel electrophoresis in the presence of SDS, the molecular weights determined for the two cardiac light chains were 27,000 (cardiac 1) and between 18,000 and 20,000 (cardiac 2) (Lowey and Risby, 1971; Sarkar et al., 1971). Sequence similarities have been demonstrated between the alkali light chains of fast muscle myosin and the cardiac 1 light chain in that they both contain the same thiol sequence, but cardiac 1 contains in addition two new thiol sequences not previously found in myosin (Table 4) (Weeds and Pope, 1971). However, the relationship between the two cardiac light chains remains unclear. Are they related in the same manner as the alkali 1 and alkali 2 light chains, or are they chemically different, in which case they may be compared with the alkali and DTNB light chains of fast muscle myosin? Neither light chain is removed by DTNB, nor have we been able to selectively remove one of them by other chemical techniques. However, we have found recently that subfragment-1 from beef cardiac myosin prepared by brief papain digestion contains only a single light chain (cardiac 1) when examined on polyacrylamide gels in the presence of SDS (Fig. 1). Since the catalytic center activity of the subfragment-1 was similar to that of the myosin, it appears that cardiac 2 is not required for ATPase activity and may therefore be

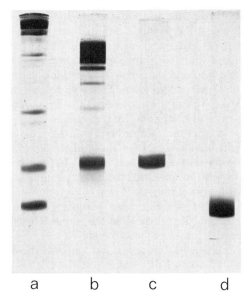

Figure 1. 10% polyacrylamide gel electrophoresis of cardiac proteins run in the presence of 0.1% SDS (Weber and Osborn, 1969). a) Cardiac myosin from bovine muscle; b) subfragment-1 from bovine cardiac myosin; c) cardiac 1 light chain; d) cardiac 2 light chain.

comparable in function to the DTNB light chain of fast muscle myosin. Nevertheless we have purified the two cardiac light chains from beef myosin for chemical characterization.

Light chains were dissociated from beef heart myosin with urea or guanidine hydrochloride as described previously and the two components separated on DEAE-cellulose (Weeds and Lowey, 1971). Figure 1 shows that the separated light chains were homogeneous. The proteins were reacted with [¹⁴C]iodoacetic acid as described previously (Weeds and Lowey, 1971), and scintillation counting indicated that cardiac 1 contained 3.0 moles of CMCys per 21,000 daltons, whereas cardiac 2 contained 0.3 moles per 18,000 daltons, suggesting some cross-contamination. Further purification of cardiac 2 was achieved by gel filtration on Sephadex G-100 in the presence of 0.1% SDS, since the purified fraction contained less than 0.16 moles CMCys per 18,000 daltons. The amino acid compositions of the two light chains are given in Table 3, confirming that they differ in their content of nearly all the amino acids.

These light chains have been further compared by peptide mapping using the methods described previously (Weeds and Lowey, 1971). Figure 2 shows the tryptic maps: peptides staining selectively for histidine, arginine, and tyrosine have been marked as well as the radioactive CMCys peptides. The numbers of unique peptides staining for these residues in cardiac 1 are in good agreement with those predicted on the basis of the composition

(Table 3), giving a minimum mol wt of 21,000 daltons, which is the same as the alkali 1 light chain and considerably lower than the value measured from the polyacrylamide gels. In the case of cardiac 2, the agreement between the composition predictions and the observed numbers of peptides is less good, since two histidine sequences should be present; nevertheless the results are consistent with a minimum mol wt of 18,000 daltons, which is similar to the measured value. The peptides maps are clearly different, from which we conclude that the two proteins are chemically unrelated. Nor is there any obvious similarity between cardiac 2 and the DTNB light chain in their peptide maps, although the compositions show certain similarities (compare with Weeds and Lowey, 1971).

These experiments confirm that cardiac myosin contains two chemical classes of light chain, one of which shows a homology of thiol sequence with the alkali light chains of fast muscle myosin and is associated with the subfragment-1 head of the molecule. The cardiac 2 light chain contains no thiol groups and is chemically different; it appears to be degraded during papain digestion since polyacrylamide gel electrophoresis in the presence of SDS shows it to be absent from subfragment-1. (Polyacrylamide gels of performic acid oxidized subfragment-1 run in the absence of SDS gave a strong band corresponding to cardiac 1 and a weak band which may indicate some residual cardiac 2.)

Lowey and Risby (1971) have determined the stoichiometry of cardiac light chains in chicken heart muscle myosin by densitometry of acrylamide gels. Assuming that bovine and chicken

Table 3. Amino Acid Compositions of Cardiac Light Chains

	Cardiac 1	Cardiac 2	Diff.
Lys	19.9	16.5	3
His	3.0	2.1	1
Arg	5.0	4.9	—
CMCys	3.0	<0.16	3
Asp	17.7	21.9	4
Thr	10.8	8.5	2
Ser	5.1	3.3	2
Glu	27.6	23.3	4
Pro	17.3	7.0	10
Gly	13.7	10.2	4
Ala	18.0	12.6	5
Val	8.8	7.9	1
Met	6.8	5.7	1
Ile	7.0	9.8	3
Leu	13.7	9.9	4
Tyr	3.0	2.0	1
Phe	9.7	13.4	3
Mol Wt	21,000	18,000	

Analyses were carried out in triplicate as described previously (Weeds and Lowey, 1971). Samples were hydrolyzed for 24 hr and 72 hr; threonine and serine values were extrapolated to zero hydrolysis time, whereas those for valine and isoleucine were extrapolated to infinite time. The results for cardiac 1 represent the mean of two different preparations; those for cardiac 2 come from a single preparation.

Table 4. Amino Acid Sequences of the Thiol Peptides from Cardiac 1

Peptide A: Leu-Met-Ala-Gly-Gln-Glu-Asp-Ser-Asn-Gly-CMCys-Ile-Asn-Tyr-Glu-Ala-Phe-Val-Lys
Peptide B: Ile-Thr-Tyr-Gly-Gln-CMCys-Gly-Asp-Val-Leu-Arg
Peptide C: Asp-Arg-Thr-Pro-Lys-CMCys-Glu-Met-Lys

Peptide A is the thiol sequence common to both the alkali light chains of fast skeletal myosin and the cardiac 1 light chain. The sequences given are for tryptic peptides, though in the case of peptide C, a slightly larger sequence has been given from knowledge of the chymotryptic sequence (Weeds and Pope, 1971).

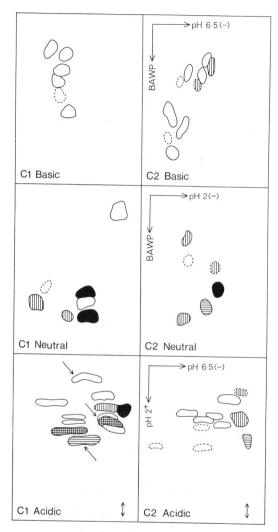

Figure 2. Two-dimensional peptide maps of tryptic digests of the two cardiac light chains (labeled with [14C]iodoacetic acid). The peptides were separated initially by electrophoresis at pH 6.5 and two-dimensional maps prepared: acidic peptides further separated by electrophoresis at pH 2; basic peptides further separated by descending chromatography in butan-1-ol-acetic acid-water-pyridine (15:3:12:10 by vol) (BAWP); the neutral band from pH 6.5 electrophoresis was submitted to further electrophoresis at pH 2, then to descending chromatography in butan-1-ol-acetic acid-water-pyridine. Solid outline, strongly staining ninhydrin peptides; dotted outline, weakly staining ninhydrin peptides; solid, peptides staining for histidine; horizontal shading, peptides staining for tyrosine; vertical shading, peptides staining for arginine. The arrows mark the radioactive CMCys peptides.

cardiac light chains have identical molecular weights, their results can be recalculated on the basis of the minimum molecular weight values estimated above, giving 1.8 moles of cardiac 1 and 1.7 moles of cardiac 2 per mole myosin. We have determined the stoichiometry of cardiac 1 using the three thiol sequences as markers in a radioisotope dilution experiment carried out as described previously (Weeds and Lowey, 1971). The three tryptic peptides (Table 4) were isolated from [14C]iodoacetate-labeled cardiac 1 and from a mixture of [14C]iodoacetate-labeled cardiac 1 and [12C]iodoacetate-labeled myosin. Comparison of the specific activities of the labeled peptides from the pure light chain and the mixture gives the molar yield of these peptides in the added [12C]-iodoacetate-labeled myosin. Assuming that the molecular weight of cardiac myosin is similar to that of skeletal myosin (470,000 daltons), the mean value for the stoichiometry of cardiac 1 determined from each of the three thiol peptides was 2.0 ± 0.3. This result taken together with the densitometry measurements of Lowey and Risby (1971) shows that cardiac myosin, like fast skeletal muscle myosin, contains four light chains in two pairs, consistent with the bipartite structure of the molecule.

Recent observations by Kendrick-Jones (1972) have shown that smooth muscle myosin from chicken gizzards also contains two classes of light chains, one of which appears to be absent in subfragment-1 and may therefore be inessential for ATPase activity. Thus it may be a common feature of mammalian myosins that they contain two classes of light chain, one of which does not seem to be required for enzymic activity, though it is conjectural from the presence of this class of light chains in such different myosins that some function must exist. The other class of light chain in mammalian myosins is essential for ATPase activity and may be responsible, at least in part, for the regulation of the contractile response (Bárány and Close, 1971).

Acknowledgment

One of us, G. F., wishes to express his thanks to the European Molecular Biology Organization for a Senior Fellowship.

References

Bárány, M. and R. I. Close. 1971. The transformation of myosin in cross-innervated rat muscles. *J. Physiol.* **213**: 455.

Dow, J. and A. Stracher. 1971a. Identification of the essential light chains of myosin. *Proc. Nat. Acad. Sci.* **68**: 1107.

———. 1971b. Changes in the properties of myosin associated with muscle development. *Biochemistry* **10**: 1316.

Dreizen, P., L. C. Gershman, P. P. Trotta, and A. Stracher. 1967. Myosin subunits and their interactions. *J. Gen. Physiol.* **50**: 85.

Dreizen, P. and L. C. Gershman. 1970. Relation of structure to function in myosin. II. Salt denaturation and recombination experiments. *Biochemistry* **9**: 1688.

Gazith, J., S. Himmelfarb, and W. F. Harrington. 1970. Studies on the subunit structure of myosin. *J. Biol. Chem.* **245**: 15.

Gershman, L. C., A. Stracher, and P. Dreizen. 1969. Subunit structure of myosin. III. A proposed model for rabbit skeletal myosin. *J. Biol. Chem.* **244**: 2726.

Holt, J. C. and S. Lowey. 1972. Light chains of myosin from chicken breast muscle. *Fed. Proc.* **31**: 866.

Kendrick-Jones, J. 1972. The subunit structure of a vertebrate smooth muscle myosin. *Proc. Roy. Soc. (London) B* In press.

Kominz, D. R., W. R. Carroll, E. N. Smith, and E. R. Mitchell. 1959. A subunit of myosin. *Arch. Biochem. Biophys.* **79**: 191.

Locker, R. H. and C. J. Hagyard. 1967a. Small subunits of myosin. *Arch. Biochem. Biophys.* **120**: 454.

———. 1967b. Variations in the small subunits of different myosins. *Arch. Biochem. Biophys.* **122**: 521.

Lowey, S. and D. Risby. 1971. Light chains from fast and slow muscle myosins. *Nature* **234**: 81.

Lowey, S. and L. A. Steiner. 1972. An immunochemical approach to the structure of myosin and the thick filament. *J. Mol. Biol.* **65**: 111.

McLachlan, A. D. 1972. Repeating sequences and gene duplication in proteins. *J. Mol. Biol.* **64**: 417.

Offer, G. W. 1965. The N-terminus of myosin. I. Studies on N-acetyl peptides from a pronase digest of myosin. *Biochim. Biophys. Acta* **111**: 191.

Perrie, W. T. and S. V. Perry. 1970. An electrophoretic study of the low molecular weight components of myosin. *Biochem. J.* **119**: 31.

Sarkar, S., F. A. Sreter, and J. Gergely. 1971. Light chains of myosins from white, red and cardiac muscles. *Proc. Nat. Acad. Sci.* **68**: 946.

Stracher, A. 1969. Evidence for the involvement of light chains in the biological functioning of myosin. *Biochem. Biophys. Res. Comm.* **35**: 519.

Tsao, T-C. 1953. Fragmentation of the myosin molecule. *Biochim. Biophys. Acta* **11**: 368.

Weber, K. and M. Osborn. 1969. The reliability of molecular weight determinations by SDS-polyacrylamide gel electrophoresis. *J. Biol. Chem.* **244**: 4406.

Weeds, A. G. 1967. Small subunits of myosin. *Biochem. J.* **105**: 25c.

———. 1969. Light chains of myosin. *Nature* **223**: 1362.

Weeds, A. G. and S. Lowey. 1971. Substructure of the myosin molecule. II. The light chains of myosin. *J. Mol. Biol.* **61**: 701.

Weeds, A. G. and B. Pope. 1971. Chemical studies on the light chains from cardiac skeletal muscle myosins. *Nature* **234**: 85.

Stoichiometry and Sequential Removal of Light Chains of Myosin

Satyapriya Sarkar

Dept. of Muscle Research, Boston Biomedical Research Institute,
and
Dept. of Neurology, Harvard Medical School, Boston, Mass. 02114

The experiments described below are designed to answer the following questions: (a) What is the stoichiometry of the light chains per mole of myosin? and (b) Is it possible to remove sequentially one light chain at a time from the myosin molecule, and, if possible, to correlate the effect of such sequential removal with the ATPase activities of myosin?

In order to determine the stoichiometry of the light chains, we have isolated the individual light chains of rabbit white myosin. These isolated pure light chains were used as standards to calibrate the densitometric scannings of SDS-polyacrylamide gel runs of highly purified myosin. Using the mol wts of 21,000; 19,000; and 17,000 obtained by amino acid analysis of the three light chains LC_1 (A1), LC_2 and LC_3 (A2), respectively, we have estimated the following stoichiometry of the light chains per mole of rabbit skeletal myosin: 2.00 moles of LC_2 or DTNB-light chain; 1.35 moles of LC_1 or alkali light chain 1 (A1), and 0.65 moles of LC_3 or alkali light chain 2 (A2) (for nomenclature and properties of light chains see Sarkar et al., 1971; Weeds, this volume).

Other workers (Weeds and Lowey, 1971; Lowey and Risby, 1971) have previously estimated that the three light chains LC_1, LC_2, and LC_3 are present in a 1:2:1 stoichiometry per mole of myosin. Since the LC_2 (or DTNB) light chain is not required for myosin ATPase activity (Gazith et al., 1970; Weeds and Lowey, 1971), it was concluded by Weeds and Lowey that each myosin head should contain an alkali light chain as an "essential" light chain to give an ATPase active site (Weeds and Lowey, 1971). (For a discussion on essential and nonessential light chains, see Weeds and Lowey, 1971.) According to our results the two alkali light chains are present in a non-equimolar stoichiometry. One of them, the LC_1 of A_1 subunit, is present in twice the molar amount of the other, LC_3 or A_2.

Figure 1. Densitometric scannings of SDS-polyacrylamide gel runs of myosin samples treated sequentially with dissociating reagents. *Panel A:* Myosin treated with 4.7 M NH_4Cl pH 7.0 in the presence of 2 mM DTT and 1 mM EDTA at 0° for 10 min (Dreizen and Gershman, 1970). The myosin was reisolated and NH_4Cl was removed from the myosin solution by dialysis as described by Dreizen and Gershman. Aliquots of the myosin were subjected to SDS-gel electrophoresis (——) Control, 45 μg myosin; (– – – –) NH_4Cl-treated myosin, 45 μg. *Panel B:* Myosin treated as described above was reisolated and treated with DTNB at 0° for 20 min according to the procedure of Weeds and Lowey (1971). After reisolation of the myosin, removal of excess DTNB, and regeneration of thiol groups, the myosin was analyzed by SDS-gel electrophoresis. (——) Control, 30 μg; (– – –) NH_4Cl- and DTNB-treated. *Panel C:* Myosin treated with DTNB as described in B was exposed at pH 11.0 for 20 min at 25° and reisolated (Weeds and Lowey, 1971). (——) Control, 30 μg; (– – –) treated at pH 11.0, 30 μg. Three μg of rabbit skeletal tropomyosin was used in each run as a marker. All gels contain 12.5% polyacrylamide. Heavy chains remained on top of the gels (not shown in the scanning).

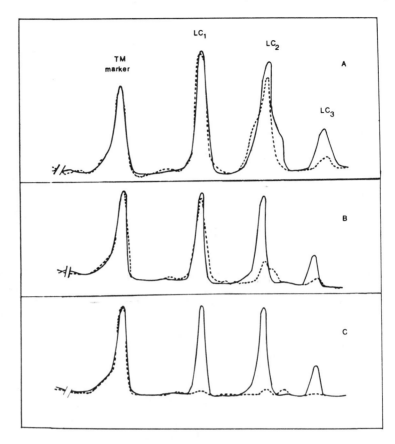

In order to achieve the sequential removal of light chains from myosin, we have treated myosin with 4.7 M NH_4Cl at pH 7.0 according to the procedure of Dreizen and Gershman (1970). The NH_4Cl-treated myosin, after reisolation and removal of NH_4Cl, was treated with 5,5'-dithiobis-(2-nitrobenzoic acid) (DTNB) according to the procedure of Weeds and Lowey (1971). Finally the myosin after reisolation, removal of excess DTNB, and regeneration of thiol groups was exposed to pH 11. The conditions used for these treatments are described in the legend to Fig. 1. Figure 1 shows the densitometer scanning of SDS-polyacrylamide gel runs of the treated myosins. By running the same amount of untreated myosin in a parallel gel run, we estimated the amount of each light chain dissociated at each step as a result of the treatments described above. As shown in a panel A, about 75% of LC_3 and 11% of LC_2 subunits were removed by NH_4Cl treatment. Sequential treatment of the reisolated myosin with DTNB showed a loss of 75% LC_2, 80% LC_3, and about 5% LC_1 from myosin (panel B). Finally treatment at pH 11 for 20 min at 25° showed a loss of almost all of the residual light chains (panel C). The conclusions derived from the gel runs of the isolated myosins are confirmed by electrophoresis of the supernatants derived at each treatment. Figure 2 shows that the

supernatants obtained from these three treatments consisted mainly of LC_3, LC_2, and LC_1 light chains, respectively.

The myosin samples obtained at each stage of the sequential treatments as described above were tested for ATPase activity. As shown in Fig. 3 the NH_4Cl-treated myosin subunit showed only an 18% loss of K^+-EDTA-activated ATPase activity. After subsequent treatment with DTNB, the myosin showed a loss of 27% ATPase activity. Finally, treatment of this myosin with alkali at pH 11 causes about 95% loss of the ATPase activity. Similar results were obtained when the Ca^{++}-activated ATPase activities of the treated myosins were determined, except that the ATPase activities were about 40% of the K^+-EDTA-activated values. These results indicate that, as a result of sequential removal of light chains, when about 80% of LC_2 and LC_3 subunits are dissociated from myosin, the resulting molecule on the average loses about 30% of the ATPase activity. Since the LC_2 or DTNB subunit is not required for ATPase activity (Weeds and Lowey, 1971; Weeds and Frank, this volume), these results suggest that the relative contribution of the two essential light chains, LC_1 and LC_3, to the ATPase activities of the myosin molecule are about 70% and 30%, respectively. These results are in agreement with

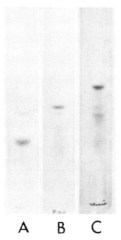

A   B   C

Figure 2. SDS-polyacrylamide gel runs of supernatants obtained from myosin samples treated sequentially with dissociating agents as described in the legend to Fig. 1. The supernatants containing the released light chains were dialyzed against a large excess of 0.05 M Tris-HCl pH 7.5, 5 mM β-mercaptoethanol, and 1 mM EDTA and were lyophilized. The lyophilized material was dissolved in 0.01 M Tris-HCl pH 8.0 and was analyzed by SDS-gel electrophoresis (Sarkar et al., 1971). A, 4 μg of supernatant light chains obtained from myosin after treatment with 4.7 M NH_4Cl as described in the legend to Fig. 1. B, Three μg of supernatant light chains obtained from myosin after treatment with DTNB as described in the legend to Fig. 1. C, Five μg of supernatant light chains obtained from myosin exposed to pH 11 at 25° for 20 min as described in the legend to Fig. 1.

the non-equimolar stoichiometry of the two essential light chains, LC_1 and LC_3, described above.

These results further suggest that in skeletal adult white (fast) muscle different populations of myosin molecules may exist, each having its own complement of light chains. Thus the light chain complement of three kinds of possible myosin molecules, which may be present in rabbit adult white muscle, are designated as follows: $LC_1(LC_2)_2LC_1$ (type I); $LC_1(LC_2)_2LC_3$ (type II);

$LC_3(LC_2)_2LC_3$ (type III). Our results suggest that the combinations containing LC_1 subunit in one or both heads represent the majority (about 70%) of the populations. It remains to be experimentally determined whether a "hybrid"-type myosin, as shown in type II in which the two heads are non-identical, really exist. The non-equimolar stoichiometry of the two essential light chains, as described above, may also result if the type I and type III molecules are present in a 70:30 ratio.

This non-equimolar stoichiometry of the two essential light chains, giving rise to different populations of myosin molecules, may arise from the independent translation of myosin subunits observed by us in an in vitro system (Sarkar and Cooke, 1970). Recently Brivio and Florini (1971) have demonstrated that there is no coordination of the in vivo synthesis of heavy and light chains. Thus it is quite likely that the individual light chains of adult white (fast) muscle myosin are non-coordinately translated, giving rise to this non-equimolar stoichiometry of the two essential light chains.

Acknowledgments

This work was supported by grants from the National Institutes of Health (AM-13238), the Massachusetts Heart Association (#1032 and 1097), and the American Heart Association (#71-915). I am grateful to Dr. John Gergely for many helpful discussions.

References

Brivio, R. P. and J. R. Florini. 1971. Independent synthesis of small and large subunits of myosin *in vivo*. *Biochem. Biophys. Res. Comm.* **44**: 628.

Dreizen, P. and L. C. Gershman. 1970. Relationship of structure to function in myosin. II. Salt denaturation and recombination experiments. *Biochemistry* **9**: 1688.

Gazith, J., S. Himmelfarb, and W. F. Harrington. 1970. Studies on the subunit structure of myosin. *J. Biol. Chem.* **245**: 15.

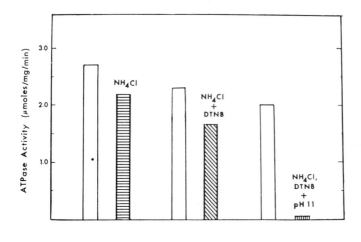

Figure 3. K+-EDTA-activated ATPase activities of myosins treated sequentially with dissociating agents as described in the legend to Fig. 1. Controls are shown in white bars, treated samples in shadowed bars. For details see the legend to Fig. 1.

Lowey, S. and D. Risby. 1971. Light chains from fast and slow muscle myosins. *Nature* **234**: 81.

Sarkar, S. and P. H. Cooke. 1970. *In vitro* synthesis of light and heavy polypeptide chains of myosin. *Biochem. Biophys. Res. Comm.* **41**: 918.

Sarkar, S., F. A. Sreter, and J. Gergely. 1971. Light chains of myosins from white, red and cardiac muscles. *Proc. Nat. Acad. Sci.* **68**: 946.

Weeds, A. G. and S. Lowey. 1971. Substructure of the myosin molecule. II. The light chains of myosin. *J. Mol. Biol.* **61**: 701.

A Phosphorylated Light Chain Component of Myosin from Skeletal Muscle

W. T. Perrie, L. B. Smillie, and S. V. Perry

Department of Biochemistry, University of Birmingham, Birmingham, England

It is now widely accepted that the light chain fraction of myosin from rabbit skeletal muscle migrates as three bands on electrophoresis in the presence of sodium dodecyl sulfate (Paterson and Strohman, 1970; Starr and Offer, 1971; Weeds and Lowey, 1971). Our determinations indicate that these bands correspond to mol wts of 15,500, 18,500, and 22,500. Upon electrophoresis in 8 M urea at pH 8.6, however, the light chain fraction can be separated into four bands (Fig. 1). Similar observations with myosin from other species indicated that this was a general phenomenon for vertebrate skeletal muscle (Perrie and Perry, 1970).

The relative amounts of the two bands of intermediate electrophoretic mobility, Ml_2 and Ml_3, (nomenclature of Perrie and Perry, 1970) varied according to the myosin extraction conditions. In general the addition of phosphate, pyrophosphate, fluoride, or ATP to the extraction medium increased the proportion of Ml_2 at the expense of Ml_3. Ml_2 was absent in myosin extracted with 0.6 M KCl to which none of these substances had been added.

A fractionation of Ml_2 and Ml_3 from Ml_1 and Ml_4 was achieved by ethanol precipitation of the whole light chain fraction at pH 8.0. The mixture was then separated into pure Ml_2 and Ml_3 by a modification of the DEAE-Sephadex chromatography method previously described by Perrie and Perry

(1970). The amino acid analyses of Ml_2 and Ml_3 were identical and in essential agreement with those of the 5,5'-dithiobis-(2-nitrobenzoic acid) (DTNB) light chain fraction (Weeds and Lowey, 1971). The peptides obtained on tryptic digestion of each component gave similar electrophoretic patterns as did those obtained after cyanogen bromide cleavage. The only gross analytical difference that could be detected between Ml_2 and Ml_3 in these preliminary studies was the presence of approximately one mole per mole of inorganic phosphate in Ml_2, with less than 10% of this amount in Ml_3. These experiments suggested that the myosin light chain component of mol wt 18,500 daltons (the DTNB light chain fraction) could exist in a phosphorylated (Ml_2) and nonphosphorylated (Ml_3) form (Perrie et al., 1972).

This relationship was confirmed by the conversion of Ml_2 to Ml_3 by incubation with the purified alkaline phosphatase of *E. coli*. It was also demonstrated that Ml_3 could be converted to Ml_2 by incubation of purified Ml_3 with a crude protein kinase preparation from skeletal muscle (32.5% ammonium sulfate precipitate, Walsh et al., 1968) in the presence of ATP, 3',5'-c-AMP, Mg^{++} and EGTA (Fig. 1). With γ-labeled [^{32}P]ATP high levels of radioactivity were found only in Ml_2 when the kinase was incubated with the complete light chain fraction of myosin.

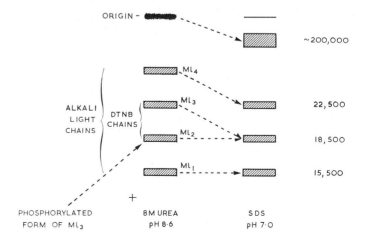

Figure 1. Scheme showing the relationship between the components of the light chain fraction of rabbit skeletal myosin observed on polyacrylamide gel electrophoresis in 8 M urea, 20 mM Tris-glycine buffer pH 8.6, and in 0.1% SDS, 85 mM Tris-borate buffer pH 7.0. Figures on the extreme right are the approximate molecular weights of the bands as judged by their electrophoretic mobility in sodium dodecyl sulfate.

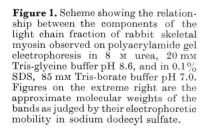

High voltage paper electrophoresis revealed the presence of one major radioactive peptide and two minor labeled peptides when ^{32}P-labeled Ml$_2$ was digested with chymotrypsin. Amino acid analysis of the major peptide after hydrolysis in N HCl at 110°C for 20 hr showed the presence of serine phosphate.

When muscle taken from an animal within one minute of death was homogenized in 6 M guanidine hydrochloride and the light chain fraction isolated by precipitation of the myofibrillar proteins with ethanol, all of the 18,500 dalton component was present in the phosphorylated form (Ml$_2$). Muscle tissue stored in ice for 2 hr before homogenization contained a significant proportion of the non-phosporylated component (Ml$_3$).

The available evidence suggests that a serine residue in the DTNB light chain component of skeletal myosin can be phosphorylated by a protein kinase system present in muscle. In resting tissue this component is fully phosphorylated but during myosin extraction is rapidly dephosphorylated by a phosphatase. As yet no role can be ascribed to the phosphorylated component. Preliminary experiments (Perrie and Perry, 1970) indicate no gross difference in the ability of the phosphorylated and nonphosphorylated forms of myosin to hydrolyze ATP. The possibility is being currently explored that the system may be involved in the regulation of myofibrillar ATPase either in addition to or in association with the troponin and the tropomyosin complex.

Acknowledgments

The work was supported in part by research grants from the Medical Research Council from the Muscular Dystrophy Associations of America, Inc., and during the tenure of a Medical Research Council of Canada Visiting Scientist award by L.B.S.

References

PATERSON, B. and R. C. STROHMAN. 1970. Myosin structure as revealed by simultaneous electrophoresis of heavy and light subunits. *Biochemistry* **10**: 4094.

PERRIE, W. T. and S. V. PERRY. 1970. An electrophoretic study of the low molecular weight components of myosin. *Biochem. J.* **119**: 31.

PERRIE, W. T., L. B. SMILLIE, and S. V. PERRY. 1972. A phosphorylated light chain component of myosin. *Biochem. J.* In press.

STARR, R. and G. OFFER. 1971. Polypeptide chains of intermediate molecular weight in myosin preparations. *FEBS Letters* **15**: 40.

WALSH, D. A., J. P. PERKINS, and E. G. KREBS. 1968. An adenosine 3',5'-monophosphate-dependent protein kinase from rabbit skeletal muscle. *J. Biol. Chem.* **243**: 3763.

WEEDS, A. G. and S. LOWEY. 1971. Substructure of the myosin molecule II. The light chains of myosin. *J. Mol. Biol.* **61**: 701.

An Immunochemical Approach to the Interaction of Light and Heavy Chains in Myosin

SUSAN LOWEY AND JOHN C. HOLT

Children's Cancer Research Foundation, Boston, Mass. 02115
and
The Rosenstiel Basic Medical Sciences Research Center, Brandeis University,
Waltham, Mass. 02154

Myosin from vertebrate skeletal muscles has two large polypeptide chains, each with a mass of about 200,000 daltons, and four smaller subunits in the range of 20,000 daltons (Lowey et al., 1969; Gershman et al., 1969; Gazith et al., 1970; Weeds and Lowey, 1971; Lowey and Risby, 1971). The heavy chains are folded into a rodlike α-helical conformation that extends over a length of about 1400 Å and terminates in two globular regions, each about 100 Å in diameter (Lowey et al., 1969). The light chains are located in the globular portion of the molecule where they are believed to be involved in determining the enzymic activity of myosin (Stracher, 1969; Driezen and Gershman, 1970).

Electrophoresis of myosin from fast skeletal muscles in the presence of sodium dodecyl sulfate (SDS) shows three low molecular weight bands on polyacrylamide gels (Fig. 1). The 25,000 and 16,000 dalton subunits are related in amino acid sequence, but the 18,000 dalton subunit is chemically different (Weeds, 1969; Weeds and Frank, this volume). The 18,000 subunit can be removed from myosin by reaction with the sulfhydryl reagent, 5,5′-dithio-bis-2-nitrobenzoic acid (DTNB) (Gazith et al., 1970; Weeds, 1969). The residual myosin retains full ATPase activity, implying that the "DTNB light chain" is not essential for hydrolytic activity. The two other light chains cannot be removed without total loss of ATPase activity. Among the rather harsh solvent conditions required for their dissociation is titration to pH 11: thus, the usage of the term "alkali light chains" to denote the 25,000 (A1) and 16,000 (A2) subunits.

The stoichiometry of the two classes of light chains from rabbit skeletal muscle myosin has been determined by the radioisotope dilution technique (Weeds and Lowey, 1971) and by densitometry of stained SDS gels (Lowey and Risby, 1971). Both methods gave two moles of DTNB light chains and a total of two moles of alkali light chains per mole myosin. Moreover, a value of four moles of light chains was consistently found for myosins from a variety of slow and fast vertebrate muscles (Lowey and Risby, 1971).

Although light and heavy chains have been separated and subsequently recombined in 4 M lithium chloride, with concomitant loss and recovery of activity (Stracher, 1969; Driezen and Gershman, 1970), the use of strong salts can readily lead to irreversible denaturation. In order to avoid introducing any chemical modification into myosin by thiol reagents or denaturing solvents, we have prepared antibodies against both classes of light chains with the purpose of using the antigen-antibody reaction to dissociate the light chains from myosin. It was found in an earlier study that specific antisera can promote dissociation of the subunits without appearing to affect the enzymic activity of myosin (Lowey and Steiner, 1972). We find that only antibodies against the DTNB light chain will dissociate the homologous antigen from the globular subfragment of myosin (HMM S-1). Antibodies against the A1 light chain show only a limited reaction with HMM S-1 and do not dissociate the alkali light chains. HMM S-1 reacted with either of the two specific antibodies shows no loss of calcium-activated ATPase activity. These preliminary data suggest that the immunological approach may provide an alternative procedure for exploring the function of the low molecular weight subunits in myosin.

Physical-Chemical Properties of Light Chains

Preparation of light chains. Before preparing antibodies against light chains, it is important to isolate these proteins in a conformational state closely resembling that in native myosin. As mentioned above, the 18,000 dalton subunit can be dissociated from myosin by treatment with a reversible thiol reagent (Gazith et al., 1970; Weeds, 1969). This procedure is relatively mild and produces a DTNB light chain only faintly contaminated by traces of the other chains (the latter can be readily removed by ion exchange chromatography) (Fig. 1c). The alkali light chains pose more of a problem since one essentially has to expose myosin to urea or guanidine or alkaline pH in order to liberate them. To minimize denaturation

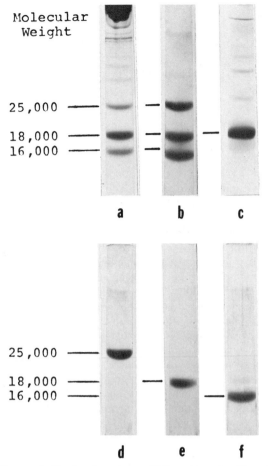

Figure 1. Electrophoresis on 10% polyacrylamide gels containing SDS. (*a*) Myosin, 80 μg; (*b*) light chains isolated from myosin by treatment with 4 M urea, 45 μg; (*c*) DTNB light chain isolated after reaction of myosin with DTNB, 20 μg. *d*, *e*, *f*: Light chains after fractionation on DEAE-cellulose: (*d*) A1, 10 μg; (*e*) DTNB light chain, 10 μg; (*f*) A2, 10 μg.

we adopted a modified version of the urea dissociation method (Gazith et al., 1970). Myosin was mixed with an equal volume of 8 M urea (in 0.05 M Tris-HCl pH 8, 5 mM dithiothreitol (DTT), 5 mM EDTA) and stirred at room temperature for 30 min. The heavy chains were precipitated with 10 vol of ice-cold water and removed by centrifugation. The supernatant (~2 liters) was applied over a period of 12–15 hours to a DEAE-cellulose column (5 × 12 cm) equilibrated with 50 mM Tris-HCl pH 8, 0.5 mM DTT. Under these conditions urea is not bound to DEAE but the light chains are. The latter can be eluted in a small volume (~70 ml) by a single step to 1 M KCl (Fig. 1b). This procedure for concentrating the light chains and separating them from urea (or DTNB in the preparation of the 18,000 subunit) has the advantage of being rapid and thereby avoiding the possible denaturation incurred by extensive dialysis and freeze-drying of large volumes (Weeds and Lowey, 1971).

Figure 2 shows a fractionation on DEAE-cellulose of an eluate obtained from the type of concentrating column described above. In this experiment all three light chains were dissociated by 4 M urea without prior removal of the DTNB light chain. The first peak to be eluted by the phosphate gradient (pH 6.0) is A1, followed by DTNB and A2 (Fig. 1d, e, f). Although this particular chromatogram gave excellent resolution of the three types of light chains, this result is unfortunately not routinely obtained. The light chains have a strong tendency to aggregate despite all precautions, and it is not uncommon to find small amounts of A1 contaminating the DTNB peak. For this reason, the DTNB light chain was usually prepared in a separate experiment by treating myosin with DTNB, rather than by urea dissociation of all three light chains.

Spectra. The ultraviolet (UV) spectra of the isolated light chains are shown in Fig. 3. The unusual appearance of the UV spectrum for A1 and A2 reflects the high content of phenylalanine relative to tyrosine (Table 1). The maxima at about 253, 259 and 265 nm are typical of the phenylalanine spectrum. The 10-fold greater extinction coefficient of tyrosine compared to phenylalanine accounts for the absence of fine structure in the 260 nm region of most proteins. Only in the rare instance of no tyrosine and tryptophan, or a very high ratio of phenylalanine to tyrosine will such fine structure be visible. Any additional aromatic groups, as in the case of the DTNB light chain, are sufficient to cancel this effect (Fig. 3 and Table 1).

The phenylalanine fine structure could also be observed in the near ultraviolet circular dichroism (CD) spectra of these proteins (Fig. 4). The 250–270 nm region was relatively insensitive to conformation, however, since the essential features of the spectrum were retained in 5 M guanidine-HCl. The far ultraviolet CD spectra of α-helical polypeptides and proteins show characteristic bands at 222 and 206 to 207 nm. From the magnitude of the trough in the 206 nm region, and from optical rotatory dispersion measurements, we estimate a value of 40–50% α-helix for all three light chains.

Molecular weight. Comparison of the electrophoretic mobilities of the light chains in SDS gels with those of proteins of known molecular weight has given values of about 25,000, 18,000, and 16,000 daltons for A1, DTNB, and A2, respectively. Minimum chemical molecular weights calculated from the amino acid composition of A1, DTNB, and A2 were 20,000, 18,500, and 16,500, respectively

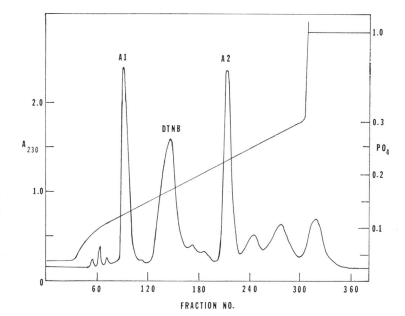

Figure 2. Fractionation of light chains. The total light chain sample, 200 mg in 50–100 ml, was applied to a column (60 × 2.5 cm) of DEAE-cellulose (Whatman DE52) equilibrated with 0.05 M potassium phosphate, pH 6.0, 0.1 mM DTT (Weeds and Lowey, 1971). The proteins were eluted by application of a linear phosphate gradient composed of 1 liter each 0.05 M and 0.35 M potassium phosphate, pH 6.0, 0.1 mM DTT. The flow rate was 35 ml/hr and the fraction size, 5 ml.

(Table 1). Although these values show good agreement for the DTNB and A2 light chains, a difference in mass of 5000 daltons exists between the two methods for the A1 light chain. A third method for determining the size of the light chains has been by equilibrium centrifugation in 5 M guanidine-HCl, 1 mM DTT. Average values obtained by the meniscus depletion procedure (Yphantis, 1964) and by the moderately high speed equilibrium method (Creeth and Holt, unpublished)

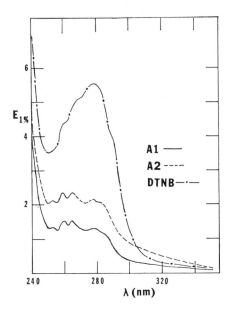

Figure 3. Ultraviolet absorption spectra of isolated light chains in 0.05 M potassium phosphate buffer, pH 6.8. Protein concentration was measured by the micro-Kjeldahl procedure, assuming a nitrogen content of 16%.

were 20,500 (A1); 19,500 (DTNB); and 17,500 (A2). It appears from these results that the A1 light chain migrates more slowly on SDS gels than is appropriate for its size. These experiments emphasize the need to determine molecular weights by a variety of methods rather than to rely on any single technique, however reliable that technique may prove to be for the majority of proteins (Weber and Osborn, 1969).

Chemical composition. As was first shown for the light chains from rabbit myosin, the A1 and A2 light chains are remarkably similar in their composition, differing primarily in the number of proline, lysine and alanine residues (Weeds and Lowey, 1971). The isolation and characterization of a fragment of 41 residues unique to A1 showed the unusual sequence of Pro_8, Ala_9 from the N-terminus of A1 (Weeds and Frank, this volume). The difference in composition between A1 and A2 light chains from chicken myosin suggests the existence of a similar sequence in chicken A1. It appears that this unusual sequence may be responsible for the anomalous migration of A1 in SDS gels. The most striking difference between the alkali light chains of chicken and rabbit lies in the presence of a single tyrosine residue in the former (Table 1). Weeds and Pope (1971) have shown that the light chains of many fast and even some slow skeletal muscles, as well as cardiac muscles, contain a thiol sequence ("Peptide A") which has a cysteine residue separated by two residues from a tyrosine residue. The single tyrosine residue found in the chicken alkali light chains is part of this conserved sequence (Weeds, pers. commun.).

The amino acid composition of the DTNB light

Table 1. Amino Acid Composition of Myosin Light Chains

	DTNB		A1		A2		A1–A2	
Lys	15.9	16.0	21.0	18.8	11.6	11.2	9	8
His	1.1	1.1	2.0	2.0	1.9	1.9	—	—
Arg	6.1	5.9	4.2	4.1	4.3	4.0	—	—
Cys	1.9	1.9	0.96	0.90	0.93	0.81	—	—
Asp	23.1	23.7	19.8	20.9	18.3	20.5	1–2	—
Thr	10.3	8.9	8.0	7.9	7.1	7.8	1	—
Ser	5.8	5.9	8.8	5.5	8.2	5.5	—	—
Glu	23.8	22.0	29.2	27.6	24.4	24.3	5	3
Pro	6.1	7.2	12.1	12.6	3.7	4.1	8	8
Gly	13.0	12.6	12.0	11.9	12.0	11.5	—	—
Ala	14.5	13.1	22.8	23.2	12.5	11.0	10	12
Val	8.9	8.8	10.6	9.8	10.2	8.8	—	1
Met	6.4	5.9	6.0	5.7	5.6	6.0	—	—
Ile	9.8	10.1	9.0	9.2	7.0	7.5	2	2
Leu	9.5	9.6	13.8	13.4	12.8	12.3	1	1
Tyr	2.0	2.0	3.0	1.0	3.0	1.0	—	—
Phe	13.0	11.8	8.1	9.9	8.5	9.8	—	—
Mol wt	19,000	18,500	21,000	20,000	17,000	16,500		

The values for rabbit myosin light chains are taken from Weeds and Lowey (1971). For chicken myosin light chains, samples were hydrolyzed for 24 or 72 hr in 6 N HCl containing 1% phenol. Cysteine was determined as cysteic acid after oxidation of the protein with performic acid. The results shown are the mean of 6 analyses including 24 hr and 72 hr hydrolysates, except that for threonine and serine 24 hr values were used, and for valine and isoleucine, 72 hr values. Tyrosine was determined only from unoxidized samples. The calculations are normalized on the basis of tyrosine residues. Preliminary spectrophotometric analysis (Beaven and Holiday, 1952) showed the absence of tryptophan in A1 and A2 and the presence of one tryptophan residue in DTNB light chain.

chain from chicken myosin looks identical to that from rabbit myosin, within the experimental error. They are probably not identical in sequence since otherwise the rabbit would be tolerant to DTNB light chain as an antigen. As discussed below, this is decidedly not the case since the DTNB light chain elicited a specific antibody titer. The extent of the sequence homology between the light chains of the two species will have to await peptide mapping and sequence studies.

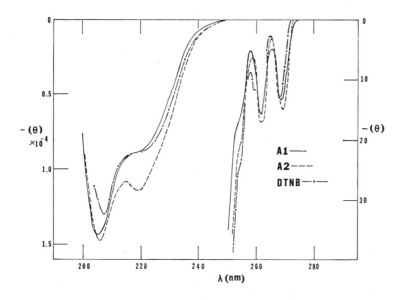

Figure 4. CD spectra of alkali 1, alkali 2 and DTNB light chains in 0.05 M potassium phosphate buffer, pH 6.8. (θ) is the mean residue ellipticity, $(\theta) = \theta_{obs} \times 100\, M/cl$ where M is mean residue weight (110), l is path length in dm, and c is concentration in g/100 ml.

Immunochemical Properties
of Light Chains

Double diffusion in agar. It was shown in an earlier study that antibodies prepared against total, unfractionated light chains react well with myosin and its proteolytic subfragment, HMM S-1 (Lowey and Steiner, 1972). Apparently, light chains prepared by alkaline dissociation of myosin (Gaetjens et al., 1968) retain sufficient native conformation to produce an antiserum half of whose antibodies are precipitable by HMM S-1 (Lowey and Steiner, 1972).

When purified A1, A2, and DTNB-light chains were reacted with this anti-light chain serum in gel diffusion, each light chain gave a single precipitin line characteristic of a pure protein (Fig. 5a). The bands for A1 and A2 fused with each other, indicative of shared antigenic determinants, and crossed with that of DTNB light chain (Horváth and Gaetjens, 1972). This gel diffusion pattern is consistent with what we know about the chemistry of the light chains of rabbit myosin: a large portion of the sequence of A1 is homologous with that of A2, and both alkali light chains are chemically distinct from the DTNB-light chain (Weeds and Lowey, 1971; Weeds and Frank, this volume). The immunochemical studies suggest that similar chemical relationships hold true for the light chains from chicken myosin.

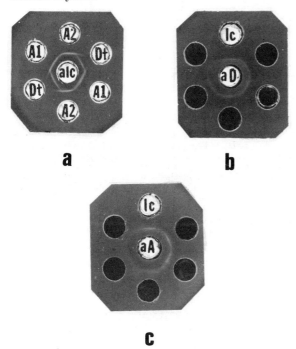

Figure 5. Immunodiffusion in 1% agar plates. (*a*) Antiserum to total light chains (alc) with A1, A2, and DTNB (Dt); (*b*) Anti-DTNB serum (aD) with a 2-fold serial dilution of total light chains (lc); (*c*) Anti-A1 serum (aA) with a 2-fold serial dilution of total light chains.

Having established the immunological purity of the isolated light chains by double diffusion in agar, we proceeded to prepare antibodies specific to the A1 and DTNB light chains (Lowey and Steiner, 1972). Figure 5(b) shows that antiserum to the DTNB light chain gave a single precipitin line when reacted with total light chains (i.e., a mixture of all three types of light chains). Similarly, antiserum to A1 reacted with total light chains gave a single band in gel diffusion (Fig. 5c). These gel diffusion patterns demonstrate the specificity of each serum for one particular type of light chain and the absence of contamination by antibodies to the heterologous light chain.

Quantitative precipitin analysis. The precipitation of alkali light chains and DTNB light chains by antisera to A1 and DTNB light chain is shown in Fig. 6 (left). The most interesting aspect of these quantitative precipitin curves is the precipitation of only 50% of the anti-A1 antibodies by A2. Since about 80% of the sequence of A1 is presumably identical to that of A2, this result implies that the conformation of A1 may be affected by the additional polypeptide chain of about 4000 daltons at the N-terminal end of the subunit. In the absence of this sequence, the A2 chain may fold in such a way as to lose antigenic determinants available in the homologous part of the A1 molecule. A small difference in conformation between the A1 and A2 light chains is suggested by their far ultraviolet CD spectra, Fig. 4.

Since the light chains were exposed to 4 M urea during their dissociation from the heavy chains, it is important to ask what effect a denaturant would have on the conformation of the light chains. The most complete unfolding of a protein can be achieved in 5 M guanidine-HCl in the presence of a reducing agent (Tanford, 1969); thus, the light chains were exposed to guanidine-HCl for a minimum of one week and then allowed to refold by dialysis against a neutral buffer. Complete recovery of the protein was achieved and, most important, complete recovery of the antigenic sites occurred (Fig. 6, filled circles). We conclude from these observations that the conformation of the isolated light chains is a very stable state; whether this state corresponds to that present in myosin will be discussed below.

Interaction of HMM S-1 with Antibodies
to Light Chains

For technical reasons it was preferable to use HMM S-1, the globular subfragment of myosin, rather than native myosin in the antibody-antigen reactions. As light chains constitute only about 15% of the weight of myosin, large amounts of

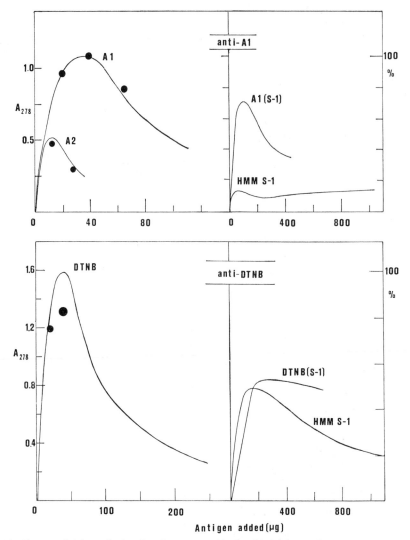

Figure 6. Quantitative precipitin analysis of antiserum specific for A1 (*above*) and DTNB light chain (*below*). Amounts of antigen indicated on the abscissa were added to 0.5 ml antiserum in 0.2 M KCl, 0.05 M Tris HCl pH 7.9, and the mixture incubated 16–30 hr at 4°C. The precipitate was collected by centrifugation for 30 min at 1300 g, washed once with 0.05 M Tris HCl pH 7.9 and twice with water, and redissolved in 0.5 ml, 0.01 M HCl. The absorbance (A_{278}) of these solutions is shown as the left ordinate; the right ordinate is a relative scale on which the maximum absorbance recorded for the appropriate immunogen is defined as 100%. The upper left-hand quadrant shows the reactivity of anti-A1 serum with its immunogen A1 and the related light chain A2. The lower left-hand quadrant shows the reactivity of anti-DTNB light chain serum with its immunogen DTNB light chain. Filled circles are values obtained for the proteins indicated after reversible denaturation with 5 M GuCl, 1 mM DTT (see text). There was no reaction between anti-DTNB and A1 or between anti-A1 and DTNB light chain, demonstrating the specificity of the antisera. The right-hand quadrants show the reactivity of HMM S-1 with the two antisera. The curves labeled A1(S-1) and DTNB(S-1) refer to total light chains isolated from HMM S-1 by treatment with 5 M GuCl, 1 mM DTT. Since the concentration of immunologically reactive species in this mixture of light chains is unknown, the position of the curves on the abscissa is arbitrary.

protein are required in the precipitin reaction and the high viscosity of concentrated solutions of myosin hinders the formation and isolation of a precipitate. HMM S-1 contains 30–40% light chains in a mass of 110,000 daltons; this factor plus a low viscosity made it a more suitable material for these immunological experiments.

HMM S-1 was prepared by papain cleavage of myosin under ionic conditions where myosin is

insoluble. The insoluble rod fragment and undigested myosin were removed by centrifugation and the water-soluble HMM S-1 in the supernatant was purified by ion exchange chromatography on DEAE-cellulose (Lowey et al., 1969). SDS gel electrophoresis of HMM S-1 isolated by this procedure shows a major band of about 90,000 daltons, flanked by two minor components, and three low molecular weight bands (Fig. 7b). The A1

Figure 7. Electrophoresis in 10% acrylamide gels containing SDS. (*a*) Myosin, 80 μg; (*b*) HMM S-1, 25 μg; (*c*) antigen-antibody precipitate from reaction of 0.5 ml anti-DTNB serum with 25 μg HMM S-1; (*d*) HMM S-1, 9 μg; (*e*) antigen-antibody precipitate from reaction of 0.5 ml anti-A1 serum with 200 μg HMM S-1. Antigen-antibody precipitates were freeze-dried from 0.01 M HCl (see legend to Fig. 6), dissolved in 8 M urea, 20 mg/ml SDS, 0.15 M 2-mercaptoethanol, boiled 2 min and applied to the gel.

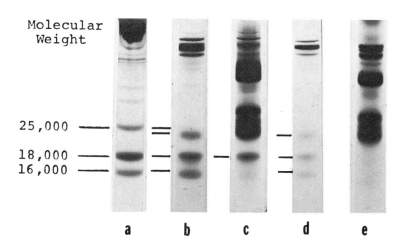

Molecular Weight

25,000 ——

18,000 ——
16,000 ——

a   b   c   d   e

light chain migrates slightly faster in HMM S-1 than in myosin; the mobility of the other two light chains is unchanged.

Anti-DTNB light chains. When HMM S-1 is added in increasing amounts to a constant volume of antiserum specific for the DTNB light chain, the precipitin curve shown in the lower right-hand quadrant of Fig. 6 is obtained. The maximum precipitate is formed when 50% of the antibodies against DTNB light chain have reacted. Evidently HMM S-1 is deficient in antigenic determinants originally present in the immunogen. This loss of antigenic sites may mean that the conformation of the bound light chains is different from that of the dissociated chains used to prepare the antisera or, alternatively, the light chains in HMM S-1 may have suffered structural changes during the proteolytic degradation of myosin. To resolve this question light chains were isolated from the same preparation of HMM S-1 as was used in the precipitin analysis. Since HMM S-1 is water soluble, light chains cannot be separated from heavy chains by the usual methods which rely on solubility differences between the chains. Instead we made use of the fact that 5 M guanidine will cause the irreversible denaturation of the heavy chains, while leaving the light chains intact (see Fig. 6, filled circles). Light chains isolated in this manner (labeled DTNB (S-1) in Fig. 6) gave a maximum precipitate with anti-DTNB serum which practically coincided with that obtained with the parent HMM S-1 molecule. These experiments show that antigenic sites were lost in the HMM S-1 molecule during proteolysis, either from a change in the conformation of the DTNB light chain or from the loss of a specific population of light chain molecules, but not from a change in conformation due to the association of light with heavy chains.

It had been reported previously that the calcium-ATPase activity of HMM S-1 is unaffected by its reaction with antibodies against the light chains (Lowey and Steiner, 1972). Consistent with this observation is the finding that anti-DTNB added in excess to antigen does not inhibit the activity of HMM S-1 (Table 2). Furthermore, SDS gel electrophoresis of the precipitate formed by this reaction shows it to contain all the DTNB light chains present in the added HMM S-1 (Fig. 7c). The amount of heavy chain in the precipitate was negligible compared to that present in HMM S-1 (Fig. 7b). (Photographic reproductions of the gels are misleading in that the high background stain due to immunoglobulins and serum proteins gives the impression of more heavy chain material than is actually present.)

Although the method described here is successful in dissociating the light chains, it has the disadvantage of simultaneously contaminating the residual HMM S-1 with a high concentration of immunoglobulins and serum proteins. Use of an immunoadsorbent would avoid this difficulty. Preliminary experiments with anti-DTNB coupled to Sepharose (Cuatrecasas and Anfinsen, 1971) indicate that it will be possible to prepare a pure HMM S-1 free of DTNB light chains. A detailed comparison of the activity of this material versus

Table 2. Effects of Antisera on Ca-ATPase Activity of HMM S-1

Antiserum (0.5 ml)	Antigen HMM S-1 (μg)	ATPase Activity at 25° (μmoles P$_i$/min/mg S-1)	
		Suspension	Supernatant
Anti-DTNB	50	1.15	1.03
Anti-A1	50	1.02	1.02
Anti-S-1	50	0	0
Normal	50	1.05	1.11

Enzyme activity was determined after antigen and antiserum had been incubated at 4° for 34 hours.

unreacted HMM S-1 may provide a clue as to the function of the 18,000 molecular weight subunit in myosin.

Anti-A1 *light chains*.   The reaction of anti-A1 with HMM S-1 proved somewhat disappointing in so far as the antibodies did not dissociate the alkali light chains from HMM S-1 (Fig. 7e). Some A1 material may be hidden by the immunoglobulin band in that region of the gel, but the complete absence of A2 implies that practically no reaction took place. Light chains from as little as 9 μg of HMM S-1 are clearly resolved on SDS gels (Fig. 7d), and the precipitin mixture leading to the precipitate in Fig. 7(e) contained as much as 200 μg of HMM S-1! The lack of reactivity of HMM S-1 with anti-A1

is further illustrated by the precipitin curves in Fig. 6 (upper right-hand quadrant). Despite the precipitation of over 75% of the anti-A1 antibodies by A1 light chains isolated from HMM S-1 (curve labeled A1 (S-1)), the subfragment precipitated less than 20% of these antibodies. The most likely interpretation of these results is that the antigenic sites of the alkali light chains are buried in HMM S-1, or that the light chains assume a different conformation when bound in HMM S-1 than in the free, dissociated state. The possibility must be considered, however, that anti-A1 and HMM S-1 do actually interact but form soluble complexes rather than the more usual highly cross-linked network leading to a precipitate. This alternative explanation will be tested by the use of

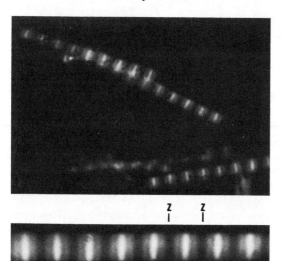

anti-A1

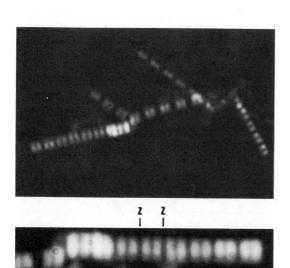

anti-DTNB

Figure 8. Fields of myofibrils stained with fluorescein-labeled immune globulin from anti-A1 light chains (*above*) and anti-DTNB light chains (*below*). Magnification: × 1875. Single myofibrils are included at a higher magnification: × 3750.

an immunoadsorbent to bind any available HMM S-1 molecules. The constant ATPase in the presence of anti-A1 antibodies is not surprising if less than 20% of the antibodies are involved in the reaction (Table 2). The small amount of HMM S-1 removed by this precipitate out of 50 μg of enzyme would be hard to detect as a significant drop in activity. If soluble complexes could be demonstrated, this measurement would assume much more importance.

Fluorescent Staining of Myofibrils

Immune globulin was prepared from antisera by ammonium sulfate precipitation followed by chromatography on DEAE-cellulose (Lowey and Steiner, 1972). Fluorescein-isothiocyanate was coupled to the immune globulin by the method of Wood et al. (1965). Myofibrils, prepared from glycerinated strips of chicken breast muscle, were incubated overnight (4°C) with an approximately equal volume of fluorescein-labeled immune globulin (2–3 mg/ml) (Pepe, 1966). Figure 8 shows typical fields of stained myofibrils examined in a standard Zeiss fluorescent microscope. It is particularly interesting to see that the pattern of fluorescence obtained with anti-A1 is almost the exact opposite of that observed with anti-DTNB light chains. While anti-A1 fluorescence was brightest in the central region of the sarcomere, with much weaker staining elsewhere in the A-band, anti-DTNB antibodies stained the A-band uniformly *except* for the central region where stain was completely excluded. We can offer no interpretation of these unique staining patterns, other than to say that they reinforce the current belief that light chains are an integral part of the myosin molecule and the thick filament. Although the existence of the alkali light chains, or "essential light chains" as they are sometimes described, is seldom disputed, the DTNB light chain still tends to be dismissed as a contaminant or not acknowledged at all. The chemical studies have clearly demonstrated that 2 moles of DTNB light chain are associated with each mole of myosin (Weeds and Lowey, 1971), and the present fluorescent studies minimize the possibility that these light chains become nonspecifically associated with myosin during its isolation. The anti-DTNB fluorescence is entirely confined to the A-band and no stain can be detected in the I-band. Moreover, the absence of any fluorescence in the I-band and the complete lack of immunological reactivity of anti-DTNB with proteins of the relaxing system (unpublished data), makes it highly unlikely that DTNB light chains are in any way related to the regulatory proteins of the thin filament.

Summary and Future Work

The major conclusions to be drawn from this study can be summarized as follows.

1. The reactivity of the DTNB light chains in HMM S-1 is very similar to that of the free light chains, implying that no significant change in conformation occurs in binding to the heavy chains. Moreover, this class of light chains can be dissociated from the heavy chains by reaction with the homologous antibodies. By use of a suitable immunoadsorbent, it should be possible to obtain a preparation of HMM S-1 entirely devoid of DTNB light chains. This material, by virtue of not having been exposed to thiol reagents or denaturants, would provide an excellent test system for studying the effect, if any, the removal of DTNB light chains has on the interactions of HMM S-1 with nucleotides, actin, and the regulatory proteins.

2. Antibodies against alkali light chains react very poorly with HMM S-1. This can be interpreted to mean that the conformation of bound alkali light chains is very different from that of the dissociated chains, or that the antigenic sites of the alkali light chains are buried within the HMM S-1 molecule. If the light chains merely had a few antigenic sites hidden by the presence of the heavy chains, one might expect the remaining available sites to react with the antibodies, thereby leading to the precipitation of the whole HMM S-1 molecule or the light chains. The absence of a sizable precipitate suggests the formation of soluble complexes or little interaction of any kind between HMM S-1 and antibodies directed against A1 light chains. The preparation of an immunoadsorbent consisting of anti-A1 or anti-A2 antibodies should help to clarify questions of this kind related to the conformation and function of the alkali light chains in HMM S-1.

3. The extent of the reaction between A2 and antibodies to A1 was smaller than might have been predicted on the basis of sequence studies on the alkali light chains from the rabbit (Weeds and Frank, this volume). One possible explanation is a difference in conformation between the A1 and A2 light chain. An alternative explanation would be that a disproportionately high fraction of the total antibodies in the A1 serum were directed against the proline-rich fragment unique to the A1 light chain and absent in A2. These hypotheses will be tested by preparing an antiserum to A2. A poor reaction of A1 light chain with anti-A2 would strengthen the argument for a conformational difference between the two alkali light chains.

4. We have seen that anti-A1 will stain all the myofibrils in a field despite the weak affinity displayed towards HMM S-1. One important

application of the fluorescent antibody technique would be to show how A1 and A2 light chains are distributed within a given population of myofibrils. This might be accomplished by preparing fluorescent antibodies specific for the proline-rich N-terminal sequence found in A1 but missing in A2. In order that this experiment be feasible, two important conditions have to be fulfilled: first, the 4000 dalton fragment must retain enough of the native conformation to produce antibodies which will be recognized by myosin; second, that region in the myosin molecule containing the N-terminus of A1 must be available for staining in the myofibril. If these conditions are fulfilled, one of two situations may be observed: All the myofibrils may fluoresce, in which case both A1 and A2 light chains are present in a given myofibril, probably as two populations of myosin molecules. If, however, only a fraction of the myofibrils fluoresce, a microheterogeneity in myofibril type will have been demonstrated, with the important corollary that each myofibril must then contain a homogeneous population of myosin molecules with one type of alkali light chain.

Acknowledgments

We thank Ms. Angela Holt for careful technical assistance, and Dr. Hans Meienhofer and Ms. Kienyin Lee for invaluable aid with amino acid analyses.

This work was supported by Public Health Service grant AM-04762, Public Health Service Research Career Program Award K3-AM-10630, National Science Foundation grant GB-8616, and a grant from the Muscular Dystrophy Associations of America, Inc. J.C.H. is Carolyn H. Rankin Fellow of the Massachusetts Heart Association, Inc.

References

BEAVEN, G. H. and E. R. HOLIDAY. 1952. Ultraviolet absorption spectra of proteins and amino acids. *Adv. Protein Chem.* **7**: 319.

CUATRECASAS, P. and C. B. ANFINSEN. 1971. Affinity chromatography. *Annu. Rev. Biochem.* **40**: 259.

DREIZEN, P. and L. C. GERSHMAN. 1970. Relationship of structure to function in myosin. I. Subunit dissociation in concentrated salt solutions. *Biochemistry* **9**: 1677.

GAETJENS, E., K. BÁRÁNY, G. BAILIN, H. OPPENHEIMER, and M. BÁRÁNY. 1968. Studies on the low molecular weight protein components in rabbit skeletal myosin. *Arch. Biochem. Biophys.* **123**: 82.

GAZITH, J., S. HIMMELFARB, and W. F. HARRINGTON. 1970. Studies on the subunit structure of myosin. *J. Biol. Chem.* **245**: 15.

GERSHMAN, L. C., A. STRACHER, and P. DREIZEN. 1969. Subunt structure of myosin. III. A proposed model for rabbit skeletal myosin. *J. Biol. Chem.* **244**: 2726.

HORVÁTH, B. Z. and E. GAETJENS. 1972. Immunochemical studies of the light chains from skeletal muscle myosin. *Biochem. Biophys. Acta* **263**: 779.

LOWEY, S. and D. RISBY. 1971. Light chains from fast and slow muscle myosins. *Nature* **234**: 81.

LOWEY, S. and L. A. STEINER. 1972. An immunochemical approach to the structure of myosin and the thick filament. *J. Mol. Biol.* **65**: 111.

LOWEY, S., H. S. SLAYTER, A. G. WEEDS, and H. BAKER. 1969. Substructure of the myosin molecule. I. Subfragments of myosin by enzymic degradation. *J. Mol. Biol.* **42**: 1.

PEPE, F. A. 1966. Some aspects of the structural organization of the myofibril as revealed by antibody-staining methods. *J. Cell Biol.* **28**: 505.

STRACHER, A. 1969. Evidence for the involvement of light chains in the biological functioning of myosin. *Biochem. Biophys. Res. Comm.* **35**: 519.

TANFORD, C. 1969. Protein denaturation. *Adv. Protein Chem.* **23**: 121.

WEBER, K. and M. OSBORN. 1969. The reliability of molecular weight determinations by dodecyl sulfatepolyacrylamide gel electrophoresis. *J. Biol. Chem.* **244**: 4406.

WEEDS, A. G. 1969. Light chains of myosin. *Nature* **223**: 1362.

WEEDS, A. G. and S. LOWEY. 1971. Substructure of the myosin molecule. II. The light chains of myosin. *J. Mol. Biol.* **61**: 701.

WEEDS, A. G. and B. POPE. 1971. Chemical studies on light chains from cardiac and skeletal muscle myosins. *Nature* **234**: 85.

WOOD, B. T., S. H. THOMPSON, and G. GOLDSTEIN. 1965. Fluorescent antibody staining. III. Preparation of fluorescein-isothiocyanate-labeled antibodies. *J. Immunol.* **95**: 225.

YPHANTIS, D. A. 1964. Equilibrium ultracentrifugation of dilute solutions. *Biochemistry* **3**: 297.

Studies on the Role of Light and Heavy Chains in Myosin Adenosine Triphosphatase

Paul Dreizen and Dennis H. Richards

Department of Medicine and Program in Biophysics, State University of New York, Brooklyn, New York 11203

The release of light and heavy components from myosin by urea, alkali, and heat denaturation was early reported (Tsao, 1953; Locker, 1956; Kominz et al., 1959). After long controversy, there now seems to be general agreement that myosin (470,000 mol wt) comprises an axial core of two heavy polypeptide chains (205,000 mol wt) that terminate in a globular region containing approximately three light chains of average mol wt about 20,000 (Gershman et al., 1966, 1969; Dreizen et al., 1967; Locker and Hagyard, 1967a, b; Lowey et al., 1969; Gazith et al., 1970; Weeds and Lowey, 1971). The globular head of myosin appears to contain symmetric halves of subfragment-1 (Slayter and Lowey, 1967; Trotta et al., 1968; Lowey et al., 1969), each of which contains one light chain and a remnant of one heavy chain (Trotta et al., 1968).

The light chains are heterogeneous with respect to net electrophoretic charge (Gershman et al., 1966; Locker and Hagyard, 1967a, b; Gershman and Dreizen, 1970, 1971; Perrie and Perry, 1970) and apparent molecular weight (Locker and Hagyard, 1967a, b; Weber and Osborn, 1969; Low et al., 1971; Sarkar et al., 1971; Starr and Offer, 1971; Weeds and Lowey, 1971; Dreizen and Kim, 1971b). The total number of light chains in myosin has been disputed in that ultracentrifugal data indicate close to three light chains in rabbit skeletal myosin (Gershman et al., 1966, 1969), whereas the use of other methods for analysis or of preparations of myosin from other vertebrate muscles yields estimates of two to four light chains per myosin molecule (Gaetjens et al., 1968; Frederiksen and Holtzer, 1968; Weeds and Lowey, 1971; Dreizen and Kim, 1971b).

The functional role of light and heavy polypeptide chains in the ATPase activity of myosin has been explored in studies on the denaturation of myosin under conditions like heat, alkaline pH, and concentrated salt solutions (Gershman et al., 1968, 1969; Dreizen et al., 1967; Dreizen and Gershman, 1970), and on the properties of reconstituted myosin or its component subunits following subunit fractionation under denaturing conditions (Gershman et al., 1969; Stracher, 1969; Dreizen and Gershman, 1970). In general those denaturing conditions which result in complete dissociation of light chains from myosin also lead to irreversible inactivation of myosin ATPase. For example, myosin ATPase is irreversibly denatured on alkaline treatment above pH 10, over the same range of pH where light chains are dissociated from myosin (Gershman et al., 1966, 1969; Dreizen et al., 1967). A close relationship between subunit dissociation and irreversible loss of ATPase has also been found in a series of experiments on myosin in concentrated salt solutions at pH 7 and 4°C (Gershman et al., 1968; Dreizen and Gershman, 1970). Treatment of myosin in those salt solutions (e.g., LiCl and KSCN) which fully dissociate light chains from the intact heavy chain core is accompanied by rapid irreversible loss of ATPase at or slightly below the transition salt concentration for light chain dissociation. More direct evidence for a dependence of myosin ATPase on subunit composition has been obtained by experiments involving dissociation of myosin in 4 M LiCl, subunit fractionation by LiCl-citrate salting out, and reconstitution of subunits in different proportion (Dreizen and Gershman, 1970) or hybridization of subunits (Dow and Stracher, 1971; Kim and Mommaerts, 1971).

There is also evidence that at least some of the light component is not essential for myosin ATPase, as indicated by selective fractionation of light chains from myosin by NH_4Cl (Gershman and Dreizen, 1970, 1971; Dreizen and Gershman, 1970) and DTNB (Gazith et al., 1970; Weeds, 1969; Weeds and Lowey, 1971), without significant change in ATPase activity of the residual myosin. The studies are somewhat ambiguous in that ATPase activity must be measured following reconstitution of the residual myosin in nondenaturing solvents. Moreover there is some confusion in that the NH_4Cl-dissociated light component contains the most anionic light chain (Gershman and Dreizen, 1970, 1971), whereas the DTNB-dissociated light chain seems to be of intermediate charge (Weeds, 1969; Weeds and Lowey, 1971).

The present paper describes work in three major areas, as follows: (1) studies on salt and heat denaturation of myosin, focusing on selective subunit dissociation and a stabilizing effect of nucleotide on myosin ATPase and subunit interactions; (2) evidence for interaction of Mg^{++} and Ca^{++} with specific light chains; and (3) studies on subunit composition and hybridization of myosin from red and white skeletal muscles.

Methods

Myosin was prepared from rabbit skeletal muscle, as described previously (Dreizen et al., 1966). Longissimus dorsi was used for white muscle; soleus, semitendinosus, biceps medial head, and intertransversarius were used for red muscles; and several thigh muscles, including gluteus medius, vastus lateralis, biceps, and adductor magnus, were used for intermediate muscles. This grouping of muscles follows that of Locker and Hagyard (1968) and is based on the proportion of red and white fibers in each muscle, ranging from approximately 90% white fibers in longissimus dorsi, to a somewhat lower (but still predominant) proportion of white fibers in "intermediate" muscles, to predominantly red fibers in "red" muscles.

Light chains were fractionated from myosin by the following procedures: (1) alkali dilution (Gershman and Dreizen, 1970); (2) ammonium chloride-ammonium sulfate salting out (Gershman and Dreizen, 1970), with modifications as noted; (3) heat dilution, in which myosin at 10–20 mg/ml was incubated at 40°C for 1 hr, in 0.4 M KCl, 0.005 M $NaHCO_3$ (or 0.05 M Tris-maleate), 1 mM dithiothreitol (DTT) at pH 7, with added nucleotides, chain phosphates, and bivalent cations, as noted. The heat-denatured protein was diluted tenfold with water and centrifuged at 10,000 rpm for 10 min. The supernatant fractions from alkali dilution and heat dilution were lyophilized and stored at −30°C until use. Supernatants from NH_4Cl-$(NH_4)_2SO_4$ fractionation were dialyzed against water and then lyophilized, or concentrated by dialysis against polyvinylpyrrolidone in some of the early experiments.

In studies involving LiCl and NH_4Cl denaturation of myosin ATPase, the protein was incubated under denaturing conditions for a specified interval, after which 2–3 ml aliquots were removed for dialysis against 0.5 M KCl, 0.05 M Tris, 1 mM DTT at pH 7.6, with repeated changes of dialysate over two days. In studies involving heat denaturation of myosin ATPase, the protein was incubated in 0.5 M KCl, 0.05 M Tris, 1 mM DTT at pH 7.6, with added nucleotides and cations, as noted; and aliquots were removed at specified times for immediate assay of ATPase.

Hybridization experiments on white, intermediate, and red muscle myosin are based on LiCl-citrate fractionation (Gershman and Dreizen, 1970;

Dreizen and Gershman, 1970), using a denaturing solution of 8 M LiCl, 0.1 M Tris-maleate (or none), 2 mM DTT at pH 7 (solution A); and a salting out suspension (B) that was freshly prepared by addition of solid potassium citrate and water to a part of solution A, so as to exceed saturation (as indicated by crystal formation) at a concentration of 4 M LiCl at 4°C. Samples of white, intermediate, and red muscle myosin in 0.4 M KCl, 0.005 M $NaHCO_3$ pH 7, were diluted with equivalent volumes of 8 M LiCl (solution A); and an appropriate volume of well-mixed citrate-LiCl suspension (B) was promptly added through a wide-bore pipette to the LiCl-treated myosin until salting out of myosin at approximately 60% saturation of citrate-LiCl. The samples were centrifuged at 10,000 rpm for 10 min; decanted into supernatant and precipitate fractions; and the fractions of supernatant, precipitate, and supernatant + precipitate of reconstituted myosin or hybridized myosin were immediately dialyzed against repeated changes of 0.4 M KCl, 0.005 M $NaHCO_3$, 1 mM DTT pH 7, over two days. The samples were then centrifuged at 5,000 rpm for 10 min to remove insoluble aggregates and used for ATPase assay. The precipitate fractions were also used for sedimentation equilibrium experiments at pH 7 and at pH 11 (after overnight dialysis against 0.4 M KCl, 0.1 M Na_2CO_3 pH 11.2). The initial fractionation procedures required a team of two, and usually three, carefully prepared participants; and all procedures were done according to a precisely arranged schedule in a cold room at 4°C, with total time 20–30 min between onset of denaturation and onset of dialysis.

Procedures used for sedimentation velocity, sedimentation equilibrium, and cellulose acetate electrophoresis are described elsewhere (Gershman et al., 1969; Gershman and Dreizen, 1970). Electrophoresis on SDS-polyacrylamide gel was done by the method of Weber and Osborn (1969).

ATPase assays are based routinely on production of inorganic phosphate during five minutes reaction of protein with 5 mM $CaCl_2$ (or EDTA), 5 mM ATP, 0.5 M KCl, 0.05 M Tris, 1 mM DTT at pH 7.6, at 25°C (or other temperature as specified). Duplicate samples and a blank were used for each assay, and inorganic phosphate was determined on TCA-supernatants, using a Technicon Autoanalyzer (Dreizen and Kim, 1971a).

Relationship between ATPase Inactivation and Selective Subunit Dissociation

LiCl denaturation. As shown in previous ultracentrifugal and electrophoretic experiments (Gershman and Dreizen, 1970), the light component of myosin is fully dissociated from the intact heavy

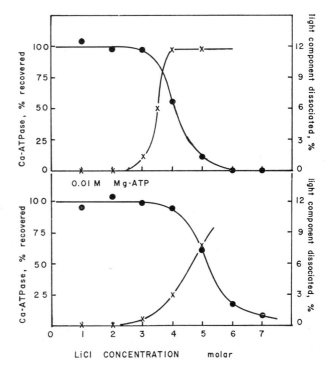

Figure 1. Denaturation of rabbit skeletal myosin in 1–7 M LiCl, 0.2 M KCl, 1 mM DTT pH 7.0, 4°C. Dissociation of light component ($\times$) as percentage total protein was determined from sedimentation velocity experiments initiated promptly after mixture of LiCl and myosin solutions. Ca-ATPase (●) was determined on samples treated in LiCl solution for 5 min prior to dialysis against 0.4 M KCl, 0.005 M NaHCO$_3$, 1.0 mM DTT pH 7 over 2 days; data are reported as percentage of original Ca-ATPase prior to LiCl treatment. 0.01 M MgATP was present during LiCl denaturation in the experiments shown in *lower* graph.

chain core in concentrated LiCl solutions; and there is complete, or nearly complete, reassociation following dialysis of LiCl-dissociated myosin against 0.4 M KCl. The transition for light chain dissociation occurs at LiCl concentrations between 3 M and 4 M at pH 7 and 4°C (Fig. 1). Subunit dissociation of myosin in LiCl solutions is accompanied by irreversible denaturation of myosin ATPase (Dreizen and Gershman, 1970). For example, after 5 min treatment of myosin in LiCl solutions prior to dialysis against 0 4 M KCl, there is almost complete recovery of Ca-ATPase after exposure to LiCl at concentration below 3 M, but negligible recovery of Ca-ATPase after exposure to LiCl at concentration above 5 M (Fig. 1). The transitions for light chain dissociation and irreversible ATPase denaturation overlap at about 3.5 M LiCl (Fig. 1).

The LiCl denaturation profile of myosin is significantly affected by MgATP in that only 3% light component is dissociated from myosin in 4 M LiCl-0.01 M MgATP, and only 8% light component is dissociated from myosin in 5 M LiCl-0.01 M MgATP (Fig. 1). After 5 min treatment of myosin in LiCl solutions containing 0.01 M MgATP, there is nearly 60% recovery of Ca-ATPase after exposure to 5 M LiCl and about 15% recovery of Ca-ATPase after exposure to 6 M LiCl (Fig. 1). Thus the transitions for subunit dissociation and irreversible denaturation of Ca-ATPase are increased from 3.5 M LiCl (in the absence of MgATP) to 5.0 M LiCl in the presence of 0.01 M MgATP (Fig. 1).

The effect of MgATP in stabilizing myosin ATPase against LiCl denaturation is clearly shown in experiments on the time-dependence of irreversible ATPase inactivation during 2 hours storage of myosin in 4 M LiCl (Fig. 2). In the absence of MgATP, the reconstituted myosin exhibits a sharp decrease in recovery of Ca-ATPase and EDTA-ATPase after prolonged treatment of myosin in 4 M LiCl, and less than 10% of total ATPase activity is recovered after 40 min storage of myosin in 4 M LiCl, with comparable rates of denaturation for EDTA-ATPase and Ca-ATPase (Fig. 2). However in identical experiments done with MgATP (0.1 mM and 1 mM) present during LiCl denaturation, there is protection of Ca-ATPase and EDTA-ATPase, with relatively slow denaturation of ATPase during 2 hours LiCl treatment (Fig. 2).

In order to determine whether nucleotide stabilization of myosin to LiCl denaturation involves the entire ATP molecule, or some part of ATP, a series of sedimentation velocity experiments were done on myosin in LiCl solutions at 1 M to 5 M LiCl, in the presence of different nucleotides and inorganic chain phosphates (Table 1). In the presence of 0.01 M Mg tripolyphosphate (PPP$_i$), the transition for subunit dissociation is increased to 5.0 M LiCl. However, 0.01 M adenosine, 0.01 M adenine, and 0.01 M MgCl$_2$ have no effect on the subunit dissociation of myosin in LiCl solutions. There is some stabilization of the myosin structure to LiCl

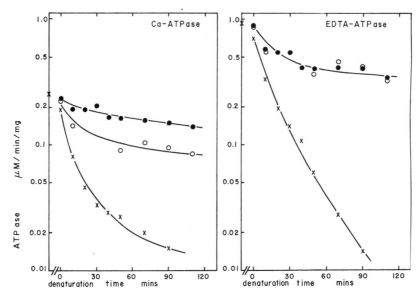

Figure 2. Irreversible denaturation of Ca-ATPase (*left*) and EDTA-ATPase (*right*) during storage of myosin in 4.0 M LiCl, 0.25 M KCl, 0.025 M Tris, 1 mM DTT pH 7.6 at 4°C. Aliquots were withdrawn at times noted and dialyzed against 0.5 M KCl, 0.05 M Tris, 1 mM DTT pH 7.6 over 2 days. ATPase activities are normalized to initial protein concentration; control values are shown along ordinate. LiCl treatment was done in the presence of 1.0 mM MgATP (●), 0.1 mM MgATP (○), and no MgATP (×). ATPase values plotted on a logarithmic scale.

denaturation in the presence of MgADP and Mg pyrophosphate (PP$_i$), but no significant stabilization in the presence of AMP or inorganic phosphate. Inosine nucleotides have approximately the same effect as adenosine nucleotides on the stability of myosin in LiCl solutions.

Under conditions in which nucleotide and chain phosphate binding is eliminated, that is, by substitution of 1 mM EDTA for 0.01 M MgCl$_2$, myosin undergoes subunit dissociation at 3.5 M LiCl, without evidence of stabilization by nucleotides or chain phosphates (Table 1). It would thus appear that stabilization of myosin to LiCl denaturation depends on the interaction of myosin with the terminal phosphate groups of ATP. The tetrahedrally arranged σ-bonding orbitals of phos-

Table 1. Molar LiCl Concentration for Dissociation of Myosin into Heavy Chain Core and Light Chains

	Phosphate	Di-phosphate	Tri-phosphate
0.01 M MgCl$_2$			
—	3.5	3.5	4.0	5.0
Adenosine	3.5	2.5	4.0	5.0
Inosine	3.5		4.0	4.5
0.001 M EDTA			
—	3.5	3.5	3.5	3.5
Adenosine	3.5		3.5	3.5

From sedimentation velocity of myosin at 5–10 mg/ml in LiCl solutions, 0.2 M KCl, 1 mM DTT, with specified chain phosphates and nucleotides at 0.01 M, pH 7, 4°C.

phate result in a nonlinear sequence of —O—P—O— groups along the phosphate chain, so that lengthening or shortening of the chain not only alters the net electrostatic charge of the nucleotide, but also changes the possible interactions between peptide groups and negatively charged oxygens or positively charged axial groups of the phosphate chain. Hence the interaction of nucleotide with myosin may be considered a kind of lock-and-key arrangement that is closely linked to integrity of the subunit structure of myosin. The linkage may derive from direct bridging of the phosphate chain between light and heavy subunits of myosin, or, alternatively, an allosteric effect of nucleotide binding on subunit interactions of myosin. In conjunction with previous evidence that ATPase activity and ADP binding depends on the presence of light and heavy polypeptide chains (Dreizen and Gershman, 1970), we would tentatively favor a direct bridging hypothesis.

NH$_4$Cl denaturation. Treatment of myosin in 4.7 M NH$_4$Cl at 4°C is accompanied by dissociation of light component, ranging from 2–4% during brief NH$_4$Cl treatment to 6–7% during NH$_4$Cl treatment for 24 hr (Gershman and Dreizen, 1970). The dissociated light component may be isolated, at least in part, in the supernatant fraction obtained during salting out with 50% saturated (NH$_4$)$_2$SO$_4$. There appears to be selective dissociation of light chains in that the supernatant fraction obtained

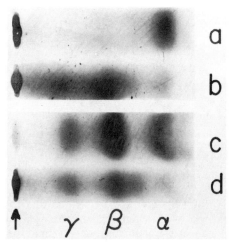

Figure 3. Cellulose acetate electrophoresis showing selective NH₄Cl dissociation of α chain from myosin (rabbit white skeletal muscle) during storage in 4.7 M NH₄Cl, 0.1 M KCl, 1 mM DTT pH 7 at 20°C and subsequent salting out with 50% saturated (NH₄)₂SO₄. *a*, Light component in (NH₄)₂SO₄ supernatant fraction after 15 min NH₄Cl treatment; *b*, light alkali component isolated from (NH₄)₂SO₄ precipitate fraction after 15 min NH₄Cl treatment; *c*, light component in (NH₄)₂SO₄ supernatant fraction after 24 hr NH₄Cl treatment; *d*, light alkali component isolated from (NH₄)₂SO₄ precipitate fraction after 24 hr NH₄Cl treatment. Total protein recovered in each sample was diluted to equivalent volume. Arrow indicates origin; anode to right in this and subsequent experiments.

after brief NH₄Cl treatment of myosin and (NH₄)₂SO₄ salting out contains predominantly fast component on cellulose acetate electrophoresis (Gershman and Dreizen, 1970, 1971). The (NH₄)₂SO₄ precipitate fraction contains the residual myosin, which on further alkali dissociation yields most of the remaining fast component as well as the other electrophoretic bands of the myosin light component. On occasion, notably in stored preparations of myosin and during NH₄Cl—(NH₄)₂SO₄ fractionation at temperatures above 4°C, there is almost complete removal of the fast electrophoretic component from myosin. In one

such experiment (Fig. 3), a sample of myosin was stored at 4°C for two weeks prior to NH₄Cl—(NH₄)₂SO₄ fractionation at 20°C. After 15 min NH₄Cl treatment the supernatant fraction (Fig. 3a) contains the fast electrophoretic component, termed the α band, and some aggregated material at the origin. The residual precipitate was subjected to fractionation by the alkali-dilution method (Gershman et al., 1966, 1969) and found to contain the two slower bands (β and γ) and negligible α band (Fig. 3b). After 24 hr storage in 4.7 M NH₄Cl at 20°C, the NH₄Cl—(NH₄)₂SO₄ supernatant fraction contains α, β, and γ bands (Fig. 3c); and the residual precipitate fraction still contains a small proportion of β and γ bands (Fig. 3d).

Although brief treatment of myosin in 4.7 M NH₄Cl is not accompanied by significant loss of ATPase, prolonged NH₄Cl treatment results in appreciable denaturation of ATPase, with loss of 40–60% of ATPase activity during 24 hr storage in 4.7 M NH₄Cl (Dreizen and Gershman, 1970). NH₄Cl denaturation of myosin ATPase is first order, with $k_{denat} = 0.16 \times 10^{-4}$ sec⁻¹ (Fig. 4). As in the case of LiCl denaturation of myosin, MgATP stabilizes myosin to NH₄Cl denaturation, and there is negligible denaturation of Ca-ATPase and EDTA-ATPase during 24 hr storage of myosin in 0.01 M MgATP-4.7 M NH₄Cl at pH 7 and 4°C. In addition there is little, if any, dissociation of light component from myosin during brief NH₄Cl treatment in the presence of 0.01 M MgATP, as shown by the virtual absence of distinct electrophoretic bands in the supernatant obtained from subsequent (NH₄)₂SO₄ fractionation of myosin.

The initial studies on NH₄Cl denaturation of myosin showed selective dissociation of the fast (α) band and minimal loss of Ca-ATPase, and it was inferred that the NH₄Cl-dissociated light component is not essential for Ca-ATPase of myosin (Dreizen and Gershman, 1970). However in subsequent experiments in which appreciable

Figure 4. Effect of MgATP on irreversible ATPase denaturation during storage of myosin (rabbit white skeletal muscle) in 4.7 M NH₄Cl, 0.05 M Tris, 1 mM EDTA, 1.0 mM DTT pH 7.5 at 4°C. Aliquots were withdrawn at times noted and dialyzed over 2 days against 0.5 M KCl, 0.05 M Tris, 1 mM DTT pH 7.6. ATPase values are normalized to original protein concentration; control values are shown along ordinate. NH₄Cl treatment was done in the presence of 10 mM MgATP (●, ▲), 1 mM MgATP (○, △), and no MgATP (×, +), for EDTA-ATPase and Ca-ATPase, respectively. ATPase values plotted on a logarithmic scale.

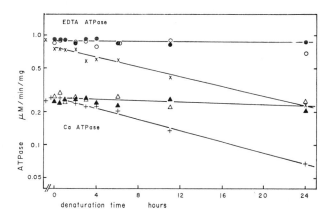

amounts of α chain are separated from myosin (Fig. 3, for example), there is concomitant decrease in the Ca-ATPase activity of the recovered protein; and it would appear somewhat uncertain from the more recent experiments whether the NH_4Cl dissociated light chain could truly be considered nonessential for Ca-ATPase. Quantitative study of subunit dissociation and ATPase inactivation is complicated by the need to employ a salting out procedure for subunit fractionation and remove the myosin from the $NH_4Cl—(NH_4)_2SO_4$ reagents prior to ATPase assay. In any event the question of essential subunits is examined further in experiments based on heat denaturation of myosin.

Heat denaturation-selective dissociation.

Heat treatment of myosin results in the dissociation of low molecular weight components (Locker, 1956; Yasui et al., 1960) that have been identified with the light chains of myosin (Gershman et al., 1968). Sedimentation velocity and sedimentation equilibrium experiments indicate extensive and almost complete dissociation of light chains from the heavy chain core of myosin during 24 hr incubation at 40°C (Gershman et al., 1969). However during 1–2 hr denaturation at 35–40°C, subunit dissociation is less than complete; and electrophoretic experiments demonstrate selective dissociation of α and γ chains from myosin. In a representative series of experiments (Fig. 5), fractionation of myosin by the alkali-dilution method (Gershman et al., 1966, 1969) yields α, β, and γ chains, with very little γ component in this preparation of myosin. After incubation of myosin

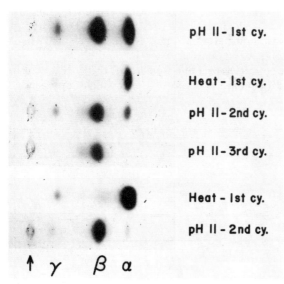

Figure 5. Cellulose acetate electrophoresis demonstrating selective dissociation of light chains during heat denaturation of myosin (rabbit white skeletal muscle) in 0.5 M KCl, 0.005 M NaHCO₃, 1 mM DTT pH 6.5.

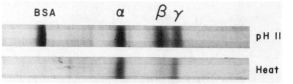

Figure 6. Electrophoresis in SDS-9% polyacrylamide gel on light component dissociated from myosin (rabbit white skeletal muscle) at pH 11 and during heat treatment (40°C, 1 hr). Fractions labeled α, β, and γ have apparent mol wts of 25,000, 17,000, and 16,000, respectively, and correspond with fast, middle, and slow bands, respectively, on cellulose acetate electrophoresis. Band labeled BSA represents bovine serum albumin.

at 40°C for 1 hr, followed by tenfold dilution with water, the water-soluble supernatant fraction contains mostly α chain and some γ chain; patterns from two independent fractionations are shown (heat, 1st cycle). The residual water-insoluble precipitate of heat-treated myosin was subjected to a cycle of alkali-dilution fractionation and yielded supernatant with predominantly β chain and small amounts of α and γ chains (pH 11, 2nd cycle). The residual precipitate of heat- and alkali-treated myosin was subjected to another cycle of alkali-dilution fractionation and yielded supernatant with β chains and little, if any, α or γ chains (pH 11, 3rd cycle).

Electrophoresis on SDS-polyacrylamide gel (Fig. 6) shows the characteristic pattern of three bands, with apparent mol wt 25,000, 17,000, and 16,000, that is obtained for the light component released from myosin by alkali-dilution. After incubation of myosin at 40°C for 1 hr, the water-soluble supernatant contains a predominant band of 25,000 apparent mol wt and a smaller band of 16,000 apparent mol wt. Thus the 25,000 mol wt band on SDS-gel electrophoresis may be identified with the α band on cellulose acetate electrophoresis; the 17,000 mol wt band on SDS-gel electrophoresis may be identified with the β component on cellulose acetate electrophoresis; and the 16,000 mol wt band on SDS-gel electrophoresis may be identified with the γ component on cellulose acetate electrophoresis.

The dissociation of light chains during heat treatment of myosin is significantly modified by chain phosphates or nucleotides. In a typical series of experiments (Fig. 7), a small amount of light component, presumably dissociated from myosin during storage at 4°C, is recovered in the water-soluble supernatant obtained immediately prior to heat treatment of myosin. The residual precipitated myosin was redissolved in 0.5 M KCl, 1 mM DTT, 0.05 M Tris-maleate pH 6.5, and one of the following: 10^{-6} M MgATP (to simulate trace amounts of nucleotide bound to myosin), 10 mM NaPPP_i, 10 mM CaPPP_i, 10 mM MgPPP_i, and

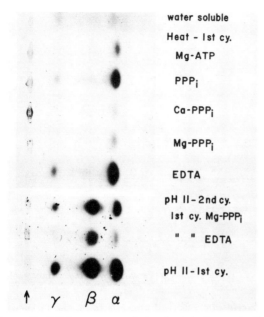

water soluble

Heat - Ist cy.

Mg-ATP

PPP$_i$

Ca-PPP$_i$

Mg-PPP$_i$

EDTA

pH II - 2nd cy.

Ist cy. Mg-PPP$_i$

"   " EDTA

pH II - Ist cy.

↑ γ β α

Figure 7. Cellulose acetate electrophoresis demonstrating stabilization of myosin structure by MgPPP$_i$ and CaPPP$_i$ during heat denaturation of one-week-old myosin (rabbit white skeletal muscle) in 0.4 M KCl, 0.005 M NaHCO₃ pH 6.5. Protein was diluted 10/1 with water, yielding some water-soluble material. The residual myosin was diluted in 0.4 M KCl, 1 mM DTT, in the presence of 10⁻⁶ M MgATP, 10 mM NaPPP$_i$, 10 mM CaPPP$_i$, 10 mM MgPPP$_i$ or 1 mM EDTA at pH 6.5. Samples were incubated simultaneously at 40°C for 1 hr, diluted 10/1 with water, yielding supernatant (*heat-1st cy.*) and precipitate fractions. The water-insoluble precipitates were subjected to alkali-dilution fractionation, yielding supernatant (*pH 11–2nd cy.*). Control 1st cycle pH 11-dilution fractionation of the same myosin preparation is shown. All samples were lyophilized, dialyzed against 0.4 M KCl, 0.005 M NaHCO₃, 1 mM D'TT pH 6.5 and diluted to approximately equivalent volumes; the loading concentration is doubled for the pH 11-1st cycle sample.

1 mM EDTA. In the samples denatured in the presence of NaPPP$_i$ and EDTA, there was dissociation of considerable α chains and some γ chains, but in the samples containing MgATP, CaPPP$_i$, and MgPPP$_i$, there was dissociation of only small amounts of α chain. These differences are confirmed by the light chain composition of the residual heat-treated myosin. The water-insoluble precipitate fractions of heat-treated myosin were subjected to a cycle of alkali-dilution fractionation and yielded supernatants containing mostly β chains in the case of prior heat denaturation in the presence of EDTA, but considerable α, β, and γ chains in the case of prior heat denaturation in the presence of 10 mM MgPPP$_i$. Other experiments demonstrate that CaCl₂ and MgCl₂ alone have no significant effect on dissociation of subunits during heat denaturation of myosin. Thus, as with LiCl and NH₄Cl denaturation of myosin, the myosin structure is stabilized to heat denaturation by ATP or PPP$_i$ in the presence of Mg⁺⁺ or Ca⁺⁺; in

particular the stabilization appears to involve interaction of α and γ chains with the residual myosin.

Heat denaturation-ATPase. Heat treatment of myosin also results in irreversible inactivation of ATPase (Ouellet et al., 1952; Yasui et al., 1960). For example on incubation of myosin at 40°C in the absence of MgATP, EDTA-ATPase diminishes over 30 min to about 10% of original activity (Fig. 8). Activation enthalpy (55 kcal/mole) and activation entropy (100 e.u.) are comparable for ATPase inactivation and light chain dissociation (Gershman et al., 1969), suggesting that the two processes may be related to one another or to some common denaturing effect.

The effect of MgATP in stabilizing myosin structure is paralleled by stabilization of myosin ATPase to heat denaturation. Figure 8 shows data on the time-dependence of denaturation during incubation of myosin at 34°C and 40°C, in the presence of MgATP at concentrations from 10⁻⁶ M to 2 mM. EDTA-ATPase is protected against heat denaturation, at least for 2 hr, in the presence of MgATP at concentrations of 0.1 mM or greater. Assays of Ca-ATPase on the same samples of heat-treated myosin show identical time-dependence of denaturation, with protection by MgATP. There is also an inhibition of EDTA-ATPase (and Ca-ATPase) with increasing concentration of MgATP (Fig. 8), presumably due to inhibition of myosin ATPase by Mg⁺⁺. The effect of MgATP on heat denaturation of myosin is analyzed further in plots of the initial reaction velocity, V_i, and the rate constant for denaturation, $k_{denat} = d \log V/dt$, against the molar ratio of MgATP to myosin, expressed on a logarithmic scale. Analysis of the data from Fig. 8 shows a sharp transition in k_{denat} from values of 2×10^{-4} sec⁻¹ (at 34°C) and 15×10^{-4} sec⁻¹ (at 40°C) to zero at molar ratios of MgATP to myosin in excess of 100 (Fig. 9).

A similar experiment on the heat denaturation of myosin ATPase in the presence of MgCl₂, at concentrations up to 1 mM, indicates inhibition of initial reaction velocity but no effect on the rate constant for irreversible denaturation of EDTA-ATPase (Fig. 9) or Ca-ATPase. There is comparable dependence of V_i on the molar concentration of MgCl₂, irrespective of the presence of ATP, confirming that inhibition of V_i is related to the direct action of Mg⁺⁺ on myosin ATPase and occurs independently of the stabilization of myosin ATPase to heat denaturation by Mg⁺⁺ and nucleotide.

MgADP exerts effects roughly similar to those of MgATP on initial reaction velocity and denaturation rate during heat treatment of myosin (Fig. 10),

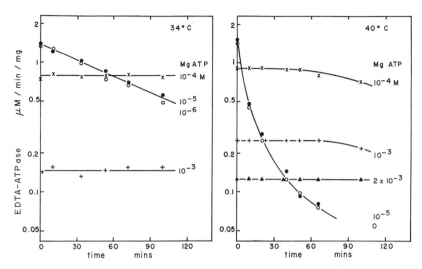

Figure 8. Effect of MgATP on heat denaturation of myosin ATPase. Samples of myosin (rabbit white skeletal muscle) in 0.5 M KCl, 0.05 M Tris, 1 mM DTT pH 7.6 were incubated at 34°C (*left*) and 40°C (*right*), in the presence of MgATP at concentrations from 10^{-6} M to 2 mM, as noted on graphs. Aliquots were removed at specified times and promptly diluted with reaction mixture (5 mM EDTA-ATP, 0.5 M KCl, 0.05 M Tris, 1 mM DTT pH 7.6) for assay of EDTA-ATPase at the same temperature employed for denaturation. ATPase values are normalized to original protein concentration and are plotted on a logarithmic scale against the duration of heat denaturation prior to ATPase assay.

although complete stabilization of myosin ATPase to heat denaturation is not achieved until concentrations of MgADP of 1 mM or greater. In contrast during heat treatment of myosin in the presence of ADP but not $MgCl_2$, there is no protection against denaturation, at least for molar ratios of ADP to myosin approaching 1000 (Fig. 10). In still another experiment, myosin-1 mM ADP was incubated at 34°C in the presence of $MgCl_2$ at concentrations up to 1 mM; there is a progressive decrease in the rate constant for denaturation as the concentration of $MgCl_2$ is increased, with protection of myosin ATPase at 0.1 M $MgCl_2$ (Fig. 10). The effect of ATP alone on the heat denaturation of myosin was not examined, due to K^+ activation of myosin ATPase

during incubation of myosin in 0.5 M KCl at temperatures above 30°C.

The present studies demonstrate that Mg^{++} and ATP (or ADP), but not cation or nucleotide alone, stabilize EDTA-ATPase and Ca-ATPase against heat denaturation at 34°C and 40°C, and similar findings are obtained during heat denaturation of myosin in the presence of Mg^{++} and inorganic chain phosphates. The evidence for protection of myosin ATPase against heat denaturation is of interest in conjunction with the observation that heat treatment of myosin (in the absence of nucleotide or chain phosphate) results in selective dissociation of α and γ chains from myosin, whereas identical heat treatment of myosin in the

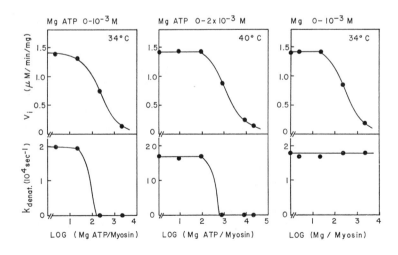

Figure 9. Effects of MgATP and $MgCl_2$ on initial reaction velocity, V_i in μM P_i/min/mg, and denaturation rate, k_{denat}, during heat denaturation of myosin at 34°C and 40°C. Values are plotted against logarithm (base 10) of the molar ratio of MgATP and $MgCl_2$ to myosin. *Left* and *middle*, values of V_i and k_{denat}, as determined from data shown in Fig. 6, at concentrations of MgATP from 0 to 2 mM; myosin at 0.1 mg/ml. *Right*, myosin at 0.2 mg/ml was incubated in 0.5 M KCl, 0.05 M Tris, 1 mM DTT pH 7.6 in the presence of 0, 10^{-6}, 10^{-5}, 10^{-4}, 10^{-3} M $MgCl_2$ at 34°C.

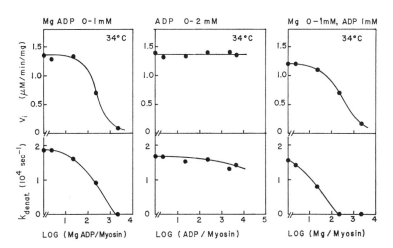

Figure 10. Effects of MgADP and ADP on initial reaction velocity, V_i, and denaturation rate, k_{denat}, during heat denaturation of myosin at 34°C. Values are plotted against logarithm (base 10) of the molar ratio of ligand to myosin. *Left*, myosin at 0.2 mg/ml in the presence of 0, 10^{-6}, 10^{-5}, 10^{-4}, 10^{-3} M MgADP. *Middle*, myosin at 0.2 mg/ml in the presence of 0, 10^{-6}, 10^{-5}, 10^{-4}, 10^{-3} M, and 2 mM ADP. *Right*, myosin at 0.1 mg/ml, 1 mM ADP, in the presence of 0, 10^{-6}, 10^{-5}, 10^{-4}, and 10^{-3} M MgCl$_2$.

presence of MgATP, MgADP, or MgPPP$_i$ is accompanied by little, if any, dissociation of α and γ chains. The evidence would suggest that binding of MgATP (among other nucleotides and chain phosphates) to myosin is closely linked with interaction between α-γ light chains and the main core of myosin.

One other feature of this work seems of general interest. One might reasonably expect the myosin molecule to remain structurally intact and functionally active at 37°C in vivo, and in this respect the evidence that myosin undergoes irreversible denaturation of ATPase and dissociates into light and heavy chains during incubation at 37°C is disconcerting. However the present evidence for nucleotide stabilization of myosin to heat denaturation would seem to answer this dilemma, in that sufficient Mg^{++} and ATP (or ADP) is available within the myofibril to saturate myosin binding sites and thus stabilize the subunit structure and ATPase site of myosin against heat denaturation at in vivo temperatures.

Interaction of Light Chains with Ca^{++} and Mg^{++}

In the presence of Mg^{++} approximately 1.5 moles of ATP, ADP, and PP$_i$ are bound per mole of myosin, with equilibrium constant about 10^6 (Kiely and Martonosi, 1968, 1969; Schliselfeld and Bárány, 1968; Nauss et al., 1969; Lowey and Luck, 1969). Studies on LiCl-citrate-fractionated subunits of myosin indicate that purified light chains and heavy chains alone do not bind MgADP, at least at molar concentrations at which MgADP is bound to native myosin or LiCl-dissociated and reconstituted myosin (Dreizen and Gershman, 1970). However, this evidence does not exclude specific interactions of Mg^{++} and nucleotide with light or heavy chains, or both, with binding energies less

than those obtained for interaction of Mg-nucleotide with native myosin or reconstituted myosin. These considerations and the opposite effects of Mg^{++} and Ca^{++} on myosin ATPase prompted studies on the possible interaction of Mg^{++} and Ca^{++} with the myosin light chains. The initial work has been done by electrophoretic analysis, rather than equilibrium dialysis, for several reasons. It was anticipated that cationic interactions with the light chains might be weak and hence difficult to analyze by equilibrium dialysis, a procedure that could be effectively used to study protein-ligand interactions at equilibrium constants above 10^4. Also, complete fractionation of light chains requires the use of two or more denaturing cycles by different methods, and one or more procedures might possibly destroy or even create weak binding sites on the light chains.

Figure 11 shows the overall changes in electrophoretic behavior of the light chains on addition of CaCl$_2$ to the electrophoretic buffer system. In the presence of 1 mM–10 mM CaCl$_2$, there is a striking decrease in the resolution of α, β, and γ chains and increase in average mobility of the light chains. The changes in mobility have a fairly complex basis, including increased conductance, electrophoretic and electrokinetic effects that are associated with movement of solvent counter to the direction of protein migration, and modification of the ionic atmosphere of the protein, with decrease in effective charge of the protein (see Cann, 1970). Thus the electrophoretic changes in Fig. 11 cannot be interpreted merely in terms of protein-ligand interaction.

Accordingly experiments were carried out in which concentrations of CaCl$_2$, MgCl$_2$, and NaCl were varied at constant increment of anion concentration or ionic strength; the concentration of electrophoretic buffer (sodium barbital) was varied

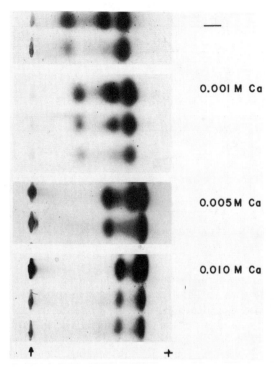

Figure 11. Cellulose acetate electrophoresis of light chains from rabbit white skeletal myosin in 0.75 M Na barbital, 1 mM DTT pH 8.6 with additional CaCl₂ at concentrations specified. Protein samples were obtained by alkali-dilution (top pattern of group) and heat (40°C, 1 hr)-dilution (bottom pattern of group).

so as to maintain approximately the same average mobility. In experiments in which the concentration of added $CaCl_2$ (or $MgCl_2$) and NaCl were varied, keeping molar anion concentration constant (1 mM, 5 mM, or 10 mM), there was an increase in mobility of the β and γ chains and a slight decrease in mobility of the α chains on replacement of NaCl by $CaCl_2$ or $MgCl_2$. Although these experiments would suggest an interaction of α chain with Ca⁺⁺ (or Mg⁺⁺), the replacement of NaCl with $CaCl_2$ (or $MgCl_2$) at constant anion concentration is accompanied by slight increase of ionic strength, thereby complicating interpretation of the changes in mobility.

Figure 12 shows electrophoretic patterns of the light chains from experiments in which different proportions of NaCl and $CaCl_2$ (or $MgCl_2$) were added to the electrophoretic solvent, at constant ionic strength increment of 0.015. In the presence of 15 mM NaCl (top pattern), there are characteristic α, β, and γ bands. In the presence of 1.5 mM $CaCl_2$ + 10.5 mM NaCl, there is no change in mobility of the β or γ chain, but the mobility of the α chains is clearly decreased. An identical change in mobility of the α chains is evident in the presence of 1.5 mM $MgCl_2$ + 10.5 mM NaCl, indicating that the α chain undergoes some kind of selective

interaction with Ca⁺⁺ and Mg⁺⁺ at levels of 1.5 mM or less. Presumably the solvated α chain has accumulated more positive charge (Mg⁺⁺ or Ca⁺⁺), or less likely although thermodynamically indistinguishable, the solvated α chain excludes more negative charges on addition of $CaCl_2$ or $MgCl_2$. On further increase of $CaCl_2$ or $MgCl_2$ concentration to 5 mM, there is no significant change in mobility of α or β chains, but some decrease in mobility of the γ chain. Identical changes in electrophoretic mobility consequent to variation in concentration of $MgCl_2$ and $CaCl_2$ were obtained on heat-dissociated light component (α and γ chains) and 3rd cycle alkali-dissociated light component (predominantly β chain), indicating that the changes in electrophoretic mobility occur independently of possible interactions among the different light chains.

With respect to possible differences between interactions of Mg⁺⁺ and Ca⁺⁺ with the light chains, the average data from a number of experiments would suggest that $CaCl_2$ has somewhat greater effect in retarding the mobility of the α chain, whereas $MgCl_2$ has somewhat greater effect in retarding the mobility of the γ chain; however the observed differences in these experiments are minor. In addition a small band with mobility intermediate between β and γ bands appeared consistently in electrophoretic experiments done in the presence of 5 mM $MgCl_2$ (Fig. 12); the minor band was not seen in electrophoretic experiments on the same protein samples in the presence of 5 mM $CaCl_2$.

Similar electrophoretic experiments were carried out at other ionic strength increments in order to determine the minimal cation concentration required for changes in mobility of the light chains.

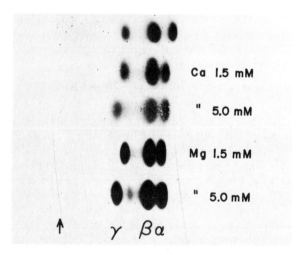

Figure 12. Cellulose acetate electrophoresis of myosin light chains (1st cycle alkali-dilution) in 0.75 M Na barbital, 1 mM DTT pH 8.6 with ionic strength increment of 0.015 due to (*from top down*) 15 mM NaCl; 1.5 mM $CaCl_2$ + 10.5 mM NaCl; 5.0 mM $CaCl_2$; 1.5 mM $MgCl_2$ + 10.5 mM NaCl; 5.0 mM $MgCl_2$.

The data would indicate that the α chain undergoes selective interaction with Ca^{++} and Mg^{++} at 0.5–1.0 mM concentration, and the γ chain undergoes selective interaction with Ca^{++} and Mg^{++} at 3–5 mM concentration.

Electrophoretic patterns are also shown for experiments done at ionic strength increments of 0.030 (Fig. 13) and 0.045 (Fig. 14). Although α, β, and γ bands are more closely spaced at the higher ionic strengths, reflecting general solvent effects on electrophoretic mobility, there is marked decrease in mobility of the γ chain and some decrease in mobility of the α chain on replacement of NaCl by MgCl$_2$ or CaCl$_2$. Comparable changes in mobility were obtained on protein fractions with either predominantly β chains (Fig. 13) and predominantly α-γ chains, confirming that the differences in mobility reflect specific interactions of cation with α and γ chains. There is little difference in the effect of Mg^{++} or Ca^{++} on the electrophoretic patterns obtained at ionic strength increments of 0.030 and 0.045, suggesting that the cationic interacting sites on α and γ chains are nearly saturated at the lower salt concentrations.

Values of the apparent mobility of α, β, and γ chains under control conditions (0.75 M Na barbital) would be consistent with the presence of two more net acidic groups in the β chain than the γ chain, and three more net acidic groups in the α chain than the γ chain. The cationic interacting sites are presumably acidic or imidazole residues on the light chains, and one might question whether the observed differences in mobility represent nonspecific interactions of Ca^{++} and Mg^{++} with polypeptide chains of different net charge. However this explanation seems unlikely in that the γ chain reacts almost as strongly with Mg^{++} and Ca^{++} as

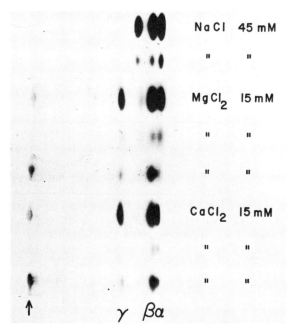

Figure 14. Cellulose acetate electrophoresis of myosin light chains in 0.45 M Na barbital, 1 mM DTT pH 8.6 with ionic strength increment of 0.045 due to 45 mM NaCl, 15 mM MgCl$_2$, or 15 mM CaCl$_2$. Same fractions as in Fig. 13; 1st cycle alkali-dilution, except for 5th and 8th patterns from the top, which are 3rd cycle alkali-dilution.

the α chain, despite the occurrence of at least three fewer acidic groups in the γ chain than the α chain. Moreover the net change in mobility of α and γ chains in the presence of Mg^{++} and Ca^{++} would be consistent with interaction of one bivalent cation per light chain; and the presence of distinct electrophoretic bands at Ca^{++} and Mg^{++} concentrations of 10 mM and 15 mM would suggest a single cationic interacting site per light chain, rather than several very weakly interacting sites. Assuming that each light chain contains a single Mg^{++}-Ca^{++} interacting site, the apparent coefficients for protein-cation interaction may be crudely approximated as 2×10^3 for the α chain and 3×10^2 for the γ chain.

A crucial question is whether the Ca^{++}-Mg^{++} interacting sites on the light chains correspond with the Mg^{++}-interacting site involved in binding of nucleotide to myosin. H. Katzeff has been exploring this problem in our laboratory by means of ultracentrifugal studies using ultraviolet absorption methods; and the preliminary evidence would indicate interaction of MgATP with the α chain.

Hybridization Experiments on Red, White, and Intermediate Muscle Myosin

In general ATPase activity is irreversibly lost under conditions in which light chains are fully dissociated from myosin. However during denaturation in concentrated salt solutions at neutral pH

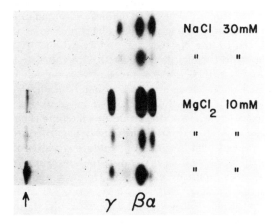

Figure 13. Cellulose acetate electrophoresis of myosin light chains in 0.60 M Na barbital, 1 mM DTT pH 8.6 with ionic strength increment of 0.30, due to 30 mM NaCl or 10 mM MgCl$_2$. All fractions are 1st cycle alkali-dilution, except for 2nd and 5th patterns from the top, which are 3rd cycle alkali-dilution.

and 4°C, there is a close overlap in the molar salt concentrations at which light chains are dissociated and ATPase is irreversibly inactivated, so that myosin may be reconstituted with appreciable recovery of ATPase activity after brief salt treatment (Dreizen and Gershman, 1970). This phenomenon permits direct investigation of the effect of changes in subunit composition on myosin ATPase, and studies have been reported involving dissociation of rabbit skeletal myosin by 4 M LiCl, subunit fractionation by LiCl-potassium citrate salting out, and subunit recombination in different proportions (Dreizen and Gershman, 1970). As there is no significant change in the amount or distribution of light component released during prolonged LiCl treatment, it would appear that subunit dissociation in 4 M LiCl occurs almost immediately, that is, within a time scale of several minutes. This would imply that time-dependent irreversible inactivation of ATPase in 4 M LiCl involves denaturation of dissociated subunits, rather than slow dissociation of subunits from myosin.

Although recombination experiments are simple in principle, they are difficult in practice, largely because of two major problems, namely, irreversible loss of myosin ATPase in 4 M LiCl (see Fig. 2 for example), and a fairly crude salting out procedure, which does not completely fractionate light and heavy chains (see Table 3 for example). Consequently LiCl-citrate fractionation of myosin requires rigid control of experimental procedures throughout, with care to complete the procedure expeditiously, denature all samples under absolutely identical conditions, and analyze the subunit composition of fractions obtained in each experiment. In experiments performed as carefully as possible with these considerations in mind, there was quantitative dependence of Ca-ATPase on the proportion of light component in reconstituted myosin, and we have interpreted this evidence to indicate an interaction of some kind between light and heavy chains in myosin ATPase (Dreizen and Gershman, 1970).

The same procedure might be sensibly used to explore the role of light and heavy chains in the regulation of myosin ATPase, and we have approached this matter through hybridization studies on preparations of myosin from white, intermediate, and red rabbit skeletal muscles. ATPase activity is two- to threefold greater in myosin isolated from white skeletal muscle than in myosin isolated from red skeletal muscle (Bárány et al., 1965; Sreter et al., 1966; Locker and Hagyard, 1968). There are evident differences in the polypeptide chains of red and white skeletal myosin, as shown by the greater susceptibility of white muscle

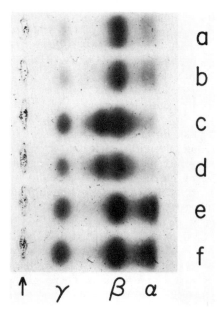

Figure 15. Cellulose acetate electrophoresis of light chains (1st cycle alkali-dilution) from preparations of myosin isolated from red (*a, b*), intermediate (*c, d*), and white (*e, f*) rabbit skeletal muscle.

myosin to tryptic digestion (Bárány et al., 1965), and differences in the electrophoretic patterns of light chains isolated from different muscles (Locker and Hagyard, 1967c, 1968; Samaha et al., 1970; Perrie and Perry, 1970; Gershman and Dreizen, 1971; Sarkar et al., 1971; Lowey and Risby, 1971). Although some workers have surmised that light chain heterogeneity assumes distinct patterns characteristic of red and white muscle, our own findings would support the original contention of Locker and Hagyard (1967c, 1968) that the electrophoretic patterns of light chains differ within the broad categories of red and white muscle myosin. For example, Fig. 15 shows electrophoretic patterns on preparations of skeletal muscle myosin isolated from a New Zealand white rabbit. Myosin from white skeletal muscle exhibits the usual α, β, and γ bands, whereas myosin from red skeletal muscle contains α, β, and γ bands in different proportions from those observed in white muscle myosin. Myosin isolated from intermediate skeletal muscle contains an additional electrophoretic component between β and γ bands, suggesting that intermediate muscle myosin cannot be considered as simply a mixture of myosins from red and white skeletal muscle. Studies on a series of myosin preparations from several strains of rabbits indicate obvious variation in the relative proportions of electrophoretic bands in red and intermediate muscle myosin; but even in preparations of myosin from a single white muscle (longissimus dorsi), the proportion of α to γ chains varies

Table 2. Sedimentation Equilibrium Experiments on Rabbit Skeletal Myosin

Prep.	Rotor Speed (rpm)	White Muscle			Intermediate Muscle			Red Muscle		
		Light Component		Heavy Component mol wt	Light Component		Heavy Component mol wt	Light Component		Heavy Component mol wt
		mol wt	%		mol wt	%		mol wt	%	
I	44,000	20,300	11.7		19,300	11.3				
	13,000			424,000			413,000			
II	40,000	20,200	14.5		21,300	11.9		18,600	13.3	
	13,000			421,000			421,000			
II	36,000	21,300	11.2		19,400	12.3		21,100	14.4	
	44,000	21,500	12.0		20,500	11.6		20,100	14.1	
	13,000									429,000
III	36,000	24,600	12.2		19,900	14.5		20,100	13.9	
	13,000			416,000			416,000			421,000
III	36,000	21,000	14.7		18,700	13.3		21,700	13.4	
	44,000	18,000	12.3		18,100	13.8		21,500	11.8	
IV	36,000	18,800	11.7		20,600	14.8				
Average		20,700	12.5	420,000	19,700	12.9	417,000	20,500	13.5	425,000

Conditions: 0.1 M Na_2CO_3, 0.4 M KCl, pH 11.

significantly in different preparations of myosin, in a somewhat reciprocal relationship (see Table 5).

Ultracentrifugal studies were done on myosin isolated from red, white, and intermediate skeletal myosin, as done previously on myosin from mixed rabbit skeletal muscle (Gershman et al., 1966, 1969). The relevant aspect of this work involves sedimentation equilibrium experiments on alkaline-dissociated myosin in 0.4 M KCl, 0.1 M Na_2CO_3 pH 11.2 (Table 2). The data indicate mol wt values about 420,000 for the intact heavy chain core of white, red, and intermediate skeletal myosin. The light component comprises 12.5–13.5% of the total protein, with mol wt approximately 20,000 in white, red, and intermediate muscle myosin. The stoichiometric composition of light and heavy components is essentially the same in preparations of myosin from red, white, and intermediate rabbit muscles. Thus, despite marked differences in electrophoretic patterns among the three kinds of myosin, the electrophoretic heterogeneity appears to represent variation in the distribution of specific light chains, the total number of light chains per myosin molecule being constant.

Data on LiCl-citrate fractionation of myosin from white, intermediate, and red rabbit skeletal muscle are summarized in Table 3. The supernatant fraction from LiCl-citrate fractionation contains light component predominantly, and the precipitate fraction contains heavy chains and 4–6% light component, most of which appears to be associated with the heavy chain core on dialysis to 0.4 M KCl, 0.005 M $NaHCO_3$ pH 7. The residual light component in the precipitate fraction represents about 30–50% of the total light component of native myosin, and there are no significant differences in the proportion of residual light chains in the precipitate fractions of white, intermediate, and red muscle myosins. Control values of Ca-ATPase are approximately 0.7 μM/min/mg for white and intermediate muscle myosin, and 0.3 μM/min/mg for red muscle myosin. Following LiCl-citrate fractionation with reconstitution of supernatant and precipitate fractions in their original proportion, specific Ca-ATPase is approximately

Table 3. Recovery of Ca-ATPase following LiCl-Citrate Fractionation of Rabbit Skeletal Myosin

Muscles	Prep.	Ca-ATPase* (μM P_i/min/mg)			% Light Component in Precipitate Fraction[†]	
		Initial Myosin	Precipitate + Supernatant	Precipitate Alone	at pH 7.0	at pH 11.0
White	I	0.71	0.38	0.25	0.5	5.4
	II	0.65	0.36	0.20	0.5	4.5
	III	0.75	0.43	0.10		4.0
	III	0.65	0.38	0.27		6.0
Intermediate	I	0.74	0.38	0.15	0.4	4.3
	II	0.70	0.46	0.21		5.8
	III	0.68	0.41	0.19		4.8
Red	I	0.28	0.20	0.07	0.4	4.1
	II	0.25	0.14	0.09	0.6	5.4
	III	0.33	0.16	0.12		5.2

* Reaction in 5 mM Ca-ATP, 0.05 M Tris, 0.5 M KCl, 1 mM DTT pH 7.6 at 37°C. Values normalized to soluble protein before and after fractionation.
† From high-speed sedimentation equilibrium.

0.4 μM/min/mg for white and intermediate muscle myosin and 0.17 μM/min/mg for red muscle myosin, that is, about 50% of the original activity. The values of Ca-ATPase in reconstituted protein have been normalized to final concentration of soluble protein, and the total recovery of Ca-ATPase after LiCl-citrate fractionation is only 25–35% of the original activity, due to loss of some protein during the course of the fractionation procedures and removal of denatured aggregated protein following reconstitution in 0.4 M KCl. Following LiCl-citrate fractionation with removal of supernatant fraction, all of the precipitate fractions retain some ATPase activity; however, Ca-ATPase is only 0.2 μM/min/mg for white and intermediate muscle myosin, and 0.1 μM/min/mg for red muscle myosin. There is considerable scatter about average values (Table 3), with some correlation with the proportion of residual light component in the precipitate fraction, but it seems preferable to consider average results for each group rather than individual experiments. In any event the data are consistent with earlier studies on myosin from mixed rabbit skeletal muscle, showing that Ca-ATPase of LiCl-citrate fractionated protein is significantly less in samples reconstituted with fewer light chains than present in native myosin.

Hybridization experiments were done simultaneously with the recombination experiments described above, in which supernatant and precipitate fractions of white, intermediate, and red skeletal myosin were hybridized immediately after centrifugation and subsequently treated identically as the reconstituted myosin and the precipitate fractions alone. The data show some dependence of Ca-ATPase on the specificity of light chains in myosin (Table 4); that is, protein reconstituted with white or intermediate muscle myosin precipitate and red muscle myosin supernatant has lower Ca-ATPase than fully reconstituted white or intermediate muscle myosin. Moreover protein reconstituted with red muscle myosin precipitate and supernatant from white or intermediate muscle myosin has somewhat higher Ca-ATPase than fully reconstituted red muscle myosin. This evidence would seem to indicate a role for light chains in determining specific Ca-ATPase of

myosin. The question remains whether heavy chains are also involved in determining ATPase specificity. Values of Ca-ATPase are greater in samples containing precipitate from white or intermediate muscle myosin than in samples containing precipitate from red muscle myosin, irrespective of the nature of the supernatant fraction (Table 4). However this finding is difficult to interpret since the observed differences in recovery of ATPase could be attributed to either the residual light chains (4–6%) or the heavy chains, which are both present in the precipitate fractions (see Table 3). The question might possibly be resolved through improvement of the fractionation procedure to yield pure light and heavy chains that could be reconstituted with complete recovery of ATPase activity; but our own experience has not yet been satisfactory in this respect.

Considerations of Structure-Function Relationships in Myosin

In previous ultracentrifugal studies on rabbit skeletal myosin, we have reported that the light chains have a weight average mol wt of 20,000 and comprise approximately 12% of the total myosin, the intact molecule having a mol wt of 470,000 (Gershman et al., 1966, 1969; Dreizen et al., 1967). This evidence would imply the presence of three light chains (on the average) per myosin molecule. In studies based on isotope dilution and analysis of gel patterns, Weeds and Lowey (1971) have concluded that myosin contains four light chains, including two "DTNB light chains" of *apparent* mol wt 18,000 on SDS-gel electrophoresis and two "alkali light chains" of *apparent* mol wt 17,000 and 25,000 on SDS-gel electrophoresis. The presence of four light chains in myosin would require a total mol wt of 72,000–80,000 (depending on assigned molecular weight values) for the light component, and this would imply that the light chains represent as much as 15–17% of the total myosin. We are disinclined to believe that the ultracentrifugal values could be misleading, albeit well aware of mishaps involving other ultracentrifugers in the analyses of globular proteins. Table 5 presents some calculations attempting to reconcile data from ultracentrifugal measurements on light chain stoichiometry and densitometric measurements on electrophoretic patterns for myosin isolated from rabbit white skeletal muscle (longissimus dorsi). The calculations are done for 12.5% light component (the *average* value for white muscle myosin in Table 2) and 14.7% light component (the *maximal* value determined for white muscle myosin in Table 2). The densitometric data indicate approximately 30–40% of α component, 35–45% of β component, and 15–30% of γ component in different preparations. Assuming 12.5% light component

Table 4. Recovery of Ca-ATPase following LiCl-Citrate Fractionation and Hybridization of Rabbit Skeletal Myosin

Precipitate fraction	None	Supernatant fraction		
		White	Intermediate	Red
White	0.21*	0.39*	0.38	0.24
Intermediate	0.19*	0.35	0.42*	0.23
Red	0.10*	0.26	0.24	0.17*

Ca-ATPase in μM/min/mg, under same conditions as Table 3.
* Averages from data in Table 3.

Table 5. Estimates of Molar Distribution of Electrophoretic Components in Light Alkali Component of Rabbit White Skeletal Muscle Myosin

| | Observed Ranges | | | Estimated Molar Ratio | |
	α	β	γ	$\alpha + \gamma$	$\alpha + \beta + \gamma$
Observed densitometric distribution					
cellulose acetate electrophoresis	30–42%	38–46%	15–27%		
SDS-gel electrophoresis	30–38%	35–45%	20–30%		
Average mol wt from sedimentation equilibrium	20,000	20,000	20,000		
Assume 12.5% light chains	0.9–1.2	1.0–1.3	0.5–0.8	1.7	2.9
Assume 14.7% light chains	1.0–1.4	1.2–1.5	0.7–1.0	2.0	3.3
Apparent mol wt from SDS-gel	25,000	17,000	16,000		
Assume 12.5% light chains	0.7–0.9	1.2–1.6	0.6–1.1	1.7	3.1
Assume 14.7% light chains	0.8–1.1	1.4–1.8	0.7–1.2	1.9	3.5

and 20,000 mol wt for each light chain, the stoichiometry would indicate approximately 1.0 α chain, 1.2 β chain, and 0.7 γ chain, with a total of 2.9 light chains per myosin molecule. In the presence of 14.5% light component, there would be 2.0 α-γ chains and 1.5 β chains. The estimates are not very different using the apparent molecular weight values obtained from SDS-gel electrophoresis, or any other set of molecular weight values in the neighborhood of 20,000. The calculations would suggest the following conclusions: (1) The number of β chains per myosin molecule appears to be 1.0 to 1.6, depending on assumed molecular weight of the β chain and the total stoichiometry for light component; (2) the total number of α and γ chains per myosin molecule seems to be two, or slightly less; (3) the molar ratio of α to γ chains varies from 1.0 to 2.3 in different preparations of white skeletal myosin.

Most of us like neat conclusions, and we are somewhat displeased over the ambiguity. It is possible that some β chains were lost during purification of myosin, although this would not be consistent with the observation that α and γ chains are those more easily dissociated from myosin by heat and NH$_4$Cl. Thus we would employ some caution in the interpretation of the β-chain population of myosin, at least until its functional role in myosin is established. Another point of ambiguity relates to the α-γ chain population of myosin. Although there seems to be a total of two α-γ chains per myosin molecule, the evidence by no means implies that each myosin molecule has one

α and one γ chain. The molar ratio of α to γ chains is variable, and in at least some preparations, the excess of α to γ chains implies the existence of myosin molecules with α-α chains. A simple explanation would be that myosin molecules contain two α chains or two γ chains, the exact proportion of which may vary in different preparations, although the occurrence of α-γ hybrids remains possible.

Previous studies involving denaturation of myosin by alkali (Dreizen et al., 1967; Gershman et al., 1969; Weeds and Lowey, 1971), heat (Gershman et al., 1969), and concentrated salt solutions at neutral pH and 4°C (Dreizen and Gershman, 1970) indicate a close relationship between ATPase inactivation and light chain dissociation. According to the present evidence (Table 6), heat denaturation of myosin ATPase is accompanied by selective dissociation of α and γ chains from myosin. The myosin molecule is stabilized to heat and salt denaturation by nucleotides and chain phosphates (in the presence of Mg^{++}), with protection of ATPase activity and little, if any, dissociation of α and γ chains. The α and γ chains appear to interact with Mg^{++} and Ca^{++}, and there is preliminary evidence for interaction of MgATP with the α chain. The findings clearly suggest a linkage between MgATP binding and subunit interactions involving the α-γ chains and the heavy chains of myosin. Moreover the recombination experiments involving LiCl-citrate fractionation of myosin suggest that light and heavy chains are both required for ATPase activity, with involvement of

Table 6. Light Chains of Rabbit Skeletal Myosin

	α	β	γ
Apparent mobility (cm/v/sec)	2.3×10^{-5}	2.05×10^{-5}	1.65×10^{-5}
Apparent molecular weight (SDS-gel)	25,000	17,000	16,000
Denaturing conditions			
pH 11	dissoc.	dissoc.	dissoc.
2 M Guanidine	dissoc.	dissoc.	dissoc.
4 M LiCl	dissoc.	dissoc.	dissoc.
4.7 M NH$_4$Cl	dissoc.	—	$\pm$ dissoc.
Heat	dissoc.	—	dissoc.
MgPPP$_i$ stabilization	+	—	+
Affinity sites			
Mg^{++}, Ca^{++} (Electrophoresis)	$K \sim 2 \times 10^3$	—	$K \sim 3 \times 10^2$

light chains (and possibly heavy chains) in determining the specific ATPase activity of myosin.

Figure 16 presents a schematic diagram of the results, with MgATP interacting with α-γ light chains and heavy chains in symmetric protomers of myosin. All the evidence would be consistent with a direct nucleotide bridge between light and heavy chains, although one cannot exclude a strong allosteric relationship between subunit interactions and nucleotide binding to myosin.

With respect to the functional role of the β light chain (or two β light chains) in myosin, Weeds and Lowey (1971) have reported that DTNB selectively dissociates the β chain from myosin without affecting ATPase activity. Our own experience in use of DTNB has not been so clear-cut, in that dissociation of β chain is somewhat variable and usually incomplete, at least in our hands. We do obtain greater (although less selective) dissociation of β chains from DTNB-reacted myosin in dilute urea solutions at room temperature, but then ATPase activity is also diminished. So we are not yet willing to commit the β chain to oblivion, especially in view of the following experiments.

In seeking simple alternatives to the rather intricate hybridization experiments on myosin, studies were done on the effect of light chains in considerable excess (at molar ratios up to 100/1) on the ATPase activity of native myosin. As shown in Fig. 17, addition of alkali-dissociated light component (α, β, γ chains) is associated with *decreased* EDTA-ATPase, ranging from 15% immediately to about 20–25% over 3–5 hr of incubation. Addition of excess light chains does not show any significant effect on Ca-ATPase of the same samples, nor is there any comparable effect of heat-dissociated light component (α-γ chains) on the ATPase activity of native myosin. The differences observed are fairly minor, but seem reasonably reproducible; and the observation could be interpreted in several very different ways: (1) The excess light chains might have been denatured by alkali and undergo exchange with the light chains of native myosin, with replacement of functionally active light chains by inactive light chains at high molar ratios; (2) the excess light chains (β chain?) might be involved in direct inhibition of myosin ATPase. These alternatives, and others, are open to experimental inquiry, and

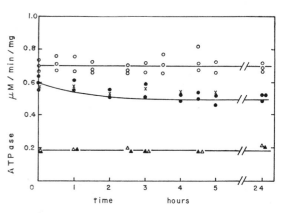

Figure 17. Effect of excess light component on myosin ATPase. Alkali-dissociated light component (α, β, and γ chains) from white muscle myosin was incubated with native myosin (white muscle) in 0.5 M KCl, 0.05 M Tris, 1 mM DTT pH 7.5 at 4°C and ATPase assays (at 25°C) were done on aliquots removed at times specified. EDTA-ATPase was determined at molar ratios of light chain to myosin of 100 (●), 20 (×), and without excess light chains (○). Ca-ATPase was determined at molar ratio of light chain to myosin of 100 (▲) and without excess light chains (△).

we will forbear obvious speculation on a possible subunit regulatory system for myosin ATPase.

Summary

There is a close relationship between ATPase inactivation and the dissociation of light chains during salt and heat denaturation of myosin. Heat denaturation of myosin is accompanied by selective dissociation of two light chains (α and γ), with concomitant loss of myosin ATPase. Under identical denaturing conditions, MgATP (or MgPPP$_i$) protects myosin ATPase and stabilizes subunit interactions involving the α-γ light chains and the heavy chains. Electrophoretic evidence is reported for specific interaction of Ca^{++} and Mg^{++} with the α and γ light chains.

Studies on the subunit composition of myosin from red, white, and intermediate rabbit skeletal muscle indicate comparable stoichiometry for light and heavy chains (from ultracentrifugal experiments), but different electrophoretic distributions of α, β, and γ light chains, and an additional electrophoretic component in intermediate muscle myosin. Recombination experiments involving LiCl-citrate fractionation of myosin show a dependence of ATPase recovery on the presence of light and heavy chains, and hybridization experiments would suggest a role for light chains (and possibly also heavy chains) in determining specific myosin ATPase.

The overall results suggest a model for myosin in which α-γ light chains interact with MgATP and heavy chains in symmetric protomers of myosin. The role of the β light chain remains uncertain.

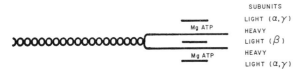

Figure 16. Schematic model for subunit structure of myosin and its interaction with MgATP.

Acknowledgments

This work was supported in part by grant AM 06165 from the National Institutes of Health and by grants from the New York Heart Association and the Health Research Council of New York City. We gratefully acknowledge the technical assistance of Mrs. Z. Capulong.

References

BÁRÁNY, M., K. BÁRÁNY, T. RECKARD and A. VOLPE. 1965. Myosin of fast and slow muscles of the rabbit. *Arch. Biochem. Biophys.* **109**: 185.

CANN, J. R. 1970. *Interacting macromolecules.* Academic Press, New York.

DOW, J. and A. STRACHER. 1971. Changes in the properties of myosin associated with muscle development. *Biochemistry* **10**: 1316.

DREIZEN, P. and L. C. GERSHMAN. 1970. Relationship of structure to function in myosin. II. Salt denaturation and recombination experiments. *Biochemistry* **9**: 1688.

DREIZEN, P. and H. D. KIM. 1971a. Contractile proteins of a benthic fish. I. Myosin ATPase. *Amer. Zool.* **11**: 513.

———. 1971b. Contractile proteins of a benthic fish. III. Subunit composition of myosin. *Amer. Zool.* **11**: 531.

DREIZEN, P., D. J. HARTSHORNE, and A. STRACHER. 1966. The subunit structure of myosin. I. Polydispersity in 5M guanidine. *J. Biol. Chem.* **241**: 443.

DREIZEN, P., L. C. GERSHMAN, P. P. TROTTA, and A. STRACHER. 1967. Myosin. Subunits and their interactions. *J. Gen. Physiol.* **50**, part 2: 85.

FREDERIKSEN, D. W. and A. HOLTZER. 1968. The substructure of the myosin molecule. Production and properties of the myosin subunits. *Biochemistry* **7**: 3935.

GAETJENS, E., K. BÁRÁNY, G. BAILIN, H. OPPENHEIMER, and M. BÁRÁNY. 1968. Studies on the low molecular weight protein components in rabbit skeletal myosin. *Arch. Biochem. Biophys.* **123**: 82.

GAZITH, J., S. HIMMELFARB, and W. F. HARRINGTON. 1970. Studies on the subunit structure of myosin. *J. Biol. Chem.* **245**: 15.

GERSHMAN, L. C. and P. DREIZEN. 1970. Relationship of structure to function in myosin. I. Subunit dissociation in concentrated salt solutions. *Biochemistry* **9**: 1677.

———. 1971. Structure and function of myosin, p. 177. In *Conference on cardiac hypertrophy*, ed. N. R. Alpert. Academic Press, New York.

GERSHMAN, L. C., P. DREIZEN, and A. STRACHER. 1966. The subunit structure of myosin. II. Heavy and light alkali components. *Proc. Nat. Acad. Sci.* **56**: 966.

GERSHMAN, L. C., A. STRACHER, and P. DREIZEN. 1968. Subunit interactions of myosin, p. 150. In *Symposium on fibrous proteins*, Australia, 1967, Butterworths.

———. 1969. Subunit structure of myosin. III. A proposed model for rabbit skeletal myosin. *J. Biol. Chem.* **244**: 2726.

KIELY, B., and A. MARTONOSI. 1968. Kinetics and substrate binding of myosin ATPase. *J. Biol. Chem.* **243**: 2273.

———. 1969. The binding of ADP to myosin. *Biochim. Biophys. Acta* **172**: 158.

KIM, H. D. and W. F. H. M. MOMMAERTS. 1971. On the reconstitution of rabbit myosin from fast and slow muscle. *Biochim. Biophys. Acta* **245**: 230.

KOMINZ, D. R., W. R. CARROLL, E. N. SMITH, and E. R. MITCHELL. 1959. A subunit of myosin. *Arch. Biochem. Biophys.* **79**: 191.

LOCKER, R. H. 1956. The dissociation of myosin by heat coagulation. *Biochim. Biophys. Acta* **20**: 514.

LOCKER, R. H. and C. J. HAGYARD. 1967a. A correlation of various subunits of myosin. *Arch. Biochem. Biophys.* **120**: 241.

———. 1967b. Small subunits in myosin. *Arch. Biochem. Biophys.* **120**: 454.

———. 1967c. Variations in the small subunits of different myosins. *Arch. Biochem. Biophys.* **122**: 521.

———. 1968. The myosin of rabbit red muscle. *Arch. Biochem. Biophys.* **127**: 370.

LOW, R. B., J. N. VERNAKIS, and A. RICH. 1971. Identification of separate polysomes active in the synthesis of the light and heavy chains of myosin. *Biochemistry* **10**: 1813.

LOWEY, S. and S. M. LUCK. 1969. Equilibrium binding of ADP to myosin. *Biochemistry* **8**: 3195.

LOWEY, S. and D. RISBY. 1971. Light chains from fast and slow muscle myosins. *Nature* **234**: 81.

LOWEY, S., H. S. SLAYTER, A. G. WEEDS, and H. BAKER. 1969. Substructure of the myosin molecule. I. Subfragments of myosin by enzymic degradation. *J. Mol. Biol.* **42**: 1.

NAUSS, K. M., S. KITAGAWA, and J. GERGELY. 1969. Pyrophosphate binding and ATPase activity of myosin and its proteolytic fragments. *J. Biol. Chem.* **244**: 755.

OUELLET, L., K. J. LAIDLER, and M. F. MORALES. 1952. Molecular kinetics of muscle adenosine triphosphatase. *Arch. Biochem. Biophys.* **39**: 37.

PERRIE, W. T. and S. V. PERRY. 1970. An electrophoretic study of the low molecular weight components of myosin. *Biochem. J.* **119**: 31.

SAMAHA, F. J., L. GUTH, and R. W. ALBERS. 1970. Differences between slow and fast muscle myosin. *J. Biol. Chem.* **245**: 219.

SARKAR, S., F. A. SRETER, and J. GERGELY. 1971. Light chains of myosins from white, red, and cardiac muscles. *Proc. Nat. Acad. Sci.* **68**: 946.

SCHLISELFELD, L. H. and M. BÁRÁNY. 1968. The binding of adenosine triphosphate to myosin. *Biochemistry* **7**: 3206.

SLAYTER, H. S. and S. LOWEY. 1967. Substructure of the myosin molecule as revealed by electron microscopy. *Proc. Nat. Acad. Sci.* **58**: 1611.

SRETER, F. A., J. C. SEIDEL, and J. GERGELY. 1966. Studies of myosin from red and white skeletal muscle of the rabbit. I. Adenosine triphosphatase activity. *J. Biol. Chem.* **241**: 5772.

STARR, R. and G. OFFER. 1971. Polypeptide chains of intermediate molecular weight in myosin preparations. *FEBS Letters* **15**: 40.

STRACHER, A. 1969. Evidence for the involvement of light chains in the biological functioning of myosin. *Biochem. Biophys. Res. Comm.* **35**: 519.

TROTTA, P. P., P. DREIZEN and A. STRACHER. 1968. Studies on subfragment-1, a biologically active fragment of myosin. *Proc. Nat. Acad. Sci.* **61**: 659.

TSAO, T. C. 1953. Fragmentation of the myosin molecule. *Biochim. Biophys. Acta* **11**: 368.

WEBER, K. and M. OSBORN. 1969. The reliability of molecular weight determinations by dodecyl-sulfate polyacrylamide gel electrophoresis. *J. Biol. Chem.* **244**: 4406.

WEEDS, A. G. 1969. Light chains of myosin. *Nature* **233**: 1362.

WEEDS, A. G. and S. LOWEY. 1971. Substructure of the myosin molecule. II. The light chains of myosin. *J. Mol. Biol.* **61**: 701.

YASUI, T., Y. HASHIMOTO, and Y. TONOMURA. 1960. Physico-chemical studies on denaturation of myosin-adenosine triphosphatase. *Arch. Biochem. Biophys.* **87**: 55.

Myosin-Linked Regulatory Systems: The Role of the Light Chains

John Kendrick-Jones, Eva M. Szentkirályi, and Andrew G. Szent-Györgyi

MRC Laboratory of Molecular Biology, Cambridge, England
and
Department of Biology, Brandeis University, Waltham, Mass. 02154

Muscular contraction is regulated by the interaction of calcium with the contractile elements. In vertebrate muscles the specific regulatory proteins, tropomyosin, and the troponin complex (Ebashi and Kodama, 1965, 1966) are associated with the thin filaments (Ohtsuki et al., 1967). Tropomyosin is required for relaxation, its role being to mediate the changes on troponin which affect the active sites of actin (Ebashi and Endo, 1968; Weber and Bremel, 1971). In molluscan muscles calcium regulation is associated with myosin; troponin and its components are absent, and tropomyosin, although present on the thin filaments, is not required for regulation. Purified molluscan myosin preparations bind calcium, and their ATPase activity depends on calcium when combined with pure actin (Kendrick-Jones et al., 1970).

In this paper we define those components of the myosin preparations which are responsible for calcium regulation. The removal of a particular light chain from myosin with EDTA leads to a loss of calcium sensitivity, which can be quantitatively regained when the light chain is recombined with the myosin in the presence of divalent cations. Therefore, the presence of this light chain on myosin is required for regulation by calcium, and it functions as a regulatory subunit.

The light chain inhibits the actin-activated ATPase activity of scallop myosin in the absence of calcium by blocking those sites on myosin which combine with actin. In the presence of calcium the inhibitory effect of the light chain is removed. Possible mechanisms describing how the light chain may control interactions between actin and myosin will be discussed. A detailed description of these observations will be published elsewhere (Szent Györgyi et al., 1973).

In these studies we focused our attention on the myosins isolated from the striated adductor muscles of two species of scallops, *Aequipecten irradians* and *Placopecten magellanicus*. These scallop myosins were prepared by the procedures previously outlined (Szent-Györgyi et al., 1971) and were found to have identical behavior.

Subunit Composition of Scallop Myosin

The basic structural features and hydrodynamic parameters of the myosins from rabbit and scallop muscles are rather similar, which indicate that they have approximately the same size and overall shape. Although the thick filaments in scallop muscles are larger than in rabbit muscles, scallop myosin forms synthetic thick filaments with characteristic bare zones and projections which are indistinguishable from those of rabbit myosin. The dimensions of the heavy chains and proteolytic fragments of scallop myosin indicate that this myosin is also composed of two chains with globular heads and a rodlike tail. The unique regulatory properties of scallop myosin are therefore not due to gross differences in the structure of this myosin.

Scallop myosin has a simple subunit structure. Polyacrylamide gel electrophoresis of scallop myosin preparations in the presence of sodium dodecyl sulfate (SDS) shows a major component with a chain weight slightly below 200,000 daltons, similar in size to the heavy chains of rabbit myosin, and a single diffuse band corresponding to a chain weight of 18,000 daltons which represent the light chains of myosin (Table 1; Fig. 1). Although the light chain appears as a single component on acrylamide gels, both in the presence and the absence of SDS and on Sephadex and ion exchange columns, two different types of light chains, very similar in size and net charge, can be prepared. EDTA releases one kind of light chain from scallop myosin; the other is removed subsequently by SDS. These light chains are the only low chain weight components of washed scallop muscles and can be prepared directly from myofibrils without the isolation of myosin. The myosin preparations contain some actin and tropomyosin, usually in amounts less than 2% of the myosin. Paramyosin,

Table 1. Quantitation of Components of Scallop Myofibrils and Myosin Preparations by Gel Densitometry

	Chain Weight	Myofibrils (percentage)	Myosin (percentage)
Myosin heavy chain	186,000	54.6	79
Paramyosin	117,000	3.8	7
(Diffuse bands) ca.	105,000	4.6	?
Actin	46,000	23.8	<2
Tropomyosin	35,000	5.2	<1
Myosin light chain	18,000	7.9	12–13
(Diffuse band) ca.	15,000		<2

Myofibrils were washed 2–3 times with 40 mM NaCl, 1 mM $MgCl_2$, 5 mM phosphate pH 7.

47

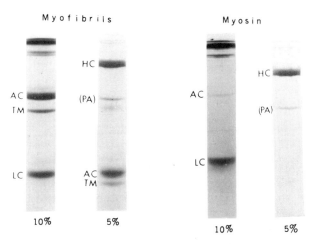

Figure 1. SDS-acrylamide gel electrophoresis of scallop myosin. AC = actin; TM = tropomyosin; LC = light chain; HC = heavy chain; PA = paramyosin. Heavy chains are about 186,000 daltons and appear as a single band. Paramyosin, actin, tropomyosin, heavy and light chains account for the components of the myofibrils. Myosin contains a trace of actin and about 5–7% paramyosin impurity; otherwise, it consists only of heavy and light chains. Troponin is not present in myofibrils or in myosin. Myofibrils and myosin were obtained from the striated adductors of *Aequipecten irradians*. Myofibrils were washed twice with a solution containing 40 mM NaCl, 5 mM phosphate pH 7.0, 1 mM MgCl₂.

with an apparent chain weight of 117,000 daltons, is the major contaminant and is present in amounts varying between 5–7% of the myosin (Table 1). Paramyosin, however, does not bind calcium and there is no evidence for its involvement in calcium regulation (Kendrick-Jones et al., 1970). The major difference, therefore, between rabbit and scallop myosin lies in their light components. In general, the light chains are the most variable features of various types of myosins. Differences in the number and size of the myosin light chain components isolated from a variety of skeletal fast and slow and cardiac muscles have previously been reported (Locker and Hagyard, 1968; Gazith et al., 1970; Samaha et al., 1970; Paterson and Strohman, 1970; Lowey and Risby, 1971; Sarkar et al., 1971; Weeds and Pope, 1971).

Role of Light Chains in Regulation

The conclusion that a light chain is the regulatory subunit is based on experiments which demonstrate that the regulatory properties of myosin are selectively lost when this light chain is removed and are quantitatively regained when the residual myosin is recombined with the light chain.

Treatment of scallop myosin with 1–10 mM EDTA removes one kind of light chain (EDTA light chain) (Fig. 2). The loss of the EDTA light chain from myosin alters its actin-activated

ATPase activity in a very specific manner by removing the calcium requirement for ATPase activity. EDTA treated myosin is a "desensitized" myosin which hydrolyzes ATP equally well both in the presence and the absence of calcium (Table 2). The myosin type ATPase (measured in 10 mM calcium) is not affected by the loss of the EDTA light chain. The removal of the EDTA light chains from scallop myofibrils is also accompanied by a loss of calcium sensitivity. The removal of the EDTA light chain reduces the amount of calcium bound by the myosin by about 40% (Table 2; Fig. 3). Scatchard plots indicate that this decrease in calcium binding is not due to an altered calcium affinity, but rather to a loss in the number of calcium binding sites. The isolated EDTA light chain, however, does not bind calcium.

Addition of the EDTA light chain to the "desensitized" myosin in the presence of divalent cations leads to a restoration in calcium sensitivity and calcium binding (Table 2; Fig. 3). These regulatory properties of the untreated myosin are regained when EDTA light chains and desensitized myosin are mixed in their original proportions. The removal of free EDTA light chain from the supernatant after recombination indicates the strong

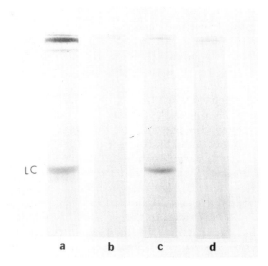

Figure 2. Release and recombination of EDTA light chain with scallop myosin. *a*) Initial myosin preparation. *b*) Control: supernatant of control myosin suspended in 40 mM NaCl, 10 mM phosphate pH 7.0. *c*) Release of EDTA light chain: supernatant of myosin to which 10 mM EDTA was added in 40 mM NaCl, 10 mM phosphate pH 7.0 for 10 min at 0° before centrifugation. Note the band of EDTA light chain. *d*) Recombination of the EDTA light chain with "desensitized" myosin: To the supernatant of (*c*) 10 mM MgCl₂ was added, the pH was readjusted to 7.0 and it was recombined with the precipitate of (*c*) dissolved in 0.6 M NaCl in the original proportions. The mixture was dialyzed against 40 mM NaCl, 1 mM MgCl₂ and 5 mM pH 7.0 phosphate overnight and the precipitated myosin was removed by centrifugation. Note the disappearance of the EDTA light chain band from the supernatant solution.

Table 2. Effect of EDTA Light Chain on ATPase Activity and on Calcium Binding of Scallop Myosin

| | Actin-activated ATPase[a] | | Ratio EGTA/Ca++ | Calcium Binding[b] |
	0.1 mM Ca++	0.1 mM EGTA		
Aequipecten				
Untreated myosin	0.22	0.05	0.23	1.67
"Desensitized" myosin	0.24	0.20	0.83	0.78
"Desensitized" myosin recombined with EDTA light chain in original proportions	0.22	0.07	0.32	1.59
EDTA light chain	none	none		none

[a] Values are given as μmoles ATP/mg/min at 23°.
[b] Values are in mμmoles Ca++/mg protein at 5×10^{-7} M Ca++.
Myosin was "desensitized" with 10 mM EDTA at low ionic strength. The removal of the EDTA light chain, the recombination of the EDTA light chain with "desensitized" myosin followed the procedure described in the legend of Fig. 2.

affinity of the light chains for myosin in the presence of divalent cations (Fig. 2). Low concentrations of calcium or magnesium ions prevent the dissociation of the EDTA light chain; half of the light chains dissociate at about 2×10^{-7} M Ca++ or at about 1×10^{-5} M Mg++. The EDTA light chain and the desensitized myosin are stable and may be stored separately for one or two days without diminishing

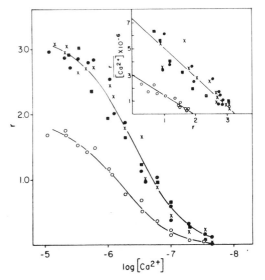

Figure 3. Calcium binding of scallop myosin (*Aequipecten*). (●——●) Untreated myosin; (○——○) myosin "desensitized" by removal of EDTA light chain; (×——×) "desensitized" myosin recombined with EDTA light chain; (■——■) untreated myosin in 1 mM MgCl$_2$. Calcium binding was measured in the presence of 4 mM MgCl$_2$ in 40 mM NaCl, 10 mM imidazole-HCl pH 7.0 and [^{3}H]glucose. 5–10 mg myosin was suspended in 8 ml 50–100 μM [^{45}Ca]-EGTA solution with enough EGTA added to achieve the desired free calcium ion concentration. After about 5 min equilibration, myosin was collected by 5 min centrifugation at about 4000 g. The myosin precipitate was resuspended in a new 8 ml aliquot of the same solution and equilibrated and recentrifuged again. The precipitate was resuspended by a few sec sonication in 1.7 ml 50 mM NaCl, 4 mM MgCl$_2$, 10 mM imidazole-HCl pH 7.0, and radioactivity and protein were measured. Three different myosin preparations. *Inset:* Scatchard plots. The μmoles of calcium bound per gram protein are denoted by r.

their ability to recombine and restore calcium sensitivity.

The correlation of calcium sensitivity with the presence of the light chain on the myosin demonstrates that the EDTA light chain is a regulatory subunit. The loss in the number of calcium binding sites upon the removal of EDTA light chain suggests that calcium may directly contribute to the binding of the light chain to the myosin molecule in such a way that it controls the interaction between actin and myosin. For example, calcium could fix the light chain in the "off" (noninhibitory) position. Calcium action, however, may be more indirect, and its effect may be mediated by conformational changes on the light or heavy chains which lead to the unblocking of actin binding sites on myosin. During inhibition the light chain is linked to the heavy chain by divalent cations, most likely magnesium under physiological conditions.

Stoichiometry of Light Chains

Quantitation of the myosin components by Sephadex chromatography and by densitometry of SDS acrylamide gels indicate that there are about three moles of light chains for each mole of myosin (Table 3). A similar ratio of light chain to myosin was obtained from washed myofibrils showing that the light chains are not selectively lost during myosin preparation. EDTA treatment removes one mole of light chain from myosin, and the loss of this single light chain leads to complete desensitization. The light chains which remain on the myosin after EDTA treatment are released by SDS and since they contain cysteine will be called thiol light chains. The Ellman reagent (DTNB) or mersalyclic acid removes about half of the light chains of intact myosin, and the material obtained is a mixture of the EDTA and the thiol light chains. The ATPase activity and the calcium binding ability of the myosin is immediately and completely lost by the treatments necessary to remove the thiol light chains. Since the loss of

Table 3. Light Chain Content of Scallop (*Aequipecten*) Myosin

	% LC[a]	Moles LC in 425,000 g Myosin
Sephadex chromatography		
Untreated myosin	11.3	2.66
"Desensitized" myosin	7.9	1.87
Densitometry of SDS gels		
Myofibrils	12.9	3.05
Untreated myosin	13.3	3.14
"Desensitized" myosin	9.4	2.22

$$^a \frac{LC}{HC + LC} \times 100$$

Chromatography was on G-200 Sephadex in 0.1% SDS. Protein determinations indicate that about 3–4% of the protein is released from myosin by two successive EDTA treatments. The value is in agreement with the estimate that there is about 0.7–1 mole of EDTA light chain in 425,000 g myosin. Note that the LC to HC ratio is the same in washed myofibrils as in untreated myosin preparations.

these functions is totally irreversible, the exact role of the thiol light chains remains unresolved.

The amino acid compositions of the EDTA light chains and thiol light chains are significantly different (Fig. 4). The most striking difference is the absence of half cystine residues in the EDTA light chain, whereas the thiol light chain contains two half cystine residues per 18,000 g. The absence of half cystine residues on the EDTA light chain allows an independent estimate of the ratio of EDTA light chain to thiol light chain in myosin and myofibrils by measuring the number of these residues in light chain preparations. The thiol groups were measured both by alkylation with radioactive iodoacetic acid and by performic acid oxydation. The values obtained by these independent procedures are similar. Two half cystine residues were found in the thiol light chains, and 1.3 to 1.4 in the total light chain preparations from myosin or myofibril, which is consistent with a

ratio of two thiol light chains to one EDTA light chain in a myosin molecule. This ratio is in agreement with that obtained by densitometry of SDS gels and gel filtration of untreated and desensitized myosin preparations.

The light chains were further characterized by preparing two-dimensional peptide maps from tryptic digests of the light chains by the procedure of Weeds and Hartley (1968) (Fig. 5). The peptide maps were selectively stained for arginine and tyrosine residues and radioautographed to detect the radioactive thiol peptides. The number of these selectively stained tyrosine- and arginine-containing peptides observed in the peptide maps of the EDTA and thiol light chain are in agreement with their amino acid composition. The absence of the radioactive thiol peptides in the EDTA light chain, even after long radioautographic exposure, indicates that the level of contamination by the thiol light chain is low. The lack of an obvious overlap between the peptides of the EDTA and thiol light chain suggests that there is little overall chemical similarity between the two types of light chains.

Speculations on Mechanism of Regulation

The results indicate that there is a single EDTA light chain and two identical thiol light chains in a myosin molecule. A significant feature of the myosin-linked regulation in scallop is that the removal of a single EDTA light chain completely desensitizes myosin. How does a single EDTA light chain regulate the activity of a double-headed myosin molecule? If only one of the S-1 subfragments of scallop myosin was an actin-activated ATPase, then the regulatory light chain would be in a one-to-one molar ratio with the functional subfragment. The mechanism of regulation could then be discussed in terms of how the relationship of the light chain to the single active subfragment is

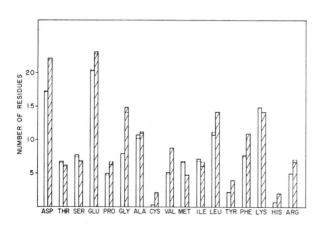

Figure 4. Amino acid analysis of light chains. Amino acid residues per molecule. The cross-hatched bars represent the thiol light chains and open bars the EDTA light chains. The dotted lines at the top of the bars are the amino acid compositions of light chains obtained from myofibrils. Note the close agreement between the amino acid compositions of the light chains obtained from the myosin and myofibrillar preparations. These analyses are the mean values for 24 and 72 hr hydrolysis times, except that the values of threonine and serine were obtained by extrapolation to zero time hydrolysis, and those of valine and isoleucine were taken from the 72 hr hydrolysates. Cysteine values were obtained as cysteic acid from the 20 hr hydrolysates of performic acid oxidized protein (Weeds and Hartley, 1968).

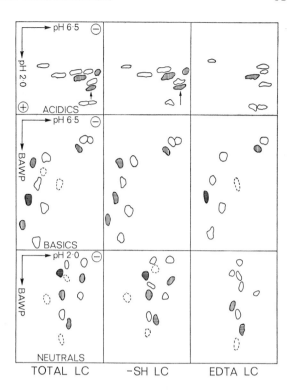

Figure 5. Two-dimensional tryptic peptide maps of the light chains of *Aequipecten irradians*. The two-dimensional peptide maps were produced by separating the peptides in the first dimension by electrophoresis at pH 6.5. The acidic peptides were rerun in the second dimension by electrophoresis at pH 2.1 and the basic peptides by descending chromatography in butan-1-ol, acetic acid, water, pyridine (BAWP). The neutral peptides from the pH 6.5 electrophoresis were separated by electrophoresis at pH 2.1 and then by descending chromatography in BAWP. Solid outlines, peptides stained with the ninhydrin-cadmium reagent. Dotted outlines indicate peptides present in trace amounts. Horizontal shading, peptides staining for tyrosine; vertical shading, peptides staining for arginine. The arrows indicate peptides containing radioactive carboxymethyl cysteine residues, which were detected by radioautography.

controlled by calcium. We consider it unlikely, however, that only one of the myosin heads is an ATPase. The observation that the phosphate burst in the early rapid phase of ATP hydrolysis amounts to about 1.6 mole of phosphate per mole of rabbit myosin (Lymn and Taylor, 1970) argues against an inactive myosin head. Nevertheless, this possibility has not yet been rigorously excluded.

If both heads participate in the actin-activated ATPase activity, then some type of cooperativity between heads is required to block the sites which interact with actin. It should be stressed that such a proposal does not assume cooperativity during ATPase activity in the presence of calcium. One possible simple model to explain this type of cooperative inhibition requires two binding sites on

Figure 6. SDS-acrylamide gel electrophoresis of the proteolytic fragments of scallop myosin. S-1 obtained from *Aequipecten* myosin by 4 min digestion at 22° in the presence of 1 mM MgCl$_2$ with insoluble papain at a papain to myosin ratio of 1:800. S-1 was purified by cosedimenting with rabbit actin at 40,000 rpm for 3 hr. Actin was removed by differential centrifugation in the presence of ATP.

HMM obtained from *Aequipecten* myosin by 4 min digestion at 22°, pH 6.5 at a trypsin to myosin ratio of 1:1000. HMM was purified by cosedimenting with rabbit actin at 40,000 rpm for 3 hr. Actin was removed by differential centrifugation in the presence of ATP. HMM is extensively degraded, although a considerable portion of the preparation contains fragments between 130,000 and 150,000 daltons. S-1 is well preserved and the major component of S-1 preparations has a chain weight of about 95,000 daltons. Note the presence of light chains in both HMM and S-1 shown on 10% gels. In

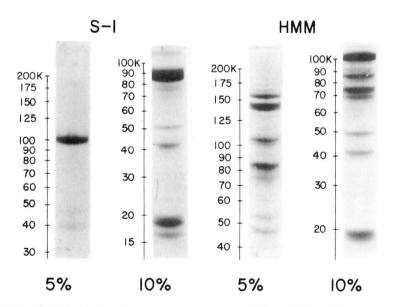

two S-1 preparations, quantitative densitometry showed that the light chains amounted to 24% and 27% of the combined light and heavy chain components, indicating the presence of about 1.5 moles of light chain in a mole of S-1.

Table 4. ATPase Activity and Calcium Binding of the Proteolytic Fragments of Scallop Myosin

| | Digestion Conditions | Actin-activated ATPase[a] | | Ratio EGTA/Ca++ | Calcium Binding[b] 1×10^{-6} M Ca++ |
		0.1 mM Ca++	0.1 mM EGTA		
HMM	Trypsin	0.36	0.11	0.30	2.0
S-1	Sol. papain + 2 mM EDTA	0.72	0.85	1.18	0.1
S-1	Sol. papain + 2 mM Mg	0.81	0.77	0.95	3.0

[a] Values are given as μmoles ATP/mg/min.

[b] Values are given as mμmoles Ca++/mg protein. The number of calcium binding sites were estimated to about five calcium per mg protein in the best S-1 preparations.

HMM was obtained by digestion with a trypsin-myosin ratio of 1:800 for 8 min in 0.6 M NaCl, 10 mM phosphate pH 7.3 at 23°. HMM was purified by centrifugation of the acto-HMM complex. 0.1 mg/ml HMM and 0.3 mg/ml actin in 10 ml for ATPase measurements.

Myosin was digested with soluble papain at a papain-myosin ratio of 1:400, in 0.2 M NaCl, 5 mM phosphate pH 7.3, 5 mM DTT and 2 mM EDTA at 0° for 20 min. Papain was activated and digestion stopped as described by Lowey et al. (1969). Preparation was dialyzed against 20 mM NaCl, 5 mM phosphate pH 7.0 and centrifuged. The supernatant consisted mostly of S-1, which was purified by centrifugation of the acto-S-1 complex. 0.36 mg/ml actin and 0.1 mg/ml S-1 in 10 ml for ATPase measurements.

S-1 was obtained by 30 min digestion at 0° with soluble papain as previous preparation but in the presence of 2.6 mM MgCl$_2$. 0.03 mg/ml S-1 and 0.5 mg/ml actin in 10 ml for ATPase measurements.

the EDTA light chain: In the absence of calcium, the EDTA light chain binds to both S-1 subfragments, fixing them in such a position that they are unable to react with actin; calcium alters the binding to allow interaction with actin. A model based on cooperative inhibition predicts that S-1 subfragment preparations should not be calcium sensitive, but heavy meromyosin, which contains both heads connected by a short tail, should require calcium for ATPase activity. One notes that in the troponin regulated systems no qualitative difference is expected in the regulation of HMM and of S-1. In the presence of troponin and tropomyosin both rabbit HMM and S-1 are similarly regulated by calcium in their interaction with actin (Eisenberg and Kielley, 1970; Weber and Bremel, 1971).

Scallop myosin is extremely susceptible to tryptic digestion which rapidly leads to a nonspecific degradation of the whole molecule (Bárány and Bárány, 1966). Under well-defined digestion conditions, however, using a very low trypsin concentration, it is possible to prepare an HMM which on SDS acrylamide gels consists mainly of fragments with chain weights between 130,000 and 150,000 daltons and apparently intact light chains (Fig. 6). This HMM has a calcium-sensitive actin-activated ATPase activity and binds considerable amounts of calcium (Table 4). A striking feature is that, despite the excessive fragmentation produced by a more prolonged digestion, the resulting low actin-activated ATPase is always calcium sensitive and the light chain components remain relatively unfragmented. HMM prepared from desensitized myosin under the same conditions has a calcium-insensitive actin-activated ATPase.

S-1 subfragment preparations isolated under a variety of conditions always lack calcium sensitivity (Table 4). These S-1 subfragments show a high degree of homogeneity and appear as a single main band of 95,000 daltons on SDS acrylamide gels (Fig. 6). Light chains are retained in a ratio of about 1.5 moles per mole of S-1, provided magnesium ions are present during papain digestion. The light chains are bound to the S-1 subfragment; they are not separated from S-1 by DEAE-cellulose chromatography and cosediment with an S-1-actin complex. The S-1 subfragment has a high calcium binding. The complete absence of calcium sensitivity in these S-1 preparations is in marked contrast to their high ATPase activities and calcium binding abilities. Calcium sensitivity cannot be restored by the addition of excess EDTA light chains prepared from intact scallop myosin. In the presence of rabbit troponin and tropomyosin, however, the actin activated ATPase activity of scallop S-1 is calcium regulated. The S-1 subfragment prepared in the presence of EDTA has no intact light chains and does not bind calcium.

The absolute lack of calcium sensitivity of S-1 subfragment preparations excludes schemes in which EDTA light chains would be functionally present on only one of the two active myosin heads. Although the differences in calcium sensitivity displayed by S-1 and HMM are consistent with a model invoking cooperative interaction between the two heads, such an interaction remains to be directly established.

Acknowledgment

This work was supported by Public Health Service Grant GM 14675 and AM 15963.

References

BÁRÁNY, M. and K. BÁRÁNY. 1966. Myosin from the striated adductor muscle of scallop (*Pecten irradians*). *Biochem. Zeit.* **345**: 37.

EBASHI, S. and M. ENDO. 1968. Calcium ion and muscular contraction. *Prog. Biophys. Mol. Biol.* **18**: 123.

EBASHI, S. and A. KODAMA. 1965. A new protein factor promoting aggregation of tropomyosin. *J. Biochem.* (Tokyo) **58**: 107.

———. 1966. Interaction of troponin with F-actin in the presence of tropomyosin. *J. Biochem.* (Tokyo) **59**: 425.

EISENBERG, E. and W. W. KIELLEY. 1970. Native tropomyosin: Effect on the interaction of actin with heavy meromyosin and subfragment-1. *Biochem. Biophys. Res. Comm.* **40**: 50.

GAZITH, J., S. HIMMELFARB, and W. F. HARRINGTON. 1970. Studies on the subunit structure of myosin. *J. Biol. Chem.* **245**: 15.

KENDRICK-JONES, J., W. LEHMAN, and A. G. SZENT-GYÖRGYI. 1970. Regulation in molluscan muscles. *J. Mol. Biol.* **54**: 313.

LOCKER, R. H. and C. J. HAGYARD. 1967. Variations in the small subunits of different myosins. *Arch. Biochem. Biophys.* **122**: 521.

LOWEY, S. and D. RISBY. 1971. Light chains from fast and slow muscle myosins. *Nature* **234**: 81.

LOWEY, S., H. S. SLAYTER, A. G. WEEDS, and H. BAKER. 1969. Substructure of the myosin molecule. I. Subfragments of myosin by enzymic degradation. *J. Mol. Biol.* **42**: 1.

LYMN, R. W. and E. W. TAYLOR. 1970. Transient state phosphate production in the hydrolysis of nucleoside triphosphates by myosin. *Biochemistry* **9**: 2975.

OHTSUKI, I., T. MASAKI, Y. NONOMURA, and S. EBASHI. 1967. Periodic distribution of troponin along the thin filament. *J. Biochem.* (Tokyo) **61**: 817.

PATERSON, B. and R. C. STROHMAN. 1970. Myosin structure as revealed by simultaneous electrophoresis of heavy and light subunits. *Biochemistry* **9**: 4094.

SAMAHA, F. J., L. GUTH, and R. W. ALBERS. 1970. Differences between slow and fast muscle myosin. *J. Biol. Chem.* **245**: 219.

SARKAR, S., F. A. SRETER, and J. GERGELY. 1971. Light chains of myosins from white, red and cardiac muscles. *Proc. Nat. Acad. Sci.* **68**: 946.

SZENT-GYÖRGYI, A. G., C. COHEN, and J. KENDRICK-JONES. 1971. Paramyosin and the filaments of molluscan "catch" muscles. II. Native filaments: isolation and characterization. *J. Mol. Biol.* **56**: 239.

SZENT-GYÖRGYI, A. G., E. M. SZENTKIRALYI and J. KENDRICK-JONES, 1973. The light chains of scallop myosin regulatory subunits. *J. Mol. Biol.* In press.

WEBER, A. and R. D. BREMEL. 1971. Regulation of contraction and relaxation in the myofibril. In *Contractility of muscle cells and related processes*, ed. R. F. Podolsky. Prentice-Hall, Englewood Cliffs, N.J.

WEEDS, A. G. and B. S. HARTLEY. 1968. Selective purification of the thiol peptides of myosin. *Biochem. J.* **107**: 531.

WEEDS, A. G. and B. POPE. 1971. Chemical studies on light chains from cardiac and skeletal muscle myosins. *Nature* **234**: 85.

Studies on the Helical Segment of the Myosin Molecule

N. A. Biró, L. Szilágyi, and M. Bálint*

Biochemistry Group, Eötvös Loránd University, Budapest, Hungary

Some sort of cyclic interaction of the globular units, the "heads" of the myosin molecule, with ATP and with actin of the thin filaments is accepted by most of the authors to be the primary force-generating event in muscle activity. This explains the wide interest devoted to HMM and HMM-S-1, the fragments containing the "heads." The functional role of the long, helical "tail" of the myosin molecule, however, can not be regarded as secondary. Aggregation of the myosin molecules to form the thick filaments secures the exact geometry of the heads relative to the thin filaments and thereby the transfer and "integration" of the mechanical force generated in the elementary cycles. This aggregation is based principally or perhaps uniquely on structural properties of the tail segment. Proteolytic fragmentation derives its interest from the fact that by this means one can make a sort of "anatomy" of this part of the myosin molecule.

Table 1. Molecular Parameters of HMM-S-2 and HMM-S-3

	HMM-S-2		HMM-S-3
	(trypsin)	(papain)	
Molecular weight			
Yphantis's method	72,000	73,000	49,500
SDS electrophoresis	75,400	——	54,000
Helix content (%)	79	99	70
Intrinsic viscosity (dl/g)	0.45	0.45	0.25

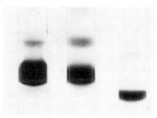

Disc electrophoretic pattern (native)

Data from Bálint et al., 1972a.

Besides the "classical" LMM known for about 20 years, more recently the rodlike part of HMM, HMM-S-2 (Lowey et al., 1967, 1969) and later the "total rod," a sort of "beheaded" myosin molecule

* Present address: Boston Biomedical Research Institute, Department of Muscle Research, Boston, Mass.

obtained by papain digestion (Lowey et al., 1969), were described. Our studies added three new fragments of LMM to these, LF-1 to F-3 (Bálint et al., 1968). They arise by successive stepwise

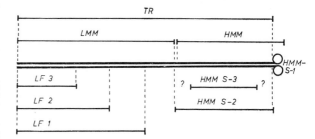

Figure 1. A sketch on the proteolytic dissection of the myosin molecule. The lengths of the helical fragments are drawn in proportion to molecular weights (see Table 3).

curtailing of LMM by prolonged tryptic digestion. In a paper about to be published (Bálint et al., 1972a) we described a fragment which is formed similarly from HMM-S-2 by the loss of some 20,000 daltons peptide material (see Table 1).

In Fig. 1 we give a sketch of the relation of all the seven helical fragments to the myosin molecule.

Summarizing the results of published works from our laboratory (Bálint et al., 1970; Szilágyi et al., 1972) and from other workers (McCubbin and Kay, 1968; Harrison et al., 1971), as well as some experiments to be described here, we can state that all these fragments are built up of two identical, continuous peptide chains. (By identical we mean within the limits of resolution of disc electrophoresis in urea and SDS.) This does not exclude a frayed-out ending, extending to a few amino acid residues only, which may well be present on the cut end(s) of the fragments. It can be stated, however, that the subunit chains of all these fragments do not give less sharp bands in urea and SDS gels than those of any intact protein.

All these fragments are highly helical; this means that they represent shorter or longer pieces of the α-helical coiled-coil of the myosin tail with essentially unaltered primary and secondary structure.

Generality of Fragmentation Pattern

A point deserving some attention is the fact that under a wide variety of proteolytic conditions one

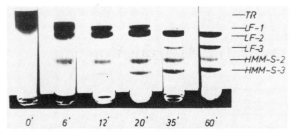

Figure 2. Digestion of total rods by trypsin. The disc electrophoretic pictures of digests obtained at different times are shown. Aliquots corresponding to equivalent amounts of *undigested* protein were electrophoresed. Total rods were prepared by dissolved papain as described by Bálint et al. (1972a) with precipitation of the helical fragments from the crude digest by acetone. Digestion by trypsin was done at pH 8.2 in 0.02 M KCl, 0.03 M Tris-HCl at 25° with a trypsin to protein ratio of 1:120 (w/w). Proteolysis was terminated by the addition of soybean trypsin inhibitor, twice the amount of trypsin present.

never obtains fragments not included among these seven. Changing the ionic milieu, e.g., can change the rate of splitting of LMM by trypsin considerably (Biró and Bálint, 1966) but not the nature of the fragments obtained (Bálint et al., 1968). Splitting of myosin by dissolved papain (Bálint et al., 1972b) results in exactly the same total rod as that obtained by the original procedure of Lowey et al. (1969), which uses papain insolubilized by attachment to cellulose. Tryptic digestion of total rods furnishes LMM and its fragments plus HMM-S-2 and HMM-S-3, identical to those obtained by other procedures (Fig. 2; Bálint et al., 1972b).

Digestion of myosin in free state, attached to actin, or built in the A filaments of myofibrils yields the same helical fragments. This holds true irrespective of the contracted or relaxed state of the myofibrils (Bálint et al., 1971).

The only case in which the influence of digestion conditions on the nature of the fragments formed is well established is the effect of Ca++ or Mg++ ion traces on the tryptic splitting of myosin and HMM (Bálint et al., 1971). Inclusion of EDTA in the digestion milieu increases the rate of splitting by trypsin of the connections between tail and heads. As a result, when myosin is digested in the presence of EDTA, the long-known splitting to LMM (and its fragments) and to HMM is greatly changed. Instead of HMM, HMM-S-2, and HMM-S-1, the tail and heads of HMM are formed (Fig. 3). As the HMM-S-2 formed directly from myosin is identical with that prepared by traditional procedures, here again conditions of digestion influence the relative rates of splitting rather than the nature of fragments.

A counterpart to this pertinacious constancy of the fragments formed under widely differing conditions of digestion is the observation that

prolonged digestion under "unappropriate" conditions does not yield all these fragments. When digestion of LMM by chymotrypsin (Bálint et al., 1968) or by papain (Bálint et al., 1972b) is forced, first a component roughly corresponding to LF-1 is formed, but later on all protein material is converted to smaller peptides.

It follows from this constancy of fragmentation pattern that the fragments are in some sense real structural parts of the molecule.

The structural relevance of this fragmentation pattern is stressed further by the results of some comparative studies made in our laboratory (Fekete et al., 1972). From rabbit red and hog cardiac myosin we obtained LMM and HMM-S-2, which were practically undistinguishable by electrophoresis in benign solvent, as well as in SDS, from those obtained from white muscle. Digestion of these LMM's by trypsin gave the same characteristic pattern described for mixed rabbit muscle. The direct formation of HMM-S-1 from red and cardiac myosin in the presence of EDTA adds a further item to the analogies with the white protein.

After all six helical components from red and heart myosin were identifiable with the corresponding white fragments, only one hitherto unknown helical component was found when digestion was done in the presence of EDTA. The relation of this

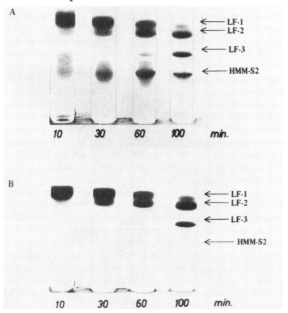

Figure 3. The effect of EDTA on the proteolysis of myosin by trypsin. The disc electrophoretic pictures of the alcohol-resistant fraction of digests obtained at different times are shown. *A*, 0.01 M EDTA included in the milieu; *B*, without EDTA. Digestion at pH 8.0 in 0.02 M KCl, 0.03 M Tris-HCl, with a trypsin to protein ratio of 1:120 (w/w). Digestion was terminated by the addition of soybean trypsin inhibitor, twice the amount of trypsin present. (From Bálint et al., 1971.)

latter fragment to the parent molecule is unelucidated till now. A further difference observed relative to white myosin was the formation of total rods in tryptic digests. When heart and red myosins were digested in the presence of EDTA, a helical component of 200,000 daltons was observed in the early digests. The well-known relative resistance of these myosins to tryptic cleavage at the LMM-HMM junction (Mueller et al., 1964; Gergely et al., 1965) together with the enhanced rate of splitting off of the heads in the presence of chelators mentioned above makes this finding quite plausible.

Some results in the literature on myosins other than from white rabbit muscle fit into this picture. Harrison et al. (1971) with chicken muscle myosin found LMM's and total rods giving segment-like aggregates very similar to the ones of the corresponding rabbit myosin fragments, and Kendrick-Jones et al. (1971) found the same with total rods obtained from chicken gizzard myosin.

Some recent results of Nakamura et al. (1971), however, showed a clear difference of the banding pattern in negatively and positively stained paracrystals formed by white LMM, on one hand, and by red or cardiac on the other. In spite of the differences in fine structural detail, a main periodicity of 43 nm was common to all three kinds of fragments. This points to some structural trait common to all three types of LMM's.

While a far-reaching analogy of the features of proteolytic fragmentation emerges from these observations, it is quite obscure what structural peculiarities of the primary structure and/or of stability relations of the coiled-coil direct proteolysis to the formation of just these fragments. If we want to turn the dissection of the myosin molecule into a sort of "functional anatomy," this question should be elucidated. One possible approach to this could be the detailed physical characterization of the fragments themselves. In the two following paragraphs we describe the results of our investigations in this direction.

Molecular Weights of Fragments

In our paper first describing the fragments of LMM (Bálint et al., 1968) we pointed out that the stretches digested away consecutively in an all-or-one fashion were all of equal size, about 30,000 daltons. The molecular weight of LMM-F-3, the shortest of the fragments, was roughly double this value. This has led us to the tentative assumption of LMM being a periodic structure built up from units of ca. 30,000 daltons. In this hypothesis L-F-3 was assumed to be built up of two, F-2 of three, F-1 of four, and LMM of five such units. The first published molecular weight of HMM-S-2

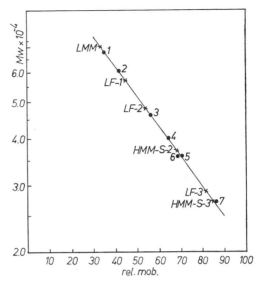

Figure 4. Determination of the subunit molecular weights of the helical fragments. Essentially the method of Weber and Osborn (1969) was used with a 10% gel, 3% crosslinking. The diagram represents log molecular weight plotted against relative mobility (referred to RNase). For calibration bovine serum albumin (1), catalase (2), actin (3), aldolase (4), phosphoglycerinaldehyde dehydrogenase (5), tropomyosin (6), and chymotrypsinogen (7) were used. All calibrating proteins were commercial preparations except actin and tropomyosin. The myosin fragments, represented by crosses, were prepared by methods usual in our laboratory (Bálint et al., 1968, 1972a, b).

(Lowey et al., 1967) promised that eventually this concept will be valid for the entire helical part of the myosin molecule.

To test this hypothesis we carried out a rather extensive series of molecular weight determinations of all the helical fragments by disc electrophoresis in SDS essentially as described by Weber and Osborn (1969). In Fig. 4 we show a representative experiment for the molecular weight determinations of all the fragments (except total rods). The average values of subunit weights obtained in all experiments of this type were used for the evaluation of the experiment, shown on Fig. 5. In this experiment the fragments of the LMM family, plus HMM-S-2, were used for calibration of the least concentrated gel used in order to obtain considerable mobility for total rods. These experiments gave a subunit weight for the total rod of 100,000–110,000 daltons.

In Table 2 we summarize the subunit weights obtained for the fragments. In Table 3 we united all the molecular weight data obtained earlier by ultracentrifugation methods (Bálint et al., 1968), the data accepted by the Lowey group (Lowey et al., 1969), and the molecular weights obtained from the data of Table 2. The agreement between values obtained by classical methods and by SDS electrophoresis, as well as between our data and

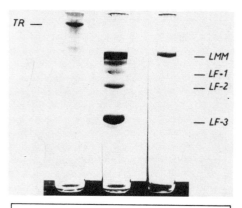

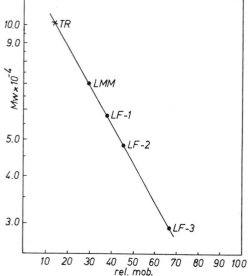

Figure 5. Determination of the subunit molecular weight of total rods. *Above:* The gels shown are, from *left to right:* Total rods, LMM plus its fragments, and LMM. *Below:* Logarithm of molecular weight vs. relative motility derived from the gels of *A*. Essentially the methods referred to in the legend of Fig. 4 were used, but with a 7.5% gel with 3% cross-linking. For calibration the fragments of the LMM family plus HMM-S-2 were used. Total rods were prepared by insoluble papain as described by Lowey et al. (1969).

those accepted by Lowey's group, is excellent. The only discrepancy that might be significant is in Lowey's and our molecular weight for HMM-S-2. Based on the accord of our high-speed equilibrium and SDS electrophoresis values, we may believe our results to be, perhaps, more confirmed.

Table 2. Subunit Molecular Weights of Helical Fragments

Fragment	Av. Mol Wt	SD	No. of Exp.
LMM	70.0	1.6	8
LF-1	57.5	0.6	12
LF-2	46.7	2.2	14
LF-3	29.5	0.5	10
HMM-S-2	37.7	1.2	7
HMM-S-3	27.0	0.6	4

Values (in thousands of daltons) derived by electrophoresis in the presence of SDS. See Fig. 4 legend for experimental details.

Table 3. Cumulative Data on Molecular Weights of Helical Fragments

Fragment	Ultracentrifugation Methods	Electrophoresis Method
Total rods	220 ± 20[a]	200[b]
LMM	150 ± 5[a]	———
	143[d]	140 ± 3.6[e]
LF-1	112[d]	115 ± 1.2[e]
LF-2	84[d]	93 ± 4.4[e]
LF-3	56[d]	59 ± 1.0[e]
HMM-S-2	62 ± 2[a]	75.4 ± 2.4[e]
	$72 \times$[c]	
HMM-S-3	48[c]	54 ± 1.2[e]

Values are in thousands of daltons.
[a] Lowey et al., 1969; [b] Harrison et al., 1971; [c] Bálint et al., 1972a—high speed equilibrium; [d] Bálint et al., 1968, apparent molecular weight values measured at 7 mg/ml concentration by the Archibald method. [e] See Table 2, this paper.

We regard the molecular weight data obtained by electrophoresis as extremely reliable. This is supported by the close agreement of the several (four to fourteen) independent measurements for each different fragment. A close scrutiny of the data shows that there are small, but perhaps not insignificant, deviations from the expectation of our hypothesis. Taking the mean deviations obtained on their face value, the consecutive "steps" digested away are between 30 to 20, 27 to 15, and 40 to 30, in thousands of daltons. For the step between HMM-S-2 and HMM-S-3, the difference lies between 25 to 18. The only generalization which can be maintained is that the degradation of these double coiled-coils goes on in all-or-one steps of a few 10,000's of daltons. This behavior is to some extent unexpected, as every such step contains some 40–50 bonds (subunit chains) falling under the specificity of trypsin. It is worth mentioning that tropomyosin (Ooi, 1967) shows a similar behavior against tryptic attack. It is well possible that the explanation lies in general principles of conformational stability.

Charge Distribution along Myosin Tail

Charges may play a predominant role in aggregation, the most important functional aspect of the myosin tail. We obtained a higher electrophoretic mobility for both HMM-S-2 and HMM-S-3 when compared to LF-3 in spite of their higher (or, in the case of HMM-S-3, roughly equal) size. This is in accord with amino acid analyses (Lowey et al., 1969) for HMM-S-2 which show that this stretch of the tail is more negatively charged in comparison to LMM.

The higher charge density of the stretch of the tail participating in the cross-bridge structure may have a functional role (Huxley, 1969).

At present we have no amino acid analyses for the fragments of LMM. Nevertheless, the relative electrophoretic mobilities in urea of these fragments

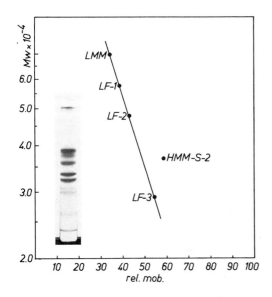

Figure 6. Electrophoresis of the LMM fragments and of HMM-S-2 in urea. The diagram shows the log molecular weight vs. relative mobility (referred to a marker dye). The inset shows a gel containing all the components concerned. Electrophoresis was in 6.6 M urea as described by Szilágyi et al., 1972.

allow the conclusion that specific charge along the LMM part of the tail is in first approximation constant. In Fig. 6 we show the disc electrophoresis in 6.6 M urea of a mixture containing all components of the L family, plus HMM-S-2. As can be seen on the diagram showing logarithm of molecular weight (obtained by SDS-electrophoresis, see Table 1) vs. relative mobility, the components of LMM and its fragments fall exactly on the same straight line, whereas HMM-S-2 falls outside the line. Urea does not suppress the individual charges of the components as SDS does. Thus the linearity of the log molecular weight-mobility relationship may be interpreted as reflecting a constancy of specific charge.

Cyanogen Bromide Degradation of LMM

Whereas molecular weight data discouraged, the findings of the charge distribution encouraged, to some extent, our speculations on a quasi-periodic structure of LMM.

If this alleged periodicity were the expression of a periodic primary structure, it could be most convincingly shown by methods of peptide chemistry. Cyanogen bromide cleavage seemed to be ideal to this end, as LMM contains only 28–30 methionines, i.e., 14–15 per chain.

After a systematic series of trials, we adapted the following procedure: The proteins were reduced and carboxymethylated in the presence of GuHCl, as described by Adelstein and Kuehl (1970), and

treated by cyanogen bromide following essentially the procedure of Elzinga (1970)—treatment in 70% formic acid, with a fivefold excess of the reagent, at 25°C for some 16 hr. The reaction mixture was lyophilized and the peptides separated in the gel system of Swank and Munkers (1971). This procedure uses excessively cross-linked and concentrated gels with urea and SDS in the solvent and separates peptides in the molecular weight range from some thousands to ca. 20,000. Staining was made by a methanolic solution of Coomassie brilliant blue.

Amino acid analyses of the peptide mixtures obtained by this procedure gave from zero to ten percent of methionines remaining; hence we regard the gel picture obtained by the finally adopted treatment as reflecting the practically complete splitting of methionines.

Figure 7 shows the gel electrophoresis pattern of the peptides obtained from LMM and from LF-3. In the case of LMM we obtained some 14–15 peptides, as expected for two identical chains, with about 30 methionines for a molecular weight of 140,000. In the case of LF-3 only four peptides were obtained, all four being represented also in the peptide spectrum of LMM. These results show clearly that LMM is built up of two identical peptide chains. There is no periodicity of the primary structure.

According to preliminary experiments the peptides obtained can be well separated by ion exchange chromatography on different exchange celluloses, monitoring being done by gel electrophoresis. We intend to carry on experiments of this sort with the two other subfragments of LMM, with HMM-S-2, as well as with total rods.

In principle these studies open up the way to the determination of the complete sequence of the

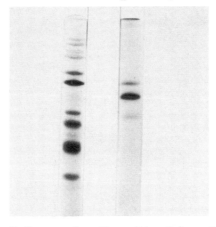

Figure 7. Cyanogen bromide peptides. *Left:* peptides of LMM; *right:* peptides of LF-3. The electrophoresis procedure of Swank and Munkers (1971) was used. Methods used for cyanogen bromide cleavage are described in the text.

myosin tail. This would be an enormous task involving the sequencing of around 200 peptides, with an average length of six residues. Besides this, some more realistic aims can be set as, e.g., the final decision to the question whether there is a substantial peptide stretch in the tail being the constituent of neither LMM nor HMM. Further, one can go deeper in comparative studies of myosins from different origins by a comparison of selected cyanogen bromide peptides.

Reversible Denaturation of Helical Fragments

Whereas in the case of myosin the reconstitutions of the native structure from subunits is still a somewhat controversial problem, the fragments of the tail can be denatured and refolded to the native state easily. McCubbin and Kay (1968) by osmometry and Bálint et al. (1970) by viscosity measurements and disc electrophoresis gave preliminary evidence for this in the case of LMM. Recently (Szilágyi et al., 1972) we substantiated and extended these findings to most of the helical fragments (total rod and HMM-S-3 not tested).

The helical fragments were transferred first to 6 M GuHCl in order to destory trypsin contaminations (Bálint et al., 1970). This solution was dialyzed against 6.6 M urea and finally transferred from urea to benign solvent by dialysis. The intrinsic viscosity and helix content (assessed by the 233 nm Cotton effect) showed a complete unfolding and refolding of LMM according to conditions. The return of helix content and intrinsic viscosity shows beyond doubt the regain of the double-helical rope structure. If the high helix content would mean the formation of single helices, intrinsic viscosity should show an increase of axial ratio.

The viscosity and ORD parameters of a mixture containing the four fragments of the LMM family plus HMM-S-2 also showed complete reversibility. In the case of a mixture the return of an additive property like viscosity to the original value does not allow an unambiguous interpretation. Disc electrophoresis carried out with the renatured mixture, however, showed that all five fragments had been exactly reformed, and nothing else is detectable in the gels. Figure 8 shows an experiment of this kind carried out with the mixture of the components of

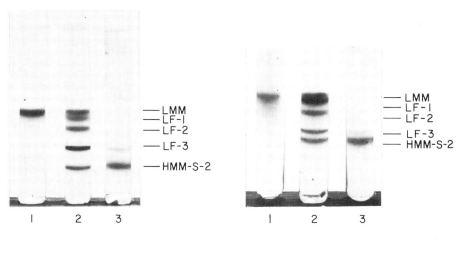

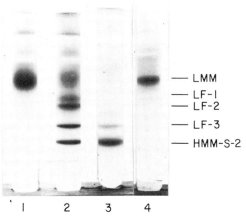

Figure 8. Disc electrophoresis of some helical fragments in native, urea-denatured, and renatured states A, Untreated samples; B, electrophoresis in the presence of 6.6 M urea; C, after elimination of urea by 48-hr dialysis. (1) LMM; (2) a mixture of the "LMM family" of fragments plus HMM-S-2; (3) HMM-S-2 contaminated with LF-3; (4) LMM as reassociated by 7 days of dialysis. (From Szilágyi et al., 1972.)

the LMM family plus HMM-S-2. The disc electrophoretic pictures for the reassociated samples were practically identical with those of the native mixture. The only difference found is the rather blurred character of the LMM band. If, however, a more prolonged dialysis is carried out, the disc of reassociated LMM becomes as sharp as that of the native one.

The unchanged number of components of the helical mixture in denaturing solvent substantiates the findings in the SDS electrophoresis experiments, showing that all these fragments are built up of a pair of equal chains; the reversibility of denaturation would hardly be possible if these chains were interrupted randomly by proteolysis.

The exact specificity of the reassociation according to chain length is somewhat unexpected. A considerable part of the peptide sequence of the shorter chains of the LMM fragments is identical with that of all longer chains present. If the aggregation of the two chains were governed uniquely by specific side chain interactions, there would be no reason for two chains of unequal length not to form a hybrid molecule, i.e., one containing at the end of a double coiled-coil stretch an unpaired polypeptide sequence (Fig. 9). Regarding the great sensitivity and resolving power of disc electrophoresis, such hybrids when present would certainly be detected. Since until now no single helical *protein* rod free in solution has been described, we may assume that in the course of reassociation the α-helical and coiled-coil structures develop simultaneously. We assume that hybrid molecules are not formed, because if they were, the part of the longer subunit not participating in the coiled-coil structure could not take the α-helical conformation either. Such a structure would be less stable thermodynamically than a

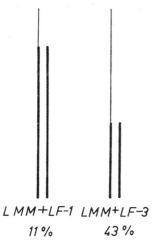

L MM +LF-1 LMM +LF-3

11 %   43 %

Figure 9. Schematic representation of two extremes of hybrid double coil molecules. Percents refer to the weight proportion of the unpaired peptide stretch.

perfect double rope because of the considerable number of hydrophobic side chains exposed to the solvent. Hence in a system in equilibrium such hybrids were eliminated.

Conclusions: Functional Outlooks

The results on the proteolytic fragments of the myosin tail allow the conclusion that these are in some sense real structural parts of the molecule. In spite of their proteolytic origin, each can in first approximation be regarded as definite a molecule as any "natural" protein is. Hence a detailed structural study of each fragment is justified and has its own merit. Beyond that we would like to look briefly into the possible functional relevance of all these findings.

The most obvious function of the tail of the myosin molecule is the building up of the A-filaments. The insolubility of myosin at low ionic strength can be regarded as a very crude expression of the forces bringing about this assembly. From the fragments only total rods and LMM share the solubility characteristics of the parent molecule. All shorter fragments are completely soluble at any ionic strength.

At ionic strength near physiological, rather uniform filamentous aggregates resembling A-filaments are formed by myosin (Noda and Ebashi, 1960; Huxley, 1963). According to our experiments (Biró, 1971, unpublished) the flow birefringence properties of total rods are strikingly similar to those observed with myosin. Under comparable conditions LMM also forms birefringent solutions, but these are extremely unstable. In the course of birefringence measurement, aggregation proceeds rapidly till reaching macroscopic dimensions (Biró and Wolf, 1971, unpublished experiments).

Recently very interesting studies were published on a special kind of aggregation of total rods and LMM (Cohen et al., 1970; Harrison et al., 1971; Kendrick-Jones et al., 1971). The fragments solubilized by the inclusion of KSCN among the constituents of a low ionic strength milieu were precipitated by low concentrations of Ca^{++} ions. Electron microscopy of the remarkably regular aggregates, "segments" formed under these conditions, gave important information on the length of the fragments, and the symmetry relations of their assembly. As the soluble LMM fragments and HMM-S-2 are precipitable by low concentrations of alkaline earth ions, the extension of these kinds of experiments to the smaller fragments seems to be feasible.

Somewhat unexpectedly our studies on the helical fragments gave some results which may have some bearing on the functioning of the heads and on the mechanism of cross-bridge movement.

The effect of EDTA on the rate of splitting-off of the heads by trypsin referred to above (Bálint et al., 1971) could be traced to the withdrawal of alkaline earth ion contaminations present in a rather important concentration (10^{-5} M) in ordinary salines. By metal buffer experiments it can be shown that Ca^{++} in a concentration of 10^{-7} M, or Mg^{++} in a concentration of 5.10^{-4} M, nearly completely suppresses the severing of the molecule by trypsin between tail and heads. We interpret these results as showing the presence at this region of the molecule of an ion-binding site or sites, the complexing of which by Ca^{++} or Mg^{++} confers some sort of stiffness to the structure. Under the ionic conditions assumed to prevail in vivo, these sites should be constantly occupied by Mg^{++} ions; hence this region should be in the rigid state irrespective of contraction or relaxation.

Nevertheless we feel that this influence of Ca^{++} and Mg^{++} ions on the structure of the stretch participating in the formation of the cross-bridges may have some functional importance. This assumption is further supported by the fact that this effect of EDTA on proteolysis seems to be shared by all kinds of myosins. With ageing the "EDTA effect" is lost, as is the case with respect to other important functional properties of myosin. This again points to a possible role of these alkaline earth-sensitive sites of myosin in the contractile mechanism.

The second fact of a possible functional importance emerges from our studies on the reversibility of denaturation of the helical fragments. We interpret the observed specificity according to length of reassociation as an expression of the instability which would prevail in the case of a hybrid formed from two subunits of unequal length.

Provided that hydrophobic (and perhaps other) interactions between the twin chains are needed to secure sufficient stability to the helical conformation, mechanical strain on one of the heads tearing open some turns of the helix belonging to this head would result in the uncoiling of the corresponding stretch of the second chain too. This in turn could influence the freedom of movement of the second head of the same myosin molecule. In vitro studies seem to indicate that myosin can attach to actin with one head at a time only (Young, 1967; Rizzino et al., 1970). In contrast to this, with isolated HMM-S-1's all actin monomers can be complexed (Moore et al., 1970). If our hypothesis on this kind of interaction between the heads via the double coil of the tail is accepted, it is possible that under mechanical strain the situation can come nearer the one found with free heads added to actin.

References

ADELSTEIN, R. S. and W. M. KUEHL. 1970. Structural studies on rabbit skeletal actin. I. Isolation and characterization of the peptides produced by cyanogen bromide cleavage. *Biochemistry* **9**: 1355.

BÁLINT, M., A. SCHAEFER, and N. A. BIRÓ. 1970. The subunit structure of light-meromyosin. *Acta Biochim. Biophys. Acad. Sci. Hung.* **5**: 45.

BÁLINT, M., GY. FEKETE, L. SZILÁGYI, M. BLAZSÓ, and N. A. BIRÓ. 1968. Studies on proteins and protein complexes of muscle by means of proteolysis. V. Fragmentation of LMM by trypsin. *J. Mol. Biol.* **37**: 317.

BÁLINT, M., A. SCHAEFER, N. A. BIRÓ, L. MENCZEL, and E. FEJES. 1971. Studies on proteins and protein complexes of muscle by means of proteolysis. VII. The presence of an alkaline earth metal sensitive site or sites in the heavy meromyosin part of myosin as revealed by proteolysis. *Physiol. Chem. Phys.* **3**: 455.

BÁLINT, M., A. SCHAEFER, L. MENCZEL, E. FEJES, and N. A. BIRÓ. 1972a. Studies on proteins and protein complexes of muscle by means of proteolysis. VIII. Characterization of the helical fragments derived from heavy meromyosin. *Acta. Biochim. Biophys. Acad. Sci. Hung.* In Press.

BÁLINT, M., L. MENCZEL, E. FEJES, and L. SZILÁGYI. 1972b. Studies of proteins and protein complexes of muscle by means of proteolysis. IX. Digestion of myosin by dissolved papain. *Acta Biochim. Biophys. Acad. Sci. Hung.* **4**: 88.

BIRÓ, N. A. and M. BÁLINT. 1966. Studies on proteins and protein complexes of muscle by means of proteolysis. I. Influence of ionic milieu on the proteolysis of myosin. *Acta Biochim. Biophys. Acad. Sci. Hung.* **1**: 13.

COHEN, C., S. LOWEY, R. G. HARRISON, J. KENDRICK-JONES, and A. G. SZENT-GYÖRGYI. 1970. Segments from myosin rods. *J. Mol. Biol.* **47**: 605.

ELZINGA, M. 1970. Amino acid sequence studies of rabbit skeletal muscle actin. Cyanogen bromide cleavage of the protein and determination of the sequence of seven of the resulting peptides. *Biochemistry* **9**: 1365.

FEKETE, GY., I. WOLF, and N. A. BIRÓ. 1972. A study of the tryptic fragmentation of white, red, and heart muscle myosins. *Symposium on Structure and Function of Normal and Diseased Muscle and Peripheral Nerve.* Kazimierz, Poland.

GERGELY, J., D. PRAGAY, A. F. SCHOLZ, I. C. SEIDEL, F. R. SRETER, and N. M. THOMSON. 1965. Comparative studies on white and red muscle. In *Molecular biology of muscular contraction*, p. 145. Igaku Shoin, Tokyo.

HARRISON, R. S., S. LOWEY, and C. COHEN. 1971. Assembly of myosin. *J. Mol. Biol.* **59**: 531.

HUXLEY, H. E. 1963. Electron microscope studies on the structure of natural and synthetic protein filaments from striated muscle. *J. Mol. Biol.* **7**: 281.

———. 1969. The mechanism of muscular contraction. *Science* **164**: 1356.

KENDRICK-JONES, J., A. G. SZENT-GYÖRGYI, and C. COHEN. 1971. Segments from vertebrate smooth muscle myosin rods. *J. Mol. Biol.* **59**: 527.

LOWEY, S., L. GOLDSTEIN, C. COHEN, and S. M. LUCK. 1967. Proteolytic degradation of myosin and the meromyosins by a water-insoluble polyanionic derivate of trypsin. Properties of a helical subunit isolated from heavy meromyosin. *J. Mol. Biol.* **23**: 287.

LOWEY, S., H. S. SLAYTER, A. G. WEEDS, and H. BAKER. 1969. Substructure of the myosin molecule. I. Subfragments of myosin by enzymic degradation. *J. Mol. Biol.* **42**: 1.

McCubbin, W. D. and C. M. Kay. 1968. The subunit structure of fibrous muscle proteins as determinated by osmometry. *Biochim. Biophys. Acta* **154**: 239.

Moore, P. B., H. E. Huxley, and D. J. DeRosier. 1970. Three-dimensional reconstruction of F-actin, thin filaments and decorated thin filaments. *J. Mol. Biol.* **50**: 279.

Mueller, H., M. Theiner, and R. E. Olson. 1964. Macromolecular fragments of canine cardiac myosin obtained by tryptic digestion. *J. Biol. Chem.* **239**: 2153.

Nakamura, A., F. Sreter, and J. Gergely. 1971. Comparative studies of light meromyosin paracrystals derived from red, white and cardiac muscle myosins. *J. Cell Biol.* **49**: 883.

Noda, H. and S. Ebashi. 1960. Aggregation of myosin. *Biochem. Biophys. Acta* **41**: 386.

Ooi, T. 1967. Tryptic hydrolysis of tropomyosin. *Biochemistry* **6**: 2433.

Rizzino, A. A., W. W. Barouch, E. Eisenberg, and C. Moos. 1970. Actin-heavy meromyosin binding. Determination of binding stoichiometry from adenosine triphosphatase kinetic measurements. *Biochemistry* **9**: 2402.

Swank, R. T. and K. D. Munkers. 1971. Molecular weight analysis of oligopeptides by electrophoresis in poly-acrylamide gel with sodium dodecyl sulphate. *Anal. Biochem.* **39**: 462.

Szilágyi, L., M. Bálint, and N. A. Biró. 1972. Specific reassociation of the polypeptide subunit chains of helical myosin fragments. *FEBS Letters* **21**: 149.

Young, M. 1967. Studies on the structural basis of the interaction of myosin and actin. *Proc. Nat. Acad. Sci.* **58**: 2393.

Weber, K. and M. Osborn. 1969. The reliability of molecular weight determinations by dodecyl sulphate—polyacrylamide gel electrophoresis. *J. Biol. Chem.* **244**: 4406.

Studies on the Structure and Assembly Pattern of the Light Meromyosin Section of the Myosin Rod

Michael Young, Murray Vernon King, Donald S. O'Hara, and Peter J. Molberg

Departments of Biological Chemistry and Medicine, Harvard Medical School 02115, and
Massachusetts General Hospital, Boston, Mass. 02114

It is now clearly established that several proteases of widely different specificity are capable of severing the myosin molecule into two main fragments (Mihályi and Szent-Györgyi, 1953; Gergely et al., 1955; Middlebrook, 1959; Kominz et al., 1965). These large pieces—heavy and light meromyosins—arise by transverse cleavage of the parent molecule somewhere near the middle of its rodlike tail section.

One of the main problems associated with preparation of the meromyosins by enzymic proteolysis is that both of these fragments are extensively cleaved internally as they are formed. For example, kinetic studies on the tryptic (Mihályi 1953; Mihályi and Harrington, 1959) and chymotryptic (Segal et al., 1967) hydrolysis of myosin reveal that the overall proteolytic process may be separated into two phases. The first and more rapid phase reflects cleavage of myosin into its main fragments, the meromyosins. The second and slower phase corresponds to hydrolysis of susceptible residues which are distributed throughout both HMM and LMM. Thus both meromyosins are more or less heterogeneous fragments of myosin, and they contain internal peptide bond interruptions.

In an attempt to study large intact sections of myosin and to avoid the heterogeneity problem discussed above, we have turned to nonenzymic cleavage with cyanogen bromide. This reagent specifically hydrolyzes methionyl peptide linkages (Gross and Witkop, 1961), and earlier studies have shown that it releases a large fragment from the C-terminus of myosin whose properties are closely similar to those of tryptic LMM (Young et al., 1968). Unlike enzymically produced LMM, however, this new fragment (LMM-C) contains no detectable internal peptide bond interruptions (Young et al., 1968). Thus BrCN slices the myosin rod within a very narrow region and releases the LMM section intact. This highly asymmetric fragment forms the bulk of the rod section of the myosin molecule, and its structure is largely a coiled-coil of α-helices (Crick, 1953; Pauling and Corey, 1953). In this study we have examined the myosin-BrCN reaction in detail, together with the molecular properties of LMM-C and some of its ordered assemblies.

Cleavage of Myosin by Cyanogen Bromide

We now have information from several sources that BrCN cleaves most compact globular proteins only at low pH values where susceptible methionyl linkages are exposed to solvent (see Witkop, 1968, for a summary of present evidence). On the other hand simple methionine-containing peptides are readily hydrolyzed by BrCN in dilute aqueous solution at *neutral* pH (Gross, 1967). Consequently in initial studies we have treated myosin with BrCN at neutral pH in an attempt to isolate both the HMM and LMM sections of the molecule. Although BrCN does split myosin under these conditions, extensive particle aggregation supervenes, and we have been unable to isolate a soluble macromolecular fragment. Similar results were obtained with myosin dissolved in 1×10^{-3} N HCl. However at pH 1 the cleavage reaction proceeds smoothly and inter-particle aggregation is minimized (Young et al., 1968). Fortunately, as discussed below, the helical structure of the LMM section of myosin remains virtually intact under these strongly acidic solvent conditions (cf. Lowey, 1965).

To study the kinetics of peptide bond cleavage solutions of myosin were first treated with DTNB (5,5'-dithiobis-(2-nitrobenzoic acid)) to block all sulfhydryl groups. (In addition to methionine BrCN also reacts slowly with cysteinyl residues, and formation of the mixed disulfide with DTNB prevents this reaction [Young et al., 1968].) The DTNB-blocked myosin dissolved in 0.1 N HCl was then treated with BrCN as described in the legend to Fig. 1. Aliquots of this solution were withdrawn as a function of time and analyzed for residual methionine. Figure 1 presents a plot of bonds hydrolyzed versus time and illustrates that extensive cleavage occurs during the first 3 hr. The half-life of this process is close to 120 min; and, although the reaction is complete after 22 hr, approximately 35% of the total methionine residues of myosin remain resistant to BrCN under the solvent conditions employed.

The rate and extent of production of LMM-C from myosin were also measured with the ultracentrifuge. For this purpose a 10 mg/ml solution of myosin in 0.1 N HCl was treated with 0.4 mg BrCN/mg protein at 25°. After several time periods aliquots were withdrawn and mixed with NaHSO₃ (equimolar to BrCN) to stop the reaction. (In acidic aqueous solvents NaHSO₃ rapidly and quantitatively reduces BrCN according to the reaction $H_2SO_3 + BrCN + H_2O \rightarrow HCN + HBr + H_2SO_4$, [Chattaway and Wadmore, 1902; Dixon and Taylor, 1913; Lang, 1925].) These mixtures were then each placed in a double-sector capillary-type synthetic boundary cell and examined with the ultracentrifuge. Under these solvent conditions HMM and undigested myosin are strongly aggregated, and they both sediment rapidly (<5–10 min at 60,000 rpm) to the base of the cell. In contrast LMM-C migrates as a single sharp boundary (with $S_{20,w}$ in the range 2–3 S at these protein concentrations), and its apparent concentration would be expected to be virtually unaffected by the Johnston-Ogston effect (Schachman, 1959). Values for LMM-C concentrations were calculated from area measurements of the LMM-C boundary, and these data were corrected for radial dilution and compared to the boundary area exhibited by undigested myosin at the same total protein concentration. Figure 2 displays the concentration of LMM-C (calculated as weight fraction of original myosin) as a function of time of digestion.

Examination of Fig. 2 reveals that the process of BrCN digestion is strikingly different from that of tryptic hydrolysis of myosin. For example, following liberation of LMM and HMM from myosin, trypsin continues to hydrolyze both meromyosins at a significant rate. In fact after

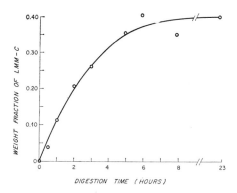

Figure 2. Kinetics of formation of LMM-C from myosin. A 10 mg/ml solution of myosin in 0.1 N HCl was treated with 0.4 mg BrCN per mg protein at 25°. At the indicated times samples were withdrawn, mixed with an equimolar (to BrCN) amount of NaHSO₃ to stop the reaction, and examined in a double-sector cell at 20° and 60,000 rpm in the ultracentrifuge. Area measurements of the LMM-C Schlieren boundary relative to undigested myosin yielded weight-fraction LMM-C.

approximately 2 hr at 25°, virtually all of the 250 lysine plus arginine bonds of LMM have been cleaved by trypsin (Young et al., 1964).

In contrast the apparent half-life of the BrCN reaction (Fig. 2) is about 120 min, and the amount of LMM-C liberated remains relatively constant 6–24 hr after digestion. Thus the data of Fig. 2 suggest that LMM-C is resistant to further fragmentation by BrCN and that it comprises close to 38% of the mass of the myosin molecule. Further evidence for the marked resistance of LMM-C to BrCN cleavage will be presented below.

Isolation and Molecular Properties

LMM-C was isolated and purified as follows. Solutions of myosin, 20 mg/ml in 0.5 M KCl, 0.01 M Tris-HCl, pH 8.0 were treated at 25° with 0.0594 mg DTNB per mg protein to block all sulfhydryl groups. The sulfhydryl-blocking reaction was followed kinetically by increase in absorption (410 nm) due to formation of 5-mercapto-2-nitrobenzoic acid (Ellman, 1959). Reaction was complete (7.5 moles of cysteinyl residues reacted/ 10^5 g of myosin) after 1.5 hr at 25°, and the protein solution was dialyzed for 48 hr against 0.1 N HCl at 4°. After dialysis the acidic solution was warmed to 25° in a constant temperature bath, protein concentration was adjusted to 10 mg/ml with 0.1 N HCl, and 0.4 mg BrCN (dissolved in 0.1 N HCl) was added per mg myosin. After 45 min the hydrolytic reaction was stopped by addition of an equimolar amount of NaHSO₃ (with respect to BrCN) dissolved in water.

To isolate LMM-C the acidic reaction mixture was first dialyzed exhaustively against 0.1 N HCl at 4° to remove HCN and then against 0.5 M KCl,

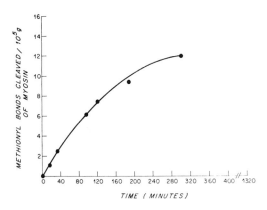

Figure 1. Cleavage of myosin with BrCN. Myosin was treated with DTNB to block all sulfhydryl groups, following which the protein (10 mg/ml, dissolved in 0.1 N HCl) was digested with 0.4 mg BrCN per mg protein at 25°. Samples were withdrawn and immediately lyophilized at the times indicated. The dry protein was hydrolyzed with 6 N HCl in vacuo at 110° and analyzed for methionine with a Beckman 120B amino acid analyzer. (From Young et al., 1968.)

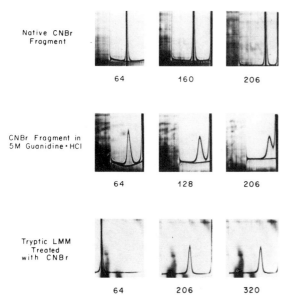

Native CNBr Fragment

64   160   206

CNBr Fragment in 5M Guanidine·HCl

64   128   206

Tryptic LMM Treated with CNBr

64   206   320

Figure 3. Sedimentative velocity studies of purified LMM-C. Time intervals (min) are shown beneath each frame. *Top:* Synthetic boundary cell run; 9.4 mg/ml protein; 0.5 M KCl, 0.01 M Tris-HCl pH 8.0; 44,770 rpm; temperature 3°. *Middle:* Reduced carboxymethylated LMM-C, synthetic boundary cell, 9.1 mg/ml protein in 5.11 M guanidine-HCl; 44,770 rpm at 24°. *Bottom:* Tryptic LMM Fr. I, prepared by the method of Young et al. (1964) and treated with 0.4 mg BrCN per mg protein for 24 hr at 25°. Solvent 0.1 N HCl; 44,770 rpm, protein concentration 10 mg/ml; temperature 25°.

0.1 M Tris-HCl pH 8.0. At this point a large precipitate (consisting of undigested and denatured myosin together with denatured HMM) is removed by centrifugation for 30 min at 48,000 g. To purify LMM-C from other hydrolysis products of myosin we have taken advantage of the fact that this fragment (like tryptic-LMM) is resistant to denaturation by high concentrations of ethanol (Szent-Györgyi et al., 1960). Three volumes of 95% ethanol was added to the supernatant solution from the preceding step. The precipitate was collected by centrifugation at 48,000 g for 10 min, dissolved in 0.5 M KCl, 0.01 M Tris-HCl pH 8.0 and dialyzed against this solvent at 4°. LMM-C was further purified by gel-filtration chromatography exactly as described earlier (Young et al., 1968).

The sedimentation properties of LMM-C were studied at high protein concentration with a double-sector synthetic boundary cell to detect small amounts of either faster or slower sedimenting components. Figure 3 depicts a single boundary.

The most compelling evidence for homogeneity of the protein stems from the fact that it contains virtually no homoserine. Table 1 presents results of amino acid analyses of LMM-C isolated after both 45 and 90 min of digestion with BrCN. With the exception of homoserine these two analyses are identical within experimental error, and they closely resemble that for tryptic LMM (Lowey and Cohen, 1962). Of particular significance is the low level of homoserine, especially that arising from the 45 min preparation where the value of 0.05 moles per 10^5 g protein represents less than 0.1 residue per molecule. Thus although BrCN splits many methionyl peptide bonds in *myosin* (Fig. 1), solutions of purified LMM-C are uncontaminated by any of these homoserine-containing fragments. Moreover as we have pointed out elsewhere (Young et al., 1968), the lack of homoserine within LMM-C establishes that this fragment arises from and comprises the C-terminal end of the myosin molecule.

Before leaving this section it is pertinent to consider another unusual property. As mentioned above the data of Figs. 1 and 2 taken together suggest that whereas myosin is extensively cleaved by BrCN, LMM-C is relatively resistant to further digestion. Table 2 lists values of the homoserine content of LMM-C isolated after several time periods of digestion, and it will be appreciated that after about 5 hr, only two moles of bonds have been cleaved per 10^5 g even though LMM-C contains 16 potentially susceptible moles of methionine residues per 10^5 g protein (Table 1). Thus although the LMM rod is readily severed from myosin by BrCN, it remains relatively resistant to further cleavage in aqueous 0.1 M HCl. We assume that this feature stems from a polypeptide chain packing arrangement wherein methionyl peptide bonds are shielded

Table 1. Summary of Amino Acid Analyses of LMM-C

| | Moles/10^5 g Protein | |
	45 min	90 min
	Moles/10^5 g	Moles/10^5 g
Lys	96	96
His	19	19
NH₃	111	107
Arg	58	58
Asp	81	81
Thr	32	32
Ser	40	40
Glu	224	226
Pro	0	0
Gly	17	15
Ala	81	82
Cys	4.6	4.9
Val	36	35
Met	17	16
Ile	33	33
Leu	94	96
Tyr	6.8	6.0
Phe	6.3	5.9
Hse	<0.05	0.3

Homoserine lactone was quantitatively converted to homoserine with pyridine acetate (Ambler, 1965; Young et al., 1968). Values for cysteine were obtained from cysteic acid analyses of performic acid-oxidized protein (Moore, 1963). Data are not corrected for hydrolytic destruction and tryptophan was not determined. Twenty-four hour 6 N HCl hydrolysis at 110° in vacuo.

Table 2. Homoserine Content of LMM-C Isolated after Varying Digestion Intervals

Digestion Time (Minutes)	Moles Homoserine/10^5 g Protein
45	<0.05
90	0.6
240	1.1
270	2.0

Myosin (10 mg/ml in 0.1 N HCl) was treated with 0.4 mg BrCN per mg protein at 25° and LMM-C isolated as described in the text. Twenty-four hour 6 N HCl hydrolysis at 110° in vacuo.

from solvent. The fact that 95–98% of the methionyl linkages of LMM-C are cleaved by BrCN in the presence of 70% formic acid supports this idea. The sedimentation patterns of Fig. 3 (bottom) further illustrate that tryptic LMM is relatively resistant to BrCN even after 20 hr at room temperature. Thus as shown in Fig. 3, preparations of tryptic LMM still yield a single boundary after prolonged treatment with BrCN in 0.1 M HCl. The sedimentation coefficient of the peak shown in Fig. 3 (bottom) is $S_{20,w} = 2.1S$—a value close to that of tryptic LMM at the same protein concentration ($S_{20,w} = 2.2S$) (Young et al., 1964). We do not mean to imply here that tryptic LMM is totally unaffected by BrCN. Indeed the single sedimenting boundary shown in Fig. 2 (bottom), although symmetrical, is too broad to represent a single component. Yet the absence of more slowly moving material together with the data of Table 2 indicate that tryptic LMM survives BrCN relatively intact.

The optical rotatory properties of LMM-C reveal that this fragment, like tryptic LMM, contains a large number of residues in α-helical conformation. Values for helix content were calculated from the Moffitt (1956) treatment over the spectral range $\lambda = 500$–300 nm. Data were corrected for the refractive index dispersion of water, and λ_0 was taken to be 212 nm. Table 3 summarizes values for

Table 3. Apparent Helix Content of LMM-C in Several Solvent Systems

Solvent	b_0	App. % Helix
0.5 M KCl, 0.01 M Tris pH 8.0	−670°	100
0.1 M HCl	−630°	94
70% HCOOH	−550°	82
1.0 M Guanidine-HCl	−300°	45
5.0 M Guanidine-HCl	−30°	0

Values of b_0 were calculated from least-squares analyses of plots of $[m']_\lambda(\lambda^2 - \lambda_0^2)/\lambda_0^2$ versus $\lambda_0^2(\lambda^2 - \lambda_0^2)$ according to the Moffitt (1956) expression. The mean residue weight of LMM-C is 118 nd λ_0 was taken to be 212 nm. Values of apparent helix content were estimated on the assumption that $b_0 = -670°$ reflects 100% α-helices. Data were recorded ($\lambda = 500$–300 nm) with a Cary Model 60 spectropolarimeter at 20° with a 1-cm jacketed quartz cell.

b_0 and apparent helix content of LMM-C in various solvent systems. It will be appreciated that b_0 changes little, if at all, when the protein is transferred from a neutral pH solvent to 0.1 M HCl. As judged by this criterion the helical backbone of LMM-C remains stable under these strongly acidic conditions. The observation that LMM-C is devoid of proline (Table 1) residues is consistent with its high helix content. On the other hand the absolute value of b_0 decreases in the presence of 70% formic acid, and, as we have already noted above, this solvent favors cleavage of all methionyl peptide bonds by BrCN.

Molecular weight and hydrodynamic properties. Values for the apparent partial specific volume (ϕ') (Casassa and Eisenberg, 1961) of LMM-C in several solvent systems are presented in Table 4. In the presence of 0.5 M KCl, the value $\phi' = 0.704$ ml/g is in good agreement with that determined previously for tryptic LMM ($\phi' = 0.701$ ml/g) (Young et al., 1964). When the protein is transferred to 5 M guanidine-HCl the mean value for ϕ' decreases by about 1%, and this change parallels that observed with other proteins in concentrated guanidine-HCl (Hade and Tanford, 1967).

A rather unexpected and more interesting finding is that ϕ' for LMM-C changes profoundly from 0.704 to 0.747 ml/g when the protein is transferred from KCl to 0.5 M phosphate (Table 4). Similar changes in ϕ' have been found with myosin (Godfrey and Harrington, 1970). The basis for this effect will be appreciated if we consider the relationship between the true thermodynamic partial specific volume ($\bar{v}$) and the apparent specific volume (ϕ') (Casassa and Eisenberg, 1961).

$$\phi' = \bar{v}_p - \gamma\left(\frac{1}{\rho} - \bar{v}_s\right)$$

In this expression $\bar{v}_p$ and $\bar{v}_s$ represent true partial specific volumes of protein and salt, respectively, ρ is the solvent density and γ is the number of grams of salt bound per g of macromolecule in the multicomponent system. (If γ is negative, then the salt component is being "rejected" by the macromolecule.) For tryptic LMM dissolved in KCl the true partial volume has been measured to be $\bar{v}_p = 0.711$ ml/g (Young et al., 1964). If we assume that this value holds for LMM-C in both KCl and phosphate solvents, then $\gamma_{KCl} = 0.02$ g/g and $\gamma_{PO_4} = -0.058$ g/g. Consequently LMM-C binds about 0.02 g of KCl when 0.5 M KCl is the solvent, yet it "rejects" phosphate (i.e., is preferentially hydrated) in the presence of 0.5 M potassium phosphate, pH 8.0. The reason for this effect is not

Table 4. Apparent Partial Specific Volume of LMM-C in Various Solvent Systems

Prep.	Method	Solvent	Protein Conc. (mg/ml)		ϕ' (ml/g)
64	D$_2$O/H$_2$O	KCl	2.88		0.702 ± 0.007
70	pycnometric	KCl	14.0		0.706 ± 0.008
				Av.	0.704
70	pycnometric	phosphate	14.7		0.757 ± 0.007
70	pycnometric	phosphate	13.1		0.737 ± 0.007
				Av.	0.747
64	pycnometric	5.11 M guanidine-HCl	10.4		0.686 ± 0.011
82	pycnometric	5.03 M guanidine-HCl	17.4		0.702 ± 0.018
				Av.	0.694
82	pycnometric	6.06 M guanidine-HCl	18.1		0.711 ± 0.018

Values for ϕ' were calculated according to the theory of Casassa and Eisenberg (1961) from pycnometric measurements and by sedimentation in H$_2$O and D$_2$O as solvents according to Edelstein and Schachman (1967). Protein concentration was determined spectrophotometrically (Young et al., 1968), and all density measurements were in duplicate. For measurements with guanidine-HCl, LMM-C was fully S-carboxymethylated (Crestfield et al., 1963). KCl refers to 0.5 M KCl, 0.01 M Tris-HCl pH 8.00. Phosphate refers to 0.5 M potassium phosphate pH 8.00.

clear, but it may depend upon the high density of polar amino acid residues within the coiled-coil backbone of LMM.

Values for the molecular weight of several LMM-C preparations have been measured over a wide range of protein concentrations and rotor velocities by high-speed equilibrium centrifugation (Yphantis, 1964). In particular the absorption optical system (operating at $\lambda = 230$ nm) was used with extremely low protein concentrations, and the numbers obtained with this system are essentially those which would apply at infinite dilution of protein.

Table 5 presents data for number- and weight-average molecular weights obtained with the absorption optical system, and the close agreement between these two numbers again reflects a high degree of homogeneity. It will also be appreciated that when the appropriate value for ϕ' is used (0.747 ml/g), the mass of LMM-C dissolved in phosphate is virtually identical to that found for KCl. When all results are taken together, the weight of LMM-C is close to 173,000 g/mole (Table 5). This number fits well with the molecular weight

calculated from the weight fraction of myosin released as LMM-C (38%, Fig. 2). For example if we take the mass of myosin to be 458,000 g/mole (Godfrey and Harrington, 1970), then 38% × 458,000 = 174,000.

Subunit structure. Amino acid analyses presented above demonstrate that LMM-C contains virtually no internal peptide bond splits. Therefore this molecule should dissociate to yield large and intact subunit polypeptide chains. Figure 3 demonstrates that when reduced S-carboxymethylated LMM-C (SCM-LMM-C) dissolved in 5 M guanidine-HCl is examined with a synthetic boundary cell, only a single sedimenting component is detected.

In light of the high degree of physical and chemical homogeneity of the LMM chains discussed above, we have measured their mass. Figure 4

Table 5. Molecular Weight of LMM-C

Solvent	RPM	Initial Conc. (mg/ml)	M_n	M_w
KCl	16,000	0.024	165,000	168,000
Phosphate	16,000	0.027	170,000	172,000
Phosphate	18,000	0.027	182,000	181,000
		Mean values:	172,000	173,000

Sedimentation equilibrium measurements utilized a photoelectric scanning system operating at $\lambda = 230$ nm and very dilute protein solutions. Three-mm columns of protein solutions were layered over perfluoromethyldecalin as a base fluid, which is transparent at 230 nm. Equilibrium traces were taken after 48 hr and temperature was maintained close to 4°. KCl and phosphate solvents are the same as in Table 4.

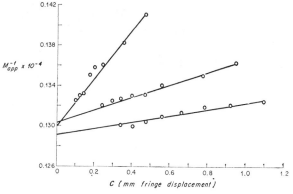

Figure 4. Plots of $1/M_{app}$ versus concentration for S-carboxymethylated LMM-C in 5.0 M guanidine-HCl. Sedimentation equilibrium patterns (Rayleigh optics) were photographed after 48 hr at 26,000 rpm and 25°. Initial protein concentration, 0.1 mg/ml. All data were analyzed by computer and top, middle, and lower plots refer to z, weight, and number average molecular weights, respectively.

illustrates typical plots of $1/M_{app}$ (number, weight, and z-average) for SCM-LMM-C in 5 M guanidine-HCl. The unfolded chains in a strongly denaturing solvent demonstrate measurable thermodynamic nonideality, and as shown in Fig. 4 the slopes of the $1/M_{app}$ vs. c plots increase in the order $n < w < z$. This behavior is just that which would be expected on theoretical grounds (Fujita, 1962), and the fact that these plots converge to closely similar values for $1/M$ at $c = 0$ again reflects a high degree of particle-size homogeneity for the LMM-C subunit.

From studies on three preparations as a function both of protein concentration and rotor speed, we find M_n and M_w are essentially identical with a mean value of 83,000 g/mole.

Length of the LMM-C rod. If we adopt the theoretical coordinates of Crick (1953) for a two-stranded coiled-coil of pure α-helices, then the length:mass ratio for LMM-C is 0.006 Å/dalton. If we further assume that LMM-C is a fully helical rod (and that its mass lies somewhere between $2 \times 83,000$ and 173,000 g/mole), then its end-to-end length must be 990–1040 Å.

The contour lengths of a total of 110 shadow cast particles were measured with the electron-microscope, and from these numbers we calculate that the mean, median, and mode of the length distribution for the LMM rod are 1000, 970, and 930 Å, respectively. The standard deviation was 160 Å.

Measurements on a different type of object provide an independent estimate of the length of the LMM-C molecule. LMM-C associates at low ionic strength to form a wide variety of ordered aggregates. Among these we have observed ribbon-like structures (Fig. 5) which resemble the segment phases obtained by Cohen et al. (1970) from myosin rods prepared by papain digestion. We have obtained two such phases (King and Young, 1972a) that throw light on the molecule length of LMM-C. One of these is illustrated in Fig. 5. Like certain of the segment structures of Cohen et al. (1970) these phases exhibit a more densely packed central strip bordered by fringes of equal width on both sides, and they show a transverse banding pattern having apparent bilateral symmetry. Thus they can be interpreted as laterally packed LMM-C molecules with alternate molecules in antiparallel

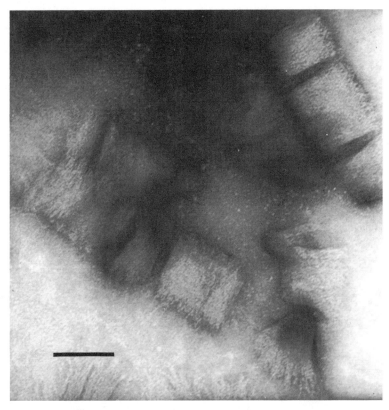

Figure 5. A phase of LMM-C prepared by dialysis with gradual dilution to a final concentration of 0.59 M ammonium acetate, stained with uranyl oxalate. Note that the objects are fringed on both sides (unsupported end of LMM-C molecules protruding from the central overlap region where the molecules make contact with neighbors on both sides). These aggregates show both transverse striations (indicating the molecular directions) and longitudinal striations (indicating alignment of uranylbinding groups or voids). Bar = 1000 Å. (From King and Young, 1972a.)

Table 6. Summary of Some Molecular Properties of LMM-C

Mass (native)	173,000 g/mole
Mass (subunit)	83,000 g/mole
b_0	$-670°$
$S_{20,w}^\circ$	3.25 S
Length of ribbon structures (from EM)	1009 ± 47 Å
Length of individual molecules (EM)	$930 - 1000$ Å
Length (calculated for 2-stranded supercoil of mass 173,000)	1000 Å

orientations. However these two phases differ from each other, and also from those prepared by Cohen et al. (1970), in the widths of the central overlap strips as well as in the details of their banding patterns.

Seven independent measurements on micrographs of the phase shown in Fig. 5 give widths of the center strip of 890 Å (standard deviation 17 Å). The corresponding width of the fringe was 119 Å (standard deviation 44 Å). Adopting the sum of widths of the center strip and one fringe as the molecular length of LMM-C, we get a value of 1009 Å, with a standard deviation of 47 Å. When taken together with measurements of individual molecules, these data place the length of LMM-C very close to 1000 Å. Table 6 lists several of the molecular properties of LMM-C outlined above.

On the Idea of a "Hinge" Section Joining LMM and HMM

Perhaps the most interesting difference between LMM-C and tryptic LMM lies in the fact that whereas trypsin hydrolyzes all of the susceptible peptide bonds of LMM quite rapidly, LMM-C remains highly resistant to BrCN even though it contains many potentially susceptible methionyl residues. A plausible explanation for this difference between trypsin and BrCN arises when we consider models for the coiled-coil class of fibrous proteins (Crick, 1953; Pauling and Corey, 1953). In these models the polypeptide chains are assumed to be arranged in such a way that the relatively nonpolar residues pack between the chains and lie shielded from contact with solvent. Thus we infer that the methionyl residues (with their long nonpolar side chains) are protected from solvent contact, whereas the more polar arginyl and lysyl residues are readily accessible to trypsin. In this connection it is interesting that the amino acid composition of LMM-C conforms rather closely to the distribution of amino acids required by a coiled-coil structure. For example as Crick (1953) pointed out, the greatest energetic stabilization of a multistranded cable would occur if all nonpolar residues tended

to pack together between the chains. Based upon the structural coordinates of a coiled-coil, this would mean that approximately 2/7 of all residues should contain hydrophobic side chains. From the data of Table 1 we calculate that the sum of alanine, methionine, valine, leucine, isoleucine, and phenylalanine is close to 2/7 of all residues. Sequence information will be necessary before we know whether these numbers are really meaningful.

Since the α-rope structure is relatively resistant to hydrolysis by BrCN, it is somewhat surprising that the LMM rod is so much more rapidly liberated from myosin. This feature suggests that there must be something rather different about the methionyl residues which *are* cleaved by BrCN. As outlined above kinetic studies on the tryptic digestion of myosin have shown that LMM and HMM are liberated rapidly by a reaction whose rate is much faster than that attributed to internal cleavage of the meromyosins themselves. Chymotrypsin displays similar kinetic behavior toward myosin (Segal et al., 1967). Two ideas have been suggested to account for this effect. The first is that there is a narrow cluster of residues which joins HMM and LMM together whose three-dimensional folded structure is more open and less compact than the adjacent tightly packed helical regions of the myosin rod (Mihályi and Harrington, 1959; Young et al., 1964; Segal et al., 1967; Young, 1969). The other possibility is that this belt of residues is preferentially rich in amino acids susceptible to attack by several proteolytic enzymes (Lowey et al., 1967; Lowey et al., 1969). In view of the facts (1) that cleavage of only two methionyl residues is necessary to cut the myosin rod in two and (2) that this reaction is much faster (at least by an order of magnitude) than the internal cleavage of the LMM-C rod by BrCN, it would appear that the local environment of *at least* two methionyl bonds in the region linking HMM and LMM together is less compactly shielded (or is more unfolded) than the remaining methionyl bonds of LMM-C. Thus although these results cannot be used to prove that a local section of the myosin rod has a more open and flexible structure, they indicate that this is a likely possibility—in agreement with kinetic studies on the proteolytic digestion of myosin referred to earlier. In no other way can we easily explain the rapid scission of the rod together with the marked resistance of LMM to further digestion by BrCN.

The structure, function, and size of this connecting region are unknown. Yet among several possibilities, one which is rather more interesting is that it might be a more or less flexible joint which serves to link HMM and LMM together (Mihályi and Harrington, 1959; Pepe, 1967). In fact the gross packing arrangement of myosin to form

Table 7. Summary of pH-Stat Studies on Proteolytic Digestion of LMM-C

Preparation	Slow Bonds	Fast Bonds	$k \times 10^2$ (slow)	$k \times 10^2$ (fast)
Myosin	260	69	1.7	15
LMM-C	240	50	1.6	20

LMM-C (40 mg) dissolved in 0.5 M KCl was treated with 0.5 mg 3 × recrystallized trypsin (Worthington) and proton liberation followed with a pH-stat (pH = 7.85) at 20° under a CO_2-free nitrogen barrier. The titrant was a standardized solution of 0.01 N NaOH. Graphic analysis of bonds cleaved versus time revealed the presence of 2 parallel first-order reactions, and the number of bonds cleaved in the "fast" and "slow" reactions were calculated as described by Young et al. (1964). Rate constants (k) were calculated from natural logarithms and time in minutes; bonds cleaved are given in moles bonds per mole of LMM-C or myosin. Data for the tryptic cleavage of myosin are taken from Mihályi and Harrington (1959), corrected for a molecular weight of myosin of 458,000 g/mole (Godfrey and Harrington, 1970).

thick filaments (Huxley, 1963; Pepe, 1967; Huxley and Brown, 1967) would appear to require some sort of deformable section of the rod portion of the molecule. Specifically we now know that at least one of the functions of LMM is to make up the backbone of thick filaments. But the HMM cross-bridges project laterally out from this backbone, and more importantly the X-ray evidence indicates that they change their angle of tilt and azimuth during contraction (Reedy et al., 1965; Huxley et al., 1965; Huxley and Brown, 1967; Huxley, 1968). It may be that a flexible HMM-LMM joint participates in the structural basis of this arrangement.

Whatever the function of this highly enzyme-sensitive region may be, the following studies reveal that the bulk of it is retained within LMM-C. Preparations of LMM-C dissolved in 0.5 M KCl were treated with trypsin at 25°C and the kinetics of proteolysis was measured with the pH-stat. Results presented in Table 7 reveal that of the 69 rapidly cleaved bonds present in myosin, at least 50 of them reside within LMM-C. The rate constants for both the "slow" and "fast" reactions are also closely similar for myosin and LMM-C. What this probably means is that BrCN severs LMM-C from myosin on the subfragment 2 (Lowey et al., 1966) side of the proteolytic enzyme-sensitive region.

Structure of Thick Filament

One particular aggregate of LMM-C has provided us with the information that the tail region of the myosin molecule is flexible and that a flexed structure can be the equilibrium conformation under certain conditions of aggregation.

This aggregate (King and Young, 1972b) grows from ammonium chloride solution in long, tubular structures. Although these tubes are presumably cylindrical when suspended in solution, they collapse and flatten upon mounting for electron microscopy and split easily along a set of parallel cleavage paths inclined to the axes of the cylinders, as shown in Fig. 6. This process results in ribbons,

some of which are much narrower than the length of the LMM-C molecule. This suggests that the LMM-C molecules lie along the cleavage paths, i.e., along helical arcs in the original cylindrical aggregate. This observation is further strengthened by the fact that the tubes show a striation pattern inclined to the edges of the tubes (with a period of 426 Å) but roughly normal to the cleavage direction. The best specimens, like a transparent barber pole, show two sets of striations inclined at equal and opposite angles to the edges, presumably arising from front and back of the specimen. However a tube that has suffered a single cleavage and has then become flattened into a single-thickness ribbon now shows striations normal to the edges, i.e., normal to the length of the molecules.

Although the curvature and torsion of the molecules varied considerably among the specimens, we did not find any specimens with straight molecules lying along the generators of the cylinders. This in itself suggests that the molecules are preferentially twisted, with fluctuation about equilibrium values of the curvature and torsion.

The import of these observations is that, if the tail region of myosin is actually flexible along its length, then it becomes feasible to build models of the thick filament based on twisting of the myosin tails into compact, ropelike structures.

Let us now examine a possible model for the thick filament of muscle that can be devised on the basis of symmetry considerations together with the notion that the tail of the myosin molecule is flexible. The postulates are: (1) Different regions of the thick filament can differ in local symmetry. The presented model is intended to refer only to the untapered part of the lateral regions of the filament where myosin cross-bridges are made. (2) The model conforms to a 6_2 or 6_4 screw axis with a 429-Å axial repeat distance (Huxley and Brown, 1967). (3) All myosin molecules in the model are related to one another by symmetry operations. (4) The myosin tails are flexible and are interwound in such a way as to maintain a close packing

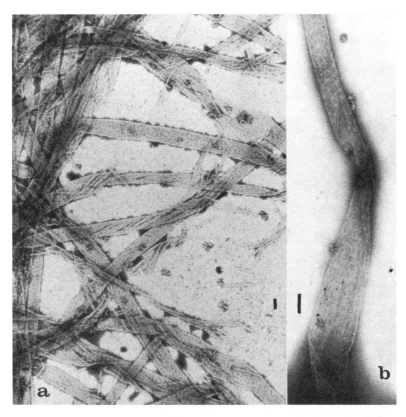

Figure 6. A phase of LMM-C prepared by dialysis with gradual dilution to a final concentration of 0.2 M ammonium chloride, pH about 8. Bars = 1000 Å. (*a*) A field containing many specimens, fixed with glutaraldehyed and stained with sodium tungstate. Many of the objects have undergone repeated splits along parallel inclined paths to give rise to ribbons that are often very narrow. These features suggest that the objects are flattened tubes that possess a direction of easy cleavage inclined to the axis and that the molecules of LMM-C lie parallel to the lines of cleavage. (*b*) A specimen that has been fixed with glutaraldehyde and stained with ammonium molybdate. The intact part of the tube shows two superposed striation patterns inclined at equal and opposite angles to the tube edges, whereas the portion that has been cleaved to give a ribbon shows striations normal to the edges. The striation patterns are rather faint, and can be seen most clearly by holding the micrograph at arm's length and inclining it to the line of sight. (Adapted from King and Young, 1972b.)

arrangement, while gradually emerging from the interior of the filament as one proceeds from tail to head end.

In detail we assume that a particular close-packed arrangement of the tails of myosin molecules must recur in transverse cutting planes every 143 Å along the filament, with a rotation $\pm 60° \pm 180\, n°$ between the patterns existing on successive levels. (Here the $\pm 60°$ rotations arise from the postulated 6_2 or 6_4 axis, and the added $180\, n°$ of rotation involves both the facts that a 6_2 or 6_4 axis contains an embedded twofold rotation axis and that there are many ways of connecting the nodes of a helical net.) Further the embedded twofold rotation axis implies that all cross sections of the filament will possess twofold rotor symmetry. Thus for a chosen cross-sectional pattern of the filament model, there will be an infinite number of ways of tracing out the molecular tails with varying amounts of twist, and several of these will prove to be physically plausible.

For concreteness we have adopted the close-packed cross section of the filament shown in Fig. 7. It satisfies our criteria and represents a physically reasonable way for the molecules to work their way from the inside to the outside of the filament. The numbers from 1 to 6 represent the height at which the respective molecular tails are cut by the sectioning plane. Number 1 represents the site of a molecular tail end, and the successive numbers represent molecules cut at successive positions spaced at 143 Å intervals along the direction of the axis of the filament. The solid lines connect sets of molecules related by this sequential offset.

Figure 8 shows the way in which the tail of a particular myosin molecule is threaded through the positions 1–6 of the cross-sectional patterns. The diagram shows how the choice of rotation angle between successive section planes is reflected in the way that the molecular tail twists as it emerges from the interior of the filament. The

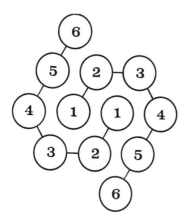

Figure 7. Cross section of a model of the thick filament. The molecules marked 1 are sectioned at their tail ends, and those marked 2 through 6 are sectioned at positions spaced at successive 143 Å unit intervals along the axis direction. The solid lines connect sets of molecules that form part of a continuous twisted sheet in which each molecule is offset by the unit spacing with respect to its neighbors.

whole model is then constructed of symmetry-related myosin molecules whose tails are twisted in exactly the same way as they emerge.

Upon examining the details of such a model we note two distinctive features: (1) The molecules of each set numbered 1 through 6 (and connected by solid lines in Fig. 7) form a continuous twisted sheet of molecular tails in which each molecule is related to its immediate neighbors by an offset of 143 Å. The molecules follow the folding of the sheet and are slightly inclined to its edges, so as to permit gradual emergence toward the head ends. (2) If we section the model in some transverse plane between

the depicted close-packed sectioning planes, then continuity and smooth bending of the molecular tails require that the molecules in position 1 must work out some distance toward the corresponding position 2, leaving a small gap in the close packing in the middle of the filament. As we proceed along the filament toward the next close-packed sectioning plane, the gap widens until it can just accommodate another pair of tail ends of molecules in position 1.

We have considered in devising this model that the thick filament probably differs in symmetry in different regions and concomitantly will show differing packing patterns of the molecular tails. Figure 9 shows how the pictured type of packing can be easily rearranged by sliding some of the molecules past one another. (Such a feature might play a role in transition regions, as for example, in the transition from the pseudo-H zone to the A zone.) The pattern at the left center is the one that we have postulated, and the patterns at top and bottom are 3-4-3-2 arrangements with triangular profiles (Pepe, 1967; Pepe and Drucker, 1972) that can be generated from it by sliding some of the molecules past one another. The pattern at the right, however, shows the same shape of cross section as the model that we have postulated, but it represents a physically less easy way of getting the molecules from the inside to the outside.

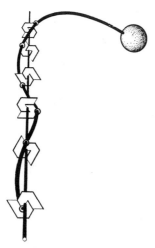

Figure 8. Configuration of the tail of a myosin molecule in the model when a rotation angle of $+60°$ between successive symmetry-related planes is adopted. The geometric figures in the transverse section planes correspond to the pattern in Fig. 7. The molecule passes successively through positions 1 to 6 in consecutive section planes.

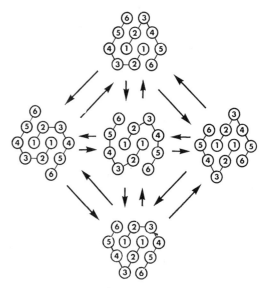

Figure 9. Possible rearrangements of the molecular packing that can give rise to different cross-sectional patterns. The pattern at the left corresponds to Fig. 7; those at top and bottom are 3-4-3-2 triangular profiles (Pepe, 1967) (since they do not have twofold rotor symmetry, they do not conform to the postulated screw axis). The pattern at the right shows the same type of cross section as in Fig. 7, but it corresponds to a physically less plausible molecular packing.

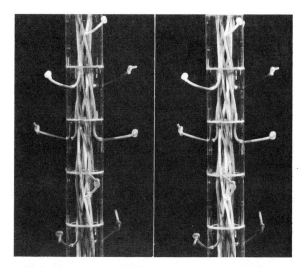

Figure 10. Stereo view of a section of the model built in plastic. The rotation angle here is −120°. The molecules gradually emerge from the middle of the filament in the top-to-bottom direction and form two separate continuous twisted sheets related by the twofold rotation axis.

Figure 10 is a stereoscopic closeup view of a short section of the model built in plastic that shows the details of packing. In presenting this model, we emphasize that we have only demonstrated that a physically plausible model can be built that conforms to the adopted symmetry principles and allows extensive contact between molecules. The fact that the model is far from unique is shown by the great variety of configurations that can arise within the framework of our postulates by different choices of rotation angles. It seems evident that neither the central pseudo-H region nor the tapered ends of the thick filament could conform to the assumed symmetry in any case, and thus they are not represented by this model.

Acknowledgments

This study was supported by a grant from the John A. Hartford Foundation, Inc.; by National Institutes of Health grant #AM 09404; by grant #70870 from the American Heart Association; and by Career Development Award #K3 AM 18,565 (M.Y.). D. S. O'Hara was supported by a post-doctoral fellowship from the National Institutes of Health (1 F2 AM19, 184), and P. J. Molberg was aided by a National Science Foundation Fellowship to the Program in Biochemical Sciences of Harvard University (GY 6082). This paper also includes material submitted by Mr. Molberg to Harvard University for the Baccalaureate Degree with Honors in the Biochemical Sciences, 1969. The authors are grateful to Dr. Jerome Gross for extended use of his electron microscope and to Drs. Dennis E. Roark and David A. Yphantis for kindly allowing us to use their computer program and to read a preprint of their paper prior to publication. We also gratefully acknowledge the expert technical assistance of Mrs. Muriel H. Blanchard and Mrs. Sylvia Mizrahi.

References

AMBLER, R. P. 1965. The behaviour of peptides formed by cyanogen bromide cleavage of proteins. *Biochem. J.* **96**: 32P.

CASASSA, E. F. and H. EISENBERG. 1961. Partial specific volumes and refractive index increments in multicomponent systems. *J. Phys. Chem.* **65**: 427.

CHATTAWAY, F. D. and J. M. WADMORE. 1902. The constitution of hydrocyanic, cyanic, and cyanuric acids. *J. Chem. Soc.* **81**: 191.

COHEN, C., S. LOWEY, R. G. HARRISON, J. KENDRICK-JONES, and A. G. SZENT-GYÖRGYI. 1970. Segments from myosin rods. *J. Mol. Biol.* **47**: 605.

CRESTFIELD, A. M., S. MOORE, and W. H. STEIN. 1963. The preparation and enzymatic hydrolysis of reduced and S-carboxymethylated proteins. *J. Biol. Chem.* **238**: 622.

CRICK, F. H. C. 1953. The packing of α-helices: simple coiled-coils. *Acta Cryst.* **6**: 689.

DIXON, A. E. and J. TAYLOR. 1913. Cyanogen bromide and cyanogen. *J. Chem. Soc.* **103**: 974.

EDELSTEIN, S. J. and H. K. SCHACHMAN. 1967. The simultaneous determination of partial specific volumes and molecular weights with microgram quantities. *J. Biol. Chem.* **242**: 306.

ELLMAN, G. L. 1959. Tissue sulfhydryl groups. *Arch. Biochem. Biophys.* **82**: 70.

FUJITA, H. 1962. *Mathematical theory of sedimentation analysis*, p. 246. Academic Press, New York.

GERGELY, J., M. A. GOUVEA, and D. KARIBIAN. 1955. Fragmentation of myosin by chymotrypsin. *J. Biol. Chem.* **212**: 165.

GODFREY, J. E. and W. F. HARRINGTON. 1970. Self-association in the myosin system at high ionic strength. I. Sensitivity of the interaction to pH and ionic environment. *Biochemistry* **9**: 886.

GROSS, E. 1967. The cyanogen bromide reaction. *Meth. Enzymol.* **11**: 238.

GROSS, E. and B. WITKOP. 1961. Selective cleavage of the methionyl peptide bonds in ribonuclease with cyanogen bromide. *J. Amer. Chem. Soc.* **83**: 1510.

HADE, E. P. K. and C. TANFORD. 1967. Isopiestic compositions as a measure of preferential interactions of macromolecules in two-component solvents. Application to proteins in concentrated aqueous cesium chloride and guanidine hydrochloride. *J. Amer. Chem. Soc.* **89**: 5034.

HUXLEY, H. E. 1963. Electron microscope studies on the structure of natural and synthetic protein filaments from striated muscle. *J. Mol. Biol.* **7**: 281.

———; 1968. Structural difference between resting and rigor muscle; evidence from intensity changes in the low-angle equatorial X-ray diagram. *J. Mol. Biol.* **37**: 507.

HUXLEY, H. E. and W. BROWN. 1967. The low-angle X-ray diagram of vertebrate striated muscle and its behaviour during contraction and rigor. *J. Mol. Biol.* **30**: 383.

HUXLEY, H. E., W. BROWN, and K. C. HOLMES. 1965. Constancy of axial spacings in frog sartorius muscle during contraction. *Nature* **206**: 1358.

KING. M. V. and M. YOUNG. 1972a. Electron microscopy of side-by-side arrays of myosin and light meromyosin-C. *J. Mol. Biol.* **63**: 539.

———. 1972b. Evidence for flexibility of the helical rod section of the myosin molecule. *J. Mol. Biol.* **65**: 519.

KOMINZ, D. R., E. R. MITCHELL, T. NIHEI, and C. M. KAY. 1965. The papain digestion of skeletal myosin A. *Biochemistry* **4**: 2373.

LANG, R. 1925. Jodometrische Bestimmung von Cyanverbindungen. *Z. anal. Chem.* **67**: 1.

LOWEY, S. 1965. Comparative study of the α-helical muscle proteins. Tyrosyl titration and effect of pH on conformation. *J. Biol. Chem.* **240**: 2421.

LOWEY, S. and C. COHEN. 1962. Studies on the structure of myosin. *J. Mol. Biol.* **4**: 293.

LOWEY, S., L. GOLDSTEIN, and S. LUCK. 1966. Isolation and characterization of a helical subunit from heavy meromyosin. *Biochem. Z.* **345**: 248.

LOWEY, S., L. GOLDSTEIN, C. COHEN, and S. M. LUCK. 1967. Proteolytic degradation of myosin and the meromyosins by a water-insoluble polyanionic derivative of trypsin. Properties of a helical subunit isolated from heavy meromyosin. *J. Mol. Biol.* **23**: 287.

LOWEY, S., H. S. SLAYTER, A. G. WEEDS, and H. BAKER. 1969. Substructure of the myosin molecule. I. Subfragments of myosin by enzymic degradation. *J. Mol. Biol.* **42**: 1.

MIDDLEBROOK, W. R. 1959. Individuality of the meromyosins. *Science* **130**: 621.

MIHÁLYI, E. 1953. Trypsin digestion of muscle proteins. II. The kinetics of the digestion. *J. Biol. Chem.* **201**: 197.

MIHÁLYI, E. and W. F. HARRINGTON. 1959. Studies on the tryptic digestion of myosin. *Biochim. Biophys. Acta* **36**: 447.

MIHÁLYI, E. and A. G. SZENT-GYÖRGYI. 1953. Trypsin digestion of muscle proteins. I. Ultracentrifugal analysis of the process. *J. Biol. Chem.* **201**: 189.

MOFFITT, W. 1956. Optical rotatory dispersion of helical polymers. *J. Chem. Phys.* **25**: 467.

MOORE, S. 1963. On the determination of cystine as cysteic acid. *J. Biol. Chem.* **238**: 235.

PAULING, L. and R. B. COREY. 1953. Compound helical configurations of polypeptide chains: structure of proteins of the α-keratin type. *Nature* **171**: 59.

PEPE, F. 1967. The myosin filament. I. Structural organization from antibody staining observed in electron microscopy. *J. Mol. Biol.* **27**: 203.

PEPE, F. A. and B. DRUCKER. 1972. The myosin filament. IV. Observation of the internal structural arrangement. *J. Cell Biol.* **52**: 255.

REEDY, M. K., K. C. HOLMES, and R. T. TREGEAR. 1965. Induced changes in orientation of the cross-bridges of glycerinated insect flight muscle. *Nature* **207**: 1276.

SCHACHMAN, H. K. 1959. *Ultracentrifugation in biochemistry*, p. 116. Academic Press, New York.

SEGAL, D. M., S. HIMMELFARB, and W. F. HARRINGTON. 1967. Composition and mass of peptides released during tryptic and chymotryptic hydrolysis of myosin. *J. Biol. Chem.* **242**: 1241.

SZENT-GYÖRGYI, A. G., C. COHEN, and D. E. PHILPOTT. 1960. Light meromyosin fraction I: A helical molecule from myosin. *J. Mol. Biol.* **2**: 133.

WITKOP, B. 1968. Chemical cleavage of proteins. *Science* **162**: 318.

YOUNG, M. 1969. The molecular basis of muscle contraction. *Ann. Rev. Biochem.* **38**: 913.

YOUNG, M., M. H. BLANCHARD, and D. BROWN. 1968. Selective nonenzymic cleavage of the myosin rod: isolation of the coiled-coil α-rope section from the C-terminus of the molecule. *Proc. Nat. Acad. Sci.* **61**: 1087.

YOUNG, M., S. HIMMELFARB, and W. F. HARRINGTON. 1964. The relationship of the meromyosins to the molecular structure of myosin. *J. Biol. Chem.* **239**: 2822.

YPHANTIS, D. A. 1964. Equilibrium ultracentrifugation of dilute solutions. *Biochemistry* **3**: 297.

Association of Myosin to Form Contractile Systems

William F. Harrington, Morris Burke, and Janice S. Barton

Department of Biology, The Johns Hopkins University, Baltimore, Maryland 21218

The association of myosin molecules at low ionic strength to form filamentous structures similar in general topology to the thick filaments of muscle has been under study in several laboratories in recent years (Jakus and Hall, 1947; Noda and Ebashi, 1960; Zobel and Carlson, 1963; Huxley, 1963; Kaminer and Bell, 1966a, b; Josephs and Harrington, 1966, 1968). Electron micrographs of negatively stained preparations of these macrostructures, prepared by lowering the ionic strength of myosin solutions near neutral pH, reveal particles of 100–150 Å diameter, with irregular surface projections except for a transverse zone devoid of such corrugations of about 0.15–0.2 μ in width which is always situated near the center of the filament. Under these ionic conditions, Huxley (1963) found the lengths of filament particles to vary between 0.25 and 2 μ with a mean length near 1.2 μ. Filaments generated by dialysis near pH 8 have similar structural properties, but they are appreciably shorter and exhibit a much sharper size distribution with mean particle length near 6300 Å.

Velocity sedimentation studies of the filament systems generated in the low or neutral pH range show rather diffuse sedimenting boundaries consistent with the broad size distribution observed in electron micrographs, but in the pH range near 8, a single, well-defined, hypersharp filament boundary is observed ($S_{20,w} = 150$ S). Previous work has shown that the filament is in rapid, reversible equilibrium with a slower sedimenting species presumed to be monomeric myosin, and it has been further established that the equilibrium is displaced in favor of monomer with increasing pH, salt, and hydrostatic pressure. For example, at pH 8.3, 0.12 M KCl only a single, rapidly sedimenting peak is observed with sedimentation coefficient characteristic of the myosin filament. At higher salt concentrations a slower sedimenting peak is detected, and as the salt concentration is increased at constant pH the bimodal boundary shows (see Fig. 1) a gradual depletion of polymer and elevation in area of the slower sedimenting boundary over a relatively narrow range of salt concentration. The transformation is complete at 0.22–0.24 M KCl, and at salt concentrations above this threshold only a single boundary is observed with sedimentation coefficient characteristics of the slower species.

Although a single sedimenting boundary is observed in high salt, it does not necessarily follow that the species under study is monomeric myosin. According to the well-documented Gilbert theory (1959) of mass transport, a rapidly reversible, self-associating monomer-dimer or monomer-dimer-trimer system can migrate in the centrifuge as a single, nonbimodal gradient peak. The presence of such an interaction in the myosin system was first observed by Godfrey and Harrington (1970a) in differential sedimentation velocity experiments in which one limb of a double-sector ultracentrifuge cell was filled with myosin at pH 6.4 and the companion limb with an identical concentration of myosin in the same solvent, but at pH 7.8. A difference in sedimentation coefficient in the two sectors was detected, by means of the Rayleigh interference optical system, which was strongly dependent on the salt concentration. Results are summarized in Fig. 2 in which the %ΔS is plotted as a function of myosin concentration. Each symbol is a data point from one differential velocity experiment, and in all cases the myosin at pH 6.4 exhibited the larger sedimentation rate. A pH-dependent self-association of myosin to one or more larger species is consistent with the data, and the suppressing effect of both low myosin concentration and high ionic strength on the differential sedimentation rate suggests the presence of a reaction leading to associated species in which electrostatic interactions play a major role. Results indicate that even in 0.5 M KCl the tendency to associate is not entirely depressed, and the sedimenting particles, though they migrate as a single hypersharp boundary in a conventional ultracentrifuge run, are in rapid, reversible equilibrium with a larger species.

The most convincing evidence for the presence of an associating system in high salt comes from the high speed sedimentation equilibrium results presented in Fig. 3. The three profiles shown here represent the reciprocal of the number average, weight average and "Z" average molecular weights as a function of protein concentration obtained from 21 high-speed sedimentation equilibrium studies of myosin in a high ionic strength solvent

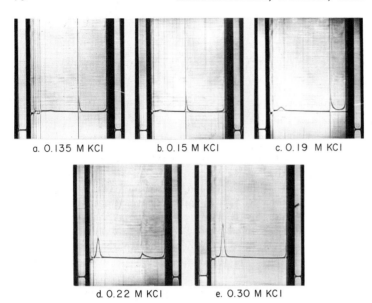

a. 0.135 M KCl   b. 0.15 M KCl   c. 0.19 M KCl

d. 0.22 M KCl   e. 0.30 M KCl

Figure 1. Effect of increasing salt concentration on the myosin-filament equilibrium system. Protein concentration is constant at 0.44% in all experiments. In all centrifuge runs a double-sector, 12 mm cell was employed at a speed of 24,000 rpm, bar angle 70°, temperature 5–7°C. Solvent was 0.002 M Tris buffer, pH 8.4. A filament system was prepared in 0.135 M KCl, 0.002 M Tris, pH 8.4 and various aliquots were dialyzed overnight vs. the buffer containing salt concentrations shown above.

system. When the results from 21 runs are averaged at ten concentrations within the 0.09–0.50 mg/ml range, they give the three sets of points shown. The solid curves are simulated 1/M vs. concentration distributions expected for a rapidly reversible monomer-dimer equilibrium defined by the three molecular parameters, the molecular

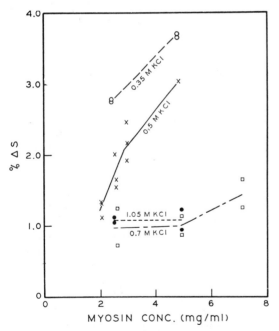

Figure 2. Percent ΔS of myosin at pH 6.4 and 7.8 as a function of myosin and KCl concentration. Each symbol is a result from one differential sedimentation velocity experiment. All runs were conducted at 6°, 24,000–32,000 rpm; additional solvent ingredients: 0.01 M EDTA, 0.025 M bis-Tris, 0.025 M Tris. In all cases, the myosin at pH 6.4 exhibited the larger sedimentation rate. (From Godfrey and Harrington, 1970a.)

weight of the monomer, the monomer-dimer association constant, and the virial coefficient. These plots show a dramatic drop in the reciprocal of molecular weight at low concentration as a result of dimerization, then increase with increasing concentration above about 0.03% because of the increasing effect of nonideality. The presence of a rapidly reversible monomer-dimer equilibrium has been difficult to detect in the myosin system as a result of the striking magnitude of nonideality in this system which overshadows the association reaction except in the very low (<0.1%) concentration range. It is therefore comforting that Herbert and Carlson (1971) have recently reached similar conclusions and have reported almost identical association parameters from their studies of spectral broadening and intensity measurements of laser light scattered from solutions of myosin under comparable ionic conditions.

The high speed sedimentation equilibrium experiments as well as the laser light scattering studies were carried out on myosin in a solvent system consisting of 0.5 M KCl, 0.2 M PO_4^{2-}, 0.01 M EDTA, pH 7.3. We have seen that reduction of the salt concentration of such myosin solutions by dilution or dialysis results in the formation of filamentous structures similar in a number of respects to the natural thick filament of muscle. Thus it seems possible that the dimeric species present in the high salt medium could be the initiating and possibly the building unit used in assembly of the thick filaments. The differential velocity sedimentation experiments of Fig. 2 demonstrate an increasing tendency for particle-particle interaction as the salt concentration is lowered, suggesting that the monomer $\leftrightharpoons$ dimer

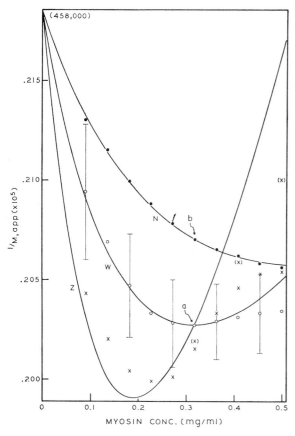

Figure 3. Averaged reciprocal moments vs. concentration from 21 high-speed sedimentation equilibrium runs with simulated moments obeying parameter values in Table 1. Points are experimental: ●, $1/M_{n,app}$; ○, $1/M_{w,app}$; ×, $1/M_{z,app}$; (×), from 6 runs at high-speed, high-loading conditions. Vertical bars are 95% confidence limits for $1/M_{w,app}$ mean values. Solvent was 0.5 M KCl, 0.2 M PO_4^{2-}, 0.01 M EDTA, pH 7.3; 6°, 9,000–12,000 rpm. Curves: reversible M-D association, $M_1 = 458,000$, $B = 6.60 \times 10^{-6}$ mole dl/g², $k_2 = 9.98$ dl/g. Myosin chromatographed on A-50 DEAE Sephadex. (From Godfrey and Harrington, 1970b.)

equilibrium is displaced toward dimer as the threshold for filament formation is approached.

Our recent high speed sedimentation equilibrium results support this view. Figure 4 shows reciprocal molecular weight vs. concentration plots of myosin obtained in a solvent system consisting of 0.2 M KCl, 0.005 M Veronal, 0.001 M EDTA, pH 8.3, i.e., at an ionic strength just above the filament threshold. The two lower profiles are the reciprocal number and reciprocal weight average molecular weight obtained from seven meniscus-depletion sedimentation experiments, and the solid curves are simulated 1/M vs. concentration distributions expected for a rapidly reversible monomer-dimer equilibrium. The number and weight average profiles obtained in the high salt medium are reproduced on the reduced ordinate scale of Fig. 4

for comparison. In the low salt solvent a much deeper minimum is observed compared to the reciprocal moment vs. concentration plot of the high salt medium suggesting that the equilibrium constant for dimerization increases significantly with decreasing ionic strength. It will be seen that all of the reciprocal moments in the figure converge at infinite dilution, irrespective of rotor speed or cell loading concentration, consistent with the presence of reversible association rather than contamination of the system by low molecular weight material.

The defining parameters for the association reaction derived from the experimental data obtained on myosin in low salt and high salt solvents are presented in Table 1. In the low salt system (0.2 M KCl, pH 8.3) under study in Fig. 4 the equilibrium constant for dimerization, $K_2 = 19$ dl/g, decreases in magnitude to a value of about 10 dl/g in the 0.5 M KCl—0.2 M PO_4^{2-} solvent employed in both the sedimentation equilibrium experiments as well as the laser light scattering studies of Herbert and Carlson (1971). It has been reported in earlier work that increasing amounts of phosphate ion shifts the monomer-dimer ratio toward monomer (Godfrey and Harrington, 1970a), and a quantitative analysis of the association reaction in 0.5 M KCl—0.5 M PO_4^{2-} has been made by Herbert and Carlson (1971). Under these ionic conditions the value of $K_2 = 1.30$ dl/g.

Thus our present evidence suggests that at the ionic strength threshold for self-assembly of the

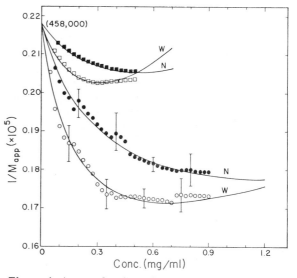

Figure 4. Averaged reciprocal moments vs. concentration plots obtained from 7 high speed sedimentation equilibrium runs with simulated moments obeying parameter values in Table 1. Solvent is 0.2 M KCl, 0.005 M Veronal, 0.001 M EDTA, pH 8.3. Myosin chromatographed on A-50 DEAE Sephadex. Number and weight average data reproduced from Fig. 3 shown in upper two curves.

Table 1. Monomer-Dimer Association Parameters for Myosin in Various Solvent Systems (5°C) .

Solvent	Mol Wt Monomer	Virial Coeff. (B) (mole dl/g^2)	Equil. Const. (K_2) (dl/g)	Method of Analysis
0.2 M KCl, pH 8.3	458,000	2.7×10^{-6}	19.3	Sed. equil.[a]
0.5 M KCl, 0.2 M PO$_4^{2-}$ 0.01 M EDTA, pH 7.3	458,000	6.60×10^{-6}	10.0	Sed. equil.[b]
0.5 M KCl, 0.2 M PO$_4^{2-}$ 0.01 M EDTA, pH 7.3	441,000	7.29×10^{-6}	10.6	Laser light scattering[c]
0.5 M KCl, 0.5 M PO$_4^{2-}$ 0·01 M EDTA, pH 8.5	476,000	7.20×10^{-6}	1.3	Laser light scattering[c]

[a] This work. [b] Godfrey and Harrington (1970b). [c] Herbert and Carlson (1971).

filament a major fraction of the protein exists in the dimer state. In a 1 % solution of myosin over 75 % is dimerized. As the salt concentration of the reacting system is lowered below the filament threshold, the rapidly sedimenting polymer peak is detected in velocity sedimentation patterns and, as noted earlier, this peak grows in size at the expense of the slower peak with decreasing salt concentration. This information strongly suggests that the slower sedimenting peak in the bimodal-Schlieren pattern is a monomer-dimer reaction boundary, since the concentration of the polymer species at salt concentrations near the filament threshold represents only a small fraction of the protein present as dimer. At any fixed salt concentration below the filament threshold, the polymer peak increases with increasing myosin concentration, and it therefore seems probable that the dimer species present in solution serves as the building block for filament assembly.

The fact that no intermediate species are detected in velocity sedimentation experiments indicates that the nucleus, once formed, is immediately transformed into filament so that at any time its concentration will be negligible. Thus we consider that the rate-limiting step for filament assembly is the formation of the nucleating species from the initiating dimer units. With this assumption we may now write the overall filament assembly reaction as a two-step process in terms of the species seen in the ultracentrifuge:

$$2M_1 \rightleftharpoons M_2 \quad ; \quad K_2 = \frac{[M_2]}{[M_1]^2} \tag{1}$$

$$nM_2 \rightleftharpoons P \quad ; \quad K_p = \frac{[P]}{[M_2]^n} \tag{2}$$

The overall filament reaction can be represented as

$$K_p(K_2)^n = \frac{[P]}{[M_1]^{2n}} \tag{3}$$

and the equilibrium constant for this process can now be seen to be the product of the equilibrium

constant for polymer formation from dimer and the equilibrium constant for dimerization raised to the nth power.

In an earlier treatment of this system (Josephs and Harrington, 1968) it was assumed that filament formation occurred through self-assembly of monomer units, leading to an expression for the equilibrium process in terms only of the "monomer" and polymer concentrations. It will be seen that the equilibrium constant of the earlier formulation is equivalent to $K_p(K_2)^n$ of Eq. 3, where the exponent n in the present treatment refers to the number of dimers associating to form filament. The present mechanism suggests that as the concentration of salt is reduced through the threshold value for filament formation, the free energy for insertion of the dimer into the nucleus changes very sharply from a positive to a dominant negative value, thereby shifting the equilibrium favorably towards polymer.

Geometry of the Myosin Dimer

There is strong reason to believe that the bare central region of the myosin synthetic filament is formed early in the self-assembly process and growth of the filament proceeds by addition of myosin particles at the two ends. In general, filaments formed by lowering the ionic strength in the pH range 7–8.5 show a rather broad distribution in lengths, but the bare central zone is always situated near the center of the filament, never at one end (Huxley, 1963; Kaminer and Bell, 1966a; Josephs and Harrington, 1966). The bare central zone clearly has a different type of organization than the corrugated surface regions, and if growth were starting near one end, it would not be expected that the systematic assembly pattern would always develop in vitro in such a way as to place the region devoid of cross-bridges in the center of the filament. Our present evidence that the predominating species of low ionic strength is a dimer leads to some interesting questions on the mechanism of assembly of the filament. If this species has antiparallel geometry, that is, with the heads of the two myosin molecules pointing in opposite directions, then the bare central region of the filament could be constructed through lateral assembly of this dimer unit. However, once this section of the filament is completed, linear extension of the structure through accretion of the same dimer unit is precluded, since this mechanism for growth would lead to a cross-bridge array incompatible with the presently accepted distribution of these projections along the native thick filament surface (Huxley and Brown, 1967). If the dimer has parallel geometry, then the bare central zone could be constructed by antiparallel association of the dimer species. In this case the construction scheme is simpler in that growth of the filament can proceed subsequent to assembly of the bare central zone through addition of the same dimer species as that used for initiating self-assembly. It seems likely to us that a specific dimer geometry is employed in the construction of the synthetic filament since assembly of the growing aggregate from an ensemble of myosin dimers of differing geometry would require a complicated and improbable sequential addition of the various dimer species. Since the geometry of the myosin dimer in solution is clearly crucial to any proposed mechanism for filament formation, it seemed worthwhile to attempt to establish its gross three-dimensional structure. We have approached this problem through an examination of the hydrodynamic behavior of LMM and rod ("headless" myosin).

It is well known that these proteolytic fragments show a remarkable tendency to associate in ordered arrays at low salt concentrations and exhibit solubility properties similar to myosin. The interaction sites responsible for association of myosin are therefore likely to be located along the rod segment of the myosin molecule and very probably within the LMM segment, since the recent work of Lowey et al. (1969) indicates that subfragment II (the segment of the rod lying between LMM and the globular head) shows no apparent tendency to associate either with itself or with LMM over a wide span of ionic strengths near neutral pH.

When preparations of LMM and rod, previously fractionated from low molecular weight contaminants on G-200 Sephadex, are examined in high speed sedimentation equilibrium experiments in the high salt solvent system 0.5 M KCl, 0.2 M PO_4^{2-}, 0.01 M EDTA, pH 7.3, they show association behavior similar to myosin. Reciprocal moment-concentration profiles of both LMM and rod exhibit a minimum in the low concentration range and converge to a common infinite dilution value consistent with the requirements of a nonideal associating system. Computer-calculated reciprocal

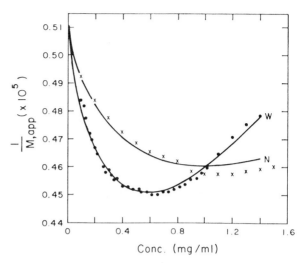

Figure 5. 1/M vs. concentration plot for myosin rod in 0.5 M KCl, 0.2 M PO_4^{2-}, 0.01 M EDTA, pH 7.3, at 5°. The circles and crosses represent weight and number averages, respectively (average for 4 runs at 14,000 and 17,000 rpm). The smooth curves represent computer simulated 1/M vs. concentration plots for a monomer ⇌ dimer reversible association with parameters $M_1 = 198,000$ g/mole, $B = 8.44 \times 10^{-6}$ mole dl/g² and $k_2 = 8$ dl/g. (From Harrington and Burke, 1972.)

apparent molecular weight moments averaged over four meniscus-depletion runs for the rod preparation and assuming a monomer-dimer equilibrium are shown as solid curves in Fig. 5. The simulated 1/M vs. concentration profiles plotted here fit the experimental data within 2.5% over the concentration range of interest, i.e., 0.1–1.6 mg/ml. Similar results were obtained with the LMM system. The simple monomer-dimer mode of association deduced from these studies is consistent with the sedimentation behavior of these two proteolytic fragments in that they exhibit single hypersharp boundaries in velocity sedimentation experiments, as expected from Gilbert (1959) theory for rapidly reversible monomer ⇌ dimer equilibria. The finding that both LMM and rod exhibit association behavior similar to that of myosin suggests a method for determining the three-dimensional geometry of the dimer.

Anticipating the viscosity results to be presented below, it is instructive to consider how this information allows us to deduce the geometry of the parent myosin dimer. Consider the two general types of myosin dimers shown schematically in Fig. 6, in which the rod segment lying between LMM and the head (subfragment II) is cross-hatched for identification. For purposes of discussion we assume the LMM and rod segments of the myosin molecules to have lengths of 860 Å and 1290 Å, respectively, and to be displaced in the various geometries by 430 Å. The method which

Anti – Parallel

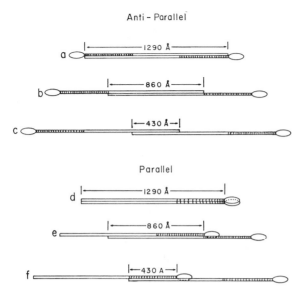

Figure 6. Schematic representation of various geometries for the myosin dimer. The rod section of the myosin molecule is subdivided into its two components subfragment II (cross-hatched) and LMM (unmarked) for convenience. For purposes of discussion LMM and rod segment lengths are assumed to be 860 and 1290 Å, respectively. (From Harrington and Burke, 1972.)

we will use, however, can be shown to be independent of the precise length of either the monomer or dimer structure.

An examination of the models reveals that the overlap distance of LMM and that of rod, when taken together, uniquely define the parent dimer. For example, a 430 Å overlap between LMM segments occurs in models (a), (c) and (e), and a 1290 Å overlap between rod segments is seen in (a) and (d). However, the only structure with both restrictions is found in model (a) of the antiparallel group. Similarly, an 860 Å overlap for LMM occurs in (b) and (d). If this restriction is taken together with a 1290 Å overlap for rod, the only structure which conforms is model (d) of the parallel group. Looking at the system of dimers in another way, it will be seen that *the length* of the LMM dimer, when taken in conjunction with *the length* of the rod dimer, establishes the geometry of the parent species. The assumption in such an argument is that the geometry of the myosin dimer is preserved in the dimerization of both LMM and rod segments.

With increasing protein concentration the relative amount of dimer in a solution of either the LMM or rod particle will increase; thus the contribution of the dimer to the overall viscosity of the solution will increase. Depending on the particle asymmetry of the dimer being formed, the viscosity increment will be either larger or smaller than that of the monomer, and the reduced viscosity vs. concentration curve would be expected to reflect this difference. The reduced viscosity-concentration

plot of a nonassociating particle should be linear with slope $k[\eta]^2$ where k is the dimensionless Huggins constant and $[\eta]$, the intrinsic viscosity. An associating system, however, would be expected to show either upward or downward curvature in the low concentration range with increasing concentration depending on the asymmetry of the associated species and the magnitude of the equilibrium constant. Figure 7 presents plots of reduced viscosity vs. concentration of LMM, rod and myosin in 0.5 M KCl, 0.2 M PO_4^{2-}, 0.01 M EDTA, pH 7.3. All of the plots show downward curvature at low concentration as expected for an associating system in which the associated species is more asymmetric than monomer. In earlier studies of the viscosity-concentration dependence of these particles (Lowey et al., 1969; Young et al., 1964) downward curvature was not detected since measurements were generally made above a concentration of 0.5 mg/ml.

Assuming a monomer-dimer equilibrium, the reduced viscosity of the mixture will be an additive

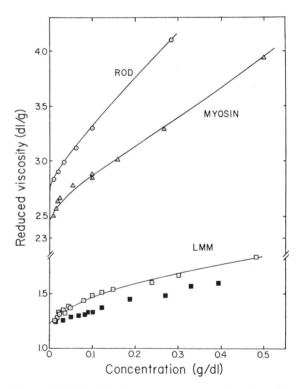

Figure 7. Reduced viscosity vs. concentration plots for (△) myosin, (□) LMM, and (○) rod in 0.5 M KCl, 0.20 M PO_4^{2-}, 0.01 M EDTA, pH 7.3, at 5°. (■) LMM in 0.5 M KCl, 0.5 M PO_4^{2-}, 0.01 M EDTA, pH 8.5. The latter has been normalized to an apparent specific volume (ϕ') = 0.710 cm³/g to account for the ϕ' change occurring when LMM is placed in the solvent containing 0.5 M PO_4^{2-}. The smooth curves represent computer simulated η_{sp}/c vs. concentration plots for a monomer ⇌ dimer reversible association reaction with parameters given in Table 2. (From Harrington and Burke, 1972.)

function of the specific viscosities of each of the species present at a total protein concentration, c:

$$(\eta_{sp}/c)_{mixture} = ([\eta]_m + k[\eta]_m^2 c_m) \frac{c_m}{c}$$

$$+ ([\eta]_d + k[\eta]_d^2 c_d) \frac{c_d}{c} \quad (4)$$

The experimental data shown in Fig. 6 have been fitted by a least squares analysis computer program in which η_{sp}/c, c, k, and the intrinsic viscosity of the monomer, $[\eta]_m$, have been used to estimate the intrinsic viscosity of the dimer, $[\eta]_d$, and the equilibrium constant for dimerization, K_2, over the concentration range of the measurements. A constant value for the Huggins constant, $k = 0.45$, was assumed in the calculations; in nonassociating systems this coefficient is generally found in the range 0.4–0.5.

Simulated η_{sp}/c vs. concentration profiles for a monomer-dimer equilibrium system generated from the derived values for $[\eta]_d$ and K_2 which best fit the experimental data are shown as the solid curves in Fig. 7. In the case of myosin, a value of 10 dl/g was assigned to K_2 since this value has been established with high precision from the sedimentation equilibrium and laser light scattering studies employing an identical solvent. It should be emphasized that the only unknown parameter required for analysis of this plot is $[\eta]_d$, since the intrinsic viscosity of monomer is obtained by extrapolation of the experimental data to infinite dilution.

As we have noted above, the equilibrium constant for dimerization of myosin is depressed in the solvent 0.5 M KCl—0.5 M PO_4^{2-} to about 1.3 dl/g. It will be seen that the curvature reflecting association behavior of the reduced viscosity vs. concentration plot of LMM is reduced in this solvent system (lower, solid squares). The virtually linear viscosity plot observed here reflects minimal self-association of LMM in accord with the laser light scattering experiments.

The viscosity parameters for monomer and dimer derived from the experimental results of Fig. 7 are shown in Table 2. Also shown are the free energies of dimerization obtained from the apparent equilibrium constant, K_2. All experiments employed the same solvent (0.5 M KCl, 0.2 M PO_4^{2-}, 0.01 M EDTA, pH 7.3) and were carried out at the same temperature (5° ± 0.02°). It will be seen that the free energies of dimerization derived from the viscosity data are, within experimental error, in the same range as those calculated from the high speed sedimentation equilibrium experiments. It will also be seen that the free energy of dimerization of LMM is apparently lower than that of myosin. This may be the result of additional charges on this

Table 2. Monomer ⇌ Dimer Reversible Association Parameters for LMM, Rod and Myosin

| | $[\eta]_m$ (dl/g) | $[\eta]_d$ (dl/g) | ΔF Dimerization (cal/mole) | |
			viscosity data	sed. equil. data
Myosin	2.45	3.11	−6900	−6860
Rod	2.65	3.44	−6960	−6300
LMM	1.23	1.90	−5800	−6100

Solvent: 0.5 M KCl, 0.2 M PO_4^{2-}, 0.01 M EDTA, pH 7.3 (5°C).

segment induced by the proteolytic treatment, or it may reflect a contribution to the interaction from the heavy meromyosin segment. In any case, the molar free energies of dimerization derived from the two independent studies are consistent in assigning most of the binding sites for association of myosin to the LMM segment.

In order to find the geometry of the dimer structure which would be consistent with the derived $[\eta]_d$ values of the LMM and rod particles shown in Table 2, we have investigated the hydrodynamic behavior of macromodels of differing dimer geometry. Scaled-up models of stainless steel rod were constructed assuming 860 Å by 20 Å geometry for LMM and 1290 Å by 20 Å for the myosin rod. Monomeric as well as dimeric models of various degrees of overlap were rotated in a viscous fluid about an axis through the center of gravity, perpendicular to the long axis of the model and the torque required to rotate the model with unit angular velocity determined (see Kuhn and Kuhn, 1952; Haltner and Zimm, 1959; Broersma, 1960; Reisler and Eisenberg, 1970). The rotational frictional coefficients derived from these measurements on single rods showed excellent agreement with that expected from the theoretical expression of Broersma (1960) which relates the torque to the axial ratio of a rigid cylinder. In the case of the rod dimer structure, for which there is no theoretical relationship between the torque expected and the particle geometry, we observed a very high degree of correlation between the torque required and a cylinder of the same volume and length as that of the dimer.

Two approaches are now possible to estimate the geometry of the dimer species which is generated on self-association of LMM and the myosin rod structures. If the geometry of the dimer is designated by a fraction of overlap, f, defined as the ratio of the length of dimer congruent to both monomer units divided by the length of monomer, then it can easily be shown (Burke and Harrington, 1972) that the intrinsic viscosity of the dimer will be given by

$$[\eta]_d = \left[\frac{(2-f)^{3/2}}{2^{1/2}} \right]^{1.8} \cdot [\eta]_m \quad (5)$$

Thus, provided that the intrinsic viscosity of monomer is known, the intrinsic viscosity of dimer for any given degree of overlap is readily deduced. If both $[\eta]_m$ and $[\eta]_d$ are known, the degree of overlap in the dimeric structure can be obtained with high precision independent of the absolute length of the monomer.

We can also estimate the length of the dimer species from its molecular weight established in the sedimentation equilibrium experiments and the intrinsic viscosity of the dimer with considerable confidence from the equation (see Yang, 1961):

$$L = 2a = 6.82 \times 10^{-8}([\eta]M)^{1/3}(p_e^2/\nu_e)^{1/3} \quad (6)$$

where p_e is the axial ratio of the equivalent ellipsoid and ν_e is the viscosity increment. In both of these treatments, we assume the usual approximation of $L_{\text{ellipsolid}} \cong L_{\text{cylinder}}$ for large p. We also assume that the dimer is hydrodynamically equivalent to a cylinder of length and volume equal to that of the dimer, in agreement with the modelling studies.

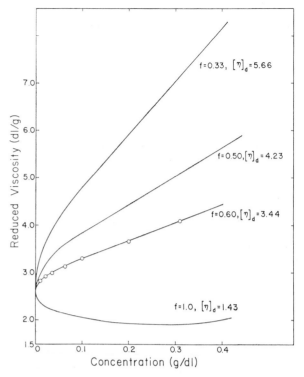

Figure 9. Viscosity vs. concentration dependence of rod (○) at 5.0°C. The solid curves represent theoretical reduced viscosity vs. concentration plots for rod in a monomer ⇌ dimer equilibrium based on Eq. 4. (f) denotes the fraction of monomer length overlapped in the dimer state. The intrinsic viscosity of the monomer, $[\eta]_m$, is assigned a value of 2.65 dl/g and the Huggins constant, (k), is assumed to be 0.45 for both monomer and dimer.

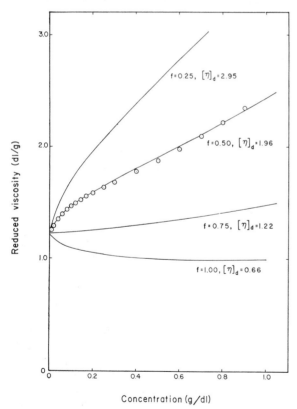

Figure 8. Viscosity vs. concentration dependence of light meromyosin (○) at 5.0°C. The solid curves represent theoretical reduced viscosity vs. concentration plots for LMM in a monomer ⇌ dimer equilibrium with respect to the specified dimer geometry based on Eq. 4. (f) denotes the fraction of monomer length overlapped in the dimer state. The intrinsic viscosity of the monomer, $[\eta]_m$, is assigned a value of 1.23 dl/g and the Huggins constant (k) is assumed to be 0.45 for both monomer and dimer.

Figure 8 presents simulated reduced viscosity vs. concentration profiles for a monomer-dimer equilibrium system of LMM expected for various degrees of overlap based on Eqs. 4 and 5. It will be seen that the expected viscosity-concentration plots show a wide distribution, allowing determination of the overlap distance with a high degree of precision. The experimental points fall along the profile expected for a 50% overlap (that is, 50% of the length of the monomeric unit is in contact with its dimer partner) out to 1% protein concentration. This behavior strongly supports the monomer-dimer mode of association. Large deviations from the plot would be expected if species larger in molecular weight than dimer were contributing to the equilibrium reaction, particularly with increasing protein concentration, unless these species had identical asymmetry to the dimer.

When similar plots are made for myosin rod (Fig. 9), the experimental results are consistent with a monomer-dimer equilibrium in which 60% of the length of the rod is overlapped in the dimer state. Thus, in view of the restrictions in the geometry of myosin dimers imposed by the overlap distance of their LMM and rod segments, the

Table 3. Geometry of LMM, Rod and Myosin Dimers Based on Viscosity, Sedimentation Equilibrium and Laser Light Scattering Studies

	Length Monomer (Å)	Length Dimer (Å)	Overlap Distance (Å)
Rod[a]	1410	1960	840
LMM[a]	895	1350	455
Myosin[b]	1480	2120	840

[a] Harrington and Burke (1972); Burke and Harrington (1972).
[b] Herbert and Carlson (1971); Carlson and Herbert (1972).

viscosity measurements strongly favor a parallel structure in which one myosin molecule is displaced 400–500 Å with respect to its neighbor. The lengths of the LMM and rod dimers estimated from Eq. 6 are in agreement with this conclusion. The length of the LMM dimer, 1320–1380 Å, depending on the assumed molecular weight of monomer, gives an overlap very close to 50% (445–465 Å) of the monomer particle length. Similarly, the length of the rod dimer, 1910–2015 Å, is consistent with an overlap near 60% (820–865 Å) of its monomer length.

Results of these calculations are summarized in Table 3, which includes the lengths of monomeric and dimeric myosin estimated by Herbert and Carlson (1971) from their laser light scattering studies. The Z-averaged radius of gyration of myosin has been recently calculated by these workers (Carlson and Herbert, 1972) for both parallel and antiparallel dimer and compared with experimentally measured values of this parameter. Their results also favor a parallel dimer with approximately 840 Å overlap.

Although our results and those of Herbert and Carlson are consistent with a parallel-type myosin dimer, variant modes of dimerization may be favored in solvent systems containing other ionic constituents. Indeed the electron microscope studies of Cohen et al. (1970) on aggregates of chicken myosin rods formed at low ionic strength in the presence of thiocyanate and Ca^{++} ions reveal both polar and antipolar ribbon-like structures. Thus, it seems essential to establish the dimer geometry under ionic conditions favoring thick filament formation. We do believe, however, that the self-assembly of myosin dimers of the parallel mode offers an attractive and rational construction scheme which would appear to be compatible with our present knowledge of the thick filament structure.

References

BROERSMA, S. 1960. Rotational diffusion constant of a cylindrical particle. *J. Chem. Phys.* **32**: 1626.

BURKE, M. and W. F. HARRINGTON. 1972. Geometry of the myosin dimer in high-salt media. II. Hydrodynamic studies on macromodels of myosin and its rod segments. *Biochemistry* **11**: 1456.

CARLSON, F. D. and T. E. HERBERT. 1972. Study of the self-association of myosin by intensity fluctuation spectroscopy. *J. Physique* **33** (suppl.): C. 1 by Fluids, Paris (in press).

COHEN, C., S. LOWEY, R. G. HARRISON, J. KENDRICK-JONES, and A. G. SZENT-GYÖRGYI. 1970. Segments from myosin rods. *J. Mol. Biol.* **47**: 605.

GILBERT, G. A. 1959. Sedimentation and electrophoresis of interacting substances. I. Idealized boundary shape for a single substance aggregating reversibly. *Proc. Roy. Soc. (London)* A **250**: 337.

GODFREY, J. E. and W. F. HARRINGTON. 1970a. Self-association in the myosin system at high ionic strength I. Sensitivity of the interaction to pH and ionic environment. *Biochemistry* **9**: 886.

———. 1970b. Self-association in the myosin system at high ionic strength II. Evidence for the presence of a monomer ⇌ dimer equilibrium. *Biochemistry* **9**: 894.

HALTNER, A. J. and B. H. ZIMM. 1959. Rotational friction coefficients of models of tobacco mosaic virus and the size of the virus particle. *Nature* **184**: 265.

HARRINGTON, W. F. and M. BURKE. 1972. Geometry of the myosin dimer in high-salt media. I. Association behavior of rod segments from myosin. *Biochemistry* **11**: 1448.

HERBERT, T. J. and F. D. CARLSON. 1971. Spectroscopic study of the self-association of myosin. *Biopolymers* **10**: 2231.

HUXLEY, H. E. 1963. Electron microscope studies on the structure of natural and synthetic protein filaments from striated muscle. *J. Mol. Biol.* **7**: 281.

HUXLEY, H. E. and W. BROWN. 1967. The low-angle X-ray diagram of vertebrate striated muscle and its behavior during contraction and rigor. *J. Mol. Biol.* **30**: 383.

JACKUS, M. A. and C. E. HALL. 1947. Studies of actin and myosin. *J. Biol. Chem.* **167**: 705.

JOSEPHS, R. and W. F. HARRINGTON. 1966. Studies on the formation and physical chemical properties of synthetic myosin filaments. *Biochemistry* **5**: 3473.

———. 1968. On the stability of myosin filaments. *Biochemistry* **7**: 2834.

KAMINER, B. and A. L. BELL. 1966a. Synthetic myosin filaments. *Science* **151**: 323.

———. 1966b. Myosin filamentogenesis: Effects of pH and ionic concentration. *J. Mol. Biol.* **20**: 391.

KUHN, H. and W. J. KUHN. 1952. Effects of hampered draining solvent on the translatory and rotatory motion of statistically coiled long-chain molecules in solution. Part II. Rotatory motion, viscosity, and flow birefringence. *J. Polymer Science* **9**: 1.

LOWEY, S., H. S. SLAYTER, A. G. WEEDS, and H. BAKER. 1969. Substructure of the myosin molecule. I. Subfragments of myosin by enzymic degradation. *J. Mol. Biol.* **42**: 1.

NODA, H. and S. EBASHI. 1960. Aggregation of myosin A. *Biochim. Biophys. Acta* **41**: 386.

REISLER, E. and H. EISENBERG. 1970. Studies on the viscosity of solutions of bovine liver glutamate dehydrogenase and on related hydrodynamic models; effect of toluene on enzyme association. *Biopolymers* **9**: 877.

YANG, J. T. 1961. The viscosity of macromolecules in relation to molecular conformation. *Advances in Protein Chemistry* (Anfinsen et al., eds.) vol. 16, p. 323. Academic Press, New York.

YOUNG, D. M., S. HIMMELFARB, and W. F. HARRINGTON. 1964. The relationship of the meromyosins to the molecular structure of myosin. *J. Biol. Chem.* **239**: 2822.

ZOBEL, C. R. and F. D. CARLSON. 1963. An electron microscope investigation of myosin and some of its aggregates. *J. Mol. Biol.* **7**: 78.

C-Protein and the Periodicity in the Thick Filaments of Vertebrate Skeletal Muscle

GERALD OFFER

Department of Biophysics, King's College, London, WC2, England

A new protein component of skeletal muscle fibrils, which is responsible for many of the transverse stripes of the A-band, has been isolated and characterized. This protein, which we call C-protein, was discovered in the following way (Starr and Offer, 1971). Myosin preparations of conventional quality were examined by SDS polyacrylamide gel electrophoresis. In addition to the bands expected from the heavy and light chains, other bands were apparent in the 80,000 to 150,000 molecular weight range and were shown to be due to impurities. These bands were termed B, C, D, etc., according to their mobility, the chief impurities being the B, C, and F-proteins.

Preparation of C-Protein

We obtain pure preparations of C and of F-protein as by-products of the purification of myosin (Table 1). Myosin is extracted from rabbit muscle and the extract diluted in the usual way. The precipitate is redissolved and subjected to ammonium sulfate fractionation. Myosin and C-protein precipitate in the 35–40% saturation fraction, but the F-protein precipitates in the 45–55% saturation fraction and can thus be removed. Myosin and C-protein may now be separated on a diethylaminoethyl Sephadex column with a phosphate buffer. C-protein passes through the column unretarded, and the myosin (together

Table 1. Scheme for Preparation of C-Protein

RABBIT MUSCLE

0.13 M KCl
0.15 M KP

EXTRACT

dilute

PRECIPITATE (CRUDE MYOSIN)

dissolve
AS fractionation

AMMONIUM SULFATE PURIFIED MYOSIN   F-PROTEIN

DEAE column

C-PROTEIN   COLUMN-PURIFIED MYOSIN

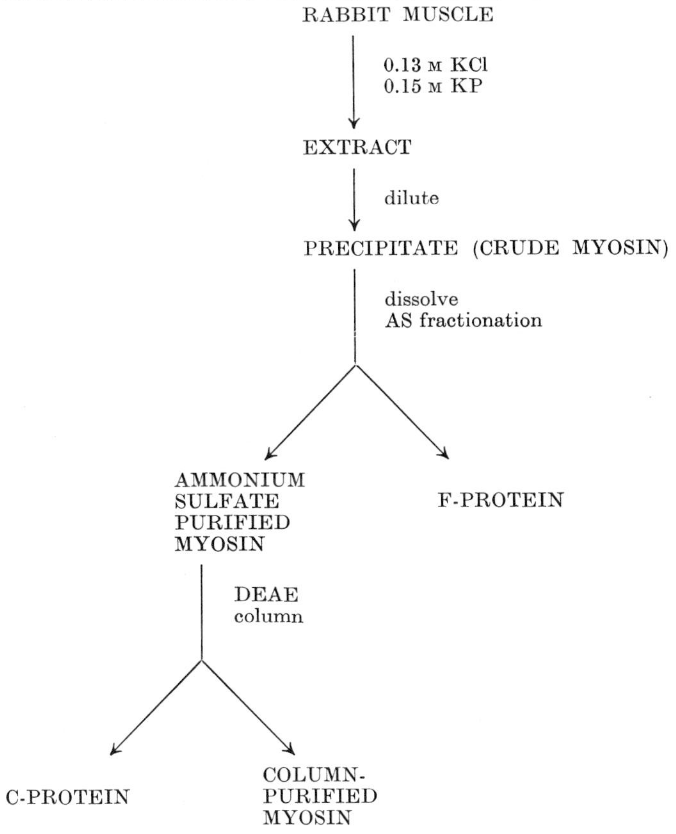

with the B-protein impurity) is eluted with a chloride gradient (Richards et al., 1967; Godfrey and Harrington, 1970). Approximately 4.7 % of the protein applied to the column emerges in the C-protein peak. At this stage the C-protein preparation is dilute and not quite pure. It is concentrated by precipitation with ammonium sulfate and purified by precipitation at very low ionic strength (10 mM imidazole chloride, pH 6.5). The impurities remain in the supernatant, and the precipitate is our pure C-protein (Offer, Moos, and Starr, in preparation).

Evidence That C-Protein is Component of Myofibril

SDS dissolves myofibrils and the polypeptide components may be separated by SDS polyacrylamide gel electrophoresis (Sender, 1971; Scopes and Penny, 1971). One of the components (approximately 2 % of the total myofibril) has a mobility identical to that of the C-protein (Fig. 1). This component is extracted from myofibrils with Guba-Straub solution (with added Mg^{++} and ATP), the solvent which is known to extract A-band material from myofibrils (Hanson and Huxley, 1955). Moreover C-protein is a persistent impurity of myosin preparations despite repeated precipitations

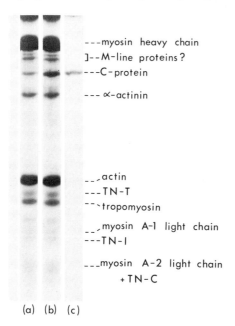

(a) (b) (c)

Figure 1. SDS electrophoresis of rabbit myofibrils and C-protein. (*a*) Rabbit myofibrils; (*b*) rabbit myofibrils with added C-protein; (*c*) C-protein only. Electrophoresis was performed in 5% gels. Note the identical mobility of C-protein and a component of the myofibril. Other bands were identified by comparison with purified muscle proteins. The tentative assignment of the two bands close to the heavy chains of myosin as M-line components is based on references quoted in the text. Other M-line components may be obscured by the bands of α-actinin and actin.

at ionic strengths where C-protein is soluble. All this suggests that C-protein is a component of the thick filaments of myofibrils. Further evidence for this conclusion will be discussed below.

Characterization of C-Protein

C-protein preparations give a single band on SDS gel electrophoresis and a single symmetrical peak in the analytical ultracentrifuge at high ionic strength. The main physical parameters of C-protein are presented in Table 2.

The similarity of the chain weight of C-protein (determined from the mobility on SDS polyacrylamide gel electrophoresis) and the molecular weight (determined by sedimentation equilibrium) in a solvent of high ionic strength (0.5 M KCl, 10 mM imidazole chloride, pH 7.0) shows that C-protein under these conditions exists as a single polypeptide chain. At a lower ionic strength (0.1) in the analytical ultracentrifuge, the schlieren peak is skewed and $S_{20,w}$ increases with concentration, suggesting dimerization. At still lower ionic strengths further association occurs. This property of association, which is characteristic of several muscle proteins, may be physiologically relevant.

The intrinsic viscosity of C-protein (13.6 ml/g) is considerably greater than for typical globular proteins. This suggests that the C-protein molecule is elongated. For example, assuming a reasonable value for the solvation (0.4 g/g) and an ellipsoidal shape for the molecule, the axial ratio would be about 9. However, C-protein is quite unlike the fibrous muscle proteins such as light meromyosin, tropomyosin, and paramyosin, since the α-helical content calculated from its mean residue ellipticity at 208 nm is essentially zero.

All these properties distinguish C-protein from the well-characterized myofibrillar proteins (myosin, actin, tropomyosin, the components of troponin, α-actinin and the components of the M-line). The distinction between C-protein and M-line proteins is particularly important. The chain weights of suspected M-line components are approximately

Table 2. Molecular Properties of C-Protein

$S^\circ_{20,w}$	4.65 S
Molecular weight	140,000 ± 10,000
Chain weight from SDS gel electrophoresis	135,000 ± 15,000
$[\eta]$	13.6 ml/g
Length of hydrodynamically equivalent ellipsoid	350 Å
$[\theta]_{208}$	−4150 deg cm²/decimole
% α-helix	0
$E^{1\%}_{278}$	10.9

45,000, 95,000, and 150,000, and other physical characteristics are different (Pepe, this volume; Etlinger and Fischman, this volume; Masaki and Takaiti, 1972; Morimoto and Harrington, 1972; personal communications from Palmer, Stromer, Goll, and Tabatabai and from Lowey). Thus we consider C-protein to be a newly characterized myofibrillar protein.

Location of C-Protein in Thick Filament

C-protein binds very strongly to myosin at an ionic strength of 0.1 (Moos et al., 1972), as we might expect from the fact that C-protein persists in myosin preparations which have been precipitated several times. Very importantly, C-protein binds to light meromyosin and to myosin rod. (The stoichiometry of binding is approximately 1 mole per mole, but too much should not be made of a stoichiometry measured on insoluble aggregates.) From this we suppose that C-protein is associated with the backbone of the thick filament rather than with the projections, but we have yet to see whether C-protein binds to subfragments 1 or 2, and we cannot rule out the possibility that C-protein interacts with the projections as well as with the backbone.

In order to locate the C-protein in the thick filament, I have collaborated with Dr. Frank Pepe in using antibody techniques. Antibodies to the rabbit proteins have been elicited in goats. We first noted that rabbit myosin of conventional quality (i.e., containing B, C, and F impurities) elicits a substantial amount of antibody to C-protein. When antiserum to such crude myosin is diffused against C-protein and against preparations of myosin at the three stages of purification—(a) crude myosin (containing B, C, and F impurities), (b) ammonium sulfate-purified myosin (containing B and C impurities), and (c) column-purified myosin (containing only the B impurity)—two precipitin lines are formed with (a) and (b). One of these precipitin lines fuses with the single line formed with C-protein, the other forms a spur with the single line formed with column-purified myosin. The presence of antibody to C-protein in antiserum elicited by ordinary myosin preparations means that we have to be very careful in interpreting electron micrographs of muscle labeled with such antiserum, because we would get information on the location of C-protein superimposed on that of myosin.

We have also elicited antiserum to purified C-protein. When this antiserum is diffused against the same antigens, crude myosin and ammonium sulfate-purified myosin give a precipitin line which fuses with the major line from C-protein, but pure myosin gives no reaction. (It should be noted that C-protein cannot be completely pure because two faint precipitin lines are seen in addition to the main line.) Thus C-protein and myosin are antigenically distinct. This disposes of any remaining suspicion that C-protein may in some way be derived from myosin.

In collabration with Dr. E. Rome, Mr. R. W. Craig, and Dr. F. A. Pepe, I have used antiserum to C-protein to find the location of C-protein using both X-ray diffraction and electron microscopy. The technique is to soak bundles of glycerinated rabbit psoas muscle in whole serum for several hours. As a control we use serum obtained from the same goat before injections of C-protein were started. The fibers are then washed in 0.1 M KCl, 1 mM MgCl$_2$, 6 mM potassium phosphate, pH 7.0, to remove unbound protein. The fibers soaked in antiserum show a very considerably enhanced X-ray meridional reflection at about 442 Å, and it appears that another meridional reflection at about 418 Å is also enhanced (Rome, this volume). The remainder of the pattern is unchanged. Control muscles soaked in normal serum give a normal diffraction pattern. Thus C-protein is partly or wholly responsible for the 442 Å reflection observed by Huxley and Brown (1967). This is the first time, to our knowledge, that X-ray diffraction has been successfully combined with an immunological technique.

Although X-ray diffraction gives information on the periodicity with which the C-protein is arranged, it does not tell us where in the thick filament C-protein is located. Craig has examined sections of the antibody-labeled muscle in the electron microscope. In each half of the A-band nine stripes separated by approximately 430 Å are emphasized (Fig. 2). (The error in measurement and the unknown amount of shrinkage do not allow us to distinguish between a 430 and a 440 Å repeat.) A similar set of stripes is produced with antiserum to rabbit myosin of conventional quality (Pepe and Offer, unpublished experiments); this antiserum, we have just seen, contains antibodies to C-protein. The outer seven stripes correspond in position to the seven antibody-labeled stripes emphasized in *chicken* muscle labeled with antiserum to *chicken* myosin (Pepe, 1967a, b). Pepe and I (see Pepe, this volume) have now shown that this antiserum also contains antibody to C-protein, and it now seems that the antibody-labeled stripes in chicken muscle also represent the location of C-protein.

Once the sites of C-protein on the thick filament have been pinpointed by the antibody label, finer structural information is best obtained by examining unlabeled muscle. As Huxley (1966, 1967) noted, there are transverse stripes at approximately 440 Å intervals in the A-bands of sectioned muscle. Because thin filaments overlap the ends of the thick

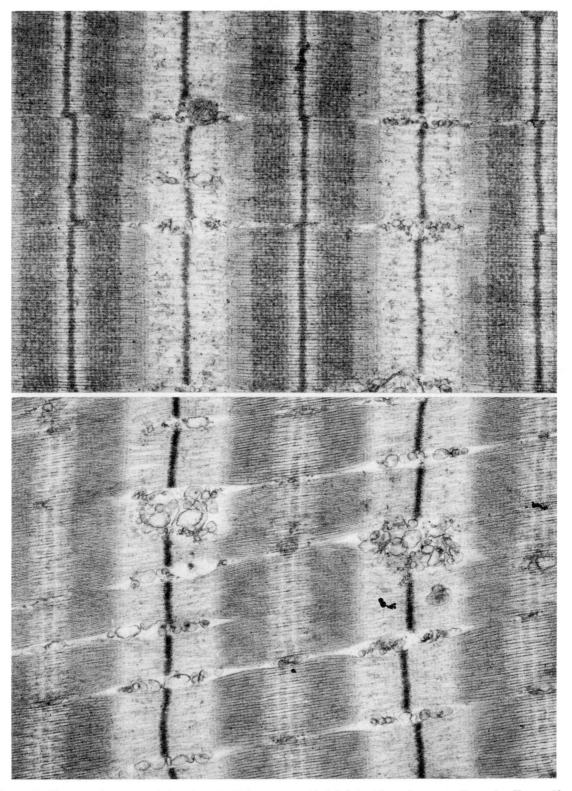

Figure 2. Electron microscopy of glycerinated rabbit psoas muscle labeled with antiserum to C-protein. *Upper:* Glycerinated muscle soaked in antiserum to C-protein for 24 hr. *Lower:* Control soaked in normal serum under identical conditions. The muscle was fixed in glutaraldehyde, post-fixed in osmium, dehydrated in ethanol (with staining in 2% uranyl acetate), and embedded in Araldite. Sections were stained in uranyl acetate/lead citrate.

filaments, it is difficult to see how far along the A-band the stripes continue; but they start closer to the center of the A-band than the set of nine antibody-labeled stripes, so that there are two stripes of material close to the bare zone which do not appear to take up antibody to C-protein. This may be discerned in Fig. 2. Structural information from sectioned muscle is limited, and Hanson et al. (1971) have shown that more detail is obtained in A-segments (from frog) examined by negative contrast. It has not so far been possible to isolate A-segments from the rabbit, but Craig (personal communication) has obtained comparable detail from intact rabbit myofibrils, also examined by negative contrast. As in frog A-segments there are eleven unstained stripes, approximately 60–100 Å wide, the outer nine of which correspond in position to the nine antibody-labeled stripes. We conclude that C-protein is responsible for the outer nine unstained stripes of the rabbit A-band and that the inner two unstained stripes are probably due to another, as yet unidentified, protein (or to C-protein which is prevented in some way from reacting with antibody).

It will be noted that in the section of antibody-labeled muscle in Fig. 2, the M-line is stained much more intensely than in the control. However, for the following reasons, we do not think that C-protein is a component of the M-line. To label the glycerinated fiber bundles, a considerable excess of antiserum was used, and small quantities of antibodies to impurities in C-protein preparations (such as M-line protein) could produce a disproportionate effect. It is well known that M-line protein is highly immunogenic (Pepe, 1966). Craig and I have shown that the nine lines and the M-line are also labeled when antiserum to C-protein is applied to myofibril preparations which are then examined by negative contrast. However, prior absorption of the antiserum with small quantities of C-protein removes the labeling of the stripes but not that of the M-line. Moreover when myofibrils are treated with fluorescently labeled antibody to myosin of conventional quality (which contains antibody to C-protein), there is little staining of the M-line (Pepe and Offer, unpublished experiments). All these observations suggest that C-protein is present in the nine stripes but not in the M-line. Attempts will be made to make a more specific antiserum to C-protein by using lower doses of C-protein to elicit antibody.

The dark stripes seen in the section of antibody-labeled muscle in Fig. 2 and the unstained stripes seen in negatively contrasted myofibrils are both narrow compared with the spacing between the stripes. If C-protein molecules were lying end-to-end along the thick filament in this region, we would expect antibody to combine with sites all along them, and no stripes would be seen. Thus we think the long axis of the C-protein molecule is arranged either radially or circumferentially relative to the filament axis rather than axially. Moreover the viscosity and molecular weight data suggest that the C-protein is not quite long enough to span the period. Thus we think it probable that the spacing

of C-protein at these intervals is dictated by the packing of the myosin molecules (and in particular the light meromyosin) to which it binds. Since the periodicity of light meromyosin aggregates is 429 Å (Szent-Györgyi et al., 1960), the periodicity of C-protein should also be 429 Å.

But how can C-protein be located at intervals of 429 Å if it is responsible for the 442 Å X-ray meridional reflection? Rome (this volume) has suggested a possible explanation. The Fourier transform of a limited periodic array of molecules in each half of the A-band would consist of a series of broad peaks with subsidiary peaks; this would be sampled by a much narrower set of peaks (cosine fringes) due to the interference of the two halves of the A-band. The result would be a large number of meridional reflections which would not immediately appear to be related. In particular, what would be a broad reflection centered at 429 Å if diffraction occurred from one-half of the A-band only, might become sampled by the interference function to give sharper reflections, for example at 442 and 418 Å. (A similar argument was used by O'Brien et al. (1971) to explain the splitting of the meridional reflection arising from the thin filaments in optical diffraction patterns of sections of frog muscle.)

These ideas may be tested by optical diffraction analysis of electron micrographs of A-segments and sectioned muscle. The optical diffraction pattern of one-half of an A-segment reveals a single periodicity of approximately 430 Å (Hanson et al., 1971). A demonstration of the effect of interference between the two halves of the A-band has recently been obtained by analysis of sections of antibody-labeled muscle (Craig, O'Brien, and Bennett, unpublished experiments). Using a mask which included the whole of the A-band, sharp meridional reflections were observed at about 420 Å and 450 Å. If now either half of the A-band was screened off leaving the mask untouched, these two sharp reflections were replaced by a single broad reflection centered at about 435 Å.

A detailed analysis is difficult because the diffraction of C-protein cannot be treated independently of that of myosin so that more structural information is required to predict the transform of the complete A-band. But it seems to be a very attractive proposition that C-protein and myosin might share a related periodicity in the thick filament so that every C-protein molecule could interact in the same way with the tails of the myosin molecules. Additional evidence that this might be so is that when C-protein is added to light meromyosin paracrystals, the transverse striations separated by approximately 400 Å are very much enhanced (Moos et al., 1972; Moos, this volume). Now electron microscopy cannot easily distinguish between 429 and 442 Å, but in the paracrystals labeled with C-protein there appears to be only one set of striations, and preliminary optical diffraction analysis shows only one periodicity (Bennett, Moos, Offer, and Starr, unpublished observations). Thus we think C-protein is adding on to these paracrystals with the same periodicity as light meromyosin and thus enhances the striation.

We should now ask how much C-protein is associated with each stripe on the thick filament. We have already noted that the fraction of C-protein in crude myosin preparations is about 4.7%. This corresponds to one C-protein molecule to 5.7 myosin molecules. Now Guba-Straub solution, with added Mg^{++} and ATP, extracts most of the myosin and C-protein from myofibrils. If therefore we can assume that when minced muscle is extracted, myofibrils in superficial layers are completely extracted but underlying myofibrils are not extracted at all, then the proportions of these two proteins in the thick filament would be similar to their proportions in the extract. (This is tentatively supported by a comparison by SDS gel electrophoresis of crude myosin preparations and of extracts of myofibrils.) Since there are about 200 myosin molecules per filament (Huxley and Brown, 1967), there would be approximately 34 C-protein molecules. With a total of 18 C-protein stripes in the A-band, this corresponds to two C-protein molecules per stripe in each filament. Further work will be required to check this figure. If it is two, this would presumably mean that C-protein is attached to every third pair of myosin molecules in the Huxley and Brown model (Fig. 3). This raises the question of why C-protein is not attached to the other myosin molecules; the answer may be that the light meromyosin backbone does not have helical symmetry but repeats axially at 429 Å, as in paracrystals. It should be noted that because neighboring thick filaments are rotated with respect to each other (Huxley and Brown, 1967), the C-protein would also be so distributed.

One major question is whether the C-protein molecules are confined to a single thick filament or

form bridges between neighboring thick filaments. No such bridges have been seen in transverse sections of muscle, and the optical diffraction analysis of A-segments suggested that the stripes do not span the gaps between filaments (Hanson et al., 1971). At present therefore, it seems likely that each C-protein molecule is confined to a single thick filament. If the C-protein molecule does indeed prove to be elongated, a reasonable possibility would be that C-protein molecules are wrapped around the circumference (very approximately 500 Å) of the backbone of the thick filament. This arrangement would allow two C-protein molecules to make contact forming a collar around the filament.

Possible Function of C-Protein

So far we have been discussing the structure and location of C-protein. As yet very little is known of its function. Considering first purely mechanical roles, I think we can rule out the possibility of C-protein being a core protein because it can react with antibodies and cannot therefore be buried. From its restricted location in one part of the A-band, it also seems unlikely that it could be a length-determining protein; if it does pack with the same periodicity as myosin, the hypothetical vernier mechanism of Huxley and Brown (1967), which demands two components of different periodicities, would in any case be inappropriate.

Another possible function that we have already mentioned is that the C-protein might form bridges between neighboring thick filaments so that neither rotation nor slip of the filaments can occur.

Finally the C-protein might act as a clamp to prevent disruption of the backbone by the radial component of the tension in the cross-bridge.

Alternatively C-protein might have a regulatory function.

Since the C-protein is not present all the way along the thick filament, it would seem unlikely that it is *directly* involved in myosin cross-bridge action. A more attractive possibility is that one C-protein molecule can regulate the behavior of a number of myosin molecules. For example, Haselgrove (1970) and Huxley (this volume) suggested that upon activation of muscle there is a conformational change in the backbone of the thick filament which allows the cross-bridges to swing out and make contact with the thin filaments. It seemed possible that C-protein by binding Ca^{++} could regulate this movement, but no Ca^{++} binding to C-protein has been detected by equilibrium dialysis against 10^{-5} M $CaCl_2$, 0.1 M KCl, 1 mM $MgCl_2$, 10 mM imidazole chloride, pH 7.0. I have also investigated whether C-protein has any effect on the interaction of F-actin and myosin. If pure myosin is added as

<--429 Å-->

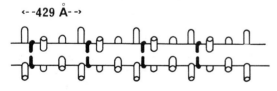

Figure 3. A possible arrangement for the C-protein molecules in the thick filament. Two C-protein molecules are shown associated with every third pair of myosin molecules of the Huxley and Brown model. They have been schematically depicted as semicircular in shape, with the long axis of the molecule arranged around the circumference of the filament.

preformed filaments to F-actin in an ATPase assay medium, the timecourse of ATP splitting is approximately linear. If, however, the myosin is added to a mixture of F-actin and a small quantity of C-protein (approximately one C-protein molecule to six myosin molecules), the rate decreases in the first few minutes to about half the rate without C-protein. This effect is independent of the presence of Ca^{++}. It is difficult to decide whether this effect is important because C-protein tends to upset the regularity of the structure of the thick filament (Moos et al., 1972; Moos, this volume), and this decreased rate may only reflect the decreased accessibility of F-actin to the globular heads of myosin. It certainly shows the need to use pure myosin in kinetic studies of actin-myosin systems.

So while we know that C-protein is a component of the thick filaments and know quite a bit about its location, its function remains unknown. But understanding more about its structure and function cannot fail to tell us more about the thick filament as a whole.

References

GODFREY, J. E. and W. F. HARRINGTON. 1970. Self-association in the myosin system at high ionic strength. II. Evidence for the presence of a monomer $\rightleftarrows$ dimer equilibrium. *Biochemistry* 9: 894.

HANSON, J. and H. E. HUXLEY. 1955. The structural basis of contraction in striated muscle. *Symp. Soc. Exp. Biol.* 9: 228.

HANSON, J., E. J. O'BRIEN, and P. M. BENNETT. 1971. Structure of the myosin-containing filament assembly (A-segment) separated from frog skeletal muscle. *J. Mol. Biol.* 58: 865.

HASELGROVE, J. C. 1970. X-ray diffraction studies on muscle. Ph.D. thesis, University of Cambridge.

HUXLEY, H. E. 1966. The fine structure of striated muscle and its functional significance. Harvey Lectures, series 60, p. 85. Academic Press, New York.

———. 1967. Recent X-ray diffraction and electron microscope studies of striated muscle. *J. Gen. Physiol.* 50: 71.

HUXLEY, H. E. and W. BROWN. 1967. The low-angle X-ray diagram of vertebrate striated muscle and its behaviour during contraction and rigor. *J. Mol. Biol.* 30: 383.

KUNDRAT, E. and F. A. PEPE. 1971. The M-band. Studies with fluorescent antibody staining. *J. Cell Biol.* 48: 340.

MASAKI, T. and O. TAKAITI. 1972. Purification of M-protein. *J. Biochem.* 71: 355.

MORIMOTO, K. and W. F. HARRINGTON. 1972. Isolation and physical-chemical properties of an M-line protein from skeletal muscle. *J. Biol. Chem.* 247: 3052.

MOOS, C., G. W. OFFER, and R. L. STARR. 1972. A new muscle protein which affects myosin filament structure. *Abstr. Amer. Biophys. Soc.* p. 208a.

O'BRIEN, E. J., P. M. BENNETT, and J. HANSON. 1971. Optical diffraction studies of myofibrillar structure. *Phil. Trans. Roy. Soc. London* B 261: 201.

PEPE, F. A. 1966. Some aspects of the structural organization of the myofibril as revealed by antibody staining methods. *J. Cell Biol.* 28: 505.

———. 1967a. The myosin filament. I. Structural organization from antibody staining observed in electron microscopy. *J. Mol. Biol.* 27: 203.

———. 1967b. The myosin filament. II. Interaction between myosin and actin filaments observed using antibody staining in fluorescent and electron microscopy. *J. Mol. Biol.* 27: 227.

RICHARDS, E. G., C. S. CHUNG, D. B. MENZEL, and M. S. OLCOTT. 1967. Chromatography of myosin on diethyl-aminoethyl-Sephadex A-50. *Biochemistry* 6: 528.

SENDER, P. M. 1971. Muscle fibrils: Solubilization and gel electrophoresis. *FEBS Letters* 17: 106.

SCOPES, R. K. and I. F. PENNY. 1971. Subunit sizes of muscle proteins as determined by sodium dodecyl sulphate gel electrophoresis. *Biochem. Biophys. Acta* 236: 409.

SZENT-GYÖRGYI, A. G., C. COHEN, and D. E. PHILPOTT. 1960. Light meromyosin fraction. I. A helical molecule from myosin. *J. Mol. Biol.* 2: 133.

STARR, R. and G. OFFER. 1971. Polypeptide chains of intermediate molecular weight in myosin preparations. *FEBS Letters* 15: 40.

Discussion: Interaction of C-Protein with Myosin and Light Meromyosin

CARL MOOS

Biochemistry Department, State University of New York, Stony Brook, N.Y. 11790

In connection with the study of C-protein reported here by Gerald Offer, I wish to report briefly some observations of the effects of C-protein on the electron microscopic appearance of artificial myosin filaments and light meromyosin paracrystals. Upon dialysis to low ionic strength, myosin which has been purified by DEAE-Sephadex chromatography forms very long filaments (up to 10–20 microns) with an exceptionally uniform diameter and regular appearance, as illustrated in Fig. 1A. In many cases an axial periodicity of about 14.5 nm can be seen in the micrographs and measured by optical diffraction. It is usually impossible to find a bare zone in these filaments, although short filaments formed from the same myosin by rapid dilution of the ionic strength do show a normal bipolar structure.

When C-protein is added to the column-purified myosin before dialysis, or when myosin is used which has not been through the DEAE column, the

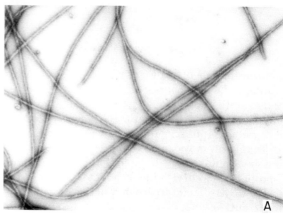

A

Figure 1. Effect of C-protein on myosin filaments. *A*, Column-purified myosin; *B*, same myosin plus C-protein in a ratio of 5 to 1 by weight. Both samples dialyzed into 0.06 M KCl-0.15 M K-phosphate-0.01 M EDTA, pH 7.6, and then into 0.1 M KCl-5 mM imidazole-HCl, pH 7. Samples applied to carbon-coated collodion-film grids and stained with 1% uranyl acetate for negative contrast. The bar is 1 micron.

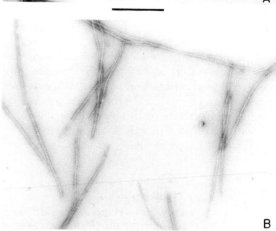

B

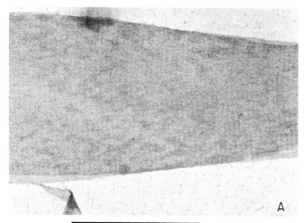

A

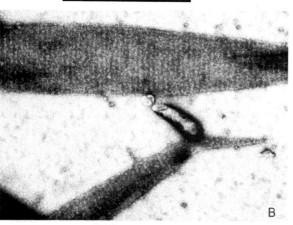

B

Figure 2. Pattern of C-protein binding to light meromyosin paracrystals. Light meromyosin was prepared from column-purified myosin by digestion with trypsin (11 min, 23°, trypsin/myosin = 1/150) and purified by ethaonl precipitation. Paracrystals were formed by dialysis from 0.5 to 0.03 M KCl in 10 mM phosphate, pH 7, at 4°. Samples were stained for negative contrast with 1% uranyl acetate. *A*, Control; *B*, paracrystals treated on the grid with a drop of C-protein solution for about 1 min before staining. The bar is 1 micron.

appearance of the filaments is noticeably different, as illustrated in Fig. 1B. In the sample shown here the amount of added C-protein was about three times that normally present in unpurified myosin, and one can easily see that considerable portions of the filaments are reduced in diameter and show a sparse and irregular arrangement of projections. The ends are particularly affected: With C-protein the filaments usually taper to a very thin tip in contrast to the rather blunt ends characteristic of pure myosin. With smaller amounts of C-protein the effect is qualitatively similar although less pronounced.

These effects of C-protein on myosin filaments suggest that it interacts with the light meromyosin part of the myosin molecule, and this was confirmed by direct measurements of the binding of C-protein to low ionic strength precipitates of myosin and its fragments. The amount of C-protein carried down with a myosin precipitate showed saturation at a maximum ratio near one mole C-protein per mole myosin, and precipitates of papain-produced myosin rod or trypsin-produced light meromyosin also bound C-protein with about the same molar stoichiometry.

The binding of C-protein to paracrystals of light meromyosin has a characteristic effect on their appearance in the electron microscope, as illustrated in Fig. 2. The typical banding at about 43 nm intervals is only barely visible in the pure light meromyosin (Fig. 2A), but addition of C-protein strongly accentuates this periodicity (Fig. 2B). The same effect is seen when C-protein is added to the light meromyosin solution before the ionic strength is reduced. As mentioned by Offer in the accompanying paper, optical diffraction measurements on micrographs such as these show no change in the repeat period when C-protein is added. It appears, therefore, that the attachment of C-protein to light meromyosin at low ionic strength is not random but bears a special relation to the structure of the paracrystalline assembly.

This work was done while the author was a visiting scientist at the MRC Muscle Biophysics Unit, King's College, London, supported by NIH Special Fellowship 1 FO3 AM5188.

The Myosin Filament: Immunochemical and Ultrastructural Approaches to Molecular Organization

Frank A. Pepe

Department of Anatomy, Medical School, University of Pennsylvania, Philadelphia, Pa. 19104

The use of antibody staining in fluorescence and electron microscopy has proved to be very useful for studying the myofibril and the structural organization of the myosin filament (Pepe 1967a,b, 1968, 1971a). The major emphasis of this work has been on the myosin-containing filaments of the A-band. The antimyosin used in all of this work was prepared against myosin which was purified by repeated reprecipitation of the myosin. Starr and Offer (1971) and Offer (this volume) have shown that myosin prepared in this way contains other proteins, especially C-protein which has been characterized by Offer (this volume). As will be seen in the work presented here, the antimyosin used previously contained antibodies to the impurities, including C-protein, as well as antibody specific for the myosin. It will also become clear that, in spite of this finding, there is no change either in the basic conclusions arrived at previously or in the model derived for the myosin filament (Pepe, 1967a,b, 1971a). All of the antibody staining work done in this paper and in previous publications has been done using the pectoralis muscles of white roosters.

Immunodiffusion

Myosin was prepared as we have done previously (Pepe, 1966, 1967a,b), and 950 mg was put onto a 2.5×90 cm DEAE Sephadex A-50 column in 0.15 M PO_4, 10 mM EDTA pH 7.5 (Richards et al., 1967; Godfrey and Harrington, 1970). After the unretarded protein came through, the purified myosin was eluted using a linear gradient of KCL in the PO_4-EDTA buffer. The elution profile is shown in Fig. 1. I will henceforth refer to the column-purified myosin as "CP-myosin" and to that before column purification as "I-myosin."

The antibody used in all our previous work (Pepe, 1966, 1967a,b) was prepared in rabbits against I-myosin. The gamma globulin fraction of the antiserum was used. I will henceforth refer to this as AM. Antiserum has also been prepared against CP-myosin. Goats were used in this case for immunization instead of rabbits, since goats proved to be more convenient to work with. The antiserum to CP-myosin was used without fractionation. I will henceforth refer to this as ASCM. All the antigens and antisera used in the immunodiffusion studies

were in 0.4 M KCL, 0.03 M PO_4 buffer pH 7.3. These studies were made in collaboration with Dr. Offer. The Ouchterlony plates were prepared using 1% Agar in the same buffer. The plates were allowed to develop for 48 hr in the cold (0–$4°C$); they were washed overnight in 0.4 M KCL, 0.03 M PO_4 buffer pH 7.3 at room temperature; they were stained for 2 hr with 1% Alizarin red in water and were destained with 1% acetic acid.

Note in Fig. 2a and b that both the I-myosin and the CP-myosin give a single line in immunodiffusion against AM. This is generally characteristic for a single antigenic species. However in Fig. 2c and 2g it can be seen that the concentrated impurities, when diffused against AM, give two lines. Note that the impurities line closest to the central antibody-containing well in Fig. 2c crosses over cleanly with the line formed with CP-myosin. This impurities line also crosses over the line formed with I-myosin, but in this case there is some deflection of the impurities line. In Fig. 2d and e note that ASCM gives a single line with I-myosin and with CP-myosin, as can also be seen in Fig. 2f. Note in Fig. 2f that ASCM gives a faint line near the well containing impurities, but there is no trace of the heavy line seen with AM (Fig. 2c). Therefore the ASCM does contain a small amount of antibody to some component of the impurities.

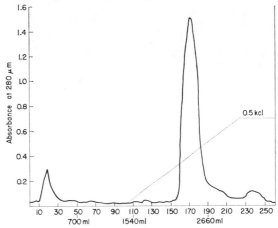

Figure 1. Elution profiles of normally purified myosin (I-myosin) on DEAE Sephadex A-50 in 0.15 M PO_4, 10 mM EDTA pH 7.5. The I-myosin (950 mg) was put onto a 2.5×90 cm column. After the unretarded peak containing impurities came through, a linear gradient of KCL was added to the buffer to elute the myosin.

Starr and Offer (1971) have shown that the impurities removed by DEAE-Sephadex A-50 chromatography consist mainly of C and F proteins which they identified on SDS gel electrophoresis. In collaboration with Dr. Offer we have been able to show that the impurities removed from our I-myosin contain C-protein as one of the constituents. C-protein was isolated (Offer, this volume) from the I-myosin. Antiserum was prepared against the C-protein in goats (henceforth ASC). As can be seen in Fig. 2i, the ASC forms a single line with the impurities, verifying that the impurities contain C-protein as one of the constituents. Also, if purified C-protein is diffused against AM

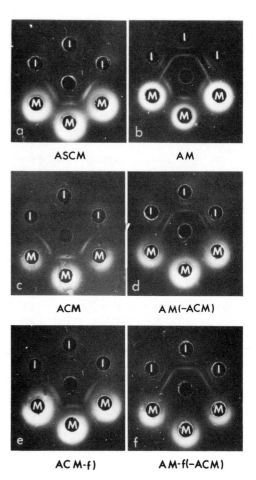

Figure 3. Immunodiffusion studies using antibody specific for column-purified myosin (*ACM*) isolated from the antimyosin used in previous work (*AM*). Notations are as follows: *ASCM*, antiserum to column-purified myosin; *AM*, the antimyosin used in previous work; *ACM*, antibody specific for column-purified myosin, isolated from AM using column-purified myosin coupled to PAB-cellulose; *ACM-f*, same as ACM only obtained from AM previously labeled with fluorescein; *AM(-ACM)*, the antibody left after ACM has been removed; *AM-f(-ACM)*, same as AM(-ACM) only fluorescein-labeled. For description of reactions, see text.

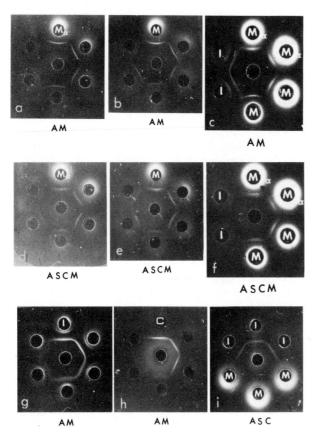

Figure 2. Immunodiffusion studies. *a, b, c:* Each center well contains the antimyosin (*AM*) used in all our previous work. This is the γ-globulin fraction of the antiserum at 29 mg/ml. $M\alpha$, normally purified myosin (I-myosin) at 5 mg/ml; *M*, column-purified myosin (CP-myosin) at 5.5 mg/ml; *I*, unretarded impurities (0.5 mg/ml) obtained on column purification of the myosin. The outer wells in (*a*) and (*b*) show serial twofold dilutions of $M\alpha$ and *M*, respectively. *d, e, f:* Each center well contains antiserum to column purified myosin (*ASCM*). The solutions in the peripheral wells are identical to those in the corresponding top three patterns. *g, h, i:* The notations *AM*, *I*, and *M* are identical to those above. The first two peripheral wells in (*g*) contain impurities at 0.5 mg/ml, the next two wells at 0.25 mg/ml, and the last two at 0.125 mg/ml. In (*h*) the *C* represents chicken C-protein at 1 mg/ml with serial twofold dilutions in the other wells. In (*i*) the central well contains antiserum to C-protein (*ASC*).

(Fig. 2h), a single line is formed, verifying that the AM contains antibody to C-protein.

In spite of column purification of the myosin prior to using it as an antigen, the ASCM still contained a small amount of antibody to some impurity. Antibody specific for CP-myosin and giving no sign of any reaction with impurities was isolated from AM (see Appendix, this paper). This was done by coupling CP-myosin to PAB cellulose. The PAB-myosin was then suspended in the AM. In this way antibody specific only for the CP-myosin was bound to the PAB-myosin. The supernatant was filtered off and the PAB-myosin-antibody complex was washed. The specific antibody was then dissociated from the PAB-myosin by bringing the pH to 2.9, and after centrifuging out the PAB-myosin, the specific antibody in the supernatant was brought to pH 7 immediately and recovered by precipitation at 50% saturated ammonium sulfate. Henceforth I will refer to antimyosin obtained in this way from fluorescein-labeled AM as ACM-f and that from unlabeled AM as ACM. The antibody left after removal of the ACM-f or the ACM will be designated AM-f(-ACM) and AM(-ACM), respectively.

In Fig. 3 are the immunodiffusion patterns obtained using these antibodies. In each pattern the top three wells contain impurities and the bottom three wells contain CP-myosin. Note in Fig. 3a that ASCM gives a single heavy line with the CP-myosin. There is a very faint line near each well containing the impurities but it is hard to see. With AM (Fig. 3b) the single heavy line with CP-myosin is evident and the two lines with impurities are also clearly seen. In Fig. 3c, ACM was used; note that only the heavy line with CP-myosin is present. There is no indication of any reaction with the impurities. In Fig. 3d, AM(-ACM) was used; note that there is no reaction with CP-myosin but that the two lines with impurities are clearly shown. Similarly in Fig. 3e and f, the same antibodies as those in Fig. 3c and d were used except that in this case they were fluorescein-labeled. The results are identical to those obtained with the unlabeled anti-

bodies. Therefore we have succeeded in isolating antibody specific only for CP-myosin, i.e., ACM and ACM-f.

Fluorescence and Electron Microscopy

Using the ACM-f obtained as described above, the staining patterns shown in Fig. 4 were obtained. These staining patterns are identical to those previously obtained with AM-f which contained antibodies to the impurities in addition to antibodies specific for CP-myosin (Pepe, 1966, 1967b, 1968, 1971a). Note at 2.3 μ sarcomere length (Fig. 4a), the bright staining in the middle of the A-band and the staining in the A-band along each A-I junction. Also note that the staining in the middle of the A-band at 2.3 μ sarcomere length is not present as a doublet. As previously reported (Pepe, 1967b, 1971a), the doublet shows up at longer sarcomere lengths. At 1.8 μ sarcomere length (Fig. 4b) where there is complete overlap of the thin and thick filaments, the bright staining in the middle of the A-band is not present. The staining in the A-band along each A-I junction has broadened. This has previously been described as consisting of both the LMM-specific staining in the A-band along the A-I junction, and the β-specific staining in the middle one-third of each half of the A-band (Pepe, 1966, 1967b, 1968, 1971a). The total antimyosin-staining pattern with ACM-f at different

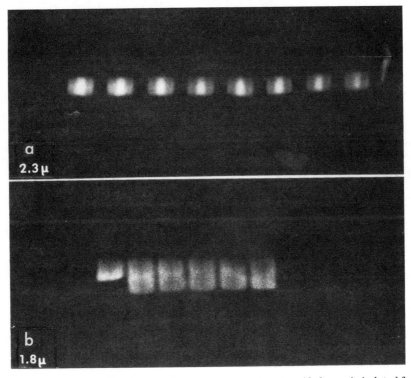

Figure 4. Fluorescent antimyosin staining using antibody specific for column-purified myosin isolated from the antimyosin used in previous work. (a) At 2.3 μ sarcomere length there is staining in the A-band along each A-I junction and staining in the middle of the A-band. (b) At 1.8 μ sarcomere length the staining in the middle of the A-band is absent and that along each A-I junction is wider and brighter.

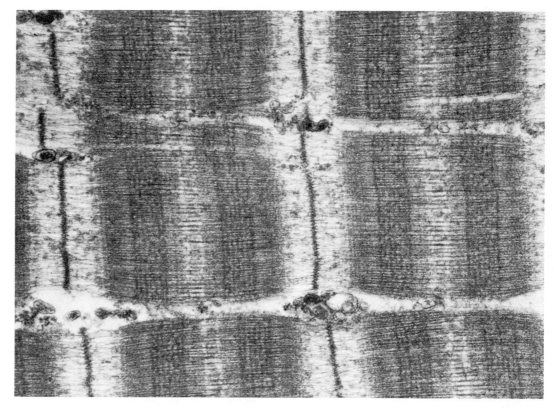

Figure 5. Staining of the A-band as observed in electron microscopy using the antimyosin used in previous work. Note seven lines in the middle one-third of each half of the A-band as previously reported.

sarcomere lengths is therefore identical to that previously obtained with fluorescent AM. The conclusion from these earlier studies that the non-uniformity of staining in the A-band is due to involvement of some of the myosin-antigenic sites in either interaction with surrounding actin filaments or myosin–myosin interactions involved in the aggregation of the myosin molecules in the filament is still valid. Such interactions would make those antigenic sites involved unavailable for antibody staining. Changes in the staining pattern therefore reflect changes in these interactions.

As has been described in previous work (Pepe, 1967a, 1968, 1971a), myofibrils stained with AM show seven lines in the middle of each half of the A-band in electron microscopy (Fig. 5). The spacing between these lines as measured from electron micrographs was 430 Å, and there was no change in either the position of the lines or the spacing between the lines with change in sarcomere length (Pepe, 1967a). In Fig. 6 are examples of sarcomeres at different sarcomere lengths which were stained with ACM. Note the absence of the seven lines in each half of the A-band. In fluorescence microscopy, staining was observed in this region of the A-band (the β-staining) with ACM

(Fig. 4). Therefore, we must conclude that the β-staining observed with AM involves *both* antibodies specific for the CP-myosin molecule and antibodies to one or more of the impurities removed by column purification of the myosin, the antibodies to the impurity being responsible for the seven lines observed in each half of the A-band. In collaboration with Dr. Offer we have found that antibody to rabbit C-protein (Offer, this volume) and also antibody to chicken C-protein gives seven to nine lines in each half of the A-band, seven of the lines corresponding in position to those seen with AM-staining of chicken fibrils. Therefore we conclude that the antibody in the AM responsible for the seven lines observed in the middle of each half of the A-band is antibody to C-protein. From X-ray diffraction studies of fibers stained with antibody to rabbit C-protein and optical diffraction studies of electron micrographs of antibody-stained fibrils, it can be concluded that the C-protein is distributed at 430 Å intervals along the backbone of the myosin filament (Rome, this volume). This reflects the underlying repeat periodicity of the myosin molecules in the myosin filament to which the C-protein binds. In the original work with AM-staining of chicken muscle fibrils, when these lines

Figure 6. Staining of the A-band as observed in electron microscopy using antibody specific for column-purified myosin isolated from the antimyosin used in previous work. The seven lines in each half of the A-band are absent (see text). There is a single line in each half of the A-band which changes in position with change in sarcomere length. The arrow indicates the line in the right half of each A-band.

were considered to be due to specific staining of myosin, the repeat periodicity was also taken to be 430 Å as measured on electron micrographs (Pepe, 1967a). Therefore all of the conclusions previously arrived at based on this 430 Å repeat periodicity are still valid. Also, since neither the periodicity nor the position of the seven lines changed with sarcomere length in the work with AM, it was concluded that the antigenic sites responsible for the periodicity had to be on the backbone of the filament.

This conclusion likewise still holds: the antigenic sites on the backbone of the filament being due to C-protein which is located along the backbone of the filament (Offer, this volume; Rome, this volume) and which reflects the organization of the underlying myosin molecules.

In the previous work with AM it was possible to show that the antibody staining the middle region of the A-band could be absorbed with the HMM fragment of myosin prepared from I-myosin. Also the antibody staining the A-band along each A-I junction could be absorbed by the LMM fragment of myosin prepared from I-myosin. The antibody staining the middle one-third of each half of the A-band (β region of the A-band) could be absorbed by the LMM fragment obtained from a short digestion of I-myosin, whereas it could not be absorbed by the LMM fragment obtained from a longer digestion of I-myosin. The antibody staining of the β region of the A-band was always absorbed by the HMM fragment regardless of the time of digestion used. From these studies it was concluded that the myosin molecule has three antigenically specific regions: the LMM-specific portion, the HMM-specific portion, and a portion in the middle having β specificity. The β specificity on the early LMM fragments was sensitive to destruction by trypsin, whereas the β specificity on the HMM fragment was not. Similar experiments will now have to be performed using ACM-f and fragments of myosin obtained from CP-myosin to be sure whether or not these conclusions are still valid. Recently Lowey and Steiner (1972) have shown that antibody prepared against the rod and S1

fragments of myosin have distinct antigenicities, which is consistent with the idea that there are antigenically specific regions on the myosin molecule.

In the previous observations using fluorescent AM, the staining observed in the LMM and β regions of the A-band increased in brightness as the sarcomere length decreased (Pepe, 1966, 1967b). This increase in availability of LMM antigenic sites was interpreted as representing a bending of the myosin molecules out from the core of the filament to accommodate the increase in distance between the filaments resulting from decrease in sarcomere length (Elliott et al., 1965). This was the first evidence obtained for bending of the myosin molecules away from the core of the filament. Later X-ray diffraction studies supported this conclusion (Huxley and Brown, 1967). Whether the increase in brightness in the β region of the A-band with decrease in sarcomere length is due to increased exposure of both the myosin-specific and C-protein-specific antigenic sites in this region, or only one of them, remains to be determined.

In the sarcomeres shown in Fig. 6, which are

Figure 7. Staining of the A-band as observed in electron microscopy using antiserum to column-purified myosin. A single line is observed in each half of the A-band exactly as seen in Fig. 6. The arrow indicates the line in the right half of each A-band.

from muscle stained with ACM, a single line can be seen in each half of the A-band. Note that the line changes in position with change in sarcomere length. The position of the line is at a constant distance of approximately 3000 Å from the free ends of the thin filaments, taking the length of the thin filaments as 1 μ. It is highly unlikely that this line could be due to staining of one of the impurities removed on chromatography of the myosin, since the ACM showed no reaction with the impurities in immunodiffusion (Fig. 3). Starr and Offer (1971) found that there is one impurity in the myosin, which they labeled B-protein and which was not removed even by DEAE-Sephadex chromatography. Therefore it is possible that the ACM may have antibody to this component. If the line seen with ACM is due to staining of B-protein, it would mean that the position of the B-protein in the A-band changes with change in sarcomere length. Although we can not presently exclude this possibility, I think it is unlikely. It is more likely that this line is due to staining specific for myosin. The appearance of a single line in each half of the A-band would then result from the orientation of the antibody molecules in such a way that they are precisely superimposed, leading to high protein density in a narrow line across the A-band. The orientation of the antibody molecules would reflect the orientation of the antigenic sites on the myosin molecules to which the antibody is bound. Since the fibrils are stained in rigor, this would mean that in rigor muscle there is only one position along the thin filament where the attached myosin cross-bridges are oriented such that some of the antibody molecules bound to them are oriented properly to be superimposed precisely. This is what would be expected if, in rigor, the attached cross-bridges are accommodating to a repeat shorter than the 430 Å repeat of the bridges on the myosin filament (Pepe, 1971b). In this case each successive bridge between an actin and a myosin filament would have to be a little longer than the preceding bridge on going from the end of the actin filament to the edge of the A-band. If the change in distance between the actin and myosin filaments which occurs with change in sarcomere length has a negligible effect on the length of the bridge compared to the differences in length of successive bridges attached to the actin filaments, then the length of a bridge at a given distance from the end of the actin filament, and presumably the orientation of that bridge, would be the same regardless of the sarcomere length. The orientation of the myosin bridges resulting in proper superimposition of antibody molecules attached to them would then be at a constant distance from the end of the actin filaments, regardless of sarcomere length. No differences

could be detected in electron microscopy on staining with ACM (Fig. 6) or with ASCM (Fig. 7).

Model for the Myosin Filament

A model has been derived for the molecular organization of the myosin filament (Pepe, 1967a, 1971a). This is the only model available which describes in detail the packing of the myosin molecules in the filament. The model was derived by relating the 430 Å periodicity seen with AM staining in electron microscopy to the structural features of the A-band and of single myosin filaments. As has already been discussed, the 430 Å periodicity seen with AM in electron microscopy is now known to be due to staining of the C-protein which is associated with the backbone of the filament. Since this periodicity reflects that of the underlying myosin to which it is attached, the original assumption that this periodicity reflects the repeat periodicity of myosin molecules along the myosin filament is still valid.

The derivation of the model has been described in detail (Pepe, 1967a, 1971a). I would like to outline here the main features of the model and the most significant observations or assumptions which led to the derivation of these features. In the model the myosin molecules are arranged as parallel linear aggregates, with the molecules in one-half of the filament oriented opposite to those in the other half of the filament. In the middle of the filament where tail-to-tail interaction of myosin molecules occurs and where the M-band protein attaches to the filaments, the linear aggregates are helically arranged. The parallel arrangement of molecules was assumed on the basis of very clear, triangular cross-sectional profiles of the filament seen in electron microscopy (Fig. 8a). It was reasoned that this was the most likely arrangement since a helical twist would be expected to give a circular cross-sectional profile. The helical aggregation in the region of the filament corresponding to the position of the M-band was suggested by observations of clear, hollow, circular cross-sectional profiles of the filament in cross sections through the M-band (Fig. 8b). Since the myosin molecules in one-half of the filament are oriented opposite to those in the other half, as first proposed by Huxley (1963), the helical portion corresponds to the region of the filament in which there is tail-to-tail aggregation of myosin molecules with head-to-tail aggregation occurring along the rest of the filament.

In each linear aggregate, 860 Å (the LMM portion) of the myosin molecule is in the backbone of the filament. The next 430 Å (the S2 portion) of the myosin molecule overlaps the next myosin molecule. The rest (the S1 portion) of the myosin molecule projects from the filament. In the filament there are three sets of four linear aggregates arranged in parallel. Within a set of four there are two pairs where the stagger within a pair is $\frac{1}{3}$ (430 Å) and the stagger between pairs is 430 Å. See the diagram in Fig. 9. The stagger relationship was derived mainly from the following two assumptions: (1) In the middle of the filament, for each linear aggregate there is tail-to-tail abutment of oppositely oriented myosin molecules (Fig. 12a). (2) The M-band material attaches to the point of tail-to-tail abutment of myosin molecules on the surface of the filament and bridges between two filaments when the tail-to-tail abutments on neighboring filaments are directly opposite. It was also assumed that there are six myosin cross-bridges every 430 Å along the filament, each coming from a different linear aggregate. Using the above assumptions and measurement of the width of the M-band as 2(430 Å), it was concluded that there are twelve parallel linear aggregates (three sets of four) forming the backbone of the filament.

The myosin filaments are not symmetrical about the middle of the filament; i.e., one end of the filament is distinguishable from the other. The myosin filaments must be stacked in the A-band in groups of three in parallel (similarly oriented), with these groups of three antiparallel (oppositely oriented) to other groups of three (Pepe, 1971a). The arrangement of the filaments in the A-band

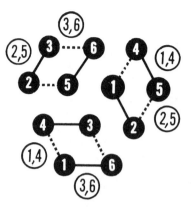

Figure 9. A model for the detailed molecular organization of the myosin filament. In the diagram the solid circles represent the LMM portion and the open circles represent the overlapping S2 portion of the molecules. A difference of one unit represents a stagger of 1/3 (430 Å) between linear aggregates. For full description of the model, see text.

was derived from the necessity to obtain a maximum of five lines in the M-band as observed in longitudinal sections through the fibril (Fig. 8c).

The model for the myosin filament (Fig. 9) predicts the following: (1) The subunit structure of the core of the myosin filament, as seen in cross sections, should consist of twelve subunits, close packed and arranged so that three are centrally located. As can be seen in Fig. 10 this subunit

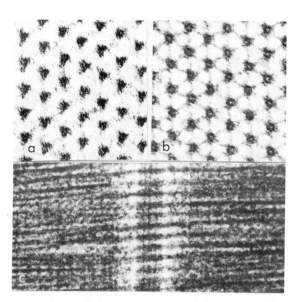

Figure 8. The thick filaments and the M-band. (a) Cross sections of the thick filaments in the pseudo-H-zone just adjacent to the M-band. (b) Cross sections of the thick filaments through the M-band region. (c) Longitudinal section through the M-band. Note the five lines of the M-band, the three central lines being more distinct than the other two (Pepe, 1971a). These sections are of the lateral muscles of the freshwater killifish (*Fundulus diaphanus*).

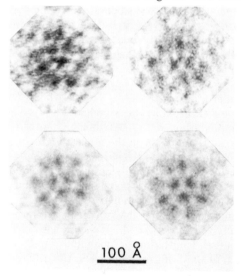

100 Å

Figure 10. Observation of the predicted subunit structure of the myosin filament. (Reprinted from Pepe, 1972). The top two micrographs are of highly magnified cross sections of thick filaments. The bottom two micrographs were printed with the same total exposure as the corresponding top micrographs except that one-third of the total exposure was given in each of three positions differing by 120° rotation about the center of the image. Note that the subunit structure visible in the top micrographs has been enhanced in the bottom micrographs. This subunit structure corresponds to the solid circles in Fig. 9 and is therefore as predicted by the model.

structure has recently been observed in cross sections of the filament (Pepe and Drucker, 1972). (2) The pattern of lines in the M-band seen in longitudinal sections should vary as the plane of the longitudinal section is rotated on the long axis of the fibril. The model predicts that, in addition to the pattern consisting of the maximum of five lines, we should see patterns consisting of four lines, only the three central lines, only two of the three central lines with the middle one missing, and only two of the three central lines including the middle one (Pepe, 1971a). All of these patterns have been observed (Pepe, 1971a) in longitudinal sections through both chicken and fish muscles (Fig. 11). The observation of these different M-band patterns cannot be explained by the model for the structure of the M-band proposed by Knappies and Carlsen (1968).

After stacking the filaments in the A-band as required to obtain a maximum of five lines in the M-band (Pepe, 1971a), we find that the orientation of the myosin cross-bridges on one filament is exactly equivalent to the orientation of the myosin cross-bridges on all neighboring filaments. That is, the filaments are all in equivalent rotational positions along the long axis of the filament. Huxley and Brown (1967) have concluded from X-ray diffraction studies of frog sartorius muscle that neighboring filaments are rotated on the long axis by 120° relative to each other. This finding contradicts what is predicted by the model. In the model, an M-bridge forms between two filaments when one end of the M-bridge attaches to a tail-to-tail abutment on one filament and the other end attaches to a tail-to-tail abutment on a neighboring filament. One end of the M-bridge therefore attaches

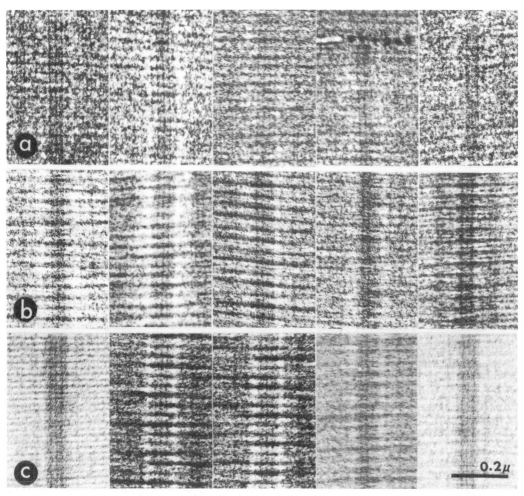

Figure 11. Observation of the predicted variation in the pattern of lines in the M-band, as the plane of longitudinal section is rotated on the long axis of the fibril. (a) Patterns obtained from the pectoralis muscles of white roosters. (b) Patterns obtained from the lateral muscles of the black molly (*Mollienesia sphenops*). (c) Patterns obtained from the lateral muscles of the freshwater killifish (*Fundulus diaphanus*). From left to right in each case there are the three middle lines, the maximum of five lines, four lines, two of the three middle lines including the central line, and two of the three middle lines with the central line missing. These different M-band patterns are predicted when the model in Fig. 9 is arranged in the A-band in such a way as to give the maximum of five lines in the M-band. (From Pepe, 1971a.)

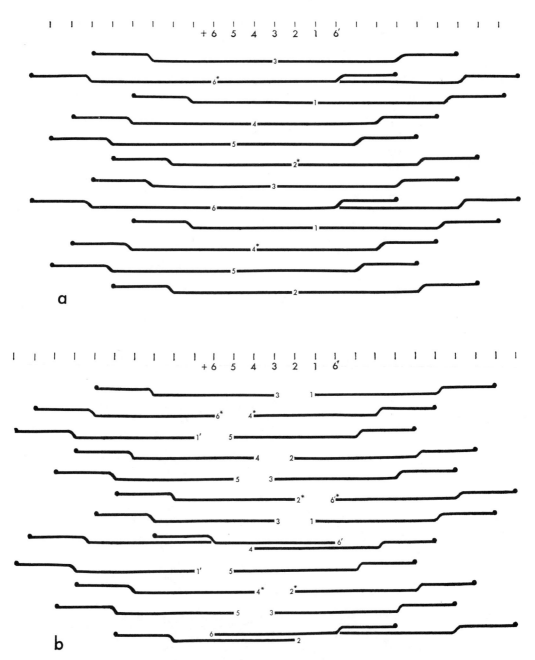

Figure 12. (*a*) Diagrammatic representation of the linear aggregates of myosin molecules in the middle of the filament where tail-to-tail aggregation occurs. On placing these filaments in the A-band as required for M-band formation (Pepe, 1971a), neighboring filaments are found to be in equivalent rotational orientation on the long axis of the filament. (*b*) By putting a gap of 2/3 (430 Å) between the ends of the myosin molecules in the region of tail-to-tail aggregation and requiring that an M-bridge can only bridge between the tails of oppositely oriented molecules on the surface of neighboring filaments, it is found that neighboring filaments are rotated, with respect to each other, by 120° on the long axis of the filament.

to the junction between two oppositely oriented myosin molecules in the middle of the myosin filament (See Fig. 12a). This is the basic information required for M-bridge formation in the model. This basic information can be shared between two neighboring filaments without changing the model in any other way by doing the following: (1) Put a

gap of $\frac{2}{3}$ (430 Å) between the tails of the oppositely oriented myosin molecules in the middle of the filament (See Fig. 12b). (2) Require that one end of the M-bridge attach to the tail of a myosin molecule on the surface of the filament and the other end of the M-bridge attach to the tail of a myosin molecule oriented in the opposite direction and on the surface of a neighboring myosin filament. When this is done without alteration of any other aspects of the model as previously derived, it is found that the relative orientation of the myosin cross-bridges on neighboring filaments is as deduced by Huxley and Brown (1967) from X-ray diffraction studies of the frog sartorius muscle. The details of this modification of the model are being prepared for publication.

Summary

The antimyosin previously used in studies of fluorescent antimyosin staining of the myofibrils (Pepe, 1966, 1967b) contained antibodies to impurities in the myosin preparations. Antibody specific only for the myosin was isolated from this antibody preparation and used to stain myofibrils. The staining pattern (Fig. 4) was identical to that previously obtained with antimyosin containing antibodies to impurities. The conclusion from the previous studies that the nonuniformity of antimyosin staining in the A-band (Fig. 4) is due to involvement of myosin antigenic sites, either in interactions with the surrounding actin filaments or in myosin–myosin interactions in the myosin filament, is therefore unchanged. The previous conclusion that there are three antigenically specific regions on the myosin molecule (Pepe, 1967b) has not yet been verified. In order to do this the antibody specific for myosin will have to be absorbed with fragments of myosin obtained by proteolytic digestion of the highly purified myosin.

When used to stain fibrils for electron microscopy, antibody specific only for myosin and isolated from the antimyosin previously used (Pepe, 1967a, 1971a) showed a single line in each half of the A-band. The position of the line changes with change in sarcomere length. It is concluded that visibility of the line results most likely from precise orientation and superimposition of antibody molecules, resulting in high protein density in a single line across the A-band. Such an alignment of antibody molecules would have to reflect the orientation of the antigenic sites to which the antibody is bound.

The seven lines in each half of the A-band previously seen with antimyosin staining (Pepe, 1967a, 1971a) have now been identified as due to the presence of antibody specific for C-protein (Offer, this volume). It can be concluded that the C-protein is distributed at 430 Å intervals along a portion of the backbone of the myosin filament (Rome, this volume), reflecting the underlying repeat periodicity of the myosin molecules in the myosin filament. In the earlier work (Pepe, 1967a, 1971a) when this periodicity was thought to represent specific staining of the myosin, it was concluded from measurements on electron micrographs that the

periodicity was 430 Å, that it reflected the repeat periodicity of the myosin molecules in the myosin filament, and that there was something different about the aggregation of the myosin molecules in the portion of the filament giving rise to the periodicity (more precise packing of myosin molecules in this region of the filament was suggested). These conclusions are still valid since the C-protein is reflecting the organization of the myosin molecules to which it is bound.

The 430 Å repeat periodicity seen with antimyosin (Pepe, 1967a, 1971a) and now known to be due to the presence of C-protein on the backbone of the filament (Offer, this volume; Rome, this volume), was related to the structural characteristics of the A-band and of separated thick filaments seen in electron microscopy. This led to derivation of the only available detailed model for the molecular organization of the myosin molecules in the myosin filament (Pepe, 1967a, 1971a). The model predicts that the subunit structure in the core of the filament should consist of three centrally located subunits and nine peripherally located, all close packed; and that the pattern of lines in the M-band seen in longitudinal sections should vary as the plane of the longitudinal section is rotated on the long axis of the fibril. Both the predicted subunit structure (Pepe and Drucker, 1972) and the patterns of lines predicted for the M-band (Pepe, 1971a) have been observed.

Acknowledgment

This work was supported by USPHS Grant RO1 AMO4806 and Career Development Award KO4 AM07342.

Appendix

The procedure used for coupling column-purified myosin to diazotized PAB-cellulose is similar to that used by Lowey et al. (1969) for preparing insoluble papain. The procedure is as follows: (1) To 5 g of PAB-cellulose (Gallard-Schlesinger Mfg. Corp., Garden City, New York, 0.32 meq/gm) 40 ml of cold 50% acetic acid was added. While stirring in ice, 5 ml of ice-cold 2 N HCL was added. Then 8 ml of ice-cold 0.5 M sodium nitrite was added in drops over approximately 15 min. This was left to stir on ice for 3 hr. (2) Ice-cold 50% NaOH was added in drops, bringing the pH to 7 and keeping the temperature below 4°C by slow addition. Addition of 2 N NaOH was used to raise the pH to 8. (3) The suspension was filtered on a Buchner funnel and washed with 200 ml of ice-cold 10% sodium acetate and then with 200 ml of ice-cold 0.15 M PO₄, 10 mM EDTA pH 7.5 (henceforth PO₄ buffer). (4) The diazotized PAB-cellulose was suspended in 50 ml of PO₄ buffer and 120 mg of CP-myosin in the same buffer was added. The mixture was stirred slowly overnight in a tightly closed polyethylene bottle in the cold room. (5) The PAB-myosin was filtered on a Buchner funnel, then washed with 200 ml of PO₄ buffer saturated with β-naphthol until a change in color was observed. It was then washed with 200 ml of PO₄ buffer alone, followed by 200 ml of 0.4 M KCL, 0.03 M PO₄ pH 7.3 (henceforth 0.43 KCL-PO₄). (6) The PAB-myosin was suspended in approximately 40 ml of 0.43 KCL-PO₄ and 90 mg of AM in 0.43 KCL-PO₄ was added. The mixture was stirred for

3 hr and was then filtered on a Buchner funnel. The filtrate was collected and precipitated in 50% saturated ammonium sulfate, dissolved in 0.43 KCL-PO$_4$ and dialyzed against 0.43 KCL-PO$_4$. This is the AM(-ACM) used in this work. (7) The PAB-myosin-antibody complex was washed with 400 ml of 0.43 KCL-PO$_4$, then suspended in 45 ml of ice-cold pH 2.9 glycine-HCL buffer, and the pH of the suspension was readjusted to 2.9. After 20 min stirring on ice, the mixture was centrifuged for 20 min at 29,000 rpm in a Beckman model L ultracentrifuge. The supernatant was brought to pH 7.0 immediately and precipitated in 50% saturated ammonium sulfate. The precipitate was dissolved in 0.43 KCL-PO$_4$ and dialyzed against 0.43 KCL-PO$_4$. This is the ACM used in this work.

References

ELLIOTT, G. F., J. LOWY, and B. M. MILLMAN. 1965. X-ray diffraction from living striated muscle during contraction. *Nature* **206**: 1357.

GODFREY, J. E. and W. F. HARRINGTON. 1970. Self association in the myosin system at high ionic strength. II. Evidence for the presence of a monomer-dimer equilibrium. *Biochemistry* **9**: 894.

HUXLEY, H. E. 1963. Electron microscope studies on the structure of natural and synthetic protein filaments from striated muscle. *J. Cell Biol.* **7**: 281.

HUXLEY, H. E. and W. BROWN. 1967. The low-angle X-ray diagram of vertebrate striated muscle and its behavior during contraction and rigor. *J. Mol. Biol.* **30**: 383.

KNAPPEIS, G. and F. CARLSEN. 1968. The ultrastructure of the M-line in skeletal muscle. *J. Cell Biol.* **38**: 202.

LOWEY, S. and L. STEINER. 1972. An immunochemical approach to the structure of myosin and the thick filament. *J. Mol. Biol.* **65**: 111.

LOWEY, S., H. S. SLAYTER, A. G. WEEDS, and H. BAKER. 1969. Substructure of the myosin molecule. I. Subfragments of myosin by enzymic degradation. *J. Mol. Biol.* **42**: 1.

PEPE, F. A. 1966. Some aspects of the structural organization of the myofibril as revealed by antibody staining methods. *J. Cell Biol.* **28**: 505.

———. 1967a. The myosin filament. I. Structural organization from antibody-staining in electron microscopy. *J. Mol. Biol.* **27**: 203.

———. 1967b. The myosin filament. II. Interaction between myosin and actin filaments observed using antibody-staining in fluorescent and electron microscopy. *J. Mol. Biol.* **27**: 227.

———. 1968. Analysis of antibody-staining patterns obtained with striated myofibrils in fluorescent microscopy and electron microscopy. In *International review of cytology* (ed. Bourne and Danielli) vol. 24, p. 193. Academic Press, N.Y.

———. 1971a. The structure of the myosin filament of striated muscle. In *Progress in biophysics and molecular biology*, (ed. J. A. V. Butler and D. Noble) vol. 22, p. 77. Pergamon Press, New York.

———. 1971b. The structural components of the striated muscle fibril. In *Biological macromolecules series, subunits in biological systems* (ed. S. N. Timasheff and G. D. Fasman) vol. 5, part A, p. 323. Marcel Dekker, New York.

PEPE, F. A. and B. DRUCKER. 1972. The myosin filament. IV. Observations of the internal structural arrangement. *J. Cell Biol.* **52**: 255.

RICHARDS, E. G., C. S. CHUNG, D. B. MENZEL, and M. S. OLCOTT. 1967. Chromatography of myosin on diethylaminoethyl-Sephadex A-50. *Biochemistry* **6**: 528.

STARR, R. and G. OFER. 1971. Polypeptide chains of intermediate molecular weight in myosin preparations. *FEBS Letters* **15**: 40.

Filaments from Purified Smooth Muscle Myosin

Apolinary Sobieszek and J. V. Small

Institute of Biophysics, Aarhus University, 8000 Aarhus C, Denmark

As one approach to the elucidation of the molecular architecture of the thick filaments of striated and other muscle types, various recent studies have been directed towards defining the assembly properties of the myosin molecule and its rod-shaped subfragments (see review by Lowey, 1971). In this paper (see also Sobieszek, 1972) we describe the structural features of filaments assembled from myosin obtained from vertebrate smooth muscle and present evidence which suggests how these filaments may be formed.

In the course of a study of actomyosin extracts from smooth muscle, it was found (Sobieszek, 1972) that for efficiently fragmented muscle, myosin was readily extracted in large amounts at low ionic strength in the presence of ATP (see also Needham and Williams, 1963). Furthermore, this myosin manifested itself, as seen in the microscope, by assembling into filaments of a very regular structure. This assembly occurred if the extract, which had been freed from actin (Sobieszek, 1972; Sobieszek and Small, 1972), was either left to stand for 24 hr in the cold or if the pH of the solution was lowered to about 6.0. By collecting these filaments by centrifugation, it was shown (Sobieszek, 1972) that they were composed predominantly, if not entirely, of myosin; this led to the idea that smooth muscle myosin could be purified, at least in part, via precipitation in these filamentous aggregates. Using this approach it was found that by repeated filament formation, through cycles of dissolving the precipitated filaments in a solution of high ionic strength and then reforming them by dialysis to low ionic strength, myosin from chicken gizzard could be purified with a yield 2–3 times greater than obtained by other methods (Bárány et al., 1966).

Filaments formed from this smooth muscle myosin (Figs. 1 and 2) could be varied in length according to the conditions used, although the various factors influencing the final length of the filaments have yet to be precisely defined. In experiments in which the ionic strength of a solution of myosin in 0.6 M KCl was decreased, either by dialysis or by dilution with buffer to give a final KCl concentration of 0.1 M, it was found that the filament length depended on the rate at which the ionic strength was lowered (cf.

Huxley, 1963). Thus if this occurred within minutes, rather short filaments were formed (Fig. 1) ranging from about 0.45 μ–1.0 μ in length (mean value about 0.6 μ). Conversely, when the ionic strength was decreased over a long period of time (about 12 hr), very long filaments were formed (Fig. 2) which ranged up to about 6 μ in length.

The striking feature of these filaments is the presence of a clear transverse periodicity of about 140 Å (Fig. 3j) which may be observed along the entire length of the filaments (Figs. 2 and 3a) and which arises from a regular arrangement of projections or cross-bridges (Sobieszek, 1972). Thus, there is no central bare zone (cf. Huxley, 1963). Preliminary optical diffraction analysis (Sobieszek, 1972) has indicated that the cross-bridges are arranged on a three-stranded helix of repeat about 720 Å and pitch about 3×720 Å.

Together with the longer filaments, shorter filaments are also observed in these preparations (Fig. 3b,c) which clearly represent the form of the myosin filaments at an early stage of assembly. These filaments show a characteristic asymmetric appearance which arises from an asymmetrical distribution of cross-bridges such that the bridges on opposite ends of a filament lie on opposite sides. The other diametrically opposed sides of these filaments are devoid of projections and exhibit smooth edges of about 2200 Å in length. That these filaments are indeed earlier stages of assembly of the larger filaments was indicated by the presence of the same smooth edge at the ends of the longer filaments (Fig. 3a, d–i). Such edges presumably represent the sites at which filament growth proceeds by the attachment of molecules available in solution. Furthermore, from the appearance and length of these edges it would seem most likely that the building unit is a bipolar aggregate of myosin (e.g., an antiparallel dimer) with a molecular overlap (taking the length of the rod of smooth muscle myosin as 1560 Å; Kendrick-Jones et al., 1971) of about 900 Å.

As yet it is not possible to exclude the possibility that these filaments contain small amounts of protein other than myosin. From recent results on striated muscle myosin (Moos, this volume) we may, however, exclude the presence of a component like the newly identified C-protein (Starr and

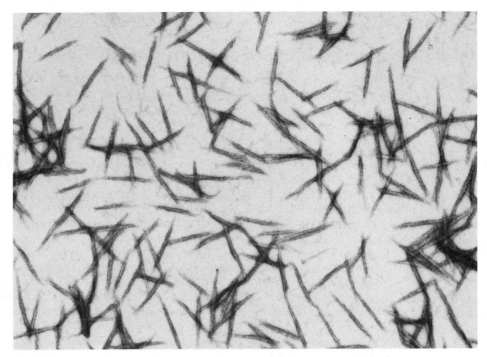

Figure 1. Filaments formed by the dilution (within 10–15 min) of gizzard myosin (10–20 mg/ml) in 0.6 M KCl (histidine buffer 0.117 M pH 6.8 with added 1 mM ATP, 0.1 mM $MgCl_2$, 1 mM cysteine) by the addition of buffer to give a final concentration of 0.1 M KCl × 20,000.

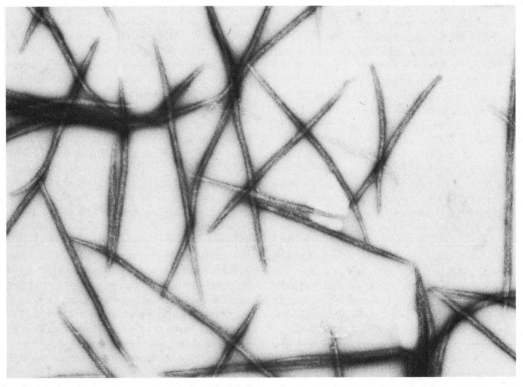

Figure 2. Filaments formed by a very slow decrease of the ionic strength of a solution of gizzard myosin (about 2 mg/ml) by dialysis. The dialysis bag containing myosin in 0.6 M KCl (buffer as Fig. 1) was placed in a small volume of buffer containing 0.6 M KCl and then the ionic strength of the external solution was decreased by the dropwise addition of KCl-free buffer over a period of about 12 hr. Note presence on many of the filaments of a regular periodicity (about 140 Å) which may be recognized along the entire filament length. × 24,000.

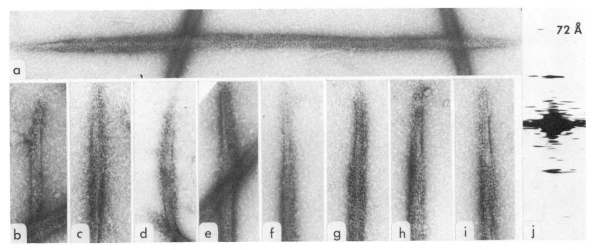

Figure 3. Fine structural details of filaments from gizzard myosin. (*a*) Filament showing the presence of cross-bridges along its entire length. × 54,000. The optical diffraction pattern from the center part of this filament (*j*) shows the presence of a strong meridional reflection at a spacing of about 144 Å. and a weaker second order of this at about 72 Å (marked). (*b*) and (*c*). Short filaments found in the preparation which evidently represent early stages of assembly of the larger filaments. These are asymmetric in appearance (see text) showing the presence of smooth cross-bridge-free regions on one side of each half of the filament. (*d*)–(*i*). The ends of several long filaments showing the presence of the same smooth edge as seen on the short asymmetric filaments (*b*) and (*c*). (*b*)–(*i*) × 67,000.

Offer, 1971; Offer, this volume) which binds strongly to myosin, since this protein acts to impair filament growth (Moos, this volume). As already described, the final length of the filaments from smooth muscle myosin was apparently unlimited, being determined only by the conditions used for filament formation. In this respect the primary factor influencing the final length of the filaments was the rate at which the ionic strength of the milieu was decreased, the effects of other factors such as pH changes (between 6.0 and 7.5) being comparatively small.

It is very interesting to note that the filaments described by Moos also possess projections along their entire length and thus lack a central bare zone. In fact, the existence of a bare zone in filaments synthesized from striated muscle myosin has not been consistently demonstrated (Kaminer and Bell, 1966). How the filaments of Moos are related to the thick filaments of striated muscle or to those filaments which do show a central bare zone (Huxley, 1963) remains to be shown. In the case of vertebrate smooth muscle the available evidence suggests that such filaments do not exist in vivo (see Small and Squire, 1972; Sobieszek and Small, 1972; Small and Sobieszek, this volume), but they clearly represent a structurally significant form of assembly of smooth muscle myosin.

Acknowledgments

We thank the Danish Statens Naturvidenska-belige Forskningsråd for their finanical support and for purchasing the Siemens 101 electron microscope. We also thank Mrs. L. Nychel and Miss K. Eskesen for technical assistance and Mr. P. Boldsen and Mrs. I. Lunde for photography.

References

BÁRÁNY, M., K. BÁRÁNY, E. GAETJENS, and G. BAILIN. 1966. Chicken gizzard myosin. *Arch. Biochem. Biophys.* **113**: 205.

HUXLEY, H. E. 1963. Electron microscope studies on the structure of natural and synthetic protein filaments from striated muscle. *J. Mol. Biol.* **7**: 281.

KAMINER, B. and A. L. BELL. 1966. Myosin filamentogenesis: Effects of pH and ionic concentration. *J. Mol. Biol.* **20**: 391.

KENDRICK-JONES, J., A. G. SZENT-GYÖRGYI, and C. COHEN. 1971. Segments from vertebrate smooth muscle myosin rods. *J. Mol. Biol.* **59**: 527.

LOWEY, S. 1971. Myosin: molecule and filament, p. 201. In *Biological macromolecules*, ed. S. N. TIMASHEFF and G. D. FASMAN, vol. 5, part A. Marcel Dekker, New York.

NEEDHAM, D. M. and J. M. WILLIAMS. 1963. Proteins of the uterine contractile mechanism. *Biochem. J.* **89**: 552.

SMALL, J. V., and J. M. SQUIRE. 1972. The structural basis of contraction in vertebrate smooth muscle. *J. Mol. Biol.* **67**: 117.

SOBIESZEK, A. 1972. Cross-bridges on self-assembled smooth muscle myosin filaments. *J. Mol. Biol.* In press.

SOBIESZEK, A. and J. V. SMALL. 1972. The assembly of ribbon-shaped structures in low ionic strength extracts obtained from vertebrate smooth muscle. *Phil. Trans. Roy. Soc. London B.* In press.

STARR, R. and G. OFFER. 1971. Polypeptide chains of intermediate molecular weight in myosin preparations. *FEBS Letters* **15**: 40.

Inhibition of Heavy Meromyosin by Purine Disulfide Analogs of Adenosine Triphosphate

RALPH G. YOUNT, JAMES S. FRYE, AND KELLY R. O'KEEFE

Graduate Program in Biochemistry, Department of Chemistry, Washington State University, Pullman, Washington 99163

The interactions of a wide variety of ATP analogs with myosin and actomyosin have given valuable insight into the role of nucleotides in contraction and relaxation. These studies have been with naturally occurring nucleoside triphosphates (e.g., Hasselbach, 1956) and more recently with a large number of synthetically prepared analogs of ATP in which principally the purine and ribose rings have been modified (Tonomura et al., 1965). The pioneering work of Moos et al. (1960) used the first phosphate-modified analog of ATP, adenylylmethylenediphosphonate, to study myosin-actin-nucleotide interactions. This analog, while very useful in studying a number of enzymes, has been of little use in myosin systems since it is sufficiently different structurally from ATP (Larsen et al., 1969) that characteristic interactions (e.g., dissociation of actomyosin at high ionic strengths) do not occur.

To overcome this difficulty we synthesized a similar analog, adenylylimidodiphosphate (AMP-PNP), in which an imido (—NH—) group replaced the β-γ bridge oxygen of ATP (Yount et al., 1971a,b). This analog, while not a substrate, has proven to be remarkably similar to ATP in its interaction with myosin and heavy meromyosin (HMM). It is the most potent competitive inhibitor of HMM ATPase known in that the K_i values in the presence of various divalent metal ions approximate the K_m values for ATP. This property of AMP-PNP, i.e., tight binding to myosin without hydrolysis, has shown that it is the binding of an ATP-like molecule alone which is responsible for relaxation of myofibrils (Dos Remedios et al., 1972) or dissociation of actomyosin (Yount et al., 1971b).

The analogs described above all interact reversibly with myosin and do not tell the nature of the amino acids involved in binding or hydrolysis. For this "affinity labels" or active-site-directed analogs are needed[1] (Singer, 1967). The only such analog currently known for myosin has resulted from the important work of Murphy and Morales (1970) with the compound 6-thioinosine triphosphate (SH-TP). In this case it was hypothesized that the purine sulfur formed a disulfide bond with a cysteine at the purine ring binding site since the nucleotide label could be removed with β-mercaptoethanol. Two such residues were found per molecule of inactive myosin, agreeing with the best estimate of the number of subunits. In addition the change in the ultraviolet spectrum of SH-TP binding to myosin could be used to show a 2:1 nucleotide: enzyme stoichiometry.

However, it was not possible to show that the SH-TP was a true active-site-directed reagent in that the inactivation was too slow (2 days at 0° to reach > 95% inactivation) to do meaningful kinetic measurements. In addition, because the analog was a substrate, a large excess of reagent had to be used (e.g., 200/1 molar ratio) and inactivation conditions chosen to minimize enzyme activity (e.g., 0.5 M NaCl, 0°). To overcome this problem the analog 6-thioinosinyl-imidodiphosphate (Fig. 1) which incorporates the mercaptopurine ring of SH-TP and the PNP linkage of AMP-PNP was synthesized (Yount, O'Keefe, and Frye, in prep.) This analog, abbreviated SHP-PNP, proved to be a good inhibitor of HMM ATPase in that the inhibition took hours rather than days and only a 20/1 excess of reagent was needed. However, the inhibition was erratic and varied from one preparation of SHP-PNP to another. Subsequent to these findings Tokiwa and Morales (1971) found that the true inactivating form of SH-TP was probably the disulfide since SH-TP solutions which had stood overnight inactivated more rapidly than freshly prepared solutions. Thus we oxidized 2 mM SHP-PNP overnight by gently bubbling oxygen through the solution. Oxygen was removed by repeated freezing and thawing of the solution on a vacuum line. A control solution using nitrogen gas was similarly treated and the effects of these two preparations in the absence of oxygen on the Ca^{++}-ATPase of HMM were determined (Fig. 2). It can be seen that the oxygen-treated SHP-PNP was a much more effective inhibitor but the nitrogen-treated sample never completely inhibited

---

[1] Our attempts to make a phosphate-modified reactive ATP analog have thus far been unsuccessful. Hence, an analog in which fluorine replaces an OH group on the γ-phosphate of ATP binds to the active site of HMM but does not react irreversibly (Haley and Yount, 1972).

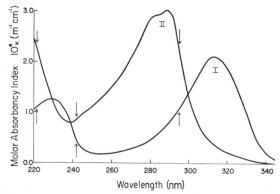

Figure 1. Structure of 6-thioinosinyl-imidodiphosphate.

HMM. The partial inhibition seen with the N₂-treated sample undoubtedly was the result of prior air oxidation of SHP-PNP, coupled with the presence of some oxygen introduced via the enzyme solution.

Attempts were made to isolate the presumed disulfide of SHP-PNP after O₂ oxidation by paper and DEAE-cellulose chromatography. No product with the appropriate properties was ever found and, in general, only the monosulfide was ever detected. After many false starts the disulfide was prepared quantitatively by oxidation with sodium triiodide solutions and isolated as the octasodium salt by precipitation with methanol and acetone. The product, abbreviated S₂P-PNP, is extremely sensitive to disulfide cleavage by nitrogen nucleophiles including Tris buffer and primary amines in general. Bicine buffer, being less nucleophilic than Tris, was used in the majority of the studies reported here.[2] Neutral water solutions of S₂P-PNP, however, may be stored frozen for weeks with

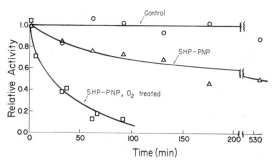

Figure 2. Effect of oxygen treatment on rate of SHP-PNP inactivation of heavy meromyosin Ca⁺⁺·ATPase. Conditions: 2 mM SHP-PNP was gently bubbled with oxygen gas for 16 hr at 25°. The oxygen was removed by repeated freezing and thawing under vacuum. A similar sample of SHP-PNP was bubbled with nitrogen gas. Inactivation reaction solutions contained 5.7 μM HMM, 50 mM Tris-HCl pH 8.0, 1.0 mM treated SHP-PNP, 20°. Assays were performed at 25° using 5.0 mM CaCl₂, 50 mM Tris-HCl pH 8.0, and 10 mM ATP. Serial samples (five) were taken for each assay and analyzed for Pᵢ using the method of Rockstein and Herron (*Anal. Chem.* **23**, 1500, 1951) using a Gilford 300-N rapid sampling spectrophotometer.

Figure 3. Ultraviolet spectra of SHP-PNP and S₂P-PNP Conditions: pH 8.0, 10 mM Tris. I = SHP-PNP; II = S₂P-PNP. ε_M S₂P-PNP = 3.02 × 10⁴ at 288 nm. The arrows indicate the isosbestic points for the reduction of S₂P-PNP to SHP-PNP.

little decomposition. The state of S₂P-PNP preparations can be readily monitored by checking the ultraviolet absorption spectrum (Fig. 3). The molar extinction coefficient of S₂P-PNP at 288 nm is 3.0 × 10⁴ and as expected the spectrum of S₂P-PNP is pH independent, unlike that of the monosulfide analogs (Murphy and Morales, 1970).

Disulfide Inhibition Studies

S₂P-PNP was shown to rapidly inactivate HMM ATPase at low concentrations (10/1 molar ratios) in minutes, instead of hours or days as required by the monosulfide preparations (Fig. 4). Figure 4 shows that both the Ca⁺⁺ and NH₄⁺-EDTA-stimulated ATPase are inactivated with first-order kinetics for at least two half-lives. This similar mode of inactivation is in marked contrast to the effect of p-CMB which first stimulates, then inhibits Ca⁺⁺-ATPase activity (Kielley and Bradley, 1956). In these studies the K⁺-EDTA activity was inhibited at all concentrations of p-CMB. In addition, Sekine et al. (1962) observed that when only one or two equivalents of NEM reacted per subunit of myosin, the Ca⁺⁺-ATPase activity was elevated while EDTA-ATPase and Ca⁺⁺-ITPase activity were lost. Again our results indicate no differences in the rates of loss of either activity under a variety of reaction conditions. Because the NH₄⁺-EDTA activity of HMM is approximately ten times the Ca⁺⁺ activity, most of our assays were done using this assay. This added

[2] A systematic study of the effects of common reaction constituents on the hydrolysis of S₂P-PNP and its monophosphate counterpart, S₂MP, has been determined (Fohndahn, Frye and Yount, unpublished results). For example, even divalent metals promote the hydrolytic cleavage of disulfide bond in S₂P-PNP. The half-life of S₂P-PNP in 10 mM Tris-HCl, pH 8.0, 25° is 460 min. Addition of 1 mM Mn⁺⁺, Ca⁺⁺, or Mg⁺⁺ decreases the half-lives to 19, 36, and 63 min, respectively.

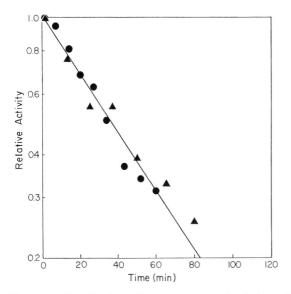

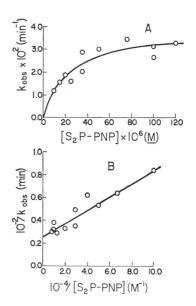

Figure 4. Inactivation of heavy meromyosin Ca^{++} and EDTA-ATPase by S_2P-PNP. Inactivation conditions: $0°$, 5.0 mM Bicine pH 8.0, 50 mM KCl, 20 μM S_2P-PNP, 1.6 μM HMM. Assay: (▲—▲) 4.6 mM Ca^{++}, 86 mM Tris-HCl pH 8.0, 10 mM ATP, $25°$. (●—●) 41 mM EDTA, 0.41 M NH_4^+, 86 mM Tris-HCl pH 8.0, 10 mM ATP, $25°$.

Figure 5A and 5B. Effect of S_2P-PNP concentration on rate of inactivation of heavy meromyosin ATPase. Inactivation conditions: 1.6 μM HMM, $0°C$, 50 mM KCl, 5 mM Bicine pH 8.0. Assays were with 4.6 mM $CaCl_2$ as given in Fig. 4. Nine to ten assays were taken for each S_2P-PNP concentration and the slope of the inactivation curve used to calculate k_{obs}.

sensitivity allowed much lower HMM concentrations to be used while at the same time the $NH_4^+(NH_3)$ and EDTA promoted the hydrolysis of excess S_2P-PNP. In addition the use of EDTA allowed the inclusion of Mg^{++}, an important variable, in the inactivation solutions.

One of the principal criteria for an affinity label is to show that the compound in question demonstrates saturation kinetics (Shaw, 1970; Kitz and Wilson, 1962). Thus if S_2P-PNP is a true site-specific reagent, it should first form a central complex with HMM as shown in Eq. 1 before inactivation

$$\boxed{HMM} \!\!\underset{}{\overset{|}{}}\!\! + S_2P\text{-PNP} \underset{k_{-1}}{\overset{k_1}{\rightleftharpoons}}$$
$$| \qquad\qquad\qquad\qquad\qquad (1)$$
$$SH$$

$$\left[\boxed{HMM} \cdot S_2P\text{-PNP}\right] \overset{k_0}{\longrightarrow} \boxed{HMM} + SHP\text{-PNP}$$
$$\quad | \qquad\qquad\qquad\qquad\qquad |$$
$$\quad SH \qquad\qquad\qquad\qquad\qquad S$$
$$\qquad\qquad\qquad\qquad\qquad\qquad |$$
$$\qquad\qquad\qquad\qquad\qquad\qquad SP\text{-PNP}$$

occurs. Such an inactivation should exhibit saturation effects similar to Michaelis-Menten kinetics. If so, then Eq. 2 should describe the

$$\frac{1}{k_{obs}} = \frac{1}{[I]}\frac{K_i}{k_2} + \frac{1}{k_2} \qquad (2)$$

inactivation (Kitz and Wilson, 1962) where k_{obs} is the first-order rate constant for inactivation by various concentrations of the inhibitor $[I]$, K_i is the dissociation constant for the EI complex, and k_2 is the maximal rate of inactivation at saturating inhibitor.

Inactivation studies with S_2P-PNP showed, indeed, that saturation occurs (Fig. 5A) and gave a $K_i = 22$ μM and $k_2 = 0.039$ min^{-1} on replotting according to Eq. 2 above (Fig. 5B). The above experiments were done with no added divalent metal ions and using $Ca^{++} \cdot ATP$ assays. The addition of Ca^{++} (1.0 mM) had no effect on the rate of S_2P-PNP inactivation. For purposes of comparison, the disulfide analog (S_2-MP) of thioinosinic acid (SH $\cdot$ MP) was prepared by a method similar to that used for S_2P-PNP and its inactivating properties tested (Fig. 6). It, too, showed saturation kinetics with a K_i value of 350 μM and a $k_2 = 0.042$ min^{-1}. Thus, S_2·MP binds much less tightly than S_2P-PNP but, as might be predicted, shows within experimental error the same rate of inactivation as S_2P-PNP at saturation.[3] In both

---

[3] S_2P-PNP has obvious uses with other enzymes. For example, it specifically inactivates the $(Na^+ + K^+)$ ATPase from trout gill preparations while having no effect on the accompanying Mg^{++} ATPase activity (Pfeiler and Yount, in prep.). The inactivation shows saturation kinetics, again indicating a central complex is formed before inactivation occurs.

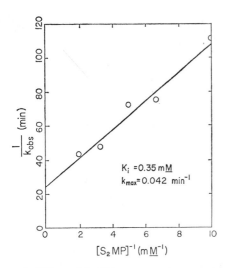

Figure 6. Effect of S_2-MP concentrations on the rate of inactivation of heavy meromyosin. Inactivation conditions were the same as Fig. 5A and 5B. Ca^{++}·ATPase assays (Fig. 4) were used.

cases ten assays were performed on reaction samples taken to span at least two half-lives of activity at each disulfide concentration. The similarity of the inactivating effects of S_2P-PNP and S_2-MP conflict with studies of these two compounds with single muscle fibers (Dos Remedios et al., 1972). These latter studies show that treatment with 5 mM S_2P-PNP (as well as with S_2-TP, the disulfide analog of ATP) acts to relax fibers *irreversibly*, whereas treatment with 5 mM S_2-MP has little or no effect on subsequent ATP-induced contraction and relaxation. One answer may be that the actin in the fibers normally protects the critical sulfhydryl on myosin unless the analog in question contains sufficient phosphates to dissociate the myosin-actin linkage and allow labeling to occur. Thus AMP-PNP and ATP (Yount et al., 1971b) will dissociate the actomyosin complex but AMP will not.

AMP-PNP Effects on S_2P-PNP Inactivation

A further criterion of S_2P-PNP's being a true active-site-directed reagent would be protection against inactivation by ATP or by the closely related substrate analog, AMP-PNP. Most surprisingly, AMP-PNP not only did not protect, it markedly accelerated the rate of inactivation (Fig. 7). This effect depends on Mg^{++} since in the presence of 10 mM EDTA, even 1.0 mM AMP-PNP is without effect. However, inactivation per se by S_2P-PNP occurs without Mg^{++} but goes faster in its presence. In the presence of 1.0 mM Mg^{++}, 5 μM AMP-PNP is saturating since doubling the concentration of AMP-PNP is without effect. In experiments not

shown IMP-PNP (inosinyl-imidodiphosphate) in place of AMP-PNP showed either no effect on S_2P-PNP inactivations, or at higher concentrations of IMP-PNP (e.g., 100 μM) some protection. Since the IMP-PNP protection appears to compete with the AMP-PNP-promoted inactivation (J. S. Frye, unpublished), it would appear that the 6-amino group on the purine ring is crucial in promoting the conformation change needed for the more rapid inactivation seen in Fig. 7.

The above experiments pose a dilemma. Since HMM in these experiments is 1.6 μM (or 3.2 μM in active sites), this means AMP-PNP is reacting essentially stoichiometrically with HMM. This binding presumably occurs at the active site since AMP-PNP is a potent competitive inhibitor of HMM ATPase (Yount et al., 1971b) and there is no evidence that equally tight-binding nonhydrolytic sites exist. If S_2P-PNP binds specifically to an inactivating site (and kinetic analysis indicates it does), it must be to a site other than the active site since AMP-PNP should block reaction there. However, an alternate explanation exists. AMP-PNP might be acting cooperatively on the heads of HMM in such a way that binding of AMP-PNP to one head would induce a conformation change in the adjacent head such that a more rapid reaction with S_2P-PNP would occur. Subsequent diffusion of AMP-PNP from site 1 would allow a second

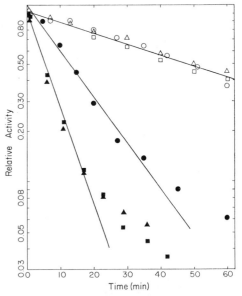

Figure 7. Effect of AMP-PNP and Mg^{++} on rate of inactivation of heavy meromyosin. Inactivation conditions: 20 μM S_2P-PNP; 20 mM Bicine pH 8.0; 1.1 μM HMM; 67 mM KCl; 0°. Further additions were (○) 10 mM EDTA; (□) 10 mM EDTA, 10 μM AMP-PNP; (△) 10 mM EDTA, 1.0 mM AMP-PNP; (●) 1 mM $MgCl_2$; (■) 1 mM $MgCl_2$ + 5 μM AMP-PNP; (▲) 1 mM $MgCl_2$ + 10 μM AMP-PNP. Assays were the EDTA assays as given in Fig. 4.

molecule of S_2P-PNP to react to give the 2 to 1 stoichiometry found previously for myosin·SH-TP inactivations (Murphy and Morales, 1970; Tokiwa and Morales, 1971). This model then proposes that the "heads" of HMM talk to each other. Fortunately it is possible to test this hypothesis since it is possible to isolate the heads of HMM as active ATPases after further proteolytic degradation. This preparation, called subfragment one (SF1), was prepared by the method described by Nauss et al. (1969). It gave a single peak in the analytical ultracentrifuge with an S value of 5.4, the same as found previously (Nauss et al., 1969). On exposure to S_2P-PNP, SF1 was rapidly inactivated (Fig. 8). Most importantly when either 0.1 mM or 0.5 mM AMP-PNP was added, the rate of inactivation was again markedly enhanced (Fig. 8). The high concentrations of AMP-PNP were purposely picked to see if any protection of the active site occurred under these conditions. The experiments indicate AMP-PNP induces the same conformation change with SF1 as it does with HMM and the implication is that *each* head has two nucleotide binding sites.

It is still possible that only one site exists (the active site) and that the presence of AMP-PNP simply induces a conformation change which exposes a new and more reactive sulfhydryl group which when blocked inactivates the enzyme. This hypothesis predicts no binding site for S_2P-PNP when AMP-PNP is present, and saturation phenomena as shown in Fig. 5A and B would not occur. In this case, S_2P-PNP is simply an elegant sulfhydryl reagent reacting in a second-order fashion with a new but very reactive sulfhydryl group. To test this possibility we determined the

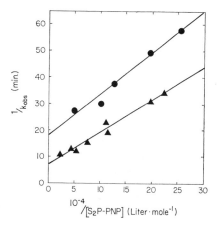

Figure 9. Effect of S_2P-PNP concentration on rate of inactivation of Mg^{++}·HMM in the presence and absence of AMP-PNP. Inactivation conditions were the same as in Fig. 8 except varying concentrations of S_2P-PNP were used. (●) No AMP-PNP, (▲) 8 μM AMP-PNP. Assay conditions: 16 mM ATP, 0.31 M NH_4Cl, 31 mM EDTA, 94 mM Tris-HCl pH 8.0, 25°. $K_i = 9$ μM and $k_2 = 5.6 \times 10^{-2}$ min^{-1}, no AMP-PNP added. $K_i = 16$ μM and $k_2 = 14.3 \times 10^{-2}$ min^{-1} with 8 μM AMP-PNP added.

rates of inactivation of HMM with various concentrations of S_2P-PNP in the presence of 8 μM AMP-PNP and 1 mM Mg^{++}. As shown in Fig. 9, S_2P-PNP inactivation exhibits saturation kinetics *both* in the presence and absence of AMP-PNP. Thus the results demonstrate that even in the presence of saturating AMP-PNP, a second site exists which binds S_2P-PNP (albeit somewhat weaker, $K_i = 16$ μM in the presence of AMP-PNP as opposed to $K_i = 9$ μM in its absence) and which allows a faster inactivation of HMM ($t_{1/2} = 4.9$ min vs. 12.4 min in the absence of AMP-PNP). If such a site did not exist, the plot with added AMP-PNP in Fig. 9 should go through the origin (Kitz and Wilson, 1962) since at infinite S_2P-PNP the rate of inactivation should be infinitely fast. Clearly this does not happen and the conclusion is that two nucleotide binding sites exist on *each* head of HMM. Furthermore, the sites must be coupled since binding of AMP-PNP to one site (the active site?) leads to a more rapid reaction at the second site, which in turn inactivates the enzyme by an as yet unknown mechanism. It is tempting to call the second site a regulatory or relaxation site since the idea of two sites being required for a contraction-relaxation cycle is a well established concept (see Weber and Portzehl, 1954; Stewart and Levy, 1970; Kominz, 1971 and references therein). However, these suggestions are concerned principally with actomyosin systems and it may be that the two sites discussed represent the two active sites of myosin acting with different functions at different times

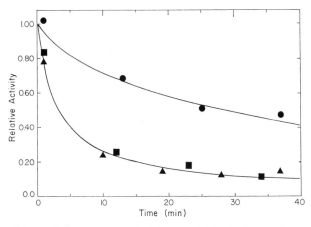

Figure 8. Inactivation of subfragment-1 by S_2-PNP in the presence and absence of AMP-PNP. Inactivation conditions: 22 μM S_2P-PNP, 0°, 50 mM KCl, 5 mM Bicine pH 8.0, 0.2 mg/ml SF-1, 1.0 mM $MgCl_2$. Further additions were (●) none; (▲) 100 μM AMP-PNP; (■) 500 μM AMP-PNP. Assays were with NH_4·EDTA as given in Fig. 4.

in the contraction cycle. Kiely and Martonosi (1968, 1969), however, have postulated on the basis of circumstantial evidence both a regulatory and a catalytic site on myosin. Our experiments give what would appear to be the first direct evidence that such additional sites exist.

Clearly these results raise many questions about the S_2P-PNP inactivation. Our results suggest S_2P-PNP reacts at a relaxation-type site which when blocked prevents myosin-actin interaction and inhibits ATP hydrolysis. It is known S_2P-PNP (as well as S_2-TP and S_2-DP) will cause single muscle fibers to relax in a fashion *not* reversible by ATP and Ca^{++} addition (Dos Remedios et al., 1972). Also actin has no effect on spin-labeled myosin after it has reacted with SH-TP (Tokiwa, 1971). However, Duke (quoted in Dos Remedios et al., 1972) has found that myosin inactivated with S_2-TP yields only SH-DP after release of the nucleotide with β-mercaptoethanol. This implies that the label was originally at the active site and was cleaved after or immediately before inactivation had occurred. This descrepancy remains to be explained; it may mean that in the absence of other nucleotides S_2-TP (or S_2P-PNP) goes to the active site, but reacts elsewhere when ATP (or AMP-PNP) is present. Kielley et al. (1967) have reported that in myosin partially reacted with N-ethylmaleimide (NEM) further reaction in the presence or absence of ATP leads to different sulfhydryl groups being blocked. It should be possible to test this possibility with HMM and S_2P-PNP by determining if different cysteines become labeled by S_2P-PNP in the presence and absence of AMP-PNP.

Summary

A purine disulfide analog of ATP has been prepared by mild oxidation of 6-thioinosinyl-imidodiphosphate. This analog rapidly inactivates either Ca^{++} or EDTA-activated heavy meromyosin ATPase, presumably by reaction at an essential sulfhydryl group. The inactivation shows saturation kinetics either in the presence or absence of Mg^{++}, indicating a central complex is formed before inactivation occurs. Surprisingly, the substrate analog, adenylyl-imidodiphosphate, promotes a *faster* inactivation of heavy meromyosin or subfragment one ATPase. It is hypothesized that the disulfide analog is reacting on a secondary site (a relaxation site?) which in turn inactivates the active site.

Acknowledgment

This research was funded by Grant AM05195 from the National Institutes of Health, U.S. Public Health Service.

References

Dos Remedios, C., R. G. Yount, and M. F. Morales. 1972. Individual states in the cycle of muscle contraction. *Proc. Nat. Acad. Sci.* **69:** 2542.

Haley, B. and R. G. Yount. 1972. γ-Fluoroadenosine triphosphate: Synthesis, properties and interaction with myosin and heavy meromyosin. *Biochemistry* **11:** 2863.

Hasselbach, W. 1956. Die Wechselwirkung verscheidner Nukleosidtriphosphate mit Actomyosin im Gelzustand. *Biochim. Biophys. Acta* **20:** 355.

Kielley, W. W., and L. Bradley. 1956. The relationship between sulfhydral groups and the activation of myosin adenosine triphosphatase. *J. Biol. Chem.* **218:** 653.

Kielley, W. W., T. Yamashita and J. P. Cooke. 1967. Structural relationships in the interaction of myosin and nucleoside polyphosphates. *Abst. 7th Int. Cong. Biochem. Tokyo* **2:** 313.

Kiely, B., and A. Martonosi. 1968. Kinetics and substrate binding to myosin adenosine triphosphatases. *J. Biol. Chem.* **243:** 2273.

———. 1969. The binding of ADP to myosin. *Biochim. Biophys. Acta* **172:** 158.

Kitz, R., and I. B. Wilson. 1962. Esters of methanesulfonic acid as irreversible inhibitors of acetylcholinesterase. *J. Biol. Chem.* **237:** 3245.

Kominz, D. 1971. The regulatory ATP-binding site of myosin, p. 253. In *Contractile proteins and muscle*, ed. K. Laki. M. Dekker, New York.

Larsen, M., R. Willett, and R. G. Yount. 1969. Imidodiphosphate and pyrophosphate: Possible biological significance of similar structures. *Science* **166:** 1510.

Moos, C., N. R. Albert, and T. C. Myers. 1960. Effects of a phosphoric acid analog of adenosine triphosphate on actomyosin systems. *Arch. Biochem. Biophys.* **88:** 183.

Murphy, A. J. and M. Morales. 1970. Number and location of adenosine triphosphatase sites of myosin. *Biochemistry* **9:** 1528.

Nauss, K. M., S. Kitagawa and J. Gergely. 1969. Pyrophosphate binding to and adenosine triphosphatase activity of myosin and its proteolytic fragments. *J. Biol. Chem.* **244:** 755.

Sekine, T., L. M. Barnett, and W. W. Kielley. 1962. The active site of myosin adenosine triphosphatase. *J. Biol. Chem.* **237:** 2769.

Shaw, E. 1970. Chemical modification by active-site-directed reagents, p. 91. In *The enzymes*, vol. 1 (ed. P. D. Boyer), Academic Press, New York.

Singer, S. J. 1967. Covalent labeling of active sites. *Advanc. Protein Chem.* **22:** 1.

Stewart, J. M. and H. M. Levy. 1970. The role of the calcium-troponin-tropomyosin complex in the activation of contraction. *J. Biol. Chem.* **245:** 5674.

Tokiwa, T. 1971. EPR spectral observations on the binding of ATP and F-Actin to spin-labeled myosin. *Biochem. Biophys. Res. Comm.* **44:** 471.

Tokiwa, T. and M. Morales. 1971. Independent and cooperative reactions of myosin heads with F-Actin in the presence of adenosine triphosphate. *Biochemistry* **10:** 1722.

Tonomura, T., S. Kubo and K. Imamura. 1965. A molecular model for the interaction of myosin with adenosine triphosphate, p. 11. In *Molecular biology of muscular contraction*, ed. S. Ebashi et al. Elsevier, New York.

Weber, H. H. and H. Portzehl, 1954. The transference of muscle energy in contraction cycle. *Prog. Biophys.* **4:** 60.

Yount, R. G., D. Babcock, W. Ballantyne and D. Ojala. 1971a. Adenylylimidodiphosphate, an adenosine triphosphate analog containing a P-N-P linkage. *Biochemistry* **10**: 2484.

Yount, R. G., D. Ojala, and D. Babcock. 1971b. Interaction of P-N-P and P-C-P analogs of ATP with heavy meromyosin, myosin and actomyosin. *Biochemistry* **10**: 2490.

Inhibition of Myosin ATPase with a Disulfide Analog of ATP

H. J. Mannherz

Max-Planck-Institut für Medizinische Forschung, Heidelberg, Germany

R. S. Goody and F. Eckstein

Max-Planck-Intitut für experimentelle Medizin, Göttingen, Germany

In an attempt to label the active site of myosin we have tried to make use of the disulfide of adenosine 5'-(O-3-thiotriphosphate) (ATPγS). Myosin ATPase is irreversibly inhibited by this compound

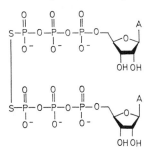

Figure 1. Formula of disulfide ATPγS.

following first-order rate kinetics. The inhibition is reversible by DTT. At 25°, 5 mM MgCl$_2$, 20 mM imidazole buffer pH 6.9, the half-time of inactivation is $t_{1/2} = 20$ min with a $K_i = 0.7$ mM (Fig. 2). This value indicates a considerably weaker affinity of this disulfide to the protein than ATP which

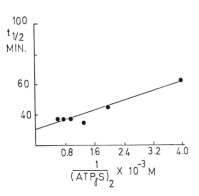

Figure 2. Kinetics of inhibition of myosin ATPase at 25°, preincubation time 45 min.

has $K_m \simeq 10^{-7}$ M (Schlisselfeld and Bárány, 1968). In order to find out whether the disulfide was binding to the active site, we carried out protection experiments with a few nucleotides.

The K_i values obtained from these experiments are approximately

$$K_i \text{ (ATP)} \simeq 1 \text{ mM}$$
$$K_i \text{ (ADP)} \simeq 10 \text{ mM}$$
$$K_i \text{ (AMPCH}_2\text{PP)} \simeq 10 \text{ mM}$$

We could not detect any protection by AMPPNP under these conditions. The large difference of the K_m and K_i values of ATP and ADP (Lowey and Luck, 1969), respectively, obtained from kinetic experiments and those values obtained from the protection experiments indicates that the inactivation does not occur at the active site.

Inspection of SDS-gels of myosin and subfragment S$_1$, which can be inactivated under the same conditions, after inhibition with the disulfide seems to show a covalent linkage of light to heavy chains. This could be explained by an SH-exchange reaction with loss of ATPγS leading to a disulfide bridge between heavy and light chains. It is, however, not clear at this time whether these are the only disulfides formed in this reaction.

References

Goody, R. S. and F. Eckstein, 1971. Thiophosphate analogs of nucleoside di- and triphosphates. *J. Amer. Chem. Soc.* **93**: 6252.

Lowey, S. and S. M. Luck. 1969. Equilibrium binding of adenosine diphosphate to myosin. *Biochemistry* **8**: 3195.

Schlisselfeld, L. H. and M. Bárány. 1968. The binding of adenosine triphosphate to myosin. *Biochemistry* **7**: 3206.

An ATP-Binding Peptide

GEORGE BARANY AND R. B. MERRIFIELD

The Rockefeller University, New York, New York 10021

It is generally accepted today that the hydrolysis of ATP, mediated by the contractile protein actomyosin, is the direct source of energy for muscle contraction. Recently, we reported in collaboration with Dr. Michael Bárány (G. Barany et al., 1972), a differential labeling of the ATPase site of frog actomyosin with *N*-ethylmaleimide (NEM). Since the ATPase and actin-binding sites of myosin overlap (M. Bárány and K. Bárány, 1959), actin exerts a specific protective effect on the ATPase site of myosin. Thus it was possible to titrate frog actomyosin with as many as 15 moles nonradioactive NEM/500,000 daltons protein before the first drop in ATPase activity could be detected. Excess reagent was quenched with cysteine and removed by several reprecipitations of the actomyosin. The ATPase site was then "opened up" by the substrate MgATP, at a concentration of 2 mM. A 5-fold molar excess of [^{14}C]NEM was added, resulting in abolition of most of the Ca^{++} and Mg^{++}-activated ATPase activities. Heavy meromyosin was isolated directly from this labeled actomyosin by a new procedure involving (1) digestion with trypsin; (2) dialysis against 0.04 M KCl and centrifugation; (3) dissociation of the clear supernatant and middle layer with 0.01 M MgATP in the ultracentrifuge; (4) (NH$_4$)$_2$SO$_4$ fractionation; (5) chromatography on Sephadex G-200. An incorporation of 1.3 moles [^{14}C]NES/350,000 daltons heavy meromyosin (theoretical = 2) provided strong evidence for a specific labeling of the essential cysteine residue at the ATPase site. After peptic, tryptic, and chymotryptic digestion of 35 mg (0.1 μmole) labeled heavy meromyosin, five radioactive peptides were isolated, purified by conventional methods, and the amino acid compositions determined. From the overlap, a consistent scheme was formulated in which all five peptides could be derived from a single region in the polypeptide chain. In particular, the following tentative amino acid sequence was deduced for the ATPase site:

Thr-Ala-Cys-Gly-Gln-Lys-Ser-Pro

Figure 1 presents the structure of this peptide and a proposed mode of binding to the substrate MgATP. The folded conformation of ATP is based on theoretical considerations of Szent-Györgyi (1957) and Boyd and Lipscomb (1969); the metal is shown coordinated to the oxygens of the β and γ phosphates and bridged to the 7-N of the adenine ring, as demonstrated by nuclear magnetic resonance studies (Cohn and Hughes, 1962; Glassman et al., 1971). The model suggests as many as five possible points of interaction between ATP and the peptide: the OH of serine with the 2'-OH of the ribose or the 3-N of the adenine; the CONH$_2$ of glutamine with the 1-N of the adenine; the SH of cysteine with the 6-NH$_2$ of the adenine; and the OH of threonine with the terminal P=O of the triphosphate chain.

Synthesis

A peptide with the structure shown in Fig. 1 was never directly isolated from the natural source, nor its sequence determined by standard degradative procedures. Only a limited amount of material was used, and we recognize that work on a larger scale is required to rigorously confirm this result. Nevertheless, we have actually synthesized this peptide in high yield and purity and thus been able to study its properties.

We have used the general strategies and methods of solid phase peptide synthesis (Merrifield, 1963; 1969). All amino acids were protected on the α-amino position by the *t*-butyloxycarbonyl (Boc) group; functional side chain-protected derivatives were *O*-benzyl-serine, *O*-benzyl-threonine, *S*-benzyl-cysteine, and *N*$^{\epsilon}$-carbobenzoxy-lysine. Boc-Pro (946 mg; 4.4 mmoles) was esterified to 5.00 g polystyrene-1%-divinylbenzene resin (8.75 mmoles chloromethyl sites) by refluxing with 0.55 ml (3.96 mmoles) triethylamine in 50 ml ethyl acetate for 50 hr. A sample of Boc-Pro-resin was hydrolyzed in 12 N HCl:acetic acid:phenol (2:1:1), and proline was determined by amino acid analysis to be 0.61 mmole/g resin (77% yield).

Boc-Pro-resin (2.00 g; 1.22 mmoles Pro) was placed in the reaction vessel of the Beckman Model 990 synthesizer, and the protected peptide chain was assembled stepwise on the resin starting from the carboxyl terminal. One cycle consisted of: (1) deprotection with 20% trifluoroacetic acid-CH$_2$Cl$_2$ (2 min + 20 min); (2) neutralization with 10% diisopropylethylamine-CH$_2$Cl$_2$ (2 min + 5

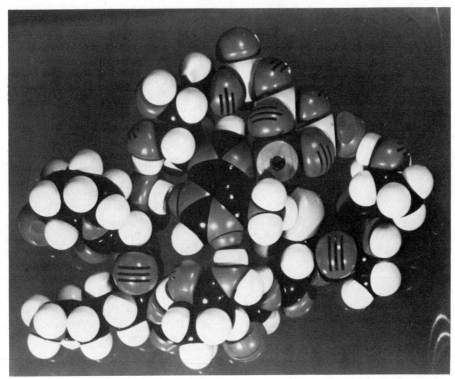

Figure 1. *Above:* CPK space-filling model visualizing the binding of MgATP by a peptide at the ATPase site of frog actomyosin. The peptide starts from the upper right (amino terminal) and stretches across the bottom. *Below:* Schematic diagram of the same complex. The proposed binding interactions are indicated by dotted lines. Note that the role of water molecules in the coordination sphere of Mg^{++} is not depicted.

min); (3) coupling of Boc-amino acid (2.5-fold excess, 3.1 mmoles) in CH_2Cl_2, mediated by N,N'-dicyclohexylcarbodiimide (3.0 mmoles) (2 hr); (4) "double coupling" program—repeat of steps (2) and (3). Between chemical steps the resin was thoroughly washed with appropriate solvents, so that one round of the above "double coupling" cycle took 7 hr 45 min on the machine. Glutamine was introduced as the Boc-nitrophenyl ester

(5-fold excess, 6.2 mmoles) in dimethylformamide (25 hr) in place of steps (3) and (4). At step (3) for the next residue, glycine, Boc-2-[^{3}H]Gly (0.6 mmole, 1 mCi) was incorporated in 83% yield, after activation with an equal molar amount of dicyclohexylcarbodiimide. The reaction was chased to completion by running step (4) twice. The final protected peptide-resin contained 0.40 mmole peptide/gram. As estimated by the radioactive

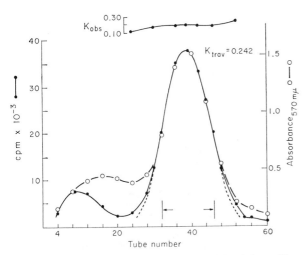

Figure 2. Counter-current distribution of S-benzyl peptide. 200 transfers were run on a 60 tube train (3 ml each phase). The solvent system consisted of the two phases from n-butanol: pyridine: 10% acetic acid (5:3:11). 10 μl lower phase was taken for liquid scintillation counting (left ordinate, ●—●), and 100 μl lower phase was developed by the ninhydrin method (right ordinate, ○—○). The dashed line is the theoretical curve, calculated according to Craig and Craig (1956). The observed k is the ratio of radioactivity in the upper and lower phase; the k of travel is based on the center of the peak.

method, 8% of the peptide chains were lost from the resin during the synthesis by gradual cleavage of the anchoring ester bond by trifluoroacetic acid.

Peptide-resin (1.00 g) was suspended in a mixture of 15 ml trifluoroacetic acid and 5 ml anisole (as a scavenger for benzyl cations) and HBr was bubbled through for 1 hr at 25°. This treatment cleaved the peptide off the resin in 88% yield and simultaneously removed all protecting groups except for the S-benzyl on cysteine. The crude S-benzyl peptide eluted in the void volume of Sephadex G-10 in 50% acetic acid and was further purified by counter-current distribution (Fig. 2). A cut containing 60% of the radioactivity (tubes 32–46) was taken. The observed partition coefficient was constant across that portion of the peak, a criterion for homogeneity. The recovery of S-benzyl peptide from this purification step was 46% by weight. Two impurities were resolved by the counter-current distribution and eventually isolated, identified, and characterized: (1) a "deletion" peptide (tubes 4–16), lacking S-benzyl-cysteine; and (2) a "truncated" peptide (tubes 12–28) terminating in a pyrollidone carboxylic acid residue derived from cyclization of a newly deprotected glutamine at the end of the peptide chain on the resin. In addition, the small amount of material which had run off the counter-current distribution train was collected. In hydrolysates of these pooled fractions, all amino acids could be

recovered except for any one of cysteine, cystine, cysteic acid, or S-benzylcysteine. Instead, a new peak appeared near ammonia on the amino acid analyzer, which we suspect to be the dibenzyl sulfonium derivative of cysteine derived by alkylation during the HBr cleavage step.

To remove the final protecting group and liberate the free sulfhydryl group, $\sim$50 μmoles purified S-benzyl peptide was dissolved in 1 ml trifluoroacetic acid and transferred to the reaction vessel of the HF cleavage apparatus. The acid was evaporated and 0.5 ml anisole was added. About 4.5 ml HF was distilled into the reaction vessel and the cleavage was run for 1 hr at 25°. After cooling and evaporation of the HF, the peptide was dissolved again in trifluoroacetic acid, and then concentrated on a rotary evaporator. It was triturated with benzene to extract anisole degradation products and finally suspended in 5.0 ml H_2O. Dithiothreitol (500 μmoles) was added and the pH was brought to 7.2 with M NH_4HCO_3. This was incubated overnight at 0° to reverse any possible $N \rightarrow O$ acyl shift at the serine residue. The peptide was finally purified on Sephadex G-15 in 50% acetic acid and lyophilized. Altogether, 150 μmoles were isolated with a specific activity of 8.7×10^5 cpm/μmole. The overall yield, based on Boc-Pro on the starting resin, was 34%. The purified peptide was homogenous as evidenced by amino acid analysis, paper chromatography and electrophoresis, and end group determination with 1-dimethylaminonaphthalene-5-sulfonyl chloride.

Binding Studies

We investigated the interaction of the synthetic peptide with ATP by several approaches. Qualitative evidence for binding was obtained by UV differential spectroscopy. In the presence of the peptide there was a significant increase in the absorbance maximum of ATP at 259 nm. In addition, we made a preliminary observation that ATP, when applied on a column of Sephadex LH-20 jointly with the peptide, eluted slightly earlier than in the control chromatogram.

In order to quantitate the binding, a new method was developed based on changes in the extraction properties of the ligands upon complex formation. Table 1 gives a typical result by this method. The two-phase system was designed so that the peptide partitioned mainly in the lower phase (tube 1), while ATP remained almost exclusively in the upper phase (tube 3). When the peptide and ATP were allowed to interact (tube 5), there was a synchronous movement of peptide into the upper phase and ATP into the lower phase. This concerted effect is the criterion for binding.

Table 1. Changes in Partition Coefficients upon Formation of Peptide-ATP Complex

Tube No.	Additions	Lower Phase (total cpm)	Upper Phase (total cpm)	c
1	[³H]peptide	333,681	43,996	0.117
2	[³H]peptide + Mg⁺⁺	328,051	43,844	0.115
3	[¹⁴C]ATP	2,328	65,688	0.966
4	[¹⁴C]ATP + Mg⁺⁺	37,201	38,787	0.484
5	[³H]peptide +	320,511	49,974	0.135
	[¹⁴C]ATP	5,771	61,967	0.915
6	[³H]peptide +	327,982	45,101	0.121
	[¹⁴C]ATP + Mg⁺⁺	40,629	32,179	0.442

A pre-equilibrated system of two phases was prepared from 5.3 g (17.5 mmoles) 4-morpholine N,N'-dicyclohexylcarboxamidine (DCHM), 10 ml 10% acetic acid, 500 ml n-butanol, and 200 ml H_2O; pH both phases = 7.5 (± 0.3). The DCHM base was prepared according to Moffatt and Khorana (1961) and apparently helped solubilize the ATP in the upper phase.

Substances listed under "Additions" were first freeze-dried and then freshly dissolved in the appropriate phase. Binding assay mixtures of final volume 1.8 ml upper phase and 0.2 ml lower phase were vigorously shaken on a Vortex, and the phases separated by centrifugation. The upper phase was removed and aliquots from both phases were taken for liquid scintillation counting. The data are presented in the final column as c, the asymmetric partitionco efficient, which gives the fraction of material found in the upper phase. The reproducibility of parallel determinations of c was within 0.003.

In this experiment, additions where noted were as follows: 0.425 μmole [³H]peptide (spec. act. = 8.7×10^5 cpm/μmole); 0.145 μmole [¹⁴C]ATP (spec. act. = 4.7×10^5 cpm/μmole); 0.1 μmole MgCl₂. All tubes contained 8.0 μmoles dithiothreitol as a reducing agent and also to chelate possible contaminating heavy metal ions.

Table 1 also shows how the well-documented formation of the MgATP complex was reflected by this method (compare tubes 3 and 4). However, this large movement into the lower phase was enchanced even further upon binding of MgATP by the peptide (tube 6).

The apparent movement of the peptide into the upper phase, ΔP_U, can be directly correlated to the amount of peptide-ATP complex, PA, in terms of the measured assymmetric partition coefficient of the peptide, c_P, (see legend Table 1) and the unknown assymmetric partition coefficient of the complex, c_{PA}.

$$\Delta P_U = PA(c_{PA} - c_P) \tag{1}$$

A similar equation holds for the change in distribution of ATP, ΔA_L, this time into the lower phase.

$$\Delta A_L = PA(c_A - c_{PA}) \tag{2}$$

Since ΔP_U, ΔA_L, c_P, and c_A are measured experimentally, Eqs. (1) and (2) can be solved simultaneously to give the unknowns of interest, PA and its distribution c_{PA}.

$$c_{PA} = \frac{\Delta P_U \cdot c_A + \Delta A_L \cdot c_P}{\Delta P_U + \Delta A_L} \tag{3}$$

$$PA = \frac{\Delta P_U + \Delta A_L}{c_A - c_P} \tag{4}$$

The affinity constants in both phases can now be calculated; that for the upper phase is given by:

$$K_U = \frac{PA}{(P_T - PA)(A_T - PA)} \cdot \frac{c_{PA}}{c_P \cdot c_A} \cdot v_U \tag{5}$$

where P_T and A_T are, respectively, total peptide and total ATP, v_U is the volume of the upper phase; c_P and c_A are experimentally determined as described previously, and c_{PA} and PA are given by Eqs. (3) and (4).

The results of a large number of experiments under varying conditions have been calculated according to Eq. (5) and are summarized in Table 2. Native myosin has an affinity for MgATP 250-fold greater than that of the isolated ATPase-site peptide; for free ATP the ratio of affinities is 25. When the peptide was oxidized with performic acid, converting a cysteine to a cysteic acid residue, the binding dropped 10- to 20-fold. This finding points to the key role of the sulfhydryl group in the binding.

As controls, neither reduced glutathione nor the pentapeptide Phe-Asp-Ala-Ser-Val significantly altered the extraction properties of ATP, even when present in large excesses. Similarly, there was no evidence that the ATPase-site peptide binds Mg⁺⁺.

We also compared the relative binding affinities of the peptide to various adenine derivatives (Table 3). The data clearly show that all parts of the ATP molecule are involved in the binding. Thirty percent of the original affinity of the peptide for ATP is retained in free adenine.

Table 2. Affinity Constants for ATP Binding

	K (M^{-1})	
	MgATP	ATP
Myosin	1.2×10^6	5.0×10^4
Peptide	4.5×10^3	2.1×10^3
Peptide, $[\text{CySO}_3\text{H}]$	1.9×10^2	1.9×10^2
Glutathione, reduced		<1
Phe-Asp-Ala-Ser-Val		<40

The affinity constants of myosin for MgATP and ATP are taken, respectively, from Schliselfeld and Bárány (1968) and M. Bárány et al. (1969). The remaining values were determined by the extraction method for the upper phase of the system described in the legend of Table 1. The affinity constants in the lower phase, K_L, were also appreciable, in all cases approximately half of K_U.

Since there are two calculated quantities in the numerator of Eq. (5), the absolute value of affinity constants by the extraction method bears an error of approximately $\pm 30\%$. For further explanation, see text.

We are now in a position to evaluate the key concepts emphasized by the model proposed at the beginning of this paper (Fig. 1). (1) The concerted effect of multiple points for binding interaction, properly oriented to achieve a large affinity constant, is visualized. Examples for this have been provided at the structural levels of both the peptide (modification of sulfhydryl group, Table 2) and ATP (see Table 3). (2) The main binding specificity of this peptide occurs for the adenine ring. The peptide is also able to discriminate between free ATP and MgATP by a mechanism possibly involving the sulfhydryl group. Obviously, this peptide consisting of eight amino acid residues cannot provide all the specificity of myosin, and other regions of the protein are undoubtedly involved in MgATP binding and catalysis.* The protein may also provide a low dielectric medium to strengthen the natural affinity of this ATPase-site peptide for ATP. (3) The peptide structure itself, in the absence of ATP, is quite flexible, and a number of other orientations favorable to binding are conceivable with other nucleoside triphosphates. This accounts both for the general specificity of the myosin NTPase activity, as well as our own preliminary observation that this peptide binds GTP and CTP (although apparently less strongly than ATP).

We wish to conclude that the results presented here represent a novel approach for the investigation of the mechanism of action of actomyosin.

Acknowledgment

We thank Drs. Michael Bárány, Lyman C. Craig, William A. Gibbons, John D. Gregory, Bernd Gutte, T. P. King, and George Némethy for helpful advice and discussions.

* For example, Tonomura et al. (1965) proposed a model, based on studies with ATP analogs, which suggested the functional involvement of two short sequences of myosin thought to be at the active site.

Table 3. Relative Binding Affinities of Peptide to Adenine Derivatives

ATP	100%
MgATP	212
ADP	81
MgADP	130
AMP	61
MgAMP	61
Adenosine	32
Adenosine + Mg^{++}	32
Adenine	30
Adenine + Mg^{++}	30

The relative values shown above (normalized to affinity peptide-ATP = 100%) apply in both phases. In fact, for precise comparative data, a "weighted" K_E, defined by

$$K_E = \frac{PA \cdot (V_U + V_L)}{(P_T - PA)(A_T - PA)} \quad (6)$$

was used. K_E is proportional to both K_U and K_L but can be determined with greater absolute accuracy ($\pm 10\%$).

The relative affinities in the presence of Mg^{++} must be taken as lower bounds, since neither ATP nor ADP were fully saturated under the conditions used for these determinations.

References

BARANY, G., M. BÁRÁNY, and R. B. MERRIFIELD. 1972. Isolation and synthesis of an ATP-binding peptide. *Fed. Proc.* **31**: 431.

BÁRÁNY, M. and K. BÁRÁNY. 1959. Studies on "active centers" of L-myosin. *Biochim. Biophys. Acta* **35**: 293.

BÁRÁNY, M., G. BAILIN, and K. BÁRÁNY. 1969. Reaction of myosin with 1-fluoro-2,4-dinitrobenzene at low ionic strength. *J. Biol. Chem.* **244**: 648.

BOYD, D. B. and W. N. LIPSCOMB. 1969. Electronic structures for energy-rich phosphates. *J. Theoret. Biol.* **25**: 403.

COHN, M. and T. R. HUGHES, JR. 1962. Nuclear magnetic resonance spectra of adenosine di- and triphosphate. II. Effect of complexing with divalent metal ions. *J. Biol. Chem.* **237**: 176.

CRAIG, L. C. and D. CRAIG. 1956. Extraction and Distribution. In *Technique of organic chemistry*, vol. 3, ed. A. WEISSBERGER. Interscience, New York.

GLASSMAN, T. A., C. COOPER, L. W. HARRISON, and T. J. SWIFT. 1971. A proton magnetic resonance study of metal ion—adenine ring interactions in metal ion complexes with adenosine triphosphate. *Biochemistry* **10**: 843.

MERRIFIELD, R. B. 1963. Solid phase peptide synthesis. I. The synthesis of a tetrapeptide. *J. Amer. Chem. Soc.* **85**: 2149.

———. 1969. Solid phase peptide synthesis. *Advanc. Enzymol.* **32**: 221.

MOFFATT, J. G. and H. G. KHORANA. 1961. Nucleoside polyphosphates. X. The synthesis and some reactions of nucleoside-5' phosphoromorpholidates and related compounds. Improved methods for the preparation of nucleoside-5' polyphosphates. *J. Amer. Chem. Soc.* **83**: 649.

SCHLISELFELD, L. H. and M. BÁRÁNY. 1968. The binding of adenosine triphosphate to myosin. *Biochemistry* **7**: 3206.

SZENT-GYÖRGYI, A. 1957. *Bioenergetics*. Academic Press, New York.

TONOMURA, Y., S. KUBO and K. IMAMURA. 1965. A molecular model for the interaction of myosin with adenosine triphosphate. In *Molecular biology of muscular contraction*, ed. S. Ebashi et al. Elsevier, Amsterdam.

Transient Kinetic Studies of the Mg++-dependent ATPase of Myosin and Its Proteolytic Subfragments

C. R. Bagshaw, J. F. Eccleston, D. R. Trentham, and D. W. Yates

Molecular Enzymology Laboratory, Department of Biochemistry, University of Bristol, U.K.

R. S. Goody

Max-Planck-Institut für Experimentelle Medizin, Göttingen, West Germany

Transient kinetic studies of the magnesium ion-dependent ATPase of heavy meromyosin have been described using the chromophoric analog of ATP, 6-mercapto-9-β-ribofuranosylpurine 5'- triphosphate (thioATP) (Trentham et al., 1972). We report here an extension of these studies using subfragment 1 and heavy meromyosin and a wider range of chromophoric ATP analogs modified both in the adenine and triphosphate moieties.

At present subfragment 1 is the smallest component of myosin known containing ATPase activity and there are several reasons for wanting to understand the mechanism of this ATPase in detail. This work describes three aspects: (1) further delineation of elementary processes of the mechanism and description of the rate constants of formation and decay of the intermediates, (2) increasing the sensitivity of the spectrophotometric probes, and (3) kinetic analysis of analogs which have been used in physiological and physical measurements on muscle. Increasing the probe sensitivity is an important step in extending these studies to more organized systems such as myosin filaments and actomyosin.

ATP Analogs

ThioATP and 2-amino-6-mercapto-9-β-ribofuranosylpurine 5'-triphosphate, (thioGTP, I), both give large absorption changes on interacting with subfragment 1. The origin of this spectral change depends principally on the fact that the pK ($= 8$) of the thiol on the purine moiety increases on binding to subfragment 1. When thioGTP binds to

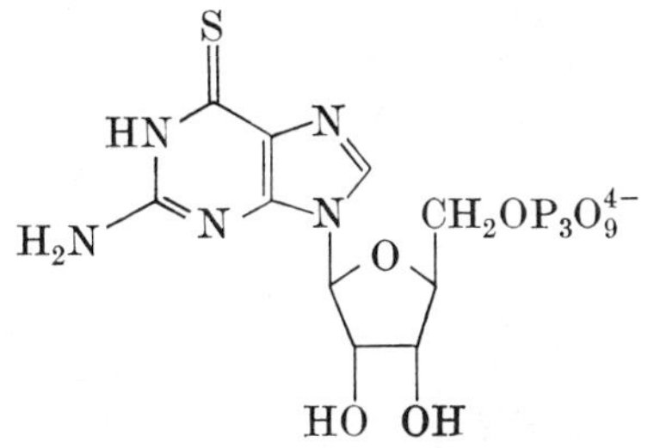

I

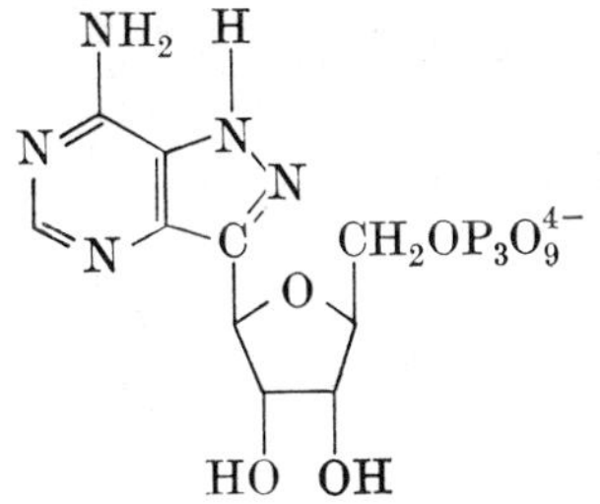

II

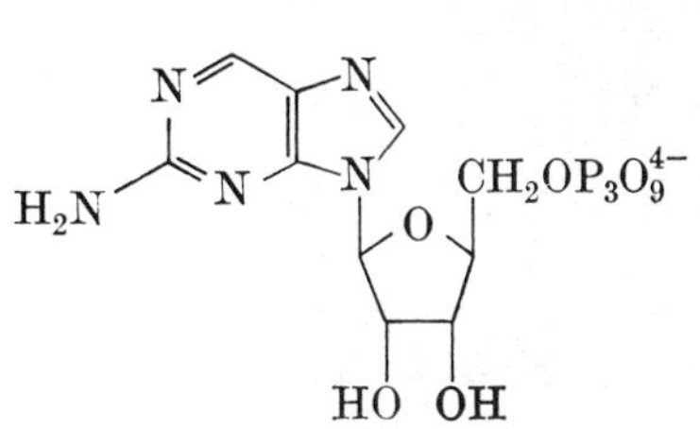

III

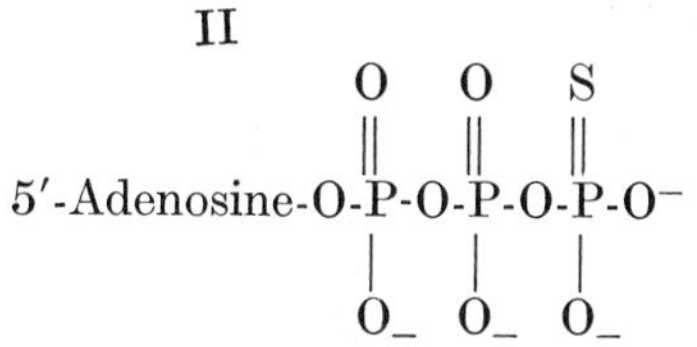

IV

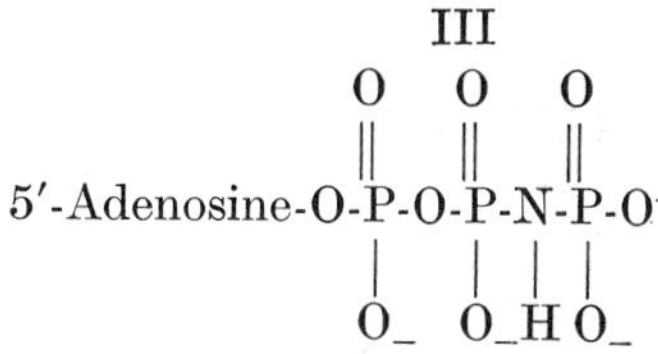

V

127

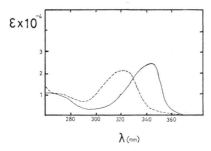

Figure 1. The UV spectrum of thioGTP in the presence of 5 mM MgCl₂ at pH 4.9 (———) and 10.0 (– – – –).

subfragment 1 at pH 8, the entire spectral change in the 340 nm region is clear of protein absorption, which makes this analog a particularly good probe for monitoring the chemistry of the active site. Figure 1 illustrates the spectral characteristics of thioGTP. In addition, both these analogs quench protein fluorescence strongly. This not only extends the range of pH over which they can usefully be used, but also increases the sensitivity of the signal by which their interaction with subfragment 1 can be detected. Formycin 5′-triphosphate (II) and 2-aminopurine-9-β-ribofuranosyl 5′-triphosphate (III) are two fluorescent analogs of ATP (Ward et al., 1969a). When excited at 310 nm the fluorescence of formycin 5′-triphosphate is enhanced on binding to subfragment 1 or heavy meromyosin, whereas that of 2-aminopurine-9-β-ribofuranosyl 5′-triphosphate is quenched. Figure 2 illustrates the spectral changes observed on the interaction of formycin 5′-triphosphate with heavy meromyosin. The reaction profile is resolved into two approximately exponential phases as noted previously with thioATP (Trentham et al., 1972). Finally, protein fluorescence is enhanced when ATP reacts with subfragment 1 or heavy meromyosin (Fig. 3). Modifications to the triphosphate moiety

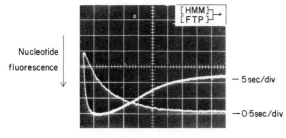

Figure 2. Stopped-flow spectrophotometric record of nucleotide fluorescence during the reaction of formycin triphosphate and heavy meromyosin (excitation at 306 nm). One syringe contained 14 μM formycin triphosphate (reaction chamber concentration) and the other 14 μM heavy meromyosin ([subfragment 1 heads]); both syringes contained 100 mM KCl, 10 mM MgCl₂ and 100 mM triethanolamine adjusted to pH 8.0 with HCl.

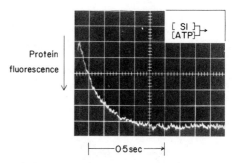

Figure 3. Stopped-flow spectrophotometric record of protein fluorescence during the reaction of ATP and subfragment 1 (excitation at 300 nm). One syringe contained 10 μM ATP (reaction chamber concentration) and the other 2.4 μM subfragment 1. Both syringes contained 100 mM KCl, 10 mM MgCl₂ and 100 mM triethanolamine adjusted to pH 8.0 with HCl.

are of two kinds, illustrated by adenosine 5′-(3-thiotriphosphate) (ATPγS, IV) (Goody and Eckstein, 1971) and 5′-adenylylimidodiphosphate (AMP·PNP, V) (Yount et al., 1971a). Chromophoric purine analogs can be used to replace the adenine in these compounds.

Materials and Methods

Preparation of proteins. Subfragment 1 and heavy meromyosin were prepared from myosin from rabbit skeletal muscle essentially as described by Lowey et al. (1969). Both proteins were purified by ion exchange chromatography on DEAE-cellulose, except for heavy meromyosin used in experiments described in Fig. 9. Concentrations of subfragment 1 are quoted as μM sites on the basis of mol wt 115,000 and $E_{280}^{1\%} = 7.9$ cm⁻¹. Concentrations of heavy meromyosin are quoted as μM subfragment 1 heads on the basis of mol wt 340,000 (equiv. wt. 170,000) and $E_{280}^{1\%} = 6.47$ cm⁻¹ (Young et al., 1965). Small corrections for light scattering were made by measuring the extinction between 320 nm and 400 nm and extrapolating back to 280 nm.

Nucleotide synthesis. Nucleotide syntheses will be described in detail elsewhere; a summary of methods is included here. ThioATP was prepared as described by Trentham et al. (1972). ThioADP was obtained from thioATP hydrolysis catalyzed by subfragment 1. ThioGTP was prepared as described by Darlix, Fromageot, and Reich (personal communication). Formycin 5′-monophosphate was prepared as described by Ward et al. (1969b) and converted to the triphosphate enzymically using phosphoenolpyruvate, pyruvate kinase, and myokinase. 2-Amino-purine-9-β-riboside was prepared by reduction of thioguanosine (Fox et al., 1958) and converted to the monophosphate by the method of Darlix et al. (1967). The triphosphate was prepared by method 1 used by Murphy et al. (1970) for thioATP synthesis. ATPγS was synthesized as described previously (Goody and Eckstein, 1971). [8-³H]ATPγS was synthesized as described by Goody et al. (1972). ThioATPγS and thioGTPγS were prepared in the same way as ATPγS, except that the protected thiophosphate intermediate was partially purified on DEAE cellulose. AMP·PNP and [2-³H]AMP·PNP were synthesized enzymically by the method of Rodbell et al. (1971), scaled up 1000-fold and characterized as described by Yount et al. (1971a). The nucleotides were purified on DEAE cellulose using triethylammonium bicarbonate buffers at pH 7.5. The nucleotides were characterized and shown to be pure by

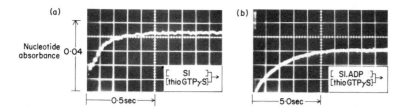

Figure 4. Displacement of ADP from subfragment 1 by thioGTPγS. Stopped-flow spectrophotometric record of the extinction change at 350 nm during the reaction of thioGTPγS with, in (*a*), subfragment 1 and, in (*b*), an equilibrium mixture of subfragment 1 and ADP. One syringe contained 11 μM subfragment 1 (reaction chamber concentrations) and in (*a*) no ADP and in (*b*) 50 μM ADP (generated from ATP in situ) and the other 75 μM thioGTPγS. Both syringes contained 100 mM KCl, 5 mM MgCl$_2$, 0.5 mM dithiothreitol and 100 mM Tris adjusted to pH 8.0 with HCl.

UV spectroscopy and standard electrophoretic, chromatographic, and enzymic techniques as outlined in publications describing the nucleotide syntheses above.

Rapid reaction equipment. Reference to stopped-flow and quenched-flow equipment appropriate to a particular experiment is made in the text except for the fluorescent stopped-flow built by Prof. H. Gutfreund and Dr. D. W. Yates which has not been described previously. In this machine, light from a Wotan XBO 250 W Xenon Arc Lamp with a current stabilized power supply (Thorn Bendix Type 481) is passed through a Farrand Foci UV monochromator, and monochromated light with a 10 nm bandwidth is used to excite fluorescence in the observation chamber of a stopped-flow machine similar to that described by Gutfreund (1972). 90° fluorescence from the 2 mm × 2 mm observation chamber was detected by an EMI 9526 B photomultiplier through Schott filters type 21611 and WG 345. The exciting light passing through the observation chamber was collected by a quartz light guide and used to excite fluorescence in a tryptophan standard cuvette viewed by a similar photomultiplier and filters, and this reference signal was subtracted from the sample fluorescence signal electronically. The resulting fluorescence difference signal was displayed on a Tektronix storage oscilloscope type 564. A Wratten 18B filter was used to reduce stray light in the exciting beam. Reactions were studied at room temperature (21° ± 2°C) with careful control of temperature when relative rates were measured.

Kinetics of ADP Binding to Subfragment 1

Analysis of the kinetics of the displacement of ADP and thioADP from subfragment 1 can be used to illustrate these spectrophotometric probes. As a demonstration of their sensitivity Fig. 4 shows the absorption change when thioGTPγS reacts with subfragment 1 in the presence and absence of bound ADP (generated from ATP in situ), and

Fig. 5 shows the same phenomenon using thioATPγS and monitoring the protein fluorescence change.

In order for the observed rate constant of ADP displacement to equal the true rate constant of ADP dissociation, it is important that the displacing agent should compete effectively kinetically with ADP for subfragment 1 and that the displacing agent should itself not be dissociated rapidly from the protein. A good diagnostic test that the observed rate constant equals the dissociation rate constant is that the observed rate constant does not change when the concentration of the displacing agent is altered. Figure 6 illustrates that when the ATP concentration is changed from 75 μM to 470 μM the observed rate constant is invariant and therefore this rate, 1.4 sec^{-1}, is the ADP dissociation rate constant from subfragment 1 at pH 8. This work is a continuation of similar studies using linked assay systems and thioATP under somewhat different salt conditions in which it was also concluded that ADP dissociates much faster than the catalytic center activity, 0.04 sec^{-1}, of the Mg^{++}-dependent ATPase (Trentham et al., 1972).

When thioADP binds to subfragment 1, protein fluorescence is quenched and this provides a spectrophotometric probe to analyze the binary complex formation, which is described most simply by:

$$\text{S1} + \text{thioADP} \underset{k_\text{d}}{\overset{k_\text{a}}{\rightleftharpoons}} \text{S1 thioADP}$$

where S1 represents subfragment 1, S1 thioADP

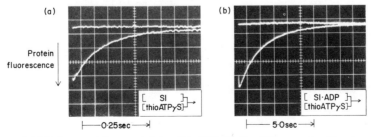

Figure 5. Displacement of ADP from subfragment 1 by thioATPγS. Conditions as for Fig. 4 except that thioATPγS replaced thioGTPγS and the protein fluorescence is recorded (excitation at 300 nm). The oscilloscope was triggered a few seconds after the initial reaction trace to give the horizontal trace.

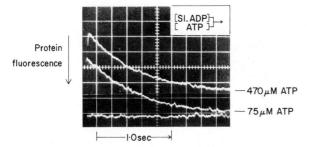

Protein fluorescence

—470 μM ATP

—75 μM ATP

|⊢—— 1·0sec ——⊣|

Figure 6. Displacement of ADP from subfragment 1 by ATP. Stopped-flow spectrophotometric record of protein fluorescence change (excitation at 300 nm) during the reaction of ATP with an equilibrium mixture of subfragment 1 and ADP. One syringe contained 11 μM subfragment 1 (reaction chamber concentrations) and 50 μM ADP (generated from ATP in situ) and the other ATP (470 μM in the upper trace and 75 μM in the middle trace). Both syringes contained 100 mM KCl, 5 mM MgCl₂, 0.5 mM dithiothreitol and 100 mM Tris adjusted to pH 8.0 with HCl. The oscilloscope was triggered a few seconds after the middle reaction trace to give the horizontal trace.

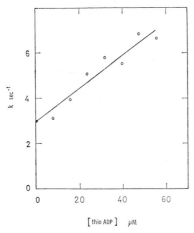

Figure 7. Kinetics of the binary complex formation between thioADP and subfragment 1. (○) First-order rate constants, k_{obs}, of the protein fluorescence change when thioADP was mixed with subfragment 1 in a stopped-flow apparatus (excitation at 300 nm). One syringe contained thioADP and the other 3.5 μM subfragment 1 (reaction chamber concentrations). Both syringes contained 160 mM KCl, 10 mM MgCl₂, 0.4 mM dithiothreitol and triethanolamine adjusted to pH 8 with HCl. (●) k_{obs} of the protein fluorescence change when 40 μM thioADP was displaced from 3.5 μM subfragment 1 on mixing with 1 mM ATP. Solvent conditions were otherwise unchanged from the association reaction (○).

the binary complex, and k_a and k_d are the rate constants. The observed rate constant of the fluorescence change when thioADP is mixed with subfragment 1 = k_a[thioADP] + k_d (for [thioADP] ≫ [S1]) and, when thioADP is displaced from subfragment 1 by excess ATP, the observed rate of fluorescence enhancement = k_d. Figure 7 describes a study of this reaction from which $k_a = 7 \times 10^4$ M⁻¹ sec⁻¹ and $k_d = 3$ sec⁻¹. The association rate of thioADP is 25% that of thioATP and the dissociation rate is five times the catalytic center activity of the thioATPase (Trentham et al., 1972).

Table 1 indicates the sensitivity of the probes by listing the lowest concentrations at which the probes can be used and still give satisfactory signal to noise ratios.

Initial Binding and Bond Cleavage of ATP

Using the quenched-flow technique, Lymn and Taylor (1970) and Tokiwa and Tonomura (1965) have studied the cleavage of ATP by heavy meromyosin and observed a transient phase of P_i production. Equation 1 describes a two-step mechanism for this process:

$$\text{M} + \text{ATP} \underset{k_{-1}}{\overset{k_1}{\rightleftharpoons}} \text{M·ATP*} \underset{k_{-2}}{\overset{k_2}{\rightleftharpoons}} \text{M·ADP·P}_i^* \quad (1)$$

where M represents myosin proteolytic subfragments, M·ATP* the binary complex, and M·ADP·P$_i^*$ the cleaved complex (water is not necessarily implicated in this step). The asterisk denotes forms of myosin with enhanced fluorescence, and the basis of this assignment is discussed below.

The fluorescence change (Fig. 3) appears to represent the binding process because as ATP is increased to large concentrations this process becomes too fast to measure (>700 sec⁻¹ the dead time of the stopped-flow equipment). If the fluorescence change corresponded only to the cleavage step, then the rate of this process would have a plateau at $k_2 + k_{-2}$, the rate of the cleavage step (160 sec⁻¹, Lymn and Taylor, 1971, Fig. 3).

The conclusion that the fluorescence change reflects the binding process is unambiguously solved using the analog ATPγS. When ATPγS reacts with myosin subfragments, a fluorescent

Table 1. Sensitivity Limits of Spectrophotometric Probes

Analog Type	Spectroscopic Method	Example	Limiting conc. (μM)
6-mercaptopurine	Absorption	Fig. 4	2
6-mercaptopurine	Protein fluorescence quenching	Fig. 5	0.03
ATP	Protein fluorescence enhancement	Fig. 3 and 6	0.1

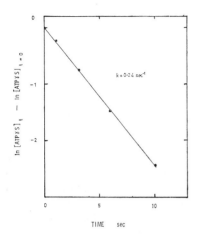

Figure 8. Rate of [8-³H] ATPγS cleavage by heavy meromyosin. The exponential decay of [ATPγS] during the reaction of 3.4 μM [8-³H] ATPγS with heavy meromyosin (5.8 μM subfragment 1 head concentration) in a solvent of 100 mM KCl, 50 mM MgCl₂, 0.5 mM dithiothreitol and 50 mM triethanolamine adjusted to pH 8 with HCl. The quenched-flow apparatus was used as a rapid mixing device to initiate the reaction between ATPγS and heavy meromyosin. The reaction was then quenched by manual addition from a well-clamped syringe of an equal volume of ice-cold 10% HClO₄ containing 1 mM ATP to prevent nucleotide adsorption onto denatured protein. In this way time points ⩾1 sec were obtained. The zero time point was obtained by quenching the reactants directly into the 10% HClO₄. The precipitated protein was separated by use of a bench centrifuge for 1 min and the solution brought to pH 3.5 by addition of 4 M sodium acetate. The radioactive nucleotides were separated using PEI-cellulose thin-layer chromatography in the presence of nonradioactive ATPγS and ADP as markers in a solvent of 0.75 M KH₂PO₄ at pH 3.4. The ATPγS and ADP bands were eluted in equal volumes of 1 M HCl overnight. 0.5 ml of the eluate was added to 15 ml of butyl-PBD scintillator and the tritium counted using a liquid scintillation counter. The butyl-PBD scintillation fluid contained 3 liters toluene, 2 liters 2-methoxyethanol, 400 g naphthalene and 30 g butyl-PBD (Koch-Light). After background corrections the difference between ln ([ATPγS]) at time t and time zero was calculated from the ratio of ADP counts to total counts.

change is observed occurring at the same rate as for ATP at equivalent concentration, although the rate of cleavage of ATPγS is reduced to 0.24 sec⁻¹ (Fig. 8).

There is no measurable fluorescence change in going from M·ATP* to M·ADP·P_i^*.

Two extreme kinetic models of Eq. 1 can be visualized, either $k_2 + k_{-2} \gg k_{-1}$ or $k_{-1} \gg k_2 + k_{-2}$. Measurement of the amplitude and rate of the fluorescence change when ATP binds to subfragment 1 at [ATP] ≫ [subfragment 1] distinguishes these models. In the former the total fluorescence change will occur in a single exponential process at a rate equal to $k_1[\text{ATP}] + \dfrac{k_{-1}k_{-2}}{k_2 + k_{-2}}$. In the latter the reaction profile will have two exponential phases; first, a rapid equilibration step between

M + ATP and M·ATP* involving a fraction of the total fluorescence change equal to $(1 + k_{-1}/k_1[\text{ATP}]^{-1})$ and occurring at a rate of $k_1[\text{ATP}] + k_{-1}$, followed by the remainder of the fluorescence change occurring at a rate of $\dfrac{k_1k_2[\text{ATP}]}{k_1[\text{ATP}] + k_{-1}} + k_{-2}$. When subfragment 1 was mixed with ATP over a wide ATP concentration range ([ATP] > [subfragment 1]), the appearance of the fluorescence change was monophasic and its amplitude remained constant until it became too fast to measure because of the dead time of the stopped-flow machine. This result suggests that $k_{-1} < k_2 + k_{-2}$ (at least a factor of 10 less).

A distinction between these two extreme kinetic models can also be made if the rate of M·ADP·P_i^* production is followed by the quenched-flow methods described by Lymn and Taylor (1970). When $k_{-1} \gg k_2 + k_{-2}$, and [ATP] ≫ $[M_0]$, the total protein concentration, M·ADP·P_i^* formation is an exponential process with a rate of $\dfrac{k_1k_2[\text{ATP}]}{k_1[\text{ATP}] + k_{-1}} + k_{-2}$. In contrast when $k_2 + k_{-2} \gg k_{-1}$, M·ADP·P_i^* formation occurs with a lag phase and its concentration at time t is given by:

$$[\text{M·ADP·P}_i^*] = \frac{k_2[\text{M}_0]}{k_2 + k_{-2}} 1\{ - e^{-(k_2+k_{-2})t} - (k_2 + k_{-2})te^{-(k_2+k_{-2})t}\} \quad (2)$$

provided that $k_1[\text{ATP}] = k_2 + k_{-2}$. Equation 2 becomes more complex if $k_1[\text{ATP}] \neq k_2 + k_{-2}$. The lag phase would in fact be most discernable when $k_1[\text{ATP}] = k_2 + k_{-2}$, and the quenched flow experiment was therefore designed to meet this requirement of [ATP].

In practice this leads to technical difficulties. [M·ADP·P_i^*] is calculated from measurements of [P_i] or [ADP] at different times during the reaction. The most important part of the reaction from the point of view of distinguishing the two models is during the first 30% of the transient formation when the levels of [P_i] or [ADP] produced as nonenzymic hydrolysis products of ATP during the acidic work-up are significant relative to [P_i] or [ADP] produced enzymatically. In addition the value of $k_2 + k_{-2}$ (= 160 sec⁻¹, Lymn and Taylor, 1971, Fig. 3), sets the time range over which the experiment can be performed and the portion of the reaction for the lag phase model in which $d[\text{M·ADP·P}_i^*]/dt \to 0$ as $t \to 0$ is over by the time the first measurement can be made. It does not help matters to lower the temperature since Lymn and Taylor's results indicate that the second-order association rate is reduced proportionately more

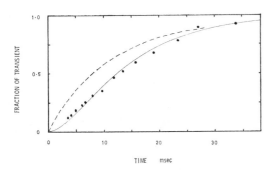

Figure 9. The rate of transient formation of $M\cdot ADP\cdot P_i^*$ followed by measurement of $^{32}P_i$ production in the reaction of $[\gamma\text{-}^{32}P]$ ATP and heavy meromyosin at $21°$. 67 μM $[\gamma\text{-}^{32}P]$ATP was mixed with heavy meromyosin (30 μM in subfragment 1 heads; the concentration was measured as described below) in a quenched-flow apparatus. Both syringes contained 50 mM KCl, 5 mM $MgCl_2$, 50 μM dithiothreitol and 20 mM Tris adjusted to pH 8.0 with HCl. The reaction was quenched into an equal volume, 3.3 ml, of ice-cold 10% $HClO_4$ containing 200 μM P_i to act as a carrier and to prevent P_i adsorption onto denatured protein. The separation of $^{32}P_i$ was based on the method of Martin and Doty (1948) except that a double extraction was performed. All the extraction steps were carried out at $0°$ and within 5 min to prevent $^{32}P_i$ production arising from ATP hydrolysis in the acid medium. 5 ml of the final isobutanol-benzene extract were mixed with 10 ml of the butyl-PBD scintillator (Fig. 8) and $^{32}P_i$ counted using a liquid scintillation counter. The dashed and solid lines are explained in the text. $[\gamma\text{-}^{32}P]$ATP was prepared by the method of Glynn and Chappell (1964). Immediately prior to use it was re-purified by ion exchange chromatography on DEAE-cellulose using as eluant a KCl salt gradient in 50 mM Tris at pH 8. The zero time point of the reaction was obtained by adding first heavy meromyosin to the $HClO_4$ followed by $[\gamma\text{-}^{32}P]$ATP. $[^{32}P_i]$ at zero time was 2% of total [ATP] and probably arose principally in the acidic work up. ATP hydrolysis at pH 8 was so slow that its effect was negligible. One g of heavy meromyosin was used in each of these experiments so that no purification was performed once the undigested myosin and light meromyosin had been removed by centrifugation following low salt dialysis. The active site concentration of heavy meromyosin was calculated from measurement of the product concentration in the 1–10 sec time range (see legend to Fig. 8), extrapolating back to zero time and multiplying the product concentration by $\dfrac{k_2 + k_{-2}}{k_2}$. This estimate, which had an error of up to 20%, was then normalized (a 5% adjustment was necessary) to give the closest fit to the solid line. Each point in the figure is the mean of a number of experiments and has an estimated error range up to 10%. More points were taken in the time range up to 10 msec to compensate for the greater error induced by nonenzymic $[\gamma\text{-}^{32}P]$ATP hydrolysis.

than $k_2 + k_{-2}$, so that a higher [ATP] is required to make $k_1[ATP] = k_2 + k_{-2}$ and the problem of background P_i or ADP production becomes worse.

Nevertheless despite these technical difficulties it is important to tackle the problem of the kinetics of ATP interaction with myosin proteolytic sub-fragments by as many independent routes as possible. The problem of background ATP hydrolysis products was reduced by making [heavy meromyosin] as large as possible relative to [ATP]. Using Lymn and Taylor's rate constants (1971, Fig. 3) an analog computer was used to simulate

the rate of $M\cdot ADP\cdot P_i^*$ production. The solid line of Fig. 9 corresponds to the condition $k_2 + k_{-2} \gg k_{-1}$ and the dashed line to the condition $k_{-1} \gg k_2 + k_{-2}$, due account being taken of the fact that [ATP] was estimated to be only 2.2 [heavy meromyosin]. This heavy meromyosin concentration was calculated by multiplying the total transient product concentration by $(k_2 + k_{-2})/k_2$ (see legend to Fig. 9) and therefore required values for k_2 and k_{-2}. Measurement of $k_2 + k_{-2}$ is known from the observed rate of transient formation at high [ATP] (Lymn and Taylor, 1971, Fig. 3), and k_2 and k_{-2} can then be calculated since k_{-2}/k_2, the equilibrium constant of the transformation $M\cdot ATP^* \rightleftharpoons M\cdot ADP\cdot P_i^*$, can be measured as follows. If ATP is mixed with a large molar excess of heavy meromyosin and the reaction is quenched after 1 sec when the initial binding is complete (controlled by k_1) but before product release occurs (controlled by k_3, Eq. 4), ATP will be predominantly equilibrated between $M\cdot ATP^*$ and $M\cdot ADP\cdot P_i^*$. Detailed experiments to determine k_{-2}/k_2 are in progress and will be reported elsewhere, although our preliminary results used here indicate that $k_{-2}/k_2 = 0.14$ (± 0.04). For the purpose of distinguishing the two kinetic models, the analog computed lines derived for Fig. 9 are not significantly altered by variation of k_{-2}/k_2 between 0 and 0.2. Lymn and Taylor (1970) have found transient amplitudes corresponding to 0.8 mole of ATP cleaved per mole of heavy meromyosin subfragment 1 heads. Stoichiometries less than 1 could arise if $k_2/(k_2 + k_{-2})$ is significantly less than 1 or if the heavy meromyosin was not pure.

The results of the quenched-flow experiments when P_i production was followed are shown in Fig. 9. The same result was obtained if ADP production was monitored using $[2\text{-}^3H]$ATP as starting material and separating the radioactive product on PEI-cellulose strips as described in Fig. 8.

The fit of the experimental points to the lag phase curve is not perfect; in particular there is a tendency for a systematic drift from points to go from above to below the analog computed solid line with increasing time. It should be noted that the technique is being pushed to its analytical limits with each point having a sizable error bar and is perhaps revealing a systematic instrumental error not revealed during the course of conventional testing. However, the points deviate considerably more from the dashed line and do not have the exponential character of that curve. The quenched-flow experiment provides significant support for the conclusion of the fluorescent stopped-flow experiments that $k_{-1} < k_2 + k_{-2}$, and fits in with the results which follow showing that the dissociation rate constants of both ATPγS and AMP·PNP are <0.2 sec^{-1}.

Kinetic Parameters of ATPγS

We investigated more fully the kinetic parameters of the reaction between ATPγS and myosin subfragments. The K_m was determined in 100 mM KCl at pH 8 and shown to be less than 1 μM, which was the limit of the sensitivity of our method using the proton indicator phenol red to monitor proton release in a split beam stopped-flow apparatus (Gutfreund, 1972). Using identical conditions to Fig. 8 but increasing [heavy meromyosin] fourfold to ascertain the concentration dependence of the reaction, the observed rate of the cleavage step was unaltered at 0.25 sec⁻¹ as would be expected from the small K_m value of ATPγS. When 25 μM ATPγS was treated with 5 μM subfragment 1 and the rate of ADP production followed, no measurable transient release of ADP was observed (Fig. 10) showing that cleavage is predominantly rate determining in the ATPγS hydrolysis catalyzed by subfragment 1 and heavy meromyosin. This result is also supported by the observation that the rate of the fluorescent transient decay in the reaction between ATPγS and excess subfragment 1 is 0.23 sec⁻¹ (Fig. 11).

Two simple kinetic models consistent with these results are described by Eq. 3 using similar nomenclature to Eq. 1

$$M + ATP\gamma S \underset{k'_{-1}}{\overset{k'_1}{\rightleftharpoons}} M \cdot ATP\gamma S^* \underset{k'_{-2}}{\overset{k'_2}{\rightleftharpoons}} M \cdot ADP \cdot P_s^*$$
$$\overset{k'_3}{\longrightarrow} \text{lower fluorescent} \qquad (3)$$
$$\text{forms of subfragment 1}$$

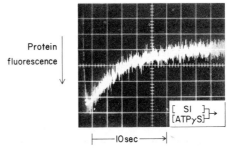

Figure 11. Stopped-flow spectrophotometric record of protein fluorescence during the reaction of ATPγS and subfragment 1 (excitation at 300 nm). One syringe contained 2 μM ATPγS (reaction chamber concentrations) and the other 4 μM subfragment 1. Both syringes contained 150 mM KCl, 50 mM MgCl₂, 0.5 mM dithiothreitol and 100 mM triethanolamine adjusted to pH 8 with HCl. The rapid initial protein fluorescence enhancement, which appears as an almost vertical faint line, is associated with the binding of ATPγS to subfragment 1.

In the first $k'_3 \gg k'_2$ and k'_{-2} so that $k'_2 = 0.25$ sec⁻¹. In the second $k'_{-2} \gg k'_2$ and k'_3 so that the observed cleavage rate (0.25 sec⁻¹) = $k'_2 k'_3 / k'_{-2}$. The essential feature of the second model is that the equilibrium between M·ATPγS* and M·ADP·P$_s^*$ is rapidly established and favors M·ATPγS*. In both models the steady-state complex of the ATPase with ATPγS as substrate is M·ATPγS*.

Given the above kinetic characteristics of ATPγS in its reaction with subfragment 1, the observed rate of the fluorescence change when an excess of ATPγS reacts with subfragment 1 is given by

$$k'_{obs} = k'_1[ATP\gamma S] + k'_{-1} + k'_2.$$

(In this paragraph the equations are written for

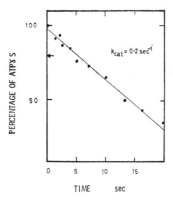

Figure 10. Steady-state rate of [8-³H]ATPγS cleavage by subfragment 1. The remaining [ATPγS] was measured during the reaction of 25 μM [8-³H]ATPγS with 5 μM subfragment 1 in a solvent of 150 mM KCl, 5 mM MgCl₂, 0.5 mM dithiothreitol and 100 mM triethanolamine adjusted to pH 8 with HCl. The techniques used are described in Fig. 8. [ATPγS] was calculated from the ratio of [8-³H]ADP counts to total counts. The arrow indicates where the line would have intersected the ordinate if there had been a rapid transient cleavage of one mole of ATPγS per mole of subfragment 1 well resolved from the steady-state rate of ATPγS hydrolysis.

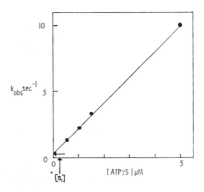

Figure 12. Dependence of the rate of ATPγS binding to subfragment 1 in ATPγS concentration. Rates were measured from stopped-flow spectrophotometric records of protein fluorescence enhancement (excitation at 300 nm). One syringe contained ATPγS and the other 0.25 μM subfragment 1 ([subfragment 1] was 2 μM for the point at 5 μM [ATPγS]). Both syringes contained 100 mM KCl, 50 mM MgCl₂, 0.5 mM dithiothreitol and 50 mM triethanolamine adjusted to pH 8 with HCl. The arrow on the ordinate corresponds to a rate of $k_2 = 0.25$ sec⁻¹ and on the abscissa to [subfragment 1].

the first model when $k_2' = 0.25$ sec^{-1}. If the second model is valid $k_2'k_3'/k_{-2}'$ should replace k_2'.) It is therefore possible by varying [ATPγS] to obtain a value or at least an upper limit of k_{-1}', since the intercept on the ordinate axis of this graph is given by $k_{-1}' + k_2'$. The results of this experiment are shown in Fig. 12 and show that the intercept = 0.25 sec^{-1}. Since $k_2' = 0.25$ sec^{-1}, this gives $k_{-1}' = 0$. In practice we conclude that $k_{-1}' < 0.2$ sec^{-1} since the intercept value is controlled by the experimental sensitivity.

Kinetic Parameters of AMP·PNP

The discovery by Yount et al. (1971b) that AMP·PNP can dissociate actomyosin and is a competitive inhibitor with respect to ATP has stimulated physical measurements on muscle relaxed with this ATP analog. It is important therefore to understand the kinetics of its interaction with myosin and its proteolytic subfragments in detail.

When AMP·PNP reacts with subfragment 1, an enhancement of protein fluorescence is observed similar to that of the ATP reaction (Fig. 3) except that the rate is reduced 20-fold to 6×10^4 M^{-1} sec^{-1}. 1.0 μM AMP·PNP is displaced from 0.85 μM subfragment 1 by either 90 or 180 μM thio-ATPγS with an observed rate constant of 0.02 sec^{-1} (solvent conditions as described in Fig. 12). This rate is probably close to the dissociation rate constant of AMP·PNP since thioATPγS binds to subfragment 1 at a rate of 2.5×10^5 M^{-1} sec^{-1} with a catalytic center activity of 0.09 sec^{-1}, and so will compete effectively with AMP·PNP for subfragment 1. The dissociation constant K_D of AMP·PNP given by these figures is 3×10^{-7} M and is in satisfactory agreement with the competitive inhibition constant of 5×10^{-7} M determined from a study of AMP·PNP acting as competitive inhibitor of subfragment 1 with respect to 2-aminopurine-9-β-ribofuranosyl 5'-triphosphate ($K_m = 13$ μM).

In view of the comparisons that are being made of X-ray diffraction patterns of muscle relaxed with AMP·PNP and ATP, it is important to establish whether AMP·PNP is reversibly cleaved but not hydrolyzed by myosin or its proteolytic subfragments. We investigated this by incubating [2-^{3}H]AMP·PNP with molar excess of heavy meromyosin at pH 8 overnight at room temperature, quenching the mixture in 10% perchloric acid, adjusting the solution to pH 8 with ammonia, and analyzing the radioactive products using paper chromatography according to Yount et al. (1971a, Table 1, solvent A). No cleavage of AMP·PNP was observed above background levels (5%). The experiments with AMP·PNP provide additional proof that the fluorescence change when ATP reacts with myosin is associated with the binding process.

Discussion

As an outline for the discussion the kinetic scheme recently proposed as an extension of Lymn and Taylor's (1971) model is used (Trentham et al., 1972):

$$M + ATP \underset{k_{-1}}{\overset{k_1}{\rightleftharpoons}} M \cdot ATP^* \underset{k_{-2}}{\overset{k_2}{\rightleftharpoons}} M \cdot ADP \cdot P_i^* \overset{k_3}{\underset{k_{-3}}{\rightleftharpoons}}$$

$$M \cdot ADP \cdot P_i \underset{k_{-4}}{\overset{k_4}{\rightleftharpoons}} M \cdot ADP + P_i \underset{k_{-5}}{\overset{k_5}{\rightleftharpoons}} M + ADP \quad (4)$$

where the asterisks denote forms of subfragment 1 with enhanced fluorescence. The fluorescence of $M \cdot ADP \cdot P_i$ was not measured in these experiments since its concentration is small relative to other intermediates of the reaction. The more definite value for k_5 supports this scheme. Martonosi and Malik report a direct measurement of k_{-5} elsewhere in this volume.

It is premature at the present time to interpret the structure of the ATP binding site from a comparison of the kinetics of the different analogs. However certain points emerge from these studies. k_1 for ATP, thioATP, and thioGTP are unaffected by substitution of sulfur on the γ-P atom (Table 2).

Table 2. Observed Rates

Substrate	Association Rates (k_1 or k_1') M^{-1} sec^{-1}	Dissociation (k_{-1} or k_{-1}') sec^{-1}	Bond Cleavage sec^{-1}	Catalytic Center Activity sec^{-1}	Diphosphate Dissociation sec^{-1}
ATP	1–2×10^6	<16	160	0.041	1.4
ThioATP	2–4×10^5			0.55	3
ThioATPγS	1–3×10^5			0.09	
ThioGTPγS	1–3×10^5				
ATPγS	1–2×10^6	<0.2	0.24	0.2	
AMP·PNP	6×10^4	0.02	0		
Formycintriphosphate	7×10^4				

Rate constants describing elementary processes of the Mg^{++}-dependent ATPase using ATP or its analogs were derived from Lymn and Taylor (1971), Trentham et al. (1972), the text and Figs. 6, 7, 8, 10, 12. Ranges of values of association rate constants are listed since these were obtained from a number of different experiments and showed significant variation with quite small perturbations of temperature and solvent, which was typically 100 mM KCl, 5 mM MgCl$_2$ and 50 mM Tris or triethanolamine at pH 8.

This contrasts with the marked effect on k_1 of N-substitution in AMP·PNP. Since AMP·PNP appears to be structurally similar to ATP (Yount et al., 1971a), it could be that the low value of k_1 indicates its binary complex with magnesium ions is stereochemically different from magnesium ATP. The low value of k_1 for formycin 5'-triphosphate (7×10^4 M^{-1} sec^{-1}) could reflect the anomalously high percentage of the synconformation probable in this compound (Koyama et al., 1966; Ward and Reich, 1968). It is interesting that although the overall rate of ATPγS cleavage is 650 times slower than ATP and the binary complex of ATPγS with the protein is the steady state complex, the rate constant k_3 of the ATPase has increased at least 10-fold with ATPγS as substrate. In contrast the catalytic center activity and K_m with thioATPγS as substrate are less than those of thioATP, although k_1 for these two compounds is the same, so that thioATPγS is a better chromophoric displacing agent than thioATP.

A significant fact for X-ray diffraction studies is that in resting muscle at least two of the states of the myosin ATPase (M·ATP*, M·ADP·P_i^*, and possibly M·ADP at low temperature) can be induced as the steady state complex. These results are reported elsewhere in this volume (Barrington Leigh et al.; Lymn and Huxley; Martonosi and Malik).

Acknowledgments

We are grateful to the Science Research Council, the Wellcome Trust, the European Molecular Biology Organisation and Deutsche Forschungsgemeinschaft (grant to Dr. F. Eckstein) for financial support. We thank Dr. A. G. Weeds, Dr. F. Eckstein and Prof. H. Gutfreund for helpful discussions and the latter for making available much newly developed rapid reaction equipment. Miss G. Witzel provided valuable technical assistance. Gifts of formycin from Dr. T. A. Khwaga, I. C. N. Corporation and from Prof. H. Umezawa are gratefully acknowledged.

References

DARLIX, J. L., H. P. M. FROMAGEOT, and P. FROMAGEOT. 1967. An improved method for the preparation of 5'-monophosphoderivatives of the common ribonucleosides. *Biochim. Biophys. Acta* **145**: 517.

FOX, J. J., I. WEMPEN, A. HAMPTON, and I. C. DOERR. 1958. Thiation of nucleosides. Synthesis of 2-amino-6-mercapto-9-β-D-ribofuranosyl purine (Thioguanosine) and related purine nucleosides. *J. Amer. Chem. Soc.* **80**: 1669.

GLYNN, I. M. and J. B. CHAPPELL. 1964. A simple method for the preparation of ^{32}P-labelled adenosine triphosphate of high specific activity. *Biochem. J.* **90**: 147.

GOODY, R. S. and F. ECKSTEIN. 1971. Thiophosphate analogs of nucleoside di- and tri-phosphates. *J. Amer. Chem. Soc.* **93**: 6252.

GOODY, R. S., F. ECKSTEIN, and R. SCHIRMER. 1972. The enzymatic synthesis of thiophosphate analogs of nucleotide anhydrides. *Biochim. Biophys. Acta* **276**: 157.

GUTFREUND, H. 1969. Rapid mixing: continuous flow. *Methods in Enzymology* (ed. K. Kustin) vol. 16, p. 229. Academic Press, New York.

GUTFREUND, H. 1972. *Enzymes: physical principles*, p. 180. Wiley-Interscience.

KOYAMA, G., K. MAEDA, and H. UMEZAWA. 1966. The structural studies of formycin and formycin B. *Tetrahedron Letters* **6**: 597.

LOWEY, S., H. S. SLAYTER, A. G. WEEDS, and H. BAKER. 1969. Substructure of the myosin molecule. Subfragments of myosin by enzymatic degradation. *J. Mol. Biol.* **4**: 21.

LYMN, R. W. and E. W. TAYLOR. 1970. Transient state phosphate production in the hydrolysis of nucleoside triphosphates by myosin. *Biochemistry* **9**: 2975.

LYMN, R. W. and E. W. TAYLOR. 1971. Mechanism of adenosine triphosphate hydrolysis by actomyosin. *Biochemistry* **10**: 4617.

MARTIN, J. B. and D. M. DOTY. 1948. Determination of inorganic phosphate. Modification of isobutyl alcohol procedure. *Anal. Chem.* **21**: 965.

MURPHY, A. J., J. A. DUKE, and L. STOWRING. 1970. Synthesis of 6-mercapto-9-β-D-ribofuranosyl purine 5'-triphosphate, a sulphydryl analog of ATP. *Arch. Biochem. Biophys.* **137**: 297.

RODBELL, M., L. BIRNBAUMER, S. L. POHL, and H. M. J. KRAUS. 1971. The glucagon sensitive adenyl cyclase system in plasma membranes of rat liver. *J. Biol. Chem.* **246**: 1877.

TOKIWA, T. and Y. TONOMURA. 1965. The pre-steady state of the myosin adenosine triphosphatase system. *J. Biochem. (Tokyo)* **57**: 616.

TRENTHAM, D. R., R. G. BARDSLEY, J. F. ECCLESTON, and A. G. WEEDS. 1972. Elementary processes of the magnesium ion-dependent adenosine triphosphatase activity of heavy meromyosin. *Biochem. J.* **126**: 635.

WARD, D. C. and E. REICH. 1968. Conformational properties of polyformycin: a polyribonucleotide with individual residues in the syn conformation. *Proc. Nat. Acad. Sci.* **61**: 1494.

WARD, D. C., E. REICH, and L. STRYER. 1969a. Fluorescence studies on nucleotides and polynucleotides. *J. Biol. Chem.* **244**: 1228.

WARD, D. C., A. CERAMI, E. REICH, G. ACS, and L. ALTWERGER. 1969b. Biochemical studies on the nucleoside analogue, formycin. *J. Biol. Chem.* **244**: 3243.

YOUNG, D. M., S. HIMMELFARB, and W. F. HARRINGTON. 1965. On the structural assembly of the polypeptide chains of heavy meromyosin. *J. Biol. Chem.* **240**: 2428.

YOUNT, R. G., D. BABCOCK, W. BALLANTYNE, and D. OJALA. 1971a. Adenylimido-diphosphate, an ATP analog containing a P-N-P linkage. *Biochemistry* **10**: 2484.

YOUNT, R. G., D. OJALA, and D. BABCOCK. 1971b. Interaction of P-N-P and P-C-P analogs of adenosine triphosphate with heavy meromyosin, myosin and actomyosin. *Biochemistry* **10**: 2490.

Actin Activation of Heavy Meromyosin and Subfragment-1 ATPases; Steady State Kinetics Studies

CARL MOOS

Biochemistry Department, State University of New York, Stony Brook, N.Y. 11790

The energy for muscle contraction is derived from the hydrolysis of ATP, and one of the principal properties of the contractile proteins in vitro is the marked activation of the myosin ATPase by actin at low ionic strength in the presence of magnesium. Since ATP hydrolysis, unlike contraction, can be quantitatively measured in a homogeneous system of purified proteins, the study of ATPase kinetics constitutes one of the principal approaches to the molecular mechanism of muscle contraction. It is difficult to apply the classical methods of steady state enzyme kinetics to the actomyosin ATPase, however, because the conditions necessary for the actin-activated ATPase also cause aggregation of myosin into filaments and precipitation of actomyosin. If simple mass-action kinetics are to be applied to the actin-myosin interaction, at least one of the proteins must be present in solution as separate molecules which can react independently with the active sites of the other. This condition can be met by the use of heavy meromyosin (HMM) or subfragment-1 (S-1) in place of myosin since these myosin fragments are soluble in virtually any ionic milieu of interest, and they appear to react independently with sites on F-actin (in the absence of tropomyosin). For some years we have been investigating the steady state kinetics of the actin-activated HMM ATPase, and in this paper I shall briefly review the principal results of that work, along with some new data on acto-S-1 ATPase and acto-HMM ITPase, in the context of the present state of the field.

Actin Activation Kinetics at High Substrate: V_{max} and K_{app}

The first step in our investigation of acto-HMM ATPase kinetics was the demonstration (Eisenberg and Moos, 1968) that activation of the HMM ATPase by actin shows a simple hyperbolic dependence on actin concentration, as illustrated in Fig. 1A. In double-reciprocal coordinates, the same data give a straight line (Fig. 1B) which, as we shall see, is quite characteristic of the system under a very wide range of conditions. Two points should be particularly noted in this experiment. First, the system is substrate saturated since

reducing the ATP concentration does not change the measured rates (Eisenberg and Moos, 1968) and the pH-stat records are linear; i.e., the reaction rate does not change with time as the substrate is depleted. Second, the curve in Fig. 1A does not represent a titration of available actin binding sites because even the lowest actin concentration is well above the molar concentration of HMM.

In the double-reciprocal plot (Fig. 1B) the intercepts of the line give two parameters by which the actin activation kinetics can be characterized. The reciprocal of the ordinate intercept is the extrapolated maximum activity at infinite actin concentration, V_{max}, which can be regarded empirically as the steady state turnover rate of fully activated HMM. The reciprocal of the abscissa intercept gives a parameter, with the dimensions of actin concentration, which is analogous to a dissociation constant or a Michaelis constant. We shall refer to it as K_{app}. Operationally it is the actin concentration required to reach half-maximal activation, and it may be regarded, in a sense, as an inverse measure of the apparent affinity of the system for actin as an activator. The interpretation of V_{max} and K_{app} in terms of specific steps in the acto-HMM ATPase reaction cycle depends on the kinetic model which is postulated, and this matter will be considered later. However, the two parameters have also proved useful empirically for analyzing the effects of various agents on the acto-HMM ATPase. This approach is illustrated by our studies of the effects of ionic strength and temperature.

Effects of ionic strength and temperature. It has long been known that actomyosin ATPase is very sensitive to ionic strength, the activity decreasing markedly as the salt concentration is increased. Such a decrease in ATPase activity could result from either an inhibition of the activity of the actomyosin complex or a decrease in the extent of interaction of actin with myosin in the mixture, and we were able to distinguish between these two possibilities by means of reciprocal plots of acto-HMM ATPase vs. actin concentration at various ionic strengths. We have consistently found that V_{max} is unaffected by ionic strength in the range from nearly zero up to 0.1, and that the

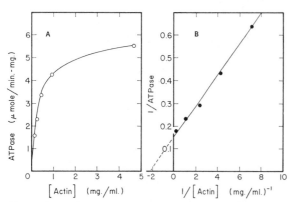

Figure 1. Actin activation of HMM ATPase. The ATPase rates, measured on a pH-stat at pH 7 and 25°, are expressed as μmoles substrate hydrolyzed per minute per mg HMM. A, ATPase rate as a function of actin concentration; B, same data plotted in double-reciprocal form. Conditions: 2.4 mM $MgCl_2$, 2.0 mM ATP, 2.33 mM KCl, 7 mM imidazole-HCl buffer, 0.039 mg/ml HMM.

decrease in ATPase activity at higher salt is due entirely to decreased actin-HMM interaction as reflected in an increased K_{app} (Eisenberg and Moos, 1968, 1970a; Rizzino et al., 1970). This is true at various temperatures (Barouch and Moos, 1971), and also, as we shall see below, with S-1 in place of HMM or with ITP as substrate. The strong ionic strength dependence of K_{app} under all these conditions suggests that ionic forces play an important role in the actin-myosin interaction. This result fits well with the work of April et al. (1968), which showed that the decreasing contractile force in intact muscle fibers with increasing extracellular osmotic pressure was principally due to the increased intracellular ionic strength caused by removal of water rather than the volume change per se. The effect of ionic strength on tension has also been directly confirmed by studies with skinned fibers (Gordon et al., 1970).

Actomyosin is also well known to be very temperature sensitive, and the analysis of the effect of temperature on acto-HMM ATPase (Barouch and Moos, 1971) is another example of the usefulness of double-reciprocal plots of ATPase against actin concentration. The Arrhenius activation energy of actomyosin ATPase is unusually high, about 25–30 kcal/mole (Levy et al., 1959; Bárány, 1967), which is about twice that of myosin ATPase, and one suggested explanation of this difference was that low temperature not only decreased the activity of the actomyosin complex but also dissociated the actin from the myosin. This idea was tested in the acto-HMM system by measuring the effect of temperature on V_{max} and K_{app} separately, and when a set of reciprocal plots at different temperatures was constructed (Barouch and Moos, 1971), it was immediately clear that the

hypothesis is wrong. Contrary to the postulated dissociating effect of reduced temperature, the actin-HMM interaction is stronger at low temperature (i.e., K_{app} is smaller). An Arrhenius plot of V_{max} gave an activation energy of 29 kcal/mole, equivalent to that of actomyosin ATPase, showing that the high temperature coefficient of actomyosin ATPase could indeed be accounted for entirely in terms of the activity of the actomyosin complex itself. It is also noteworthy that this Arrhenius plot was linear from 6° to 30° with no sign of the break often seen with actomyosin or myofibrils (Levy et al., 1959; Bárány, 1967); therefore the change in slope in these more complex systems must arise from some feature which is absent in the acto-HMM system, such as the cooperative character of filament-filament interactions or an effect of the tropomyosin-troponin regulatory system (Hartshorne et al., 1972).

These two examples serve to illustrate the usefulness of V_{max} and K_{app} as parameters for characterizing the activating effect of actin under various conditions. Another application of this approach was the demonstration that the tropomyosin-troponin system acts only on K_{app} and not on V_{max} (Eisenberg and Kielley, 1970), and it is expected that the method will be found useful in the future in other studies as well.

Inosine triphosphatase of acto-HMM.
Recently we have begun an investigation of the acto-HMM kinetics with ITP as substrate instead of ATP, and actin activation curves are shown in Fig. 2A for both substrates measured under

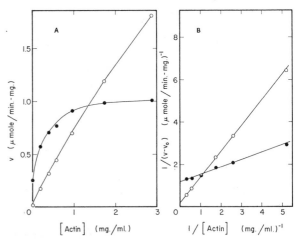

Figure 2. Comparison of ITPase and ATPase of acto-HMM. A, ATPase ($\bigcirc$) or ITPase ($\bullet$) rates, v, measured as in Fig. 1 at various actin concentrations; B, same data plotted in double-reciprocal form after subtracting v_0, the rate in the absence of actin. Conditions: 6 mM ATP or ITP, 7 mM $MgCl_2$, 25 mM KCl, 2 mM imidazole-HCl buffer; HMM varied between 0.03 and 0.30 mg/ml pH 7, 25°. The actin used in both systems contained IDP as its bound nucleotide.

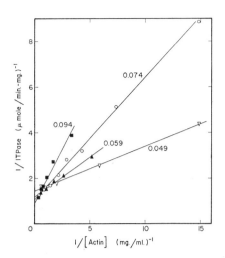

Figure 3. Effect of ionic strength on acto-HMM ITPase. Double-reciprocal plots under the same conditions as in Fig. 2B except for KCl concentration, which was varied to give the total ionic strength indicated on each line.

identical conditions. As is well known for myosin (Levy et al., 1962), the ITPase rate in the absence of actin (v_0) is some tenfold higher than the ATPase, and this is shown by the experimental points on the ordinate in Fig. 2A. However, starting from this rate as a baseline, the actin activation curve again looks hyperbolic, as is verified by the linearity of the reciprocal plot of $v - v_0$ in Fig. 2B. The ATPase plot is also hyperbolic as usual, but its curvature is only barely perceptible in the range of actin concentrations used in Fig. 2A. This brings out another striking difference between the two substrates: K_{app} for the ITPase system is some 20-fold smaller than for the ATPase, or, in other words, a much lower actin concentration is required to activate the ITPase. The effect of salt on the acto-HMM ITPase is shown in Fig. 3. It is evident that, as in the ATPase system, K_{app} is very sensitive to ionic strength and is entirely responsible for the fall in measured ITPase rates with increasing salt. Actually, in Fig. 3, the V_{max} of acto-HMM ITPase increased slightly with increasing KCl concentration, but this effect is too small to be accepted with confidence at this stage. Table 1 summarizes

Table 1. Comparison of Actin Activation Parameters for ATPase and ITPase

	Ionic Strength	ATPase	ITPase
v_0* (μmole/min-mg HMM)	—	0.02	0.24
V_{max} (μmole/min-mg HMM)	0.059 0.074	6	1.1 1.2
K_{app} (mg/ml actin)	0.059 0.074	7 12	0.3 0.5

* v_0 = activity in the absence of actin.

the differences between the ITPase and ATPase parameters of the acto-HMM system for two different salt concentrations, derived from the intercepts in Figs. 2 and 3 and from previous ATPase studies in this laboratory. It is noteworthy that, despite the large difference in the corresponding values of K_{app} between the two substrates, the change in ionic strength affects both in about the same proportion.

Actin activation of subfragment-1 ATPase. The experiments described so far were all done with HMM, but since HMM has two sites for ATPase and actin binding, it is clearly important to carry out kinetic studies with S-1 as well. With the development by Lowey et al. (1969) of a simple procedure for making S-1 by direct papain digestion of myosin, several laboratories have begun kinetic studies with S-1 in place of HMM. Figure 4 shows double-reciprocal plots of the actin activation of HMM and S-1 at two different ionic strengths. It is clear that, as Margossian and Lowey have previously shown (Lowey, 1971), acto-S-1 also gives a linear plot, and its V_{max} is about half that of acto-HMM, as might be expected if each HMM has two S-1 heads. The K_{app} for acto-S-1 is about twice that of acto-HMM, and it changes with salt in about the same proportion in the two systems (Table 2). These data have particular implications for the evaluation of kinetic models, which will be discussed later, but the important point to note here is that these results are obtained with a myosin subfragment which is well established to have only one site for ATP (Eisenberg and Moos, 1970b; Young, 1967). This site is the hydrolytic site, and Fig. 4 shows that its ATPase is activated

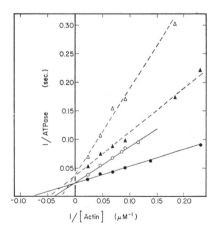

Figure 4. Comparison of acto-S-1 and acto-HMM ATPases. Rates expressed in sec^{-1}, i.e., moles ATP hydrolyzed per mole HMM or S-1 per sec. (– – –) S-1; (——) HMM. ($\triangle$, $\bigcirc$) ionic strength 0.035;($\blacktriangle$, $\bullet$) ionic strength 0.020. Conditions: 3 mM MgCl$_2$, 2 mM ATP, 4 mM imidazole-HCl buffer, 5 or 20 mM KCl pH 7, 25°.

Table 2. Comparison of Actin Activation Parameters for HMM and S-1

	Ionic Strength	HMM	S-1
V_{max} (sec^{-1})	—	28	44
K_{app} (μM actin)	0.020	21	13
	0.035	40	27

by actin to quantitatively the same extent in S-1 as it is in HMM. It is well known also that acto-S-1 is dissociated by ATP, so it seems clear that dissociation must be caused by ATP binding at the hydrolytic site. This is strong evidence against the view that a nonhydrolytic "relaxing site" for ATP binding is responsible for the dissociating effect of ATP in actomyosin.

Substrate dependence of acto-HMM ATPase. To provide further insight into the acto-HMM ATPase kinetics, we have studied the dependence of the reaction rate on substrate as well as actin concentrations, using creatine kinase to maintain low ATP concentrations in a steady state (Eisenberg and Moos, 1970a). Figure 5A shows a family of

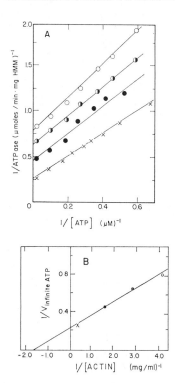

Figure 5. Substrate dependence of acto-HMM ATPase. *A,* Lineweaver-Burk plots of rate vs. ATP at various actin concentrations; *B,* ordinate intercepts from *A,* i.e., reciprocal of ATPase rate at infinite substrate, plotted against 1/actin. Conditions: 2.5 mM MgCl$_2$, 2.5 mM phosphocreatine, 1 mg/ml creatine kinase, 0.036 mg/ml HMM pH 7, 25°. Actin concentrations: (○) 0.24 mg/ml; (◖) 0.36 mg/ml; (●) 0.65 mg/ml; (×) 3.3 mg/ml. (From Eisenberg and Moos, 1970a.)

Lineweaver-Burk plots of acto-HMM ATPase at various actin concentrations. These plots are linear and can be extrapolated to the ordinate to obtain the ATPase activity at infinite substrate for each actin concentration. The values obtained in this way, by extrapolation from micromolar ATP concentrations, correspond well with the values measured directly at millimolar substrate in the previous graphs and can be plotted in double reciprocal form against actin concentration (Fig. 5B) to give V_{max} and K_{app} values which are the same as those obtained directly from measurements at high ATP. This means that over the entire ATP concentration range up to several millimolar, only one kind of ATP binding site is reacting, and these sites react independently without any allosteric interaction between them. This is further evidence that binding of ATP at the hydrolytic site is responsible for dissociation of acto-HMM.

There is no evidence in the acto-HMM kinetics of the substrate inhibition which is a familiar feature of actomyosin systems. One can speculate that substrate inhibition in actomyosin could be due to an internal allosteric property of myosin which is destroyed by proteolysis when HMM is made; or it might reflect a cooperativity between myosin molecules along the myosin filament when it reacts with an actin filament; or it might involve cooperativity between the sites on an actin filament caused by tropomyosin, which is absent in our system (cf. Bremel et al. this volume).

The data of Fig. 5 can also be plotted as 1/v vs. 1/actin, as shown in Fig. 6A, and these plots too are linear at all ATP concentrations. Extrapolation to the ordinate now yields a new set of limiting rates representing the activity at infinite actin for each ATP concentration, and these values can be plotted in Lineweaver-Burk fashion against ATP concentration, as in Fig. 6B. The ordinate intercept of this plot naturally gives the same V_{max} as Fig. 5B for saturating ATP and actin concentrations, but the abscissa intercept yields a new parameter of the system, an apparent Michaelis constant for the maximally actin-activated HMM ATPase. We have shown previously (Eisenberg and Moos, 1970a) that, unlike K_{app}, this Michaelis constant of acto-HMM ATPase is quite unaffected by ionic strength. Its value from Fig. 6B is about 6 μM, which will be significant in the discussion of kinetic models in the next section.

To summarize the main observations, the actin activation of HMM ATPase at saturating substrate concentration shows a simple hyperbolic dependence on actin concentration, i.e., linear double-reciprocal plots of rate against actin under a wide range of conditions: at ionic strengths from near zero up to 0.1, from temperatures near 0° up to 30°, and with

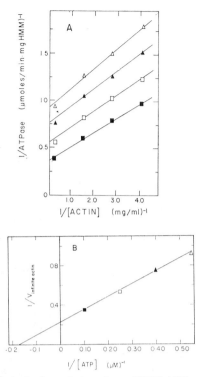

Figure 6. Actin dependence of acto-HMM ATPase at low substrate concentration. A, Double-reciprocal plots of rate vs. actin at various ATP concentrations (points taken from the curves in Fig. 5A). B, Ordinate intercepts from A, plotted against 1/ATP, i.e., Lineweaver-Burk plot at infinite actin. ATP concentrations: ($\triangle$) 1.8 μM; ($\blacktriangle$) 2.5 μM; ($\square$) 4 μM; ($\blacksquare$) 10 μM. (From Eisenberg and Moos, 1970a.)

ITP as substrate in place of ATP. The activation can be described in terms of two parameters, V_{max} and K_{app}, derived from the two intercepts of the double-reciprocal plots, and these parameters are useful for comparing the activation kinetics under different conditions. The dependence on substrate also follows a simple Michaelis relationship from less than 2 μM up to several millimolar ATP, suggesting the presence of only a single kind of substrate site; and this conclusion is strongly supported by the finding that S-1, which has only one ATP-binding site, shows both an actin-activated ATPase and an ATP-induced dissociation of its actin complex.

Evaluation of a Kinetic Model

We turn now to a consideration of the kinetic mechanism of the actin-activated ATPase and the interpretation of constants such as V_{max} and K_{app} in terms of steps in the reaction cycle. The kinetics discussed above were first analyzed in terms of a simple enzyme-substrate-modifier type of scheme as shown in Fig. 7, in which actin is treated as a

general enzyme modifier (Eisenberg and Moos, 1968, 1970a). The scheme includes a classical Michaelis pathway for the ATPase of HMM alone (reactions 1 and 5), a reversible reaction of HMM with actin in the absence of substrate (reaction 2), and a ternary acto-HMM-ATP complex as an intermediate in the activated ATPase. The actin activation of the ATPase means that $k_6 > k_5$, and the dissociation of acto-HMM by ATP means that the dissociation constant of actin from the acto-HMM-ATP complex to form HMM-ATP and actin ($K_3 = k_{-3}/k_3$) is larger than that of acto-HMM without ATP ($K_2 = k_{-2}/k_2$). At saturating substrate concentration the steady state rate equation for this model is Eq. 1 (Botts, 1958),

$$\frac{1}{v - k_5} = \frac{1}{k_6 - k_5} + \frac{K_3}{(k_6 - k_5)} \cdot \frac{1}{[\text{Actin}]} \quad (1)$$

which predicts a linear relation between $1/(v - k_5)$ and 1/actin. This equation is quite consistent with the ATPase data where $1/v$ was plotted against 1/actin, however, because k_5 is equivalent to v_0, the ATPase of HMM alone, and we have seen that it is negligible compared to v throughout most of the measured range. In Fig. 3 we saw that in the ITPase system, where v_0 is a larger fraction of v, $1/(v - v_0)$ is linear in 1/actin, and Fig. 8 shows that a distinctly nonlinear plot is obtained if v_0 is not subtracted from the ITPase rate. It should be recognized, of course, that this result does not prove the model; other more complex schemes, including that of Lymn and Taylor (1971), also predict a linear dependence of $1/(v - v_0)$ on 1/actin at saturating substrate.

To determine what our constants K_{app} and V_{max} would mean in our model, we can match Eq. 1 to our plots of 1/rate against 1/actin at high substrate. We find that V_{max} from the ordinate intercept of the plot corresponds in the model to k_6, the rate constant for substrate hydrolysis and product release from the ternary acto-HMM-ATP complex, and that K_{app} from the abscissa intercept is the true dissociation constant, K_3, of the ternary

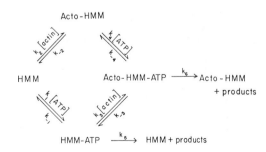

Figure 7. The acto-HMM ATPase as a simple enzyme-substrate-modifier system.

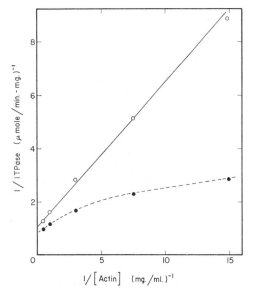

Figure 8. Reciprocal plots of acto-HMM ITPase. (●)
Reciprocal of measured ITPase activity, v; (○) reciprocal
of $(v - v_0)$, where v_0 is the rate in the absence of actin.
Conditions as in Fig. 2, with 40 mM KCl.

complex to actin and HMM-ATP. If the model
were correct, therefore, quantitative analysis of the
effects of temperature, ionic strength, and other
agents on K_{app} could give thermodynamic information about the binding of actin to HMM-ATP;
however, from the discussion below and also from
observations reported in other papers in this
volume, it seems unlikely that this simple model is
in fact adequate, and consequently such analyses
are unjustified.

The substrate dependence of the rate in this type
of kinetic system has been analyzed in a very
useful manner by London (1968), and the results
presented here are internally consistent with that
analysis (Eisenberg and Moos, 1970a). For our
system, however, London's treatment also predicts
a definite relation among the parameters of the
system, shown in Eq. 2:

$$\frac{K_3}{K_2} = \frac{K_m \text{ of acto-HMM ATPase}}{K_m \text{ of HMM ATPase}} \qquad (2)$$

Qualitatively, the prediction is that if ATP
dissociates acto-HMM, the Michaelis constant of
acto-HMM ATPase must be larger than that of
HMM, and indeed the value of about 6 μM for the
apparent Michaelis constant at infinite actin,
obtained from Fig. 6B, is about ten times larger
than that reported for HMM alone (Schliselfeld
and Bárány, 1968; Eisenberg and Moos, 1970b).
However, when the relation in Eq. 2 is examined
quantitatively, we encounter difficulty. The binding
of actin to HMM in the absence of ATP at ionic
strength about 0.1 has been measured in the

ultracentrifuge by Margossian and Lowey (quoted
in Lowey, 1971) and by Eisenberg et al. (1971),
and both groups obtained a dissociation constant,
K_2 in our model, less than 0.5 μM. Our ATPase
measurements at this ionic strength gave a K_{app}
of the order of 400 μM (Eisenberg and Moos, 1968).
Thus the ratio on the left side of Eq. 2 is at least
800, which is clearly not equal to the ratio of
about 10 for the Michaelis constants.

There is another reason for doubting that this
simple kinetic model is adequate. The model
treats HMM as a single unit which reacts one-to-
one with both substrate and actin, whereas it is
well established that HMM has two subunits,
each of which can probably react with actin and
ATP. The comparison of the actin-activation of
S-1 and HMM in Fig. 4 indicated that both subunits
of HMM are fully activated at infinite actin. It can
be shown that, according to the model in Fig. 7,
the activation of both heads must mean that they
both bind to actin simultaneously in the presence of
ATP, and in this case the binding energy of HMM-
ATP to actin should be twice that of S-1-ATP. We
have seen that the dissociation constant for this
interaction in the kinetic model is given by K_{app},
so K_{app} for HMM should be the square of that for
S-1. However, in Fig. 4 they were shown to differ
by no more than a factor of two. This paradox
might be resolved by postulating either that the two
subunits of HMM are not identical and only one
has ATPase activity, or that there is a strong
interaction between them such that actin binding
at one head activates both; but it seems more likely
that these data, as well as the binding data in the
absence of substrate, are actually telling us that
K_{app} is not really the thermodynamic dissociation
constant for the binding of actin to HMM-ATP.
This would mean that the simple model in Fig. 7
is not adequate to account for all the properties of
the system.

Still further reasons to doubt the model arise
from the transient kinetic studies of Taylor's group
(Lymn and Taylor, 1971), especially their studies
of the rate of dissociation of acto-HMM upon
addition of ATP. As they have pointed out, this
dissociation appears to be too fast to be accounted
for by dissociation to free HMM followed by binding
of ATP; it is more likely that it involves a direct
attack of ATP on the acto-HMM complex. Furthermore, our model does not include the intermediate
containing bound hydrolyzed substrate which both
Tonomura's and Taylor's laboratories have demonstrated. Such an intermediate could fit into the
model if it were in very rapid equilibrium with one
of the species in our ring, but in fact neither its
formation from bound substrate nor its dissociation
to free products is as fast as the ATP-induced

dissociation of acto-HMM (Lymn and Taylor, 1971). It therefore seems clear that, as tempting as it was to explain everything in terms of the simple four-species model in Fig. 7, the actin-activated ATPase mechanism is actually more complex than that.

Various more complex kinetic models for the acto-HMM ATPase have been introduced recently, and some of these are discussed in other papers in this volume (cf. Eisenberg and Kielley; Tonomura et al.; Koretz et al.). They would certainly avoid some of the difficulties which have led us to abandon the simple model in Fig. 7 above, but since general steady state rate equations for these models have not yet been analyzed in detail, it cannot yet be determined whether they are compatible with all the data summarized here.

Acknowledgment

This work was supported by Research Grant GM-10249 from the National Institute of General Medical Sciences, U.S. Public Health Service.

References

APRIL, E., P. W. BRANDT, J. P. REUBEN, and H. GRUNDFEST. 1968. Muscle contraction: The effect of ionic strength. *Nature* **220**: 182.

BÁRÁNY, M. 1967. ATPase activity of myosin correlated with speed of muscle shortening. *J. Gen. Physiol.* (No. 6, Part 2) **50**: 197.

BAROUCH, W. W. and C. MOOS. 1971. Effect of temperature on actin activation of heavy meromyosin adenosine triphosphatase. *Biochim. Biophys. Acta* **234**: 183.

BOTTS, J. 1958. Typical behavior of some simple models of enzyme action. *Trans. Faraday Soc. London* **54**: 593.

EISENBERG, E. and W. W. KIELLEY. 1970. Native tropomyosin. Effect on the interaction of actin with heavy meromyosin and subfragment-1. *Biochem. Biophys. Res. Comm.* **40**: 50.

EISENBERG, E. and C. MOOS. 1968. The adenosine triphosphatase activity of acto-heavy meromyosin. A kinetic analysis of actin activation. *Biochemistry* **7**: 1486.

——. 1970a. Actin activation of heavy meromyosin adenosine triphosphatase. Dependence on adenosine triphosphate and actin concentrations. *J. Biol. Chem.* **245**: 2451.

——. 1970b. Binding of adenosine triphosphate to myosin, heavy meromyosin, and subfragment-1. *Biochemistry* **9**: 4106.

EISENBERG, E., L. DOBKIN, and W. W. KIELLEY. 1971. Binding of actin to heavy meromyosin in the absence of ATP. *Fed. Proc.* **30**: 1310 (Abstr.).

GORDON, A. M., R. E. GODT, and J. W. WOODBURY. 1970. Ionic strength as a determinant of calcium-activated tension in skinned muscle fibers in various salt solutions. *Fed. Proc.* **29**: 656 (Abstr.).

HARTSHORNE, D. J., E. M. BARNS, L. PARKER, and F. FUCHS. 1972. The effect of temperature on actomyosin. *Biochim. Biophys. Acta* **267**: 190.

LEVY, H. M., N. SHARON, and D. E. KOSHLAND, JR. 1959. Purified muscle proteins and the walking rate of ants. *Proc. Nat. Acad. Sci.* **45**: 785.

LEVY, H. M., N. SHARON, E. M. RYAN, and D. E. KOSHLAND, JR. 1962. Effect of temperature on the rate of hydrolysis of adenosine triphosphate and inosine triphosphate by myosin with and without modifiers. Evidence for a change in protein conformation. *Biochim. Biophys. Acta* **56**: 118.

LONDON, W. P. 1968. Steady state kinetics of an enzyme reaction with one substrate and one modifier. *Bull. Math. Biophys.* **30**: 253.

LOWEY, S. 1971. In *Subunits in Biological Systems*, part A, p. 201, ed. S. N. TIMASHEFF and G. D. FASMAN. Marcel Dekker, New York.

LOWEY, S., H. S. SLAYTER, A. G. WEEDS, and H. BAKER. 1969. Substructure of the myosin molecule. I. Subfragments of myosin by enzymatic degradation. *J. Mol. Biol.* **42**: 1.

LYMN, R. W. and E. W. TAYLOR. 1971. Mechanism of adenosine triphosphate hydrolysis by actomyosin. *Biochemistry* **10**: 4617.

RIZZINO, A. A., W. W. BAROUCH, E. EISENBERG, and C. MOOS. 1970. Actin-heavy meromyosin binding. Determination of binding stoichiometry from adenosine triphosphatase kinetic measurements. *Biochemistry* **9**: 2402.

SCHLISELFELD, L. H. and M. BÁRÁNY. 1968. The binding of adenosine triphosphate to myosin. *Biochemistry* **7**: 3206.

YOUNG, M. 1967. On the interaction of adenosine diphosphate with myosin and its enzymically active subfragments. Evidence for three identical catalytic sites per myosin molecule. *J. Biol. Chem.* **242**: 2790.

Evidence for a Refractory State of Heavy Meromyosin and Subfragment-1 Unable to Bind to Actin in the Presence of ATP

E. Eisenberg and W. Wayne Kielley

*Section on Cellular Physiology, Laboratory of Biochemistry, National Heart and Lung Institutes,
N.I.H., Bethesda, Maryland 20014*

The sliding filament theory, first proposed in 1954 (Huxley and Niedergerke, 1954; Huxley and Hanson, 1954) is now generally accepted as the overall basis for muscle contraction, and the elegant structural studies of H. E. Huxley and his co-workers suggest that the key event in the sliding process is the cyclic interaction of myosin bridges with F-actin and ATP (Huxley, 1969). Kinetic studies on the interaction of actin, myosin, and ATP in vitro should provide useful information about the cyclic interaction occurring in vivo; but unfortunately, because myosin occurs as insoluble filaments at low ionic strength, these studies have been difficult to interpret in a quantitative manner (Eisenberg and Moos, 1967). On the other hand, heavy meromyosin (HMM), a tryptic digestion product of myosin which retains the two-headed (two-site) structure of the myosin, and subfragment-1 (S-1), a further proteolytic digestion product which consists of single myosin heads (Lowey et al., 1969), are both soluble at low ionic strength and are therefore quite amenable to quantitative kinetic analysis of their interaction with actin and ATP.

Numerous studies on the actin activation of the HMM and S-1 ATPases have clearly demonstrated that, over a wide range of temperature and ionic strength, there is a simple hyperbolic relationship between ATPase activity and F-actin concentration (Eisenberg and Moos, 1968, 1970; Eisenberg et al., 1968; Barouch and Moos, 1971). The simplest interpretation of such a hyperbolic relationship is that in the presence of ATP the binding of HMM to F-actin follows a normal adsorption isotherm, so that at infinite actin concentration where the actin-activated ATPase equals V_{max}, all of the HMM is bound to the F-actin. Whether this is in fact the case can only be tested by directly measuring the binding of actin to HMM in the presence of ATP. In recent studies we have used the analytical ultracentrifuge to make such measurements with HMM (Eisenberg et al., 1972), and we now describe further studies with both HMM and S-1.

Unexpectedly we find that, under conditions where the HMM and S-1 ATPases are almost maximally activated by actin, more than 50% of the HMM and more than 75% of the S-1 are dissociated from the actin. ATPase studies comparing the myosin, HMM, and S-1 suggest that very little if any of the protein is denatured (Eisenberg et al., 1972). It thus appears that during the cycle of actin-HMM interaction in the presence of ATP many of the HMM heads are in a refractory state during which, even at very high actin concentration, they cannot bind to actin. ATPase studies with HMM and S-1 confirm this view and further suggest that the two HMM heads act independently. Therefore, in contrast to the situation in the absence of ATP where both heads of the HMM bind simultaneously to the actin (Eisenberg et al., 1971; Margossian and Lowey, 1972), in the presence of ATP it appears that, simply on a random basis, only one HMM head at a time binds to actin while the other head remains refractory.

Methods

Myosin was prepared by the method of Kielley and Harrington (1960) and HMM was prepared from the myosin as previously described (Eisenberg and Moos, 1967). S-1 was prepared from the myosin using soluble papain as described by Lowey et al. (1969). In the earlier experiments described actin was prepared with a Sephadex G-200 column (Adelstein et al., 1963), whereas in the later experiments it was prepared by the method of Spudich and Watt (1971). No difference was noted in the properties of the actin prepared by the two different methods. All protein concentrations were determined by UV absorption at 280 nm (Eisenberg and Moos, 1967; Eisenberg et al., 1968). ATPase was measured at pH 7.0 and 0.5° with a Radiometer pH-stat (Eisenberg and Moos, 1967), and temperature was monitored during the reaction by a Yellowstone thermister and a No. 427 Teflon-coated probe. For all ATPase rates given in this paper, the ATPase of the HMM or S-1 alone was subtracted from the measured ATPase rate in the presence of actin, as described previously (Eisenberg and Moos, 1968). The centrifuge experiments were performed in a model E analytical

ultracentrifuge equipped with a photoelectric
scanner as described previously (Eisenberg et al.,
1972). For purposes of calculation, the molecular
weights of HMM, S-1, and actin were taken to be
350,000 (Mueller, 1964), 120,000 (Young et al.,
1965; Lowey et al., 1969), and 45,000 (Rees and
Young, 1967), respectively. The molecular weight
of a HMM "head" was considered to be 175,000
(one-half of the HMM molecular weight).

Results

Binding of HMM to actin. The optimal
conditions for determining the binding of actin to
HMM in the presence of ATP using the analytical
ultracentrifuge are: (1) as high an actin concentration and as low an ionic strength as possible so
that a significant fraction of the HMM is bound
(Rizzino et al., 1970; Eisenberg and Moos, 1970);
(2) sufficient ATP and as low a temperature as
possible so that the actin sediments to a significant
degree before all of the ATP is hydrolyzed (Barouch
and Moos, 1971); and (3) as high an HMM concentration as possible so that the free HMM present
can be accurately measured. The conditions we
actually chose are shown in Fig. 1, where the

Table 1. Binding of HMM to Actin in the Presence of ATP

	HMM Bound	V_{max}
Standard condition	41%	85%
[HMM] halved	44%	85%
ATP all hydrolyzed	96%	—

Standard condition: 11.4 μM HMM head; 44 μM actin,
3 mM MgATP; 2 mM P_i; 3 mM imidazole pH 7, temperature
0.5°; rotor speed = 30,000 rpm.

ATPase activity as a function of actin concentration is plotted both directly and in reciprocal
form. The ATPase shows a simple hyperbolic
dependence on actin concentration and at 45 μM
actin, the ATPase activity is 85% of V_{max}.

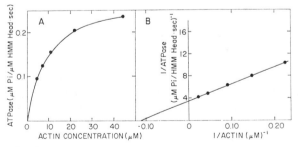

Figure 1. Dependence of actin-HMM ATPase on actin
concentration. A, Direct plot; B, double-reciprocal plot. All
samples contained 3 mM $MgCl_2$, 3 mM ATP, 3 mM imidazole pH 7.0, 2 mM P_i and 11.6 μM HMM heads.
Temperature = 0.5°.

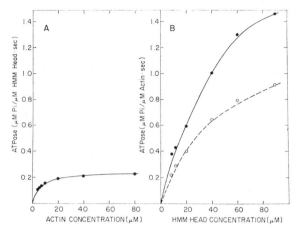

Figure 2. Dependence of actin-HMM ATPase on actin and
HMM concentration. All samples contained 2.5 mM Mg^{++},
2 mM ATP, 1.2 mM imidazole pH 7.0, and 1.5 mM P_i.
Temperature = 0.5°. A, 8.5 μM HMM heads, actin
concentration varied. B, (●) 1 μM actin; (○) 4 μM actin,
HMM concentration varied.

The results of the centrifuge experiments are
shown in Table 1. When the ATPase activity is
85% of V_{max} (line 1), only 41% of the HMM
sediments with the F-actin. As would be expected
if all of the unbound protein were HMM, when the
HMM concentration is halved (line 2), the amount
of protein in the supernatant and correspondingly
the amount bound to the actin also halves. Furthermore after all of the ATP is hydrolyzed to ADP
and P_i (line 3), essentially all of the HMM sediments with the F-actin showing that ATP is
required for the effect we observe.

ATPase studies also suggest that a large fraction
of the HMM is unable to bind to actin. As can be
seen by comparing Figs. 2A and 3A with Figs. 2B
and 3B, the turnover rate per mole of HMM head at

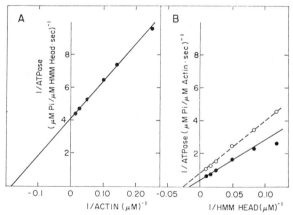

Figure 3. Double-reciprocal plots of data shown in Fig. 2.
A, Double-reciprocal plot of Fig. 2A; B, double-reciprocal
plot of Fig. 2B. (●) 1 μM actin; (○) 4 μM actin.

high actin concentration is much less than the turnover rate per mole of actin monomer at high HMM concentration. A complicating factor in this data is the observation shown in Figs. 2B and 3B that the ATP turnover rate per mole of actin itself depends on the actin concentration, an effect which we observe even at very high HMM concentration and do not completely understand at the present time. Nevertheless, the basic qualitative observation shown in these Figures is most easily explained if, in fact, only a small fraction of the HMM is actually bound to actin at high actin concentration, whereas at high HMM concentration almost all of the actin is complexed with HMM.

Binding of S-1 to actin. To determine whether this same behavior would be shown by S-1 which in contrast to HMM has only a single head, we repeated both the centrifuge and ATPase studies using S-1 prepared with soluble papain. Figure 4 shows that at infinite actin concentration the ATP turnover rate per mole of S-1 is nearly equal to the turnover rate per mole of HMM head. This confirms the work of Margossian and Lowey (1972) and shows that under the conditions of our experiments, the two HMM heads apparently act independently.

Although Fig. 4 shows that the binding of S-1 to F-actin in the presence of ATP is somewhat weaker than that of HMM, it is still possible to measure this binding using the analytical ultra-centrifuge. As can be seen in Fig. 5, the S-1 ATPase shows a typical hyperbolic dependence on actin concentration. At 45 μM actin, the ATPase is 78% of V_{max} and as shown in line 1 of Table 2, under this condition only 22% of the S-1 sediments with

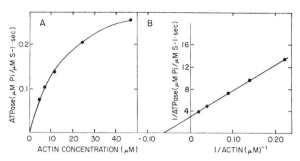

Figure 5. Dependence of acto-S-1 ATPase on actin concentration. *A*, Direct plot; *B*, double-reciprocal plot. All samples contained 3 mM MgCl$_2$, 3 mM ATP, 3 mM imidazole pH 7.0, 2 mM P$_i$ and 12.6 μM S-1. Temperature = 0.5°.

the F-actin. Furthermore, as for HMM, when the S-1 concentration is halved (line 2), the amount of S-1 bound to the actin also halves; and after all of the ATP is hydrolyzed to ADP and P$_i$ (line 3), all of the S-1 sediments with the F-actin. The ATPase studies with S-1 are also very similar to those with HMM. As shown in Figs. 6 and 7, the

Table 2. Binding of S-1 to Actin in the Presence of ATP

	S-1 Bound	V_{max}
Standard condition	22%	78%
[S-1] halved	19%	78%
ATP all hydrolyzed	95%	—

Standard condition: 12.5 μM S-1; 44 μM actin; 3 mM MgATP, 2 mM P$_i$; 3 mM imidazole pH 7, temperature 0.5°; rotor speed = 30,000 rpm.

turnover rate per mole of actin at high S-1 concentration is markedly higher than the turnover rate per mole of S-1 at high actin concentration, and as with the HMM, the turnover rate per mole of actin depends on the actin concentration. Therefore the two-headed structure of HMM does

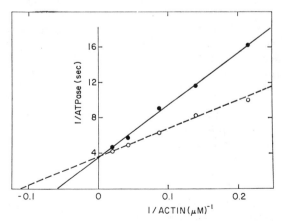

Figure 4. Comparison of actin-HMM and actin-S-1 ATPases. Conditions same as Fig. 1. (○) 12 μM HMM heads, 1/ATPase = (μmole P$_i$/μmole HMM head sec)$^{-1}$; (●) 12.3 μM S-1, 1/ATPase = (μmole P$_i$/μmole S-1 sec)$^{-1}$.

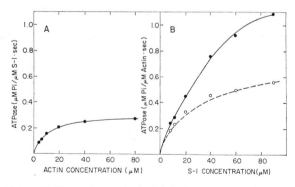

Figure 6. Dependence of acto-S-1 ATPase on actin and S-1 concentrations. All samples contained 2.5 mM Mg^{++}, 2 mM ATP, 1.2 mM imidazole pH 7.0, and 1.5 mM P$_i$. Temperature = 0.5°. *A*, 8.2 μM S-1, actin concentration varied; *B*, (●) 1 μM actin; (○) 4 μM actin, HMM concentration varied.

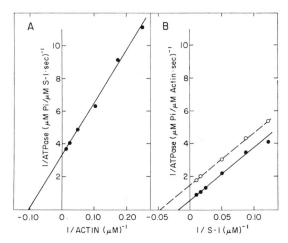

Figure 7. Double-reciprocal plots of data shown in Fig. 6.
A, Double-reciprocal plot of Fig. 6A. *B*, Double-reciprocal
plot of Fig. 6B. (●) 1 μM actin; (○) 4 μM actin.

not seem to be involved in the effects we observe—
very similar effects also occur with S-1.

Although our results with HMM and S-1 agree
qualitatively, an important quantitative difference
does seem to occur. Table 3 shows that, as might

Table 3. Comparison of HMM and S-1 Binding to Actin at
Varied Actin Concentration

Actin Conc. (μM)	HMM		S-1	
	Protein Bound	V_{max}	Protein Bound	V_{max}
11	19%	56%	11%	41%
22	29%	73%	16%	59%
46	45%	85%	22%	78%

Conditions: 11.5 μM HMM or 12.5 μM S-1; 3 mM
MgATP, 2 mM P_i; 3 mM imidazole pH 7, temperature 0.5°;
rotor speed 30,000 rpm.

be expected for both HMM and S-1, as the actin
concentration is increased, there is a corresponding
increase in the amount of actin-bound protein.
However, at 46 μM actin where the ATPases of
both the HMM and S-1 are approaching V_{max}, it
appears that only about half as much S-1 as HMM
sediments with the F-actin, suggesting, as will be
discussed below, that perhaps only one of the two
HMM heads interacts with actin at a time.

Discussion

Ultracentrifuge studies. Using the ultracentrifuge to measure the binding of a slowly
sedimenting molecule to a rapidly sedimenting one
is a standard technique (Steinberg and Schachman,
1966), which in the case of the muscle proteins has
been used successfully to measure the binding of
HMM to F-actin in the absence of ATP (Eisenberg

et al., 1971; Margossian and Lowey, 1972). The
ultracentrifuge studies presented in this paper show
that in the presence of ATP a large fraction of both
HMM and S-1 does not sediment with F-actin under
conditions where kinetic studies indicate that the
actin-activated HMM and S-1 ATPases have nearly
reached their maximum value. The question arises
as to whether some artifact could be causing this
result. It does not appear that any of the protein
remaining in the supernatant is actin, since at
constant actin concentration the amount of free
protein halves when the added HMM or S-1
concentration is halved. Furthermore Table 3
shows that rather than increasing, the free protein
decreases as the actin concentration is increased.
It also appears unlikely that pressure is causing
dissociation of the acto-HMM, because at 30,000
rpm, the pressure head near the meniscus where we
make our measurement is very low (Eisenberg et
al., 1972). Finally, after all of the ATP is hydrolyzed
to ADP and P_i, all of the HMM and S-1 sediment
with the F-actin. Therefore the effect we are
observing must be due to the presence of ATP, and
it is difficult to see what centrifuge artifact could
be caused by ATP but not by ADP. We therefore
conclude from our data that in the presence of
ATP, at any one time, a large fraction of both the
HMM and S-1 are unable to bind to F-actin.
Parenthetically it should be noted that whereas our
control experiments tend to rule out the possibility
that the fraction of free HMM or S-1 is much smaller
than we observe, it may well be larger.

ATPase studies. Unless all of this free protein
is denatured, which seems very unlikely (Eisenberg
et al., 1972), it would appear that during the cycle
of interaction of HMM with actin, the HMM goes
through a relatively long-lived "refractory" state
during which it cannot bind to the F-actin. On the
other hand, an F-actin monomer should always be
able to bind HMM so that, in a sense, an actin
monomer should be able to "juggle" several
activated HMM molecules. Therefore the ATP
turnover rate per mole of actin monomer at very
high HMM or S-1 concentration should be much
faster than the ATP turnover rate per mole of
HMM head at high actin concentration. In fact,
qualitatively this is what we observe. However the
ATPase studies are complicated by the fact that at
high HMM concentration, the turnover rate per
mole of actin itself depends on the actin concentration, increasing with decreasing actin concentration. This is not necessarily unexpected in a
system with more than one kinetic intermediate of
HMM present. However, it precludes quantitative
analysis of the kinetic data at the present time.
It also makes determination of stoichiometry from

this type of data as was previously attempted by one of the authors (E. E., Rizzino et al., 1970) essentially impossible, since the ratio of $V_{\max}$ at infinite actin to that at infinite HMM will vary markedly, depending on the actin concentration and the conditions of the experiment. Despite the quantitative complexity of this data, however, it does show that the ATP turnover rate per mole of actin can be much higher than the turnover rate per mole of HMM head and therefore, at least qualitatively, these ATPase studies are consistent with the results of our centrifuge experiments.

Comparison of HMM and S-1. Comparison of the behavior of S-1 with HMM also suggests that the HMM heads may go through a refractory state during which they cannot bind to F-actin. Figure 4 showed that the ATP turnover rate per mole of HMM was twice that per mole of S-1, suggesting that the HMM heads act independently and can therefore bind simultaneously to the actin. If this is the case, however, it might be expected that the binding of HMM to actin would be manyfold stronger than the binding of S-1 to actin, whereas in fact the apparent binding constants in the presence of ATP only differ by about a factor of two. Therefore the ATPase data suggest that perhaps only one of the HMM heads binds to actin at a time.

This conclusion is supported by our centrifuge data. As was shown in Table 3, at a given actin concentration, roughly twice as much HMM as S-1 sedimented with the F-actin, which is what might be expected if, while one of the two HMM heads interacted with F-actin, the other head remained refractory and was carried down by the actin without actually interacting with it. With S-1 of course, no extra heads in the refractory state would sediment, and therefore for a given level of ATPase less protein would bind. In summary then, we need a kinetic model which explains the following: (a) even when the ATPase nearly equals $V_{\max}$, only a small fraction of the HMM or S-1 binds to F-actin; (b) the heads of the HMM act independently; and (c) apparently only one of the HMM heads interacts with actin at a time.

Kinetic models. In developing a kinetic model to explain this data several lines of evidence must be considered. First, the steady state kinetic studies of Eisenberg and Moos (1970) and Eisenberg et al. (1968) clarified the dual role of ATP, which not only acts as a substrate for the actin-activated myosin ATPase, but is also able to dissociate the actomyosin complex. Their studies showed that, rather than there being separate hydrolytic and "dissociating" sites for ATP on each myosin head, there is a single site, and ATP binding to this site

plays a dual role acting both as a substrate and as a modifier of the actin binding to the myosin.

Second, the pre–steady state studies of Tonomura et al. (1969) showed that ATP bound to myosin can exist in at least two forms, ATP itself and a hydrolyzed intermediate. In an elegant series of experiments Taylor and his associates (Finlayson et al., 1969; Finlayson and Taylor, 1969; Lymn and Taylor, 1970; Taylor et al., 1970) expanded on this work and proposed that ATP on myosin is rapidly hydrolyzed to products in an initial hydrolysis of the ATP, and then in a rate-limiting step, the products are very slowly released.

$$\text{M} + \text{S} \rightleftarrows \text{MS} \xrightarrow{\text{fast}} \text{M} - \text{Product} \xrightarrow{\text{slow}} \text{M} + \text{Product}$$

By then determining with light scattering measurements that the dissociation of acto-HMM by ATP is faster than the initial ATP hydrolysis, Lymn and Taylor (1971) proposed that ATP dissociates actin from HMM, and then, following the rapid hydrolysis of the ATP, actin dissociates the ADP and P_i from the HMM. Therefore the Lymn and Taylor model explains the dual role of ATP by proposing that each of its functions occurs when the ATP exists in a different intermediate state. However the transition between these states is so rapid at $0°$ (more than 15 sec^{-1}) (Lymn and Taylor, 1970) compared to the steady state rate (0.5 sec^{-1}) that the model predicts that, when the acto-HMM ATPase $= V_{\max}$, more than 95% of the HMM will be complexed with actin. To explain our data we must modify this model so that a rate-limiting step occurs prior to the interaction of the HMM with actin. One possibility is that the rate of the initial hydrolysis of ATP is much lower in the steady state than during pre–steady state measurements. In this case we would have:

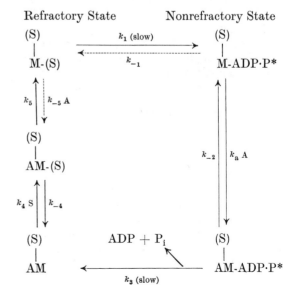

where M = HMM, S = ATP, A = actin, M-(S) = one head of the HMM in the refractory state, and M-ADP·P* = one head of the HMM in the nonrefractory state. Another possibility is that, in fact, a rapid initial hydrolysis of ATP does occur but there is a slow step following this hydrolysis:

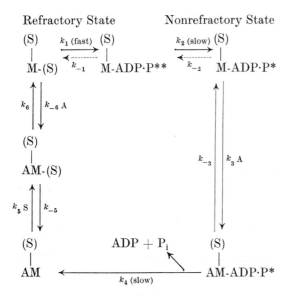

Both of these models are consistent with our findings. At saturating actin concentration both models predict that two species of HMM will be present in significant concentrations: the refractory state of the HMM that cannot bind to actin and the complex of the nonrefractory state with actin. Furthermore both models assume that the two HMM heads act independently. However since an HMM head in the refractory state transforms to the nonrefractory state in a first-order process, as one head of the HMM goes through the cycle the other head will remain refractory simply on a random basis, so that in most cases only one head of the HMM will bind to actin at a time.

In addition to postulating a rate-limiting step occurring on the HMM prior to its interaction with actin, these models also differ from that of Lymn and Taylor in that M-ADP·P* is not considered to be simply myosin with bound product. Presumably in the absence of Ca^{++} with the troponin-tropomyosin complex present, the M-ADP·P* complex would not bind to actin (Eisenberg and Kielley, 1970), whereas if it were simply myosin with bound product, it would form a rigor complex with actin. Therefore M-ADP·P* must differ in some way from myosin in equilibrium with free product. The possibility that myosin intermediates other than bound substrate and bound product occur has in fact been suggested by the pre–steady state ATPase studies of Trentham et al. (1972).

At the present time we have no way of determining which of the two models we have proposed is correct. However if the second model is correct, it suggests that ATP analogs which cannot be hydrolyzed may not be able to dissociate actomyosin completely. Of course it is also quite possible that neither model is correct and the refractory state is due to some conformational change in the myosin which is relatively long-lived and is independent of the nucleotide species present on the myosin. Clearly much more work will be required before the true nature of the refractory state is determined.

Implications for muscle contraction. The concept of a refractory state can easily be adapted to give the simple model of muscle contraction shown in Fig. 8. The key point in this model is that at any one time most of the bridges are off of the actin. Occasionally a bridge in the refractory state (M-ATP) transforms to the nonrefractory state (M-ADP*). It then goes through a cycle of interaction with actin, but after it dissociates, it again remains off of the actin for a relatively long period of time. In this model we have not linked the transition from the refractory to the nonrefractory state with a change in the orientation of the myosin bridge, but of course such a change might well occur.

A model such as this can provide possible explanations for several puzzling in vivo findings. First, it may explain X-ray diffraction studies which show that, in contrast to the situation in rigor muscle, in contracting muscle at any one

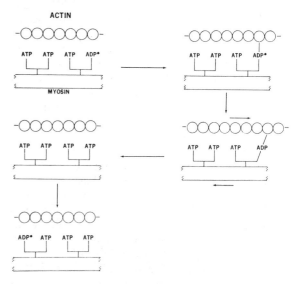

Figure 8. A possible model for muscle contraction based on the assumption that a refractory state occurs in vivo.

time only a small fraction of the myosin bridges are actually binding to actin (Huxley and Brown, 1967; Miller and Tregear, 1970). Second, it may explain the puzzling result that, when muscle contracts at high velocity, it appears that the myosin bridges move along the actin filament much more rapidly than one would expect from the ATP turnover rate measured in vitro (Podolsky and Nolan, 1971). Clearly if a myosin bridge spends a large part of its cycle off of the actin, then the length of time it spends on the actin filament will be much shorter than the total turnover time of the ATP on myosin as measured in vitro. Finally this model may be related to the force-velocity relationship observed for muscle (Hill, 1938). As the velocity increases and the filaments travel past each other at a faster rate, a bridge in the refractory state will, on the average, travel a longer distance between each interaction with actin, and therefore it will exert less force per unit distance. Furthermore because the duration of the refractory state will depend on the myosin ATPase, this model provides a possible explanation for the link observed between myosin ATPase activity and the force-velocity curves of slow and fast muscles (Bárány, 1967). Of course at the present time this type of model building is pure speculation. We only have evidence for a refractory state in vitro at 0°, and much more work will be required before we can extend our findings to the situation in vivo with any confidence whatever.

Acknowledgment

We wish to thank Louis Dobkin for excellent technical assistance.

References

ADELSTEIN, R., J. GODFREY, and W. W. KIELLEY. 1963. G-actin: Preparation by gel filtration and evidence for a double stranded structure. *Biochem. Biophys. Res. Comm.* **12:** 34.

BÁRÁNY, M. 1967. ATPase activity of myosin correlated with speed of muscle shortening, p. 197. In *Contractile processes in macromolecules*, ed. A. Stracher. Little Brown and Co., Boston.

BAROUCH, W. W. and C. MOOS. 1971. Effect of temperature on actin activation of heavy meromyosin ATPase. *Biochim. Biophys. Acta* **234:** 183.

EISENBERG, E. and W. W. KIELLEY. 1970. Native tropomyosin: Effect on the interaction of actin with heavy meromyosin and subfragment-1. *Biochem. Biophys. Res. Comm.* **40:** 50.

EISENBERG, E. and C. MOOS. 1967. The interaction of actin with myosin and heavy meromyosin in solution at low ionic strength. *J. Biol. Chem.* **12:** 2945.

———. 1968. The adenosine triphosphatase activity of acto-heavy meromyosin. A kinetic analysis of actin activation. *Biochemistry* **7:** 1486.

———. 1970. Actin activation of heavy meromyosin adenosine triphosphatase: Dependence on ATP and actin concentration. *J. Biol. Chem.* **245:** 2451.

EISENBERG, E., L. DOBKIN, and W. W. KIELLEY. 1971. Binding of actin to heavy meromyosin in the absence of ATP. *Fed. Proc.* **30:** 1310 (Abstr.).

———. 1972. Heavy meromyosin: Evidence for a refractory state unable to bind to actin in the presence of ATP. *Proc. Nat. Acad. Sci.* **69:** 667.

EISENBERG, E., C. R. ZOBEL, and C. MOOS. 1968. Subfragment-1 of myosin: ATPase activation by actin. *Biochemistry* **7:** 3186.

FINLAYSON, B. and E. W. TAYLOR. 1969. Hydrolysis of nucleoside triphosphates by myosin during the transient state. *Biochemistry* **8:** 802.

FINLAYSON, B., R. W. LYMN, and E. W. TAYLOR. 1969. Studies on the kinetics of formation and dissociation of the actomyosin complex. *Biochemistry* **8:** 811.

HILL, A. V. 1938. The heat of shortening and the dynamic constants of muscle. *Proc. Roy. Soc. (London) B* **136:** 1967.

HUXLEY, A. F. and R. NIEDERGERKE. 1954. Interference microscopy of living muscle fibres. *Nature* **173:** 971.

HUXLEY, H. E. 1969. The mechanism of muscular contraction. *Science* **164:** 1356.

HUXLEY, H. E. and W. BROWN. 1967. The low-angle X-ray diagram of vertebrate striated muscle and its behaviour during contraction and rigor. *J. Mol. Biol.* **30:** 383.

HUXLEY, H. E. and J. HANSON. 1954. Changes in the cross-striations of muscle during contraction and stretch and their structural interpretation. *Nature* **173:** 973.

KIELLEY, W. W. and W. F. HARRINGTON. 1960. A model for the myosin molecule. *Biochim. Biophys. Acta* **41:** 401.

LOWEY, S., H. S. SLAYTER, A. G. WEEDS, and H. BAKER. 1969. Substructure of the myosin molecule. I. Subfragments of myosin by enzymic degradation. *J. Mol. Biol.* **42:** 1.

LYMN, R. W. and E. W. TAYLOR, 1970. Transient state phosphate production in the hydrolysis of nucleoside triphosphates by myosin. *Biochemistry* **9:** 2975.

———. 1971. Mechanism of adenosine triphosphate hydrolysis by actomyosin. *Biochemistry* **10:** 4617.

MARGOSSIAN, S. S. and S. LOWEY. 1972. Substructure of the myosin molecule. IV. Interactions of myosin and its subfragments with ATP and F-actin. *J. Mol. Biol.* In press.

MILLER, A. and R. T. TREGEAR. 1970. Evidence concerning crossbridge attachment during muscle contraction. *Nature* **226:** 1060.

MUELLER, H. 1964. Molecular weight of myosin and meromyosin by Archibald experiments performed with increasing speed of rotation. *J. Biol. Chem.* **239:** 797.

PODOLSKY, R. J. and A. C. NOLAN. 1971. Cross-bridge properties derived from physiologic studies of frog muscle fibers, p. 247. In *contractility of muscle cells and related processes*, ed. R. J. Podolsky. Prentice-Hall, Englewood Cliffs, New Jersey.

REES, M. K. and M. YOUNG. 1967. Studies on the isolation and molecular properties of homogeneous globular actin: Evidence for a single polypeptide chain structure. *J. Biol. Chem.* **242:** 4449.

RIZZINO, A. A., W. W. BAROUCH, E. EISENBERG, and C. MOOS. 1970. Actin-heavy meromyosin binding. Determination of binding stoichiometry from ATPase kinetic measurements. *Biochemistry* **9:** 2402.

SPUDICH, J. A. and S. WATT. 1971. The regulation of rabbit skeletal muscle contraction. I. Biochemical studies of the interaction of the tropomyosin-troponin complex with actin and the proteolytic fragments of myosin. *J. Biol. Chem.* **246:** 4866.

STEINBERG, I. Z. and H. K. SCHACHMAN. 1966. Ultra-centrifugation studies with absorption optics. V. Analysis of interacting systems involving macromolecules and small molecules. *Biochemistry* **12:** 3728.

TAYLOR, E. W., R. W. LYMN, and G. MOLL. 1970. Myosin-product complex and its effect on the steady-state rate of nucleoside triphosphate hydrolysis. *Biochemistry* **9:** 2984.

TONOMURA, Y., H. NAKAMURA, N. KINOSHITA, H. ONISHI, and M. SHIGEKAWA. 1969. The pre-steady state of the myosin-ATP system. X. The reaction mechanism of the myosin-ATP system and a molecular mechanism of muscle contraction. *J. Biochem.* **66:** 599.

TRENTHAM, D. R., R. G. BARDSLEY, J. F. ECCLESTON, and A. G. WEEDS. 1972. Elementary processes of the magnesium ion-dependent adenosine triphosphatase activity of heavy meromyosin: A transient kinetic approach to the study of kinases and adenosine triphosphatases and a colorimetric inorganic phosphate assay in situ. *Biochem. J.* **126:** 635.

YOUNG, D. M., S. HIMMELFARB, and W. F. HARRINGTON. 1965. On the structural assembly of the polypeptide chains of heavy meromyosin. *J. Biol. Chem.* **240:** 2428.

Possible Differentiation between Rigor Type Interactions of the Two Myosin Heads with Actin

M. C. Schaub and J. G. Watterson

Institute of Pharmacology, University of Zurich, 8006 Zurich, Switzerland

It is now generally accepted that the myosin molecule consists in part of two globular units, the heads (Slayter and Lowey, 1967; Lowey et al., 1969), both of which contain one hydrolytic site (Schliselfeld and Barany, 1968; Nauss et al., 1969; Murphy and Morales, 1970). These units are thought to be equivalent and independent in their hydrolytic activity since the measured specific activity of systems in which the heads, heavy meromyosin subfragment-1, are detached from the rest of the molecule is twice that of systems containing myosin or heavy meromyosin (Lowey et al., 1969). In contrast some differences between the two heads might be expected since analytical investigations on the myosin molecule indicate variations in amino acid composition of both the heavy and light chains, although it has not yet been conclusively demonstrated that these variations are to be found within the one molecule or from one molecule to another (Johnson and Perry, 1970; Hardy et al., 1970; Weeds and Lowey, 1971).

For contraction however, where actin directly influences the Mg^{++}-activated ATPase, there must exist interactions between the two myosin heads and actin. These interactions may be different from, and even dependent on, one another at least at some time during the contraction cycle. It has been suggested that in order for the superprecipitation reaction to take place, interdependent reactions of the two myosin heads with actin are necessary (Tokiwa and Morales, 1971). These workers used a nucleotide affinity label which binds covalently with myosin at the active sites, producing systems containing a mixture of the three unlabeled, half, and totally labeled, molecular types. The statement that half-labeled molecules alone cannot initiate the process was not unambiguously demonstrated by their experiments. However it is known that similar ATP analogs in conjunction with divalent metal ions are able to produce dissociation of actomyosin (AM), so that superprecipitation would not be expected to occur if both heads were labeled (Yount et al., 1971). On the other hand it has been reported that the reaction does proceed when the enzymic sites of both heads have been blocked with thiol reagents such as N-ethylmaleimide (NEM) and p-chloro-

mercuribenzoate (Sekine and Yamaguchi, 1966). This may indicate that a rigor type interaction between actin and myosin, which persists even in a nucleotide-free high salt medium, is first necessary before the superprecipitation reaction can be initiated.

We investigated this rigor state of AM with the help of NEM to detect possible differences between the interactions of the two myosin heads with actin. Since thiol groups are known to influence the enzymic sites of myosin, such differences would be indicated by changes in the myosin and actin-modified myosin ATPase activities. The desensitized actomyosin used in this study was extracted from myofibrils as the intact complex, which has been purified to remove the regulatory proteins (Schaub et al., 1967), and is known to have an actin to myosin stoichiometric ratio of near 2:1 on the basis of 3-methylhistidine analyses (S. V. Perry and C. I. Harris, personal communication). Although a considerably higher ratio than this was found to be needed for attachment of both myosin heads to actin when heavy meromyosin and actin were combined in vitro (Young, 1967; Szentkiralyi and Oplatka, 1969; Tawada, 1969), it will be shown here that both heads are interacting with actin in desensitized actomyosin.

Methods

Purified, desensitized actomyosin free of regulatory proteins (Schaub et al., 1967; Schaub and Perry, 1969) and myosin (Trayer and Perry, 1966) were prepared from mixed rabbit skeletal muscle. Alkylation was carried out on 3.0–3.5 mg of protein per ml in 50 mM Tris-HCl pH 7.6 at 0°C for 10 min, under various conditions and NEM concentrations as indicated in the text. To remove excess reagents the protein was diluted 10 times with cold water, centrifuged for 20 min at 25,000 g and resuspended in 10 mM Tris-HCl pH 7.6. ATPase activity was assayed (Schaub and Perry, 1969) on 0.15–0.30 mg of protein per ml at 25°C in 50 mM Tris-HCl pH 7.6, 2.5 mM Tris-ATP in the presence of 2.5 mM $MgCl_2$ or 1 M KCl and 10 mM EDTA as indicated. On the average the specific activity of different AM preparations was for the Mg^{++}-activated ATPase 0.52 and for the K^+-activated ATPase 0.86 μmoles of phosphorus released per mg of protein per min. Viscometric determinations were made in an Ubbelohde type viscometer with an outflow time for 1 M KCl at 8°C of 2 min 10 sec. Superprecipitation was recorded at 360 nm automatically under identical conditions as used for the assessment of Mg^{++}-activated ATPase of AM. Protein concentrations were determined by a micro-method involving nesslerization (Strauch, 1965).

Results

The influence of the presence of salt and Mg-pyrophosphate during incubation with NEM is shown in Fig. 1. Mg-pyrophosphate and salt dissociate the AM complex, and as can be seen, the smallest concentration of NEM used causes a decrease in the K^{+}-activated ATPase. The dependence of this decrease with increasing NEM concentrations exhibits the same behavior as that of the K^{+}-activated ATPase of myosin alone, pretreated in the presence of 1 M KCl. Little effect on the hydrolysis rate of ATP is found when AM is treated in the absence of salt for this concentration range of NEM. On the other hand the presence of salt allows the destruction of the active site during NEM treatment with concentrations above 50 μM. However the role played by this medium is in some degree specific since the transition from the fully active to inactivated state of the enzyme does not indicate a direct dependence on NEM concentration. In fact the shape of the curve would suggest a leveling out in the range of 80–150 μM NEM.

The effect of increasing amounts of NEM on both enzymic activities of AM was examined under

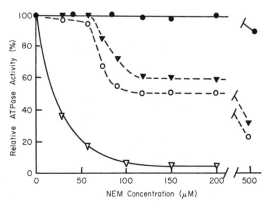

Figure 2. Effect of pretreatment with NEM of actomyosin on the Mg^{++}-activated and the K^{+}-activated ATPase activities. ATPase assays were carried out under standard conditions with 2.5 mM ATP at pH 7.6 in 2.5 mM MgCl$_2$ or in 1 M KCl and 10 mM EDTA after pretreatment with NEM under various conditions and subsequent removal of excess reagents as described in methods. (●) Pretreatment with NEM in 5 mM CaCl$_2$ only; (▼, ○) pretreatment with NEM in 1 M KCl and 5 mM CaCl$_2$; (▽) pretreatment with NEM in 1 M KCl and 5 mM Ca-pyrophosphate. Full symbols = Mg^{++}-activated ATPase; open symbols = K^{+}-activated ATPase.

various conditions. 10 mM EDTA instead of MgCl$_2$, 10 mM EDTA and 5 mM Na-pyrophosphate in the presence of 1 M KCl, or salt alone produced a similarly shaped curve, again suggesting a stepwise inactivation of both types of ATPase activities. High salt concentration, however, proved to be essential to produce this effect. The most clearly defined plateau occurred when 5 mM CaCl$_2$ and 1 M KCl were employed during the alkylation (Fig. 2). For NEM concentrations higher than 60 μM the K^{+}-activated ATPase drops sharply from its full activity to a level of 50 % and remains constant for increasing NEM concentrations to above 200 μM. The Mg^{++}-activated ATPase again behaves in a closely parallel manner. As with Mg^{++}, Ca^{++} in conjunction with pyrophosphate and salt brings about full dissociation of AM, indicated by the immediate drop in the K^{+}-activated ATPase. Figure 2 also shows that the Mg^{++}-activated ATPase is not affected by the alkylating agent in the absence of salt.

Under dissociating conditions the AM solution has virtually the same viscosity as myosin alone in 1 M KCl (Fig. 3). However in the absence of pyrophosphate, 5 mM of Ca^{++} or Mg^{++} in the presence of 150 μM NEM and 1 M KCl in no way affects the higher viscosity of undissociated AM in 1 M KCl. Furthermore no drop in the viscosity of the AM solution was detected by the additional presence of either 10 mM EDTA or 10 mM EDTA + 5 mM Na-pyrophosphate. Hence all those conditions which led to a stepwise inactivation of the enzymic activities of AM by alkylation do not produce a detectable physical dissociation.

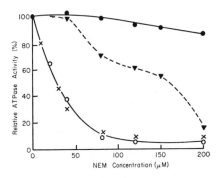

Figure 1. Effect of pretreatment with NEM of actomyosin and myosin on the K^{+}-activated ATPase activity. ATPase assays were carried out under standard conditions in 1 M KCl and 10 mM EDTA with 2.5 mM ATP at pH 7.6 after pretreatment with NEM under various conditions and subsequent removal of excess reagents as described in methods. (●) AM pretreated with NEM in the presence of 0.5 mM MgCl$_2$ only; (▼) AM pretreated with NEM in 1 M KCl and 0.5 mM MgCl$_2$; (○) AM pretreated with NEM in 1 M KCl and 5 mM Mg-pyrophosphate; (×) myosin alone pretreated with NEM in 1 M KCl.

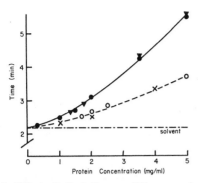

Figure 3. Effect of alkylating conditions on viscosity of actomyosin solutions. Outflow times were measured at different protein concentrations in 1 M KCl and 50 mM Tris-HCl pH 7.6 at 8°C. (●) AM alone; (▼) AM in the presence of 5 mM $CaCl_2$ and 150 μM NEM; (○) AM in the presence of 5 mM Mg-pyrophosphate; (×) myosin alone.

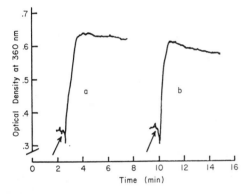

Figure 4. Time course of superprecipitation of control and of partially alkylated actomyosin. Superprecipitation was carried out on 0.32 mg of AM per ml in 2.5 mM $MgCl_2$ and the reaction started by the addition of 2.5 mM ATP, indicated by the arrows. (a) AM, the K^+-activated ATPase of which was reduced to 50% by pretreatment with 150 μM NEM in 1 M KCl and 5 mM $CaCl_2$ and subsequent removal of excess reagents; (b) control AM.

The effect of various conditions of NEM treatment on the superprecipitation reaction of AM was also investigated. The first plot of Fig. 4 illustrates the behavior of an AM pretreated under conditions corresponding to the plateau region of Fig. 2, and the second illustrates that of a similarly treated, but not alkylated, AM. The extent of the superprecipitation is seen to be almost identical, although the rate of the reaction in the case of the alkylated sample is somewhat slower than that of the control.

Discussion

The NEM treatment of myosin alone, or of AM under conditions when actin and myosin are fully dissociated from one another, leads to a considerable decline in the enzymic activity at the smallest NEM concentration used. Calculations based on molecular weight values of 500,000 and 45,000 for myosin and actin, respectively, show that this concentration corresponds to about 2 moles of NEM per mole of myosin or one NEM molecule per head. This is in agreement with reports that blocking of only two thiol groups per myosin molecule causes inactivation of the K^+-activated ATPase, while the Ca^{++}-activated ATPase of myosin is simultaneously activated (Sekine et al., 1962; Sekine and Kielley, 1964). We also observed a threefold activation of this latter reaction under conditions where there is no K^+-activated ATPase. Treatment of AM in the gel state, i.e., in the absence of salt, reveals that actin protects the important thiol group from reaction with NEM. That the Mg^{++}-activated ATPase also remains unaffected implies that the same thiol group or groups are also necessary for the functioning of the Mg^{++}-activated reaction when the ionic forces involved in the overall interaction between actin and myosin are in effect.

In the presence of salt and Ca^{++}, however, treatment with increasing NEM concentrations produces a stepwise inactivation of both ATPases, as indicated by the well-defined plateau. The protein unit of one myosin and two actins may react first with about eight NEM molecules without effect. Although the ionic interactions are suppressed in the presence of salt, as known from the lack of activation by Mg^{++} ions, the hydrophobic interaction of the heads with actin generate a conformation of the protein in which a number of thiol groups become more readily available for reaction with NEM than those which influence both ATPase reactions. In contrast, under conditions where both types of interaction are absent, namely, AM in a fully dissociating medium or myosin alone, the latter essential thiol groups represent the most reactive ones.

The drop to the 50% activity level is brought about by only 2–4 more molecules of NEM in addition to those which had no effect. The three following alternatives are possible explanations of this intermediate level: (1) A fraction of about half the myosin molecules are totally inactivated; (2) the turnover rate of each site is reduced by a factor of about half; (3) one site of each myosin molecule is inactivated. As myosin is totally inactivated under dissociating conditions, alkylation of half the myosin molecules, as suggested in the first alternative, would cause a detectable drop in viscosity, which was not found. Considering the remaining alternatives, the third is favored by the independent findings that first, each head has one hydrolytic site, and second, the K^+-activated ATPase of this site is inactivated by the blocking of only one thiol group.

Assuming that one site only per myosin molecule is destroyed implies that the corresponding thiol group of the other site remains protected over the whole plateau region where the unit of one myosin and two actins is subjected to an additional 16–18 NEM molecules. It may therefore be concluded that under these conditions of NEM treatment the two heads show a difference in their rigor type interaction with actin, and that both heads are attached as shown by the lack of effect of the first eight NEM molecules.

References

HARDY, M. F., C. I. HARRIS, S. V. PERRY, and D. STONE. 1970. Occurrence and formation of the N^ε-methyllysines in myosin and the myofibrillar proteins. *Biochem. J.* **120**: 653.

JOHNSON, P. and S. V. PERRY. 1970. Biological activity and the 3-methylhistidine content of actin and myosin. *Biochem. J.* **119**: 293.

LOWEY, S., H. S. SLAYTER, A. G. WEEDS, and H. BAKER. 1969. Substructure of the myosin molecule. I. Subfragments of myosin by enzymic degradation. *J. Mol. Biol.* **42**: 1.

MURPHY, A. J. and M. F. MORALES. 1970. Number and location of adenosine triphosphatase sites of myosin. *Biochemistry* **9**: 1528.

NAUSS, K. M., S. KITAGAWA, and J. GERGELY. 1969. Pyrophosphate binding to an adenosine triphosphatase activity of myosin and its proteolytic fragments. Implications for the substructure of myosin. *J. Biol. Chem.* **244**: 755.

SCHAUB, M. C. and S. V. PERRY. 1969. The relaxing protein system of striated muscle. Resolution of the troponin complex into inhibitory and calcium ion-sensitizing factors and their relationship to tropomyosin. *Biochem. J.* **115**: 993.

SCHAUB, M. C., D. J. HARTSHORNE, and S. V. PERRY. 1967. The adenosine-triphosphatase activity of desensitized actomyosin. *Biochem. J.* **104**: 263.

SCHLISELFELD, L. H. and M. BARANY. 1968. The binding of adenosine triphosphate to myosin. *Biochemistry* **7**: 3206.

SEKINE, T. and W. W. KIELLEY. 1964. The enzymatic properties of N-ethylmaleimide-modified myosin. *Biochim. Biophys. Acta* **81**: 336.

SEKINE, T. and M. YAMAGUCHI. 1966. Superprecipitation of actomyosin reconstructed with F-actin and NEM-modified myosin. *J. Biochem.* (Tokyo) **59**: 195.

SEKINE, T., L. M. BARNETT, and W. W. KIELLEY. 1962. The active site of myosin adenosine-trophosphatase. 1. Localization of one of the sulphydryl groups. *J. Biol. Chem.* **237**: 2769.

SLAYTER, H. S. and S. LOWEY. 1967. Substructure of the myosin molecule as visualized by electron microscopy. *Proc. Nat. Acad. Sci.* **58**: 1611.

STRAUCH, L. 1965. Ultramikro-Methode zur Bestimmung des Stickstoffes in biologischem Material. *Z. klin. Chem.* **3**: 165.

SZENTKIRALYI, E. M. and A. OPLATKA. 1969. On the formation and stability of the enzymically active complexes of heavy meromyosin with actin. *J. Mol. Biol.* **43**: 551.

TAWADA, K. 1969. Physicochemical studies of F-actin-heavy meromyosin in solutions. *Biochim. Biophys. Acta* **172**: 311.

TOKIWA, T. and M. F. MORALES. 1971. Independent and cooperative reactions of myosin heads with F-actin in the presence of adenosine triphosphate. *Biochemistry* **10**: 1722.

TRAYER, I. P. and S. V. PERRY. 1966. The myosin of developing skeletal muscle. *Biochem. Z.* **345**: 87.

WEEDS, A. G. and S. LOWEY. 1971. Substructure of the myosin molecule. II. The light chains of myosin. *J. Mol. Biol.* **61**: 701.

YOUNG, M. 1967. Studies on the structural basis of the interaction of myosin and actin. *Proc. Nat. Acad. Sci.* **58**: 2393.

YOUNT, R. G., D. OJALA, and D. BABCOCK. 1971. Interaction of P-N-P and P-C-P analogs of adenosine triphosphate with heavy meromyosin, myosin and actomyosin. *Biochemistry* **13**: 2490.

A Proposal for the Mechanism of Contraction in Intact Frog Muscle

Michael Bárány and Kate Bárány

Departments of Contractile Proteins and Physical Chemistry, Institute for Muscle Disease,
New York, New York 10021

In the past decade considerable knowledge has been accumulated about the structure and enzymic activity of myosin in vitro. Identification of the globular head of the myosin molecule as the cross-bridge in the structure of muscle made possible studies on how myosin functions in the living muscle. Huxley and Brown (1967) and Tregear and Miller (1969) showed by X-ray diffraction that the cross-bridges change their position during muscle contraction. Furthermore, Huxley (1968) has found that an amount of material equal to about 30% of the original mass of the thick filaments is transferred to the vicinity of the thin filaments when a muscle passes into rigor. These findings give strong support to the idea that during contraction the myosin and actin filaments interact with each other.

The resolution of the X-ray data is not high enough to detect changes within the individual cross-bridges or within the various parts of the myosin molecule in the muscle. Therefore, a chemical method has been worked out in this laboratory to study structural changes in myosin during contraction of frog muscle. This is based on a change in the reactivity of myosin with 1-fluoro-2,4-dinitro-[³H]benzene (FDNB) between contracting and resting muscle (Bárány and Bárány, 1970). The incorporation of the reagent into myosin was decreased during isotonic contraction; in contrast, no difference was found in the incorporation into the sarcoplasmic proteins of contracting and resting muscle. Subsequent studies revealed that other reagents were capable of reproducing the FDNB effect (Bárány et al., 1971). On account of its high incorporation into myosin, N-ethyl-1-[¹⁴C]maleimide ([¹⁴C]NEM) was used to determine the distribution of label in the substructure of myosin during muscle contraction. The results indicated that the change in reactivity during contraction is localized to the cross-bridges, whereas the backbone of the filament and its linkage with the cross-bridges remain unaffected. Analysis of radioactive peptides, derived from myosin of contracting and resting muscle, gave results which were compatible with the idea that during contraction a con-

formational change occurs in a certain area of the cross-bridges rendering this area inaccessible for reaction with [¹⁴C]NEM.

This paper describes the incorporation of [¹⁴C]NEM into myosin of functionally different muscles. The interaction of actin, myosin, and ATP was modified in the intact muscle: (1) Actin was prevented from interacting with myosin by stretching the muscle to 140% rest length, so that no tension was developed upon stimulation. (2) ATP was eliminated by inducing rigor with iodoacetate, thus allowing maximal combination of the actin filaments with the cross-bridges. (3) Both actin and ATP were blocked by treatment of the muscle at 140% rest length with iodoacetate. By comparing the reactivity of myosin with [¹⁴C]NEM in the modified muscles with that of normal muscles, the requirements for the changed reactivity of myosin during contraction could be established. These experiments naturally raised the question of what kind of nucleotide is bound to myosin in the resting muscle. Finally, we describe the reactivity of [¹⁴C]NEM with various proteins, other than myosin, during contraction. These studies have led to a proposal for the mechanism of contraction in the intact frog muscle.

Reactivity of [¹⁴C]NEM with Myosin of Functionally Different Muscles

Normal contractions. Figure 1 shows the experimental set-up. For isotonic twitches the semitendinosus muscles were stimulated 25 times per minute with a duration time of 1 msec at 25 V (as indicated by the dial giving the open circuit output of the Model 104-A stimulator of the American Electronic Laboratories Inc.), under a load of 1.5 g, in oxygenated Ringer's solution at 1°. Under these conditions about 0.5 sec was required to reach the maximal shortening; thus, the stimulated muscles have spent about 21% of the total time in the shortening phase of the contraction cycle. After equilibration with normal Ringer's solution, the muscles were treated with 70 ml of a Ringer's solution containing 0.20 mM [¹⁴C]NEM (630,000 cpm/μmole) under oxygen at

Figure 1. The general set up for the muscle experiments. This picture illustrates the paired semitendinosus muscles of the frog (*Rana pipiens*). The two pelvic tendons of the muscles were knotted with suture and tied to a hook attached to a lever. The tibial tendon was tied to a plastic block, and the block was screwed to the chamber so that both heads of the muscle were facing the two platinum stimulating electrodes (Sandow, 1947). The length of the muscle could be changed with the fine screw adjustment of the Palmer stand connected to the lever. The weight of the semitendinosus muscles used in these experiments ranged from 120–150 mg, and the length varied from 28–33 mm. For other details see Bárány and Bárány (1970).

1° for 30 min. During the [^{14}C]NEM treatment one muscle was stimulated, as described above, while the other muscle was resting. At the end of the experiment, unreacted [^{14}C]NEM was removed from the muscles by washing with 5 batches of 70 ml normal Ringer's solution at 1°. Finally, myosin was isolated from the muscles and the incorporated N-ethyl-1-[^{14}C]succinimide ([^{14}C]NES) determined.

The first experiment in Table 1 shows typical results for the incorporation of [^{14}C]NEM into the myosin of these muscles. The incorporation in the stimulated muscle was decreased to 81% that of the resting one. In order to account for the total myosin, the myofibrils of the same muscles were isolated. This was based on the fact, to be described later, that [^{14}C]NEM reacts insignificantly with actin. As may be seen from the data of experiment

Table 1. Incorporation of [^{14}C]NEM into Myosin and Myofibrils of Frog Semitendinosus Muscle in Various Functional States

Expt. No.	Type of Experiment	Length of Muscle (%)	Myosin Experimental	Resting	Myofibrils Experimental	Resting
			(moles [^{14}C]NES/500,000 g Protein)			
1	Stimulated vs. resting	100	0.94	1.16	0.75	0.94
2	Stimulated vs. resting	140	1.23	1.22	1.00	1.02
3	Iodoacetate-treated vs. resting	100	1.16	1.25	0.92	0.97
4	Iodoacetate-treated vs. resting	140	1.58	1.19	1.27	1.03

Myosin and myofibrils were prepared from the same muscles. To isolate enough protein the corresponding muscles (about 130 mg each) from two experiments were pooled. The washed muscles were chopped with scissors and extracted with 100 ml of 0.04 M KCl, pH 7.0, in the cold room for 1 hr. After centrifugation at 33,000 g, the supernatant was discarded and the residue was blended with additional 100 ml of 0.04 M KCl at 4° for 3 min. The residue obtained after centrifugation was found to consist essentially of myofibrils. An aliquot of the residue was dissolved in 0.02 M HCl by homogenization and was used to determine protein by the biuret method and radioactivity by liquid scintillation counting. The remaining part of the residue was extracted with 50 ml of a solution containing 0.3 M KCl, 0.15 M potassium phosphate buffer, pH 6.6, and 0.01 M ATP at 4° for 20 min, then centrifuged, and the myosin in the supernatant was precipitated by dialysis against 200 vol 0.02 M KCl pH 7 overnight. The myosin precipitate was collected, dissolved in a solution of 0.6 M KCl, 0.05 M Tris-HCl pH 7.4, 0.01 M MgSO$_4$, and 0.01 M ATP at 2.0–2.5 mg protein/ml, and centrifuged at 226,400 g for 2 hr. The myosin in the supernatant was precipitated with 5% trichloroacetic acid (TCA), washed with 0.1% TCA, and dissolved in 0.02 M HCl; protein and radioactivity was determined from this solution.

No. 1 in Table 1, the incorporation into the fibrils followed the same pattern as that into the myosin; i.e., the [14C]NES content of fibrils from contracting muscle was 80% that from resting muscle.

On the average, a 21% decrease was found in the incorporation of [14C]NEM into the myosin of isotonically contracting semitendinosus muscles, as compared to the resting state. This is shown in the first row of Table 2.

Ringer's solution in the bath was exchanged for a Ringer's solution containing 0.20 mM [14C]NEM. Both muscles were left in this solution under nitrogen at 1° for 30 min; subsequently, the muscles were washed with normal Ringer's solution to remove excess [14C]NEM.

The full development of rigor, after treatment with iodoacetate for 60 min, was evidenced by the absence of mechanical response upon tetanic

Table 2. Changes in Incorporation of [14C]NEM into Myosin of Functionally Different Semitendinosus Muscle as Compared to Resting Muscle

Expt. No.	Type of Experiment	Length of Muscle (%)	Inducer of Change	% of Change	No. of Determinations
1	Stimulated vs. resting	100	Actin and ATP	−20.9 ± 3.4	45
2	Stimulated vs. resting	140	ATP	−0.3 ± 2.9	46
3	Iodoacetate-treated vs. resting	100	Actin	−6.7 ± 1.5	44
4	Iodoacetate-treated vs. resting	140	None	+19.7 ± 1.9	45

The 100% incorporation value, expressed as mole [14C]NES/500,000 g myosin of the resting muscle, was in the range 0.95–1.48. Percentage change is given with ± standard error of means.

In several experiments, we measured the incorporation into myosin of muscles which contracted isometrically. There was a 9% decrease in the incorporation during this contraction. The isometric contraction time of these muscles was about 0.2 sec; thus, with a stimulation frequency of 25 times per minute, these muscles spent about 8% of the total time in contraction.

Stretched and stimulated muscle. The muscles were stretched to 140% rest length, so they did not develop tension upon stimulation. It is known from the work of Gordon et al. (1966a,b) that in semitendinosus muscle fibers at 140% length there is no significant overlap between actin and myosin filaments. The stretched muscles were treated with [14C]NEM, as described before under "Normal contractions"; during the treatment one of the paired muscles was stimulated 25 times per minute.

Myosin and myofibrils isolated from such stretched and stimulated muscle incorporated [14C]NEM to the same extent as did those from stretched and resting muscle (Expt. No. 2 in Tables 1 and 2). Therefore, these data indicate that actin plays a role in eliciting the change in the reactivity of myosin with [14C]NEM.

Rigor. One of the paired muscles was incubated in 70 ml of Ringer's solution containing 1.0 mM iodoacetate under nitrogen at 25° for 60 min, while the other muscle was incubated under identical conditions but in the absence of the reagent. To remove iodoacetate the muscle was washed 5 times; the control muscle was washed identically. Then both muscles were chilled to 1° and the normal

stimuli, by the tension produced, and by the stiffness of the muscle. Furthermore, nucleotide determinations showed that the ATP content of the iodoacetate-treated muscles was reduced to 6% that of control muscles (cf. Expt. No. 2 with No. 1 in Table 3), and that the sum of IMP and inosine in the iodoacetate-treated muscles was 77% of the total nucleotide and nucleoside (Expt. No. 2 in Table 3).

Although iodoacetate is an SH group reagent, it does not react appreciably with myosin in the muscle. Thus, treatment of the muscle with [3H]-iodoacetate at 25° for 60 min resulted only in 0.03–0.04 mole of 3H-label per 500,000 g of myosin. In contrast, the same amount of sacroplasmic proteins contained 1.60–1.76 moles of [3H]acetate.

The incorporation of [14C]NEM into myosin and myofibrils of muscle in rigor was slightly decreased, as compared to the resting state (Expt. No. 3 in Table 1). The difference was found to be 7% (Expt. No. 3 in Table 2). The ATP content of the iodoacetate and NEM-treated muscles was 4.6% that of muscles treated with NEM alone (cf. Expts. No. 3 with No. 4 in Table 3). Therefore, these results show that in the absence of ATP, a decrease in reactivity of myosin with [14C]NEM takes place.

Stretched and ATP-depleted muscle. Both actin and ATP were eliminated from the interaction with myosin by treatment of the muscle at 140% rest length with 1.0 mM iodoacetate. The experimental conditions were otherwise the same as described above for the muscle in rigor. The ATP content of the stretched and iodoacetate-treated or iodoacetate and NEM-treated muscles was 5% and

Table 3. Nucleotide and Nucleoside Content of Semitendinosus Muscles after Various Treatments

Expt. No.	Treatment	Length of Muscle (%)	μmole/g muscle						
			ATP	ADP	AMP	Adenosine	IDP	IMP	Inosine
1	None	100	2.25 ± 0.12	0.46 ± 0.03	0.17 ± 0.01	0.06 ± 0.01	0.03 ± 0.01	0.16 ± 0.02	0.12 ± 0.03
2	Iodoacetate	100	0.13 ± 0.02	0.27 ± 0.02	0.19 ± 0.02	0.09 ± 0.02	0.07 ± 0.01	1.43 ± 0.19	1.14 ± 0.15
3	Iodoacetate and NEM	100	0.10 ± 0.03	0.25 ± 0.02	0.20 ± 0.02	0.08 ± 0.02	0.08 ± 0.01	1.50 ± 0.18	1.08 ± 0.16
4	NEM	100	2.18 ± 0.14	0.45 ± 0.02	0.15 ± 0.01	0.07 ± 0.01	0.02 ± 0.01	0.20 ± 0.02	0.16 ± 0.03
5	None	140	2.29 ± 0.10	0.49 ± 0.03	0.14 ± 0.02	0.05 ± 0.01	0.01 ± 0.01	0.18 ± 0.03	0.14 ± 0.02
6	Iodoacetate	140	0.11 ± 0.03	0.23 ± 0.01	0.16 ± 0.02	0.08 ± 0.02	0.06 ± 0.01	1.56 ± 0.17	1.11 ± 0.16
7	Iodoacetate and NEM	140	0.09 ± 0.03	0.23 ± 0.02	0.18 ± 0.02	0.07 ± 0.01	0.07 ± 0.02	1.48 ± 0.20	1.17 ± 0.18
8	NEM	140	2.20 ± 0.11	0.45 ± 0.03	0.18 ± 0.02	0.06 ± 0.01	0.04 ± 0.01	0.15 ± 0.02	0.10 ± 0.02

Values are expressed as μmole/g muscle and give the average of 4 determinations with ± standard error of means. All treatments were performed under nitrogen; 1.0 mM iodoacetate was used at 25° for 60 min and 0.20 mM NEM at 1° for 30 min. At the end of the treatment the muscles were washed 3 times with Ringer's solution at 1°, then dropped into 25 ml of 1-butanol at 0° and blended in this solvent, in ice, with the turbo-shear assembly of the Virtis homogenizer Model 45. Four pairs of muscles were used for each determination; the homogenates were kept at 0° until all four muscles had been worked up. The butanol was then removed at reduced pressure in a Virtis freeze-dryer and the residue extracted with 15 ml of 2% perchloric acid at 0°. The extract was centrifuged, the supernatant adjusted to pH 10.0–10.5 with 5 M KOH, and the residue extracted with 40 ml 0.05 M NaOH at 25° overnight to determine the noncollagenous protein content of the muscles (Lilienthal et al., 1950). The supernatant was left at 0° for 1 hr, centrifuged to remove the potassium perchlorate, then applied on a Dowex 1 column and chromatographed (Bárány et al., 1960). From the noncollagenous protein content the weight of the muscles was calculated, using a determined amount of 180.7 mg noncollagenous protein per gram of fresh semitendinosus muscle.

4%, respectively, compared to that of control muscles (cf. Expts. No. 6 with No. 5 and No. 7 with No. 8 in Table 3). Furthermore, the accumulation of IMP and inosine was as high as 80% of the total nucleotide and nucleoside (Expts. No. 6 and 7 in Table 3).

Unexpectedly, the incorporation of [¹⁴C]NEM into the myosin or myofibrils of the stretched and ATP-depleted muscle was increased, compared to that of the ATP-containing muscle (Expt. No. 4 in Table 1). Quantitation of this increase with myosin gave a value of 20% (Expt. No. 4 in Table 2). The reason for the increased incorporation will be described subsequently.

Nature of Nucleotide Bound to Myosin in Resting Frog Muscle

Reaction of frog myosin with [¹⁴C]NEM in vitro. In order to clarify the reason for the increased reactivity of myosin in the iodoacetate-treated stretched muscle, experiments were carried out with isolated frog myosin. This myosin was prepared as described previously (Bárány et al., 1968) with the exception that 0.2 mM dithiothreitol (DTT) was present throughout the preparation. The myosin was reacted with [¹⁴C]NEM in an ionic milieu resembling that of frog muscle (Dubuisson, 1954); before initiating the reaction, the DTT was blocked by carboxymethylation with a 2-fold excess of iodoacetate.

Figure 2 shows a typical experiment for the incorporation of [¹⁴C]NEM into myosin in the presence of MgADP or MgATP. With MgADP there

was no change in the incorporation when the nucleotide concentration was varied from 5×10^{-5} M to 5×10^{-3} M, whereas the same variation in MgATP concentration decreased the incorporation to 70% that without any nucleotide. Results similar to those shown in Fig. 2 for MgADP were obtained for MgAMP, MgIDP, MgIMP, Mg⁺⁺

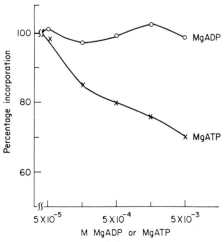

Figure 2. The incorporation of [¹⁴C]NEM into frog myosin as a function of the MgADP or MgATP concentration. Reaction conditions: 1.0 mg myosin per ml, 100 mM KCl, 25 mM Tris-HCl pH 7.0, 12.6 mM MgCl₂, ADP or ATP as indicated on the abscissa, and 0.02 μmole [¹⁴C]NEM (630,000 cpm/μmole) per ml. The final volume was 5.0 ml, the temperature 1°, and the reaction time 2.5 min. The reaction was stopped by the addition of 5 ml 10% TCA and the incorporation determined after 5 washes with 10 ml batches of 2% TCA. The value for the 100% incorporation was 1.44 moles [¹⁴C]NES/500,000 g myosin.

plus adenosine, and Mg^{++} plus inosine (the nucleotides and nucleosides found in Expts. No. 6–8, in Table 3). The decrease in the incorporation of [^{14}C]NEM into myosin is, thus, specifically caused by MgATP. These results led to the conclusion that in the stretched resting muscle, myosin contains bound MgATP which decreases its reactivity with [^{14}C]NEM, and upon irreversible loss of ATP, as in the stretched iodoacetate-treated muscle, more [^{14}C]NEM will be incorporated into the myosin.

ADP content of resting frog muscles.

Marston and Tregear (1972) reported recently that myosin contains bound ADP in water-glycerol-extracted rabbit psoas and *Lethocerus* flight muscle fibers, relaxed with MgATP in the presence of EGTA at 1.5°. They suggested that this myosin-ADP complex is the intermediate in the ATP hydrolysis catalyzed by myosin in the fibers, as shown previously by Taylor and his collaborators (1970) in myosin systems. The apparent discrepancy could conceivably be rationalized by assuming that such a MgADP intermediate, trapped after hydrolysis, has properties different from that of a "normal" complex formed through the binding of free MgADP to myosin. Any noncovalently bound ADP, as the intermediate, should be released from myosin in the muscle after denaturation of the protein, and therefore, determination of the ADP content of the muscle should reveal whether MgADP or MgATP is responsible for the decreased incorporation of [^{14}C]NEM into myosin.

Since in the muscle actin contains bound ADP, it is necessary to know how much actin and myosin is present in the frog muscle. With this knowledge the ADP found in the resting muscle can be assigned either to actin alone, or to both actin and myosin. Table 4 shows the actin and myosin content of the semitendinosus muscle. Two methods were used for quantitating actin: (1) direct determination of the actin protein, and (2) a specific procedure through the actin-bound nucleotide. The first method gave a somewhat lower value than the latter one, possibly due to incomplete extraction with 0.6 M KI of actin from the acetone-dried powder. Nevertheless, the average value of the two different methods yielded an actin content of about 0.5 μmole/g muscle, which is reasonable (Seraydarian et al., 1962). Myosin was determined as the difference of contractile proteins and actin. In the four experiments performed, the percentage deviation between individual analyses was less than 10%.

Parallel with the determination of actin and myosin content, we determined the distribution of nucleotide in the resting frog semitendinosus

Table 4. Actin and Myosin Content of Frog Semitendinosus Muscle

Expt. No.	Actin (mg protein/g fresh muscle)		Myosin
	(a)	(b)	
1	21.2	24.4	103.6
2	18.1	24.7	106.5
3	18.5	25.4	109.0
4	19.0	22.8	100.4

Determination of actin: Two procedures were used. In *a*, the direct protein determination, about 0.5 g sample of freshly minced muscle was weighed on the analytical balance in centrifuge tubes. The mince was washed at 25° 10 times with 5 ml of 0.04 M NaCl pH 7 each time to remove the soluble proteins from the muscle. The residue was washed 4 times with 10 ml acetone to denature the myosin and air-dried. The powder was extracted 3 times with 10 ml of 0.6 M KI pH 7; the combined supernatants were dialyzed against 0.05 M phosphate buffer pH 7, then fractionated with (NH$_4$)$_2$SO$_4$. The precipitate, between 0–40% saturation, was dialyzed against distilled water and the proteins quantitated by the biuret method. This fraction was found to contain only actin on SDS-polyacrylamide gel electrophoresis.

In procedure *b*, the bound nucleotide method, the weighed sample of muscle mince was washed 10 times with 0.04 M NaCl; the residue was chilled in ice and then extracted with 7.0 ml of cold 4% perchloric acid with vigorous shaking for 5 min. By this procedure only the nucleotide bound to actin is released. After centrifugation the nucleotide was determined by spectrophotometry in the supernatant fluid. The actin content of the muscle was calculated, assuming that 1.0 mole of nucleotide is bound per 45,000 dalton of actin.

Determination of myosin: The myosin content of the muscle was determined as the difference between the contractile proteins and actin. To determine the contractile proteins, the weighed muscle sample was washed 10 times with 0.04 M NaCl, the residue extracted with 50 ml of 0.05 M NaOH at 25° overnight, and the total protein in the supernatant was measured by the biuret method. The protein fraction extracted with 0.05 M NaOH contains, in addition to the contractile, the regulatory proteins also. These latter proteins amount to about 10 mg/g frog muscle, according to our preliminary results. The myosin content of the muscle was calculated from the difference of the value of the contractile proteins, corrected for the regulatory proteins, and from the average value of actin, determined by the two independent methods.

(Table 5). To obtain maximal ATP content, the muscles were oxygenated before disintegration for one hour. Furthermore, Mg^{++} and Ca^{++} were removed from the perchloric acid extract before nucleotide analysis, since we have found that these metals tend to reduce the ATP content and increase the ADP content of the muscle in Dowex 1 chromatography. Under these conditions, 77–82% of the total nucleotide and nucleoside was ATP, and the ratio of ATP and ADP varied from 5.2 to 5.8.

Table 5 compares the ADP content, 0.45–0.50 μmole/g muscle, with the amount of nucleotide binding sites in actin and myosin, 0.45–0.51 and 0.40–0.44 μmole/gram of muscle, respectively. These results show clearly that the ADP content of the muscle can be accounted for exclusively as bound to

Table 5. Nucleotide Binding Sites of Contractile Proteins and ADP Content of Resting Semitendinosus Muscle

Expt. No.	Protein	Nucleotide Binding Site (μmole/g muscle)	Nucleotide and Nucleoside Content (μmole/g muscle)		
			ADP	ATP	Other
1	Actin	0.51	0.47	2.73	0.12
	Myosin	0.41			
2	Actin	0.48	0.50	2.62	0.27
	Myosin	0.43			
3	Actin	0.49	0.45	2.60	0.23
	Myosin	0.44			
4	Actin	0.46	0.48	2.51	0.19
	Myosin	0.40			

The nucleotide binding sites of actin and myosin were calculated from the data of Table 4, assuming 1.0 mole of binding site per 45,000 dalton of actin and 250,000 dalton of myosin. To determine the nucleotide and nucleoside content, the muscles were oxygenated at 100% rest length in normal Ringer's solution at 25° for 1 hr. Four muscles were used for each experiment. These were simultaneously homogenized in butanol and subsequently analyzed as described in the legend of Table 3. There was one additional step; namely, the perchloric acid extract was passed through a Dowex 50 column in the cold room to remove Mg^{++} and Ca^{++} before the Dowex 1 chromatography. Other nucleotide and nucleoside in the table refers to the sum of AMP, adenosine, IDP, IMP, and inosine. The muscles used in these experiments were from frogs kept several months in the cold without food.

actin. If myosin would contain bound ADP as well, the ADP found in the muscle would have been almost twice as high. The results from both [^{14}C]-NEM incorporation and direct nucleotide determination, thus, support the view that ATP is the nucleotide bound to myosin in the resting muscle.

Literature. Various data of the literature support our findings that the ADP content of resting muscle is in the range of the ADP bound to actin and that most of the adenine nucleotide is in the form of ATP. Mommaerts and Wallner (1967) found 0.42 μmole ADP and 2.55 μmole ATP per gram of frog sartorius. Cain et al. (1963) found 0.47 μmole ADP and 2.75 μmole ATP per gram of frog rectus. Newbold and Scopes (1971) reported 0.4 μmole ADP and 2.6 μmole ATP per gram of minced ox skeletal muscle in the presence of 100 mM P_i. Mommaerts and Rupp (1951) analyzed 12.7% ADP and 85.1% ATP in rabbit psoas. The ATP:ADP ratio was 8.2:1 in the rectus muscle of rat (Hohorst et al., 1962), or 8.9:1 in the same muscle and 9.2:1 in the breast muscle of the pigeon (Arese et al., 1965), and 5.4:1 in the beating and KCl-arrested rabbit heart (Liu and Feinberg, 1971).

Change in Incorporation of [^{14}C]NEM into Various Proteins during Muscle Contraction

Preparation of various proteins. The use of gastrocnemius, instead of semitendinosus, allowed us to isolate six different protein fractions from three pooled muscles, totaling 3–3.5 grams.

1. *Sarcoplasmic proteins.* The muscle mince was extracted with 100 volumes of 0.04 M KCl, pH 8, at 4°, centrifuged, and the supernatant fluid saved. The residue was blended with additional 100 volumes of 0.04 M KCl for 3 min, the homogenate centrifuged, and the supernatant combined with the previous one.

2. *Myosin.* The residue obtained after extraction of sarcoplasmic proteins was extracted with a solution containing 0.3 M KCl, 0.15 M phosphate buffer, pH 6.6, and 0.01 M ATP at 4° for 20 min, then centrifuged. The myosin in the supernatant was purified by dialysis against 200 volumes of 0.02 M KCl, followed by clarification at 226,400 g in 0.6 M KCl, pH 7.4, and 0.01 M MgATP^{2-} for 2 hr. The yield was about 30 mg myosin per gram of fresh muscle.

3. *Light chains.* These were released from myosin, in 0.04 M KCl, pH 7, either by (a) urea or (b) ethanol-ether treatment. (a) In the urea procedure, the myosin, 10 mg/ml, was stirred in 5.0 M urea, pH 8.7, at 25° for 2 hours. The solution was then diluted with 3 volumes of distilled water and the heavy chains precipitated with 1.0 M potassium citrate (Gaetjens et al., 1968). After centrifugation at 48,000 g, the supernatant containing the light chains was saved; the yield was 6–7% of the weight of myosin. (b) In the ethanol-ether procedure, the myosin was treated three times with 5 volumes of 95% ethanol containing 1 mM DTT and twice with 3 volumes of ether and 1 mM DTT. The dried powder was extracted twice with 50 volumes each of a solution consisting of 1.0 M KCl, 1 mM Tris-HCl, and 1 mM DTT, pH 7.4, at 4° and the supernatants combined. The yield of the light chains was slightly less than that obtained by the urea procedure.

4 and 5. *Actin* and *"actin supernatant."* The KI procedure of A. G. Szent-Györgyi (1951) was used to prepare actin from the residue obtained after extraction of myosin. The KI solution contained additional 1 mM ATP. During purification of this actin by centrifugation at 226,400 g for 2 hr, a well-defined protein fraction referred to as "actin supernatant" was isolated. The actin in the pellet was further purified by repeated centrifugation. The yield was about 10 mg actin per gram of muscle. The protein of the "actin supernatant" was about 2 mg per gram of muscle.

6. *Troponin-tropomyosin*. The residue obtained after extraction of actin was washed with distilled water, then treated 3 times with 5 volumes of 95% ethanol and twice with ether (always in the presence of 1 mM DTT). The dried powder was extracted overnight with 10 ml of a solution containing 1.0 M KCl, 1 mM DTT, and 1 mM Tris-HCl, pH 7.4, and reextracted once more with 10 ml of the same solution. The yield of troponin-tropomyosin was about 1 mg protein per gram of muscle.

Change in incorporation during contraction.

The gastrocnemius muscles were stimulated as described previously (Bárány et al., 1971) in the set up of Fig. 1; the [^{14}C]NEM concentration was 0.25–0.35 mM.

Table 6 shows a typical experiment for comparing specific incorporation of [^{14}C]NEM into various proteins of resting and contracting gastrocnemius muscles. The bound [^{14}C]NES content per 100,000 g protein was the highest for sacroplasmic proteins, closely followed by the myosin head. The lowest

Table 6. Comparison of Specific Incorporation of [^{14}C]-NEM into Various Proteins of Resting and Contracting Gastrocnemius Muscle

Protein	Resting	Contracting
	(mole [^{14}C]NES/100,000 g protein)	
Sarcoplasmic	0.482	0.459
Myosin head	0.441	0.350
Light chains	0.178	0.168
Actin	0.028	0.022
Actin supernatant	0.184	0.177
Troponin-tropomyosin	0.160	0.151

The preparation of various proteins is described in the text. For the isolation of myosin head see Bárány et al. (1971). The incorporation was determined after precipitating and washing the proteins with TCA and dissolution in 0.02 M HCl.

incorporation was obtained for actin, whereas for light chains, actin supernatant, and troponin-tropomyosin the values were similar. The decrease in incorporation during contraction was evident in the case of the myosin head, but a decrease was observable with other proteins as well.

The extent of change in the incorporation during contraction is shown in Table 7. For these analyses,

we did not isolate the myosin head but compared only the incorporation into the intact myosin molecule, since it was shown previously (Bárány et al., 1971) that the entire difference in incorporation between resting and contracting muscles occurs in the globular head of myosin. Table 7 shows 79% incorporation of [^{14}C]NEM into myosin of contracting muscle, compared to that of resting

Table 7. Change in Incorporation of [^{14}C]NEM into Various Proteins of Contracting Gastrocnemius Muscle as Compared to Resting Muscle

Protein	Percentage Incorporation	No. of Determinations
Sarcoplasmic	97.0 ± 4.5	23
Myosin	79.1 ± 3.4	45
Light chains	93.6 ± 5.3	31
Actin	80.2 ± 2.5	37
Actin supernatant	94.7 ± 5.6	28
Troponin-tropomyosin	93.8 ± 5.0	26

The 100% incorporation values, expressed as mole [^{14}C]NES/100,000 g protein, were in the following range: 0.44–0.60 for sarcoplasmic proteins; 0.19–0.30 for myosin; 0.12–0.18 for light chains; 0.02–0.04 for actin; 0.11–0.18 for actin supernatant; and 0.10–0.18 for troponin-tropomyosin. Results are given with ± standard error of means.

muscle. The incorporation into actin was 80% which is similar to that into myosin. The troponin-tropomyosin fraction of contracting muscle gave 94% incorporation of the resting muscle, while the incorporation into the light chains was 94% and into the actin supernatant 95%. Under these conditions the incorporation into the sarcoplasmic proteins of contracting muscle was 97% compared to these proteins of resting muscle.

The data of Table 7 reveal a change in the reactivity of actin during contraction synchronous to that of myosin. It is the merit of the radioactive method that this 20% change in actin can be detected when only 0.01 mole of label is bound per actin globule (as may be calculated from the data of Table 6). In contrast to the well-defined changes in the incorporation of [^{14}C]NEM into the contractile proteins during contraction, there is only a 5–6% change in the incorporation into the light chains of myosin, into the protein remaining in the supernatant after ultracentrifugation of KI-extracted actin, and into troponin-tropomyosin (Table 7). Because of the relatively large standard error which accompanies these percentage changes, it remains inconclusive whether or not there is a concrete decrease in the incorporation of [^{14}C]NEM into these three protein fractions during contraction.

The wide range of specific incorporation of [^{14}C]NEM into various proteins of frog muscle (Table 6) deserves comment. According to our observation, the cysteine content of these proteins does not differ from each other significantly; thus, the variation in the incorporation can not be caused by the composition of the proteins. It is known from the work of Tsao and Bailey (1953) that NEM readily reacts with half of the total SH groups both in G-actin and F-actin. Therefore, the lack of reactivity of NEM with the cysteine residues of actin in the muscle suggests that these groups are masked, perhaps through some noncovalent protein-protein interactions. The analogous incorporation of [^{14}C]NEM into the myosin head and sarcoplasmic proteins of resting muscle invites speculation that the head has similar mobility, "floating" in the cytoplasm where it is readily accessible to the reagent. It is of interest that such proteins as the light chains, actin supernatant, and troponin-tropomyosin, isolated by such widely different routes, possess essentially the same incorporation. A preliminary report from this laboratory (Bárány et al., 1972) described the resemblance of troponin-tropomyosin to the light chains and actin supernatant in frog muscle.

Proposal for Mechanism of Contraction

The present work establishes the involvement of both actin and ATP in the decreased reactivity of myosin during contraction. The data of Table 2 show no change in the reactivity of myosin in stretched and stimulated muscle, where there is no overlap between actin and myosin filaments. Furthermore, there is only a 7 % decrease for muscle in rigor, which is depleted of its normal ATP content. In contrast, isotonically contracting muscle, spending only 21 % of the total time in the shortening phase of the mechanical activity, exhibits a 21 % decrease in the incorporation. It follows that muscle in rigor, or stretched and stimulated, essentially lacks the fundamental change in the reactivity of myosin with [^{14}C]NEM found in the contracting muscle. Therefore, it appears that the concerted interaction of the myosin filaments, actin filaments, and ATP is necessary to induce a full decrease in the reactivity of myosin with [^{14}C]NEM.

As shown in Tables 6 and 7, there is no change in the incorporation of [^{14}C]NEM into various proteins during contraction, with the specific exception of myosin and actin. Due to the very low absolute levels of [^{14}C]NES content in actin of resting and contracting muscles (Table 6), we did not attempt to analyze the nature of the change in the incorporation into actin during contraction, and therefore this effect will not be discussed further. A detailed study was carried out previously to distinguish possible reasons for the changed reactivity of myosin (Bárány et al., 1971). Experiments on the ATPase activity and actin-binding ability of myosin, prepared from [^{14}C]NEM-treated resting

and contracting muscles, showed clearly that the changed reactivity does not occur at the active sites of myosin. Furthermore, the large number of [^{14}C]NEM-labeled peptides, prepared from four different proteolytic digests of the myosin, proved that the reaction is not restricted to a specific cysteine residue, but rather a general reaction takes place involving eight to twelve SH groups. The distribution of radioactivity in peptides isolated from myosin of resting and contracting muscles exhibited a manifold difference, suggesting that several parts of the myosin head are changing their configuration during contraction.

The increased incorporation of [^{14}C]NEM into myosin of stretched and ATP-depleted muscle (Tables 1 and 2) has led to the conclusion that in the resting muscle MgATP protects a part of myosin from reaction with NEM. From quantitation of the ATP content further evidence was obtained that myosin contains bound ATP in the resting muscle; at 12.6 mM Mg^{++} concentration in the intracellular water (Dubuisson, 1954), this ATP occurs as MgATP complex. In view of the overwhelming evidence that in myosin the head part is the locus for binding of ATP, it follows that in the resting muscle the cross-bridges (the myosin heads) contain bound MgATP.

In consideration of these findings, the following mechanism of muscle contraction emerges: In the resting muscle interaction between actin and myosin filaments is inhibited by the troponin-tropomyosin system (Ebashi et al., 1969). The calcium released from the sarcoplasmic reticulum by the stimulus inactivates the regulatory proteins; this makes possible the combination of the cross-bridges with the actin filaments, initiating the hydrolysis of bound MgATP. During the concerted interaction of actin and MgATP with the myosin head, the head undergoes a conformational change while it pulls the actin filament. This is schematized in Fig. 3.

We propose that this conformational change is the driving force for generating tension and movement. An intramolecular change in the cross-bridge, while attached to actin, will develop tension, as in isometric contraction, or will allow shortening, as in isotonic contraction. The conformational change is related to the hydrolysis of ATP, and this may be the explanation for the relationship between ATPase activity of myosin and speed of muscle contraction (Bárány, 1967). Accordingly, the mechanochemistry of muscle contraction is based on the coupling of ATP hydrolysis with deformation of the myosin head.

Relaxation involves the rearranged structure of the myosin head, which facilitates detachment from the actin filament and the release of MgADP.

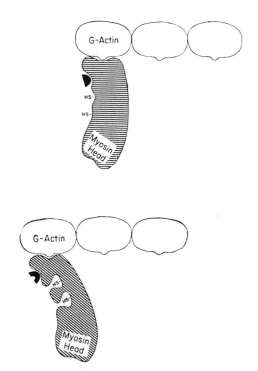

Figure 3. A model for the mechanism of contraction. The upper part of the figure depicts the attachment of the actin monomer to the head of myosin containing bound MgATP (solid semicircle) and two readily available SH groups. The lower part shows the changed configuration of the head containing bound MgADP (solid crescent) and the buried SH groups. As a result of these changes the actin chain was moved with one monomer unit. (Note that actually about 8–12 SH groups are involved in the conformational change, as indicated in the text.)

Thus, another function of the actin and ATP-induced conformational change is to make possible unloading of actin and the reaction products of hydrolysis, after which MgATP is rebound to regenerate the resting state.

This proposal differs from that of Lymn and Taylor (1971), which was deduced from their kinetic studies on the mechanism of ATP hydrolysis catalyzed by the heavy meromyosin-actin (acto-HMM) complex. These authors suggest that in the resting muscle myosin exists as its "long half-life" complex with ADP and P_i, and that during contraction actin displaces the hydrolysis products from the active sites of myosin. A key point in their suggestion is the finding that the rate of acto-HMM dissociation is extremely fast, from which it follows that the dissociation of acto-HMM precedes the hydrolysis of ATP. It was shown in this work that in the intact resting muscle myosin does not contain bound ADP, and therefore this part of their suggestion appears to be excluded from the molecular events of contraction in living muscle. It is known that acto-HMM is dissociated by ATP

at virtually any ionic strength (Szent-Györgyi, 1953); in contrast, there is no direct evidence for such a dissociation of actomyosin at low ionic strength. Therefore the results obtained with the simple acto-HMM system may not be extrapolated to the environment of the muscle cell. Trentham and associates (1972) reported that the dissociation rates of the products ADP and P_i from heavy meromyosin are faster than the rate controlling process, which occurs after the initial bond cleavage of ATP. This conclusion varies from that of Lymn and Taylor. Furthermore, Trentham and associates raised the possibility that the slow step after the hydrolysis involves a conformational adjustment of the protein.

There are theories of muscle contraction which are based on conformational changes. Volkenstein (1969) gave the mathematical analysis of how the transconformation of cross-bridge produces work and liberates heat. If the transconformation of the cross-bridge moves the actin filament by the distance of a G-actin monomer, the energy liberated (work and heat) is 2.6×10^{-13} erg per cross-bridge. This energy corresponds to 40% of the ATP energy (10 kcal/mole), assuming that the transconformation is the result of the hydrolysis of one molecule of ATP per cross-bridge per interaction with actin. Hill (1969) discussed various kinetic schemes in his proposal, relating conformational change of the cross-bridge to tension development on the actin filament and to movement. Chapman and Gibbs (1972) formulated a transduction model for the energetics of muscle contraction, consisting of a conformational change which occurs within the cross-bridge-actin-MgATP complex, when tension is generated with simultaneous breakdown of ATP. This model is adaptable to different muscles and accounts entirely for the total energy balance of contraction, after the heat due to the Ca^{++} pump has been subtracted.

A molecular mechanism of muscle contraction by operation of a cycle, in which transconformation of the myosin molecule takes place when it interacts with ATP and actin, was first proposed by Tonomura et al. (1961). Since then, fluorescence, spin-label, chromophoric probe, and transient kinetic methods showed that the globular head of myosin undergoes conformational changes upon interaction with substrates or modifiers (Morales, 1970). It was also recognized, some time ago, that the site of myosin associated with its ATPase activity differs from that involved in its interaction with actin (Bárány and Bárány, 1959; Perry and Cotterill, 1964; Stracher, 1964). Thus, the architecture of the myosin head provides a rational background for the concerted interaction of actin and ATP (Fig. 3); the simultaneous interaction of several groups in the head with actin and ATP, during the catalytic process, results in significant changes in conformation.

Summary

The incorporation of [^{14}C]NEM into myosin of frog muscle decreased during isotonic and isometric contraction, as compared to resting muscle. However, modification of the interaction of myosin, actin, and ATP in the muscle resulted in different patterns of incorporation. Actin was prevented from interacting with myosin by stretching the semitendinosus muscle to 140% rest length, so that no tension was developed upon stimulation. Myosin of such stretched and stimulated muscle showed no decrease in incorporation. ATP was eliminated by inducing rigor with iodoacetate; the incorporation into myosin of this muscle was slightly decreased. Both actin and ATP were blocked by treatment of the muscle at 140% rest length with iodoacetate; this treatment increased the incorporation.

A study was undertaken to clarify the nature of the increased reactivity of myosin in the iodoacetate-treated stretched muscle. In vitro experiments with frog myosin revealed that in Mg^{++}-containing media, omission of ATP resulted in enhanced incorporation of [^{14}C]NEM into myosin. Thus MgATP decreased the reactivity of myosin with [^{14}C]NEM, an effect not reproduced by ADP nor any other nucleotide. These results suggest that in the resting muscle, myosin contains bound MgATP, which decreased its reactivity with NEM, and upon irreversible loss of ATP, as in the iodoacetate-treated muscle, more [^{14}C]NEM will be incorporated into the myosin. From quantitation of nucleotides, further evidence was obtained for the occurrence of a myosin-ATP complex in resting muscle.

The incorporation of [^{14}C]NEM into six different proteins of frog muscle was determined. Only actin was able to duplicate the specific change in the incorporation during contraction observed for myosin.

These data, along with our previous findings, which show a manifold difference in distribution of radioactivity of peptides isolated from myosin of resting and contracting muscles (Bárány et al., 1971), suggest the following mechanism for contraction: Combination of the actin filaments with the myosin cross-bridges initiates the hydrolysis of the bound MgATP. During this concerted interaction, the cross-bridges undergo a conformational change while they pull the actin filaments. It appears that this conformational change is the driving force for generating the tension and the movement.

Acknowledgments

We thank Mr. Mu-Ju Shen, Robert Haire, Ronald Silverman, and Mrs. Maria Topal for their expert assistance.

This work was supported by grants from the Muscular Dystrophy Associations of America, Inc., and a grant from the Muscular Dystrophy Association of Canada.

NOTE ADDED IN PROOF. The nucleotide distribution in resting semitendinosus muscle was also determined according to the procedure of Dydynska and Wilkie (1966). The muscles were frozen in isopentane and then powdered in a Teflon mortar, generously donated by Prof. Douglas R. Wilkie. The results were the same as those obtained by the butanol procedure shown in Table 5. Through the use of trace [^{14}C]ADP as an internal standard, the recovery of nucleotides from the muscles by either method was determined to be 95–97%.

References

ARESE, P., R. KIRSTEN, and E. KIRSTEN. 1965. Metabolitgehalte und -gleichgewichte nach tetanischer Kontraktion des Taubenbrustmuskels und des Rattenskeletmuskels. *Biochem. Z.* **341**: 523.

BÁRÁNY, M. 1967. ATPase activity of myosin correlated with speed of muscle shortening. *J. Gen. Physiol.* **50**: (suppl. part 2) 197.

BÁRÁNY, M. and K. BÁRÁNY. 1970. Change in the reactivity of myosin during muscle contraction. *J. Biol. Chem.* **245**: 2717.

———. 1959. Studies on "active centers" of L-myosin. *Biochim. Biophys. Acta* **35**: 293.

BÁRÁNY, M., K. BÁRÁNY, and G. BAILIN. 1968. Reactivity of actomyosin and myosin with 1-fluoro-2,4-dinitrobenzene in vivo and in vitro. *Biochim. Biophys. Acta* **168**: 298.

BÁRÁNY, M., K. BÁRÁNY, and E. GAETJENS. 1971. Change in the reactivity of the head part of myosin during contraction of frog muscle. *J. Biol. Chem.* **246**: 3241.

BÁRÁNY, M., K. BÁRÁNY, und W. TRAUTWEIN. 1960. Die Hemmung der Aktin-L-Myosin Interaktion in lebenden und extrahierten Muskeln durch Urea. *Biochim. Biophys. Acta* **45**: 317.

BÁRÁNY, M., K. BÁRÁNY, E. GAETJENS, and B. Z. HORVÁTH. 1972. Resemblance of troponin-tropomyosin to other protein fractions in frog muscle. *Abstr. 16th Annu. Meet. Biophys. Soc.*, Toronto, Canada, page 284a.

CAIN, D. F., M. J. KUSHMERICK, and R. E. DAVIES. 1963. Hypoxanthine nucleotides and muscular contraction. *Biochim. Biophys. Acta* **74**: 735.

CHAPMAN, J. B. and C. L. GIBBS. 1972. An energetic model of muscle contraction. *Biophysical J.* **12**: 227.

DUBUISSON, M. 1954. *Muscular contraction.* Charles C. Thomas, Springfield, Illinois.

DYDYNSKA, M. and D. R. WILKIE. 1966. The chemical and energetic properties of muscle poisoned with fluorodinitrobenzene. *J. Physiol.* **184**: 751.

EBASHI, S., M. ENDO, and I. OHTSUKI. 1969. Control of muscle contraction. *Quart. Rev. Biophys.* **2**: 351.

GAETJENS, E., K. BÁRÁNY, G. BAILIN, H. OPPENHEIMER, and M. BÁRÁNY. 1968. Studies on the low molecular weight protein components in rabbit skeletal myosin. *Arch. Biochem. Biophys.* **123:** 82.

GORDON, A. M., A. F. HUXLEY, and F. J. JULIAN. 1966a. Tension development in highly stretched vertebrate muscle fibers. *J. Physiol.* **184:** 143.

———. 1966b. The variation in isometric tension with sarcomere length in vertebrate muscle fibers. *J. Physiol.* **184:** 170.

HILL, T. I. 1969. A proposed common allosteric mechanism for active transport, muscle contraction, and ribosomal translocation. *Proc. Nat. Acad. Sci.* **64:** 267.

HOHORST, H. J., M. REIM, and H. BARTELS. 1962. Studies on the creatine kinase equilibrium in muscle and the significance of ATP and ADP levels. *Biochem. Biophys. Res. Comm.* **7:** 142.

HUXLEY, H. E. 1968. Structural difference between resting and rigor muscle; evidence from intensity changes in the low-angle equatorial X-ray diagram. *J. Mol. Biol.* **37:** 507.

HUXLEY, H. E. and W. BROWN. 1967. The low-angle X-ray diagram of vertebrate striated muscle and its behaviour during contraction and rigor. *J. Mol. Biol.* **30:** 383.

LILIENTHAL, J. L., JR., K. L. ZIERLER, B. P. FOLK, R. BUKA, and M. J. RILEY. 1950. A reference base and system for analysis of muscle constituents. *J. Biol. Chem.* **182:** 501.

LIU, M. S. and H. FEINBERG. 1971. Incorporation of adenosine-8-^{14}C and inosine-8-^{14}C into rabbit heart adenine nucleotides. *Amer. J. Physiol.* **220:** 1242.

LYMN, R. W. and E. W. TAYLOR. 1971. Mechanism of adenosine triphosphate hydrolysis by actomyosin. *Biochemistry* **10:** 4617.

MARSTON, S. B. and R. T. TREGEAR. 1972. Evidence for a complex between myosin and ADP in relaxed muscle fibres. *Nature New Biol.* **235:** 23.

MOMMAERTS, W. F. H. M. and J. C. RUPP. 1951. Dephosphorylation of adenosine triphosphate in muscular contraction. *Nature* **168:** 957.

MOMMAERTS, W. F. H. M. and A. WALLNER. 1967. The break-down of adenosine triphosphate in the contraction cycle of the frog sartorius muscle. *J. Physiol.* **193:** 343.

MORALES, M. F. 1970. Conformation and displacement in muscle contraction. *Proc. Nat. Acad. Sci.* **66:** 236.

NEWBOLD, R. P. and R. K. SCOPES. 1971. Post-mortem glycolysis in ox skeletal muscle: Effect of adding nicotinamide-adenine dinucleotide to diluted mince preparations. *J. Food Sci.* **36:** 215.

PERRY, S. V. and J. COTTERILL. 1964. The action of thiol inhibitors on the interaction of F-actin and heavy meromyosin. *Biochem. J.* **92:** 603.

SANDOW, A. 1947. Latency relaxation and a theory of muscular mechanochemical coupling. *Ann. N.Y. Acad. Sci.* **47:** 895.

SERAYDARIAN, K., W. F. H. M. MOMMAERTS, and A. WALLNER. 1962. The amount and compartmentalization of adenosine diphosphate in muscle. *Biochim. Biophys. Acta* **65:** 443.

STRACHER, A. 1964. Disulfide-sulfhydryl interchange studies on myosin A. *J. Biol. Chem.* **239:** 1118.

SZENT-GYÖRGYI, A. G. 1953. Meromyosins, the subunits of myosin. *Arch. Biochem. Biophys.* **42:** 305.

———. 1951. A new method for the preparation of actin. *J. Biol. Chem.* **192:** 361.

TAYLOR, E. W., R. W. LYMN, and G. MILL. 1970. Myosin-product complex and its effect on the steady-state rate of nucleoside triphosphate hydrolysis. *Biochemistry* **9:** 2984.

TONOMURA, Y., K. YAGI, S. KUBO, and S. KITAGAWA. 1961. A molecular mechanism of muscle contraction. *J. Res. Inst. Catalysis* (Hokkaido U.) **9:** 256.

TREGEAR, R. T. and A. MILLER. 1969. Evidence of cross-bridge movement during contraction of insect flight muscle. *Nature* **222:** 1184.

TRENTHAM, D. R., R. G. BARDSLEY, J. F. ECCLESTON, and A. G. WEEDS. 1972. Elementary processes of the magnesium ion-dependent adenosine triphosphatase activity of heavy meromyosin. *Biochem. J.* **126:** 635.

TSAO, T. C. and K. BAILEY. 1953. The extraction, purification and some chemical properties of actin. *Biochim. Biophys. Acta* **11:** 102.

VOLKENSTEIN, M. V. 1969. Muscular contraction. *Biochim. Biophys. Acta* **180:** 567.

Formation and Decomposition of the Myosin-Phosphate-ADP Complex in the Myosin-ATPase Reaction

Yuji Tonomura, Yutaro Hayashi, and Akio Inoue

Department of Biology, Faculty of Science, Osaka University, Toyonaka, Osaka

To clarify the molecular mechanism of muscle contraction, it is essential to elucidate the mechanism of the reaction of myosin with ATP, which is the key process in muscle contraction.

From kinetic studies on the myosin- and actomyosin-ATPase reactions, we proposed that the intermediates of the myosin-ATPase reaction were the myosin-ATP complex, phosphoryl myosin, and the myosin-phosphate-ADP complex. Since these studies have recently been treated in considerable detail by one of the authors (Tonomura, 1972; Tonomura and Oosawa, 1972), only a brief outline of that will be given here. The existence of the myosin-ATP complex was originally proposed from the finding that the rate of the myosin-ATPase reaction in the steady state conforms to the Michaelis-Menten equation. Its existence was later supported by comparison of the UV-spectral change induced by ATP and the rate of the ATPase reaction in the steady state. The formation of phosphoryl myosin and the myosin-phosphate-ADP complex were proposed from the transient kinetics of the myosin-ATPase reaction. We showed that the amount of the initial burst of P_i liberation is 1 mole per mole of myosin under various conditions of pH, ionic strength, and temperature, and concluded that the initial burst of P_i liberation is due to the formation of the myosin-phosphate-ADP complex, E_P^{ADP}. The initial burst was confirmed by Sartorelli et al. (1966) and Lymn and Taylor (1970). We also observed that the liberation of a stoichiometric amount of H^+ and the change in the UV spectrum occur just before the formation of E_P^{ADP}. The formation of phosphoryl myosin was proposed from the findings that p-nitrothiophenyl myosin, which shows no initial burst of P_i liberation, is formed by the reaction of myosin and p-nitrothiophenol only in the presence of MgATP, and that a P-exchange reaction occurs between the intermediate and the terminal phosphate of ATP during the initial phase of the reaction. However, phosphoryl myosin is unstable and occurs transiently during the initial phase of the reaction, and its amount is so small that it can be neglected in kinetic analyses under usual conditions. On the other hand, E_P^{ADP} is a stable intermediate, and many kinetic features of myosin-ATPase can be explained from the kinetic properties of E_P^{ADP}. This paper is on our recent studies on the structure-function relationship of myosin, and particularly on the kinetically important intermediate, E_P^{ADP}.

Heterogeneity of Two Heads of Myosin Molecule

It is appropriate to mention briefly our recent studies on the substructure of the myosin molecule, since the interpretation of kinetic results on myosin-ATPase depends largely on how many active sites of the ATPase are present in the myosin molecule and in which parts of the molecule they are. Tonomura and Morita (1959) studied the binding of PP_i to myosin by equilibrium dialysis and light-scattering and concluded that 2 moles of PP_i bind to 1 mole of myosin and that only 1 mole of these takes part in the dissociation of actomyosin. Morita (1967; 1971) recently showed that 2 moles of ADP can bind to 1 mole of H-meromyosin and that only the one of these two moles of ADP which binds the more strongly induces the change in the UV spectrum of H-meromyosin. We (Kanazawa and Tonomura, 1965; Hayashi and Tonomura, 1970; Inoue et al., 1972) showed that the amounts of the initial burst of P_i liberation from the myosin, H-meromyosin, and S-1-ATP systems, i.e., the amounts of the myosin-, H-meromyosin- and S-1-phosphate-ADP complexes, were 1, 1, and 0.5 mole per mole of protein, respectively, when the molar concentration of ATP was higher than that of protein. This stoichiometric burst of P_i liberation disappeared when 1 mole of p-nitrothiophenol was bound to 1 mole of myosin (Imamura et al., 1965; Tonomura and Kanazawa, 1965; Kinoshita et al., 1969a), but this specific p-nitrothiophenylation did not affect myosin-ATPase activity in the steady state (Kinoshita et al., 1969a).

It is generally agreed that the bulk of the myosin molecule with a molecular weight of 4.6–4.8 × 10^5 is composed of two heavy subunits, each with a mol wt of about 2.0 × 10^5, and that these two subunits have very similar chemical

structures. Studies on myosin by electron microscopy (Slayter and Lowey, 1967; Lowey et al.,
1969) and proteolysis (Trotta et al., 1968; Lowey
et al., 1969) showed that one side of the molecule is
separated into two globular units and that two
S-1's are produced from the bipartite head by
enzymic degradation. The capacities for myosin-
ATPase activity (Nauss et al., 1969; Hayashi and
Tonomura, 1970) and for formation of the myosin-
phosphate-ADP complex (Hayashi and Tonomura,
1970) per mole of myosin or H-meromyosin were
preserved in 2 moles of S-1. Thus, the problem of
whether the myosin molecule contains two identical
head parts or not must be solved by studying the
substructure of the two S-1 molecules.

Stracher (1969) and Dreizen and Gershman
(1970) showed that the light subunits, as well as
the heavy subunits of myosin, are essential for
expression of the ATPase activity. One of the
authors (Hayashi, 1972) showed by SDS-gel
electrophoresis that myosin contained three kinds
of light subunits with molecular weights of 2.5, 1.8,
and 1.4×10^4 (Fig. 1A). We will call these three
subunits g_1, g_2, and g_3, respectively, in order of
decreasing molecular weight. Recently, Weeds and
Lowey (1971) and Lowey and Risby (1971)
showed that the myosin molecule from skeletal
white muscle contained 1 mole of g_1, 2 moles of g_2,
and 1 mole of g_3; that the chemical structures
around the SH groups of g_1 and g_3 were identical

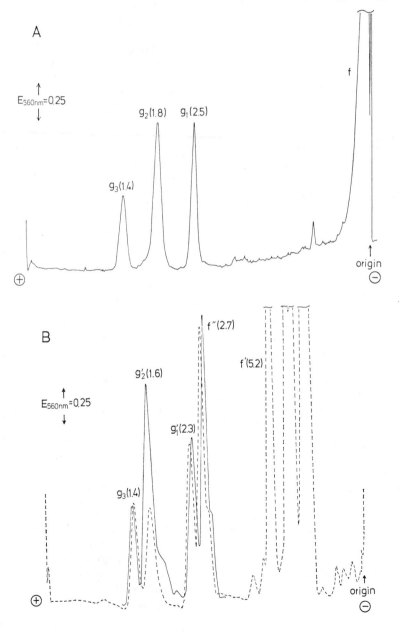

with each other; and that g_2 could be almost completely removed from myosin by treatment with 5,5′-dithio-(bis-2-nitrobenzoic acid) without loss of ATPase activity in the steady state. We also specifically removed about a half of g_2 and its derivative, g_2' (see below), from myosin and H-meromyosin, respectively, by treatment with p-chloromercuribenzoate (PCMB) without loss of ATPase activity (Fig. 1B).

S-1 prepared by tryptic digestion of H-meromyosin was composed of five components with molecular weights of 5.2–5.5, 2.7, 2.1, 1.6, and 1.4×10^4, as shown in Fig. 1C (Hayashi, 1972). The changes in the components in myosin and H-meromyosin during tryptic digestion were followed by SDS-gel electrophoresis. The results indicated that the components with mol wts of 5.2–5.5 (f′) and 2.7×10^4 (f″) are derived from the heavy subunits, that with a mol wt of 2.1×10^4 (g_1'') is derived from g_1, that with a mol wt of 1.6×10^4 (g_2') is derived from g_2, and the component with a mol wt of 1.4×10^4 is g_3 itself. Judging from the areas under the peaks in densitometer tracings, the molar ratio of $g_1'':g_2':g_3$ in S-1 was 1.0:0.4:0.3, although this ratio in myosin and H-meromyosin was 1.0:2.1:1.1. These results strongly suggest that the light subunit essential for ATPase is g_1 alone, since, although the contents of g_2' and g_3 are much less than the stoichiometric amounts, all the activity of myosin-ATPase was retained in the two S-1's, and since about half of g_2 was removed without affecting myosin-ATPase activity. Thus, it is concluded that the one of the two heads of the myosin molecule containing g_1

has an active site for ATPase, but the other head does not.

However, the results so far obtained by chemical modifications of myosin-ATPase are not conclusive with regard to the number of active sites. When 1 mole of a specific residue per mole of myosin was chemically modified with reagents such as trinitrobenzene sulfonate (Tokuyama et al., 1966), diazonium-1 H tetrazole (Shimada, 1970), or monoiodoacetamide (Ohe et al., 1970), the maximum change in enzymic activity occurred, indicating the presence of one active site in the myosin molecule. On the other hand, chemical modification of myosin with a nitroxide derivative of iodoacetamide (Seidel et al., 1970) or an analog of ATP (Murphy and Morales, 1970) supported the presence of two identical active sites. The latter results can be reasonably expected assuming that myosin contains two similar heavy subunits and two sets of light subunits with similar chemical structures, as mentioned above.

ATPase Activity in Steady State

The dependence of the rate of myosin-ATPase in the steady state on the ATP concentration was measured over a wide range of ATP concentrations from 0.1 to 5 μM in 0.5 M KCl, 2.5 mM MgCl$_2$ and 50 mM Tris-maleate, pH 7.8 at 0°C (Inoue et al., 1972). Figure 2 shows the results on four myosin preparations. At ATP concentrations above 1 μM, the values of K_M and V_m were 1 μM and 0.44 min^{-1}, respectively. At ATP concentrations below 0.3 μM, the rate deviated from the straight line obtained at higher ATP concentrations and became almost

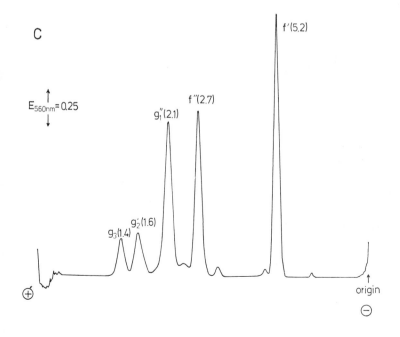

C

$E_{560nm} = 0.25$

$g_3(1.4)$ $g_2'(1.6)$ $g_1''(2.1)$ f″(2.7) f′(5.2)

⊕ origin ⊖

Figure 1. Densitometer tracings of electrophoretograms of myosin, H-meromyosin with and without PCMP-DTT-treatment, and S-1 on SDS gel. Electrophoreses were performed on 10% acrylamide gel in the presence of 0.1% SDS, essentially by the method of Weber and Osborn (1969). The electrophoretograms were scanned with a Model 2410-S Linear Transport Scanner on a Gilford Model 240 spectrophotometer. The figures in parentheses are molecular weights ($\times 10^4$). See text for notations. *A.* Myosin purified on phosphocellulose. *B.* H-meromyosin: (——) intact; (– – – –) treated with PCMB, subjected to gel filtration on Sephadex G-200, then treated with dithiothreitol (DTT) to remove the PCMB. *C.* S-1 treated with alkali and purified by combination with actin.

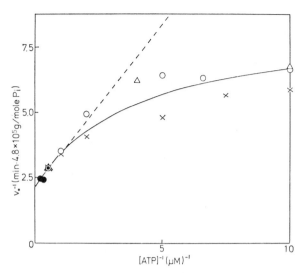

Figure 2. Double reciprocal plot of the rate of the ATPase reaction in the steady state, v_0, against the concentration of ATP. The mixture contained 0.003 mg/ml myosin, 0.5 M KCl, 2.5 mM MgCl$_2$, and 50 mM Tris-maleate at pH 7.8 and 0°C. ($\bigcirc$, $\bullet$, $\times$, $\triangle$) different preparations of myosin. The rates of each preparation were plotted relative to that at 2 μM ATP. The rates at 2 μM were 0.36, 0.36, 0.35, and 0.33 min^{-1}, and the average value of 0.35 min^{-1} was used as the rate at 2 μM ATP. The solid line indicates

$$v_0(\text{min}^{-1}) = 0.11 + \frac{0.33}{1 + \dfrac{1\ \mu\text{M}}{[\text{S}]}}.$$

independent of the ATP concentration. The dependence of the rate of the overall reaction, v_0, on the ATP concentration was given by

$$v_0(\text{min}^{-1}) = 0.11 + \frac{0.33}{1 + \dfrac{1\ \mu\text{M}}{[\text{S}]}}.$$

Thus, the value of V_m obtained in a higher ATP concentration range was four times that in a lower ATP concentration range.

The following three mechanisms may be considered to interpret this phenomenon. At high concentrations of ATP (i) the number of ATPase active sites increases, (ii) the rate of decomposition of intermediate increases, and (iii) hydrolysis of ATP occurs via a route different from that at lower ATP concentrations. Taylor and coworkers (Lymn and Taylor, 1970; Taylor et al., 1970) reported curvature of a Scatchard plot of the rate of ATPase in the steady state at ATP concentrations of about 0.1 μM. They proposed mechanisms for (i) and (ii), reporting an increase in the amount of the initial burst of P$_i$ liberation from 1 to 2 moles/mole of myosin at ATP concentrations of around 20 μM, and acceleration of decomposition of E$_P^{ADP}$ by high concentrations of ATP. However, we (Inoue et al.,

1972) have recently shown that the maximum amount of the initial burst was constant (1 mole/ mole of myosin) and was independent of the ATP concentration. Moreover, we observed no acceleration of decomposition of E$_P^{ADP}$ at high concentrations of ATP, as will be mentioned later.

On the other hand, mechanism (iii) seems most probable, since the dissociation constant of the binding of ATP with myosin measured by UV-spectroscopy (Sekiya and Tonomura, 1967) or the luciferin-luciferase method (Nanninga and Mommaerts, 1960) was a few μM, which was similar to the K_M value of myosin-ATPase in the steady state in a high ATP concentration range; but no ATP binding was detected at ATP concentrations below 0.3 μM. Thus, the Michaelis intermediate of ATPase at low ATP concentrations seems to be free myosin which does not bind nucleotide, whereas the intermediate at high ATP concentrations seems to be a myosin-nucleotide complex.

Formation of Myosin-Phosphate-ADP Complex

The rate of E$_P^{ADP}$ formation, v_f, was previously measured by us (Onishi et al., 1968) and later by Lymn and Taylor (1970) at only high concentrations of ATP. Therefore, we measured the value

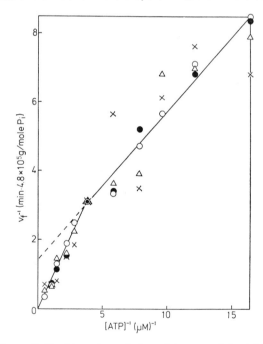

Figure 3. Double reciprocal plot of the rate of the initial burst of P$_i$ liberation, v_f, against the concentration of ATP. The mixture contained 0.03 mg/ml myosin, 0.5 M KCl, 2.5 mM MgCl$_2$, and 25 mM Tris-maleate at pH 7.8 and 0°C. ($\bigcirc$, $\bullet$, $\times$, $\triangle$) different preparations of myosin. The rates of each preparation were plotted relative to that at 0.26 μM ATP. The rates at 0.26 μM ATP were 0.28, 0.35, 0.34, and 0.31 min^{-1}, and the average value of 0.32 min^{-1} was used as the rate at 0.26 μM ATP.

of v_f over a wide range of ATP concentrations from 0.06 to 5 μM in the presence of 0.03 mg/ml myosin, 0.5 M KCl, 2.5 mM MgCl$_2$, and 25 mM Tris-maleate, pH 7.8 at 0°C (Inoue et al., 1972). A double reciprocal plot of v_f against the ATP concentrations is shown in Fig. 3. It gave two straight lines intersecting at 0.3 μM ATP. At ATP concentrations below 0.3 μM the maximum value of v_f, V_f, and the Michaelis constant, K_f, were 0.7 min^{-1} and 0.3 μM, respectively. At ATP concentrations above 0.3 μM the rate of E_P^{ADP} formation increased with increase in the ATP concentration, and the values of V_f and K_f could not be measured accurately with our simple mixing apparatus, whereas Lymn and Taylor (1970) reported that the values of V_f and K_f at 20°C were in the order of 3000 min^{-1} and 50 μM, respectively. From these and other results we proposed the following mechanism for formation of E_P^{ADP}:

$$E + S \rightleftharpoons ES \underset{\underset{\substack{\uparrow \\ \text{acceleration by S}}}{}}{\xrightarrow{\hspace{3cm}}}$$

(phosphoryl myosin) $\rightarrow E_P^{ADP}$

The mechanism of acceleration of E_P^{ADP} formation by high ATP concentrations still remains to be clarified, but it is noteworthy that formation of phosphorylated intermediates of the Na$^+$-K$^+$-dependent ATPase and the Ca^{++}-Mg^{++}-dependent ATPase of the sarcoplasmic reticulum are also accelerated by high concentrations of ATP (Tonomura, 1972).

Liberation of P_i and ADP from Myosin-Phosphate-ADP Complex

The rate of P_i liberation from E_P^{ADP} was measured by Taylor et al. (1970) using the gel filtration method. They showed that the rate is comparable with that of ATPase in the steady state under various conditions of pH, ionic strength, and temperature. We have also measured the rate by the rapid flow-dialysis method (Colowick and Womack, 1969). To measure fast reactions, a Millipore filter with a pore size of 0.22 μ was used as a membrane, and the pressure on both sides of the membrane was maintained at the same level by applying a negative pressure to the efflux tube from the chamber. The diffusion of myosin across the membrane could be neglected. The rate was calculated by comparing the time course of appearance of P_i in the lower chamber with that observed when P_i was added to myosin in a predetermined manner in the upper chamber. The reaction was started by adding 0.2 ml of 55 μM [^{32}P]ATP solution to 2 ml of 2.75 mg/ml myosin

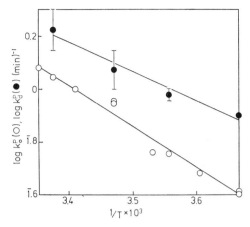

Figure 4. Dependence on temperature of the rate of P_i liberation from E_P^{ADP} and the rate of the myosin-ATPase reaction in the steady state. The mixture contained 0.5 M KCl, 2.5 mM MgCl$_2$, and 50 mM Tris-HCl at pH 7.8 and 0°C. (○) Rate constant of ATPase in the steady state (k_o^P); (●) rate constant of P_i liberation from E_P^{ADP} (k_d^P), measured by the rapid flow-dialysis method.

solution in the upper chamber. The rate of P_i liberation from E_P^{ADP} was measured under several conditions of pH, ionic strength, and temperature. Figure 4 shows the dependence on temperature of the rate constant of P_i liberation and that of ATPase in the steady state in the presence of 0.5M KCl, 2.5 mM MgCl$_2$, and 50 mM Tris-HCl, pH 7.8 at 0°C. The rate of P_i liberation from E_P^{ADP} was in the same order of magnitude as that of ATPase in the steady state at high ATP concentrations, but the two did not coincide. Furthermore, P_i liberation was not accelerated by adding 0.2 mM non-radioactive ATP 15 sec after starting the reaction.

The rate of ADP liberation was first observed by us (Kinoshita et al., 1969a) by measuring the liberation of free ADP coupled with the pyruvate kinase system or by measuring the rate of recovery of the change in the UV spectrum. The rate was found to be similar to that of ATPase in the steady state at high ATP concentrations. This was later confirmed by Taylor et al. (1970) using the gel filtration method. They also reported that the rate of ADP liberation from E_P^{ADP} was accelerated by adding nonradioactive ATP and that the E-ADP complex formed from free enzyme and ADP was decomposed at a rate comparable with that of decomposition of E_P^{ADP}.

We have recently remeasured the time course of ADP liberation from E_P^{ADP} using the rapid flow-dialysis method. The reaction was started by adding 0.2 ml of 55 μM [^{3}H]ATP solution to 2 ml of solution containing 2.75 mg/ml myosin in the upper chamber. Under the conditions used in the experiment shown in Fig. 5, i.e., in 0.5 M KCl,

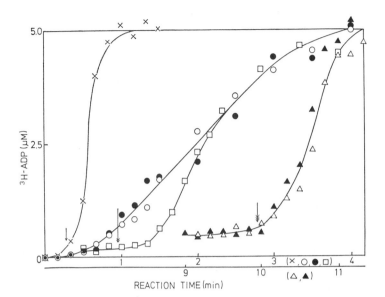

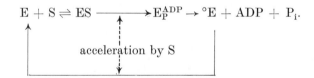

Figure 5. Time-course of ADP liberation from E_P^{ADP} measured by the rapid flow-dialysis method. The mixture contained 0.5 M KCl, 2.5 mM MgCl$_2$, and 50 mM Tris-HCl at pH 7.8 and 0°C. ($\times$) A mixture of 0.2 ml of 55 μM [^{3}H]ATP, 2 ml of 2.75 mg/ml myosin, and 0.1 ml of 4 mM ADP was kept at 0°C for 10–20 min, then poured into the upper chamber at time 0. (○, □, △) The reaction was started by adding 0.2 ml of 55 μM [^{3}H]ATP to 2 ml of 2.75 mg/ml myosin at time 0, and 0.1 ml of 4.4 mM nonradioactive ADP was added at the times indicated by the symbols ↓, ⇓, and ⇟, respectively. (●, ▲) 0.1 ml of 4.4 mM nonradioactive ATP was added at the times indicated by the symbols ↓ and ⇓, respectively.

2.5 mM MgCl$_2$, and 50 mM Tris-HCl, pH 7.8 at 0°C, almost all the [^{3}H]ADP produced by the ATPase reaction remained to be bound to myosin, and the rate of [^{3}H]ADP liberation was measured by adding 0.2 mM nonradioactive ADP or ATP. When nonradioactive ADP was added 17 sec after the start of the reaction, [^{3}H]ADP was liberated slowly from E_P^{ADP}, and the rate was almost the same as that of ATPase in the steady state at high ATP concentrations. On the other hand, when nonradioactive ADP was added 10 min after the start of the reaction, [^{3}H]ADP was liberated too rapidly to be estimated by the present method. The rate of liberation was unaffected by adding nonradioactive ATP instead of ADP. The liberation of ADP from the myosin-ADP complex formed by mixing free enzyme with ADP also occurred very rapidly. It was also reported recently by Trentham et al. (1972) that the rate of liberation of thio-ADP from the simple H-meromyosin-thio-ADP complex is much higher than that of the reaction intermediate, the H-meromyosin-phosphate-thio-ADP complex, using the UV-spectral change due to binding of H-meromyosin with a thio-nucleotide.

Thus, the rate of decomposition of E_P^{ADP} was even higher than that of ATPase in the steady state at higher ATP concentrations (cf. Fig. 4), and was about four times that in the steady state at lower ATP concentrations. Therefore, we concluded that in a low ATP concentration range E_P^{ADP} is decomposed via two steps: $E_P^{ADP} \rightarrow {}^\circ E +$ ADP + P$_i$ → E + ADP + P$_i$, where $^\circ$E is myosin, which differs in conformation from E and does not form E_P^{ADP} on adding ATP (see below). Furthermore, the rate-determining step is that for the conversion of $^\circ$E to E. This mechanism is consistent with the results that in a low ATP concen-

tration range the most stable intermediate of ATPase does not contain bound nucleotide, and that the net liberation of a stoichiometric amount of H$^+$ (Tonomura et al., 1969) and the recovery of the initial burst of P$_i$ liberation (Nakamura and Tonomura, 1968) after adding a stoichiometric amount of ATP to myosin is several times slower than that of the ATPase reaction in the steady state at high ATP concentrations.

However, the result that the rate of decomposition of E_P^{ADP} is of the same order of magnitude as that of the ATPase reaction in the steady state in a high ATP concentration range can be most easily explained by the following mechanism:

$$E + S \rightleftharpoons ES \longrightarrow E_P^{ADP} \rightarrow {}^\circ E + ADP + P_i.$$

acceleration by S

But the following results could not be explained by this mechanism: (a) *p*-Nitrothiophenylation of myosin decreased the amount of the initial burst of P$_i$ liberation to zero, but did not affect the rate of the ATPase reaction in the steady state (Kinoshita et al., 1969a). (b) The amount of initial burst of P$_i$ liberation of usual S-1 was 0.5 mole/mole, as described above. However, the amounts of initial burst for S-1 obtained by drastic proteolysis were much less than 0.5 mole/mole, whereas the ATPase activities in the steady state were almost the same as that of usual S-1 (Yagi et al., 1972). For example, the amount decreased to 0.1 mole/mole by treatment of usual S-1 with Nagarse. (c) The rate of recovery of the initial burst was much lower than that of decomposition of E_P^{ADP} as mentioned above.

(d) Furthermore, Bárány and Bárány (this volume) have shown that in resting muscle the nucleotide bound to myosin is ATP but not ADP.

Taylor et al. (1970) assumed that the enzyme-ADP complex has the same kinetic properties as the E_P^{ADP} complex. However, the UV spectrum (Morita, 1967; Sekiya and Tonomura, 1967) and ESR spectrum of spin-labeled myosin (Seidel et al., 1970; Seidel and Gergely, 1971) in the presence of ATP were different from those of the E-ADP complex. The rate of ADP liberation from E_P^{ADP} was also very different from that of ADP liberation from the myosin-ADP complex, as mentioned above. Furthermore, we have recently observed that a typical initial burst occurs after adding 20 μM ATP to myosin under conditions in which all the active sites of ATPase were occupied by ADP, i.e., with 1.9 mg/ml myosin, 15 μM ADP, and 0.5 M KCl at pH 7.8 and 0°C. Therefore it is evident, at least, that a simple enzyme-ADP complex cannot be the most stable intermediate of the ATPase reaction. On the other hand, all our results can be explained by the following mechanism (Inoue et al., 1972; Tonomura, 1972)—that the ATPase reaction in the steady state in a high concentration range of ATP is simple hydrolysis of ATP catalyzed by °E or E (route 2), different from the main route for hydrolysis in a low concentration range of ATP (route 1):

$$E + S \underset{K_1}{\rightleftharpoons} ES \xrightarrow{k_2} E_P^{ADP} \xrightarrow{k_3} {}^{\circ}E + ADP + P_i \quad (1)$$
$$\underset{k_4}{\big\uparrow}$$

$$^{\circ}E(E) + S \underset{K_5}{\rightleftharpoons} {}^{\circ}ES(ES) \xrightarrow{k_6} {}^{\circ}E(E) + ADP + P_i$$
$$(2)$$

The rate-limiting steps of these two routes are steps (4) and (6), respectively. The values of k_3, k_4, and k_6 were 0.44, 0.18, and 0.33 min^{-1}, respectively, in 0.5 M KCl at pH 7.8 and 0°C. The K_M value of route (2) was 1 μM, while that of route (1) was calculated to be 0.05 μM, which was consistent with the experimental value of less than 0.1 μM.

Acceleration of Decomposition of Myosin-Phosphate-ADP Complex by F-Actin

Since earlier studies showed two types of binding ratio of myosin to F-actin, indicating that the myosin molecule forms a complex with an actin monomer or dimer, we (Takeuchi and Tonomura, 1971) recently remeasured the binding ratios of H-meromyosin and S-1 to F-actin by light-scattering and ultracentrifugal methods. These studies clearly showed that H-meromyosin bound to an actin dimer and that S-1 bound to an actin monomer, indicating that each head part of the

myosin molecule can bind with an actin monomer. It is well known that splitting of ATP is accelerated by binding of F-actin with myosin and that actomyosin-ATPase is directly coupled with muscle contraction. Allosteric effects may be considered to be involved in the effect of F-actin on myosin-ATPase, since ADP bound to F-actin is not required for the accelerating effect (Bárány et al., 1966; Tokiwa et al., 1967) and since the effect can be simulated by trinitrophenylation of myosin (Tokuyama and Tonomura, 1967), although the site for trinitrophenylation is different from that for F-actin binding (Kubo et al., 1960).

The following results obtained by us clearly showed that E_P^{ADP} is a reaction intermediate in ATP hydrolysis catalyzed by actomyosin and that F-actin accelerates the decomposition of the intermediate into E, ADP, and P_i. First, as soon as F-actin was added to a mixture of myosin and ATP, the ATPase activity became approximately that of actomyosin in the steady state. There was no initial burst of P_i liberation or any lag phase (Kinoshita et al., 1969b). Second, no initial burst of P_i liberation could be observed with p-nitrothiophenyl myosin, but the rate of the myosin-ATPase reaction in the steady state was only slightly altered by this modification (Kinoshita et al., 1969b). Acto-p-nitrothiophenyl myosin reconstituted from p-nitrothiophenyl myosin and F-actin showed only myosin-ATPase activity under conditions where actomyosin-ATPase activity was observed with untreated myosin. The effects of pH and temperature on actomyosin-ATPase also supported the mechanism described above (Onishi et al., 1968).

Reconstituted actomyosin was dissociated, at least partially, into myosin and F-actin by addition of ATP under the conditions used for studies on actomyosin-ATPase. This dissociation complicated quantitative kinetic analyses of the actomyosin-ATPase reaction. We (Tonomura and Yoshimura, 1960; Tonomura, Yoshimura and Kanazawa, 1961) previously showed that the ATPase activity of actomyosin reconstituted from F-actin and myosin which had been treated with PCMB and then with β-mercaptoethanol to remove the PCMB was not inhibited by excess substrate. Recently, myosin or H-meromyosin was treated with PCMB, isolated by gel filtration, and then treated with dithiothreitol (DTT). This treatment, as described above, removed about half of g_2 and g_2' from myosin and H-meromyosin, respectively, without loss of myosin-ATPase activity (cf. Fig. 1B). When ATP was added to acto-H-meromyosin reconstituted from the PCMB-DTT-treated H-meromyosin and F-actin prepared by extraction at room temperature, H-meromyosin

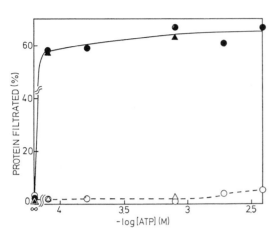

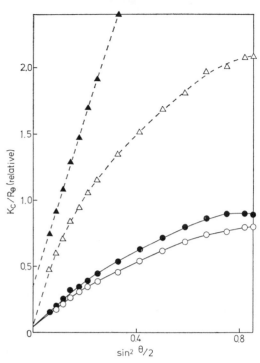

Figure 6. Rapid filtration of acto-H-meromyosin through a Millipore filter in the presence of ATP. The mixture contained 0.71 mg/ml H-meromyosin, 0.40 mg/ml F-actin, 2 mM MgCl$_2$, 0.05 M KCl, and 0.02 M Tris-maleate at pH 7.0 and 18–20°C. H-Meromyosin was prepared from untreated myosin ($\bullet$, $\blacktriangle$) or from PCMB-DTT-treated myosin ($\bigcirc$, $\triangle$), and F-actin was prepared by extraction at room temperature. ($\bigcirc$, $\bullet$) 0.1 mM CaCl$_2$; ($\triangle$, $\blacktriangle$) 1 mM EGTA. The pore size of the Millipore filter was 0.45 μ, and filtration was performed under vacuum within 15 sec after adding ATP.

Figure 7. Effect of treatment of H-meromyosin with PCMB and DTT on the change in angular distribution of light-scattering of acto-H-meromyosin on adding ATP. The mixture contained 0.25 mg/ml H-meromyosin and 0.14 mg/ml F-actin. The ionic environments were as for Fig. 6. H-Meromyosin was prepared from untreated myosin ($\bullet$, $\blacktriangle$) or from PCMB-DTT-treated myosin ($\bigcirc$, $\triangle$). Light-scattering intensities were measured within 3 min after adding ATP. ($\bigcirc$, $\bullet$) No ATP; ($\triangle$, $\blacktriangle$) 0.6 mM ATP.

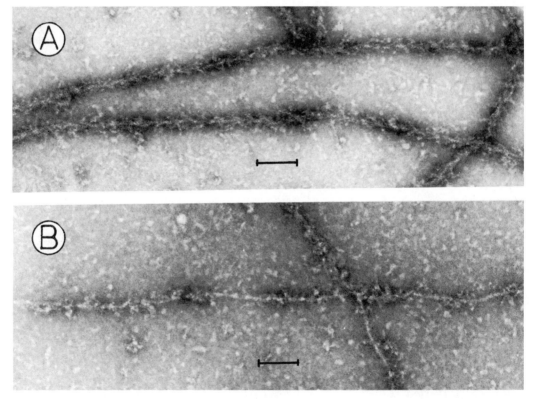

Figure 8. Electron micrographs of PCMB-DTT-treated acto-H-meromyosin in the presence and absence of ATP. 0.13 mg/ml PCMB-DTT-H-meromyosin, 0.07 mg/ml F-actin extracted at room temperature, 0.04 M KCl, 2 mM MgCl$_2$, 40 μM CaCl$_2$, 0.02 M Tris-maleate, pH 7.0, 20°C. *A.* No ATP; *B.* 0.6 mM ATP. Bar equals 0.1 μ.

176

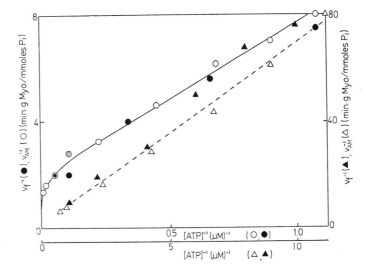

Figure 9. Double reciprocal plots of the rate of the actomyosin-ATPase reaction, v_{AM}, and the rate of the initial burst of P_i liberation of myosin-ATPase, v_f, against the ATP concentration. The ionic environments were as for Fig. 6; temperature was 24–25°C. ($\bigcirc$, $\triangle$) ActomyosinATPase, 0.015 mg /ml of myosin pretreated with PCMB and β-mercaptoethanol, 0.015 mg/ml F-actin; ($\bullet$, $\blacktriangle$) initial burst of P_i liberation, 0.1 mg/ml of myosin pretreated with PCMB and β-mercaptoethanol.

did not pass through a Millipore filter (pore size, 0.45 μ), whereas most of the H-meromyosin of acto-H-meromyosin reconstituted from untreated H-meromyosin and F-actin could pass through the filter (Fig. 6). The angular distribution of light-scattering intensity of acto-H-meromyosin reconstituted from the PCMB-DTT-treated H-meromyosin and F-actin showed that the weight-average molecular weight was scarcely changed by adding ATP, but that the radius of gyration increased to about twice as large as that in the absence of ATP (Fig. 7). Thus, it was concluded that the complex of F-actin with PCMB-DTT-treated H-meromyosin did not dissociate, but expanded in size on adding ATP, whereas the complex of F-actin with untreated H-meromyosin almost completely dissociated into F-actin and H-meromyosin. Figure 8 shows the electron micrographs of negatively stained PCMB-DTT-treated acto-H-meromyosin in the presence and absence of ATP. Thus, ATP induced remarkable changes in the ultrastructure of PCMB-DTT-treated acto-H-meromyosin: ATP changed the "arrowhead" structure of acto-H-meromyosin into a new structure, in which H-meromyosin remained bound to F-actin and formed projections oriented more perpendicularly to the F-actin axis than in the absence of ATP.

We (Inoue et al., 1972) compared the rate of formation of E_P^{ADP} with the rate of the actomyosin-ATPase reaction in the steady state using myosin treated with PCMB and then with β-mercaptoethanol. As shown in Fig. 9, the two rates were equal over a wide range of ATP concentrations. These results also strongly support the conclusions that E_P^{ADP} is a reaction intermediate in actomyosin-ATPase and that actin greatly accelerates the

step(s) of regeneration of myosin from E_P^{ADP} without affecting the rate of its formation.

The molecular mechanism of translocation of the head part of the myosin molecule, which is assumed to be coupled with ATP-hydrolysis, remains to be clarified. Therefore, several types of molecular mechanism of muscle contraction can be presented on the basis of the studies on the structure and function of myosin described in this report, and they are discussed in our monograph (Tonomura, 1972).

References

BÁRÁNY, M., A. F. TUCCI, and T. E. CONOVER. 1966. The removal of the bound ADP of F-actin. *J. Mol. Biol.* **19:** 483.

COLOWICK, S. P. and F. C. WOMACK. 1969. Binding of diffusible molecules by macromolecules: Rapid measurement by rate of dialysis. *J. Biol. Chem.* **244:** 774.

DREIZEN, P. and L. C. GERSHMAN. 1970. Relationship of structure to function in myosin. II. Salt denaturation and recombination experiments. *Biochemistry* **9:** 1688.

HAYASHI, Y. 1972. Submolecular structure of subfragment-1 of the myosin molecule. *J. Biochem.* **72:** 83.

HAYASHI, Y. and Y. TONOMURA. 1970. On the active site of myosin A-adenosine triphosphatase. X. Functions of two subfragments, S-1, of the myosin molecule. *J. Biochem.* **68:** 665.

IMAMURA, K., T. KANAZAWA, M. TADA, and Y. TONOMURA. 1965. The pre-steady state of the myosin-adenosine triphosphate system. III. Properties of the intermediate. *J. Biochem.* **57:** 627.

INOUE, A., K. SHIBATA-SEKIYA, and Y. TONOMURA. 1972. The pre-steady state of the myosin-adenosine triphosphate system. XI. Formation and decomposition of the reactive myosin-phosphate-ADP complex. *J. Biochem.* **71:** 115.

KANAZAWA, T. and Y. TONOMURA. 1965. The pre-steady state of the myosin-adenosine triphosphate system. I. Initial rapid liberation of inorganic phosphate. *J. Biochem.* **57:** 604.

KINOSHITA, N., S. KUBO, H. ONISHI, and Y. TONOMURA. 1969a. The pre-steady state of the myosin-adenosine triphosphate system. VIII. Intermediate formation and activation of myosin by ATP. *J. Biochem.* **65:** 285.

KINOSHITA, N., T. KANAZAWA, H. ONISHI, and Y. TONOMURA. 1969b. The pre-steady state of the myosin-adenosine triphosphate system. IX. Effect of F-actin on the myosin-ATP system. *J. Biochem.* **65:** 567.

KUBO, S., S. TOKURA, and Y. TONOMURA. 1960. On the active site of myosin A-adenosine triphosphatase. I. Reaction of the enzyme with trinitrobenzenesulfonate. *J. Biol. Chem.* **235:** 2835.

LOWEY, S. and D. RISBY. 1971. Light chains from fast and slow muscle myosins. *Nature* **234:** 81.

LOWEY, S., H. S. SLAYTER, A. G. WEEDS, and H. BAKER. 1969. Substructure of the myosin molecule. I. Subfragments of myosin by enzyme degradation. *J. Mol. Biol.* **42:** 1.

LYMN, R. W. and E. W. TAYLOR. 1970. Transient state phosphate production in the hydrolysis of nucleotide triphosphates by myosin. *Biochemistry* **9:** 2975.

MORITA, F. 1967. Interaction of heavy meromyosin with substrate. I. Difference in ultraviolet absorption spectrum between heavy meromyosin and its Michaelis-Menten complex. *J. Biol. Chem.* **242:** 4501.

————. 1971. Interaction of heavy meromyosin with substrate. V. Heterogeneity in the binding of ADP. *J. Biochem.* **69:** 517.

MURPHY, A. J. and M. F. MORALES. 1970. Number and location of adenosine triphosphatase sites of myosin. *Biochemistry* **9:** 1528.

NAKAMURA, H. and Y. TONOMURA. 1968. The pre-steady state of the myosin-adenosine triphosphate system. V. Evidence for a phosphate exchange reaction between adenosine triphosphate and the "reactive myosin-phosphate complex". *J. Biochem.* **63:** 279.

NANNINGA, L. B. and W. F. H. M. MOMMAERTS. 1960. Studies on the formation of an enzyme-substrate complex between myosin and adenosinetriphosphate. *Proc. Nat. Acad. Sci.* **46:** 1155.

NAUSS, K. M., S. KITAGAWA, and J. GERGELY. 1969. Pyrophosphate binding to and adenosine triphosphatase activity of myosin and its proteolytic fragments. Implications for the substructure of myosin. *J. Biol. Chem.* **244:** 755.

OHE, M., B. K. SEON, K. TITANI, and Y. TONOMURA. 1970. On the active site of myosin A-adenosine triphosphatase. IX. Chemical modification of the enzyme by monoiodoacetamide. *J. Biochem.* **67:** 513.

ONISHI, H., H. NAKAMURA, and Y. TONOMURA. 1968. The pre-steady state of the myosin-adenosine triphosphate system. VI. Effect of ATP concentration, pH and temperature. *J. Biochem.* **63:** 739.

SARTORELLI, L., H. J. FROMM, R. W. BENSON, and P. D. BOYER. 1966. Direct and ^{18}O-exchange measurement relevant to possible activated or phosphorylated stage of myosin. *Biochemistry* **5:** 2877.

SEIDEL, J. C. and J. GERGELY. 1971. The conformation of myosin during the steady state of ATP hydrolysis: Studies with myosin spin labeled at the S_1 thiol groups. *Biochem. Biophys. Res. Comm.* **44:** 826.

SEIDEL, J. C., M. CHOPEK, and J. GERGELY. 1970. Effect of nucleotides and pyrophosphate on spin labels bound to S_1 thiol groups of myosin. *Biochemistry* **9:** 3265.

SEKIYA, K. and Y. TONOMURA. 1967. Change in ultraviolet absorption spectrum of H-meromyosin induced by its binding with substrate and competitive inhibitor. *J. Biochem.* **61:** 787.

SHIMADA, T. 1970. On the active site of myosin A-adenosine triphosphatase. VIII. Modification of the enzyme with diazonium-1H-tetrazole. *J. Biochem.* **67:** 185.

SLAYTER, H. S. and S. LOWEY. 1967. Substructure of the myosin molecule as visualized by electron microscopy. *Proc. Nat. Acad. Sci.* **58:** 1611.

STRACHER, A. 1969. Evidence for the involvement of light chains in biological functioning of myosin. *Biochem. Biophys. Res. Comm.* **35:** 519.

TAKEUCHI, K. and Y. TONOMURA. 1971. Formation of acto-H-meromyosin and acto-subfragment-1 complexes and their dissociation by adenosine triphosphate. *J. Biochem.* **70:** 1011.

TAYLOR, E. W., R. W. LYMN, and G. MOLL. 1970. Myosin-product complex and its effect on the steady state rate of nucleotide triphosphate hydrolysis. *Biochemistry* **9:** 2984.

TOKIWA, T., T. SHIMADA, and Y. TONOMURA. 1967. Role of ADP of F-actin in superprecipitation and enzymatic activity of actomyosin. *J. Biochem.* **61:** 108.

TOKUYAMA, H. and Y. TONOMURA. 1967. On the active site of myosin A-adenosine triphosphatase. VII. Effect of trinitrophenylation of myosin on the decomposition of phosphoryl myosin. *J. Biochem.* **62:** 456.

TOKUYAMA, H., S. KUBO, and Y. TONOMURA. 1966. Molecular properties of fraction S-1 from a trypsin digest of myosin. *Biochem. Z.* **345:** 57.

TONOMURA, Y. 1972. Muscle proteins, muscle contraction and cation transport. University of Tokyo Press, Tokyo.

TONOMURA, Y. and T. KANAZAWA. 1965. Formation of a reactive myosin-phosphate complex as a key reaction in muscle contraction. *J. Biol. Chem.* **240:** PC4110.

TONOMURA, Y. and F. MORITA. 1959. The binding of pyrophosphate to myosin A and myosin B. *J. Biochem.* **46:** 1367.

TONOMURA, Y. and F. OOSAWA. 1972. Molecular mechanism of contraction. *Ann. Rev. Biophys.* **1:** 159,

TONOMURA, Y. and Y. YOSHIMURA. 1960. Inhibition of myosin B-adenosine triphosphatase by excess substrate. *Arch. Biochem. Biophys.* **90:** 73.

TONOMURA, Y., Y. YOSHIMURA, and S. KITAGAWA. 1961. On the active site of myosin A-adenosine triphosphatase. III. Effect of pretreatment and metal ions on the clearing response of actomyosin to adenosine triphosphate. *J. Biol. Chem.* **236:** 1968.

TONOMURA, Y., H. NAKAMURA, N. KINOSHITA, H. ONISHI, and M. SHIGEKAWA. 1969. The pre-steady state of the myosin-adenosine triphosphate system. X. The reaction mechanism of the myosin-ATP system and a molecular mechanism of muscle contraction. *J. Biochem.* **66:** 599.

TRENTHAM, D. R., R. G. BARDSLEY, J. F. ECCLESTON, and A. G. WEEDS. 1972. Elementary processes of the magnesium ion-dependent adenosine triphosphatase activity of heavy meromyosin. *Biochem. J.* **126:** 635.

TROTTA, P. P., P. DREIZEN, and A. STRACHER. 1968. Studies on subfragment-1, a biological active fragment of myosin. *Proc. Nat. Acad. Sci.* **61:** 659.

WEBER, K. and M. OSBORN. 1969. The reliability of molecular weight determinations by dodecyl sulfate-polyacrylamide gel electrophoresis. *J. Biol. Chem.* **244:** 4406.

WEEDS, A. G. and S. LOWEY. 1971. Substructure of the myosin molecule. II. The light chain of myosin. *J. Mol. Biol.* **61:** 701.

YAGI, K., Y. YAZAWA, F. OHTANI, and Y. OKAMOTO. 1972. The enzymatically active portion of myosin. In *Organization of energy-transducing membranes*, ed. M. Nakao and L. Packer. University of Tokyo Press, Tokyo.

Studies on Mechanism of Myosin and Actomyosin ATPase

J. F. Koretz,* T. Hunt, and E. W. Taylor*

Department of Biophysics, University of Chicago and MRC Muscle Biophysics Unit, Kings College, London WC2, England

Studies of the structure and of the mechanical and thermal properties of striated muscle have led to the general acceptance of a model of contraction which could be described somewhat awkwardly as the sliding filament—moving bridge model. The evidence in support of the model has been summarized in numerous articles (H. E. Huxley, 1968; Taylor, 1972) and is presented in some detail in this volume. The essential point is that the macroscopic properties are a summation of the molecular processes occurring at a single myosin cross-bridge. The contraction cycle involves (1) detachment of the cross-bridge from the actin-containing thin filament, (2) movement of the free cross-bridge, (3) reattachment of the cross-bridge in a new orientation, and (4) movement of the cross-bridge, probably by rotation, to its original orientation with sliding of the thin filament. This cyclic process is accompanied by the hydrolysis of one or more molecules of ATP.

The problem to be solved by studies of the chemistry of actomyosin can be stated with reasonable clarity. The contraction cycle described above must include at least four steps. The hydrolysis of ATP by actomyosin in solution must involve intermediate steps, the number being as yet unknown. The purpose of kinetic studies is to determine the sequence of intermediate steps and to establish a correspondence between the steps in the enzyme mechanism and the postulated contraction cycle.

Two kinds of difficulties stand in the way of carrying out this program. In muscle the reaction occurs subject to the restrictions imposed by a highly ordered lattice structure. It could be argued that the dominant reaction pathway in muscle is quite different from that in solution. There is no simple way to get around this objection, other than the careful application of common sense. Properties of the homogeneous system in solution cannot be immediately translated into the properties of muscle. Conclusions based on enzyme studies have to be validated whenever possible by a study of ordered structures of increasing complexity, namely thick filaments, myofibrils, and glycerinated muscles. A second problem is the coupling of the

reaction to the performance of work. Again it can be argued that coupling will alter the rate of the reaction in such a way as to favor an alternate reaction pathway in muscle. This is a serious difficulty, and perhaps the hardest property to explain is not that muscle contracts, but that it does so with an efficiency of 50 percent. Studies of the enzyme mechanism cannot directly determine which is the mechanochemically coupled step. However introducing a coupled step does not present a conceptual problem, and it may be hoped that eventually the enzyme mechanism and the contraction cycle will fit together in a "natural" way so that there is only one logical choice for the coupled step.

Previous studies have established some important aspects of the mechanism. For myosin under physiological conditions (10^{-3} M $MgCl_2$, 0.1 M KCl) the steady state rate is very slow, but the first molecule of ATP is hydrolyzed with a rate constant the order of 100 sec^{-1} (Lymn and Taylor, 1970). As there is no evidence for a phosphorylated intermediate state and both reaction products dissociate slowly, the rate-limiting step is the breakdown of some intermediate complex into enzyme plus products (Taylor et al., 1970). With acto-HMM, while the steady state rate is much faster, the early phase is essentially described by similar rate processes. The rate of dissociation of acto-HMM on mixing with ATP appears to be a faster step than the hydrolysis of ATP, which suggests that dissociation occurs into actin plus myosin-substrate complex. The myosin-intermediate complex will recombine with actin with displacement of products. (Lymn and Taylor, 1971). On the basis of this evidence the following provisional scheme was advanced

$$S + AM \leftrightarrows AMS \quad (AM, Pr) \xrightarrow[(5)]{\text{fast}} AM + ADP + P_i$$

(1)

where (M, Pr) refers to the intermediate complex formed by hydrolysis of ATP. Various lines of

* Present address: MRC Muscle Biophysics Unit, Kings College London.

evidence suggest that the (M, Pr) intermediate is not identical with the complex generated by adding ADP and P_i to myosin (this complex is written $M \cdot ADP \cdot P_i$). The difference spectrum (Morita, 1967), spin-labeled spectrum of myosin (Seidel and Gergely, 1971), and fluorescence emission spectrum (Szent-Geörgyi, 1972) are different for $M \cdot ADP$ than for the complex generated by adding ATP. Furthermore, indirect measurements of the rate of dissociation of $M \cdot ADP$ gave values some 20 to 50 times larger than the steady state rate of hydrolysis (Malik and Martonosi, 1972; Trentham et al., 1972). The intermediate complex may pass through a $M \cdot ADP \cdot P_i$ state before the products dissociate or it may undergo an ordered dissociation. In either case the environment of bound ADP and therefore presumably the configuration of myosin is different in the (M, Pr) and $M \cdot ADP$ complexes.

The work presented here will deal with two aspects of the mechanism, the liberation of a proton during the early phase and the control of the rate of ATP hydrolysis by Ca ion.

Materials and Methods

Rabbit myosin and heavy meromyosin were prepared as described previously (Finlayson and Taylor, 1969). Actin was obtained free of relaxing proteins by the method of Spudich and Watt (1971); subfragment 1 was prepared by papain digestion using either soluble or insoluble papain as described by Lowey et al. (1969). Natural tropomyosin (the complex of tropomyosin plus troponins) was made by the method of Hartshorne (1969). Proton liberation was measured by the change in optical density of orthocresol sulfonethaleine ($\lambda = 572$ nm) at pH 8.0 employing a modified Aminco-Morrow stopped-flow apparatus. Turbidity was measured to follow the association and dissociation of actin and myosin using the same apparatus at $\lambda = 400$ nm.

The Proton Early Burst

At pH 8 there is a stoichiometric release of a H^+ in the hydrolysis of ATP and such a release might be expected to accompany the phosphate early burst. Previous studies (Finlayson and Taylor, 1969) did show a H^+ burst, but the magnitude was less than the phosphate burst and the data under some conditions did not appear to fit a single-rate process. Tonomura and collaborators (Kokiwa and Tonomura, 1965) reported an absorption rather than a liberation of a proton during the early phase.

The problem required a more thorough investigation over a range of ionic strengths and temperatures. The possibility of two rate processes prompted a comparison of subfragment-1 with myosin or HMM under the same conditions. The results of a preliminary study are reported here.

Typical oscilloscope traces for myosin and subfragment 1 in 1 M KCl are shown in Fig. 1. The recordings clearly show a rapid H^+ release at the beginning of the reaction. The steady state rate at this ionic strength is sufficiently slow to give an almost horizontal trace at this sweep rate (50 msec per major division). Under all conditions employed (0.1 M to 1.0 M KCl, 10° and 20°, pH 8, 2×10^{-5} M to 2×10^{-3} M ATP, 10^{-3} to 5×10^{-3} M $MgCl_2$) a proton liberation was obtained during the early phase for myosin, HMM, and S-1.

The simple two-step mechanism

$$M + ATP \xrightleftharpoons{k_1} M \cdot ATP \xrightleftharpoons{k_2} (M, Pr) + H^+ \quad (2)$$

predicts that the time course of H^+ liberation should fit two exponential terms. One term is a small early lag which is difficult to distinguish

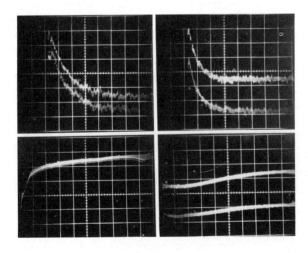

Figure 1. *Upper:* Oscilloscope traces, in duplicate showing early proton burst phase for myosin (*left*) and subfragment-1 (*right*). Experimental conditions: 0.5 M KCl pH 8, 20°C, 8×10^{-5} M ATP, 10^{-3} M $MgCl_2$. Time scale is 50 msec per major division.

Lower: Recombination of actin-troponin-tropomyosin complex with heavy meromyosin-product intermediate in the presence (*left*) and absence (*right*) of Ca^{++} as measured by increase in turbidity at $\lambda = 400$ nm. Turbidity change for complete recombination of actin with heavy meromyosin was approximately five major divisions. Experimental conditions: 0.1 M KCl, 10^{-3} M $MgCl_2$, 10°C, 0.8 moles ATP added per heavy meromyosin site plus creatine phosphate and creatine kinase. *Left:* 2 sec per major division; *right* duplicate traces, 5 sec per major division.

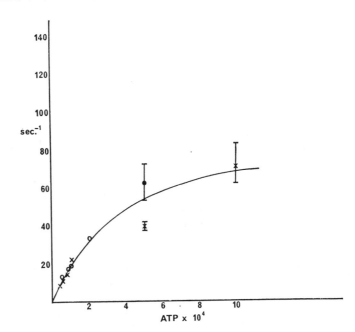

Figure 2. Variation of the apparent rate constant for proton liberation with ATP concentration for myosin ($\bigcirc$) and subfragment-1 ($\times$). Experimental conditions: 1 M KCl, 20°C, 10^{-3} M $MgCl_2$ pH 8.

from the perturbations occurring when drive stops. The time course of most of the process should follow a single exponential, and analysis of the records by plotting the log of voltage difference versus time gave a good fit to a straight line. The steady state rate was determined from the slope of a slow sweep (5 sec per major division) and this contribution to the voltage change was subtracted before plotting the data.

The variation in the apparent rate constant with concentration is shown for myosin and S-1 in 1.0 M KCl (Fig. 2). Each point is the average of at least four determinations; it is evident that the rate is the same for both preparations within experimental error. Although studies are not completed, the rates appear approximately the same for myosin, HMM, or S-1 at the same ionic strength at 20°.

Denoting the forward and reverse rates for the mechanism (Eq. 2) by k_1, k_{-1} and k_2, k_{-2}, it is easily shown, subject to the condition $k_{-1} \ll k_2$ that the apparent rate is equal to k_1S if $k_1S \ll k_2$ and reaches a maximum value of k_2 at high S such that $k_1S \geq k_2$. If $k_{-1} \gg k_2$, the rate at low S is $\dfrac{k_1}{k_{-1}} k_1S$. Since further evidence is needed to distinguish between these two possibilities, we will refer to the rate constant determined from the slope of the plot at low S as an apparent second-order rate constant for substrate binding.

While this work was in progress, a preliminary report was published by Pembrick and Walz (1972) who found a single proton step for S-1 but apparently two rate processes for myosin. In total these studies have raised more problems than they have answered, although the following points appear to be clearly established. A proton is released during the early phase with a time course at least approximately the same as the phosphate formation measured by Lymn and Taylor (1970). The apparent first-order rate constant obtained from the concentration dependence of the rate is also in good agreement for the recent proton measurements and phosphate measurements. This is to be expected since both measurements should give the apparent rate constant for substrate binding.

For subfragment-1 a single-rate process describes the data, but for myosin there is a disagreement: the present work appears to be fitted by a single rate constant at a given concentration, whereas our earlier studies at 20° and recent work of Pembrick and Walz requires two rate constants. This is a crucial problem since it would suggest interaction between substrate sites or a change in configuration when one substrate site is occupied. The answer depends on the accuracy of fitting data to a known function, and it can be resolved by proper statistical analyses and improved methods of data collection. As further studies are in progress the problem is best put aside for the present.

The H^+ liberation studies raise further questions about the mechanism proposed in Eq. 1. As can be seen in Fig. 2, the rate appears to attain a maximum value of about 80 sec^{-1} at $1-2 \times 10^{-3}$ M ATP (i.e., $k_2 \simeq 80$ sec^{-1}). As the ionic strength is reduced, k_2 is larger and consequently more difficult to measure. In 0.5 M KCl k_2 is 200–250 sec^{-1}. In previous studies of the early phosphate production a value of $k_2 = 60$ sec^{-1} was obtained in 0.5 M KCl and 100–150 sec^{-1} was obtained in 0.05 M KCl. Thus the maximum rate for the H$^+$ step appears to be faster than for phosphate formation, but both values should be the same according to the simple mechanism, Eq. 2. However for technical reasons measurements of phosphate formation were made up to a concentration of only 2.5×10^{-4} M ATP. At this concentration, although the slope of the plot is decreasing in the case of H$^+$ measurements, expansion of the concentration scale by a factor of ten indicates that the actual maximum rate is higher than would have been estimated from data taken out to only 2.5×10^{-4} M. It appears likely that the figures quoted for the maximum rate from phosphate measurements are low by a factor of at least two, and further studies are being attempted to extend the concentration range.

In summary Eq. 2, which attributes the early H$^+$ burst to the very fast hydrolysis of ATP, is an approximate description of the mechanism. However the accuracy of the rate constant measurements is sufficiently poor so that other possibilities are not yet excluded. For the overall reaction the H$^+$ would arise from the ionization of the phosphate moiety, but the product remains bound to the enzyme in the early phase. As the environment of the product is unknown, the pK could be lower than that of free inorganic phosphate, and the occurrence of ^{18}O exchange (Sartorelli et al., 1966) could indicate that other phosphate-containing moieties such as metaphosphate PO_3^- may contribute. In any case the magnitude of the H$^+$ is smaller than the phosphate burst, and the magnitude decreases with decreasing ionic strength.

Mechanism of Relaxation

Activation of actomyosin ATPase requires the binding of Ca ion to the troponin sites on the thin filament. Examination of the kinetic scheme (Eq. 1) suggests that the step

$$(M, Pr) + A \frac{k_4}{k_{-4}} (AM, Pr)$$

or the dissociation of products (step 5) or both should be blocked by removal of Ca ion. Studies by Eisenberg and Kielley (1970) and by Parker et al. (1970) support the explanation that removal of Ca weakens the interaction of actin with myosin in the presence of ATP. Their results are also consistent with the Eisenberg and Moos mechanism (1968) as well as our Eq. 1 and do not aid in distinguishing between the two mechanisms. A direct approach is to measure the rate constants in the transient state for the various steps in Eq. 1. Preliminary studies have been made of the forward

rates of the reactions $A + M \rightleftharpoons AM$, $AM + ATP \rightleftharpoons A + M \cdot ATP$, $A + (M, Pr) \rightleftharpoons (AM, Pr)$ in the presence and absence of Ca ion. For the third reaction the (M, Pr) state must be maintained while the free ATP concentration is kept sufficiently low so that dissociation of AM by ATP does not mask the initial recombination.

One syringe of the stop-flow was loaded with a solution containing creatine phosphate, creatine kinase, HMM, and ATP at ratios of ATP to HMM heads of 0.8, 1.0, and 1.2. Sufficient CP was added to maintain a constant rate of hydrolysis for at least ten minutes. The second syringe contained actin-troponin-tropomyosin and in some experiments excess creatine to partially block ATP generation after mixing. Experiments were performed at 10° or 20° in 0.1 M KCl. Aliquots of the myosin-CP-ATP-CP kinase solution containing [^{32}P]CP and [^{3}H-γ-^{32}P]ATP at the same ^{32}P specific activity were subject to rapid column chromatography at 10° to verify that a myosin-intermediate complex had been formed. The starting mixture was found to contain roughly 1–1.2 moles of products per mole of HMM; i.e., on the average approximately one site per HMM was occupied by products.

Removal of Ca ion reduced the rate of the $A + M \rightleftharpoons AM$ reaction by a factor of two or three; it had little effect on the rate of dissociation by ATP and essentially blocked the recombination of actin with the (M, Pr) complex. A typical $A + (M, Pr)$ recombination experiment is shown in Fig. 1. In the presence of Ca^{++} there is a rapid rise in turbidity after mixing. The curve is similar to that obtained for the mixing of $A + M$, although the second-order rate constant is reduced by roughly a factor of ten. In the absence of Ca^{++} there is essentially no turbidity increase over a period of ten seconds, followed by a slow increase amounting to about 20% of the change which accompanies the formation of the AM complex at the same concentration. No attempt will be made to analyze the results in terms of rate processes as the nature of the secondary turbidity change is unknown. However the general result is fairly clear. Recombination of actin with the (M, Pr) intermediate is blocked by the removal of Ca ion. Even the rate of binding of myosin alone to actin is affected by Ca^{++}. Thus the binding of Ca^{++} to troponin affects the configuration of the site on actin to which myosin attaches.

Discussion

The occasion of a general symposium on muscle contraction provides a strong stimulus to present a model of the molecular mechanism. Considerable progress has indeed been made in the past few years in our understanding of the properties of the various protein components, the mechanism of ATP hydrolysis by actomyosin, and the control of enzyme activity by Ca^{++} and the relaxing protein system.

A preliminary mechanism has been proposed, based largely on work in this laboratory, which appears to be in accord with much of the available data and which fits in a natural way with the properties expected of the contraction cycle. But at best such a scheme is incomplete; it represents a first attempt at ordering the steps in the cycle. Other intermediates remain to be described; the nature of the (M, Pr) state and its relation to the products complex $M \cdot ADP \cdot P_i$ is unclear; and little is known about the (AM, Pr) complex. The question of interaction between the substrate binding sites has not been satisfactorily answered.

It does not appear reasonable to expend much effort on formulating theoretical models which attempt to deduce the macroscopic properties of muscle from an enzyme scheme, although models can be of some use if their failure to work suggests to us the kind of information that is missing.

If we begin with an enzyme scheme such as Eq. 1 and assume that all of the rate constants are known, we could generate a contraction model by making a few assumptions. First, the rate of combination of actin with a myosin-intermediate state in solution is a second-order reaction, but in muscle the rate is determined by the constraints introduced by the lattice, and some assumption must be made. Second, at least one step in the cycle must be coupled to the load in such a way that increasing the load reduces the rate of reaction. This problem is probably not serious, since the effect of load could be taken into account by introducing an equilibrium constant for the coupled step which varies with the load. The recent model of Huxley and Simmons (1971) provides an example of this type of treatment.

If we proceed in this way, it becomes clear that the available evidence falls far short of what is required for a realistic model. Kinetic studies hopefully can determine the sequence of steps for the major pathway of ATP hydrolysis in solution. However nothing is known as yet about the rates of configurational changes of myosin which might occur with binding or hydrolysis of substrate or interaction with actin. Indirect evidence for configurational changes has been discussed here and in other papers in this volume, but it is not the kind of evidence that can define a contraction model.

What is needed is a detailed description of the various configurational states and a knowledge of the rates of the possible transitions. The con-

traction cycle is a closed pathway consisting of several steps. The overall efficiency is at least 50% at optimal load, yet there are a number of steps at which the pathway could branch to bypass the mechanochemically coupled step. For example, if a myosin bridge can combine with actin, rotate and do work, it can also rotate, combine and do no work. A satisfactory model must not only enumerate the steps which do occur in the contraction cycle, but also give evidence why alternate reaction pathways don't occur. The slow decay of the (M, Pr) intermediate provides an example of how branching at a particular step can be ruled out on the basis of kinetic evidence. For other steps in the presumed mechanism, particularly those concerned with bridge movement, little is known about the chemical or configurational states, and there is no evidence that would allow us to eliminate alternate pathways.

Acknowledgments

This work was supported by grants from the National Institutes of Health, General Medical Research, Muscular Dystrophy Association of America, and Life Insurance Medical Research Fund.

References

EISENBERG, E. and W. W. KIELLEY. 1970. Native tropomyosin: Effect on the interaction of actin with heavy meromyosin and subfragment 1. *Biochem. Biophys. Res. Comm.* **40:** 50.

EISENBERG, E. and C. MOOS. 1968. The adenosine triphosphatase activity of acto-heavy meromyosin. A kinetic analysis of actin activation. *Biochemistry* **7:** 1468.

FINLAYSON, B. and E. W. TAYLOR. 1969. Hydrolysis of nucleoside triphosphate by myosin during the transient state. *Biochemistry* **8:** 802.

HARTSHORNE, D. J., M. THEINER, and H. MUELLER. 1969. Studies on troponin. *Biochim. Biophys. Acta* **175:** 301.

HUXLEY, A. F. and R. M. SIMMONS. 1971. Proposed mechanism of force generation in striated muscle. *Nature* **233:** 533.

HUXLEY, H. E. 1969. The mechanism of muscle contraction. *Science* **164:** 1356.

LYMN, R. W. and E. W. TAYLOR. 1970. Transient state phosphate production in the hydrolysis of nucleoside triphosphates by myosin. *Biochemistry* **9:** 2984.

————. 1971. Mechanism of adenosine triphosphate hydrolysis by actomyosin. *Biochemistry* **10:** 4617.

LOWEY, S., H. S. SLAYTER, A. G. WEEDS, and H. BAKER. 1969. Substructure of the myosin molecule. I. Subfragments of myosin by enzymic degradation. *J. Mol. Biol.* **42:** 1.

MALIK, M. N. and A. MARTONOSI. 1972. The regulation of the rate of ATP hydrolysis by H-meromyosin. *Arch. Biochem. Biophys.* **152:** 243.

MORITA, F. 1967. Interaction of HMM with substrate. I. Difference in ultraviolet absorption spectrum between HMM and its Michaelis-Menten constant. *J. Biol. Chem.* **242:** 4501.

PARKER, L., M. Y. PYUN, and J. HARTSHORNE. 1970. The inhibition of adenosine triphosphate activity of the subfragment-1 actin complex by troponin plus tropomyosin, troponon B plus tropomyosin and troponin B. *Biochim. Biophys. Acta* **223**: 453.

PEMBRICK, S. M. and F. G. WALZ. 1972. Initial rapid proton liberation during hydrolysis of adenosine triphosphate by myosin subfragment-1. *J. Biol. Chem.* **247**: 2959.

SARTORELLI, L., H. J. FROMM, R. W. BENSON, and P. D. BOYER. 1966. Direct and ^{18}O-exchange measurements relevant to possible activated or phosphorylated states of myosin. *Biochemistry* **5**: 2877.

SEIDEL, J. C. and J. GERGELY. 1971. The conformation of myosin during the steady state of ATP hydrolysis. Studies with myosin spin labeled at the S_1 thiol groups. *Biochem. Biophys. Res. Comm.* **44**: 826.

SPUDICH, J. A. and S. WATT. 1971. The regulation of rabbit skeletal muscle. I. Biochemical studies of the interaction of tropomyosin-troponin complex with actin and proteolytic fragments of myosin. *J. Biol. Chem.* **247**: 4866.

SZENT-GYÖRGYI, A. G. 1972. *Biochemistry.* In Press.

TAYLOR, E. W. 1972. Chemistry of muscle contraction. *Annu. Rev. Biochem.* **41**: 577.

TAYLOR, E. W., R. W. LYMN, and G. MOLL. 1970. Myosin-product complex and its effect on the steady-state rate of nucleoside triphosphate hydrolysis. *Biochemistry* **9**: 2984.

TOKIWA, T. and Y. TONOMURA. 1965. The pre-steady state of the myosin adenosine triphosphate system. II. Initial rapid absorption and liberation of hydrogen ion followed by a stopped-flow method. *J. Biochem.* (Japan) **57**: 616.

TRENTHAM, D. R., R. G. BARDSLEY, J. F. ECCLESTON, and A. G. WEEDS. 1972. Elementary processes of the magnesium ion-dependent adenosine triphosphatase activity of heavy meromyosin: A transient kinetic approach to the study of kinases and adenosine triphosphatases and a colorimetric inorganic phosphate assay in situ. *Biochem. J.* **126**: 635.

Kinetics of Formation and Dissociation of H-Meromyosin-ADP Complex

A. MARTONOSI AND M. N. MALIK

Department of Biochemistry, St. Louis University School of Medicine, St. Louis, Missouri 63104

Previous work by Taylor et al. (1970) and by our laboratory (Malik and Martonosi, 1971) established that the rate-limiting step of Mg^{++}-moderated ATP hydrolysis by H-meromyosin (HMM) at 6–10°C may be the release of ADP from the active site.

There are indications for a temperature-dependent transition in the kinetics of myosin ATPase at 16°C, which is reflected in a change of the activation energy of ATP hydrolysis (Levy et al., 1959) and in altered responses to various modifiers of ATPase activity (Perry, 1967). These observations raise the possibility that with increasing temperature, the rate-limiting step shifts from the ADP dissociation mechanism to another unidentified step of the hydrolytic process.

The purpose of the present studies was to compare the rate of dissociation of ADP from HMM-ADP complex, with the steady-state rate of ATP hydrolysis at 6° and 23°C. The dissociation rate constant (k_2) of HMM-ADP complex was calculated from the equilibrium constant (K) and association rate constant (k_1) on the basis of the general equation:

$$ \text{HMM} + \text{ADP} \underset{k_2}{\overset{k_1}{\rightleftharpoons}} \text{HMM-ADP} $$

where $K = k_1/k_2$ and therefore $k_2 = k_1/K$. The equilibrium constant (K) was determined by equilibrium dialysis and the association rate constant (k_1) by rapid kinetic techniques as described earlier

Table 1. Dissociation Rate Constant (k_2) of HMM-ADP Complex and Its Relation to Turnover Number of ATP Hydrolysis

	Control HMM		Treated HMM[a]	
	6°	23°	6°	23°
K (M^{-1})	7.04×10^5 $\pm 0.5 \times 10^5$	7.0×10^4 $\pm 1.76 \times 10^4$	1.4×10^6 $\pm 0.5 \times 10^6$	2.7×10^5 $\pm 0.83 \times 10^5$
k_1 (M^{-1} sec^{-1})	1.71×10^4 $\pm 0.035 \times 10^4$	5.10×10^4 $\pm 0.13 \times 10^4$	0.44×10^4 $\pm 0.006 \times 10^4$	5.30×10^4 $\pm 0.3 \times 10^4$
k_2 (sec^{-1})	0.024	0.73	0.003	0.196
Turnover number of HMM ATPase (sec^{-1})	0.0144 ± 0.007	0.034 ± 0.001	0.012 ± 0.001	0.175 ± 0.0025

The measurements were carried out in a medium of 0.6 M KCl, 0.05 M Tris-HCl buffer pH 8.0 using 1 mM $MgCl_2$ as activator, essentially as described earlier (Malik and Martonosi, 1971).

[a] Treated with 2.25 mole NEM per 3×10^5 g protein.

(Malik and Martonosi, 1971). The behavior of untreated HMM was compared with HMM reacted with 2.25 moles of N-ethylmaleimide (NEM) per 3×10^5 g protein, which causes activation of ATP hydrolysis at 23°C and a slight inhibition at 6°C.

In Table 1 the average values of K and k_1 derived from several experiments are presented together with calculated values of k_2 and the turnover number of ATP hydrolysis obtained under identical conditions with 1 mM $MgCl_2$ as activator at pH 8.0.

The striking increase in the value of k_2 upon increasing the temperature from 6°C to 23°C is noteworthy.

A comparison of k_2 with the turnover number of ATP hydrolysis permits the following conclusions.

At 6°C both with control and NEM-treated HMM, the values of k_2 and the turnover number are of comparable magnitude, supporting the possibility that under these conditions dissociation of ADP from the active site may limit the rate of ATP hydrolysis.

At 23°C with untreated HMM the dissociation rate constant of HMM-ADP complex (k_2) is much greater than the turnover number of ATP hydrolysis. This suggests, in agreement with the independent observations of Trentham et al. (1972), that release of ADP from the active site does not limit the rate of ATP hydrolysis at 23°C.

Treatment of HMM with 2.25 moles of NEM per 300,000 g protein activates the ATPase activity at 23°C and decreases the rate of dissociation of ADP from the active site. As a result the difference between the values of k_2 and the turnover number of ATP hydrolysis decreases, and maximum activation of ATPase activity by NEM may be defined by the rate of dissociation of ADP from the active site.

It is implicit in these conclusions that the HMM-ADP complex formed by the binding of MgADP to the enzyme is kinetically equivalent to the HMM-product complex generated by ATP hydrolysis. This assumption is supported by the essential agreement between our observations and those of Trentham et al. (1972) obtained by entirely different techniques. Nevertheless further work is required to ascertain whether the kinetics of ADP release from the active site is significantly influenced by the preceding hydrolytic step.

References

Levy, H. M., N. Sharon, and D. E. Koshland, Jr. 1959. A mechanism for the effects of dinitrophenol and temperature on the hydrolytic activity of myosin. Biochim. Biophys. Acta 33: 288.

Malik, M. N. and A. Martonosi. 1971. Equilibrium and rapid kinetic studies of the effect of N-ethylmaleimide on the binding of ADP to myosin and H-meromyosin. Arch. Biochem. Biophys. 144: 556.

Perry, S. V. 1967. The structure and interactions of myosin. Prog. Biophys. Mol. Biol. 17: 325.

Taylor, E. W., R. W. Lymn, and G. Moll. 1970. Myosin-product complex and its effect on the steady-state rate of nucleoside triphosphate hydrolysis. Biochemistry 9: 2984.

Trentham, D. R., R. G. Bardsley, J. F. Eccleston, and A. G. Weeds. 1972. Elementary processes of the magnesium ion-dependent adenosine triphosphatase activity of heavy meromyosin. Biochem. J. 126: 635.

Investigation of Conformational Changes in Spin-Labeled Myosin: Implications for the Molecular Mechanism of Muscle Contraction

J. C. Seidel and J. Gergely

Department of Muscle Research, Boston Biomedical Research Institute, Boston, Mass. 02114
Department of Neurology, Massachusetts General Hospital, Boston, Mass. 02114
Departments of Neurology and Biological Chemistry, Harvard Medical School, Boston, Mass. 02115

Spin labels, stable compounds that contain an unpaired electron, have become important tools in the investigation of biological systems. These probes, which are analogous to optical probes or reporter groups, are usually attached covalently to a molecule of biological interest and can be used for the study of individual protein molecules or to obtain information on the structure of more complex systems, e.g., membranes (Hamilton and McConnell, 1968; McConnell and McFarland, 1970; McConnell, 1971). When these stable radicals are placed in a steady magnetic field (H_0), they absorb the energy of an oscillating electromagnetic field the same way as chromophores absorb light. On varying the strength of H_0 one obtains the so-called electron spin resonance (ESR) spectrum which is usually displayed as the derivative of the absorption.

Nitroxide radicals are useful probes because their spectrum depends on the orientation of the $2p\pi$ orbital containing the unpaired electron with respect to the magnetic field. When these radicals are rapidly tumbling, this anisotropy is averaged out and three sharp lines are observed in the spectrum. As the rate of tumbling is reduced, the spectrum gradually broadens. Thus the ESR spectrum will reflect changes in the tumbling rates or mobility of the labels, and in the case of spin labels attached to macromolecules, changes in the mobility may indicate changes in the environment of the probe. Thus, for example, McConnell and his colleagues were able to show that a spin label attached to hemoglobin undergoes changes in mobility when oxygen combines with the heme which, in turn, can be attributed to changes in the conformation of protein (McConnell, 1971).

From the point of view of being able to utilize changes in the mobility of the spin label attached to a macromolecule, two points have to be borne in mind. First, if the spin label is attached to a specific group or groups within the macromolecule, preferably having some relation to the active center, the information derived is more useful than in the case where the spin labels are attached at random. Second, the labeling process itself may change the properties of the macromolecule so that caution is needed in interpreting the data.

Since it is generally believed that muscle contraction involves some changes in the conformation of the macromolecules that are the components of the contractile machinery, spin labeling has obvious potentialities among other techniques such as optical spectroscopy, light scattering, fluorescence polarization, etc. This paper deals with some studies on myosin and its proteolytic fragments and their interaction with nucleotides and with actin.

Results and Discussion

Selective spin labeling at SH groups. It has been known for some time that myosin contains two classes of functionally important SH groups. Blocking of the most reactive thiol groups (S_1) results in activation of Ca-activated ATPase and inhibition of the K-EDTA-stimulated ATPase (Kielley and Bradley, 1956). Subsequent blocking of the second class (S_2), whose reactivity to thiol reagents is increased in the presence of ATP, ADP or pyrophosphate (Sekine and Yamaguchi, 1963), results in a loss of ATPase activity (Kielley and Bradley, 1956; Sekine and Yamaguchi, 1963). Although a variety of reagents can produce this pattern of activation and inhibition, it is clear from the comparison of the effect of various reagents that the selectivity of the reaction depends on the choice of the reagent. If one uses the piperidinyl nitroxide analog of iodoacetamide,

the inhibition of K-ATPase takes one mole per heavy chain, and this modification of myosin ATPase is accompanied by the appearance of a highly immobilized electron spin resonance spectrum. As the amount of reagent is increased other groups react, as is indicated by the appearance of peaks in the spectrum characteristic of more mobile spin labels (Seidel et al., 1970).

These results have two implications: (1) Taking into account the heavy chain structure of myosin, the stoichiometry of labeling suggests the presence of one S_1 group per heavy chain; and (2) the linear relationship between ATPase activity and the amount of label bound indicates that the label reacts quite selectively with S_1 groups.

SH groups of the second type, S_2, will react with thiol reagent in the presence of ATP and ADP. If one first blocks the S_1 thiol groups with NEM, a maleimide spin label combining with S_2 groups is also immobilized and the same stoichiometry can be demonstrated as shown for S_1 groups (Seidel, 1972).

Effects of ATP, ADP, and PP_i on ESR spectrum.

Although blocking of the rapidly reacting thiol groups modifies ATPase activity of myosin or heavy meromyosin (HMM), spin labels attached to these groups serve as indicators of the interaction of myosin with ATP and its analogs. As one adds increasing amounts of ATP, ADP or

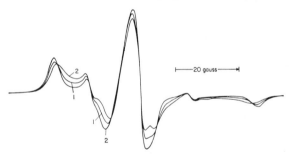

Figure 1. Effect of pyrophosphate on ESR spectrum of S_1-spin-labeled HMM. HMM was labeled with 2 moles of label A/mole of HMM. Spectra were recorded in a solution containing 0.08 M Tris pH 7.5, 1 mM MgCl$_2$, and 39 mg of HMM per ml (1.1×10^{-4} M). Curves were traced twice and were found to be exactly reproducible. Unlabeled curve, no pyrophosphate; curve 1, 10^{-4} M potassium pyrophosphate; curve 2, 2×10^{-4} M and 3×10^{-4} M potassium pyrophosphate. Myosin and HMM were prepared and spin labeled as described previously (Seidel et al., 1970). All spectra in this paper were recorded at room temperature. (From Seidel et al., 1970.)

PP$_i$, the spectra indicate an increased mobility of the spin label. Titration with PP$_i$ suggests a ratio of 2 moles of ligand per mole of HMM (Fig. 1). Thus, when these compounds combine with the ATP binding site, there is a change in mobility of the spin label, presumably arising from a change in the conformation of myosin.

Changes in conformation associated with ATP hydrolysis.

If one does these experiments with ATP without regard for the time elapsed after adding the nucleotide and compares the effect with

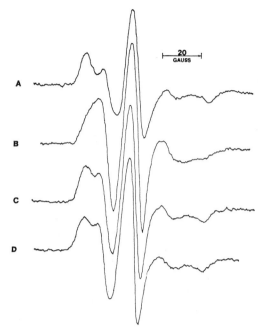

Figure 2. Effect of MgATP and MgADP on ESR spectra of spin-labeled myosin. Selective spin labeling at the S_1 thiol groups with N-(1-oxyl-2,2,6,6-tetramethyl-4-piperidinyl) iodoacetamide was carried out as previously described (Seidel et al., 1970). Spectra of solutions containing S_1-labeled myosin, 14 mg/ml, 0.4 M KCl and 0.04 M Tris pH 7.5 were recorded at room temperature. A, no further addition; B, 5 mM MgCl$_2$ and 5 mM ATP, recorded 2 min after addition of ATP; C, as B, recorded 10 min after addition of ATP; D, 5 mM MgCl$_2$ and 5 mM ADP. (From Seidel and Gergely, 1971.)

that of ADP or PP$_i$, there seems to be no difference (Seidel et al., 1970). It also appears that, provided these compounds are added in sufficiently high amounts, there is no requirement for any divalent cation. If, however, one adds MgATP to myosin spin-labeled at the S_1 thiol groups and records the first spectrum within 2 min after mixing, one finds a large change in the ESR spectrum, indicating a much greater increase in mobility than that previously seen (Fig. 2). This spectrum does not change for a period of time, but then in a relatively

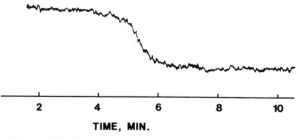

Figure 3. Time course of change in ESR spectrum of spin-labeled myosin following addition of MgATP. To a solution of myosin containing 0.4 M KCl, 0.04 M Tris pH 7.5, 5 mM MgCl$_2$ and 14 mg protein/ml, 5 mM ATP was added at 0 time. (From Seidel and Gergely, 1971.)

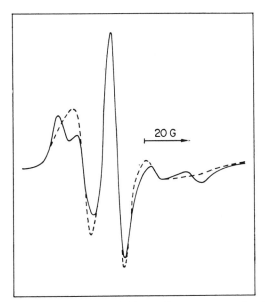

Figure 4. Effect of MgATP on the spectrum of spin-labeled subfragment-1. Spectra were recorded in 0.4 M KCl, 0.04 M Tris pH 7.5, protein 6.5 mg/ml. (———) Spectrum of subfragment-1 alone; (– – –) after addition of 5 mM MgATP. The latter spectrum was recorded within 2 min after the addition of MgATP. For details on preparation and labeling see Seidel et al. (1970).

short time it changes (Fig. 3) into the spectrum that one gets with ADP or pyrophosphate. By measuring the rate of hydrolysis of ATP it was found that ATP must be completely hydrolyzed when this latter spectral change occurs. Thus it appears that this spectrum is associated with the steady state of ATP hydrolysis, and as the substrate is used up a change takes place. If one uses Ca^{++}

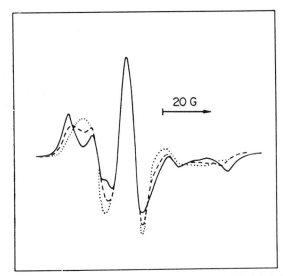

Figure 5. Effect of AMPPNP on spectrum of S_1-labeled myosin. Spectra were recorded as described in the legend to Fig. 2 except that the pH was 8.0. (———) Spectrum of myosin alone; (– – –) in the presence of 5 mM $MgCl_2$ + 1 mM AMPPNP; (. . . .) in the presence of 5 mM MgATP.

instead of Mg^{++} essentially the same effect is obtained, but one has to use an ATP-regenerating system to record the spectral change because of the higher rate at which ATP is hydrolyzed. If the divalent metal is omitted, then no significant hydrolysis takes place, and no matter how early one begins the observation, the ESR spectrum is identical with that obtained at the end of the ATPase reaction or on adding ADP or PP_i. These results show that during the hydrolysis of ATP the myosin molecule is in a state which is different from that obtained on adding the products of the reaction.

In view of the various suggestions made concerning the interaction between the two subfragment-1 heads of the myosin molecule (Tokiwa and Morales, 1971; Lymn and Taylor, 1970), it was of interest to compare the effect of ATP on spin-labeled subfragment-1, which contains only one ATP binding site. It is clear that the change in mobility of the spin label does not require interaction of the heads, since the spectrum of subfragment-1 is identical with that of HMM or of myosin following the addition of MgATP (Fig. 4).

To clarify the question whether hydrolysis of ATP is required for the appearance of the steady state spectrum, we have done further experiments under conditions where hydrolysis does not take place, although the nucleotide does combine with the protein as shown by the formation of the ADP-type spectrum. On addition of ATP to enzymically inactive myosin in which both S_1 and S_2 groups are blocked, the spectral change is the same as that produced by ADP. The most direct evidence establishing a relation between the hydrolytic step and the formation of the steady state spectrum has been obtained with the use of 5′-adenylyl imidodiphosphate (AMPPNP) which, as shown by Yount and his colleagues (Yount et al., 1971a, b), combines with myosin and competes with ATP but is not hydrolyzed. This compound produces the ADP-type spectrum, but again no steady state spectrum is observed (Fig. 5).

As shown by Taylor and his colleagues (Lymn and Taylor, 1970), the hydrolysis of ATP by native myosin involves a rapid initial pre–steady state step leading to the formation of an intermediate complex, whose rate of breakdown determines the steady state rate and characterized by the appearance of P_i on adding trichloroacetic acid. If the same is true for S_1-modified myosin, one could interpret the steady state ESR spectrum as being due to this complex. With the use of $[\gamma\text{-}^{32}P]ATP$ we could show that, in fact, there is an initial rapid burst with modified myosin (Fig. 6).

The simplest interpretation of this burst is the formation of a complex consisting of the hydrolytic

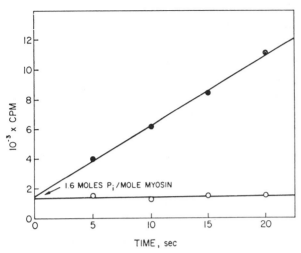

Figure 6. Effect of spin labeling of S_1 groups on rate of hydrolysis of ATP by myosin. Hydrolysis of ATP was determined by measurement of radioactivity of inorganic phosphate released from γ-^{32}P-labeled ATP by the method of Lymn and Taylor (1970). Mixing was done with automatic pipettes of the Eppendorfer type. Conditions of incubation were 0.5 M KCl, 0.04 M Tris pH 8.0, 40 μM ATP, 10 mM MgCl$_2$ and 2 mg myosin/ml. Temp. 22°C. ATP was added at zero time, trichloroacetic acid at times indicated on the abscissa. ($\bullet$) Spin-labeled myosin; ($\bigcirc$) native myosin.

products bound to myosin, as originally proposed by Taylor and his colleagues (Lymn and Taylor, 1970; Taylor et al., 1970), although the possibility of an acid-labile phosphorylated intermediate (Tokiwa and Tonomura, 1965) or of ATP itself being labilized in the complex cannot be completely ruled out. The ESR data suggest that the conformation of the protein in the intermediate complex is different from that in the complex formed with added ADP and P_i. When we first noticed this difference in the two ESR spectra, we also realized that the complex formed in the pre-steady state step described by Lymn and Taylor for unmodified myosin cannot be identical with the complex formed with added ADP and P_i in view of the previously reported differences between the ultraviolet difference spectra during the steady state of hydrolysis of ATP and that in the presence of added ADP (Morita, 1967). The same interpretation would apply to the extrinsic fluorescence data of Cheung (1969) and also to recent intrinsic fluorescence studies (Werber et al., 1972).

These studies then suggest that one may think of myosin combined with the products prior to their dissociation as being in a distinct conformation characterized both by a different absorption spectrum, suggesting a new environment for some of the protein chromophores, and one in which the mobility of a spin label attached to the S_1 group is altered in a characteristic way. It

would thus seem that myosin has at least three different detectable conformational states: that of the free protein, the complex with products of the reaction or some ATP analogs, and the protein in the intermediate complex (see also Botts et al., this volume). On the basis of results with AMPPNP one might speculate that the ATP myosin complex would be similar to the ADP complex, although direct evidence is lacking.

Trentham et al. (1972), using a transient kinetic approach, and Malik and Martonosi (1972), combining the measurements of the rate of binding of ADP to HMM with determination of the binding constant, also came to the conclusion that the intermediate complex, whose breakdown is rate limiting, is different from the complex produced by combination with ADP. Trentham et al. (1972) suggest that in the kinetic sequence the intermediate complex changes to a complex identical with that formed on adding ADP before dissociation of the products takes place. It seems to us that the present evidence is equally compatible with the assumption that the intermediate complex breaks down, perhaps in a sequential fashion, to form free myosin and the products, which in turn then form a complex with myosin as ATP is used up:

$$M + S \rightleftharpoons M*S \rightarrow M**Pr$$
$$\uparrow \qquad \swarrow \qquad \searrow$$
$$\rule{1cm}{0.4pt}M + Pr \rightleftharpoons M*Pr$$

Interaction of myosin with actin. The combination of myosin and actin, clearly one of the important steps in the molecular scheme of contraction, can also be studied with the use of spin labels. Combination of actin and spin-labeled myosin produces a slight effect opposite to that of ATP, resulting in a further restriction in the motion of the spin label attached to myosin (Fig. 7). This decreased mobility has also been observed by Tokiwa (1971). A clearer manifestation of the interaction of myosin and actin is the reduction in the amplitude of the ESR signal at higher microwave power upon addition of actin. This effect, which is confined to those spectral peaks that have contributions from immobilized labels bound to S_1 groups, arises because the ESR signal saturates at lower microwave power in the presence of actin. This saturation effect has been used to titrate spin-labeled HMM with actin; the effect becomes maximal when two actin monomers are added per mole of HMM (Fig. 8), suggesting that each myosin head can independently interact with an actin monomer. The same conclusion was reached by Tokiwa (1971) and is consistent with studies by other methods (Lowey, 1970; Eisenberg et al., 1971; Takeuchi and Tonomura, 1971).

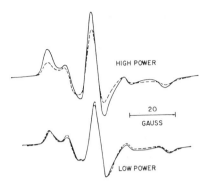

Figure 7. Effect of actin on ESR spectrum of S_1-labeled myosin. Myosin concentration, 5 mg/ml; actin concentration, 2.5 mg/ml; in 0.25 M KCl, 0.02 M Tris pH 7. The high microwave power was chosen so as to produce a maximal difference between myosin and actomyosin. (——) Myosin; (– – –) actomyosin. Actin free of tropomyosin and troponin were prepared essentially as described elsewhere (Lehrer et al., 1972) but DTT was omitted. Two cycles of polymerization were carried out with 0.6 mM $MgCl_2$.

Table 1. Effect of Spin Labeling S_1 Groups of Myosin on the Ca^{++} Sensitivity of the Actomyosin ATPase Activity

Treatment of Myosin	ATPase Activity (μmoles/mg/min)		
	Myosin	Actomyosin +Ca	+EGTA
None	0.01	0.40	0.04
Spin label	0.09	0.13	0.19
Spin label + DTT	0.08	0.42	0.14

ATPase activity was assayed in a system containing 0.05 M KCl, 0.04 M Tris pH 7.5, 2 mM $MgCl_2$, 2 mM ATP, myosin (0.4 mg/ml) and, where added, actin (0.2 mg/ml), troponin (0.066 mg/ml), tropomyosin (0.06 mg/ml), and 10^{-4} M $CaCl_2$ or 10^{-4} M EGTA. Myosin ATPase was measured in the absence of troponin, tropomyosin, $CaCl_2$ or EGTA. Troponin (unfractionated) and tropomyosin were prepared as described by Greaser and Gergely (1971). Assays for rows 1 and 3 (none and spin label + DTT) were carried out in the presence of 2×10^{-4} M DTT. Spin labeling was carried out as described (Seidel et al., 1970) and terminated either with 2×10^{-3} M DTT or by overnight dialysis. Other details are given elsewhere (Seidel et al., 1970).

It was also of interest to study the effect of actin on spin-labeled myosin in the presence of ATP under conditions where actin increases the ATPase activity. The activity of spin-labeled myosin is increased by actin and the actin-activated ATPase is the same as that of native myosin. This, to return to an earlier point, suggests that spin labeling did not profoundly alter the myosin molecule.

It was somewhat disappointing that actin had no observable effect on the ESR spectrum during the steady state of ATP hydrolysis with either myosin or HMM. This could mean either that when actin combines with the intermediate complex the

spectrum does not change or, if the interaction with actin does affect the spin label, the concentration of the actomyosin or acto-HMM product complex must be too low for this change to be reflected in the ESR spectrum. Eisenberg et al. (1972) have shown by means of ultracentrifugation at low temperature that, during the hydrolysis of ATP even at very high actin concentrations, a large fraction of the HMM molecules are not combined with actin.

Although myosin labeled at the S_1 groups is capable of interacting with actin, the question arises whether blocking of the S_1 group may be responsible for the loss of Ca^{++} sensitivity of actomyosin produced by the treatment of myosin with NEM (Hartshorne and Daniel, 1971). It appears that reconstituted actomyosin containing spin-labeled S_1 groups does show Ca^{++} sensitivity when tropomyosin and troponin are added. Thus the SH group(s) whose blocking has been shown to abolish Ca^{++} sensitivity are not the S_1 groups (Table 1). It should be pointed out that in order to observe Ca^{++} sensitivity it is important to add a reducing agent to make sure that thiol groups other than S_1 are not oxidized in the course of the experimental steps leading to blocking.

Conclusions

The studies here described have shown that the use of spin labels selectively attached to groups likely to be involved in, or at least having an effect on, the ATPase reaction makes it possible to distinguish various states, or conformations, of the myosin molecule. Of particular interest is the conformation that appears associated with the

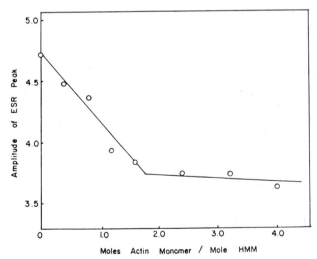

Figure 8. ESR titration of spin-labeled heavy meromyosin with actin at high microwave power. Final HMM concentration, 5.0 mg/ml. F-actin was added to achieve the ratio indicated on the abscissa. Other conditions and setting of microwave power as described in the legend to Fig. 7.

hydrolysis of ATP and is likely to correspond to the (M, ADP, P) intermediate complex. The mobility of the label in this state is maximal; the lowest mobility is shown by unliganded myosin, whereas in the complex formed with ADP or P_i, or with the unhydrolyzable ATP analog AMPPNP, the label has an intermediate mobility.

The existence of these states raises two kinds of question. One has to do with the mechanism of the movement of the cross-bridges, the other with the energetics of muscle contraction. Both are too complicated to be solved in the present state of our knowledge, but some speculation may be worthwhile.

Concerning the first question, it would seem that the conformational change observed on forming the intermediate complex with myosin may not be immediately involved in cross-bridge movement. The latter takes place after attachment of the myosin head, presumably as the intermediate complex, to actin. X-ray work on the effect of ATP and various ATP analogs promises to be a powerful tool for establishing whether myosin heads in the form of the intermediate complex have an intrinsically different orientation than those combined with unhydrolyzed ATP or an unhydrolyzable analog, even when not attached to actin. Various articles in this volume point to the multiplicity of states during the contraction-relaxation cycle; and the mapping of states defined by one technique, e.g., analysis of force transients, onto those defined by another, e.g., analysis of ATPase kinetics, will require a great deal of work and caution.

Nor can one unequivocally suggest that the conformational change associated with ATP hydrolysis is concerned with the conservation of the negative free energy of hydrolysis. It seems that not only the combination of myosin with actin is exergonic, but there are strong indications that the movement of the attached subfragment-1 moiety into a position that could correspond to a more stable attachment in terms of the model recently proposed by Huxley and Simmons (1971) is also exergonic. The latter conclusion is based on the way in which the subfragment-1 molecules are aligned on an actin filament and on the angle of cross-bridges observed in insect muscle (Moore et al., 1970). Ultimately, of course, the work carried out during the full cycle of bridge attachment, movement, and detachment is done at the expense of breaking down an ATP molecule. On this view no single step could be looked upon as being singled out for the utilization of the free energy of ATP. Whether the distinctive conformation of the intermediate complex plays a role in some special sequentially ordered attachment of myosin to actin is only a matter of speculation at present.

Acknowledgments

We thank Mr. Michael Punchekunnel for skilled technical assistance, Mr. Stephen Anderson for collaboration with the radioisotope experiments, and Dr. M. L. Greaser for collaboration in the experiments with troponin. We also thank Dr. R. G. Yount for a gift of AMPPNP.

This work was supported by grants from the National Institutes of Health (HE-5949), the National Science Foundation, and the Muscular Dystrophy Associations of America, Inc.

References

CHEUNG, H. C. 1969. Conformation of myosin: Effects of substrate and modifiers. *Biochim. Biophys. Acta* **194:** 478.

EISENBERG, E., L. DOBKIN, and W. W. KIELLEY. 1971. Binding of actin to HMM in the absence of ATP. *Fed. Proc.* (Abstr.) **30:** 1501.

———. 1972. Heavy meromyosin. Evidence for a refractory state unable to bind to actin in the presence of ATP. *Proc. Nat. Acad. Sci.* **69:** 667.

GREASER, M. L. and J. GERGELY. 1971. Reconstitution of troponin activity from three protein components. *J. Biol. Chem.* **246:** 4226.

HAMILTON, C. L. and H. M. McCONNELL. 1968. Spin labels, p. 115. In *Structural chemistry and molecular biology*, ed. A. RICH and N. DAVIDSON. Freeman, San Francisco.

HARTSHORNE, D. J. and J. L. DANIEL. 1971. The importance of sulfhydryl groups for the calcium-sensitive response of natural actomyosin. *Biochim. Biophys. Acta* **223:** 214.

HUXLEY, A. F. and R. M. SIMMONS. 1971. Proposed mechanism of force generation in striated muscle. *Nature* **233:** 533.

KIELLEY, W. W. and L. B. BRADLEY. 1956. The relation between sulfhydryl groups and the activation of myosin adenosinetriphosphatase. *J. Biol. Chem.* **218:** 653.

LEHRER, S. S., B. NAGY, and J. GERGELY. 1972. The binding of Cu^{2+} to actin without loss of polymerizability: the involvement of the rapidly reacting -SH group. *Arch. Biochem. Biophys.* **150:** 164.

LOWEY, S. 1970. Structure and interaction of myosin properties. (Abstr.) *8th Int. Cong. Biochem.*, p. 28.

LYMN, R. W. and E. W. TAYLOR. 1970. Transient state phosphate production in the hydrolysis of nucleoside triphosphates by myosin. *Biochemistry* **9:** 2975.

MALIK, M. N. and A. MARTONOSI. 1972. The regulation of the rate of ATP hydrolysis by colorimetric inorganic phosphate assay in situ. *Fed. Proc.* (Abstr.) **31:** 1622.

McCONNELL, H. M. 1971. Spin label studies of cooperative oxygen binding to hemoglobin. *Annu. Rev. Biochem.* **40:** 227.

McCONNELL, H. M. and B. G. McFARLAND. 1970. Physics and chemistry of spin labels. *Quart. Rev. Biophys.* **3:** 91.

MOORE, P. B., H. E. HUXLEY, and D. J. DeROSIER. 1970. Three-dimensional reconstruction of F-actin, thin filaments and decorated thin filaments. *J. Mol. Biol.* **50:** 279.

MORITA, F. 1967. Interaction of HMM with substrate. I. Difference in ultraviolet absorption spectrum between HMM and its Michaelis-Menten constant. *J. Biol. Chem.* **242:** 4501.

SEIDEL, J. C. 1972. Conformation of myosin spin labeled at the S_2 thiol groups. *Arch. Biochem. Biophys.* In press.

SEIDEL, J. C. and J. GERGELY. 1971. The conformation of myosin during the steady state of ATP hydrolysis: Studies with myosin spin labeled at the S_1 thiol groups. *Biochem. Biophys. Res. Comm.* **44:** 826.

SEIDEL, J. C., M. CHOPEK, and J. GERGELY. 1970. Effect of nucleotides and pyrophosphate on spin labels bound to S_1 thiol groups of myosin. *Biochemistry* **9:** 3265.

SEKINE, T. and M. YAMAGUCHI. 1963. Effect of ATP on the binding of *N*-ethylmaleimide to SH groups in the active site of myosin ATPase. *J. Biochem.* **54:** 196.

TAYLOR, E. W., R. W. LYMN, and G. MOLL. 1970. Myosin-product complex and its effect on the steady-state rate of nucleoside triphosphate hydrolysis. *Biochemistry* **9:** 2984.

TAKAEUCHI, K. and Y. TONOMURA. 1971. Formation of acto-H-meromyosin and acto-subfragment-1 complexes and their dissociation by adenosine triphosphate. *J. Biochem.* **70:** 1011.

TOKIWA, T. 1971. EPR spectral observations on the binding of ATP and F-actin to spin labeled myosin. *Biochem. Biophys. Res. Comm.* **44:** 471.

TOKIWA, T. and M. F. MORALES. 1971. Independent and cooperative reactions of myosin heads with F-actin in the presence of adenosine triphosphate. *Biochemistry* **10:** 1722.

TOKIWA, T. and Y. TONOMURA. 1965. Presteady state of the myosin adenosine triphosphate system. II. Initial rapid absorption and liberation of hydrogen ion followed by a stopped flow method. *J. Biochem.* **57:** 616.

TRENTHAM, D. R., R. G. BARDSLEY, J. F. ECCLESTON, and A. G. WEEDS. 1972. Elementary processes of the Mg ion dependent ATPase activity of HMM. A transient kinetic approach to the study of kinases and ATPases. *Biochem. J.* **126:** 635.

WERBER, M. M., A. G. SZENT-GYÖRGYI, and G. D. FASMAN. 1972. Fluorescence studies on heavy meromyosin: Substrate interaction. *Biochemistry* **11:** 2872.

YOUNT, R. G., D. BABCOCK, W. BALLANTINE, and D. OJALA. 1971a. Adenyl imidodiphosphate, an ATP analog containing a P-N-P. *Biochemistry* **10:** 2484.

YOUNT, R. G., D. OJALA, and D. BABCOCK. 1971b. Interaction of P-N-P and P-C-P analogs of ADP with HMM, myosin and actomyosin. *Biochemistry* **10:** 2490.

Does a Myosin Cross-Bridge Progress Arm-over-Arm on the Actin Filament?

Jean Botts, Roger Cooke, Cristobal dos Remedios, Joseph Duke, Robert Mendelson,
Manuel F. Morales, Tomonubo Tokiwa and Gustavo Viniegra
Cardiovascular Research Institute, University of California, San Francisco, California 94122

Ralph Yount
Department of Chemistry, Washington State University, Pullman, Washington 99163

The title question is not answered in this paper; nevertheless the question motivates a great deal of our research, and some of our results are undoubtedly *components* of the answer. Other of our results—on fiber fluorescence and fiber labeling—are broader than the question, but are relevant means of unraveling some of the issues in this Symposium. At any rate, multiplicity of subjects in one paper is a price to be paid for inviting but one person from a large laboratory.

It is now generally accepted that the myosin molecule is structurally duplex, but even the princes of our subject continue to ignore this odd feature when they think about how myosin *functions*. We have asked whether an individual half of a myosin molecule is functionally equivalent to half of a whole molecule; this question was posed in two ways: Cooke and Morales (1971) tested the ability of myosin moieties, S-1 and HMM, to bind—more accurately to immobilize—G-actins. It turned out that S-1 binds *one* actin, but in HMM each S-1 moiety binds *two*. Cooke's results were obtained in the region of excess G-actin. They are not to be confused with what happens in the equivalence zone, where the bound S-1 concentration approaches that of actin monomers. The point is that in at least one zone the two S-1 moieties of myosin cooperate to grasp actin. Tokiwa and Morales (1971) compared the ability of doubly native, half-native, and doubly dead myosin molecules in regard to whether F-actin could stimulate their ATPase, and also in regard to their effectiveness in increasing the superprecipitation *rate* of the actomyosin complex. In the test of ATPase enhancement molecules were effective in proportion to the number of native S-1 moieties; e.g., a doubly native molecule was twice as effective as a half-native molecule. However, *only doubly native molecules contributed to the superprecipitation rate*. It is not our purpose to prove that superprecipitation = contraction; the point here is that superprecipitation is an operationally well-defined process for which both moieties must be native—strongly implying that in this process *the two moieties work together*. Tokiwa's work revealed incidentally certain new features of superprecipitation; viz., the extent of the process does *not* go hand in hand with rate. For example, the extent achieved by a very few doubly native molecules and many doubly dead molecules is indistinguishable from the extent achieved by doubly native molecules, even though the rates exhibited by the two systems are vastly different.

In pondering Cooke's and Tokiwa's results one is naturally led to wonder about the meaning of intermoiety cooperativity when actin is present (as contrasted with no cooperativity when actin is absent [Murphy et al., 1970]). In this speculative vein Viniegra and Morales (1972) suggested that perhaps a duplex myosin molecule progresses along a thin filament by an arm-over-arm movement. In such a movement the two S-1 moieties of a myosin might maximally straddle the number of actins which Cooke observed to be immobilized by HMM; furthermore, Tokiwa's observation would find a natural explanation in that it is hard for a no-armed or one-armed person to pull in a rope.

Viniegra's speculation prompted many questions, of which the simplest and most directly answerable question was: Can the S-1 moieties of myosin move relative to the stem (LMM moiety) in something of a rope-drawing manner? As is well known, there are many inferential reasons for thinking myosin to be flexible—for instance, from enzyme-susceptibilities or because of features in fiber organization. However, the flexibility of myosin molecules free in solution has never been studied. Our attempt to do so follows closely the technique employed by Stryer and associates (Yguerabide et al., 1970) in studying the segmental flexibility of immunoglobulin. Single photon-counting circuits receive the output of a photomultiplier tube trained on the parallel-polarized component of fluorescence emitted by a solution and (in our case simultaneously) of a tube trained on the perpendicular component. The time-dependence of single photon emission is followed for many hundreds of nanoseconds after the solution has been illuminated by a very brief (ca. 2 nsec) flash of plane-polarized exciting light. The excitation process is repeated at about 75 KHz, and the data accumulated in a computer is processed to yield $r(t) \equiv [(I_\parallel(t) - I_\perp(t)]/[(I_\parallel(t) + 2I_\perp(t)]$, the "polarization anisotropy" as a function of time. Phenomenologically, $r(t)$ is well described by a single decaying exponential, i.e., $r(t) = A \exp(-t/\phi)$, where ϕ and A are related to parameters of molecular *rotation* and to parameters relating the absorption and emission dipoles to the geometric axes of the rotating segment that bears the fluorophore. The fluorescence from the intrinsic fluorophores of myosin decays too rapidly to be useful for our purposes, so we have to use "labels." A very good label for myosin has been developed by Dr. Bud Hudson (personal communication); it has fluorescence characteristics like those of "ANS" and a chemical reactivity toward proteins like that of "IAA." Upon reaction with myosin this label selectively and strongly stimulates the Ca^{++}-ATPase, indicating that it binds to the two rapidly reacting SH groups of myosin—one group on each S-1 moiety. From the parent labeled myosin are prepared heavy meromyosin (HMM) and subfragment one (S-1). Then all three systems are purified by chromatography. The excited-state lifetime for Hudson's dye is about 20 nsec, considerably longer than that of tryptophanyl residues, but short relative to the time over which $r(t)$ must be followed; nevertheless, instruments like Yguerabide's (1970) and ours (Mendelson et al., 1972) have the sensitivity and stability to measure ϕ values of 2000 nsec. If the structure bearing the fluorophore is "internally rigid," then $r(t)$ decays (following a flash of polarized exciting light) because the optic axes initially defined by the flash are randomized by rotational diffusion of the structure. In principle, if internal rigidity is assured, it is possible to write a *theoretical* expression for $r(t)$. Although recently experts in the field seem to have agreed on such an expression (Belford et al., 1972) for macromolecules having the shape of an ellipsoid with no two axes alike, no one has derived (or is likely to derive) $r(t)$ for an ellipsoid attached to a larger structure by means of a hinge. Accordingly for myosin and HMM we have to be satisfied with qualitative inferences (Mendelson et al., 1972 and Table 1), realizing that the relaxation time, ϕ, is an empirical parameter. Subfragment 1 data, however, can be treated more precisely by adapting the equation of Belford et al. to the case of a prolate ellipsoid of revolution.

Values of ϕ of the order of 10^2 nsec are too slow o arise from completely free rotation of covalently attached dye. Furthermore, aggregation of the *non*labeled ends of myosin molecules (Experiment 1, Table 1) increases ϕ over what it is in solution (Experiments 2, 3, 4, Table 1). These observations suggest that there is an acceptable internal rigidity where the dye molecules attach; i.e., depolarization arises not because the dye moves relative to the member, but because the member bearing the dye moves. With this proviso our results immediately show that S-1 is quite elongate, not spherical. Comparison of our results with the appropriate theoretical prediction for a hydrated ellipsoid indicates that the axial ratio of S-1 exceeds 3.8 (e.g., that S-1 is 170 Å × 45 Å); were S-1 spherical, its ϕ value would be 70 nsec, which is about one-third of the observed relaxation time. It is interesting that an S-1 moiety approximately 170 Å long would almost span the interfilament distance. Our principal conclusion arises from comparing Experiments 6–8 (ϕ value for S-1, ca. 240 nsec) and Experiment 5 (ϕ value for HMM, ca. 405 nsec). The mass of HMM (3.4×10^5 daltons) is considerably greater than that of two S-1 moieties (2.3×10^5 daltons), and this difference alone would tend to make ϕ(HMM) greater than ϕ(S-1). However, even if this difference is ignored, it is clear that if HMM were a perfectly rigid structure its motions would be equivalent to moving two S-1 moieties (each with solvent flowing around it); on these accounts ϕ(HMM) should exceed twice ϕ(S-1). In fact this expectation is not realized, indicating that the HMM structure must allow its S-1 moieties motions in addition to tumbling as parts of one rigid body; in other words, *HMM must be flexible.* The foregoing experimental indication that the S-1 moieties of myosin enjoy considerable motion relative to the remainder of the myosin structure does not prove that myosin can draw in an actin filament by an "arm-over-motion," but it is *consistent* with the hypothesis. Of course current experimentation with this system is exploiting these initial findings. For example, it has already been found that complexation of labeled S-1 with G-actin significantly lengthens the excited lifetime

Table 1. Values of ϕ, Relaxation Time of Fluorescence Polarization Anisotropy, for Myosin Systems in Different Physical States

System	pH	[KCl]	Physical State	ϕ (in nsec)
Myosin	7.0	0.10	filaments	1800 ± 300
Myosin	8.0	0.60	solution	450 ± 75
Myosin	7.0	0.60	solution	460 ± 40
Myosin	6.5	0.60	solution	450 ± 75
HMM	7.0	0.15	solution	405 ± 25
S-1	6.5	0.10	solution	240 ± 20
S-1	7.0	0.15	solution	220 ± 20
S-1	6.5	0.60	solution	250 ± 35

of the labels; this observation bears on the proximity of the actin-binding site to the reactive thiol and ATPase sites, and indicates that the effect will be useful in equilibrium binding studies.

We turn now to an attempt at detecting cross-bridge motion in the organized fiber system, using a different fluorescence polarization effect. The seminal observation was that of Aronson (Aronson and Morales, 1969), who showed that P, the polarization of the fluorescence emitted by glycerol-extracted rabbit psoas muscle fibers excited at 300 nm by plane-polarized light, is sensitive to physiological state if the plane is perpendicular ($\perp$) to the fiber axis (and insensitive if the plane is parallel, $\parallel$). Subsequently we (dos Remedios et al., 1972a) have collected evidence indicating that this effect, present in insect fibers and in living frog fibers as well as in rabbit psoas, arises from trypto-phanyl residues in the S-1 moieties: This evidence is that $P_\perp$ has a consistent relationship to mechanical parameters such as tension or the dynamic stiffness, as well as to filament overlap; moreover, there appears to be a concentration of tryptophanyl in the S-1 moieties (Shimizu et al., 1971). Our working hypothesis is that the $\perp$-orientation of the exciting plane illuminates *many* (just a few would not be detected) tryptophanyls rigidly imbedded in S-1 moieties, and that the two components of the intensity of fluorescence (therefore the $P_\perp$ that they constitute) change because these tryptophanyls move coordinately in space as the *attitude* of the S-1 moieties changes. It is perhaps worth emphasizing that this effect could not result from the more familiar depolarization due to rotational diffusion, for relative to the excited lifetime of tryptophanyl (ca. 10^{-9} sec) the cross-bridge movements (ca. 10^{-2} sec) are negligibly slow. Since $P_\perp$ data can be acquired at rates much higher than those of cross-bridge movement (dos Remedios et al., 1972a), $P_\perp(t)$ could conceivably reveal kinematic cycles of the S-1 moieties; however, it is now widely thought that there are only random phase relations among such cycles—i.e., the cycles, especially in psoas, are not synchronized. A provisional way around this difficulty is to decide (on the basis of independent information) what states occur to a significant extent during the *chemical* cycle, and by various tricks attempt to force all S-1 moieties into each of these states for a time long enough to characterize the state positionally (e.g., by $P_\perp$ or by some X-ray diffraction parameter). If now we imagine an active process in the steady state (e.g., isometric tension development or isovelocity shortening), during which the S-1 moieties are passing through the chemical-kinematic states, we believe that the measured value of the positional parameter, say

$P_\perp$ (active process), will have the form,

$$P_\perp(\text{active process}) = \frac{\sum_j P_\perp(j)\bar{C}_j}{\sum_j \bar{C}_j} \qquad (1)$$

where $P_\perp(j)$ is the statically measured value for state j, and $\bar{C}_j$ is the steady state concentration of species j. The $\bar{C}_j$ are functions of velocity constants defined by the chemical scheme. Elsewhere (dos Remedios et al., 1972b) we have attempted a beginning of this kind of analysis; here we repeat its salient features.

Experimentally, the analysis depends on the measurement of $P_\perp$ and also on the availability of two kinds of ATP analogs. Since one of us is discussing these analogs elsewhere in the volume (Yount et al., this volume), we list here only their properties relevant to this application.

1. At pH 8.0 the form, electrical state, and cation binding properties of AMP-PNP (an ATP analog in which the terminal —O— bridge is replaced by —N(H)—) are very similar to those of ATP, but of course myosin does not catalyze the hydrolysis of this analog; it is, so to speak, a permanent or unsplittable ATP (Yount et al., 1971a,b).

2. The 6-SH substituted analogs of the adenine nucleotides, e.g., HS-X, where X can be the triphosphate (TP) or diphosphate (DP), are functionally very similar to the corresponding parent compounds, but have two special properties of interest to contractility research—(a) strong absorbance peaks at 322 nm which are generally reduced when the molecule binds to the active site (e.g., of myosin ATPase), and (b) through the formation of a disulfide intermediate, S_2X, the capacity to affinity label *reversibly* the active sites of myosin (Tokiwa and Morales, 1971; Yount, Frye, and O'keefe, unpublished; Yount et al., this volume); X can be the tri-, di-, or monophosphate, or even -P-PNP. It is important to add, however, that if a myosin system is labeled with S_2TP, the nucleotide later removed (by adding a thiol) for analysis under conditions which discourage hydrolysis (e.g., 0.6 M NaCl, no activator), is HS-DP. Furthermore, in the case of fibers, the mechanical consequences of labeling with S_2TP or with S_2DP are exactly the same, so it may be that when labeling is done with S_2TP the actual label is nevertheless the *di*phosphate; i.e., the enzyme may be quite capable of catalyzing the hydrolysis of covalently attached substrate.

Selection of myosin complexes which are in the cycle is not straightforward, but from the work of many investigators over the past decades it is obvious that (letting "M" stand for a single S-1 moiety of myosin and "Act" stand for a monomer of actin in F-actin) M·ATP, M·ADP, and M·Act are in the main sequence. On the other hand, there is nothing to indicate long lifetimes for any complexes with orthophosphate, P_i. There is, however, mounting evidence for the existence of a complex occurring in between M·ATP and M·ADP. We postulated (Viniegra and Morales, 1972) a complex of this kind because added ADP or IDP did not alter the time course of proton liberation in pre–steady state ATPase and ITPase (Imamura et al., 1970) and because of Ulbrecht and Ruegg's (1971), as well as Hotta's (1971), observation indicating that while actomyosin ATPase is in the steady state it catalyzes the "back incorporation" of $^{32}P_i$ into ATP. It seems clear now that the existence of this complex would also explain catalysis, during ATPase, of "intermediate" exchange between ^{18}O atoms of the water and of P_i (Swanson and Yount, 1966; Boyer, 1967). Finally, it also seems necessary to postulate this complex because the EPR signal (from spin-labeled myosin) (Seidel and Gergely, 1971) and fluorescence signal (from tryptophanyl in myosin) (Werber et al., 1972) obtained during ATPase are different from those emitted by M·AMP-PNP (Werber et al., 1972; D. B. Stone, unpublished) or by M·ADP (Seidel and Gergely, 1971; Werber et al., 1972; Stone, unpublished). We have called the complex "M * ADP," understanding that because of the uncertainty of P_i complexing, it could as well be "M * ADP, P_i"; moreover, we believe that an exergonic process exists wherein the energy in M * ADP is redistributed in such a way that M is deformed to M*, and that simultaneously the affinity of myosin for ADP is reduced to that obtaining in the usual complex of M and ADP; i.e., there is a process, M * ADP → M*·ADP. As will be shown shortly, the properties of M·ADP and M-S-S-DP are very different, and we have considered the possibility that M-S-S-DP may closely resemble M * ADP.

Another question that bedevils scheme-writers is whether ternary complexes of myosin, actin, and nucleotide exist. It is generally conceded that this is not the case if the nucleotide is ATP, and Kiely and Martonosi's work (1969) suggests that this is also true if it is ADP; but the properties of M * ADP in this regard are unknown. If M-S-S-DP is like M * ADP, then Tokiwa's observation (1971) that the EPR spectrum of M-S-S-DP is unaltered by F-actin, as well as a result cited below, suggests that also in this case no ternary complexes exist; however, since some aspects of the matter are unsettled, the existence of a complex such as M * ADP·actin will not be excluded, as it is an attractive concept in accounting for certain observations.

It is a central fact of muscle biochemistry that, under the ionic conditions of contraction, actin powerfully activates myosin ATPase; so actin should enter the ATPase scheme in such a way that it diverts the pathway around the process which otherwise would rate-limit myosin ATPase. Until recently, the observations and suggestions of Blum many years ago (Blum, 1955), and the recent studies of Taylor and his associates (Taylor et al., 1970), made it seem likely that ADP desorption (M·ADP → M + ADP) was rate-limiting. Were this so, the aforementioned Martonosi observation (1969) (actin + M·ADP → M·actin + ADP) would complete the explanation of why actin activates myosin ATPase. The recent results of Trentham et al. (1972) and of Malik and Martonosi (1972) dispute Taylor's conclusion and suggest that at room temperature rate-limitation occurs prior to ADP desorption. Because it entails a redistribution of energy and a change in conformation, we suppose that rate-limitation resides in M * ADP → M*·ADP; this will be assumed in one of the schemes below.

Returning now to our experimental results, we list the chemical and mechanical states in which $P_\perp$ has been statically measured (in all cases the SD of the measurement is between 0.001 and 0.004); the first three on the list are very sure; the last two are harder to get, therefore less sure.

1. A fiber treated with a conventional "relaxing solution" that includes ATP, or that includes AMP-PNP (but at pH 7.9), or affinity labeled with S_2-PNP, or affinity labeled with S_2-TP: in each case mechanical parameters such as tension or dynamic stiffness assume values characteristic of "relaxation," and $P_\perp$ assumes the value 0.127. It is assumed that in this state the myosin moieties bear a ligand (in at least one case the ligand is certainly a *tri*phosphate), and there is no mechanical union with actin.

2. A fiber affinity labeled with S_2-DP; the mechanical parameters are those of "relaxation", and $P_\perp$ is again 0.127. The ligand is the *di*phosphate and presumably there is no contact with actin.

3. A fiber at normal length is treated with a nonnucleotide solution; the mechanical parameters are those of "rigor" and $P_\perp$ is 0.095. If the length of the fiber is oscillated with amplitude of $\pm 1\%$ (about 100 Å/bridge), the tension fluctuates but $P_\perp$ does not, suggesting that potential energy is being exchanged by virtue of flexibility in the myosin molecules, but that the attitude of the S-1 moieties is constant.

4. A fiber in which the myosin moieties are equilibrated with ADP, but actin has been withdrawn (by stretching beyond interfilament overlap) to avoid disturbing the M·ADP complex. The $P_\perp$ value is 0.110.

Table 2. Possible Reaction Cycles of an S-1 Moiety Progressing on Actin

State	$P_\perp$	Sequence A	Sequence B
(1)	0.127	M(s)·ATP	M(s)·ATP
(2)	0.127	M(s) * ADP	M(s) * ADP
(3)	0.110	M*$(s + \Delta s)$·ADP	M(s) * ADP·actin
(4)	0.095	M*$(s + \Delta s)$·actin	M*$(s + \Delta s)$·ADP·actin
(5)	0.095	M(s)·actin	M*$(s + \Delta s)$·actin

The bottom entry is assumed to convert into the top entry. The kinematic cycle is specified by $P_\perp$ (see text). The chemical cycle (Additions or deletions of small molecules are omitted for clarity; no attempt has been made to place P$_i$ in the cycle.) is specified by Sequence A if the power stroke is $s + \Delta s \rightarrow s$, and by Sequence B if it is $s \rightarrow s + \Delta s$. During isometric contraction $P_\perp$ has the value 0.116, which is regarded as a time average of the state values.

5. A fiber in which the myosin moieties are equilibrated with a nonnucleotide ("rigor") solution, but actin has been withdrawn (by stretch) to prevent M·Act formation. The $P_\perp$ value is 0.110.

In addition to states (1)–(5), we have measured many times the value of $P_\perp$ when a fiber held isometrically at normal length is developing full active tension in a solution containing ATP, Mg^{++}, and Ca^{++}. This $P_\perp$ value is 0.116 ± 0.003.

There is a final point to consider before assembling the foregoing states in reaction sequences. Many investigators (e.g., the Viniegra proposal [Viniegra and Morales, 1972]) have discussed the momentum transfer in terms of a single site on an S-1 moiety successively attaching → pulling (or pushing) → detaching → reaching out → etc.; appropriate to such a cycle is a linear coordinate, x, describing the axial distance of the S-1 site from some reference point on the core of the thick filament, and undergoing a cycle, $x \rightarrow x + \Delta x \rightarrow x \rightarrow$, etc. From this point of view the recent proposal of A. F. Huxley and Simmons (1971) is not different except that it is an angular coordinate which would most appropriately describe the cycle, $\theta \rightarrow \theta + \Delta\theta \rightarrow \theta \rightarrow$, etc. If we let s be in general a space coordinate (linear or angular), it is, however, very important whether the momentum transfer ("power stroke") is assumed to occur during $s \rightarrow s + \Delta s$, or $s + \Delta s \rightarrow s$. In Table 2 we have assembled the states (1)–(5) in two ways (A and B), depending upon this choice.

Both sequences A and B have virtues and defects. For example, sequence A includes no ternary complexes and is very suitable if indeed myosin ATPase is rate-limited by ADP desorption. Sequence B has the possibility of explaining the Ulbrecht-Ruegg effect, and is consistent with a rate-limiting step prior to ADP desorption. Neither sequence calls for a large positional change as a result of hydrolysis; such changes would appear to accompany the transition (2) → (3). The large positional change in (5) → (1) would seem more consistent with sequence B, which calls for the reversal of the deformation of myosin in that step. In sequence B it is the binding of the next ATP that resets the system to begin the next cycle. Obviously experiments of various kinds have yet to be done, but the outline of some connection between position, chemistry, and kinetics seems at hand.

Acknowledgment

The research reported in this paper was supported by grants from the U.S.P.H.S., the N.S.F., and the American Heart Association. The authors gratefully acknowledge the assistance of Kathleen Ue in making the protein preparations.

References

ARONSON, J. F. and M. F. MORALES. 1969. Polarization of tryptophane fluorescence in muscle. *Biochemistry* **8:** 4517.

BELFORD, G. G., R. L. BELFORD, and G. WEBER. 1972. Dynamics of fluorescence polarization in macromolecules. *Proc. Nat. Acad. Sci.* **69:** 1392.

BLUM, J. J. 1955. The enzymatic interactions between myosin and nucleotides. *Arch. Biochem. Biophys.* **55:** 486.

BOYER, P. D. 1967. ^{18}O and related exchanges in enzymic formation and utilization of nucleoside triphosphates. In *Current topics in bioenergetics* (ed. D. R. Sanadi) vol. 2, p. 99. Academic Press, New York.

COOKE, R. and M. F. MORALES. 1971. Interaction of globular actin with myosin subfragments. *J. Mol. Biol.* **60:** 239.

DOS REMEDIOS, C. G., R. G. C. MILLIKAN, and M. F. MORALES. 1972. Polarization of tryptophane fluorescence from single striated muscle fibers. *J. Gen. Physiol.* **59:** 103.

DOS REMEDIOS, C. G., R. G. YOUNT, and M. F. MORALES. 1972. Individual states in the cycle of muscle contraction. *Proc. Nat. Acad. Sci.* **69:** 2542.

HOTTA, K. and Y. FUJITA. 1971. On the intermediate complex of actomyosin ATPase. *Physiol. Chem. Phys.* **3:** 196.

HUXLEY, A. F. and R. M. SIMMONS. 1971. Proposed mechanism of force generation in striated muscle. *Nature* **233:** 533.

IMAMURA, K., J. A. DUKE, and M. F. MORALES. 1970. Studies on myosin catalysis and modification. *Arch. Biochem. Biophys.* **136:** 452.

KIELY, B. and A. MARTONOSI. 1969. The binding of ADP to myosin. *Biochim. Biophys. Acta* **172:** 158.

MALIK, M. N. and A. MARTONOSI. 1972. The regulation of the ATPase activity of H-meromyosin. *Arch. Biochem. Biophys.* **152:** 243.

MENDELSON, R. M., J. BOTTS, P. C. MOWERY, and M. F. MORALES. 1972. The segmental flexibility of the S-1 moiety of myosin. *Abstr. 16th Meet. Biophys. Soc.*, p. 281a.

MURPHY, A. J. and M. F. MORALES. 1970. Number and location of adenosine triphosphatase sites of myosin. *Biochemistry* **9:** 1528.

MURPHY, A. J., J. A. DUKE, and L. STOWRING. 1970. Synthesis of 6-mercapto-9-β-D-ribofuranosyl-purine 5′-triphosphate. *Arch. Biochem. Biophys.* **137:** 297.

SHIMIZU, T., F. MORITA, and K. YAGI. 1971. Tyrosine and tryptophane contents in heavy meromyosin and subfragment-1. *J. Biochem.* **69:** 447.

SEIDEL, J. C. and J. GERGELY. 1971. The conformation of myosin during the steady state of ATP hydrolysis: Studies with myosin spin-labeled at the S_1 thiol groups. *Biochem. Biophys. Res. Comm.* **44:** 826.

SWANSON, J. F. and R. G. YOUNT. 1966. The properties of heavy meromyosin and myosin catalyzed "medium" and intermediate ^{18}O-phosphate. *Biochem. Z.* **345:** 395.

TAYLOR, E. W., R. W. LYMN, and G. MOLL. 1970. Myosin-product complex and its effect on the steady-state rate of nucleoside triphosphate hydrolysis. *Biochemistry* **9:** 2984.

TOKIWA, T. 1971. EPR spectral observations on the binding of ATP and F-actin to spin-labelled myosin. *Biochem. Biophys. Res. Comm.* **44:** 471.

TOKIWA, T. and M. F. MORALES. 1971. Independent and cooperative reactions of myosin heads with F-actin in the presence of adenosine triphosphate. *Biochemistry* **10:** 1722.

TRENTHAM, D. R., R. G. BARDSLEY, J. F. ECCLESTON, and A. G. WEEDS. 1972. Elementary processes of the magnesium ion-dependent adenosine triphosphatase activity of heavy meromyosin. *Biochem. J.* **126:** 635.

ULBRECHT, M. and J. C. RÜEGG. 1971. Stretch induced formation of ATP-^{32}P in glycerinated fibers of insect flight muscle. *Experientia* **27:** 45.

VINIEGRA, G. and M. F. MORALES. 1972. Toward a theory of muscle contraction. *J. Bioenerget.* **3:** 101.

WERBER, M. M., A. G. SZENT-GYÖRGYI, and G. D. FASMAN. 1972. Fluorescence studies on heavy meromyosin: Substrate interaction. *Biochemistry.* **11:** 2872

YOUNT, R. G., D. BABCOCK, W. BALLANTYNE, and D. OJALA. 1971a. Adenylyl imidodiphosphate, an adenosine triphosphate analog containing a P-N-P linkage. *Biochemistry* **10:** 2484.

YOUNT, R. G., D. OJALA, and D. BABCOCK. 1971b. Interaction of P-N-P and P-C-P analogs of adenosine triphosphate with heavy meromyosin, myosin, and actomyosin. *Biochemistry* **10:** 2490.

YGUERABIDE, J., H. F. EPSTEIN, and L. STRYER. 1970. Segmental flexibility in an antibody molecule. *J. Mol. Biol.* **51:** 573.

NOTE: After this paper had been prepared and submitted, it was called to our attention that in 1966 Young, Himmelfarb, and Harrington (*J. Biol. Chem.* **240:** 2428) considered the possible rotational mobility of the S-1 moieties in the HMM structure, employing fluorescence effects. In some respects their approach (labeling with the sulfonyl chloride of a dye and measuring steady state flourescence) and their conclusions (motion of the S-1 moieties in HMM is the same as motion of free S-1's) differed significantly from ours, but they inferred S-1 dimensions (250 Å × 30 Å) which resemble those deduced by us. We are pleased to call attention to this earlier work by Professors Young and Harrington, and by Miss Himmelfarb.

Links between Mechanical and Biochemical Kinetics of Muscle

D. C. S. WHITE

Department of Biology, University of York, York YO1 5DD

Muscle is highly organized structurally to convert chemical energy into mechanical energy, enabling it to develop tension or shorten depending upon the external load. An understanding of the mechanisms involved in muscle contraction requires several types of information. These fall into three main classes: structural, mechanical, and biochemical.

At the present time our understanding of what might be termed the "statics" of striated muscle is fairly good for all three classes. We know the basic structure of the contractile elements from detailed X-ray diffraction and electron microscopic work; the mechanical properties of vertebrate striated muscle in the steady state (i.e., under conditions in which either the length or the tension is held constant) has been known for many years from the work of, in particular, A. V. Hill; the relationships between ATPase activity and chemical conditions have been continually better understood as the proteins actin and myosin have been obtainable in purer forms.

Much of the present-day work is directed towards the kinetics. The problems here are that the time constants involved are very short, and this necessitates the ability to resolve experimental variables in comparable times. The catalytic site activity of active muscle is of the order of 10 sec^{-1} at 0°C (derived from 4–6 μmole/min/mg HMM, Eisenberg and Moos, 1968); thus, under the common assumption that mechanical changes in crossbridges are directly linked to the hydrolysis of ATP, the slowest chemical rate constants in active muscle may be of the order of 10 sec^{-1}, and presumably there will be much faster ones.

The mental picture that many people have is of a population of cross-bridges in an ordered array performing a cycle of activity of some kind. The cycle involves the hydrolysis of one (or possibly two) molecule of ATP, which takes place as a succession of biochemical steps. Coupled with this in some way is a progression through a series of mechanical states. If held at a constant length the muscle develops a certain tension and hydrolyzes ATP at a certain rate.

Whether or not this view is correct, it provides the framework for the experiments described in this paper.

Two basic approaches can be used to study a system of this kind. The first involves the rapid mixing of two solutions containing the reactants involved in one step of the cycle and recording the change in some parameter associated with the reaction, say release of a product or spectroscopic change associated with the reacted form of the reaction. It is possible to mix two solutions within a few milliseconds, and then with optical methods, to follow the resulting changes continuously with a time resolution of better than one millisecond. Such experiments are known as stopped-flow or quenched-flow experiments, depending upon the precise details, and have been used very successfully by Taylor and his coworkers (Lymn and Taylor, 1971 is the latest work) and Trentham et al. (1972). Most of the results so far have been performed on the interaction of ATP with myosin in the absence of actin.

The second way is to perturb suddenly the steady state system in which the cycle is undergoing its normal activity and to follow the manner in which a new steady state is reached after application of the perturbation. This second way is the only one suitable for most mechanical and structural work on intact muscle, since the delay involved in diffusing chemicals into the filament array precludes the rapid initiation of reactions by chemical changes. How the new steady state is reached is dependent upon the rate constants of the various stages in the cycle and upon mechanical factors. The problem with this method is to identify the various steps in the cycle. In biochemical solutions (in which the proteins are disorganized) this problem may not be acute, since the individual stages may each have their own characteristics. However, such systems have the serious drawback that mechanical constraints cannot be readily applied, and a full understanding of the kinetics of the contractile cycle *must* require that the dependence of the parameters of the cycle upon the obtainable mechanical distortions of the molecules involved be measured. How, for example, do the rate constants that the biochemists measure in solution vary with changes in the angle that the cross-bridge makes with the filaments? Mechanically it is much less easy to identify the various steps that give rise to the kinetics, since so far the only parameter that has been measured with the necessary time resolution is tension, and it is possible that several stages in the biochemically determined cycle may contribute to this.

A second problem concerns the way in which the steady state system is to be perturbed. In solutions one possibility is to change the temperature of the solution extremely rapidly; this is known as the temperature jump method. With intact fibers a rapid change in either the tension or the length of the fibers can be used. This results, presumably, in a change in the conformation of the cross-bridges, which therefore exert a new tension. The resulting tension changes are dependent upon the way in which the tension exerted by a cross-bridge changes with this distortion and upon the rate constants of attachment and detachment of the bridges. Insect flight muscle has one advantage over vertebrate striated muscle for this kind of experiment in that one or more of the rate constants appear to depend upon the strain of the sarcomere (Thorson and White, 1969).

Structural measurements during perturbations of this type have been made on insect flight muscle (Tregear and Miller, 1969; Miller and Tregear, 1970) in which cross-bridge conformational changes have been demonstrated from X-ray diffraction pictures obtained during sinusoidal oscillation of the length of the fibers.

Mechanical experiments designed to study the kinetics have been made on both vertebrate striated muscle (Civan and Podolsky, 1966; Sugi, 1969; Huxley and Simmons, 1970, 1971a) and insect fibrillar flight muscle (Jewell and Rüegg, 1966; Abbott, 1969; White and Thorson, 1972). This paper is concerned with the interpretation of the latter.

Methods

The experimental results illustrated in Figs. 1, 3, and 4 were obtained by Dr. John Thorson and myself. The techniques and a full account of the results are given in White and Thorson (1972). In these figures the top line is of muscle length and the bottom line of muscle tension. The apparatus was capable of applying length changes and measuring the resulting tension with a time resolution of about 1 msec. Glycerol-extracted fibers from the giant waterbug *Lethocerus cordofanus* were used within 2–3 weeks after glycerol extraction. They were immersed in one of two solutions: (A) phosphate solution (10 mM ATP, 12 mM MgCl₂, 5 mM EGTA, 3 mM CaCl₂, 20 mM KCl, 20 mM orthophosphate, pH 7.0); (B) nonphosphate solution (10 mM ATP, 10 mM MgCl₂, 5 mM EGTA, 3 mM CaCl₂, 45 mM KCl, 20 mM histidine, pH 7.0). Relaxing solutions were made up identically but without calcium. The reasons for choosing the particular concentrations

of chemicals in the solutions are given in White and Thorson (1972).

Figure 2 shows a set of earlier experimental results that were obtained with a tension transducer which limited the time resolution to about 3–4 msec. The length changes were applied as ramps with duration of 5 msec.

The units used for tension are μdyn/A filament. The values were calculated using the data from Chaplain and Tregear (1966) for the number of A filaments per fiber (6.4×10^5).

"Rest length" is defined as that length at which the tension in the relaxed fibers just reaches zero. Because of the high stiffness of the relaxed insect flight muscle, this is a well-defined length.

Properties of Glycerol-extracted Insect Fibrillar Flight Muscle

1. In the resting state, insect flight muscle is very stiff by comparison with vertebrate striated muscle (Machin and Pringle, 1959). This property is maintained in the glycerol-extracted state (White, 1967).

2. If a rapid length increase of less than a few percent of the length of the fiber is applied to the active muscle, the resulting tension change, in the presence of orthophosphate ions (White and Thorson, 1972), shows an initial very fast transient rise in tension, which decays in less than 5 msec, followed by a delayed rise in tension, whose time course is approximately exponential with a time constant of about 30 msec (Fig. 1B, 2). It is this delayed rise in tension, analogous to a negative viscosity, which enables the muscle to do cyclic work, as in flight, on a suitable load.

3. In the absence of orthophosphate ions the response is strikingly dissimilar for step heights greater than a few tenths of a percent in amplitude (Fig. 1A). There is an initial rapid transient response, followed by a delayed tension change; but this delayed tension change no longer rises monotonically to a new equilibrium level. There is now a delayed increase in tension which rises to a high tension but which then decays to a final equilibrium level, approximately that which would have been reached in the presence of phosphate for that particular step height. This we have named the phosphate starvation transient (PST). At very low amplitudes the response is more similar to that in the presence of phosphate (Jewell and Rüegg, 1966; White and Thorson, 1972).

Strikingly, the response to the step decrease in length is no longer even qualitatively the inverse of that to the step increase. Provided the initial length is sufficient (at least 1–2 % above rest length) there is a delayed *increase* in tension in response to the *decrease* in length (Figs. 1A(i), 3, and 4). This is still

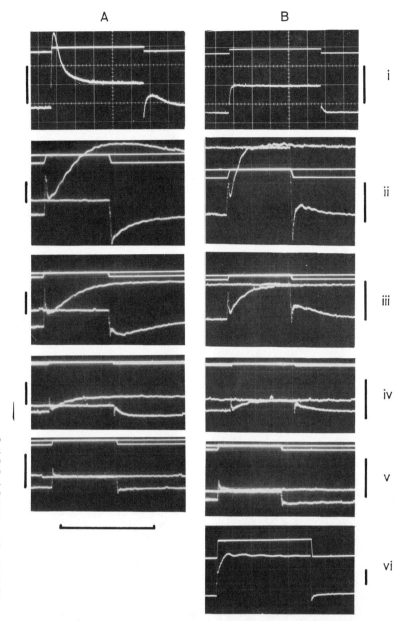

Figure 1. Comparison of the response of the muscle to step length changes in (A) nonphosphate and (B) phosphate solutions. Vertical bars are for the tension records and equal 10 μdyn/A filament. Horizontal bar is for the time axis and represents: A(i) 5 sec, B(i) 1 sec, B(vi) 500 msec, the rest, 100 msec. Step length change is (i) 0.5%, (ii) 1%, (iii) 0.5%, (iv) 0.2%, (v) 0.5%, (vi) 2%. Brackets enclose records from the same set of fibers: [A(i), B(i)], [A(ii, iii, iv, v), B(ii, iii, iv, v)], [B(vi)]. Zero tension is approximately at the bottom of each record, but slow drift in the tension transducer did not allow an accurate estimate of this.

seen even when the duration of the step change in length is made very short (50 msec in Figs. 3B and 4), and under these circumstances is in fact larger, following a pulse decrease in length, than a pulse increase.

4. At high amplitudes of step change of length (greater than about 2%), the response has, superimposed upon that described in sections 2 and 3, a series of damped oscillations (Schadler et al., 1969; White and Thorson, 1972), See Fig. 1B(vi).

5. The rate of hydrolysis of ATP by the active muscle fibers increases markedly as the length of the fibers is increased from rest length for the first 3–4% (Rüegg and Tregear, 1966; Rüegg and Stumpf,

1969a). This is the same range of length changes over which the mechanical effects described in paragraphs 2 and 3 above are best observed. The increase is of the order of 100 pmoles/cm fiber/min/ 1% length increase. In order to calculate the change in catalytic site activity, we must make one of a number of assumptions. The two extremes are to assume either that every heavy meromyosin (HMM) molecule is equivalent, or that only one ATPase site per cross-bridge can be active at any one time. Chaplain and Tregear (1966) calculate the myosin content of 1 cm fiber to be 1.95 μg (equivalent 3.9 pmoles) and the cross-bridge content to be 1.3 pmoles. With two HMM per myosin the

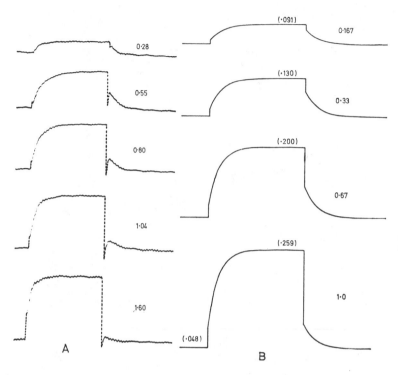

Figure 2. *A.* Effect of varying the step height on the response in phosphate solutions. Step length change (%) indicated on the records. Due to the slower response time of the apparatus used for this experiment, the initial transient changes in tension observable in the other records are not obtained. Vertical: 25 μdyn/A filament/cm. Step duration, 500 msec. *B.* Predicted response of the two-state model including cross-bridge elasticity (Model 2) as specified by Eqs. 1, 2, 5, and 6. Values of parameters: $p_d = 20$ sec^{-1}, $Q = 6$ (sec · % strain)$^{-1}$, $x_0 = 12$ nm; steps of strain are 0.167, 0.33, 0.67, 1%. (1% corresponds to 12 nm.) Tension units are arbitrary. The figures in brackets on the curves indicate the number of attached bridges at equilibrium at the various tensions.

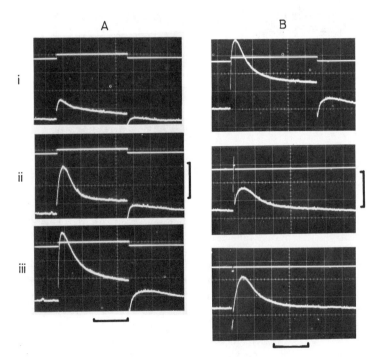

Figure 3. *A* and *B.* Experiments in nonphosphate solution. *A.* effect of varying initial fiber length on response to a 0.5% applied step length change. Starting lengths: (*i*) 0.5%, (*ii*) 1.5%, (*iii*) 2.5% above rest length. Starting tensions are approximately 2, 8, and 15 μdyn/A filament, respectively. *B.* Comparison of response to a long 0.5% step (*i*) with that to 50 msec upward (*ii*) and downward (*iii*) pulses. Vertical bars, 10 μdyn/A filament. Horizontal bars, 1 sec.

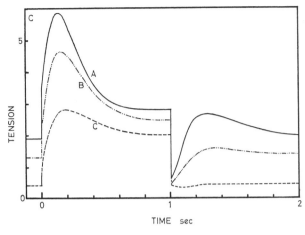

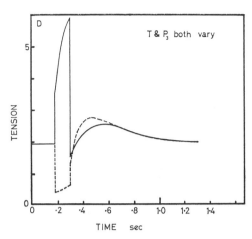

Figure 3. C and D. Predictions from model 3 with both p_3 and tension varying with distortion. Values of parameters given in Table 1. Tension units arbitrary. In all the graphs of model 3 predictions, the tensions are drawn to the same scale. C. Effect of varying starting length. D. Effect of applying short upward (solid line) and downward (dashed line) perturbations in p_1.

first assumption gives the change in catalytic site activity as $0.21 \text{ sec}^{-1}/\%$ length change; the second assumption gives a value of $1.3 \text{ sec}^{-1}/\%$ length change.

6. Rüegg et al. (1971) have demonstrated that the addition of orthophosphate ions to the solutions causes both a reduction in the steady-state tension of the fibers and an increase in the speed of the response, measured either as the frequency of the isometric oscillations at high amplitude, or as the frequency at which the maximal amount of work

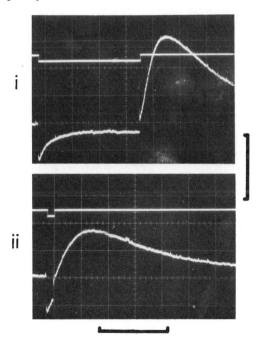

Figure 4. Effect of step length on experimental response in nonphosphate solution. Vertical bar: 10 μdyn/A filament. Horizontal bar: 500 msec.

per cycle is obtained from the fibers when the length is oscillated sinusoidally at a peak-to-peak amplitude of 2% (see also White and Thorson, 1972). Abbott (this volume) found that the rate constant at low amplitude is also usually increased by the addition of P_i, although White and Thorson (1972) found no change. The P_i concentration required for half the maximal effect in all these experiments was between 1 and 5 mM P_i.

7. Considerable work has been done experimentally to determine the relationship between the amount of oscillatory work performed by the fibers and the amount of ATP hydrolyzed (Rüegg and Tregear, 1966; Rüegg and Stumpf, 1969b; Steiger and Rüegg, 1969; Pybus and Tregear, this volume). These studies all demonstrate that a bundle of fibers, oscillated about a mean length, hydrolyze ATP at a faster rate, the greater the power output from the muscle. The effect is very marked, a trebling of the ATPase activity being possible comparing the oscillating muscle with the static, but active, muscle held at the same mean length (Pybus and Tregear, this volume).

Theoretical Treatment of Cross-Bridge Cycle

Dr. John Thorson and I have, over the past few years, been doing experiments to answer the question: Can we exclude the possibility that specific models of the cross-bridge cycle can explain the detailed mechanical responses observed in insect flight muscle and the dependence of ATP hydrolysis upon mechanical parameters? We have been hoping to establish a dialogue in which biochemical results suggest critical mechanical experiments and vice versa.

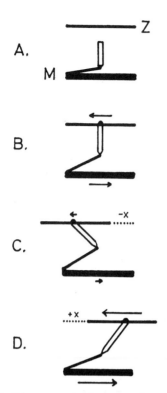

Figure 5. Diagrammatic demonstration of terminology, based upon Huxley (1969). *A.* Detached bridge. Directions of Z line and M line indicated. *B.* Attached, undistorted bridge. Arrows represent magnitude of shear force exerted. *C.* Shortening of the muscle results in negative distortion of undistorted bridges. *D.* Lengthening results in positive distortion.

It is convenient to define at the outset what is meant by the various terms that will be used. To begin with we shall consider two-state cycles in which a cross-bridge can be either *detached* from the I filament or *attached* to it (Figs. 5 and 7A). If detached, it has a certain probability per unit time (p_a) of attaching. We assume that at the instant of attachment the cross-bridge has some preferred orientation or conformation. Such a bridge is undistorted. While it is attached to the I filament, it is possible that there will be a sliding of the I filament relative to the A filament. In this case the cross-bridge will be *distorted*, and we denote the degree of *distortion* (x) by the sliding of the I filament relative to the A filament since the instant of attachment. An attached bridge will have some probability per unit time of detaching (p_d). The tension exerted by an attached cross-bridge will depend upon the shape of its curve of potential energy (PE) vs. distortion. In the actual process of attachment we consider that this curve has a minimum at $x = 0$ and thus exerts no force. Some subsequent event causes a shift of the PE curve along the filament axis, and the cross-bridge

now exerts a shear force between the two sets of filaments. This subsequent event may be very rapid; in fact, in all our mathematical formulations the two steps are treated as one, governed by the "attachment" rate, p_a.

In "one-bridge" models one assumes that the filaments are very stiff and that there can be no movement of attached cross-bridges in the absence of length changes at the end of the muscle (cf. the effects of distributed elasticity in Thorson and White, 1969). Whether or not the filaments slide relative to one another depends upon the summed effects of all cross-bridges relative to the external loading on the muscle.

Model 1 (*Thorson and White, 1969*). The critical mechanical property which enables work to be produced in flight is that imposed length changes induce *delayed tension* changes. Similarly in any hypothetical cycle of attachment and detachment, the fraction of bridges attached must undergo delayed readjustment following an alteration of the rate constants. We naturally proposed therefore, that the critical delay in tension was the delay in readjustment of this attached fraction.

In order to test this idea quantitatively we had to make certain assumptions as to the means by which stretch perturbed the rate constants. Since fibrillar muscle is very stiff and A filament-Z line links are suspected (White and Thorson, 1973), it was natural to assume that the (necessarily maintained) perturbation was mediated by strain in the A filaments.

The exponentially *delayed tension* arises naturally from the simplest equation for the two-state cycle:

$$\frac{dn}{dt} = p_a(1 - n) - p_d n = p_a - (p_a + p_d)n \qquad (1)$$

for if one treats the rate constant p_a as a forcing function, the response obtained to a step change in p_a from p_{a1} to p_{a2} is:

$$n(t) = n_0 + (n_\infty - n_0)(1 - e^{-(p_{a2}+p_d)t}) \qquad (2)$$

where

$$n_0 = \frac{p_{a1}}{p_{a1} + p_d} \qquad (3)$$

$$n_\infty = \frac{p_{a2}}{p_{a2} + p_d} \qquad (4)$$

The response is exponential, with rate constant $p_{a2} + p_d$.

Note that with the extremely small length changes that can be applied experimentally (less than 0.1%) attached cross bridges cannot undergo appreciable *distortion*, and thus the unknown dependencies of bridge force and p_d upon distortion, severe sources of arbitrariness in large-signal analyses of the cyclic hypothesis, did not affect our tests.

Although treatment of the simple two-state cycle as in Equation 1 serves to illustrate the hypothesis, it is clear that any critical test must take account of the mutual feedback between rate constant changes and the *strain-activation* mechanism, as well as the crucial contributions to tension of the viscoelastic properties (White, 1967). Equation 1 in various forms and the assumption that tension is proportional to n have been used to examine potential nonlinear properties of the cyclic hypothesis (see models 2 and 3 below), and in some discussions of the relationships of certain measurements (see Abbott, this volume). But it must be borne in mind that in the muscle n need not be proportional to, or in phase with, tension and that cycling rate vs. tension is also sensitive to these considerations (see, e.g. White and Thorson, 1972). In particular note that the cycle alone predicts no passive tension which in fibrillar muscle is on the order of the active tension.

Accordingly in order to test the hypothesis, we (Thorson and White, 1969) calculated the length-tension dynamics of a representative nonhomogeneous viscoelastic sarcomere with a continuum of strain-activated cross-bridge cycles in the overlap region. Not only did the predictions agree well with Abbott's (1968) mechanical data but the estimated average cycling rate of bridges in the sarcomere (assuming one ATP is hydrolyzed per cycle) agreed well with ATP hydrolysis measurements.

At the time we wrote that paper there was insufficient evidence to test this description of the behavior of muscle any further. Obvious experiments to test it included measurements of the dynamics of the muscle at higher amplitudes. These have been described in the experimental section.

Model 2 (*White and Thorson, 1972*). In a more recent paper (White and Thorson, 1972) we investigated the effect of removing the low-amplitude restriction. Under these conditions a mathematical description of the cross-bridge dynamics must include an expression for the way in which (a) tension and (b) p_d vary with cross-bridge *distortion*. In that paper we showed that if

$$p_d = \text{constant} \qquad (5)$$

(i.e., independent of distortion, x) and that if the tension generated by a cross-bridge was related linearly to the *distortion*

$$T_x = k(x_0 + x) \qquad (6)$$

in which x_0 is the displacement from the point of attachment for a cross-bridge to give zero tension, then the response is as shown in Fig. 2B. If the initial very rapid transients are ignored, then the model including distortion describes the results obtained in the presence of phosphate ions fairly well. In particular, notice that the elastic asymmetry of the experimental response is obtained in the model (due to the greater number of attached bridges at the higher tension), and that as the amplitude of the step is increased the delayed rise in tension becomes faster than the delayed fall.

We have, however, been unable to account for the high-amplitude results obtained in the absence of phosphate with a two-state model.

Model 3. From a biochemical viewpoint a two-state model incorporating ATP hydrolysis is absurd. The simplest complete scheme probably includes at least seven steps (Fig. 6). Thus the states of a two-state mechanical model must include several biochemical stages. Restricting the number of mechanical states to fewer than the number of biochemical states will be valid only if the transitions between each biochemical state within a mechanical state are sufficiently rapid and if the mechanical effects of each of the biochemical stages are identical. Good arguments can be put forward therefore for investigating the properties of a three-state mechanical cycle.

A three-state cycle is much more complex than a two-state cycle. Strictly, the transitions between each pair of states should be reversible; the sensible possibility exists that either one or two of those states shall be tension generating, and that if two states are attached and capable of generating tension, they may do so with different parameters. Unfortunately, not much help is obtained by looking for values of the relevant rate constants from solution biochemistry since of those likely to be involved only the ones indicated in Fig. 6A with heavy arrows have been measured. The options available are therefore numerous.

Since one major object of the extended model is to find a testable role for orthophosphate in the cycle, an obvious step to include is one with a backward reaction whose rate would be proportional to phosphate concentration. This is the only backward step that we shall consider. The three-state cycle is therefore as depicted in Fig. 7B, and we shall consider for the moment that states B and C are separated at the point in the biochemical cycle where

$$\text{AM·(ADP)·P} \rightleftharpoons \text{AM·(ADP)} + \text{P} \qquad (7)$$

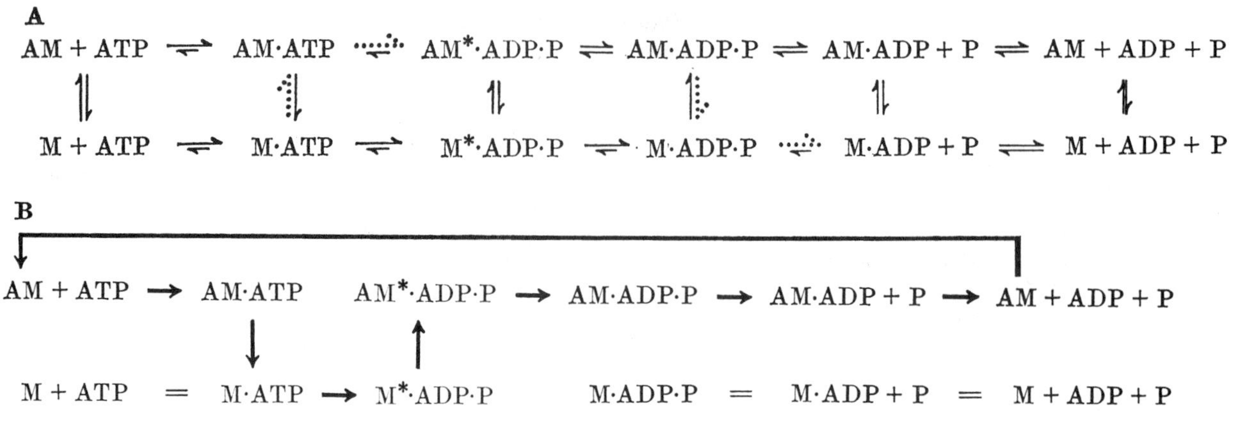

A

$$\text{AM + ATP} \rightleftharpoons \text{AM·ATP} \cdots\!\rightarrow \text{AM*·ADP·P} \rightleftharpoons \text{AM·ADP·P} \rightleftharpoons \text{AM·ADP + P} \rightleftharpoons \text{AM + ADP + P}$$

$$\text{M + ATP} \rightleftharpoons \text{M·ATP} \rightleftharpoons \text{M*·ADP·P} \rightleftharpoons \text{M·ADP·P} \cdots\!\rightarrow \text{M·ADP + P} \rightleftharpoons \text{M + ADP + P}$$

B

$$\text{AM + ATP} \rightarrow \text{AM·ATP} \quad \text{AM*·ADP·P} \rightarrow \text{AM·ADP·P} \rightarrow \text{AM·ADP + P} \rightarrow \text{AM + ADP + P}$$

$$\text{M + ATP} = \text{M·ATP} \rightarrow \text{M*·ADP·P} \quad \text{M·ADP·P} = \text{M·ADP + P} = \text{M + ADP + P}$$

Figure 6. Suggested biochemical reactions relevant to muscle contraction. *A.* Full scheme. Long, solid arrows indicate that rate constants have been measured; long dotted arrows that rate constants can be inferred, or are known approximately; short arrows, rate constants not known (to me!) *B.* The pathway indicated is that suggested as the likely sequence of biochemical events taking place in active muscle. The schemes are discussed fully by White and Thorson (in press).

The ADP is included in brackets as it is not certain whether ADP is released before or after the phosphate.

As before we will make one of the rate constants, p_1, dependent upon the overall strain applied to the muscle. (Notice the distinction between strain, which is a measure of the length changes applied to the muscle, and distortion, which applies to an individual cross-bridge.) At any constant degree of strain, the number of bridges in distorted states decays to zero as they detach from their distorted configurations and reattach at the undistorted configuration. p_1 is the equivalent of p_a in the two-state model.

At first sight it seems that both states B and C should be tension generating in that they both include biochemical states in which actin and myosin are associated. However, if the non-phosphate results are to be described by this scheme, state B must be the only state in which significant tension is generated by the undistorted bridge. The equations governing the performance of this model are as follows:

$$\frac{dA}{dt} = p_3C - p_1A \tag{8}$$

$$\frac{dB}{dt} = p_1A - (p_qB - q_2C) \tag{9}$$

$$C = 1 - A - B \tag{10}$$

$$p_1 = QL \tag{11}$$

$$\text{Tension} = T_1B \tag{12}$$

$$q_2 = q_2^*[\text{phosphate}] \tag{13}$$

As they stand, Eqs. 8–13 do not include any effect of cross-bridge distortion. With the values for the constants indicated in Fig. 8, however, they reproduce one of the major qualitative differences between the phosphate and nonphosphate results, the PST following the upward step (Fig. 9). Quantitatively the behavior is entirely wrong.

In particular the predicted magnitude of the response is much larger, the rate constants to the peak tensions appear slower, and the initial tension is much greater in the phosphate than the non-phosphate case. However, if the forward rate constant (p_2) is also increased in the presence of phosphate, all these discrepancies disappear, and for the step up the three-state model can account for the major features of both phosphate and non-phosphate responses (Fig. 9). I can think of no obvious reason why p_2 should be affected by phosphate. Abbott (quoted in Rüegg et al., 1971) suggested that all Rüegg's results presented in paragraph 6 of the experimental section could be accounted for by an increase in p_d of our two-state model. The increase in p_2 with phosphate is the equivalent suggestion for the three-state model.

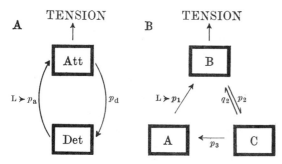

Figure 7. The two-state (*A*) and three-state (*B*) models showing the rate constants used. Length (*L*) perturbs p_a and p_1, and tension is generated as described in the text by the states indicated. *Det.* = detached; *Att.* = attached.

The reason for the large delayed transient rise in tension is easy to understand. At the initial length p_1 is small, and about $\frac{3}{4}$ of the bridges are in state A. The large increase in p_1 causes these to transfer to state B very rapidly. If p_2 is not sufficiently large, then there is a build-up of bridges in this state which then decay more slowly to state C. If p_2 is sufficient, then the extra build-up does not occur. It is not sufficient, however, just to increase p_2 by

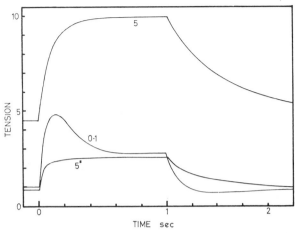

Figure 9. Predicted responses of the three-state model for various values of the rate constants between stages B and C, under conditions in which neither tension generation or p_3 are affected by distortion. The values of the parameters are given in Table 1. The concentrations indicated on the graphs give the values of q_2 in the table for a value of q_2^* (Eq. 13) of 10^4 (msec)$^{-1}$. The curves labeled 0.1 and 5 mM have the same value of p_2, that labeled 5* mM has a much higher value of p_2.

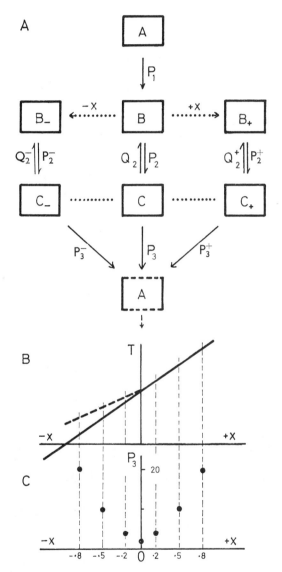

Figure 8. A. The three-state model of Fig. 7A, but including the effects of distortion as described in the text. B_- and C_- are the states occupied by bridges with negative distortion and B_+ and C_+ those occupied by bridges with positive distortion. This model, with discrete distortion states, is suitable only for predicting responses of the model to step length changes. B. Effect of distortion upon the tension generated by a cross-bridge. The solid line is that described by Eq. 6. Figure 12 uses the dashed line for negative distortions and the solid line for positive distortions. C. Variation in p_3 with distortion used to obtain Fig. 12.

the addition of phosphate. The magnitudes of the final tensions and the values of the rate constants also require that the rate of the back reaction (q_2) also be increased. The important feature is the increase in the rate at which equilibrium between states B and C is reached.

It is obvious that account must be taken in this, as in the two-state model, of the effects of cross-bridge distortion, both on the tension generated by a bridge and upon the rate constants. If the step between states B and C is that at which phosphate is released, then state C must include biochemical states in which actin and myosin are associated. Although the possibility arises that such bridges may have a finite stiffness (and thus contribute tension when distorted), this effect has not been included in the predictions in this paper. Figure 10 shows the effect of making the assumption that the tension exerted by a cross-bridge in state B is related linearly to its distortion, as in Eq. 6 and Fig. 8B (cf. Figs. 1A and 3A). The fit remains.

Thus far the PST following an initial step increase in length has been accounted for, but that following a step decrease in length (as in Fig. 1Ai, 3A, B) has not. Neither has any account been given of perhaps the most striking experimental result, the PST obtained in response to a brief pulse change of length of either sign (Fig. 3B, 4).

It is probable that the PST is basically the same phenomenon in all these cases for the following reason. Consider the response to a long downward pulse length change (Fig. 4). The PST at the end of this pulse must have the same cause as that for the

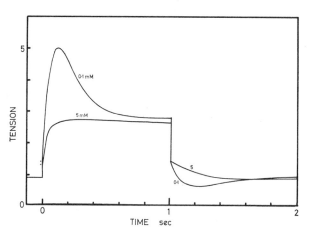

Figure 10. Predicted response of model **3** under the same conditions as Fig. 9, but with the addition of cross-bridge tension dependence upon distortion.

PST following an initial step increase of length since they are the same experiment at different starting lengths; i.e., both are the response to a step increase in length from an equilibrium situation. If the duration of the downward pulse is now made shorter and shorter, there is a gradual decrease in amplitude of the PST, until at very short durations (less than about 10 msec) the PST disappears. The PST to a brief downward pulse thus also has the same cause. The similarity of time course of the response to a short upward length pulse suggests strongly that this is also the same phenomenon, in which case so is the PST following the step decrease in length at the end of a long upward step change of length.

The cause of the PST following a step increase in length on the three-state model is a transient build-up of bridges in state B. This can occur because state A becomes transiently overpopulated for the new value of the attachment probability p_1. Hence all the PST's may occur because of an over-population of state A. This can occur if the effect of negative bridge distortions is to cause rapid detachment of bridges from states B and C into state A.

Figure 8 illustrates the way in which these effects have been computed. Since all the experimental results have been obtained with steps of equal magnitude, the addition of distortion-dependent parameters is straightforward. The states A, B, and C are as before, with the same rate constants describing the transitions. If a step increase in length is applied, then all attached bridges are distorted by an amount equal to the extent of the applied length change. Thus all bridges in states B and C become distorted and are represented by states B^+ and C^+. Their transitions are governed by the rate constants indicated, and they will return

to state A and then reenter the cycle. Note that bridges in the detached state A can only attach at the undistorted configuration. A similar argument holds for negative steps of strain. Of course if, say, a negative step is applied before the system has regained equilibrium after a positive step, there will still be bridges in states B^+ and C^+. These will then revert to the undistorted configuration, having had positive and negative distortions of equal magnitude applied to them during their lifetime as attached bridges. The only way in which an attached cross-bridge can change its distortion is by means of length changes. There are no chemical rate constants between state B^+ and B, etc.

It is not necessarily the case that state C represents attachment. If the necessary biochemical steps included in mechanical state C proceed rapidly to a detached state, then it might be more appropriate to consider transitions between B^+ and C and between B^- and C, and to exclude states C^+ and C^-. The justification for not doing this is two-fold: (1) the effect of addition of phosphate no longer appears, and (2) the agreement with the effects of applying pulsed length changes of short duration is no longer maintained.

For these reasons the notion that large negative distortions cause bridges to repopulate the detached state A has been examined by making p_3 dependent upon distortion in the manner illustrated in Fig. 8C.

The simulation of the phosphate results obtained by making p_2 and q_2 large holds as well here as for the two-state model. The predictions made by the model with the added effects of cross-bridge distortion are shown, for the nonphosphate case, in Figs. 3C, 3D, and 11. The main features of the observed response, apart from the very rapid

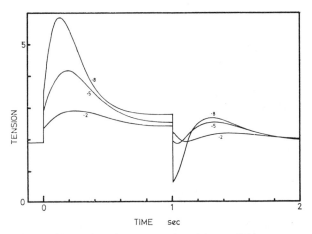

Figure 11. Predicted response of model **3** to different step perturbations of p_1 using the parameters indicated in Table 1. Both p_3 and tension generation vary with distortion as indicated in Fig. 8B, C. The figures on the graphs are of the ratio of the length change to x_0.

transients, are now present:

1. The PST following the step-down now arises naturally. At low starting lengths (Figs. 3A, 3C) it is absent. Its magnitude increases as the starting length is increased.

2. At the same starting length, but for different amplitudes of length change, the magnitude of the PST relative to the final equilibrium level is greater following the increase in length at greater step amplitudes. Further, following the step decrease the positive PST is almost absent for very low-amplitude steps, and there is an initial negative delayed tension change. At higher amplitudes the negative change disappears and the delayed positive transient change in tension is the only observed effect.

3. The proper responses to very short pulses are now predicted in a qualitative manner, and in particular the rate of rise in tension after the pulse decrease in length is faster than that after the pulse increase in length (Figs. 3B, D).

There is a further intriguing property of the nonphosphate response. With the experimental phosphate results the "elastic" tension change in response to a length increase was always smaller than that to the ensuing step decrease. With the nonphosphate results (see Fig. 1Aiii) this is not the case. The theory, however, predicts that there will be a greater number of bridges attached at the greater tension, as in the phosphate case. Thus the linear relationship between tension and distortion (as depicted by the solid line in Fig. 8B) cannot account for this. One possible resolution of this problem is that the distortion-tension diagram of a cross-bridge in the absence of phosphate is non-linear. The dashed line in Fig. 8B has been used for negative bridge distortions in Fig. 11, and illustrates the degree of nonlinearity required to produce the experimental asymmetry.

No account has been given of two important findings. The first concerns the very rapid initial transient changes. Two different interpretations of these have been made for both vertebrate and insect muscle: that they represent a very rapid detachment of tension-generating bridges (Podolsky et al., 1969; Julian, 1969), and that they are some passive "viscous" effect within the cross-bridge itself (Huxley and Simmons, 1970, 1971a, 1971b). I have tacitly assumed that the second interpretation is correct. If, however, there is a very rapid detachment of bridges during this time, the above discussion would need extensive modification. The only good experimental evidence on this point is that of Huxley (Fig. 6 in Huxley, 1971), which indicates (as Huxley discusses) that the assumptions Podolsky makes to fit the initial rapid changes are invalid.

The second finding is that the magnitude of the tension change in relaxed muscle is large (the initial changes may be larger than in active muscle). The model described here takes no account of this. Either the effects responsible for the elastic changes in relaxed muscle are in effect not observable in active muscle under the experimental conditions used—as might occur, for example, if there were small length changes in the A or I filaments, or if the cross-bridges were responsible for the elasticity of relaxed muscle and changed their properties upon activation—or the effects of relaxed muscle do contribute to the tension changes in active muscle, in which case the ideas behind the work reported here will probably still give rise to mathematical expressions with similar solutions, but the values of the parameters, in particular of cross-bridge elasticity, will need changing. In any case the cross-bridge cycle described here must be tested in a viscoelastic half-sarcomere (as was done for Model 1) in order to determine the effect on the predicted response of the distribution of strain activation along the length of the half-sarcomere, and in order to provide a model accounting for both relaxed and activated properties of the muscle.

Comparison of Biochemical and Mechanical Kinetics

The three models that have been presented in this paper form a logical succession of ideas. All the findings of the first model are true of the second model when it is constrained to small perturbations; and likewise the third model simplifies to the second when the equilibrium between states B and C is achieved very rapidly by increasing the values of the rate constants involved. Thus the three models are not separate. It would be possible to present the third model alone, but in so doing much of the insight that one gains from the two simpler models is lost. In its turn the third model is either a useful simplification of some more complete understanding, or else it is based upon the wrong assumptions. Our best hope is to find experiments that will establish that the assumptions are invalid. In this sense the models are satisfactory in that they predict mechanical responses under a much wider range of conditions than those used to obtain them.

Although the models are based upon the mechanics of cross-bridge activity alone, their eventual usefulness must be the ease with which the biochemical events can be fitted into the same framework. If this can be achieved, then the range of predictions becomes very much more extensive, and such a model should be able to account for both the biochemical and energetic performance of

Table 1. Values of Parameters Used to Obtain Predicted Responses of Model 3

Figure	Line	p_{1a}	p_{1b}	p_2	q_2	p_3^-	p_3	p_3^+	k	L
3C	A	2	6	12	0.2	20	2	20	10	8
	B	1	4	12	0.2	20	2	20	10	8
	C	0.2	2	12	0.2	20	2	20	10	8
3D	solid	2	6	12	0.2	20	2	20	10	8
	dashed	2	0.4	12	0.2	20	2	20	10	8
9	0.1	0.6	6	12	0.2	2	2	2	0	5
	5	0.6	6	12	10	2	2	2	0	5
	5*	0.6	6	72	10	2	2	2	0	5
10	0.1	0.6	6	12	0.2	2	2	2	10	5
	5	0.6	6	72	10	2	2	2	10	5
11	0.2	2	3	12	0.2	4	2	4	10	2
	0.5	2	4	12	0.2	10	2	10	10	5
	0.8	2	6	12	0.2	20	2	20	10	8

In all the figures: $p_2^- = p_2^+ = p_2$; $q_2^- = q_2^+ = q_2$. p_{1a} indicates the initial and final values of p_1; p_{1b} indicates the value of p_1 during the step perturbation of this parameter. The cross-bridge stiffness (k) is given the value 0 for the situation in which there is no dependence of cross-bridge tension upon distortion, and 10 otherwise. L indicates the values of the applied length change. x_0 has the value 10 in these arbitrary units.

muscle. The simplest method of introducing biochemical events is to make the assumption that one ATP molecule is hydrolyzed per complete mechanical cycle. This will then enable comparison to be made with the results mentioned in paragraph 7 of the experimental section.

However, a more complete identification of the biochemical events underlying the mechanical events is obviously desired. Figure 6A is my summary of the likely biochemical events, derived from the work of Trentham et al. (1972), Finlayson and Taylor (1969), Finlayson et al. (1969), Lymn and Taylor (1970, 1971), Eisenberg et al. (1968), Eisenberg and Moos (1968, 1970). In Fig. 6B I have indicated the sequence of biochemical reactions that seem, on present evidence, to be the dominant ones in active muscle. In Fig. 12 I have redrawn this set of reactions to indicate one possible way in which they might fit into the three-state mechanical cycle. The transition within the tension-producing state is an obvious candidate for the biochemical event underlying the shift in the potential energy versus distortion curve of a cross-bridge subsequent to attachment mentioned in the previous section. In this case the intermediate state AM·ADP·P will not generate tension, and I am therefore assuming that the transition to AM·ADP·P is very rapid. The scheme requires that actomyosin in the absence of

phosphate bound to the hydrolytic site be very different mechanically from that in which phosphate is bound. Making the third mechanical transition be that which occurs as ATP is bound to the actomyosin (rather than the, at first sight, attractive possibility of making it the next step at which actin dissociates) enables the three-state model to account for the increase in mechanical rate constant observed at higher ATP concentrations (Thorson and White, 1972).

Acknowledgments

I should like to thank Dr. John Thorson and Prof. J. D. Currey for helpful criticism.

References

ABBOTT, R. H. 1968. The mechanism of oscillatory contraction of insect fibrillar flight muscle. D. Phil. thesis, University of Oxford.

CHAPLAIN, R. A. and R. T. TREGEAR. 1966. The mass of myosin per cross-bridge in insect fibrillar flight muscle. *J. Mol. Biol.* **21:** 275.

CIVAN, M. M. and R. J. PODOLSKY. 1966. Contraction kinetics of striated muscle fibres following quick changes in load. *J. Physiol.* **184:** 511.

EISENBERG, E. and C. MOOS. 1968. The adenosine-triphosphatase activity of acto-heavy meromyosin. A kinetic analysis of actin activation. *Biochemistry* **7:** 1486.

———. 1970. Actin activation of heavy meromyosin adenosine triphosphatase. *J. Biol. Chem.* **245:** 2451.

EISENBERG, E., C. R. ZOBEL, and C. MOOS. 1968. Subfragment 1 of myosin: Adenosine triphosphatase activation by actin. *Biochemistry* **7:** 3186.

FINLAYSON, B. and E. W. TAYLOR. 1969. Hydrolysis of nucleoside triphosphatase by myosin during the transient state. *Biochemistry* **8:** 802.

FINLAYSON, B., R. W. LYMN, and E. W. TAYLOR. 1969. Studies on the kinetics of formation and dissociation of the actomyosin complex. *Biochemistry* **8:** 811.

HUXLEY, A. F. 1971. The activation of striated muscle and its mechanical response. *Proc. Roy. Soc. (London)* B **178:** 1.

HUXLEY, A. F. and R. M. SIMMONS. 1970. A quick phase in the series-elastic component of striated muscle,

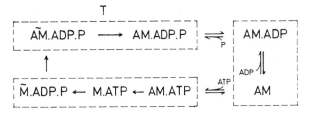

Figure 12. One possible relationship between the biochemical states of Fig. 6B and the three mechanical states of the three-state model (indicated by the dashed boxes).

demonstrated in isolated muscles from the frog. *J. Physiol.* **208**: 52.

———. 1971a. Mechanical properties of the cross-bridges of frog striated muscle. *J. Physiol.* **218**: 52.

———. 1971b. Proposed mechanism of force generation in striated muscle. *Nature* **233**: 533.

HUXLEY, H. E. 1969. The mechanism of muscular contraction. *Science* **164**: 1356.

JEWELL, B. R. and J. C. RÜEGG. 1966. Oscillatory contraction of insect fibrillar muscle after glycerol-extraction. *Proc. Roy. Soc. (London)* B **164**: 428.

JULIAN, F. J. 1969. Activation in a skeletal muscle contraction model with a simplification for insect fibrillar muscle. *Biophys. J.* **9**: 547.

LYMN, R. W. and E. W. TAYLOR. 1970. Transient state phosphate production in the hydrolysis of nucleoside triphosphates by myosin. *Biochemistry* **9**: 2975.

———. 1971. Mechanism of adenosine triphosphate hydrolysis by actomyosin. *Biochemistry* **10**: 4617.

MACHIN, K. E. and J. W. S. PRINGLE. 1959. The physiology of insect fibrillar muscle. *Proc. Roy. Soc. (London)* B **151**: 204.

MILLER, A. and R. T. TREGEAR. 1970. Evidence concerning cross-bridge attachment during muscle contraction. *Nature* **226**: 1060.

PODOLSKY, R. J., A. C. NOLAN, and S. A. ZAVELER. 1969. Cross-bridge properties derived from muscle isotonic velocity transients. *Proc. Nat. Acad. Sci.* **64**: 504.

RÜEGG, J. C. and H. STUMPF. 1969a. Activation of the myofibrillar ATP-ase activity by extension of glycerol-extracted insect fibrillar muscle. *Pflugers Arch.* **305**: 34.

———. 1969b. The coupling of power-output and myofibrillar ATPase activity in glycerol-extracted insect fibrillar muscle at varying amplitudes of ATP-driven oscillation. *Pflugers Arch.* **305**: 21.

RÜEGG, J. C. and R. T. TREGEAR. 1966. Mechanical factors affecting the ATPase activity of glycerol-extracted insect fibrillar flight muscle. *Proc. Roy. Soc. (London)* B **165**: 497.

RÜEGG, J. C., M. SCHADLER, G. T. STEIGER, and G. MULLER. 1971. Effects of inorganic phosphate on the contractile mechanism. *Pflugers Arch.* **325**: 359.

SCHADLER, M., G. STEIGER, and J. C. RÜEGG. 1969. Tension transients in glycerol-extracted fibres of insect fibrillar muscle. *Experientia* **25**: 942.

STEIGER, G. J. and J. C. RÜEGG. 1969. Energetics and efficiency in the isolated contractile machinery of an insect fibrillar muscle at various frequencies of oscillation. *Pflugers Arch.* **307**: 1.

SUGI, H. 1969. The mode of tension development by stretch in active frog muscle fibres. *Proc. Jap. Acad.* **45**: 413.

THORSON, J. and D. C. S. WHITE. 1969. Distributed representations for actin-myosin interaction in the oscillatory contraction of muscle. *Biophys. J.* **9**: 360.

TREGEAR, R. T. and A. MILLER. 1969. Evidence of cross-bridge movement during contraction of insect flight muscle. *Nature* **222**: 1184.

TRENTHAM, D. R., R. G. BARDSLEY, J. F. ECCLESTON, and A. G. WEEDS. 1972. Elementary processes of the magnesium-ion-dependent adenosine triphosphatase activity of heavy meromyosin. *Biochem. J.* **126**: 635.

WHITE, D. C. S. 1967. Structural and mechanical properties of insect fibrillar flight muscle in the relaxed and rigor states. D. Phil. thesis, University of Oxford.

WHITE, D. C. S. and J. THORSON. 1972. Phosphate starvation and the nonlinear dynamics of insect fibrillar flight muscle. *J. Gen. Physiol.* **60**: 307.

WHITE, D. C. S. and J. THORSON. 1973. The kinetics of muscle contraction. *Proq. Biophys.* In press.

Regulatory Proteins of Muscle with Special Reference to Troponin

S. Ebashi, I. Ohtsuki, and K. Mihashi*

Department of Pharmacology, Faculty of Medicine, University of Tokyo, Hongo, Tokyo

The contraction-relaxation cycle of muscle under physiological conditions is regulated by the intracellular concentration of free Ca ion, which is under the control of the intracellular membraneous system, i.e., the sarcoplasmic reticulum (cf. Weber, 1966; Ebashi and Endo, 1968). The responsiveness to Ca ion requires the participation of a protein system in the interaction of myosin and actin (Ebashi, 1963). The protein system, first called "native tropomyosin" (Ebashi and Ebashi, 1964), consists of tropomyosin and a new protein, troponin (Ebashi and Kodama, 1965), which is the real Ca-receptive protein (Ebashi et al., 1967, 1968; Yasui et al., 1968). In the absence of Ca ion, troponin in collaboration with tropomyosin exerts a particular effect on F-actin to depress its interaction with myosin; Ca ion removes this depression through its binding to troponin. Thus, Ca ion behaves like a kind of de-repressor (Ebashi and Endo, 1968; Ebashi et al., 1968, 1969).

Ohtsuki et al. (1967) have shown that the antibody against troponin binds to thin filaments at intervals of approximately 400 Å. Based on this finding and related observations, we have proposed a model of the thin filament (Ebashi et al., 1969) in which tropomyosin molecules are located in the grooves of double strands of actin helix, and one tropomyosin molecule, having one troponin molecule at a specified site, covers seven actin molecules. This stoichiometry among actin, tropomyosin, and troponin molecules has been substantiated by optical diffraction studies (Ohtsuki and Wakabayashi, 1972; Spudich et al., 1972).

Concomitantly with the discovery of troponin and of the function of troponin-tropomyosin system, some new structural proteins, α-actinin (Ebashi and Ebashi, 1965), β-actinin (Maruyama, 1965a,b), and M-protein (Masaki et al., 1968) have been isolated and shown to participate in the structural organization of myosin and actin filaments (cf. Ebashi and Nonomura, 1972). Based on these facts Maruyama and Ebashi (1970) have proposed a concept of "regulatory proteins," the proteins which serve the muscle function by controlling the myosin-actin interaction or by arranging the steric correlation of myofilaments to form the ordered structure of the sarcomere.

Components of Troponin

Troponin was first thought to be a single protein because of its behavior in ultracentrifugation and electrophoresis (Ebashi and Kodama, 1965; Wakabayashi and Ebashi, 1968). However Hartshorne and Mueller (1968) first showed that it could be separated into two components, troponin A and troponin B. This was followed by several papers to identify the components of troponin. There is now general agreement that troponin consists of three components, of which the mol wts are 37,000–40,000, 22,000–24,000, and 17,000–18,000 respectively (Schaub and Perry, 1969, 1971; Hartshorne and Pyun, 1971; Ebashi et al., 1971; Greaser and Gergely, 1971; Drabikowski et al., 1971; Staprans et al., 1972; Ebashi, 1972) (Fig. 1). The outline of the method to separate troponin into its three components is shown in Fig. 2.

In view of the quantitative analysis of SDS (sodium dodecyl sulfate)-polyacrylamide gel electrophoretic patterns and of the yields of these components, it is probable that their molar ratio in the original troponin is 1:1:1.

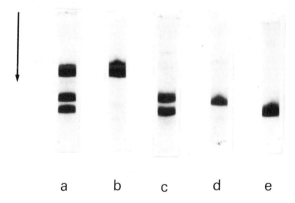

a   b   c   d   e

Figure 1. SDS polyacrylamide gel electrophoresis of troponin and its components. *a*, Original troponin; *b*, troponin I; *c*, troponin II plus troponin A (eluate from SE-Sephadex column); *d*, Troponin II; *e*, troponin A.

* On leave of absence from Faculty of Science, Nagoya University; the Exchange Scientist of Japan Society for the Promotion of Science.

Functional Reconstitution of Troponin

All three components are necessary for the reconstitution of physiological function of the original troponin, as illustrated in Fig. 3. The 17,000 component (troponin A) is certainly the Ca-receptive protein as first indicated by Hartshorne and Mueller (1968). This was confirmed utilizing the different affinities of skeletal and cardiac troponins for Sr ion (Ebashi et al., 1968); i.e., troponin A derived from cardiac troponin affords a high Sr-sensitivity and that from skeletal troponin a low Sr-sensitivity, irrespective of the sources of other two troponin components (Ebashi, 1972). Essentially the same observation was made by Staprans et al. (1972).

The 22,000 component, troponin II, exerts a strong inhibition on actin-myosin interaction in accordance with the effect of troponin B of Hartshorne and Mueller (1968) and the "inhibitory factor" of Schaub and Perry (1969, 1971). As shown in Fig. 4 saturation of the inhibitory effect is attained when the molar ratio of actin to troponin II becomes around 6, in accordance with the proposal already mentioned that one troponin molecule would exist at every seven actin molecules. (This relationship has been well demonstrated with the original troponin—Ebashi et al., 1968; Spudich and Watt, 1971; Weber and Bremel 1972—this is

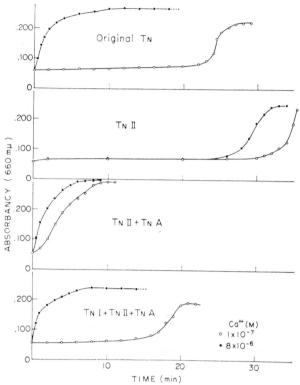

Figure 3. Reconstitution of troponin function from its components. Reaction mixtures contained: 0.04 M KCl, 1 mM MgCl$_2$, 0.02 M Tris-maleate pH 6.8, 0.1 mM GEDTA (glycoletherdiaminetetraacetic acid)-Ca which gives free Ca ion concentrations mentioned in the figure, 0.5 mM ATP, 0.5 mg/ml of desensitized myosin B, 25 μg/ml of tropomyosin and a specified amount of troponin or troponin component. Troponin, 20 μg/ml; troponin I, 12.5 μg/ml; troponin II or troponin A, 10 μg/ml. Temperature 20°C. For other details see Ebashi et al., 1968.

the crucial point in judging between a phenomenon related to a physiological mechanism and one in vitro.)

The addition of troponin A to troponin II removes the depressing action of troponin II. Since different kinds of polyanions, i.e., SDS, dextran sulfate derivatives, heparin, polyamino acids, etc. (Fig. 5), show a similar action, this effect of troponin A seems to be due to the acidic nature of troponin A (see Table 1). It should be noted that the removal of the inhibitory action is slightly but significantly less in the absence of Ca ion than in the presence of Ca ion. In other words the combination of troponin II and troponin A can provide a weak Ca-sensitivity to the myosin-actin interaction.

However the Ca-sensitizing effect of the original troponin cannot be restored unless troponin I, the tropomyosin aggregating factor, is combined with troponin II and troponin A. Thus the function of the original troponin requires the cooperation of all three components, in agreement with the report of Greaser and Gergely (1971).

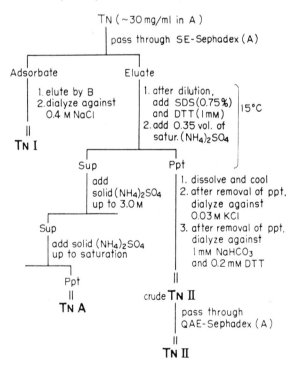

Figure 2. Flow diagram of the method for separation of troponin into its three components.

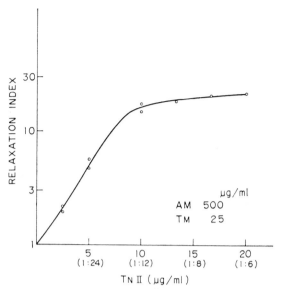

Figure 4. Inhibitory effect of troponin II on super-precipitation. Reaction mixture contained: 0.03 M KCl, 1 mM MgCl$_2$, 0.02 M Tris-maleate pH 6.8, 0.5 mM ATP, 0.5 mg/ml of desensitized myosin B, 25 μg/ml of tropomyosin and a specified concentration of troponin II at 19°C. Relaxation index: the ratio of the time required for half-maximum increase in absorbancy in the presence of an indicated concentration of troponin II to that in the absence of troponin II. For others see legend for Fig. 3.

Role of Each Component in Troponin Mechanism

The necessity of all three components for the reconstitution of troponin function is thus established, but their precise roles in the troponin mechanism have remained to be the subject of further investigation.

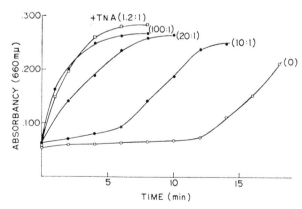

Figure 5. Effect of SDS on the inhibitory effect of troponin II on superprecipitation. Experimental conditions were very similar to those in Figs. 3 and 4. Figures in parentheses refer to the molar ratio of SDS or troponin A to troponin II. SDS in these concentrations does not show any accelerating effect on the superprecipitation of pure actomyosin plus tropomyosin, so that the above effect of SDS is mainly due to its binding to troponin II (this does not mean that SDS is entirely bound to troponin II).

Table 1. Amino Acid Composition of Troponin and Its Components

	Troponin	Troponin I	Troponin II	Troponin A
Asp	77	66	76	129
Thr	22	19	22	23
Ser	33	26	42	35
Glu	154	178	150	146
Pro	22	27	37	7
Gly	39	25	40	69
Ala	72	81	68	68
Val	35	32	32	37
Met	30	15	39	43
Ile	31	26	26	45
Leu	67	62	82	55
Tyr	9	10	10	8
Phe	21	14	16	51
Lys	115	138	115	57
His	17	19	20	5
Arg	72	79	74	39

As shown in the previous section, troponin II and troponin A by themselves cannot significantly sensitize the myosin-actin interaction to Ca ion as long as the physiological ratio between actin and two components is kept. However as illustrated in Fig. 6, the use of much larger amounts of

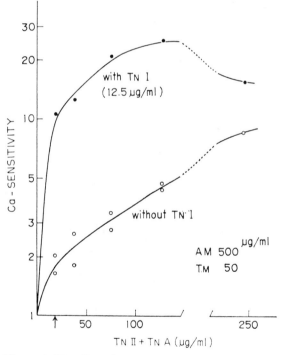

Figure 6. The effect of varying concentrations of troponin II-troponin A complex on superprecipitation. Experimental conditions were essentially the same as those in Fig. 3 except that 50 μg/ml tropomyosin was used. The first eluate from SE-Sephadex in Fig. 2 was used; the ratio of troponin A to troponin II by weight was estimated about 1.2. Ca-sensitivity is defined as the ratio of the time required for half-maximum increase in absorbancy in a low concentration of Ca ion, 1×10^{-7} M, to that in a high concentration, 8×10^{-6} M. Arrow indicates the point where the molar ratio of troponin II or troponin A to actin is about 1:7.

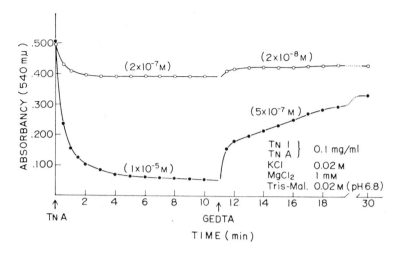

Figure 7. Reversible turbidity change of troponin I plus troponin A induced by the change in Ca ion concentration. Troponin I was suspended in a mixture containing 0.02 M KCl, 1 mM MgCl₂, 0.02 M Tris-maleate pH 6.8 and Ca ion of a concentration indicated in the figure. Troponin I, 100 μg/ml; troponin A, 50 μg/ml. Temperature 23°C.

troponin A and troponin II exhibits a marked Ca-sensitizing action fairly comparable to that of the original troponin or of reconstituted troponin.

It is conceivable that troponin A in the absence of Ca ion would dissociate, though only slightly, from troponin II, and actin-bound tropomyosin would possess a higher affinity for free troponin II than the troponin II-troponin A complex. As a result the amount of dissociated troponin II in the absence of Ca ion would increase with increase in the amount of troponin II-troponin A complex and consequently could exert its inhibitory action. The results so far obtained seem to support this interpretation, but a conclusion should be reserved until further information is obtained.

In this connection it should be mentioned that even in the absence of tropomyosin, high concentrations of troponin II can depress the myosin-actin interaction to some extent. It has also been reported that a protease-digested fragment of troponin II exerts an inhibitory action irrespective of the presence of tropomyosin (Wilkinson et al., 1971). These inhibitory actions require a higher molar ratio of troponin II to actin for its maximum activity, nearly one to one. In this respect they have some resemblance to the inhibitory action of a large amount of troponin II-troponin A complex with tropomyosin in the absence of Ca ion (Fig. 6). However this resemblance may be only apparent, since the extent of inhibition by the former is far less than the latter, and without tropomyosin the troponin II-troponin A complex cannot exhibit an inhibitory effect in the absence of Ca ion.

The next question to be raised is what is the role of troponin I in the troponin mechanism. Except for its strong affinity for tropomyosin, troponin I apparently does not show a property which can fit in with the Ca-sensitizing mechanism. However the following finding may deserve attention as suggesting its physiological role.

The solubility of troponin I decreases with decrease in ionic strength; this low solubility at low ionic strengths is counteracted by troponin A in the presence of Ca ion but not in the absence of Ca ion (Fig. 7). The removal of Ca from a solution of troponin I and troponin A deprives troponin A of solubilizing activity and restores the turbid state of troponin I (almost all polyanions so far tested can also solubilize troponin I more or less, but their action cannot be reversed by the removal of Ca ion). In the precipitated as well as solubilized state of troponin I, troponin A is firmly bound to troponin I. This means that the troponin I-troponin A complex assumes quite different structures according to the presence or absence of Ca ion. If we plot the final turbidity of the troponin I-troponin A complex against pCa, the correlation is exactly a mirror image of that between pCa and contractile responses (Fig. 8). This reminds us of the Ca-dependent quaternary structure change of troponin shown by ultracentrifugation and electrophoresis examinations (Wakabayashi and Ebashi, 1968).

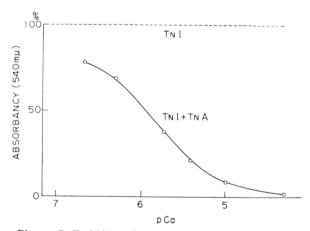

Figure 8. Turbidity of troponin I plus troponin A at varied pCa. The turbidity of troponin I 10 min after addition of troponin A (see Fig. 7) is plotted against varied pCa.

It may be bold to deduce any conclusion from this simple finding. However it is attractive to assume that troponin A would alter its steric relationship with troponin I in accordance with the change in the Ca ion concentration and that this alteration would induce functional association and dissociation of troponin A with and from troponin II, thus giving rise to the contraction-relaxation cycle.

Steric Correlation of Troponin in Myofibrils and Crystalline Structures of Tropomyosin

The work of Ohtsuki et al. (1967) demonstrating the periodic distribution of troponin in thin filaments at intervals of 400 Å has stimulated an

investigation to decide the fine structural correlation between tropomyosin and troponin. Higashi and Ooi (1968) have shown that troponin binds to the middle of the longer arm of the unit network of tropomyosin crystals. According to Nonomura et al. (1968) troponin binds to the middle of the broad light band of Cohen-Longley type paracrystals (1966) (troponin I has the same binding site as the original troponin).

In view of these findings, if the end of the tropomyosin molecule in the crystalline structure is known, we can determine the steric correlation between troponin and tropomyosin. On the basis of the observation of the ends or broken portion of paracrystals (Fig. 9a,b), we have reached a

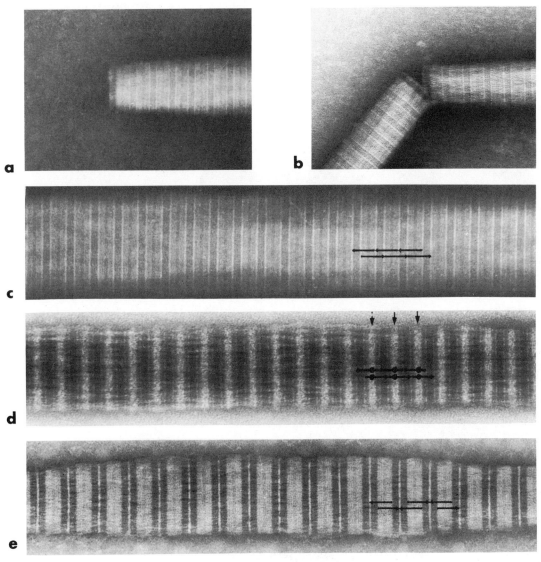

Figure 9. Arrangement of tropomyosin molecules in paracrystal structure. *a*, Edge of tropomyosin paracrystal. Paracrystal terminates at the extreme edge of narrow dark band. × 121,000. *b*, Broken portion of tropomyosin paracrystal. × 153,000. *c*, Tropomyosin paracrystal. Arrangement of tropomyosin molecules, represented as bars with arrowheads, is shown schematically. × 160,000. *d*, Troponin-tropomyosin paracrystal. Troponin is localized in the middle of the wide bright band (indicated by arrows). × 160,000. *e*, Paracrystals made from tropomyosin treated with a derivative of dextran sulfate. × 160,000.

conclusion that filamentous tropomyosin molecules, the length of which is 400 Å, are arranged in such a way that two boundary lines of each band (either wide or narrow) in the paracrystal correspond to the positions of end-to-end bonding of tropomyosin molecules. Since tropomyosin molecules behave like polyelectrolytes, having electric charges of different signs at both ends of the molecule, respectively, it is reasonable to assume that the conjugation of two identical ends having the electric charge of the same sign is not allowed in the paracrystal. From these observations and considerations it may be concluded that troponin binds to a region about 130 Å apart from one end of the tropomyosin molecule (Fig. 9c,d).

The addition of a polyanion, a derivative of dextran sulfate, produces an interesting paracrystal (Fig. 9e). It may not be bold to postulate that the polyanion would combine with the cationic part of tropomyosin, and consequently the conjugation of cationic ends of two tropomyosin molecules becomes possible. If this assumption is right, troponin binds to a site 270 Å from the cationic end of the tropomyosin molecule.

Antitroponin-stained thin filament brush shows that the first troponin is located at about 250 Å from the top of the thin filament. If we assume that the end of the tropomyosin molecule starts from the top of thin filament, troponin is located at 250 Å from the end on the H-band side, viz., the cationic end, of the tropomyosin molecule. However this conclusion is dependent on some assumptions which require further confirmations.

The separation of troponin into three components has stimulated experiments in which myofibrils are stained with each antibody against each of three components. The results are consistent with what we had expected (Fig. 10); each antibody stains thin filaments with about 380 Å periodicity in the same way as does the antibody against the original troponin (Ohtsuki et al., 1967). However there is one point to be noted. The band produced by antitroponin I is definitely broader than other cases. The width of the band reaches nearly 150 Å. The H-band and Z-band sides of the first band due to antitroponin I are 250 Å and 400 Å from the top of the thin filament; the former coincides with the site of antibodies against the

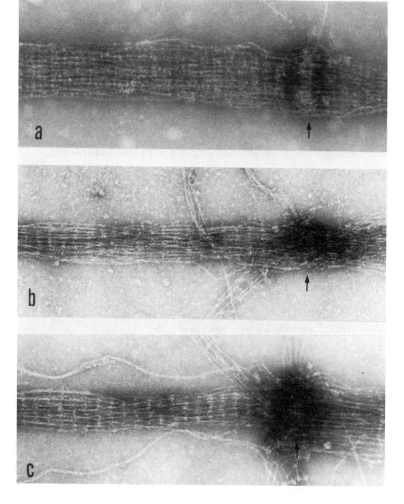

Figure 10. Separated thin filaments stained with (*a*) antitroponin I, (*b*) antitroponin II, (*c*) antitroponin A. Arrows indicate the Z-line structure. × 88,000.

other two components as well as the original troponin, whereas the latter is fairly close to the end on the Z-band side of the first tropomyosin molecule if the above assumption is true. The reason why we could not see this feature with the antibody against the original troponin might be that the antibodies against troponin II and troponin A contained in that antitroponin preparation would have hindered sterically the binding of antitroponin I to its proper position. Indeed according to immunofluorescent studies each antibody completely inhibits the binding of the other two antibodies to myofibrils.

Mechanism of Troponin Action

We have postulated that the troponin-tropomyosin system would bring about an inhibitory mechanism in the interaction of myosin and actin in the following way (Ebashi and Endo, 1968; Ebashi et al., 1969): (1) The effect of Ca ion on troponin is mediated to F-actin through tropomyosin. (2) In the absence of Ca ion, troponin exerts an inhibitory effect on F-actin. (3) This depression is removed by Ca ion.

Assumptions (2) and (3) have been fully substantiated by later findings. The fact that troponin contains "inhibitory factor," troponin II, as a component (Hartshorne and Mueller, 1968; Schaub and Perry, 1970) of which the action is counteracted by another component, troponin A, in the presence of Ca ion may be considered as direct evidence for these assumptions. The finding of Ishiwata and Fujime (1972), using a quasi-elastic light scattering technique, that in the absence of Ca ion troponin together with tropomyosin induces a rigid or less flexible state of F-actin, which cannot be seen under usual conditions, has added strong support to this idea from a physicochemical standpoint.

To substantiate the point that tropomyosin plays an indispensable role in modifying the conformation of F-actin, considerable evidence has been furnished (Tonomura et al., 1969; Ishiwata and Fujime, 1972; Mihashi, personal communication). However it is still a matter of conjecture whether troponin would exert its effect on actin only through tropomyosin or in part directly on actin itself.

Troponin I binds to F-actin in the absence of tropomyosin and enables F-actin to form unique ordered aggregates (Fujii et al., 1972; the binding of the original troponin to F-actin shown by Drabikowski and Nonomura, 1968, is largely due to the property of troponin I), but this may not be considered as having a physiological significance since this effect is intensely hindered by tropomyosin. Troponin II and troponin A also can bind to F-actin according to observations using a fluorescent probe, but this binding may not be of a specific nature. Troponin II has some affinity for tropomyosin in view of ultracentrifugation and other physicochemical observations, but this is also not so marked as that of troponin I for tropomyosin, and no clear stoichiometry has been found yet.

It is shown, however, from the excitation spectrum of actin-bound FMA (fluorescein mercuric acetate), that tropomyosin and troponin II could form a stoichiometric complex together with F-actin (Fig. 11). Tropomyosin alone does not show

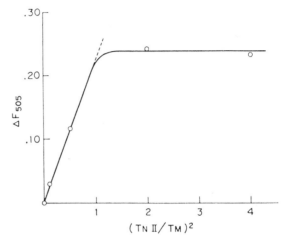

Figure 11. Excitation spectrum of FMA bound to F-actin at varied molar ratios of troponin II to tropomyosin. Ordinate indicates the fractional increase in the excitation spectrum at 505 mμ. Emission was observed at 530 mμ. Molar ratio of tropomyosin to actin was 1:3.

any significant effect on the excitation spectrum of actin-bound FMA. When an increasing amount of troponin II is added, the intensity at 505 mμ is enhanced in such a way that the fractional increase in F_{505}: not linear, but proportional to the second power of the molar ratio of troponin II to tropomyosin. Moreover the increase in F_{505} reaches a plateau at the point where the molar ratio of troponin II to tropomyosin becomes unity.

Thus a clear stoichiometric correlation between troponin II and tropomyosin emerges when F-actin is present. There might be two explanations for this. One is that tropomyosin makes some change in its conformation due to its binding to actin and as a result can form stoichiometric combination with troponin II. The other is that troponin II would bind also to actin to form a stable complex with tropomyosin (this may imply that troponin II would exert its inhibitory action by bridging a gap between tropomyosin and contralateral actin strand). At present we could not decide between these two alternatives.

The next question might be at what step of myosin-actin-ATP interaction the troponin-tropomyosin system in the absence of Ca ion would exert its inhibitory action. This may be the step of reaction 1 in the following scheme for myosin-actin interaction:

$$A + M{:}_P^{ADP} \rightarrow AM{:}_P^{ADP} \tag{1}$$

$$AM{:}_P^{ADP} \rightarrow AM^* + ADP + P_i \tag{2}$$

$$AM^* + ATP \rightarrow AM^* \cdot ATP \tag{3}$$

$$AM^* \cdot ATP \rightarrow A + M \cdot ATP \tag{4}$$

$$M \cdot ATP \rightarrow M{:}_P^{ADP} \tag{5}$$

Miscellaneous

The presence of α-actinin and M-protein in the Z-band and M-line, respectively, and their entity as a protein have been established (Table 2). (This does not, of course, exclude the possibility that some other unidentified proteins will

Table 2. Properties of α-Actinin, β-Actinin and M-Protein

	Mol Wt	Localization
α-Actinin[a]	95,000–100,000	Z-band
β-Actinin[b]	70,000	I-band region, close to A-I junction
M-Protein[c]	155,000	M-line

[a] Masaki et al., 1968; Goll et al., 1969, 1971; Masaki, personal communication.
[b] Masaki et al., 1972; Maruyama, personal communication (complex of 33,000 and 37,000 components).
[c] Masaki and Takaiti, 1972.

be isolated from the Z-band as well as the M-line.) Although its exact role has not yet been settled, β-actinin, discovered by Maruyama (1965a,b) as the actin-dispersing factor, is shown to have an interesting but somewhat puzzling localization (Table 2).

Table 3. Turnover Rates of Muscle Proteins of Rabbits Estimated from Rates of Amino Acid Incorporation

	Days
Myosin	29
Actin	74
Troponin	12
Tropomyosin	23
α-Actinin	24
M-Protein[a]	12
10S Component of crude α-actinin	44
Soluble proteins	20

From T. Koizumi, personal communication; derived from determinations with five kinds of amino acids.

[a] The value of the complex with 94,000 component (Masaki and Takaiti, 1972).

This suggests its important role in contractile processes, perhaps arranging the ordered distribution of thin filaments at the entrance of the tunnel made of thick filaments.

The discovery of several regulatory proteins has awakened an interest in the question of whether or not the onset of synthesis of myofibrillar proteins in the embryonic stage would be simultaneous. This interest may be increased by the fact that myofibrillar proteins have entirely different turnover rates (Table 3). Using fluorescent antibody techniques, Masaki and Yoshizaki (1972) have shown that in chick embryo no appreciable difference is detected in the time of the first appearance of all myofibrillar proteins except for actin, i.e., myosin, troponin, tropomyosin, α-actinin, β-actinin, and M-protein. Thus it is very likely that the synthesis of all regulatory proteins starts at the same time as that of myosin. Unfortunately Masaki and Yoshizaki could not give information as to actin because of the difficulty in preparing its antibody. It is an interesting question at this point whether or not actin would preexist before the appearance of other myofibrillar proteins, in view of the fascinating work of Ishikawa et al. (1969) about the omnipresence of actin-like filaments even in noncontractile tissues.

Acknowledgments

This work was supported in part by the research grants of the U.S. Public Health Service, AM 04810, Muscular Dystrophy Association of America, Inc., Ministry of Health and Welfare, Japan, No. 216, the Iatrochemical Foundation, Toray Science Foundation, and Mitsubishi Foundation.

References

COHEN, C. and W. LONGLEY. 1966. Tropomyosin paracrystals formed by divalent cations. *Science* **152:** 794.

DRABIKOWSKI, W. and Y. NONOMURA. 1968. The interaction of troponin with F-actin and its abolition by tropomyosin. *Biochim. Biophys. Acta* **160:** 129.

DRABIKOWSKI, W., R. DABROWSKA, and B. BARYLKO. 1971. Separation and characterization of the constituents of troponin. *FEBS Letters* **12:** 148.

EBASHI, S. 1963. Third component participating in the superprecipitation of "natural actomyosin." *Nature* **200:** 1010.

———. 1972. Separation of troponin into three components. *J. Biochem.* **72:** 787.

EBASHI, S. and F. EBASHI. 1964. A new protein component participating in the superprecipitation of myosin B. *J. Biochem.* **55:** 604.

———. 1965. α-Actinin, a new structural protein from striated muscle. I. Preparation and action on actomyosin-ATP interaction. *J. Biochem.* **58:** 7.

EBASHI, S. and M. ENDO. 1968. Calcium ion and muscle contraction. *Prog. Biophys. Mol. Biol.* **18:** 123.

EBASHI, S. and A. KODAMA. 1965. A new protein factor promoting aggregation of tropomyosin. *J. Biochem.* **58:** 107.

EBASHI, S. and Y. NONOMURA. 1972. *Proteins of the myofibril. Muscle* III, ed. G. H. Bourne. Academic Press, New York. In press.

EBASHI, S., F. EBASHI, and A. KODAMA. 1967. Troponin as the Ca^{++}-receptive protein in the contractile system. *J. Biochem.* **62**: 137.

EBASHI, S., A. KODAMA, and F. EBASHI. 1968. Troponin. I. Preparation and physiological function. *J. Biochem.* **64**: 465.

EBASHI, S., M. ENDO, and I. OHTSUKI. 1969. Control of muscle contraction. *Quart. Rev. Biophys.* **2**: 351.

EBASHI, S., T. WAKABAYASHI, and F. EBASHI. 1971. Troponin and its components. *J. Biochem.* **69**: 441.

FUJII, T., M. KAWAMURA, K. YAMAMOTO, and K. MARUYAMA. 1972. Interaction of troponin I component and F-actin. *J. Biochem.* In press.

GOLL, D. E., A. SUZUKI, and I. SINGH. 1971. Some properties of purified α-actinin. *Biophys. Soc. Abstr.*, p. 107a.

GOLL, D. E., W. F. H. M. MOMMAERTS, M. D. REEDY, and K. SERAYDARIAN. 1969. Studies on α-actinin like proteins liberated during trypsin digestion of α-actinin and myofibrils. *Biochim. Biophys. Acta* **175**: 174.

GREASER, M. L. and J. GERGELY. 1971. Reconstitution of troponin activity from three protein components. *J. Biol. Chem.* **246**: 4226.

HARTSHORNE, D. J. and H. MUELLER. 1968. Fractionation of troponin into two distinct proteins. *Biochem. Biophys. Res. Comm.* **31**: 647.

HARTSHORNE, D. J. and H. Y. PYUN. 1971. Calcium binding by the troponin complex, and the purification and properties of troponin A. *Biochim. Biophys. Acta* **229**: 698.

HIGASHI, S. and T. OOI. 1968. Crystals of tropomyosin and native tropomyosin. *J. Mol. Biol.* **34**: 699.

ISHIKAWA, H., R. BISCHOFF, and H. HOLTZER. 1969. Formation of arrowhead complexes with heavy meromyosin in a variety of cell types. *J. Cell Biol.* **43**: 312.

ISHIWATA, S. and S. FUJIME. 1972. Effect of calcium ions on the flexibility of reconstituted thin filaments of muscle studied by quasielastic scattering laser light. *J. Mol. Biol.* **67**: 1.

MARUYAMA, K. 1965a. A new protein-factor hindering network formation of F-actin in solution. *Biochim. Biophys. Acta* **94**: 208.

———. 1965b. Some physico-chemical properties of β-actinin, "actin-factor," isolated from striated muscle. *Biochim. Biophys. Acta* **102**: 542.

MARUYAMA, K. and S. EBASHI. 1970. Regulatory proteins of muscle, p. 373. In *The physiology and biochemistry of muscle as a food*, ed. E. J. Briskey et al. Univ. Wis. Press, Madison, Wisconsin.

MASAKI, T. and O. TAKAITI. 1972. Purification of M-protein. *J. Biochem.* **71**: 355.

MASAKI, T. and C. YOSHIZAKI. 1972. The onset of myo-fibrillar protein synthesis in chick embryo in vivo. *J. Biochem.* **71**: 755.

MASAKI, T., O. TAKAITI, and S. EBASHI. 1968. "M-substance," a new protein constituting the M-line of myofibrils. *J. Biochem.* **64**: 909.

MASAKI, T., O. TAKAITI, H. HAMA, M. KAWAMURA, and K. MARUYAMA. 1972. A study on the localization of β-actinin in chicken myofibril. *J. Biochem.* In press.

NONOMURA, Y., W. DRABIKOWSKI, and S. EBASHI. 1968. The localization of troponin in tropomyosin paracrystals. *J. Biochem.* **64**: 419.

OHTSUKI, I. and T. WAKABAYASHI. 1972. Optical diffraction studies on the structure of troponin-tropomyosin-actin paracrystals. *J. Biochem.* **72**: 369.

OHTSUKI, I., T. MASAKI, Y. NONOMURA, and S. EBASHI. 1967. Periodic distribution of troponin along the thin filament. *J. Biochem.* **61**: 817.

SCHAUB, M. C. and S. V. PERRY. 1969. The relaxing protein system of striated muscle-resolutions of the troponin complex into inhibitory and calcium ion sensitizing factors and their relationship to tropomyosin. *Biochem. J.* **115**: 993.

———. 1971. The regulatory proteins of the myofibril—characterization and properties of the inhibitory factor (troponin B). *Biochem. J.* **123**: 367.

SPUDICH, J. A. and S. WATT. 1971. The regulation of rabbit skeletal muscle contraction. I. Biochemical studies of the interaction of the tropomyosin-troponin complex with actin and the proteolytic fragments of myosin. *J. Biol. Chem.* **246**: 4866.

SPUDICH, J. A., H. E. HUXLEY, and J. T. FINCH. 1972. The regulation of skeletal muscle contraction. II. Structural studies of the interaction of the tropomyosin-troponin complex with actin. *J. Mol. Biol.* In press.

STAPRANS, I., H. TAKAHASHI, M. P. RUSSELL, and S. WATANABE. 1972. Skeletal and cardiac troponins and their components. *J. Biochem.* **72**: 723.

TONOMURA, Y., S. WATANABE, and M. MORALES. 1969. Conformational changes in the molecular control of muscle contraction. *Biochemistry* **8**: 2171.

WAKABAYASHI, T. and S. EBASHI. 1968. Reversible change in physical state of troponin induced by calcium ion. *J. Biochem.* **64**: 731.

WEBER, A. 1966. Energized calcium transport and relaxing factors, p. 203. In *Current topics in bioenergetics*, ed. D. R. SANADI. Academic Press, N.Y.

WEBER, A. and R. D. BREMEL. 1972. In *Contractility of muscle cells and related processes*, ed. R. J. Podolsky. Prentice-Hall, Englewood Cliffs, N.J. In press.

WILKINSON, J. M., S. V. PERRY, H. COLE, and I. P. TRAYER. 1971. Characterization of components of inhibitory factor (TN-B) preparations of the myofibril. *Biochem. J.* **124**: 55.

YASUI, B., F. FUCHS, and F. N. BRIGGS. 1969. The role of the sulfhydryl group of tropomyosin and troponin in the calcium control of actomyosin. *J. Biol. Chem.* **243**: 735.

Studies on the Subunit Composition of Troponin

D. J. HARTSHORNE

Departments of Biological Sciences and Chemistry, Carnegie-Mellon University,
Pittsburgh, Pennsylvania 15213

P. DREIZEN

Department of Medicine and Program in Biophysics, State University of New York,
Brooklyn, New York 11203

Troponin was discovered several years ago by Ebashi and Kodama (1965, 1966a) and as yet its molecular weight has not been unequivocally established. This is not due to a lack of experimental evidence but rather to the considerable variation of the published values. An initial value of 80,000 was given by Ebashi and Endo (1968) although the same laboratory later showed a preference for a lower estimate of 50,000 (Wakabayashi quoted by Ebashi et al., 1968). Arai and Watanabe (1968) using gel filtration demonstrated that treatment with dithiothreitol caused a variation of the molecular weight between 150,000 and 44,000. These authors also calculated a value of 86,000 based on the sedimentation constant and intrinsic viscosity of troponin. Drabikowski et al. (1971a) using chromatography on Sephadex G-100 suggested a value of 70,000. Based on the amount of the troponin-tropomyosin complex required to bring about 50% inhibition of the ATPase activity of desensitized actomyosin, Schaub and Perry (1969) proposed 35,000 as the upper limit of the molecular weight. Although it is often convenient to be given such a generous range of values, one should be more selective, and we have endeavored to establish more precisely the molecular weight of troponin. To do this the technique of high speed sedimentation equilibrium has been used, a method which has not previously been applied to troponin.

The need for a reliable estimate of the molecular weight is obvious when considering, for example, the distribution of troponin on the thin filament or the activity and molecular stoichiometry of the troponin-tropomyosin complex. Also it would serve as a target for the models of troponin built up from the subunits in that a framework would be available into which the pieces could be more accurately fitted and the stoichiometry determined.

It was thought originally that troponin was a homogeneous protein. That this was not so was first demonstrated by Hartshorne and Mueller (1968) (see also Hartshorne et al., 1969) who separated troponin into two functional components, A and B.

Troponin A is the component which binds Ca^{++} and has a mol wt of approximately 18,000 (Hartshorne and Pyun, 1971). Troponin B, however, is more complex and its properties of inhibition of the ATPase activity of actomyosin and binding to tropomyosin have not been associated with a single component. Subsequent to the discovery of troponin A and B many reports have appeared confirming the heterogeneity of troponin. Greaser and Gergely (1971) were the first to apply the technique of SDS polyacrylamide gel electrophoresis to troponin, and they concluded that of four components of troponin three were essential; these had mol wts of 21,000 (equivalent to troponin A), 24,000, and 35,000. The fourth component had a mol wt of 14,000. Similar subunits have been found by Murray and Kay (1971) and by Perry and coworkers (Schaub et al., 1972; Wilkinson et al., 1972).* However Ebashi et al. (1971) found three classes of subunits, 17,000, 22,000, and 40,000, of which only the latter two were considered to be essential. In order to add our ideas to the models for troponin we determined the molecular weights of troponins A and B by sedimentation equilibrium and compared these to the subunit molecular weights given by SDS gel electrophoresis.

A disturbing feature of this work, however, has been the possibility that proteolytic enzymes were present. Wilkinson et al. (1972) have found some proteolytic activity in the low ionic strength extract of myofibrils, and they suggested that the 14,000 component is derived by catheptic activity from one of the other components. We observed, during the ultracentrifuge study to be described, that the amount of low molecular weight component increased as the preparation aged, which is consistent with the presence of proteases. One is obviously cautious in accepting the validity of the subunits until it can be shown that these are not the products of limited proteolysis. In this regard it is interesting to note that Ebashi and Kodama

---

* The names used in various laboratories for these components are different, and a table given by Wilkinson et al. (1972) should help to clarify the situation.

(1966b) showed that troponin was extremely sensitive to tryptic digestion. Thus we have examined our preparations for proteolytic enzyme activity and in particular we have assayed various inhibitors, the object being to isolate troponin under conditions of minimal proteolytic degradation.

Further physical characterization of the subunits was hampered by the lack of success we had with the existing separation techniques. To help us design a method of separation we determined the isoelectric points of the troponin subunits and then used this information as a guide to the subsequent ion exchange chromatography. The method of choice yielded the 18,500 and 24,000 components in one fraction and the 39,000 component in the other. Some results with these fractions are presented.

Experimental Procedures

Protein preparations. The troponin used for the molecular weight determinations was prepared by the method of Hartshorne and Mueller (1969). For the cathepsin experiments this method and that of Ebashi et al. (1971) was used. Troponin A and B were prepared as described by Hartshorne and Mueller (1968). Troponin A was further purified using DEAE-cellulose chromatography (Hartshorne and Pyun, 1971). Small amounts of troponin A which frequently contaminated troponin B were removed by chromatography on SE-Sephadex under conditions similar to those described by Schaub and Perry (1969).

SDS polyacrylamide gel electrophoresis. The methods of Weber and Osborn (1969) and Laemmli (1970) were used. The gels were stained with Coomassie blue and after destaining were scanned on a Gilford 2000 with a model 2410 linear transport attachment.

Analytical isoelectric focusing. The conditions which were used were: 1% ampholyte, 5% acrylamide, 0.27% bisacrylamide, 12.5% sucrose, and 6 M urea: The protein was usually applied to the gel mixture before polymerization by ammonium persulfate. Focusing was achieved in gel columns 10 cm × 0.4 cm over a period of about 5 hr. The anode solution was 2% phosphoric acid and the cathode solution 4% ethylenediamine. The protein bands were visualized either by precipitation with 12% trichloroacetic acid, 3.6% sulfosalicylic acid, 30% methanol (Vesterberg, 1971) or by staining with Coomassie blue using the method of Riley and Coleman (1968). The pH gradients were determined by soaking 0.5-cm segments of a gel in 0.5 ml distilled water overnight and then measuring the pH with a Radiometer combination electrode (GK2321C).

Cathepsin assay. Acid-denatured hemoglobin was used as a substrate at pH values below 5 and casein at values above 5 (Anson, 1938; Kunitz, 1947). The conditions of assay were: 20 mg acid-denatured hemoglobin or casein, 0.067 M glycine HCl, 0.067 M sodium acetate-acetic acid, 0.067 M Na_2HPO_4-NaH_2PO_4 (the pH's were varied between 2 and 8), enzyme and distilled water to 3 ml. After the desired incubation time at 37°C, the reaction was stopped by the addition of 1 ml 25% trichloroacetic acid. The precipitated protein was removed by filtration and the absorption of the filtrate was read at 280 nm.

ATPase assays. These were carried out under the conditions described by Hartshorne and Mueller (1969) using desensitized actomyosin (Schaub et al., 1967).

Ultracentrifuge experiments. The experiments were done at 4°C in a Beckman Model E ultracentrifuge equipped with electronic speed control. High-speed sedimentation equilibrium experiments were done by the method of Yphantis (1964) as modified for multicomponent analysis based on successive equilibria at two or more rotor speeds (Gershman et al., 1966, 1969). The molecular weight and concentration of low molecular weight component were determined from interference patterns at the higher speed(s); and the molecular weight of the heavier component was determined from interference patterns at the lower speed(s), correcting for the presence of known low molecular weight component. The proportion of different components in the preparation is based on estimation of total protein concentration from synthetic boundary experiments (at 2–5 mg/ml protein).

Results and Discussion

Molecular weight determinations. The results of a representative sedimentation equilibrium experiment on troponin are shown in Fig. 1. Equilibrium was attained at two speeds. The nonlinearity of the plots indicated that the sample was heterogeneous with a contaminant of low molecular weight. At the higher speed (ω_2) the low molecular weight component was better resolved than at the lower speed and it was possible to estimate its contribution to the concentration distribution. Using this value the lower speed run could then be corrected and the value of the major heavy component estimated. Table 1 summarizes the results obtained from several estimations. The low molecular weight component represents about 18% of the total protein and has a mol wt of approximately 18,500. The residual heavy component has a mol wt of approximately 88,000, and this component may be identified with troponin.

Figure 1. Fringe displacement plotted against square of radial distance (r^2) from sedimentation equilibrium experiment on crude troponin at 0.7 mg/ml in 1 M KCl, 10 mM Tris-HCl pH 7.6, 1 mM dithiothreitol at 4°C. Graph at 31,410 rpm (ω_2) shows observed fringe displacement ($\bullet$) and troponin component ($\bigcirc$) as determined from observed data less low molecular weight component. Proportion of low molecular weight component is expressed as percent of total protein. Graph at 24,630 rpm (ω_1) shows observed data ($\bullet$), and troponin component ($\bigcirc$), corrected for contribution from low molecular weight material as determined from ω_2 data. m, meniscus; b, bottom; M, molecular weight.

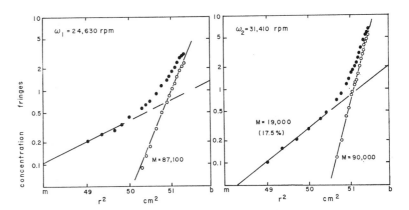

Similar molecular weight values for troponin were obtained at rotor speeds from 24,000–44,000 rpm and at different initial concentrations of protein. These results may be considered reasonably accurate to within ±5000, despite the presence of considerable, possibly heterogeneous, low molecular weight material.

Two samples of troponin were stored for one month at 4°C prior to ultracentrifugal analysis. The stored samples contained more low molecular weight material (31–40% of the protein) than was obtained in fresh preparations of troponin (see Table 1). However the residual troponin fraction showed essentially the same molecular weight as observed initially. This finding would suggest that no one component of troponin is preferentially degraded, since selective degradation of troponin would enrich one or two components and cause a

marked change in the apparent molecular weight of troponin.

As shown in a representative sedimentation equilibrium experiment on troponin A (Fig. 2), this protein is fairly homogeneous, with a linear plot of log (concentration) against r^2, at different rotor speeds. There is no significant change in molecular weight of troponin A after dialysis against 5 M guanidine HCl, confirming the presence of a single polypeptide chain. The ultracentrifuge data on troponin A are summarized in Table 2 and indicate an average mol wt of 18,300.

Data from a sedimentation equilibrium experiment on troponin B are shown in Fig. 3. Again the system is heterogeneous, with a low molecular component of mol wt about 14,000 that represents about 7% of the total protein. The residual heavy component has a mol wt about 72,000 and may be

Table 1. Crude Troponin: High-Speed Sedimentation Equilibrium

Preparation	Initial Conc. (mg/ml)	Rotor Speed (rpm)	Troponin Mol Wt	Low Molecular Weight Component (%)	Low Molecular Weight Component (Mol Wt)
I	0.70	32,000	87,200	17.0	18,400
		24,000	88,200	17.0	18,400
I	0.42	44,000	87,810	15.5	18,400
		32,000	87,700	15.5	18,400
		24,000	87,050	15.5	18,400
II	0.69	31,400	89,970	17.5	18,800
		24,630	87,100	17.5	18,800
II	0.46	31,400	90,000	24.4	16,100
		24,630	83,700	24.4	16,100
III	0.70	31,400	87,400	17.5	19,100
		24,630	89,900	17.5	
III	0.45	32,000	90,500	14.6	20,200
		24,000	89,300	14.6	
Average ± SD			88,100	17.6	18,500
			±1,840	±3.3	±1,230
Stored one month at 4°C					
I	0.42	32,000	87,100	30.9	18,200
		24,000	89,800	30.9	18,200
II	0.43	31,400	88,100	40.4	29,600
		24,600	92,100	40.4	30,700

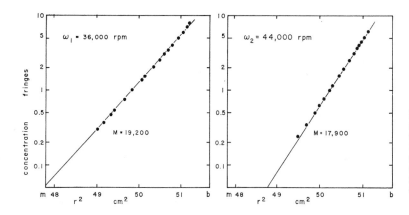

Figure 2. Sedimentation equilibrium experiment on troponin A at 0.9 mg/ml, in 1 M KCl, 10 mM Tris-HCl pH 7.6, 1 mM dithiothreitol at 4°C. Successive equilibria at 36,000 rpm (ω_1) and 44,000 rpm (ω_2) with fringe displacement plotted against square of radial distance (r^2). Symbols as in Fig. 1.

Table 2. Troponin A: High-Speed Sedimentation Equilibrium

Expt.	Solvent	Initial Conc. (mg/ml)	Rotor Speed (rpm)	Mol Wt
1	1.0 M KCl	0.9	44,000	17,900
			36,000	19,200
2	1.0 M KCl	0.6	35,600	18,100
3	1.0 M KCl	0.45	42,000	19,400
			35,600	15,850
4	1.0 M KCl	0.5	44,000	17,800
			36,000	18,800
5	5 M Guanidine	0.33	36,000	18,770
6	5 M Guanidine	0.17	42,040	17,600
			37,020	19,800
7	5 M Guanidine	0.33	36,000	18,100
Average ± SD				18,300 ±1,080

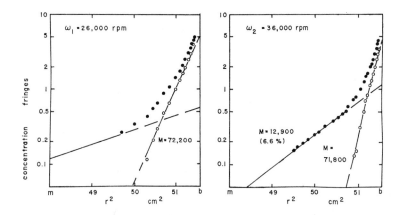

Figure 3. Sedimentation equilibrium experiment on troponin B at 1.1 mg/ml, in 1 M KCl, 10 mM Tris-HCl pH 7.6, 1 mM dithiothreitol at 4°C. Graph at 36,000 rpm (ω_2) shows observed fringe displacement (●) and troponin B component (○) as determined from observed data less low molecular weight component. Graph at 26,000 rpm (ω_1) shows observed data (●), and troponin B component (○), corrected for contribution from low molecular weight material as determined from ω_2 data. Symbols as in Fig. 1.

Table 3. Troponin B: High-Speed Sedimentation Equilibrium

Preparation	Initial Conc. (mg/ml)	Rotor Speed (rpm)	Component 1 Mol Wt	Component 2 %	Mol Wt
I	1.1	36,000	71,800	6.6	12,900
		26,000	72,200	6.6	12,900
I	0.80	44,000	63,300	7.2	12,100
		34,000	73,500	7.2	12,100
		26,000	73,200	7.2	12,100
I	0.55	44,000	72,130	7.2	11,900
		28,000	71,300	7.2	11,900
		24,000	73,800	7.2	11,900
II	0.84	35,600	70,100	15.1	13,400
		27,690	71,900	15.1	15,000
II	0.42	40,000	69,900	12.0	12,900
		26,000	70,600	12.0	14,800
III	0.88	36,000	71,800	14.0	16,600
		28,000	68,300	14.5	17,700
III	0.44	27,690	68,050	16.0	17,900
IV	0.60	44,770	69,700	8.5	13,710
		33,450	71,730	8.5	13,710
		26,000	69,800	8.5	13,710
Average ± SD			70,730 ±2,470	10.0% ±3.5%	13,720 ±1,940

Troponin B stored for 2–4 weeks at 4°C

Preparation	Initial Conc. (mg/ml)	Rotor Speed (rpm)	Component 1 Mol Wt	Component 2 %	Mol Wt
I	0.57	29,500	73,660	16.4	13,000
		24,600	70,800	16.4	13,000
III	0.4	36,000	71,840	22.5	13,800
		26,000	70,600	22.5	13,800
IV	0.4	44,770	69,700	32	11,600
		37,020	71,300	27	13,200
		26,000	70,700	27	13,200
Average ± SD			71,230 ±1,260		13,090 ±740

identified with troponin B. Table 3 summarizes data from a series of experiments on preparations of troponin B. There is a variable proportion (7–16%) of low molecular weight protein (14,000 mol wt), but the major fraction of troponin B has a mol wt approximately 71,000, and this value is unchanged at different rotor speeds and different initial protein concentrations. Although prolonged storage of troponin B for 2–4 weeks at 4°C is accompanied by a marked increase in the extent of low molecular weight material (to 16–32% of total protein), there is no significant change in the molecular weight of the residual troponin B.

Comparison of ultracentrifuge results with those from SDS gel electrophoresis. Our results with the SDS gel electrophoresis method are essentially the same as those first reported by Greaser and Gergely (1971). The molecular weights we obtained were 18,500, 24,000, and 39,000 (see Fig. 4). The low molecular weight contaminants were shown not as a discrete component, but as a mixture of components. This is in agreement with the results of Drabikowski et al. (1971b). If a 1:1:1 stoichiometry is assumed for each of the major components, a mol wt of 81,500 is obtained, which

is in reasonable agreement with the figure of 88,000 obtained by sedimentation equilibrium. A similar value may be computed, again assuming a 1:1:1 ratio, from the results of Greaser and Gergely (1971), Perry (1971), and Murray and Kay (1971).

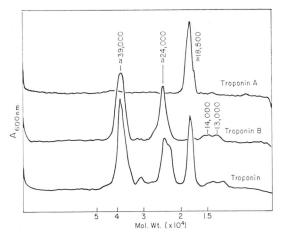

Figure 4. Scans of SDS polyacrylamide gels after staining with Coomassie blue. *Upper trace,* troponin A; *middle trace,* troponin B; *lower trace,* troponin. The approximate molecular weights are indicated.

These results are not compatible with those of Ebashi et al. (1971).

Troponin A, which is the Ca++-binding protein (Hartshorne and Pyun, 1971), has been shown to have a mol wt of about 18,000, and this is confirmed in the present study. Troponin B, which is a Ca++-insensitive inhibitor, was found to have a mol wt of about 71,000, which is consistent with a composition of the 24,000 and 39,000 subunits plus lower molecular weight contaminants.

Presence of proteases in troponin preparations.

For most of the work to be described below troponin was prepared according to Ebashi's method (Ebashi et al., 1971). The presence of proteolytic enzymes in this preparation is illustrated in Fig. 5. Activity is maximum at about pH 4 and probably indicates the presence of a cathepsin. The Ebashi procedure for the preparation of troponin involves two ammonium sulfate fractionations. A precipitate is formed at about 48% saturation (P48) and is normally discarded; troponin is then collected at a higher ammonium sulfate concentration (approximately 65% saturation). Both of these fractions are shown in Fig. 5. A neutral protease is evident in the P48 fraction as well as the catheptic type protease (Busch et al. [1972] have recently reported that a proteolytic enzyme may be prepared from rabbit skeletal muscle, and one of the enrichment procedures was the precipitation of the enzyme at about 40% ammonium sulfate saturation). In the troponin fraction the neutral protease was absent. Since the cathepsin activity represented the bulk of the

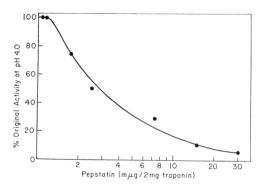

Figure 6. Inhibition of catheptic activity in a preparation of troponin by pepstatin. Acid-denatured hemoglobin was used as a substrate at pH 4.0.

proteolytic activity, we concentrated initially on finding an inhibitor to block the acidic proteases. (It should be stressed that most of the preparative procedures for troponin involve an isoelectric fractionation around pH 4.6.) Many compounds were tested and most were without effect. Partial inhibition to about 50% was obtained with dithiothreitol (2 mM), NaCl (0.3 M), and EGTA (2 mM). Trasylol (aprotinin) was without effect. Fortunately one compound was found which gave complete inhibition, and this was pepstatin (see Barrett and Dingle, 1972). At a weight ratio of approximately 1:50,000 (pepstatin:troponin) virtually all catheptic activity was inhibited (Fig. 6). Another way of completely removing the catheptic activity was by acidifying the troponin. Exposure of troponin for 30 min at a pH between 1.0 and 2.5 inactivates the cathepsin without altering the activity of the troponin. This latter method is probably of greater practical use than the addition of pepstatin and should be used for example when doing chromatography at acidic pH's.

The amount of catheptic activity varied considerably in troponins made by different methods. In general troponin prepared by the Ebashi procedure contained the highest level of acid proteases, and that prepared following an ethanol-ether step (Hartshorne and Mueller, 1969) contained the lowest level of proteolytic activity.

No inhibitors for the neutral proteases were found. Among the compounds tested were: pepstatin, trasylol, soybean trypsin inhibitor, EGTA, phenylmethylsulfonylfluoride, and sulfhydryl reagents.

The point of this brief study of proteases was to establish conditions where proteolytic degradation was minimized, and under these conditions to prepare troponin and examine the subunit profile. Before continuing with the study of the troponin subunits, we wished to reduce the possibility that they might be products of proteolysis. This

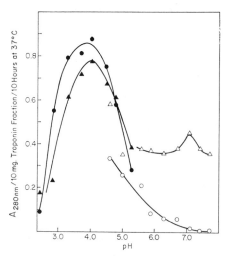

Figure 5. pH profile of the proteolytic activity found in troponin preparations. (△, ▲) Fraction precipitated at 48% saturated ammonium sulfate; (○, ●) fraction precipitated at 65% saturated ammonium sulfate, i.e., troponin. Solid symbols indicate the use of acid-denatured hemoglobin as a substrate. Open symbols indicate the use of casein as a substrate.

was done by incorporating pepstatin into each of the preparative steps and, of course, by completing the isolation as quickly as possible. Myofibrils were made by the Perry and Zydowo (1959) method, with the modification that pepstatin was added to the borate buffers (50 $\mu g/100$ ml). The myofibrils were treated with ethanol-ether and the troponin isolated according to the procedure of Hartshorne and Mueller (1969). Pepstatin (10 $\mu g/100$ ml) was present during the isoelectric fractionation at pH 4.6. The troponin so prepared behaved identically to that produced under the usual conditions. Thus it is unlikely that the subunits are artifacts, although until possible effects of the neutral proteases can be eliminated we cannot be certain.

Isoelectric points of components.

The technique of isoelectric focusing in polyacrylamide gels was applied to troponin. In addition to the ampholyte (1%) the gels contained 6 M urea to dissociate the subunits and 12.5% sucrose to stabilize the pH gradient (Doerr and Chrambach, 1971). The band pattern was quite complicated and is represented schematically in Fig. 7. The correlation of the bands to the different subunits was done either by focusing purified components or by cutting out the stained bands from the isoelectric focusing gel, homogenizing in 2% SDS, and then applying the homogenate to an SDS polyacrylamide gel. The most acidic component was troponin A. Usually two bands were seen corresponding to isoelectric points of approximately 4.1 and 4.4. In the presence of EGTA the isoelectric points were reduced by about 0.2 pH units. The 24,000 component usually appeared as a triplet of close bands between pH 5.0 and 5.2. The 39,000 component was the most basic protein. It was not possible to assign it a distinct isoelectric point as its banding pattern was very varied and ranged over a considerable pH range. The reason for this complexity is not understood; a possible complication could arise through the reaction of cyanate with the side chain amino groups or by reaction of the proteins with the ampholyte. However for the purposes of ion exchange chromatography it was adequate to know that the 39,000 component was the most basic component.

The results of isoelectric focusing of tropomyosin are also shown in Fig. 7. Two bands of approximately equal staining intensity were usually seen at pH's 5.4 and 5.6.

Cation exchange chromatography.

For the purpose of this study we chose the cation exchanger SP-Sephadex. The function of troponin A we felt was understood sufficiently to allow us to concentrate on the 24,000 and 39,000 components. Therefore the emphasis of this study was to separate

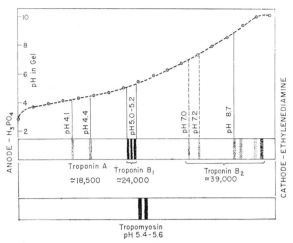

Figure 7. Schematic representation of the isoelectric focusing done in polyacrylamide gels on troponin (*upper diagram*) and tropomyosin. The 3–10 ampholyte was used. The pH along the gels is indicated by the broken line. The two subunits of troponin B are termed B_1 and B_2.

these two subunits. The 18,500 and 24,000 subunits were collected as one fraction and the 39,000 as another. The presence of troponin A with the 24,000 component was also a practical advantage in that the mixture of components was water soluble. The most suitable pH to separate the subunits was judged from a pilot study done at many pH values. In general the isoelectric points given by the focusing method were lower than those indicated by ion exchange chromatography. The results are shown in Fig. 8. Within the pH range

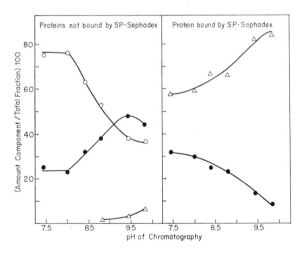

Figure 8. Elution of the troponin subunits from SP-Sephadex at different pH values. The proteins not bound by the SP-Sephadex were eluted with the equilibration buffer (8 M urea, 20 mM Tris-HCl for the pH values of 7.5–9.0 *or* 8 M urea, 20 mM sodium borate for the pH values of 9.0–9.8). The protein bound by the SP-Sephadex was eluted with the above buffers adjusted to 1 M KCl. The composition of each fraction with regard to the troponin subunits is shown. (○) 18,500 component; (●) 24,000 component; (△) 39,000 component.

which was chosen the protein not held by the SP-Sephadex was composed mostly of the 18,500 and 24,000 subunits, and the point here was to determine at which pH the 39,000 subunit began to be eluted. This was at about pH 9.0. The protein bound by the SP-Sephadex was mostly the 39,000 and should be eluted free from either of the smaller subunits. The most suitable pH value to obtain separation through one column procedure was at about pH 9.8, and this was used for the work to be described below. At this pH one accepts slight contamination of the major components (in the order of 5%).

Properties of subunits. Other workers (Greaser and Gergely, 1971; Wilkinson et al., 1972) have suggested that the 24,000 component is the inhibitor of the Mg^{++}-activated ATPase activity of actomyosin. However the mixture of the 24,000 and 18,500 components should not be inhibitory because of the influence of the 18,500 subunit. This was shown to be the case using the first

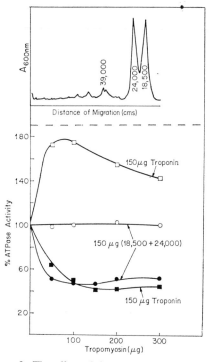

Figure 9. The effect of the 18,500 and 24,000 components with tropomyosin on the ATPase activity of desensitized actomyosin. The mixture of the 18,500 and 24,000 components was kept constant (150 μg) and tropomyosin varied ($\bigcirc$, $\bullet$). ATPase activity was determined at 25° in 2.5 mM $MgCl_2$, 2.5 mM ATP, 25 mM Tris-HCl pH 7.6 (open symbols) and in 2.5 mM $MgCl_2$, 2.5 mM ATP, 25 mM Tris-HCl pH 7.6, 1 mM EGTA (closed symbols). The same amount of troponin was assayed under the same conditions ($\square$, $\blacksquare$). Desensitized actomyosin 0.2 mg/ml. The upper part of the figure shows the tracing of the SDS polyacrylamide gel done on the fraction containing the 18,500 and 24,000 components and indicates only a slight contamination by the 39,000 subunit.

fraction obtained by SP-Sephadex chromatography at pH 9.8. The major components of this fraction were the 18,500 and 24,000 subunits. As shown in Fig. 9 inhibition occurred only when tropomyosin was added and in the absence of Ca^{++}. Since the Ca^{++} sensitivity of troponin is thought to be carried by troponin A, this would implicate the 24,000 component with the inhibition of ATPase activity and confirms the reports cited above. The involvement of another protein component with either the sensitization to Ca^{++} or the inhibition of ATPase activity is unlikely since the preparation which was used contained principally the 24,000 and 18,500 subunits. (A scan of the SDS gel done on this fraction is shown in Fig. 9.)

However there is a difference between the control troponin plus tropomyosin curves and those of the troponin subunits plus tropomyosin mixture, and that is in the effect produced on the Mg^{++}-activated ATPase activity of actomyosin in the presence of Ca^{++} (i.e., in the absence of EGTA). In the control troponin-tropomyosin curves a marked activation of ATPase activity was observed. This was not the case for the troponin minus the 39,000 component, where no activation was seen. It should be pointed out that the 39,000 subunit by itself had little effect on the ATPase activity of actomyosin. Thus the 39,000 subunit does appear to affect the ATPase activity of actomyosin and causes an activation. Whether this is a reflection of its in vivo function is not clear as the physiological equivalent of the activation of ATPase activity has not been established.

How the activation is accomplished is of some interest. It is reasonable to suppose that since the ATPase effects are manifest only in the presence of tropomyosin an interaction with tropomyosin occurs. This assumption is confirmed by the sedimentation boundary patterns of various mixtures of components (Fig. 10). The troponin-tropomyosin boundary is characteristic (see Hartshorne and Mueller, 1967) and the sedimentation profile of the 39,000 subunit plus tropomyosin is essentially the same. As the amount of tropomyosin was increased a double boundary appeared, which was due to an excess of tropomyosin. Similar patterns were not observed with mixtures of 18,500, 24,000, and tropomyosin. These ultracentrifuge studies were done at high ionic strength, and the conditions were therefore not the same as in the ATPase assay media. At low ionic strength tropomyosin is polymeric and a hypersharp sedimentation boundary is obtained, which is not suitable for binding studies of this type. Greaser and Gergely (personal communication) have also found that the largest troponin subunit binds to tropomyosin.

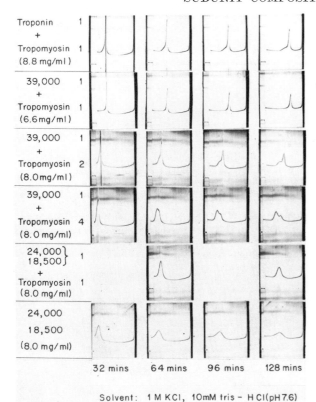

Troponin 1
+
Tropomyosin 1
(8.8 mg/ml)

39,000 1
+
Tropomyosin 1
(6.6 mg/ml)

39,000 1
+
Tropomyosin 2
(8.0 mg/ml)

39,000 1
+
Tropomyosin 4
(8.0 mg/ml)

24,000
18,500   1
+
Tropomyosin 1
(8.0 mg/ml)

24,000
18,500
(8.0 mg/ml)

32 mins   64 mins   96 mins   128 mins

Solvent: 1 M KCl, 10mM tris - HCl(pH 7.6)

Figure 10. Sedimentation velocity diagrams of mixtures of various components of troponin and tropomyosin. The composition of the solutions and the weight ratios used are shown on the left. The total protein concentration is also shown. Photographs were taken at the indicated times after reaching the speed of 60,000 rpm. Temperature 20°. The solvent in each case was 1 M KCl, 10 mM Tris-HCl pH 7.6.

The observation that the 39,000 subunit binds to tropomyosin does not preclude the binding of the 24,000 subunit to tropomyosin. Indeed since the inhibition of ATPase activity by the 24,000 subunit is enhanced by tropomyosin, one might propose that it does interact with tropomyosin. However if this interaction occurs it is not apparent at high ionic strength. An alternative, although less likely explanation for the effect of tropomyosin on the inhibition by the 24,000 subunit, is that the primary effect is that of tropomyosin on actin, which then allows a more effective interaction of actin with the 24,000 subunit. It is difficult to see how this latter suggestion could work without assigning a more active role to tropomyosin. A lot more work must be done with this system as the sites of interaction of the various components is the key towards a better understanding of the regulatory mechanism.

Acknowledgments

This work was supported by grants to D. J. H. HE-09544 and GM-46407 from the National Institutes of Health and GB-8388 from the National Science Foundation; and to P. D. by grants from the National Institutes of Health (AM-06165), the Health Research Council of New York City, and the New York Heart Association. We gratefully acknowledge the technical assistance of Mrs. Z. Capulong and Mrs. L. Abrams.

References

Anson, M. L. 1938. The estimation of pepsin, trypsin, papain and cathepsin with hemoglobin. *J. Gen. Physiol.* **22:** 79.

Arai, K. and S. Watanabe. 1968. A study of troponin, a myofibrillar protein from rabbit skeletal muscle. *J. Biol. Chem.* **243:** 5670.

Barrett, A. J. and J. T. Dingle. 1972. The inhibition of tissue acid proteinases by pepstatin. *Biochem. J.* **127:** 439.

Busch, W. A., M. H. Stromer, D. E. Goll, and A. Suzuki. 1972. Ca²⁺-specific removal of Z lines from rabbit skeletal muscle. *J. Cell Biol.* **52:** 367.

Doerr, P. and A. Chrambach. 1971. Anti-estradiol antibodies: Isoelectric focusing in polyacrylamide gel. *Anal. Biochem.* **42:** 96.

Drabikowski, W., R. Dabrowska, and B. Barylko. 1971a. Separation and characterization of the constituents of troponin. *FEBS Letters* **12:** 148.

Drabikowski, W., U. Rafalowska, R. Dabrowska, A. Szpacenko, and B. Barylko. 1971b. The effect of proteolytic enzymes on the troponin complex. *FEBS Letters* **19:** 259.

Ebashi, S. and M. Endo. 1968. Calcium ion and muscle contraction. *Prog. Biophys. Mol. Biol.* **18:** 123.

Ebashi, S. and A. Kodama. 1965. A new protein factor promoting aggregation of tropomyosin. *J. Biochem.* **58:** 107.

———. 1966a. Interaction of troponin with F-actin in the presence of tropomyosin. *J. Biochem.* **59:** 425.

———. 1966b. Native tropomyosin-like action of troponin on trypsin-treated myosin B. *J. Biochem.* **60:** 733.

Ebashi, S., A. Kodama, and F. Ebashi. 1968. Troponin. *J. Biochem.* **64:** 465.

Ebashi, S., T. Wakabayashi, and F. Ebashi. 1971. Troponin and its components. *J. Biochem.* **69:** 441.

Gershman, L. C., P. Dreizen, and A. Stracher. 1966. Subunit structure of myosin. II. Heavy and light components. *Proc. Nat. Acad. Sci.* **56:** 966.

Gershman, L. C., A. Stracher, and P. Dreizen. 1969. Subunit structure of myosin. III. A proposed model for rabbit skeletal myosin. *J. Biol. Chem.* **244:** 2726.

Greaser, M. L. and J. Gergely. 1971. Reconstitution of troponin activity from three protein components. *J. Biol. Chem.* **246:** 4226.

Hartshorne, D. J. and H. Mueller. 1967. Separation and recombination of the ethylene glycol bis (β-aminoethyl ether)-N,N'-tetraacetic acid-sensitizing factor obtained from a low ionic strength extract of natural actomyosin. *J. Biol. Chem.* **242:** 3089.

———. 1968. Fractionation of troponin into two distinct proteins. *Biochem. Biophys. Res. Comm.* **31:** 647.

———. 1969. The preparation of tropomyosin and troponin from natural actomyosin. *Biochim. Biophys. Acta* **175:** 301.

Hartshorne, D. J. and H. Y. Pyun. 1971. Calcium binding by the troponin complex, and the purification and properties of troponin A. *Biochim. Biophys. Acta* **229:** 698.

HARTSHORNE, D. J., M. THEINER, and H. MUELLER. 1969. Studies on troponin. *Biochim. Biophys. Acta* **175**: 320.

KUNITZ, M. 1947. Crystalline soybean trypsin inhibitor. *J. Gen. Physiol.* **36**: 291.

LAEMMLI, U. K. 1970. Cleavage of structural proteins during the assembly of the head of bacteriophage T4. *Nature* **227**: 680.

MURRAY, A. C. and C. M. KAY. 1971. Separation and characterization of the inhibitory factor of the troponin system. *Biochem. Biophys. Res. Comm.* **44**: 237.

PERRY, S. V. 1971. The relaxing protein system and the regulation of the myofibrillar adenosine triphosphatase. *Biochem. J.* **125**: 83P.

PERRY, S. V. and M. ZYDOWO. 1959. The nature of the extra protein fraction from myofibrils of striated muscle. *Biochem. J.* **71**: 220.

RILEY, R. F. and M. K. COLEMAN. 1968. Isoelectric fractionation of proteins on a microscale in polyacrylamide and agarose matrices. *J. Lab. Clin. Med.* **72**: 714.

SCHAUB, M. C. and S. V. PERRY. 1969. The relaxing protein system of striated muscle. Resolution of the troponin complex into inhibitory and calcium ion-sensitizing factors and their relationship to tropomyosin. *Biochem. J.* **115**: 993.

SCHAUB, M. C., D. J. HARTSHORNE, and S. V. PERRY. 1967. The adenosine triphosphatase activity of desensitized actomyosin. *Biochem. J.* **104**: 263.

SCHAUB, M. C., S. V. PERRY, and W. HÄCKER. 1972. The regulatory proteins of the myofibril. Characterization and biological activity of the calcium-sensitizing factor (troponin A). *Biochem. J.* **126**: 237.

VESTERBERG, O. 1971. Staining of protein zones after isoelectric focusing in polyacrylamide gels. *Biochim. Biophys. Acta* **243**: 345.

WEBER, K. and M. OSBORN. 1969. The reliability of molecular weight determinations by dodecyl sulfate-polyacrylamide gel electrophoresis. *J. Biol. Chem.* **244**: 4406.

WILKINSON, J. M., S. V. PERRY, H. A. COLE, and I. P. TRAYER. 1972. The regulatory proteins of the myofibril. Separation and biological activity of the components of inhibitory factor preparations. *Biochem. J.* **127**: 215.

YPHANTIS, D. A. 1964. Equilibrium ultracentrifugation of dilute systems. *Biochemistry* **3**: 297.

Troponin Subunits and Their Interactions

M. L. Greaser, M. Yamaguchi, and C. Brekke

Muscle Biology Laboratory, University of Wisconsin, Madison, Wisconsin 53706

J. Potter and J. Gergely

Department of Muscle Research, Boston Biomedical Research Institute
and
Department of Biological Chemistry, Harvard Medical School, Boston, Massachusetts 02115

The activation of muscle contraction is believed to be mediated by the interaction of calcium with troponin in the myofibril (Ebashi et al., 1967; Fuchs and Briggs, 1968). To understand this interaction, the location of the calcium binding sites on troponin and the intra- and intermolecular changes which accompany the Ca^{++} binding and release must be determined. Toward this end we have separated the protein components that are usually present in troponin preparations and studied their properties. The purpose of this report is to describe some of the characteristics of the individual troponin subunits and the protein-protein interactions which occur both among themselves and with tropomyosin.

Subunit Separation and Purification

Troponin was prepared from extracts of ethanol-ether powder as described previously (Greaser and Gergely, 1971). It was fractionated into four major components by chromatography on DEAE-Sephadex in 6 M urea (Fig. 1). Each of these components had a different molecular weight (as determined by polyacrylamide gel electrophoresis in sodium do-

decyl sulfate) with values of 14,000, 24,000, 37,000, and 20,000 for the successive fractions from the DEAE-Sephadex column. The latter three components have been designated TN-I, TN-T, and TN-C because they *inhibit* actomyosin ATPase activity, bind to *tropomyosin*, and bind *calcium*, respectively.

The TN-I and TN-C were nearly homogeneous, while the 14,000 dalton component and TN-T showed significant contamination with other protein species. TN-T could be purified to give a single band by rechromatography on DEAE-Sephadex or by chromatography of troponin on SP-Sephadex at pH 8.5 in 6 M urea. The gel electrophoresis patterns for all four components are shown in Fig. 2.

The tests for troponin activity were conducted with fractions obtained directly after the first DEAE-Sephadex separation; all other results reported here were obtained by using preparations which were essentially homogeneous as determined by their gel electrophoresis patterns.

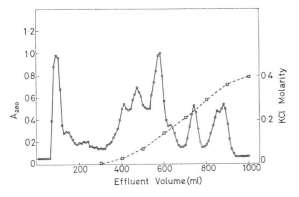

Figure 1. DEAE-Sephadex chromatography of troponin in 6 M urea. Sixty ml of troponin (18.6 mg/ml) were dialyzed against a solution containing 6 M urea, 50 mM Tris HCl pH 8.0 and 1 mM DTT and then applied to a 30 cm × 4 cm column of DEAE-Sephadex A-50 which had previously been washed with the same solution. The protein was eluted with a linear gradient (600 ml of the solution described above and an equal volume of the same solution containing 0.6 M KCl). (○) Absorbance at 280 nm; (□) KCl concentration of effluent. (From Greaser and Gergely, 1971.)

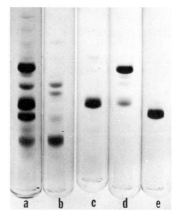

Figure 2. Gel electrophoresis of troponin fractions. Fractions collected in the experiment depicted in Fig. 1 were pooled as shown below and a portion containing the amount of protein indicated was subjected to SDS polyacrylamide gel electrophoresis. The method of Weber and Osborn (1969) was used with the modifications described previously (Greaser and Gergely, 1971). (*a*) Original troponin, 33 μg; (*b*) effluent volume 90–110 ml, 6 μg; (*c*) effluent volume 400–420 ml, 8 μg; (*d*) effluent volume 580–600 ml, 7 μg; (*e*) effluent volume 750–770 ml, 8 μg. (From Greaser and Gergely, 1971.)

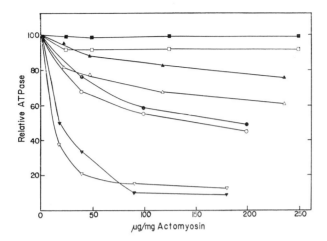

Figure 3. Effect of separated troponin components on actomyosin ATPase. Assays were conducted at 25° and pH 7.5 in a medium containing 25 mM Tris, 25 mM KCl, 5 mM MgCl$_2$, 5 mM ATP, and 0.01 mM CaCl$_2$ or 1 mM EGTA. Reactions were terminated with acid and the ATPase activity determined by measuring the inorganic phosphate liberated. Actomyosin was prepared by mixing purified myosin and actin in a 4:1 weight ratio. Troponin components dialyzed free of urea were added in the amounts indicated on the abscissa. Tropomyosin was added in an amount approximately equal to that of the fraction tested. (■, ▼, ▲, ●) Ca^{++} medium; (□, ▽, △, ○) EGTA medium. (●, ○) 14,000 dalton fraction; (▼, ▽) TN-I; (▲, △) TN-T; (■, □) TN-C. (From Greaser and Gergely, 1971.)

Tests for Troponin Activity

The question of which of the troponin fractions are required for activity could be answered once the methods for their separation had been developed. Each of the individual fractions as well as the various combinations were tested in a system containing actin, myosin, and tropomyosin. Unfractionated troponin strongly inhibits actomyosin ATPase activity in the absence of Ca^{++} and activates ATP splitting when Ca^{++} is present. Figure 3 depicts the effects of the individual troponin components on the actomyosin system. TN-T, the 14,000 dalton fraction, and TN-I all inhibited the ATPase activity. The inhibition by TN-I was much stronger than that of any of the other fractions. However, no fraction showed a significant change in response to Ca^{++}. Therefore no single fraction appeared to account for the original troponin activity.

Mixtures of two of the components also did not appear to reconstitute troponin (Fig. 4). TN-I plus

TN-T produced a calcium-independent inhibition similar to TN-I alone. TN-T plus TN-C had little effect on the actomyosin system. A mixture of TN-C with TN-I gave less inhibition than TN-I alone and only a small calcium dependence. Thus troponin activity could not be obtained from any mixture of two components.

A combination of TN-I, TN-T, and TN-C gave full activity (Fig. 5). The 14,000 dalton component could not replace any of the other fractions, and recent evidence suggests that it is a proteolytic fragment of one of the other components (Drabikowski et al., 1971; Wilkinson et al., 1972).

The reconstitution of troponin activity depended on mixing the required components before the removal of urea by dialysis. If the urea in each protein was dialyzed away before the fractions were

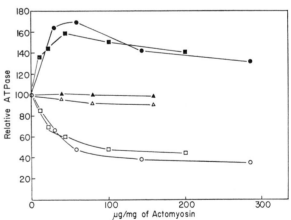

Figure 5. Effect of unfractionated and reconstituted troponin on actomyosin ATPase. Assays were conducted as described in the legend to Fig. 3. (▲, ■, ●) Ca^{++} medium; (△, □, ○) EGTA medium. (■, □) original troponin; (▲, △) 14,000 dalton fraction, TN-I, TN-T, and TN-C were dialyzed separately in order to remove urea and then mixed in equal proportions by weight; (●, ○) TN-I, TN-T, and TN-C were mixed before removal of urea by dialysis. (From Greaser and Gergely, 1971.)

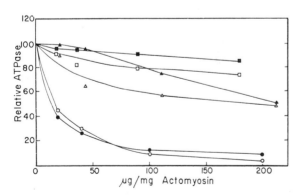

Figure 4. Effect of different pairs of troponin components on actomyosin ATPase. Assays were conducted as described in the legend to Fig. 3. (▲, ●, ■) Ca^{++} medium; (△, ○, □) EGTA medium. (●, ○) TN-I plus TN-T; (▲, △) TN-I plus TN-C; (■, □) TN-T plus TN-C. (From Greaser and Gergely, 1971.)

mixed, very little activity was recovered. However, Eisenberg and Kielley (1972) have been able to reconstitute an active troponin from mixtures of the three urea-free components. The reasons for this apparent discrepancy are under study.

Amino Acid Composition

The amino acid compositions of the three major troponin components are shown in Table 1. Each

Table 1. Amino Acid Composition of Troponin Components

	TN-I	TN-T	TN-C
Lysine	108	123	53.1
Histidine	21.4	20.5	5.9
Arginine	62.8	75.2	37.6
Aspartic acid	84.1	67.5	127
Threonine	29.7	20.5	30.6
Serine	48.6	29.7	36.7
Glutamic acid	138	194	190
Proline	38.7	41.0	10.9
Glycine	46.5	27.6	69.0
Alanine	73.2	85.1	69.9
Valine	50.4	38.1	38.6
Methionine	29.9	14.4	50.2
Isoleucine	29.3	27.8	49.4
Leucine	85.1	65.8	49.0
Tyrosine	15.8	14.4	10.0
Phenylalanine	21.0	17.4	50.3
Cysteic acid	12.0	<0.1	5.0
Tryptophan	8.2	5.6	0

Residues per 10^5 g of protein.

protein has a high proportion of charged amino acids, accounting for over 40% of the total residues. The sulfhydryl groups are found only in TN-I and TN-C with 3 and 1 moles per mole, respectively. TN-C has a high methionine and phenylalanine content but lacks typtophan. Both TN-I and TN-T have high proline contents.

The possibility that one or more of the three major troponin components could have arisen from proteolysis of another can be ruled out by comparing their amino acid compositions. Neither the 24,000 dalton TN-I nor the 20,000 dalton TN-C could have arisen from the 37,000 dalton TN-T since the latter contains no cysteine (Table 1). The fact that the phenylalanine content of TN-C is twice that of TN-I also eliminates the possibility of a TN-I to TN-C conversion.

Calcium Binding

Unfractionated troponin bound 2.28 moles of calcium per 10^5 g of protein (Table 2) which was in agreement with previous work (Fuchs and Briggs, 1968; Yasui et al., 1968; Hartshorne and Pyun, 1971). Of the four troponin components only TN-C had high-affinity calcium-binding sites (Greaser and Gergely, 1970). A mean value of 4.68 moles of calcium bound per 10^5 g of protein was consistent with a 1:1 molar binding ratio (Table 2).

Table 2. Calcium Binding of Troponin Subunits

	n (moles/10^5 g protein)	K (M^{-1})
Troponin	2.28	9×10^5
14,000 dalton fraction	0	—
TN-I	0	—
TN-T	0	—
TN-C	4.58	1.4×10^6

Ca†† binding was determined using Chelex-100 (Fuchs and Briggs, 1968) in a medium containing 0.1 M KCl, 4 mM MgCl$_2$, and 10 mM Tris-HCl (pH 7.5). The number of binding sites (n) and the binding constant (K) were calculated from Scatchard plots. Experiments with TN-I and TN-T were carried out in 0.3 M KCl because of their limited solubility at lower salt concentrations.

These results are in disagreement with the report of Ebashi et al. (1971) which indicated calcium-binding ability in two different troponin components. One of their fractions (fraction III) clearly corresponded to TN-C whereas the other (fraction II) appeared to be like TN-I, as judged by the molecular weight and amino acid composition. High-affinity calcium-binding ability of TN-I has never been detected in our preparations in spite of attempts to assay the activity as soon as possible after dialysis or Sephadex G-25 chromatography to remove urea. It seems most likely that the calcium-binding ability of Ebashi et al.'s (1971) fraction II arose from a contamination with their fraction III material.

A binding of one mole of Ca^{++} per mole of TN-C could not account for the approximately two moles of Ca^{++} which were bound by the unfractionated troponin complex. Since TN-I did not bind Ca^{++}, the question arose as to whether mixtures of the separated components would restore full Ca^{++}-binding ability. Table 3 shows the Ca^{++}-binding of TN-C and its mixtures with TN-I and TN-T. TN-T did not affect the calcium-binding ability of TN-C. A 1:1 molar mixture of TN-I and TN-C

Table 3. Calcium Binding of Mixtures of Troponin Subunits

Sample	Moles Ca^{++}/mole of TN-C
TN-C	1.06
TN-T + TN-C	1.15
TN-I + TN-C	2.19
TN-T + TN-I + TN-C	2.20

The components separated by chromatography in urea were mixed, dialyzed against 0.3 M KCl, 10 mM Tris-HCl (pH 7.5), 0.1 mM DTT to remove urea, and then dialyzed against large volumes of 60 mM KCl, 2 mM MgCl$_2$, 1 μM CaCl$_2$, and 10 mM Tris-HCl (pH 8.1). After dialysis the amount of calcium bound was determined by atomic absorption spectroscopy.

bound two moles of Ca++ per mole of TN-C. A similar value was obtained with molar mixtures of TN-I, TN-T, and TN-C. Thus the calcium-binding ability of the unfractionated troponin could be regenerated with a mixture of the three required components. Equilibrium dialysis experiments indicated that the affinity constants for the separate sites were approximately equal and in the range of 10^6 M^{-1}.

The calcium-binding ability of TN-C can be affected by factors in addition to TN-I (Potter and Gergely, 1972). If 1 mM EDTA is added to TN-C after separation on DEAE-Sephadex and before urea removal, the protein only binds 0.5 mole of Ca++ per mole. However, DEAE-Sephadex chromatography with urea buffers containing 1 mM EDTA and subsequent addition of 2 mM $CaCl_2$ to the TN-C solution before dialysis to remove urea resulted in a calcium-binding capacity of 2 moles per mole. These experiments suggest that Ca++ affects the refolding of TN-C. They also indicate that both calcium-binding sites on the troponin complex may potentially be located on the TN-C, although there is presently no way to distinguish whether the additional binding site that appears when TN-C and TN-I are mixed is situated on either component.

Troponin Subunit Stoichiometry

The problems with proteolysis during troponin preparation (Ebashi et al., 1971; Drabikowski et al., 1971) give considerable variation in the ratios of the

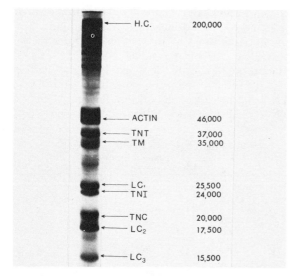

Figure 6. SDS polyacrylamide gel electrophoresis of myofibrils. Myofibrils were prepared from rabbit back muscle and dissolved in sodium dodecyl sulfate. Electrophoresis was conducted with 10% acrylamide gels which were 15 cm in length, using the procedure described by Weber and Osborn (1969). Total protein added was 150 µg and the gel was stained with fast green. H.C. and LC_1, LC_2, LC_3 refer to the myosin heavy chain and light chains respectively.

troponin subunits found in the isolated complex. Therefore experiments were conducted to determine the stoichiometry of the TN-I, TN-T, and TN-C in the myofibril. Myofibrils from rabbit back muscle were dissolved in sodium dodecyl sulfate and subjected to electrophoresis on polyacrylamide gels. The gels were stained with fast green and the staining intensity of the various bands was determined by scanning with a densitometer. In order to insure that the measured stoichiometry was correct, it was necessary to determine the relative staining intensities of the purified troponin components to account for any difference in dye binding. TN-I and TN-C gave the same absorbance per unit weight, but the staining intensity of TN-T was approximately twice as high. An SDS gel of a myofibril preparation is shown in Fig. 6. The troponin subunits are clearly resolved from the other myofibrillar protein components. Using the staining intensity correction factors for each of troponin components, the observed molar stoichiometry was close to 1:1:1 for TN-I, TN-T, and TN-C. Preliminary results also suggested that one mole of tropomyosin and 7 moles of actin were present per mole of each troponin component.

Interactions of Troponin Components with Tropomyosin

Ultracentrifuge experiments. An interaction between troponin and tropomyosin can easily be demonstrated by the large increase in viscosity which occurs when the two proteins are combined at low ionic strength (Ebashi and Kodama, 1965; Ebashi and Endo, 1968). This complex migrates as a single peak in the ultracentrifuge with a sedimentation constant greater than that of either protein alone (Hartshorne and Mueller, 1967). Troponin B (which contains components whose molecular weights approximately correspond to TN-I and TN-T) also interacted with tropomyosin, but troponin A did not (Hartshorne et al., 1969).

We have used the ultracentrifuge approach to look at the possible interactions between purified troponin components (individually and complexed) and tropomyosin. Figure 7 shows the sedimentation patterns of tropomyosin, tropomyosin with TN-I plus TN-C, and tropomyosin with TN-T. Tropomyosin had a sedimentation constant of approximately 1.85 in the concentration range used. Mixtures of tropomyosin with TN-I plus TN-C gave a single boundary with an S value of 1.7. Separate ultracentrifuge runs indicated that TN-I plus TN-C had an S value of 2.1. Thus there was no evidence of complex formation.

In contrast to these results was the striking change in the sedimentation pattern when TN-T was mixed with tropomyosin. A new faster boundary appeared with an S value of 2.9 (Fig. 7). The sedimentation constant of TN-T alone was about 2.25. The 2.95 boundary was hypersharp and similar to the boundaries which are characteristic of tropomyosin-troponin and tropomyosin-troponin B complexes (Hartshorne and Mueller, 1967; Hartshorne et al., 1969).

Ultracentrifugation of mixtures of TN-C with tropomyosin gave no evidence of interaction, in agreement with previous findings using troponin A (Hartshorne et al., 1969). The mixture of TN-I with tropomyosin also gave no indication of interaction. These results therefore point to TN-T as the link between tropomyosin and the troponin complex.

Experiments have been conducted to determine the stoichiometry of the TN-T–tropomyosin interaction. The two proteins were mixed in various ratios and their sedimentation patterns observed. When mixed in a 1 : 1 molar ratio (37,000 daltons of TN-T to 70,000 daltons of tropomyosin), the hypersharp boundary was nearly symmetrical (Fig. 6). Mixtures of 1.5 to 2 moles of tropomyosin to 1 mole of TN-T and vice versa gave a pronounced trailing edge, indicating uncomplexed protein. Preliminary results obtained using the high speed equilibrium technique gave a molecular weight of approximately 100,000 with molar mixtures of the two proteins.

Thus the data suggest that the optimum combining ratio is one mole of TN-T per mole of tropomyosin.

Electron microscopy. Solutions of tropomyosin can be readily transformed into several forms of highly ordered aggregates (Huxley, 1963; Cohen and Longley, 1966). These aggregates may provide information about the tropomyosin-actin associations in the thin filament and the troponin-tropomyosin interaction. We have examined the effects of adding the individual troponin components and certain mixtures of these components to preexisting tropomyosin crystals and tactoids. Also tropomyosin in solution has been mixed with the troponin subunits prior to crystallization. Tropomyosin was purified by hydroxylapatite chromatography (Eisenberg and Kielley, 1972).

A tropomyosin tactoid is shown in Fig. 8. It consists of light staining bands (width of 250 Å) and dark staining bands (width of 75 Å) which alternate to give a period of about 400 Å. This structure has been described previously (Cohen and Longley, 1966; Casper et al., 1969). Addition of TN-T to preexisting tropomyosin tactoids results in the

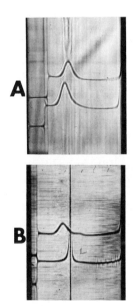

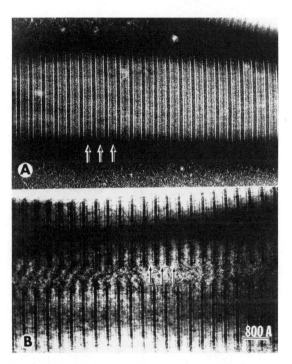

Figure 7. Ultracentrifugation of mixtures of tropomyosin with troponin components. Both runs were made at 6° with a rotor speed of 52,640 rpm and a bar angle of 65°. The solvent was 1 M KCl, 5 mM Tris (pH 7.5). Sedimentation was from left to right. *A*, 156 min at full speed. *Upper,* 5.0 mg/ml of tropomyosin; *lower,* a mixture containing 2.5 mg/ml of tropomyosin and 2.08 mg/ml of TN-I plus TN-C. *B,* 184 min at full speed. *Upper,* 5.76 mg/ml of tropomyosin; *lower,* a mixture containing 5.76 mg/ml of tropomyosin and 2.93 mg/ml of TN-T.

Figure 8. Magnesium tactoids of tropomyosin and tropomyosin plus TN-T. *A,* A tropomyosin solution was dialyzed versus a solution containing 50 mM MgCl$_2$ and 50 mM Tris-HCl (pH 8.2) as described by Casper et al., 1969. *B,* Tropomyosin tactoids formed as above were mixed with TN-T solution (in 1 M KCl, 5 mM Tris HCl, pH 7.5). This mixture was then dialyzed one to two days against the MgCl$_2$-Tris HCl solution. Both preparations were negatively stained with uranyl acetate. The TN-T appeared to bind to the center of the light band of the tropomyosin tactoids (see arrows).

appearance of a lighter staining zone in the center of the light band with a 400 Å periodicity (Fig. 8). This binding pattern is similar to that which occurs when the troponin complex is added to tropomyosin paracrystals (Nonomura et al., 1968).

An apparent transition in form was observed to occur at the ends of the tactoids (Fig. 9). A hex-

agonal net pattern emerged with a spacing of 400 Å between intersection points. If the periodicity of the TN-T binding to the tropomyosin tactoid is followed into the region of the hexagonal nets, it suggested that the TN-T was binding at the intersection points.

Mixing TN-T and tropomyosin solutions before

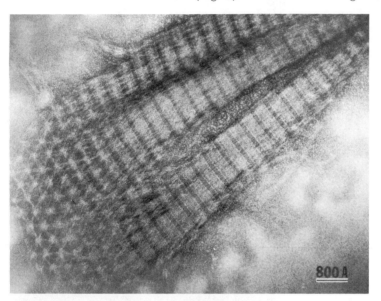

Figure 9. Magnesium tactoid of tropomyosin plus TN-T. TN-T solution was added to preformed magnesium tactoids of tropomyosin as described in Fig. 8B. The end of the tactoid has transformed into a hexagonal net pattern. If the TN-T binding periodicity is followed through the transformation region, it suggests that the TN-T is located at the intersection points of the hexagonal net pattern.

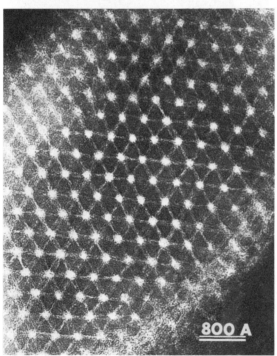

Figure 10. Hexagonal net aggregate formed from a mixture of tropomyosin and TN-T. Solutions of tropomyosin and TN-T were mixed in 1 M KCl, 5 mM Tris-HCl pH 7.5, and dialyzed versus the standard MgCl₂-Tris HCl solution (see Fig. 8B). The distance between intersection points in the lattice was 400 Å.

dialysis against a solution containing $MgCl_2$ resulted in the appearance of the majority of the protein in the hexagonal net structure (Fig. 10). Note that the arms between the intersection points are straight with a length of 400 Å. Observation of the edges of the hexagonal net crystals revealed that the ends of the tropomyosin molecules terminated at the intersection points (Fig. 11). This result, plus the periodicity in binding of TN-T (Fig. 9), suggests that TN-T binds to the end of the tropomyosin molecule.

Other types of tropomyosin–TN-T aggregation also occur (Fig. 12). In addition to the hexagonal nets and tactoids, two other forms were commonly observed. The first is a double-stranded net. The second appeared to be a tactoid-like structure which resulted from the collapse of the double-stranded net. This second type can be clearly differentiated from the normal tactoid because of its different staining pattern. It does, however, have a 400 Å periodicity. A clearer picture of the latter two forms is shown in Fig. 13.

It is interesting to note that all three of the new TN-T plus tropomyosin crystals have been observed previously. The hexagonal net was described by Higashi-Fujime and Ooi (1969) as their tropomyosin crystal form III. The double-stranded net and the collapsed form of this structure was also accidently found by these same workers (Higashi-Fujime and Ooi, 1969).

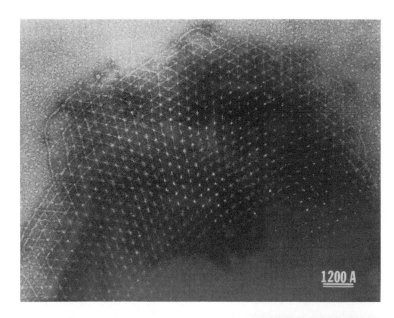

Figure 11. Edge of a hexagonal net aggregate formed from a mixture of tropomyosin and TN-T. A tropomyosin solution was dialyzed versus 0.2 M KCl, 0.01 M sodium acetate (pH 5.6) to form the usual crystal lattices with kite-shaped net patterns (Huxley, 1963; Higashi-Fujime and Ooi, 1969; Casper et al., 1969). The crystallized tropomyosin was mixed with a TN-T solution and redialyzed versus KCl-sodium acetate. The resulting hexagonal nets formed were similar to those formed in MgCl$_2$ (see Fig. 10). This micrograph shows that the ends of the tropomyosin molecules apparently terminate at the intersection points.

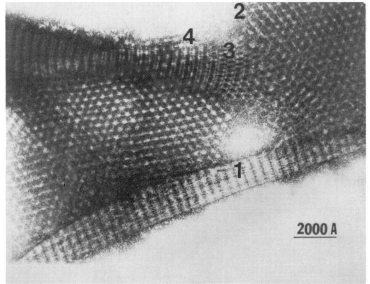

Figure 12. Polymorphism of tropomyosin–TN-T aggregates. TN-T was added to tropomyosin tactoids formed with magnesium as described in Fig. 8B. Four different forms were commonly observed: (1) tropomyosin tactoids with TN-T binding; (2) hexagonal nets; (3) double-stranded nets; (4) the collapsed form of the double-stranded nets.

Addition of TN-I, TN-C, or TN-I plus TN-C to tropomyosin tactoids (formed with MgCl$_2$) caused no change in the appearance of the banding patterns. This suggests that either the specificity of binding of these components was too weak to be observed or else no binding occurred. Our present data support the latter case.

Mixing TN-T with a previous mixture of TN-I plus tropomyosin resulted in a new type of tactoid form (Fig. 14). The center of the light staining band appeared much lighter, indicating additional protein binding. The width of this band was very constant and much narrower than that which can be observed with whole troponin or TN-T binding to tropomyosin, suggesting a greater specificity of

attachment. The width of the dark staining band also decreases from approximately 75 Å to 15 Å. We as yet do not have an adequate model for this striking transformation. The problem remains to relate the structural changes observed to the troponin-tropomyosin transformations which occur during contraction and relaxation.

The results obtained with TN-T binding to tropomyosin suggest that this interaction may occur at the end of the tropomyosin molecule. This is in conflict with previous postulations. It has been suggested that tropomyosin molecules end in the short arms of the conventional tropomyosin crystal net (Higashi-Fujime and Ooi, 1969). The position of the troponin binding has been clearly shown to be

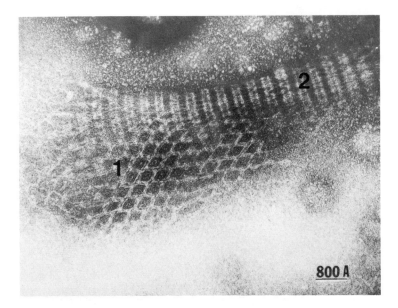

Figure 13. Aggregates of tropomyosin and TN-T. This structure was formed using the standard MgCl₂-Tris HCl solutions for dialysis (Fig. 8B). The double-stranded nets (*1*) appeared to collapse into a tactoid-like structure (*2*).

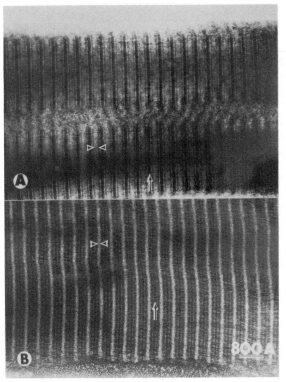

Figure 14. Magnesium tactoids of tropomyosin plus TN-T and tropomyosin plus TN-I and TN-T. *A*, Tropomyosin plus TN-T tactoids formed in MgCl₂-Tris HCl (see Fig. 8B). *B*, Tropomyosin tactoids formed in MgCl₂-Tris HCl were mixed with a solution of TN-I and redialyzed against MgCl₂-Tris HCl for 3 days. Observation of the mixture revealed only the normal tropomyosin tactoid patterns. TN-T was then added to the tropomyosin tactoid plus TN-I solution mixture and dialyzed against MgCl₂-Tris HCl for 2 more days. The resulting tactoid (*B*) shows protein binding in the center of the light band. The width of this band was much narrower than when only TN-T was present (see double arrows). Also the width of the dark band was reduced (see vertical arrows) in tactoids formed from tropomyosin, TN-I and TN-T compared to those from tropomyosin and TN-T.

in the center of the long arms of the crystal lattice (Higashi and Ooi, 1968; Fukazawa et al., 1970) and would therefore be in the center of the tropomyosin molecule. It seems unlikely that there are multiple binding sites on tropomyosin for troponin or TN-T. The experimental evidence for both the previous and present models appears to be insufficient to settle this important question of the site of troponin and tropomyosin interaction.

Different forms of tropomyosin. The hydroxylapatite purification of tropomyosin (Eisenberg and Kielley, 1972) was used to separate tropomyosin with subunits of 35,000 daltons from tropomyosin which contained 35,000 and 37,000 dalton subunits, as determined by SDS gel electrophoresis. The amino acid composition of both preparations was nearly identical, indicating that the 37,000 dalton material could not be TN-T contamination. Figure 15 is an electron micrograph of tropomyosin which

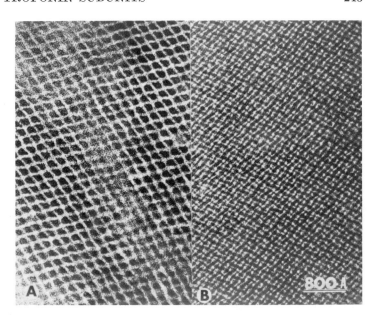

Figure 15. Crystals from single band and double band tropomyosin. Tropomyosin was purified by hydroxylapatite chromatography and separated into two fractions. One fraction gave a single band with a mol wt of 35,000 on SDS gels (S-TM); the other had approximately equal proportions of 35,000 and 37,000 dalton proteins (D-TM). The two protein solutions were dialyzed versus a solution containing 0.2 M KCl, 0.01 M sodium acetate pH 5.6. *A*, The D-TM formed the normal kite-shaped nets. *B*, The majority of the S-TM crystals gave a new type of aggregation pattern. The S-TM also formed kite-shaped nets.

had been crystallized at pH 5.6 and which contained both 35,000 and 37,000 dalton subunits. The usual crystal lattice with kite-shaped net patterns (Huxley, 1963; Higashi-Fujime and Ooi, 1969; Casper et al., 1969) was observed. In contrast the homogeneous 35,000 dalton tropomyosin formed a completely different structure when crystallized under the same conditions (Fig. 15). It also produced some of the normal kite-shaped crystal patterns. Comparison of the two types of tropomyosin under a variety of other crystallization and tactoid formation conditions has revealed no other significant differences. The interactions of the two types with troponin subunits using the conditions described (see Figs. 8 to 14) have also given identical results to date. The significance of two types of tropomyosin remains to be assessed.

Cardiac Troponin

The application of the techniques used with skeletal muscle for preparing troponin from cardiac muscle yielded extremely heterogeneous preparations when examined by SDS gel electrophoresis. We have developed a new procedure for preparing bovine cardiac troponin which utilizes a 1 M KCl extraction of the residue remaining after myosin extraction. This extract was fractionated with ammonium sulfate, and the protein, which precipitated between 30 and 50% saturation, was redissolved and dialyzed versus a solution containing a 1 M KCl, 1 mM KPO_4 (pH 6.8), and 5 mM β-mercaptoethanol. The protein was then applied to a hydroxylapatite column and eluted with a phosphate gradient (Eisenberg and Kielley, 1972). The elution pattern usually gave three or four

poorly resolved peaks (Fig. 16). The highest troponin activity appeared in the most retarded peak. Polyacrylamide gels of the bovine cardiac and rabbit skeletal troponins and their mixtures revealed clear differences in molecular weight (Fig. 17). The cardiac troponin had two major and one minor bands, with approximate molecular weights of 40,000, 30,000, and 20,000. Only the minor 20,000 dalton band migrated in a similar position to the corresponding skeletal troponin component. Rabbit cardiac troponin contained components with molecular weights which corresponded to the bovine cardiac preparations. It remains to be established which of these components are necessary

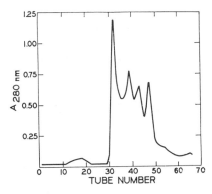

Figure 16. Hydroxylapatite chromatography of cardiac troponin. A solution of cardiac troponin (26 ml of 7.1 mg/ml) was applied to a 30 cm × 1.9 cm hydroxylapatite column which had been equilibrated with a buffer containing 1 M KCl, 1 mM potassium phosphate pH 6.8, and 5 mM β-mercaptoethanol. The protein was eluted with a linear gradient (200 ml of starting buffer and 200 ml of 1 M KCl, 0.2 M potassium phosphate pH 6.8, 5 mM β-mercaptoethanol). The highest troponin activity was found in the most retarded peak (tubes 46–48).

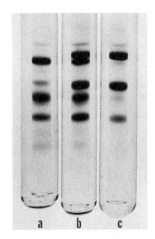

Figure 17. SDS gel electrophoresis of skeletal and cardiac troponin. (*a*) Rabbit skeletal troponin (4 μg); (*b*) a mixture of rabbit skeletal troponin (4 μg) and bovine cardiac troponin (4 μg); (*c*) bovine cardiac troponin (4 μg). Rabbit cardiac troponin components gave similar gel patterns to the bovine cardiac preparations.

for troponin activity and whether their properties are similar to those found in the skeletal troponin system.

Summary

Skeletal muscle troponin consists of three essential subunits. TN-I, with a mol wt of 24,000, strongly inhibits actomyosin ATPase activity. TN-T, with a mol wt of 37,000, interacts with tropomyosin as evidenced by the appearance of a new hypersharp peak in the ultracentrifuge and the drastic alterations in tropomyosin aggregation as viewed in the electron microscope. TN-C, with a mol wt of 20,000, is the only component which has been found to bind calcium. TN-T appears to bind to tropomyosin in a mole to mole stoichiometry, and electron microscopic observations suggest the binding may be at the end of the tropomyosin molecule. Cardiac troponin contained subunits which differed markedly in molecular weight from those obtained from skeletal troponin.

Acknowledgments

The authors would like to acknowledge the assistance of Marshall Elzinga and John Collins in performing the amino acid analyses. This research was supported by grants from the National Institutes of Health (HL 5949, TO 1-HL 05811, AM 14776) and the National Science Foundation.

References

CASPER, D. L. D., C. COHEN, and W. LONGLEY. 1969. Tropomyosin: crystal structure, polymorphism, and molecular interactions. *J. Mol. Biol.* **41**: 87.

COHEN, C. and W. LONGLEY. 1966. Tropomyosin paracrystals formed by divalent cations. *Science* **152**: 794.

DRABIKOWSKI, W., U. RAFALOWSKA, R. DABROWSKA, A. SZPACENKO, and B. BARYLKO. 1971. The effect of proteolytic enzymes on the troponin complex. *FEBS Letters* **19**: 259.

EBASHI, S. and M. ENDO. 1968. Calcium ion and muscle contraction. *Prog. Biophys. Mol. Biol.* **18**: 123.

EBASHI, S. and A. KODAMA. 1965. A new protein factor promoting aggregation of tropomyosin. *J. Biochem.* **58**: 107.

EBASHI, S., F. EBASHI, and A. KODAMA. 1967. Troponin as the Ca^{++} receptive protein in the contractile system. *J. Biochem.* **62**: 137.

EBASHI, S., T. WAKABAYASHI, and F. EBASHI. 1971. Troponin and its components. *J. Biochem.* **69**: 441.

EISENBERG, E. and W. W. KIELLEY. 1972. Reconstitution of active troponin-tropomyosin complex in the absence of urea from its four column-purified components. *Fed. Proc.* **31**: 502 (Abstr.).

FUCHS, F. and F. N. BRIGGS. 1968. The site of calcium binding in relation to the activation of myofibrillar contraction. *J. Gen. Physiol.* **51**: 655.

FUKAZAWA, T., E. BRISKEY, and W. F. H. M. MOMMAERTS. 1970. A new form of native tropomyosin. *J. Biochem.* **67**: 147.

GREASER, M. L. and J. GERGELY. 1970. Calcium binding component of troponin. *Fed. Proc.* **29**: 463 (Abstr.).

———. 1971. Reconstitution of troponin activity from three protein components. *J. Biol. Chem.* **246**: 4226.

HARTSHORNE, D. J. and H. MUELLER. 1967. Separation and recombination of the ethylene glycol bis (β-aminoethyl ether)-N, N'-tetraacetic acid-sensitizing factor obtained from a low ionic strength extract of natural actomyosin. *J. Biol. Chem.* **242**: 3089.

HARTSHORNE, D. J. and H. V. PYUN. 1971. Calcium binding by the troponin complex, and the purification and properties of troponin A. *Biochim. Biophys. Acta* **229**: 698.

HARTSHORNE, D. J., M. THEINER, and H. MUELLER. 1969. Studies on troponin. *Biochim. Biophys. Acta* **175**: 320.

HIGASHI, S. and T. OOI. 1968. Crystals of tropomyosin and native tropomyosin. *J. Mol. Biol.* **34**: 699.

HIGASHI-FUJIME, S. and T. OOI. 1969. Electron microscopic studies on the crystal structure of tropomyosin. *J. de Microscopie* **8**: 535.

HUXLEY, H. E. 1963. Electron microscope studies on the structure of natural and synthetic protein filaments from striated muscle. *J. Mol. Biol.* **7**: 281.

NONOMURA, Y., W. DRABIKOWSKI, and S. EBASHI. 1968. The localization of troponin in tropomyosin paracrystals. *J. Biochem.* **64**: 419.

POTTER, J. and J. GERGELY. 1972. Calcium binding of unfractionated and reconstituted troponin. *Fed. Proc.* **31**: 501 (Abstr.).

WEBER, K. and M. OSBORN. 1969. The reliability of molecular weight determinations by dodecyl sulfate polyacrylamide gel electrophoresis. *J. Biol. Chem.* **224**: 4406.

WILKINSON, J. M., S. V. PERRY, H. A. COLE, and I. P. TRAYER. 1972. The regulatory proteins of the myofibril. Separation and biological activity of the components of inhibitory-factor preparations. *Biochem. J.* **127**: 215.

YASUI, B., F. FUCHS, and F. N. BRIGGS. 1968. The role of the sulfhydryl groups of tropomyosin and troponin in the calcium control of actomyosin contractility. *J. Biol. Chem.* **243**: 735.

Troponin—Its Composition and Interaction with Tropomyosin and F-Actin

W. Drabikowski, Ewa Nowak, Barbara Baryłko, and Renata Dąbrowska

Department of Biochemistry of Nervous System and Muscle,
Nencki Institute of Experimental Biology, Warsaw, Poland

Ebashi and coworkers (for review see Ebashi and Endo, 1968) were the first to show that the complex of troponin and tropomyosin is required to confer Ca^{++} sensitivity to the actomyosin system. As it has been found in our and other laboratories, preparations of troponin from rabbit skeletal muscle consist of four components (Drabikowski et al., 1971a, b; Greaser and Gergely, 1971; Wilkinson et al., 1972; Schaub et al., 1972). The average molecular weights of the components obtained by us with the use of SDS-polyacrylamide gel electrophoresis are: 13,000, 18,000, 24,000 and 39,000 daltons, respectively. Occasionally an additional small band corresponding to mol wt of about 30,000 daltons is also observed.

It is already well established that the 24,000 component inhibits Mg^{++}-stimulated ATPase activity of synthetic actomyosin regardless of the concentration of free calcium ions (Schaub and Perry, 1971; Greaser and Gergely, 1971; Drabikowski et al., 1971b), whereas the 18,000 component binds Ca^{++} tightly and neutralizes the inhibitory effect of the 24,000 component (Greaser and Gergely, 1970; Drabikowski and Baryłko, 1971; Drabikowski et al., 1971a; Hartshorne and Pyun, 1971; Schaub et al., 1972).

Greaser and Gergely have recently proposed (this volume) calling the 24,000 component TN-I (I for inhibition) and the 18,000 component TN-C (C for calcium), according to their properties. Both components seem to be essentially sufficient for restoration in the presence of tropomyosin of the physiological role of troponin, although Greaser and Gergely (1971) postulated the necessity of the 39,000 component for full reconstitution of troponin activity. Our observations (Drabikowski et al., 1971b) indicate that the latter protein is directly involved in the activation of ATPase of actomyosin in the presence of Ca^{++} often caused by some troponin preparations in the presence of tropomyosin. Since the 39,000 component interacts with tropomyosin, the TN-T has been proposed for it by Greaser and Gergely. We have adopted in this paper the terminology proposed by these authors for TN-I and TN-C, but, in view of the evidence presented below that the 39,000 component interacts specifically not only

with tropomyosin but also with F-actin, we propose to call it TN-B (B for binding).

We have shown recently that both TN-I and TN-B are very susceptible to proteolytic enzymes, forming protein fragments with molecular weights within the range of 10,000–13,000 daltons (Drabikowski et al., 1971b). We have also suggested that the 13,000 daltons component present in troponin is formed during preparation as the result of action of endogenous proteases. Figure 1 shows that this is indeed the case. During incubation of the tropomyosin-troponin complex at pH 4.5—a necessary step for fractionation of the complex—splitting of both TN-I and TN-B takes place. SDS-gel electrophoresis reveals the appearance, as an intermediate step, of protein fragments with mol wt 10,000–13,000 daltons (Fig. 1). When the whole mixture is preincubated at 100° before incubation at pH 4.5, no splitting occurs. All these results clearly show that the lower molecular weight components are

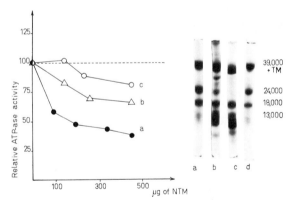

Figure 1. Changes in composition and properties of tropomyosin-troponin complex during incubation at pH 4.5. Two mg/ml of the tropomyosin-troponin complex (a protein fraction salting out between 40 and 60% (NH$_4$)$_2$SO$_4$ saturation from 1 M KCl extract from alcohol-ether muscle powder [Drabikowski et al., 1969], denoted on the abscissa as NTM) was incubated at pH 4.5 in 1 M KCl at 20° for (*a*) 10 min, (*b*) 6 hr, and (*c*) 24 hr. (*d*) Sample heated 5 min at 100° before 24 hr incubation. ATPase activity was measured at 25° for 5 min in a medium containing 10 mM Tris-HCl buffer pH 7.5, 10 mM KCl, 2 mM ATP, 2 mM MgCl$_2$, 2 mM EGTA, 0.30 mg synthetic actomyosin (reconstituted from actin and myosin mixed in proportion 1:3, w/w) and tropomyosin-troponin complex as indicated on the abscissa. SDS-polyacrylamide gel electrophoresis was performed as previously described (Drabikowski et al., 1971b).

the products of proteolytic splitting caused by cathepsins of lysosomal origin, active at acid pH and contaminating the preparations of the tropomyosin-troponin complex.

Figure 1 also shows that splitting of TN-I and appearance of the 13,000 daltons component are correlated with a decrease of the inhibitory effect of the complex on actomyosin ATPase in the absence of Ca^{++}. As shown earlier (Drabikowski et al., 1971a) the 13,000 component also inhibits Mg^{++}-stimulated ATPase of actomyosin, although the inhibitory activity is weaker than that of TN-I. All these results indicate that the 13,000 component is the degradation product of TN-I. This is supported by the fact that the 13,000 component usually contains sulfhydryl groups and that only TN-I, but not TN-B, contains cysteine residues (see Table 1).

A protein of mol wt of about 30,000 daltons, present usually in small amounts in troponin preparations from mixed rabbit skeletal muscle, seems to be the first product of splitting of TN-B, caused by contamination of endogenous proteases probably active at neutral pH range (Dąbrowska et al., 1972). Protein of the same molecular weight is formed from TN-B as the result of an extremely mild treatment with trypsin.

Several kinds of interaction seem to exist between individual components of troponin. TN-C interacts with both TN-I and TN-B, changing their solubility properties. This interaction seems to involve Ca^{++}, since for full dissociation of TN-C from the complexes on a DEAE-Sephadex column, besides urea, the presence of EDTA is necessary (Drabikowski et al., 1971a). The complex between TN-I and TN-B requires guanidine for dissociation (Hartshorne and Dreizen, 1971). Consequently all three constituents of troponin are tightly bound to each other and move on Sephadex G-100 as a single peak of mol wt equal to about 85,000 daltons (Fig. 2).

Chromatography on a Sephadex G-200 column of the mixture of tropomyosin and troponin reveals

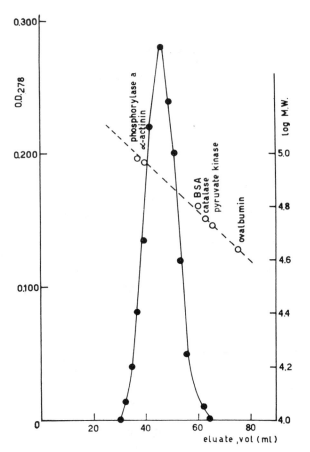

Figure 2. Sephadex chromatography of troponin. Troponin was chromatographed on Sephadex G-100 column, calibrated with standard proteins indicated on the figure. Elution medium 0.3 M KCl and 20 mM Tris-HCl pH 7.5.

Table 1. Thiol Group Content of Troponin Constituents

Expt.	TN-B	TN-I	Degradation Products (13,000 component)	TN-C
1	0	15.4	13.0	4.2
2	0	—	7.5	5.4
3	0	16.2	7.9	5.3

Number of SH equivalents per 10^5 g protein was determined as previously described (Drabikowski and Nowak, 1970) after preincubation with 0.1 M DTT and 6 M urea and removal of DTT on Sephadex G-25.

in a nonretarded peak tropomyosin and three constituents of troponin: TN-C, TN-I, and TN-B. (Fig. 3A). The second retarded peak contains no TN-B, some TN-C and TN-I, and all the 13,000 component. When, however, the mixture of tropomyosin and the preparation of troponin devoid of TN-B is chromatographed, the nonretarded peak contains only tropomyosin (Fig. 3B). These results clearly indicate that tropomyosin forms a complex with all three constituents of troponin, but TN-B is the only component which interacts directly with tropomyosin. They also show a lack of binding of the 13,000 component with tropomyosin. That the affinity of TN-B to tropomyosin is very strong, perhaps even stronger than that to TN-I and TN-C, is indicated by the fact that at pH 4.5, in the presence of 1 M KCl, TN-B is the troponin constituent which is the most difficult to separate from precipitated tropomyosin.

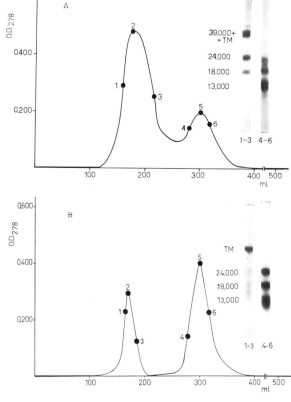

Figure 3. Interaction of tropomyosin with the constituents of troponin. Sephadex G-200 column chromatography of the mixture of tropomyosin with troponin (*A*) or tropomyosin and troponin devoid of TN-B (*B*) was carried out as previously described (Drabikowski et al., 1969). The sample contained 10 mg tropomyosin and 20 mg troponin. Elution medium 0.3 M KCl containing 20 mM Tris-HCl pH 7.5.

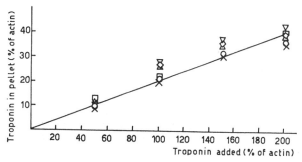

Figure 4. Binding of troponin to F-actin. Samples containing various amounts of troponin labeled with [^{14}C]NEM or ^{125}I and F-actin (1.0–1.2 mg/ml) were centrifuged in the preparative ultracentrifuge or were precipitated by myosin in the presence of various amounts of tropomyosin: (×) 0, (○) 10, (△) 20, (▽) 40, (□) 60, and (◇) 100% of actin.

Figure 4 shows the binding of troponin to F-actin. The extent of binding is rather small and the curve shows no sign of saturation. The Scatchard plot from the values obtained is almost parallel to the abscissa, indicating a very weak binding. It is remarkable, however, that tropomyosin does not enhance virtually the extent of binding of troponin.

It has been earlier observed (Drabikowski and Nonomura, 1968) that troponin causes precipitation of F-actin. Some of the troponin preparations obtained by us in recent years also show this phenomenon. Analysis of their composition indicates that precipitation of F-actin is caused by preparations of troponin rich in TN-B and poor in TN-C. Figure 5 shows that precipitated F-actin contains troponin in amounts up to about 30% by weight. This amount corresponds roughly to the molar ratio of one troponin per 6–7 actin monomers.

The experiments with individual troponin constituents show that TN-B interacts directly with F-actin. In order to show the binding, the mixture

of F-actin labeled with [^{14}C]ADP and TN-B in 0.5 M KCl is dialyzed against 0.1 M KCl. The radioactivity found in the precipitate indicates the presence of F-actin, which forms, as revealed by negative staining, paracrystals very similar to those previously found with whole troponin (Drabikowski and Nonomura, 1968). The formation of F-actin paracrystals by TN-B is abolished by TN-C.

The binding of tropomyosin by F-actin (Drabikowski and Nowak, 1968) is of different character than that of troponin, and the Scatchard plot indicates two sets of binding sites (Fig. 6). The

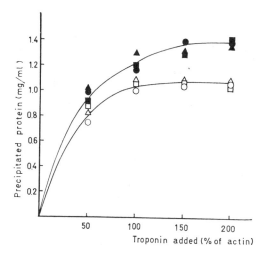

Figure 5. Precipitation of F-actin by troponin. Samples of F-actin labeled with [^{14}C]ADP (1.0 mg/ml) and various amounts of troponin, as indicated on the abscissa, were centrifuged, and in the supernatant radioactivity and protein content were measured. (○, □, △) F-actin; (●, ■, ▲) total protein in the precipitate. Various symbols denote different experiments.

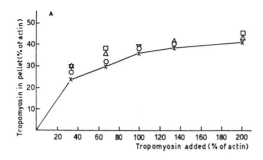

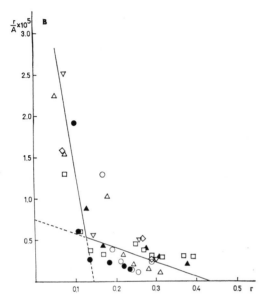

Figure 6. Binding of tropomyosin to F-actin. *A.* Samples containing various amounts of tropomyosin labeled with [^{14}C]NEM or ^{125}I and F-actin (1.0–1.2 mg/ml) were centrifuged in the preparative ultracentrifuge or were precipitated by myosin in the presence of various amounts of troponin: ($\times$) 0, ($\bigcirc$) 33, ($\triangle$) 66, ($\triangledown$) 100, and ($\square$) 133% of actin. *B.* Scatchard plot giving the values of tropomyosin bound to F-actin in the absence of troponin obtained in various experiments, as denoted by different symbols. ($\bigcirc$, $\square$, $\triangle$, $\triangledown$) centrifugation of F-actin; ($\bullet$, $\blacktriangle$) precipitation with myosin. r = moles of tropomyosin bound per mole of actin monomer; A = concentration of free tropomyosin in M.

higher affinity binding constant is equal to 3.3 $\times 10^6$ M^{-1} and the maximum number of sites is 0.135 moles of tropomyosin per mole of actin monomer. This value corresponds to one tropomyosin molecule per seven actin monomers, and agrees well with the model of thin filament, originally proposed by Hanson and Lowy (1964), in which two strands of tropomyosin lie in the grooves of a double-stranded F-actin helix. Troponin has no effect on the extent of binding of tropomyosin to F-actin but may increase somewhat the value of the binding constant of tropomyosin to F-actin, which is consistent with the cementing role of

troponin in the binding of the former proteins observed earlier (Drabikowski et al., 1968).

All the results presented in this work seem to indicate that the role of TN-B might be the formation of a link between the other components of troponin, i.e., TN-C and TN-I, tropomyosin and F-actin. The role of tropomyosin, apart of its significance in the regulatory protein system, may consist in preventing F-actin from chaotic interaction with TN-B and enabling the binding of this protein, and consequently other constituents of troponin, only at regular (about 400 Å) periods, as previously postulated (Drabikowski, 1968). Besides tropomyosin, TN-C with its unique highly charged character seems to be also involved in the proper arrangement of all proteins, forming a multiprotein assembly of thin filament.

Acknowledgment

This work was supported in part by a Foreign Research Agreement No. 05-015-1 of N.I.H. under P.L. 480.

References

DĄBROWSKA, R., M. DYDYŃSKA, A. SZPACENKO, and W. DRABIKOWSKI. 1972. Comparative studies of the composition and properties of troponin from fast, slow and cardiac muscle. *Comp. Biochem. Physiol.* (In press.)

DRABIKOWSKI, W. 1968. Interaction of F-actin with structural muscle proteins other than myosin. Symposium on functional and structural aspects of the myofibrillar proteins, Balatonboglar Hungary. Abstracts of Communications p. 2, –1; Muscle Notes, 42.

DRABIKOWSKI, W. and B. BARYŁKO. 1971. Calcium binding by troponin. *Acta Biochim. Polon.* **18**: 353.

DRABIKOWSKI, W. and Y. NONOMURA. 1968. The interaction of troponin with F-actin and its abolition by tropomyosin. *Biochim. Biophys. Acta* **160**: 129.

DRABIKOWSKI, W. and E. NOWAK. 1968. Studies on the interaction of F-actin with tropomyosin. *Europe. J. Biochem.* **5**: 376.

———. 1970. Thiol group content in tropomyosin and troponin. *Acta Biochim. Polon.* **17**: 221.

DRABIKOWSKI, W., R. DĄBROWSKA, and B. BARYŁKO. 1971a. Separation and characterization of the constituents of troponin. *FEBS Letters* **12**: 148.

DRABIKOWSKI, W., R. DABROWSKA, and E. NOWAK. 1969. Comparative studies on the composition and properties of EGTA-sensitizing factors. *Acta Biochim. Biophys. Acad. Sci. Hung.* **4**: 112.

DRABIKOWSKI, W., D. R. KOMINZ, and K. MARUYAMA. 1968. Effect of troponin on the reversibility of tropomyosin binding to F-actin. *J. Biochem.* (Tokyo) **63**: 802.

DRABIKOWSKI, W., U. RAFAŁOWSKA, R. DĄBROWSKA, A. SZPACENKO, and B. BARYŁKO. 1971b. The effect of proteolitic enzymes on the troponin complex. *FEBS Letters* **19**: 259.

EBASHI, S. and M. ENDO. 1968. Calcium ion and Muscle Contraction. *Prog. Biophys. Mol. Biol.* **18**: 123.

GREASER, M. L. and J. GERGELY. 1970. Calcium binding component of troponin. *Fed. Proc.* **29**: 463.

——. 1971. Reconstitution of troponin activity from three protein components. *J. Biol. Chem.* **246**: 4226.

HANSON, J. and J. LOWY. 1964. The structure of actin filaments and the origin of the axial periodicity in the I-substance of vertebrate striated muscle. *Proc. Roy. Soc. (London)* B **160**: 449.

HARTSHORNE, D. J. and P. DREIZEN. 1971. Quoted by P. Dreizen in Structure and function of the myofibrillar contractile proteins. *Annu. Rev. Med.* **22**: 365.

HARTSHORNE, D. J. and H. Y. PYUN. 1971. Calcium binding by the troponin complex, and the purification and properties of troponin A. *Biochem. Biophys. Acta* **229**: 698.

SCHAUB, M. C. and S. V. PERRY. 1971. The regulatory proteins of the myofibril: Characterization and properties of the inhibitory factor (troponin B). *Biochem. J.* **123**: 367.

SCHAUB, M. C., S. V. PERRY, and W. HÄCKER. 1972. The regulatory proteins of the myofibril: Characterization and biological activity of the calcium-sensitizing factor (troponin A). *Biochem. J.* **126**: 237.

WILKINSON, J. M., S. V. PERRY, H. A. COLE, and I. P. TRAYER. 1972. Separation and biological activity of the components of inhibitory factor preparation. *Biochem. J.* **127**: 215.

Localization and Mode of Action of the Inhibitory Protein Component of the Troponin Complex

S. V. Perry, H. A. Cole, J. F. Head and F. J. Wilson

Department of Biochemistry, University of Birmingham, Birmingham, England

It is now well established that the troponin complex consists of three distinct protein components that differ in molecular weight, amino acid composition, and function. A number of names for these components have been used by different investigators (Table 1), and in this article we shall use the terms inhibitory protein and calcium-binding protein to describe the proteins with inhibitory activity on the Mg^{++}-stimulated ATPase of actomyosin and with a high affinity for calcium, respectively. In the absence of a clearly defined function for the third component of the troponin complex, it will be identified by its molecular weight and referred to as the 37,000 component.

The inhibitory protein is a highly specific inhibitor of the Mg^{++}-stimulated ATPase of "synthetic" actomyosin, desensitized actomyosin, acto-heavy-meromyosin (acto-HMM), and the actin subfragment 1 complex. It is identical with the inhibitory factor originally detected and isolated from extracts of myofibrils (Perry et al., 1966; Perry, 1967; Hartshorne et al., 1967) and later prepared in a more

highly purified form from the troponin complex (Hartshorne and Mueller, 1968; Schaub and Perry, 1969).

By chromatographic procedures in the presence of dissociating agents such as strong urea solutions, and preferably in the presence of calcium chelating agents to break down the complex formed with the calcium-binding protein, the inhibitory protein can be isolated from the troponin complex in a pure form which moves as a single band corresponding to a mol wt of 23,000 on electrophoresis in sodium dodecyl sulfate (SDS). The electrophoretic properties of the inhibitory protein suggests that it is a rather basic protein with an isoelectric point close to neutrality. It is presumably for this reason that it is almost insoluble at physiological pH values in media of low ionic strength. Its solubility is increased by complexing with the calcium binding protein, in the presence of polyanions and by partial proteolytic digestion. The inhibitory protein has a characteristic amino acid composition clearly distinct from that of the acidic calcium-binding protein (Fig. 1). Its amino

Table 1. Protein Components of Troponin Complex

Calcium-Binding Protein	Inhibitory Protein	37,000 Component	Reference
Troponin A	Troponin B		Hartshorne and Mueller, 1968
Calcium-sensitizing factor	Inhibitory factor	37,000 component	Schaub and Perry, 1969; Wilkinson et al., 1971, 1972
Troponin 4	Troponin 2	Troponin 3	Greaser and Gergely, 1971
Troponin III	Troponin II	Troponin I	Ebashi et al., 1971
Component III	Component II	Component I	Murray and Kay, 1971
Troponin C	Troponin I	Troponin T	Potter and Gergely, 1972

Under each of the names adopted in this article for the components of the troponin complex are listed the names used by other authors to describe fractions of similar biological activity and amino acid composition.

Figure 1. Representation of the amino acid composition of the components of the troponin complex. The figures are formed by plotting the number of residues per mole of each amino acid along the appropriate radii and joining up the points obtained. (*a*) Calcium-binding protein; (*b*) inhibitory protein. The unshaded area in the center of this figure represents the composition of the 14,000 dalton component with inhibitory activity which has been isolated from some troponin preparations; (*c*) 37,000 component. The amounts of the hydrophilic amino acids are plotted in the upper half of the figure and the nonpolar amino acids in the lower half so that the shape gives an indication of the relative proportions of these two classes of amino acids.

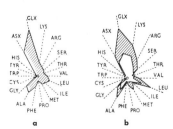

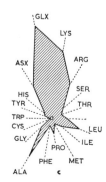

251

acid profile is similar to that of the 37,000 component
which is also slightly basic and very insoluble when
isolated in the pure form. Despite the broad similarities in composition the inhibitory protein differs
from the 37,000 component in containing three
cysteine residues per mole, whereas the latter has
none.

In resting muscle the inhibitory protein is fully
effective, and contraction cannot take place as the
hydrolysis of ATP by the myofibril is reduced to a
low level. On stimulation the calcium concentration
of the sarcoplasm rises to a concentration of about
10^{-5} M, under which conditions the calcium-binding
protein neutralizes the action of the inhibitory protein in some way not yet understood. ATP is immediately hydrolyzed at a rapid rate and contraction
ensues.

Localization of Inhibitory Protein
in Myofibril

The properties of the inhibitory protein and its
mode of action suggest that it is closely associated
with actin in the myofibril. Localization of the inhibitory protein in the I filament is implied by the
report of Ebashi et al. (1969) that the troponin complex is distributed along the I filament with a
periodicity of 400 Å. This conclusion is valid only if
the troponin complex acts as a single antigen and
produces an antibody specific to the complex as a
whole rather than the individual components. To
clarify this point we have studied the antigenic properties of the troponin complex and its purified
components.

When guinea pigs were immunized with the troponin complex prepared from rabbit skeletal muscle,
by the method of Schaub and Perry (1971), a complex response was invariably obtained. On examination of the immune serum by the Ouchterlony gel
diffusion procedure at least four precipitin lines were
seen (Fig. 2a). Purified preparations of the inhibitory
protein and the 37,000 component, which gave single
bands on electrophoresis in SDS, each gave a single
precipitin line with the antibody to the troponin
complex. The precipitin lines obtained with the
purified components appeared to correspond with
those seen in the complex response of the antitroponin to the troponin preparation used for immunization. A highly purified sample of the rabbit
calcium-binding protein reacted less consistently
with guinea pig antitroponin. This was probably due
to the low antigenicity in the guinea pig of the
rabbit calcium-binding protein either injected
alone or as the troponin complex (Hirabayashi,
personal communication). All the evidence obtained
strongly suggests that when guinea pigs were immunized with the rabbit troponin complex, the

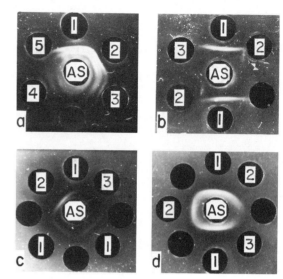

Figure 2. Immunochemical analyses of antisera to rabbit
troponin complex and antisera to rabbit inhibitory protein.
(*a*) Center well (*AS*) contains guinea pig antiserum to unfractionated troponin complex; peripheral wells contain the
troponin complex used as antigen at (1) 2800 μg/ml, (2)
1000 μg/ml, (3) 330 μg/ml, (4) 110 μg/ml, and (5) tropomyosin, 4200 μg/ml. Gel matrix is 1.25% agarose in 150
mM NaCl and 15 mM Tris pH 7.6. In this experiment and in
those illustrated below the well capacity is 0.05 ml. (*b*)
Center well (*AS*) contains guinea pig antiserum to rabbit
inhibitory protein; peripheral wells contain (1) inhibitory
protein preparation used as antigen, 600 μg/ml; (2) tropomyosin, 500 μg/ml; (3) calcium-binding protein, 100 μg/ml.
Gel matrix as in (*a*). (*c*) Center well (*AS*) contains guinea pig
antiserum to rabbit inhibitory protein; peripheral wells
contain (1) inhibitory protein used as antigen, 130 μg/ml;
(2) inhibitory protein, 200 μg/ml; (3) 37,000 component,
200 μg/ml. Gel matrix as in (*a*). (*d*) Center well (*AS*) contains guinea pig antiserum to rabbit inhibitory protein;
peripheral wells contain the 14,000 dalton component at
(1) 50 μg/ml; (2) 25 μg/ml; (3) inhibitory protein, 50 μg/ml.
Gel matrix is 1.25% agarose in 0.01% CdCl$_2$, 150 mM NaCl
and 15 mM Tris pH 7.6. Note the inhibitory protein used as
antigen contained, in addition to the 14,000 dalton component, the 37,000 component. It is referred to as the
inhibitory protein antigen to distinguish it from the inhibitory protein, which is a pure component of 23,000 mol
wt.

antigenic response was to the individual components of the complex and to any impurities present
in the preparation rather than to the complex as a
single antigen.

To determine the localization of the inhibitory
protein in the myofibril, an antibody to an inhibitory
protein preparation, obtained by the method of
Schaub and Perry (1971) from rabbit skeletal muscle,
was raised in the guinea pig. When injected at low
dose levels, i.e., three injections each of 50 μg over a
period of one month, a single precipitin line response
was obtained on gel diffusion over a range of antigen
concentrations and ionic conditions. At the time
when the immunization studies were started, it was
considered that the inhibitory protein preparation
consisted of a single component of 23,000 mol wt.

Indeed the single precipitin line obtained when the antibody was tested with different concentrations of antigen tended to confirm this view (Fig. 2b,c). Nevertheless subsequent study (Wilkinson et al., 1972) indicated that the inhibitory protein preparation made by the procedure of Schaub and Perry (1971) often contained the proteolytic fragment of the inhibitory protein of mol wt 14,000 and the 37,000 component. On further testing with samples of the components of the troponin complex, the antiserum reacted only with a highly purified sample of the inhibitory protein, mol wt 23,000, which showed a single band on electrophoresis in SDS (Wilkinson et al., 1972). There was no reaction with calcium-binding protein, tropomyosin (Fig. 2b), or with the 37,000 component (Fig. 2c). A faint band was observed with some 37,000 component preparations when used at high concentration. It was considered that this band was due to contamination of the 37,000 component with inhibitory protein, which was responsible for the very low inhibitory activity often present in 37,000 component preparations and which could be removed on repeated purification.

Apart from the inhibitory protein the only other protein that gave a pronounced reaction with the antiserum from guinea pigs immunized against inhibitory protein antigen was the 14,000 component, which has been shown to be present in some troponin preparations but not in fresh myofibrils (Wilkinson et al., 1971, 1972; Greaser and Gergely, 1971; Drabikowski et al., 1971). Previous investigations (Wilkinson et al., 1972) indicated that the 14,000 component is probably produced by the action of muscle cathepsins on the inhibitory protein during preparation of the troponin complex. Such a relationship was confirmed by the fact that the precipitin lines obtained on reaction of the antibody with the inhibitory protein and with the 14,000 component were continuous, indicating similarity of antigenic sites on the two proteins (Fig. 2d). The observation strongly suggests that the degradation of the inhibitory protein to the 14,000 component by the action of muscle proteases, probably cathepsin D, preserves both its antigenic site and the part of the molecule essential for inhibitory activity.

It was noted that, although the preparation used for immunization probably contained some 37,000 component, there was little antigenic response to this protein. The reason for this was not clear since other studies have suggested that some preparations of the 37,000 component are good antigens. The lack of response to this component in these experiments may have been due to the presence of excess inhibitory protein and the antigenically similar proteolytic fragment derived from it and the low levels of total protein injected.

When myofibrils at rest length prepared from strips of glycerinated rabbit psoas were treated with antibody to the inhibitory protein made fluorescent by labeling with fluorescein isothiocyanate (FITC) (Coons and Kaplan, 1950) and observed in ultraviolet light, two intense bands of fluorescence were seen in the I band on either side of the Z line region (Fig. 3a,b). The nonfluorescent region in the I band appeared to be broader than the width of the Z line as judged by its appearance in phase contrast. It was of interest that when the myofibril was treated with fluorescent antibody to the complete troponin complex, a similar distribution of fluorescence in the I band was observed. There was, however, a noticeable difference in that the nonfluorescent region at the Z line was distinctly narrower than was the case when the myofibril was stained with the fluorescent antibody to the inhibitory protein (Fig. 3e,f; cf. 3c,d). This implied that the immunoglobulin prepared against the troponin complex contained an antibody to material lying closer to the Z line than the inhibitory protein. In addition to the strong fluorescence observed in the I band of myofibrils treated with the fluorescent antibody to the inhibitory protein, fainter fluorescence was apparent in the A band which extended from the A-I junction to the edge of the H space (Fig. 3a,b).

The results were consistent with localization of the inhibitory protein in the I filament. The faint fluorescence in the regions of overlap of the I with the A filaments was considered to be due to the reduced accessibility of the inhibitory protein for combination with the antibody. This interpretation was confirmed by the changes in fluorescent pattern observed in stretched and contracted myofibrils and by the increase in length of the fluorescent bands in myofibrils stretched so that the I filaments were largely withdrawn from the A band (Fig. 4a,b; cf. 3a,b). In myofibrils that had shortened to 65% of the resting length, the fluorescence zones in the I band were much narrower, and in addition a strong fluorescence appeared in the center of the A band (Fig. 4c,d; cf. Fig. 3c,d). Such a fluorescent zone in the middle of the A band was only observed in myofibrils that had narrow I bands and were obviously contracted. It was considered that the fluorescence in the middle of the A band represented a region where access to the inhibitory protein was not obstructed by the myosin heads, as in other parts of the A band. Such a region would exist in the center of the A band in the shortened myofibril due to I filaments extruding into the central region free of myosin cross-bridges, and particularly in regions where the I filaments overlapped.

The pattern of fluorescence obtained with the unabsorbed immunoglobulin was essentially the same as that obtained with the immunoglobulin that had been absorbed with the 37,000 component.

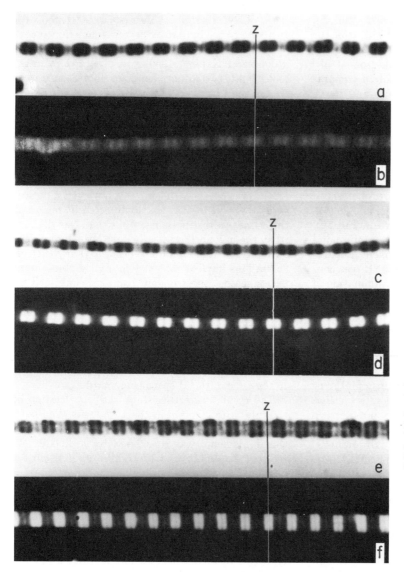

Figure 3. Immunofluorescence of myofibrils at rest length stained with antibody to rabbit inhibitory protein and with antibody to rabbit troponin complex. Magnification 6750. (*a*) Phase contrast photomicrograph of a myofibril stained directly with FITC guinea pig γ globulin to rabbit inhibitory protein. (*b*) The same myofibril shown in (*a*) seen in fluorescence. In this example and in those shown below the sarcomeres are in register and the Z line is marked. (*c*) Phase contrast photomicrograph of a myofibril stained by the indirect method with guinea pig antiserum to rabbit inhibitory protein. (*d*) The myofibril shown in (*c*) now seen in fluoresence. (*e*) Phase contrast photomicrograph of a myofibril stained by the indirect method with guinea pig antiserum to rabbit troponin complex. (*f*) The myofibril seen in (*e*) shown in fluorescence.

When the antibody was absorbed with the inhibitory protein, the specific I band fluorescence was removed.

A fluoresence pattern indicating the presence of inhibitory protein in the I band was also obtained when the myofibrils were treated with unfractionated anti-inhibitory protein serum which was then localized by subsequent treatment with FITC-labeled rabbit antibody to guinea pig γ globulin (Fig. 3c,d; 4c,d). Band patterns observed under these conditions were essentially similar to those obtained by the direct staining procedure, i.e., when the antibody to the inhibitory protein itself was fluorescent.

It was therefore concluded that the inhibitory protein is localized on the I filament but does not extend right up to the Z line, or is possibly masked

in some way in this region so that it does not react with the antibody.

Preliminary studies with guinea pig antisera to rabbit 37,000 component have indicated that this component is localized in the I filament in a manner similar to the inhibitory protein.

Biological Activity of Inhibitory Protein and the Effects of Tropomyosin

The localization of the inhibitory protein in the I filament implies that its action on the actomyosin ATPase is a consequence of the special nature of its interaction with actin. Such a relationship is also indicated by the fact that it is the Mg^{++}-stimulated ATPase and not the Ca^{++}-stimulated ATPase of

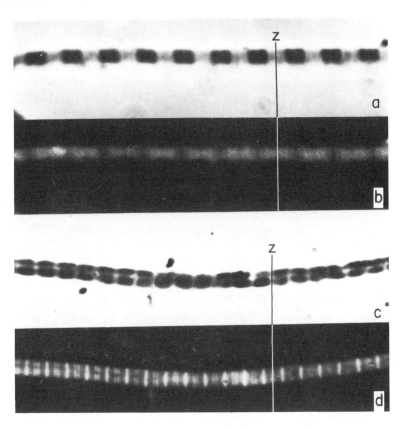

Figure 4. Immunofluorescence of stretched and contracted myofibrils stained with antibody to rabbit inhibitory protein. Magnification 6750×. (*a*) Phase contrast photomicrograph of a myofibril stretched to 120% rest length and stained with FITC guinea pig γ globulin to rabbit inhibitory protein. (*b*) The same myofibril seen in fluorescence. For a comparison with a myofibril at rest length see Fig. 3(a) and (b). (*c*) Phase contrast photomicrograph of a myofibril contracted to 65% rest length and stained by the indirect method with guinea pig antiserum to rabbit inhibitory protein. (*d*) The same myofibril in fluorescence; compare with Fig. 3(c) and (d).

actomyosin which is inhibited by the inhibitory protein. Whereas ATPase activity in the presence of Ca^{++} takes place at a high level whether actin is present or not, the modifying effect of actin on the enzymic activity of myosin is essential for significant rates of hydrolysis of ATP in the presence of Mg^{++}. Although it is highly specific so far as the divalent cation and the enzymes are concerned, the inhibitory protein inhibits the hydrolysis of all the nucleotide triphosphates so far tested, namely, ATP, CTP, and ITP.

The Mg^{++}-stimulated ATPase activity of actomyosin and acto-HMM systems that have been purified to free them of the other components of the troponin complex is significantly reduced by the inhibitory protein in the absence of tropomyosin. The amount of inhibitory protein required to bring about 50% of the maximum inhibition of the ATPase of a given actomyosin or acto-HMM system obtainable with inhibitory protein was, however, decreased several times by the addition of tropomyosin (Fig. 5). The increased effectiveness of the inhibitory protein depended on the molar ratio of actin to tropomyosin. In the absence of tropomyosin 50% of the maximum inhibition of the ATPase was obtained with a molar ratio of actin to inhibitory protein of about 2:1, whereas at optimum tropomyosin concentrations the maximum

value of the ratio obtained was 6–8:1. Under the latter conditions there was about one molecule of tropomyosin present in the system for every 3–4 actin monomers.

When all the components are organized in relation to one another with greater precision than can be obtained with in vitro studies, as for example they are in the I filament of the myofibril, it may be valid to assume that the amount of inhibitory factor required to bring about maximum inhibition is twice that required for 50% inhibition. The following conclusions can then be drawn from the in vitro enzymic experiments. In the absence of tropomyosin one inhibitor molecule is required per actin monomer for maximum inhibition of the Mg^{++}-stimulated ATPase. When optimum amounts of tropomyosin were present, it was demonstrated that under some conditions one molecule of inhibitory protein is able to exert its effect on 3–4 actin monomers and prevent the interaction with myosin that is essential for a high level of Mg^{++}-stimulated ATPase.

These results suggest that although tropomyosin is not essential for the action of the inhibitory protein, its presence is required if one molecule of the inhibitory protein is to be effective on more than one actin monomer. In the absence of other components of the regulatory protein system it would appear that the inhibitory protein interacts with

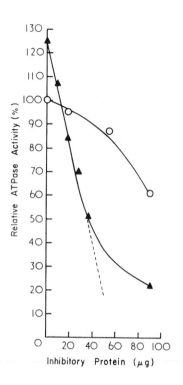

Figure 5. Effect of inhibitory protein on the Mg^{++}-stimulated ATPase of acto-HMM in the presence and absence of tropomyosin. Enzymic assays carried out on 300 μg actin free of troponin components and 600 μg HMM incubated for 5 min at 25°C in 2.5 mM MgCl$_2$, 2.5 mM Tris ATP and 25 mM Tris HCl, pH 7.6. ATPase activity expressed as a percentage of that obtained with acto-HMM alone. (○) No added tropomyosin; (▲) 200 μg tropomyosin added.

actin in such a way that the Mg^{++}-stimulated hydrolysis of ATP by the actomyosin system is much slower than in the absence of inhibitory protein. Indeed in the presence of excess inhibitory protein the rate of ATP hydrolysis by actomyosin in the presence of Mg^{++} is more comparable to that of myosin itself. This effect of the inhibitory protein may involve a conformational change in the actin which, in the presence of tropomyosin, could be transmitted by cooperative action to adjacent actin monomers, themselves not in direct contact with the inhibitory protein. The fact that the studies with acto-HMM suggest that one molecule of inhibitory protein can block the effect of one actin monomer in the absence of tropomyosin indicates that the tropomyosin is not involved as an essential component of the inhibitory process, but in some way transmits the effect of an inhibitory protein molecule on one actin monomer to neighboring monomers. A possible representation of the arrangements of the components which satisfies these facts is given in Fig. 6b.

If, as is generally supposed, tropomyosin lies in the groove of the actin double-helical filament, one molecule of 400 Å length would extend along a linear array of about seven actin monomers (Fig. 6). If all the actin monomers associated in this way with one tropomyosin molecule could act cooperatively, it would follow that in the presence of tropomyosin one inhibitory protein molecule could influence seven actin monomers. The in vitro studies suggest that one inhibitor molecule influences three or four actin monomers and, rather strikingly, one tropomyosin molecule is associated with a similar number of actin monomers to give the maximum cooperative effect. Whether this discrepancy between the in vitro results and what might be considered theoretically possible is simply a feature of the relatively poor organization of the in vitro system or a real property of the system must await further work. If one molecule of inhibitory protein can, at a maximum, influence three or four actin monomers, then clearly two molecules will be required to influence seven actin monomers.

The maximum extent of the inhibition obtained with the inhibitory protein is of some interest. In resting muscle the level of myofibrillar ATPase activity is probably no greater than 1 or 2% of the maximum obtained during contraction. In the presence of EGTA the ATPase of isolated myofibrils is reduced by 90–95% although sometimes the inhibition obtained is less than this. The reasons for variation in the extent of inhibition of the ATPase of the isolated myofibril in the absence of Ca^{++} are not clear but may reflect changes occurring in the protein components and their structural organization during preparation and storage of the myofibrils. Likewise with natural and reconstituted actomyosin systems the extent of inhibition obtained with the regulatory protein system is often

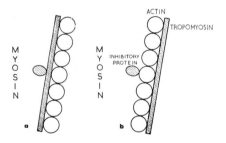

Figure 6. Diagrammatic representation of possible arrangement of the inhibitory protein in relation to the actin and tropomyosin in the I filament. The scheme shown in (b) represents the inhibitory protein acting directly on an actin monomer and its effect being transmitted to neighboring actin monomers by the tropomyosin molecule lying alongside them.

much less than that obtained in the best conditions with myofibrils. The extent of inhibition obtained with the inhibitory protein alone on actomyosin systems was similar to that obtained with the whole regulatory system on similar enzyme preparations. In some cases up to 90% inhibition could be obtained with excess inhibitory protein although on some occasions it did not exceed 60–70%. Such a variation was obtained with a single inhibitory protein preparation when tested on different actomyosin suspensions and on a given actomyosin suspension tested after different periods of storage. This would suggest that the condition of the actomyosin system is an important factor in determining the extent of inhibition obtained.

Some troponin preparations contain in addition to the main components a protein of mol wt about 14,000 which also possesses inhibitory activity (Wilkinson et al., 1971, 1972; Greaser and Gergely, 1971; Drabikowski et al., 1971). This protein is not present in freshly prepared myofibrils, and in view of this observation, its amino acid composition and immunochemical behavior, it is now clear that the protein of 14,000 mol wt is produced by partial proteolysis of the original inhibitory protein by the endogenous proteases of muscle, probably cathepsin D. Degradation in this way removes a substantial part of the molecule containing all the cysteine residues present in the inhibitory protein but nevertheless preserves the biological activity of the original molecule (Wilkinson et al., 1972). The specific inhibitory activity of the 14,000 component of the ATPase of an acto-HMM system is about the same on a molar basis as that of the intact inhibitory protein. In contrast to the behavior of the latter, however, the inhibitory activity of the 14,000 component is not significantly increased by tropomyosin. This suggests that the part of the inhibitory protein molecule which, in association with tropomyosin, is essential for inducing cooperativity between actin monomers is lost. Nevertheless the ability to interact with calcium-binding protein is retained insofar as the latter neutralizes inhibitory activity in about the same molar ratio as with the whole intact inhibitory protein.

Interaction of the Inhibitory Protein and Calcium-Binding Protein

The calcium-binding protein has the property of neutralizing the inhibitory action of the inhibitory protein on the Mg++-stimulated ATPase of actomyosin. As has been pointed out elsewhere (Schaub et al., 1972; Perry, 1972) two aspects of this interaction can be distinguished.

1. In the absence of tropomyosin the neutralization of the action of the inhibitory protein is independent of Ca++ concentration. It also occurs in

the absence of the 37,000 component and is not significantly affected by its presence. This activity is a stable property of the calcium-binding protein and is shown by all preparations so far tested. It is not lost on carboxymethylation of the calcium-binding protein (Schaub et al., 1972). So far as can be estimated from enzymic studies with highly purified samples of the components in the absence of tropomyosin, approximately one molecule of calcium-binding protein is required to fully neutralize one molecule of inhibitory protein (Fig. 7). Over the initial stages of the titration when the neutralization of inhibitory activity is more linear in response, the preliminary data suggests that calcium-binding protein is more effective, and the molar ratio of calcium-binding protein to inhibitory protein required for neutralization is closer to 1:2 (Fig. 7).

2. In the presence of tropomyosin the neutralization of the effect of inhibitory protein by the calcium-binding protein becomes sensitive to the calcium concentration, and neutralization only occurs when the Ca++ is > 10^{-5} M. When the Ca++ concentration falls to 10^{-7} or lower values, the calcium-binding protein is unable to neutralize the inhibitory protein which consequently exerts its effect on the Mg++-stimulated ATPase of actomyosin. Although not essential for calcium sensitivity

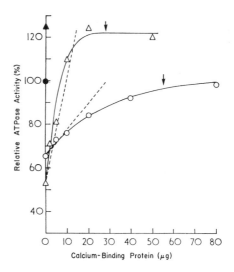

Figure 7. Neutralization of the action of the inhibitory protein on the Mg++-stimulated ATPase of acto-HMM by the calcium-binding protein. Enzymic assays carried out on 700 μg of desensitized actomyosin under the same conditions as for Fig. 5. ATPase activity expressed as a percentage of that obtained with desensitized actomyosin alone. Arrows represent points on curves where the molar ratio of inhibitory to calcium-binding protein is 1:1. Dotted lines represent the response if one molecule of calcium-binding protein neutralizes 2 inhibitory protein molecules. (▲) 100 μg tropomyosin; (△) 100 μg tropomyosin + 35 μg inhibitory protein and increasing amounts of calcium-binding protein; (●) no additions; (○) 70 μg inhibitory protein and increasing amounts of calcium-binding protein.

(Wilkinson et al., 1972; Drabikowski et al., 1971) there are suggestions that the 37,000 component may play a part in this system. Neutralization of inhibitory action in the presence of tropomyosin is at least as effective as in its absence. Complete neutralization is obtained when equimolar amounts of the two proteins are present. Likewise as in the absence of tropomyosin, neutralization over the initial stage of the titration curve is more effective.

With the data available and the variations obtained between preparations it is impossible to be precise at this stage of investigation, but there does not appear to be any pronounced effect of tropomyosin on the neutralizing action of the calcium-binding protein comparable to its effect on the action of the inhibitory protein. Also the stoichiometry is such over the initial stages of the titration curve that one molecule of calcium-binding protein may be able to neutralize more than one and possibly two molecules of inhibitory protein.

Whereas the simple neutralization of the action of the inhibitory protein by calcium-binding protein is a relatively stable effect consistently observed with all preparations of the protein, the property of the reversible neutralization that is sensitive to Ca^{++} concentration appears to be less easy to obtain experimentally by recombination of inhibitory protein, calcium-binding protein, and tropomyosin alone. On some occasions, in the absence of 37,000 component, simple mixing of pure inhibitory pro-

tein and calcium-binding protein produces a preparation which restores calcium sensitivity to the Mg^{++}-stimulated ATPase of desensitized actomyosin in the presence of tropomyosin. In our experience the most consistent demonstration of the restoration of calcium sensitivity to the Mg^{++}-stimulated ATPase of actomyosin, by the inhibitory and calcium-binding proteins in the presence of tropomyosin, has been when the two troponin components have been isolated without separation from each other (Fig. 8). Restoration of calcium sensitivity with such mixtures has been obtained with desensitized actomyosin and acto-HMM, both of which can be demonstrated to be free of significant contamination with the 37,000 component. Thus under some circumstances calcium sensitivity can be restored by two of the three components of the troponin complex. There is, however, evidence that the mixture of inhibitory protein and calcium-binding protein alone is not so effective on a unit weight basis as the whole troponin complex in restoring calcium sensitivity. Some authors (Greaser and Gergely, 1971; Eisenberg and Kielley, 1972) consider the 37,000 component to be essential for restoration of calcium sensitivity to actomyosin. The reason for this discrepancy between findings is not clear, but it may be explained by the recent report of Ebashi (personal communication, 1971) that the effectiveness of the inhibitory and calcium-binding proteins alone in restoring calcium sensitivity is considerably increased by the addition of the 37,000 component to the system.

Clearly the existence of the troponin complex indicates that the three components interact with each other and remain associated during extraction from muscle, despite exposure to conditions of high ionic strength. The marked effects on the Mg^{++}-stimulated ATPase of the inhibitory protein alone and in the presence of the calcium-binding protein in the absence of the 37,000 component suggest that certain special features are involved in the interaction between the former two proteins.

It was noted in earlier studies that the separation of inhibitory protein from the calcium-binding protein was much improved if separation was carried out by chromatography either in citrate buffer or in other buffers containing EGTA (Schaub and Perry, 1969; Schaub et al., 1972). The fact that the two proteins were more readily separated in the absence of Ca^{++} implies that in the presence of this cation they interact more strongly than when it is absent. It was also noted that the electrophoretic mobility of preparations of calcium-binding protein, presumed at that time to be pure, were changed markedly when Ca^{++} was reduced to a low level by the addition of EGTA (Schaub and Perry, 1969; Schaub et al., 1972).

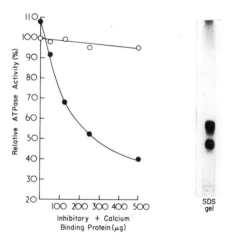

Figure 8. Calcium sensitization of the Mg^{++}-stimulated ATPase of acto-HMM on addition of a mixture of the calcium-binding and inhibitory proteins in the presence of tropomyosin. Enzymic assays carried out on 210 μg actin free of troponin components, 630 μg HMM under the same conditions as for Fig. 5. ATPase activity expressed as in Fig. 5. (○) No EGTA; (●) mM EGTA. On the right of the diagram is illustrated an electrophoretogram carried out in SDS (as in Fig. 9) of 40 μg of the mixture of calcium-binding and inhibitory proteins used for the enzymic tests. This preparation was isolated from troponin as a mixture free of 37,000 component.

Evidence for the formation of a stable complex between inhibitory protein and calcium-binding protein was obtained from electrophoretic studies in acrylamide gels in 6–8 M urea, 25 mM Tris, 80 mM glycine buffer pH 8.6, using defined mixtures of purified samples of the two proteins. When the mixture contained a molar excess of inhibitory protein, only one band moved into the gel but protein remained at the origin; whereas if the calcium-binding protein was in excess, all the protein migrated towards the anode as two bands. In the presence of EGTA a single fast band that had an identical mobility to that of pure calcium-binding protein in the presence of EGTA moved into the gel, and a protein band which could be correlated in intensity with the amount of inhibitory protein in the mixture remained at the origin (Fig. 9).

These results indicate that in the presence of Ca^{++} the inhibitory protein and the calcium-binding protein form a complex which migrates as a single band in 6–8 M urea pH 8.6. From the band patterns obtained with mixtures of different molar ratios of the two proteins, the complex appeared to contain equimolar proportions of the two components (Fig. 10). A similar picture was obtained on electrophoresis in which the 6 M urea was replaced with 40% glycerol, suggesting that the stoichiometry of the interaction was unchanged by the high urea concentration. When the Ca^{++} concentration is reduced by EGTA, the complex dissociates and the strongly negatively charged calcium-binding pro-

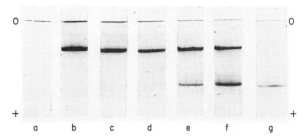

Figure 10. Electrophoresis of mixtures containing known molar ratios of the calcium-binding and inhibitory proteins. Conditions: 8% acrylamide, 6 M urea, 25 mM Tris, 80 mM glycine pH 8.6, 5 mM CaCl$_2$. The molar ratios of inhibitory calcium-binding proteins in the mixtures applied were as follows: (a) 1:0, (b) 3:1, (c) 2:1, (d) 1:1, (e) 1:2, (f) 1:3, (g) 0:1. The electrophoretic mobilities of the components are as for Fig. 9b and c; i.e., the inhibitory protein stays at the origin (10a), the complex has an intermediate mobility (10d), and the calcium-binding protein has the highest mobility (10g).

tein migrates to the anode with a faster mobility than the complex. The inhibitory protein with an isoelectric point of pH 7–8 remains close to the origin under these conditions, as has been reported previously (Schaub and Perry, 1969).

Gel filtration experiments on Sephadex G-200 under conditions similar to those used for electrophoresis with mixtures of the two proteins confirmed the existence of a complex containing one molecule of each component. In addition to more retarded peaks representing the excess inhibitory or calcium-binding proteins, a less retarded peak was eluted corresponding to the molecular weight to be expected if one mole of each were present as a complex.

The electrophoretic mobility of pure samples of calcium-binding protein was slightly lower in the presence of EGTA. This difference is the reverse from what would be expected from the difference in net charge, and the change in mobility may be due to a conformational change resulting from removal of Ca^{++} from the protein.

Complex formation between the inhibitory protein and the calcium-binding protein of the type indicated by the electrophoretic experiments was observed whenever the preparations were mixed. Nevertheless even though simple mixing of the components formed the complex, it did not usually produce a system which effectively restored calcium sensitivity to the Mg^{++}-stimulated ATPase of actomyosin. Also a mixture of the two proteins that moved as a single component on gel filtration in urea and from which the urea had been removed by subsequent dialysis restored only slight Ca^{++} sensitivity to actomyosin in the presence of tropomyosin. Thus it would appear that the ability of the two proteins to form a complex did not indicate that the

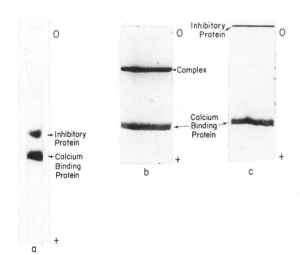

Figure 9. Effect of EGTA on the electrophoresis of a mixture of the inhibitory and calcium-binding proteins. (a) Gel electrophoresis (on 10% acrylamide, 100 mM Na phosphate buffer pH 7.0, 1% 2-mercaptoethanol, 1% SDS) of mixture of inhibitory and calcium-binding proteins. (b) Mixture used in (a) electrophoresed in an 8% acrylamide gel, in 6 M urea, 25 mM Tris, 80 mM glycine pH 8.6. (c) As in (b) but 2 mM GETA added to sample.

system was able to restore Ca^{++} sensitivity to acto-myosin ATPase, even though the existence of the complex itself required the presence of Ca^{++}.

In the absence of tropomyosin, neutralization of the inhibitory action could be brought about by the calcium-binding protein in the presence of EGTA. It therefore can be concluded that formation of a complex between the inhibitory protein and the calcium-binding protein, such as can be demonstrated in urea or on gel filtration, was not essential for neutralization of the inhibitory activity. This conclusion assumes that EGTA has the same effect on the complex under the conditions of electrophoresis and the enzymic tests.

The reason why the mixtures of calcium-binding and inhibitory proteins did not always restore sensitivity to the actomyosin system in the presence of tropomyosin is obscure. From the preliminary experiments carried out, it would appear that the formation of the complex between the two proteins in the presence of Ca^{++} is not the only requirement for the restoration of calcium sensitivity to the enzymic activity of actomyosin. Possibly the two components require reassociation in some subtle way which is facilitated by the presence of the 37,000 component, itself not essential for restoration of calcium sensitivity. Another explanation could be that during preparation some change occurs in the structure of the calcium-binding protein or of the inhibitory protein. This modification, although it does not destroy the ability of the two proteins to complex in the presence of Ca^{++}, may involve a change in the conformation of one of the proteins from the form that is required for restoration of calcium sensitivity to the actomyosin ATPase. Allosteric sites on proteins sometimes show similar instability and such a site may be involved in this system.

Amounts of Calcium-Binding and Inhibitory Proteins in the Myofibril

The evidence available indicates that the troponin complex is localized on the I filament, and that it acts by modifying in some way the interaction between actin and myosin which is responsible for the high level of ATP hydrolysis in the presence of Mg^{++}. Because every actin monomer is potentially a point of interaction with the myosin, each monomer must be subject to the influence of the troponin complex. Although the evidence from the yields obtained on isolation from whole muscle is not very precise, it suggests that there is not enough of the troponin complex present in the myofibril for one molecule each of the calcium-binding protein, inhibitory protein, and 37,000 component to be associated with every actin monomer. Indeed if one molecule of each of the three components were

present for every two actin monomers, about 15% of the total myofibrillar protein would consist of the troponin complex. Although a drastic revision of the analytical data for the myofibril might accommodate such a large contribution, it is unlikely that there is so much of the troponin complex present. Electron microscope investigations also indicate a period localization of part, at least, of the troponin complex along the I filament (Ebashi et al., 1969).

These facts suggest that some type of cooperative effect occurs in which the action of the troponin complex is transmitted to actin monomers themselves not in direct contact with the complex. Studies with isolated inhibitory protein indicate that the inhibitory action of one molecule of the inhibitory protein can be exerted over several actin monomers in the presence of tropomyosin. Thus the tropomyosin may act by transmitting the effect of a single molecule of the inhibitory protein to other actin monomers themselves probably not in direct contact with the inhibitory protein.

In such a scheme in which the effect of the inhibitory protein determines the state of activity of the myofibril, the calcium-binding protein localized in close proximity to the inhibitory protein could in turn regulate its action in response to changes in calcium concentration. The evidence available indicates that, unlike the activity of the inhibitory protein, the neutralizing activity of the calcium-binding protein is not significantly increased by the presence of tropomyosin. This supports the view that the mechanism involves control by the calcium-binding protein of the inhibitory protein, the effect of which is transmitted to a number of actin monomers.

To substantiate the indications from the in vitro experiments and as a first approach to this problem, we have determined the amounts of calcium-binding protein present in the myofibril, using the isotope dilution technique. This procedure is much facilitated in the case of the calcium-binding protein compared to the other proteins of the troponin complex because its electrophoretic behavior in the absence of Ca^{++} enables it to be readily separated from the whole muscle extracts on electrophoresis. A sample of pure calcium-binding protein was carboxyamidomethylated by a procedure similar to that used by Weeds and Lowey (1971) and shown to migrate as a single band with a mobility indistinguishable from that of untreated protein. A known amount of the labeled protein was then added to a sample of whole muscle that had been dispersed in urea. After equilibration, which was shown to be complete in less than 3 hr at room temperature, a pure sample of calcium-binding protein was isolated from the solution of the muscle in urea. Isolation was carried out by preparative electrophoresis in

6 M urea, Tris-glycine buffer pH 8.6. When the sample of muscle was applied to the gel in 4 mM EGTA, the calcium-binding protein ran as a fast band of pure protein, well separated from other muscle proteins, and was readily eluted free of contaminating proteins and could be rerun without change in specific activity. From the dilution of the radioactivity the average value for the content of the calcium-binding protein of rabbit psoas muscle is 0.67 ± 0.023 of the total protein nitrogen of muscle (Table 2). Assuming the myofibril represents 60% of the total protein nitrogen, this means that 1.1% of the myofibril is calcium-binding protein. If the myofibril contains 20% actin of mol wt 45,000, the ratio of moles of actin monomer to calcium-binding protein of mol wt 18,000 is 7.3:1.

Dr. Hirabayashi in our laboratory has shown, using the indirect staining method with an antibody raised in the guinea pig to a very pure sample of chicken calcium-binding protein, that in chicken skeletal muscle the calcium-binding protein is localized in the I band in a similar manner to the inhibitory protein and the 37,000 component. It follows from the analytical data that, if the calcium-binding protein is evenly distributed along the length of the I filament, there will be one molecule for every seven actin monomers. The evidence from the in vitro studies is that the calcium-binding protein does not act directly on the actin but regulates its function through the inhibitory protein. As yet precise figures for the amounts of the inhibitory protein in the myofibril are not available, but from the results of the enzymic and electrophoretic studies some estimate can be made as to the limits within which must fall its molar ratio with respect to the calcium-binding proteins.

The electrophoretic studies on mixtures of the inhibitory and calcium-binding proteins indicate

that, at pH 8.6 in 6 M urea, one molecule of calcium-binding protein is complexed with one molecule of inhibitory protein. When there is molar excess of calcium-binding protein, the latter protein appears as an electrophoretic component in addition to the complex. Whenever whole extracts of rabbit psoas were examined by the electrophoretic procedures described in the presence of Ca^{++}, the calcium-binding protein could not be detected as a band corresponding to the pure protein. It migrated as the complex with the inhibitory protein, and only when EGTA was added to the system did it appear as the band of high mobility characteristic of the calcium-binding protein. This observation taken in conjunction with the results of the electrophoretic studies of the complex suggests that in the myofibril, in the presence of Ca^{++}, there is always at least one molecule of inhibitory protein available to complex with each molecule of the calcium-binding protein.

In most troponin preparations the inhibitory protein was usually present in slight molar excess over the calcium-binding protein. This observation may be a consequence of selective extraction during preparation rather than actual differences in the relative amounts of the components in the myofibril. Nevertheless there are suggestions from the preliminary studies with the reconstituted systems that a plausible mechanism of function could be proposed which involves more than one molecule of inhibitory protein per molecule of calcium-binding protein.

For example, the neutralization experiments indicate that, at least under conditions where the inhibitory protein is present in excess, one calcium-binding protein molecule can neutralize the effect of more than one, possibly of two, inhibitory protein molecules. Further, if the stoichiometry of the action of the inhibitory protein alone on acto-HMM can be extended to the myofibril, at least two molecules would be required to be fully effective on seven actin monomers.

These speculations are open to the criticism that they do not include a role for the 37,000 component, which as several reports indicate, is able to render mixtures of the inhibitory and calcium-binding proteins more effective in restoring calcium sensitivity. Without more information about its biological function and until careful comparisons are available between efficacy of the inhibitory and calcium-binding protein mixture in the presence and absence of 37,000 component, it is difficult to formulate a role for this component. It is possible that its role in the calcium-sensitizing system may be a passive one, rendering the components more effective by facilitating attachment and localization within the contractile system. Whatever its role may be, the evidence presently available suggests that the

Table 2. Calcium-Binding Protein Content of Rabbit Psoas Muscle

Urea Conc. (M)	Equilibration Time (hr)	Calcium-Binding Protein (% total protein N)
6	3	0.70
6	12	0.66
6	24	0.63
6	72	0.68
6	12	0.66
8	12	0.69
8	12	0.67
		Av. 0.67 ± 0.023

Samples of rabbit psoas muscle (200–400 mg) were dispersed in the urea solution, a known amount of radioactive calcium-binding protein added and the total calcium-binding protein isolated by preparative electrophoresis of the whole muscle extract on 8% acrylamide, 25 mM Tris 80 mM in glycine pH 8.6, 4 mM EGTA, and urea as indicated. The specific radioactivity of the calcium-binding protein so isolated was determined.

37,000 component is not an essential requirement for restoration of some degree of calcium sensitivity to an actomyosin system.

Acknowledgments

The work was supported in part by research grants from the Medical Research Council and the Muscular Dystrophy Associations of America, Inc.

References

Coons, A. H. and M. H. Kaplan. 1950. Localization of antigen in tissue cells. II. Improvements in a method of the detection of antigen by means of fluorescent antibody. *J. Exp. Med.* **91**: 1.

Drabikowski, W., U. Rafalowska, R. Dabrowska, A. Szpacenko, and B. Barylko. 1971. The effect of proteolytic enzymes on the troponin complex. *FEBS Letters* **19**: 259.

Ebashi, S., M. Endo, and I. Ohtsuki. 1969. Control of muscle contraction. *Quart. Rev. Biophys.* **2**: 351.

Ebashi, S., T. Wakabayashi, and F. Ebashi. 1971. Troponin and its components. *J. Biochem. (Tokyo)* **69**: 441.

Eisenberg, E. and W. W. Kielley. 1972. Reconstitution of active troponin-tropomyosin complex in the absence of urea from its 4 column-purified components. *Fed. Proc.* **31**: 502.

Greaser, M. L. and J. Gergely. 1971. Reconstitution of troponin activity from 3 protein components. *J. Biol. Chem.* **246**: 4866.

Hartshorne, D. J. and H. Mueller. 1968. Fractionation of troponin into two distinct proteins. *Biochem. Biophys. Res. Comm.* **31**: 647.

Hartshorne, D. J., S. V. Perry, and M. C. Schaub. 1967. A protein factor inhibiting the magnesium-activated adenosine triphosphatase of desensitized actomyosin. *Biochem. J.* **104**: 907.

Murray, A. C. and C. M. Kay. 1971. Separation and characterization of the inhibitory factor of the troponin system. *Biochem. Biophys. Res. Comm.* **44**: 237.

Perry, S. V. 1967. Factors involved in the interaction of actin myosin and ATP. *Abstr. 7th Int. Cong. Biochem.*, Tokyo. p. 332.

————. 1972. Regulatory protein system of skeletal muscle. *PAABS Revista.* In press.

Perry, S. V., V. Davies, and D. Hayter. 1966. Natural tropomyosin and the factor sensitizing actomyosin adenosine triphosphatase to ethylenedioxy bis-(ethyleneamino) tetraacetic acid. *Biochem. J.* **99**: 1C.

Potter, J. and J. Gergely. 1972. Calcium binding of unfractionated and reconstituted troponin. *Fed. Proc.* **31**: 501 (Abstr.).

Schaub, M. C. and S. V. Perry. 1969. The relaxing protein system of striated muscle. Resolution of the troponin complex into inhibitory and calcium-sensitizing factors and their relationship to tropomyosin. *Biochem. J.* **115**: 993.

————. 1971. The regulatory proteins of the myofibril. Characterization and properties of the inhibitory factor (Troponin B). *Biochem. J.* **123**: 367.

Schaub, M. C., S. V. Perry, and W. Hacker. 1972. The regulatory proteins of the myofibril. Characterization and biological activity of the calcium-sensitizing factor (Troponin A). *Biochem. J.* **126**: 237.

Weeds, A. G. and S. Lowey. 1971. Substructure of the myosin molecule. II. The light chains of myosin. *J. Mol. Biol.* **61**: 701.

Wilkinson, J. M., S. V. Perry, H. A. Cole, and I. P. Trayer. 1971. Characterization of components of inhibitory factor (Troponin B) preparations of the myofibril. *Biochem. J.* **124**: 44P.

————. 1972. Separation and biological activity of the components of inhibitory-factor preparations. *Biochem. J.* **127**: 215.

Phosphorylation and Dephosphorylation of the Inhibitory Component of Troponin

J. T. STULL, P. J. ENGLAND, C. O. BROSTROM, AND E. G. KREBS

Department of Biological Chemistry, School of Medicine, University of California, Davis, California 95616

Protein phosphorylation and dephosphorylation is a mechanism by which the activities of a number of enzymes are known to be regulated in the cell. In particular, the enzymes of glycogen metabolism are in this category (see Krebs, 1972, for references). We have considered the possibility that muscle contraction is also influenced by such a mechanism, and present evidence for the phosphorylation of the inhibitory component of troponin (TN-I; Potter and Gergely, 1972) by phosphorylase kinase and its dephosphorylation by phosphorylase phosphatase.

Phosphorylase kinase catalyzes the conversion of phosphorylase b to the physiologically active a form in the presence of Ca^{++} and ATP-Mg^{++}. The formation of phosphorylase a then causes mobilization of stored carbohydrate to meet the real and potential energy needs of the muscle. Phosphorylase kinase in resting muscle is in the nonactivated form which has a high K_m for phosphorylase b at pH 7 or below. However, this form of the enzyme has sufficient catalytic activity in vivo for the rapid transformation of phosphorylase b to a during electrical stimulation or with low concentrations of catecholamines (Drummond et al., 1969; Stull and Mayer, 1971). When cyclic AMP formation is stimulated, phosphorylase kinase is converted to the activated form by a cyclic AMP-dependent protein kinase and this form has a low K_m for phosphorylase b at pH 7.0. Phosphorylase kinase has an absolute requirement for Ca^{++}, the activated form having a higher affinity, and it is thought that Ca^{++} may be the common agent linking muscle contraction to glycogen breakdown. Phosphorylase phosphatase catalyzes the formation of phosphorylase b from phosphorylase a. This enzyme has been studied in less detail, and various conflicting reports exist as to its control (see Krebs, 1972, for references).

The previous papers of this section have described in detail the composition and probable functions of troponin, and the reader is referred to these for detailed information. Briefly, troponin is composed of three subunits (Potter and Gergely, 1972): TN-I (mol wt 22,000), which inhibits actomyosin adenosine triphosphatase activity equally in the presence or absence of Ca^{++}; TN-C (mol wt 18,000), which binds Ca^{++} and blocks the inhibition by TN-I in the presence or absence of Ca^{++}; and

TN-T (mol wt 38,000), which restores the original Ca^{++}-sensitizing activity of troponin when added to TN-I and TN-C. In our investigations of phosphorylation and dephosphorylation we have used a troponin B preparation as described by Hartshorne and Mueller (1969). This contains both TN-I and TN-T; in the preparation used for these experiments TN-I accounted for 44% of the protein. Full experimental details are given elsewhere (Stull et al., 1972; England et al., 1972).

Figure 1 shows the incorporation of [^{32}P]phosphate into the protein of troponin B from [^{32}P]ATP in the presence of phosphorylase kinase. After 30 min of incubation the phosphate incorporated was 0.64 moles per 10^5 g protein. Troponin B, which had been incubated for 3 hr with phosphorylase kinase and [^{32}P]ATP, was subjected to electrophoresis on sodium dodecyl sulfate polyacrylamide gels, followed by radioautography of the gel (Fig. 2). The radioautogram demonstrated that the radioactivity was essentially associated with TN-I. Quantitation of the radioactivity in slices of the gel showed that 95% was associated with TN-I, and only 5% with TN-T. The total phosphate in troponin B before incubation with phosphorylase kinase was 1.7 moles

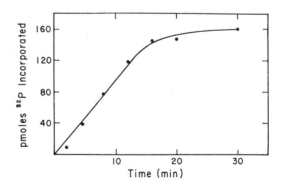

Figure 1. The phosphorylation of troponin B by phosphorylase kinase. Incorporation of phosphate into troponin B was measured in a reaction mixture containing: 50 mM Tris pH 8.5; 3 mM [γ-^{32}P]ATP; 10 mM magnesium acetate; 10 mM 2-mercaptoethanol; 3 μg/ml phosphorylase kinase; 0.5 mg/ml troponin B. The incubation was performed at 30°; 50 μl aliquots were removed at the indicated times and protein-bound ^{32}P determined. (Reproduced from Stull et al. (1972) with permission of the American Society of Biological Chemists, © 1972.)

263

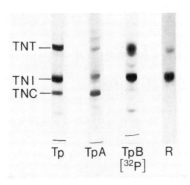

Figure 2. Analysis of troponin fractions by SDS polyacrylamide gel electrophoresis. Troponin (Tp), troponin A (TpA) and troponin B (TpB) were prepared as described by Hartshorne and Mueller (1969), and [^{32}P]troponin B as described in Fig. 1, except the pH was 7.8 and the incubation time 3 hr. Electrophoresis of 20 μg samples of protein was performed on 10% polyacrylmide gels in the presence of 0.1% SDS. R is the radioautogram of the [^{32}P]troponin B gel. (Reproduced from Stull et al. (1972) with permission of the American Society of Biological Chemists, © 1972.)

per 10^5 g protein. After additional phosphate (as ^{32}P) was incorporated with phosphorylase kinase, the total phosphate was 2.5 moles per 10^5 g protein. If it is assumed that all of the phosphate (both endogenous and that incorporated from [^{32}P]ATP) is present in TN-I, approximately 1 mole phosphate per mole TN-I is present.

Further evidence for the endogenous phosphate and that incorporated by phosphorylase kinase being on the same site in TN-I was obtained using phosphorylase phosphatase. Figure 3 shows that

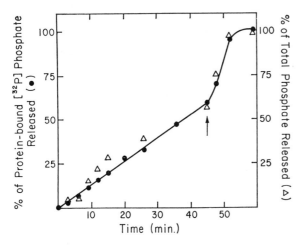

Figure 3. Time course of release of ^{32}P$_i$ and total phosphate from [^{32}P]troponin B by phosphorylase phosphatase. A solution of 2.1 mg/ml [^{32}P]troponin B in 0.2 M KCl, 0.1 M tris, 10 mM 2-mercaptoethanol, pH 7.5, was incubated with 10^{-4} units/ml phosphatase at 30°. Samples were taken at the times shown and acid-soluble ^{32}P$_i$ and total phosphate determined. The arrow indicates the addition of 10^{-3} units/ml phosphatase. (Reproduced from England et al. (1972) with permission of the American Society of Biological Chemists, © 1972.)

incubation of the phosphatase with troponin B, which had been maximally phosphorylated with [^{32}P]ATP and phosphorylase kinase, released both total phosphate and ^{32}P$_i$ at the same rate. All of the protein-bound phosphate was released and corresponded to 1 mole per mole TN-I. After dephosphorylation it was found that 1 mole phosphate per mole TN-I could be reincorporated by phosphorylase kinase. Phosphorylase phosphatase has high substrate specificity (Graves et al., 1960), and it is therefore very probable that both the endogenous phosphate and that incorporated by phosphorylase kinase are at the same site on TN-I.

If a physiological role for this phosphorylation and dephosphorylation is to be postulated, it was important to obtain estimates of the K_m and V_{max} values for the kinase and phosphatase when TN-I is a substrate. Comparison of these values with those when phosphorylase is a substrate enables the significance of the reactions in vivo to be evaluated. The K_m of nonactivated phosphorylase kinase for the phosphorylase b monomer at pH 7.0 is 370 μM, whereas with TN-I the K_m is 5 μM. The V_{max} is approximately 11-fold greater when phosphorylase b is the substrate. The activated kinase has a much lower K_m for phosphorylase b at pH 7.0, approximately 20 μM, whereas the K_m for TN-I is unchanged at 5 μM, this still being fourfold less than the K_m for phosphorylase b. For phosphorylase phosphatase the K_m for phosphorylase a is 6 μM, for TN-I it is 20 μM. The V_{max} is four times greater when TN-I is the substrate.

The low K_m values of phosphorylase kinase and phosphatase when TN-I is the substrate, coupled with the appreciable V_{max} rates obtained, give the potential in vivo for phospho-dephospho interconversions of TN-I in response to physiological stimuli. That such a transformation may occur is supported by isolation of troponin from skeletal muscle with protein-bound phosphate, and the incorporation of one mole phosphate per mole TN-I. Since phosphorylase kinase is a Ca^{++}-requiring enzyme, stimuli that produce Ca^{++} translocations in muscle may alter contractility primarily by stimulation of phosphorylase kinase activity with subsequent phosphorylation of TN-I. The functional significance of the phosphorylation and dephosphorylation of the inhibitory subunit of troponin is under current investigation.

Acknowledgments

This work was supported by grants from the Muscular Dystrophy Association of America, Inc. and the American Heart Association, and by Grant AM 12842 from the National Institutes of Health. J.T.S. holds a Damon Runyon Cancer Research

Fellowship. P.J.E. holds a Sir Henry Wellcome Travelling Fellowship of the Medical Research Council of Great Britain.

References

DRUMMOND, G. I., J. P. HARWOOD, and C. A. POWELL. 1969. Studies on the activation of phosphorylase in skeletal muscle by contraction and by epinephrine. *J. Biol. Chem.* **244**: 4235.

ENGLAND, P. J., J. T. STULL, and E. G. KREBS. 1972. Dephosphorylation of the inhibitor component of troponin by phosphorylase phosphatase. *J. Biol. Chem.* **247**: 5275.

GRAVES, D. J., E. H. FISCHER, and E. G. KREBS. 1960. Specificity studies on muscle phosphorylase phosphatase. *J. Biol. Chem.* **235**: 805.

HARTSHORNE, D. J. and H. MUELLER. 1969. The preparation of tropomyosin and troponin from natural actomyosin. *Biochim. Biophys. Acta* **175**: 301.

KREBS, E. G. 1972. Protein kinases. *Current Topics Cell. Reg.* **5**: 99.

POTTER, J. and J. GERGELY. 1972. Calcium binding of unfractionated and reconstituted troponin. *Fed. Proc.* **31**: 501.

STULL, J. T. and S. E. MAYER. 1971. Regulation of phosphorylase activation in skeletal muscle *in vivo*. *J. Biol. Chem.* **246**: 5716.

STULL, J. T., C. O. BROSTROM, and E. G. KREBS. 1972. Phosphorylation of the inhibitor component of troponin by phosphorylase kinase. *J. Biol. Chem.* **247**: 5272.

Manifestations of Cooperative Behavior in the Regulated Actin Filament during Actin-Activated ATP Hydrolysis in the Presence of Calcium

ROBERT D. BREMEL,* JOHN M. MURRAY,† AND ANNEMARIE WEBER†

Department of Biochemistry, St. Louis University School of Medicine, St. Louis, Missouri 63104

In the presence of sufficient ATP intact myofibrils respond to changes in the calcium concentration of the surrounding medium by contracting or relaxing. Troponin senses the level of calcium, and the actin molecules in the thin filament respond either by being "turned off," that is, by becoming inaccessible to ATP-activated myosin (Weber and Bremel, 1971) so that the actin and the myosin filaments remain dissociated from each other, or by being "turned on," that is, by permitting contraction to take place. The mechanism by which troponin in conjunction with tropomyosin controls the behavior of actin is not yet understood. We know, however, that the proteins of the regulated actin filament, i.e., containing troponin + tropomyosin, are assembled into repeating units (Ohtsuki et al., 1967; Ebashi et al., 1968a; O'Brien et al., 1971; Spudich, Huxley, and Finch, in press) consisting of seven actin monomers, one tropomyosin, and one troponin (Bremel and Weber, 1972). It is likely that these morphological units represent functional entities, as first suggested by Ebashi and his colleagues several years ago (1968b). If this is correct, one would expect protein-protein interactions within such units. Evidence for cooperation between proteins of the regulated actin filament does exist. In several instances manipulation of a single protein has been shown to alter the behavior of the complete filament. First, when calcium is removed from troponin (Fuchs and Briggs, 1968; Ebashi et al., 1968b) all of the actin monomers are "turned off" (Weber and Bremel, 1971; Bremel, 1972). Second, when a fraction, less than 50%, of the actin monomers are combined with myosin in rigor complexes[1] all of the monomers in the filament are "turned on" in the complete absence of calcium (Bremel and Weber, 1972; Bremel, 1972) and at the same time the affinity of troponin for calcium is increased.

We shall now describe some further manifestations of cooperative behavior within the actin filament. Apparently protein-protein interactions are revealed not only by transitions between the "on" and "off" state but also by modulations of the "on" state. In the following experiments, troponin was always saturated with calcium and the actin filaments were "on."

We used in our experiments subfragment 1 (S-1) (Lowey et al., 1969) instead of complete myosin molecules in order to insure that each molecule had an equal statistical chance to interact with any actin monomer in the filament. The regulated actin filaments were prepared by polymerizing pure monomeric actin (single band on SDS gel electrophoresis) with an excess of an extract containing the troponin-tropomyosin complex. When the polymerized filaments were separated from the mixture, they were found to be of fairly reproducible composition, containing one tropomyosin and one troponin for seven actin monomers; they appeared to be fully regulated since 90–96% of the actin molecules were turned off when calcium was removed (Bremel and Weber, 1972; Bremel, 1972). Actin-myosin interaction was evaluated by measuring the rate of actin-activated ATP hydrolysis. It is obtained by correcting the total rate of ATP hydrolysis for that of S-1 alone as determined in control experiments. In Fig. 1 the cofactor activity of regulated filaments is compared with that of pure actin. There is no evidence for cooperative behavior in filaments of pure actin: The kinetic experiments by Eisenberg and Moos (1970) indicate that rate constants for interaction between pure actin, myosin, and ATP remain constant under all conditions. In accordance, the rate of hydrolysis is a simple hyperbolic function of substrate concentration. By contrast, with regulated filaments as cofactor, the function becomes biphasic giving the appearance of substrate inhibition.[2] Furthermore, at optimal ATP the rate of ATP hydrolysis is considerably higher than the maximal velocity obtained with pure actin. Only in the range of

Present address: * Department of Anatomy, Duke University Medical Center, Durham, North Carolina; † University of Pennsylvania, Department of Biochemistry, Philadelphia, Pennsylvania 19104.

[1] Rigor complexes are formed from actin and myosin free of bound nucleotide. The association constant for this complex is high for regulated actin even in the complete absence of calcium.

[2] Eisenberg and Kielley (1970) observed a similar phenomenon when they found that the rate of ATP hydrolysis became markedly activated as the ATP was hydrolyzed, especially at high ratios of HMM to actin.

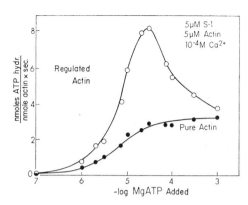

Figure 1. The effect of regulatory proteins on the cofactor activity of actin. ATPase activity determinations were made by measuring the rate of creatine liberation (cf. Weber, 1969) in a 2.0 ml incubation mixture containing the actin and S-1 concentrations indicated. The other constituents in the medium were 2 mg/ml creatine kinase; 5.0 mM creatine phosphate; 1 mM $MgCl_2$; 10 mM imidazole pH 7.0; 100 μM Ca^{++}; I = 0.035; 25°C. The reaction was initiated by addition of S-1 to the remainder of the incubation mixture.

apparent substrate inhibition, e.g., at 1.0 mM MgATP, does the rate of ATP hydrolysis with regulated actin as a cofactor approach that with pure actin.

Let us first consider the biphasic behavior. ATP does not bind to any component of the regulated filament; it binds only to S-1. All evidence indicates that S-1 molecules possess a single binding site for ATP, namely, the hydrolytic site (Morita, 1969; Maruyama and Weber, 1972; Kiely and Martonosi, 1968; Schliselfeld and Bárány, 1968; Lowey and Luck, 1969). This makes it very difficult to explain the biphasic behavior on the basis of interaction between two ATP binding sites. How, under these circumstances, can biphasic behavior result from monotonic changes in substrate saturation of S-1? For an explanation one must take into account the properties of nucleotide-free S-1 molecules. S-1 molecules not containing bound nucleotide have a very high affinity for actin with a K_D of 10^{-7} M (unpublished observations) or lower. Therefore, they are not free in solution but are combined with actin as rigor complexes. Thus at lower ATP concentrations and therefore higher concentrations of nucleotide-free S-1, there is a higher occupancy of actin filaments by rigor complexes. With increasing ATP and increasing substrate saturation of S-1, the concentration of rigor complexes falls, ultimately becoming zero at infinite ATP.

Taking these facts into consideration, it is possible to explain the biphasic behavior with regulated actin filaments as cofactor on the basis of a few simple assumptions: (1) In the presence of troponin + tropomyosin strong interactions between actin molecules are induced which are

absent in filaments of pure actin. (2) A functional unit within a filament responds in a cooperative manner to the presence of rigor complexes on a fraction of the actin monomers causing the cofactor activity of all actin molecules to be potentiated over that of pure actin. (3) Potentiation of a single functional unit is an all-or-none transition without intermediate stages; i.e., in the presence of calcium a functional unit is either in a potentiated or non-potentiated state. (4) Rigor complexes are randomly distributed[3] on the actin filaments. It follows that potentiation appears as soon as threshold occupancy of the filament with rigor complexes has been attained, i.e., a critical ratio of rigor complexes to total actin. Potentiation is not altered on further increasing this ratio, but cofactor activity of a functional unit switches to nonpotentiated as soon as the ratio falls below the threshold value.

With these assumptions, the biphasic behavior may be described in the following way. The rate of ATP hydrolysis increases with increasing saturation, as long as all functional units remain potentiated by sufficient rigor complexes. On further increase in substrate saturation the concentration of rigor complexes is so reduced that more and more units return to the nonpotentiated state because their occupancy by rigor complexes has become less than threshold. The rate of ATP hydrolysis declines when the increase in substrate saturation of the potentiated units no longer compensates for the fall in the number of the potentiated units. From this it follows that the cofactor activity of regulated actin should become equal to that of pure actin when most of the functional units are not combined with myosin.

This was tested by the experiment of Fig. 2. Under the conditions of this experiment the presence of the regulatory proteins depressed rather than potentiated cofactor activity, showing that troponin + tropomyosin are not simple activators of actomyosin ATPase activity as may have been concluded from Fig. 1 (cf. also Stewart and Levy, 1970; Sekiya and Tonomura, 1971); instead they can be inhibitory. Furthermore the curve with regulated actin is a simple hyperbolic function of ATP concentrations just as the curve with pure actin. However over the whole range of ATP concentrations the ATPase activity with regulated actin was nearly 50% lower than with pure actin. Experiment 1 and 2 differ by the molar ratio of actin monomers to S-1 (as well as absolute concentrations of each protein). It was 1:1 in

---

[3] Since it is likely that the cooperative effect also increases the formation of rigor complexes (see Fig. 4 for the mechanism of potentiation), one would expect two binomial distributions: one for potentiated and one for nonpotentiated units.

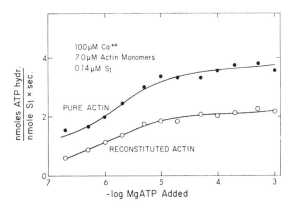

Figure 2. The absence of biphasic behavior at ratios of actin/S-1 of 50/1. ATPase activity measurements were made under essentially the same conditions as in Fig. 1.

experiment 1, whereas in the second experiment actin was present in considerable excess over S-1 (50:1). This means that even in the absence of ATP, i.e., with S-1 fully desaturated, only one unit (= seven monomers) in three units could possibly contain one rigor complex and an even smaller fraction of units more than one rigor complex. In other words, in the second experiment the possibility for significant cooperative interactions is very much lower than in the first. Therefore on the basis of our view on cooperativity, one would expect biphasic behavior and potentiation of the rate of ATP hydrolysis to be greatly diminished.

Figure 3 describes the effect of increasing the ratio S-1 to actin. Actin was kept constant at 2 μM, and the rate of hydrolysis as a function of ATP

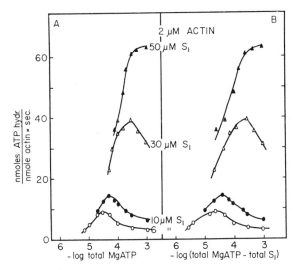

Figure 3. The shift in optimal MgATP concentration with increasing S-1 concentration at a constant actin concentration. ATPase activity measurements made as in Fig. 1.

concentration was determined for increasing concentrations of S-1 ranging from 6 to 50 μM; that is, a ratio of S-1 to actin from 3 to 25. Only regulated actin was used. The abscissa of Fig. 3A indicates the concentration of total ATP, i.e., the highest possible value for free ATP if none of it had been bound to S-1. The abscissa of Fig. 3B gives the lowest possible values for free ATP, assuming saturation of all ATP binding sites of S-1 at all ATP concentrations. The true concentration of free ATP lies between these two limiting values. Since S-1 was present in excess, actin-activated ATPase activity is expressed as nmoles ATP hydrolyzed per second per nmole actin. Figure 3 shows two effects of increasing S-1 concentration. First, over the whole range of ATP concentrations the rate of actin-activated ATP hydrolysis increased with S-1 concentration. This, of course, is not specific for regulated actin but also happens with pure actin as cofactor, as shown by Eisenberg and Moos (1970), because the steady state concentration of the actin-substrate-S-1 complex increases with S-1 concentration. Second, the optimal ATP concentration for ATPase activity was shifted to higher values; at 50 μM S-1, a declining phase of ATPase activity was not seen even at 1.0 mM MgATP (see below).

How do we interpret the shift of the ATP optimum? The ATP concentration defines, in addition to substrate saturation, the ratio of rigor complexes to total actin, at constant total S-1 (S_0). Assuming a random distribution of rigor complexes on the actin filament, the fractions of functional units containing one and more, up to seven, rigor complexes can be calculated for each ratio of rigor complexes to total actin. At ATP concentrations above optimal, a ratio of potentiated to nonpotentiated units has been attained where the decline in the number of potentiated units with increasing ATP is no longer compensated by increased saturation of each unit with enzyme-substrate complexes; therefore, the ATPase activity falls. The ATP concentration at which a given ratio of potentiated to nonpotentiated units (i.e., a given ratio of rigor complexes to total actin) is reached is shifted to higher values with increasing S-1 concentrations, independent of the pathway of rigor complex formation; that is, there is no qualitative difference with respect to the direction of the shift when rigor complexes (= AS) are formed primarily by either the reaction A + S → AS or the reaction A $\sim$ S-ADP-P → AS[4] (A = free actin; S = S-1 free of bound nucleotide;

---

[4] We realize that it has not yet been proved that the same complexes are formed by the two reactions. Nevertheless, the identity of the two complexes seems to us so very likely that we proceed on that assumption.

A $\sim$ S-ADP-P = enzyme-substrate-cofactor complex). We used the second reaction to derive the expression which shows this right shift (see appendix to this paper), since preliminary modeling, using the scheme of Lymn and Taylor (1971) and a combination of constants taken from their work and our own data suggest that at most ATP concentrations the second reaction prevails. In addition to the ratio potentiated to nonpotentiated units, the value for optimal ATP concentration also depends on the occupancy of the potentiated units with enzyme-substrate complex, i.e., the substrate saturation of the system S-1-potentiated actin. Substrate saturation is also shifted to higher ATP concentrations with increasing S-1 as shown by the expression given in the appendix.

It appears then that the observed right shift in ATP optimum would be expected on the basis of the postulated cooperative effects of rigor complexes.

So far the results of all three experiments confirm our view that a cooperative response was elicited by the formation of rigor complexes. What is the mechanism of the ensuing potentiation of actin cofactor activity? Is ATP hydrolyzed faster because the cofactor-enzyme-substrate complex is degraded more rapidly? Or is the apparent affinity of actin for S-ADP-P increased so that at finite S-1 concentrations a greater fraction of the total actin is shifted into the enzyme-substrate-cofactor complex, increasing its steady state concentration? In Fig. 4 the rate of actin-activated ATP hydrolysis is plotted as a function of increasing S-1 concentration, using a double-reciprocal plot and comparing the cofactor effect of pure and regulated actin. Cofactor activity was identical for both actins at infinite S-1 when the rate of ATP

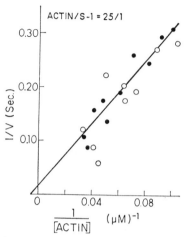

Figure 5. Double-reciprocal plot of the cofactor activity of pure ($\bullet$———$\bullet$) and regulated ($\circ$———$\circ$) actin filaments with the actin concentration in excess of S-1 at a constant ratio of 25/1. Determinations of ATPase activity as in Fig. 4.

hydrolysis was limited by the rate constant for degradation of the cofactor-substrate-enzyme complex because the formation of the complex is infinitely fast. Therefore the cooperative response does not change the rate of degradation of the cofactor-substrate-enzyme complex. It seems that, instead, it causes the apparent affinity of regulated actin for S-ADP-P to increase since maximal velocity was approached at lower S-1 concentrations than with pure actin as a cofactor. Whereas with pure actin, in confirmation of the experiments of Eisenberg and Moos (1970), the double-reciprocal plot was a straight line, it was curved with regulated actin. This indicates that, with increasing S-1 concentrations, the rate constants governing the formation and dissociation of the actin-substrate-S-1 complex were constantly changing in such a way—as indicated by the direction of the curvature—as to shift the equilibrium towards formation of this complex.

In the experiment described by Fig. 5, ATPase activity was measured as a function of increasing actin concentrations, comparing again the cofactor activity of regulated and pure actin. The ratio actin to S-1 was kept constant at 25/1 in order to minimize cooperative behavior induced by myosin binding to actin filaments. In accordance with our expectations, the double-reciprocal plot with regulated actin as a cofactor was linear and possibly superimposable over that with pure actin (the scattering of the points is still too great to be sure whether the lines superimpose).

In summary, the results of all five experiments agree with our view that rigor complexes induce a cooperative response which results in an increase of the apparent affinity of actin for S-ADP-P.

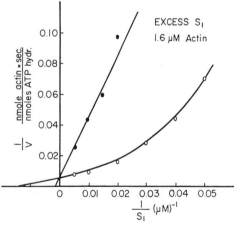

Figure 4. Double-reciprocal plot of an experiment comparing the cofactor activity of pure ($\bullet$———$\bullet$) and regulated ($\circ$———$\circ$) actin filaments with the S-1 concentration in excess of that of actin. ATPase activity was determined as in Fig. 1 except in this case the MgATP concentration was held constant at 1 mM.

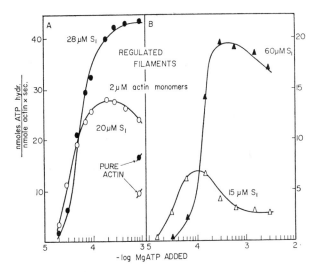

Figure 6. The disappearance of biphasic behavior at high concentrations of S$_1$. (A) and (B) were carried out with different protein preparations and at a slightly different ionic strength which accounts for the differences in activity. I(A) = 0.035; I(B) = 0.040; other conditions as in Fig. 1.

Figures 6A and B suggest that with high concentrations of S-1, potentiation may be maintained in spite of the continuous decline in the number of rigor complexes, presumably by cooperative interactions induced by enzyme-substrate-cofactor complexes. With 28 μM S-1 at low ionic strength and with 60 μM S-1 at higher ionic strength, ATPase activity does not fall off significantly after a maximal velocity has been reached (the decline in Fig. 6B is only 8% with a tenfold increase in ATP concentration), although at the ATP concentrations greater than necessary for maximal velocity the concentration of rigor complexes must have declined continuously to very low values. That the rate with regulated actin at plateau was much higher than with pure actin can be seen on Fig. 6A where the velocity with pure actin is indicated by one point at each highest ATP concentration. By comparison, with a lower concentration of S-1 the ATPase activity of regulated actin was biphasic, although with both S-1 concentrations the number of units occupied above threshold with rigor complexes must have become small at high ATP levels. However, if enzyme-substrate complexes should be capable of substituting for rigor complexes and also induce potentiation through cooperative interactions, the difference between the monotonic and biphasic curves in each figure may be explained: Since the S-1 concentration needed to saturate actin filaments with enzyme-substrate complexes (S-ADP-P) is much higher ($K_M = 10^{-4}$) than the S-1 concentration needed to saturate with rigor complexes ($K_D = 10^{-7}$), only 60 μM S-1, but not

15 μM S-1, may have been sufficiently high to permit threshold occupancy of all units with enzyme-substrate complexes. By contrast, potentiation of all units with rigor complexes (at the appropriately low ATP) was obtained at both S-1 concentrations. Therefore at 15 μM S-1 the number of potentiated units decreased at high ATP, giving rise to the biphasic curve; whereas with 60 μM S-1 rigor complexes were replaced by a sufficient number of enzyme-substrate complexes to keep the number of potentiated units maximal.

It does not seem implausible that both kinds of complexes with myosin induce similar cooperative behavior in regulated actin units. We conclude that systems containing regulated actin and S-1, HMM, or myosin exhibit biphasic behavior with respect to ATP concentration, that is, apparent substrate inhibition, when interactions with myosin-ADP-P are insufficient to produce as much potentiation as with rigor complexes. That happens when the S-1 or HMM concentrations are low (Weber and Bremel, 1971; Shigekawa and Tonomura, 1972) and it was observed with certain low-activity myosin preparations (Muir et al., 1971). With well-regulated myofibrils (95% relaxation) at ionic strength close to 0.1 M, apparently the potentiation by enzyme substrate complexes is equal to that by rigor complexes because we observed no substrate inhibition with skeletal myofibrils up to 10 mM MgATP (Bremel and Weber, unpublished observations). Differences in the occurrence of substrate inhibition with different actomyosin preparations may result from variations in the structural arrangement of the two filaments, i.e., the number of S-1 molecules capable of interacting with the same actin unit, and from variations in the rate constants governing the formation of enzyme-substrate-cofactor complexes.

Figures 7 and 8 indicate that cooperative behavior does not depend on the presence of troponin: It is induced by tropomyosin alone. For the experiment in Fig. 7 rabbit tropomyosin was used. Tropomyosin from chicken breast muscles was used for the experiment in Fig. 8. The tropomyosin from chicken showed only one band in gel electrophoresis in both SDS and urea systems as compared to two bands with rabbit tropomyosin. At low ATP, in the concentration range where rigor complexes are present, the rate of ATP hydrolysis was accelerated as with troponin + tropomyosin, and at high ATP the rate returned towards that with pure actin. Similar observations were made by Shigekawa and Tonomura (1972). Troponin modified the cooperative behavior induced by tropomyosin: The ATP optimum is shifted to higher ATP concentrations. Furthermore the ATPase activity may possibly remain more

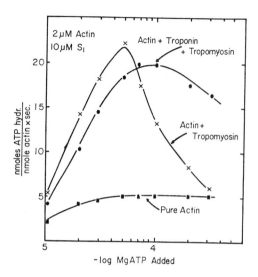

Figure 7. A comparison of the effects of tropomyosin with that of a combination of troponin and tropomyosin on the cofactor activity of actin. For this experiment rabbit tropomyosin (0.75 μM) was used that showed the typical two bands in SDS gel electrophoresis. Assay conditions as in Fig. 1.

potentiated at very high ATP concentrations when the concentration of rigor complexes is negligible and potentiation depends on enzyme-substrate complexes only.

Lehman and Szent-Györgyi (1972), using *Limulus* myosin as enzyme, found that tropomyosin

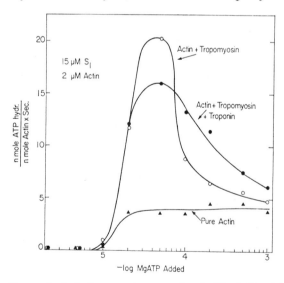

Figure 8. A similar experiment to that in Fig. 7. For this experiment actin filaments were mixed with 2 mg of chicken tropomyosin for each mg of actin and the complex collected after centrifugation at 105,000 g was used as cofactor. This tropomyosin preparation showed a single band in SDS and urea gel electrophoresis, contained no proline, no tryptophan (determined by J. Vanderkooi using phosphorescence analysis), and was approximately 100% α-helical as determined by optical rotatory dispersion. (Amino acid composition and ORD measurements were kindly performed by J. Morrissey.)

potentiated actin-activated ATP hydrolysis about fourfold even at high ATP concentrations where the concentration of rigor complexes was minimal. This is not observed with rabbit actomyosin (Shigekawa and Tonomura, 1972). By very neat experimentation, Lehmann and Szent-Györgyi were able to show that potentiation is due to an increase in the apparent affinity between tropomyosin-actin and the myosin-substrate complex. Their observations resemble our data with very high concentrations of rabbit S-1 and regulated actin. This leads us to suggest that in the presence of ATP an actomyosin system with *Limulus* myosin contains a higher steady state concentration of enzyme-substrate-cofactor complexes than a system with rabbit myosin, a concentration sufficiently high to initiate maximal cooperative interactions within the actin filaments. This increase in the steady state concentration of enzyme-substrate-cofactor complex may be the result of its much lower rate of degradation, since *Limulus* actomyosin hydrolyzes ATP at a maximal rate of 0.1 μmoles min^{-1} mg^{-1} as compared to a rate of 1.0 μmole min^{-1} mg^{-1} by rabbit myofibrils.

In conclusion, it seems that when actin is assembled together with tropomyosin and troponin into filaments, the actin monomers lose the ability to act independently from each other as they act in filaments of pure actin; they become cooperative in the regulated filament. As a result, the presence of sufficient rigor complexes on the filament causes all actin molecules to be "turned on" in a calcium-free medium or, with calcium present, to be potentiated in their cofactor activity. The same constants are affected in both cases. In the absence of calcium the ratio of the rate constants for association and dissociation of the enzyme-substrate-cofactor complex is altered greatly from a value favoring complete dissociation to one permitting about 50% association at 100 μM S-1 excess. In the presence of calcium the change in the value of the rate constants is smaller since the same final ratio is reached as in the absence of calcium, but the starting point is already a value favoring association.

It is not known how the proteins interact. We are inclined to speculate that actin molecules polymerized in the presence of tropomyosin are associated in a special manner, allowing the propagation of conformational changes from one active site of actin to that of the next molecule. By contrast, in the absence of tropomyosin conformational changes in one actin molecule cannot affect the active sites of neighboring molecules within the filament. In support one may cite previous observations indicating that tropomyosin modifies some interactions between actin molecules, as manifested by changes in extent and rate of polymerization (Grant, 1967; Pragay and Gergely, 1968) or the rigidity of the actin chain as measured by laser light scattering (Fujime and Ishiwata, 1971). These conformational changes probably involve troponin (since rigor complexes change its calcium affinity) and possibly tropomyosin (Tonomura et al., 1969).

Summary

1. When we compared the cofactor activity of pure and regulated actin in the presence of calcium, with S-1 as enzyme, we found regulated actin to be a more potent cofactor under the following conditions. First, cofactor activity of regulated actin was potentiated at low (but not at high) ATP concentrations when the concentration of S-1 was equal to that of actin. It declined again to that of pure actin on raising the ATP concentration to values above optimal. This means that under these conditions ATPase activity was a biphasic function of ATP concentration when regulated actin was the cofactor.

Cofactor activity of regulated actin was not potentiated even at low ATP when actin was present in large excess over S-1.

Second, cofactor activity of regulated actin was potentiated at high as well as low ATP concentrations when the concentration of S-1 was high, and in excess of actin, in the range of 30 to 60 μM, depending on ionic strength.

2. Our data indicate that the potentiation of cofactor activity is caused by cooperative effects induced in the regulated actin filament by rigor complexes at low ATP, and at high ATP by complexes with nucleotide-containing myosin, i.e., enzyme-substrate-cofactor complexes. Since the latter complexes have a much lower apparent association constant than the rigor complexes, potentiation by enzyme-substrate-cofactor complexes occurs only at fairly high S-1 concentrations. We assume that upon threshold occupancy by S-1 of a functional unit (i.e., 7 actin monomers controlled by 1 troponin and 1 tropomyosin) all actin molecules within the units are potentiated in their cofactor activity (threshold value is probably not more than 3–4 actin molecules occupied in a total of 7).

3. Potentiation was shown to be due to an alteration of the constants involved in the formation and/or dissociation of the enzyme-substrate-cofactor complex. The rate of degradation of this complex (= V_{max} at infinite S-1) was not significantly altered.

4. Cooperative interactions were induced in the presence of tropomyosin alone. Troponin was not necessary for cooperativity but modified it further.

Acknowledgments

This work was supported by NIH grants 9 R01 HE 13637-06 and 5-R01 HE 13637-07; training grant GM 44612 and clinical training grant HE 5672-07.

Appendix

The derivation of an expression showing the ratio [AS]/[A]$_0$ *as a function of* [S]$_0$ *and* [ATP]. Abbreviations: [AS], rigor complexes; [A]$_0$, total actin; [S]$_0$, total S-1; [A], free actin; [A $\sim$ S], enzyme-substrate-cofactor complex formed from actin and S-ADP-P; [S], nucleotide-free S-1; [S-ATP], complex between S-1 and ATP; [S-ADP-P] intermediate of S-1 formed from S-ATP according to Lymn and Taylor (1970, 1971), which interacts with actin.

We base the calculations on the scheme proposed by Lymn and Taylor (1971)

with the following simplifying assumptions. First, we assume that the concentration of S-ADP-P is essentially determined by the equilibrium reactions

$$S + ATP \underset{k_{-3}}{\overset{k_3}{\rightleftharpoons}} S\text{-}ATP \underset{k_{-6}}{\overset{k_6}{\rightleftharpoons}} S\text{-}ADP\text{-}P \quad (1)$$

and second, that AS is primarily formed by A $\sim$ S $\xrightarrow{k_5}$ AS.

According to the scheme above

$$k_1[ATP][AS] = k_5[A \sim S] \quad (2)$$

$$k_1[ATP][AS] = k_4[S\text{-}ADP\text{-}P][A] \quad (3)$$

$$[A]_0 = [AS] + [A] + [A \sim S] \quad (4)$$

Therefore we may write

$$[A]_0 = [AS]\left(1 + \frac{k_1[ATP]}{k_4[S\text{-}ADP\text{-}P]} + \frac{k_1[ATP]}{k_5}\right) \quad (5)$$

$$\frac{[AS]}{[A]_0} = \frac{1}{k_1[ATP]\left(\frac{1}{k_5} + \frac{1}{k_4} \times \frac{1}{[S\text{-}ADP\text{-}P]}\right) + 1} \quad (6)$$

[S-ADP-P] may be derived from expression (1) and, since S-1 was in considerable excess over actin by assuming

$$[S]_0 = [S] + [S\text{-}ATP] + [S\text{-}ADP\text{-}P] \quad (7)$$

In that case and from (1)

$$[\text{S-ATP}] = \frac{k_{-6}}{k_6} [\text{S-ADP-P}] \qquad (8)$$

$$[\text{S}] = \frac{k_{-3}[\text{S-ATP}]}{k_3[\text{ATP}]} \qquad (9)$$

Combining (7), (8), and (9) one obtains

$$\frac{1}{[\text{S-ADP-P}]} = \left(1 + \frac{k_{-6}}{k_6} + \frac{k_{-6}k_{-3}}{k_6 k_3 [\text{ATP}]}\right) \times \frac{1}{[\text{S}]_0} \qquad (10)$$

By defining $1 + \dfrac{k_{-6}}{k_6} = \alpha$ and $\dfrac{k_{-6}k_{-3}}{k_6 k_3} = \beta$, (10) is simplified to

$$\frac{1}{[\text{S-ADP-P}]} = \frac{1}{[\text{S}_0]} \times \left(\alpha + \frac{\beta}{[\text{ATP}]}\right) \qquad (11)$$

Substituting (11) into (6) we obtain

$$\frac{[\text{AS}]}{[\text{A}]_0} = \frac{1}{k_1[\text{ATP}]\left(\dfrac{1}{k_5} + \dfrac{\alpha}{k_4[\text{S}_0]} + \dfrac{\beta}{k_4[\text{S}_0][\text{ATP}]}\right) + 1} \qquad (12)$$

The derivation of an expression showing the ratio $[\text{A} \sim \text{S}]/[\text{A}]_0$ *as a function of* $[\text{S}]_0$ *and* $[\text{ATP}]$. Signs and assumptions are the same as above. Combining (2), (3), (4) we obtain

$$\frac{[\text{A} \sim \text{S}]}{[\text{A}]_0} = \frac{1}{1 + \dfrac{k_5}{k_1[\text{ATP}]} + \dfrac{k_5}{k_4[\text{S-ADP-P}]}} \qquad (13)$$

By substituting (11) into (13) we obtain

$$\frac{[\text{A} \sim \text{S}]}{[\text{A}]_0} = \frac{1}{\dfrac{k_5}{[\text{ATP}]}\left(\dfrac{\beta}{k_4[\text{S}_0]} + \dfrac{1}{k_1}\right) + 1 + \dfrac{k_5\alpha}{k_4[\text{S}_0]}}$$

which may be rewritten as

$$\frac{[\text{A} \sim \text{S}]}{[\text{A}]_0} = \frac{k_1 k_4 [\text{ATP}][\text{S}_0]}{k_1[\text{ATP}](k_4[\text{S}_0] + \alpha k_5) + k_4 k_5 [\text{S}_0] + k_1 k_5 \beta}$$

NOTE ADDED IN PROOF

Preliminary modeling, using a PDP 12 computer with a floating point processor and the program written by David J. Bates which numerically integrates differential (rate) equations and allows computation of final steady state values, confirmed the conclusions drawn from our calculations, namely, that the plateau at high S-1

concentrations results from the cooperative effect of actin-S-ADP-P complexes.

We should like to thank Dr. C. Frieden for the use of the computer and Mr. David Bates for giving us the program and his kind help.

References

BREMEL, R. D. 1972. Protein-protein interactions and the regulation of actomyosin activity. Ph.D. thesis, St. Louis University.

BREMEL, R. D. and A. WEBER. 1972. Cooperative behavior within the functional unit of the actin filament in vertebrate skeletal muscle. *Nature New Biology* **238**: 97.

EBASHI, S., M. ENDO, and I. OHTSUKI. 1968a. Control of muscle contraction. *Quart. Rev. Biophys.* **2**: 351.

EBASHI, S., A. KODAMA, and F. EBASHI. 1968b. Troponin. *J. Biochem.* (Tokyo) **64**: 465.

EISENBERG, E. and W. W. KIELLEY. 1970. Native tropomyosin: effect on the interaction of actin with heavy meromyosin and subfragment I. *Biochem. Biophys. Res. Comm.* **40**: 50.

EISENBERG, E. and C. MOOS. 1970. Actin activation of heavy meromyosin adenosine triphosphatase. *J. Biol. Chem.* **245**: 2451.

FUCHS, R. and F. N. BRIGGS. 1968. The site of calcium binding in relation to the activation of myofibrillar contraction. *J. Gen. Physiol.* **51**: 655.

FUJIME, S. and S. ISHIWATA. 1971. Dynamic study of F-actin by quasielastic scattering of laser light. *J. Mol. Biol.* **62**: 251.

GRANT, R. J. 1967. The reversibility of G-actin-ADP polymerization: Physical and chemical characterization. Ph.D. thesis, Columbia University, New York.

KIELY, B. and A. MARTONOSI. 1968. Kinetics and substrate binding of myosin adenosine triphosphatase. *J. Biol. Chem.* **243**: 2273.

LEHMAN, W. and A. G. SZENT-GYÖRGYI. 1972. Activation of the adenosine triphosphatase of *Limulus polyphemus* actomyosin by tropomyosin. *J. Gen. Physiol.* **59**: 375.

LOWEY, S. and S. M. LUCK. 1969. Equilibrium binding of adenosine diphosphate to myosin. *Biochemistry* **8**: 3195.

LOWEY, S., H. S. SLAYTER, A. G. WEEDS, and H. BAKER. 1969. Substructure of the myosin molecule. *J. Mol. Biol.* **42**: 1.

LYMN, R. W. and E. W. TAYLOR. 1970. Transient state phosphate production in the hydrolysis of nucleoside triphosphate. *Biochemistry* **9**: 2975.

———. 1971. Mechanism of adenosine triphosphate hydrolysis by actomyosin. *Biochemistry* **10**: 4617.

MARUYAMA, K. and A. WEBER. 1972. The binding of adenosine triphosphate to myofibrils during contraction and relaxation. *Biochemistry* **11**: 2990.

MORITA, F. 1969. Interaction of heavy meromyosin with substrate. *Biochim. Biophys. Acta* **172**: 319.

MUIR, J. R., A. WEBER, and R. E. OLSON. 1971. Cardiac myofibrillar ATPase: a comparison with that of fast skeletal actomyosin in its native and in an altered conformation. *Biochim. Biophys. Acta* **234**: 199.

O'BRIEN, E. J., P. M. BENNETT, and J. HANSON. 1971. Optical diffraction studies of myofibrillar structure. *Phil. Trans. Roy. Soc. London* B **261**: 201.

OHTSUKI, I., T. MASAKI, Y. NONOMURA, and S. EBASHI. 1967. Periodic distribution of troponin along the thin filament. *J. Biochem.* (Tokyo) **61**: 817.

PRAGAY, D. A. and J. GERGELY. 1968. Effect of tropomyosin on the polymerization of ATP-G-actin and ADP-G-actin. *Arch. Biochem. Biophys.* **125**: 727.

SCHLISELFELD, L. H. and M. BÁRÁNY. 1968. The binding of adenosine triphosphate to myosin. *Biochemistry* **7**: 3206.

SEKIYA, K. and Y. TONOMURA. 1971. Regulation of the acto-H-meromyosin-ATP system by calcium ion and treatment of H-meromyosin with p-chloromercuri-benzoate. *J. Biochem.* (Tokyo) **69**: 935.

SHIGEKAWA, M. and Y. TONOMURA. 1972. Activation of actomyosin-ATPase by tropomyosin. *J. Biochem.* (Tokyo) **71**: 147.

SPUDICH J. A., HUXLEY H. E. and FINCH, J. T. 1972. *J. Mol. Biol.* In press.

STEWART, J. M. and H. M. LEVY. 1970. The role of the calcium-troponin-tropomyosin complex in the activation of contraction. *J. Biol. Chem.* **245**: 5764.

TONOMURA, Y., S. WATANABE, and M. MORALES. 1969. Conformational changes in the molecular control of muscle contraction. *Biochemistry* **8**: 2171.

WEBER, A. 1969. Parallel response of myofibrillar contraction and relaxation to four different nucleoside triphosphates. *J. Gen. Physiol.* **53**: 781.

WEBER, A. and R. D. BREMEL. 1971. Regulation of contraction and relaxation in the myofibril. In *Contractility of muscle cells and related processes*, ed. R. J. Podolsky. Prentice-Hall, Englewood Cliffs, New Jersey.

Dynamic Property of F-Actin and Thin Filament

Fumio Oosawa, Satoru Fujime, Shin'ichi Ishiwata, and Koshin Mihashi

Institute of Molecular Biology and Department of Physics, Nagoya University, Nagoya, Japan

The purpose of this study is to investigate whether or not any change takes place in the structure of F-actin or the thin filament during contraction and relaxation in muscle. In most of the theories of muscle contraction the thin filament is treated as if it is a rigid rod on which myosin cross-bridges can walk (A. F. Huxley, 1957; H. E. Huxley, 1969). Regulation of the contraction of the striated muscle, however, is believed to be performed by the change of state of the thin filament induced by binding of calcium ions (Ebashi and Endo, 1968). It may also be probable that in the process of force generation or sliding a conformational change is caused in the thin filament by myosin charged by the energy of ATP (Oosawa et al., 1961).

The overall dynamic property of F-actin and the thin filament in solution has been studied by the method of the quasi-elastic light scattering under various environmental conditions corresponding to contraction and relaxation. The local structure of the filament has been studied by fluorescent probe. In this paper are briefly described the experimental results obtained to date which suggest a possible role of the conformational change of the thin filament in contraction and its regulation. Details of experimental procedures are found in several papers from this laboratory.

Flexibility of F-Actin

The quasi-elastic scattering of laser light in a solution of F-actin gives information about the dynamic property of F-actin. The optical homodyne method employed here for the spectral analysis of scattered light is convenient for the measurement of the frequency shift in the range of $10–10^3$ Hz; that is, the Brownian motion which produces the displacement of almost the order of the wavelength in 10^{-1} to 10^{-3} sec can be extracted.

Figure 1 is an example of the spectral density obtained in an F-actin solution (Fujime, 1970). The spectrum fits well with a single Lorentzian, of which the half height width Γ increased linearly with the square of the scattering vector K ($|K| = (4\pi/\lambda')\sin(\theta/2)$), where λ' is the wavelength in the solution and θ is the scattering angle, as shown in Fig. 2. The width can be written as

$$\Gamma = 2DK^2 + 1/\tau' \tag{1}$$

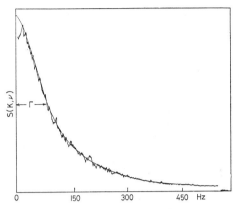

Figure 1. The spectral density of scattered light from a solution of F-actin. F-actin concentration, 3 mg/ml; solvent conditions, 0.1 M KCl pH 8.3 at room temperature; scattering angle θ, 60°.

The coefficient D must give the translational diffusion constant. This width Γ was independent of the concentration of F-actin from 1 to 5 mg/ml. The value of D was about 0.03×10^{-7} cm²/sec, which is consistent with the calculated diffusion constant of a filament of the length of a few μm. Therefore the constant width $1/\tau'$ in the above equation is considered to be due to the internal

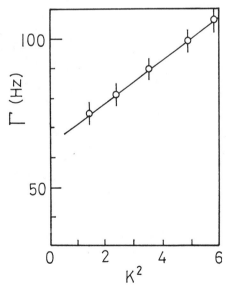

Figure 2. The relation between the half height width Γ and the square of the scattering vector K^2 ($\times 10^{10}$ cm^{-2}) in a solution of F-actin. Conditions the same as in Fig. 1

motion of a single F-actin filament. If F-actin is a semiflexible filament, its bending motion is expected to contribute to the frequency broadening. The dynamic property of a semiflexible filament is represented by the flexural rigidity ε or the statistical length λ^{-1}. Then, the relaxation time τ of the bending motion of the lowest mode is related to the parameters ε and λ by the equation

$$\varepsilon = 3kT/4\lambda \qquad (2)$$

$$\tau = (\tfrac{4}{3})(2/3\pi)^4(\Theta/kT)L^2(\lambda L) \qquad (3)$$

under the condition $\lambda L < 1$, where Θ is the frictional constant and L is the length of the filament. Since F-actin polymerized in vitro has a broad exponential length distribution (Oosawa and Kasai, 1962), the length L in the above equation must be its $(z + 1)$ average in the present homodyne method. The relaxation time τ is obtained from the experimentally determined value of τ' by the relation

$$1/\tau' = N/\tau \qquad (4)$$

where N is an even integer, which was tentatively made 4 in the case of long F-actin filaments. For F-actin of the average length of 2.5 μm, the value of τ' was found to be 9.0 msec, and $\lambda L \doteq 0.42$; $\lambda \doteq 0.17$ μm^{-1}.

The rigidity or the statistical length is an intrinsic property of the filament independent of the total length. The relation between the contour length L and the mean square of the end-to-end distance R is given by

$$R^2\rangle = [\exp{(-2\lambda L)} - 1 + 2\lambda L]/2\lambda^2 \qquad (5)$$

Electronmicrographs of F-actin negatively stained show that it assumes a curved shape. By measuring the end-to-end distance of a large number of F-actin filaments having different lengths in the electronmicroscopic field, Kawamura found that the relation between L and R fits very well in Eq. 5 with $\lambda = 0.17$ μm^{-1} (Fujime, 1970). This agreement between the result of spectral analysis and electronmicroscopic observation supports the interpretation that the frequency broadening of $1/\tau'$ is due to the bending motion of F-actin filaments in the solution.

The rotational relaxation time of F-actin around the minor axis is too long; the internal motion of the higher modes is too small and its relaxation time is too short to be detected in the present space-time range.

Thus F-actin is not a rigid rod but is regarded as a semiflexible filament. The mean end-to-end distance of F-actin of contour length 2.5 μm is

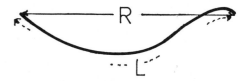

Figure 3. The average shape of a F-actin filament in solution. The mean end-to-end distance of F-actin of contour length 2.5 μm is about 2.2 μm.

about 2.2 μm and that of contour length 1.0 μm is about 0.9 μm (Fig. 3). The mean transverse amplitude is about 1300 Å and 400 Å, respectively (Ishiwata and Fujime, 1972). That is, the amplitude is of the same order as the distance between neighboring filaments in striated muscle.

The curved form of F-actin is a result of thermal motion. The extensibility of F-actin under tension K can be estimated by the relation

$$\partial R/\partial K = \langle \delta R^2 \rangle / kT \qquad (6)$$

where $\langle \delta R^2 \rangle$ is the mean square fluctuation of the end-to-end distance. The fluctuation of the end-to-end distance in electronmicrographs was of the same order as the difference between the contour length and the mean end-to-end distance (Kawamura; private communication). This means that the end-to-end distance is easily extended by a weak tension. Without load F-actin fluctuates to assume a curved form. However if a tension of the order of that developed in activated muscle is applied, F-actin is fully extended to a straight form.

The value of the elastic modulus for bending ε of F-actin determined above was about 1.7×10^{-17} g-cm^3 sec^{-2}, which corresponds to about one-hundredth of the elastic modulus of a steel wire of the same radius. The elastic modulus for stretching, Young's modulus Y, of a thin wire of radius a is related to its elastic modulus for bending through the equation

$$Y(\pi a^4/4) = \varepsilon \qquad (7)$$

If tension K is applied to stretch the wire, the change of its length ΔL is given by

$$\Delta L = (LK/Y)/\pi a^2$$

$$= LKa^2/4\varepsilon \qquad (8)$$

Under a tension of 1×10^{-5} dyne, which corresponds to the tension per thin filament in muscle at isometric contraction, the above equation gives $\Delta L = 3 \times 10^{-7}$ cm for $\varepsilon = 1.7 \times 10^{-17}$ g cm^3 sec^{-2}, $L = 10^{-4}$ cm, and $a = 15$ Å; and $\Delta L = 10^{-6}$ cm for the same values of ε and L and $a = 30$ Å. In the case of F-actin, which is a two-stranded helical polymer of globular protein molecules, Eqs. 7 and 8 have no theoretical basis. However the above calculation may give a clue for estimation of stretching of F-actin under tension.

Effect of Regulatory Proteins

Quasi-elastic scattering measurements have been made on solutions of the complex of F-actin with tropomyosin and troponin. With addition of tropomyosin to F-actin the half height width of the spectrum increased. The translational diffusion constant remained constant, while the relaxation time of the internal motion decreased. That is, the flexibility of F-actin decreased with increasing amount of bound tropomyosin (Fujime and Ishiwata, 1971). It attained a minimum at a weight ratio of tropomyosin to actin of from 1:6 to 1:3.

The addition of troponin to the complex of F-actin and tropomyosin decreased the translational diffusion constant because of formation of a loose network structure. However the addition of EGTA to decrease the concentration of free calcium ions increased the diffusion constant to the original value. Moreover it was found that when free calcium ions were almost completely removed by EGTA, the width $1/\tau'$ increased appreciably, suggesting a decrease of the flexibility of the complex of F-actin, tropomyosin, and troponin (Ishiwata and Fujime, 1972).

By using Ca-EGTA buffer, it was confirmed that with decreasing Ca ion concentration, the increase of the half height width independent of the scattering angle occurred at a micromolar range of the concentration. Figure 4 gives the half height width and the calculated mean transverse amplitude of the F-actin-tropomyosin-troponin complex of length 1.0 μm as a function of Ca ion concentration (Ishiwata and Fujime, 1972).

Both tropomyosin and troponin are necessary for this sensitivity of the flexibility of F-actin to Ca ions. The weight ratio of tropomyosin and troponin to actin of 1:1:6 is enough to induce the sensitivity. The Ca ion concentration where the flexibility change occurs just corresponds to its critical value for contraction or superprecipitation. The F-actin-tropomyosin-troponin complex is in a most rigid state at the condition of relaxation of muscle. Upon activation it is made more flexible by Ca ions. Ca ions bound to troponin induce some structural change in F-actin in such a way that the influence of tropomyosin and troponin on F-actin is removed.

By measuring the flow birefringence it was found by Tanaka (1972, private communication) that in a dilute solution of urea the binding between F-actin and tropomyosin is stronger in

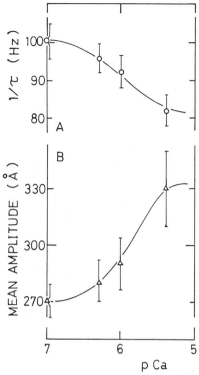

Figure 4. The relation between the free Ca^{++} concentration and the half height width $1/\tau'(A)$ or the mean transverse amplitude of F-actin of the length 1 μm (B). F-actin concentration, 3 mg/ml; the weight ratio of actin, tropomyosin, and troponin, 6:1:0.5; KCl, 0.1 M; Tris-maleate pH 6.8, 10 mM; at room temperature. The Ca^{++} concentration was adjusted by using a Ca-EGTA buffer.

the presence of troponin without Ca ions than with Ca ions. This is consistent with the above result.

If the thin filament in striated muscle is a straight rod and the distance between thin and thick filaments is larger than the length of the cross-bridge of myosin, not all of the cross-bridges can interact with the thin filament. Flexibility of the thin filament is needed for interaction. Recent experiments by Endo (1972) using a skinned muscle fiber showed that in micromolar or lower Ca ion concentrations the isometric tension is not always proportional to the length of the overlapping region of two filaments: With decreasing Ca ion concentration the tension at maximum overlap begins to drop below the tension at shorter overlap. This suggests that at low Ca ion concentrations the increase of the interfilament distance prevents the cross-bridge from interaction with the thin filament. Therefore it may be probable that the flexibility increase of the thin filament directly plays a part in the initiation of contraction.

In solution the most rigid state of F-actin or the thin filament can neither activate the Mg-ATPase activity of myosin nor cause superprecipitation in the millimolar range of Mg-ATP concentration. Some local change in the structure of actin molecules for interaction with myosin must be associated with the change of the overall dynamic property of the filament observed above. A change sensitive to Ca ions was detected in the ESR signal from labeled actin having tropomyosin and troponin (Tonomura et al., 1969). The fluorescent probe attached to actin also gives a similar result, as will be described later.

Interaction with Heavy Meromyosin in Absence of ATP

With increasing amount of heavy meromyosin added to F-actin, the straight line between the half height width of the spectrum Γ and the square of the scattering vector K^2 shifted to smaller values of the width without changing the slope. That is, the relaxation time of the internal motion increased without a change of the translational diffusion constant (Fujime and Ishiwata, 1971).

Heavy meromyosin molecules have various effects on F-actin solution. The network structure formed by heavy meromyosin molecules linked to two F-actin filaments causes a large increase of viscosity and a rubber-like elasticity of the solution. Such large aggregates, if existing, might give long relaxation times. The present experiment of quasi-elastic light scattering, however, picked up a motion having the relaxation time of the order of 10^{-1} to 10^{-3} sec and showed that the translational diffusion constant was not changed by heavy meromyosin. The diffusion constant of long filaments such as F-actin is mainly determined by their lengths, although bound heavy meromyosin makes F-actin thicker. Two quantities D and $1/\tau'$ in Eq. 1 must come from the movement of the same structure. Therefore it is reasonable to consider that the decrease of $1/\tau'$ implies an increase of the flexibility of F-actin.

This interpretation does not exclude formation of the network structure in the solution, but simply says that its contribution was out of the range measurable by the present method. An experiment carried out by using the heterodyne method supported the above interpretation (Fujime et al., 1972).

The flexibility of F-actin reaches a maximum at a molar ratio of heavy meromyosin to actin of about 1:6 and then decreases again to the initial value at saturation of bound heavy meromyosin on actin, as shown in Fig. 5 (Fujime and Ishiwata, 1971). The flexibility change is large; the mean end-to-end distance of F-actin of contour length 1.0 μm was about 0.9 μm without heavy meromyosin, but becomes about 0.8 μm with bound heavy meromyosin at the ratio 1:6.

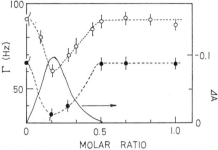

Figure 5. The relation between the half height width Γ and the molar ratio of heavy meromyosin to actin. F-actin, 3 mg/ml; KCl, 60 mM; pH 8.3 at room temperature. Scattering angle, 90° ($\bigcirc$) and 0° ($\bullet$). The solid line is the difference spectrum $\triangle$Å at 280 nm.

Heavy meromyosin seems to induce loosening of the polymer structure of F-actin. Such an effect of heavy meromyosin was previously suggested by Tawada by measuring flow birefringence and ultraviolet absorption (Tawada, 1969; Nakaoka, 1972). Parallel to the flexibility change, a change of ultraviolet absorption near 280 nm was brought about by heavy meromyosin in the same direction as in the case of depolymerization of F-actin (Higashi and Oosawa, 1965). Electric birefringence measurements also showed that a large dipole moment of F-actin perpendicular to the long axis is eliminated by heavy meromyosin at a molar ratio 1:6 (Kobayashi et al., 1964; Kobayashi, 1964).

The increase of flexibility of F-actin is not caused by binding of the subfragment of myosin S-1 (Fujime and Ishiwata, 1971), which has a single head for hydrolysis of ATP and binding with actin. The flexibility change requires the concerted interaction of two heads of heavy meromyosin with actin monomers in F-actin. A possible mechanism may be distortion in F-actin imposed by two heads-two monomers binding, owing to a small misfit between the distance of two neighboring monomers in F-actin and the distance of two heads of heavy meromyosin molecules (Fujime et al., 1972). The important function of the two heads-two monomers interaction has been suggested by using copolymers of native and modified actin monomers (Tawada and Oosawa, 1969a) and partially modified heavy meromyosin (Tokiwa and Morales, 1971).

In the case of F-actin having bound tropomyosin also, a similar change of the half height width Γ was observed with the addition of heavy meromyosin. In this case the maximum flexibility appeared at a molar ratio of heavy meromyosin to actin of about 1:2 (Fujime and Ishiwata, 1971).

As described previously F-actin with tropomyosin and troponin is in the most rigid state in the absence of calcium ions. This state is maintained even after the addition of heavy meromyosin.

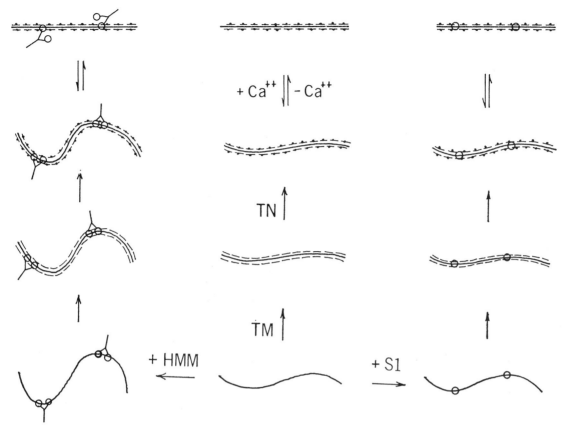

Figure 6. Illustration of various states of F-actin interacting with other proteins: (—) tropomyosin (TM); (●) troponin (TN); $\left(\begin{array}{c} \circ\,\,\,\,\circ \\ \diagdown\!\diagup \\ | \end{array} \right)$ heavy meromyosin (HMM); (○) subfragment S-1.

Under this condition heavy meromyosin can bind to F-actin but cannot change its flexibility. This depression by the tropomyosin-troponin system of the effect of heavy meromyosin, however, is completely released by a micromolar concentration of Ca ions. There is a small difference in the flexibility of the F-actin-tropomyosin-troponin complex in the absence and presence of calcium ions. This difference is greatly amplified by binding of heavy meromyosin (Ishiwata and Fujime, 1971). Heavy meromyosin acts as a negative tension on F-actin or the complex.

Figure 6 illustrates the experimental results described above.

Interaction with Heavy Meromyosin in Presence of ATP

When ATP is added to the complex of F-actin and heavy meromyosin, it is dissociated, and after hydrolysis of ATP by heavy meromyosin, the complex is formed again. Viscosity and turbidity of the solution first drops and later increases to the initial value or higher. Spectral analysis of scattered light has been carried out during this process. A high concentration of Mg-ATP was added and a solvent condition was chosen in order to sufficiently slow down the rate of ATP hydrolysis for frequency scanning at every measurement.

As shown in Fig. 7 where the molar ratio was 1:6, the spectrum at the beginning was similar to that of a simple solution of F-actin, suggesting complete dissociation of the two proteins. Then the concentration of ATP was gradually decreased by the ATPase. At about 4 to 3 millimolar Mg-ATP, suddenly a large decrease of the spectrum width occurred. This decrease was independent of the scattering angle; thus it is likely due to an increase of the relaxation time of the internal motion of the F-actin filament. After this decrease the width was kept constant until all the ATP was hydrolyzed.

Figure 8 shows the result obtained at a higher molar ratio of heavy meromyosin to actin (1:2). A drop of width Γ similar to the previous case was observed at about the same concentration of Mg-ATP. The final value of the width depended

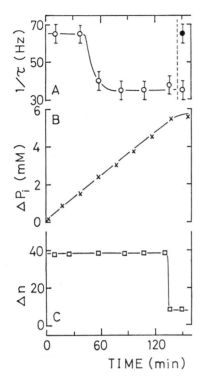

Figure 7. The change of the half height width at $\theta = 0°$, $1/\tau$ ($\bigcirc$); the inorganic phosphate liberation $\triangle P_i$ ($\times$); and the degree of flow birefringence $\triangle n$ ($\square$) after the addition of ATP to a solution of heavy meromyosin and F-actin at a molar ratio of about 1:6. F-actin, 0.5 mg/ml; heavy meromyosin, 0.5 mg/ml; KCl, 0.1 M; MgCl$_2$, 6 mM; Tris-HCl pH 8.0, 10 mM; at room temperature. The initial concentration of ATP was 6 mM. ($\bullet$ F-actin only)

on the molar ratio, as expected from the result in the absence of ATP.

A high concentration of pyrophosphate can dissociate heavy meromyosin and actin. However with decreasing concentration of pyrophosphate, the change of the spectrum occurred only once at a very low concentration of pyrophosphate (Fujime et al., 1972).

Under the solvent conditions in these experiments the rate of ATP hydrolysis was about one mole of ATP per mole of heavy meromyosin per sec. At the low molar ratios of heavy meromyosin to actin chosen here, in the presence of millimolar Mg-ATP, only a very small number of heavy meromyosin molecules are expected to be interacting with actin at one time. In the absence of ATP, such a small number of heavy meromyosin could not change the flexibility of F-actin. Therefore the observed sudden drop of width $1/\tau'$ requires heavy meromyosin molecules charged with the energy of ATP.

During this change of the spectrum no appreciable change has been found in other properties, such as turbidity, flow birefringence, and the rate

of ATP hydrolysis, as shown in Fig. 7. At present it is difficult to derive a definite interpretation of the decrease of the width $1/\tau'$. However it seems most probable that it means the increase of the flexibility of F-actin, or more generally, the increase of the relaxation time of F-actin to recover the average conformation from a deviated one. Such a deviation may be caused not only by thermal energy, but also by specific interaction with heavy meromyosin molecules storing the energy of ATP. Too high a concentration of Mg-ATP inhibits this specific interaction.

When heavy meromyosin is replaced by myosin, ATP induces superprecipitation, which has been considered to correspond to muscle contraction. The above experiment indicates that a remarkable change takes place in the structure of F-actin interacting with myosin charged with ATP in the clearing phase before superprecipitation. An attractive idea is that this structural change may be more directly connected with the mechanism of muscle contraction (Oosawa et al., 1961).

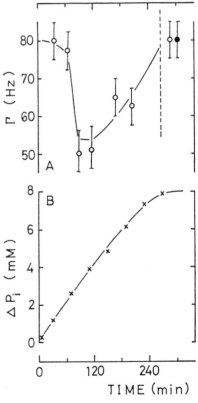

Figure 8. The change of the half height width Γ ($\bigcirc$) at $\theta = 60°$ and the inorganic phosphate liberation $\triangle P_i$ ($\times$) after the addition of ATP to a solution of heavy meromyosin and F-actin at a molar ratio of about 1:2. F-actin, 0.45 mg/ml; heavy meromyosin, 1.12 mg/ml; KCl, 0.1 M; MgCl$_2$, 8 mM; Tris-maleate pH 6.8, 20 mM; at room temperature. The initial concentration of ATP was 8 mM. ($\bullet$ F-actin only)

In a solution of myosin and the F-actin-tropo-myosin-troponin complex superprecipitation occurs soon after the addition of Mg-ATP in the presence of calcium ions. It is very much delayed by removal of free calcium ions by EGTA (Ebashi and Endo, 1968). Mg-ATP is necessary for super-precipitation, but its concentration must not be too high. This upper limit of the Mg-ATP concentration is very low in the absence of calcium ions.

When the F-actin-tropomyosin-troponin complex was used instead of F-actin, after the addition of heavy meromyosin and Mg-ATP, the spectrum width changed similarly to the case of Fig. 7 in the presence of calcium ions. On the other hand, when calcium ions were removed by EGTA, the width $1/\tau'$ stayed constant at a large value, corresponding to the most rigid state of F-actin, until the Mg-ATP concentration became very low. Just before ATP was consumed, a drop of the width at a low scattering angle was observed. Finally, it recovered to the original value, which is consistent with the result described in the previous section.

This result suggests that the structure change of F-actin or the thin filament interacting with heavy meromyosin having bound ATP, represented by the large drop of the spectrum width is an important phenomenon necessary for superpre-cipitation. There may be a close correlation between the concentration range of Mg-ATP for muscle contraction and that for this structural change of the thin filament.

Effect of Heavy Meromyosin on Actin Studied by Fluorescent Probe

A study complementary to the above has been developed by using a fluorescent probe attached to actin, which must be sensitive to the local structure of F-actin or the thin filament. The probe employed was fluorescein mercuric acetate (FMA) at a 1:1 or smaller ratio to actin. When the emission intensity was measured at a fixed wave-length of 525 nm, the excitation spectrum was found to depend on the state of actin, as shown in Fig. 9. A most remarkable change associated with polymerization of actin appeared at 505 nm, whereas the spectrum at shorter wavelengths did not change. When heavy meromyosin was added to F-actin having this probe, a change of the excitation spectrum occurred in the same range of wavelength. The ratio of the emission intensities at excitation at 505 nm and at 470 nm has been adopted as the best indicator of this change of fluorescence.

With increasing amount of heavy meromyosin bound to actin, the value of this ratio begins to increase at a very low molar ratio, and the change

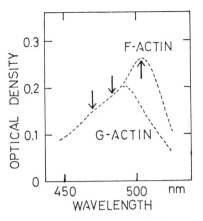

Figure 9. Absorption spectra of fluorescein mercuric acetate (FMA) bound to F-actin and to G-actin. FMA was bound to F-actin in the presence of 0.1 M KCl, 1 mM MgCl₂, and 20 mM Tris-acetate pH 7.0 at a molar ratio to actin equal to 0.6. G-actin having FMA was obtained by dialysis against cold distilled water containing 100 μM ATP and 0.5 mM bicarbonate. Concentration of F-actin and G-actin was 1.5 mg/ml. Solvent conditions: KCl, 0.1 M; MgCl₂, 1 mM; bicarbonate, 0.5 mM for F-actin; ATP, 100 μM; bicarbonate, 0.5 mM for G-actin. The excitation spectra of FMA bound to F-actin and G-actin where the emission was measured at 525 nm were approximately proportional to the optical densities. Arrows indicate the wavelengths 470 nm, 485 nm, and 505 nm.

is saturated at a molar ratio very much smaller than 1:2. Further addition of heavy meromyosin did not change the fluorescence. Data in Fig. 10 were replotted in Fig. 11 as a relation between the logarithm of the molar ratio and the relative magnitude of the fluorescence change. This result suggests that actin molecules in F-actin make a transition between two states under the influence of bound heavy meromyosin, and each heavy meromyosin molecule extends its influence over a large number of actin molecules.

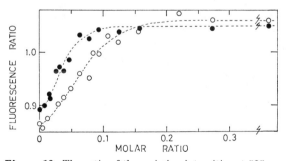

Figure 10. The ratio of the emission intensities at 525 nm of FMA bound to F-actin with excitation at 505 nm and 470 nm as a function of the molar ratio of heavy mero-myosin to F-actin. F-actin, 0.1 mg/ml; KCl, 0.1 M; MgCl₂, 1 mM; Tris-acetate pH 7.0, 20 mM; at room temperature. (○) Pure F-actin; (●) F-actin decorated with tropomyosin and troponin at a weight ratio of about 4:1:1. F-actin having tightly bound magnesium ions was used in these experiments.

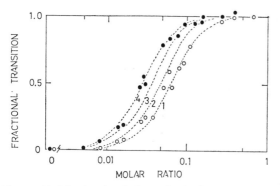

Figure 11. The fractional change in the fluorescence ratio as a function of the molar ratio of heavy meromyosin to F-actin. (○) Pure F-actin; (●) F-actin decorated with tropomyosin and troponin. The dotted lines were theoretical curves calculated on the basis of the model proposed in the text. The parameter a assumed was 25 for curve 1, 35 for curve 2, 45 for curve 3, and 60 for curve 4. The parameter n was 2 for all curves.

In order to analyze the above data it is assumed here that a certain length of F-actin containing a actin molecules is a unit to make transition, and binding of n heavy meromyosin molecules to this unit is required for the transition. In other words, if an arbitrarily chosen length of F-actin composed of a actin molecules has bound more than n heavy meromyosin molecules, all of these actin molecules are assumed to be in the second state, giving the larger value of fluorescence. Calculation showed that the curve of pure F-actin in Fig. 11 can be explained very well under the assumption that $n = 2$ and $a = 25$, over the whole range of the molar ratio. The approximate values of a and n can be estimated simply from the slope of the curve and the value of the molar ratio where half of the actin molecules finished the transition.

Thus the fluorescence study succeeded in showing that a very small number of heavy meromyosin molecules can induce a cooperative change of state of all actin molecules in F-actin or the thin filament. Extension of the effect of bound heavy meromyosin to several actin molecules was suggested previously by the ESR study (Cooke and Morales, 1971). The present experiment shows that the effect is of extremely long range.

A similar behavior of the fluorescence from the probe was observed with the addition of heavy meromyosin to F-actin decorated by tropomyosin and troponin in the presence of calcium ions. In this case, as shown in Fig. 11, the fluorescence increase took place at lower molar ratios than in the case of pure F-actin. That is, a larger value of a (about 50) was obtained although the value of n was the same. The effect of bound heavy meromyosin extends over a longer range in the F-actin-tropomyosin-troponin complex. The cooperativity

was enhanced by tropomyosin (and troponin) on F-actin.

When calcium ions were removed by EGTA, the fluorescence from the probe on the F-actin-tropomyosin-troponin complex was not changed by heavy meromyosin. Here again it was confirmed that F-actin having bound tropomyosin and troponin in the absence of calcium ions is in an insensitive state to heavy meromyosin. The observed fluorescence may reflect a local structure of actin important for contraction and its regulation.

The effect of ATP on the fluorescence from the same probe has been investigated in solutions of heavy meromyosin and F-actin or the F-actin-tropomyosin-troponin complex under various conditions. According to preliminary experiments the degree of fluorescence depolarization increased upon the addition of ATP. This increase was not observed in the case of the F-actin-tropomyosin-troponin complex in the absence of calcium ions.

Discussion

All of these experimental results suggest a structural change of F-actin or the thin filament not only in the regulation of muscle contraction, but also during the sliding or force-generation process in muscle.

The thin filament is in a rigid state under the condition of relaxation and is made flexible by binding of calcium ions to troponin. Then, with binding of heavy meromyosin, a specific structure is prepared in each actin molecule in F-actin, and later a remarkable change occurs in the overall dynamic property of F-actin. The polymer structure is distorted or loosened by heavy meromyosin.

In the presence of ATP heavy meromyosin is not permanently bound to F-actin but is in a cycle of binding and dissociation. In the experimental condition adopted here the rate of binding is rather slow. Nevertheless a large change was induced in the structure of F-actin by a small amount of heavy meromyosin. Heavy meromyosin molecules charged by ATP (ADP and inorganic phosphate) left a strong after effect on F-actin. Such a specific action of the heavy meromyosin plus ATP system to regulate the actin-actin binding has been suggested previously by Tawada and Oosawa (1969b). It is to be noted here that in activated muscle only a small fraction of cross-bridges seems to be in the state of binding with actin, according to recent structural studies.

Various experiments have been undertaken to determine whether there is any condition where F-actin undergoes a conformational change. The increase of the rate of exchange of nucleotides and divalent cations bound to F-actin has been an indicator of conformational change. In the case of pure F-actin sonic vibration or high temperature was found to increase the exchange of ADP and the accompanying splitting of ATP (Asakura et al., 1963; Asai and Tawada, 1966). In a mixture of F-actin and myosin a large and fast exchange of nucleotides and divalent cations was found under the conditions of superprecipitation (Szent-Györgyi and Prior, 1966; Kasai and Oosawa, 1969). Later experiments, however, have not always proved a close connection between this exchange and superprecipitation or contraction. Thus not enough attention has been paid to the possibility of a conformational change of F-actin in muscle contraction.

The experimental result reported here shows that this possibility must be reexamined. Structural analysis of the change observed in quasi-elastic light scattering and fluorescence must be made by various methods. Careful design of experimental conditions is needed to approximate in vivo conditions. The fluorescence study suggests that the dynamic property of F-actin or the thin filament may depend on the species of divalent cations bound to actin. The thin filament in muscle has magnesium ions instead of calcium ions (Kasai, 1969). It is also useful to apply the experimental technique used here directly to the muscle fiber. Another important problem is to compare the result on muscle actin with that on non-muscle actin. In the case of actin from plasmodium, its polymer was found to undergo a large conformational change depending on the concentration of ATP and divalent cations (Hatano, 1972; Fujime and Hatano, 1972). In cells other than muscle, the range of the conformational change of F-actin or the thin filament may be larger than in striated muscle.

References

ASAI, H. and K. TAWADA. 1966. Enzymic nature of F-actin at high temperature. *J. Mol. Biol.* **20**: 403.

ASAKURA, S., M. TANIGUCHI, and F. OOSAWA. 1963. Mechanochemistry of F-actin. *J. Mol. Biol.* **7**: 55.

COOKE, R. and M. MORALES. 1971. Interaction of globular actin with myosin subfragments. *J. Mol. Biol.* **60**: 249.

EBASHI, S. and M. ENDO. 1968. Calcium ions and muscle contractions. *Prog. Biophys. Mol. Biol.* **18**: 123.

ENDO, M. 1972. Stretch-induced increase in activation of skinned muscle fibres by calcium. *Nature* **237**: 211.

FUJIME, S. 1970. Quasi-elastic light scattering from solutions of macromolecules. II. Doppler broadening of light scattered from solutions of semi-flexible polymers, F-actin. *J. Phys. Soc. Japan* **29**: 751.

FUJIME, S. and S. HATANO. 1972. Plasmodium actin polymers studied by quasielastic scattering of laser light. *J. Mechanochem. Cell Motil.* **1**: 81.

FUJIME, S. and S. ISHIWATA. 1971. Dynamic study of F-actin by quasielastic scattering of laser light. *J. Mol. Biol.* **62**: 251.

FUJIME, S., S. ISHIWATA, and T. MAEDA. 1972. F-actin-heavy meromyosin complex studied by optical homodyne and heterodyne methods. *Biochim. Biophys. Acta* In press.

HATANO, S. 1972. Conformational changes of plasmodium actin polymers formed in the presence of Mg^{2+}. *J. Mechanochem. Cell Motil.* **1**: 75.

HIGASHI, S. and F. OOSAWA. 1965. Conformational changes in actin molecules associated with polymerization and nucleotide binding. *J. Mol. Biol.* **12**: 843.

HUXLEY, A. F. 1957. Muscle structure and theories of contraction. *Prog. Biophys.* **7**: 255.

HUXLEY, H. E. 1969. The mechanism of muscular contraction. *Science* **164**: 1356.

ISHIWATA, S. and S. FUJIME. 1971. The effect of Ca^{2+} on dynamic properties of muscle proteins studied by quasielastic light scattering. *J. Phys. Soc. Japan* **31**: 1601.

————. 1972. Effect of calcium ions on the flexibility of reconstituted thin filaments of muscle studied by quasielastic scattering of laser light. *J. Mol. Biol.* **68**: 511.

KASAI, M. 1969. The divalent cations bound to actin and thin filament. *Biochim. Biophys. Acta* **172**: 171.

KASAI, M. and F. OOSAWA. 1969. Behavior of divalent cations and nucleotides bound to F-actin. *Biochim. Biophys. Acta* **172**: 300.

KOBAYASHI, S. 1964. Effect of electric field on F-actin oriented by flow. *Biochim. Biophys. Acta* **88**: 541.

KOBAYASHI, S., H. ASAI, and F. OOSAWA. 1964. Electric birefringence of actin. *Biochim. Biophys. Acta* **88**: 528.

NAKAOKA, Y. 1972. Effects of tropomyosin, troponin and Ca^{2+} on the interaction between F-actin and H-meromyosin. *Biochim. Biophys. Acta* **267**: 558.

OOSAWA, F. and M. KASAI. 1962. A theory of linear and helical aggregation of macromolecules. *J. Mol. Biol.* **4**: 10.

OOSAWA, F., S. ASAKURA, and T. OOI. 1961. Physical chemistry of muscle protein "actin". *Prog. Theoret. Phys. (Kyoto)* (Suppl.) **17**: 14.

SZENT-GYÖRGYI, A. G. and G. PRIOR. 1966. Exchange of ADP bound to actin in superprecipitated actomyosin and contracted myofibrils. *J. Mol. Biol.* **15**: 515.

TANAKA, H. 1972. The effect of troponin on the binding of tropomyosin and F-actin. *Biochim. Biophys. Acta.* In press.

TAWADA, K. 1969. Physicochemical studies of F-actin-heavy meromyosin solutions. *Biochim. Biophys. Acta* **172**: 311.

TAWADA, K. and F. OOSAWA. 1969a. Activation of H-meromyosin ATPase by polymers of actin and carboxymethylated actin. *J. Mol. Biol.* **44**: 309.

————. 1969b. Effect of the H-meromyosin *plus* ATP system on F-actin. *Biochim. Biophys. Acta* **180**: 199.

TOKIWA, T. and M. F. MORALES. 1971. Independent and cooperative reactions of myosin heads with F-actin in the presence of ATP. *Biochemistry* **10**: 1722.

TONOMURA, Y., S. WATANABE, and M. MORALES. 1969. Conformational changes in molecular control of muscle contraction. *Biochemistry* **8**: 2171.

Tropomyosin-Troponin Assembly

C. COHEN, D. L. D. CASPAR, J. P. JOHNSON, K. NAUSS, S. S. MARGOSSIAN, D. A. D. PARRY

Laboratory of Structural Molecular Biology, Children's Cancer Research Foundation, Boston, Mass. 02115
and the
Rosenstiel Basic Medical Sciences Research Center, Brandeis University, Waltham, Mass. 01254

Tropomyosin crystals illustrate dynamic interactions of this molecule in muscle (Cohen et al., 1971). The very properties which make these crystals biologically interesting, however, limit what can be seen by X-ray crystallography. The 400 Å long tropomyosin molecules are firmly bonded end-to-end to form an open mesh of cross-connected supercoiled filaments (Caspar et al., 1969). These filaments occupy only about 5% of the volume of the crystal, and their flexibility results in disorder of the crystal lattice. Moreover, the bonds which cross-connect the filaments are weak, and slight rotations at these bonds produce large changes in lattice dimensions. So far we have only been able to visualize the molecules in one projection to a resolution of about 20 Å. We can see changes in intermolecular interactions without being able to distinguish details of molecular conformation.

Poor as these crystals are, they provide images of the tropomyosin filaments which are clearer than those obtained from muscle. The tropomyosin filaments in the crystal do correspond to those which are coiled in the grooves of the actin helix in muscle. Tropomyosin is the structural link which positions the troponin every 400 Å along the thin filaments. Troponin can bind to tropomyosin in the crystals as in the thin filaments: The binding to the crystal produces striking changes in cell dimensions which are due to small changes in the cross-connections of the filaments. Images of these crystals reveal the location and form of the troponin. The molecular rearrangements of tropomyosin in the crystal lattice produced by troponin may illustrate motions of tropomyosin in the thin filaments of muscle.

Chemistry of Tropomyosin and Troponin

The structures formed by tropomyosin depend on the composition of the preparations. The purification is complicated by the heterogeneity of the molecule and the tight binding of the troponin components to the tropomyosin. This molecular heterogeneity also adds to the problems in the crystallography. In any case, knowledge of the chemical composition is essential for interpreting the structures formed by tropomyosin and troponin.

Rabbit skeletal tropomyosin is heterogeneous (Cummins and Perry, 1972; Hodges and Smillie, 1972a, b). The SDS gel patterns show two bands: one of mol wt about 37,000, the other of mol wt about 33,000 in proportions of about one to three, respectively. The molecule is a two-chain α-helical coiled-coil of mol wt about 65,000. The SDS gels show that there are at least two chemically distinct polypeptide chains in the preparations from rabbit skeletal muscle. There are three cysteine residues for a weight of 65,000 (Table 1), and Cummins and Perry (1972) have reported that the subunit present in the smaller amount has twice the SH content as the other. The analysis of the peptide fragments (Hodges and Smillie, 1972a, b) suggests that the two polypeptide chains of each kind of molecule may have similar but nonidentical sequences.

The chemical heterogeneity of rabbit tropomyosin may be related to the presence of both red and white fibers in this skeletal muscle. Two components have also been observed in the tropomyosins from cow, sheep, and pig skeletal muscles (Cummins and Perry, 1972). We have observed only one band on SDS gels of tropomyosin from chicken breast muscle; however, there are 2.5 SH groups per 65,000 mol wt (Table 1), indicating a heterogeneity of the tropomyosin from this muscle which also has mixed fiber types. In contrast, fish skeletal muscle has a very low proportion of red fibers. Carp muscle tropomyosin may be homogeneous since it shows a single band on SDS gels and has two cysteines per 65,000 mol wt (Table 1). Rabbit and veal heart tropomyosins also show a single band on SDS gels, but veal heart tropomyosin appears to have 2.5 SH groups per 65,000 mol wt. Our observations and those of Cummins and Perry (1972) suggest that chemically homogeneous tropomyosins may be obtained from muscles with a single fiber type. Moreover measurement of an even number of SH groups per molecule may be an indication of a homogeneous preparation with similar subunits.

These chemical results are consistent with the picture of tropomyosin as a two-stranded coiled-coil molecule (Crick, 1953). The sequences recently reported by Hodges and Smillie (1972a) from rabbit tropomyosin show that nonpolar side chains often

Table 1. Amino Acid Compositions of Tropomyosins Isolated from Muscle

	Moles AA/10⁵ g				
	Veal Cardiac	Carp Skeletal	Chicken Skeletal	Rabbit Skeletal	Rabbit Skeletal[g]
Lys	116.3 ± 2.9	129.6 ± 2.2	119.3 ± 4.4	119.0 ± 3.3[a]	116
His	7.0 ± 1.1	7.4 ± 0.5	8.5[e]	5.8 ± 0.8	5.7
Arg	44.6 ± 2.3	45.2 ± 1.5	42.6 ± 4.8	40.1 ± 5.6	42
Asp	89.5 ± 1.6	88.9 ± 1.0	86.2 ± 1.2	89.3 ± 2.3	89
Thr[b]	24.1	30.2	26.2	23.5	24
Ser[b]	43.9	31.4	37.6	39.0	39
Glu	229.1 ± 2.4	213.6 ± 5.4	226.2 ± 0.7	226.6 ± 4.1	216
Pro	0	0.1	0	0	1.4
Gly	9.0 ± 0.7	14.4 ± 0.6	10.6 ± 0.3	10.9 ± 0.3	11
Ala	110.4 ± 2.7	103.6 ± 2.0	105.8 ± 1.9	108.9 ± 2.8	106
Val	25.8 ± 1.5	32.4 ± 3.1	25.6 ± 2.1	30.0[c]	30
Met	16.5 ± 1.7	19.0 ± 0.5	20.1 ± 0.4	21.6 ± 1.8	19
Ile	29.4 ± 2.2	26.6 ± 0.6	32.3 ± 1.4	34.8[c]	34
Leu	100.4 ± 1.0	104.9 ± 2.4	101.3 ± 1.8	101.0 ± 2.7	96
Tyr	15.0 ± 0.9	16.5 ± 1.1	16.4 ± 1.3	16.5 ± 0.8	16
Phe	3.4 ± 0.3	2.7 ± 0.4	2.6[e]	3.0[e]	4.3
Cys[d]	3.5 ± 0.5	3.2	3.9	4.7	4.1
Trp[f]	—	0.2	<1.0	0.2	—

[a] Unless otherwise indicated, all values are duplicate analyses of 24, 48, and 72 hr hydrolysates and are expressed ± SD.
[b] Serine and threonine values extrapolated to zero times of hydrolysis.
[c] Value of 72 hr hydrolysate.
[d] Cysteine determined as cysteic acid according to the method of Hirs (1967). Average of two determinations.
[e] Value of 24 hr hydrolysate.
[f] Determined by the method of Beaven and Holiday (1952).
[g] Determined by Hodges and Smillie (1970).

occur every three or four residues, which would be expected for the regular pattern of interactions in an α-helical coiled-coil. The preliminary chemical data show no large-scale repeating unit in the molecule, but Parry and Squire (pers. commun.) have suggested that the strong 28 Å reflection from tropomyosin aggregates (Caspar et al., 1969) may indicate a sequence periodicity in the molecule related to the 55 Å repeat of the actin helix.

Rabbit troponin preparations consist of three major components of mol wts about 37,000, 24,000, and 18,000 (Ebashi et al., 1971; Hartshorne and Pyun, 1971; Schaub et al., 1972; Greaser and Gergely, 1971) designated as TN_T, TN_I, and TN_C, respectively (Fig. 1). The heaviest component binds strongly to tropomyosin and has about the same mobility on SDS gels as the slower rabbit tropomyosin component. The overlapping of bands in the SDS gels has complicated determinations of the amount of troponin in tropomyosin preparations. Amino acid analyses of purified rabbit tropomyosin and troponin components (Ebashi et al., 1971; Greaser and Gergely, pers. commun.; Schaub et al., 1972; Wilkinson et al., 1972) show that the proteins can be distinguished by analysis of key residues (Table 2). Tropomyosin contains no proline or tryptophan in contrast to the heaviest troponin component (TN_T) which has about 14 proline residues and 2 tryptophans. Furthermore, no cysteic acid is detected in the amino acid analysis of this component. The other two troponin components, which are essential for inhibition and calcium

sensitization of actomyosin ATPase, are readily detected on SDS gels. Thus, chemical and physico-chemical criteria can be used to measure the troponin content of different tropomyosin preparations. In muscles other than rabbit skeletal, the troponin components can be readily distinguished from tropomyosin by SDS gels.

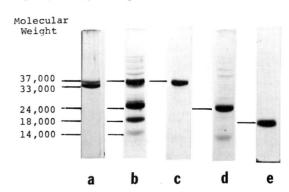

Figure 1. SDS polyacrylamide gel electrophoresis of tropomyosin, troponin and its components. Rabbit muscle troponin prepared as described by Greaser and Gergely (1971) was dialyzed against 0.05 M Tris-HCl pH 7.9, 1 mM EGTA, then against 6 M urea in the same buffer (Eisenberg and Kielley, 1972). The troponin components were fractionated by chromatography on a DEAE-cellulose column (41 × 2.5 cm) equilibrated with 6 M urea, 1 mM DTT, 0.05 M Tris-HCl pH 7.9. The protein was eluted by applying a linear KCl gradient. The fractions off the column were pooled and monitored by SDS gel electrophoresis. (a) Rabbit tropomyosin (4 μg); (b) unfractionated rabbit troponin (25 μg); (c) tropomyosin-binding component of troponin TN_T (6 μg); (d) the inhibitory component of troponin TN_I (7 μg); (e) the calcium-binding component of troponin TN_C (8 μg).

Table 2. Amino Acid Analysis of Rabbit Skeletal Troponin Components (Moles AA/10^5 g)

	TN_T 37,000 Mol Wt[a]		TN_I 24,000 Mol Wt		TN_C 18,000 Mol Wt	
	This Study	Wilkinson et al., 1972	This Study	Wilkinson et al., 1972	This Study	Schaub et al., 1972
Lys	126.9	115.2	101.0 ± 4.6	105.3	49.7 ± 2.6	56.2
His	18.2	17.4	19.1 ± 1.8	17.0	6.0 ± 0.5	10.0
Arg	81.5	73.9	63.8 ± 3.8	66.1	38.3 ± 1.5	42.3
Asp	68.5	71.8	83.1 ± 1.8	81.8	126.0 ± 2.6	111.0
Thr[b]	19.5	19.2	29.2	17.4	32.5	29.5
Ser[b]	26.2	29.7	49.0	45.2	38.6	37.2
Glu	203.8	205.7	143.0 ± 7.6	163.6	184.7 ± 4.0	194.0
Pro	37.8	30.0	36.3 ± 4.1	24.4	6.1 ± 1.0	19.4
Gly	26.4	28.4	53.8 ± 1.2	42.2	71.3 ± 2.5	62.3
Ala	84.0	91.0	74.8 ± 3.3	74.4	71.2 ± 1.6	79.0
Val	33.6	38.9	45.4 ± 3.0	37.8	36.5 ± 2.2	43.9
Met	12.4	15.3	24.5 ± 3.9	40.0	53.8 ± 1.0	40.5
Iso	24.8	25.2	30.8 ± 1.1	22.6	49.1 ± 2.1	44.4
Leu	62.7	64.2	79.7 ± 1.4	87.0	51.6 ± 1.8	56.2
Tyr	11.5	12.9	15.1 ± 2.7	10.9	10.0 ± 3.5	9.4
Phe	14.5	15.3	21.6 ± 5.2	14.4	53.5 ± 1.8	42.4
Cys[c]	0.4	1.6	10.2	13.9	3.8	8.9
Trp[d]	—	—	7.7	—	1.3	Nil

Unless otherwise indicated, all values are duplicate analyses of 24, 48, and 72 hr hydrolysates and are expressed ± SD.

[a] Values for this component are the average of four 24 hr hydrolysates except for tyrosine, which was determined from the two non-performate oxidized samples.

[b] Serine and threonine values extrapolated to zero times of hydrolysis.

[c] Cysteine determined as cysteic acid according to the method of Hirs (1967). Average of two determinations.

[d] Determined by the method of Beaven and Holiday (1952).

Tropomyosin Nets and Fibers

The simple structure of the tropomyosin molecule contrasts strikingly with the variety of assemblies it can form. The ordered aggregates that can be produced in vitro depend on solvent conditions, muscle source, and troponin content. Bailey (1948) grew crystals of tropomyosin from many vertebrate muscles and found that they had the same remarkable form. More than 95% of the crystal volume is water. Analysis of the crystal structure has shown that the very open lattice is built from cross-connected molecular filaments (Caspar et al., 1969). Under crystallizing conditions, a variety of net forms have been observed with rabbit tropomyosin. The period along the strands of all the nets is about 400 Å, but they are cross-connected in different ways. The simplest pattern is a 400 Å square net (Fig. 2). All the other forms are characterized by cross-connections between the filaments which are separated by the same long (225–235 Å) and short (165–175 Å) intervals observed in the crystal lattice. A striking feature of the polymorphic nets is that wavy and straight filaments occur. The crystal lattice, which is the most stable form, shows the most accentuated filament bending. In the meshes of different shape and symmetry the same bond sites appear to be connected in different combinations.

Bailey crystals have not been obtained from chicken skeletal muscle tropomyosin, although they are readily grown from other vertebrate striated muscle tropomyosins. Ebashi and Ohtsuki (pers. commun.) observed a distinctive square net from chicken tropomyosin precipitated in 0.1 M KCl at pH 5.4. We have reproducibly grown this aggregate (Fig. 2a) and it appears to be the only net formed by this tropomyosin. The mesh is different from any of the square nets formed from rabbit tropomyosin. The repeat along the filaments in all of the forms is 400 Å, but the edge length of the unit cell is 2 × 400 Å. The filaments are directed along the diagonal of the square lattice and the crossovers between them form nodes which appear alternately as large bright spots and as a square cluster of four smaller spots. This pattern arises from pairs of filaments which are in contact at the bright nodes and which are separated at the open crossovers. The filaments are wavy with an amplitude comparable to that in the Bailey crystal lattice, and the crossovers are separated by similar long-short intervals. Moreover the negatively stained chicken nets appear as highly ordered as the Bailey crystals.

The 400 Å square rabbit net, in contrast, appears to distort easily (Fig. 2c), although there is a regular periodicity along the filaments. The quartered-square rabbit net (Fig. 2b) has thick strands separated by 400 Å with a pair of closely spaced thin filaments midway between. The thick filaments transform into pairs of thin filaments and vice versa; therefore, the net is locally disordered although the lattice retains a regular square form. The accentuation of the bright nodes in the chicken net is similar to that seen in thicker portions of negatively stained

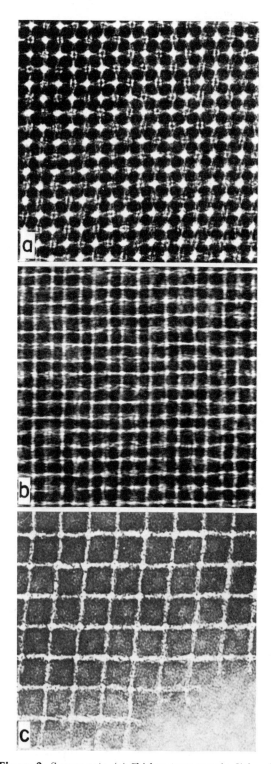

Figure 2. Square nets. (a) Chicken tropomyosin dialyzed against 0.1 M KCl, 0.01 M sodium acetate pH 5.4. (b) "Quarter-square" net formed by rabbit tropomyosin dialyzed against 0.1 M KCl, 0.01 M sodium acetate pH 5.4. (c) Rabbit tropomyosin dialyzed against 0.12 M $(NH_4)_2SO_4$, 0.01 M sodium acetate pH 5.4. Negatively stained with uranyl acetate. × 200,000.

Bailey crystals. Since the optical density in the image is an exponential function of stain thickness, stain-excluding regions will appear accentuated when they are regularly superposed. The appearance of the nodes and the regularity of the lattice in thicker portions of the chicken nets indicate that they are proper three-dimensional crystals. Chicken tropomyosin may thus provide another crystal form suitable for X-ray analysis.

In addition to the various nets, tropomyosin also forms compact fibrous aggregates. Precipitation with divalent cations produces highly ordered tactoids with a period of about 400 Å (Cohen and Longley, 1966; Millward and Woods, 1970). All vertebrate striated muscle tropomyosins, including chicken, form identically banded nonpolar tactoids when precipitated with magnesium or calcium. The band pattern has a broad light-staining region 250 Å long bounded by narrow stain-excluding lines. Different dihedral band patterns can be produced by precipitation with divalent cations, and a polar pattern has been observed with lead precipitation (Caspar et al., 1969).

Tactoids with flat ends show the position of the molecules in the structures. The flat end in the magnesium form is at a position corresponding to one of the narrow stain-excluding bands. These narrow bands therefore mark the ends of the molecules. The tactoid is formed from oppositely directed arrays of polar filaments, and the positions of the dyads relative to the ends of the molecules define the shift between the arrays of filaments. In the magnesium tactoid one of the dyads is 75 Å from a molecular end. If this dyad were at an end, the oppositely directed arrays would be in register. The 400 Å period is determined by the head-to-tail bonding of the polar filaments which build the structure. The narrow stain-excluding band may represent an overlap of 20–30 Å in the head-to-tail bonding. It could also correspond to crossover points in a three-dimensional network of molecular filaments in the tactoid (Cohen et al., 1971). Further observations on the relation between the net and fibrous forms have been obtained from studies of complexes with troponin components described below.

Troponin in the Crystal Lattice

Troponin can cocrystallize with rabbit tropomyosin in the Bailey lattice (Higashi and Ooi, 1968) (Fig. 3). X-ray diffraction measurements have been made on crystals of "pure" tropomyosin and on some crystals grown from "relaxing factor" preparations which contain about equal amounts of tropomyosin and troponin (Fig. 4). Comparing the diffraction patterns of crystals with and without

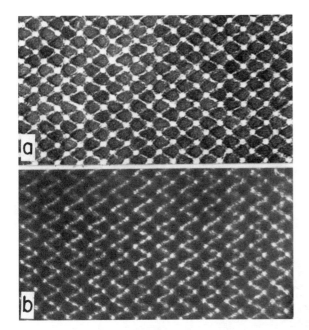

Figure 3. Electron micrographs of crystals: (a) tropomyosin, (b) tropomyosin-troponin complex. Crystals are negatively stained with uranyl acetate. The view in these electron micrographs corresponds to the [100] projection of the crystal (see Fig. 4). The troponin is seen at the middle of the long arm of the kite-shaped mesh in (b). From the symmetry of the crystal, each strand of the net consists of a pair of oppositely directed molecular filaments. Correspondingly, the thickening identified with troponin consists of a pair of units related by a dyad. × 200,000.

troponin, there are significant differences in unit cell dimensions and in the intensities of some of the reflections. The patterns show that the symmetry of the molecular packing is unchanged, and that the form of the tropomyosin molecule is not greatly altered when some troponin is incorporated. The magnitude of the intensity changes indicates that there is only limited binding of troponin to tropomyosin in these crystals.

The three-dimensional packing arrangement of the tropomyosin molecules in the lattice has been inferred from the crystal symmetry together with the image of the a-axis projection (Cohen et al., 1971). Electron micrographs show the view down the a-axis, but images of the crystal viewed in other directions have not yet been obtained. The 400 Å long molecules are connected in continuous filaments which are directed along the body diagonals of the cell. Each molecule has eight nearest neighbors: two at either end in the filament and six forming cross-connections in the three-dimensional meshwork. Four of these connections are seen as crossovers in the a-axis projection and are separated by the long-short intervals. The unequal separation of these contacts leads to a bending of the filaments and the appearance of the kite-shaped mesh.

The unit cell dimensions of crystals containing troponin differ significantly in the length of the b- and c-axes from those of pure tropomyosin (Table 3). In contrast, the body diagonal is very close to 402 Å in all crystals. The period along the diagonally oriented filaments is determined by the invariant head-to-tail bonding of the tropomyosin molecules. The body diagonal of the different lattices is constant because the strands are stiff and are specifically connected at their pivot points. When troponin is complexed in the lattice, the X-ray diffraction diagrams (Fig. 4a, b) show that the rectangular cell in the a-axis projection becomes less elongated. Correspondingly, the diagonally oriented spikes on the X-ray diagram become more nearly perpendicular reflecting the pivoting of the filaments at the cross-connections. The large changes in unit cell dimensions result from small cooperative motions of the molecules relative to each other (Fig. 4c, d).

Electron microscope observations show troponin as a broad mass attached near the middle of the long arms in the lattice (Higashi and Ooi, 1968).

Table 3. Unit Cell Parameters of Some Rabbit Tropomyosin Crystals Space Group P22$_1$2

Crystal	a-axis (Å)	b-axis (Å)	c-axis (Å)	Unit Cell Diagonal	Ratio of c/b
"Pure" tropomyosin					
I	124	—	299		
II	—	238	303		
III	123	236	295		
Average	123	237	299	402	1.27
Tropomyosin with bound troponin					
V	122	254	286		
VI	120	249	291		
VII	122	255	286		
VIII	120	257	286		
Average	121	254	287	402	1.13

"Pure" tropomyosin crystals were shown to be free of troponin. Parameters are averaged from three crystals. Tropomyosin crystals containing troponin were grown from relaxing factor preparations in which the amounts of tropomyosin and troponin were comparable. Parameters are averaged from four crystals grown from three different preparations. Crystallizing medium 0.1 M NaAc, 0.05 M (NH$_4$)$_2$SO$_4$ pH 5.8, 1 mM DTT and 1 mM EDTA were also included in the crystallization medium for pure tropomyosin crystals.

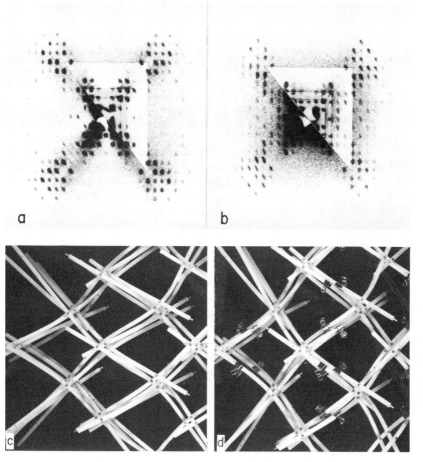

Figure 4. Tropomyosin diffraction patterns and models. (*a*) and (*b*) 3° precession photographs of [100] projection. A composite of weaker exposures is superimposed to the right of the diagonal to show the very strong low-order reflections. (*a*) Crystal of "pure" tropomyosin (crystal III in Table 3). *b*-axis 236 Å, *c*-axis 295 Å. (*b*) Crystal of tropomyosin containing troponin (crystal VIII in Table 3). *b*-axis 257 Å, *c*-axis 286 Å. (*c*) Model of tropomyosin crystal viewed down the *a*-axis; the *b*-axis is vertical and the *c*-axis is horizontal. The bonding relations in the model are described by Cohen et al., 1971. (*d*) Same view of the tropomyosin-troponin crystal illustrating change in lattice dimensions. The troponin is represented by clips attached near the middle of the long arm.

Correspondingly the electron density maps calculated from the X-ray patterns show that there is a concentration of matter near the middle of the long arm in crystals with troponin. Structural analysis of the diffraction patterns of crystals with and without troponin (Fig. 4) is complicated by the disorder in the crystals. The data are available for quantitative comparison of these structures to a resolution of about 20 Å if criteria for scaling intensities can be established. Comparison of key reflections with very different intensities on the two photographs provides a direct measure of the separation of the pair of troponin complexes related by a dyad. The binding site of each globular troponin unit appears to be about 50 Å from the dyad at the middle of the long arm.

Locating the troponin near the middle of the long arms in the lattice does not show how it alters the bonding at the crossover points. Simple mechanical models could be constructed to illustrate the structural changes, but such models are apt to be misleading. The form of the lattice is determined both by short-range interactions at the crossovers and by long-range forces between the highly charged rod-shaped molecules. The crystal is a highly hydrated

gel and its soft texture demonstrates the weakness of the forces stabilizing the lattice. The troponin complex has about the same mass as the rod-shaped molecule to which it binds. Attachment of even a small amount of troponin to the open tropomyosin mesh will obviously alter the long-range intermolecular interactions. It is striking that the lattice is merely modified rather than restructured by the attachment of this bulky component.

Crystals grown from tropomyosin preparations containing small amounts of troponin have lattice constants intermediate between the two states in Table 2: The states with intermediate axial ratio may indicate a graded response of the molecular packing to the troponin content. The modification of the lattice is cooperative since the crystal is a single phase; this suggests that each troponin unit in the crystal may have an effect on the conformation of all the tropomyosin molecules in the lattice.

Troponin Components in Nets and Fibers

Rabbit tropomyosin complexed with troponin forms tactoids when precipitated with magnesium (Nonomura et al., 1968), and chicken preparations show the same structure (Huxley and Hitchcock) pers. commun.). The troponin appears as a stain-excluding region about 100 Å wide located at the center of the broad light band of the negatively stained Mg tactoid (Fig. 5a). Varying amounts of troponin can be complexed in the tactoid. As Nonomura et al. (1968) described, binding a small amount of troponin "fills in" the thin dark band at

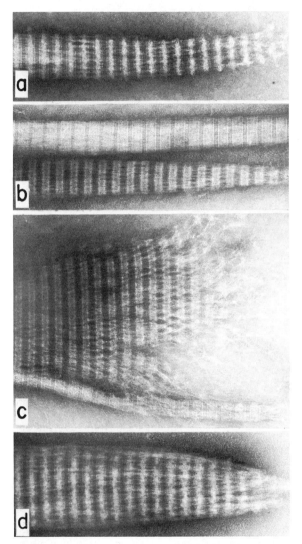

Figure 5. Tropomyosin-troponin tactoids. (*a*) Paracrystal obtained by complexing rabbit tropomyosin and troponin in a 1:2 weight ratio and precipitating against 0.01 M TES, 0.01 M EDTA, 0.05 M magnesium acetate, pH 7.0. (*b*) Paracrystal obtained by complexing rabbit tropomyosin and troponin-binding component in 2:1 molar ratio. Precipitation by dialysis against 0.05 M Tris, 0.05 M MgCl₂, pH 7.9. (*c*) Paracrystal obtained by complexing rabbit tropomyosin and troponin-binding component in 1:1 molar ratio. Precipitation by dialysis against 0.05 M Tris, 0.05 M MgCl₂, pH 7.9. (*d*) Paracrystal obtained by complex of rabbit tropomyosin, troponin-binding component and calcium-binding troponin subunit. Molar ratio 1:1:1. Precipitation by dialysis against 0.1 M sodium acetate, 0.05 M (NH₄)₂SO₄, 1 mM DTT, 1 mM EDTA, pH 5.1. Negatively stained with uranyl acetate. × 100,000.

the middle of the light-staining region. The central region of the broad band becomes progressively lighter as more troponin is incorporated, corrugation is observed at the edges of the tactoid, and fraying at the ends may occur. The characteristic Mg-band pattern becomes indistinct in tactoids saturated with troponin, and the troponin bands dominate the image. The globular troponin units act as spacers separating the tropomyosin filaments which are observed as light longitudinal striations in the negatively stained tactoids.

The troponin band consists of pairs of units connected to oppositely directed pairs of tropomyosin filaments. The dyad relating the troponin units is 125 Å from a molecular end. The 100 Å width of the troponin band suggests that the center of individual troponin units may be about 25 Å from the center of the dyad. Therefore the troponin is bound either at about 150 Å or 100 Å from a molecular end.

The organization of the individual components of troponin in the relaxing factor can now be investigated since methods for their purification have been developed (Greaser and Gergely, 1971; Eisenberg and Kielley, 1972). Remarkable assemblies can be produced by combining the binding component of troponin (Fig. 1) with tropomyosin. Precipitation of this complex with magnesium produces tactoids (Fig. 5b, c), some of which resemble those obtained from the relaxing factor, and a new form, strikingly different from the usual Mg tactoid. The dihedral symmetry and 400 Å period in the new form are the same as in other divalent cation tactoids, but the band pattern is very different. The period consists of a pair of bright stain-excluding bands and a 200 Å wide, very darkly staining region. This form is in fact identical to the unusual tactoid observed by Higashi-Fujime and Ooi (1969), but conditions for producing this structure were not established. Their micrograph showed this tactoid being generated by the collapse of a net with a distinctive diamond mesh. What appears to be critical for forming this tactoid is the presence of the binding component. A consistent feature of the tactoids we have observed is the light longitudinal striations in the dark band; some of the tactoids are in fact opened out, revealing strands as in the diamond mesh (Fig. 5c).

In order to obtain the open diamond mesh, we precipitated the complex as in the Bailey crystallization procedure. The diamond lattice was indeed produced (Fig. 6), displaying the whole range of transitions from undistorted mesh to compact tactoid. This lattice is distinct from that of the Bailey crystal. The net has plane group symmetry *cmm* in projection (Fig. 7). The length of the unit cell diagonal is 800 Å, but since the net is centered, the period along the strands is 400 Å as in other net forms. The crossovers are separated by about 100 Å and 300 Å

Figure 7. Enlarged view of double diamond mesh. The superposed drawing accentuates the molecular filaments and distinguishes the two types of cross connections. The dotted lines indicate an orthogonal pair of mirrors of the plane group *cmm*. These mirrors correspond to twofold axes of the crystal lattice seen in projection. × 200,000.

to produce the mesh of large and small diamonds. These long-short intervals of the double diamond mesh are different from the crossover intervals of 230 Å and 170 Å characteristic of other net forms. Filament crossovers appear as bright spots in the negatively stained image. The node at the acute diamond vertex is distinctly brighter than that at the obtuse vertex. This feature can be attributed to a difference in mass at the vertices. Crossovers of the filaments generally appear enhanced in negatively stained images; but if the filaments are uniform, these intersections should look similar at the two kinds of vertices. The bright node at the acute vertex therefore shows the location of the globular troponin-binding component.

The regularity of the double diamond lattice in projection and its thickness indicate that the net seen in the electron microscope is the projection of a three-dimensional crystal. There are four enantio-morphic orthorhombic space groups which would show *cmm* symmetry in projection: $C222_1$ $C222$, $I222$, $I2_12_12_1$. These four space groups all have mirror lines in projection corresponding to pairs of superposed dyads in the lattice. Designating the direction of view in the electron micrograph as the c-axis, the view down the a- and b-axes, parallel to the plane of projection, will show dyads which relate structural elements in the lattice "face-to-

Figure 6. Double diamond nets. Rabbit tropomyosin and troponin-binding component mixed in a molar ratio of 2:1 and precipitated by dialyzing against 0.1 M sodium acetate, 0.05 M $(NH_4)_2SO_4$, 1 mM DTT, 1 mM EDTA, pH 5.6. Negatively stained with uranyl acetate. × 100,000.

face" and "back-to-back." As seen in the electron micrograph, pairs of tropomyosin molecules cross at or near the position of the troponin binder. The observed symmetry indicates that in the three-dimensional lattice the tropomyosin filaments cross with the binding components facing each other.

Comparison of the structures formed by tropomyosin with troponin components provides information about their interactions. Observations on the relaxing factor show that the troponin complex fits into aggregates which can be formed by tropomyosin alone. In contrast the interaction of the binding component with tropomyosin appears to generate new cross-connections between the tropomyosin filaments.* Almost certainly the troponin binding component bonds to the same site on the tropomyosin molecule whatever may be bonded elsewhere. The binding component interacts strongly and specifically with tropomyosin and it must also link the other two troponin components. Absence of these components could expose a new site for bonding. The location of the binding component in the double diamond net indicates that it may actually cross-connect the tropomyosin filaments. The symmetry of the net implies that the cross-connection is formed by a dimer link between pairs of troponin-binding components.

The fact that the double diamond net is formed when troponin binder is present, and not with the whole relaxing factor, implies that the regulatory components of troponin might block the dimer interaction on the binding component. We have added calcium-binding component to the tropomyosin-troponin binder and precipitated under conditions where the tactoid or double diamond net would be produced. Paracrystals were formed (Fig. 5d) which are very similar to those obtained with the complete troponin complex (Fig. 5a). This finding is consistent with the idea that the calcium-binding subunit may indeed prevent dimer linking between binding components.

The transformations of the double diamond lattice (Fig. 6) vividly illustrate how tropomyosin tactoids may be fabricated. The net compacts as the filaments pivot cooperatively at the cross-connections. Just as woven fabrics stretch on the bias, so these tropomyosin nets extend on the diagonal. The extreme collapsibility of the double diamond lattice compared to other tropomyosin nets may be due to the troponin swivels at the joints. This range of lattice transitions is not generally observed in the

---

* In addition to the double diamond lattice, the complex of the troponin-binding component and tropomyosin also produces a hexagonal lattice with a 400 Å spacing between nodes. We have also observed this form which is described in this volume by Ebashi and Ohtsuki and by Greaser and Gergely. The nodes in the hexagonal net presumably represent the location of the troponin-binding component.

formation of other fibrous aggregates of tropomyosin. The tropomyosin filaments may, however, be organized in a compact network in all tactoids (Cohen et al., 1971). Features of the mesh can be observed in Mg tactoids when troponin spacers pry apart the filaments (Fig. 5). Just as there are many polymorphic forms of the nets (Caspar et al., 1969), there are a variety of band patterns in the tactoids. Supercoiling of the tropomyosin filaments would occur in compacting nets forming fibers, and this plaiting may be characteristic of the assembly of protein fibers.

Motions of Tropomyosin in Muscle

The tropomyosin-troponin crystals display aspects of the interaction between these molecules in muscle. In the lattice the binding of troponin induces small cooperative changes in the bonds which cross-connect the tropomyosin filaments. These small changes in bonding lead to the large differences in unit cell dimensions. We have suggested (Cohen et al., 1971) that these changes in the crystal may provide an analogy for the structural changes in the thin filament related to regulation of contractile activity. In muscle the tropomyosin filaments are coiled in the groove of the actin helix: The bending of the tropomyosin molecules is comparable to that observed in the supercoiled filaments in the crystal lattice. In the lattice the molecules pivot or slip at the crossover contacts when troponin is bound. By analogy the tropomyosin filaments might slip or twist in the actin groove under different physiological states. The specific end-to-end bonding would be conserved, whereas links to actin would be modified. The change in the tropomyosin actin link could correspond to the pivoting of the tropomyosin molecules at cross-connections in the lattice.

Recent X-ray diffraction studies on a variety of muscles have provided the first direct evidence for motion of tropomyosin in the thin filaments during contraction (Huxley, 1970; Lowy, 1971; Vibert et al., 1971, 1972; see also this volume: Haselgrove, H. Huxley, Lowy). The intensity of the second layer line of the actin pattern increases significantly when the muscles are activated. This finding indicates that there is a structural change in the thin filaments during contraction. There is a single tropomyosin filament in each of the two grooves of the actin helix (Spudich et al., 1972). O'Brien et al. (1971) have shown that intensity of the second layer line in optical diffraction patterns of actin paracrystals depends on the tropomyosin content. The three-dimensional reconstruction of the actin-tropomyosin-troponin complex by Spudich et al. (1972) shows that each tropomyosin filament lies to one side of the center of the actin grooves. Model calculations by

Parry and Squire (pers. commun.) indicate that the intensity of the second actin layer line is very sensitive to the position of tropomyosin in the groove of the actin helix. It is obvious that if the tropomyosin were nearer the middle of the groove, its diffraction would be in phase with the second layer line of the actin. Placing it to one side would reduce the diffracted intensity of this layer line.

The observed change in the diffraction pattern on contraction of muscle can therefore be interpreted to indicate a change in the position of the tropomyosin in the groove of the thin filament. The fact that a similar intensity change occurs during contraction of molluscan muscles (Vibert et al., 1971, 1972) indicates that motion of the tropomyosin molecules may also occur in thin filaments which do not have troponin. Since calcium regulation is associated with myosin in molluscan muscle, the observed change in the diffraction pattern suggests that the molecular rearrangements in the thin filaments in this case may be following the linking of actin and myosin, rather than preceding it. These comparative studies on vertebrate and molluscan muscles indicate that although the calcium regulation is different in these muscles, in both cases motions of tropomyosin may occur in the thin filaments. Comparable displacements of tropomyosin are observed in different states of the crystal lattice. The crystals thus illustrate the motions of the tropomyosin molecule.

Regulation by calcium in vertebrate muscles involves structural rearrangements in the thin filaments which are triggered by the binding of calcium to troponin. The stoichiometry of ATPase inhibition of actomyosin by tropomyosin-troponin (Weber and Bremel, 1971; Spudich and Watt, 1971) and the periodicities of the thin filament complex (Ohtsuki et al., 1967; Huxley and Brown, 1967) indicate that there are about seven actin monomers related to each tropomyosin molecule in the thin filament (Ebashi and Endo, 1968). It is very unlikely that each actin is related in the same way to successive portions of the tropomyosin molecule (Cohen et al., 1971). If the ratio were in fact exactly seven to one, each tropomyosin molecule would be identically related to every seventh actin monomer. In this case the two tropomyosin strands in a thin filament would be shifted by an odd multiple of 27 Å relative to one another. However there may not be a simple integral ratio in the axial periodicities of tropomyosin and actin. The experimental measurements cannot rule out the possibility that the ratio of actin to tropomyosin is as much as 1% greater than an integral value of seven. There would then be an exact repeat after 100 or more actin monomers and a vernier relation between the actin helix and the tropomyosin filaments (Huxley and Brown, 1967; Cohen et al., 1971). In this case the tropomyosin

filaments should be regarded as interacting with the whole actin helix. On this assumption, the displacement of one strand in relation to the other is not fixed by the symmetry of the actin helix; the two tropomyosin strands could be in register. The critical specificity in the tropomyosin-actin bonding might be the initial combination at the base of the thin filament and could involve interactions with Z-line proteins. Although the actin helix is periodic, the combination with tropomyosin and troponin may be structurally different along the length of the thin filament.

If there were an integral combining ratio between actin and tropomyosin, the troponin might act independently on each tropomyosin molecule to control the seven actins in a functional "unit." Alternatively, whether the combining ratio is integral or not, the troponin may act on the whole set of tropomyosin molecules which are connected end-to-end in a filament. There are two tropomyosin strands in the thin filament, but only one actin helix. Each actin monomer is connected to four actin neighbors —one above, one below, and one to each side. Throwing one troponin switch in the thin filament may have an effect on all the interconnected molecules. In this case, as with a crystal lattice, the thin filament could act as a single phase.

An essential pattern of connections is maintained in the thin filaments, providing the framework which unites the moving parts. Conserved connections include the head-to-tail linkage of tropomyosin and the bonding of each actin unit with its four neighbors. In the thin filament variable linkages appear to involve changes in the connection between the tropomyosin strands and the actin helix. The molecular displacements may involve breaking some short-range contacts and forming others. The structural results on the troponin binding component illustrate again the importance of both conserved and variable linkages in the muscle machine (Caspar and Cohen, 1969). The linkage between the troponin-binding component and tropomyosin appears to be another invariant connection. The functioning of the other two troponin components, one of which responds to calcium and the other which blocks ATPase activity, could require variable connections. Just as tropomyosin is the link between troponin and actin, the troponin-binding component is the link between the regulatory subunits of the complex and the thin filament.

Acknowledgments

We thank Drs. S. Ebashi and I. Ohtsuki for communicating unpublished results; Dr. Hans Meienhofer and Ms. Jean Lee for invaluable aid with amino acid analyses; Drs. A. G. Szent-Györgyi and Susan Lowey for discussion; Ms. Isa Bernardini,

Mr. Charles Ingersoll, Mr. Ray Fronk, Ms. Marjorie Kasac, Mr. Paul Norton, and Mr. Larry Tobacman for technical assistance. This work was supported by Grants AM 02633 from the National Institute of Arthritis and Metabolic Diseases, CA 04696 from the National Cancer Institute, and the National Science Foundation. This work was done during the tenure of a Research Fellowship of the Muscular Dystrophy Associations of America awarded to D. A. D. P. and a Massachusetts Heart Association Fellowship to J. P. J.

References

BAILEY, K. 1948. Tropomyosin: A new asymmetric protein component of the muscle fibril. *Biochem. J.* **43**: 271.

BEAVEN, G. H. and E. R. HOLIDAY. 1952. Ultraviolet absorption spectra of proteins and amino acids. In *Advances in protein chemistry* (ed. M. L. Anson et al.) vol. 7, p. 319. Academic Press, New York.

CASPAR, D. L. D. and C. COHEN. 1969. Polymorphism of tropomyosin and a view of protein function. In *Nobel Symposium* (ed. A. Engstrom and B. Strandberg) vol. 2, p. 393. Wiley, New York.

CASPAR, D. L. D., C. COHEN, and W. Longley. 1969. Tropomyosin: Crystal structure, polymorphism and molecular interactions. *J. Mol. Biol.* **41**: 87.

COHEN, C. and W. LONGLEY. 1966. Tropomyosin paracrystals formed by divalent cations. *Science* **152**: 794.

COHEN, C., D. L. D. CASPAR, D. A. D. PARRY, and R. M. LUCAS. 1971. Tropomyosin crystal dynamics. *Cold Spring Harbor Symp. Quant. Biol.* **36**: 205.

CRICK, F. H. C. 1953. The packing of α-helices: simple coiled: coils. *Acta Cryst.* **6**: 689.

CUMMINS, P. and S. V. PERRY. 1972. Subunit structure and biological activity of tropomyosin B from different muscle types. *Biochem. J.* **128**: 106.

EBASHI, S. and M. ENDO. 1968. Calcium ion and muscle contraction. *Prog. Biophys. Mol. Biol.* **18**: 123.

EBASHI, S., T. WAKABAYASHI, and F. EBASHI. 1971. Troponin and its components. *J. Biochem.* **69**: 441.

EISENBERG, E. and W. W. KIELLEY. 1972. Reconstruction of active troponin-tropomyosin complex in the absence of urea from its 4 column-purified components. *Fed. Proc.* **31**: 1630.

GREASER, M. L. and J. GERGELY. 1971. Reconstitution of troponin activity from three protein components. *J. Biol. Chem.* **246**: 4226.

HARTSHORNE, D. J. and H. Y. PYUN. 1971. Calcium binding of the troponin complex and the purification and properties of troponin A. *Biochim. Biophys. Acta* **229**: 698.

HIGASHI, S. and T. OOI. 1968. Crystals of tropomyosin and native tropomyosin. *J. Mol. Biol.* **34**: 699.

HIGASHI-FUJIME, S. and T. OOI. 1969. Electron micro-scopic studies on the crystal structure of tropomyosin. *J. Microscopie* **8**: 535.

HIRS, C. H. W. 1967. Determination of cystine as cysteic acid. In *Methods in enzymology*, ed. C. H. W. Hirs, vol. 11, p. 59. Academic Press, New York.

HODGES, R. S. and L. B. SMILLIE. 1970. Chemical evidence for chain heterogeneity in rabbit muscle tropomyosin. *Biochem. Biophys. Res. Comm.* **41**: 987.

———. 1972a. The histidine and methionine sequences of rabbit skeletal tropomyosin. *Can. J. Biochem.* **50**: 312.

———. 1972b. Cysteine sequences of rabbit skeletal tropomyosin. *Can. J. Biochem.* **50**: 330.

HUXLEY, H. E. 1970. Structural changes in muscle and muscle proteins during contraction. *8th Int. Cong. Biochem. Interlaken:* 23.

HUXLEY, H. E. and W. BROWN. 1967. The low angle X-ray diagram of vertebrate striated muscle and its behavior during contraction and rigor. *J. Mol. Biol.* **30**: 383.

LOWY, J. 1971. Bulletino di Zoologia, *Symposium on evolution of fibrous proteins.* 11th Ital. Zool. Cong., Garda. In press.

MILLWARD, G. R. and E. F. WOODS. 1970. Crystals of tropomyosin from various sources. *J. Mol. Biol.* **52**: 585.

NONOMURA, Y., W. DRABIKOWSKI, and S. EBASHI. 1968. The localization of troponin in tropomyosin paracrystals. *J. Biochem.* **64**: 419.

O'BRIEN, E. J., P. M. BENNETT, and J. HANSON. 1971. Optical diffraction studies of myofibrillar structure. *Phil. Trans. Roy. Soc.* London B **261**: 201.

OHTSUKI, I., T. MASAKI, T. NONOMURA, and S. EBASHI. 1967. Periodic distribution of troponin along the thin filament. *J. Biochem.* **61**: 817.

SCHAUB, M. C., S. V. PERRY, and W. HÄCKER. 1972. The regulatory proteins of the myofibril, characterization and biological activity of the calcium-sensitizing factor (troponin A). *Biochem. J.* **126**: 237.

SPUDICH, J. A. and S. WATT. 1971. The regulation of rabbit skeletal muscle contraction. *J. Mol. Biol.* **246**: 4866.

SPUDICH, J. A., H. E. HUXLEY, and J. T. FINCH. 1972. The regulation of skeletal muscle contraction. II. Structural studies of the interaction of the tropomyosin-troponin complex with actin. *J. Mol. Biol.* In press.

VIBERT, P. J., J. LOWY, J. C. HASELGROVE, and F. R. POULSEN. 1971. Changes in the low-angle X-ray diffraction patterns of smooth muscles during contraction. 1st *Europe. Biophys. Congr.*, Vienna.

VIBERT, P. J., J. C. HASELGROVE, J. LOWY, and F. R. POULSEN. 1972. Structural changes in actin-containing filaments of muscle. *Nature New Biol.* **236**: 182.

WEBER, A. and R. D. BREMEL. 1971. Regulation of contraction and relaxation in the myofibril, p. 37. In *Contractility of muscle cells and related processes*, ed. R. J. Podolsky. Prentice-Hall, Englewood Cliffs, New Jersey.

WILKINSON, J. M., S. V. PERRY, H. A. COLE, and J. P. TRAYER. 1972. The regulatory proteins of the myofibril: separation and biological activity of the components of inhibitory factor preparations. *Biochem. J.* **127**: 215.

Tropomyosin: Amino Acid Sequence and Coiled-Coil Structure

R. S. Hodges,* J. Sodek, L. B. Smillie, and L. Jurasek

Department of Biochemistry, University of Alberta, Edmonton, Alberta, Canada

Tropomyosin preparations of mol wt of about 70,000 daltons are known to dissociate in the presence of reducing agents and denaturing media to monomers of about 35,000 daltons (McCubbin et al., 1967; Holtzer et al., 1965; Woods, 1967, 1969; Weber and Osborne, 1969; Olander et al., 1967). The highly helical (>90%) polypeptide chains are believed to be arranged in a two-stranded coiled-coil (Crick, 1953; Cohen and Holmes, 1963) whose structure is stabilized by hydrophobic interactions between the two strands of α-helix. The relative unimportance of electrostatic interactions in the stabilization of the structure is indicated by the unusual stability of the molecule in acid and alkali (Lowey, 1965; Riddiford and Scheraga, 1962), even though the content of charged amino acid residues is remarkably high (Hodges and Smillie, 1970). In the structure proposed by Crick (1953), two α-helices are packed side-by-side in a "knobs-into-holes" manner. In order to interlock two helices with their axes parallel one to the other, the number of residues per turn must be $n/2$, where n is an integer. In the α-helix the number of residues per turn is 3.6. If nonpolar groups occur every seventh residue, then optimum nonpolar side-chain interactions occur when the α-helices pack at an angle of about 20° to each other. In order for the α-helices to remain in contact indefinitely, it is necessary to deform them slightly along their entire length. This results in the two helices slowly winding around each other to form the coiled-coil.

That this simple coiled-coil concept required modification was suggested by Elliott et al. (1968) on the basis of X-ray diffraction studies of paramyosin. Thus the ratio between the meridional reflection at about 5.1 Å and the projected residue translation of about 1.5 Å was not exactly the expected value of 7/2. An alternative structure, the segmented rope originally suggested by Fraser and MacRae (1961a,b), has been proposed by Parry (1970) to account for anomalies in the X-ray diffraction patterns of α-fibrous proteins not predicted by simple theory. This model consists of short lengths of straight α-helix inclined to one another, approximating the coiled-coil over a short range with abrupt bends in the axes of the α-helices at intervals of 20–30 Å. Caspar et al. (1969), however, consider that the principal features of the α-protein X-ray diagrams are in agreement with the physically significant features of Crick's model for the coiled-coil. Thus the supercoiling of tropomyosin molecules in the crystals studied by these workers (Cohen et al., 1971) may possibly account for the puzzle posed by Elliott et al. (1968). Electron microscopic and X-ray diffraction studies of these crystals and of fibrous aggregates or tactoids have also led these workers to the interesting conclusions that the tropomyosin molecules are associated head-to-tail in polar filaments with a 400 Å period, and that the individual chains in each tropomyosin molecule are arranged in a parallel manner. Further, there appeared to be no discontinuity along the filaments which would mark the ends of molecules. This indicated that the coiled-coil might be continuous along the length of the tropomyosin filaments.

Heterogeneity of Rabbit Skeletal Tropomyosin

Although the postulate of a coiled-coil structure for tropomyosin, paramyosin, the rod portion of myosin, and other fibrous proteins was made many years ago, chemical and sequence data supporting such a structure have been fragmentary (Hodges and Smillie, 1970; Fraser et al., 1971). Since tropomyosin is the smallest and simplest of the fibrous proteins, it was the logical choice for amino acid sequence analysis. Rabbit skeletal muscle was the tissue source since this tropomyosin was the most fully characterized by chemical and physical methods. Initial efforts were directed to the question of identity or nonidentity of the polypeptide chains of the protein. Elucidation of the numbers of unique sequences about the cysteine residues was found to be inconsistent with a homogeneous preparation of two chemically identical polypeptide chains (Hodges and Smillie, 1970, 1972a). Similarly the results of polyacrylamide gel electrophoresis in solutions of concentrated urea and/or sodium dodecyl sulfate indicated chain heterogeneity (Weber and Osborne, 1969; Sender, 1971). Cummins and Perry (1972) have demonstrated, in fact, that of the two bands observed by

* Present address: The Rockefeller University, New York, N.Y. 10021.

them on gel electrophoresis of ^{14}C-labeled S-carboxymethylated tropomyosin, the slower and minor component contained twice the specific radioactivity of the faster, indicative of a higher cysteine content. However the demonstration that there are only two unique histidine sequences in tropomyosin and likely not more than seven or eight unique methionine sequences predicts that these different polypeptide chains are highly similar in amino acid sequence (Hodges and Smillie, 1972b; Hodges et al., 1972).

In the present work, attempts by several techniques to separate these chains in adequate amounts for sequence analysis have been unsuccessful. Consequently in an attempt to extend the sequence data, cyanogen bromide cleavage of the unfractionated ^{14}C-labeled S-carboxymethylated tropomyosin was undertaken.

Fractionation of Cyanogen Bromide Fragments

The protein preparation was treated with cyanogen bromide at 22–25°C for 20 hr as described by Steers et al. (1965). The molar ratio of cyanogen bromide to methionine was 100:1 in 70% formic acid at a protein concentration of 1%. The reaction was terminated by dilution with 10 volumes of water and lyophilization. This treatment resulted in a loss of 94% of the methionines and a 91% recovery of these residues as homoserine (Table 1).

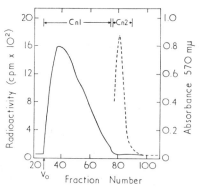

Figure 1. Fractionation of cyanogen bromide fragments of ^{14}C-labeled CM-tropomyosin by gel filtration on Sephadex G-50 in 0.05 N acetic acid. A sample of 500 mg in 25 ml was applied to the top of a Sephadex column (100 × 5 cm) and developed at a flow rate of 40 ml/hr. Fractions of 20 ml were collected and aliquots monitored by manual ninhydrin assays (– – –) and radioactive counting (——).

There was no significant destruction or conversion of other amino acids.

Separation of the fragments into two fractions, representing large and small components, was achieved by gel filtration on Sephadex G-50 in 0.05 N acetic acid (Fig. 1). Better separations were achieved in solutions of 8 M urea on either Sephadex G-75 or G-50, but because of the difficulties of recovering the small peptides from urea solutions, separation with acetic acid was used for this purpose. Fraction Cn2, representing the small cyanogen bromide fragments, was further fractionated by ion exchange chromatography and high voltage electrophoresis, and the sequences of the purified peptides elucidated by conventional methods. These have been reported elsewhere (Hodges and Smillie, 1972c).

Fraction Cn1, at a concentration of 6.6 mg/ml in 1.1 M KCl, 0.025 M phosphate buffer pH 7.0, sedimented as a single symmetrical peak in the ultracentrifuge, indicating that if more than one component were present, they were of similar molecular size. Upon polyacrylamide gel electrophoresis in 8 M urea at pH 9.5, the fraction separated into two poorly resolved bands consistent with a high degree of similarity in size and charge of the components. A low-speed sedimentation equilibrium run at 13,000 rpm was performed to determine the approximate molecular weight of the components in benign medium (1.1 M KCl, 0.025 M phosphate pH 7.0). Similarly, high-speed sedimentation equilibrium at 36,000 rpm was performed in 8 M urea (0.2 M KCl, 0.025 M phosphate buffer pH 7.0). The ln y vs. r^2 plots were linear in both cases, indicating a relative mass homogeneity of the components. The calculated molecular weights were approximately 30,000 in

Table 1. Amino acid Composition of Cyanogen Bromide-Treated CM-Tropomyosin and of Fragment Cn1B (residues per mole)

	CM-Tropomyosin[a]	CNBr-Treated CM-Tropomyosin[a,b]	Cn1B[b]	
Lys	40.8	40.9	15.8	16
His	2.0	2.1	1.8	2
Arg	14.8	14.7	5.6	6
Asp	30.1	30.3	13.4	13
Thr	7.8	7.7	4.4	5
Ser	12.1	11.6	7.5	8
Glu	77.2	76.6	37.6	38
Pro	0.5	0.4	—	
Gly	4.0	4.0	1.8	2
Ala	36.4	37.1	15.7	16
Val	9.7	9.6	7.8	8
Met	6.3	0.4	—	
Ile	10.8	11.1	7.3	7
Leu	33.3	33.3	17.0[c]	17
Tyr	5.9	5.6	4.6	5
Phe	1.4	1.4	1.0	1
CM-cys	1.3	1.2	0.9[d] (1.0)[e]	1
Hse	—	5.7	1.0	1
			Total residues	146

[a] Analyses converted to residues per 35,000 daltons; Hodges and Smillie (1970).
[b] All values for 22 hr hydrolysates at 110°.
[c] Leucine taken as 17.0.
[d] Value from amino acid analysis.
[e] Value from radioactive content.

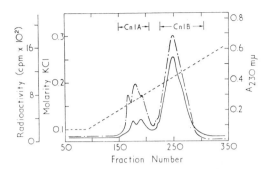

Figure 2. QAE-Sephadex A-50 chromatography of fraction Cn1. Approximately 500 mg of sample in 250 ml of starting buffer (0.10 M KCl, 8 M urea, 0.05 M Tris-HCl buffer pH 7.5) was applied to a Sephadex column of dimensions 5 × 100 cm. A linear gradient was generated with 2.5 liters of starting buffer and 2.5 liters of 0.30 M KCl, 8 M urea, 0.05 M Tris-HCl buffer pH 7.5. Flow rate was 35 ml/hr and 20 ml fractions were collected; absorbancy at 230 mμ (— — —) and radioactivity (———). The salt gradient is indicated by the broken line. Fractions A and B were pooled as indicated.

the benign medium and 17,000 in the denaturing solvent. A circular dichroism spectrum of the fraction Cn1 (in 1.1 M KCl, 0.025 M phosphate buffer pH 7.0) exhibited the normal pattern observed for α-helical proteins with two negative dichroic peaks located at 209 and 221 mμ. Their magnitude indicated an α-helical content of approximately 70 %. Optical rotatory dispersion data gave somewhat lower estimates of 50–60 %.

The above data indicated that fraction Cn1 was composed of a limited number of fragments, possibly only two, of roughly equal size and charge properties. After a number of attempts, separation into two major fractions was achieved by ion exchange chromatography on QAE-Sephadex (Fig. 2). The material used in this fractionation was derived from fraction Cn1 collected from Sephadex G-50 chromatography in 0.05 N acetic acid, and for this reason probably contained a small percentage of uncleaved polypeptide chains. This may account in part for the apparent heterogeneity of fraction Cn1A. To date this fraction has not been further characterized. Fraction Cn1B, containing the majority of the radioactivity and of the material absorbing at 230 mμ, was collected, dialyzed exhaustively against distilled water, and freeze-dried.

Amino acid analysis of this component (Table 1) gave a single homoserine residue and a calculated minimal mol wt of 17,000 daltons. Interestingly, it contained the bulk of the tyrosine and phenylalanine residues of tropomyosin, a fact consistent with the relatively high absorbancy of fraction Cn1B compared to Cn1A. Its S-carboxymethylcysteine content of one residue per mole, as measured by both radioactive content and amino acid analysis, indicated that it contained the two

highly homologous cysteine sequences previously described (Hodges and Smillie, 1970, 1972a), a fact recently confirmed from sequence analysis in our laboratory (Hodges et al., 1972). In this respect the fragment may be considered as heterogeneous in that it is composed of two or more different polypeptide chains corresponding to the intact tropomyosin preparation from which it was derived, but to be homogeneous in that it is free of cyanogen bromide fragments arising from other regions of the tropomyosin molecule(s). Finally, and of most significance, both of the two histidine residues of the tropomyosin subunit (35,000 daltons) were present. Since one of these histidines was known from our previous studies to be near the COOH-terminus of the protein (Hodges and Smillie, 1972b), this fragment of mol wt 17,000 daltons must be derived from the COOH-terminal half of the molecule.

Physical Properties and Amino Acid Sequence of Fraction Cn1B

The molecular weight of fraction Cn1B was estimated in benign medium (0.6 M KCl, 0.05 M phosphate pH 7.0) by both high- and low-speed sedimentation equilibrium and by high-speed sedimentation equilibrium in 8 M urea (0.2 M KCl, 0.05 M phosphate pH 7.0). The plots of ln y vs. r^2 are shown in Fig. 3 and provided molecular weights of

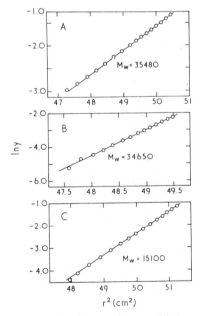

Figure 3. Typical sedimentation equilibrium data for fraction Cn1B in the absence and presence of 8 M urea. A, Low speed (16,000 rpm) in 0.6 M KCl, 0.05 M phosphate buffer pH 7.0; 0.92 mg/ml. B, High speed (32,000 rpm) in same solute as A; 0.92 mg/ml. C, High speed (40,000 rpm) in 0.2 M KCl, 8 M urea, 0.05 M phosphate pH 7.0; 1.0 mg/ml. The plots represent the natural logarithms of the fringe displacement with respect to the square of the radial distance.

1	2	3	4	5	6	7	8	9	10	11	12	13	14
(GLX,	ILE,	GLX,	GLX,	ILE,	GLX)	[LEU]	LYS -	GLU -	(ALA LEU)	LYS -	HIS -	ILE	[ALA]

15	16	17	18	19	20	21	22	23	24	25	26	27	28
GLU -	ASP -	(ALA)	ASP -	ARG -	LYS -	[TYR]	GLX -	GLX -	(VAL)	ALA -	ARG -	LYS -	[LEU]

29	30	31	32	33	34	35	36	37	38	39	40	41	42
VAL -	ILE -	(LEU ILE)	GLX -	SER GLY	ASX GLX	[LEU]	GLX -	ARG -	(ALA SER)	GLX -	GLX -	ARG	[ALA]

43	44	45	46	47	48	49	50	51	52	53	54	55	56
GLX -	LEU VAL	(SER ALA)	GLX -	GLY SER	LYS -	[CYS]	ALA GLY	GLX ASX	(LEU)	GLX -	GLX -	GLX -	[LEU]

57	58	59	60	61	62	63	64	65	66	67	68	69	70
LYS -	THR ILE LYS	(VAL)	THR -	ASN -	ASX -	[LEU]	LYS -	SER -	(LEU)	GLX -	ALA -	GLX VAL	[ALA]

71	72	73	74	75	76	77	78	79	80	81	82	83	84
GLX -	LYS -	(TYR)	SER -	GLX -	LYS -	[GLX]	ASX -	LYS -	(TYR)	GLX -	GLX -	GLX -	[ILE]

85	86	87	88	89	90	91	92	93	94	95	96	97	98
LYS -	VAL LEU	(LEU)	SER -	ASX -	LYS -	[LEU]	LYS -	GLX -	(ALA)	GLX -	THR -	ARG -	[ALA]

99	100	101	102	103	104	105	106	107	108	109	110	111	112
GLX -	PHE -	(ALA)	GLX -	ARG -	SER -	[VAL]	ALA THR	LYS -	(LEU)	GLX -	LYS -	SER -	[ILE]

113	114	115	116	117	118	119	120	121	122	123	124	125	126
ASX -	ASX -	(LEU)	GLX -	ASX -	GLX -	[LEU VAL]	TYR -	ALA -	(GLX)	LYS -	LEU -	LYS -	[TYR]

127	128	129	130	131	132	133	134	135	136	137	138	139	140
LYS -	ALA -	(ILE)	SER -	GLU -	GLU -	[LEU]	ASP -	HIS -	(ALA)	LEU -	ASN -	ASP -	[MET]

141	142	143
THR -	SER -	(ILE)

Figure 4. Tentative amino acid sequence of COOH-terminal half of rabbit skeletal tropomyosin. Cyanogen bromide fragment Cn1B is from residues 1–140. The COOH-terminal sequence (residues 128–143) was determined previously (Hodges and Smillie, 1972b). The residues (approximately six) at the NH$_2$-terminal end have not been sequenced. An unambiguous overlap is still required in the region of residues 124–129. A regular pattern of nonpolar amino acid residues occurs in two series, I and II. The residues in each of these series repeat at every seventh residue in the sequence. Series I positions are squared; Series II positions are circled. Amino acid substitutions have been observed in 15 positions and indicate a minimum of 4 different polypeptide chains in rabbit skeletal tropomyosin.

approximately 35,000 in benign medium and of about 15,000 in the denaturing solvent. The latter value is in reasonable agreement with 17,000 from the amino acid composition. As with the mixed cyanogen bromide fraction Cn1, the circular dichroism spectrum was typical of a protein with high α-helical content (53–61%).

Amino acid sequence analysis of this fragment has now reached an advanced stage and we are able to report what is, we believe, a tentative but essentially correct primary structure for the C-terminal half of the tropomyosin molecule (Fig. 4). The only major uncertainty is in the NH$_2$-terminal region of the fragment where we have been unable as yet to assign a definite sequence to the terminal six or more residues. The overlap for the region of residues 124–129 is also unsatisfactory. Full details of this sequence analysis will be reported elsewhere.

Examination of the sequence reveals that a regular pattern of hydrophobic residues occurs in two series, I and II, indicated by the open squares and circles in Fig. 4. The residues in each of these series repeat at every seventh position in the sequence. Thus the residues of series I occur at positions 7, 14, 21, 28, . . . , and the residues of series II occur at positions 10, 17, 24, 31, ... In this pattern an amino acid in series I is always found 3 residues on the NH$_2$-terminal side and 4 residues on the COOH-terminal side of a residue in series II. Thus, considering the two series together, there is an alternating 3 to 4 pattern of repeating hydrophobic residues which appears to be essentially continuous throughout the sequence of

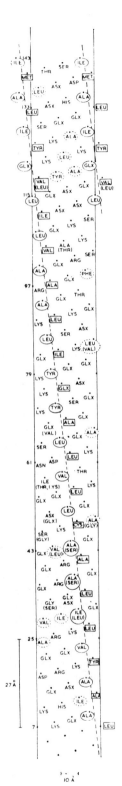

Figure 5. Amino acid sequence of the COOH-terminal half of tropomyosin (residues 7–143 of Fig. 4) plotted on an α-helical net. The radius of the α-helix is taken as 5 Å with 3.6 residues per turn and a residue translation of 1.5 Å. Residues in positions of series I are squared and those in series II circled. Other hydrophobic residues are enclosed in broken circles.

this half of the tropomyosin molecule. The pattern clearly provides on the average a nonpolar amino acid at every 3.5 residues and is consistent with the coiled-coil model.

At several positions in either series I or II, there occur amino acid residues which are not ordinarily classified as hydrophobes. Thus glutamic acid or glutamine occurs at positions 77 and 122, serine as a replacement of alanine is found at positions 38 and 45, and cysteine is located at position 49. Four tyrosines are found at positions 21, 73, 80, and 126. However the polar or nonpolar nature of these residues, with the exception of glutamic acid, is ambivalent and their presence in certain circumstances has been reported in the interior of globular proteins. On the other hand, it is possible that their occurrence at certain positions in the hydrophobic pattern leads to a local disruption of the regularity of the coiled-coil. This may be of importance in the interaction of tropomyosin with actin or one or more members of the troponin complex. Irrespective of their significance, it is clear that their presence does not lead to a disruption of the long-range pattern of repeating hydrophobes in the two series.

Nonpolar residues are not confined exclusively to the positions of series I and II. This is best illustrated by the projection of the sequence on a helical net as illustrated in Fig. 5. Here a regular right-handed α-helix with 3.6 residues per turn is represented as a cylinder which has been opened up and laid flat on the paper. The position of each amino acid residue in the sequence is then plotted on the cylinder in a position corresponding to its position in the α-helix. It can be seen that a line can be drawn midway between the positions occupied by the residues in the two series. This line is not parallel to the axis of the cylinder because of the nonintegral nature of the α-helix. Nonpolar residues which do not occur in the positions of either series I or II are circled by broken lines. It is clear that the great majority of the hydrophobic residues in the sequence are found along the band formed by the residues of series I and II and that the majority of residues outside this region are polar and charged residues.

If two such projected cylinder or helical nets are superimposed face-to-face at an angle, so that the hydrophobic band of one helix lies over that of the other, it is found that the side chains of one helix fall between those of the other helix. This can only be done if the side chains of residues in series I or II are aligned along the hydrophobic line with residues of the same series, as illustrated in Fig. 6. Thus in the coiled-coil, the two identical or very similar α-helices can only be packed together in a knobs-into-holes arrangement in a manner such

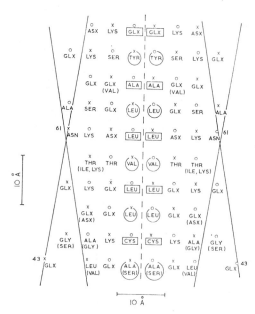

Figure 6. Face-to-face superposition of two α-helical nets at an angle so that the hydrophobic band of one helix lies over the other. Positions of amino acid residues of one helix are indicated by X and of the other helix by O. Residues in series I positions are squared and those in series II positions are circled. The regions of sequence shown are from residues 43–75 and the two chains are in register.

that the two chains are in register or out of register by seven amino acid residues or some multiple of it.

Model Building and a Staggered Parallel Coiled-Coil

Inspection of the sequence shown in Figs. 4 and 5 shows that there appears to be an alternation of groups with bulky and less bulky side chains in the hydrophobic pattern. Thus three alanines are grouped together at positions 10, 14, and 17; three alanines and cysteine occur at positions 38, 42, 45, and 49 and a further three alanines at positions 94, 98, and 101. These regions of small hydrophobes are separated by regions rich in bulky side chains such as leucine, valine, and isoleucine. The significance of this pattern may be to bring the chains into appropriate alignment. Thus if the parallel chains are in register, there would be interaction between the bulky hydrophobes of one chain with the same bulky hydrophobes on the other. Alternatively, the parallel chains may be staggered by 7, 14, 21, or 28 residues such that bulky hydrophobes of one chain interact with the less bulky of the other. In an attempt to test which of these two possibilities seemed more likely, various arrangements were tested with α-helical models of the sequence built from CPK space-filling atoms.

The case of two parallel and identical chains in register results in nine valine to valine or isoleucine

to isoleucine interactions. The minimum radius of the coiled-coil at these positions occurs when the γ-carbons and hydrogens of the side chains of these residues are in Van der Waal's contact. The major systematic hydrophobic bonding would then result from γ- to γ-carbon and hydrogen interaction. However such an arrangement would prevent interaction between alanine residues unless the radius of the coiled-coil was variable and constricted in the regions of these residues.

A more attractive arrangement would appear to be one in which the radius of the coiled-coil is intermediate between a situation in which two valines or isoleucines interact and that in which two alanines are paired through β- to β-carbon and hydrogen interactions. This will occur, for example, when an alanine is paired with an isoleucine through β- to γ-carbon and hydrogen interaction, or when two leucines interact through their β- and δ-carbons and hydrogens. To represent this intermediate situation, a coiled-coil was constructed in which all hydrophobic positions were occupied by leucines and in which the two α-helices were running in the same direction. Pairs of leucines were removed from positions corresponding to series I and II as illustrated in Fig. 7 and replaced with all possible combinations of amino acid pairs. It was necessary to do this for positions in both series I and II since residues in the two series are not stereochemically equivalent in the coiled-coil. Each of the amino acid pairs was tested for its stereochemical fit without disruption of the coiled-coil established by the leucine to leucine interactions in the remainder of the structure. In this way it was possible to construct matrices for all possible combinations of amino acids for both series. These are shown in Figs. 8 and 9. It is clear that amino acid pairs fall into one of four categories in the case of series I and into one of three categories in the case of series II positions. In the case of series I, the situation is more restricted since residues involving β-carbon branched side chains (i.e., isoleucine, valine, and threonine) cannot be sterically accommodated with themselves without an increase in the radius of the coiled-coil established by the leucine to leucine interactions. However each of these residues can be paired with any other

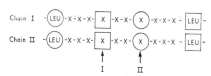

Figure 7. A coiled-coil was constructed of CPK space-filling components in which all positions of Series I and II were occupied by leucine residues. All possible combinations of amino acid pairs were then tested for stereochemical fit in each of positions I and II.

	Lys	Arg	Glu	His	Asp	Gly	Ala	Ser	Cys	Phe	Tyr	Asn	Gln	Met	Thr	Val	Ile	Leu
Lys	○	○	○	○	○	—	○	○	○	○	○	○	○	○	○	○	○	○
Arg	○	○	○	○	○	—	○	○	○	○	○	○	○	○	○	○	○	○
Glu	○	○	○	○	○	—	○	○	○	○	○	○	○	○	○	○	○	○
His	○	○	○	—	—	—	—	—	—	—	—	—	○	○	○	○	○	○
Asp	○	○	○	—	—	—	—	—	—	—	—	—	○	○	○	○	○	○
Gly	—	—	—	—	—	—	—	—	—	—	—	—	—	—	—	—	—	—
Ala	○	○	○	—	—	—	—	—	—	—	—	—	●	●	●	●	●	●
Ser	○	○	○	—	—	—	—	—	—	—	—	—	●	●	●	●	●	●
Cys	○	○	○	—	—	—	—	—	—	—	—	—	●	●	●	●	●	●
Phe	○	○	○	—	—	—	—	—	—	—	—	—	●	●	●	●	●	●
Tyr	○	○	○	—	—	—	—	—	—	—	—	—	●	●	●	●	●	●
Asn	○	○	○	—	—	—	—	—	—	—	—	—	●	●	●	●	●	●
Gln	○	○	○	○	○	—	●	●	●	●	●	●	●	●	●	●	●	●
Met	○	○	○	○	○	—	●	●	●	●	●	●	●	●	●	●	●	●
Thr	○	○	○	○	○	—	●	●	●	●	●	●	●	●	□	□	□	●
Val	○	○	○	○	○	—	●	●	●	●	●	●	●	●	□	□	□	●
Ile	○	○	○	○	○	—	●	●	●	●	●	●	●	●	□	□	□	●
Leu	○	○	○	○	○	—	●	●	●	●	●	●	●	●	●	●	●	●

Figure 8. Amino acid pairs in series I. (—) Stereochemical fit; no β- to γ- or δ-carbon and hydrogen Van der Waal's contact. (□) Not stereochemically permissible. (○) Van der Waal's contact between β- and γ- or δ-carbon and hydrogen but involve charged residue(s). (●) Van der Waal's contact between β- and γ- or δ-carbons and hydrogens and do not involve charged residues.

	Lys	Arg	Glu	His	Asp	Gly	Ala	Ser	Cys	Phe	Tyr	Asn	Gln	Met	Thr	Val	Ile	Leu
Lys	○	○	○	○	○	—	○	○	○	○	○	○	○	○	—	○	○	○
Arg	○	○	○	○	○	—	○	○	○	○	○	○	○	○	—	○	○	○
Glu	○	○	○	○	○	—	○	○	○	○	○	○	○	○	—	○	○	○
His	○	○	○	—	—	—	—	—	—	—	—	—	○	○	—	○	○	○
Asp	○	○	○	—	—	—	—	—	—	—	—	—	○	○	—	○	○	○
Gly	—	—	—	—	—	—	—	—	—	—	—	—	—	—	—	—	—	—
Ala	○	○	○	—	—	—	—	—	—	—	—	—	●	●	—	●	●	●
Ser	○	○	○	—	—	—	—	—	—	—	—	—	●	●	—	●	●	●
Cys	○	○	○	—	—	—	—	—	—	—	—	—	●	●	—	●	●	●
Phe	○	○	○	—	—	—	—	—	—	—	—	—	●	●	—	●	●	●
Tyr	○	○	○	—	—	—	—	—	—	—	—	—	●	●	—	●	●	●
Asn	○	○	○	—	—	—	—	—	—	—	—	—	●	●	—	●	●	●
Gln	○	○	○	○	○	—	●	●	●	●	●	●	●	●	—	●	●	●
Met	○	○	○	○	○	—	●	●	●	●	●	●	●	●	—	●	●	●
Thr	—	—	—	—	—	—	—	—	—	—	—	—	—	—	—	—	—	—
Val	○	○	○	○	○	—	●	●	●	●	●	●	●	●	—	●	●	●
Ile	○	○	○	○	○	—	●	●	●	●	●	●	●	●	—	●	●	●
Leu	○	○	○	○	○	—	●	●	●	●	●	●	●	●	—	●	●	●

Figure 9. Amino acid pairs in series II. (—) Stereochemical fit; no β- to γ- or δ-carbon and hydrogen Van der Waal's contact. (○) Van der Waal's contact between β- and γ- or δ-carbons and hydrogens but involve charged residues. (●) Van der Waal's contact between β- and γ- or δ-carbons and hydrogens and do not involve charged residues.

residue which does not have a branched β-carbon side chain. Of the remaining three categories in the matrices, the one most favorable to the stabilization of the coiled-coil would be that which provides β- to γ- or δ-carbon and hydrogen Van der Waal's contacts and does not involve charged residues. We will refer to this as the favorable pairing category.

It is probably significant that if we stagger the two chains of tropomyosin sequence by 14 residues as illustrated in Fig. 10, then the regions of small hydrophobes tend to pair with the bulkier hydrophobes in a most satisfactory manner. All pairs now fall into the favorable pairing category with only two exceptions. Thus at position 94 of chain I, the pairing of alanine with tyrosine provides no β- to γ- or δ-interactions but is stereochemically permissible without a change in the radius of the

coiled-coil structure. It has also been observed that the aromatic ring of tyrosine (or phenylalanine) can interact with other side chains in the immediate vicinity to perhaps compensate for the absence of β- to γ- or δ-interactions. A more serious discrepancy occurs at position 119 where the pairing of two valines in series I is stereochemically impermissible without an increase in radius of the coiled-coil. However the valine at position 119 of chain I was found only as a substitution for leucine and in much lower yield. Thus it probably arises from the minor tropomyosin component, and it is possible that a compensating substitution for valine occurs in chain II at this position but has not yet been detected in our sequence analysis. The staggering of the two chains by 7, 21, or 28 residues leads to a less satisfactory packing arrangement since 8, 6, and 10 amino acid pairs, respectively,

	7	8	9	10	11	12	13	14	15	16	17	18	19
Chain I	LEU	LYS	GLU	ALA LEU	LYS	HIS	ILE	ALA	GLU	ASP	ALA	ASP	ARG

	20	21	22	23	24	25	26	27	28	29	30	31	32
Chain I	LYS	TYR	GLX	GLX	VAL	ALA	ARG	LYS	LEU	VAL	ILE	LEU ILE	GLX
Chain II		LEU	LYS	GLU	ALA LEU	LYS	HIS	ILE	ALA	GLU	ASP	ALA	ASP

	33	34	35	36	37	38	39	40	41	42	43	44	45
Chain I	SER GLY	ASX GLX	LEU	GLX	ARG	ALA SER	GLX	GLX	ARG	ALA	GLX	LEU VAL	SER ALA
Chain II	ARG	LYS	TYR	GLX	GLX	VAL	ALA	ARG	LYS	LEU	VAL	ILE	LEU ILE

	46	47	48	49	50	51	52	53	54	55	56	57	58
Chain I	GLX	GLY SER	LYS	CYS	ALA GLY	GLX ASX	LEU	GLX	GLX	GLX	LEU	LYS	THR ILE LYS
Chain II	GLX	SER GLY	ASX GLX	LEU	GLX	ARG	ALA SER	GLX	GLX	ARG	ALA	GLX	LEU VAL

	59	60	61	62	63	64	65	66	67	68	69	70	71
Chain I	VAL	THR	ASN	ASX	LEU	LYS	SER	LEU	GLX	ALA	GLX VAL	ALA	GLX
Chain II	SER ALA	GLX	GLY SER	LYS	CYS	ALA GLY	GLX ASX	LEU	GLX	GLX	GLX	LEU	LYS

	72	73	74	75	76	77	78	79	80	81	82	83	84
Chain I	LYS	TYR	SER	GLX	LYS	GLX	ASX	LYS	TYR	GLX	GLX	GLX	ILE
Chain II	THR	VAL	THR	ASN	ASX	LEU	LYS	SER	LEU	GLX	ALA	GLX VAL	ALA

	85	86	87	88	89	90	91	92	93	94	95	96	97
Chain I	LYS	VAL LEU	LEU	SER	ASX	LYS	LEU	LYS	GLX	ALA	GLX	THR	ARG
Chain II	GLX	LYS	TYR	SER	GLX	LYS	GLX	ASX	LYS	TYR	GLX	GLX	GLX

	98	99	100	101	102	103	104	105	106	107	108	109	110
Chain I	ALA	GLX	PHE	ALA	GLX	ARG	SER	VAL	ALA THR	LYS	LEU	GLX	LYS
Chain II	ILE	LYS	VAL LEU	LEU	SER	ASX	LYS	LEU	LYS	GLX	ALA	GLX	THR

	111	112	113	114	115	116	117	118	119	120	121	122	123
Chain I	SER	ILE	ASX	ASX	LEU	GLX	ASX	GLX	LEU VAL	TYR	ALA	GLX	LYS
Chain II	ARG	ALA	GLX	PHE	ALA	GLX	ARG	SER	VAL	ALA THR	LYS	LEU	GLX

	124	125	126	127	128	129	130	131	132	133	134	135	136
Chain I	LEU	LYS	TYR	LYS	ALA	ILE	SER	GLU	GLU	LEU	ASP	HIS	ALA
Chain II	LYS	SER	ILE	ASX	ASX	LEU	GLX	ASX	GLX	LEU VAL	TYR	ALA	GLX

	138	137	139	140	141	142	143			
Chain I	LEU	ASN	ASP	MET	THR	SER	ILE			
Chain II	LYS	LEU	LYS	TYR	LYS	ALA	ILE	SER	GLU	etc.

Figure 10. Staggered arrangement of the two helices in coiled-coil in which the two chains are out of register by 14 residues to maximize hydrophobic interaction and regularity of the coiled-coil.

would then not fall within the favored pairing category as detailed in the matrices of Figs. 8 and 9. In these comparisons the residues at positions 77 and 122 have been assumed to be glutamines rather than glutamic acids. However even if this is not so, the conclusion that a staggering of 14 residues is the most favorable arrangement would not be invalidated, since all staggering arrangements would be affected to the same extent.

Photographs of a CPK space-filling model of a portion of the staggered coiled-coil are shown in Fig. 11A and B. The region shown is approximately from residues 28–86 of chain I and from residues 14–72 of chain II (Fig. 10). Several points emerge from an inspection of the details of the structure. First, it is clear that portions of the bulkier residues forming the hydrophobic core of the structure are turned to the outside and will be exposed at the surface, creating patches or streaks of hydrophobicity. A good example of this is seen in Fig. 11A, where below the exposed cysteine sulfhydryl in the middle of the photograph one observes in

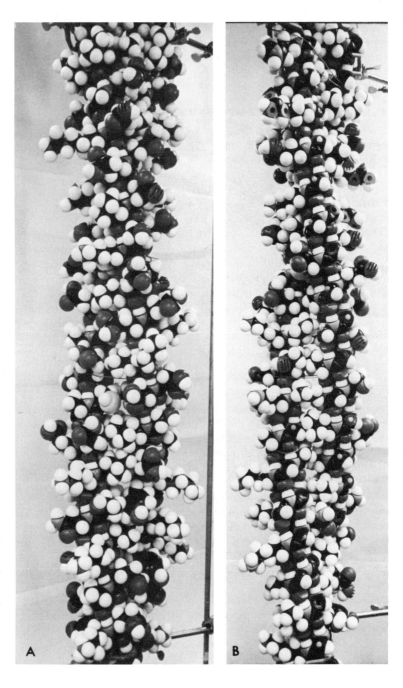

Figure 11. Photographs of a CPK space-filling model of a portion of the staggered coiled-coil of tropomyosin. The region shown is approximately from residues 28–86 of chain I and from residues 14–72 of chain II (Fig. 10). The direction of the chains (NH_2-terminal to COOH-terminal) is from top to bottom. *A*, the sulfhydryl of cysteine-49 of chain II can be seen just below the middle of the photograph. A hydrophobic ring or collar (made up of residues leucine-28, valine-29, isoleucine-30, and leucine-31 of chain I and isoleucine-13 and alanine-14 of chain II) occurs at the top of the model. *B*, taken from the opposite side of the molecule, many of the side chains have been removed to illustrate the pairing of the hydrophobic side chains.

descending order: leucine, leucine, tyrosine, leucine, and tyrosine. A further feature of the packing arrangement is that where ambivalent hydrophobes, such as tyrosine or cysteine, occur in the positions of either series I or II, their reactive groups are still available at the surface for ionization or chemical reaction. Thus the model is consistent with the titration data of Lowey (1965), even though four of the five tyrosines of the cyanogen bromide fragment occur in the positions of either series I or series II. A similar situation exists with the glutamine or glutamic acid residues which occur at two positions (77 and 122) in the hydrophobic pattern. Their side chains are sufficiently long that the acidic or amidated moieties can be turned towards the exterior and still provide the necessary β- to γ-carbon Van Der Waal's interaction with their paired amino acid residues.

Possible Interaction Sites

Physical-chemical studies of tropomyosin in solution have indicated that the aggregation behavior of tropomyosin at low ionic strength is explicable in terms of end-to-end association of the molecules (Tsao and Bailey, 1953; Kay and Bailey, 1960). Electron microscopic and X-ray diffraction studies of tropomyosin crystals have shown that there is a specific head-to-tail bonding of tropomyosin molecules in the filaments of these kitelike structures (Cohen et al., 1971; Casper et al., 1969). The electron microscopic observations of Higashi-Fujime and Ooi (1969) on the edges of negatively stained crystals suggest that the molecules end near the middle of the short arms of these kites. However there appears to be no discontinuity or appreciable overlap in this region between the ends of the two-stranded coiled-coil molecules. Our observation that the amino acid sequence of the COOH-terminal half of the tropomyosin polypeptide chains may be best accommodated in a coiled-coil structure in which the two α-helices are staggered by 14 residues may be of some relevance to this question. If the two polypeptide chains are of equal length, then at each end of the molecule there will be a 14 residue region of single-stranded structure which in monomeric tropomyosin in solution would presumably exist in a nonhelical or disordered form. In an aggregated state or in tactoids and crystals however, these "sticky ends" may well overlap and provide the end-to-end attachment sites such that the coiled-coil structure could be continuous along the length of the tropomyosin filaments. Although our knowledge of the amino acid sequence of tropomyosin at the NH_2-terminal region is still fragmentary (Hodges and Smillie, 1972c), available evidence is consistent with such a hypothesis. Thus the pattern of hydro-

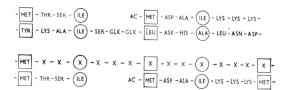

Figure 12. Postulated head-to-tail junction of tropomyosin molecules. Only the first eight residues of the NH_2-terminal region of the chains have been sequenced (Hodges and Smillie, 1972c).

phobes in two series, I and II, at an average distance of 3.5 residues is present and can be aligned with the C-terminal sequence (Fig. 12), leaving only a gap of three amino acid residues of single-stranded structure at the junction points. There would be no interruption in the hydrophobic pairing pattern. It is probably also significant that the NH_2-terminal ends of the tropomyosin polypeptide chains are N-acetylated, reducing the local disruption that might otherwise occur to the order of the structures with a charged NH_2-terminal group.

The binding sites of one or more members of the troponin complex to the tropomyosin molecule can only be a matter for conjecture at this time. The observation that p-chloromercuribenzoate appears to bind at a pair of sites, one of which may be near the troponin binding site (Cohen et al., 1971), may be of relevance to the location of the cysteine residues in the COOH-terminal cyanogen bromide fragment reported in this work. Thus the two homologous sulfhydryl sequences reported in an earlier study (Hodges and Smillie, 1970, 1972a) have been located 95 residues, or roughly one-third of the distance, from the COOH-terminus of the protein. If a relationship between the binding of troponin and the sulfhydryl groups of tropomyosin is confirmed (Parker and Kilbert, 1970), it would establish the site of troponin binding to the COOH-terminal half of the molecule and fix the head-to-tail direction of the tropomyosin molecules in the crystal filaments.

The nature of the tropomyosin to actin monomer interaction is at this time quite obscure. As discussed by Cohen et al. (1971) it is not possible for there to exist an identical relation between each actin monomer and its portion of the tropomyosin molecule unless there were identical areas of sequence distributed approximately sevenfold along the tropomyosin polypeptide chains. The present work establishes that this is not the case. However a similar but not identical spatial arrangement of charged and nonpolar groups on the surface of the coiled-coil, repeated along the length of the molecule, cannot be ruled out. An inspection of the structure for this possibility must await the completion of the entire amino acid sequence and the assignment of all amides and acidic groups.

Although a structure in which the two α-helices are staggered by fourteen residues provides an attractive explanation for the head-to-tail association of tropomyosin molecules in solution, in crystals, and in the thin filaments of muscle, the possibility that the two chains are in register cannot be excluded at this time. Indeed, the more irregular structure provided by changes in the radius of the coiled-coil arising from the two chains in register could well be of importance in the interaction of tropomyosin with actin and the troponin complex in the thin filament of the myofibril.

The repeating hydrophobic pattern observed with rabbit skeletal tropomyosin in the present work undoubtedly has implications for the structures of other fibrous proteins, and in particular for those of paramyosin and the rod portion of myosin. Both of these proteins probably exist as two-stranded coiled-coils in which the two polypeptide chains run in the same direction (Lowey et al., 1969). It seems that the only gross differences in the molecular architecture of these molecules are their overall lengths. However there is a significant difference in the association properties of tropomyosin on the one hand and the myosin-paramyosin proteins on the other. Thus while tropomyosin aggregates in a specific polar end-to-end fashion, the others tend to form specific staggered side-to-side associations. Although one would predict a similar regular repeat of hydrophobic amino acid residues in myosin and paramyosin as we observe in tropomyosin, the longer-range patterns of bulky and small hydrophobes and the distribution of these and ionic and polar residues on the surface of the coiled-coil may be quite different.

Summary and Conclusions

The isolation and amino acid sequence analysis of a cyanogen bromide fragment from rabbit skeletal muscle tropomyosin has permitted us to report a tentative primary structure for the COOH-terminal half of the molecule. These studies have confirmed our previous conclusion that this tropomyosin consists of several different, but highly similar, polypeptide chains. In the sequence, nonpolar amino acid residues occur in two series at intervals of every seven residues. Amino acid residues in series I are three residues on the NH_2-terminal side, and four residues on the COOH-terminal side of residues in series II. Thus on the average, nonpolar residues occur every 3.5 residues. The occasional finding of a charged or ambivalent hydrophobic residue in the positions of series I or II does not lead to a disruption of this long-range

pattern. The majority of residues located between the nonpolar residues are charged or polar amino acids.

Two highly similar or identical α-helices with the reported sequence can be packed together in parallel in a coiled-coil structure. These may be in register or staggered by seven residues or some small multiple of it. The observation that groups of small hydrophobic side chains appear to alternate with groups of bulky side chains suggests that a staggered arrangement of the two α-helices would maximize the regularity and hydrophobic interactions of the coiled coil. Model-building considerations show that this would occur with a stagger of 14 residues. Such an arrangement could account for the end-to-end aggregation of tropomyosin molecules in solution and in crystal and tactoid filaments. However a structure in which the two polypeptide chains are in register cannot be ruled out at this time. A more irregular structure might be of importance, for instance, in the interactions of tropomyosin with actin and the several members of the troponin complex. It is predicted that a similar pattern of repeating hydrophobic residues occurs in the rod region of myosin and in paramyosin, but that the distribution of small and bulky hydrophobes and of ionic and polar amino acid residues on the surface of the coiled-coils is different in these proteins in view of their different associating properties.

Acknowledgments

We are indebted to Mr. M. Natriss and Mr. M. Carpenter for their expert technical assistance. We wish to acknowledge the support of this research by the Canadian Heart Foundation and the Medical Research Council of Canada. R. S. Hodges was a recipient of a Medical Research Council of Canada Studentship and J. Sodek holds a Post-doctorate Fellowship from the same organization.

References

CASPAR, D. L. D., C. COHEN, and W. LONGLEY. 1969. Tropomyosin: Crystal structure, polymorphism and molecular interactions. *J. Mol. Biol.* **41**: 87.

COHEN, C. and K. C. HOLMES. 1963. X-ray diffraction evidence for α-helical coiled-coils in native muscle. *J. Mol. Biol.* **6**: 423.

COHEN, C., D. L. D. CASPAR, D. A. D. PARRY, and R. M. LUCAS. 1971. Tropomyosin crystal dynamics. *Cold Spring Harbor Symp. Quant. Biol.* **36**: 205.

CRICK, F. H. C. 1953. The packing of α-helices; simple coiled-coils. *Acta Cryst.* **6**: 689.

CUMMINS, P. and S. V. PERRY. 1972. Subunit structure and biological activity of tropomyosin B from different muscle types. *Biochem. J.* **128**: 106P.

ELLIOTT, A., J. LOWY, D. A. D. PARRY, and P. J. VIBERT. 1968. Puzzle of the coiled coils in the α-protein paramyosin. *Nature* **218**: 656.

FRASER, R. D. B. and T. P. MACRAE. 1961a. α-Configuration of fibrous proteins. *Nature* **189**: 572.

——. 1961b. The molecular configuration of α-keratin. *J. Mol. Biol.* **3**: 640.

FRASER, R. D. B., T. P. MACRAE, G. R. MILLWARD, D. A. D. PARRY, E. SUZUKI, and T. A. TULLOCH. 1971. The molecular structure of keratins. *Applied Polymer Symp.* **18**: 65.

HIGASHI-FUJIME, S. and T. OOI. 1969. Electron microscopic studies on the crystal structure of tropomyosin. *J. de Microscopie* **8**: 535.

HODGES, R. S. and L. B. SMILLIE. 1970. Chemical evidence for chain heterogeneity in rabbit muscle tropomyosin. *Biochem. Biophys. Res. Comm.* **41**: 987.

——. 1972a. Cysteine sequences of rabbit skeletal tropomyosin. *Can. J. Biochem.* **50**: 330.

——. 1972b. The histidine and methionine sequences of rabbit skeletal tropomyosin. *Can. J. Biochem.* **50**: 312.

——. 1972c. Cyanogen bromide fragments of rabbit skeletal tropomyosin. *Can. J. Biochem.* (in press).

HODGES, R. S., J. SODEK and L. B. SMILLIE. 1972. The cyanogen bromide fragments of rabbit skeletal tropomyosin. *Biochem. J.* **128**: 102P.

HOLTZER, A., R. CLARK, and S. LOWEY. 1965. The conformation of native and denatured tropomyosin B. *Biochemistry* **4**: 2401.

KAY, C. M. and K. BAILEY. 1960. Light scattering in solutions of native and guanidinated rabbit tropomyosin. *Biochim. Biophys. Acta* **40**: 149.

LOWEY, S. 1965. Comparative study of the α-helical muscle proteins: Tyrosyl titration and effect of pH on conformation. *J. Biol. Chem.* **240**: 2421.

LOWEY, S., H. S. SLAYTER, A. G. WEEDS, and H. BAKER.

1969. Substructure of the myosin molecule. I. Subfragments of myosin by enzymic degradation. *J. Mol. Biol.* **42**: 1.

MCCUBBIN, W. D., R. F. KOUBA, and C. M. KAY. 1967. Physicochemical studies on bovine cardiac tropomyosin. *Biochemistry* **6**: 2417.

OLANDER, J., M. F. EMERSON, and A. HOLTZER. 1967. On the dissociation and reassociation of the polypeptide chains of tropomyosin and paramyosin. *J. Amer. Chem. Soc.* **89**: 3058.

PARKER, C. J. and L. H. KILBERT. 1970. A study of the role of sulphydryl groups in the interaction of troponin and myofibrils. *Arch. Biochem. Biophys.* **140**: 326.

PARRY, D. A. D. 1970. A proposed conformation for α-fibrous proteins. *J. Theoret. Biol.* **26**: 429.

RIDDIFORD, L. and H. SCHERAGA. 1962. Structural studies of paramyosin. I. Hydrogen ion equilibria. *Biochemistry* **1**: 95.

SENDER, P. M. 1971. Muscle fibrils: Solubilization and gel electrophoresis. *FEBS Letters* **17**: 106.

STEERS, E., G. R. CRAVEN, and C. B. ANFINSEN. 1965. Evidence for non-identical chains in the β-galactosidase of *Escherichia coli* K12. *J. Biol. Chem.* **240**: 2478.

TSAO, T.-C. and K. BAILEY. 1953. Aspects of polymerization in proteins of the muscle fibril. *Disc. Faraday Soc.* **13**: 145.

WEBER, K. and M. OSBORN. 1969. The reliability of molecular weight determinations by dodecyl sulfate-polyacrylamide gel electrophoresis. *J. Biol. Chem.* **244**: 4406.

WOODS, E. F. 1967. Molecular weight and subunit structure of tropomyosin B. *J. Biol. Chem.* **242**: 2859.

——. 1969. Comparative physicochemical studies on vertebrate tropomyosins. *Biochemistry* **8**: 4336.

Structure of the Actin-containing Filaments in Vertebrate Skeletal Muscle

JEAN HANSON, V. LEDNEV, E. J. O'BRIEN, AND PAULINE M. BENNETT

Medical Research Council Muscle Biophysics Unit, King's College, London, England

In the skeletal muscles of vertebrates the chain of processes coupling excitation with contraction terminates in the release of the actin subunits in the thin filaments from an inhibited state. While the muscle is relaxed, the actin subunits are prevented from interacting with myosin: Cross-bridges cannot be formed, and tension cannot be generated. The state of the actin subunits is regulated by a multi-component protein system which is believed to be situated in the thin filaments and which is itself controlled by the level of free Ca^{++} ions in the myofibrillar compartment of the muscle cell. On excitation Ca^{++} ions enter the myofibril and are captured by one component of the regulatory protein system, with the result that the actin subunits are released from inhibition. The regulatory system consists of tropomyosin (TM) and at least two other components, collectively known as troponin (TP), one of which binds the triggering Ca^{++} ion (review by Ebashi and Endo, 1968). In this paper we shall consider the structural basis of the mechanism by which the regulatory proteins affect the state of the actin subunits.

The main evidence that the regulatory proteins are situated together with actin in the thin myofilament can be summarized as follows: (1) The filaments contain another structure besides F-actin (Hanson and Lowy, 1963, 1964; Moore et al., 1970); (2) the Mg ATPase activity of myosin in the presence of natural thin filaments separated from myofibrils is inhibited when the Ca^{++} ion level is reduced (Kendrick-Jones et al., 1970); (3) antibodies to components of the regulatory system bind to the thin filaments in the myofibril (review by Ebashi and Endo, 1968).

Following the suggestion by Hanson and Lowy (1963) that TM may be situated as two long strands in the two long-pitch grooves of the F-actin double helix, Ebashi et al. (1969) proposed further that TP, pictured as having a globular shape, is attached to the TM strands and spaced at intervals of about 400 Å along the actin-TM filament. This model is supported by the following evidence: (1) In vitro under near physiological conditions, TM binds to F-actin, and TP binds to TM but not to actin (review by Ebashi and Endo, 1968). (2) TM molecules aggregate end-to-end in vitro, and the shape of the molecule, as observed in crystals of TM, is

curved so that it could follow the long-pitch helical groove in the F-actin structure (Cohen et al. 1971). (3) The three-dimensional image reconstructed from an electron micrograph of a thin myofilament shows extra material lying mainly between the two long-pitch actin strands (Moore et al., 1970). (4) When TP is incorporated in the various crystalline structures that TM forms in vitro, it is located at intervals of about 400 Å along the strands of TM (see Cohen et al., 1971). (5) The thin myofilaments, examined in the intact muscle by X-ray diffraction or in sections in the electron microscope, show an axial periodicity that could be accounted for if some material were attached at sites spaced about 400 Å apart (review by Hanson, 1968a). (6) Ferritin-labeled antibody to TP binds to the thin myofilaments at intervals of about 400 Å (Ohtsuki et al., 1967).

Here we shall present additional evidence that supports this model and has been obtained by optical and X-ray diffraction studies on synthetic thin filaments. In particular our results strengthen the evidence that TP is located at intervals of about 400 Å along the filaments, and they provide the first good evidence that TM does indeed lie in the long-pitch grooves of the F-actin helix. The changes that we observe in the X-ray and optical diffraction patterns when TM (with or without TP) is added to F-actin provide a basis for interpreting published X-ray diffraction data from the intact muscle in the resting, contracting, and rigor states. Some of our results have been reported briefly elsewhere (Hanson, 1967a, b, 1968b; O'Brien et al., 1971).

Methods

Our work on the structure of synthetic filaments was started in 1965 and has continued while characterization of the regulatory proteins has made rapid progress. Therefore we have used many different protein preparations (Table 1), with the advantage that our conclusions are based on a variety of experiments, all giving results that are consistent with one another. Details of the methods used for preparing the proteins are given in Hanson (1972).

Crystalline aggregates of synthetic filaments (Hanson, 1967a, b, 1968b) were formed by raising the Mg^{++} ion concentration to 0.07 M or higher

311

Table 1. Cross-Striation in Sectioned Crystals Correlated with TP

Preparation	Composition (SDS electrophoresis)	Ca++-Sensitizing Activity	Cross-Striation
Purified actin[a]	A	Absent (Greaser and Gergely, 1971)	Absent
KI actin[b]	A, TM, TP	Present (J. Spudich, pers. commun.)	Present
20°-extracted actin[c]	—	Present (Katz, 1966)	Present
0°-extracted actin[c]	A, trace TM	Very low (Katz, 1966)	Absent
Purified actin + Mueller TM[d]	—	Present (Mueller, 1966)	Present
Purified actin + TM. TP[e]	A, TM, TP[e]	Present (Spudich and Watt, 1971)	Present
Purified actin + purified TM[f]	A, TM	Absent (Schaub and Perry, 1969)	Absent
Purified actin + TM + TP[e]	A, TM, TP[e]	Present (Spudich and Watt, 1971)	Present

[a] Most preparations were purified by a method based on the work of Martonosi (1962); others were purified according to Rees and Young (1967).

[b] Prepared by the method Szent-Györgyi (1951). The polymerized actin was collected by centrifugation and resuspended in 0.1 M KCl before the crystals were formed. SDS gel electrophoretograms show that α-actinin is absent from the washed polymers and that the proportions of A, TM, and TP in the polymers are about the same as in myofibrils.

[c] Prepared by the Straub method (Katz and Hall, 1963). The G-actin was extracted from the acetone-dried residue either at room temperature (about 20°) or at 0°. In each case the actin was polymerized in 0.1 M KCl, collected by centrifugation, and resuspended in 0.1 M KCl before the crystals were formed.

[d] The TM was prepared in the presence of dithiothreitol by the method of Mueller (1966) except that it was precipitated only once at pH 4.3 and only once in 70% saturated $(NH_4)_2SO_4$.

[e] Preparations generously donated by Dr. J. Spudich (see Spudich and Watt, 1971).

[f] Two preparations were used, one generously donated by Professor S. V. Perry, the other prepared according to Noelken (1962). SDS gel electrophoretograms of the latter material show the usual two bands at positions corresponding to chain weights of about 35,000 and 37,000 (see Cohen et al., 1971).

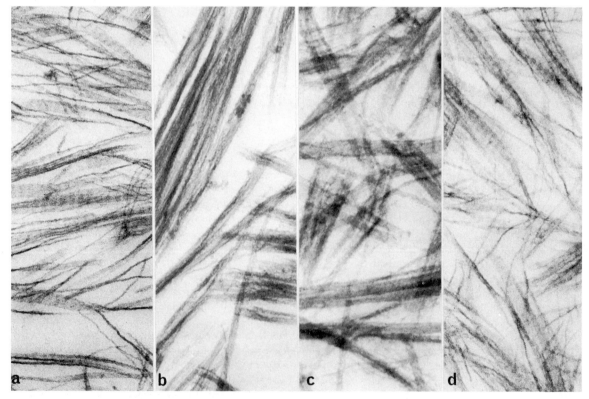

Figure 1. Sections of crystalline aggregates of synthetic filaments formed from (a) KI actin, (b) purified actin, (c) purified actin combined with purified TM, (d) purified actin combined with TM and TP. Note axial periodicity in (a) and (d) which contained TP and absence of periodicity in (b) and (c) which did not contain TP (see Table 1). Magnifications: (a) 63,000, (b) 53,000, (c) and (d) 69,000.

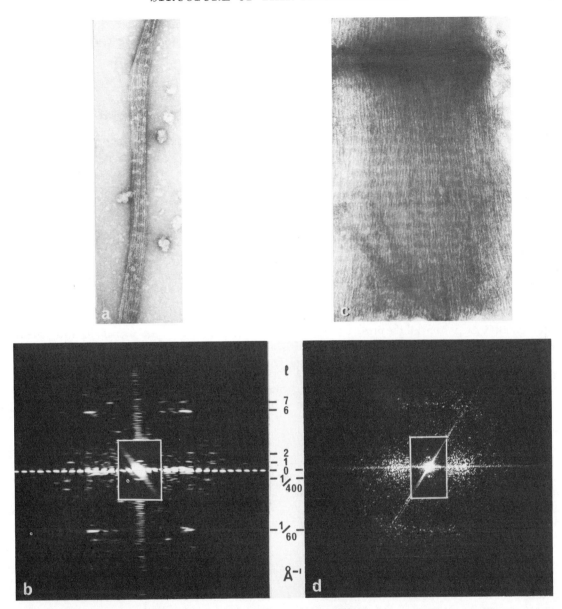

Figure 2. Negatively stained preparations of (*a*) a crystalline aggregate of filaments synthesized from KI actin (see Table 1) and (*c*) an I-segment separated from a relaxed fibril of frog skeletal muscle (only part of the segment is shown). In each case the cross-striation is attributed to TP located at intervals along the filaments. (*b*) and (*d*) are optical diffraction patterns of parts of the micrographs shown in (*a*) and (*c*), respectively. The central area of each pattern, which has been exposed less than the rest, shows a meridional first-order reflection from the cross-striation, and (*d*) shows higher orders also. In (*b*) note the off-meridional layer lines arising from the helical structure of the filaments. Because the filaments are not packed as regularly in the I-segment as in the crystal, the pattern in (*d*) does not show clear layer lines, but note the intensity in the region of the 6th layer line. Magnifications: (*a*) and (*c*) 80,000.

while holding the pH at 6.6. Sections of the crystals were prepared by standard procedures: Centrifuge pellets were fixed in OsO_4, dehydrated in ethanol, stained with phosphotungstic acid, and embedded in Araldite (Hanson, 1972). Other crystals were deposited on carbon-filmed copper grids, fixed with glutaraldehyde, and negatively contrasted with either uranyl acetate or potassium phosphotungstate (Hanson, 1972). I-segments were prepared as

described by Hanson et al. (1971). The optical diffraction methods we have used are outlined by O'Brien et al. (1971).

Specimens for X-ray diffraction studies were prepared as follows. A soft translucent pellet of the synthetic filaments was formed by centrifugation and then placed in the wide end of a Pantak capillary tube. The gel was forced down the capillary and the tube sealed. The specimens were stored at 4° and the diffraction pattern examined at intervals. Some of the best-oriented gels were concentrated by breaking the seal and allowing slow evaporation (monitored by taking diffraction patterns). A toroidal camera (Elliott, 1965), focal length about 10 cm, and a mirror-monochromator camera, focal length about 20 cm, were used. The cameras were filled with helium, and exposure times were 20–30 hr, using a Hilger microfocus or an Elliott rotating anode X-ray source. The temperature of the specimens was held at about 2° during exposure.

Location of Troponin

As shown in Table 1 an axial periodicity was observed in sectioned crystals of synthetic filaments only when TP was present in addition to TM and actin (Hanson, 1972) (Fig. 1). Correspondingly crystals containing TP and prepared for microscopy by negative contrast methods show bands of unstained amorphous material crossing the filament assembly at fairly regular intervals (O'Brien et al., 1971) (Fig. 2a). Crystals from all the protein preparations listed in Table 1 were examined using negative contrast methods, and the bands were only observed when TP was present; but the results were less reproducible than those from sectioned preparations (Hanson, 1972). A different axially repetitive structure, accounted for by the helical arrangement of the subunits in the filaments and by the packing of the filaments in the crystal, is observed in negatively contrasted crystals where actin is the only component (Hanson, 1967a, b, 1968b; Moore et al., 1970; O'Brien et al., 1971). The conformation of the helix is not resolved in sections, and therefore pure actin crystals show no axial periodicity in sections (Fig. 1b).

The cross-striation in the sectioned crystals that contain TP appears similar to the well-known cross-striation of the thin filament assembly in sections of myofibrils, though in the crystals the stained bands seem to be broader than in the fibril (this is presumably due to differences in packing). Similarly, negatively contrasted I-segments (Z-discs plus attached filament assemblies) separated from relaxed myofibrils show regularly spaced transverse bands that resemble those in the negatively contrasted crystals which contain TP (Hanson et al., 1971; O'Brien et al., 1971) (Fig. 2).

The spacing of the stripes in the sectioned crystals is the same in all the different preparations that have been measured, and the mean of all the results is 378 Å (range 333–410 Å) (Hanson, 1972). (Only crystals that lay with their long axes approximately parallel to the edge of the knife while the sections were cut were measured.) The corresponding stripes in the negatively contrasted crystals are

spaced at intervals of about 370 Å (Hanson, 1972), and similar measurements were obtained from the I-segments. The amount of shrinkage that had occurred in each of these three types of preparation is not known. Dr. S. G. Page (pers. commun. quoted in O'Brien et al., 1971) has measured the repeat of the stripes in the thin filaments in sections of muscle prepared so as to minimize shrinkage: Her result was 382 Å ± 10 Å.

In optical diffraction patterns from electron micrographs of negatively contrasted crystals and I-segments, the cross-striation is represented by reflections on the meridian (Fig. 2b, d). Also present in these patterns are off-meridional layer lines arising from the helical structure of the filaments (Figs. 2b and 3d); the same layer lines are present in the patterns from crystals that do not contain TP and are not striped, but the low-angle meridional reflections are absent (Fig. 3a) (O'Brien et al., 1971). In some of the X-ray diffraction patterns obtained from oriented gels of synthetic thin filaments containing TM and TP (KI actin, Table 1), reflections on the meridian at spacings of about 192 Å and 128 Å (orders of 385 Å) have been observed; the central scatter in these patterns obscures the region where the first-order reflection would appear. The optical diffraction pattern from the cross-striated I-bands in sectioned muscle (the micrographs were masked to exclude all regions except the I-bands) has a reflection on the meridian at about 385 Å (O'Brien et al., 1971; G. Borisy, H. E. Huxley, and S. G. Page, pers. commun.). There seems little doubt that the meridional X-ray reflection at 385 Å in the diffraction pattern from the intact living muscle is the first-order reflection from periodicity in the thin filaments (Huxley and Brown, 1967).

All these results taken together show that we have synthesized thin filaments which resemble the natural ones in having an axial periodicity of about 385 Å that appears in the electron microscope as a cross-striation of the filament assembly and is represented in diffraction patterns by a set of low-angle reflections on the meridian. Because this periodicity is reproduced in the synthetic systems only when TP is present, this component of the regulatory system is clearly responsible for this periodicity in the muscle. The micrographs suggest that the TP is located at intervals along the filaments, rather than being helically arranged, and the diffraction patterns confirm this model because the first-order reflection is on the meridian. The spacing of this reflection in the X-ray diffraction pattern from the intact living muscle (385 Å according to Huxley and Brown, 1967) is the best available measurement of the separation of the TP sites along the thin myofilaments.

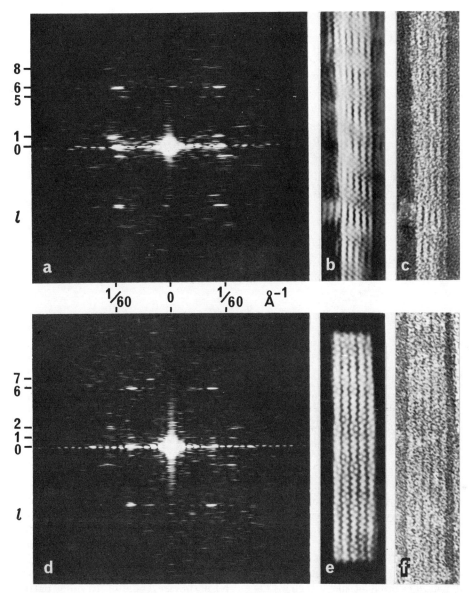

Figure 3. Optical diffraction analysis of negatively stained crystals of purified actin (a–c) and of purified actin combined with Mueller TM (see Table 1) (d–f). The first layer line is strong in the pattern from pure actin (a); the second layer line is strong, relative to the first one, in the pattern from the TM-containing crystal (d). The filtered image of the pure actin crystal (b) shows filaments with the typical F-actin structure. The filaments in the filtered image from the TM-containing crystal (e) show evidence of four-stranded helical structure (see text). The original micrographs of (b) and (e) are shown in (c) and (f), respectively, Magnifications: (b), (c), (e), and (f) 270,000.

Location of Tropomyosin

The first evidence that TM lies in the two long-pitch grooves of the F-actin double helix was obtained by comparing electron micrographs of negatively contrasted crystals assembled from actin alone and from actin, TM, and TP (O'Brien et al., 1971). (The various protein preparations we used in these studies are listed in Table 1; mostly we used purified actin, KI actin, and a mixture of purified actin and "Mueller TM.") The first layer line in all

the good optical diffraction patterns obtained from micrographs of pure actin crystals is strong and the second layer line weak (Fig. 3a). In contrast in many of the good patterns from crystals that contain TM and TP, the second layer line is strong and the first one weak (Fig. 3d); in other patterns from such crystals both layer lines are weak. ("Good" patterns are those showing well-defined layer lines.) Apart from the different relative intensities of the first two layer lines and the different row-line distribution of the maxima (depending

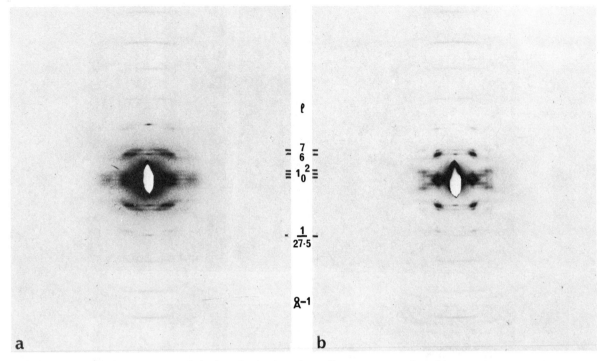

Figure 4. X-ray diffraction patterns of oriented gels of synthetic filaments formed from (a) purified actin and (b) KI actin (see Table 1). Note that the second layer line in (b) is much stronger than in (a). In each case the gel had been concentrated by controlled evaporation of the solvent until the filaments became packed in near crystalline arrays. The nature of the crystal lattice differs in (a) and (b), but the detailed indexing of the diffraction patterns is still under investigation. The meridional reflections on the 6th layer line in (a) and on the 6th and 7th layer lines in (b) are observed only in patterns from concentrated gels and are probably features of the increased crystallinity of these preparations. Each pattern was taken with a toroidal camera, focal length about 10 cm, filled with helium; a Hilger microfocus X-ray source was used, with an exposure time of about 20 hr.

on how the filaments are packed), the patterns from the two types of crystals (with and without TM) are similar.

The intensification of the second layer line relative to the first one indicates that, to a first approximation, the TM-containing filament is a four-stranded helix (O'Brien et al., 1971). "Filtered" images have been reconstituted from the optical diffraction patterns after masking out noise spectra (Figs. 3b, e). The filtered image of a filament that contains TM is clearly different from that of a pure actin filament: Instead of appearing narrow where the two long strands of actin subunits cross over one another, the profile of the complex filament widens in this region in such a way that the axial repeat appears to have been halved (O'Brien et al., 1971). Furthermore the filtered images show that the TM-containing filaments are staggered in the crystal in the manner expected of four-stranded helices arranged with "ridge-and-groove" packing; pure actin filaments pack differently (O'Brien et al., 1971).

A pronounced intensification of the second layer line relative to the first one is also observed in X-ray

diffraction patterns from oriented gels of TM-containing synthetic filaments, compared with patterns from gels of pure F-actin. The two layer lines are only resolved in the patterns from concentrated gels (Fig. 4), and as yet such patterns have only been obtained from purified actin and from KI actin (see Table 1). In other patterns from less concentrated gels of TM-containing filaments (including some formed from purified actin mixed with purified TM), although the first two layer lines are not resolved, it is apparent from the shape and orientation of the broad reflection present in this region that the second layer line is considerably stronger than in comparable patterns from pure F-actin.

Considering that the optical diffraction patterns were obtained from filaments that had been aggregated into crystals by Mg^{++} precipitation and then fixed, stained and dried, it is remarkable that they give the same main result as the one obtained from the X-ray diffraction patterns of untreated filaments.

The finding that the most pronounced effect of TM addition is on the first two layer lines in the actin diffraction pattern, rather than also on the 6th or higher-order layer lines, indicates that the TM strands are smooth; i.e., the diffraction patterns do not distinguish subunits in these strands.

We conclude that in synthetic thin filaments, TM forms two smooth helices situated in the two long-pitch grooves of the F-actin double helix. It should be noted that enough calcium was present in all the specimens to prevent inhibition of the actin subunits, so that the structure observed should correspond to that in the activated muscle. Attempts (by V. Lednev and T. Wakabayashi) to obtain well-oriented gels of synthetic thin filaments in an EGTA-containing medium (where the actin subunits would be in an inhibited state) have not been successful.

Interpretation of Diffraction Patterns from Intact Muscles

The second actin layer line has not been observed in X-ray diffraction patterns from vertebrate skeletal muscle in the resting state, but the first one is visible (Huxley and Brown, 1967). When the muscle goes into rigor, the second actin layer line appears in the pattern (Vibert et al., 1972; Lednev, unpublished); moreover Huxley (1970) has briefly reported its appearance in patterns recorded during isometric contraction. These observations suggest that when the actin subunits are capable of interacting with myosin the strands of TM lie in the grooves of the actin helix in approximately the same positions as in the synthetic thin filaments we have studied (in the presence of Ca^{++} ions). In the resting muscle, on the other hand, it seems that the TM strands are situated elsewhere in the filament. If each TM strand lay closer to one of the long-pitch strands of actin subunits than to the other actin strand, the filament would approximate to a two-stranded helix and give a diffraction pattern with the first layer line stronger than the second one, as observed in the pattern from resting muscle.

The Regulatory Mechanism

According to our interpretation of the diffraction patterns from the intact muscle, the strands of TM change their positions relative to the actin subunits when the regulatory system switches the actin off and the muscle relaxes. It seems that when the actin subunits are inhibited each TM strand may lie nearer to one of the long-pitch actin strands than to the other one, and that when the inhibition is released the TM moves away from that strand. Does this signify that TM molecules block the myosin-combining sites on the actin subunits while the muscle is relaxed and then move out of the way on activation? Alternatively does TM switch the

actin subunits off by changing their conformation or their relationship to one another in the polymer? In this connection it is noteworthy that G-actin is an extremely poor activator of the ATPase of myosin subfragment 1 (Offer et al., 1972). This might be due to a difference in tertiary structure between G-actin and the subunit in F-actin. A more attractive possibility for the present discussion is that more than one subunit in F-actin might participate in the site which binds S-1 and accelerates the release of the products of ATPase activity. On removal of Ca^{++} the regulatory protein system might cause a small change in the arrangement of the actin subunits (not necessarily involving a change in the pitch of the helix), thereby disrupting the S-1 binding site. It must be stressed that we do not suppose that F-actin in its inhibited state is functionally equivalent to G-actin. Indeed G-actin appears to be functionally inadequate both with respect to the rate at which it binds to the S-1-products complex and to the rate at which the actin-S-1-ADP·P_i complex breaks down (Offer et al., 1972), whereas in the case of F-actin in the inhibited state only the former rate is affected (Eisenberg and Kielley, 1970; Koretz et al., this volume).

Acknowledgments

This work was done with the technical assistance of Mr. H. Baker, Mrs. L. Baker, Mrs. M. Blythe, Miss M. Castle, Miss D. Howard, and Mrs. S. Weindling. Mr. Z. Gabor assisted with photography, and Mr. A. Fasoli and Mr. C. McCarthy with apparatus construction. The X-ray cameras were designed by Drs. A. Elliott, E. Rome, and M. Spencer. We are grateful to Dr. G. Offer for the electrophoretic analyses and to Drs. Offer and Rome for discussion. We thank Professor S. V. Perry and Dr. J. Spudich for gifts of proteins. The electron microscope was bought with a grant (AM 06166-01) from the U.S. Public Health Service.

References

COHEN, C., D. L. D. CASPAR, D. A. D. PARRY, and R. M. LUCAS. 1971. Tropomyosin crystal dynamics. *Cold Spring Harbor Symp. Quant. Biol.* **36**: 205.

EBASHI, S. and M. ENDO. 1968. Calcium ion and muscle contraction. *Prog. Biophys. Mol. Biol.* **18**: 123.

EBASHI, S., M. ENDO, and I. OHTSUKI. 1969. Control of muscle contraction. *Quart. Rev. Biophys.* **2**: 351.

EISENBERG, E. and W. W. KIELLEY. 1970. Native tropomyosin: Effect on the interaction of actin with heavy meromyosin and subfragment 1. *Biochem. Biophys. Res. Comm.* **40**: 50.

ELLIOTT, A. 1965. The use of toroidal reflecting surfaces in X-ray diffraction cameras. *J. Sci. Instr.* **42**: 312.

GREASER, M. L. and J. GERGELY. 1971. Reconstitution of troponin activity from three protein components. *J. Biol. Chem.* **246**: 4226.

HANSON, J. 1967a. Axial period of actin filaments. *Nature*
213: 353.
――――. 1967b. The structure of actin filaments, p. 327.
7th Int. Congr. Biochem. (Abstr.).
――――. 1968a. Recent X-ray diffraction studies of muscle.
Quart. Rev. Biophys. **1**: 177.
――――. 1968b. In *Symposium on Muscle* p. 99, ed. E.
Ernst and F. B. Straub. Akademiai Kiado, Budapest.
――――. 1972. Evidence from electron microscope studies
on actin crystals concerning the origin of the cross-
striation in the thin filaments of vertebrate skeletal
muscle. *Proc. Roy. Soc.* (*London*) B. In press.
HANSON, J. and J. LOWY. 1963. The structure of F-actin
and of actin filaments isolated from muscle. *J. Mol.
Biol.* **6**: 46.
――――. 1964. The structure of actin filaments, and the
origin of the axial periodicity in the I-substance of
vertebrate striated muscle. *Proc. Roy. Soc.* (*London*)
B **160**: 449.
HANSON, J., E. J. O'BRIEN, and P. M. BENNETT. 1971.
Structure of the myosin-containing filament assembly
(A-segment) separated from frog skeletal muscle. *J.
Mol. Biol.* **58**: 865.
HUXLEY, H. E. 1970. Structural changes in muscle and
muscle proteins during contraction. *8th Int. Cong.
Biochem.*, Interlaken.
HUXLEY, H. E. and W. BROWN. 1967. The low-angle X-ray
diagram of vertebrate striated muscle and its be-
haviour during contraction and rigor. *J. Mol. Biol.* **30**:
383.
KATZ, A. M. 1966. Purification and properties of a tropo-
myosin-containing protein fraction that sensitises
reconstituted actomyosin to calcium-binding agents.
J. Biol. Chem. **241**: 1522.
KATZ, A. M. and E. J. HALL. 1963. Actin from heart muscle:
isolation, purification and physicochemical properties.
Circulation Res. **13**: 187.
KENDRICK-JONES, J., W. LEHMAN, and A. G. SZENT-
GYÖRGYI. 1970. Regulation in molluscan muscles. *J.
Mol. Biol.* **54**: 313.

MARTONOSI, A. 1962. Studies on actin. VII. Ultracentrifugal analysis of partially polymerised actin
solutions. *J. Biol. Chem.* **237**: 2795.
MOORE, P. B., H. E. HUXLEY, and D. J. DE ROSIER. 1970.
Three-dimensional reconstruction of F-actin, thin
filaments and decorated thin filaments. *J. Mol. Biol.*
50: 279.
MUELLER, H. 1966. EGTA-sensitising activity and molecular properties of tropomyosin prepared in presence
of a sulfhydryl protecting agent. *Biochem. Zeit.* **345**:
300.
NOELKEN, M. E. 1962. The denaturation of paramyosin
and tropomyosin and attempts to isolate actin. Ph.D.
thesis, Washington Univ.
O'BRIEN, E. J., P. M. BENNETT, and J. HANSON. 1971.
Optical diffraction studies of myofibrillar structure.
Phil. Trans. Roy. Soc. London B **261**: 201.
OFFER, G. W., H. F. BAKER, and L. BAKER. 1972. Interaction of monomeric and polymeric actin with myosin
subfragment 1. *J. Mol. Biol.* **66**: 435.
OHTSUKI, I., T. MASAKI, Y. NONOMURA, and S. EBASHI.
1967. Periodic distribution of troponin along the thin
filament. *J. Biochem.* **61**: 817.
REES, M. K. and M. YOUNG. 1967. Studies on the isolation
and molecular properties of homogeneous globular
actin. *J. Biol. Chem.* **242**: 4449.
SCHAUB, M. C. and S. V. PERRY. 1969. The relaxing protein
system of striated muscle. Resolution of the troponin
complex into inhibitory and calcium ion-sensitising
factors and their relationship to tropomyosin. *Biochem.
J.* **115**: 993.
SPUDICH, J. A. and S. WATT. 1971. The regulation of rabbit
skeletal muscle contraction. I. Biochemical studies of
the interaction of the tropomyosin-troponin complex
with actin and the proteolytic fragments of myosin.
J. Biol. Chem. **246**: 4866.
SZENT-GYÖRGYI, A. G. 1951. A new method for the preparation of actin. *J. Biol. Chem.* **192**: 361.
VIBERT, P. J., J. C. HASELGROVE, J. LOWY, and F. R.
POULSEN. 1972. Structural changes in actin-containing
filaments of muscle. *Nature New Biol.* **236**: 182.

Myosin-linked Regulatory Systems: Comparative Studies

William Lehman, John Kendrick-Jones, and Andrew G. Szent-Györgyi

Department of Biology, Brandeis University, Waltham, Mass. 02154
and
MRC Laboratory of Molecular Biology, Hills Road, Cambridge, England

The mechanism of regulation in molluscan muscles differs from that in vertebrate muscle. In molluscan muscles calcium triggers contraction by interacting with myosin (Kendrick-Jones et al., 1970), whereas in vertebrate and arthropod muscles, calcium reacts with the components of the thin filaments (Ebashi and Kodama, 1965, 1966). In this paper we will compare these two control systems, pointing out similarities as well as some of the fundamental differences. We will show that the myosin-linked regulation is not restricted to molluscan muscles and will report on our other preliminary comparative studies. Finally we will discuss the possible changing role of the components involved in regulation during evolution.

The myosin-dependent regulatory system has only recently been discovered in molluscan muscles, and there is currently only a single publication on the subject (Kendrick-Jones et al., 1970). Much of the data presented here and their interpretation are based on comparative studies which will be published in detail elsewhere.

Criteria

The regulatory systems are best identified by directly testing isolated thin filaments and myosin preparations. The availability of these preparations from different muscles, without a significant loss of regulatory function, in a sufficiently high yield is an important prerequisite of comparative studies.

We have prepared native thin filaments from a number of different muscles, including rabbit, chicken, frog, horseshoe crab, insect, molluscan, and annelid muscles. The method of thin filament separation is relatively simple and avoids the use of organic solvents or solutions of high ionic strength (Szent-Györgyi et al., 1971; Kendrick-Jones et al., 1970; Hardwicke and Hanson, 1971). The muscle is homogenized in the presence of EDTA, ATP and magnesium ions, at an ionic strength between 0.05 and 0.15 and at pH 6–7. In this solution interaction between thin and thick filaments is reduced, and the thick filaments tend to form aggregates which can be removed at a relatively low centrifugal force (30,000–50,000 g). The yield of thin filaments may be as high as 80% of the actin (the smooth red adductor muscles of *Mercenaria*) or as low as 0.5–1.0% of the actin (rabbit or chicken striated muscles). The yield appears to depend on the ease with which the thin filaments may be removed from the structures to which they are attached (Z-lines, dense bodies). The procedures were not applicable to some crustacean muscles, such as lobster and crayfish.

The thin filaments can be analyzed by various techniques. SDS-acrylamide gel electrophoresis reveals the various components present and may be used to establish their stoichiometry. These gel patterns show similarities in related phyla and may help in the classification of the muscles. If a thin filament preparation contains all the neccessary components of regulation, it will confer calcium dependence to the actin-activated ATPase activity of rabbit myosin. Calcium binding by thin filaments at low calcium concentrations in the presence of magnesium ions will show the presence of specific calcium-binding proteins.

In order to test the regulatory function of myosin, one requires a myosin preparation which is free of actin and tropomyosin contaminations. Since the light chains of myosins vary extensively even among myosins prepared from muscles which have similar regulatory systems, SDS gels of myosin are not decisive in detecting myosin-linked regulatory systems.

Although thin filaments and pure myosin are the most suitable materials to determine experimentally the type of regulatory system present in a particular muscle, it is difficult in some instances to obtain thin filaments or myosin because of proteolysis, an extreme lability of the pure myosin or the limited amounts of muscle tissue available. In such cases the identification of the regulatory system is based on less direct methods. These alternative approaches will be discussed later.

Evidence for Two Types of Regulation

Rabbit muscle is a typical example for actin-linked regulation, whereas molluscan muscle exemplifies myosin-linked regulation. Molluscan myosin binds calcium, and when combined with pure rabbit actin has a calcium-sensitive ATPase. In contrast rabbit myosin does not bind calcium, and its ATPase activity is not influenced by calcium (Table 1). The results obtained with the thin filaments complement the findings on myosin.

Table 1. Comparison of Myosins

Source	% Inhibition of ATPase in absence of Ca++ with pure actin	μmoles Ca/g 5×10^{-6} M Ca++
Rabbit	0	<0.3
Limulus	0	<0.2
Aequipecten (striated)	92	2.8
Aequipecten (smooth)	>95	2.4
Mercenaria (red)	84	2.3
Glottidia (peduncle)	70	2.5

ATPase activity was measured in a pH-stat by following proton liberation at pH 7.5 at 23°C. 0.5–2 mg myosin was mixed with rabbit actin in 0.4 M NaCl at a ratio of 2–3 g to 1 g actin. The rabbit actin used was free of tropomyosin and troponin contamination and moved as a single band in SDS-acrylamide gel electrophoresis. The actomyosin formed was stirred into 10 ml 25 mM NaCl (final concentration), 1 mM MgCl$_2$, 0.5 mM ATP which additionally contained either 0.1 mM CaCl$_2$ or 0.1 mM EGTA. Frequently calcium was added directly to the EGTA-containing sample once the ratio of proton liberation was established. Inhibition of ATPase activity equals

$$\left(1 - \frac{\text{ATPase in EGTA}}{\text{ATPase in Ca++}}\right) \times 100.$$

Calcium binding was measured as described by Kendrick-Jones et al. (1970).

Molluscan thin filaments do not bind calcium and do not sensitize the ATPase of rabbit myosin; thin filaments of vertebrates, arthropods and annelids, on the other hand, both bind calcium and confer calcium sensitivity to the ATPase activity of rabbit myosin (Table 2). One can conclude, therefore, that in molluscan muscles regulation occurs by modifying myosin and in vertebrates and arthropods by altering actin.

Composition of Thin Filaments

Molluscan thin filaments. Molluscan thin filaments contain only actin and tropomyosin in significant amounts; components corresponding to troponin are absent (Fig. 1). This simple composition characterizes the thin filaments of a number of different molluscan muscles, e.g., the smooth and striated adductor muscles of scallop (*Aequipecten irradians, Placopecten magellanicus*), the anterior byssus retractor of *Mytilus edulis*, the red adductor muscle of *Mercenaria mercenaria*.

The protein composition of molluscan muscles is also remarkably simple. Most of the bands seen on SDS-acrylamide gel electrophoresis of washed scallop myofibrils can be accounted for by the thin filaments and myosin. The outstanding feature of washed myofibrils is the paucity of low molecular weight components. There appears to be only a single small component present in scallop myofibrils, having a chain weight of about 18,000 daltons (Fig. 2). This component is also present in myosin preparations and represents the light chains of scallop myosin. The rest of the major protein bands of scallop myofibrils are accounted for by actin, tropomyosin, paramyosin, and the heavy chains of myosin. In vertebrate muscle troponin consists of several different components (Hartshorne and Mueller, 1968; Schaub and Perry, 1969)

Table 2. Comparison of Thin Filaments

Source	% Inhibition of ATPase in absence of Ca++ with rabbit myosin	μmoles Ca/g at 5×10^{-6} M Ca++
Rabbit	50–93	~4.0
Chicken	82	~2.0
Frog	50	~2.0
Limulus	76–97	0.6
Lethocerus	78	0.8
Nereis	51	0.5
Glycera	62	0.5
Lumbricus	60	1.3
Aequipecten (striated)	<3	<0.1
Mercenaria (pink)	<3	<0.1
ABRM	<3	—
Glottidia	0	—
Mercenaria and rabbit relaxing proteins	70	~6.0

Thin filament preparations were mixed with purified rabbit myosin in various proportions (1 g to 5–20 g). ATPase measurements and calcium binding were as described in the legend of Table 1.

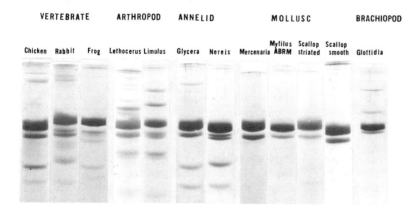

VERTEBRATE   ARTHROPOD   ANNELID   MOLLUSC   BRACHIOPOD

Chicken   Rabbit   Frog   Lethocerus   Limulus   Glycera   Nereis   Mercenaria   Mytilus ABRM   Scallop striated   Scallop smooth   Glottidia

Figure 1. Thin filaments from different muscles. 10% SDS polyacrylamide gels. All the thin filaments show a main actin band with a tropomyosin band immediately beneath. The vertebrate, arthropod, and annelid preparations show additional bands which correspond to the components of troponin. Molluscan and brachiopod preparations show one of these troponin components.

Chicken   Scallop striated

Myf.   T. F.   Myf.   T. F.

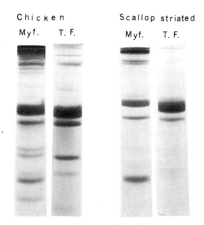

Figure 2. Comparison of thin filaments and myofibrils from chicken and scallop. Chicken myofibrils contain the components of the troponin complex and the light chains of myosin. The troponin components are also present on the thin filaments. A 40,000 dalton component is not resolved from actin on these gels. Scallop myofibrils only show a single low molecular weight component which corresponds to the myosin light chain. No troponin is observed.

that are individually responsible for calcium binding, ATPase inhibition, and binding to tropomyosin (Greaser and Gergely, 1971); none of these components are present in scallop muscles.

Vertebrate and arthropod thin filaments. The presence of troponin on the thin filament preparations of vertebrate and arthropod muscles can be easily demonstrated (Fig. 1). Thin filaments from muscles of rabbit, chicken, frog, horseshoe crab, waterbug, clam worm, and bloodworm combine with purified rabbit myosin and form an ATPase which requires calcium. The inhibition of the ATPase activity in the absence of calcium varies from 50 to 90% (Table 2). The thin filaments from vertebrate muscles contain several components in addition to actin and tropomyosin, which correspond to the components of purified troponin preparations from rabbit and chicken (Greaser and Gergely, 1971; Schaub and Perry, 1971). Troponin-like proteins differing in chain weights from vertebrate troponins can also be obtained from horseshoe crab and lobster, and these components are also present on arthropod thin filaments. Although thin filaments from muscles of vertebrates and invertebrates both contain troponin, there are differences between these troponins. Arthropod thin filaments contain components which differ in number and in chain weight from those of vertebrate thin filaments. Thin filaments from vertebrates bind 3–5 times more calcium than those from arthropod muscles (Table 2). Nearly full calcium sensitivity is obtained with arthropod thin filaments although considerably less than one mole of calcium is bound for each mole of tropomyosin.

Myosin-linked or actin-linked regulatory systems are not tissue or organ specific and are not restricted to a particular muscle type. The striated and smooth adductor muscles of scallop and the anterior byssus retractor of *Mytilus* (a "catch" muscle) all possess myosin-linked regulation. The thin filaments of rabbit obtained from mixed muscles and those obtained from the breast muscle of chicken are rather similar.

The stoichiometry of the components of the thin filaments may be estimated by the densitometry of SDS gels. These ratios can also be compared with corresponding ratios found in myofibrils. We could not obtain the stoichiometry of all components from experiments in a single species since some of these are not fully separated on SDS gels; particularly the highest chain weight component of troponin (troponin T), which Greaser and Gergely (1972) found to interact with tropomyosin, remains unresolved. In rabbit this component stays close to tropomyosin and in chicken close to actin. The combined results of the rabbit and chicken preparations indicate the following molar ratios: five to six actins: one tropomyosin: one 25,000 component (troponin B or I): one-half 18–20,000 component (troponin A or C). Where a comparison is feasible, the indications are that the same stoichiometry exists in washed myofibrils. The reliability of our technique was further checked by determining the stoichiometry of the light chains of chicken myosin on SDS acrylamide gels of chicken myofibrils. The stoichiometry obtained agrees very well with those reported by Lowey and Risby (1971) for purified chicken myosin. This ratio of troponin components, found on thin filament preparations showing high calcium sensitivity, differs from studies using purified troponin, in which a simpler mole to mole ratio was found for troponin I and C (Hartshorne, this volume; Greaser and Gergely, this volume).

Tropomyosin is retained by molluscan thin filament preparations and the ratio of tropomyosin to actin can be obtained both from molluscan thin filaments and myofibrils. This ratio is similar to the one observed in vertebrate muscles and suggests that there are 5–6 actin monomers for each tropomyosin molecule.

Comparison of Regulatory Systems

In molluscan, vertebrate, and arthropod muscles regulation is achieved by controlling the interaction between actin and myosin. The basic difference between the two regulatory systems lies in the location of the sites where the control is exerted and the mechanism by which interaction between actin and myosin is prevented. In vertebrate and arthropod muscles, active sites on actin are blocked by the regulatory proteins. In this system calcium interacts with troponin (Ebashi et al., 1967; Hartshorne, et al., 1969; Schaub and Perry, 1969; Greaser and Gergely, 1971). The troponin effect is mediated by tropomyosin and alters the reactivity of actin (Weber and Bremel, 1971; Spudich and Watt, 1971). In these muscles calcium sensitivity requires the presence of tropomyosin. The tropomyosin requirement is a key feature of actin-linked regulations. In myosin-linked regulatory systems, for example in molluscs, the interaction between actin and myosin is prevented by blocking sites on myosin, and tropomyosin is not needed for regulation. In our comparative studies the lack of a tropomyosin requirement is viewed as strong evidence that the control system is linked to myosin. Regulatory proteins similar to those found in vertebrate muscles do not play a role in molluscan regulation. Regulation in molluscs depends on the presence of a particular light chain on myosin. This light chain, for example, can be removed from scallop myosin by reducing the concentration of divalent cations with the aid of EDTA. The dissociation of the light chain is reversible, and it readily recombines with the residual myosin when the magnesium ion concentration is raised above 10^{-5} M. The actin-activated ATPase of scallop myosin depends in a very characteristic manner on the presence of the EDTA light chain. The removal of this light chain abolishes the calcium sensitivity of myosin; its ATPase activity when combined with pure rabbit actin does not require calcium any longer. Scallop myosin, produced by recombination of the EDTA light chain and "desensitized" EDTA-treated myosin, regains its calcium sensitivity. The reconstituted myosin behaves as an untreated preparation and its ATPase activity is again depressed in the absence of calcium. Calcium binding by scallop myosin is reduced by about 40% when the EDTA light chain is removed; the isolated light chain itself, however, does not bind calcium. Calcium binding is restored when myosin is reconstituted from the EDTA light chain and the "desensitized" myosin (Szent-Györgyi et al., 1972; Kendrick-Jones et al., this volume). On the basis of these findings, we conclude that the EDTA light chain functions as a regulatory subunit. In the absence of calcium it interferes with sites on myosin which combine with actin, and this inhibition is relieved by calcium ions. Since control is exerted by modifying myosin sites and calcium regulation does not require tropomyosin, the mechanism for this regulation is quite different from the one existing in actin-linked systems.

It appears that, in molluscan as in vertebrate

regulation, the formation of cross-links between actin and myosin is prevented in the absence of calcium. For this reason thin filaments separate from thick filaments and can be isolated in the presence of calcium chelators and ATP. It is possible, however, to demonstrate in a more direct fashion that, in the absence of calcium, molluscan myosin does not combine with actin.

If molluscan myosin and pure actin cannot combine in a calcium-free solution containing ATP, the actin present should be uncomplexed and therefore available to react with added rabbit myosin since, in contrast to the molluscs, rabbit myosin does not require calcium to combine with pure actin. We find that in a calcium-free medium, the ATPase activity of rabbit myosin is activated to the same extent by rabbit actin, whether scallop myosin is present or absent (Fig. 3). These results show that the sites on actin remained free to react with rabbit myosin, i.e., in the presence of ATP, scallop myosin requires calcium to combine with pure actin.

Using a similar technique, the lack of interaction between rabbit myosin and rabbit actin, complexed with regulatory proteins, can also be demonstrated.

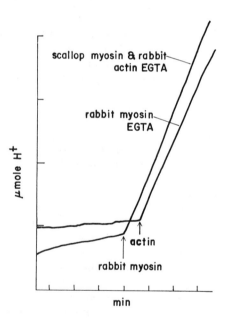

Figure 3. Calcium requirement for the cross-link formation of scallop myosin with rabbit actin. ATPase activity followed by proton liberation using a pH-stat set at pH 7.5. Curves represent two sets of experiments. To 2 mg scallop myosin and 0.36 mg rabbit actin in 10 ml 0.1 mM EGTA, 25 mM NaCl, 1 mM MgCl$_2$, 0.5 mM ATP, and 1.1 mg rabbit myosin were added (indicated by arrow) in one experiment. In the other experiment 0.36 mg rabbit actin was added (arrow) to 1.1 mg rabbit myosin. The rate of proton liberation is similar when rabbit myosin and actin are mixed in the presence or absence of scallop myosin. This indicates that in the absence of calcium scallop myosin does not combine with actin.

In this case the sites on actin are blocked in the absence of calcium, and the myosin sites remain free. The availability of the myosin sites can be tested in a calcium-free medium by adding pure rabbit actin to a system containing actin complexed with troponin-tropomyosin. Addition of pure actin to the mixture activates the ATPase immediately, indicating that the rabbit myosin is free to react with pure rabbit actin and that the thin filaments cannot compete with the actin. The results confirm that the troponin-tropomyosin system acts by reducing the affinity of actin to myosin (Parker et al., 1970; Eisenberg and Kielley, 1970). We will use this technique later to characterize the regulatory system of a muscle, without the isolation of myosin (cf. Table 5).

The similarity between the myosin- and actin-linked regulatory systems is that both suppress the ATPase activity in the absence of calcium and both require about 10^{-6} to 10^{-5} M calcium ions to reverse this inhibition. The mechanism of regulation, however, is entirely different in the two systems. Scallop muscle is regulated by a single regulatory light chain which does not bind calcium by itself, although it influences the calcium binding of the whole myosin. This relatively simple system is distinct from the vertebrate regulatory system in which regulation appears to be a result of an interaction among several different components. The troponin complex consists of several subunits and each component is individually responsible for the ATPase inhibition, calcium binding, and binding of the complex to tropomyosin (Greaser and Gergely, 1971). The light chain differs from the components of troponin and it is not a "troponin-like" molecule which combines with myosin. The sites on myosin are not the same as those on actin, and it is reasonable to suppose that they are regulated differently.

Although we have previously considered that in vertebrate muscles the calcium-binding component may be present on myosin at rest and translocated during activation (Kendrick-Jones et al., 1970), the evidence does not support a "flip-flop" mechanism.

First, ATPase sensitivity depends on the ratio of troponin-tropomyosin to actin and not on the ratio of troponin to myosin (i.e., myosin subfragment-1) when S-1 or actin are respectively in excess, which indicates that all the components of troponin are associated with actin when the ATPase activity is inhibited (Weber and Bremel, 1971; Spudich and Watt, 1971). Significantly one also finds all the components of the troponin-tropomyosin system on the thin filaments and none on myosin, even though the thin filaments tested were isolated in relaxing conditions. Finally in the molluscs the regulatory light chain does not interact with components of the thin filaments.

All of the regulatory components of molluscan muscles are confined to myosin, and molluscan myosin has a calcium-dependent ATPase when mixed with pure rabbit actin. In contrast all of the regulatory components of vertebrate and arthropod muscles are confined to the thin filaments, and the thin filaments of these muscles sensitize the ATPase of purified rabbit myosin. These thin filaments also sensitize an EDTA-treated desensitized scallop myosin preparation even though this myosin is obtained from a muscle which does not contain troponin (Table 3). Similarly S-1 subfragment of scallop myosin, which is not calcium sensitive, forms calcium-requiring complexes with troponin-containing thin filaments. One can infer from these results that specific light chains unique to vertebrate muscles are not required for troponin-linked regulation. The observations that pure rabbit actin can be regulated by scallop myosin and desensitized scallop myosin regulated by vertebrate thin filament proteins illustrate further that the regulatory actions of the two systems are different, that they act independently of each other, and are located on different filaments.

Cross Reactions between Components of Regulatory Systems

The actins of the different thin filaments are similar in that they are all able to combine with

Table 3. Interaction of Desensitized Scallop Myosin and S-1 with Rabbit Relaxing System

	Actin-activated ATPase (μmole ATP/mg/min)		% Inhibition of ATPase in absence of calcium
	0.1 mM Ca++	0.1 mM EGTA	
Untreated myosin + rabbit actin (2:1 w/w)	0.23	0.04	83
Desensitized myosin + rabbit actin (2:1 w/w)	0.21	0.19	10
Desensitized myosin + rabbit actin containing rabbit relaxing proteins[a] (1.5:1 w/w)	0.31	0.05	84
S-1 + rabbit actin (1:4 w/w)	0.72	0.85	0
S-1 + rabbit actin containing rabbit relaxing proteins[a] (1:5 w/w)	1.30	0.13	90

From Szent-Györgyi et al., 1972.
[a] Rabbit actin was mixed with about equal weight of crude relaxing protein extract (Ebashi and Ebashi, 1964), pelleted by centrifugation at 100,000 g and combined with myosin.

Table 4. Interaction of *Mercenaria* Tropomyosin with Rabbit Troponin

Mixture	% Inhibition of ATPase in absence of Ca++
Rabbit actin + rabbit troponin, *no tropomyosin* + *rabbit myosin*	0
Rabbit actin + rabbit troponin, *rabbit tropomyosin* + *rabbit myosin*	55
Rabbit actin + rabbit troponin, *Mercenaria* tropomyosin, *rabbit myosin*	50

Actin was premixed with tropomyosin and troponin as indicated and combined with myosin in 0.4 M NaCl. ATPase activity was measured as described previously. Mixtures contained 3.3 mg myosin, 0.5 mg actin, 0.5 mg tropomyosin, and 0.7 mg troponin.

tropomyosin or with the tropomyosin-troponin complex. Thin filaments from molluscs form a hybrid with rabbit relaxing proteins which binds calcium, regulates rabbit myosin, and behaves like rabbit thin filaments (Table 2). The explanation for the lack of troponin in molluscan thin filaments is that this regulatory protein is apparently not produced by these muscles (Fig. 2). Although tropomyosin is not required for regulatory function in molluscs, it is a component of their thin filaments. Molluscan tropomyosin combines with rabbit actin and with rabbit troponin and can substitute for rabbit tropomyosin in a synthetic rabbit regulatory system (Table 4). This remarkable cross-reaction indicates that the troponin binding sites on tropomyosin preceded the appearance of troponin during evolution; i.e., the development of troponin took advantage of the preexisting tropomyosin-actin relationship.

The apparent ease by which troponin and tropomyosin complexes with molluscan thin filaments reflects the similarity and constancy of actin during evolution (cf. Elzinga and Collins, this volume). Similarly lobster and horseshoe crab troponins form a functional complex with rabbit actin-tropomyosin. In marked contrast we have been unable to combine the regulatory light chains of molluscs with rabbit myosin either in the presence or in the absence of EDTA. Evidently myosin has varied greatly during evolution and the combining sites of the light chains and the light chains themselves have not been conserved. The actin-linked regulatory systems thus readily form hybrids with the actins of all the muscles tested; the myosin-linked regulatory units, however, are a great deal more specific. The extent of the specificity and the amount of variation in the composition and sequence of regulatory light chains remains to be explored.

Comparative Aspects

It is pertinent to ask whether the myosin-linked regulation is an evolutionary offshoot unique to molluscan muscles, or if this system has occurred more often in the mainstream of evolution. It is clear that a myosin-linked control system characterizes a number of Lamellibranch muscles. It is, however, of considerable interest to establish the presence of myosin-linked regulation in other phyla. As previously indicated a direct characterization of a muscle's regulatory system requires the isolation of purified myosin and thin filaments, a measurement of their respective calcium binding coupled with a determination of their individual contribution to calcium sensitivity, and a study of their components. We have been able to obtain direct evidence using the smooth peduncle muscle of the brachiopod, *Glottidia pyramidata*. Brachiopods, commonly called lampshells, were abundant in the Cambrian period, have evolved little since, and are thought to have considerably preceded molluscs in the evolutionary sequence. *Glottidia* muscles have a myosin-linked regulatory system (Table 5). Their thin filaments contain no components whose chain weights are lower than tropomyosin and actin (Fig. 4), and they do not form a calcium-dependent ATPase complex with pure rabbit myosin (Table 2). The myosin of *Glottidia* binds calcium, and, when mixed with rabbit actin, the resulting ATPase requires calcium (Table 1).

Although one can characterize the regulatory system of a muscle in a direct manner using isolated thin filaments and myosins, frequently these preparations are not easily made. Muscles may be difficult to obtain in sufficient quantities for preparative purposes, and even worse, the myosins of

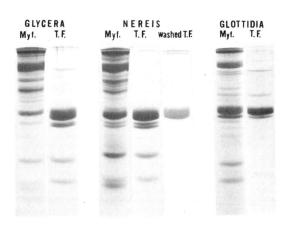

Figure 4. Washed muscle (Myf.) and thin filament (T.F.) preparations from annelid and brachiopod muscles. Tropomyosin and troponin can be removed from *Nereis* by washing in a solution consisting of 0.6 M NaCl, 5 mM Tris-HCl pH 8.0, and 0.2 mM ATP.

Table 5. Regulatory Systems in Different Animals

	Ca++ Binding		Troponin on T.F. SDS gels	ATPase Sensitization		Competition Studies Actin Activation	Type of Regulation
	My	T.F.		My + pure actin	T.F. + rabbit My		
Rabbit	−	+	+	−	+	+	Tr
Chicken		+	+		+	+	Tr
Frog		+	+	−	+	+	Tr
Mouse		+	+		+	+	Tr
Limulus	−	+	+	−	+	+	Tr
Lobster	−			−		+	Tr
Scallop (striated)	+	−	−	+	−	−	My
Scallop (smooth)	+	−	−	+	−	−	My
Mercenaria (red)	+	−	−	+	−	−	My
ABRM		−			−	−	My
Cerebratulus[a]		?			−	−	My
Golfingia[b]		?			−	−	My
Glottidia	+	−	−	+	−	−	My
Nereis	+	+	+		+	−	Tr + My
Glycera		+	+		+	−	Tr + My
Lethocerus		+	+		+	−	Tr + My
Lumbricus		+	+		+	−	Tr + My

Chicken troponin has originally been prepared by Ohtsuki et al. (1967). Lobster troponin has an inhibitory component and a calcium-binding component which also sensitizes the ATPase activity. Tropomyosin is required for the function of lobster troponin (J. M. Regenstein, personal communication).

[a] Components on thin filaments not present in myofibrils.
[b] Components on thin filaments not identified.

some invertebrate muscles are very labile and lose their ATPase activity upon removal of actin. In fact, at present *Glottidia* is the only non-molluscan species where we could directly show a myosin-linked regulation. It was, therefore, of considerable importance to devise a method which allows differentiation between a myosin-linked and an actin-linked regulatory system without making the isolation of thin filaments or myosin necessary. We have developed a simple procedure which appears to be reliable, rapid, and requires only milligram quantities of muscle tissue. The method is based on the differences in the response of molluscan and rabbit myosins to pure actin which is free of tropomyosin. Molluscan myosin cannot react with actin in the absence of calcium; therefore addition of *excess* actin to molluscan muscle or actomyosin in an EGTA-containing solution will have no effect on their ATPase activity which will remain low and still require calcium for activation. Conversely, when calcium is absent, rabbit myosin selectively combines with pure actin in a mixture containing pure actin and actin-tropomyosin-troponin complex, since the effect of troponin-tropomyosin is to reduce the affinity of actin to myosin. Therefore introduction of excess pure actin to a vertebrate actomyosin should fully activate its ATPase even in the absence of calcium. *Mercenaria* and rabbit myofibrils show the expected behavior. When calcium is absent, *Mercenaria* myofibrils are not activated by pure actin, whereas rabbit myofibrils are (Fig. 5). In practice, it is helpful to solubilize

the actomyosin of the muscle suspension in 0.6 M NaCl and 0.5 mM ATP before the experiments are performed to insure that the actin added can react with myosin.

This competitive actin-binding experiment can establish unambiguously the presence of a troponin-dependent regulation. Indeed in the absence of calcium pure actin activates the ATPase of a number of muscles which contain troponin, such as rabbit, chicken, frog, mouse, lobster, and horseshoe crab. Moreover addition of calcium ions after activation with pure actin has no further effect on the ATPase activity of these muscles.

This method can also be used to establish the presence of a myosin-dependent regulation. It is important to realize, however, that the test measures only the reactivity of myosin, and an independent study of the thin filaments is required to exclude the coexistence of a troponin-like regulation with the myosin-linked regulatory system. In principle, one may probe for a troponin-linked regulation by the addition of rabbit myosin to myofibrils. However this is not practical since different myosins vary considerably in their specific ATPase activities (Bárány, 1967), whereas actin from all species activates a particular myosin to the same extent.

The competition studies employing pure actin indicate that the myosin-linked regulation is rather widespread. The Nemertine worm, *Cerebratulus lacteus* (ribbonworm), and the Sipunculid, *Golfingia gouldi*, both respond like molluscan muscles (Table 5). Thin filaments can be prepared from these

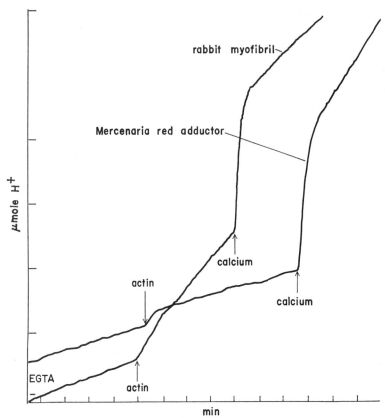

Figure 5. Response of myosin-linked and troponin-linked regulatory systems to pure rabbit actin; pH-stat records of proton liberation. To 0.8 mg rabbit myofibrils and 1.5 mg *Mercenaria* muscle 0.4 and 0.7 mg rabbit actin were added (arrow). The ATPase of rabbit myofibrils is fully activated by actin, and no further increase is observed on the addition of calcium (arrow). In the case of *Mercenaria* myofibrils, actin has no effect; calcium is needed for activation. The burst of proton liberation upon addition of calcium is due to displacement of protons on the EGTA by calcium. 0.1 mM EGTA, 25 mM NaCl, 1 mM MgCl$_2$ and 0.5 mM ATP followed by 0.2 mM CaCl$_2$.

muscles, and they do not confer calcium sensitivity to rabbit myosin. Components, however, are present in thin filament preparations in addition to actin and tropomyosin, and at this stage we are not able to establish definitively whether these muscles contain only a myosin-linked regulation.

The regulation of annelids is of particular interest since their muscles appear to contain both a myosin-linked and an actin-linked control system. Competitive actin binding studies indicate the presence of a myosin-linked regulatory system in the polychaete worms, *Glycera dibranchiata* (blood-worm), and *Nereis virens* (clam or sand worm) (Table 5). In addition their thin filaments confer about 50–60% calcium sensitivity to rabbit myosin and contain low molecular weight components which very likely correspond to troponin (Fig. 4). The thin filaments bind small amounts of calcium in quantities comparable to the thin filaments of *Limulus* and *Lethocerus* (Table 2). Both *Nereis* and *Glycera* myosins are extremely unstable once actin is removed, and our conclusion that they contain a molluscan-type myosin is based solely on competition studies with rabbit actin. However the alternative interpretations explaining the lack of activation by pure actin are improbable. The possibility that the interaction between actin and

myosin in these polychaete worms is species specific and the mode of interaction restricted within the phylum is not likely in view of the great similarity of actins obtained from different species. Significantly tropomyosin and troponin can be removed from *Nereis* thin filaments (Fig. 4) by washing in 0.6 M NaCl, a method designed by Spudich and Watt (1971) for purifying rabbit actin. This *Nereis* preparation, containing only actin, behaves exactly like rabbit actin; i.e., it does not activate the ATPase of myofibrils in the absence of calcium. The results support the conclusion that *Nereis* myosin requires calcium for actin combination in the presence of ATP.

Our observation that annelid muscles contain both the myosin-dependent and the troponin-linked regulatory systems can be interpreted in alternative ways. The two regulatory systems may be segregated in different muscles, some of the muscles containing only the troponin-linked regulation and some only the myosin-linked one. If this explanation is correct, there should be two populations of myosins, a calcium-sensitive one and a calcium-independent one, and there also should be two populations of thin filaments, one containing troponin, the other troponin free. This interpretation deserves serious consideration, since the muscle preparations used

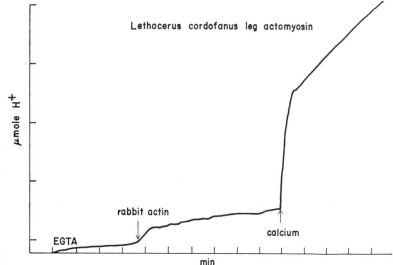

Figure 6. Competitive actin binding assay of *Lethocerus* actomyosin. 1.4 mg leg actomyosin in 0.1 mM EGTA, 25 mM NaCl, 1 mM MgCl$_2$, 0.5 mM ATP. At the first arrow 0.7 mg pure rabbit actin was added followed by 0.2 mM CaCl$_2$ at the second arrow.

were obtained by the homogenization of the whole body wall which contains both longitudinal and circular muscles. An alternative explanation is that all the muscle cells contain both regulatory systems. This interpretation predicts a single population of myosin and a single population of thin filaments. Our experiments support the latter alternative. The fact that the calcium sensitivity of the myofibrils and actomyosin is greater than 95% even in the presence of *excess* pure actin indicates that the muscles contain only calcium-sensitive myosin molecules. Furthermore rabbit myosin is calcium sensitive even with an *excess* of *Glycera* and *Nereis* thin filaments, indicating that the preparation does not contain a significant population of troponin-free thin filaments. If the thin filament preparation consisted of a mixture of troponin-free and troponin-containing actins, calcium sensitivity would decline at higher thin filament concentrations because of the higher affinity of rabbit myosin to troponin-free actins. Moreover densitometry of acrylamide gels indicates that the molar ratio of tropomyosin to actin is about 1:5.5, and the ratio of the 25,000 dalton component to tropomyosin is about 1:1, tending to support the contention that essentially all the thin filaments contain troponin. These results demonstrate that both regulatory systems coexist in the same muscle cell.

Surprisingly insect muscles show great similarity to Annelid muscles and also appear to contain both regulatory systems. Thin filament preparations of the asynchronous flight muscles and of the walking leg muscles of the waterbug, *Lethocerus cordofanus*, and *Lethocerus griseus*, sensitize rabbit myosin by 60–80%. The thin filaments contain low molecular weight components, similar in size to *Limulus*

troponin (Fig. 1, Table 2). The presence of a troponin-like regulation in these muscles was expected and has been reported previously (Maruyama et al., 1968). However actin competition studies on myofibrils and actomyosins indicate that rabbit actin does not activate the ATPase in the absence of calcium, and interaction between actin and myosin requires calcium (Fig. 6). Although preparations of myofibrils and of actomyosins are quite stable, the myosin preparations lose activity rapidly, and we have not as yet been able to consistently obtain enzymically active myosin preparations. Nevertheless we believe the competition studies provide reasonably strong evidence for a direct action of calcium on myosin. A more direct demonstration of a myosin-linked regulatory system in the waterbug would be highly desirable in view of the important X-ray evidence indicating that cross-bridges move in response to calcium prior to interaction with actin (Miller and Tregear, 1970).

Evolutionary Aspects: Speculations

Although our comparative studies on the regulatory systems of different muscles are preliminary and very few phyla have been investigated, the following points seem to emerge (cf. Fig. 7): (1) Myosin-linked regulation is not restricted to molluscs; (2) animals from phyla which evolved early have a myosin-linked regulation; (3) regulation utilizing troponin systems is the result of a more recent evolutionary development; (4) muscles of some species contain both myosin-linked and troponin-linked regulatory systems; (5) the troponins of vertebrates differ from the troponins of arthropods; (6) tropomyosin is present in all

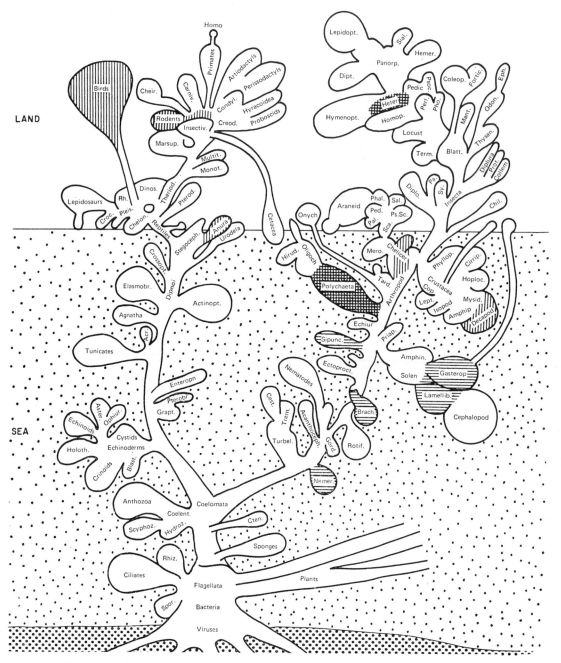

Figure 7. Evolution of regulatory systems. Phylogenetic tree: vertical lines indicate the presence of troponin, horizontal lines the presence of a myosin-linked regulatory system. Note that both regulatory systems are found in annelids and insects. From L. A. Borradaile et al., 1963.

muscles, and it arose earlier than troponin during evolution.

It would appear that the control systems have not evolved in a parallel manner and that regulation by myosin light chains preceded the troponin-tropomyosin regulation of actin. The troponins of vertebrates differ from the troponins of arthropods in the size of their constituents and in their calcium binding. These differences suggest that troponins may have evolved by convergent pathways and may only have appeared after the branch point in evolution which led separately to vertebrates and to arthropods.

Regulation of muscle is a vital function; a loss of regulation would certainly be lethal. Since the myosin-linked and the troponin-linked regulatory systems are apparently unrelated, one could not have arisen from the other simply by mutation. If the evolution of the regulatory systems were indeed sequential, one would predict that a number of species should have contained both systems and that these organisms should occupy a position somewhere between molluscs and arthropods on the phylogenetic tree (Fig. 7). Such transitional muscles may also characterize some species which lie in the pathway of the evolution of vertebrates. Animals with muscles containing both regulatory systems need not have been transient during evolution and, strikingly, appear to be common among present-day annelids.

In higher organisms the troponin system was obviously selected over the myosin-linked regulation. The selective advantage of the troponin system could be explained if mutation in troponin were less likely to be lethal than mutation in a regulatory light chain. Since troponin and tropomyosin are not part of the enzymic site, one would expect that modifications in these proteins would be more tolerated than alterations in the light and heavy chains of myosin.

Tropomyosin seems to have preceded troponin during evolution. All the muscles we studied contain tropomyosin in an approximately constant actin to tropomyosin ratio. The regulatory function of tropomyosin appears to be a more recent acquisition. A different primary role for tropomyosin may be its involvement in length determination of the thin filaments. One notes that in some thin filaments (e.g., scallop) tropomyosin and actin are the only major components present in sufficient quantities to build structures which correspond to the length of the thin filaments. Another alternative role for tropomyosin could be to increase the mechanical stability of the actin filaments, and it might be required for tension bearing. One would expect then that the only motile system lacking tropomyosin would be one in which only little tension is generated. Perhaps a contractile apparatus of this type participates in the protoplasmic streaming of some primitive motile systems.

It is conceivable that some of the light chains of higher animals are regulatory light chains which lost their regulatory function during evolution and represent vestiges. Comparative studies of the light chains may help to test the reality of such a speculation.

Acknowledgments

We thank Mr. Joe M. Regenstein for data on lobster troponin and Dr. Sarah E. Hitchcock for data on chicken troponin, Dr. Annemarie Weber for discussions, and Mrs. Ruth Hoffman for excellent assistance.

This work was supported by Public Health Service Grants GM 14675 and AM 15963 to A.G.S.G. and by a Fellowship from the Muscular Dystrophy Association of America, Inc., to W.L.

Note Added in Proof

Since the submission of the manuscript, a number of additional animals were examined. Myosin-linked regulation was found in the Echinoderm *Thyone briareus* (sea cucumber), the Cephalopod *Loligo pealei* (squid), the Lamellibranchs *Mya orenaria* (steamer clam) and *Spisula solidissima* (surf clam). Actin-linked regulation was found in the Crustaceans *Balanus nubilis* (giant barnacle) and *Libinia emarginata* (spider crab), the Teleosts *Cyprinus carpio* (carp) and *Anguila anguila* (common eel), the Elasmobranch *Raia clavata* (skate). Both myosin-linked and actin-linked regulation was found in the Olygochaete *Lumbricus terrestris* (earthworm) and the Polychaete *Eudistylia polymorpha* (feather duster).

References

BÁRÁNY, M. 1967. ATPase activity of myosin correlated with speed of muscle shortening. *J. Gen. Physiol.* **50**: no. 6, part 2: 197.

BORRADAILE, L. A., F. A. POTTS, L. E. S. EASTHAM, and T. T. SAUNDERS, revised by G. A. Kerkut. *The invertebrata:* A manual for the use of students. Cambridge University Press, 1963.

EBASHI, S. and F. EBASHI. 1964. A new protein component participating in the superprecipitation of myosin B. *J. Biochem.* (Tokyo) **55**: 604.

EBASHI, S. and A. KODAMA. 1965. A new protein factor promoting aggregation of tropomyosin. *J. Biochem.* (Tokyo) **58**: 107.

——. 1966. Interaction of troponin with F-actin in the presence of tropomyosin. *J. Biochem.* (Tokyo) **59**: 425.

EBASHI, S., F. EBASHI, and A. KODAMA. 1967. Troponin as the Ca++-receptive protein in the contractile system. *J. Biochem.* (Tokyo) **62**: 137.

EBASHI, S., A. KODAMA, and F. EBASHI. 1968. Troponin. I. Preparation and physiological function. *J. Biochem.* (Tokyo) **64**: 465.

EISENBERG, E. and W. W. KIELLEY. 1970. Native tropomyosin: Effect on the interaction of actin with heavy meromyosin and subfragment-1. *Biochem. Biophys. Res. Comm.* **40**: 50.

GREASER, M. L. and J. GERGELY. 1971. Reconstitution of troponin activity from three protein components. *J. Biol. Chem.* **246**: 4226.

HARDWICKE, P. M. D. and J. HANSON. 1971. Separation of thick and thin myofilaments. *J. Mol. Biol.* **59**: 509.

HARTSHORNE, D. J. and H. MUELLER. 1968. Fractionation of troponin into two distinct proteins. *Biochem. Biophys. Res. Comm.* **31**: 647.

HARTSHORNE, D. J., M. THEINER, and H. MUELLER. 1969. Studies on troponin. *Biochim. Biophys. Acta* **175**: 320.

KENDRICK-JONES, J., W. LEHMAN, and A. G. SZENT-GYÖRGYI. 1970. Regulation in molluscan muscles. *J. Mol. Biol.* **54**: 313.

LOWEY, S. and D. RISBY. 1971. Light chains from fast and slow muscle myosin. *Nature* **234**: 81.

MARUYAMA, K., F. W. S. PRINGLE, and R. T. TREGEAR. 1968. The calcium sensitivity of ATPase activity of myofibrils and actomyosins from insect flight and leg muscles. *Proc. Roy. Soc.* (London) B **169**: 229.

MILLER, A. and R.T. TREGEAR. 1970. Evidence concerning crossbridge attachment during muscle contraction. *Nature* **226**: 1060.

OHTSUKI, I., T. MASAKI, Y. NONOMURA, and S. EBASHI. 1967. Periodic distribution of troponin along the thin filament. *J. Biochem.* (Tokyo) **61**: 817.

PARKER, L., H. Y. PYUN, and D. J. HARTSHORNE. 1970. The inhibition of the adenosine triphosphatase activity of the subfragment 1-actin complex by troponin plus tropomyosin, troponin B plus tropomyosin and troponin B. *Biochim. Biophys. Acta* **223**: 453.

SCHAUB, M. C. and S. V. PERRY. 1969. The relaxing protein system of striated muscle. Resolution of the troponin complex into inhibitory and calcium ion-sensitizing factors and their relationship to tropomyosin. *Biochem. J.* **115**: 993.

———. 1971. The regulatory proteins of the myofibril. Characterization and properties of the inhibitory factor (Troponin B). *Biochem. J.* **123**: 367.

SPUDICH, J. A. and S. WATT. 1971. The regulation of rabbit skeletal muscle contraction. I. Biochemical studies of the interaction of the tropomyosin-troponin complex with actin and the proteolytic fragments of mysoin. *J. Biol. Chem.* **246**: 4866.

SZENT-GYÖRGYI, A. G., C. COHEN, and J. KENDRICK-JONES. 1971. Paramyosin and the filaments of molluscan "catch" muscles. II. Native filaments: Isolation and characterization. *J. Mol. Biol.* **56**: 239.

SZENT-GYÖRGYI, A. G., E. M. SZENTKIRALYI, and J. KENDRICK-JONES. 1972. The light chains of scallop myosin as regulatory subunits. *J. Mol. Biol.* In press.

WEBER, A. and R. D. BREMEL. 1971. Regulation of contraction and relaxation in the myofibril, p. 37. In *Contractility of muscle cells and related processes*, ed. R. F. Podolsky. Prentice-Hall, Englewood Cliffs, New Jersey.

Structural Studies by X-Ray Diffraction of Striated Muscle Permeated with Certain Ions and Proteins

ELIZABETH ROME

Medical Research Council Muscle Biophysics Unit, King's College, 26–29 Drury Lane, London, W.C.2, England

X-ray diffraction techniques have been used very extensively to study not only many different kinds of muscles, but also muscles in different states such as resting, contracting, and rigor. In this paper I am going to describe a new approach in which X-ray diffraction has been used to study glycerinated muscle permeated with certain ions and proteins which effect structural changes in, or labeling of the filaments. New structural information has been obtained from these studies.

Glycerinated muscle (Szent-Györgyi, 1949) is a system in which the cell membranes have been rendered permeable and the soluble enzymes and substrates leached out, but the gross structure of the contractile apparatus and its ability to contract have been unaffected. The muscle is in an inextensible (rigor) condition in which the projections from the myosin filaments are attached to the actin filaments. The unique feature of this system compared with living muscle is that the medium permeating the contractile apparatus can be directly controlled. Usually glycerinated muscle is studied in a medium whose composition is roughly similar to that of the interior of the cell in living muscle. This medium —0.1 M KCl, 1 mM MgCl$_2$, 6.67 mM phosphate buffer, pH 7.0—will be referred to as standard salt solution. I have now studied glycerinated (rabbit psoas) muscle under the following conditions: (a) in the presence of a relaxing medium, (b) after permeation with subfragment 1 molecules, and (c) after permeation with antibodies to the newly discovered protein of the thick filament, C-protein.

In all the experiments both standard glycerinated muscle preparations and "fast" glycerinated preparations have been used. To make the latter preparations the fiber bundles, immediately after excision from the rabbit, are osmotically shocked by alternating them several times between standard salt solution and glycerol solution, so that the membrane is rapidly broken down and the extraction process completed quickly. These preparations are suitable for use after 1–2 days of extraction, whereas the standard preparations require up to five weeks. There are no apparent differences in the results obtained from the two different types of preparations.

Glycerinated Muscle in Relaxing Medium

When vertebrate striated muscle passes from the living to the rigor state, X-ray diffraction patterns show that there is a structural change in the myosin filaments and the projections move out and attach to the actin filaments (Huxley and Brown, 1967; Huxley, 1968; Haselgrove, 1970). The object of the present experiments was to see whether the glycerinated (rigor) muscle could be effectively relaxed and the structure returned to that of the living muscle. Previously only glycerinated insect muscle has been effectively relaxed and corresponding changes in the X-ray pattern observed (Reedy et al., 1965). The present experiments on glycerinated rabbit psoas muscle have been reported in detail elsewhere (Rome, 1972) so only the main features will be discussed here.

The procedure used for relaxation was to soak the glycerinated muscle specimen initially in a salt solution containing EGTA but no ATP, (0.1 M KCl, 5 mM MgCl$_2$, 4 mM EGTA, 10 mM histidine-Cl, pH 7.0) for one hour and then to place it in a similar solution but containing in addition 5 mM ATP. In this relaxing solution the best specimens remained relaxed and extensible for several days; they could be stretched to sarcomere lengths up to about 4.4 μm.

The equatorial X-ray diffraction pattern of relaxed glycerinated muscle was investigated as a function of sarcomere length in the range 2.25–4.4 μm. Sarcomere lengths were determined by optical diffraction, and the equatorial patterns were obtained using Franks cameras with specimen-to-film distances of about 10 cm. Typical equatorial patterns from both rigor and relaxed glycerinated muscles are shown in Fig. 1. In general two reflections are seen: these arise from the 1,0 and 1,1 planes of the hexagonal filament lattice. The relative intensity of the reflections depends on the relative contributions from the thick (myosin-containing) and thin (actin-containing) filaments: an increase in the thin filament contribution increases the intensity of the 1,1 reflection ($I(1,1)$) relative to that of the 1,0 reflection ($I(1,0)$). In patterns from both rigor and relaxed muscles the intensity ratio $I(1,1)/I(1,0)$

Rigor   **Relaxed**

S = 2.12 μm

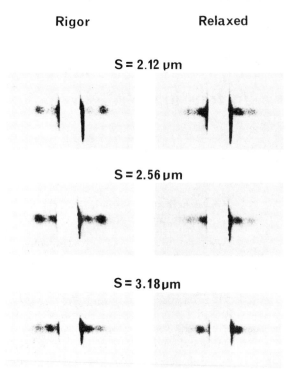

S = 2.56 μm

S = 3.18 μm

Figure 1. Comparison of equatorial X-ray diffraction patterns from glycerinated muscle in the rigor and the relaxed conditions taken on a Franks camera with specimen-to-film distance ∼10 cm. The magnification is 11.25. Each pair of patterns is from one specimen, first in the rigor condition (*left*) and then in the relaxed condition (*right*). The sarcomere lengths, S, are as shown. The inner reflection is the 1, 0 reflection and the outer reflection is the 1,1 reflection. At each sarcomere length the intensity ratio I(1, 1)/I(1, 0) is much greater in the rigor than in the relaxed condition.

decreases as the sarcomere length increases. This is because only in the A-band of the sarcomere are the thin filaments sufficiently ordered to contribute to the reflections (Elliott et al., 1963; Rome, 1967). The most significant feature of the patterns, however, is that at any given sarcomere length, I(1,1)/I(1,0) is very much less in the relaxed muscle than in the rigor muscle. Comparison with data from living and rigor muscles obtained by Huxley (1968) shows that the intensity ratio in the relaxed glycerinated muscles is similar to that in the living resting muscles. Now Huxley interpreted the increase of I(1,1)/I(1,0) in passing from the living to the rigor state as due to the attachment of the myosin projections to the actin filaments. The reverse effect observed here, the decrease of I(1,1)/I(1,0) when a muscle passes from the rigor to the relaxed state, can likewise be attributed to the detachment of the myosin projections from the actin filaments. Thus it can be seen that equatorial X-ray patterns provide a very clear test of relaxation.

Meridional X-ray patterns of relaxed glycerinated muscle were obtained on a mirror-monochromator

camera with specimen-to-film distance about 20 cm. A typical pattern showing low-angle myosin reflections is reproduced in Fig. 2, where it is compared with patterns from living resting muscle and from rigor glycerinated muscle. In the living muscle pattern layer lines indexing on a repeat of 429 Å are seen; these arise from the myosin projections which are arranged in a helix around the myosin filament axis with 143 Å axial separation and 429 (3 × 143) Å pitch (Huxley and Brown, 1967). The most prominent features of the rigor pattern are a meridional reflection at 144.4 Å arising from the axial repeat of the myosin projections, and a strong but rather diffuse layer line at about 370 Å thought to arise from the attachment of the projections to the actin filaments in an "actin-like" helix (pitch ∼370 Å). When the rigor muscle is relaxed the characteristic rigor pattern disappears and is replaced by a pattern resembling, but very much weaker and more diffuse than, the living pattern: reflections indexing

Living   **Rigor**

Å

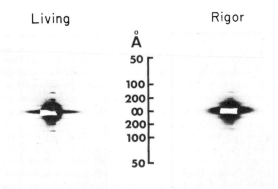

Relaxed glycerinated

Å

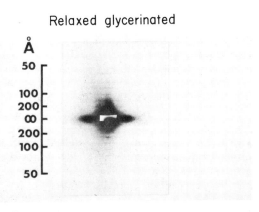

Figure 2. Comparison of meridional low-angle X-ray diffraction patterns from living resting, rigor glycerinated, and relaxed glycerinated muscle taken on a mirror monochromator camera with specimen-to-film distance ∼20 cm. The magnification is 2.25. In the living resting and relaxed glycerinated patterns, reflections indexing on a 429 Å repeat can be seen. In the rigor glycerinated pattern the predominant features are a strong layer line at about 370 Å and a meridional reflection at 144.4 Å.

on the 429 Å repeat can be measured out to the sixth order. These patterns demonstrate not only that the myosin projections are detached from the actin filaments in the relaxed glycerinated muscle, but also that the helical conformation of the projections around the myosin filaments is re-established as in the living muscle.

Both the equatorial and the meridional X-ray data therefore demonstrate that the process whereby the contractile apparatus of a muscle passes from the living to the rigor state is a reversible one, and the structure of relaxed glycerinated muscle resembles that of the living resting muscle. The importance of this type of experiment is that the processes and structural changes associated with the contractile process are being directly examined. Obviously an even more relevant experiment in this context would be to obtain X-ray patterns from glycerinated muscle in the presence of ATP and calcium, i.e., in the contracting state itself. However, all attempts so far have been unsuccessful because of the great difficulty of obtaining uniform contraction throughout the specimen.

Glycerinated Muscle Labeled with Subfragment 1

In this section and the following one I describe our most recent work in which large protein molecules have been diffused into the contractile apparatus of glycerinated muscle. This work is still in progress, and hence this is only a preliminary report.

The myosin molecule contains two globular heads which can be cleaved off by proteolytic digestion (Lowey et al., 1969); these are termed subfragment 1 (S-1). In vitro S-1 will decorate actin filaments, one S-1 molecule attaching to each actin subunit (Moore et al., 1970). In glycerinated (rigor) muscle it might be expected that S-1 molecules would attach to vacant actin subunits and therefore produce enhancement of those X-ray reflections arising from the attachment of the S-1 parts of the myosin projections to the actin filaments. If this were the case, it should be possible to observe directly which parts of the standard rigor pattern arise from the attachment of S-1 to the actin filaments and therefore to resolve the ambiguities (noted by Haselgrove, 1970) concerning the origin of certain of the rigor layer lines. There is, however, the problem that large molecules such as S-1 (size approximately $150 \times 45 \times 30$ Å³; Moore et al., 1970) might not be able to permeate into the interfilament spaces of the contractile apparatus. The purpose of the present experiments was therefore to see whether there was any change in the X-ray patterns of specimens which had been soaked in S-1 solutions, and if so, what structural information on the actin-myosin attachment could be obtained from these patterns.

Normally for X-ray diffraction of glycerinated muscle, specimens of about 1 mm diameter are used. In the present experiments in order to facilitate the diffusion of the S-1 molecules into the fibers, the specimens were usually shredded over most of their length into smaller fiber bundles (diameter $\sim$ 0.3 mm). The specimens were soaked overnight in a salt solution (0.1 M KCl, 1 mM MgCl₂, 1 mM DTT, 1 mM Tris-Cl, pH 7.0) containing 10 mg/ml S-1. To remove excess S-1 the specimens were then soaked for about 2 hr in standard salt solution, and then examined in this solution. Typical patterns taken on a mirror-monochromator camera with specimen-to-film distance about 20 cm are shown in Fig. 3 where they are compared with standard living and (glycerinated) rigor patterns. Figure 3d is from a specimen which was shredded before soaking in the S-1 solution, whereas Fig. 3c is from a specimen which was not shredded. In both of these patterns certain reflections (which are shown below to index on an "actin-like" helix) are clearly enhanced, indicating that S-1 molecules can diffuse into the contractile apparatus and label the actin filaments. The enhancement of the reflections is very much greater in the pattern from the shredded specimen, indicating that this procedure really does improve the diffusion of the S-1 molecules into the contractile apparatus. However, the shredded specimen has the disadvantage that the reflections are very much more arced and diffuse; this is because the shredding procedure tends to disorient the fiber bundles.

There has been controversy over a number of years about the exact configuration of the actin helix in both living and rigor states and in different kinds of muscle (e.g. Huxley and Brown, 1967; Lowy and Vibert, 1967; Millman et al., 1967; Haselgrove, 1970). The helix is perhaps best described as a nonintegral two-strand helix (approximately 13 subunits in six turns of the genetic helix) with subunit repeat 27.5 Å and near axial repeat about 360–385 Å. In X-ray patterns from living muscle the strongest layer line reflection observed from the actin structure is that at 59 Å (see Fig. 3a). When muscle passes into rigor, the most obvious changes are the movement of the intensity maximum of the 59 Å reflection closer to the meridian and the appearance of a very strong but rather diffuse layer line at about 370 Å (see Fig. 3b); other weak layer lines are also seen. The pattern is interpreted in terms of the attachment of the myosin projections to the actin filaments in an "actin-like" helix (Huxley and Brown, 1967; Haselgrove 1970). In the S-1 labeled patterns (Fig. 3c, d) the intensity of the 59 Å reflection is increased and is even closer to the meridian, the intensity of the 370 Å layer line is increased, and layer line reflections at 185, 86, 70 and 51 Å which can be seen only faintly or not at

Living   Rigor

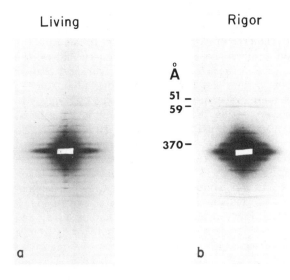

a   b

S-1 labeled rigor

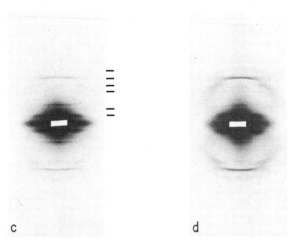

c   d

Figure 3. Comparison of meridional X-ray diffraction patterns from living, rigor, and S-1 labeled rigor muscle taken on a mirror-monochromator camera with specimen-to-film distance ~20 cm. The magnification is 2.25. Pattern (c) is from a specimen which was not shredded before soaking in the S-1 solution; pattern (d) is from a specimen which was shredded. In the living pattern only the 59 Å actin reflection can be clearly seen; in the standard rigor pattern a strong layer line at approximately 370 Å can be seen together with the 59 and 51 Å actin reflections. In the S-1 labeled rigor patterns reflections (indicated in the figure) with spacings 370, 185, 86, 70, 59, and 51 Å lying on an "actin-like" helix are intensified.

all in the standard rigor pattern are now seen clearly. The assignment of all these features when seen in the standard rigor pattern to the projections attaching to the actin filaments is therefore confirmed. Furthermore, both the spacings and radial positions of all these reflections seen in the S-1 labeled patterns are consistent with the S-1's being arranged in an

"actin-like" helix around the actin filament axis, again confirming the interpretation of the standard rigor pattern.

A further feature of the standard rigor pattern, discussed in some detail by Haselgrove (1970), is a series of weak, rather diffuse layer lines close into the meridian of the pattern and indexing on a repeat of about 2×365 Å. Haselgrove suggested that these might arise either from the parts of the myosin projections close to the myosin filament (the S-2 connecting links perhaps) arranged in a helix around the actin filament axis, or from the S-1 parts of the myosin projections attached to the actin filaments and arranged in a helix around the myosin filament axis. There is no enhancement of these layer lines in the S-1 labeled patterns, therefore suggesting that the former of Haselgrove's interpretations is the correct one. (I am most grateful to Dr. Haselgrove for pointing this aspect of the S-1 labeled patterns out to me.) No other reflections are enhanced in the S-1 labeled patterns, and the remainder of the patterns are unchanged from the standard rigor pattern.

Detailed measurements of the spacings of the S-1 "actin-like" helical reflections have not been made; as stated before the reflections are rather diffuse, and a large number of patterns would need to be measured in order to obtain statistically significant data. Approximate measurements of the radial positions of the reflections have been made where possible. The measurements on the 370, 59, and 51 Å layer lines are confused by lattice sampling and by the presence of intensity maxima due to both the actin helix and the S-1 helix. It can be seen, however, that as the quantity of matter in the S-1 helix increases, i.e., from the living to the rigor to the S-1 labeled states, the intensity maxima on the 59 and 51 Å layer lines move in very much closer to the meridian, indicating that there is increasingly more scattering matter at a greater distance from the actin filament axis. The reflections on the 185, 86, and 70 Å layer lines seen in the S-1 labeled patterns can be assigned to the S-1 helix rather than the actin helix because their radial positions correspond to a helix with point scatterers at a distance of about 75 Å from the filament axis, much too great for the actin helix. Lednev (personal communication) has also obtained a helix radius of about this value for the myosin projections attached to the actin filaments in standard rigor muscle.

It is an interesting feature of the actin helix pattern in living muscle that the intensity of the 59 Å reflection is very much greater than that of the 51 Å reflection (see Fig. 3a). Furthermore when the myosin projections or S-1 molecules label the actin helix, the relative increase of the intensity of the 59 Å reflection is far greater than that of the

51 Å reflection. This suggests that the shape of the S-1 molecules, together with their mode of attachment to the actin filaments, is such as to accentuate the 59 Å planes in the helical structure (see Fig. 4). Now Moore et al. (1970) have obtained three-dimensional reconstructions from electron micrographs of actin filaments decorated with S-1, and

Figure 4. Diagrammatic representation of the actin helix, assuming there are exactly 13 subunits in six turns of the genetic helix. The genetic helix of pitch 59 Å is indicated by the dotted lines.

their model shows S-1 molecules which are somewhat elongated and curved and which tend to lie along the 59 Å planes. The present X-ray results therefore complement the optical diffraction studies of Moore et al. and indicate that the attachment of the S-1 molecules to the actin filaments observed in the electron microscope is the same as that in the muscle itself.

Glycerinated Muscle Labeled with Antibody to C-Protein

I am now going to describe work which has been done in collaboration with Dr. G. Offer, Dr. F. A. Pepe, and Mr. R. W. Craig. This work concerns the recently discovered C-protein (Starr and Offer, 1971) which is strongly bound to myosin in the thick filament (Moos et al., 1972; Offer, this volume). In order to determine the precise location of the C-protein we have used antibody techniques. Antibody to rabbit C-protein (anti-C) has been elicited in goats and diffused into the contractile

apparatus of glycerinated rabbit psoas muscle; both X-ray diffraction and electron microscopy techniques have then been used to examine these specimens. I describe here the X-ray work.

The anti-C molecules are of approximately the same size as the S-1 molecules, and a similar procedure was used to diffuse them into the contractile apparatus: the specimens were shredded, soaked overnight in whole antiserum dialyzed against standard salt solution, soaked for a couple of hours in standard salt to remove excess protein, and then examined in just the standard salt solution. As a control the same procedure was used but with serum obtained from the goat before the C-protein injections were started; the X-ray patterns obtained from such preparations were quite normal. Typical X-ray patterns from a specimen soaked in anti-C and taken on a mirror-monochromator camera with specimen-to-film distance about 40 cm are shown in Fig. 5 where they are compared with a standard (glycerinated) rigor pattern. In the standard pattern meridional reflections at 442 and 385 Å are seen just clear of the central scatter. These reflections were first observed by Huxley (Huxley, 1967; Huxley and Brown, 1967) and attributed to repeating periodicities in the A-band and the I-band respectively; optical diffraction studies have now confirmed this assignment (O'Brien et al., 1971). In the anti-C labeled pattern the intensity of the 442 Å reflection is greatly enhanced (by ~5 times), therefore showing that the C-protein in the thick filament must be partly or wholly responsible for this reflection. The spacing appears to have increased slightly from 442 ± 1 Å in the standard rigor pattern to 445 ± 1 Å in the anti-C labeled pattern. Also, in several particularly good patterns a subsidiary, very weak reflection with a spacing of 417.5 ± 2 Å is observed. The rest of the anti-C labeled pattern is unchanged: there is no obvious intensification of any other reflections. Electron micrographs taken by Craig of anti-C labeled specimens show that the C-protein is located in nine stripes in each half of the A-band (a typical micrograph is reproduced in Offer's paper in this volume); the separation of the stripes is about 420–440 Å, and the distance between the centers of the arrays (i.e., the distance of the 5th stripe on one side of the M-line to the 5th stripe on the other side) is about 6800 Å.

C-protein binds strongly to myosin, and when added to light meromyosin (LMM) paracrystals it accentuates their approximately 430 Å periodicity (Moos, discussion to Offer, this volume). The puzzle is, therefore, that the X-ray evidence suggests that myosin and C-protein in the thick filament have different periodicities. Both Huxley and Brown (1967) and Haselgrove (1970) have made detailed measurements on the living muscle pattern showing

Standard rigor

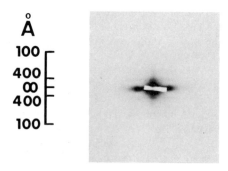

Anti−C labeled rigor

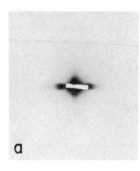

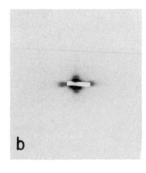

Figure 5. Comparison of meridional low-angle X-ray diffraction patterns from standard rigor and anti-C labeled rigor muscle taken on a mirror-mono-chromator camera with specimen-to-film distance ~40 cm. The magnification is 1.5. In the standard rigor pattern the 442 and 385 Å reflections can be seen just clear of the central scatter; the 144.4 Å reflection can be seen further out. Patterns (a) and (b) are different exposures from the same anti-C labeled specimen; the relative intensification of the 442 Å reflection compared with the 144.4 Å reflection can be seen in pattern (a), whereas in the shorter exposure of pattern (b) the 442 Å reflection can be seen very clear from the central scatter.

that the 442 Å reflection cannot possibly be indexed on the 429 (3 × 143) Å myosin repeat. In rigor muscle the 143 Å myosin reflection is increased by approximately 1% to 144.4 Å (Haselgrove, 1970), but my measurements suggest that there is not an equivalent increase in the 442 Å reflection; it still cannot be indexed on the myosin repeat. We believe, however, that the interpretation of the X-ray patterns may be less straightforward than has so far been envisaged and that myosin and C-protein may, in fact, have related periodicities. The clue for this new interpretation is the subsidiary reflection at 417.5 Å seen in the anti-C labeled patterns. This subsidiary reflection has been seen before in high resolution patterns (in living muscle Huxley and Brown (1967) measured it at 416 Å and Haselgrove (1970) at 417.6 Å, compared with 442 Å for the primary reflection), but the significant point here is that it is intensified along with the primary reflection in the anti-C labeled patterns and must therefore also be due to C-protein. The 385 Å reflection is also seen to be a doublet at high resolution and O'Brien et al. (1971) have evidence that this is caused by interference between the two halves of

the I-band. Similarly the 442 and 417 Å doublet could be caused by interference between the C-protein arrays in each half of the A-band. The diffraction pattern of such a system is the diffraction pattern of one array multiplied by an interference function; the interference function is a set of fringes whose separation is related to the separation of the two arrays. Thus it could be possible for the true repeat of the C-protein stripes to be the same as that of the myosin, i.e. in living muscle 429 Å, but for the interference function to sample the 429 Å peak so that it is split into two reflections, one with a spacing less than, and the other with a spacing greater than 429 Å. If this were the case, higher orders of the 442 Å reflection would not be expected. In high resolution patterns from living muscle (Huxley and Brown, 1967; Haselgrove, 1970) strong meridional reflections indexing on the 429 Å repeat are seen together with a large number of fainter, subsidiary reflections, usually close to the primary reflections. Although some of these subsidiary reflections do seem to index on a 442 Å repeat, they are of no greater (often smaller) intensity than any of the other adjacent subsidiary reflections. There is no reason to pick out a 442 Å repeat from these subsidiary reflections, and an alternative interpretation of these reflections is suggested later.

We have roughly tested the plausibility of our suggestion by assuming a model for the arrangement of C-protein in the thick filament and deducing the diffraction pattern. The procedure is illustrated diagrammatically in Fig. 6. It is assumed that the C-protein is located in nine lines of repeating periodicity 429 Å in each half of the thick filament, and that the diffraction pattern of either one of these arrays is simply that of a nine-line diffraction grating. The interference function is assumed to be the "cosine-squared" fringes arising from the separation

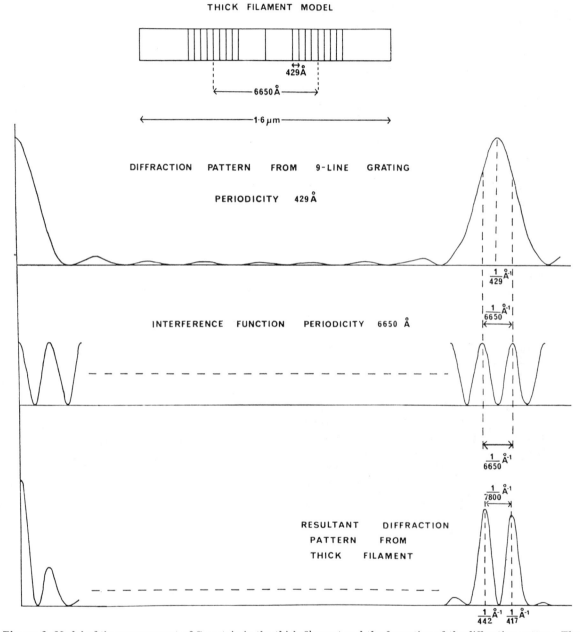

Figure 6. Model of the arrangement of C-protein in the thick filament and the formation of the diffraction pattern. The C-protein is arranged in an array of nine lines, separation 429 Å, in each half of the thick filament, and the diffraction pattern from one of these arrays is assumed to be that of a nine-line diffraction grating (the diffraction pattern is only shown out as far as the 1st principal maximum). The center-to-center separation between the two arrays is 6650 Å, and the interference function from the two arrays is assumed to be a cosine-squared function with repeat 6650 Å. The resultant diffraction pattern of the thick filament is the product of these two functions and is shown for the region of the 1st order reflection: the original broad reflection centered at 429 Å is split into two sharper reflections at 442 and 417 Å.

of two identical arrays. The periodicity of the fringes, 6650 Å, has been chosen to be consistent with the separation of the two C-protein arrays seen in the micrographs and to be such that there is a peak at $1/443$ Å^{-1}. The product of the two functions then gives, instead of the original broad 429 Å peak, a doublet with sharper peaks at 442 and 417 Å. It can be seen that the separation of these two peaks is less than the separation of the peaks in the interference function ($1/7800$ Å^{-1} compared with $1/6650$ Å^{-1}) and that this arises because of the overriding shape of the diffraction pattern from the single array. This is an effect which must be taken into account when the separation of two arrays in the real system is being deduced from the splitting of a reflection in the diffraction pattern.

Although the spacings of the reflections predicted by this model are in remarkably good agreement with the values actually observed in muscle, the intensities are not in such good agreement: in the real system the 442 Å reflection is very much more intense than the 417 Å reflection, whereas the model system predicts only a small difference in the intensities. There are, however, several features of the real system which have not been taken into account in the model system. For example, Hanson et al. (1971) have observed that isolated (frog) A-segments when examined by negative contrast in the electron microscope show eleven unstained stripes of repeating periodicity 420 ± 15 Å; eleven similar stripes are seen in Craig's micrographs of rabbit myofibrils also examined by negative contrast, but only the outer nine of these stripes are labeled by anti-C. It may be that some other protein component is responsible for the inner two stripes. The LMM backbone of the filament is also thought to have a similar periodicity since oriented gels of LMM give a series of X-ray reflections indexing on a 430 Å repeat (Szent-Györgyi et al., 1960). Since the diffraction from C-protein cannot be considered independently from that of the rest of the thick filament, these other repeating structures must be taken into account in order to predict the exact diffraction pattern. It is also necessary to consider the shape and size of the C-protein stripes, and the opposite polarity of the thick filament structure on either side of the M-line. Thus the diffraction pattern deduced above for the C-protein arrays alone could be considerably changed when these other features are included in the analysis. An example of the type of changes involved may be indicated by the observation that the 442 Å reflection seen in standard rigor patterns is shifted to 445 Å in the anti-C labeled patterns. It seems likely that this occurs because the labeling of just nine of the eleven stripes slightly changes the shape and sampling position of the 429 Å peak.

Despite the simplicity of the model used for the diffraction analysis, it does demonstrate the plausibility of our suggestion. Furthermore we feel that this model of the thick filament in which C-protein, which binds so strongly to myosin, has the same periodicity as myosin is undoubtably preferable to any other model in which they have different periodicities. We hope that further analyses of models incorporating the other periodically repeating structures of the thick filament will clarify our concept of the structure. It may also be possible to account for several of the detailed features of the living muscle pattern observed by Huxley and Brown (1967) and Haselgrove (1970). For example, it seems likely that the subsidiary meridional reflections referred to earlier may correspond to subsidiary peaks of the diffraction pattern of one-half of the thick filament sampled by the interference function. (I am most grateful to Dr. E. J. O'Brien for suggesting this to me.) Also the slight but significant difference between the measured and calculated spacings of certain of the orders of the 429 Å repeat may arise because the interference function produces a slight shift in the peak spacing. A further speculation is that the 1% increase in the spacing of the 143 Å reflection when a living muscle passes into rigor has a similar origin to that discussed above for the shift of the 442 Å reflection when rigor muscle is labeled with anti-C; i.e., a slight structural change in the thick filament (caused in this case perhaps by the myosin projections reaching further out to attach to the actin filaments) could change both the shape and the sampling position of the 143 Å peak. These features are being investigated further.

Acknowledgments

I thank Professor Jean Hanson for her constant encouragement and ready advice, Dr. G. Offer for his interest and enthusiasm and much helpful advice, Dr. E. J. O'Brien and Miss P. M. Bennett for useful discussions, and Mr. Z. Gabor for photographic assistance.

References

ELLIOTT, G. F., J. LOWY, and C. R. WORTHINGTON. 1963. An X-ray diffraction and light diffraction study of the filament lattice of striated muscle in the living state and in rigor. *J. Mol. Biol.* **6**: 295.

HANSON, J., E. J. O'BRIEN, and P. M. BENNETT. 1971. Structure of the myosin-containing filament assembly (A-segment) separated from frog skeletal muscle. *J. Mol. Biol.* **58**: 865.

HASELGROVE, J. C. 1970. X-ray diffraction studies on muscle. Ph.D. thesis, University of Cambridge.

HUXLEY, H. E. 1967. Recent X-ray diffraction and electron microscope studies of striated muscle. *J. Gen. Physiol.* **50**: No 6, Part 2, 71.

———. 1968. Structural difference between resting and rigor muscle; evidence from intensity changes in the low-angle equatorial X-ray diagram. *J. Mol. Biol.* **37**: 507.

HUXLEY, H. E. and W. BROWN. 1967. The low-angle X-ray diagram of vertebrate striated muscle and its behaviour during contraction and rigor. *J. Mol. Biol.* **30**: 383.

LOWEY, S., H. S. SLATER, A. G. WEEDS, and H. BAKER. 1969. Substructure of the myosin molecule. I. Subfragments of myosin by enzymic degradation. *J. Mol. Biol.* **42**: 1.

LOWY, J. and P. J. VIBERT. 1967. Structure and organization of actin in a molluscan smooth muscle. *Nature* **215**: 1254.

MILLMAN, B. M., G. F. ELLIOT, and J. LOWY. 1967. Axial period of actin filaments: X-ray diffraction studies. *Nature* **213**: 356.

MOORE, P. B., H. E. HUXLEY, and D. J. DeROSIER. 1970. Three-dimensional reconstruction of F-actin, thin filaments and decorated thin filaments. *J. Mol. Biol.* **50**: 279.

MOOS, C., G. W. OFFER, and R. L. STARR. 1972. A new muscle protein which affects myosin filament structure. *Abstr. Amer. Biophys. Soc.*

O'BRIEN, E. J., P. M. BENNETT, and J. HANSON. 1971. Optical diffraction studies of myofibrillar structure. *Phil. Trans. Roy. Soc. London* B **261**: 208.

REEDY, M. K., K. C. HOLMES, and R. T. TREGEAR. 1965. Induced changes in orientation of the cross-bridges of glycerinated insect flight muscle. *Nature* **207**: 1276.

ROME, E. 1967. Light and X-ray diffraction studies of the filament lattice of glycerol-extracted rabbit psoas muscle. *J. Mol. Biol.* **27**: 591.

———. 1972. Relaxation of glycerinated muscle: Low-angle X-ray diffraction studies. *J. Mol. Biol.* **65**: 331.

STARR, R. and G. OFFER. 1971. Polypeptide chains of intermediate molecular weight in myosin preparations. *FEBS Letters* **15**: 40.

SZENT-GYÖRGYI, A. 1949. Free-energy relations and contraction of actomyosin. *Biol. Bull.* **96**: 140.

SZENT-GYÖRGYI, A. G., C. COHEN, and D. E. PHILPOTT. 1960. Light meromyosin fraction 1: a helical molecule from myosin. *J. Mol. Biol.* **2**: 133.

X-Ray Evidence for a Conformational Change in the Actin-containing Filaments of Vertebrate Striated Muscle

J. C. HASELGROVE

MRC Laboratory of Molecular Biology, Hills Road, Cambridge CB2 2QH England*
and
Biofysisk Institut, Aarhus Universitat, DK8000 Aarhus C, Denmark

The control of vertebrate striated muscle from rest to activity is regulated by the combination of the regulatory proteins tropomyosin and troponin on the thin actin filament (see reviews by Ebashi and Endo, 1968; Ebashi et al., 1969). In the absence of calcium the regulatory proteins inhibit the actin-myosin interaction. Activation of muscle is caused by calcium binding to troponin which then removes the inhibition. Since there is only one troponin molecule for every seven actin molecules (Weber and Bremel, 1971) and troponin does not interact with actin, then the control of activation is presumably mediated by the tropomyosin molecules which run along the grooves in the actin filaments (Hanson and Lowy, 1963; Moore et al., 1971). It is, therefore, of importance to investigate any structural changes that occur in the thin filaments when actin-myosin interaction is allowed to take place, for such changes may give us information about the control mechanism. X-ray diffraction for the study of muscle has the great advantage that the structure can be studied at the molecular level without destruction of the specimen. The low-angle X-ray pattern (corresponding to spacings above about 20 Å) from vertebrate striated muscle consists of layer lines and meridional reflections that arise both from the myosin filaments and from the actin filaments (Huxley and Brown, 1967), so it is possible, by observing changes in the low-angle diffraction pattern, to follow changes occurring in the thin filament structure when the muscle contracts or passes into rigor. The great disadvantage of X-ray diffraction studies on muscle is that they do not give information directly about the muscle structure; the structure must first be assumed from other studies and the structure can be accepted if it gives a diffraction pattern like the observed pattern.

The earlier studies of the muscle low-angle X-ray diffraction patterns (Elliott et al., 1967; Huxley and Brown, 1967) showed the gross structure of the thin filaments was remarkably insensitive to the state of the muscle. The axial period of the helical

structure was the same whether the muscle was resting, contracting, or in rigor, and this indicated that there is little difference in the structure of the basic helix of F-actin (composed of G-actin monomers). The first evidence for a significant change in the actin X-ray diffraction pattern of contracting muscle was obtained by Huxley (1970, 1971a,b), who found that a reflection corresponding to the 2nd actin layer line was present in the patterns from muscles contracting at rest length, although this reflection was absent from the patterns from relaxed muscles. We then noticed that similar changes occur when a striated muscle is put into rigor at rest length, the 2nd layer line being distinct while no sign was seen of the 3rd layer line (Fig. 1). Much stronger patterns from live muscles (Fig. 1) showed that the 3rd layer line is visible, although it is not as strong as the 2nd layer line from rigor muscles. We also found (Vibert et al., 1972a,b) that such changes also occur in the smooth muscles anterior byssus retractor muscle of *Mytilus* (ABRM) and taenia coli of the guinea pig. The increase in the intensity of the 2nd actin layer line is presumably related to the changes that occur in the thin filaments when actin and myosin are allowed to interact. O'Brien et al. (1971) showed with optical transforms of actin paracrystals, with and without tropomyosin, that the relative intensities of the layer lines are changed by the introduction of tropomyosin—specifically, the 2nd layer line becomes stronger. Although the changes in the X-ray diffraction pattern from whole muscle are not likely to be due to the complete removal and addition of tropomyosin to the thin filaments, they could well be caused by movement of the position of the tropomyosin molecules with respect to the actin, as suggested by Huxley (1970, 1971a,b) and Spudich et al. (1972). Their work shows that the tropomyosin is not held in the center of the groove between two strands and is thus in a position where it may interact with the myosin binding sites.

We therefore made calculations (using an IBM 360/44 computer at the Institute of Theoretical Astronomy in Cambridge) of the diffraction patterns that would be given by various simple

---

* Present address.

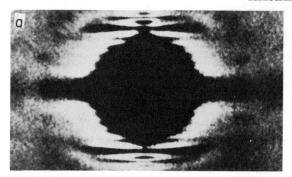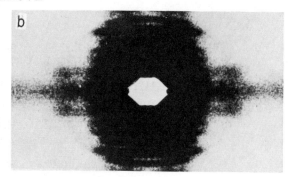

Figure 1. (a) The diffraction pattern from a live relaxed muscle at rest length. The 3rd layer line can be seen but not the 2nd. The layer lines are more easily seen if the picture is viewed at a glancing angle in the direction of the layer line. (b) The diffraction pattern of a muscle in a standard glycerol rigor at rest length. The 2nd layer line can be seen clearly but not the 3rd layer line. The same pattern is obtained from muscles in calcium-free rigor. (All the patterns were taken using a focusing X-ray camera (Huxley and Brown, 1967; Haselgrove, 1970) and have been reproduced by a background-leveling technique designed to enhance the weaker layer lines. The appearance of the pattern has thus been distorted to allow selected features to be demonstrated. We thank E. B. Neergaard for showing us this technique.)

models of the thin filament to see if we could account for the changes that occur in the observed diffraction patterns. The calculated patterns were compared with patterns obtained from live and rigor muscles. Rigor muscles were chosen for the comparison because in this state all possible myosin-actin attachments are made and we are investigating the changes associated with such interactions. However comparison of the patterns from vertebrate striated muscle contracting (patterns kindly lent by Dr. H. E. Huxley) and in rigor at rest length showed the second layer line to be of similar intensity compared with the rest of the pattern.

Calculations

The actin filament without any tropomyosin present may be considered for the purpose of calculation as a series of spherical subunits, each representing one actin monomer, joined in two

long strands which wind around each other forming a slowly turning double helix (Hanson and Lowy, 1963). Mathematically the helix can be thought of in terms of the genetic helix with a repeat length of 51 Å and two subunits in slightly more than one turn of the helix. The structure repeats again after 15 turns of the genetic helix and in this distance there are 28 subunits. This distance (764.4 Å) is the pitch length of one of the two long strands of monomers. It makes very little difference to the diffraction patterns if the whole filament is twisted slightly so that the structure does not repeat exactly after one period of the long helix; so for convenience this 28/15 genetic helix was taken as our model for the F-actin helix which gives rise to reflections only on defined layer lines with indices (l), given by the selection rule $l = 28m + 15n$ (Cochran et al., 1952). Values for l and n for the actin layer lines are shown in Table 1. The intensity on a given layer line is the square of the amplitude

Table 1. Comparison of Observed and Calculated Layer Lines

Layer Line[a]	Index (l)	Bessel Order (n)	Intensity obs.[b]	Intensity calc. ($\times 10^{-4}$)	Radial Position Å^{-1} obs.	Radial Position Å^{-1} calc.
1 Relaxed	2	2	S	760	0.011	0.016
Activated			W	460	—	0.016
2 Relaxed	4	4	—	17	—	0.030
Activated			W	93	0.024	0.024
3 Relaxed	6	6	VW	30	0.029	0.028
Activated			—	10	—	0.032
5	11	−3	W	31	0.018	0.017
6	13	−1	S	401	0.008	0.009
7	15	+1	S	209	0.007	0.009

[a] The first three layer lines were calculated using the following parameters: radius of actin monomer, 20 Å; actin helix 24, Å; tropomyosin cylinder, 10 Å; $\phi = 50°$ for relaxed pattern, 70° for activated pattern. The 5th, 6th, and 7th layer lines were calculated using a 24 Å sphere at a radius of 24 Å. Diffraction from a 28/13 helix gives two exceedingly weak layer lines in the region between the 3rd and 5th layer lines. These layer lines have not been discussed because their intensity is so low. Data for the first three layer lines for a model with 8 Å radius tropomyosin can be read from Fig. 4.

[b] Intensity classification: S, strong; W, weak; VW, very weak.

F_A calculated from the equations

$$F_A = \Phi(x)Jn(X)$$
$$\Phi(x) = \frac{3}{x^3}(\sin x - x \cos x) \quad (1)$$
$$X = 2\pi r R$$
$$x = 2\pi r_0 s$$

where $Jn(X)$ is an nth order Bessel function, n is given by the selection rule. $\Phi(x)$ is the form factor for a sphere. R = the radius in reciprocal space of the point where the diffraction intensity is considered, s = distance from the origin in reciprocal space of this point, r = radius of the helix, and r_0 = radius of the sphere. No correction was made for the Lorentz factor because no evidence was ever obtained that the actin layer lines are sampled by a lattice repeat.

The actin filament structure is often thought of in terms of the 28/13 genetic helix which has a pitch of 59 Å and a repeat (or near repeat) of the structure after 28 subunits in 13 turns of the helix. It is useful when calculating the diffraction intensities from helices to follow the conventions of Cochran et al. (1952) for describing the helices, since their equations can then be used for calculations without adjustment of signs. Following their convention the 51 Å genetic helix describes an actin filament with a right-handed, long double chain, and the 59 Å helix describes an actin filament with a left-handed long chain. Actin is now thought to be a right-handed structure (Depue and

Rice, 1965; Reedy, 1967, 1968). For the calculations described here, the results are not dependent on which model is chosen since X-ray diffraction cannot distinguish between two structures which are mirror images. However later calculations in which the angle or position of attachment of subunits is dependent on the direction of twist of the helix may well be confused if the wrong basic model is taken at this stage.

For calculations of the diffraction expected from the thin filaments the tropomyosin molecules were included in the model. The tropomyosin was represented by two continuous cylinders, of radius r_{0T} and wound into a double helix of radius r_T so that one cylinder followed each groove in the actin structure. Since the cylinder representing tropomyosin runs so nearly parallel to the helix axis, it was assumed that the form factor f_{TM} of tropomyosin could be represented by the form factor of a thin disc of radius r_{0T}. The intensity of diffraction from the thin filament on a given layer line is then the square of the amplitude F given by

$$F = F_A + W_{TM}f_{TM}Jn(2\pi r_T R)\exp\{-in\phi\}$$
$$f_{TM} = \frac{2}{2\pi r_{0T}R}J_1(2\pi r_{0T}R) \quad (2)$$

W_{TM} is the weight/unit length of tropomyosin relative to the weight/unit length of actin in the structure. ϕ is the angle subtended at the center of the filament by the tropomyosin cylinder and the actin chain with which it is associated (Fig. 2). ϕ goes from 0 to $\pi/2$ as the tropomyosin moves from the very edge of the filament towards the center of the groove. Because tropomyosin is represented by a continuous helix it makes little or no contribution to the intensities of the 5th, 6th, 7th, and 8th actin layer lines which occur near the meridian because the actin-monomer helix is discontinuous. Therefore the intensity of the first four layer lines was calculated using Eqs. 2 taking both the tropomyosin and actin components into account, and the other layer lines were calculated using Eqs. 1 assuming that tropomyosin does not contribute to the intensity of these layer lines. (For details of the helical diffraction see Cochran et al., 1952 and Klug et al., 1958.) Little is known about the position or shape of the troponin in the thin filament except that it occurs regularly every 385 Å (Ohtsuki et al., 1967). We therefore did not take the troponin into account when performing the calculations, although we recognize that it may well affect the diffraction pattern detectably because the total weight of troponin in the thin filament $(2 \times 80,000$ daltons every 385 Å (Ebashi et al.,

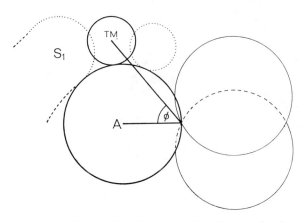

Figure 2. Cross section through a thin filament showing the relative positions of actin (A), tropomyosin (TM) and the S_1 subfragment of myosin when attached as in a decorated thin filament. The tropomyosin is shown both in the relaxed position (full circle) where $\phi = 50°$ and in the activated position (dotted circle) where $\phi = 70°$. Only one of the two tropomyosin molecules associated with the filament is shown; the other sits in a corresponding position in the other groove. For consistency with the equations of Cochran et al. the angle ϕ as shown should be negative.

1971) is about the same as the tropomyosin (2 × 70,000 daltons every 385 Å (Holtzer et al., 1965; Woods, 1967). The occurrence of similar intensity changes in the frog muscle, which contains troponin, and the ABRM (Vibert et al., 1972a,b), which is troponin-free, encouraged us to make this simplification at this stage.

The object of the calculations was to simulate the diffraction patterns expected from different models of the thin filament structure and by comparison with the observed diffraction patterns to define some of the parameters of the structures that give the best agreements. Thus we could determine the structural changes that might be expected to occur when a muscle contracts or passes into rigor and that give rise to the changes in the diffraction pattern. Intensity measurements of the actin layer lines are very difficult because the reflections are very weak and diffuse (the 2nd layer line in a contracting muscle is probably about 100 times weaker than the 1st myosin layer line), making quantitative measurements of the intensity of the weaker lines impossible. Therefore we had to make the comparison of calculated and experimentally obtained patterns using a qualitative comparison by eye of the shape, intensity, and radial position of the layer lines (Table 1). The

computer was programmed to give both the calculated intensity and a "visual" output of characters of different visual density to simulate a diffraction pattern; all the intensities below a selected level appeared blank and the others appeared more or less dark according to the intensity (Fig. 3). Comparison of the calculated and observed patterns was thus made straightforward.

In order to reduce the number of calculations that need be made (each model contains six parameters that can be varied) and to simplify the interpretation of the patterns, we imposed the following restrictions: (a) The helix of actin monomers has the same parameters in the relaxed muscle and the contracted muscle. (b) The radius of the tropomyosin cylinder is constant. (c) The tropomyosin cylinder always touches one long chain of actin monomers but the angle ϕ (Fig. 2) can vary, and with it the radius of the tropomyosin helix. (d) The values for the parameters should be physically meaningful in that the actin subunit should be about 20 Å in radius (Moore et al., 1971), and the tropomyosin cylinder should be about 10 Å radius (Holtzer et al., 1965). (e) Each 385 Å of filament length contains two tropomyosin molecules of weight 70,000 daltons (Woods, 1967)

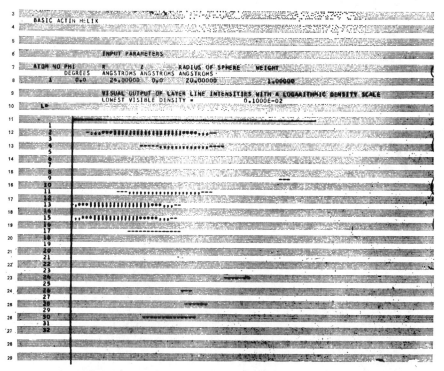

Figure 3. Computed visual output of the lower right-hand quadrant of the diffraction pattern from an actin helix consisting only of spheres 20 Å radius arranged into a helix 24 Å in radius. Points with an intensity of less that 0.001 are not shown. Each character printed represents a density level of $\sqrt{10}$ greater than the next visually less dense character. The intensity distribution along the first three layer lines is shown in Fig. 5b.

and 14 actin monomers of weight 46,000 daltons (Rees and Young, 1967). The value of W_{TM} was therefore held constant at 0.22 for all calculations.

The first parameters to fix were those of the actin-monomer helix. Calculations of models with different helix radii were compared with diffraction patterns from the frog sartorius, from the smooth muscles ABRM and taenia coli of the guinea pig. We have already found that the actin diffraction patterns from all these muscles is very similar (Vibert et al., 1972a,b). In making the comparison of calculated and observed patterns more weight was placed on the agreement of the 5th, 6th, and 7th lines, which are not affected by tropomyosin, than on the first three lines, which are affected. The best fit for all the patterns (Figs. 1, 3) came from a model with spherical subunits each 24 Å in radius at a helix radius of 24 Å. The same values were also obtained by Miller and Tregear (1971) from a study of the diffraction pattern from insect muscle. The 1st layer line of the calculated pattern is the strongest and extends as far from the meridian as any of the other lines. The 6th and 7th layer lines (the well-known layer lines at 59 and 51 Å) are also strong and are flanked nearer the equator by the 5th layer line. The 2nd layer line, although weak, is significantly stronger than the peak of the 3rd layer line which lies in the region in which the subunit transform is zero.

A helix of 24 Å radius spheres at a helix radius of 24 Å has a total diameter of 96 Å, which is slightly greater than the value of 80 Å usually accepted for the thickness of the thin filaments. In view of our restrictions to spherical subunits, a 24 Å radius sphere is an acceptable approximation to a subunit which is expected to be about 55 Å long (the repeat length of subunits along one chain) and about 40 Å thick in a cross section as found by Moore et al. (1970). Indeed if we were to reduce the cross-sectional size of the subunits to turn them into prolate ellipsoids, then the 6th layer line would extend slightly further from the meridian than it does relative to the 7th layer line, and the 5th layer line would extend slightly further still and so improve the agreement between the calculated and observed patterns. Nearer the equator of the pattern, the first few layer lines are more dependent on the cross-sectioned shape of the subunits than they are on the length of the subunit. Therefore when we calculated the intensity of these layer lines from models containing both actin and tropomyosin, we tried different models for the actin structure: (a) a 24 Å radius sphere in a 24 Å helix, (b) a 20 Å radius sphere in a 24 Å helix, and (c) a 20 Å radius sphere in a 20 Å helix.

The tropomyosin cylinder is sufficiently thin to give an appreciable contribution to the diffracted intensity in the region of the 2nd and 3rd layer lines, where the diffraction from actin is very weak, although it only has one-fifth of the weight of actin. As has already been mentioned, the changes in the 2nd and 3rd actin layer lines could be brought about by a movement of the tropomyosin in the groove of the actin helix. If we consider the structure of the thin filament at low resolution, then we can see the type of change that we would expect. With tropomyosin situated near the edge of the groove it would appear to divide the actin helix at one-third of its repeat length and correspondingly the 3rd layer line would be strong. With the tropomyosin near the center of the groove the actin repeat would appear to be half that of the actin monomer chain and so the 2nd layer line will be strong. Using the constraints described above we calculated the diffraction patterns expected from models with different diameter tropomyosin cylinders (in the region of 20 Å) at different positions in the actin groove. To determine the best models, a comparison was made of only the 2nd and 3rd layer lines of the calculated and the observed patterns. The first layer line in patterns from vertebrate striated muscle at rest length is obscured by the myosin diffraction pattern, the 4th line is too weak to see, and the 5th, 6th, and 7th layer lines are not dependent upon the position of the tropomyosin.

As expected, we found that it was possible to simulate qualitatively the changes we see in the X-ray pattern by changing the position of the tropomyosin in the actin groove. What we did not expect was that the changes in diffracted intensity could be accomplished by a relatively very small change in position, viz., a movement of the tropomyosin to alter the angle ϕ by 20° and corresponding to a movement of about 15 Å (Figs. 4, 5). The exact parameters chosen for the size of the actin and tropomyosin molecules were not very critical. All the parameters we used for the actin monomer helix gave acceptable patterns—24 Å spheres in a helix of radius 24 Å and 20 Å spheres in helices with radii of 20 or 24 Å. It was not possible to make the 3rd layer line much stronger than the 2nd if the tropomyosin was represented by a cylinder as large as 12 Å radius, but 8 and 10 Å cylinders gave patterns in which a change in the angle ϕ between the tropomyosin and actin helices of only 20° from 50° to 70° produced considerable changes in the pattern. It was not possible to choose between these six models with confidence because of the lack of quantitative data on the layer line intensities, but the models all showed the same characteristics as follows:

The 3rd layer line was strongest when the angle ϕ was about 50°, near which point the 2nd layer

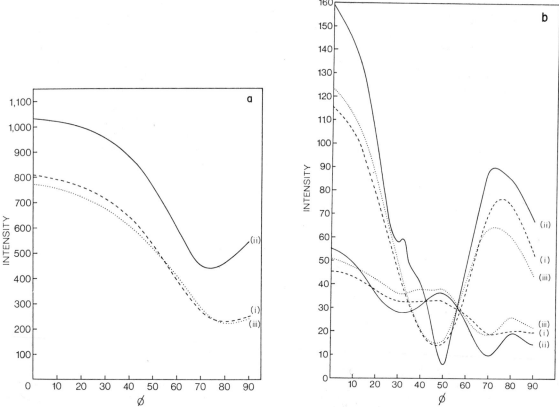

Figure 4. Graphs showing how the calculated peak intensities of the first three layer lines depend on the position of tropomyosin in the thin filament. (a) First layer line. (b) Top curves: 2nd layer line; lower curves: 3rd layer line. The position of tropomyosin is defined by the angle ϕ in Fig. 1. Tropomyosin radius = 8 Å. Curves are plotted for different models of actin: (i) 20 Å sphere in 20 Å radius helix; (ii) 20 Å sphere in 24 Å radius helix; and (iii) 24 Å sphere in 24 Å radius helix. Calculations with a tropomyosin cylinder of 10 Å radius give very similar curves showing that the exact parameters chosen for the model are not important.

line was at its weakest (Fig. 4). The ratio of the intensities of the 2nd and 3rd layer lines was then about 0.5:1. An increase of only 20° in ϕ decreased the intensity of the 3rd layer line to its weakest where it was only about one-half to one-third as strong. The same structural change strengthened the intensity of the 2nd layer line between four and ten times (depending on the model chosen) so that the relative intensities of the 2nd and 3rd layer lines was about 5:1 or more (Figs. 4, 5). The intensity of the 2nd layer line at its strongest is considerably stronger than the 3rd layer line at its strongest, in agreement with the patterns from vertebrate striated muscle. Further increase in the angle ϕ from 70° to 90° caused a very slight further increase in the intensity of the 2nd layer line from some models, but then the intensity began to drop again whereas the 3rd layer line intensity rose again slightly. The 1st layer line is worthy of note. Although the 1st layer line arising from actin alone is so strong and the weight of the tropomyosin

component so small, the strength of the 1st layer line decreases by over one-third when the angle ϕ increases from 50–70°. This change in the position of the tropomyosin molecule in the thin filament corresponds to a linear movement of only about 15 Å (Fig. 1), which is sufficiently small to be explained in terms of conformational changes of molecules within the thin filament.

Thus we have shown using calculations from a very elementary model that the large change in intensity of the 2nd layer line that occurs when vertebrate striated muscles contract or pass into rigor (Table 2) can satisfactorily be explained by a movement of the tropomyosin molecule in the groove of the filament, as suggested by Spudich et al. (1972). Moreover only a very small structural change is necessary to produce quite pronounced changes in the diffraction pattern. A most pleasing aspect of the model that we deduce is that it is essentially the same as the structure of the actin-tropomyosin-troponin filaments in the presence of

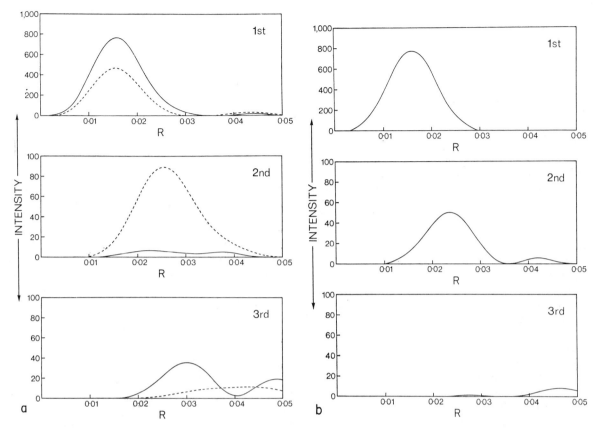

Figure 5. Intensity distribution along the first three layer lines calculated for a model with an actin monomer 20 Å radius at a helix radius of 24 Å. Note that the scale of the plot for the 1st layer line is $10\times$ smaller than that for the 2nd and 3rd layer lines. (*a*) Layer lines calculated for actin filament with a 10 Å radius cylinder of tropomyosin. (———) Tropomyosin in relaxed position with $\phi = 50°$; (– – –) tropomyosin in activated position with $\phi = 70°$. (*b*) Layer lines calculated for actin alone.

calcium found by Spudich et al. (1972). They found that the tropomyosin was associated with one actin chain and the tropomyosin and actin chains subtended an angle of 60° with the center of the filament, which is in the narrow region where we think the tropomyosin sits. We should stress here that the agreement between the observed diffraction patterns and those calculated from a model of spheres

Table 2. Strength of the 2nd Layer Line from Vertebrate Striated Muscle

	Sartorius $s = 2.2\,\mu$	Semitendinosus nonoverlap
Live resting	—	—
Contracting (Huxley, 1970)	+	
Rigor in presence of Ca++	+	+
Rigor in absence of Ca++	+	—

+ 2nd layer line distinct, 3rd layer line not visible.
− 2nd layer line weaker than 3rd layer line, which is visible but very weak.

and cylinders does not prove that the thin filament structure is the same as our model. It only shows that at low resolution the filament structure could be represented by such a model, and the agreement of our model structure with the structure deduced by other means supports the correctness of our model. We have no proof, for example, that the change in the X-ray pattern when a muscle contracts or goes into rigor is caused by a movement of the tropomyosin from a position where $\phi = 50°$ to one where $\phi = 70°$. Figure 4 shows that qualitatively a similar change would be produced if ϕ increased only to 60°, or up to 90° and the tropomyosin moved right to the center of the groove, or even moved in the opposite direction towards the edge of the filament. We choose to discuss the tropomyosin movement from $\phi = 50°$ to $\phi = 70°$ because this very small movement gives the largest change and therefore the most easily visible change in the diffraction pattern. Also the changes in the first three layer lines can all be interpreted by this one movement.

In the relaxed muscle where actin-myosin inter-action is inhibited, the 2nd actin layer line is weaker than the 3rd layer line and tropomyosin is situated towards the edge of the groove in the filament. When actin-myosin inhibition is removed and the cross-bridges interact with actin (Table 2), the 3rd layer line becomes weaker and the 2nd layer line becomes much stronger because the tropomyosin has moved, or has been moved, to a position nearer the center of the groove in the actin helix. We call these two positions of the tropomyosin the relaxed and the activated positions, corresponding to the relaxed and acti-vated (or rigor) X-ray patterns. There seems little doubt that this movement of tropomyosin is associated with the mechanism that controls actin-myosin interaction, but the immediate cause of the movement still has to be identified. Two possible mechanisms seem likely:

1. Calcium ions released into the sarcoplasm when muscles are activated bind to troponin which then undergoes a conformational change (Waka-bayashi and Ebashi, 1968), causing the tropo-myosin to move relative to the actin. This mecha-nism would be calcium sensitive and independent of the number of bound cross-bridges and would occur even if no cross-bridges were bound to actin.

2. Cross-bridges bind to actin tangentially on one side (Moore et al., 1970) so that they are projecting into the filament. The present X-ray data provide no way of distinguishing which side of the groove the tropomyosin is associated with (i.e., whether ϕ is negative as in Fig. 3 or whether it is positive); but if the tropomyosin were on the side on which the cross-bridges attach, then upon activation the cross-bridges would bind to actin and in so doing would move tropomyosin away from the point of attachment of myosin towards the center of the groove.

Experiments show that both of these mechanisms may operate.

Experimental Results

The dependence of the tropomyosin movement on the attachment of cross-bridges was investi-gated by putting muscles into rigor in conditions where the cross-bridges cannot interact with actin (Haselgrove, 1970). By stretching a semitendinosus muscle beyond overlap length (the length at which the actin and myosin filaments just overlap), muscles can be put into rigor either by soaking them in a frog Ringer solution containing 1 mM iodoacetate, or by soaking them in a glycerol solution for 24 hr followed by a frog Ringer solution or a standard salt solution (Rome, 1972). Pictures from such muscles (Fig. 6a) show a pronounced 2nd actin layer line although the cross-bridges have

been physically prevented from attaching to the thin filament. Although densitometric measure-ments were not possible, the 2nd layer line looks as strong relative to the 6th and 7th lines as it is in the rest length rigor muscle.

One advantage of studying the actin pattern from such stretched muscles is that there are no layer lines from the myosin filaments to obscure the actin pattern; the myosin layer lines seen in relaxed muscle have completely disappeared, and the rigor series arising from the attachment of cross-bridges to the actin are not generated. On the other hand, the disadvantage of working with nonoverlap muscles is that the filaments when separated do not all remain parallel and result in a disordered pattern. However even in the best-ordered patterns no trace of the third layer line could be seen. The 1st layer line was much weaker than would have been expected from the cal-culations of the thin filament in the rigor state and in many patterns was very difficult to see at all. The discrepancy may be accounted for by the simplicity of the model chosen for calculations and the omission of the effect of troponin which may well help weaken the first layer-line further than does the tropomyosin alone. The patterns did show the 5th actin layer line predicted from the calculations.

The experiment shows then that in standard conditions the change in the state of the thin filament from the relaxed to the activated state can occur even if no cross-bridges can attach to the thin filaments. The mechanism causing this change of tropomyosin position must therefore be con-tained within the thin filaments, as is the case with the first of the possible mechanisms described above. This experiment does not disprove the second mechanism since it shows only that cross-bridge attachment is not a necessary requirement in the given conditions; it does not show that the cross-bridges cannot move the tropomyosin when they do attach.

If the movement of the tropomyosin is controlled by the troponin component of the thin filaments, then we would expect the movement to be calcium sensitive. We were able to investigate the calcium dependence of the change in intensity of the first three layer lines by putting muscles into rigor at different sarcomere lengths in the complete absence of calcium. Soaking a live muscle for 24 hr in a calcium-free frog Ringer solution containing 2 mM EGTA completely destroyed the ability of the muscle to be excited electrically, presumably by binding all the available calcium. The X-ray diffrac-tion pattern was not altered by such treatment, indicating that the structure of the filaments was identical to that of a fresh live muscle. Subsequent

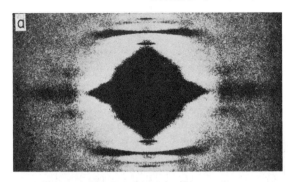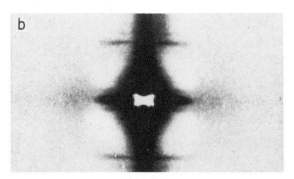

Figure 6. Diffraction patterns from semitendinosus muscles put into rigor at nonoverlap lengths. (*a*) Standard iodoacetate rigor. The 2nd layer line is visible. (*b*) Calcium-free iodoacetate rigor. The 2nd layer line is absent, but the first layer line is clearly visible. It was not possible to reproduce this pattern to show the 3rd layer line.

treatment of the relaxed muscle, either with the same Ringer solution containing 1 mM iodoacetate or by glycerination using solutions without calcium but containing 2 mM EGTA, depleted the muscle of ATP and put it into rigor. The glycerinated muscles were X-rayed either in standard salt solution containing 2 mM EGTA or in calcium-free frog Ringer containing 2 mM EGTA. Tension records were made of some muscles being put into rigor by glycerination at rest length. Muscles in calcium-containing solutions generate tensions up to about 20 g when returned from the glycerol solution to either frog Ringer solution or standard salt. Only a very small (less than 0.5 g) effect was seen when muscles were glycerinated in the absence of calcium. We thus feel confident that our treatment of muscles with calcium-free solutions has affected the calcium-dependent, tension-generating mechanism of the muscle. As with muscles put into rigor in solutions containing calcium there was no difference in the patterns from muscles put into rigor using iodoacetate or using glycerol.

Semitendinosus muscles stretched to nonoverlap lengths and then put into a calcium-free rigor give a pattern that is different from that given by muscles put into a standard rigor (Fig. 6). The first layer line is now distinct whereas the 2nd layer line is much weaker than the weak 3rd layer line. The pattern is very similar to the actin component of a relaxed muscle, indicating that little if any change had occurred in the thin filament structure although the loss of the myosin layer line pattern showed that changes had occurred in the myosin structure. The observed pattern was pleasingly similar to that calculated for the thin filament in the relaxed state, with the 1st and 6th layer lines strong, the 3rd and 5th layer lines weak, the 2nd very weak, and the 4th absent (Table 1). We have thus been able to demonstrate that the whole of the low-angle pattern predicted from the actin thin filament structure is present in vertebrate striated muscle. It therefore seems likely

that the 400 Å reflection seen in live striated muscle by Huxley and Brown (1967) does have components from both the 1st myosin layer line and the 1st actin layer line as they suggested. Furthermore it is now possible to see that the diffraction pattern from striated muscle actin is remarkably similar to the diffraction pattern from smooth muscles (Vibert et al., 1972a,b).

We conclude therefore that the tropomyosin had not moved when the muscle went into calcium-free rigor at nonoverlap length and that the thin filament was still in the relaxed state. We were worried that the lack of change in the thin filaments might be due to the muscle's being incompletely in rigor, but the complete loss of the myosin layer line pattern showed that the whole of the muscle had been affected by the rigor process. Confirmation that the thin filament structure is affected by the presence of calcium came from the following experiment. Glycerinated muscle in rigor at nonoverlap lengths were X-rayed both in calcium-containing and calcium-free solutions. No 2nd layer line was visible in the calcium-free solution but it was visible in the calcium-containing solution. Therefore when cross-bridges cannot attach to the thin filament, the position of the tropomyosin molecule is sensitive to the concentration of calcium. It is known that the combination of tropomyosin and actin is not calcium sensitive with respect to the interaction of actin and myosin, so there seems no reason to assume that these two molecules alone should be sensitive to calcium with respect to their structural relationship. We have assumed, however, that the changes we are investigating are related directly or indirectly to the mechanism controlling actin-myosin interaction. Calcium sensitivity is conferred on the relaxing system by troponin, which is situated regularly along the thin filaments (see Ebashi and Endo, 1968). It seems, therefore, that the movement of tropomyosin in the thin filament is either caused or controlled by the troponin.

Weber and Bremel (1971) have shown that at very low levels of ATP concentration the tropomyosin-troponin relaxing system is unable to suppress actin-myosin interaction which then occurs regardless of the calcium concentration. This is confirmed by Spudich et al. (1972) who have shown that actin filaments can be decorated with S, subfragments of myosin in the absence of both ATP and calcium. It is therefore expected that when a muscle at rest length is depleted of ATP it will go into rigor and the cross-bridges will attach to actin even in the absence of calcium. Indeed sartorious muscles at rest length put into a calcium-free rigor gave diffraction patterns which were indistinguishable from the usual patterns of rigor muscles (Fig. 1); viz., the 6th and 7th layer lines appear closer to the meridian than in a relaxed muscle, and also close to the meridian are the two series of layer lines associated with the attachment of cross-bridges to the actin filament. Further from the meridian the 2nd actin layer line is prominent but the 3rd layer line is not visible. On well-ordered patterns the end of the 1st layer line can be seen extending out from the meridian almost as far as the 2nd layer line, and, as predicted by the calculations (Fig. 5), are about the same intensity in this region. Although the muscle contains no calcium, the presence of a distinct 2nd layer line shows that the position of tropomyosin in the thin filament is very similar, if not identical to, the activated state that can be induced by the presence of calcium. However we would not expect the mechanism operating here to be the same as the mechanism operating in nonoverlap muscles, since the latter mechanism is sensitive to calcium. As far as we know the only difference between muscles put into calcium-free rigor at different lengths is that at nonoverlap no cross-bridges attach to actin, and the diffraction pattern indicates that the actin filament is in a relaxed state; whereas at rest length cross-bridges attach to actin and the diffraction pattern indicates that the actin filament is in an activated state. The experiments with nonoverlap muscles showed that the attachment of cross-bridges to actin is not a necessary requirement for the change if actin is present, but it did not show that the cross-bridges cannot cause the change. We must now conclude that in the absence of calcium, the attachment of cross-bridges to the thin filament can cause movement of the tropomyosin towards the center of the groove in the filament, thus supporting the second possible mechanism proposed earlier.

This demonstration of actin-myosin attachment and the enhancement of the 2nd actin layer line even in the absence of calcium allows us to demonstrate that local changes the cross-bridges make on the thin filaments are not transmitted for long distances along the thin filaments. Muscles were stretched so that the length of overlap of the thick and thin filaments was considerably reduced, but not eliminated, so that cross-bridges could interact with actin at the end of the filaments but not along the rest of their length. These muscles were then put into a calcium-free rigor; X-ray patterns from the muscles did not show a strong 2nd layer line, indicating that along most of the length of the thin filament the tropomyosin was in the relaxed state. Since we have just shown that in the region where interaction can take place the tropomyosin is moved to the activated position, we presume that this is happening at the end of the filament but that the movement is not propagated along the whole of the filaments.

Discussion

These results show then that the changes in intensity of the first three actin layer lines in X-ray diffraction patterns may be accounted for by movement of the tropomyosin molecule within the filament. The total movement necessary to affect the X-ray pattern in the way we observe is relatively small and is about 15 Å towards the center of the groove in the filament. In relaxed muscles the tropomyosin is sitting at the edge of the groove in the thin filament and is very close to the point on the actin at which S_1 subunits of myosin attach (Fig. 3; Moore et al., 1970). In this position tropomyosin may well inhibit actin-myosin interaction by physically blocking the active site for myosin attachment. On addition of calcium the tropomyosin is moved to a position nearer the center of the groove where it no longer interferes with the actin-myosin interaction. Since this interaction is not inhibited if no troponin is present, it seems likely that the normal position for tropomyosin in the thin filament is in the activated position near the center of the filament and inhibition of actin-myosin interaction by troponin in the absence of calcium is caused by troponin moving the tropomyosin to the relaxed position. In this way it would be possible for one troponin molecule to exert a controlling influence over all seven actin monomers in contact with the tropomyosin molecule controlled by that troponin. Although such a mechanism involving simple steric blocking of an active site would explain the control of activation by calcium in a live contracting muscle, the whole system is not likely to be quite so straightforward, as the following discussion shows.

Relaxation of actin-myosin interaction is dependent on the presence of ATP as well as upon the level of calcium present (Weber and Bremel, 1971).

Our X-ray patterns have shown that in the absence of ATP and calcium the thin filament is in the relaxed state if the cross-bridges are physically prevented from attaching by stretching the muscles to beyond overlap length. So at rest length the cross-bridges are able to attach firmly to the thin filament although the tropomyosin has not been moved before the attachment. Once attached, the cross-bridge is then able to move in a way that will cause local movement of the tropomyosin. It seems, therefore, unlikely that the tropomyosin covers the active site completely, but must inhibit the interactions in a way that prevents attachment of myosin loaded with ATP (or its split products; Lymn and Taylor, 1971) but allows attachment of unloaded myosin. Moreover Bremel et al. (this volume) have shown recently that the presence of tropomyosin activates the actomyosin ATPase. Sequence work on tropomyosin by Smillie, (personal communication) shows that tropomyosin does not have a repeating sequence that would enable tropomyosin to interact equivalently with each actin monomer near the active site and in this way control the actin-myosin interaction.

Smooth muscles also give X-ray diffraction patterns showing that upon activation the 2nd layer line increases in intensity with respect to the third layer line (Vibert et al., 1972a,b), although the change is less than in vertebrate striated muscle. Studies on the muscle ABRM have shown (Kendrick-Jones et al., 1970) that the thin filaments of molluscan smooth muscle contain no troponin and that it is the myosin which is calcium sensitive and controls activity. Thus the changes in the thin filament in the ABRM cannot be brought about by the action of troponin but must be caused by the attachment of cross-bridges. The cycling of cross-bridges during contraction means that not all cross-bridges are attached to actin at once, so only some regions of the thin filament are affected by cross-bridges attaching. This in turn produces the change in the X-ray pattern which is small compared with the complete change produced by the action of troponin in vertebrate striated muscle. Therefore there remains much about the causes and effect of the movement of tropomyosin in the thin filament that must still be investigated.

At the same time as the calculations described here were being made, Drs. Parry and Squire independently performed similar calculations and arrived at a similar conclusion about the movement of tropomyosin. Their work is now in press. I would like to thank Drs. Spudich, Huxley, and Finch for allowing me to study and quote from their paper, and especially Dr. H. E. Huxley for many helpful and stimulating discussions.

References

COCHRAN, W., F. H. C. CRICK, and V. VAND. 1952. The structure of synthetic polypeptides. I. The transform of atoms on a helix. *Acta Cryst.* **5**: 581.

DEPUE, R. H. and R. RICE. 1965. F. Actin is a right-handed helix. *J. Mol. Biol.* **12**: 302.

EBASHI, S. and M. ENDO. 1968. Calcium ion and muscle contraction. *Prog. Biophys. Mol. Biol.* **18**: 125.

EBASHI, S., M. ENDO, and I. OHTSUKI. 1969. Control of muscle contraction. *Quart. Rev. Biophys.* **2**: 351.

EBASHI, S., T. WAKABAYASHI, and F. EBASHI. 1971. Troponin and its components. *J. Biochem.* **69**: 441.

ELLIOTT, G. F., J. LOWY, and B. MILLMAN. 1967. Low angle X-ray diffraction studies of living striated muscle during contraction. *J. Mol. Biol.* **25**: 31.

HANSON, J. and J. LOWY. 1963. The structure of F-actin and of actin filaments isolated from muscle. *J. Mol. Biol.* **6**: 46.

HASELGROVE, J. C. 1970. X-ray diffraction studies on muscle. Ph.D. thesis, University of Cambridge.

HOLTZER, A., R. CLARK, and S. LOWEY. 1965. The conformation of native and denatured tropomyosin B. *Biochemistry* **4**: 2401.

HUXLEY, H. E. 1970. Structural changes in muscle and muscle proteins during contraction. *8th Int. Cong. Biochem.*, Interlaken.

———. 1971a. Cross-bridge movement and filament overlap. *Amer. Biophys. Abstr.*, p. 235.

———. 1971b. Structural changes during muscle contraction. *Biochem. J.* **125**: 85p.

HUXLEY, H. E. and W. BROWN. 1967. The low angle X-ray diagram of vertebrate striated muscle and its behaviour during contraction and rigor. *J. Mol. Biol.* **30**: 383.

KENDRICK-JONES, J., W. LEHMAN, and A. G. SZENT-GYÖRGYI. 1970. Regulation in molluscan muscles. *J. Mol. Biol.* **54**: 313.

KLUG, A., F. H. C. CRICK, and H. W. WYCKOFF. 1958. Diffraction by helical structures. *Acta Cryst.* **11**: 99.

MILLER, A. and R. T. TREGEAR. 1971. X-ray studies on the structure and function of vertebrate and invertebrate muscle. In *Contractility of muscle cells and related processes*, ed. R. J. Podolsky. Prentice-Hall, Englewood Cliffs, N.J.

LYMN, R. W. and E. W. TAYLOR. 1971. Mechanism of adenosine triphosphate hydrolysis by actomyosin. *Biochemistry* **10**: 4617.

MOORE, P. B., H. E. HUXLEY, and D. J. DE ROSIER. 1970. Three-dimensional reconstruction of F-actin, thin filaments and decorated thin filaments. *J. Mol. Biol.* **50**: 279.

O'BRIEN, E. J., P. M. BENNETT, and J. HANSON. 1971. Optical diffraction studies of myofibrillar structure. *Phil. Trans. Roy. Soc. London* B **261**: 201.

OHTSUKI, I., T. MASAKI, Y. NONOMURA, and S. EBASHI. 1967. Periodic distribution of troponin along the thin filament. *J. Biochem.* (Tokyo) **61**: 817.

REEDY, M. K. 1967. Cross-bridges and periods in insect flight muscle. *Amer. Zool.* **7**: 465.

———. 1968. Ultrastructure of insect flight muscle. I. Screw sense and structural grouping in the rigor cross-bridge lattice. *J. Mol. Biol.* **31**: 155.

REES, M. K. and M. YOUNG. 1967. Studies on the isolation and molecular properties of homogeneous globular actin. *J. Biol. Chem.* **242**: 4449.

ROME, E. 1972. Relaxation of glycerinated muscle: Low angle X-ray diffraction studies. *J. Mol. Biol.* **65**: 331.

SPUDICH, J. A., H. E. HUXLEY, and J. T. FINCH. 1972. The regulation of skeletal muscle contraction. II. Structural studies of the interaction of the tropomyosin-troponin complex with actin. *J. Mol. Biol.* In press.

VIBERT, P. J., J. C. HASELGROVE, J. LOWY, F. R. POULSEN. 1972a. Structural changes in actin-containing filaments in muscle. *Nature New Biol.* **236**: 182.

———. 1972b. Structural changes in actin containing filaments in muscle. *J. Mol. Biol.* In press.

WAKABAYASHI, T. and S. EBASHI. 1968. Reversible change in physical state of troponin induced by calcium ion. *J. Biochem.* **64**: 731.

WEBER, A. and R. D. BREMEL. 1971. Regulation of contraction and relaxation in the myofibril. In *Contractility of muscle cells and related processes*, ed. R. J. Podolsky. Prentice-Hall, Englewood Cliffs, N.J.

WOODS, E. F. 1967. Molecular weight and subunit structure of tropomyosin B. *J. Biol. Chem.* **242**: 2859.

Studies of the Low-Angle X-Ray Pattern of a Molluscan Smooth Muscle during Tonic Contraction and Rigor

J. LOWY AND P. J. VIBERT

Institute of Biophysics, Aarhus University, Denmark

Experiments with intact muscles by methods of low-angle X-ray diffraction provide unique information in that structural features can be studied both in the resting and activated states at or near the molecular level. But as regards the timing and the nature of the structural changes in actively contracting muscles, the techniques available at present only indicate what happens in average terms.

Concerning the attachment of cross-bridges to actin, this has been shown to occur asynchronously in all living muscles that have been investigated during contraction (Huxley and Brown, 1967; Vibert et al., 1971). It means that the information obtained with the X-ray method can only be of rather limited value unless ways are found to synchronize actin-myosin interaction or to stop this reaction at various phases of the cross-bridge cycle.

Attempts to find out what happens in the myosin structure meet with another difficulty which arises from the very strong likelihood that the crucial events during contraction occur in the globular S_1 units (Huxley, 1969). From this it follows that experiments with intact muscles cannot provide the information required about actin-myosin interaction at the molecular level because the myosin pattern at high angles is dominated by diffraction from the part of the molecule (light meromyosin) which does not interact with actin.

A few years ago the prospects of obtaining useful results from studying the actin pattern in contracting muscles did not appear particularly encouraging. As pointed out by Huxley and Brown (1967), the fact that during contraction only a relatively small number of cross-bridges are attached to actin at any given time would not lead one to expect to find a major structural change in the actin filaments. It was therefore somewhat surprising when we noted indications of such a change by observing an increase in the intensity of certain actin layer lines in two kinds of contracting smooth muscles, namely the taenia coli of the guinea pig (TCGP)* and the anterior byssus retractor of *Mytilus* (ABRM). In the latter case the intensity of the second actin layer line was found to increase more than that of the others (Vibert et al., 1971). This particular phenomenon was first reported by Huxley (1970) from experiments with contracting toad striated muscle.

* Results briefly described in Lowy and Small, 1970.

From the start, the change observed in the actin structure was very interesting because it indicated the occurrence of an increase in order rather than a decrease, as had been the case in the findings concerning cross-bridges. At about the same time, a lot more became known about the components of the actin-containing filaments and the control of actin-myosin interaction, particularly in molluscan muscles like the ABRM (Kendrick-Jones et al., 1970). These considerations, and our observation that the actin pattern could be more easily studied in such muscles, led us to undertake an intensive X-ray investigation of the ABRM. Here we report on the behavior of its low-angle pattern during tonic contraction and rigor and discuss the significance of the major change observed in the actin pattern.

Functionally, the ABRM is remarkable for its well-known property of maintaining a state of tension for a very long time with a minimal expenditure of energy (the "catch" mechanism). Structurally, the muscle is of interest because its myosin-containing filaments are very large (diameter up to 1300 Å, lengths from 10 to 50 μ; Lowy and Hanson, 1962) and contain substantial amounts of the protein paramyosin (Bailey, 1956, 1957). For many years attention was focused on the mechanism of "catch" and in particular on the question of whether or not the functioning of this mechanism could in any way be related to the presence of paramyosin.

Tonic Contraction

In our X-ray experiments with tonically contracted muscles we tried to find out if any structural changes could be detected which might provide clues about the nature of the "catch" mechanism. The methods we used are described elsewhere (Vibert et al., 1972b).

Figure 1 shows a record of the mechanical activity during a prolonged contraction of the ABRM. This is a tonic response because when the muscle is released and allowed to shorten, very little tension is redeveloped (Jewell, 1959); and when stimulation is stopped, the decay of tension follows the characteristically slow tonic time course (Lowy and Millman, 1963). Comparing patterns obtained during such a contraction with patterns from resting muscles (Figs. 2a and 2b), we note five main features:

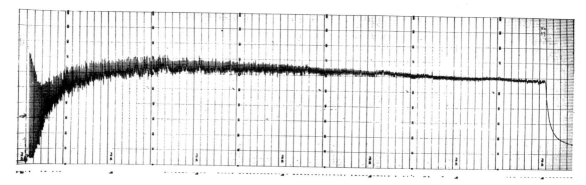

Figure 1. Record of isometric mechanical activity from partially innervated ABRM maintained in oxygenated artificial seawater of the composition given by Millman (1964). Longitudinal 50 Hz stimulation was applied for about 30 sec at intervals of 4 min. Temperature 5°C. Maximum tension developed, about 6 kg cm^{-2}. Duration of experiment 20 hr. Note the spontaneous tension development in the early stages of the experiment which leads to the establishment of a high level of tension. The slow decay of tension when stimulation ceases, and release experiments (not shown), indicate that the high tension level is largely due to "catch" tension, although a small phasic component is superimposed.

1. Within the errors of our measurements there is no change in the axial spacing of the 59 Å actin layer line or of the meridional 143 Å reflection from the myosin-containing filaments. This indicates that no appreciable length changes occur in the actin and myosin filaments. Similar findings have been reported by Millman and Elliott (1965). We consider it worth mentioning our results because in our experiments the muscles could be maintained at a much higher tension level than in the previous ones.

2. No change can be seen in the position of the intensity maximum on the 59 Å actin layer line. This is in line with what happens in activated frog muscles (Huxley and Brown, 1967) and indicates that only a relatively small number of cross-bridges are attached to actin at any given time. This interpretation can be put forward because it has been possible to demonstrate that in a state of rigor (where all the cross-bridges are presumably attached to actin) similar changes occur in the X-ray pattern of the ABRM as have been observed in rigorized frog muscles. Our experiments with rigorized ABRM preparations are described in the next section.

3. The intensity of the 190 Å actin layer line shows a striking increase (Vibert et al., 1971, 1972a, b). A similar change has been observed by Huxley (1970, 1971a, b) in contracting toad muscles, and its general significance will be discussed later.

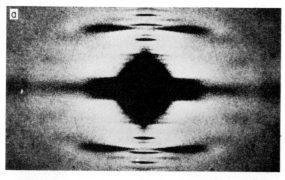

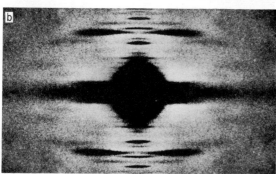

Figure 2. Low-angle X-ray diffraction patterns from living ABRM, taken on mirror-monochromator camera of length 23 cm. Elliott GX6 rotating anode X-ray generator with focus 1 mm × 0.1 mm loaded at 40 kV, 18 mA. Fiber axis vertical. (*a*) Pattern from unstimulated muscle in oxygenated artificial seawater at 5°C. (*b*) Pattern from muscle during activity of the kind shown in Fig. 1. Note the increase in the intensity of the peak on the second actin layer line. (From Vibert et al., 1972a.)

4. Cross-bridge movement during contraction is clearly indicated by a decrease in the intensity of the 143 Å meridional myosin reflection in striated frog and insect muscles (Huxley and Brown, 1967; Tregear and Miller, 1969), as well as in the smooth TCGP muscle (Lowy et al., 1970). However, in the ABRM no appreciable change can be detected in the intensity of the meridional reflection at 143 Å.

Considering the very large amounts of paramyosin in the backbone of the myosin-containing filaments of the ABRM (Kendrick-Jones et al., 1970), it seems reasonable to assume that most of the intensity of the 143 Å reflection comes from the paramyosin structure (Bear and Selby, 1956), the contribution due to the axial periodicity of the cross-bridges being relatively small. If that were so, it follows that even if all cross-bridges moved during contraction, this might still produce too small an effect to be detectable by our methods. The same explanation could also account for the fact that whereas in a state of rigor, which presumably involves some movement of all the cross-bridges, there is a decrease in the intensity of the 143 Å reflection in frog muscles (Huxley and Brown, 1967); no such change can be detected in our experiments with rigorized ABRM muscles.

It should also be noted that as regards the feature responsible for the 143 Å reflection due to paramyosin, no changes occur in the tonically contracted ABRM that can be detected by our methods.

5. In smooth muscles like the ABRM and TCGP the regular sideways packing of the actin filaments gives rise to an equatorial reflection at about 120 Å (Lowy and Vibert, 1967; Elliott and Lowy, 1968). We have not seen any change in the intensity, shape, or spacing of this reflection during tonic contraction of the ABRM.

As shown in the next section, the 120 Å reflection becomes more diffuse in the rigor ABRM where one can assume that a considerable rearrangement occurs in the disposition of the actin filaments as a result of cross-linking with myosin. It would appear, therefore, that no comparable phenomenon takes place during contraction, presumably because not enough actin filaments are interacting with myosin at any given time. This would, of course, be in accordance with the observation described under (2).

Accepting the explanation given for the absence of a change in the intensity of the 143 Å reflection, all the features that have so far been observed in the tonically contracted ABRM are comparable to those that occur in contracting frog muscles. From the viewpoint of a unitary theory for contraction, this is a very satisfying conclusion because it indicates that essentially the same sliding mechanism operates in two kinds of muscles that differ vastly in structure and function. However, our present results from experiments with tonically contracting ABRM muscles are somewhat dissappointing in that they give no clues about the nature of the "catch" mechanism.

Muscles in Rigor

The experimental procedure was as follows: The ABRM muscles were left intact in the animal; they were freed from the connective tissue surrounding them and slightly stretched by a weight tied to the byssus thread. This preparation was then placed into a glycerol solution (of a composition given by Rüegg and Weber, 1963), great care being taken to maintain the normal position of the muscles in the

Figure 3. Low-angle X-ray diffraction patterns from ABRM. (*a*) Resting muscle, fixed in 2.5% glutaraldehyde in seawater. (*b*) Muscle extracted for 48 hr in glycerol solution (see text), then fixed in 2.5% glutaraldehyde in buffered 0.1 M KCl, pH 7.5. Patterns from unfixed muscles show essentially similar features, except for some differences in the relative intensities of certain meridional and near-meridional reflections from paramyosin.

animal. Glycerination was carried out at 0°C and the muscles were examined by X-ray diffraction at that temperature after various periods. The rigor patterns showed (Fig. 3a) the following changes when compared with patterns obtained from living resting muscles (Fig. 3b).

(a) As deduced from measurements of the spacing of the 51 and 59 Å layer lines, there is a slight increase (5 to 10 Å) in the pitch of the actin helix.

(b) The 51 Å layer line increases in intensity relative to the 59 Å layer line.

(c) There is an increase in the intensity of the second actin layer line which has an axial spacing near 190 Å.

(d) Together with the intensity maximum on some other actin layer lines (see e), that on the 51 and 59 Å layer lines moves toward the meridian.

(e) Layer lines are seen with axial spacings of about 69, 85, and 107 Å. These are weak and rather diffuse and lie somewhat further from the meridian than the 51 and 59 Å layer lines, though not as far as the principal actin layer lines present in patterns from resting and contracting muscles (cf. Figs. 2 and 3). Their spacings and appearance suggest that they are comparable to one of the series of layer lines found in patterns from rigor frog muscles (Huxley and Brown, 1967; Haselgrove, 1970) and probably arise from the actin helix "decorated" with cross-bridges. In the ABRM measurement of their spacing, together with those for the 51 and 59 Å layer lines (see under a) indicates an axial repeat for the "decorated" actin helix of about 2×370 Å. If these layer lines are assumed to arise from a composite structure whose symmetry is basically that of the F-actin structure (Moore et al., 1970), then their radial positioning corresponds to a complex whose electron density is concentrated at a radius of 60 to 70 Å from the axis of the actin helix, much as expected from thin filaments "decorated" with cross-bridges (Moore et al., 1970).

(f) Very close to the meridian a rather strong layer line with an axial spacing of about 185 Å also appears, and there is a considerably weaker one at about 125 Å. These again resemble layer lines seen in patterns from rigor frog muscles. In view of the very great differences between the thick filament structures of ABRM and frog muscles, it is tempting to suggest that these layer lines originate from some feature of the "decorated" actin complex.

Except for the change mentioned under c, all the features described above occur only in frog muscles rigorized at a length where there is a substantial degree of overlap between the actin and myosin filaments; they are absent in muscles stretched to a sarcomere length of about 3.7 μ (Haselgrove, 1970). This suggests that all these features depend on the attachment of cross-bridges to actin (Haselgrove, 1970). However, the change noted under c can also be seen in patterns from frog muscles rigorized at a sarcomere length of 3.7 μ (Vibert et al., 1972b) and this will be discussed later.

The results quoted in a, b, d, e, and f establish that the effects of attachment of cross-bridges in the ABRM are comparable to those demonstrated in frog muscles. (The absence of a decrease in the intensity of the 143 Å reflection in the ABRM has already been commented on in the section on Tonic Contraction.) Taking what is probably the most straightforwardly interpretable of these effects, namely, the shift of the intensity maximum of the 59 Å actin layer line toward the meridian, its most

interesting consequence is the possible deduction of asynchronous cross-bridge attachment during contraction (Huxley and Brown, 1967). The non-appearance of the strong 185 Å layer line may perhaps be interpreted along the same lines, because we have not seen this reflection in the tonically contracting ABRM; nor has it been noted in patterns obtained from contracting frog muscles by Huxley and Brown (1967) and Haselgrove (1970).

The increase in the intensity of the 51 Å actin layer line suggests some change in the shape and/or orientation of the actin monomers, probably accompanied by the slight increase in the pitch of the helical structure. (We may note that there are some indications that the latter change also occurs in contracting frog muscles; cf. Haselgrove, 1970; Elliott et al., 1967.) At present the significance of these changes remains obscure; reasons are given later why one fairly obvious possibility is unlikely.

We now come to two phenomena seen only in the rigorized ABRM. The first concerns the equatorial reflection at about 120 Å which becomes more diffuse. This is to be expected from what is known about the arrangement of the actin and myosin-containing filaments in the glycerinated ABRM (Lowy et al., 1966). Here very many of the latter are seen surrounded by rosettes of actin filaments so that fewer of these are present in the regular sideways arrangement characteristic of the actin lattice.

The second phenomenon concerns reflections seen at higher angles. The diffuse reflection on the equator at about 20 Å (Cohen and Holmes, 1963) becomes somewhat sharper and more intense, and another diffuse reflection appears on or near the equator at a spacing of about 35 Å. The latter value is about twice that obtained for the distance between molecules in paramyosin filaments examined under a variety of conditions in the "wet" state (Elliott et al., 1968; Elliott and Lowy, 1970). But the 35 Å value is close to the spacing of the "dots" seen in electron micrographs of negatively stained paramyosin filaments (Elliott and Lowy, 1970). The appearance of a reflection at that spacing may indicate that paramyosin molecules have become grouped in pairs. It remains to be seen whether or not this has any functional significance, say for the operation of the "catch" mechanism, or is due to nonspecific effects of glycerination, changes in ionic strength, etc.

Structural Change in Actin Filaments

Hitherto the only moderate-angle actin reflections that could be recognized in patterns from various kinds of intact muscles were those at about 51 and 59 Å (Huxley and Brown, 1967; Millman et al., 1967; Lowy and Vibert, 1967), and maybe in insect flight muscles a first layer line near 390 Å

(Reedy et al., 1965). However, in smooth muscles it is possible to distinguish, in addition, layer lines with spacings of about 70, 125, 187, and 375 Å; these can be shown to arise from a structure with the symmetry of the actin helix because of their shape and axial spacing, as well as the radial position of their intensity maxima (Vibert et al., 1972b).

It should be noted here that there are several features of the diffraction pattern from relaxed and contracted ABRM muscles which are not explained by our current model of the actin helical structure. These features include three diffuse layer lines (at axial spacings of about 45, 85, and 105 Å) whose intensity maxima are all closer to the meridian than those predicted by a simple actin model. The axial spacings of the latter two layer lines are similar to those of two of the layer lines seen in patterns from the rigorized ABRM, but the layer lines in the pattern from living muscles are closer to the meridian than those in the pattern from rigorized muscles.

The known presence of tropomyosin in thin filaments indicates that it may be necessary to refine the model for such filaments to take account of the "four-strandedness' that would result from placing tropomyosin in the long grooves of the actin helix. It is as yet not known whether, and/or to what extent, tropomyosin may be specifically bound to successive actin monomers, or whether tropomyosin should be regarded as a continuous helix located within the thin filament structure. If tropomyosin turned out to be effectively discontinuous, as might be inferred from the images formed by three-dimensional reconstruction of thin filament structure (Moore et al., 1970), then higher-order actin layer lines such as the one at 59 Å could be influenced. The diffuse peak at a large radial position on approximately the 4th layer line which we see in patterns from living ABRM and frog muscles (Vibert et al., 1972b) may also contain a contribution from tropomyosin. We discuss later the likelihood that the position of tropomyosin in the grooves of the actin helix could affect the intensities of the lower-order actin layer lines.

We now wish to put forward some suggestions as to why the actin pattern in smooth muscles like the TCGP and ABRM is so much clearer than in striated muscles. To begin with, there appear to be no myosin layer lines due to the arrangement of the cross-bridges in such smooth muscles. In the ABRM all the actin layer lines except the first can easily be distinguished from those due to paramyosin. In the case of the 1st actin layer line, the situation is greatly improved following incubation of the muscle in a hypertonic Ringer solution. This has no effect on the 1st actin layer line other than to cause it to move away from the meridian; in that position it can

unambiguously be distinguished from the paramyosin layer line at about 360 Å, and therefore measurement of the axial spacing of the actin layer line is facilitated. We may note here that in the living resting state the first actin layer line is more intense in the ABRM than in the TCGP.

Another reason why the actin filaments in smooth muscles give a better pattern may be that, in comparison with those in striated muscles, they are much longer and in their regular sideways packing remain parallel for a considerable part of their length. It may be recalled that in striated muscles like those of the frog, the relatively short actin filaments are not normally in a parallel arrangement; that is, they pass from a hexagonal array in the overlap regions to a near-square lattice in the Z lines.

Having identified the various actin layer lines in the resting ABRM, it became possible to study their behavior in muscles in various states. We are here concerned with one particular set of observations, namely, the increase in the intensity of the 2nd actin layer line seen in both rigorized and contracting ABRM muscles. This will now be discussed in the light of the fact that the same observations have been made in experiments with frog muscles during contraction and rigor (Huxley, 1970, 1971a,b; Vibert et al., 1971, 1972a,b).

An increase in the intensity of the 2nd actin layer line would result from the introduction of extra electron density into the long grooves of the actin helix. This was pointed out by O'Brien et al. (1971) in an attempt to explain a similar observation in optical diffraction patterns from electron micrographs of paracrystals of F-actin and of such paracrystals which contained, in addition, tropomyosin (TM) and proteins of the troponin complex (TN). Patterns from the latter preparation showed an appreciably stronger 2nd layer line, and it was suggested that this could be due to the introduction of TM/TN material into the long grooves of the actin helix. But in the actin filaments of intact muscles TM is very likely located in approximately that position already (Hanson and Lowy, 1964; Moore et al., 1970). To explain the observed change in the actin pattern of intact muscles, one could postulate that it occurs as the result of a movement of TM, say from a less to a more central position within the grooves. This may be called the TM shift.

Extra electron density could also be introduced into the long grooves of the actin helix by some change in shape and/or orientation of the G-actin monomers. This may be called the actin monomer change. Taking the evidence available at present, the arguments for and against the two possibilities are as follows.

The occurrence of some kind of actin monomer

change in rigorized muscles is evident from the observation that the intensity of the 51 Å layer line increases. At the same time there is a slight change in the helical parameters of the actin structure. But it is unlikely that such changes could account for the relatively large change in electron density required to produce the increase in the intensity of the 2nd actin layer line. Furthermore, the latter phenomenon is still clearly seen in two situations where an actin monomer change cannot be detected, namely, in normally contracting muscles and in muscles rigorized at a sarcomere length of about 3.7 μ (Haselgrove, 1970; Vibert et al., 1972b).

From these arguments it would appear that the TM shift is the more likely possibility. Though there is no direct evidence for this structural change at present (and all further considerations must therefore be speculative), it seems worthwhile to discuss the situation in functional terms, in view of what is now known about the composition of the actin-containing filaments in relation to the mechanism involved in the control of actin-myosin interaction. There is good evidence that in vertebrate skeletal muscles contraction is regulated via the TM/TN system located in the thin filaments (see review by Ebashi and Endo, 1968). This suggests the possibility that the TM shift may be associated with the operation of such a regulatory mechanism. In favor of this is the observation that the TM shift is still seen in a frog muscle that has been stretched before being rigorized to a sarcomere length where the actin and myosin filaments no longer overlap (Vibert et al., 1972a,b).

On the other hand, in smooth molluscan muscles like the ABRM (as well as in striated molluscan muscles) the mechanism for regulation of contraction is located in the myosin molecule; TN is not present in the thin filaments and TM is not required at all for the regulation of actin-myosin interaction (Kendrick-Jones et al., 1970). Here it would appear that the TM shift could be associated with the attachment of cross-bridges to actin. In favor of this possibility argues our finding that an increase in the intensity of the 2nd actin layer line is seen both in tonically contracting muscles and in rigorized specimens. This can be explained most straightforwardly on the assumption that the change in the actin filament structure is due to cross-bridge attachment. From biochemical studies of the molluscan system (Kendrick-Jones et al., 1970) it would appear that the isolated thin filaments are in a state ready to interact with myosin. If this also applied to the intact muscle, the possibility arises that the positioning of TM might not be the same in the resting ABRM as in resting vertebrate skeletal muscle.

In the vertebrate system one expects that activation by calcium via TM/TN would cause the TM shift to occur along the whole length of the thin filaments. This need not necessarily be the case in molluscan muscles where we assume that the TM shift is due to cross-bridge attachment. The latter situation could be clarified by making measurements of the extent of the intensity change on the 2nd actin layer line in contracting and rigorized ABRM muscles and in molluscan striated muscles contracting or rigorized at various sarcomere lengths.

Preliminary studies show that the 2nd actin layer line is strong in patterns from molluscan striated muscles rigorized at rest length. Present work includes experiments with such muscles stretched to sarcomere lengths where the actin and myosin filaments no longer overlap. In this situation one would not expect to see any signs of the TM shift, nor should it be possible to bring it about by introducing calcium into the system.

As regards the functional significance of the TM shift, the most obvious suggestion is that it may serve to unmask sites on the actin monomers and/or be necessary for their complete activation.

Acknowledgments

We extend thanks to E. B. Neergaard for introducing us to the technique used for reproduction of the X-ray patterns, and to him, P. Boldsen, and Mrs. I. Lunde for carrying out the photographic work involved. We thank also Miss K. Eskesen, Mrs. L. Nychel, and Messrs. F. Marquard and J. Pedersen for excellent technical assistance. This work was supported by grants from Statens Naturvidenskabelige Forskningsråd and Statens Lægevidenskabelige Forskningråd.

References

BAILEY, K. 1956. The proteins of adductor muscles. *Publ. Staz. Zool. Napoli* **29**: 96.
————. 1957. Invertebrate tropomyosin. *Biochim. Biophys. Acta* **24**: 612.
BEAR, R. S. and C. C. SELBY. 1956. The structure of paramyosin fibrils according to X-ray diffraction. *J. Biophys. Biochem. Cytol.* **2**: 55.
COHEN, C. and K. C. HOLMES. 1963. X-ray diffraction evidence for α-helical coiled-coils in native muscle. *J. Mol. Biol.* **6**: 423.
EBASHI, S. and M. ENDO. 1968. Calcium ion and muscle contraction. *Prog. Biophys. Mol. Biol.* **18**: 123.
ELLIOT, A. and J. LOWY. 1970. A model for the coarse structure of paramyosin filaments. *J. Mol. Biol.* **53**: 181.
ELLIOTT, A., J. LOWY, D. A. D. PARRY, and P. J. VIBERT. 1968. Puzzle of the coiled-coils in the α-protein paramyosin. *Nature* **218**: 656.
ELLIOTT, G. F. and J. LOWY. 1968. Organization of actin in a mammalian smooth muscle. *Nature* **219**: 156.

ELLIOTT, G. F., J. LOWY, and B. M. MILLMAN. 1967. Low-angle X-ray diffraction studies of living striated muscle during contraction. *J. Mol. Biol.* **25**: 31.

HANSON, J. and J. LOWY. 1964. The structure of actin filaments and the origin of the axial periodicity in the I-substance of vertebrate striated muscle. *Proc. Roy. Soc. (London)* B **160**: 449.

HASELGROVE, J. C. 1970. X-ray diffraction studies on muscle. *Ph.D. thesis, University of Cambridge.*

HUXLEY, H. E. 1969. The mechanism of muscular contraction. *Science* **164**: 1356.

———. 1970. Structural changes in muscle and muscle proteins during contraction. *8th Int. Cong. Biochem.* Interlaken.

———. 1971a. Cross-bridge movement and filament overlap. *Biophys. Soc. Abstr.* ThAM-Cll.

———. 1971b. Structural changes during muscle contraction. *Biochem. J.* **125**: 85 P.

HUXLEY, H. E. and W. BROWN. 1967. The low-angle X-ray diagram of vertebrate striated muscle and its behaviour during contraction and rigor. *J. Mol. Biol.* **30**: 383.

JEWELL, B.R. 1959. The nature of the phasic and tonic responses of the anterior byssal retractor muscle of *Mytilus. J. Physiol.* **149**: 154.

KENDRICK-JONES, J., W. LEHMAN, and A. G. SZENT-GYÖRGYI. 1970. Regulation in molluscan muscles. *J. Mol. Biol.* **54**: 313.

LOWY, J. and J. HANSON, 1962. Ultrastructure of invertebrate smooth muscles. *Physiol. Rev.* **42**, (suppl. 5): 34.

LOWY, J. and B. M. MILLMAN. 1963. The contractile mechanism of the anterior byssus retractor muscle of *Mytilus edulis. Phil. Trans. Roy. Soc. (London)*, B **246**: 105.

LOWY, J. and J. V. SMALL. 1970. The organisation of myosin and actin in vertebrate smooth muscle. *Nature* **227**: 46.

LOWY, J. and P. J. VIBERT. 1967. Structure and organisation of actin in a molluscan smooth muscle. *Nature* **215**: 1254.

LOWY, J., F. R. POULSEN, and P. J. VIBERT. 1970. Myosin filaments in vertebrate smooth muscle. *Nature* **225**: 1053.

LOWY, J., J. HANSON, G. F. ELLIOTT, B. M. MILLMAN, and M. W. McDONOUGH. 1966. p. 229. In *Principles of biomolecular organisation*, ed. G. E. W. Wolstenholme and M. O'Connor. Churchill, London.

MILLMAN, B. M. 1964. Contraction in the opaque part of the adductor muscle of the oyster. *J. Physiol.* **173**: 238.

MILLMAN, B. M. and G. F. ELLIOTT. 1965. X-ray diffraction from contracting molluscan muscle. *Nature* **206**: 824.

MILLMAN, B. M., G. F. ELLIOTT, and J. LOWY. 1967. Actin: X-ray diffraction studies. *Nature* **213**: 356.

MOORE, P. B., H. E. HUXLEY, and D. J. DE ROSIER. 1970. Three-dimensional reconstruction of F-actin, thin filaments and decorated thin filaments. *J. Mol. Biol.* **50**: 279.

O'BRIEN, E. J., P. M. BENNETT, and J. HANSON. 1971. Optical diffraction studies of myofibrillar structure. *Phil. Trans. Roy. Soc. (London)* B **261**: 201.

REEDY, M. K., K. C. HOLMES, and R. T. TREGEAR. 1965. Induced changes in orientation of the cross-bridges of glycerinated insect flight muscle. *Nature* **207**: 1276.

RÜEGG, J. C. and H. H. WEBER. 1963. In *Perspectives in biology*, p. 301. ed. C. F. Cori et al. Elsevier, Amsterdam.

TREGEAR, R. T. and A. MILLER. 1969. Evidence of cross-bridge movement during contraction of insect flight muscle. *Nature* **222**: 1184.

VIBERT, P. J., J. C. HASELGROVE, J. LOWY, and F. R. POULSEN. 1972a. Structural changes in the actin-containing filaments of muscle. *Nature New Biol.* **236**: 182.

———. 1972b. Structural changes in actin-containing filaments in muscle. *J. Mol. Biol.* In press.

VIBERT, P. J., J. LOWY, J. C. HASELGROVE, and F. R. POULSEN. 1971. Changes in the low angle X-ray diffraction patterns of smooth muscles during contraction. *Proc. 1st Europe. Biophys. Congress*, Baden, vol. 5, p. 409.

Structural Changes in the Actin- and Myosin-containing Filaments during Contraction

H. E. HUXLEY

MRC Laboratory of Molecular Biology, Hills Road, Cambridge CB2 2QH, England

Earlier studies on the low-angle X-ray diffraction diagrams given by striated muscles showed that some features of the diagrams remained virtually unchanged during contraction, whereas others changed in a characteristic manner. Thus the subunit repeat and the pitch of the actin helices in the thin filaments remained apparently constant, and the subunit repeat in the myosin filaments (the 143 Å cross-bridge spacing) remained almost constant too (Elliott et al., 1965, 1967; Huxley et al., 1965; Huxley and Brown, 1967), although more extensive measurements revealed a small increase in spacing during contraction, by about 1% (Huxley and Brown, 1967; Haselgrove, 1967, 1970). On the other hand, there was a very substantial decrease in the *intensity* of the X-ray diagram given by the helical arrangement of cross-bridges on the thick filaments, especially in the off-meridional parts of the pattern, which was interpreted as showing that the cross-bridges on any given filament moved from a relatively well-ordered arrangement characteristic of live relaxed muscle to a more random arrangement in contracting muscle. These results showed that there must be some axial movement of the cross-bridges, combined with a more extensive radial and/or azimuthal disordering, and it was suggested that this was brought about by the asynchonized movement of the cross-bridges during activity as they each went through their individual contractile cycles of actin attachment and ATP breakdown (Huxley and Brown, 1967).

Observations on the equatorial reflections (which arise from the side-by-side arrangement of the actin and myosin filaments in the region of overlap) showed only very small changes in spacing when the muscle became active (Elliott et al., 1965, 1967), but observations on the changes in the relative intensities of the equatorial reflections associated with rigor (Huxley, 1968) or with activity (Haselgrove, 1970; Haselgrove and Huxley, in prep.) indicated that a very substantial sideways redistribution of mass occurred when interaction between the filaments took place. Material originally associated with the myosin filaments in live relaxed muscles became closely associated with the actin filaments; in rigor the amount of material involved corresponded approximately to the mass of all the myosin S_1 subunits, whereas in active muscle about half that amount was involved. It was suggested (Huxley, 1968) that this might represent the active end of the cross-bridge leaning out sideways to attach to actin, and it was pointed out that such a scheme provided a good way out of the difficulties which hitherto had seemed to exist with models involving direct physical interaction between actin and myosin filaments across a variable side-spacing. It was also noted that mechanical considerations made it probable that the region of active force generation was the contact area between the S_1 head subunit and the actin monomer to which it attached (Huxley, 1968, 1969).

Thus several of the X-ray observations (and also others that I do not have space to mention here but are described by Haselgrove [1970] and Haselgrove and Huxley, in prep.) gave strong support to the sliding-filament model, in which the filaments are of virtually invariant length, and force is developed by moving cross-bridges. The results also suggested some new features of the force-generating mechanism itself. However there were a number of points at which the interpretation of the X-ray patterns was far from complete or where additional data were needed; the present paper will deal with two of these.

The first part is concerned with efforts to reach a more realistic understanding of the properties and behavior of the cross-bridges. In the earlier work it quickly became apparent that in an active muscle, although the pattern characteristic of the regular resting arrangement of cross-bridges was greatly attenuated (to 30% or less of its original intensity), it was not replaced by any trace of a new pattern— for instance, there was no sign of the pattern characteristic of muscle in rigor. If cross-bridge movement outside the boundaries of the movement already present in relaxed muscle (and evidenced by the fading of the diffraction pattern beyond spacings below about 50 Å) depends on attachment to actin (to allow the myosin component to recognize that the muscle has been switched on and to allow the cycle of ATP splitting to be completed), then the results require that in an isometrically contracting muscle a high proportion of the cross-bridges are attached to actin at any given moment (so as to abolish the "relaxed" X-ray pattern) but that their angles of attachment are distributed over a range so that the characteristic rigor "labeling" of the actin

X-ray pattern is not obtained. On the other hand, the 143 Å meridional reflection is only reduced to about two-thirds its original intensity, which implies that the average extent of axial movement of the bridges is only about 15–20 Å, hardly enough to abolish the rigor pattern. Thus the evidence indicated that the assumption that attachment to actin was required for movement might be wrong and that there might be a certain amount of radial and azimuthal movement of unattached cross-bridges during contraction. It was suggested at that time that this might be the result either of some cooperative interaction of bridges or of the direct recognition, by the myosin filaments, that the muscle had been switched on, perhaps by some calcium-sensitive mechanism (Huxley and Brown, 1967)—two possibilities which it has proved surprisingly difficult to choose between.

Haselgrove (1970) explored this problem further by examining the X-ray diffraction diagrams given by frog sartorius muscles, stretched to the maximum length possible (about 3 μ) and stimulated at this length, so that less than 50% of normal overlap of the actin and myosin filaments was present during contraction. He found in most experiments that the loss of the layer line pattern was just as extensive as in a muscle at full overlap, indicating that cross-bridge movement was occurring in the region of the myosin filaments not overlapped by actin.

In the present experiments, I have taken advantage of the fact that an intact frog semitendinosus muscle can be stretched fairly readily to a length at which the sarcomere length along most of the muscle is in excess of 3.6–3.7 μ, so that overlap between actin and myosin filaments should be abolished altogether. Such muscles will shorten passively again to more normal sarcomere lengths and will then develop a large part of their original tension at that length. They therefore represent a system in which one can look for cross-bridge movement under conditions where one might expect actin-myosin interaction to have been completely eliminated. Some of these experiments have already been mentioned briefly in conference abstracts (Huxley, 1970, 1971a, b), but the present account covers a more detailed and extensive study whose implications are somewhat unexpected.

The second topic is concerned with the structure of the thin actin-containing filaments. As I have already mentioned, the helical parameters of these filaments remain unchanged during contraction, and even in rigor the subunit repeat remained constant and no decisive evidence could be found for a change in the pitch of the helix. This was a not unexpected result, in terms of a model in which myosin seemed to be the more active partner and where the role of the actin seemed to be to activate

the myosin ATPase and to provide an attachment site for the cross-bridges which themselves would undergo the structural changes necessary to change their angle of attachment. However, the X-ray result attained rather more interest when it became clear from the work of Ebashi and others that activation of contraction by calcium ions was effected in vertebrate striated muscle by changes produced in the properties of the thin filaments—changes in the myosin ATPase-activating ability of actin which were controlled by the troponin-tropomyosin system.

It was with a view to looking for additional details of the changes in the X-ray reflections from both the actin and myosin filaments during contraction that further developments in technique, especially as regards the X-ray tube itself, were pursued; and in the course of this work, new reflections from the actin-containing filament system during contraction have been picked up. These, too, have been mentioned previously, briefly, in conference reports (Huxley, 1970, 1971a, b); the present account will describe them in more detail, analyze their structural interpretation, and discuss their implications in the light of our three-dimensional electron-microscopic observations on analogous systems and of recent biochemical studies on the control mechanism of muscle, especially those of Annemarie Weber and her colleagues (e.g. Bremel and Weber, 1972). Analogous changes in the X-ray diagrams from smooth muscles during contraction have been described by Vibert et al. (1971, 1972).

Experimental Methods

X-ray techniques. The cameras used were essentially the same as those described by Huxley and Brown (1967). For some of the experiments an improved collimating system was used in which two horizontal slits of appropriate width and about 10 cm apart were placed between the monochromator (which was focusing in a horizontal plane) and the specimen. This shielded off-meridional parts of the diagram from stray radiation scattered by the mirror and the monochromator was especially useful when examining weak patterns. Most of the experiments were carried out using a prototype 18-in diameter, 3000 rpm rotating anode X-ray tube, developed by Huxley and Holmes, and operated with an Elliott gun giving a focal spot whose foreshortened dimensions were approximately 100 $\mu \times$ 100 μ, with a beam current of 50 mA, 40 kV.

Preparation of muscles. Semitendinosus muscles (posterior half) were carefully dissected from frogs and toads kept either in an outdoor pond or in a cold room. Some toads kindly provided by Dr. Richard Adrian were caught in the wild state and proved to have muscles of exceptional endurance for long contraction series experiments. For the work on very stretched semitendinosus muscles, a rather elaborate protocol had to be employed. Haselgrove (1970) had observed that when frog sartorius muscles were stretched, the myosin layer line pattern began to decrease rather sharply in intensity beyond a certain sarcomere length (around 2.8–2.9 μ); he found, however, that much stronger patterns could be produced if a delay of about 24 hr was interposed either between the dissection of the muscle and the time when it was stretched, or between the stretching of the muscle and the time when it was photographed. In the case of semitendinosus muscles stretched beyond overlap, I observed the same effect; but the loss of the pattern was often so severe that even after the waiting period it was too weak for changes in it to be recorded reliably. However I found that if the muscle was first allowed to stand for 24 hr at rest length in normal Ringer's solution, then transferred to $\frac{2}{3}$ R solution (i.e., Ringer diluted with half its volume of water) for a few hours before stretching and then kept at the stretched length for 8–12 hr before the experiment proper began, maintaining the $\frac{2}{3}$ R solution throughout the experiment, then satisfactory patterns, similar to those from muscles at rest length, could be obtained even from muscles stretched to a sarcomere length of 4.5 μ.

Sarcomere lengths were measured from the laser diffraction pattern given by the muscle. This was often recorded in situ on the X-ray camera either before or during the experiment; but the most accurate values of sarcomere length were obtained by fixing the muscle in situ in gluteraldehyde at its experimental length, cutting it by hand into sections about 100–200 μ in thickness, and measuring the very clear optical diffraction patterns given by these.

Procedure during contraction series. In the stretch experiments two rather different techniques were employed. In the first, the isometric case, the muscle was held fixed at one end and the other was attached to the tension recorder, an RCA 5734 transducer. This was held stationary during the experiment but could be moved so as to change the length of the muscle. Initially, active tension (P_0) was recorded at rest length, and the muscle was then stretched until the active tension developed was less than about 0.2 P_0, usually a stretch of 4–5 mm in muscles which measured about 25 mm in length between the attached wire and thread at the tendons. At this length, a significant resting tension (about $\frac{1}{2}$ P_0 or more) was present immediately after stretch, then gradually decayed. Laser diffraction patterns usually showed sarcomere lengths in the central zone of the muscle between 3.6 and 4.0 μ,

but no doubt shorter sarcomeres were present at the ends of the muscle, well outside the region monitored by the X-ray beam. Often a small increase in sarcomere length, of the order of a few percent, could be perceived in the central region of the muscle during stimulation.

The muscles were kept in well-oxygenated $\frac{2}{3}$ R solution at about 4°C and stimulated directly via platinum strips on either side of them, using differentiated square wave pulses with a repetition frequency of 5 or 10 per sec, which was quite adequate to give a steady response. Tetani of 6.8 sec duration, repeated once every two minutes, were found to give satisfactory results and usually the muscles maintained their activity remarkably well, perhaps because, at no overlap, much less ATP was being used. In the second type of experiment, the isotonic case, the muscle was stretched by means of a weight attached to it via a very low friction pulley. After a little experience weights could be chosen which gave a sarcomere length between 3.7 and 4.5 μ (constant to ± 0.1 μ for a given muscle) and against which only a millimeter or two of shortening occurred upon stimulation. Since the tension on the muscle was now constant during the whole experiment, i.e., equal to the load whether the muscle was stimulated or not, it was hoped to eliminate any possible effects on the "no-overlap" sarcomeres arising from tension produced by shorter sarcomeres at the ends of the muscle. (In fact, no evidence for such an effect was found.) The isotonic experiments were more difficult to carry out since often the residual shortening during contraction was very small and difficult to use as an indicator of the condition of the muscle. Also, weights adequate to prevent an excessive amount of shortening sometimes stretched the muscles excessively and gave preparations whose contractile response during and after the long periods of time involved was somewhat unpredictable.

As a check on the possible effect of applied tension on the X-ray diffraction pattern given by the no-overlap sarcomeres, several experiments were performed in which an additional load was applied to the unstimulated muscle by a relay device operating on the same protocol as the stimulator (i.e., 6.8 sec once every two minutes), the pattern being recorded only when the additional load was applied.

Procedures for recording changes in X-ray patterns from thin filaments. The parts of the actin filament pattern which were of interest here were some of the reflections rather far out from the meridian along the layer lines. These reflections are very weak and take an hour or two to record even with the fastest cameras. Several toad muscles went on contracting long enough to give measurable patterns individually, but in the case of frog sartorius muscle, I was obliged to resort to the expedient of photographing 4–6 muscles in succession on the same film.

Vibert et al. (1971, 1972) have pointed out that muscles in rigor show similar changes to active ones in the region of the diagram in question. However there is some disadvantage in the use of such preparations at normal sarcomere lengths in that the attachment of a large number of cross-bridges at the same time may itself modify the reflections given by the thin filaments in this region; whereas in contracting muscle little sign of "labeled" actin is visible, and the myosin layer lines are very weak. Following an earlier observation by Huxley and Brown (1967), I found that heating a frog muscle for 5–10 min at 42°C destroyed the myosin pattern and gave an actin filament pattern normal in all respects, except that in the fine detail of the low-order layer line reflections, it corresponded to that from an active muscle, presumably as a consequence of the release of calcium from the sarcoplasmic reticulum (heat rigor). In another preparation used, muscles were heated to 65° for 10 min, following the report by Tanaka and Oosawa (1971) that tropomyosin dissociated from actin at higher temperatures.

Recording and analysis of results. X-ray patterns were recorded on Ilford Industrial G film and their intensities measured by means of a Joyce-Loebl microdensitometer (model III C). Computations were carried out on an IBM 360/44 in the Institute of Theoretical Astronomy, Madingley Road, Cambridge, using programs generously provided by Dr. J. C. Haselgrove.

Results

Myosin Layer Line Reflections in Very Stretched Muscles

Isometric contractions. A total of about twenty satisfactory isometric contraction experiments were carried out at sarcomere lengths between 3.6 μ and 4.1 μ, and the X-ray patterns recorded during eighteen of these were analyzed densitometrically, the measurements being tabulated in Table 1. The overall result was the again surprising one that, although the muscles were at sarcomere lengths at which no overlap would be expected between thick and thin filaments, a substantial decrease occurred in the intensity of the pattern from the myosin cross-bridges. The extent of the decrease—to about 0.6 to 0.7 of the intensity in resting muscle—is about half that observed in muscles at normal degrees of overlap (Huxley and Brown, 1967) but is nonetheless a very prominent

Table 1. Isometric Contractions

Experiment No.	Sarcomere Length (μ)	Intensity during Stimulation as a Fraction of Resting Intensity		
		1st L.L.	2nd L.L.	3rd L.L.
1761	3.8	0.63	0.73	1.21
1762	3.6	0.65	0.51	0.63
1770	3.8	0.37	0.57	0.52
1771	3.9	0.48	0.63	0.46
1773	3.8	0.53	0.72	0.80
1774	3.8	0.68	0.89	0.74
1775	3.8	0.48	0.84	0.34
1777	3.9	0.66	0.81	0.80
1887	3.8	0.85	0.89	0.99
1905	3.8	0.57	0.85	0.81
1916	3.8	0.61	0.66	0.52
1917	3.7	0.97	0.67	1.06
1921	3.7	0.59	0.75	0.72
2209	4.0	0.64	0.84	0.85
2210	3.7	0.88	0.80	0.78
2220	3.9	0.68	0.64	0.65
2237	4.0	0.68	0.81	0.63
2247	3.8	0.53	0.75	0.80
Average		0.63 ± 0.11	0.74 ± 0.09	0.74 ± 0.17

effect. Considerable care was taken to make sure that the sarcomere lengths had the intended values. In many experiments, measurements included ones made during stimulation; after each experiment the muscle was fixed in situ with glutaraldehyde, the region which had been in the X-ray beam was cut into sections, and the sarcomere length in each piece measured by laser diffraction. Usually the sarcomere length distribution was very uniform, within about 0.1 μ of the average value, with occasional groups of fibers very much more stretched (i.e., up to 4.5 μ or more) than the rest. Even at these degrees of stretch, the muscles often developed a significant amount of tension (10–20% P_0), presumably arising in large part from the contraction of the shorter sarcomeres towards the ends of the fibers (A. F. Huxley and Peachey, 1961). The question therefore arose whether the X-ray pattern given by the central region of the muscle might be influenced by this applied tension. Two types of experiment were done to investigate this possibility. In the first, muscles were stretched to the required length and a normal contraction series carried out. They were then loaded with an additional 10 g weight, applied for 6 sec periods at 2 min intervals while the X-ray shutter was open. The muscles were not stimulated in this part of the experiment. This additional tension was two or three times as great as that developed during stimulation; but although clear changes were visible in the myosin layer lines during stimulation, none could be seen to result from the "passively" applied tension, so it seemed unlikely that the explanation lay along these lines. The second type of experiment is described below.

Isotonic experiments. In these, tension on the muscles was maintained constant during the whole experiment, as described in the Methods section. No shortening of the sarcomeres in the region of the muscle transversed by the X-ray beam could be detected by light diffraction. The pattern often became somewhat fainter during stimulation, perhaps due to longitudinal displacements caused by shortening at the ends of the fibers as described by Gordon et al. (1966a, b); but no indication of significant *increases* in sarcomere length were seen. A total of some 49 successful experiments were carried out. This number was large for two reasons. First, it was desired to have data from muscles over the whole range of sarcomere lengths between 3.6 μ and 4.5 μ, but the loading necessary was not easy to judge with certainty and muscles tended to "creep" during the first 2 hr of loading; too high a load would cause the eventual rupture of the muscle, usually part way through the contraction series. Second, from time to time preparations were encountered in which no change of pattern was seen during stimulation but which in other respects were normal, and so it was thought advisable to have a fairly large statistical sample. From the 49 experiments, the best 13 sets of X-ray photographs were selected for densitometric analysis, selection being on the basis of the technical quality of the photograph, not the extent of change visible in it. The results of these measurements are tabulated in Table 2, where it can be seen that a substantial decrease in the intensity of the myosin layer line pattern was once again found. The extent of the decrease, to 0.6–0.7 of the "resting" value, was about the same as in the isometric case. In the 36 experiments which were evaluated visually, most patterns showed a decrease in intensity during stimulation; but in six cases, no discernible change was visible. These were all muscles with longer than average sarcomere lengths—in two cases, they were 4.5 μ, the longest successfully employed—but there was no compelling reason to think that the activation mechanism might have been damaged at this length. Three of the muscles which were allowed to shorten after the experiment would still develop considerable tension, and several other muscles which were kept stretched to between 4.0 and 4.5 μ during experiments developed half their original tension when returned to shorter sarcomere lengths afterwards. And other muscles *did* show a change in pattern at $s = 4.5 \mu$.

These "no change" experiments, though not numerous, occurred often enough to be disturbing, and in the course of work done to clarify the situation, a new phenomenon was discovered, not necessarily related to the anomalies, but having some interest both for its own sake and because of its implications for other types of measurement. In experiments on frog sartorius muscles at rest length or slightly stretched, it was found that the decrease in intensity of the myosin layer line pattern which

occurs during stimulation persists for several seconds after stimulation is over (Fig. 1). Data from two typical experiments in which the muscle was tetanically stimulated for 1.5 sec and the exposure began 0.5 sec after stimulation ceased and terminated 3.5 sec or 5 sec later is shown in Table 3. It will be seen that the loss of the myosin pattern is almost half as great as during contraction itself, the actin reflections being unaffected. This experiment has been repeated several times successfully. By about 10 sec after contraction, the pattern usually seemed to have returned to normal. This phenomenon has not been explored in great detail—it can be done much better with an experimental arrangement different from the one we have at present—but it seems to be a real effect and whereas it does not alter the earlier conclusion that movement of the cross-bridges occurs upon contraction, it means that measurements of the extent of the change in the axial X-ray diagram need to be interpreted with caution, as we will discuss later.

Table 2. Isotonic Contractions

Experiment No.	Sarcomere Length (μ)	Intensity during Stimulation as a Fraction of Resting Intensity		
		1st L.L.	2nd L.L.	3rd L.L.
2016	4.1	0.60	0.83	—
2062	4.1	0.65	0.85	0.88
2065	3.9	0.65	0.78	—
2069	3.9	0.80	0.81	0.58
2073	3.6	0.48	0.51	0.55
2077	3.8	0.65	0.75	0.54
2099	3.8	0.51	0.80	0.77
2110	4.5	0.71	0.69	0.83
2122	4.5	1.01	1.46	0.85
2135	4.0	0.74	1.04	0.59
2143	3.9	0.63	0.86	0.64
2146	4.0	0.67	0.67	0.82
2171	4.2	0.62	0.46	0.69
Average		0.67 ± 0.09	0.80 ± 0.15	0.70 ± 0.11
(Omitting Exp. 2122)		0.64 ± 0.06	0.75 ± 0.11	0.69 ± 0.11

Table 3. Examples of Data from Delayed Exposure Experiments

Expt. 2222[a]	1st L. L.	2nd L. L.	3rd L. L.
Relaxed intensity before experiment	394	107	162
Experimental intensity	202	88	83
Relaxed intensity after experiment	437	140	150

Expt. 2241[b]	2nd Film 1st L. L.	2nd L. L.
Relaxed intensity before experiment	2037	842
Experimental intensity	1584	556
Relaxed intensity after experiment	1845	780

[a] Exposure 3.5 sec, begun 0.5 sec after end of 1.5 sec tetanus. Intensities fall on the average to 0.6 that of control value.
[b] Exposure 5 sec, begun 0.5 sec after end of 1.5 sec tetanus. Intensities fall on the average to 0.75 that of control value.

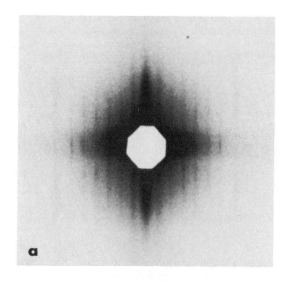

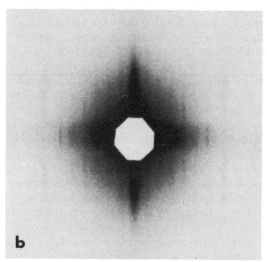

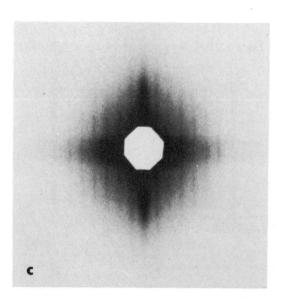

Reflections Associated with the Thin Filaments

Resting muscles. In addition to the series of actin reflections usually described (i.e., the ~59 Å and ~51 Å off-meridional layer lines, the 27.3 Å first meridional reflections and other meridional reflections indexing approximately as higher orders of a 54.6 Å repeat, and an off-meridional reflection near 400 Å which is difficult to disentangle from the outer part of the first myosin layer line, but which probably really represents the first actin layer line at ~370 Å), there are other reflections, apparently on the same layer line system, but at much larger distances from the meridian. Thus, while the very prominent 59 Å reflection has a broad maximum centered about 0.008–0.009 Å^{-1} off meridian, the reflections we are now considering have radial positions in the range 0.02–0.03 Å^{-1}. These reflections are very weak and rather diffuse, and to be visible they require exposure times about 20 times or more larger than are necessary to register the 59 Å reflection. The reflections are listed in Table 4.

The strongest of these weak reflections in live frog sartorius muscle occurs at an axial spacing of about 130 Å, i.e., not significantly different from the position of the 3rd actin layer line, and has a radial position of 0.026 Å^{-1}. Two other reflections of comparable intensity occur at axial spacings of approximately 70 Å and 51 Å (i.e., on other layer lines of the actin pattern) and at radial positions of approximately 0.025 and 0.024 Å^{-1}. There are also significant reflections on a layer line at 29.3 Å at radial positions of 0.015 Å^{-1}. No reflection is visible in the position of the 2nd actin layer line, but a weak one there (probably anything less than one-half to one-third the intensity of the 3rd layer line reflection) would be difficult to recognize over the background.

Actively contracting muscle. Although the reflections in question are weak ones, there is a clear difference visible between resting and active muscle, provided that sufficiently long exposure times are available. Contracting muscle shows a characteristic reflection at an axial spacing of approximately 190 Å (i.e., in the position of the 2nd actin layer line) and at a radial position of 0.021 Å^{-1} (Fig. 2). The reflection previously visible on the 3rd actin layer line is now not seen, i.e., its intensity

Figure 1. X-ray diffraction diagrams of live frog sartorius muscle, showing the layer line pattern given by the myosin cross-bridges, and the 59 Å actin reflection (outer horizontal lines). (*a*) Relaxed muscle, precontrol; (*b*) experimental muscle, same as (*a*), but photographed by opening shutter on X-ray camera for series of 3.5 sec exposures, each beginning 0.5 sec after end of 1.5 sec tetanus; (*c*) relaxed muscle, post-control, same as (*b*), normal exposure after experiment. Note large drop in intensity of myosin layer line pattern in (*b*) even though muscle was not stimulated during the exposures.

Table 4. Reflections Given by Actin-containing Filaments in Muscle at Larger Off-meridional Distances

	Layer Line	Axial Spacing	Radial Spacing	Intensity
Relaxed Muscle	Second	—	—	Very weak ($<\frac{1}{3}I_3$ L.L.) Not visible
	Third	~130 Å	0.026 Å⁻¹	Weak but visible
	—	70 Å	0.025 Å⁻¹	Weak but visible
	—	51 Å	0.024 Å⁻¹	Weak but visible
Contracting Muscle	Second	~190 Å	0.021 Å⁻¹	Medium—weak, visible
	Third	—	—	Very weak Not visible
	—	70 Å	0.02–0.025 Å⁻¹	Weak but visible
	—	51 Å	0.02–0.025 Å⁻¹	Weak but visible
Heated Muscle	Second	~190 Å	0.021 Å⁻¹	Medium—weak, visible
	Third	—	—	Very weak Not visible
	—	70 Å	0.019 Å⁻¹	Weak but visible
	—	51 Å	0.022 Å⁻¹	Weak but visible

must have decreased by a factor of two to three times or more, though, since it is such a weak reflection even in relaxed muscle, it is difficult to be certain about this. The 2nd layer line reflection in contracting muscle appears considerably stronger than the 3rd layer line reflection in relaxed muscle, but that may in part be due to improved orientation. The 70 Å and 51 Å reflections are also visible at approximately the same radial positions as before, but so far sufficiently strong photographs have not been obtained to enable us to say whether small changes in their positions or intensity have occurred during contraction.

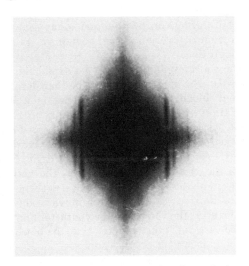

Figure 2. X-ray diffraction diagram given by actively contracting toad sartorius muscle, showing the new reflection in the position of the 2nd actin layer line, which is now stronger than the 3rd. *Note:* The relative intensities of reflections and background in Figs. 2 and 3 was artificially altered during processing to improve visibility, using the procedure described by Vibert et. al. (1971).

Heated muscle. If a frog sartorius muscle is kept at a temperature over 40°C for a short while (42°C for 10 min in this instance), it becomes white, opaque, and inextensible, as though it were in rigor. However not only does the myosin layer line pattern disappear in the X-ray diagram, but there is no sign of the "labeled actin" reflections (e.g., there is no strong ~380 Å layer line close to the meridian) characteristic of rigor, nor of the myosin meridional reflection at 143 Å, so presumably the myosin has been denatured. However such muscles still give a very good actin pattern which at higher angles (i.e., at the 59 Å reflection and beyond) is indistinguishable from that given by live muscle. At lower angles, however, the pattern is characteristically different from that given by live muscle in that the 3rd actin layer line has disappeared, and a conspicuous reflection has appeared in the 2nd actin layer line at the same radial position as that observed in active muscle, i.e., 0.021 Å⁻¹ (Fig. 3).

The reflections at the larger radial spacings on the 70 Å and 51 Å layer lines are also visible; now it can be seen that they do not differ appreciably in intensity from those visible in the diagrams of relaxed muscle, but there has been a very slight change in their radial positions which now measure approximately 0.019 and 0.022 Å⁻¹, respectively.

Heated muscles do not represent ideal experimental material, but they do have the advantage over rigor muscles that all traces of reflections from cross-bridges have been wiped out; in rigor muscle there is a much stronger reflection at a small radial spacing (0.011 Å⁻¹) on the 2nd actin layer line, which no doubt arises from the attached myosin cross-bridges, so that the presence of measurable intensity further out on the same layer line is of a

Figure 3. X-ray diffraction diagram given by frog sartorius muscle heated to 42°C for 10 min. The myosin reflections disappear, but the actin pattern remains intact and now shows a prominent 2nd layer line reflection, probably resulting from calcium release by the reticulum and uptake by thin filaments, giving them the "active" configuration.

Table 5. Measurement of Troponin-Tropomyosin Axial Repeat during Contraction

Film No.	Spacing in Å		
	Precontrol (Resting)	Experimental (Contracting)	Post-control (Resting)
1876	383	385	380
1879	387	388	381
1886	382	380	382
1892	380	381	382
1903	380	383	381
1911	383	383	385
1912	383	380	380
1915	390	386	382
1920	385	382	384
	384 ± 3 Å	383 ± 2 Å	382 ± 1 Å[a]

[a] Apparent probable error fortuitously low.

little more ambiguous significance, at normal sarcomere lengths anyway (see Haselgrove, this volume).

When muscles are heated to 60° for 10 min, all the reflections on the lower-order actin layer lines disappear, but the pattern from 59 Å outwards persists.

The ~385 Å meridional reflection.

A series of experiments were performed using a 60-cm focal length camera to investigate the behavior of the meridional reflection near 385 Å during contraction. This reflection is believed to arise primarily from the troponin component of the thin filaments, repeating with a periodicity defined by tropomyosin and probably equal to seven times the actin subunit repeat ($7 \times 54.6 = 382.2$ Å). The rather long exposure times required to record this region of the diagram (1–2 hr minimum) placed the experiment close to the limit of present techniques, but no significant change in spacing could be detected (see Table 5). In a number of cases the reflection decreased somewhat in intensity during contraction, but this may have been a consequence of impaired lateral order between the thin filaments. The 440 Å meridional reflection decreased in intensity to an even greater extent.

Discussion

Behavior of Cross-Bridges during Stimulation of Very Stretched Muscles

The finding that considerable amount of cross-bridge movement takes place in sarcomeres stretched to a length greater than the sum of the lengths of the thick and thin filaments, though it follows from Haselgrove's earlier result on partially overlapped sarcomeres, is still very surprising and puzzling. There is a great deal of biochemical evidence that activation in vertebrate striated muscle is exclusively concerned, in the first instance, with changes in the thin filaments, and that myosin "recognizes" that activation has taken place only by finding that the combining sites on actin are no longer blocked. At the concentrations of free magnesium probably present in muscle (1–5 mM), myosin binds only small amounts of calcium at the levels of free calcium adequate to switch on fully troponin-tropomyosin-controlled actomyosin ATPase (Bremel, 1972); and so far no other calcium-binding protein has been found in the thick filaments. One would therefore expect that only these cross-bridges which can make contact with actin would be affected during contraction, and so myosin outside the region of overlap should remain in its resting configuration.

The contrary finding might be explained in two ways. On the one hand, it may be that despite the present biochemical evidence there *is* a second activation mechanism (i.e., additional to the troponin system) present in vertebrate striated muscle —perhaps with the function of giving more complete relaxation—and it may be that this mechanism holds the cross-bridges away from the actin filaments in a resting muscle and releases them during contraction so that they can move about more freely and make the necessary contacts. For this to be the case, one must suppose that the biochemical evidence is wrong or misleading in some way, or that the system involved is very labile. Whereas this is a possibility, it remains no more than that unless it can be shown that there is no other possible explanation for the structural findings described here. But I think there is still a second possible way of explaining these observations.

Gordon et al. (1966a, b) found that single fibers from frog semitendinosus muscles, stretched beyond the expected no-overlap length and maintained at that length by a "length clamp", would on some occasions develop significant amounts of tension. This tension developed with a time course of a second or two, as distinct from the much more rapid tension development in a normal contraction. It was the fast component of tension which fell to zero as the muscles were stretched to the point of no overlap, whereas the slow component needed considerably longer stretches ($>3.9\,\mu$) to eliminate it. Gordon et al. suggested that this slow component of tension might arise from a few filaments, initially out of register or of greater than normal length, which might be able, by producing some local shortening or shearing, to involve progressively more of the filaments in tension development.

In the present experiments this slow component, if present, obviously could not have been responsible for producing a tension greater than the muscle as a whole was producing, and in the isometric experiments this was never more than 0.2 P_0 and often much less. Thus half the cross-bridges cannot possibly have been involved in its generation. However we have already mentioned the earlier suggestion from the work on muscle at normal degrees of overlap that cross-bridge movement might be to some extent cooperative, and the present results on the delayed disorder effect make this even more probable. The picture that emerges from this work is one in which the cross-bridges in a resting muscle are rather flexibly attached to the thick filaments—as has already been deduced on other grounds—and where the equilibrium position of the bridge is sensitive to changes in its environment and in particular to the exact position of other nearby bridges on the same filament and on neighboring filaments. The weakening of the cross-bridge pattern in passively stretched muscle and its recovery with the passage of time (through an unknown process) or with decrease in ionic strength is clear evidence that the extent of ordering of the cross-bridges can vary in the absence of any evident actin-myosin interaction. After activation of a muscle ceases and tension generation comes to an end, one would expect all the cross-bridges to return to their "chemical" resting state in a time less than the "cycling time" during contraction—say in the time necessary to charge up each myosin head with ATP or its split products, a very small fraction of a second at most. However if the equilibrium position of the bridges is weakly defined and if long range interactions do occur with other bridges, then an appreciable time may elapse before the system settles down into a fully ordered state again.

If this is the case, then one cannot exclude the possibility that cross-bridge interaction during contraction in a small region in an A-band might disturb the regularity of the arrangement of neighboring cross-bridges, even though these were not in a position to interact with actin themselves, and that this disturbance might spread until a considerable proportion of all the cross-bridges was involved. It may be that this is the effect that has been showing up in the present experiments and that the "anomalous" preparations which showed little change in pattern were ones in which local irregularities in overlap were fewer, or where very high degrees of stretch had eliminated them. Thus I do not think one can say with certainty that there *must* be some way in which the myosin filaments themselves can recognize activation "chemically" (e.g., by being sensitive directly to changes in calcium concentration), and although one should remain alert for such a possibility, there is no compelling reason for supposing the present biochemical picture to be wrong.

If cross-bridge movement is a cooperative effect with a relatively long time constant, then, whereas tension generation is still produced by and requires cross-bridge movement and the onset of contraction can still be correlated with the extent of cross-bridge disorder, no simple relationship will exist at longer times, for example, between the number of cross-bridges attached at different loads and speeds of shortening and the changes seen in the X-ray diagram; and the recovery rate of the pattern after activation will not reflect the decay of the active state but some other phenomenon.

Changes in Thin Filaments

There is now a considerable body of biochemical evidence that in vertebrate striated muscle contractile activity is controlled by changes in the actin-tropomyosin-troponin complex in response to changes in the free calcium level within the myofibrils. It appears that in a resting muscle the free calcium concentration is very low (10^{-8} M or less) and that under these conditions, actin is unable to combine with the myosin cross-bridges, so that the actomyosin ATPase is not activated and no tension is produced. Since actin on its own activates myosin ATPase whether calcium is present or not, it is clear that tropomyosin-troponin is functioning as an interaction inhibitor in the absence of calcium. Upon activation, the sarcoplasmic reticulum releases calcium so that its concentration in the myofibrils rises to 10^{-5}–10^{-4} M. Calcium then becomes bound to the troponin component, actin myosin interaction takes place, and the muscle contracts. What gives this process special interest is the fact that troponin appears to be located along the length of the thin filaments at approximately 400 Å intervals; yet it

is able to control the activity of all the actin monomers in the intervening region, keeping them all
switched off in a resting muscle and switching on
probably all of them during contraction (Weber and
Bremel, 1972). This feature of the mechanism is
borne out by in vivo studies of the stoichiometry of
the regulation process (Bremel and Weber, 1972;
Spudich and Watt, 1971). Troponin appears to be a
relatively globular molecule, but tropomyosin is a
two-chain coiled-coil α-helical structure (Cohen and
Szent-Györgyi, 1957; Caspar et al., 1969) present in
sufficient amounts to provide two continuous
strands running along the length of the actin filaments and hence able to make contact with each
actin monomer. Since the presence of tropomyosin
is necessary for the regulation mechanism to
operate, it is natural to suppose that the influence of
the troponin is transmitted to the actin monomers
via the tropomyosin strands. It is therefore of
interest to see whether structural information about
the thin filament complex can provide any clues as
to how this mechanism operates.

Hanson and Lowy (1963) suggested that tropomyosin might lie in the long-pitch grooves of the
actin double helix and this idea has been confirmed
and extended by later work. Moore et al. (1970),
applying the 3-D reconstruction techniques to
electron micrographs of negatively stained specimens, found that preparations of purified actin
showed simply a helical arrangement of relatively
simple globular units, but that preparations of thin
filaments prepared directly from muscle showed
additional material asymmetrically located in the
long-pitch grooves of the structure. The protein
composition of these filaments was not investigated
(since they were not available in purified form), but
it was noted that the additional material might
correspond to the tropomyosin-troponin which one
would expect to be present.

O'Brien et al. (1971) reported that there was
usually a characteristic difference in the optical
diffraction pattern given by micrographs of paracrystals of pure actin on the one hand and of
unpurified actin, or actin mixed with impure tropomyosin, on the other. This difference took the form
of an increase of relative intensity of the reflection
on the 2nd layer line of the actin pattern (regarding the ∼365 Å layer line as the first) in the presence
of the additional components, and they pointed out
that it was consistent with the presence of additional
material within the grooves of the structure. The
problem was taken some stages further by Spudich
et al. (1972), using preparations of highly purified
actin and of the same actin mixed with the purified
regulatory protein complex, giving a system shown
to have full calcium sensitivity in an actomyosin
ATPase assay. We confirmed the result of Moore

et al. that pure actin had the form of relatively
simple globular units arranged in the two-chain
helix, and also the strong indications from O'Brien
et al. that it was the presence of the regulatory
proteins which gave rise to the strengthening of the
2nd and 3rd layer line reflections. Moreover, we
were able to reconstruct the complex filaments
and show that a fairly continuous strand of material
ran along in each of the two grooves in the actin
helix, situated asymmetrically, so that if the actin
structure is thought of as two strings of monomers
twisted around each other, one regulatory strand
is more closely associated with one string of monomers, and the other with the other string. The
angular position of each strand, measured from the
axis of the helix, is about 60–70° relative to the associated actin monomer at the same level, as can
be seen from the end-on projection of the structure
shown in Fig. 4. In these reconstructions the strong
385 Å axial periodicity believed to arise from the
troponin component was not included, so that the
resultant structure would be expected to show the
location of the tropomyosin only. Thus the tropomyosin appears to run alongside the chains of
actin monomers in a position very suggestive of
possible interaction. However it must be realized
that the preparations were not made in a calcium-
free medium, so that even if the "switched-off"
configuration could be preserved by negative
staining, there is no reason to suppose that the
reconstructed tropomyosin configuration in these

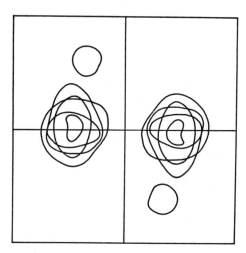

Figure 4. Arrangement of actin and tropomyosin as seen
in end-on helical projection of 3-D reconstruction of composite filament (copied directly from Plate X(a) in Spudich
et al., 1972). Each tropomyosin strand seems to be
associated with its "own" helix of actin monomers. In this
type of view, the position of the tropomyosin peak lies at
an azimuthal angle of about 60° with respect to the actin.
However, a more reliable estimate of position is given by
the difference Fourier (loc. cit. Plate X(b)) in which the
angle is 70°.

experiments corresponds to the inhibitory form. Indeed, we will see later it more probably corresponds to the "switched-on" form.

Another 3-D reconstruction result that is relevant here concerns the mode of attachment of the myosin S_1 subunits to the actin monomers, studied on so-called "decorated" actin (i.e., actin mixed with S_1 in absence of ATP) which probably corresponds to the "rigor" configuration of the cross-bridges and to their probable position at the end of their working stroke. Moore et al. (1970) found that the S_1 subunit, besides being tilted and skewed in a characteristic manner, was attached to the actin monomers in a somewhat tangential fashion, so that the end of the S_1 subunit and the contact area between it and the actin extended round into the groove in the actin structure. This can be seen in Fig. 5, showing an end-on projection of a short length of the structure. In this diagram the occurrence of additional material in the grooves of the actin structure can also be seen, apparently asymmetrically placed; but in the entire reconstruction this density is not sufficiently well resolved to specify with certainty to which of the two actin chains it is more closely associated.

It must be recalled that the formation of "rigor links" (i.e., in absence of ATP) between myosin and actin is not inhibited by the regulatory complex even in the absence of calcium, and therefore attachment of the type seen in Fig. 5 would not be inhibited by the troponin-tropomyosin complex.

Nevertheless the projection of myosin into the groove is very suggestive and it seems at least a

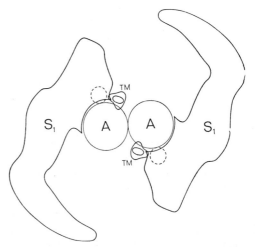

Figure 6. Composite end-on view of actin-tropomyosin-S_1 structure. The shape of the S_1 subunits and the position at which they attach to the actin is copied directly from Plate IX of Moore et al. (1970). The actin structure (24 Å diameter spheres placed at a radius of 24 Å) is that found by Haselgrove (this volume) to give the best agreement (for a simple model) with the stronger features of the X-ray diagram. The two tropomyosin (TM) positions correspond to: *dotted contours*, activated state taken directly from difference Fourier (Plate X(b)) of Spudich et al. (1972); *solid contours*, relaxed state, based on radial position of reflection on third layer line of relaxed muscle. (This assumes that the "sense" of the azimuthal position of tropomyosin with respect to the attachment site of S_1 has been chosen correctly, which has not yet been proved.) The possible way in which tropomyosin could block the attachment of a cross-bridge is very evident.

working possibility that when myosin attaches to actin in the presence of ATP (or its split products), the end of the molecule projects a little bit further into the groove and that the attachment is vulnerable to the influence of tropomyosin. For the tropomyosin to just make contact with the myosin head in the rigor attachment, it would have to lie at an angle of about 60–70° to the respective actin monomer, measured from the helix axis. Figure 6 shows the position of the rigor links again, together with the outline of the probable positions of tropomyosin given by the difference Fourier of Spudich et al. (1972) (activated state) and from the X-ray data on resting muscle (relaxed state), as discussed below.

With these observations and possibilities in mind, let us now turn to the X-ray diffraction results, both the ones described here and earlier ones. The first result we should consider is the finding that the pitch and subunit repeat of the actin helix remain unchanged in an actively contracting muscle. As was pointed out at the time, this does not rule out repetitive cyclical changes taking place in a small part of the thin filament structure at any given time; but since calcium will be tightly bound to the troponin of the thin filaments for the whole period of activity, the changes

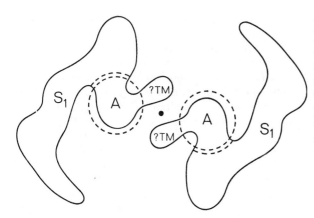

Figure 5. Arrangement of actin and S_1 subunits of myosin as seen in end-on view of 3-D reconstruction of "decorated" actin filaments (copied from Plate IX of Moore et al. (1970), using contour levels corresponding to approximate centers of subunits). The dotted circles correspond to the probable position of the actin subunits, represented as spheres 24 Å in diameter. Due to foreshortening of the structure (anisodimensional shrinkage of the negative stain over a hole in the carbon film), the diameter of the filament is probably too large.

it produces should be maintained virtually continuously throughout contraction and hence they cannot be of a kind which affects the helical parameters of the thin filaments. This view is strengthened by the observation (Huxley and Brown, 1967) that even in rigor (when the level of free calcium is normally high) the subunit repeat in the actin filaments is unchanged, and any change in the pitch of the actin helix is either relatively small or absent. Now experience with other systems (e.g., hemoglobin) indicates that even relatively small internal changes in structure inside a molecule or subunit are liable to produce appreciable changes in the way the subunits pack together, and I therefore believe that the most likely interpretation of the constancy of the actin structure is a virtually complete constancy in the internal structure of the actin monomers themselves when they are switched on or off. This strongly favors the possibility that regulation is affected by a steric blocking mechanism involving tropomyosin (Huxley 1970, 1971a,b).

Next let us consider the significance of these rather faint reflections on the 2nd and 3rd (and higher) layer lines of the X-ray pattern given by the thin filament helix, of their radial positions, and of the changes in their relative intensities which seem to be associated with the switching on of the actin structure. I have already mentioned the finding that the optical diffraction patterns from purified actin had virtually no measurable intensity on the 2nd and 3rd layer lines, whereas reflections were visible there in the presence of tropomyosin. This indicates strongly that in the X-ray case too these reflections arise from or are enhanced by the presence of tropomyosin, so that changes in them would reflect changes in the location of the tropomyosin. If the tropomyosin remained at a constant radius from the center of the thin filament helix, and if some contribution from the actin structure itself was present on the layer lines in question, then a decrease in 3rd layer line intensity and an increase in 2nd layer line intensity would require the tropomyosin to move away from a position one-third the way between actins (i.e., angle 60° as defined previously) and towards a halfway position (angle 90°). Thus, in a very crude way, the direction of movement away from the myosin attachment site seems correct.

One can take this argument a little further if, for the purposes of computation, one accepts a simplified model consisting of spherical globules to represent actin monomers and solid tubes of density appropriately placed and wound around with the actin helix to represent tropomyosin. I am indebted to John Haselgrove for designing the necessary programs. Haselgrove (private communication) has shown that very good agreement with the stronger features of the X-ray diagram from

actin in muscle can be obtained with actin monomers 24 Å in radius arranged with their centers on a helix of radius 24 Å and with the appropriate subunit and helical repeats. He observed that such a model gave very low intensity on the 2nd and 3rd layer lines, especially the latter, since the Bessel function maximum there lay close to the minimum of the actin molecular transform. The position of this minimum is not very sensitive to the exact choice of the shape of the actin monomer, so the very small contribution of actin to the 3rd layer line is likely to remain true even with a more realistic model. If the intensity of the 59 Å axial actin reflection is taken as 1000, then the intensities of the maxima of the reflections from actin alone on the 2nd and 3rd layer lines would be approximately 4 and 0.8.

If tropomyosin is represented by a solid cylinder 8.5 Å in radius (corresponding to one tropomyosin molecule of 70,000 daltons to seven F-actin monomers each of 48,000 daltons), then it gives rise to relatively strong diffractions on the 2nd and 3rd layer lines. On the 3rd layer line, for all positions of tropomyosin within the groove (i.e., at radii from the center of the helix ranging from 24 Å to about 50 Å), the maxima of the transform would have values ranging from approximately three to nine on the same scale as before, i.e., between four and ten times greater than the maximum value of the actin contribution (0.8). Moreover, the radial position of the *observed* intensity maximum on this layer line from relaxed muscle (0.026 Å^{-1}) corresponds to a region of the actin transform which is even lower still (approximately 0.01, i.e., $\frac{1}{80}$ of the *actin* maxima). Thus the reflection in question seems likely to arise essentially from the tropomyosin component alone, and accordingly the radius at which the tropomyosin must be situated can be computed quite readily and is approximately 44 Å. The contribution from the actin and tropomyosin component is illustrated in Fig. 7. Since the actin contribution is so small, interference in the region in question between its transform and that of the tropomyosin is also small, and so the intensity of the resultant reflection is not sensitive to the azimuthal position of the tropomyosin. A tropomyosin strand placed at 44 Å radius and in the center of the groove would give an almost equally strong reflection at the same position on the 3rd layer line. However in this position the tropomyosin would be lying freely suspended some 20 Å away from the surface of the actin monomers. This seems rather an unlikely arrangement and a much more plausible one would be a model in which the tropomyosin is located at this same radius and just makes contact with actin. The azimuthal position of the tropomyosin would then be 46½° (Fig. 8), and the possibilities of its interfering

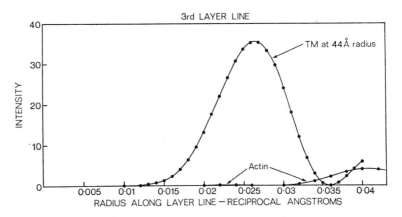

Figure 7. Calculated X-ray intensity distributions along 3rd layer line for models in which tropomyosin is represented by a solid cylinder 8.5 Å, wound into helix of same pitch as actin and radius 44 Å, actin by spherical subunits of radius 24 Å arranged on a helix of radius 24 Å, pitch 2 × 382.2 Å, subunit repeat 54.6 Å. Since actin contribution on 3rd layer line is so small, the intensity given by composite filament at all radii <0.035 Å^{-1} is essentially the same as that given by tropomyosin alone and is not sensitive to azimuthal angle, only to radius.

with myosin attachment in the active configuration are at once apparent, although it also overlaps the end of the myosin in the rigor configuration (see Fig. 6).

Let us now consider the effect of this positioning on the intensity of the reflections on the 2nd layer line. Here the situation is rather more complicated since the expected height of the tropomyosin transform is only about four times that of actin at similar radii, so appreciable interference effects will occur (Fig. 9). The maximum in the tropomyosin transform will occur at a radius of 0.018 Å^{-1} and will have a value of approximately 15, whereas that of actin occurs at a radius of 0.020 Å and has a value of approximately 4; thus the *amplitudes* of the two transforms are in the ratio of approximately 2:1 in the region of their maxima, which virtually coincide. The resultant amplitude will then depend on the

relative phases of the two contributions. They will be in phase when the tropomyosin is placed in the center of the groove (azimuthal position 90°) and exactly out of phase when the tropomyosin is at azimuth 45°. Thus there will be a minimum in the intensity on the 2nd layer line in relaxed muscle for a tropomyosin position lying quite close to that expected on the basis of its radial position, deduced from the 3rd layer line, and of the requirement that it should be in contact with actin. Thus the observed characteristics of both the 2nd and 3rd layer lines in relaxed muscle agree well with the idea of the tropomyosin strands being in contact with the actin monomers and being located at a position well placed for interfering with the attachment of myosin.

When the muscle is activated, the 3rd layer line decreases considerably in intensity, and a characteristic reflection not visible previously occurs on the 2nd layer line at a radial position of 0.022 Å^{-1}. As I have already mentioned, actin and tropomyosin make comparable contributions to the transform on this layer line, and the actin maximum is at a radius of about 0.02 Å. One could therefore get a strong reflection by putting tropomyosin in the center of the groove and at radius 44 Å; this would increase the intensity by a factor of 9 compared with the 45° azimuthal angle. However the 3rd layer line intensity would, as explained above, still remain high and the tropomyosin would not be in contact with actin. Better agreement can be obtained if the radial distance of the tropomyosin from the center of the helix is reduced. If the tropomyosin is kept on the midline of the groove (azimuth 90°) and brought into contact with actin (which would occur at a radial distance of about 24 Å), then the intensity on the 2nd layer line remains high since the actin and tropomyosin maxima still overlap and remain in the same phase. The maximum intensity would be 3–4 times that given by the relaxed configuration (46½°, 44 Å), and the position of the maxima would

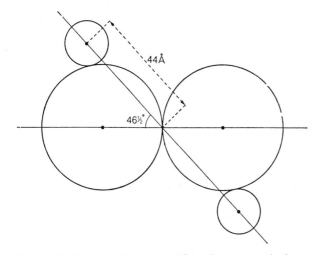

Figure 8. Diagram showing position of tropomyosin in relaxed state. The radius of 44 Å is derived from the measurements of the position of the X-ray intensity maximum on the 3rd layer line. The azimuthal angle of 46½° is derived from the assumption that tropomyosin and actin are in contact.

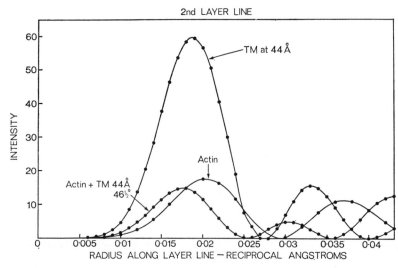

Figure 9. Calculated X-ray intensity distributions along 2nd layer line for models with tropomyosin at 44 Å radius, actin with standard dimensions, and composite filament with tropomyosin at $46\frac{1}{2}°$ azimuth. The tropomyosin and actin peaks occur at almost the same radius and almost exactly out of phase (45° would put them exactly out of phase) so that the resultant intensity is rather low, as in the case of relaxed muscle.

be at radius 0.023 Å^{-1}, very close to the observed value (the maxima lies closer to the actin maxima rather than the tropomyosin maxima (at 0.031 Å for this radius) because the actin transform goes negative now under the tropomyosin peak, whereas the tropomyosin transform is small and positive under the actin peak (Fig. 10). The intensity on the third layer line would be reduced to about two-thirds of its relaxed value.

However even this positioning is not entirely satisfactory, for three reasons. First, it does not correspond to the position of tropomyosin seen in the 3-D reconstruction from micrographs of filaments most probably in the activated configuration. These show tropomyosin at azimuth 60–70° and radius ~35 Å. Second, tropomyosin exerts an *activating* effect on actomyosin ATPase; i.e., it increases the activity over and above what it would be with actin alone, when the tropomyosin is present either without troponin, or with troponin in the

presence of calcium. This indicates that when tropomyosin is moved from the "blocking" position, it still remains in a position where it can influence actin-myosin interaction, which could plausibly be the case for the position visible in the electron micrographs. Third, the midgroove position leads to difficulties about the 3rd layer line. When tropomyosin is placed at 24 Å radius, the maximum of its transform on this layer line occurs at a radius of 0.043 Å^{-1}. The intensity of the tropomyosin reflection (10) here would be only about one-quarter its maximum value when the tropomyosin was located at radius 45 Å, but unfortunately it now almost coincides with the weak actin maximum, at radius 0.041 Å^{-1} and intensity ~4. As the two reflections are not 180° out of phase, reinforcement occurs and the resultant intensity is (as already mentioned) about two-thirds its value with tropomyosin at (44 Å, $46\frac{1}{2}°$). No doubt disorientation effects would weaken a peak at this rather large

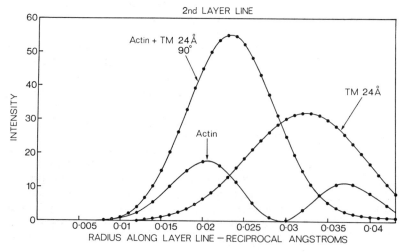

Figure 10. Calculated X-ray intensity distributions along 2nd layer line for models with tropomyosin at 24 Å radius, actin with standard dimensions, and composite filament with tropomyosin at 90° azimuth. The tropomyosin and actin peaks overlap and are exactly in phase, so that the resultant intensity is high and lies at an intermediate spacing.

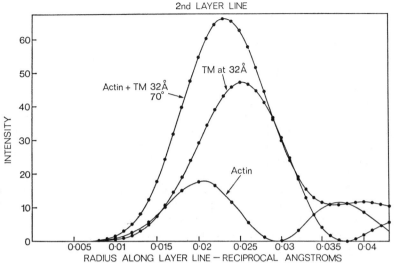

Figure 11. Calculated X-ray intensity distributions along 2nd layer line for models with tropomyosin at 32 Å, actin with standard dimensions, and composite filament with tropomyosin at 70° azimuth. (This position corresponds to that given by the 3-D reconstructions of the electron micrographs of actin plus relaxing system in presence of calcium.) The actin and tropomyosin peaks, although not exactly in phase, are sufficiently in phase in the region where they overlap to give a strong resultant peak, as is in fact observed on the 2nd layer line in activated muscle.

radial position, but the behavior of the 3rd layer line certainly provides no strong reason for wanting to put tropomyosin at this radius and on the center line.

If, on the other hand, in the active muscle tropomyosin is placed at azimuth 60–70° (as in the 3-D reconstructions) and radius 38–32 Å, the general agreement seems more satisfactory. On the 2nd layer line, the peak intensity lies between 2.7 and 4.4 times the value for the relaxed system (azimuth $46\frac{1}{2}°$ and radius 44 Å). On the 3rd layer line, the intensity lies between 0.7 and 0.4 of its relaxed value. For the 60° tropomyosin azimuth, the position of the maximum lies at radius ~ 0.0285 Å^{-1}, rather too close to the value of 0.026 for the relaxed system if radial distance is to be invoked to give a further decrease in visibility. At 70° azimuth, the peak, now only 0.4 of its maximum value, is at a radial position of 0.034 Å, which could make it very difficult to detect; and at 80° azimuth the intensity has increased again. Thus tropomyosin at azimuth 70° and radius 32 Å seems (Fig. 11), in this admittedly over-simplified model used for computation, to give reasonable agreement with the rather meager data presently available for active muscle, and at the same time corresponds quite well with its position as deduced from micrographs.

Another argument favoring the view that most of the change structure represents a movement of the tropomyosin strand, as a whole, to a different radial and azimuthal position may be based on the behavior of the weak off-meridional reflections on the 72 Å and 51 Å layer lines of the actin system. In contrast to the effects on the 2nd and 3rd layer lines, where a major change in relative intensity occurs, only relatively minor changes are seen on these higher layer lines when the muscle changes from the

relaxed state. Now if a substantial change in actin structure occurred (even if it did not affect the helical parameters), it would have to be a change which repeated with the actin subunit period, and one would therefore expect changes in the lower-order layer lines to be accompanied by ones in the higher-order ones too. That the latter are very much smaller indicates that the part of the structure that is moving has only a small component with an axial periodicity in the range of the actin repeat and is very consistent with a model involving a relatively smooth tropomyosin rod moving sideways as a whole.*

It may therefore be worthwhile to consider the steric blocking model for regulation as a serious possibility and to discuss some of its implications. One of the most attractive features of the mechanism is that it can account for the cooperative features of muscle activation in a very natural and straightforward way. Bremel and Weber (1972) interpreting earlier data (e.g., Weber, 1969; Weber et al., 1969) concluded that if a significant proportion of the actin monomers in an actin-tropomyosin-troponin filament are occupied by myosin or S_1 in the rigor linkage, the unoccupied actins are switched on even in the complete absence of calcium and are able to combine with myosin carrying ATP (or its split products) and activate the ATPase cycle. This could be simply explained if the myosin heads which had already attached were able to keep the tropomyosin strand pushed sideways even without calcium, so exposing the attachment site for myosin

---

* While this work was being written up, a communication was received from D. A. D. Parry and J. M. Squire, in which they analyze some of the data given here and of Vibert et al. by computations of model systems similar, though not identical, to the ones used here, and arrive independently at very similar conclusions.

on the neighboring actin monomers which were not yet occupied and allowing them to accept activated myosin.

Again, Bremel et al. (this volume) have found that the activating effect of tropomyosin alone is also a cooperative phenomenon, enhanced by the presence of attached myosin heads, and this could be explained along the same lines if attached myosin displaced tropomyosin to a position which generated a more attractive site on the neighboring actins.

How could the change in position or behavior of tropomyosin, required for such an activation mechanism, be effected via the troponin components? One might speculate that tropomyosin was held rather firmly in the blocking position in relaxed thin filaments and that it was free to move nearer the center of the groove in the activated case. Since tropomyosin binds to actin strongly even in the absence of troponin, there must be at least one actin attachment site per tropomyosin molecule, and it has been suggested (Spudich et al., 1972) that these occur at 385 Å intervals along a tropomyosin strand and so match up with each seventh actin monomer. (At a 44 Å radius this corresponds to a "contour length" repeat for tropomyosin of 413 Å.) The troponin binding sites on tropomyosin also repeat at a similar interval (Nonomura et al., 1968; Hitchcock and Huxley, unpublished), but are not necessarily in the same position as the actin binding sites. With actin and tropomyosin alone, the latter is not held in the blocking position, whereas troponin may supply additional bonding which holds the tropomyosin in the inhibitory position until calcium is supplied. This would suggest that calcium might influence the bonding of troponin (or parts of it) to actin or to tropomyosin, possibilities that clearly can be explored biochemically. Meanwhile, we are proceeding with our 3-D reconstruction work on the influence of troponin and its components on the actin-tropomyosin structure.

References

BREMEL, R. D. 1972. Protein-protein interactions and the regulation of actomyosin activity. Ph.D. thesis, St. Louis University.

BREMEL, R. D. and A. WEBER. 1972. Cooperative behavior within the functional unit of the actin filament in vertebrate striated muscle. *Nature* **238**: 97.

CASPER, D. L. D., C. COHEN, and W. LONGLEY. 1969. Tropomyosin: Crystal structure polymorphism and molecular interactions. *J. Mol. Biol.* **41**: 87.

COHEN, C. and A. G. SZENT-GYÖRGYI. 1957. Optical rotation and helical polypeptide chain configuration in α-proteins. *J. Amer. Chem. Soc.* **79**: 248.

ELLIOTT, G. F., J. LOWY, and B. M. MILLMAN. 1965. X-ray diffraction from living striated muscle. *Nature* **206**: 1357.

———. 1967. Low-angle X-ray diffraction studies of living muscle during contraction. *J. Mol. Biol.* **25**: 31.

GORDON, A. M., A. F. HUXLEY, and F. J. JULIAN. 1966a. Tension development in highly stretched vertebrate muscle fibers. *J. Physiol.* **184**: 143.

———. 1966b. The variation in isometric tension with sarcomere length in vertebrate muscle fibers. *J. Physiol.* **184**: 170.

HANSON, J. and J. LOWY. 1963. The structure of F-actin and of actin filaments isolated from muscle. *J. Mol. Biol.* **6**: 46.

HASELGROVE, J. C. 1970. X-ray diffraction studies on muscle. Ph.D. thesis, University of Cambridge.

HUXLEY, A. F. and L. D. PEACHEY. 1961. The maximum length of contraction for vertebrate striated muscle. *J. Physiol.* **156**: 150.

HUXLEY, H. E. 1968. Structural differences between resting and rigor muscle; evidence from intensity changes in the low-angle equatorial X-ray diagram. *J. Mol. Biol.* **37**: 507.

———. 1969. The mechanism of muscle contraction. *Science* **164**: 1356.

———. 1970. Structural changes in muscle and muscle proteins during contraction. *8th Int. Cong. Biochem*, Interlaken.

———. 1971a. Cross-bridge movement and filament overlap. *Amer. Biophys. Soc.* (Abstr.) p. 235a.

———. 1971b. Structural changes during muscle contraction. *Biochem. J.* **125**: 85.

HUXLEY, H. E. and W. BROWN. 1967. The low-angle X-ray diagram of vertebrate striated muscle and its behavior during contraction and rigor. *J. Mol. Biol.* **30**: 383.

HUXLEY, H. E., W. BROWN, and K. C. HOLMES. 1965. Constancy of axial spacing in frog sartorius muscle during contraction. *Nature* **206**: 1358.

MOORE, P. B., H. E. HUXLEY, and D. J. DEROSIER. 1970. Three-dimensional reconstruction of F-actin, thin filaments and decorated thin filaments. *J. Mol. Biol.* **50**: 279.

NONOMURA, Y., W. DRABIKOWSKI, and S. EBASHI. 1968. The localization of troponin in tropomyosin paracrystals. *J. Biochem.* (Tokyo) **64**: 419.

O'BRIEN, E. J., P. M. BENNETT, and J. HANSON. 1971. Optical diffraction studies of myofibrillar structure. *Phil. Trans. Roy. Soc. London* B **261**: 201.

SPUDICH, J. A. and S. WATT. 1971. The regulation of rabbit skeletal muscle. I. Biochemical studies of the interaction of the tropomyosin-troponin complex with actin and proteolytic fragments of myosin. *J. Biol. Chem.* **247**: 4866.

SPUDICH, J. A., H. E. HUXLEY, and J. T. FINCH. 1972. The regulation of skeletal muscle contraction. II. Structural studies of the interaction of the tropomyosin-troponin complex with actin. *J. Mol. Biol.* In press.

TANAKA, H. and F. OOSAWA. 1971. The effect of temperature on the interaction between F-actin and tropomyosin. *Biochim. Biophys. Acta* **253**: 274.

VIBERT, P. J., J. C. HASELGROVE, J. LOWY, and F. R. POULSEN. 1972. Structural changes in actin-containing filaments in muscle. *Nature New Biol.* **236**: 182.

VIBERT, P. J., J. LOWY, J. C. HASELGROVE, and F. R. POULSEN. 1971. Changes in the low-angle X-ray diffraction patterns of smooth muscles during contraction. *1st Europe. Biophys. Cong.*, Vienna.

WEBER, A. 1969. Parallel response of myofibrillar contraction and relaxation to four different nucleoside triphosphates. *J. Gen. Physiol.* **53**: 781.

WEBER, A. and R. D. BREMEL. 1971. Regulation of contraction and relaxation in the myofibril. In *Contractility of muscle cells and related processes*, ed. R. J. Podolsky. Prentice-Hall, Englewood Cliffs, N.J.

WEBER, A., R. HERZ, and I. REISS. 1969. The role of magnesium in the relaxation of muscle. *Biochemistry* **8**: 2266.

Changes in the Polarization of Tryptophan Fluorescence in the Actomyosin System of Working Muscle Fibers

G. J. Steiger and J. C. Rüegg

Institut für Zellphysiologie, Ruhr-Universität Bochum, D-4630 Bochum, Postfach 2148, Germany

K. M. Boldt and D. W. Lübbers

Max-Planck-Institut für Arbeitsphysiologie, D-4600 Dortmund, Rheinlanddamm, Germany

W. Breull

Physiologisches Institut der Universität Bonn, D-5300 Bonn, Nussallee, Germany

Electron microscope investigations and X-ray diffraction studies of muscles have established that there is a different conformation in the myosin molecule (cross-bridge position) in rigor than in relaxation (Huxley, 1968; Reedy et al., 1965).

As Aronson and Morales (1969) have shown first on skeletal muscles, the polarization of tryptophan fluorescence can also be an appropriate indicator for changes in conformation or orientation of the myosin crosslinks. Under static conditions the polarization is higher in relaxed muscle than in rigor muscle, and there is an intermediate polarization when the preparation is contracted isometrically. These results have been verified qualitatively on functional isolated actomyosin systems of insect flight muscles.

Insect flight muscles, incubated in contraction solution with MgATP and Ca ions, react with an increased contractile activity when they are stretched or, conversely, with a decreased contractile activity when they are released (Jewell and Rüegg, 1966). There are some indications that it

may be possible, at least partially, to synchronize the cross-bridge movement by means of a quick, rectangular length step (Schädler et al., 1971). Figure 1 shows the delayed tension rise after a rapid length change, followed by several rhythmical fluctuations in tension although the length remains constant. According to all appearances these isometric oscillations in tension are caused by a synchronized stroke of a number of cross-bridges. This seems not to be a unique feature of insect flight muscles because we obtain similar, but 10 times slower, oscillations in extracted rabbit heart preparations.

Furthermore insect flight muscles are able to do work on the apparatus during forced sinusoidal length changes (i.e., $2\% \ L_0$, 4 Hz). Therefore a method was developed for dynamic registration of quick changes in the polarization of tryptophan fluorescence during sinusoidal or rectangular length changes.

The exciting light (295 nm) was directed through a bichromatic, semitransparent mirror onto the

RABBIT HEART MUSCLE   **INSECT FLIGHT MUSCLE**

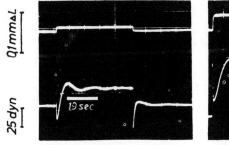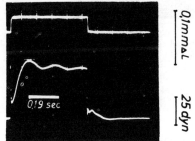

Figure 1. Isometrical oscillations in tension. *Left:* papillary heart muscle preparation of rabbit. *Right:* DLM preparation of waterbug (*Lethocerus maximus*). The upper traces show the rectangular change in length. The curve below the tension course: a quick stretch causes, besides an initial peak, a delayed tension rise followed by several rhythmical fluctuations in tension, although the length remained constant at this time. The preparations were extracted with ether and freeze-dried. The incubation solution contained 17.5 mM MgATP, 10^{-6} to 10^{-5} M Ca ions, 4 mM imidazol buffer, 10 mM Na azide at pH 6.5. The experimental temperature was 20°C. Note the different time scale: for the rabbit heart it is 10 times slower than for the insect flight muscle.

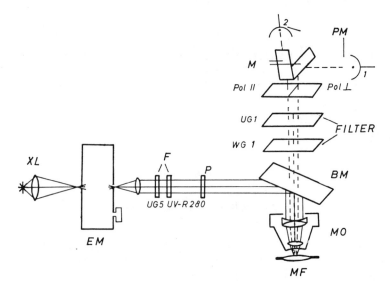

Figure 2. Schematic diagram of the experimental arrangement. Light from a xenon (or mercury) lamp (*XL*) passes through a monochromator (*EM*) (295 nm) and two filters (UG 5 and UV-R 280), then through a polarizer (*P*). It is directed upon the muscle (*MF*) preparation by a bichromatic, semitransparent mirror (*BM*). A part of the fluorescence light is sampled through a microscope objective (*MO*), separated into two polarization components (*Pol* ‖ and *Pol* ⊥) parallel and perpendicular to the fiber axis, and measured in two photomultipliers (*PM*). The apparatus used for stretching and releasing muscle fibers (not shown) was similar to that described earlier (Steiger, 1971).

muscle. The electrical vector of the exciting light was perpendicular to the fiber axis. The muscle fiber was attached between two glass rods, one of them connected to an RCA transducer valve, the other one to an electromechanical vibrator. A part of the fluorescence light was sampled through a microscope objective, separated into two polarization components, parallel and perpendicular to the fiber axis, and measured in two photomultipliers. In this way we obtained the necessary time resolution of about 1 msec. The complete experimental assembly is diagrammed in Fig. 2.

The polarization of the tryptophan fluorescence changes in phase with the delayed tension development (Fig. 3): An increase of the parallel component and a decrease of the perpendicular component means a decrease in the polarization. There are similar results during sinusoidal length changes in contraction solution.

These findings speak in favor of a cyclic conformational change of the myosin molecules during the chemo-mechanical energy transformation.

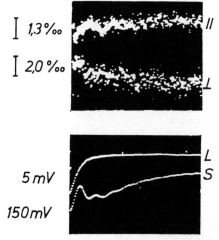

Figure 3. Changes in the intensity of the parallel and perpendicular components of the emitted tryptophan fluorescence (*upper*). The increase of the parallel component and the decrease of the perpendicular component occur in phase with the tension change, and not in phase with the length change. The lower part of the figure shows the length (*L*) and the tension recordings (*S*). Preparation in standard ATP salt solution, pH 6.5, 10^{-5} M Ca ions. The polarization of the exciting light (295 nm) was perpendicular to the fiber axis. The duration of the recorded period is 80 msec. The length signal of 5 mV corresponds to a length change of about 1% L_0. The tension signal of 150 mV corresponds to a tension response of about 15 dynes/fiber.

References

Aronson, J. F. and M. F. Morales. 1969. Polarization of tryptophan fluorescence in muscle. *Biochemistry* **8:** 4517.

Huxley, H. E. 1968. Structural differences between resting and rigor muscle; evidence from intensity changes in the low-angle equatorial X-ray diagram. *J. Mol. Biol.* **37:** 507.

Jewell, B. R. and J. C. Rüegg. 1966. Oscillatory contraction of insect fibrillar flight muscle after glycerol extraction. *Proc. Roy. Soc.* (London) B **164:** 428.

Reedy, M. K., K. C. Holmes, and R. T. Tregear. 1965. Induced changes in orientation of the crossbridges of glycerinated insect flight muscle. *Nature* **207:** 1276.

Schädler, M., G. J. Steiger, and J. C. Rüegg. 1971. Mechanical activation and isometric oscillation in insect fibrillar muscle. *Pflügers Arch.* **330:** 217.

Steiger, G. J. 1971. Stretch activation and myogenic oscillation of isolated contractile structures of heart muscle. *Pflügers Arch.* **330:** 347.

The Structure and Function of Insect Muscle

P. ARMITAGE, A. MILLER, C. D. RODGER, AND R. T. TREGEAR

A.R.C. Unit of Insect Physiology
and
Laboratory of Molecular Biophysics, Zoology Department, Oxford University

Insect muscle offers two advantages when used in structural studies on the mechanism of muscle contraction. First, the degree of order in the filament array of water bug fibrillar flight muscle is better than that of any other muscle yet examined; this means that the X-ray diffraction method can be used here to best advantage. Second, the range of physiological properties peculiar to insect muscle permits a comparison of structures with dissimilar functions and also allows experiments with time resolution on the intensities of parts of the X-ray diffraction pattern which give information about the molecular arrangement in muscle during the course of active contraction. In this paper we summarize our recent results in three projects: the molecular arrangement in water bug flight muscle, the relationship between molecular arrangement and the asynchronous property of some insect flight muscles, and the changes in molecular arrangement during oscillatory contraction.

Molecular Arrangement in Resting Water Bug Flight Muscle

The low-angle X-ray diffraction pattern from *Lethocerus* flight muscle contains a number of well-resolved reflections (Fig. 1). We have calculated the X-ray diffraction patterns predicted for various models of the molecular arrangement in this muscle and compared these with the observed patterns;

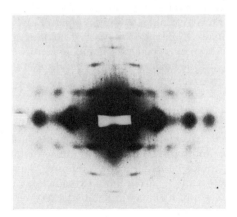

Figure 1. X-ray diffraction pattern from glycerinated water bug flight muscle in rigor.

results will be reported in detail in a forthcoming publication (Miller and Tregear, 1972). Here we will summarize our main findings.

The flight muscles of three species (*maximus, cordofanus,* and *annulipes*) of the giant water bug *Lethocerus* have been studied in the presence (relaxed muscle) and in the absence (rigor muscle) of ATP. Most of the experiments were carried out on glycerinated muscles, but we have shown (Miller and Tregear, 1971) that these give similar X-ray diffraction patterns to intact muscles.

Layer lines of spacing 38.5, 5.9, and 5.1 nm were observed in the X-ray diffraction pattern of relaxed muscle and attributed to the actin helix. Because of the meridional breadth of these reflections it is not possible to estimate the helical parameters of actin to better than 10%. In rigor muscle, however, the actin helix which is now marked by myosin heads generates a number of additional layer lines which indicate that the actin helix is of pitch 5.92 nm, with an axial subunit separation of 2.75 nm. Such a structure repeats after 28 residues or 13 turns of the helix; it can be alternatively described as a two-start helix of pitch 77.0 nm, subunit repeat 5.4 nm, with a 2.7 nm axial displacement between strands. Since we could detect no change in the spacing of the 5.9 nm reflection between relaxed and rigor muscle, it is likely that the actin helix is similar in relaxed muscle. The intensity distribution on the layer lines in relaxed muscle was fitted with an actin helix of radius 24 nm with 24 nm diameter spherical subunits.

In relaxed muscle the myosin heads are arranged helically around the surface of the thick filaments. Although we could not completely exclude multiple helices of the type discussed by Squire (1971), our observations favored a two-start helix of pitch 77.0 and axial spacing of 14.5 nm. The myosin heads are spread azimuthally to account for the low intensity of the 38.5 nm layer line (Fig. 2).

In rigor muscle the myosin heads become helically ordered around the actin filaments. The layer line intensities could be fitted when two-thirds of the actin monomers are marked with myosin heads of three times their mass. The myosin head is 11.2 nm long, 5.6 nm wide, and angled axially by 35° (to the horizontal) and azimuthally by −35°

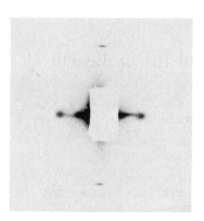

Figure 2. X-ray diffraction pattern from glycerinated water bug flight muscle relaxed.

(taking the direction of the 5.9 nm pitch helix as negative or the 77.0 nm pitch as positive). It was not necessary to change the dimensions of the myosin heads between the models for relaxed and rigor muscles. The model has the appropriate actin:myosin mass ratio and is consistent with Reedy's (1968) electron micrographs.

Reedy (1968) further deduced from electron micrographs that thin filaments were arranged around a thick filament on the locus of a helix of pitch 77.0 and with axial separation 12.7 nm; features of our X-ray diffraction patterns also required an arrangement like this, but it raised the difficulty that it is not possible to arrange such helices of actin filaments on a perfect lattice with a hexagonal unit cell of side 56.0 nm (the distance between neighboring thick filaments). However the helices of actin filaments can be accommodated on hexagonal lattices in which certain actin filaments occupy statistical positions, and the X-ray diffraction patterns suggested that a statistical structure may exist in water bug flight muscle.

The diffraction pattern of a continuous helical line of pitch P consists of layer lines spaced at integral values of $1/P$ about the origin. The distance from the meridian of the intensity maximum on the nth layer line increases with increasing n so that the pattern resembles a cross centered on the origin. A two-start helix of pitch P yields a diffraction pattern in which the odd-layer lines (odd n) are systematically absent. The diffraction pattern of a discontinuous helix consisting of points equally spaced with axial separation p along the locus of a helix of pitch P, has, in addition to the cross centered on the origin, identical crosses centered on points on the meridian axially separated by m/p, where m takes the values $\pm 1, \pm 2, \pm 3, \ldots$. The origin cross is obtained with $m = 0$.

The actin helix may be regarded as a two-start helix of pitch 77.0 nm (Miller and Tregear, 1972) with subunits axially spaced by 55.0 nm along one strand; the two strands are staggered axially by 2.75 nm and rotated azimuthally by just under 180° with respect to each other. The fact that, in rigor muscle, the layer-lines with $m = 0$ are sampled by the filament reciprocal lattice, whereas those with $m \neq 0$ are not, indicates some type of screw disorder; viewed as continuous helices the actin filaments fit on the lattice, but viewed as discontinuous helices they do not since screw disorder allows the subunits to move along the locus of the continuous helix. The nature of possible screw disorders requires consideration of the lattice and filament symmetries.

Electron micrographs of transverse sections of water bug flight muscle show that the thick filaments are arranged on a hexagonal lattice with the actin filaments midway between each pair of neighboring thick filaments. The positions of the equatorial reflections of the X-ray diffraction pattern also indicate that this unit cell has hexagonal symmetry, and it is evident that the actin filaments lie on the dyad (twofold axis) positions of a P6 unit cell.

If the actin filament is regarded as a continuous two-start helix, it has a dyad axis and thus can be accommodated in a P6 lattice. The helix cross with $m = 0$ in the diffraction pattern is represented by the 38.5 and the 19.3 nm layer lines, and these are sampled by the same reciprocal hexagonal lattice as the equator in rigor muscle. The outer actin layer lines are on helix crosses with $m \neq 0$ and are not observably sampled, though some well-resolved patterns show intensity fluctuations along the layer lines. This lack of sampling can be correlated with the fact that the actin filament, considered as a discontinuous helix, does not have a dyad axis and so does not conform to the symmetry of the P6 lattice.

When the discontinuous actin filaments are arranged on a helix around a thick filament, it is clear that this structure cannot have the symmetry of the P6 lattice indicated by the equator and $m = 0$ layer lines. If extended statistical lattices are considered however, the actin filaments can be accommodated in a hexagonal unit cell. By a statistical lattice we mean one in which the unit cells have identical size but the contents may be any of a specified number of types. In a perfect lattice there is only one type of unit cell contents and all the unit cells are identical. If the filaments had complete azimuthal disorder, the number of unit cell types could be regarded as infinite. The actin filaments can be fitted into extended lattices in which certain of the actin filaments may be in one

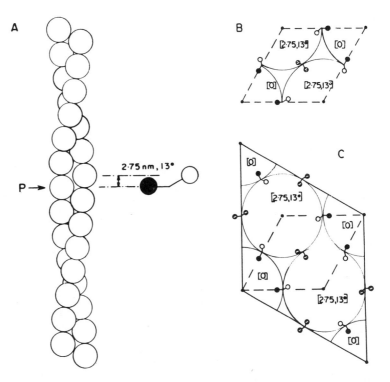

Figure 3. *A*, Diagram of the actin helix. The two-stranded rope is right-handed, of pitch 77.0 nm and with a 5.5 nm axial separation of subunits. The two strands are related by an azimuthal rotation of just under 180° and an axial translation of 2.7 nm. The genetic helix is left-handed with pitch 5.9 nm and axial subunit spacing 2.7 nm. The helix may be represented in section at level *P* by the shaded and open circles shown. *B* and *C*, Possible hexagonal unit cell containing statistically positioned actin molecules. The hatched circles mark the locations populated equally by actin molecules in two azimuthal orientations related by 180°. Different types of statistical unit cells are feasible; the simplest is clearly a cell with the size of that shown in *B* and all the actin molecule positions statistical.

of two azimuthal orientations separated by 180° (Fig. 3).

The effect of statistical unit cells on the X-ray diffraction pattern is to reduce the intensity of sharp Bragg reflections and produce layer line streaks; this is consistent with the diffuse appearance of the layer lines with $m \neq 0$. The layer lines, however, are also closer to the meridian than is predicted by models of actin filaments marked with myosin heads (Miller and Tregear, 1972). A displacement of intensity maxima towards the meridian would be brought about if the actin filaments were arranged on the postulated helix around the thick filaments. This displacement would always be operative on the $m = 0$ layer lines and it explains the otherwise puzzling near-meridional intensity maximum on the 19.3 nm layer line; but it would only be operative on the $m \neq 0$ layer lines if such large helices can extend throughout the myofibril when the actin filament is considered as discontinuous. The statistical unit cells show that this is possible and an intensity maximum displacement therefore could occur on the $m \neq 0$ layer lines.

Comparative Study of Relation between Structure and Function

Water bug flight muscle differs from vertebrate striated muscle, such as frog sartorius, in that each single muscle twitch is not synchronous with an incoming electrical impulse from the nerve. In the so-called asynchronous flight muscles of certain insects, the electrical impulse results in a rise in the interfilament calcium ion concentration (to about 10^{-5} M), but this is not accompanied by a large rise in tension or ATPase. We shall term this state "static calcium activation." If the muscle is stretched by a few percent of its length when the calcium concentration is about 10^{-5} M, an increase in tension and ATPase does occur after a delay (Rüegg and Tregear, 1966b). During flight these muscles are stretch-activated by the elastic thorax of the insect and resonance between the muscle and thorax results in oscillatory tension and the production of work.

The pitch of the actin helix in the water bug flight muscle is similar to that of frog striated muscle (77.0 nm), but there are also a number of structural differences. The positions and numbers of the thin filaments in the hexagonal lattice of thick filaments differ and the pitches of the helices of cross-bridges differ (77.0 and 85.8 nm in water bug flight and frog sartorius, respectively). We may inquire whether these structural differences are related to differences in muscle function and, in particular, ask if the close numerical relationship of pitch length between the actin and cross-bridge helices is related to the asynchronous property of water bug flight muscle.

In view of the lack of a suitable theory for muscle contraction, the hypothesis that the asynchronous property is related to numerical similarity of actin and cross-bridge helix pitches was tested by comparing the structures of a number of insect muscles. We used the flight muscles of water bug (*Lethocerus*) and elephant dung beetle (*Helicopris* and *Panchnoda*), which are asynchronous, and the leg and siphon retractor muscles of *Lethocerus* and the flight muscles of locust (*Schistocerca gregaria*) and "giant locust" (*Eutropidacris*), which are synchronous. This project is still incomplete but we can list our results at present.

Filament lattice. In water bug fibrillar flight muscle the thin filaments occupy the dyad positions of the P6 lattice, and the ratio of the number of thin filaments to the number of thick filaments is 3:1. The special positions of the thin filaments lead to a very low intensity of the $(1,1,\bar{2},0)$ reflection. In the flight muscles of elephant dung beetle and locust (*Schistocerca gregaria*) the ratio of filament numbers is also 3:1, and we have observed that the $(1,1,\bar{2},0)$ reflection is also weak in these muscles. In water bug leg and siphon retractor muscles the filament number ratio of 6:1, and the X-ray diffraction pattern of rigor retractor muscle shows a strong $(1,1,\bar{2},0)$ reflection as expected. In the giant locust flight muscle the filament number ratio is difficult to determine from electron micrographs but appears to have an average value of about 4.5:1. In relaxed giant locust muscle the $(1,1,\bar{2},0)$ reflection is present but it is replaced by the $(2,0,\bar{2},0)$ when the muscle passes into rigor. This suggests that in rigor muscle the myosin heads attach mainly to thin filaments on the dyad positions of the P6 cell. Also it is clear from electron micrographs and the X-ray results that the 3:1 filament lattice is not peculiar to asynchronous muscles.

Axial order of myosin heads in relaxed and rigor muscle. In relaxed water bug flight muscle the most intense off-equatorial reflection in the X-ray diffraction pattern is the 14.5 nm meridional reflection. It is due to the axial order of myosin heads on the surface of the thick filament. When water bug flight muscle passes into rigor, the spacing of the 14.5 nm reflection remains unchanged; but its intensity is greatly diminished (by about 80%), indicating axial movement of the myosin heads. In frog sartorius muscle the corresponding reflection has spacing 14.3 nm in relaxed muscle; this changes to 14.5 nm in rigor muscle and the intensity is decreased by about 40%.

We have found that the 14.5 nm reflection is also intense in relaxed flight muscles of elephant dung beetle, locust, and giant locust. The intensity is diminished when the muscles pass into rigor, so in these muscles also there is an axial movement of the myosin heads during the relaxed-rigor transition.

When water bug flight muscle undergoes the relaxed-rigor transition, the spacing of the 14.5 nm meridional reflection remains unchanged to within 0.3%. The pitch of the helix of myosin heads on the thick filaments in relaxed muscle is the same as the pitch of the actin helix. In contrast, when frog sartorius muscle undergoes the relaxed-rigor transition, the 14.3 nm meridional reflection changes spacing to 14.5 nm; together with the disappearance of the 42.9 nm layer-lines of relaxed muscle, this suggests that a molecular rearrangement takes place in the thick filaments.

We have measured the spacings of the 14.5 nm meridional spacings in the synchronous giant locust flight muscle and find that in relaxed and rigor muscles the spacings are identical to within 0.5%. We have not yet estimated the absolute spacing of this reflection, but the lack of shift in spacing during the relaxed-rigor transition suggests that, as in the asynchronous water bug flight muscle, no molecular rearrangement occurs in the thick filaments.

Helix pitches. In water bug flight muscle the pitch of the actin helix derived from the layer line spacings in rigor muscle is 77.0 nm. In relaxed water bug flight muscle a layer line with spacing 38.5 and intensity profile different from that in rigor muscle suggests that the pitch of the helix of myosin heads on a thick filament is 77.0 nm. This is confirmed by the fact that the 23.5 nm layer line is the $n = -1$ member of the $m = 1$ helix cross centered on the 14.5 nm meridional reflection. In our X-ray diffraction patterns of rigor muscle this 23.5 nm layer line is not visible (Fig. 1). In elephant dung beetle asynchronous flight muscle in rigor, the X-ray diffraction pattern (Fig. 4) is closely

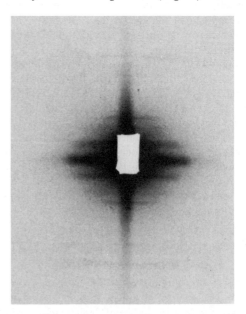

Figure 4. X-ray diffraction pattern from glycerinated elephant dung beetle flight muscle in rigor.

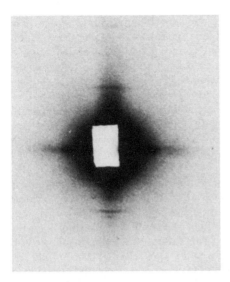

Figure 5. X-ray diffraction pattern from glycerinated elephant dung beetle flight muscle relaxed.

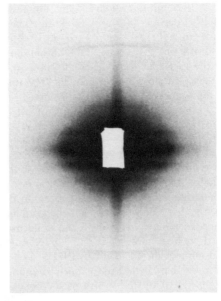

Figure 6. X-ray diffraction pattern from giant locust flight muscle in rigor.

similar to that from water bug flight muscle in rigor. The most evident feature is the sharp layer line with spacing 38.5 nm. In addition there are two layer lines of spacing 19.3 and 23.5 nm. Weaker layer lines of spacing 5.1, 5.9, 7.0, 8.6, and 11.0 nm are also evident.

In X-ray diffraction patterns from relaxed muscle (Fig. 5) the 14.5 nm reflection is more intense than in those from rigor, but the intensity of the 38.5 nm layer line is greatly diminished and it is difficult to observe. The 5.1 and 5.9 layer lines are visible.

The weakness of the layer line due to the pitch of the helix of myosin heads on the thick filament in relaxed muscle is similar to that in water bug flight muscle but in contrast to the relatively intense 42.9 nm layer line of frog sartorius muscle (Huxley and Brown, 1967). In all the insect muscles we have examined this layer line is very weak in relaxed muscle.

In Figure 5 a weak layer line of spacing 38.5 nm is visible. Since this is the most intense layer line from rigor muscle, we must rule out the possibility that it originates from a small fraction of the muscle being in rigor if we are to be able to use it to estimate the pitch of the helix of myosin heads in relaxed muscle. We have not yet been able to do this for elephant dung beetle flight muscle.

The X-ray diffraction pattern from giant locust synchronous flight muscle in rigor (Fig. 6) is also similar to that from water bug asynchronous flight muscle in rigor (Fig. 1). There are differences. In the patterns from giant locust the 19.2 nm layer line has a strong meridional component, and there is a 23.5 nm layer line; we have not observed either of these in patterns from water bug flight

muscle in rigor. (Diffraction patterns from the flight muscle of ordinary locust (*Schistocerca gregaria*) and the siphon retractor muscle of water bug also showed a meridional component on the 19.3 nm layer line.) Layer lines of spacing 5.1 and 5.9 nm streak across the meridian, and there are also layer lines of spacing 7.0, 8.5, and 11.0 nm.

Figure 7 shows the X-ray diffraction pattern from the relaxed flight muscles of giant locust. This

Figure 7. X-ray diffraction pattern from giant locust flight muscle relaxed.

shows a 14.5 nm meridional reflection and layer lines with spacing 38.5 and 23.5 nm. There is a weak meridional reflection of spacing 19.3 nm. The 5.1 and 5.9 nm reflections are visible and split across the meridian, but the other layer lines seen from rigor muscle are absent. The 23.5 nm layer line is more intense that that at 19.2 nm in relaxed muscle, whereas in patterns from rigor muscle the 19.3 nm layer line is the more intense. It follows that in relaxed flight muscle of giant locust the pitch of the helix of myosin heads on the thick filaments is 77.0 nm and equal to the pitch of the actin helix. Since these muscles are synchronous, it is clear that muscles with the helices of identical pitch do not necessarily show asynchronous behavior. In evolutionary terms the structures with identical helices preceded the development of asynchronous behavior (Cullen, 1971).

We note that it has now been demonstrated (Rüegg et al., 1970) that rabbit psoas muscle under certain conditions can also display oscillatory behavior. It would be interesting to know the thick filament structure under these conditions.

Changes in Molecular Arrangement during Active Muscle Contraction

As a result of the model computations referred to in the first section of this paper, the structural significance of most of the low-angle X-ray diffraction pattern of water bug flight muscle is known. Changes in molecular arrangement can therefore be investigated by studying changes in appropriate regions of the X-ray diffraction pattern during active muscle contraction.

Asynchronous flight muscle of insects has the useful property of developing periodically varying tension so that the X-ray diffraction pattern can be examined at various phases of the oscillatory cycle and the intensities of X-ray reflections plotted against muscle tension. In addition the diffraction pattern can be studied in the intermediate state of static calcium activation.

The main factor limiting the practicality of such experiments is the low intensity of the X-ray reflections from muscle and the relatively short life of the active muscle; the available number of scattered quanta of X rays is small. In principle, photographic films record each quantum of X rays which fall on them (until the grains coalesce), but they suffer from a constant background of "chemical fog" due to film processing. Since this background does not increase with exposure time of the film, films are ideal when sufficient quanta are available to overcome the background. Proportional X-ray counters, on the other hand, have a background which is proportional to exposure time.

Thus in our experimental conditions we have used proportional counters in preference to films. Proportional counters have the added advantage of being more suitable for recording intensities which are continuously varying. Counts can be fed into a multi-channel scaler, scaling in synchrony with the oscillating muscle so that over a number of cycles of oscillation the counts recorded in selected segments of the oscillatory cycle can be aggregated. A computer program written by Dr. Roger Abbott allowed the counter to be connected on-line to a PDP8-1 computer; aggregates of counts in each eighth of the cycle were displayed on a cathode ray oscilloscope and could be inspected or photographed at any time. We confirmed by counts on the main X-ray beam that the counts followed Poisson statistics, and this allowed us to add separate counts taken on the same spot and take the standard deviation of the sum as Poisson.

Preliminary reports of studies on two parts of the X-ray diffraction pattern have been published. These are on the 14.5 nm meridional reflection which comes from the axial order of myosin heads (Tregear and Miller, 1969) and the $(1,0,\bar{1},0)$ and $(2,0,\bar{2},0)$ equatorial reflections which give information about the distribution of mass between the filaments and hence the radial position of the myosin heads (Miller and Tregear, 1970). During active oscillation of the muscle the intensity of the 14.5 nm reflection falls in phase with the rise in tension (Fig. 8). This implies a decrease in the axial order of the myosin heads, and it may be estimated from the amplitude of the intensity variation that a minimum of 16% of the myosin heads are displaced axially; but it is not possible from these data alone to specify the nature of the axial displacement. Changes in the intensities of the two equatorial reflections indicate a lateral redistribution of mass equivalent to between 13 and 20% of the myosin heads being transferred from the surface of the thick filaments to the surface of the thin filaments when the muscle is statically activated by calcium. No further mass transfer is observed when the muscle performs oscillatory work.

However these studies cannot give decisive information about the important question of whether or not during muscle contraction the myosin heads attach to the actin filaments in the sense of marking the actin helix. The intensity changes described above are consistent with attachment, but unequivocal evidence can only be obtained by looking for changes in the regions of the X-ray diffraction pattern which would be uniquely affected by attachment to actin. We have studied the two most intense of such regions, namely, the 38.5 and 19.4 nm layer lines on the $(1,0,\bar{1},l)$ row line.

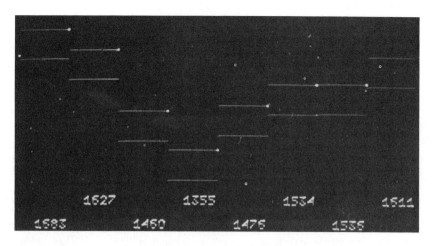

Figure 8. Photograph of cathode ray oscilloscope display of results on the proportional counter experiment on the intensity of the 14.5 nm meridional X-ray reflection during active muscle oscillation. The ordinate represents the intensity of the reflection; the distance between the horizontal lines corresponds to 2 × standard deviation. The whole abscissa corresponds to one cycle of oscillation and this is divided into eight "bins" where the counts are accumulated. The numbers along the abscissa are the total counts in each bin. This method of output was devised for us by Dr. Roger Abbott.

In the experiments on the 38.5 nm and 19.4 nm layer lines a lead mask with two holes was used to take in two quadrants of the reflection simultaneously; for the 19.4 nm layer line the reflections in all four quadrants were counted simultaneously by using a counter with a larger window. In each single experiment on the oscillating muscle the counts from these reflections in each eighth of the cycle were recorded continuously for up to 12 hours while the muscle developed work. Measurements were also made of the background count so that the observed counts could be corrected and control experiments were done by oscillating the muscle in relaxing solution for some hours and again accumulating the counts from the reflections in each eighth of the cycle.

The result of the 38.5 nm reflection is shown in Figure 9. The intensity of the reflection varies significantly in anti-phase with the tension of the muscle such that at maximum tension the intensity falls to 81% of its value in relaxed muscle. Now in relaxed muscle the intensity of this reflection is 0.17 counts per second (cps) compared with 0.6 cps in rigor. It is likely that the decrease in intensity is due to myosin heads leaving the thick filament helix; and if we assume that these myosin heads then mark the actin helix, the 19% drop in intensity of the 38.5 nm layer line indicates such an attachment of either 16% or 30% of the myosin heads. In view of the fact that the intensity changes in the meridional and equatorial reflections indicate that no more than 13–20% of the myosin heads move out of the actin filament positions, the figure of 16% seems more likely.

The intensity of the 19.2 nm reflection is zero in relaxed muscle and 1.9 cps in rigor. Because it is so weak it is very difficult to detect any change in intensity. Our results are shown in Figure 10. There is a small but just significant increase in the intensity when the muscle is bearing full tension, and the amplitude of the change indicates that at that point 15–30% of the myosin heads are marking the actin helix.

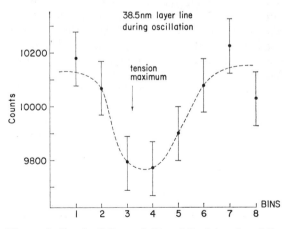

Figure 9. Graph of the variation of the intensity of the 38.5 nm layer line reflection during the cycle of active muscle oscillation. The errors in intensity are represented by vertical bars through the observed points.

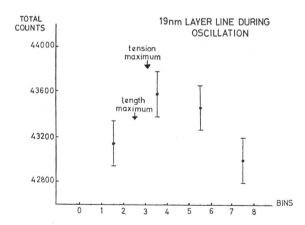

Figure 10. Graph of the variation of the intensity of the 19.3 nm layer line reflection during the cycle of active muscle oscillation.

These observations show unequivocally that the myosin heads do attach to the actin helix during active muscle contraction. All our observations on the 14.5 nm meridional, the $(1,0,\bar{1},0)$ and $(2,0,\bar{2},0)$ equatorial, and the 38.5 and 19.3 nm layer line reflections show intensity changes do not occur when muscle is oscillated in relaxing solution. The amplitudes of changes of each of these reflections in actively oscillating muscle are consistent with the same fraction of myosin heads (10–20% if we assume attachment in activity the same as that in rigor) being transferred from the thick filament helix to the actin helix when the muscle is developing maximum active tension. Since each of these different reflections indicates the same fraction of myosin heads attached, it is likely, in view of the different effects that different types of attachment would have on the intensities of these reflections, that the mode of attachment is similar to that in rigor.

If we suppose that the drop in the intensity of the 14.5 nm meridional reflection is due to a population of axially moving myosin heads with a distribution of axial displacements varying from zero to, say, 10.0 nm, then we should have to conclude that a total of more than 16% of the myosin heads were moving but less than 16% were angled at the rigor orientation. If all of these moving bridges were attached to the actin helix, the resulting shift of mass from thick to thin filaments would be larger than that indicated by the change in intensity of the equatorial reflections; if only those of the moving population in the extreme rigor orientation were attached, the mass shift would be lower than that indicated by the equatorial intensity changes. If in this last case the unattached myosin heads were

nevertheless in the environment of the thin filaments, as seems to happen in static calcium-activated muscle, then the number out at the thin filaments would be larger than that indicated by the equatorial intensity changes while the number attached would be less than that indicated by the layer-line intensity changes. It is obvious that the estimates of percentage of moving myosin heads indicated by the various parts of the X-ray diffraction pattern are subject to some uncertainty, but our present results suggest that all or most of the myosin heads which move during active muscle contraction spend most of their time marking the actin helix and in an orientation similar to that in rigor muscle.

Three additional points may be mentioned. When the muscle is statically activated by calcium, there is a small (1–2%) increase in intensity of all the reflections measured. However the ratio of the intensities of the two equatorial reflections changes with increase in calcium ion concentration, which indicates that there is a lateral mass transfer from thick to thin filaments. The extent of this mass transfer depends on the calcium ion concentrations at low concentrations ($<10^{-5}$ M), then levels off (Fig. 11) when the mass transfer is equivalent to 13–20% of the myosin heads moving from the thick to the thin filament surface. Thus at values of $\{Ca^{++}\} < 10^{-5}$ M the number of myosin heads out at the thin filaments appears to be determined by the calcium ion concentration. Second, on static calcium activation there is no decrease in intensity of the 14.5 nm reflection; nor is there a change in the intensity of the 38.5 layer line. This means that the myosin heads which move out to the environment of the thin filaments on static calcium activation do not at this stage attach to the actin helix. Finally, a feature of the phase relationship between the intensities of *all* the reflections studied during oscillation and the active tension of the muscles (Figs. 8, 9 and 10) is that the intensity variation lags slightly but perceptibly behind the tension variation. This implies that maximum muscle tension is developed before maximum displacement of the myosin heads. The intensity variation therefore contains potential information not only about the number of myosin heads in motion but also about the nature of the cycle of movement and its relation to force generation.

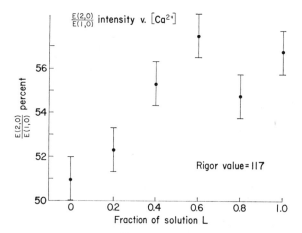

Figure 11. Graph of ratio of the intensity of the $(2,0,\bar{2},0)$ to the intensity of the $(1,0,\bar{1},0)$ reflections at different calcium ion concentrations. 1.0L corresponds to $[Ca^{++}]$ = 10^{-4}M and 0.0L to 10^{-9} M.

References

CULLEN, M. J. 1971. A Comparative study of the anatomical basis of flight: Hemiptera. D. Phil. thesis, Oxford University.

HASELGROVE, J. D. 1970. X-ray diffraction studies on muscle. Ph.D. thesis, Cambridge University.

HUXLEY, H. E. and W. BROWN. 1967. The low-angle X-ray diagram of vertebrate striated muscle and its behaviour during contraction and rigor. *J. Mol. Biol.* **30**: 383.

MILLER, A. and R. T. TREGEAR. 1970. Evidence concerning crossbridge attachment during muscle contraction. *Nature* **226**: 1060.

———. 1971. X-ray studies on the structure and function of vertebrate and invertebrate muscle, pp. 205–228. In *Contractility of muscle cells and related processes* (ed. R. J. Podolsky) Prentice-Hall, Englewood Cliffs, N.J.

———. 1972. The structure of insect flight muscle in the presence and absence of ATP. *J. Mol. Biol.* **70**: 85.

REEDY, M. K. 1968. Ultrastructure of insect flight muscle. *J. Mol. Biol.* **31**: 155.

RÜEGG, J. C. and R. T. TREGEAR. 1966. Mechanical factors affecting the ATP-ase of glycerol extracted insect fibrillar flight muscle. *Proc. Roy. Soc.* B **165**: 497.

RÜEGG, J. C., G. J. STEIGER, and M. SCHADLER. 1970. Mechanical activation of contractile system in skeletal muscle. *Pflugers Arch. ges Physiol.* **319**: 139.

SQUIRE, J. M. 1971. General model for the structure of all myosin containing filaments. *Nature* **233**: 457.

TREGEAR, R. T. and A. MILLER. 1969. Evidence of crossbridge movement during contraction of insect flight muscle. *Nature* **222**: 1184.

Intensity Fluctuation Autocorrelation Studies of Resting and Contracting Frog Sartorius Muscle

Francis D. Carlson, Robert Bonner, and Allan Fraser

Department of Biophysics, Johns Hopkins University, Baltimore, Maryland 21218

The development of the continuous wave laser, self-beating spectroscopy, and photon correlation techniques have made it possible to study the microscopic dynamics of a system by observing the fluctuations in the intensity of coherent light scattered by the system. This technique has already found applications in molecular biology in measuring the diffusion constants of several viruses and various macromolecules. We have developed light-beat techniques and used them in our laboratory to study myosin and its self-association (Herbert and Carlson, 1971) and to study TMV (Cummins et al., 1969). Fujime (1970) and Fujime and Ishiwata (1971) reported its use in the study of F-actin and its interaction with tropomyosin and troponin. In this paper we present the results of some preliminary studies of intensity fluctuation autocorrelations obtained from coherent light scattered from resting and contracting muscle. The experiments reported here provide a survey of intensity fluctuation autocorrelations of muscle. Systematic and detailed studies now in progress will be reported elsewhere. The results obtained so far lead to conjectures about the dynamics of muscular contraction.

Theoretical Background

If a collimated plane-polarized coherent beam of light, such as that produced by continuous wave laser, is incident upon a small bundle of muscle fibers, it will produce in space a scattered optical field. For striated muscle this scattered field exhibits the familiar sarcomere diffraction bands. The space-time characteristics of the field are determined by the space-time optical character-istics of the scattering muscle. If the scattering properties of the scattering system are invariant in time, then the scattered optical field in space and its intensity will also be invariant in time. If there are changes in the dimensions, the optical polariz-ability, or the positions of the elementary scatterers within the scattering volume, there will be corre-sponding fluctuations in the space-time amplitude and intensity of the scattered field. For a wide variety of systems it is possible to relate these intensity fluctuations in the scattered optical field to the dynamics of the scatterers which produce

them. To do so one must measure continuously the instantaneous intensity at a point in the optical field. This requires a photodetection device, usually a high quality photomultiplier. The use of the photomultiplier complicates the problem of meas-uring the intensity fluctuation in the scattered field, for it introduces its own statistical fluctuation into the measurements through the inherent probabilistic nature of the photoelectric effect. Consequently the statistics of the intensity fluctuations of the scattered optical field are mixed with the statistics of the photon detection process. The techniques for unmixing or unfolding these two random processes have been extensively developed. It is possible to make extremely precise measurements of the statistical parameters of a scattered optical field. From these measurements strong inferences can be made about some of the dynamic characteristics of the scatterers.

There are two light-beating techniques for measuring the intensity fluctuations of scattered optical fields of the type produced by muscle. These are photocurrent spectrum analysis and photon autocorrelation. Photocurrent spectrum analysis involves the detection of the optical field at a point by a broad-band photomultiplier-amplifier system and subsequent determination of the power spectrum of the fluctuating photo-multiplier current. From this power spectrum it is possible to determine the spectrum of the optical field which arises from the intensity fluctuations produced by the scatterers. The Fourier transform of this spectrum is the autocorrelation of the intensity fluctuations. A more precise and rapid way to measure intensity fluctuations at a point in an optical field is to detect single photon arrivals and to autocorrelate directly the arrival sequence. The direct measurement of the intensity autocorre-lation from the photon arrival times is technically complex, but its speed and precision make it preferable.

The general theory which relates the intensity of the scattered optical field to the space-time optical properties of the scatterers has been developed by Van Hove (1954), Komorov and Fisher (1963), and Pecora (1964, 1968). Cummins et al. (1969) have presented a development of this

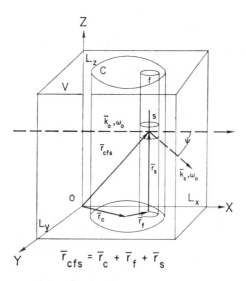

$$\overline{r}_{cfs} = \overline{r}_c + \overline{r}_f + \overline{r}_s$$

Figure 1. Scattering volume reference coordinates. An idealized cubic scattering volume is defined by the optical system. The fiber axis of the muscle is aligned with the Z coordinate axis. Within the scattering volume, the axis of a muscle cell, c, is specified by the vector, $\bar{r}_c$, from the laboratory coordinate origin. Within the cell is a myofibril whose axis is specified by the vector, $\bar{r}_f$, from the respective cell axis. Finally, a sarcomere along the myofibril is specified by the vector in the Z direction, $\bar{r}_s$. Individual myofibrillar sarcomeres and their immediate surrounds are considered the fundamental light scatterers. They scatter the incoming coherent plane wave from the laser, identified by frequency (ω_0) and the x-directed wave vector, $\bar{k}_0$. Quasi-elastically scattered light is scattered to all angles. In any one measurement an angle in the XY plane is taken for study, and scattered light with the same frequency (ω_0) and a new wave vector, $\bar{k}_s$, is observed at a point in the far scattered optical field.

theory that is appropriate for macromolecules in solution. We shall employ their format and notation in the following introduction to the theory as it applies to muscle.

Figure 1 depicts the scattering volume and the geometry of the scattering system. This scattering volume is usually less than 1 cubic mm. A plane wave of plane polarized coherent light is incident on the scattering volume in the X direction. It has a wave vector, $\bar{k}_0$, and an optical frequency, ω_0. Since the source of this light is a stabilized c.w. laser, its intensity is constant. Light scattered at an angle ψ with respect to the incident light is assigned a wave vector, $\bar{k}_s$, in the direction of scattering. A single myofibrillar sarcomere is considered the elementary scatterer, although in principle some other subcellular structures, such as the thick and thin filaments, might be taken as the elementary scatterers. The scattering volume contains many muscle cells, each of which contains many myofibrils, each of which in turn contains many sarcomeres. The position vector in the laboratory-based coordinate system of the sth

sarcomere of the fth myofibril in the cth muscle cell is $\bar{r}_{cfs}$. $\bar{r}_c$ locates the center of mass on the cth cell in laboratory coordinates; $\bar{r}_f$ is the vector from this center of mass to the center of mass of the fth fibril; $\bar{r}_s$ is the vector from the center of mass of the fibril to the center of mass of the sth sarcomere. Then

$$\bar{r}_{cfs} = \bar{r}_c + \bar{r}_f + \bar{r}_s.$$

The light scattered in the $\bar{k}_s$ direction by the sarcomere at $\bar{r}_{cfs}$ produces at a distant point a fluctuating optical electric field, $E_{cfs}(t)$. $A_{cfs}(t)$ designates the amplitude of this field. The phase of the scattered field will depend upon the position of the sarcomere relative to the origin of our laboratory coordinate system. Thus

$$E_{cfs}(t) = A_{cfs}(t) \cdot e^{i\bar{q}\cdot\bar{r}_{cfs}(t)} \cdot e^{-i\omega_0 t}$$

where $\bar{q} = \bar{k} - \bar{k}_s$. For quasi-elastic scattering where n_0 is the refractive index of the scatterer and λ_0 the wavelength of the incident light in vacuo:

$$|\bar{q}| = \frac{4\pi n_0}{\lambda_0} \sin \psi/2$$

The scattered optical field due to scattering from all of the sarcomeres present in the scattering volume is given by the sum of the fields of each of the individual sarcomeres, that is

$$E(t) = \sum_{c=1}^{C} e^{i\bar{q}\cdot\bar{r}_c(t)} \cdot \sum_{f=1}^{F} e^{i\bar{q}\cdot\bar{r}_f(t)} \cdot \sum_{s=1}^{S} A_{cfs}(t) e^{i\bar{q}\cdot\bar{r}_s(t)} \cdot e^{-i\omega_0 t}$$

where C = number of cells, F = number of myofibrils in a cell, and S = number of sarcomeres in a myofibril. The instantaneous intensity $I(t)$ of the scattered field at the point of interest is defined as

$$I(t) = E^*(t) \cdot E(t)$$

where $E^*(t)$ is the complex conjugate of $E(t)$. The time-averaged intensity is

$$\langle I(t) \rangle = \langle E^*(t) \cdot E(t) \rangle$$

where angular brackets denote the time average of the quantity in the brackets.

To characterize the intensity fluctuations of the scattered field we use the autocorrelation function of the instantaneous intensity. The autocorrelation of a time-dependent quantity provides a measure of how closely correlated the value of the function is at a particular time with its value at some earlier or later time. The normalized intensity autocorrelation function is defined as

$$g^{(2)}(\tau) = \frac{\langle I(t) \cdot I(t + \tau) \rangle}{\langle I \rangle^2}$$
$$= \frac{\langle E^*(t) \cdot E(t) \cdot E^*(t + \tau) \cdot E(t + \tau) \rangle}{\langle |E(0)|^2 \rangle^2}.$$

When the optical field is Gaussian, there is a relationship between the intensity autocorrelation, $g^{(2)}(\tau)$, and the field autocorrelation, $g^{(1)}(\tau)$, namely,

$$g^{(2)}(\tau) = 1 + g^{(1)}(\tau)^2.$$

Using this relationship it is possible to determine the field autocorrelation from measurements of the intensity autocorrelation. Since the field autocorrelation depends on the space-time dependent characteristics of the scatterers through $\bar{r}_{cfs}(t)$ and $A_{cfs}(t)$, it tells us something about these space-time properties.

An appropriate dynamic and optical model of muscle must be used to interpret the autocorrelation of a scattered field and to obtain information about the time-dependent characteristics of the sarcomeres. It is beyond the scope of the present paper to present this theory in detail. Theoretical considerations and the experimental findings which are reported below seem to indicate that the decay times of the intensity autocorrelation in the scattered field of a contracting muscle are related to the relaxation time of the continuous chaotic fluctuations in optical polarizability of the sarcomeres in the contracting muscle.

Experimental Technique

The spectrometer used in these studies was designed specifically for use with biological samples, muscle in particular. It can, however, be used with other biological material and with solutions of macromolecules. The source of coherent light was a Coherent Radiation Laboratory model #52G argon ion laser. The power in the incident beam was below 5 mW and produced no visible damage to the muscle. Calculation shows that if the entire 5 mW were absorbed, the muscle would heat up only a few degrees. The optical collecting system consisted of a rotatable optical bench which could be set at any angle to the incident beam and on which a double pinhole collimating system was mounted. Following the pinhole collimator a photomultiplier (ITT Model FW4085) was used as a single photon detector. The output of the photomultiplier was appropriately amplified and fed into a discriminator and pulse-shaping unit. This produced pulses of uniform height and

duration whose leading edge defined the arrival time of a photon detected by the photomultiplier. These uniform pulses were then fed into a real-time autocorrelator. The autocorrelator was designed by Dr. Allan Fraser and built by the Applied Physics Laboratory of The Johns Hopkins University. The details of this correlator have been described by Fraser (1971). It calculates directly the single clipped autocorrelation of the photon arrival times, which can be corrected according to the theory of Jakeman (1970) to give the normalized intensity correlation $g^{(2)}(\tau)$ or the optical field normalized autocorrelation $g^{(1)}(\tau)$ for the case of a Gaussian field. The entire spectrometer and autocorrelator are illustrated schematically in Fig. 2.

Lucite chambers were provided for clamping a frog's sartorius or sternocutaneous muscle isometrically and keeping them bathed in Ringer solution. The pelvic end of the sartorius muscle was clamped, and an isometric tension transducer of the strain gauge type was attached to the tibial tendon. Direct electrical stimulation by 0.5 msec shocks was provided through two platinum electrodes. The mounting arrangement permitted the muscle to be tilted into the incident beam and rotated in the plane normal to the incident beam. The scattered light could be observed at all angles except in the neighborhood of 90° where the muscle itself obscured the scattered radiation.

A unique feature of the autocorrelator used in these studies was its split memory. In normal operation the autocorrelator calculated the autocorrelation function at 62 lag times. The split memory feature made it possible to divide the memory into as many as eight separate memories with a correspondingly smaller number of lag times in each. Each of the separate memories could be used to measure the autocorrelation function in sequential phases of any process. In the case of contracting muscle the memory was often split into four parts, enabling the observation of the autocorrelation function during four sequential phases of a twitch or tetanus. In a typical autocorrelation measurement, a sequence of some 10–20 twitches was used with a 10–20 sec separation time between twitches. In the case of tetani, 3–10 0.63 sec tetani separated by 50 sec were used. All experiments were done between 15° and 22°C.

Throughout these studies sartorius muscles obtained from the American Bullfrog, *Rana catesbeiana*, were used. The details of this preparation are the same as those used for *Rana pipiens*, and they are described in Carlson et al. (1967).

As mentioned, single clipped photon autocorrelations were measured and corrected according to Jakeman (1970) to obtain the normalized

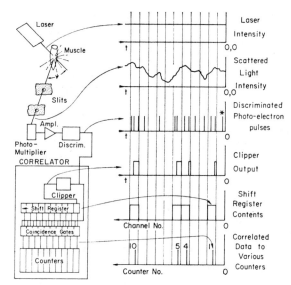

Figure 2. Data gathering scheme with indicative charts. Constant intensity laser radiation was incident upon the middle of a muscle and was scattered. The light scattered into an angle ψ from the laser beam and in the zeroth order of the sarcomere diffraction pattern was collected by a double pinhole collimator. The intensity of scattered light varied in time because of the fluctuations in the scattering properties of the muscle cells in the scattering volume. Scattered light intensity was measured with a photon detecting system consisting of a photomultiplier, an amplifier, and a discriminator. The detected photoelectron pulses arrived randomly in time with their probability of occurrence directly proportional to the instantaneous scattered light intensity. Discriminated photoelectron pulses were fed into the correlator, which, in the arrangement most often used, produced the "single-clipped autocorrelation" function of its input. The input of the correlator fed two kinds of circuits, the "clipper" and one input of each of the coincidence gates. A "clipper" is a digital threshold circuit; its output was one when a preset threshold (set at one in the diagram) was exceeded. The "clipper" output was reset to zero at the end of each correlator time interval. Correlator time intervals were set to divide time into equal contiguous intervals. Time division is indicated by the vertical lines running through the charts. The "clipper" output fed the input of the shift register. The shift register contents were a digital one in each shift register channel n if and only if the clipper output was one at the end of the nth previous time interval. The shift register was thus used as a delay line for final "clipper" outputs as it shifted its contents one channel to the left at the end of each correlator time interval. Each shift register channel had an output that went to a corresponding coincidence gate. The coincidence gates had two inputs: discriminated photoelectron pulses, and clipped and then delayed discriminated photoelectron pulses. Each coincidence gate multiplies its two inputs by the logical operation of detecting coincidences, and the multiplied outputs are added in counters corresponding to the various correlation time delays. In the sample at time zero shown in the figure, correlated data to various counters are one count, corresponding to the one asterisked discriminated photoelectron pulse, and the ones in the shift register contents, into the first, fourth, fifth, and tenth counters. The system accumulated data in up to billions of contiguous time intervals in this same fashion, and the counters continually added their input pulses over all of the intervals. The final contents of the bank of counters is the correlation function that is related to the true autocorrelation function of the scattered light intensity, which is in turn related to the fluctuations in the scattering properties of the muscle.

field autocorrelation from the clipped intensity autocorrelation. That is

$$g_{\text{k}}^{(2)}(\tau) = 1 + |g^{(1)}(\tau)|^2 \left\{ \frac{1 + k}{1 + \bar{n}} \right\}$$

where k is the clip number and $\bar{n}$ is the average count rate per correlator time interval.

The single-clipped corrections of Jakeman (1970) were developed for Gaussian-Lorentzian optical fields and affect the amplitude but not the shape of the autocorrelation function.

Uncorrected values of the amplitudes of the autocorrelations varied from 0.1 to 1 for twitches and tetani. After correction they varied between 0.4 and 0.7, except for rest muscle where the maximum value was 0.10 ± 0.09. Except for resting muscle, the corrected values of $g^{(2)}(0)$ were constant to within 10% for the same optical alignment and the same physiological state of the muscle. These high values for the corrected amplitudes of $g^{(2)}(0)$ are a strong indication that the optical field of the light scattered from the contracting muscle is Gaussian. They do not prove this point, however, and an experimental verification is planned.

Results

Resting muscle. Corrected intensity autocorrelations for rest muscles are shown in Fig. 3. The amplitude of the correlation is about 0.1, and the time dependence cannot be fitted by a single exponential, although two exponentials give a satisfactory fit. The time for the normalized autocorrelation function minus the baseline, $g^{(2)}(\tau) - 1$, to decay to one-half the maximum amplitude, $T_{1/2}$, varied from 10–25 msec.

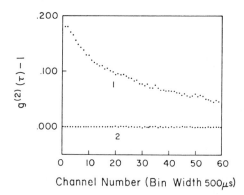

Figure 3. The normalized intensity fluctuation autocorrelation minus baseline, $g^{(2)}(\tau) - 1$, for (*1*) a rest sartorius muscle at 2.25 μm sarcomere length during a 10 sec sample period; (*2*) a glycerinated muscle in rigor at 2.6 μm. Both were at 40° scattering angle, illuminated with 4658 Å line analyzed with a vertical polarizer (VV).

Glycerol-extracted muscle. Intensity auto-correlations measured on glycerol-extracted muscle in rigor (Fig. 3) were essentially flat. This indicates that no appreciable changes in the scattering properties occurred during the course of this experiment.

Twitch contraction. Highly reproducible intensity autocorrelations were obtained from single twitch contractions. Autocorrelations were determined during the rising phase, at the peak of the twitch, during relaxation, and after the end of the twitch tension. The amplitude of the autocorrelation was virtually independent of the physiological phase of the twitch. Although the autocorrelations obtained for these different physiological phases showed widely different decay times, their shapes were virtually the same (see Figs. 4, 5, and 7). During the first 20–25% of the decay the autocorrelations showed some rounding, but for the remainder of the decay they were all approximately exponential. These results are summarized in Fig. 7. The autocorrelations decayed more rapidly during phases when the tension was changing rapidly than during phases when the tension was relatively constant or changing slowly. At rest length the autocorrelation decay times varied from 10 μsec–100 μsec. It should be noted that even when the twitch tension had fallen to zero, prominent autocorrelations with large $T_{1/2}$ were obtained.

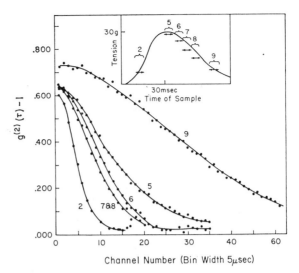

Figure 5. $g^{(2)}(\tau) - 1$ for a series of different sample periods during the fall of tension with the fastest decaying curve from the previous figure as a reference. These experiments used the same muscle under identical conditions to those of the previous figure. The time of sample for the various curves appears in the inset.

Tetanic contraction. Intensity autocorrelations obtained for various phases of a tetanus showed the same large amplitudes as for twitches. Decay times during the early rising phase of the tetanus were comparable to those found in the rising phase of a twitch. There was a steady increase in $T_{1/2}$ throughout the plateau of the tetanus (Fig. 6). Although $T_{1/2}$'s for the plateau of the tetanus, 100 μsec–1000 μsec, were considerably

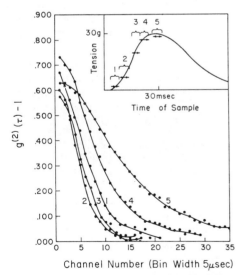

Figure 4. $g^{(2)}(\tau) - 1$ for a series of different sample periods during the rise of tension of twitches from one sartorius muscle at 2.6 μm rest sarcomere length, 40° scattering angle, illuminated with 4658 Å line and analyzed with a vertical polarizer (VV). Each curve is the average from 10–25, 5 msec samples. The time of sample of each numbered autocorrelation is indicated with respect to the tension record in the upper right inset.

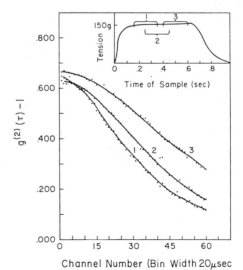

Figure 6. $g^{(2)}(\tau) - 1$ for a series of different sample periods during the plateau of tetanus tension from one sartorius muscle at 2.4 μm rest sarcomere length, 40° scattering angle, illuminated with a 4658 Å line and analyzed with a vertical polarizer (VV). Each curve is the average from 3–4200 msec samples. The time of sample for the 3 numbered curves is again indicated in the inset.

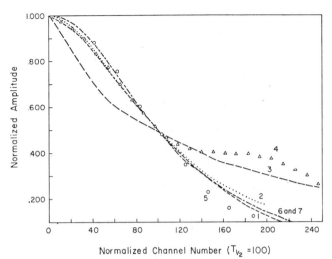

Figure 7. The amplitudes and channel numbers of $g^{(2)}(\tau) - 1$ have been normalized to facilitate comparison of shapes from various physiological states. The numbered curves were from (*1*) Twitch of sternocutaneous muscle: 100 msec sample of entire rising phase at tension, average of 7 repeats, 3.06 μm rest sarcomere length, $T_{1/2} = 530$ μsec, $g^{(2)}(0) - 1 = 0.53$, $\psi = 50°$. (*2*) Plateau of tetanus of sartorius muscle: 200 msec sample of early plateau, average of 3 repeats, 2.4 μm rest sarcomere length $T_{1/2} = 500$ μsec, $g^{(2)}(0) - 1 = 0.64$, $\psi = 40°$. (*3*) Rest sartorius muscle: 10 sec samples, 2.25 μm rest sarcomere length, $T_{1/2} = 10.5$ msec, $g^{(2)}(0) - 1 = 0.18$, $\psi = 40°$. (*4*) Vibrated glycerinated rigor sartorius muscle: 20 sec sample, 2.6 μ sarcomere length, 18 cps amplitude 0.1 mm vibration, $T_{1/2} = 1.86$ msec, $g^{(2)}(0) - 1 = 0.60$, $\psi = 50°$. (*5*) Twitch of sartorius: 5 msec sample of fastest rising phase of tension, average of 15 repeats, 2.6 μm rest sarcomere length, $T_{1/2} = 24$ μsec, $g^{(2)}(0) - 1 = 0.64$, $\psi = 40°$. (*6*) Twitch of sartorius: 5 msec sample of peak of tension, average of 19 repeats, 2.6 μm rest sarcomere length, $T_{1/2} = 68$ μsec, $g^{(2)}(0) - 1 = 0.63$, $\psi = 40°$. (*7*) Twitch of sartorius: 10 msec sample of late decay of tension, average of 10 repeats, 2.6 μm rest sarcomere length, $T_{1/2} = 136$ μsec, $g^{(2)}(0) - 1 = 0.71$, $\psi = 40°$. All of the above experiments used 4658 Å laser line analyzed with a vertical polarizer (VV).

larger than those for the twitch, the shapes of the autocorrelation functions were the same (Fig. 7).

Dependence on sarcomere length. When a muscle was stretched to 1.33 times its in situ length, all decay times for both twitch and tetanic contractions were increased by approximately one order of magnitude.

Control experiments. In order to determine whether the intensity autocorrelations observed could be the result of heterogeneous regions of the muscle being moved through the scattering beam by internal force imbalances in the muscle, autocorrelations were taken on a twitched muscle for widely different scattering volumes of linear dimensions from 200–1000 μ. These autocorrelations were approximately the same.

To determine whether simple bulk translation of the muscle through the scattering volume could produce autocorrelations with the same characteristics as those obtained from contracting muscle, a glycerol-extracted muscle was mounted rigidly on the driving coil of a loud speaker and vibrated transversely to the incident beam with various frequencies and amplitudes. Autocorrelations obtained from these vibrated preparations showed secondary maxima and did not monotonically decrease. A typical autocorrelation is shown in Fig. 7. Furthermore the decay times obtained from vibrated muscle were orders of magnitude longer than the fastest obtained from twitched muscle. For example, when the muscle was vibrated with peak to peak amplitude of 3 mm at a frequency of 60 Hz, the fastest decay time of the autocorrelation obtained was 15 times greater than the fastest

decay times obtained during a twitch. This vibrated muscle was moving at velocities in excess of 30 cm/sec. This velocity exceeds the maximum velocity of shortening of an unloaded sartorius muscle at the temperatures at which these experiments were performed and is greatly in excess of any local velocities that might be expected during an isometric contraction. The autocorrelations on vibrating muscle were independent of whether the direction of vibration was along the muscle axis or perpendicular to it. Decay times observed on vibrated glycerol-extracted muscle did not increase with the increase of the size of the scattering volume.

The sternocutaneous muscle of a bullfrog is only about 160 μm, or about 5 cells, thick. Intensity autocorrelations collected from stimulated sternocutaneous muscles were virtually the same as those from sartorius muscle (Fig. 7).

Angular dependence of decay time. A systematic study of the angular dependence of the decay time of the intensity autocorrelation was not possible at the time this work was done. However some crude experiments were done on twitches over a range of scattering angles from 0° to 70°. Decay time increased, but not indefinitely, with a decrease in scattering angle. That is, in the limit of zero scattering angle finite decay times were obtained. To a rough approximation a plot of $1/T_{1/2}$ against $\sin \psi/2$ was linear with a positive intercept.

Change in orientation of muscle. When the muscle was rotated slightly in the plane normal to the incident beam so that its axis was no longer vertical, the autocorrelation obtained showed no significant differences from that obtained with the long axis of the muscle aligned vertically.

Discussion

It is not yet possible to fully interpret these results or to develop fully their significance to the mechanism of muscular contraction. There are, however, some important conclusions that can now be drawn. The experiments described above support the following conclusions:

1. The intensity fluctuations which occur when a muscle contracts do not arise primarily from translation of the whole muscle through the scattering volume. Such movements might occur from internal axial and off-axial forces developed by the contracting muscle. The intensity fluctuations arising in contracting muscles arise from more rapid fluctuations in the scattering properties of the muscle than such bulk translation could produce.

2. The large amplitudes and the monotonic decay of the autocorrelations from contracting muscle indicate that most of the scattering material within the muscle is behaving in a chaotic fashion. In resting muscle, however, the amplitude of the intensity autocorrelation is small and suggests that a much smaller fraction of the scattering material present is responsible for the intensity fluctuations.

3. Although the decay times of the autocorrelations vary widely from one stage of contraction to another in both the twitch and the tetanus, the shape of the intensity autocorrelations are very nearly the same when linearly scaled. The intensity autocorrelation for the resting muscle is an exception, however, for its shape is not the same as that of the autocorrelations obtained from contracting muscle.

4. The intensity fluctuation autocorrelations observed are not strongly dependent on the amount of tension in the muscle, for they occur with equal prominence when the tension has decayed to less than 10% of its maximum value and persist for a short period following a contraction. There is then a gradual return to the small amplitude correlations of the resting muscle.

5. The intensity fluctuations do not vanish in the limit of zero scattering angle, and further, there is clearly a dependence of the decay times on the scattering angle. This suggests that some of the intensity fluctuation is due to small chaotic movements of the sarcomeres normal to their fiber axis. There are also changes in the scattering properties, probably associated with fluctuations in the dimensions, orientation, or optical polarizability of the scatterers that give rise to the intensity fluctuation that one sees in the limit of zero scattering angle.

As already mentioned the full interpretation of these results must await further theoretical and experimental studies. However certain conjectures are justified at this time. It is likely that these intensity fluctuations arise from rapid and chaotic fluctuations in the shape or optical polarizability of the myofibrillar sarcomeres. Further, the decay times of the intensity autocorrelations can be related to the relaxation times of this sarcomere fluctuation process. If this interpretation is correct, then even in isometric contractions sarcomeres are constantly changing their dimensional or optical properties. This would imply that even in a servo-regulated isometric contraction the dynamic change of the sarcomere dimensions would make it unlikely that a constant amount of overlap of the thick and thin filaments exists in each sarcomere throughout the course of a contraction. If one accepts this concept of the dynamic state of the myofibrillar sarcomere during contraction, then it seems likely that the interpretation of some properties of muscle will have to be modified. For example, a muscle with its interior in a dynamic state of agitated movement would not be expected to have the same heat transport properties as one which was stationary. Consequently the interpretation of muscle heat data would become more complex. Further, the steady heat production which one sees in an isometrically contracting muscle could arise from the dissipation of energy by the internal movement of the sarcomeres. The isometric tension which a muscle develops may be best regarded as a time average of the integral over the length of the muscles of the forces developed by each sarcomere, and so on.

Finally the question naturally arises: "How much of the intensity fluctuation that we see is due to the movement of cross-bridges?" No meaningful answer to this question can be given at the present time. If, as suggested, it is true that the intensity autocorrelation decay times are a measure of the relaxation times of the dimensional and optical fluctuations of sarcomeres and that these dimensional changes are due to action of cross bridges, then we might conjecture that the autocorrelation decay times correspond to changes in cross-bridges configurations. It may eventually be possible to make some statements about the magnitude of the fluctuations of the number of cross-bridges attached in a single sarcomere during the plateau of an isometric tetanus. Answers to these questions and a full understanding of the data reported here will have to wait further light-scattering studies, on both intact muscle and solutions of contractile proteins.

Acknowledgment

This work was supported by research grant #2 RO1 AM12803 from the USPHS to Dr. Francis D. Carlson.

References

CARLSON, F. D., D. HARDY, and D. R. WILKIE. 1967. The relation between heat produced and phosphorylcreatine split during isometric contraction of frog's muscle. *J. Physiol.* **189**: 209.

CUMMINS, H. Z., F. D. CARLSON, T. HERBERT, and G. WOODS. 1969. Translational and rotational diffusion constants of tobacco mosaic virus from Rayleigh linewidths. *Biophys. J.* **9**: 518.

FRASER, A. 1971. A digital pulse correlator and an advantage of double clipping. *Rev. Sci. Inst.* **539**: 1539.

FUJIME, S. 1970. Quasi-elastic light scattering from solutions of macromolecules. II. Doppler broadening of light scattered from solutions of semi-flexible polymers, F-actin. *J. Phys. Soc. Japan* **29**: 751.

FUJIME, S. and S. ISHIWATA. 1971. Effect of Ca^{++} on the F-actin-tropomyosin-troponin complex studied by quasi-elastic light scattering. *J. Phys. Soc. Japan* **30**: 303.

HERBERT, T. J. and F. D. CARLSON. 1971. Spectroscopic study of the self-association of myosin. *Biopolymers* **10**: 2231.

JAKEMAN, E. 1970. Theory of optical spectroscopy by digital autocorrelation of photon-counting fluctuations. *J. Phys.* Series A **3**: 201.

KOMOROV, L. I. and I. Z. FISHER. 1963. *J. Expt. Theor. Phys.* **16**: 1358.

PECORA, R. 1964. Doppler shifts in light scattering from pure liquids and polymer solutions. *J. Chem. Phys.* **40**: 1604.

———. 1968. Spectral distribution of light scattered by monodisperse rigid rods. *J. Chem. Phys.* **48**: 4126.

VANHOVE, L. 1954. Correlations in space and time and Born approximation scattering in systems of interacting particles. *Phys. Rev.* **95**: 249.

How Many Myosins per Cross-Bridge?
I. Flight Muscle Myofibrils from the Blowfly, *Sarcophaga bullata*

Michael K. Reedy

Department of Anatomy, Duke University, Durham, N. Carolina 27710

Gunter F. Bahr

Biophysics Branch, Armed Forces Institute of Pathology, Washington, D.C. 20315

AND

Donald A. Fischman

Department of Biology, University of Chicago, Chicago, Illinois 60637

Cross-bridges project from the myosin filaments in groupings which repeat every 14.5 nm along the filament shaft. We are going to call this repeating group of projections a "crown." Now, we share a general belief that each cross-bridge represents one myosin molecule (two S_1 subunits) and we thought that we might be able to prove it with favorable material and techniques. In vertebrate striated muscle, it is still typical to speak in terms of two cross-bridges per crown, although there seems to be enough myosin to supply four molecules per crown (Huxley, 1963). When an insect fibrillar flight muscle was clearly seen in thin sections to display four cross-bridges per crown (Reedy, 1968), it seemed attractive to consider such a grouping for vertebrate muscle. However the insect muscle story already had its own anomaly in the matter of myosin to cross-bridge correspondence. For Chaplain and Tregear (1966) estimated that the myosin content was sufficient for six molecules per crown in this muscle. Given reasonable precautions, which Chaplain and Tregear certainly took, it still appears that specimen selection requires some good luck to avoid premature or degenerating stages of an imperfectly characterized seasonal variation in water bug flight musculature, lest the cross-bridge population of an underdeveloped muscle be matched for correspondence with the myosin content of a fully mature muscle. We thought that estimates of myosin content and of cross-bridge population should be made on one and the same specimen; in the ideal case both estimates would be made on a single fibril. So we began the present study in 1968, hoping to approach this ideal experiment and hoping for a different answer.

It would have been ideal to proceed with the same material used in the initial structural and protein studies of fibrillar insect flight muscle,

namely, that from the giant water bug, *Lethocerus*. So let us explain why the present study is almost totally concerned with flight muscle from the blowfly, *Sarcophaga bullata*. First, the flies are cheap and readily available, unlike the seasonal water bug which must be imported. Second, it was easy to isolate fly flight myofibrils with normal and uniform cross banding, while *Lethocerus* fibrils almost always showed contraction bands and other disturbances of structure. Third, under the phase microscope it looked as though myosin could be rapidly and quantitatively extracted from fly fibrils, something we have never yet approached in experiments with water bug flight fibrils. Fourth, fly fibril suspensions could be quickly and completely cleared of non-fibrillar particles with a solution containing 0.1%–1.0% Triton X-100; whereas the mitochondria, etc., in water bug muscle homogenates did not dissolve and vanish with this treatment. Fifth, the cross-bridge lattice in the fly shows enough general similarities to that of the water bug to serve our purposes, even though the exact lattice pattern is evidently somewhat different. In the end, we could not rely on quantitative myosin extraction (see Bullard and Reedy, this volume) and we chose not to use the detergent, but that still leaves three of our justifications for using the blowfly.

When we began in 1968, we sought to measure the dry mass of intact and of myosin-extracted myofibrils, using quantitative electron microscopy (Zeitler and Bahr, 1962; Bahr and Zeitler, 1964). We did a rather lengthy experiment on a single fly, weighing fibrils from the homogenate of muscles of the left side and counting cross-bridges in the embedded and sectioned muscles of the right side. (Fischman et al., 1968; Reedy et al., 1969). When some uncertainties arose to cloud the value of our labor in that experiment, we put it aside, turning to an improved experimental strategy which is reported here.

We describe here our results from dry mass measurements and structural studies of fly flight myofibrils. Our arguments depend as well upon the estimate of percentage myosin content reported from the same lots of myofibrils in the accompanying paper by Bullard and Reedy (this volume). Our mass measurements have been obtained by interference microscopy (IMy henceforth) and by quantitative electron microscopy (QEMy henceforth). Both techniques were applied to fibrils from the same lot, and in one experiment 29 fibrils were each measured twice, once by IMy and once by QEMy. The rather close agreement (in the latter experiment mean results were identical) seems to indicate that either technique can give reliable results when used as we used them, for the sources of possible error are quite different, and the agreement probably indicates that we were not prey to them.

Sustained by new faith in the applicability of our methods, we finally returned to *Lethocerus*, and we report some first mass measurements from IMy of water bug flight myofibrils which are within 5% of our results from the fly. This helps to minimize our worry about the difference between the cross-bridge lattice patterns. That of the fly, which we describe here, is not so regular nor yet so clearly worked out that we can count every cross-bridge as confidently as we can for the water bug.

What is our best answer? It tends strongly to agree with Chaplain and Tregear, and to that extent accommodates the model of thick filament structure proposed by Squire (1971). Our present results indicate five or six myosins per crown, with six as the likelier value. The methods and arguments which provided this value leave us with no real expectation that later work will indicate four myosins/crown. Our greatest uncertainty concerns

the filament lattice spacing in isolated myofibrils, which we must know in order to use light microscope measurements of fibril diameter for calculating numbers of filaments (and thus cross-bridges) in a given sarcomere. In order to make clear the significance of our results and their possible errors in calculating the best myosin/cross-bridge ratio, we have provided a sort of "slide-rule" in the form of a graph (Fig. 10) and a nomogram (Fig. 11) which give an immediate visual indication of how the ratio may vary in fibrillar flight muscles for different values of filament spacing, myosin percentage, and dry mass concentration.

Materials and Methods

Table 1 lists the composition of many of the aqueous solutions used.

Specimens. The blowfly, *Sarcophaga bullata*, was chosen because we believe that it is the largest Dipteran (10–12 mm) easily available in large numbers. Initially our flies were cultured from a parent stock maintained at UCLA (Zoology) by Dr. Franz Engelmann, but later we obtained them as age-synchronized imagos from Carolina Biological Supply. Flies freshly emerged within a period of 24 hr or less were collected. We designated their age as zero days at the end of the collection period when they were shipped. Transit time in the post was 12–18 hr. They were then kept on 5% glucose solution and pork liver in screened gallon jars in the open laboratory until they reached the desired age, usually 5–7 days. We chose this age because we did not wish to do either deliberate or inadvertent research on development or senescence of flight muscles. During the first 3–5 days of adult life we observed gross flight muscle bulk to double or triple, accompanied by a proportionate increase in mean myofibrillar diameter, and by a 10-fold to 100-fold increase in the time required for A band extraction by H-S-Z solution. Occasional degenerated flight muscles were encountered at as early as 8 days of age and fly deaths were well under way by 12 days of age; a group of age-mates seldom survived $3\frac{1}{2}$ weeks.

For dissection flies were immobilized by cooling, and wings and head were removed before gluing the dorsal thorax to a plastic microscope slide using 5-min epoxy adhesive. The adhesive trapped the hairs rather than binding effectively to the chitin surface, so some care was required to avoid dislodging the thorax prematurely. Legs were cut off and the abdomen was neatly pulled off, drawing the esophagus with it. The ventral thorax was removed by cutting the lateral chitin wall and dorso-ventral muscles

Table 1. Composition of Solutions

	pH	K-Na-PO₄ buffer (mM)	KCl-KQ (M)	MgCl₂ (mM)	EGTA (mM)	Na azide (mM)	Sucrose	Glycerol	Paraformaldehyde	Na pyrophosphate
Complete rigor	7.0	20	0.1	5	5	5–10				
Rigor glycerol	7.0	20	0.1	5	5	5–10		50% v/v		
Sucrose rigor	7.0	20	0.1	5	5	5–10	10% or 30% w/v			
Simple rigor	7.0	20	0.1	5						
H-S-Z[a] (for myosin extraction)	6.5	20	1.0	1.0						10 mM
4% PF	7.0	100							4% w/v	

[a] H-S-Z solution is Hasselbach-Schneider solution as modified by Zebe (1966) and others to extract insect myosin.

with iris scissors under a dissecting microscope. The dorsal longitudinal muscles (DLM) were freed of overlying loose tissues. If the muscle was to be glycerinated or fixed for electron microscopy, the thorax was split in the midline with a razor, the cut going through the underlying adhesive and plastic support slide, to give two halves in which the muscles were reasonably well supported and oriented. With care, this cut exactly split each of the six pairs of giant fibers composing the DLMs without injury to any. Only the DLMs were used in our work. When we wished fibrils rather than whole fibers, everything lateral to the DLMs was severed by one razor stroke and peeled from the epoxy in one or two passes with fine forceps. Thus isolated, the 12 giant fibers were cut away from their chitin attachments with fine iris scissors and placed in an ice-chilled rigor solution containing 10% or 30% sucrose (Table 1). Up to 70 flies could be dissected and processed by one person in a day. Each fly yielded about 1 mg of cleaned fibrils after the following procedure. After they were collected, all muscles were homogenized by a Sorvall Omni-mixer micro attachment, monitored by phase microscope to ensure that fibrils were as long as possible compatible with complete disruption of all cells. Chilled sucrose rigor solution was used for homogenization and for the first seven cycles of centrifugal washing, as the sucrose seemed to help avoid development of contraction bands and so to ensure that variations in sarcomere length were due to variations in A band length, not in I or H band length. A bench centrifuge in a cold room was used for washing, employing 12 ml of solution for each washing; pellet volume shrank from 1.5 ml to about 0.6 ml in the course of 15 washes. This treatment of fresh DLM homogenate resulted in a microscopically clean myofibril suspension in which other material was comfortably estimated to be much less than 1% of the total mass or volume.

The phase microscope appearance of these fibrils remained constant and free of microorganisms when the suspension was stored in the refrigerator in rigor solution (with azide) for as long as a year. Material relied on in this study for mass studies or protein analysis was examined over a range of 1–60 days after preparation. Conspicuous aging changes in the suspension were not evident, though inconspicuous changes may now be watched for in future studies. The one group of truly anomalous findings, the "heavy" fibrils identified by interference microscopy, were encountered only as infrequent objects in one suspension, examined five weeks after preparation, but not in suspensions examined both earlier and later. On another occasion low mass fibrils were encountered in a year-old suspension which had apparently begun to support a yeast contaminant.

At a late stage in the project, we managed to prepare relatively acceptable fibrils of water bug flight muscle. Some glycerinated *Lethocerus maximus* which had been prepared in rigor glycerol and stored therein at −20° for 3 years was used. The secret was to homogenize in glycerol. A 2 × 2 × 13 mm bundle of flight muscle was homogenized in 0.3 ml of rigor glycerol using a Virtis micro-homogenizer with its invaluable 1-ml stainless steel microchamber. Brief high-speed homogenization was sufficient to give large numbers of very long fibrils which appeared free of the traumatic disturbances of cross band pattern common when homogenization was carried out in salt solutions.

Mounting fibrils for light microscopy. When only light microscopy was contemplated, a cover slip was lowered directly and flat onto a drop of myofibril suspension sufficiently generous to keep the cover slip afloat. We were careful to avoid sliding or squeezing the cover slip while excess fluid was being withdrawn and the preparation was sealed, in order not to squash or shear fibrils. The optical thickness of the trapped suspension was easily 15 μm at most points, giving plenty of room for fibrils whose diameter was 5 μm or less. Fibrils within a few micrometers of the cover slip always provided the best images and the clearest detail when using oil immersion lenses. Cover slips were sealed with clear fingernail lacquer or with a paraffin-

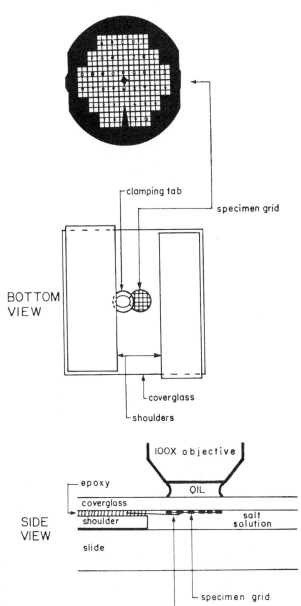

Figure 1. At the top is a Maxtaform Finder grid of the type found useful for coordinated light and electron microscope studies of identical fibrils. Below this are bottom and side views of the clamping platform used to hold electron microscope grids in an aqueous solution during light microscopy with oil immersion lenses. Fibrils adsorbed to the grids could be examined wet, before or after fixation, and later dried or fixed and embedded for electron microscopy.

vaseline-lanolin mixture. We suspect that lacquer solvents, diffusing under the cover slip, could be implicated in the "heavy" fibrils, which have not been encountered with the other sealant or with unsealed preparations.

Often we needed to observe the same fibrils by light microscopy and by electron microscopy. This key procedure in our project was relatively easy with the aid of the wet grid-clamping platform shown in Fig. 1. The basic element is a square No. 1 cover slip, with two strips of No. 1

cover slip cemented to its underside to form supporting "shoulders," leaving a 6- to 8-mm wide specimen channel between the shoulders. A slot grid becomes trapped against the ceiling of this channel by one of the shoulders before the epoxy cement sets. Wooden clothespins are used to clamp the shoulders and cover slip together during gluing and are also convenient to hold the finished platform upside down on a supporting glass slide while a specimen grid is being mounted. The slot grid provides the actual clamping tab which holds a specimen grid close against the cover slip. The entire platform is cleaned and dried before each use, and the upper surface should remain dry during grid mounting. With the platform held upside down, a drop of rigor solution was placed over the clamping tab, and a Finder grid carrying adsorbed wet fibrils was submerged face down in the drop and slid partly beneath the clamping tab until secure, as in Fig. 1. Clothespins were removed. A clean glass slide was brought near until it contacted the drop of solution. The platform at once adhered to the slide through surface tension, and the slide was turned over, bringing everything right side up for light microscope viewing. The channel was just filled with rigor solution and subsequent evaporation losses were made up every few minutes with distilled water.

The grid was surveyed by phase contrast with $10 \times$ and $40 \times$ dry lenses to locate the most suitable fibrils quickly and conveniently. Locations were noted on a map corresponding to the central 1.6 mm circular area of the Finder grid pattern which would later be the only part accessible for viewing in the electron microscope. The maps were simply Xerox copies of the pattern, taken from a photomicrographic print showing an entire grid at $125 \times$. Next the grid was viewed with $100 \times$ oil immersion objectives, the map was used to find the fibrils, and appropriate measurements were made of sarcomere length, fibril diameter and, if interference microscopy was used, of retardation through the center of the fibril.

Grids were mounted fibril side up, against the cover slip. Careful focusing was needed to be sure that fibrils chosen for measurement were adhering flat to the grid film and were not precariously adsorbed, or worse still, adhering to the cover glass surface just 5 or 10 μm above the grid film.

Before demounting grids, all immersion oil was removed with dry swabs, followed by ether-moistened swabs. Then an excess of rigor solution was added to float the platform, it was lifted and placed upside down on some dry Parafilm, and more solution was quickly added to keep the specimen grid wet while it was carefully demounted with forceps. Most fibrils had been fixed before grid mounting by floating the grid 5–6 min on buffered 4% paraformaldehyde. Those not fixed previously were fixed immediately after grid demounting. The grid was then washed 30 sec or more in distilled water to remove all salts and was air dried at room temperature. (Some grids were not washed and dried, but were embedded and sectioned at this point.)

We have yet to resolve completely some problems and anxieties regarding the metal of the grids. The clamping tab was a copper slot grid, as were most of our Finder grids. When we used nickel Finder grids for a while, light microscopy became impossible due to electrolytic formation of gas bubbles all over the grids after mounting, until this problem was solved by using simple rigor solution rather than complete rigor solution (Table 1). However fibrils mounted and examined in simple rigor solution have occasionally shown alterations in cross-banding contrast in the EM or even in light microscope appearance, which suggests that grid metals, copper and even gold, may ionize sparingly and then rebind to fibrils when EGTA is excluded. No systematic effect on dry mass measurements has been observed in comparing grid-mounted fibrils with fibrils never exposed to grids. Nevertheless some attention to the interaction of grid metals, specimens, and solutes would be advisable in further work of this kind and perhaps in any work where native composition of adsorbed biological particles on EM grids is assumed.

Electron microscopy of embedded material.

Glycerol-extracted dorsal longitudinal flight muscles from blowflies aged 5–7 days were prepared for thin sectioning. After extraction in rigor glycerol for 1–5 days, they were washed free of glycerol, fixed in buffered 5% glutaraldehyde followed by 1% osmium tetroxide, dehydrated in ethanol, embedded in Araldite, sectioned with diamond knives, stained with $KMuO_4$ followed by lead citrate, and micrographed variously in a Siemens 1a, 101, and a Phillips EM 300 with goniometer stage. This muscle was presumed to be in rigor. A variety of fixation techniques were attempted to get a good-looking cross-bridge lattice from fresh muscle or ATP-relaxed glycerinated muscle, with uniformly unsatisfying results. Very little order survived in the relaxed cross-bridge lattice and the results will not be further described here.

In order to embed single fibrils which had been examined and measured by light microscopy, the wet grid was rescued from under the coverslip, fixed 10 min each in glutaraldehyde and osmium tetroxide, block-stained 10 min in uranyl acetate, and embedded in agar to simplify further handling. Ordinary agar was too lumpy for subsequent light microscopy, so special grade agar was used. The grid was immersed in warm 2% agar solution, then gelled into a thin agar slab between two glass slides, then cut out in a square which was dehydrated through alcohols and embedded in Araldite. The embedded grid was oriented and trimmed to permit cross sectioning of the desired fibril at the point where light microscope measurements had been made. The fibril could be viewed after embedding by ordinary or phase contrast microscopy, the more easily if it had been prestained with uranyl. The exact position and angle for sectioning was determined with the aid of photomicrographs previously taken to show the fibril and grid square as originally seen under the light microscope. Liberal use of immersion oil helped to secure a view of the degree to which the trimmed block face coincided with the desired segment of fibril length. The plane of the embedded grid was parallel to a horizontal plane at the knife edge. Two or three copper grid bars were left within the block face. Glass knives were used; copper proved soft enough to be cut when sections between 0.1 and 0.2 μ thick are taken, as was done here. Sections were collected on Formvar coated slot grids to be sure the fibril would not be missed.

Two to four fibrils in different quadrants of a grid could be subjected to light microscope measurements and then sectioned. This required glutaraldehyde fixation of the agar gel after the grid was embedded, further hardening of the agar by dehydration and infiltration into liquid Araldite, and finally, halving or quartering of the grid by slicing the agar block with razor blades.

Calibration and microscopic measurement.

To a very great extent, the significance of our whole effort in the recent work depended upon the accuracy with which we could calibrate magnification and measure fibril dimensions. Primary standards were a stage micrometer cross-ruled in 10 μm intervals, used to calibrate an eyepiece reticle and the image-shearing eyepiece for light microscopy, and a grating replica cross-ruled with 2160 lines/mm used for electron microscopy to calibrate all instrumental settings and sessions.

For light microscopy each objective lens was individually calibrated, since we found that equivalent nominal magnifications did not give identical true magnification. For electron microscopy all lenses were normalized for hysteresis by switching on and off several times at the exact setting used, and the specific focal setting of coarse and medium objective lens current controls were noted during specimen microscopy and were duplicated (i.e., physical focal position in the microscope was duplicated) for exposure of magnification standards. Electron microscope lens settings which gave negligible distortion were found empirically and used throughout. Dimensions from electron micrographs were measured on the negatives using a Scherr-Tumico optical comparator, Model 22-1500.

A different approach was used in one case. In comparing dimensions of the rigor cross-bridge lattice, sections of water bug and fly flight muscle were mounted on the same grid, micrographs were taken in sequence and optical diffraction patterns were taken in sequence from these to show that the 38 nm and the 14.3 nm layer lines were identically spaced to better than ±1%.

Our estimates of cross-bridge population and dry mass concentration depended on the accuracy with which we could use the light microscope to measure sarcomere length and especially fibril diameter. Oil immersion 100× lenses for phase contrast or interference microscopy were used. In ten instances, the same fibril was measured with both objectives, giving results which were essentially identical, despite the much narrower condenser aperture used with the interference microscope. Average sarcomere length along a given fibril segment was obtained by estimating the number of sarcomeres (to the nearest 0.1 sarcomere) which could be fitted to one or two large divisions of a calibrated eyepiece reticle. Typical examples gave results such as 2.6 sarcomeres in one division, or 7.0 sarcomeres in two divisions, etc.

Fibril diameter was measured with a Vickers image-shearing eyepiece (VISE) mounted in the vertical tube of a trinocular eyepiece tube. The setting precision of this instrument is better than one-tenth of the resolution limit for a microscope objective, and it can give exact dimensions to this accuracy under some conditions (Dyson, 1960; see also Powell and Errington, 1963). Sheared fibril images as seen in the VISE are shown in Figs. 2L, 2M, and 2N. Just before the setting point, only the A bands were touching (they usually bulge slightly) so that an interrupted black line of image overlap was seen (Fig. 2M). When this overlap was just completely absent, the VISE reading was taken. Focus was crucial to this decision; the final focal search was most rapidly made by simply pressing vertically up or down on the fine focus knob rather than by rotating it. Operator constancy was important; one of us (M.K.R.) made 90% of the measurements and trained a technician carefully to make the remainder. Lack of fatigue was important; each VISE measurement was taken standing up after a 5–20 min session of seated microscopy on the same fibril. We used "double apposition," where the fibril image is sheared in both directions, and the central unsheared position is not read. Typical measurements were 10–20 divisions, representing 2–4 μm, reproducible to the nearest ±0.15 division, representing ±30 nm, and corresponding to the standard deviation of setting precision obtained by Powell and Errington (1963). When the interference microscope was used, the image was adjusted to give dark fibril edges with maximum contrast (Figs. 2F and 2L, M, N).

Early in the design of this project, 12 fibrils were embedded and cross-sectioned after VISE diameter readings (1.9–2.8 μm) to count myofilaments or filament layers across the measured diameter (see Fig. 3) and determine what thick filament separation, S, would account for the VISE measurements. The range was 50–59 nm; the mean was 55.4 ± 2.3 nm. Since the filament spacing in fiber bundles by X-ray diffraction is 53–56 nm (Reedy et al., 1965), we felt that the VISE gave fibril diameters accurate to possibly ±2% and we proceeded under this assumption. Late in the progress of the work reported here, we made a spot check and obtained rather different results. Five fibrils from our recent preparations were embedded and cross-sectioned after VISE diameter measurements, and the mean filament spacing necessary to account for the measurements was 62 nm, with a range of 59.5–67.0 nm. Our final conclusions are quite sensitive to filament spacing, and so it will be evaluated more carefully in future work, perhaps by X-ray diffraction of a fibril suspension or pellet.

Interference microscopy. The Zeiss 100X oil immersion interference equipment after Jamin-Lebedeff was used with Senarmont compensation (cf. Bennett, 1950) for measuring the retardation of the A band in flight myofibrils. Rather steady monochromatic light ($\lambda = 546$ nm) was obtained from a DC powered HBO 200W mercury arc with a glass heat filter and a Zeiss PIL interference filter. The WL microscope was elevated to permit illumination through the baseplate tube so that the pinhole used as a

luminous field stop with the 100× equipment could be normally positioned in the baseplate light exit. Köhler illumination was aligned so that adequate light was obtained after full closing of iris diaphragms in lamp housing and condenser and placement of the pinhole. The iris diaphragms were then opened very slightly. The half-angle of the illumination cone used for retardation measurements was 7° or less. Use of such low condenser apertures (NA of 0.12 or less) secured maximum contrast and setting sensitivity and prevented measurement errors due to oblique ray bundles as discussed by Ingelstam (1957).

We could not obtain optimum extinction without going beyond the directions given with the instruction manual. We have observed the same problem in several other sets of Zeiss interference equipment, so our remarks here may help others to understand and use it. We adjusted the equipment for imaging a myofibril with green light, then substituted a strain-free 10× or 100× objective lens for the interference objective. (The myofibril helped to confirm the identity of measuring and comparison beams during what follows.) The polarizer was exactly East-West. The analyzer was rotated to measure the plane of polarization of each of the two beams now visible, which turned out to be +48.5° (measuring beam) and −41.5° (comparison beam). These beams are supposed to be at +45° and −45°; they are thus 3.5° anti-clockwise from the specified azimuth in the case of our condenser. We attribute our extinction difficulty to this. Our additional procedure for obtaining best extinction, after the fixed 1/4λ compensator plate was in place and all other instructions to get extinction had been followed, was to rotate the *polarizer* anti-clockwise until the background was maximally dark; a rotation of about 7° was required.

We obtained identical measuring results with a different adjustment employing a rotating 1/4λ plate, set at 7° anti-clockwise from the position of the fixed 1/4λ plate, between accurately crossed and zeroed analyzer and polarizer. We had expected 3.5° to do it, but 7–9° proved necessary empirically for best extinction. Since the rotating 1/4λ plate is six times the price of the fixed plate, we do not recommend it.

Once the preceding adjustments were made, the analyzer could be set to zero (our preference) or nearby, and optimum extinction could be reacquired by readjusting the condenser beam tilt knob. This was the only subsequent adjustment ever needed to reacquire extinction when it was spoiled during searching or rotation of specimen slides.

Retardation of myofibrils was verified to be less than one wavelength by observing interference colors during illumination with white light. Phase retardation or optical path difference (OPD) was measured using Senarmont compensation, where the analyzer was rotated clockwise (mathematically negative) from the setting for maximum background extinction (Fig. 2E) to the setting for maximum extinction at the center of a succession of half-A bands (Fig. 2H). The image of a cylindrical fibril darkened progressively from the edges toward the center as the analyzer was neared (Fig. 2F). As the final setpoint was neared, only a minute faintly lighter center was evident against a darker background (Fig. 2G is 2° short of the setpoint for a fibril where total analyzer rotation was −24.5°), and the setpoint was read off where this central lightening just became dark. Each of the two analyzer positions was set and read off to the nearest 0.5° three or four times to establish a setting accuracy of ±0.5 for each point. Typical readings ranged from 20–50°, corresponding to total phase retardations of 40–100°, or total OPD's of 60–150 nm.

We doubt that any inadvertently flattened fibrils were included in our measurements; when they were squashed by coverslip pressure the effects were general, unmistakable and easily avoided.

Since the measuring beam was linearly polarized, some account was taken of the fact that myofibrils are birefringent (Figs. 2C,D,K). Birefringence was carefully measured in two fibrils. Two methods were used. A standard birefringence measurement was made using

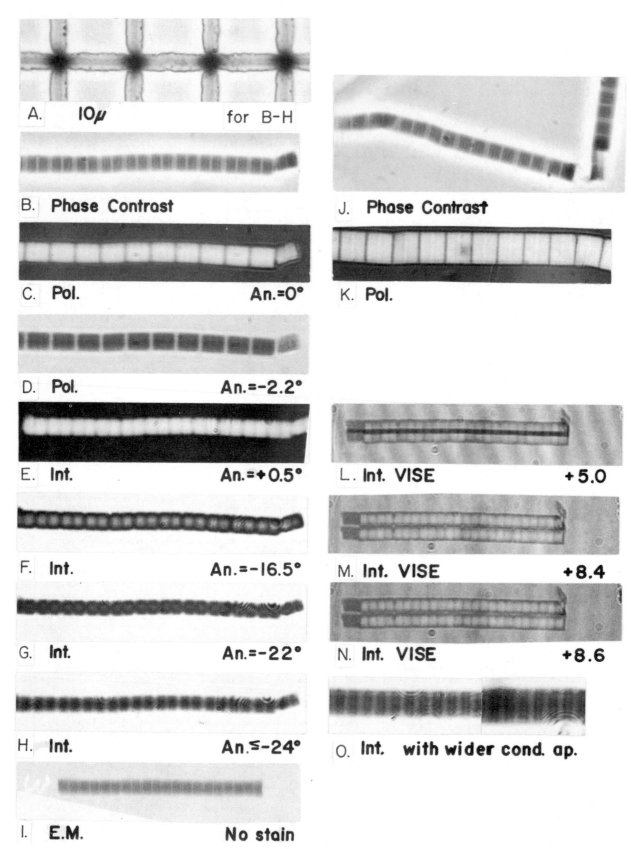

Figure 2. These show the images of *Sarcophaga* flight myofibrils by several techniques of microscopy, including phase contrast, polarized light, and interference microscopy. Analyzer settings (with the 1/4λ plate in place for Senarmont compensation) are printed beneath polarization and interference microscope pictures. The truest image of mass distribution in the fibrils is seen in *I*, an electron micrograph. The VISE images are shown with increasing shear; the difference between 8.4 and 8.6 represents 40 nm in fibril diameter. The micrometer *A* calibrates *B* and *J* exactly and the remaining light micrographs within ±2%.

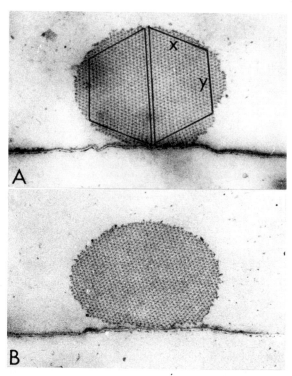

Figure 3. After viewing by light microscopy, fibrils adsorbed to grid support film may be fixed, embedded and transversely sectioned as seen here to check dimensions and filament numbers surmised from light microscope estimations. Film is visible beneath each fibril here. Elliptical form is due to section compression. Filament lattice may be variably oriented when fibril adheres to support film. Filament counting is simplified by striking off semi-hexagons, counting filaments along sides x and y, and calculating enclosed filament count $(x^2 + 2xy - x)/2$, then counting remainder of excluded peripheral filaments. ×23,000.

strain-free optics and Senarmont compensation (see Bennett, 1950). Interference optics were then substituted, and the retardation of the same fibril was measured with the fibril parallel to and then normal to the plane of the measuring beam. Both methods indicated that the retardation of green light polarized parallel to the long axis of the fibril was about 10% greater than the retardation when it was polarized normal to the fibril.

In an effort to standardize the influence of birefringence, we rotated many fibrils to lie parallel to the plane of the measuring beam. When this could not be done without superimposing ghost images of grid bars or other fibrils on the object area, some other arbitrary position was used, but the approximate angular departure from the desired azimuth was recorded in case some eventual correction were to become appropriate.

In the end we could not conveniently adapt photography or photometry to the measurements of retardation required for this stage of our study. We chose to depend upon the cylindrical form of myofibrils, measuring their horizontal diameter with the image-shearing eyepiece, then measuring the maximum OPD along the vertical diameter through the center of each half-A band. The measured diameter expressed in nm was used as the thickness, t_0, in calculating dry mass concentration in the A band from the expression $[mass] = OPD/t_0\alpha$, where OPD is 3.03 nm/ degree of analyzer rotation with Senarmont compensation at 546 nm and α is the specific refractive increment (Barer and Joseph, 1954). α is taken as 0.18 cm³/g initially here, although it has been measured as 0.19 cm³/g for

myosin (Rupp and Mommaerts, 1957; Gellert and Englander, 1963) and 0.17 cm³/g for actin (Mommaerts, 1952).

Quantitative electron microscopy. For quantitative electron microscopy, all electron micrographs and photometric readings were made in G. F. Bahr's laboratory. Procedures used at the AFIP for this technique have been previously described (cf. Zeitler and Bahr, 1962; Bahr and Zeitler, 1965; DuPraw and Bahr, 1969). The following is therefore not complete, but serves to emphasize special features of the approach to our material and some aspects of the technique not elsewhere described.

Formalin-fixed, air-dried myofibrils on Formvar-coated Maxtaform Finder grids were photographed at 940× (±1% distortion) in a Siemens Elmiskop 1a using 100 kV, a 100 µm objective aperture, 100 µm aperture condenser, and beam currents on the order of 3×10^{-6} A/cm² at the specimen. Kodak medium projector slide plates were exposed 3 sec and developed 5 min in D 72 developer, water-washed and fixed. Fibrils previously measured by light microscopy were approached on each Finder grid with the aid of a map. To minimize beaming of the fibrils, focusing was done on an adjacent square and the beam intensity was adjusted to that used in photography. The desired fibril was only then brought into the beam for the first time. Photography was usually completed within 30 sec or less. At this low beam intensity, irradiation extended for 2–4 min only reduced the measurable mass by 1–3%, as determined by photometric weighing. A 1–2 min exposure with a bright crossover beam was adequate to remove 15–20% of the mass of the irradiated segment of the fibril, as would be expected when a dose approaching or exceeding 10^{-2} C/cm² was delivered to the specimen (see Bahr et al., 1965; Stenn and Bahr, 1970). Beam current was measured using a Keithley 410 pico-ammeter to read the current incident upon an 18 mm sensor of silver (an American dime mounted just above the final image screen), as described by Bahr et al. (1963). Readings from $0.4-0.7 \times 10^{-11}$ were appropriate to give the desired optical density of 0.5 to 0.75 for the background (empty support film) of the electron microscope negative. This density range corresponds to 17–31% transmission.

The emulsion becomes significantly more sensitive to electrons as desiccation is prolonged before exposure. We followed a standard procedure used in the AFIP laboratory for the 1968 experiments. The camera was loaded with a dozen freshly opened undesiccated plates; they were forepumped for 15 min and then desiccated for about 20 min more under high vacuum until high voltage could just be switched on; the plates in a loading were then exposed rapidly in the next 10 min, taking all myofibrils as they were encountered on a specimen grid. In the present case, where specific fibrils were being sought and photographed, it sometimes took up to an hour or 90 min to find, prepare for and photograph all 10–15 fibrils of interest on a given grid, taking suitable precautions to avoid irradiation until absolutely necessary. It was therefore necessary to calibrate the sensitivity increase as time of desiccation proceeded. Typically, 0.7×10^{-11} A for 3 sec produced a negative with 25% transmission from freshly desiccated plates, whereas a lesser exposure of 0.6×10^{-11} A was used for plates desiccated 30–90 min. When plates were unavoidably desiccated for still longer periods, 0.5×10^{-11} A was used. These adjustments exposed plates to a density of 20–25% transmission with high reliability. Under other instrumental conditions these precise values will certainly not obtain, but some calibration of the "emulsion desiccation factor" will be similarly helpful.

Absolute mass calibration standards (micrographs) of mixed sizes of polystyrene latex spheres were taken with each session and processed with each rack of plates, along with magnification standards. Just as specimen magnification was chosen to fit the fibril images into standard rectangular apertures of the photometer, so the spheres were photographed at 7120× so that their images would fit the standard disc aperture sizes in the photometer. To

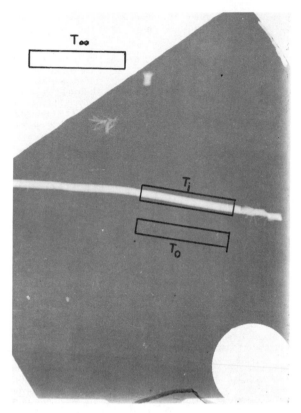

Figure 4. This duplicates the appearance of an electron microscope negative used for photometric weighing of a *Sarcophaga* myofibril segment. Rectangular aperture is used in each of the three locations shown to give each of the three readings described in text. Mass is proportional to the difference between the background reading and object image reading. Mass per sarcomere is calculated on the basis of counting (in this case) 7.15 sarcomeres within aperture. ×1200.

get the necessary exposure, this required specimen irradiation at 2×10^{-4} A/cm². Polystyrene is more resistant to radiation mass loss than many other polymers. A 2–4 min exposure of spheres at this beam intensity removed 2% or less of their mass by photometric weighing. We used only freshly made grids and photographed each group of spheres only once to avoid radiation losses in the standards.

In order to determine relative mass distribution along a fibril, a scanning densitometer (G-3 Schnellphotometer, Zeiss-Jena) was adjusted to produce a linear response to transmission, T_i, as the illuminated fibril image was drawn past a slit. The base line, T_0, was established by a scan of the darker image of adjacent empty support film, and the difference, $T_\Delta = T_i - T_0$, was treated as a direct analog display of the varying relative mass profile within the different parts of the sarcomere (Fig. 4).

Absolute mass measurements were obtained through the use of a special purpose integrating photometer, the Zeiss IPM-2, from which integrated transmission readings can be read directly. Each photometric reading, R, requires that an aperture be selected to fit around an object image and that one setting and two readings be made with this aperture. (Figure 4 illustrates this for a myofibril image; the same procedure is followed with disc apertures for sphere images in reading mass standards.)

T_∞ = an unexposed area of plate, corresponding to infinite mass; the dial or digital readout is adjusted to read 100% transmission here for each aperture used on each

negative, before reading T_i and T_0. T_i = reading of image of the object. T_0 = reading of adjacent background, corresponding to zero mass. If this reading was less than 17% or greater than 31%, the negative was not used.

The desired reading $R = (T_i - T_0) \times A$, where A is different for each aperture and corresponds to the negative area seen through each aperture. For the mass standards, R is plotted against the known mass of the spheres to obtain the proportionality factor for converting R into absolute weight, W. (Sphere mass is obtained from their specific gravity (1.05) and volume, based on individual diameter measurements.) This proportionality factor is adjusted for the difference in magnification between standards and specimens in order to convert specimen readings to absolute weights. In the case of myofibrils, a Z line or M line was fitted against the end of an aperture, the exact number of sarcomeres fitting into a standard aperture was counted (to the nearest 0.1 sarcomere), and the weight per sarcomere was calculated. Then, the sarcomere volume for this particular segment of fibril was calculated from the original light microscope measurements of sarcomere length and fibril diameter taken when the fibril was still wet, and the dry mass concentration was calculated in g/cm³ = pg/µm³.

Results

Our basic strategy in the 1972 experiments has been to use microscopy to determine the dry mass concentration for intact fibrils and the concentration of cross-bridges or crowns in the same fibrils, but to develop the myosin percentage and myosin molecular weight from protein analysis of the same fibrils (reported by Bullard and Reedy in the following paper). It should be noted that both IMy and QEMy measure the total mass of particles. Our assumption that myofibrillar total mass represents nothing but protein remains to be checked in further work. We want to describe first the appearance of the fibrils with various light microscopes, and then some results from electron microscopy of fibers and fibrils, before going on to consider the distribution and measurement of dry mass. All of the figures and general text descriptions concern fibrils from the dorsal longitudinal muscles (DLM) of *Sarcophaga*. One group of mass measurements and a few observations have been specifically identified with fibrils from the corresponding muscles in *Lethocerus*.

Light microscopy of isolated fibrils. Figure 2 shows the appearance of a single typical fibril by three different techniques of light microscopy (B to H) for comparison with a fibril at the same magnification by electron microscopy (I). By phase contrast microscopy, the typical cross striation pattern (type I pattern of Hanson, 1956; "elongated" pattern of Aronson, 1962) is remarkable for the faintness of the Z bands, the multiple "substriping" of each half A band (B, J), and the uniformity of H and I band lengths encountered over a range of sarcomere lengths from 2.5–4.1 µm. Z bands are equally faint by electron microscopy of isolated unstained fibrils (I) and are not at all visible by polarized light (C, D, K) nor by interference microscopy (E to H, O). Fibrils with visibly narrowed I or H bands or with prominent Z bands were taken as an indication of an alteration in sarcomere length which was not acceptable for mass measurements.

A band substriations were not evident by electron microscopy (Figs. 2I, 5A, 7B, 7D) or polarization microscopy (Figs. 2C, D, K). However they were quite marked in phase contrast and still obvious in interference images. This was easier to recognize by dark or positive contrast (Figs. 2B, J, H, O) than by light or negative contrast (Fig. 2E). Like Aronson (1962), we regard it as a feature of image structure but not of fibril structure, because we could not detect it by electron microscopy and because we found it to vary with numerical aperture and sarcomere length. Using a phase contrast objective with a built-in iris diaphragm, we observed that a typical half A band had two dark stripes when NA was set between 0.6 and 0.8, and three dark stripes whenever higher NA was used (Fig. 2B, sarcomere length 3.23 μ). As for the variation with sarcomere length, four dark stripes instead of three were visible per half A band, using a Zeiss phase achromat of NA 1.25, whenever sarcomere length exceeded about 3.65 μm (Fig. 2J, SL = 4.02 μm), and substriping was often of low contrast unless sarcomere length was at least 0.2 μm more or less than this threshold value.

In the interference microscope image, there were always two substripes per half A band (2H), which became much clearer but did not change to three stripes as the condenser aperture was opened. (NA for the interference objective is nominally 1.0.) We decided to disregard this complexity of interference image structure for measurement purposes, and we measured by setting extinction for the entire half A band, which almost appeared as a single stripe anyway when the condenser aperture had been stopped down to 0.10 for accurate retardation measurements (2E, F, G, H).

Lesser cross bands such as the I band and H band were found by electron microscopy (Figs. 2I, 5A, 7B, 7D) to be shorter in relation to sarcomere length than light microscope appearances indicated.

We have no notable misgivings about our assumption that the suspension of centrifugally washed fibrils would be virtually 100% protein, perhaps contaminated by less than 1% chitin from microtracheoles. However we have come to appreciate that our former routine use of uranyl block-staining eliminates and thus conceals the massive glycogen infiltration of the H band otherwise found in sections of whole (even glycerinated) fibers. (Compare Fig. 5B with Figs. 6A, 7D; the latter two were block stained.) The marked birefringence (Figs. 2C, D, K) and low interferometric retardation and low mass thickness of the H band by electron scattering all suggest that this glycogen is no longer present in the fibrils as we prepared them, but this supposition has not been definitively checked by PAS-staining and amylase digestion. The presence of such a quantity of glycogen could possibly add 2–5% nonprotein to total fibrillar dry mass.

Despite the great regularity of the myofilament lattice and the cross-bridge lattice in fibrillar flight muscle, we encountered substantial variations in sarcomere length and in fibril diameter, even apart from developmental changes, in agreement with Hanson (1956), Aronson (1962), Gregory et al. (1968), and Auber (1969). We tried to evaluate overall mean sarcomere length for the six DLM fiber pairs from flies aged 2 and 14 days by light diffraction using a laser, which was difficult even after fibers had been partially cleared by up to a week of extraction in glycerol solutions and detergent solutions. (We found that the half-sarcomere spacing dominated the diffraction pattern.) Diffraction indicated sarcomere lengths of 2.5–3.0 μm for the smallest fibers No. 1 and No. 2 (Auber, 1969) and 3.3–3.9 μm for the larger four fibers. The range of measured diameters for isolated fibrils from flies 1–7 days old is from 1.8–5.2 μ.

The mean values with standard deviations for 66 fibrils from two lots of 7-day-old flies were: sarcomere length, 3.46 ± 0.37 μm; fibril diameter, 3.20 ± 0.66 μm. Diameters were biased in that we preferred wide fibrils for interference microscopy, and 61 of these 66 selected fibrils were wider than 2.3 μm, which tends to indicate our cutoff.

Electron microscopy: cross sections of whole fibers.

The gross anatomical arrangements and the general appearance of the dorsal longitudinal flight muscles of *Sarcophaga* can be adequately understood by applying the description of *Calliphora* from the paper on myofibrillogenesis by Auber (1969). The variation with age of sarcomere length and of the fibril diameter and filament number per fibril is well documented there. We found that each giant fiber of the 2-day-old *Sarcophaga* was semi-partitioned into 50 or 60 subfibers, each subfiber with a complement of 300–500 myofibrils. In cross sections large areas looked as though the fibrils were of uniform diameter, but there was an impression, later verified, of larger fibrils at the outermost surfaces of the giant fibers. The fibers are arbitrarily numbered by Auber from 1 to 6, running dorsally to ventrally and shortest to longest, and we adopted the same numbering. In counting filaments in our single-fly experiment to determine a value for the mean number of thick

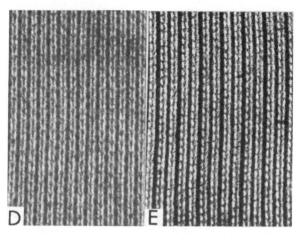

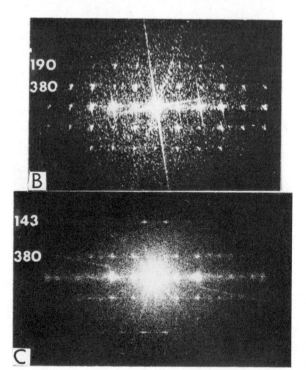

Figure 5. *A*, half sarcomere from Z band (*top*) to glycogen-crowded H band from a rigor specimen of glycerinated *Sarcophaga* flight muscle. Axial beading marking 38.8 nm period of cross-bridge lattice becomes one-sided laddering due to cross-bridge helix when section is tilted top-forwards by 45° as in *E* (74,000 ×). Although *B* and *C* were both from rigor specimens, 14.5 nm period is clear only in *C*, split across meridian. *D* (74,000 ×) is enlarged from *A* (47,000 ×) and *B* is a diffraction pattern from *A*.

filaments in a cross section of a fibril, we were anxious to obtain a random and representative sample. We cut cross sections of each whole giant fiber, a profile extending about 300 μm $\times$ 700 μm in the 2-day-old fly. Fully visible on slot grids, this large area was sampled by superimposing a square grid, using the calibrated controls for specimen stage traverse movement of the Siemens 1a electron microscope. A grid with meshes 100 μ $\times$ 100 μ was easily generated by the controls, and micrographs showing 20–30 fibrils in cross section were obtained at every nodal point. Counts were completed in all micrographs for two fibers, No. 1 and No. 5. We encountered a range of 860 to 1850 filaments/fibril. We developed our sampling grid after first finding in random micrographs of fiber No. 1 that all 23 fibrils in one EM taken at the periphery averaged 1355 filaments/fibril, whereas 64 fibrils in two EMs of the central region averaged 1114 filaments/fibril. We tried to be sure that our sampling grid was laid out so as not to sample the periphery disproportionately. Total filament counts were made on each fibril using a technique indicated in Fig. 3. For 80 such fibrils (16 micrographs, 5 fibrils each) in fiber 1, the mean filament number was 1174 $\pm$ 162. For 114 fibrils of fiber No. 5 (19 micrographs, 6 fibrils each), the mean was 1338 $\pm$ 89. The grand mean for these two fibers was 1270 $\pm$ 148; we considered that it was probably biased slightly in favor of smaller fibrils, but we used it in calculation anyway. We never made such systematic counts in older flies, but it is clear that fibril diameters and filament numbers were in a substantially higher range for 7-day-old flies.

Electron microscopy: cross sections of single fibrils.

Single fibrils were embedded and sectioned after light microscopic measurements primarily to check the relationship between VISE measurements and actual filament spacings and numbers across the measured diameter. The results of this have already been noted in the Methods section describing VISE technique. An additional point we want to make here is that 11 out of the 17 fibrils to which this has so far been applied looked considerably messier than those seen in Fig. 3; it required some care to single out the filament layers and count them accurately. However we have been relieved to trace this largely and perhaps entirely to a problem in accurate orientation of transverse sections, in that crisp, good-looking end-on views of a well-ordered filament lattice have been obtained in several of these by seeking such views with the aid of a tilting stage.

Electron microscopy of the cross-bridge lattice.

Our knowledge of cross-bridge numbers in *Sarcophaga* depends on sections from embedded rigor muscle which demonstrate that periodic structure (Fig. 5) and cross-bridge grouping (Fig. 6) are very much like what has been worked out for the rigor cross-bridge lattice of *Lethocerus*. A 38.8 nm axial period (38.0 nm due to shrinkage in embedded material) is conspicuous in relatively thick (150 nm) longitudinal sections cut or tilted to give a type I view (Reedy, 1968) of the filament lattice (Figs. 5A, D). Forty-one repeats can be counted in the half A band seen in Fig. 5A. This period is absent from the H band and the I–Z band regions. (The gap is usually equal in length to about 4, sometimes 5, repeats of the period in each of these bridge-free zones.) We have found that different sarcomere lengths involved differences in the length of the A band, of the thick and thin filaments, and thus in the number of repeats of this period, whereas the H band and I–Z region appear to remain at constant length. We have counted 39–45 repeats of this period in different fibrils and half sarcomeres of a single fiber. (For comparison, *Lethocerus* tends to have 28–31 repeats per half sarcomere.) This is strikingly important. It extends previous observations showing a natural variation in A band and thick and thin filament lengths in certain other Arthropod muscles (Franzini-Armstrong, 1970) to the most seemingly regular of all Arthropod muscles, the fibrillar insect flight muscles. Furthermore it indicates that the variations in sarcomere length detected by light microscopy or light diffraction can be confidently interpreted as variations in filament length, so long as contractures or ruptures are absent. In a way it means that the light microscope can be used to estimate the number of cross-bridges along the thick filaments of any normally cross-striated sarcomere, regardless of its length.

As in *Lethocerus* 38.8 nm evidently marks the length for a half-turn of twin helical structure along both actin and myosin filaments, marked by the projection of cross-bridges from myosins and the attachment of bridges to actins. The appearance in thick sections is consistent with this. The period is expressed in Figs. 5A and 5D as a dense beading along thick and thin filament profiles, in lateral register from thick to thick or from thin to thin profiles. The bridges projecting laterally (thin filament beading) are staggered by one-quarter turn of a helix ($\frac{1}{4}$ of 388 Å) with respect to the bridges projecting vertically, overlying the thick filament. Figure 5E represents a tilted view of the same sort of type I section. Tilt axis is horizontal, and the filaments tilt forward toward the observer by 45° as they run toward the top of the page. The cross-bridge period has become very marked and regular on the right side of all of the thick filament profiles,

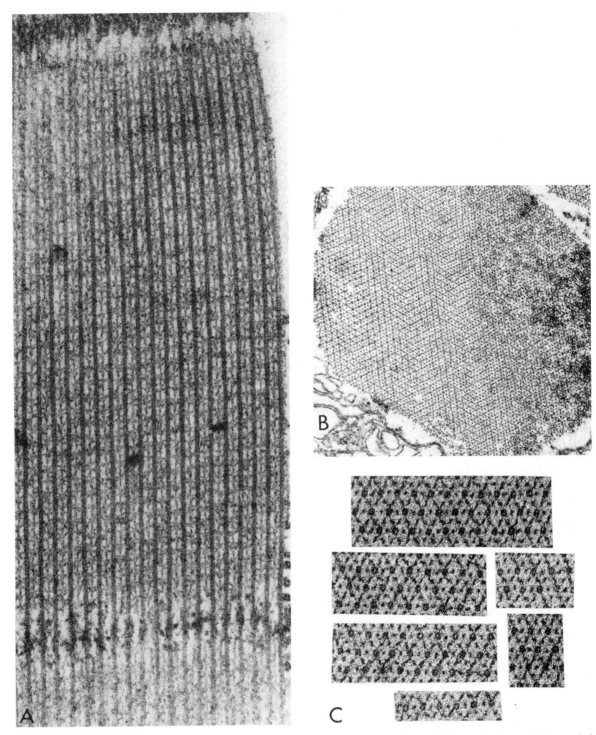

Figure 6. *A*, a single myac filament layer from *Sarcophaga*. Chevrons, often double, mark 39 repeats of cross-bridge period of 38.8 nm in this half sarcomere. Thick filament terminals appear to split and join thin filaments rather than entering Z band, evidently a modification of the usual C filament morphology. ×94,500. *B*, 19 nm thick transverse section crosses 2 μm fibril with slight obliquity, leaving Z band at right and advancing through four repeats of 38 nm axial period before reaching left edge of fibril. Resulting herringbone mosaic pattern cannot be seen in thicker sections. ×31,000. *C*, four bridge groupings of the flared-X form, seen in very thin transverse sections like *B*, are displayed here. Each thick filament is connected to four of the surrounding six thin filaments by bridges. ×92,000.

but has become quite indistinct on the left of each thick profile. This is because this tilt lets the viewer sight directly along the right-sided gyres of the cross-bridge helix, and, as in *Lethocerus* (Reedy, 1968), the helix is a double left-handed one along the myosin filaments.

In single myac filament layers, the 38.8 nm repeat is clearly associated with the transverse rows of cross-bridges (Fig. 6A). The chevron configuration of cross-bridges is the dominant regularity in the lattice. However the regularity of the chevron form, the double nature of the chevron repeat, and the overall clarity and regularity of the lattice pattern are all somehow less impressive than in corresponding views of *Lethocerus* flight muscle (Reedy, 1967, 1968, 1971). The optical diffraction patterns from micrographs of the *Sarcophaga* lattice (Figs. 5B, C) have not shown the degree of order nor been so reproducible from one rigor specimen to another as was the case with *Lethocerus*. The diffraction patterns indicate some difference in rigor lattice structure from *Lethocerus*. *Sarcophaga* patterns have not yet shown any equivalent for the *Lethocerus* layer lines corresponding to 116, 58, 23, and 12.7 nm periods (Reedy, 1967). The 14.5 nm line is variably weak, smeared, or clearly split across the meridian, in the latter case exactly as found by Elliott (1965) for the 14.5 nm and 7.3 nm periods in X-ray diffraction patterns from another blowfly, *Calliphora*. The 19.4 nm layer line is surprisingly variable in *Sarcophaga* optical transforms, and the micrographs have not yet given evidence for preservation of 7.3 nm and 4.8 nm periods in *Sarcophaga* as they have in *Lethocerus*. Inspection of the cross-bridge detail in our micrographs has not yet shown us the basis for the meridional splitting of the 14.5 nm layer line when this was present. Further studies of this have been held in abeyance because it seemed that the best evidence would come from ordered views of the relaxed *Sarcophaga* cross-bridge lattice, but our efforts to obtain such pictures have met with no success up to now.

Very thin transverse sections offer an encouraging point of further similarity with *Lethocerus*. The effect of lateral register in this lattice of cross-bridge helices is clearly detectable in the herringbone mosaic appearance of such sections when they traverse the fibril with a very slight obliquity (Fig. 6B). Such a section catches single crowns or cross-bridge groupings in successive azimuthal orientations at different levels along the axis. Upon close inspection, these sections display the same flared-X four-bridge grouping (Fig. 6C) which led Reedy (1968) to conclude that *Lethocerus* muscle had four cross-bridges per crown. That same conclusion now seems justified for *Sarcophaga*.

It may take much further work before the cross-bridge lattice of *Sarcophaga* can be as thoroughly specified on structural grounds as that of *Lethocerus*. Until then, the question may remain as to whether the less perfectly ordered lattice might indicate a different number of cross-bridges per unit length of filament or per unit volume of lattice than is the case for *Lethocerus*. It is certainly possible to compare Fig. 6A with similar views of *Lethocerus* myac layers and to receive the qualitative impression that there may be more bridges, grouped in a less orderly manner, on display in *Sarcophaga*. Whatever the differences in cross-bridge arrangement, it will be seen (below) that the number of myosins per crown is virtually identical in *Lethocerus* and *Sarcophaga*.

Dry mass distribution in the sarcomere. The mass thickness profile of a typical sarcomere was examined for two reasons. We needed to know how or whether myosin and nonmyosin could be reasonably fitted into the profile of the sarcomere in order to produce plausible calculations of myosin concentration in the cross-bridge zones. Also we wanted to know whether the cross-bridge zones of each half A band would have a significantly different dry mass concentration, $[mass_A]$, than the mean concentration for the whole sarcomere or fibril, $[mass_F]$, since we were measuring $[mass_F]$ by QEMy and our nonintegrating point-measurement approach to interference microscopy could only be used to measure OPD through the cross-bridge zones to give a value for $[mass_A]$.

Using standard micrograph negatives taken as part of our QEMy, we made densitometer scans of the transmission (mass thickness) profiles of several sarcomeres and found them qualitatively identical, with Z band and A band heights closely similar and a modest dip at H band and I band (Fig. 7C). We selected one well-qualified sarcomere for closer study (Figs. 7A, B). It was 3.42 μm long and had lengthened only 1.5% in mean sarcomere length during the switch from hydrated light microscopy to vacuum-dried electron microscopy. The fibril had straight sides, and the finer cross striations were straight and relatively undistorted by drying collapse artifacts. The negative image was scanned by a slit of size and placement shown in Figs. 7A and B to give the T_A profile. The actual height of the T_A profile in Fig. 7C was obtained by subtracting the T_o scan of the adjacent background made with the same slit.

We tried, but discarded the attempt, to decide band lengths from the tracing on internal evidence. Instead, we measured two fibrils of embedded and sectioned muscle showing what we believe is the same sarcomere pattern and same A band length,

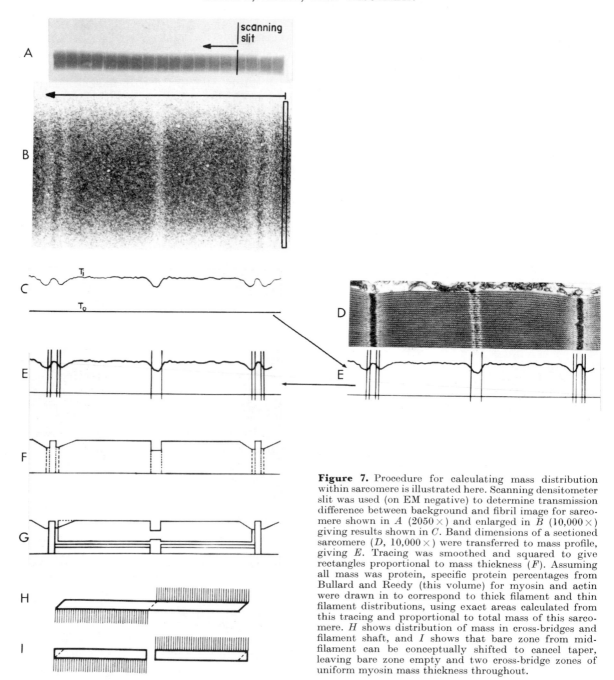

Figure 7. Procedure for calculating mass distribution within sarcomere is illustrated here. Scanning densitometer slit was used (on EM negative) to determine transmission difference between background and fibril image for sarcomere shown in A (2050×) and enlarged in B (10,000×) giving results shown in C. Band dimensions of a sectioned sarcomere (D, 10,000×) were transferred to mass profile, giving E. Tracing was smoothed and squared to give rectangles proportional to mass thickness (F). Assuming all mass was protein, specific protein percentages from Bullard and Reedy (this volume) for myosin and actin were drawn in to correspond to thick filament and thin filament distributions, using exact areas calculated from this tracing and proportional to total mass of this sarcomere. H shows distribution of mass in cross-bridges and filament shaft, and I shows that bare zone from mid-filament can be conceptually shifted to cancel taper, leaving bare zone empty and two cross-bridge zones of uniform myosin mass thickness throughout.

and we sliced the mass profile in accord with these length measurements, allocating 5.2% for H band, 5.8% for the I–Z band, and 3.2% for the Z band alone. The wavy tracing with its rounded corners was converted to straight lines and sharp corners by eye, while preserving the mean tracing heights and the taper at the ends of the A band, adjusting for the knowledge that our 50 μ slit must give a sloped tracing even while advancing over a sharp density step. The tracing with which we worked was 50 mm

high and sarcomere length was 317 mm. The area blocks resulting from the procedure just described were measured. The total summed area proportional to total mass for the single sarcomere was 15,070 mm². The sub-band areas were calculated as a percentage of the total, giving Z = 3.2%, I–Z = 5.6%, and H = 3.8%. These areas correspond to the fractions of total sarcomere mass located in each band.

We have assumed that no glycogen, lipids,

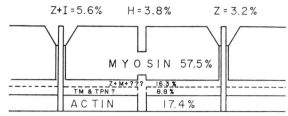

Figure 8. Sarcomere dry mass distribution in *Sarcophaga* flight myofibrils. This is a vertically expanded version of Fig. 7G, in which areas are proportional to mass. The relative sarcomeric fractions of mass located in the H band, the Z band, and the I–Z band as found in 7F are given above the diagram. Myosin quantity and myosin-actin ratio were given appropriate locations and areas. Tropomyosin and troponin are tentatively proposed on the basis of 2 tropomyosins and 2 troponins for every 14 actin subunits. Remaining unknown material includes part of the M line, the Z line, part of the I band, as well as part of the A band. Mass thickness for the highest part of the A band is $1.04\times$ greater than the mean mass thickness for the entire sarcomere. Myosin actually represents 62% of this mass thickness throughout all but the tapered ends of the A band.

nucleic acids, minerals or even significant amounts of formaldehyde were present in the fibrils. We have considered all the dry mass to be protein. Taking the values of Bullard and Reedy (this volume) for myosin as 57.5% of fibrillar protein and myosin/actin ratio of 3.31 w/w, we calculated the areas for these two proteins in our typical sarcomere mass profile and fitted them in accord with our knowledge of the length of thick and thin filaments.

An appropriate small area of the H band was given over to myosin. It was considered that the taper at the ends of the A band was due to the taper at the ends of the thick filaments (7H). The "missing mass" due to the taper at each end (indicated for one end by dotted triangle, Fig. 7G) was calculated, and a corresponding area of myosin mass (2.6% of total myosin) was placed in the H zone. The remaining myosin mass was fitted into the cross-bridge zones most simply by calculating the total area of these zones (A band minus H band), subtracting myosin, and then blocking in all the nonmyosin remainder as rectangular areas at the bottom, giving myosin all the upper portion which includes the tapered feature of the mass-thickness profile. The result gives a mass-thickness for myosin in the H band which is 43% of the mass-thickness of myosin throughout the A band plateau, in very rough agreement with the $\frac{250}{470}$ ratio of the weight of the myosin rod (LMM plus HMM-S_2) to whole myosin.

The actin area has been calculated as $(1/3.31) \times$ myosin mass and located like the thin filaments between Z and H bands. We have assumed that the thin filament extensions which penetrate the Z band are nonactin. If actin were to penetrate the Z

bands completely to the opposite surface, then actin filaments would be 1.08 times longer, and actin mass thickness would be 0.93 times the thickness actually portrayed in Figs. 7G and 8, and about 35% of the mass thickness of the Z band would represent actin.

Tropomyosin and troponin percentages have yet to be properly quantified for *Sarcophaga*. The area assigned to them in Figs. 7G and 8 is probably a maximum. It is calculated in proportion to actin mass by assuming that, as in rabbit skeletal muscle, every 38.8 nm length of the myofilament accounts for 14 actin subunits (mol wt 45,000), two tropomyosin rods (mol wt 70,000), and two troponin complexes (mol wt 89,000) (see other papers in this volume, for example, Ebashi et al., Hartshorne and Dreizen, and Cohen et al.).

The unassigned 16% of the mass diagram may be given over to the Z and M line specific proteins, to C filaments (Auber and Couteaux, 1963), and to a core material which we have found variably affecting the hollow appearance of *Sarcophaga* thick filaments in cross sections (centrofibrillar filaments look more solid) or by negative staining. (Such core material appears to be strikingly absent from all thick filaments of *Drosophila*; cf. Goode, 1972.)

The mean mass thickness for the entire sarcomere turns out to be 0.96 of the mass thickness for the plateau in the cross-bridge zone, so that one might expect that values for [mass$_A$] obtained by interference microscopy could be 4% higher than those obtained for [mass$_F$] by quantitative electron microscopy. Furthermore myosin represents 62% of the mass thickness of the plateau region of the A band, so that one would expect this amount to be lost if interference microscopy could be used to monitor the change in OPD during specific and quantitative myosin extraction under the microscope. Unfortunately the conditions for such precise extraction seem to have eluded us so far (see Bullard and Reedy in the following paper).

Weighing sarcomeres: the single fly experiment, 1968-1969. In our single fly experiment of 1968–1969, we made mass measurements by QEMy on fibrils from a homogenate of the six giant fibers of the left DLM of a 2-day-old *Sarcophaga*. The fibrils were washed with rigor solution containing 1% Triton X-100, resuspended in rigor solution, and adsorbed to Formvar-coated 100 mesh grids. Some grids were then floated 5 min on H-S-Z solution to extract myosin. All grids were finally floated 5 min on drops of buffered 4% paraformaldehyde, then washed in distilled water and air-dried. The grids were then photographed as we came to them under the standard conditions described in Methods for QEMy, rejecting only badly kinked, distorted, or frayed fibrils, but rejecting none on the basis of size. Photometric weighing followed.

We chose formaldehyde fixation over critical point drying and glutaraldehyde fixation because lateral shrinkage after these latter two treatments led to dried fibrils too dense to be adequately penetrated by 100 kV electrons. We also suspected that glutaraldehyde would add more mass than formaldehyde, and recently hemoglobin has been shown to bind enough glutaraldehyde to add $\frac{1}{3}$ to its weight (Morel et al., 1971), which would obviously be undesirable for QEMy. If formaldehyde were similarly reactive, which it is not, it could produce up to a 10% increase in weight.

We weighed 300 sarcomeres in 40 intact fibrils from our single fly, obtaining a mean weight of $2.3 \pm 0.35 \times 10^{-12}$ g/sarcomere. In another 41 extracted fibrils, the mean weight of 307 sarcomeres was $1.30 \pm 0.52 \times 10^{-12}$ g. Our interpretation then was that an average sarcomere contained 1.0 pg of myosin, all of which was extracted by the H-S-Z solution. When we last reported on that work (Reedy et al., 1969), our filament counting in fibrils of the other (embedded-sectioned) side indicated a mean of 1167 filaments/fibril, and an average cross-bridge zone length of $3.30 \mu/$ sarcomere. Assuming four bridges per crown and a crown every 14.5 nm on every filament, the cross-bridge population in an average sarcomere was calculated to be

$$\frac{3.3 \times 10^3 \text{ nm/filament}}{14.5 \text{ nm/crown}} \times 1167 \text{ filaments/ sarcomere}$$
$$\times 4 \text{ bridges/crown} = 1.06 \times 10^6 \text{ bridges/sarcomere.}$$

The weight of myosin per bridge was thus calculated as

$$\frac{1.0 \times 10^{-12} \text{ g}}{1.06 \times 10^6 \text{ bridges}} = 0.94 \times 10^{-18} \text{ g/bridge.}$$

Now, multiplying this by Avogadro's number of daltons/g gives 5.68×10^5 daltons/bridge, and we concluded that this was a reasonable weight for one insect myosin molecule. However now that Bullard and Reedy (this volume) have shown that insect myosin weighs about 4.7×10^5 daltons instead, and that the myosin content is about 57.5% of sarcomere mass, rather than 43%, the myosin estimations from our single fly experiment are chiefly of historical interest.

If our ignorance about myosin weight and percentage was a major weakness of the 1968–1969 experimental strategy, our uncertainty about filament spacing in isolated fibrils represents a corresponding weakness in the strategy of our more

recent experiments. Either strategy now seems capable of giving the desired type of answer if we can remedy the weakness in each original experimental design. The results of the single fly experiment can be used, in fact, to give the type of answer sought by the 1972 experiments. Continued counting altered the average to 1270 filaments/fibril for the single fly experiment before counting was discontinued. If we assume that these filaments were separated by 53 nm as in intact muscles, the cross-sectional area of a cylindrical fibril would be (1270) (0.866) (53 nm²) or 3.09 μm². Mean sarcomere length was estimated at 3.65 μ from optical diffraction of semi-thin sections of this fly and from optical diffraction of other 2-day-old fly fibers. The mean sarcomere volume so calculated can be matched with the mean sarcomere mass from intact fibrils in the same fly, giving

$$\frac{2.3 \times 10^{-12} \text{ g}}{11.3 \times 10^{-12} \text{ cm}^3} = 0.204 \text{ g/cm}^3 \quad \text{for} \quad [\text{mass}_F].$$

This is how we calculated the value entered for this experiment in Table 2.

Weighing sarcomeres; the single fibril experiments, 1972. The results of six recent experiments on fibrils from two lots of 7-day-old flies are summarized in Table 2. The major part of each fibril suspension was shipped on ice to England for protein analysis as reported by Bullard and Reedy (this volume). These 1972 experiments actually fall into three groups. The first group included one experiment using each technique, to see if IMy and QEMy gave approximately similar [mass] values on the same lot of fibrils. In Experiment 3-24-72, phase contrast microscopy was used to measure sarcomere length and fibril diameter in complete rigor solution of 22 formalin-fixed grid-mounted fibrils in which sarcomeres were subsequently weighed by QEMy to determine [mass$_F$]. The mean value of 0.167 includes two extremely light fibrils around 0.11 g/cm³ and two rather heavy ones around 0.23 g/cm³. In Experiment 4-9-72, 21 unfixed fibrils sealed under a cover slip in complete rigor solution with nail lacquer were measured by IMy, always with the fibril axis parallel to the plane of the polarized measuring beam. Here, the seemingly extreme individual values were all high, including 0.23, 0.23, 0.24, 0.28, and 0.31 g/cm³, and the mean was rather high as well. When values which we judged extreme were excluded from averaging, the mean values were very close, 0.171 by QEMy and 0.174 by IMy. We believe that the low values represent fibrils which had somehow experienced A band extraction in standard rigor solution. Such fibrils could at times

Table 2. Results of Dry Mass Measurements in *Sarcophaga* and *Lethocerus* Myofibrils

Exp. No.	Source of DLM Fibrils	Number of Fibrils	Technique and Fixation	Mean [mass] per Sarcomere[e] (g/cm)³	A band Overlap Zone [mass] (g/cm)³
4-24-68	2-day-old blowfly	40 (weighed), 194 (filaments counted)	QEMy[a] formalin	0.204[c]	
3-24-72	Fly, lot 3-1-72	22	QEMy formalin	0.167 ± 0.032	
4-5-72	Fly, lot 3-1-72	7	IMy[b] unfixed		0.276 ± 0.019
4-9-72	Fly, lot 3-1-72	21	IMy unfixed		0.193 ± 0.040
5-10-72	Fly, lot 4-26-72	29	IMy formalin		0.171 ± 0.016
5-10-72	Fly, lot 4-26-72	29	QEMy formalin	0.171 ± 0.0172	
5-10-72	Mean difference between IMy and QEMy of same fibril for 29 fibrils = 0.011 ± 0.010 g/cm³				
5-10-72[d]	Fly[d]	19	IMy formalin		0.174 ± 0.017
6-1-72	*L. maximus* glycerinated 3 years	16	IMy unfixed		0.166 ± 0.011

[a] QEMy refers to quantitative electron microscopy.
[b] IMy refers to interference microscopy.
[c] Mass concentration *calculated* on basis of assigned thick filament spacing of 53 nm.
[d] Oriented selection, parallel to plane of polarized measuring beam, from full group of 29 fibrils.
[e] Abbreviated [mass_F] in text.
[f] Abbreviated [mass_A] in text.

be explicitly identified during light microscopy and were never deliberately included in weighing experiments. On the other hand, the heavy fibrils included in Experiment 4-9-72 and ubiquitous in Experiment 4-5-72 have not been met during IMy of grid-mounted preparations where we could have double-checked them by embedding-sectioning or by QEMy. We include them without knowing yet whether they are artifacts or simply unusual fibrils of authentic but obscure biological significance. If they were elliptical in cross section, others should be oriented to give unusually "light" values, and a far greater ellipticity would be required than we ever recognized in seeing thousands of fibrils in sections.

The second grouping is an experiment entered as 5-10-72 and displayed graphically in Fig. 9, where we measured 29 fixed fibrils mounted on four different grids by both QEMy and IMy, measuring sarcomere length and fibril diameter during interference microscopy. The mean values and standard deviations were practically identical, regardless of whether or not the effect of birefringence on IMy readings was isolated. The disagreement between values for individual fibrils may be seen by matching [mass_A] from IMy for a given fibril (any bar in the lower graph) with the bar directly above in the upper graph which shows whether [mass_F], determined on same fibril by QEMy, turned out to be lower (bar below the line) or greater (bar above the line), and by how much (length of bar). The mean difference per fibril was on the order of 6.5% of the total weight, and the sum of all negative and

all positive differences was practically zero (−0.0016 to be exact).

In our QEMy experiments we calculated [mass_F] for each fibril. However we have also evaluated the results in summary form, as follows. From the mean values for sarcomere length (3.46 μm) and fibril diameter (3.20 μm) obtained as an average from 66 fibrils selected and measured with the intention of weighing them by QEMy, the mean volume was calculated as 27.9 μm³. Fifty-two of these fibrils from 7-day-old flies were actually weighed by QEMy, giving a mean mass/sarcomere of 4.68 × 10^{-12} g/sarcomere, or just twice what we obtained in unselected fibrils of the 2-day-old fly. The mean overall [mass_F] by QEMy determined in this way is 0.168 g/cm³ for *Sarcophaga* fibrils.

We were encouraged by overall agreement between the results of the two techniques. The major lesson seems to be that 10 to 30 weighings by either method can give quite a reproducible mean value. (The agreement we find between the two techniques certainly matches that found by Ruch and Bahr in weighing nuclei, 1970.) So the last experiment which we report is IMy of unfixed fibrils from glycerinated water bug DLM. Again these were all measured with the fibril axis parallel to the electric vector of the measuring beam. The mean [mass_A] is within 5% of the mean for such oriented IMy measurements of *Sarcophaga* fibrils.

The agreement between unfixed glass-mounted fibrils subjected to IMy and fixed grid-mounted fibrils subjected to IMy as well as QEMy suggests

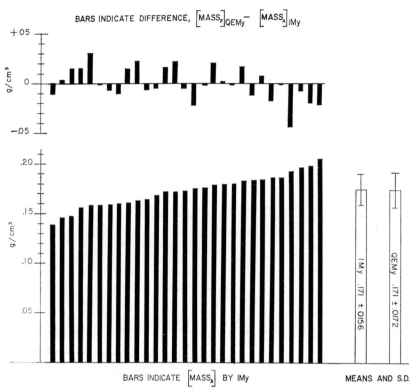

Figure 9. Graphs of 29 myofibrils, each weighed by two methods. These two bar graphs are not histograms. Each bar in the lower graph represents mass concentration in the A band of a single fibril determined by interference microscopy, and the bar directly above it in the upper graph indicates mean mass concentration in the entire sarcomere as determined by quantitative electron microscopy, in terms of whether the value was less than or more than that for the same fibril in the lower graph.

that any metal dissolved from the grids during our experiments was not binding to fibrils in such a way as to alter dry mass or lattice volume. Furthermore no effect of formaldehyde is detectable.

The two mean values with standard deviation brackets drawn in at the lower right of Fig. 9 summarize our present sense of all the results to date in the 1972 experiment series, namely, that both techniques find a concentration of dry mass around 0.17 g/cm³ in the sarcomeres and cross-bridge zones of both blowfly and water bug fibrils. The different value from the single fly experiment seems to be the consequence of an approach in which swelling of the filament lattice cannot influence the determination.

How Many Myosins per Crown?

We come back now to the question which started this work. How many myosins are there per cross-bridge? A nomogram offers a useful way to relate our data to this question. We first prepared a nomogram similar to Fig. 11, except that the middle scale showed myosin molecules per crown. Then it turned out that we didn't know enough about filament spacing in our fibril to offer such a middle

scale, because in checking five fibrils from recent material, we found thick filaments spaced at 62 nm rather than at 55 nm as had been found in a pilot experiment in 1969 (see Methods and Fig. 3). In order to display graphically the influence of different filament spacings without going to a three-dimensional nomogram, we have had to use two different figures, Figs. 10 and 11. Figure 10 is used to find the number of crowns per liter in the cross-bridge zone as a function of thick filament separation, and Fig. 11 is used to find the number of myosin molecules per liter as a function of fibrillar myosin percentage and dry mass concentration. The ratio of these two numbers is just what we are after, the number of myosin molecules per crown. After explaining the figures, we will use them to consider the significance of our best data.

Each figure displays a relationship which was first developed as an algebraic expression. We will describe these expressions now. In both cases our treatment aims to consider only the cross-bridge zone of the sarcomere and to exclude and ignore that volume of the sarcomere containing the I–Z zone and the bare zone region (coinciding exactly with H band length in fibrillar flight muscle).

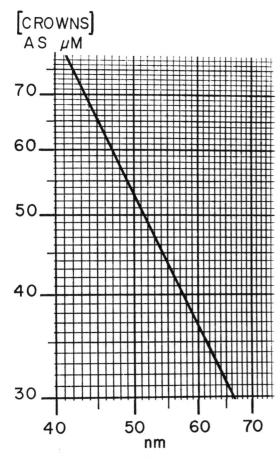

S = SPACING BETWEEN MYOSIN FILAMENTS

Figure 10. This log-log plot graphically displays an equation given in the text, and shows how sensitive the concentration of cross-bridge locations or crowns is to the variation in spacing between thick filaments. See text for various experimental values which can be projected on Xerox copies of this diagram, used in conjunction with Fig. 11.

In Fig. 10 we have expressed the number of crowns per liter as a fraction of Avogadro's number, that is, as a molarity. First, we calculate the volume per crown. If thick filaments in a hexagonal lattice are separated by a center-to-center spacing numbering S in nanometers, the cross-sectional area per filament is given by $0.866 \, S^2 nm^2$. The ringlike zone of cross-bridge origins which we have termed a "crown" recurs every 14.5 nm along each thick filament. The volume per crown is therefore $12.55 \, S^2 nm^3$. One liter contains $10^{24} \, nm^3$, so that there are

$$\frac{10^{24} \, nm^3/liter}{12.55 \, S^2 nm^3/crown} = \frac{7960}{S^2} \times 10^{19} \, crowns/liter.$$

The whole function of Fig. 10 arises because we have found some variation in S and therefore in the number of crowns per liter. To express this number as a molarity is to express the fraction of Avogadro's number of crowns present in one liter and also to invoke the curious idea of a mole of crowns, which seems inescapably useful at this juncture between morphology and biochemistry. Expressed as molarity then, the concentration of crowns is

$$\frac{7060/S^2 \times 10^{19} \, crowns/liter}{6.023 \times 10^{23} \, crowns/mole} = \frac{0.1322}{S^2} \, moles/liter$$

$$= \frac{132,200}{S^2} \, \mu M.$$

Myosin concentration turns out to be in the neighborhood of 200–250 μmoles/liter, so for convenience in comparison we express the concentration of crowns in the same units. The expression gives an absolutely straight line, when [crowns] is plotted against filament spacing S on log-log graph paper. That's all there is to Fig. 10. It works equally well for vertebrate muscle except for a 1% difference in the axial spacing of crowns (14.3 nm in frog muscle) and a range of filament spacings extending below 40 nm.

As for Fig. 11, the nomogram seen there was found to require left and right scales proportional to log x, and a middle scale proportional to log x^2 (e.g., the center diagonal on log-log paper). Our data on dry mass concentration and that on myosin percentage from the following paper (Bullard and Reedy) both refer to the whole sarcomere or the whole fibril, so we must first take account of the fact that myosin is not distributed evenly throughout the sarcomere, but is restricted to the A band, to the region of thick filaments, about 94% of the 3.5 μ sarcomere length in *Sarcophaga*. Even here the mass thickness profile for myosin is not uniform. Although it is constant at its maximum value throughout most of the A band, it is lower in the H band bare zone (no HMM) and at the tapered ends (declining LMM as the end of the filament shaft is reached). Actually these two zones between them can be summed to make a short zone (5% of sarcomere length) having the same myosin mass thickness as the rest of the A band. In other words (Figs. 7H and 7I) the bare zone cancels out the tapered ends, and we can conceive of the myosin concentration as uniform throughout the cross-bridge region. In what follows, quantities$_A$ refer to this cross-bridge region, excluding the H band (which is exactly coextensive with the bare zone in fibrillar flight muscle). Quantities$_F$ refer to the whole fibril, i.e., to the whole sarcomere.

We consider the sarcomere as a cylinder of

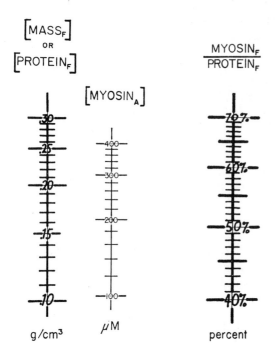

ASSUMES:

FIBRILLAR INSECT FLIGHT MUSCLE
SARCOMERE LENGTH = 3.5 μm
M.W. OF MYOSIN = 470,000

Figure 11. This nomogram can be used to plot myosin concentration in the cross-bridge zone using the values for myosin percentage from protein analysis of fibrils and the values for total dry mass concentration from interference microscopy and quantitative electron microscopy of the same fibril preparations (assuming a sarcomere length = 3.5 μm and a mol wt for myosin = 470,000). Current best values for *Sarcophaga* are 57.5% myosin and 0.17 g/cm³ for mass concentration. A reduction in sarcomere length to 2.4 μm (e.g., *Lethocerus*) might be expected to bring the myosin percentage down to 55%.

uniform diameter, regardless of variations in mass density as indicated in Figs. 7 and 8. From our length measurements (Fig. 7D) the bridge-free I–Z and H zones represent 11% of sarcomere length in a 3.5 μm sarcomere. Therefore, from the length ratio of 0.89 for A/F,

$$\frac{\text{volume}_A}{\text{volume}_F} = 0.89.$$

From the data on myosin percentage and mass concentration, we readily find mean fibrillar myosin concentration as g/cm³,

$$[\text{myosin}_F] = [\text{mass}_F] \times (\text{myosin } \%_F).$$

But as we have been noting, this quantity of myosin is distributed at higher concentration in a smaller volume.

$$[\text{myosin}_F](\text{volume}_F) = [\text{myosin}_A](\text{volume}_A).$$

Thus,

$$\begin{aligned}
\text{myosin}_A &= \frac{\text{volume}_F}{\text{volume}_A} \times [\text{myosin}_F] \\
&= 1.12\,[\text{myosin}_F] \\
&= 1.12\,[\text{mass}_F](\text{myosin } \%_F).
\end{aligned}$$

Figure 11 will be as accurate as the report by Bullard and Reedy that the weight of a micromole of myosin is 0.47 g. To finish, we convert myosin concentration so that it expresses the number of molecules/liter, as follows

$$[\text{myosin}_A \text{ as } \mu\text{moles/liter}]$$

$$= \frac{[\text{myosin}_A \text{ as g/cm}^3] \times 1000 \text{ cm}^3/\text{liter}}{0.47 \text{ g}/\mu\text{mole}}$$

$$= 1.12\,[\text{mass}_F \text{ as g/cm}^3] \times (\text{myosin } \%_F)$$

$$\times \frac{1000}{0.47} \times \frac{\text{cm}^3 \; \mu\text{moles}}{\text{liter g}}.$$

The algebraic equivalent of the nomogram, then, is the expression

$$[\text{myosin}_A \text{ as } \mu\text{moles/liter}] = [\text{mass}_F \text{ as g/cm}^3]$$

$$\times (\text{myosin } \%_F) \times 2390 \, \frac{\text{cm}^3 \; \mu\text{moles}}{\text{liter g}}.$$

How might the nomogram require a change to accommodate future results? The middle scale would have to be adjusted if a different value for myosin molecular weight were later accepted as more accurate. For example, a change from 0.47 g/μm to 0.49 g/μm would adjust myosin molarities downwards 4% from the values obtained with Fig. 11. With shorter sarcomere lengths, (e.g., *Lethocerus*), one would expect a slight drop in the length fraction of A band and thus in the myosin percentage, since I–Z and H zones remain constant at about 0.36 μm at all sarcomere lengths. For SL = 2.4 μm, we would expect A band to be 84% of sarcomere length and myosin to be 55% of fibrillar protein, so a slight readjustment of the middle scale could be made for the sake of precision, though such precision is needless for our present purposes.

We have chosen not to enter any particular set of values in Figs. 10 and 11. They are intended as calculating tools; photocopies may be made and marked up if desired. The significance of our results and error limits as they bear on a final answer can be quickly seen with these diagrams. For example using the Fig. 11 nomogram, the best mean value for myosin percentage, 57.5% (Bullard and Reedy, this volume), may be located on the right scale, and from this a straight line may be drawn to our best mean value for [mass_F], 0.171 g/cm³, crossing the middle scale to give the concentration of myosin in the cross-bridge zone as 235 μm. Our estimate of mean filament spacing in this material, 62 nm, is applied on Fig. 10 and found to indicate [crowns] at 34 μm; the ratio 235/34 is 6.8 myosins per crown. Our previously assumed value for S, 55 nm, gives 44 μm crowns, or 5.4 myosins per crown. Midway between these two filament spacings is 58.5 nm, which gives 38.6 μm crowns, indicating 6.1 myosins per crown.

Our single fly experiment gives a very similar value when analyzed with Figs. 10 and 11. We have already assumed the filament spacing S = 53 nm as given by X rays for intact *Lethocerus* flight muscle (Miller and Tregear, 1972). We find in Fig. 10 that this corresponds to 47 μm crowns. 53 nm was used in calculating the mean sarcomere volume, which gave [mass_F] = 0.204 g/cm³. Using 57.5 as myosin percentage, Fig. 11 indicates a myosin concentration of 280 μm and 280/47 gives 5.96 myosins per crown.

Finally it can be shown that four myosins per crown appears rather outside the reach of present or probable values and is attractive only because it would give exact correspondence with the number of bridges seen in micrographs. For example, if myosin concentration is 235 μm, [crowns] must be 58 μm, requiring a filament spacing of 47 nm in order to give this value. We can also calculate the necessary values for either of the other two parameters required to give four myosins per crown. Accepting 62 nm as the filament spacing, which gives 34.4 μm for [crowns], a myosin concentration of 138 μm would allow four myosins per crown. This could result from (Fig. 11) 57.5% myosin and a [mass_F] of 0.10 g/cm³, or alternatively, from a [mass_F] of 0.17 g/cm³ and a myosin percentage somewhat below 35%. In each case, the values required are too far from those experimentally determined to expect future work to favor them.

The mean [mass_A] for *Lethocerus* fibrils from glycerinated flight muscle, 0.167 g/cm³, gives 229 μm myosin in the cross-bridge zone; obviously the above conclusions apply without significant modification.

It must be obvious that we expect an integral number of myosins per crown. As we said in the introduction, we believe that each cross-bridge is constituted from the two heads of one myosin molecule, and conversely, that each myosin molecule is designed to serve as one cross-bridge. At one time there was discussion as to whether some myosin molecules were inside the shaft of the thick filament and unrepresented by the surface complement of cross-bridges. This is no longer required for plausible models, and we do not think it is

required by our results. We do indeed have strong indication for an integral number. Like Chaplain and Tregear (1966), we find that there do not seem to be four myosins per crown in insect fibrillar muscle; there seem to be more like six instead. The interesting question then becomes, not "Where did those two extra myosins come from in the protein analysis?", but "Where did those two missing cross-bridges go in the electron microscopy?" We were not previously prepared to consider this latter question seriously and now we are; that is what has changed.

Discussion

Interference microscopy, microradiography, and quantitative electron microscopy are three physical techniques which can be used to determine the distributed and integrated dry mass of cells and biological particles. Bull spermatozoa have been weighed using all three methods (microradiography and IMy by Carlson and Gledhill, 1966; QEMy by Bahr and Zeitler, 1964) and red blood cells have also been weighed with all these (microradiography and IMy by Gamble and Glick, 1960; QEMy by Bahr and Zeitler, 1962). Gravimetric weighing and interference microscopy have been used to evaluate a preparation of thymus cell nuclei by Müller (1969) and remarkably close agreement with his results were obtained by Ruch and Bahr using QEMy on the same preparation (1970). Such duplicate weighings by parallel techniques help to establish the reliability of both the techniques and the results. We felt that this would certainly be the case if we employed two techniques when weighing sarcomeres of insect flight myofibrils. This was of particular value because we felt that the likeliest source of major error tended to run in opposite directions with each technique.

With interference microscopy the minimum intensity method of extinction point matching which we used has been demonstrated to lead to maximum overestimates of retardation and therefore of dry mass (Ross and Galavazi, 1965), with errors up to 30–50% expected under some conditions. With respect to electron microscopy, electron irradiated protein films lose about 20% of their mass rather quickly under some typical instrumental conditions, and so this method seemed liable to possible error on the low side in determining fibril weights. We were concerned about this in our 1968–1969 experiments, which came to a halt partly because counting filaments in scores of fibrils was doubly tedious work if we had uncertain values for mean sarcomere dry weight, not to mention our uncertainties about myosin percentage and molecular weight at that time.

Our present results indicate sufficiently good

agreement between the IMy and QEMy to show that both are quite reliable, and we should ask why the expected errors did not turn up. We are most sure of our apology for apparent accuracy with QEMy. The radiation dose was just too low to cause significant mass loss, as we stated in Methods. According to Bahr et al., (1965) our specimen irradiation at 4×10^{-6} A/cm^2 would have required 1000 sec to induce loss of 10 % mass from a gelatin film. Greater radiation sensitivity was more recently indicated for albumin and hemoglobin, which would lose 20 % after no more than 100 sec at our beam intensity, according to Stenn and Bahr (1970). In practice we found 1.5–3 % mass loss when comparing micrographs taken of the same fibrils after 10 sec and after 150–250 sec; and we found localized 15–20 % mass losses only after using a bright crossover beam (5 μm diameter; intensity not measured) for 2 min on a few sarcomeres. Thus radiation mass losses from particulate specimens on standard copper grids seem to have proceeded much more slowly under the actual specimen conditions we used for QEMy than the losses observed from mesh-supported protein films in the special 70–75 kV electron irradiation chamber used for testing by Bahr et al. (1965) and by Stenn and Bahr (1970). The complex structure of the specimen itself may have been in our favor here, in that such mass losses from electron irradiation became less with increasing molecular weight and increased polymerization (Stenn and Bahr, 1970). In short, no underweight estimate due to specimen mass loss seems at all likely under our experimental conditions.

We are less certain of our apology for the apparent accuracy of interference microscope determinations. The subjective physiological basis for overestimating retardation is that, while object darkens and background lightens (Figs. 2D–F), the eye cannot judge the setting of actual minimum intensity in the object and chooses instead the more advanced setting where the object appears at maximum *contrast* against the bright background (Ross and Galavazi, 1965). We chose our endpoint, however, on the basis of object contrast *with itself*; as the analyzer was rotated, extinction occurred for progressively greater optical path differences, and so two darkening boundaries of extinction proceeded from both sides toward the center, permitting a fairly precise judgment (about $\pm \lambda/200$) of when they met and merged completely in the center. It is clear, of course, that photometric measurements would be the most accurate.

Both our IMy and QEMy weighings were subject to the same possibility of error from wrong estimates of fibrillar diameter or shape. We developed some faith in the absolute accuracy of the VISE when we checked it in a pilot experiment three years ago (see Methods). Our final result is sensitive to diameter measurement because the concentration for given dry mass confined in the cylinder varies as the square of the diameter, regardless of whether diameter is used to calculate sarcomere volume, as in our QEMy estimates, or to evaluate total optical path, t_0, as in our IMy measurements.

When we did our single fly experiment, we included a series of VISE measurements which were not required by our formal experimental logic, but which are now serviceable to give better statistical support to our results from embedded single fibrils regarding the errors involved in diameter measurements used in the 1972 experiments. The VISE measurements were made of 100 fibrils in rigor solution from the left DLM homogenate used in the single fly experiment, with an oil immersion phase contrast objective. These fibrils were taken strictly as we came to them, and the mean diameter was 2.24 ± 0.12 μm. Using the filament count of 1270 from embedded-sectioned fibrils of the right DLM, we find that filament spacing S $= 59.8$ nm in fibrils of that diameter. At S $= 53$ nm, 1270 filaments should make a fibril of diameter 1.98 μm. Thus we would predict that IMy of those 2-day-old fibrils, had it been done, would have indicated a dry mass concentration of 0.16 g/cm^3, rather than the value of 0.204 g/cm^3 calculated for S $= 53$ nm. We cannot tell, in fact, whether the error resides with the VISE or with our assumption about filament lattice spacing. It would require light and electron microscope measurements on identical objects not prone to swelling, such as polystyrene spheres, in order to calibrate the absolute accuracy of the VISE on spherical and cylindrical objects.

A type of error related to that concerning the accuracy of diameter measurements arises in the consideration of truly elliptical fibrils. We measured horizontal diameter D, assuming that vertical diameter H was the same. H is the actual light path in IMy, whereas in QEMy it enters into calculation of sarcomere volume. Nevertheless any error due to H $\neq$ D is shown to operate identically for the two different techniques by the following comparison (L = sarcomere length here):

$$[\text{mass}]_{\text{calculated}} = \frac{\text{weight}}{\text{LD}^2\pi} \quad \text{for QEMy, and}$$

$$= \frac{\text{OPD}}{\text{D}\alpha} \quad \text{for IMy}$$

$$[\text{mass}]_{\text{true}} = \frac{\text{weight}}{\text{LDH}\pi} \quad \text{for QEMy, and}$$

$$= \frac{\text{OPD}}{\text{H}\alpha} \quad \text{for IMy.}$$

In both cases,

$$\frac{[\text{mass}]_{\text{calculated}}}{[\text{mass}]_{\text{true}}} = \frac{H}{D}.$$

Thus elliptical fibrils will lead to both overestimates and underestimates of mass in individual cases, but if the ellipses are randomly oriented to the adsorbing support film, as indicated by our limited gallery of embedded and cross-sectioned fibrils (after correction for section compression), then such errors will be compensated by averaging results from many fibrils. Thus the errors indicated by the disagreements between duplicate measurements on identical fibrils (upper graph in Fig. 9) probably are random errors of the method and are about equally distributed between both methods.

The precise agreement between the overall mean values for $[\text{mass}_A]_{\text{IMy}}$ and $[\text{mass}_F]_{\text{QEMy}}$ on 29 fibrils in Experiment 5-10-72 is somewhat accidental. In the first place, interference microscopy measured only the cross-bridge zone of the A band, which has a mass thickness of $1.04 \times$ the mean value for $[\text{mass}_F]$ (Figs. 7 and 8). In the second place, our exact values by IMy depend on the role of fibrillar birefringence (note that all fibrils parallel to the electric vector of the measuring beam during IMy have a mean value at least 2% higher than the overall mean for 29 fibrils; cf. Table 2) and depend further on α, the specific refractive increment. We took 0.18 cm^3/g because even with published values of 0.19 and 0.17 for myosin and actin, respectively (see methods), the remaining 15% of A band protein (Fig. 8) is not evaluated yet; nor could we assess the possibility that aggregation into filaments would raise α for myoproteins just as crystallization raises α (by 5%) for chymotrypsinogen (Davies and Thornburg, 1960). To be sure, the exact value for α of solubilized insect fibrillar proteins could be determined and the remaining uncertainties could doubtless be evaluated, but we have no plans yet for such investigations.

We do not know of any previous publications reporting absolute mass determinations on isolated myofibrils from any muscle, although some estimates from interference microscopy of single whole fibers of vertebrate skeletal muscle have been made. A. F. Huxley and Niedergerke (1958) using frog muscle estimated mean whole fiber [mass] as about 0.24 g/cm^3, inferred a fibrillar value, $[\text{mass}_F]$ of 0.143 g/cm^3, and indicated (indirectly) a value for $[\text{mass}_A]$ between 0.178 and 0.201 g/cm^3. Ross and Casselman (1960) using mouse muscle estimated mean whole fiber [mass] and $[\text{mass}_F]$ as well to be 0.16 g/cm^3, and further estimated $[\text{mass}_A]$ as

0.23 g/cm^3, including dissolved ("nonfibrillar") proteins. These are certainly in the general vicinity of our values, especially considering the differences in material, and we found them very heartening when we were trying to decide whether to believe our first few measurements.

Until now quantitative evaluation of mass in isolated myofibrils has been concerned with the relative distribution of dry mass in the different cross bands, as in the classic microinterferometric study of Huxley and Hanson (1957) and the provocative study of Vajda et al. (1967), in which quantitative electron microscopy was applied to myofibrils for the first time. These studies used rabbit psoas fibrils which can vary greatly in cross sectional profile, so that volume and thus [mass] could not be estimated without additional information. Such information might now be obtained by embedding and sectioning single fibrils after interference microscopy.

If we could do our own experiments over again in more ideal form, we would include an absolute VISE calibration, a determination of filament spacing from fibril suspensions by X-ray diffraction, an assay for non-protein myofibril components, a determination of the specific refractive increment for solubilized whole fibrillar protein (0.25–0.50 M MgCl$_2$ dissolves *everything* except thinned remnants of the Z and M discs), and use of photometric methods for integrating OPD over the whole image area of selected fibril segments. And we would favor more use of *Lethocerus* fibrils. But we doubt that any or all of these improvements would significantly alter our best value of six myosins per crown, nor find support for the value we favored at the outset of four myosins per crown.

So, as we mentioned earlier, we now consider that it is worthwhile to develop experiments which might demonstrate the missing cross-bridges presumably associated with the two occult myosins in each crown. The highly ordered images of embedded and sectioned *Lethocerus* muscle remain stubborn and clear in the regularity of appearance which indicates four bridges/crown, but some further careful analysis of these images will nevertheless be attempted in beginning the search. It may be that some bridges are ill-situated to cross-link with actin in setting up the rigor cross-bridge lattice, as suggested by Squire (1971) in support of his myosin filament modeling which proposes six bridges per crown in insect fibrillar flight muscle. Reedy (1971) has noted a response of cross-bridges in the rigor lattice to the occasional absence of actin filaments which is in line with Squire's suggestion, in that the numerous adjacent cross-bridges which should have attached to the missing actin become obscure, perhaps crumpled close to the filament shaft. A few may become visible because of attachment to an azimuthally inconvenient actin, 60° away from the position of the missing filament, but there is no evidence that this happens often, and it seems that any bridge which does not attach is not preserved as a recognizable bridge.

Acknowledgments

We are indebted for valuable discussions of QEMy to Dr. E. Zeiller. For several suggestions concerning experimental and analytical logic, we are grateful to R. Tregear.

For technical help we thank M. Norris, W. Engler, B. Estes, and J. Must. M.K.R. was supported during this work by NIH Research Career Development Award 5-K3-NB-21,075, by USPHS Program Project Grant HE 11351, and by research grants from the USPHS (1-RO1-AM-14317) and the Los Angeles County Heart Association (Award 406). G.F.B. acknowledges support from American Cancer Society Research Grant V-C-53L. D.A.F. acknowledges support from an NSF Research Grant (GB-7591) and a USPHS Grant (HE 09172).

References

ARONSON, J. 1962. The elongation of myofibrils from the indirect flight muscle of *Drosophila*. *J. Cell. Biol.* **13:** 33.

AUBER, J. 1969. La myfibrillogenèse du muscle strié. I. Insectes. *J. Microscopie* **8:** 197.

AUBER, J. and R. COUTEAUX. 1963. Ultrastructure de la strie Z dans des muscles de dipteres. *J. Microscopie* **2:** 309.

BAHR, G. F. and E. ZEITLER. 1962. Determination of the total dry mass of human erythrocytes by quantitative electron microscopy. *Lab. Invest.* **11:** 912.

———. 1964. Study of bull spermatozoa. *J. Cell Biol.* **21:** 175.

———. 1965. The determination of the dry mass in populations of isolated particles. *Lab. Invest.* **14:** 955.

BAHR, G. F., F. B. JOHNSON, and E. ZEITLER. 1965. The elementary composition of organic objects after electron irradiation. *Lab. Invest.* **14:** 1115.

BAHR, G. F., O. SACKERLOTZKY, and E. ZEITLER. 1963. Electromagnetic exposure system for Siemens electron microscope. *Rev. Sci. Instr.* **34:** 1443.

BARER, R. and S. JOSEPH. 1954. Refractometry of living cells. I. Basic principles. *Quart. J. Microscop. Sci.* **95:** 399.

BENNETT, H. S. 1950. The microscopical investigation of biological materials with polarized light, p. 591. In *McClung's handbook of microscopical technique*, ed. R. McClung Jones, Hoeber, New York.

CARLSON, L. and B. L. GLEDHILL. 1966. Studies on the dry mass of bull spermatozoal heads. Comparative measurements using soft X-ray microradiography and microinterferometry. *Expt. Cell Res.* **41:** 376.

CHAPLAIN, R. A. and R. T. TREGEAR. 1966. The mass of myosin per cross-bridge in insect fibrillar flight muscle. *J. Mol. Biol.* **21:** 275.

DAVIES, H. G. and W. THORNBURG. 1960. The specific refraction increment of crystalline protein. II. α-Chymotrypsinogen. *Biochim. Biophys. Acta* **37:** 25.

DUPRAW, E. J. and G. F. BAHR. 1969. The arrangement of DNA in human chromosomes, as investigated by quantitative electron microscopy. *Acta Cytol.* **13:** 188.

DYSON, J. 1960. Precise measurement by image splitting. *J. Opt. Soc. Amer.* **50:** 754.

ELLIOTT, G. F. 1965. Low-angle X-ray diffraction patterns from insect muscle. *J. Mol. Biol.* **13:** 956.

FISCHMAN, D. A., M. K. REEDY, and G. F. BAHR. 1968. A quantitative electron microscope determination of the mass of myosin per thick filament of insect flight muscle. *J. Cell Biol.* **39:** 44a.

FRANZINI-ARMSTRONG, C. 1970. Natural variability in the length of thin and thick filaments in single fibres from a crab, *Portunus depurator*. *J. Cell Sci.* **6:** 559.

GAMBLE, C. N. and D. GLICK. 1960. Studies in histochemistry. LVII. Determination of the total dry mass of human erythrocytes by interference microscopy and X-ray microradiography. *J. Biophys. Biochem. Cytol.* **8:** 53.

GELLERT, M. F. and S. W. ENGLANDER. 1963. The molecular weight of rabbit myosin A by light scattering. *Biochemistry* **2:** 39.

GOODE, M. D. 1972. Ultrastructure and contractile properties of isolated myofibrils and myofilaments from *Drosophila* flight muscle. *Trans. Amer. Microscop. Soc.* **91:** 182.

GREGORY, D. W., R. W. LENNIE, and L. M. BIRT. 1968. An electron-microscopic study of flight muscle development in the blowfly *Lucilia cuprina*. *J. Roy. Microscop. Soc.* **88:** 151.

HANSON, J. 1956. Studies on the cross-striation of the indirect flight myofibrils of the blowfly. *Calliphora*. *J. Biophys. Biochem. Cytol.* **2:** 691.

HUXLEY, A. F. and R. NIEDERGERKE. 1958. Measurement of the striations of isolated muscle fibres with the interference microscope. *J. Physiol.* **144:** 403.

HUXLEY, H. E. 1963. Electron microscope studies on the structure of natural and synthetic protein filaments from striated muscle. *J. Mol. Biol.* **7:** 281.

HUXLEY, H. E. and J. HANSON. 1957. Quantitative studies on the structure of cross-striated myofibrils. I. Investigations by interference microscopy. *Biochim. Biophys. Acta* **23:** 229.

INGELSTAM, E. 1957. Some considerations concerning the merits of interference microscopes. *Expt. Cell Res.* (suppl.) **4:** 150.

MILLER, A. and R. T. TREGEAR. 1972. The structure of insect fibrillar flight muscle in the presence and absence of ATP. *J. Mol. Biol.* In Press.

MOMMAERTS, W. F. H. M. 1952. The molecular transformations of actin. I. Globular actin. *J. Biol. Chem.* **198:** 445.

MOREL, F. M. M., R. F. BAKER and H. WAYLAND. 1971. Quantitation of human red blood cell fixation by glutaraldehyde. *J. Cell Biol.* **48:** 91.

MÜLLER, W. 1969. Untersuchungen zur interferenzmikroskopischen Massenbestimmung an inhomogenen Objekten. Ph.D. thesis, Swiss Federal Institute of Technology, Zürich.

POWELL, E. O. and F. P. ERRINGTON. 1963. The size of bacteria, as measured with the Dyson image-splitting eyepiece. *J. Roy. Microscop. Soc.* **82:** 39.

REEDY, M. K. 1967. Cross-bridges and periods in insect flight muscle. *Amer. Zool.* **7:** 465.

———. 1968. Ultrastructure of insect flight muscle. I. Screw sense and structural grouping in the rigor cross-bridge lattice. *J. Mol. Biol.* **31:** 155.

———. 1971. Electron microscope observations concerning the behavior of the cross-bridge in striated muscle, p. 229. In *Contractility of muscle cells and related processes*, ed. R. J. Podolsky. Prentice-Hall, Englewood Cliffs, N.J.

REEDY, M. K., D. A. FISCHMAN, and G. F. BAHR. 1969. The correspondence between myosin content and cross-bridge population in sarcomeres of insect flight muscle. *Biophys. J.* **9**: A-95.

REEDY, M. K., K. C. HOLMES, and R. T. TREGEAR. 1965. Induced changes in orientation of the cross-bridges of glycerinated insect flight muscle. *Nature* **207**: 1276.

ROSS, K. F. A. and W. G. B. CASSELMAN. 1960. The total solids in the A- and I-band regions of living mammalian striated muscle-fibers. *J. Roy. Microscop. Soc.* **101**: 223.

ROSS, K. F. A. and G. GALAVAZI. 1965. The size of bacteria, as measured by interference microscopy. *J. Roy. Microscop. Soc.* **84**: 13.

RUCH, F. and G. F. BAHR. 1970. Dry mass determination by interference microscopy; agreement with quantitative electron microscopy. *Expt. Cell Res.* **60**: 470.

RUPP, J. C. and W. F. H. M. MOMMAERTS. 1957. The scattering of light in myosin solutions. II. A determination of the molecular weight. *J. Biol. Chem.* **224**: 277.

SQUIRE, J. M. 1971. General model for the structure of all myosin-containing filaments. *Nature* **233**: 457.

STENN, K. S. and G. F. BAHR. 1970. A study of mass loss and product formation after irradiation of some dry amino acids, peptides, polypeptides and proteins with an electron beam of low current density. *J. Histochem. Cytochem.* **18**: 574.

VAJDA, E., F. GUBA, and V. HARSÁNYI. 1967. Protein distribution in myofibrils of different sarcomere lengths. *Acta Biochim. Biophys. Acad. Sci. Hung.* **2**: 317.

ZEBE, E. 1966. Zur Spaltung von Adenosintriphosphat durch die Z-Scheiben der indirekten Flugmuskeln von *Phormia regina* (*Diptera*). *Experientia* **22**: 96.

ZEITLER, E. and G. F. BAHR. 1962. A photometric procedure for weight determination of submicroscopic particles, quantitative electron microscopy. *J. Appl. Phys.* **33**: 847.

How Many Myosins per Cross-Bridge?
II. Flight Muscle Myosin from the Blowfly, *Sarcophaga bullata*

Belinda Bullard

Agricultural Research Council Unit of Insect Physiology, Department of Zoology, Oxford University

Michael K. Reedy

Department of Anatomy, Duke University, Durham, North Carolina 27706

This paper describes part of a project concerned with the question of numerical correspondence between cross-bridges and myosin molecules in fibrillar insect flight muscles.

In their attempt to estimate the myosin content of Dipteran flight myofibrils, Reedy, Fischman, and Bahr (1969) relied on the assumption that their myosin extracting solution was totally specific and totally effective, and they further assumed a molecular weight for insect myosin of 500,000 to 550,000. The work reported here began with preliminary biochemical experiments which indicated enough possible error in these assumptions to require a new strategy. We decided not to use "before" and "after" mass measurements of extracted myofibrils as a means of determining myosin content. Instead, we decided to measure myosin percentage in myofibrils (and myosin molecular weight as well) by conventional biochemical methods, as reported below, and to reserve the microscopic methods solely to determine mass concentration and distribution in intact unextracted fibrils, as reported by Reedy et al. (this volume). The resulting argument concerning myosin correspondence with cross-bridge numbers is considered in that paper. Another argument, based on the myosin to actin ratio found here, is considered in this paper.

Materials and Methods

Dorsal longitudinal flight muscle myofibrils from 20 to 80 blowflies (*Sarcophaga bullata*) were prepared, centrifugally washed 12–15 times, and suspended in a standard salt solution with 10 mM Na azide as described by Reedy et al. (this volume). Analytical ultracentrifuge studies and some Kjeldahl protein analyses were done in the Department of Physiology at UCLA in Los Angeles. Gel electrophoresis and later Kjeldahl analyses were all done in Oxford on fibrils sent by air on ice from Durham, North Carolina. The standard salt with azide solution was replaced by thorough washing with 0.1 M KCl, 10 mM Na phosphate pH 7.0 before further analysis of fibrils. Each fly yielded about 1 mg of fibrillar protein.

Myosin extraction was performed according to Zebe (1966) using a modified Hasselbach-Schneider solution. This H-S-Z solution was used at 0°C, pH 6.5 (results at pH 7.0 were equivalent), and contained 1 M KCl, 10 mM Na pyrophosphate, 1 mM $MgCl_2$, and 20 mM potassium phosphate buffer.

Sedimentation measurements were made with a Spinco Model E ultracentrifuge. Molecular weights were determined by high-speed sedimentation equilibrium centrifugation with Rayleigh interference optics using the method of Yphantis (1964). Rotor speeds were about 10,000 rpm and runs were made at 4°. The partial specific volume of myosin was taken as 0.720 ml/g at 4° (Kay, 1960).

In preparation for sodium dodecyl sulfate (SDS) gel electrophoresis, myofibrillar solids and extracts were first completely dissolved by dialysis against 8 M urea with 1% SDS and 10 mM Na phosphate buffer pH 7.0. Electrophoresis was by the method of Weber and Osborn (1969) except that samples were incubated 5 min at 100°C with an equal volume of a solution containing 2% SDS, 0.02 M Na phosphate buffer pH 7.0, 20% β-mercaptoethanol, 12% glycerol, 0.07% bromophenol blue. Two types of gel were used. One was 7.5% acrylamide and the other was a double-layered gel, the bottom half 10% and the top half 5% acrylamide. The double gel had the advantage of allowing the myosin heavy subunit to migrate well into the 5% region while the actin was well resolved in the 10% region. Total protein loads ranging from 2–20 μg were put on parallel gels for each run. After staining with Coomassie brilliant blue the gels were scanned with a densitometer and areas under the bands measured with a planimeter.

Protein was estimated by the Kjeldahl procedure, assuming all the fibrillar proteins contain 16.2% nitrogen.

Dispersion of measured values around a mean had been expressed here as a standard error of the means, rather than as a standard deviation.

Characterization of Fly Myosin

The material referred to as myosin was so identified by several criteria. Under the phase microscope the H-S-Z solution uniformly removed considerable density from the A band of myofibrils within 20–60 min (for 5- to 7-day-old flies; extraction took only seconds with 1-day-old flies), leaving intact, separable I-Z brushes which did not extract further (Fig. 1) and which contained only thin filaments by electron microscopy of sections (performed by D. A. Fischman; Cf. also Zebe, 1966). Such extracts contained no actin filaments or identifiable nonmyosin material when checked by electron microscopy using negative staining. Synthetic thick filaments were produced by lowering the ionic strength of the extract to 0.1 M KCl (Huxley, 1963). Finally, the sedimentation behavior and electrophoretic behavior of the extract very closely resembled those of rabbit skeletal muscle myosin. The identification of myofibrillar proteins was aided by direct comparison with gel electrophoretic patterns obtained from rabbit myoproteins and myofibrils.

Sedimentation velocity measurements on myosin extracted up to 10 min with H-S-Z solution showed that the myosin was homogeneous and there were no large aggregates (Fig. 2). Figure 3 gives the sedimentation constant, $S_{20,W}$, as a function of protein concentration. A least squares analysis

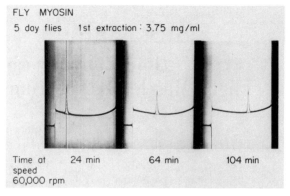

FLY MYOSIN
5 day flies   1st extraction : 3.75 mg/ml

Time at   24 min   64 min   104 min
speed
60,000 rpm

Figure 2. Sedimentation velocity of fly myosin. Solvent H-S-Z solution. Double sector cell. Speed 60,000 rpm; temperature 4°C, bar angle 70°. Five-day-old flies.

gave the line

$$10^{-13} \times \frac{1}{S_{20,W}} = 0.162 + 0.010\,c$$

where c is the concentration in mg/ml. The $S_{20,W}$ obtained by extrapolation was 6.17 S which is not significantly different from that for rabbit myosin.

The mean weight average molecular weight, $\overline{M}_W$, of fly myosin was $490,000 \pm 7000$ daltons ($n = 6$) for myosin from three lots of flies, 0, 3, and 5 days old. The molecular weight of myosin from each lot was measured at 0.6 and 0.8 mg/ml, where centrifugation time was 35–41 hr. The value obtained for rabbit myosin was $471,000 \pm 11,000$ daltons ($n = 4$) for 2 preparations, each measured at 0.6 and 0.8 mg/ml, where centrifugation time was 46 hr. Least squares analysis showed that the molecular weights of fly and rabbit myosin are not significantly different at the 10% level with these data.

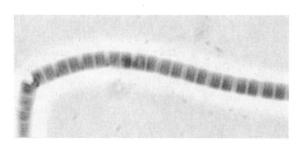

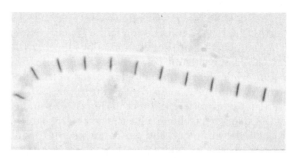

Figure 1. Oil immersion phase contrast photomicrographs of flight muscle myofibril from *S. bullata* before and after removal of A-band density by Hasselbach-Schneider-Zebe (H-S-Z) solution containing 1.0 M KCl, 10 mM Na pyrophosphate, 1 mM MgCl$_2$ and 20 mM Na phosphate buffer, pH 6.5. Two-day-old fly. Extraction was complete in 3 min.

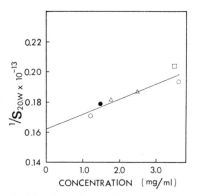

Figure 3. The dependence of the sedimentation coefficient of fly myosin on concentration. Solvent H-S-Z solution; speed 60,000 rpm; temperature 4°C. Corrections were made for solvent viscosity and density; partial specific volume taken as 0.720 at 4°C. Different symbols represent different lots of flies: (●, □) 3-day-old; (○, △) 5-day-old. Intercept of regression line gives intrinsic sedimentation coefficient of 6.17 S with correlation coefficient of 0.92.

The molecular weight of the myosin heavy subunit estimated by SDS gels was the same as the heavy subunit of rabbit myosin, which is 200,000 daltons. The light subunits were 32,000 and 15,000 daltons, apparently different in weight and in number from those of rabbit myosin.

Extraction of Myosin from Myofibrils

Reedy et al. (1969) found that 43% of myofibrillar mass was extracted by the H-S-Z solution from fibrils freshly adsorbed to EM specimen grids. Our preliminary in vitro extractions removed 50% or more of fibrillar protein, and this result was the chief reason for a change in experimental strategy as mentioned in the introduction. First however, in an effort to rescue the usefulness of the microscope techniques for quantitating myosin extraction, we tried to evaluate the specificity and completeness of H-S-Z extraction. About 6 mg of myofibrils from 7-day-old flies was washed free of standard salt and sodium azide, as detailed in Methods, and then extracted repeatedly at 0°C with about 2 volumes of H-S-Z solution for each extraction step. From two to seven successive extractions were done, and times were varied from 5 min to 16 hr. The residue

Table 1. Protein and Myosin Distribution in H-S-Z Extraction of Fly Myofibrils (lot 4-26-72)

Fraction	Total Protein[a] (mg)	Myosin[b] (%)	Myosin[c] (mg)
1. 1-hr extraction	1.52	81%	1.23
2. 4-hr additional extraction	1.14	76%	0.87
3. Residue after 5hr extraction	1.40	15%	0.21
Original fibrils, reconstructed from extraction	4.06	57%	2.31
Whole fibrils, unextracted		57±3.5%(SE)	

[a] Determined by Kjeldahl method.
[b] By SDS gels.
[c] Combining SDS gels and Kjeldahl results.

was easily spun down after each extraction, using a bench centrifuge, and consisted of separate I-Z segments or brushes by phase microscopy. The amount of protein in successive extracts and in the residue was measured by the Kjeldahl method. The percentage of myosin in each fraction was estimated by densitometry of SDS gels. Brief extraction for 5–10 min removed almost pure myosin (SDS scan as seen in Fig. 4B), although there were some high molecular weight bands associated with the myosin heavy subunit. In later extracts the light subunits could not be measured directly because they overlap with other low molecular weight proteins in the myofibrils, so the proportion of heavy subunits was measured and a correction was made for the light subunits to obtain total myosin percentage.

Table 1 displays results from a two-step 5-hr extraction of fibrils from 7-day-old flies, lot 4-26-72. Figure 4C is a gel scan showing that the residue still contained 15% myosin. The time course and end point of this H-S-Z extraction are fairly typical of our results, in that more steps and longer extractions did not procure more specific or complete extraction of myosin. Table 1 shows that 5 hr extraction removed 66% of fibrillar protein, but only 91% of fibrillar myosin.

It appears that quantitation of H-S-Z extraction under the microscope would not afford decent quantitative myosin content estimates for fly myofibrils. Recent preliminary interference microscopy has agreed with our in vitro result in showing removal of 60–75% of A-band mass by H-S-Z extraction. However when fibrils were mounted on EM grids, only 30–45% was removed, in agreement with the result from quantitative electron microscopy by Reedy et al. (1969). Pitting and notching of copper grid bars, which we observed after 5 hr exposure to H-S-Z solution, suggests that dissolved copper could well interfere with even the earliest stages of protein extraction under these conditions.

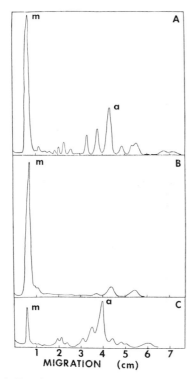

Figure 4. Densitometer traces of fly myofibrils and fractions on 7.5% acrylamide gels. *A*, Total proteins of washed myofibrils; *B*, myosin extracted from myofibrils after 10 min in H-S-Z solution; *C*, residue of myofibrils after extracting for 5 hr with myosin-extracting solution. *m* = myosin, *a* = actin. Seven-day-old flies, lot 4-26-72.

In fibrils from five different lots of flies, the total protein extracted by H-S-Z solution ranged from 66%–82%. For the same five lots the total myosin percentage, estimated from gel scan densitometry of all extracts and residue, ranged from 55%–65% of total fibrillar protein, with a mean of 59%. This myosin percentage (e.g., 57% in Table 1) was considered less accurate than the slightly lower value which was usually obtained from SDS gel scans of whole myofibrils from the same lots of flies.

SDS Gel Electrophoresis of Whole Myofibrils

Figure 4A shows a densitometer trace of electrophoresis pattern of whole myofibrils on a 7.5% acrylamide gel. There are two bands which are not seen in gel scans of vertebrate striated muscle fibrils. These have mobility slightly less than actin, corresponding to molecular weights of 56,000 and 53,000 daltons. In the 7.5% gels the areas of the myosin heavy subunit band, the actin band, and the total area of all bands were measured. For the double gels only the area of the bands for the myosin heavy subunit and for the actin were measured.

Adherence to Beer's law was checked for each set of gels (Fig. 5) and was satisfactory over the range of protein loadings used here. Figure 6 shows results for the proportion of myosin heavy subunit to total protein obtained for myofibrils from a total of about 60 7-day-old flies (lots 3-1-72 and 4-26-72) run on five sets of gels. Regression analysis gave a line corresponding to 50.6 ± 3.0 ($n = 17$) percent myosin heavy subunit. From Fig. 7 the proportion of myosin heavy subunit to actin can be estimated. The measurements are from the same gels as Fig. 6, together with values from double gels run at the same time. The regression line gave a myosin heavy subunit to actin ratio of 3.03 ± 0.16 ($n = 31$).

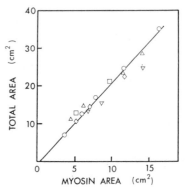

Figure 6. The proportion of myosin heavy subunit to total protein in fly myofibrils. Myofibrils run on 7.5% acrylamide gels and areas of densitometer traces measured by planimetry. Seven-day-old flies, lots 3-1-72 and 4-26-72. Myosin area is for heavy subunit only. Different symbols represent different sets of gels. Reciprocal of slope of regression line is 0.506 ± 0.030 ($n = 17$) with correlation coefficient of 0.98.

The reliability of estimates of relative protein concentration from staining intensity depends on the uptake of stain by the different proteins. It has already been shown that myosin light and heavy subunits stain equally for equal amounts of protein (Lowey and Risby, 1971). In our experiments the relative intensity of staining of insect myosin and rabbit actin was tested. Waterbug (*Lethocerus*) myosin and rabbit actin were dialyzed against 1% SDS, 10 mM Na phosphate buffer pH 7.0, the protein concentration of each solution was measured by the Kjeldahl method, and they were mixed to give a solution containing equal amounts of myosin and actin. This mixture was run on 7.5% and on double gels with loadings from 2–24 μg of total protein. Densitometer traces had minor heavy bands in addition to the bands for the myosin

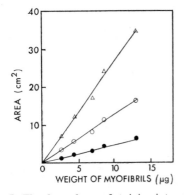

Figure 5. The dependence of staining intensity of fibrillar proteins on amount loaded onto gels. Fly myofibrils run on 7.5% acrylamide gels and areas of densitometer traces measured by planimetry. Seven-day-old flies, lot 4-26-72. (△) Total for all bands; (○) myosin heavy subunit; (●) actin.

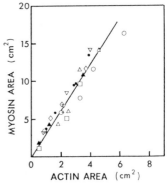

Figure 7. The proportion of myosin heavy subunit to actin in fly myofibrils. Myofibrils run on 7.5% and double gels of 5 and 10% acrylamide, areas of densitometer traces measured by planimetry. Seven-day-old flies, lots 3-1-72 and 4-26-72. Myosin area is for heavy subunit only. Different symbols represent different sets of gels; 7.5% gels are same as in Fig. 3. Slope of regression line is 3.03 ± 0.16 ($n = 31$) with correlation coefficient of 0.96.

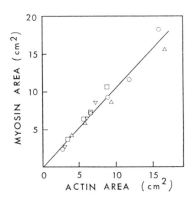

Figure 8. The relative staining intensity of myosin and actin standards. Two standard mixtures containing equal amounts of myosin (insect) and actin (rabbit) run on 7.5% and double gels of 5 and 10% acrylamide, areas of densitometer traces measured by planimetry. Myosin area is sum of all myosin bands. Different symbols represent different sets of gels. Slope of regression line is 1.07 ± 0.07 ($n = 16$) with correlation coefficient of 0.98.

heavy subunit and the light subunits; all these were measured to compute total area for myosin. Figure 8 shows results obtained from two standard mixtures. The regression line gave a value of 1.07 ± 0.07 ($n = 16$) for the relative staining intensity of myosin to actin.

If all proteins had the same relative staining intensity as myosin, then the values of 50.6% myosin heavy subunit and 3.03 as a ratio of myosin heavy subunit to actin, as given above, would need no correction. However it is clear that actin at least does not stain identically with myosin. If myosin is unique and all other myofibrillar proteins stain similarly to actin, then the 1.07 staining ratio corrects the proportions. The corrected ratio of myosin heavy subunit to actin becomes 2.83.

The light subunits must now be brought in before estimating total myosin. We chose to consider that a 10 min H-S-Z extract contained myosin of normal subunit composition. This material had two light subunits which were estimated by densitometry to represent 14.4% of the light and heavy subunits (mean of three samples). Allowing for light subunits then, the proportion of myosin to total fibrillar protein becomes $57.0 \pm 3.5\%$ and the ratio of myosin to actin becomes 3.31 ± 0.18. If actin is unique and all other myofibrillar proteins stain similarly to myosin, then the corrected myosin proportion allowing for light subunits is 58%.

Cross-Bridge Numbers Based on Myosin to Actin Ratio

As detailed in the accompanying paper by Reedy et al., the myosin percentage of 57% appears

to indicate five or six myosin molecules every 145 Å along each thick filament. The question may also be argued separately from consideration of the known structure of the actin filament and the ratio of myosin to actin in whole fibrils obtained above from gel scans. In the L_3 hexagonal filament lattice of insect flight muscle, there are three actin filaments per myosin filament. The lengths of myosin and actin are almost identical in the flight muscle sarcomere, where there is virtually total overlap of actin filaments with the cross-bridge region of the myosin filaments. So we can simply calculate the mass per unit length of myosin and actin in an L_3 cell. There is one actin every 2.75 nm along an actin filament, and each subunit weighs 45,000 daltons.

For three actin filaments, $3 \times 45,000$ daltons/ 2.75 nm = 49,000 daltons/nm. Turning to the thick filament and using 470,000 as the weight of myosin, then for one myosin filament with four myosins/ 14.5 nm, we get $4 \times 470,000$ daltons/14.5 nm = 130,000 daltons/nm; and with six myosins/14.5 nm, we get $6 \times 470,000$ daltons/14.5 nm = 194,000 daltons/nm. Myosin/actin ratio should then be 3.96 for six myosins per "crown" and 2.65 for four myosins per "crown." The measured ratio from SDS gel scans is 3.31, which corresponds exactly to five myosins per "crown." This is not a welcome result, but it is particularly difficult to dismiss because the myosin:actin ratio seems to be one of the least contestable results to emerge from the SDS gel quantitation.

It is difficult to see where our theoretical mass/ length calculation could have ignored the additional 15–20% of actin actually present in gels and responsible for bringing the ratio down from theoretical 3.96 to measured 3.31; not more than 10% could be buried in the Z band. Neither can we clearly develop the opposite sort of argument, that 15–20% of the fibrillar myosin fails to appear on the gels, or that 15–20% of the material under the actin peak is not actin but some electrophoretically identical protein.

Conclusions

In the myofibrils of 7-day-old *S. bullata* flight muscles, myosin constitutes $57.0 \pm 3.5\%$ ($n = 17$) of the total protein; the ratio of myosin to actin is 3.31 ± 0.18 ($n = 31$), estimated by quantitating SDS gels of whole myofibrils. The myosin percentage estimated by quantitating gels of fractionated myofibrils, i.e., of extracts and residue, is 55–65% with a mean of 59%, though this method is probably less accurate. H-S-Z solution extracts 66–82% of fibrillar protein, part of which is nonmyosin, and it fails to extract about 10% of fibrillar myosin. The myosin percentage agrees well with that reported by Chaplain and Tregear (1966) for flight muscle myofibrils from the water-bug, *Lethocerus cordofanus*. Our best value of 57% is more than the 43% reported by Reedy et al. (1969). The values for myosin percentage are somewhat higher than those found for the myosin of rabbit psoas by Hanson and Huxley (1957) and Huxley and Hanson (1957), where the actin filaments are relatively longer but relatively fewer in proportion to the myosin filaments.

Acknowledgments

We wish to thank Mr. Mark Norris for expert technical assistance in preparation of fibrils. This research was supported by USPHS Program Project Grant HE 11351, Los Angeles County Heart Research Award 406, NIH Research Career Development Award 5-K3-NB-21,075 (to M.K.R.), and NIH Research Grant 1-R01-AM-14317. We are indebted to Mr. Douglas Brown and Dr. Darrel Goll for valuable discussions concerning ultra-centrifugation experiments, and to Mr. Richard Tregear for the computer program used for regression analysis of gel scans.

References

CHAPLAIN, R. A. and R. T. TREGEAR. 1966. The mass of myosin per cross-bridge in insect fibrillar flight muscle. *J. Mol. Biol.* **21**: 275.

HANSON, J. and H. E. HUXLEY. 1957. Quantitative studies on the structure of cross-striated myofibrils. II. Investigations by biochemical techniques. *Biochim. Biophys. Acta* **23**: 250.

HUXLEY, H. E. 1963. Electron microscope studies on the structure of natural and synthetic protein filaments from striated muscle. *J. Mol. Biol.* **7**: 281.

HUXLEY, H. E. and J. HANSON. 1957. Quantitative studies on the structure of cross-striated myofibrils. I. Investigations by interference microscopy. *Biochim. Biophys. Acta* **23**: 229.

KAY, C. M. 1960. The partial specific volume of muscle proteins. *Biochim. Biophys. Acta* **38**: 420.

LOWEY, S. and D. RISBY. 1971. Light chains from fast and slow muscle. *Nature* **234**: 81.

REEDY, M. K., D. A. FISCHMAN, and G. F. BAHR. 1969. The correspondence between myosin content and cross-bridge population in sarcomeres of insect flight muscle. *Biophys. J.* **9**: A 95.

WEBER, K. and M. OSBORN. 1969. The reliability of molecular weight determinations by dodecyl sulfate-polyacrylamide gel electrophoresis. *J. Biol. Chem.* **244**: 4412.

YPHANTIS, D. A. 1964. Equilibrium ultracentrifugation of dilute solutions. *Biochemistry* **3**: 297.

ZEBE, E. 1966. Zur Spaltung von Adenosintriphosphat durch die Z-Scheiben der indirekten Flugmuskeln von *Phormia regina* (*Diptera*). *Experientia* **22**: 96.

Biochemical and Ultrastructural Studies on Vertebrate Smooth Muscle

ROBERT V. RICE AND ARLENE C. BRADY

Department of Biological Sciences, Mellon Institute of Science, Carnegie-Mellon University, Pittsburgh, Pa. 15213

Three sets of relatively long filaments appear in the cytoplasm of vertebrate smooth muscles. The filaments are distinguished in the electron microscope by their distinctly different diameters, by their subunit fine structure, and by their spatial arrangements. Although there is not good agreement as to the exact diameters of the three sets of filaments, we will refer to them as 8-nm, 10-nm, and 15-nm filaments. Electron microscopic observations have also consistently reported the presence of dense bodies in vertebrate (and invertebrate) smooth muscles. Dense bodies are about 50 nm by 500 nm and are best distinguished at low magnification as clumps of amorphous material. A better name would be amorphous bodies, but for the sake of uniformity we will refer to them as dense bodies.

There appears to be good agreement from the results of both electron microscopic and X-ray diffraction studies as to the identity, fine structure, and spatial arrangement of the smallest filament. This 8-nm filament is certainly actin (Hanson and Lowy, 1963; Huxley, 1963; Rice et al., 1966; Somlyo and Somlyo, 1968). Negative staining suggests the subunit structure of actin to be very similar in a wide variety of muscles (Hanson and Lowy, 1964; Rice et al., 1966). X-ray diffraction studies were the first to detect a close packing arrangement of actin filaments in smooth muscle (Elliott and Lowy, 1968) which was then subsequently found by electron microscopy (Rice et al., 1970; Lowy and Small, 1970; Heumann, 1970; Devine and Somlyo, 1971; Rice et al., 1971; Devine et al., 1972; Cooke and Fay, 1972). The X-ray diffraction results showed a close packing nearest neighbor distance of 11.5 nm (Elliott and Lowy, 1968). Electron microscopy results varied from 12 nm to 13 nm and also showed the actin filaments in a variety of packing modes (Rice et al., 1970); hexagonal, square, and concentric circle packing were all found. Occasionally rosettes of thin 8-nm filaments surrounding a thick 15-nm filament have been reported in smooth muscle, but the true in vivo arrangement is not yet agreed upon.

Thick filaments were sometimes seen in many early electron microscopic examinations of verte-brate smooth muscle (for reviews see Somlyo and Somlyo, 1968; Burnstock, 1970). Early results from this laboratory (Kelly and Rice, 1968) indicated that thick filaments were found only at lower pH; these findings were amended later to suggest that thick filaments were found only in contracted smooth muscle cells (Kelly and Rice, 1969; Rice et al., 1970). More recently we have reported their presence in both contracted and relaxed muscle (Rice et al., 1971). The main reason for the above sequence was our inability to show continued relaxation of taenia coli after addition of fixative. When rabbit portal anterior mesenteric vein smooth muscle was relaxed in Ca^{++}-free Krebs solution containing 10^{-3} M procaine and fixed in 4% formaldehyde, no evidence of contraction resulted. Such fixed relaxed smooth muscle showed excellent thick filaments (Rice et al., 1971). Many other recent studies have shown thick filaments to be present in vertebrate smooth muscles (Cooke and Chase, 1971; Devine and Somlyo, 1971; Garamvölgyi et al., 1971; Somlyo et al., 1971a,b,c, 1972; Heumann, 1970; Keyserlingk, 1970; Lowy et al., 1970; Cooke and Fay, 1972; Kelly and Arnold, 1972).

When organized thin filament (8-nm) arrays were first successfully found together with thick filaments, it was felt that at last electron microscopic techniques had succeeded in preserving the in vivo state (Rice et al., 1970). This attitude was taken since the original thin filament arrays had been found from "living" muscle (Elliott and Lowy, 1968). There remained the puzzle as to why X rays could not detect the thick filaments. The puzzle was partially resolved when X-ray diffraction results gave a 14.4 nm reflection which was assigned to myosin thick filaments (Lowy et al., 1970). However because of the very narrow dimension of the 14.4 nm reflection, Lowy et al. (1970) suggested that ribbons of myosin thick filaments were required rather than a lattice. Ribbons were found in the electron microscope when smooth muscle was severely stretched and incubated at low temperatures in nonphysiological media (Lowy and Small, 1970; Somlyo et al., 1971b). The last two studies also reported many thick filaments (15 nm) not associated with ribbons. Rice et al.

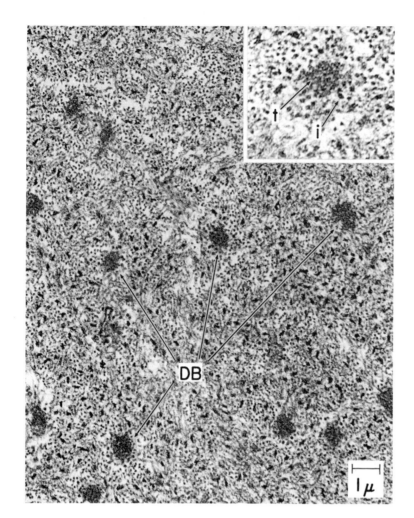

Figure 1. Electron micrograph of a thin cross section of glycerinated (24 hr) chicken gizzard smooth muscle. Fixed for 1 hr in 6.25% glutaraldehyde in Krebs-Ringer buffer at pH 7.2 at 22°C. Post-fixed in 1% OsO_4 in Palade buffer at pH 7.4 for 1 hr at 22°C. Block stained for $1\frac{1}{2}$ hr in 2% uranyl acetate at pH 7.0. Embedded in eponaraldite, sectioned with Porter-Blum MT-2 ultramicrotome with Du Pont diamond knife. Section stained with lead citrate and uranyl acetate. Dense bodies (*DB*) with intermediate filaments (*i*) are seen; the latter are more evident in the inset. Glycerination has disturbed arrays of thin filaments (*t*) and much fewer thick filaments are found than in freshly excised and fixed tissue. Dense bodies are enhanced by glycerination. ×80,000 (inset 148,000).

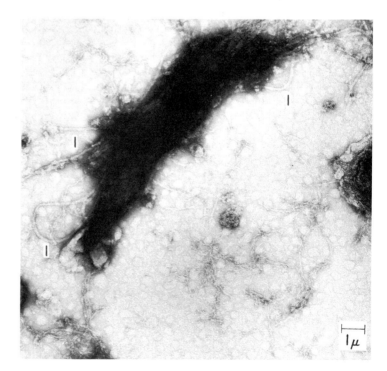

Figure 2. Electron micrograph of dense bodies isolated in the 50% sucrose gradient explained in the text. Negative stained with uranyl acetate. Smooth intermediate filaments (*I*) are seen inserting into the amorphous mass and some decorated actin filaments are present. ×65,000.

430

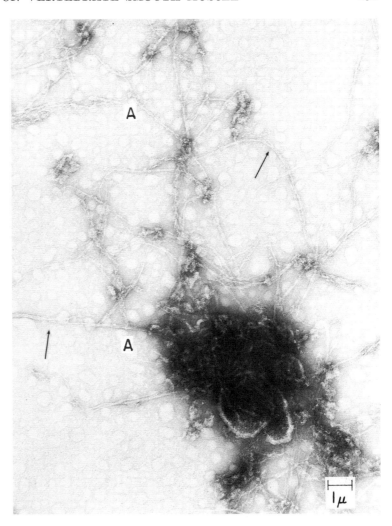

Figure 3. Dense body isolated by Cooke and Chase (1971) procedure. Extracted for 10 days with 2 mM Tris. Negatively stained with uranyl acetate. Smooth intermediate filaments are seen and a small amount of decorated actin. Frayed filaments (arrows) exhibit 3.5-nm subunit filaments. In some locations intermediate filaments appear to be attached (A) to another intermediate filament. Some contorted thicker filaments are seen within the densely stained remains of a dense body. ×70,000.

(1971) suggested that a lattice of thick filaments could account for the narrow dimension of the 14.4 nm reflection in reporting a rectangular array of thick filaments about 57 nm by 77 nm from electron microscopic studies. They suspected that X-ray diffraction on in vivo smooth muscle might find some such reflection closer to 80 or 90 nm. To date no such reflection has been reported.

The detection of 14.4 nm reflection (Lowy et al., 1970) and indications of cross-bridges on tapered thick filaments (Rice et al., 1970; Somlyo et al., 1971b, 1972; Lowy and Small, 1970) strongly suggests that they are composed of myosin. Ribbons have never been detected either by electron microscopy or X-ray diffraction under usual physiological conditions. This indicates that ribbons are aggregates of thick filaments. These ordered aggregates may be useful in deducing details of the thick filament structure, but there is no evidence for their presence in normal functioning smooth muscle.

The variation of filament diameters reported in smooth muscle studies by electron microscopy has long been part of the confusion as to whether thick filaments existed or were artifacts. Reviews should be consulted for details (Somlyo and Somlyo, 1968; Burnstock, 1970). The confusion is reduced somewhat if only studies using glutaraldehyde are considered. Kelly and Rice (1968, 1969) found thick filament diameters from 10.0–20.0 nm. Nonomura (1968) reported a similar range.

Rice et al. (1970) noted the less electron-opaque central region of 10-nm filaments; Somlyo et al. (1971b), Cooke and Chase (1971), and Cooke and Fay (1972) noted that the 10-nm intermediate filaments were closely associated with dense bodies. Large numbers of 10-nm filaments were found in developing smooth muscle (Uehara et al., 1971). Cooke and Chase (1971) isolated and partially purified dense bodies with associated intermediate 10-nm filaments after first extracting actomyosin from chicken gizzard. Prosser et al. (1960) have provided considerable evidence that dense bodies are composed of proteins. Several reports have shown thin filaments (8.0 nm) running into dense bodies (Devine and Somlyo, 1971; Devine et al., 1972). The apparent hollow nature of the 10-nm intermediate filament suggested to us that they might not be either actin or myosin in composition.

Lane (1965) apparently first noticed microtubules in smooth muscle, and Rice et al. (1970) pointed out their apparent association with cyto-

plasmic vesicles of smooth muscle. The smooth muscle cytoplasmic microtubules are readily distinguished from all three sets of myofilaments by their diameter (25.0 nm) and subunit structure typical of microtubules.

Electron Microscopy

Figure 1 is an electron micrograph of a cross section of glycerinated chicken gizzard smooth muscle. Numerous dense bodies (DB) are seen with

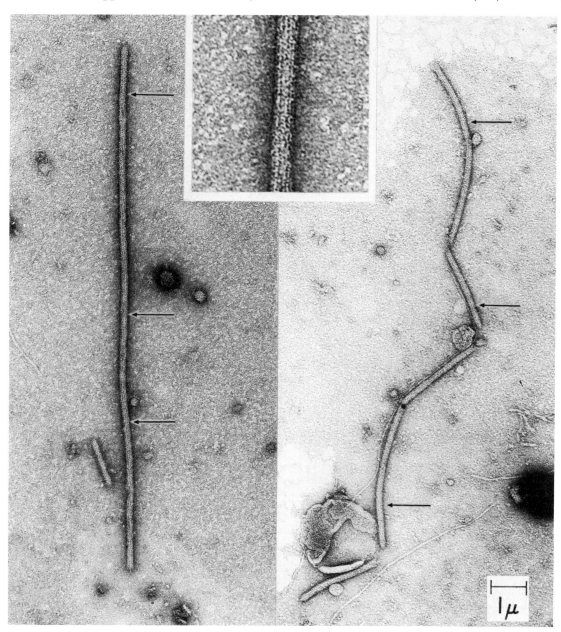

Figure 4. Dense body material after 12 days extraction with 2 mM Tris and further high speed (60,000 rpm) centrifugation (Spinco 60 Ti for 90 min in 0.5 M KCl, 2.5 mM ATP, 5 mM MgCl$_2$, 2 mM Tris pH 7.6. Negatively stained in uranyl acetate. Actin filaments and a 20-nm aggregate of 3.5-nm filaments remain. Substructure (arrows) of the aggregate show 3.5–4.0-nm filaments. ×100,000. The substructure arrangement of parallel rows of 3.5-nm filaments is clearly seen in the insert. ×300,000.

associated intermediate filaments (I) and thin filaments (t). Glycerinated gizzard enhances dense bodies for electron microscopic observation. Only fresh gizzards were used for isolation studies, however. The population of dense bodies seems to vary among smooth muscle varieties. Gizzard is a particularly rich source of dense bodies and because it can readily be obtained in relatively large quantities, it was chosen by Cooke and Chase (1971) and by us as an ideal smooth muscle for studying the structure and function of dense bodies.

We have confirmed the original observations of Cooke and Chase (1971) in that dense bodies and associated intermediate filaments remain after exhaustive extraction of gizzard smooth muscle with 0.6 M KCl. Such preparations frequently show banded collagen fibers and decorated actin filaments. We hoped to purify the intermediate filaments to remove the possible contamination. Cooke and Chase's (1971) purification procedure was to differentially centrifuge the residue from actomyosin extraction in an 8% sucrose solution.

We have found after several trials that a discontinuous sucrose gradient allows the isolation of dense bodies with intermediate filaments in the 50% sucrose portion of a gradient comprised of equal volumes of 25%–35%–50%–65% sucrose in 50 mM Tris buffered at pH 7.2. The blended residue was layered on the top sucrose solution in SW27 tubes and spun at 5200 rpm for 15 min. All operations were carried out at 4°C. Only the material in the 50% sucrose layer was used for SDS gel electrophoresis and amino acid analysis. Figure 2 is an example of isolated dense bodies with intermediate filaments. Large amorphous masses are seen with smoothly contoured intermediate filaments curving among each other and inserting into the dense mass. The electron opacity of such masses prevents identification of details within. Extensive surveys in the electron microscope have not revealed any collagen fibers, mitochondria, vesicles, or any other organelle in this fraction. Therefore this fraction contains principally the amorphous material of dense bodies and their associated intermediate filaments. However, small amounts of decorated actin filaments can be found and must be removed by high speed differential centrifugation in the presence of ATP (Weber, 1956).

Before the discontinuous gradient was used, we had found that extractions with low ionic strength Tris (2 mM) of dense bodies isolated by the Cooke and Chase procedure (1971) removed the amorphous material during a time course similar to the removal of Z lines of striated muscle (Stromer et al., 1969). These experiments were undertaken in an attempt to discern structural details of the inter-

relations of 10-nm filaments within the dense body opaque mass. Figure 3 is an example of an isolated dense body from a suspension which had been extracted for 10 days in 2 mM Tris. Much amorphous material remains, but more smoothly contoured 10-nm filaments are exposed. An aliquot of dense bodies which had been dialyzed in 2 mM Tris for 12 days was treated with 2.5 mM ATP in 0.5 M KCl which contained 5 mM $MgCl_2$ and 2 mM Tris pH 7.6. The suspension was centrifuged at 60,000 rpm in a Spinco 60 Ti rotor for 90 min. The material in the pellet was then negatively stained and examined in the electron microscope. No decorated actin was seen, only bare actin and what appears to be an ordered aggregate of intermediate (10-nm) filaments, shown in Fig. 4. These aggregates are 20-nm wide and have only been found after the ATP centrifugation. In contrast to the flexible intermediate filament, the 20-nm ordered aggregates are very straight and somewhat brittle. These filaments have a substructure of 3.5–4.0-nm filaments as seen in the inset at higher magnification. The substructure may be compared to that of an actin filament which fortuitously lies alongside the aggregate. Actin appears to have a diameter about twice that of the subfilament.

The brief (20 min) low speed (9000 rpm) centrifugation in a Sorvall SS-34 used to pellet the partially extracted dense bodies leaves a good many 10-nm filaments in the supernatant. Figure 5 is an example of smoothly contoured filaments released from dense bodies during 12 days of extraction in 2 mM Tris. Little amorphous material remains and some fraying is evident, indicating a subunit fibril with a 3.5–4.0-nm diameter. In some instances 10-nm filaments appear to be attached to one another (A). The flexibility of intermediate filaments is particularly evident as seen in Fig. 5. Thus our procedure results in great losses of 10-nm filaments as the low ionic strength extraction releases the filaments from amorphous dense body material. Such losses were permitted since our goal was to obtain a final preparation as completely free of amorphous material as possible. After 15 days extraction essentially no amorphous material remains; only jumbled masses of filaments were found in negative stained preparation.

Polypeptide Chains of Intermediate Filaments

The low ionic strength extraction of dense bodies was monitored by SDS acrylamide gel electrophoresis to determine the molecular weights of the proteins both released and remaining in association with dense bodies. The methods of Laemmli (1970) and Weber and Osborn (1969) were followed. The supernatant contained increasing amounts of two

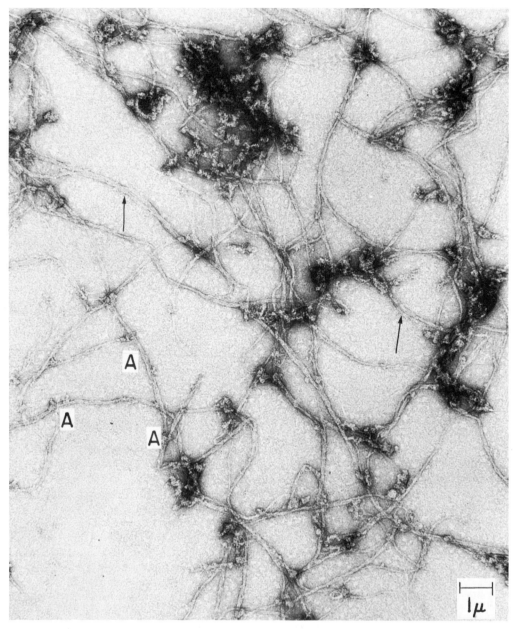

Figure 5. Smooth intermediate filaments from dense bodies isolated in 50% sucrose layer after 12 days extraction in 2 mM Tris. Supernatant of low speed (9000 rpm) centrifugation. Some filaments appear to be attached (*A*) in a specific fashion. Some fraying (arrows) of filaments shows 3.5-nm subfilaments. Negatively stained in uranyl acetate. ×90,000.

bands of about 16,000 and 18,000 daltons, but the presence of many other bands precludes positive identification. After 12–15 days two high molecular weight bands about 85,000 and 105,000 daltons remained in the pellet. Preliminary amino acid analysis of the pellet indicates an amino acid composition different from any other muscle protein reported in the literature. However, the preparation is impure and more purification is necessary.

Discussion

The preliminary evidence presented here suggests that dense bodies of gizzard smooth muscle are comprised of smoothly contoured 10-nm filaments found from negatively stained preparation. These are probably the same as the intermediate filament reported in thin sections associated with dense bodies but also found in other locations in the cytoplasm. These filaments are apparently connected to one another inside the dense body, but the way in which they are connected is not certain. One possibility is that the small (3.5–4.0 nm) subfilaments of the intermediate filament anastomoses in the dense body to form a three-dimensional structure. The three-dimensional structure within the dense body may be similar to the Z line of striated muscle. Amorphous material is removed at low ionic strength from dense bodies in a manner strikingly similar to the removal of amorphous material of Z lines. Some indication of the anastomotic fibrillar structure within dense bodies can be gathered from dense bodies partially extracted from amorphous material, but since the structure apparently collapses, it is highly distorted. The relationship between the intermediate filaments of smooth muscle and the 10-nm filament found in nonmuscle cells is yet to be determined. However their overall morphology is similar.

The intermediate filaments are distinct from thin (8-nm) actin filaments and from thick (15-nm) myosin filaments in their substructure revealed by electron microscopy, by their polypeptide chain molecular weights, and by their amino acid composition. Unpublished observations of C. Tan, D. J. Hartshorne, and R. V. Rice on the molecular weights of gizzard myosin indicate that gizzard myosin has a heavy chain of 200,000 daltons and light chains of 20,000 and 16,000 daltons. The 45,000 dalton chain of actin has also been obtained in a variety of smooth muscles. Tropomyosin bands at about 37,000 daltons. A band at 90,000 daltons has been reported from striated muscle, as has an unidentified band at 105,000 daltons (Scopes and Penny, 1971). It is possible that one or both high molecular weight bands may be one of the actinins of smooth muscle, but the published amino acid compositions do not agree with our present analysis.

We can conclude that intermediate filaments are neither myosin nor actin. The ultrastructural evidence agrees with the biochemical analysis in that three different types of filaments are present in smooth muscle. The smallest diameter filament is most certainly actin, but whether smooth muscle tropomyosin and troponin are associated with it has yet to be determined. At present we can only eliminate all known muscle proteins except possibly an actinin as a component of the intermediate (10-nm) filament. In addition to the elimination of myosin and actin we can exclude tropomyosin on the basis of SDS acrylamide gel electrophoresis. The thickest filaments of smooth muscle are most probably primarily made up of organized associations of myosin much in the same manner as has been amply demonstrated for striated muscle. Under normal physiological conditions only individual thick filaments are detected. The polarity of the heads of myosin macromolecules is not yet

known. A lattice arrangement (66 nm nearest neighbor packing) is suggested. We do not know whether some sort of "A" band and "I" band arrangement of thick (myosin) and thin (actin) filaments exists in smooth muscle under physiological conditions. The sliding filament theory (Huxley and Hanson, 1954; Huxley and Niedergerke, 1954) is certainly compatible with the present evidence that all three types of filaments are present during contraction and relaxation and whether the smooth muscle is stretched or unstretched.

Acknowledgments

We are greatly indebted to Ms. Joan Moses, Diane Raymond, and Judith Baldassare for expert technical assistance. Prof. D. J. Hartshorne provided expertise in SDS acrylamide gel electrophoresis. We also thank Dr. A. P. Somlyo for discussions. Supported in part by NIH grant AM 02809.

References

Burnstock, G. 1970. Structure of smooth muscle and its innervation, p. 1–69. In *Smooth muscle*, ed. E. Bülbring et al. Edward Arnold Publ. London.

Cooke, P. H. and R. H. Chase. 1971. Potassium chloride-insoluble myofilaments in vertebrate smooth muscle cells. *Exp. Cell Res.* **66**: 417.

Cooke, P. H. and F. S. Fay. 1972. Thick myofilaments in contracted and relaxed mammalian smooth muscle cells. *Exp. Cell Res.* **71**: 265.

Devine, C. E. and A. P. Somlyo. 1971. Thick filaments in vascular smooth muscle. *J. Cell. Biol.* **49**: 636.

Devine, C. E., A. V. Somlyo, and A. P. Somlyo. 1972. Sarcoplasmic reticulum and excitation-contraction coupling in mammalian smooth muscles. *J. Cell Biol.* **52**: 690.

Elliott, G. F. and J. Lowy. 1968. Organization of actin in a mammalian smooth muscle. *Nature* **219**: 156.

Garamvölgyi, N., E. S. Vizi, and J. Knoll. 1971. The regular occurrence of thick filaments in stretched mammalian smooth muscle. *J. Ultrastr. Res.* **34**: 135.

Hanson, J. and J. Lowy. 1963. The structure of F-actin and of actin filaments isolated from muscle. *J. Mol. Biol.* **6**: 46.

———. 1964. The structure of actin filaments and the origin of the axial periodicity in the I-substance of vertebrate striated muscle. *Proc. Roy. Soc. B* **160**: 449.

Heumann, H.-G. 1970. A regular actin filament lattice in a vertebrate smooth muscle. *Experientia* **26**: 1131.

Huxley, A. F. and R. Neidergerke. 1954. Structural changes in muscle during contraction. Interference microscopy of living muscle fibers. *Nature* **173**: 971.

Huxley, H. E. 1963. Electron microscope studies on the structure of natural and synthetic protein filaments from striated muscle. *J. Mol. Biol.* **7**: 281.

Huxley, H. E. and J. Hanson. 1954. Changes in the cross-striations of muscle during contraction and stretch and their structural interpretation. *Nature* **173**: 973.

Kelly, R. E. and J. W. Arnold. 1972. Myofilaments of the pupillary muscles of the iris fixed in situ. *J. Ultrastr. Res.* **40**: 532.

KELLY, R. E. and R. V. RICE. 1968. Localization of myosin filaments in smooth muscle. *J. Cell Biol.* **37**: 105.

———. 1969. Ultrastructural studies on the contractile mechanism of smooth muscle. *J. Cell Biol.* **42**: 683.

KEYSERLINGK, D. G. 1970. Ultrastructure of glycerinated small intestine muscle cells of the rat before and after contraction. *Z. Zellforsch.* **111**: 559.

LAEMMLI, U. K. 1970. Cleavage of structural proteins during the assembly of the head of bacteriophage T$_4$. *Nature* **227**: 680.

LANE, B. P. 1965. Alterations in the cytologic detail of intestinal smooth muscle cells in various stages of contraction. *J. Cell Biol.* **27**: 199.

LOWY, J. and J. V. SMALL. 1970. The organization of myosin and actin in vertebrate smooth muscle. *Nature* **227**: 46.

LOWY, J., F. R. POULSEN, and P. J. VIBERT. 1970. Myosin filaments in vertebrate smooth muscle. *Nature* **225**: 1053.

NONOMURA, Y. 1968. Myofilaments in smooth muscle of guinea pig's taenia coli. *J. Cell Biol.* **39**: 741.

PROSSER, C. L., G. BURNSTOCK, and J. KAHN. 1960. Conduction in smooth muscle: Comparative structural properties. *Amer. J. Physiol.* **199**: 545.

RICE, R. V., A. C. BRADY, R. H. DEPUE, and R. E. KELLY. 1966. Morphology of individual macromolecules and their ordered aggregates by electron microscopy. *Biochem. Z.* **345**: 370.

RICE, R. V., G. M. McMANUS, C. E. DEVINE, and A. P. SOMLYO. 1971. A regular organization of thick filaments in mammalian smooth muscle. *Nature* **231**: 242.

RICE, R. V., J. A. MOSES, G. M. McMANUS, A. C. BRADY, and L. M. BLASIK. 1970. The organization of contractile filaments in a mammalian smooth muscle. *J. Cell Biol.* **47**: 183.

SCOPES, R. K. and I. E. PENNY. 1971. Subunit sizes of muscle proteins as determined by sodium dodecyl sulphate gel electrophoresis. *Biochim. Biophys. Acta* **236**: 409.

SOMLYO, A. P. and A. V. SOMLYO. 1968. Vascular smooth muscle. I. Normal structure, pathology, biochemistry, and biophysics. *Pharmacol. Rev.* **20**: 197.

SOMLYO, A. P., C. E. DEVINE, and A. V. SOMLYO. 1971a. Thick filaments in unstretched mammalian smooth muscle. *Nature* **233**: 218.

SOMLYO, A. P., A. V. SOMLYO, C. E. DEVINE, and R. V. RICE. 1971b. Aggregation of thick filaments into ribbons in mammalian smooth muscle. *Nature* **231**: 246.

SOMLYO, A. P., C. E. DEVINE, A. V. SOMLYO, and S. R. NORTH. 1971c. Sarcoplasmic reticulum and the temperature-dependent contraction of smooth muscle in calcium-free solutions. *J. Cell Biol.* **51**: 722.

SOMLYO, A. P., C. E. DEVINE, A. V. SOMLYO, and R. V. RICE. 1972. Filament organization in vertebrate smooth muscle. *Proc. Roy. Soc.* (London). In press.

STROMER, M. H., D. J. HARTSHORNE, H. MUELLER, and R. V. RICE. 1969. The effect of various protein fractions on Z- and M-line reconstitution. *J. Cell Biol.* **40**: 167.

UEHARA, Y., G. R. CAMPBELL, and G. BURNSTOCK. 1971. Cytoplasmic filaments in developing and adult vertebrate smooth muscle. *J. Cell Biol.* **50**: 484.

WEBER, A. 1956. The ultracentrifuged separation of L-myosin and actin in an actomyosin sol under the influence of ATP. *Biochim. Biophys. Acta* **19**: 345.

WEBER, K. and M. OSBORN. 1969. The reliability of molecular weight determinations by dodecyl sulfate-polyacrylamide gel electrophoresis. *J. Biol. Chem.* **244**: 4406.

DISCUSSION

J. LOWY

Low-angle X-ray diffraction experiments with living contracting and relaxed guinea pig taenia coli muscle (TCGP) have lead to the discovery of a meridional reflection at about 143 Å whose presence establishes the existence of a regular assembly of myosin molecules in the form of filamentous elements (Lowy et al., 1970). I want to present evidence from further X-ray studies which indicates that the structure of the myosin elements giving rise to the 143 Å reflection persists unchanged regardless of the state of the muscle (Lowy et al., 1972).

From measurements of the shape of the 143 Å reflection, it is possible to deduce that in the axial direction the diffracting elements must be *at least* 5000 Å long; whilst in a direction perpendicular to the filament axis these elements must diffract coherently over a lateral distance of *at least* 600 Å. It has been possible to visualize such very long elements in the electron microscope (Lowy and Small, 1970) and to identify them as the myosin elements by the presence on their surface of cross-bridges which show an axial periodicity of about 140 Å (Small et al., 1971). Furthermore, these myosin elements were found to be ribbon-shaped, about 100 Å thick and with a lateral width that varies from 200 to 1100 Å (Lowy and Small, 1970). The cross-bridges are located on the two ribbon faces; they are organized laterally (with a separation of about 100 Å) in rows across each ribbon face, and the rows are spaced axially at a constant period of about 140 Å (Small et al., 1971; Small and Squire, 1972). The lateral register of the cross-bridges across the ribbon faces would easily account for the 600 Å equatorial breadth of the 143 Å reflection. We can detect no changes in the shape of this reflection when the muscle passes from a resting to a contracted state, when it is loaded with weights from 1–10 g, or when it is incubated at different temperatures in Ringer's solutions made hypertonic to various extents (up to 500 milliosmolar) by the addition of sucrose.

These observations show that the structure of the myosin elements responsible for the 143 Å reflection persists unchanged regardless of the state of the muscle, and in our view this constitutes good circumstantial evidence that the ribbons represent the normal form of the myosin elements.

Dr. Rice has argued that myosin is normally present in large round filaments (diameter about 180 Å) with irregular outlines, and that these filaments are organized in a quasi-regular lattice (nearest neighbor spacing about 700 Å). He considers that the latter feature could account for the equatorial breadth of the 143 Å reflection.

He also claims that the ribbons are artifacts resulting from the aggregation of round filaments due to two conditions, both of which are said to reduce the separation between the round filaments. These conditions are excessive stretching of the muscle and treatment with hypertonic solutions.

I have already discussed some of the evidence which indicates that the structure of the myosin elements persists unchanged under these two conditions. Two additional points may be mentioned here. As in the case of another kind of smooth muscle, namely, the anterior byssus retractor muscle of *Mytilus* (ABRM) (Elliot and Lowy, 1961), we found in our X-ray experiments with the TCGP that stretching does not affect the separation of the actin filaments. There is therefore no reason to believe that the spacing between the myosin elements will be changed under these conditions. The distance between the actin filaments is indeed decreased in hypertonic solutions in both the TCGP (Elliot and Lowy, 1968) and the ABRM (Lowy and Vibert, 1967 and unpublished results). But if this were to lead to the aggregation of myosin elements in the TCGP, one might expect that a mixture of round filaments and ribbons would be present under conditions of varying hypertonicity and that such a state of affairs could have an appreciable effect on the shape of the 143 Å reflection. However, our experiments with TCGP muscles in Ringer's solutions of varying hypertonicity showed no signs of a shape change in that reflection. I also want to point out that in our original electron microscope studies (Lowy and Small, 1970) the ribbons were seen under prefixation conditions which involved incubation of the TCGP in *isotonic* Ringer's solutions, and that further electron microscope work has demonstrated that—other things being equal—the presence of ribbons does not depend on whether the muscle has been loaded with weights of 1 or 10 g before fixation (Small and Squire, 1972).

These considerations can be interpreted to indicate that stretching or hypertonic solutions do not lead to an aggregation of myosin elements into ribbons. There remains the question of the possible significance of a lattice arrangement of the myosin elements. It can be shown (Lowy et al., 1972) that the arrangement of myosin elements in a regular lattice has no effect on the equatorial breadth of the 143 Å reflection. This is determined solely by the fact that the structures responsible for the 143 Å reflection (i.e. the cross-bridge) are in register (to within ± 12 Å axially) over a distance of at least 600 Å laterally. It is not easy to envisage such a situation in the case of the round filaments without any clear evidence for a connecting structure equivalent to the M line present in

certain striated muscles. But even if such an M line structure existed, it would undoubtedly be compressed during the assumed aggregation of round filaments into ribbons. This would upset the lateral register of the cross-bridges and hence produce an effect on the shape of the 143 Å reflection.

Rice was able to see round filaments regardless of whether the muscles were spontaneously active or relaxed, that is, respectively, in muscles incubated before fixation at room temperature in isotonic Ringer's solutions, or in calcium-free Ringer's solutions containing procaine. Under these conditions Small and Squire (1972) did not succeed in preserving adequately the structure of the myosin elements either in the form of round filaments or ribbons. Ribbons were most consistently seen in muscles relaxed in isotonic or hypertonic Ringer's solutions at 4°C (Lowy and Small, 1970). We suggest that the ribbons are seen at low temperatures because the spontaneous cycling of cross-bridges is effectively suppressed, and this somehow allows the myosin elements to be adequately preserved by chemical fixation. As regards to the state of the cross-bridges, the X-ray results show that the *intensity* of the 143 Å reflection increases markedly at the lower temperatures when spontaneous contractile activity is absent, whereas the reflection is very weak at room temperature when the muscle is active. This is to be expected on the assumption that the intensity of the 143 Å reflection is mainly due to the degree of axial order of the cross-bridges and that (as in striated muscles) this decreases in the contracting TCGP muscle (Lowy et al., 1970).

Our conclusion is that one of the main reasons why the ribbon form of the myosin elements could be preserved in a relatively intact state was because means were found to effectively suppress cross-bridges movement. In fact, cross-bridge order could be better preserved than in frog or rabbit muscles, and in the best TCGP preparations (Small et al., 1971; Small and Squire, 1972) cross-bridge order is almost as good as in insect flight muscles. This made it possible to demonstrate that in the TCGP the cross-bridges are arranged on a two-dimensional lattice, as well as to obtain good evidence that they face in opposite directions on opposite ribbon faces (Small et al., 1971; Small and Squire, 1972).

To summarize: (1) There is no evidence from our studies of the shape of the 143 Å reflection which supports the idea that the ribbons are formed by aggregation of round filaments. (2) The equatorial breadth of the 143 Å reflection can be most straightforwardly interpreted in terms of the ribbon structure. (3) The arrangement of cross-bridges on a two-dimensional lattice on the two ribbon faces argues strongly against the possibility that such extremely well-organized structures are artifacts formed by aggregation of round filaments.

References

ELLIOTT, G. F. and J. LOWY. 1961. X-ray diffraction studies of a molluscan smooth muscle. *J. Mol. Biol.* **3:** 41.

————. 1968. Organization of actin in a mammalian smooth muscle. *Nature* **219:** 156.

LOWY, J. and J. V. SMALL. 1970. The organization of myosin and actin in vertebrate smooth muscle. *Nature* **227:** 46.

LOWY, J. and P. J. VIBERT. 1967. Structure and organization of actin in a molluscan smooth muscle. *Nature* **215:** 1254.

LOWY, J., F. R. POULSEN, and P. J. VIBERT. 1970. Myosin filaments in vertebrate smooth muscle. *Nature* **225:** 1053.

LOWY, J., P. J. VIBERT, J. C. HASELGROVE and F. R. POULSEN. 1972. The structure of myosin elements in vertebrate smooth muscles. *Proc. Roy. Soc.* B (in press).

SMALL, J. V. and J. M. SQUIRE. 1972. Structural basis of contraction in vertebrate smooth muscle. *J. Mol. Biol.* **67:** 117.

SMALL, J. V., J. LOWY, and J. M. SQUIRE. 1971. The myosin ribbons of vertebrate smooth muscle. *Proc. 1st Europ. Biophys. Cong.*, Baden, vol. 5, p. 419.

The Core Component of the Myosin-Containing Elements of Vertebrate Smooth Muscle

J. V. Small and Apolinary Sobieszek

Institute of Biophysics, Aarhus University, 8000 Aarhus C, Denmark

Recent electron microscope studies of thin sections from vertebrate smooth muscle (Lowy and Small, 1970; Small and Squire, 1972) and of extracts obtained from minced muscle at low ionic strength (Sobieszek, 1970; Sobieszek and Small, 1972) have indicated that the myosin-containing filaments are ribbon-like in shape and have provided information about the detailed structure of these ribbons. In this report we summarize the evidence which has demonstrated the presence and nature of a second component in the ribbons, in addition to myosin, which forms the core of the ribbons and which is presumably responsible for their ribbon-like shape.

The ribbons were shown from longitudinal sections to be several microns in length and to bear a regular arrangement of projections (Fig. 1b), which were organized in rows across the ribbon face and with the rows spaced axially at a constant period of about 140 Å. The presence of projections on the ribbons of the same dimension and axial spacing as the "cross-bridges" observed on the myosin filaments of striated muscles (Huxley, 1957, 1969; Reedy, 1968) and there considered to represent the head part of the myosin molecule (see review by Lowey, 1971) was the first firm indication that the ribbons did indeed contain myosin. Furthermore, it was apparent that myosin molecules on one face of a ribbon were polarized in the opposite direction to those on the opposite face (Small and Squire, 1972).

From a study of transverse sections (e.g. Fig. 1a) it was apparent that the ribbons could readily be distorted and/or dissociated, presumably as a consequence of chemical fixation. Such changes produced two other filament types. First, as a result of longitudinal splitting and distortion of the ribbons, large filaments with an irregular but approximately circular cross section and average diameter of about 180 Å diameter were formed (Small and Squire, 1972). Alternatively, from images such as shown in Fig. 2a–d it was clear that the ribbons could also dissociate so as to give rise to filaments of a very regular size (Lowy and Small, 1970), these filaments having the same appearance and dimension (about 100 Å diameter) as the neurofilaments and tonofilaments found in

other cell types (see also Uehara et al., 1971). Further studies showed that these filaments were square in cross section, about 75 Å across, and composed of four filamentous subunits (Fig. 2e). From the absence in longitudinal sections of projections on these square filaments and from their resemblance to tonofilaments, it was concluded that they were not composed of myosin but of another type of structural protein, and that this protein made up the core of the ribbons.

In summary, the data available from sections thus indicated that the ribbons were composed of at least two components, a core of nonmyosin material and a surface lattice of myosin molecules.

Although there were several reasons from the studies of thin sections (Small and Squire, 1972) to consider the ribbons as the in vivo myosin-containing structures, it was not possible to confirm this unequivocally using the X-ray diffraction evidence, which had provided the first demonstration of the existence of myosin in filamentous form in the living muscle. Thus while the X-ray patterns obtained from living taenia coli muscle (Lowy et al., 1970) could readily be explained on the basis of the presence of ribbon-shaped myosin-containing elements, the existence of myosin in filaments of another form could not be excluded (Lowy et al., 1972). In this respect it has been considered by other investigators (Somlyo et al., 1971; Rice et al., 1971) that the approximately cylindrical filaments of about 180 Å average diameter observed in thin sections are the most likely form of myosin. To be consistent with the X-ray data, however, such filaments would have to be organized with a degree of longitudinal register which would require the existence of some form of interfilament link (see also Lowy et al., 1972) and, as yet, no evidence for any such link has been obtained. As indicated previously these large filaments are explained by Small and Squire (1972) as merely products of splitting and distortion of the ribbons.

As another approach to the investigation of the form of the contractile components, experiments were carried out in an attempt to isolate the filaments from the cell of several vertebrate smooth muscles and study them using the negative staining

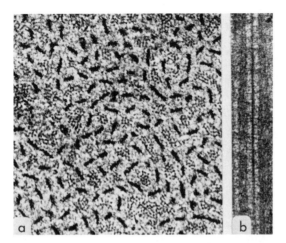

Figure 1. (*a*) Transverse section of part of a cell from the guinea pig vas deferens showing the general arrangement of the contractile elements. Ribbons containing myosin (dark lines) occur together with thin actin-containing filaments (dots), the latter being organized into small, regular lattices. × 78,000.

(*b*) Longitudinal section of a ribbon seen in "edge view" (plane of ribbon parallel to electron beam), showing cross-bridges spaced at regular intervals of about 140 Å. Guinea pig taenia coli, × 66,000.

technique (Sobieszek, 1970; Sobieszek and Small, 1972). The results of these studies showed that while intact myosin-containing filaments could not, under the conditions used, be released from the cells, a protein that was extracted in large amounts at low ionic strength in the presence of ATP could readily assemble into long, ribbon-shaped structures (Fig. 3a, b). This assembly

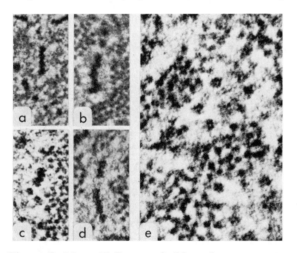

Figure 2. (*a*) to (*d*) Images of ribbons from transverse sections which indicate that the filaments of near 100 Å diameter, that have been consistently observed in smooth muscle cells, are derived from a dissociation of the ribbons. Taenia coli, × 185,000.

(*e*) The "100 Å filaments" are in fact square in cross section, 75 Å across and composed of four filamentous subunits. Middle part of micrograph shows a group of four of such filaments which are surrounded by two main groups of thin filaments. Taenia coli, × 290,000.

occurred either if $CaCl_2$ (50 mM) was added to the low ionic strength extract or if the extract was simply left to stand for one or more days in the cold. The ribbons observed in these preparations are often very wide, up to about 1000 Å across, and show a prominent transverse striation at a period of about 56 Å. From optical diffraction analysis (Sobieszek and Small, 1972) it was found that these ribbons possessed a long repeat of about 400 Å (391 Å ± 4 Å), that is, the same long repeat as observed for assemblies of tropomyosin obtained from various muscle types (Tsao et al., 1965; Cohen and Longley, 1966; Caspar et al., 1969; Millward and Woods, 1970). In addition, tropomyosin prepared from chicken gizzard according to the procedure of Bailey (1948) could be induced to form into essentially the same ribbon-shaped structures (Fig. 3c, d; and Sobieszek and Small, 1972). It was therefore apparent that the ribbon-shaped structures observed in the extracts obtained at low ionic strength were composed of smooth muscle tropomyosin.

From the dimensions of these ribbons it was immediately tempting to conclude that they corresponded to the core component or backbone of the ribbons identified in thin sections of intact muscle. That this was indeed the case was indicated by two further lines of evidence. First, it was found that the ribbons (Fig. 3a) readily dissociated into finer filaments (Fig. 4a), including some of width about 80 Å, the same dimension as the square filaments observed in thin sections, and all of which had subfilaments of about 20 Å to 30 Å diameter. This could be achieved by simply dialyzing a preparation of ribbons in buffer against distilled water. Second, if extracts containing the ribbons together with smooth muscle myosin were left to stand for a week or more in the cold, the ribbons commonly appeared to have additional material on their surfaces which obscured the fine 56 Å periodicity and which gave them a roughened appearance. In some instances this extra material gave rise to a transverse striation with an axial period of about 140 Å (Fig. 4b), the same period as observed on the ribbons in longitudinal sections and also on cylindrical filaments assembled in vitro from purified smooth muscle myosin (Sobieszek, 1972; Sobieszek and Small, this volume), and in each of these latter cases shown to arise from a regular arrangement of projections, or cross-bridges. Optical diffraction analysis of such composite ribbons (Sobieszek and Small, 1972) clearly demonstrated the 140 Å periodicity and additionally, some patterns also showed meridional reflections characteristic of tropomyosin.

The results from actomyosin extracts of smooth muscle therefore indicated that the ribbons formed

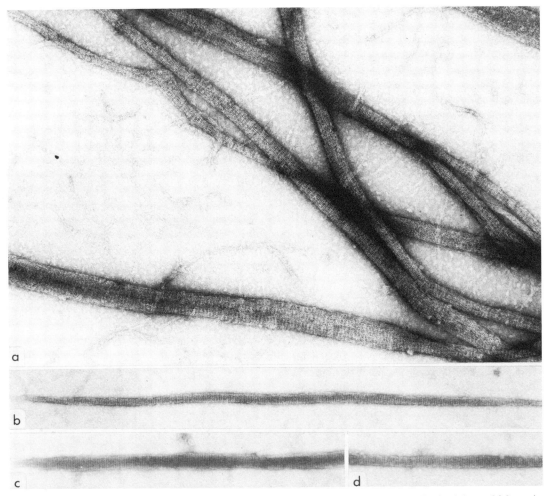

Figure 3. (a) and (b) Ribbon-shaped structures observed in a low ionic strength extract obtained from chicken gizzard. Finest period is 56 Å. (a), ×71,000; (b), ×45,000.
(c) and (d). Ribbon-shaped structures formed from purified chicken gizzard tropomyosin in the presence of Ca⁺⁺. These show the same fine structure as the ribbons in (a) and (b). ×45,000.

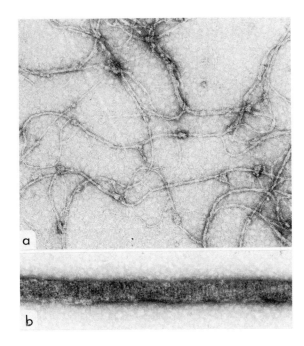

Figure 4. (a) Fine filaments formed by dialyzing ribbons shown in Fig. 3a against distilled water. Many of these are identical in size to the "100 Å filaments" observed in section material. Gizzard, ×73,000.
(b) Ribbon observed in a fraction containing ribbons together with smooth muscle myosin which was left to stand for about one week in the cold. Additional material (myosin) is present on the ribbon which gives rise to a transverse periodicity of about 140 Å. ×60,000.

in vitro were essentially the same structures as those identified in thin sections. They possess a core of nonmyosin material which apparently provides a surface suitable for interaction with myosin. Furthermore, the core component has been identified as tropomyosin.

In other muscle types, tropomyosin has been generally shown to form an integral part of the Ca^{++}-regulatory-protein system that is associated in vertebrate striated muscle with the actin filaments. The identification of the core component of the myosin ribbons as tropomyosin thus raises the problem of the nature of the regulatory mechanism in vertebrate smooth muscle. Is tropomyosin associated with actin in this muscle type; and if so, is it the same as, or different from, the protein that forms the core of the myosin ribbons? Our further experiments will be designed to answer these important questions.

Acknowledgments

We thank the Danish Statens Naturvidenskabelige Forskningsråd for their financial support and for purchasing the Siemens 101 electron microscope. We also thank Mrs. L. Nychel and Miss K. Eskesen for technical assistance and Mr. P. Boldsen and Mrs. I. Lunde for photography.

References

BAILEY, K. 1948. Tropomyosin: A new asymmetric protein component of the muscle fibril. *Biochem. J.* **43**: 271.

CASPAR, D. L. D., C. COHEN, and W. LONGLEY. 1969. Tropomyosin: Crystal structure, polymorphism and molecular interactions. *J. Mol. Biol.* **41**: 87.

COHEN, C. and W. LONGLEY. 1966. Tropomyosin paracrystals formed by divalent cations. *Science* **152**: 794.

HUXLEY, H. E. 1957. The double array of filaments in cross-striated muscle. *J. Biophys. Biochem. Cytol.* **3**: 631.

———. 1969. The mechanism of muscular contraction. *Science* **164**: 1356.

LOWEY, S. 1971. Myosin: molecule and filament, p. 201. In *Biological macromolecules*, ed. S. N. TIMASHEFF and G. D. FASMAN, vol. 5, part A. Marcel Dekker, New York.

LOWY, J. and J. V. SMALL. 1970. The organization of myosin and actin in vertebrate smooth muscle. *Nature* **227**: 46.

LOWY, J., F. R. POULSEN, and P. J. VIBERT. 1970. Myosin filaments in vertebrate smooth muscle. *Nature* **225**: 1053.

LOWY, J., P J. VIBERT, J. C. HASELGROVE, and F. R. POULSEN. 1972. The structure of the myosin elements in vertebrate smooth muscle. *Phil. Trans. Roy. Soc. London B.* In press.

MILLWARD, G. R. and E. F. WOODS. 1970. Crystals of tropomyosin from various sources. *J. Mol. Biol.* **52**: 585.

REEDY, M. K. 1968. Ultrastructure of insect flight muscle. I. Screw sense and structural grouping in the rigor cross-bridge lattice. *J. Mol. Biol.* **31**: 155.

RICE, R. V., G. M. McMANUS, C. E. DEVINE, and A. P. SOMLYO. 1971. Regular organization of thick filaments in mammalian smooth muscle. *Nature New Biol.* **231**: 242.

SMALL, J. V. and J. M. SQUIRE. 1972. The structural basis of contraction in vertebrate smooth muscle. *J. Mol. Biol.* **67**: 117.

SOMLYO, A. P., A. V. SOMLYO, C. E. DEVINE, and R. V. RICE. 1971. Aggregation of thick filaments into ribbons in mammalian smooth muscle. *Nature New Biol.* **231**: 243.

SOBIESZEK, A. 1970. Isolation of ribbon-shaped elements from vertebrate smooth muscle. *J. Ultrastr. Res.* **38**: 208.

———. 1972. Cross-bridges on self-assembled smooth muscle myosin filaments. *J. Mol. Biol.* In press.

SOBIESZEK, A. and J. V. SMALL. 1972. The assembly of ribbon-shaped structures in low ionic strength extracts obtained from vertebrate smooth muscle. *Phil. Trans. Roy. Soc. London B.* In press.

TSAO, T. C., T. H. KUNG, C. M. PENG, Y. S. CHANG, and Y. S. TSOU. 1965. Electron microscopical studies of tropomyosin and paramyosin. *Scientia Sinica* **14**: 206.

UEHARA, Y., G. R. CAMPBELL, and G. BURNSTOCK. 1971. Cytoplasmic filaments in developing and adult vertebrate smooth muscle. *J. Cell Biol.* **50**: 484.

Effects of ATP Analogs on the Low-Angle X-Ray Diffraction Pattern of Insect Flight Muscle

J. Barrington Leigh, K. C. Holmes, H. G. Mannherz and G. Rosenbaum

Max-Planck-Institut für medizinische Forschung, Heidelberg, Germany

F. Eckstein and R. Goody

Max-Planck-Institut für experimentelle Medizin, Göttingen, Germany

Some years ago Reedy, Holmes, and Tregear (1965) noted that ATP had a dramatic effect upon the low-angle X-ray diffraction pattern of glycerinated fibers of the flight muscle from the giant water bug *Lethocerus maximus*. A strong meridional reflection appeared at a spacing of 145 Å in the presence of ATP. When the ATP was washed out, this reflection disappeared and its place was taken by a 380 Å layer line. With the help of electron microscopy these changes were shown to be due to an alteration in the orientation of the cross-bridges. In the absence of ATP (rigor state) the cross-bridges are at 45° to the filaments. In the presence of ATP the bridges appear to be free from the actin and are right-angled to the thick (myosin) filament (Fig. 1).

The glycerinated fibers of an insect flight muscle are, therefore, a very favorable system for observing differences of cross-bridge orientation. We wish now to discuss the following questions which arise out of this observation: (a) What is the nature of the steady state complex observed in the presence of an excess of ATP? (b) Can we identify other steady state complexes with the help of ATP analogs?

Steady State Complex in Excess ATP

The kinetic studies of Taylor and his coworkers (Lymn and Taylor 1970, 1971; Taylor et al., 1970) show that the magnesium-activated ATPase of actomyosin resides entirely in the myosin and that the actin stimulates release of the products of hydrolysis. These results may be summarized by the following scheme where A = actin; M = myosin:

$$AM + ATP \overset{k_{+1}}{\rightleftharpoons} MATP + A \overset{k_{+2}}{\rightleftharpoons}$$
$$MADPP_i + A \overset{k_{+3}}{\rightleftharpoons} AM + ADP + P_i$$

Taylor and his associates find that k_1 is very fast, k_2 has a value of about 100 sec^{-1}, and k_3 is slow with a value around 0.05 sec^{-1} if actin is not present. The rate increases dramatically on the addition of actin. All the back rates are slow enough to be neglected. Taylor argues that since the dissociation of actomyosin (k_1) is very fast, there is no chance of actomyosin being the actual enzyme carrying out ATP hydrolysis. From the work of Trentham et al. (1971) it is clear that the Taylor's picture is oversimplified insofar as the step designated by k_3 must involve two or three conformational changes characterized by one or two more first-order rate constants. Taylor's kinetic data were taken using rabbit myosin and at pH 8, 20°C, whereas the X-ray experiments on insect flight muscle were made at pH 7, 2°C. If we assume that the changes in rate constants caused by altering muscle type, temperature, and pH are not too extreme, then we can predict that the steady state complex in glycerinated flight muscle in the presence of an excess of ATP should be the species MADP·P$_i$ because the slow rate constant k_3 will bring about an accumulation at this stage. We therefore correlate this steady state complex with the

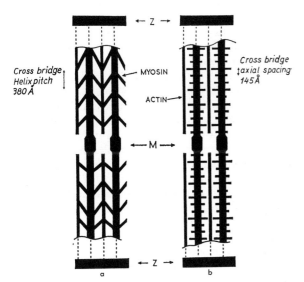

Figure 1. ATP-induced changes in the orientation of the cross-bridges in insect flight muscle: (*a*) Arrangement of angled cross-bridges in the sarcomere in rigor; (*b*) the right-angled position of the bridges in the relaxed state (Reedy et al., 1965).

right-angled cross-bridge configuration found from the X-ray and electron microscope results. The existence of the complex MADP has recently been demonstrated in relaxed insect muscle fibers by Marston and Tregear (1972).

ATP Analogs

A logical development of this program is to look for analogs or variations of conditions which bring about large changes of the k_2/k_3 ratio, thereby producing different distributions of steady state complexes. In practice we have first tried analogs on the muscle fibers to see if they gave X-ray diffraction patterns different from those given by normal ATP. We have used α,β-methylene-ATP (APCPP) (Miles Biochem), adenosine 5′-O-(3-thiotriphosphate) (ATP-γS) (Goody and Eckstein, 1971), and β,γ-imino-ATP (AMPPNP) (Yount et al., 1971).

α,β-Methylene-ATP (APCPP).

It has been shown (Mannherz et al., 1972) that APCPP is a substrate for myosin. It also relaxes actomyosin as judged from ultracentrifuge studies. As a check on the ability of this analog to relax insect actomyosin in situ, we have measured the stiffness of a glycerinated muscle fiber as a function of analog concentration (Fig. 2). All stiffness measurements were made on a simple stretching apparatus. The tension referred to in this figure is the rigor tension generated when relaxed muscle held isometrically is put into rigor. The release of this tension, which falls off more steeply than the stiffness, might be associated with small induced conformational changes in the myosin. In contrast the stiffness is a measure of the average number of bridges attached. The results in Fig. 2 indicate that at about 30 mM APCPP concentration the relaxation of insect actomyosin in the fibers is complete.

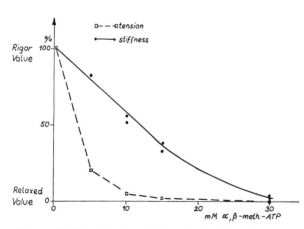

Figure 2. The behavior of rigor-induced tension and stiffness, measured relative to their relaxed state values, with varying concentrations of α,β-methylene ATP (APCPP).

Table 1. Composition of Solutions Used to Bath Muscle Fibers

Rigor Solution:	10 mM MgCl$_2$, 4 mM EGTA, 10 mM NaN$_3$, 20 mM imidazole-chloride buffer pH 6.9
Relaxing Solution:	15 mM MgATP, 4 mM EGTA, 10 mM NaN$_3$, 20 mM imidazole-chloride buffer pH 6.9
Analog Solutions:	Rigor solution + 15 or 30 mM MgAPCPP, 20 mM MgATP-γS, 10 mM Mg AMPPNP, respectively.

Using a 40-cm focusing camera and an Elliott X-ray rotating anode tube, we conducted experiments on the effect of various concentrations of APCPP. In order to avoid the possibility that the analog might destroy the structure of muscle without causing any specific changes, we ran a series of photographs in which a control rigor photograph was made, followed by the analog photograph, followed by an equal exposure in the presence of 15 mM ATP. (Details of solutions are given in Table 1.) An experimental series made with 15 mM APCPP is shown in Fig. 3. Note the strong 380 Å layer line in rigor and note also the 145 Å meridional reflection given in the presence of ATP (Fig. 3c). In the presence of 15 mM APCPP the diffraction pattern given by glycerinated insect flight muscle is virtually unchanged from that given in rigor, even though at this concentration of analog the stiffness curve (Fig. 2) shows that at least 50% of the bridges are not attached to the actin. Apparently enough connections are still being made to sustain the actomyosin lattice and give a crystal-like diffraction pattern on the 380 Å layer line. In the presence of 30 mM APCPP (Fig. 4b) there is complete disappearance of the 380 Å layer line without the appearance of the 145 Å meridional reflection. Hence the cross-bridges are detached from the actin and yet do not assume the classical right-angled configuration for unattached cross-bridges.

The strength of the 145 Å meridional reflection is an indication of the distribution of the cross-bridges between various states of orientation. If all cross-bridges are at an angle of 45° to the myosin axis, then the 145 Å meridional will be very weak. If all bridges are at right angles, then the 145 Å meridional will be strong. We must deduce from our present observations that in the presence of APCPP the cross-bridges can detach from the actin but stay in the 45° configuration or are highly disordered.

The question naturally arises as to which steady state complex of the Taylor-Lymn scheme this is. At present transient kinetic data on APCPP are not available. However the turnover rate of APCPP hydrolysis by rabbit myosin is not accelerated by the addition of actin. In contrast, in the case of normal ATP, k_3 is greatly increased, thereby accelerating the turnover rate of ATP hydrolysis. The

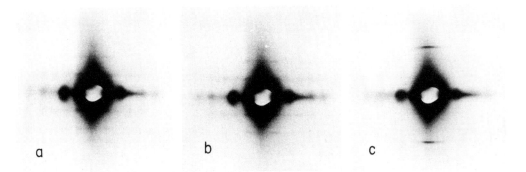

Figure 3. Low-angle X-ray diffraction photographs taken with a mirror-monochromator focusing camera and an Elliott fine focus rotating anode tube of the dorsolongitudinal flight muscles of *Lethocerus maximus*. Specimen temperature 2°C specimen-to-film distance 40 cm. $C\mu K_\alpha$ radiation. Exposure time 10 hr. The sequence is as follows: (*a*) muscle in rigor solution; (*b*) muscle in 15 mM α,β-methylene ATP; (*c*) muscle in relaxing solution (see Table 1).

Figure 4. Low-angle X-ray diffraction photographs, conditions as in Fig. 3. The camera has been altered to give better low-angle resolution. (*a*) Muscle in rigor; (*b*) muscle in 30 mM α,β-methylene ATP; (*c*) muscle in relaxing solution (see Table 1).

measured hydrolysis rates of APCPP and ATP with magnesium-activated myosin are in fact similar. The fact that the steady state rate for APCPP is actin insensitive suggests that k_3 is already faster than k_2 and that k_2 limits the reaction rate.* Thus the myosin cross-bridge in a 45° position but not bound to actin is probably the species MAPCPP.

Adenosine 5'-O-(3-thiotriphosphate) ATP-γS.

ATP-γS has been shown to be a substrate for myosin with a turnover rate somewhat higher than that of normal ATP (Bagshaw et al., this volume). It also dissociates actomyosin in the ultracentrifuge. Stiffness measurements on insect flight muscle (Fig. 5) show that the stiffness drops to a low value at 20 mM ATP-γS. Kinetic measurements show (Bagshaw et al.) that k_2 for rabbit myosin (pH 8, 22°C) is 0.25 sec^{-1}; k_3 could not be measured but is probably faster than 1 sec^{-1}. If these constants were applicable directly to the insect myosin system, we would expect to find in the steady state

an unequal mixture of MATP-γS and MADP (PS), the former being the major species. We find using low-angle X-ray diffraction (Fig. 6) a half-strength 145 Å reflection. This could be explained on a two-state model, by having 75% of the bridges *up*

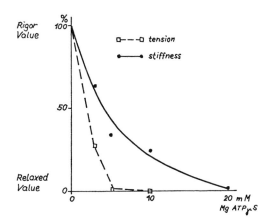

Figure 5. The behavior of rigor-induced tension and stiffness with varying concentration of adenosine-5'-O-(3-thiotriphosphate) (ATPγS).

---

* This has recently been confirmed by direct kinetic measurements by Mannherz and Trentham.

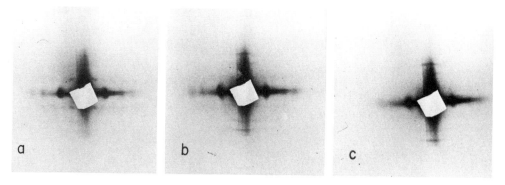

Figure 6. Low-angle X-ray diffraction photographs, conditions as in Fig. 3. (*a*) muscle in rigor; (*b*) muscle in 20 mM ATP-γS; (*c*) muscle in relaxing solution.

and 25% *down*. This is not what one would predict from the kinetic data. One explanation of the X-ray result would be that k_2 is in fact greater than k_3 for insect myosin at 2°C and pH 7. We are presently determining the kinetic constants under the relevant conditions. An alternative explanation may be that the sample is contaminated by ATP, produced by myokinase in the fibers acting on the ADP resulting from cleavage of ATP-γS.

β,γ-*Imino-ATP (AMPPNP*). This analog has been reported by Yount et al. (1971) to dissociate actomyosin. It differs from ATP in one notable respect: the pK of the γ-phosphate group is one unit higher. At pH 7.0 the state of ionization of the terminal phosphate group may therefore be different from that of ATP. Stiffness measurements (Fig. 7) indicate that it does not dissociate actomyosin very well, although the drop in rigor tension

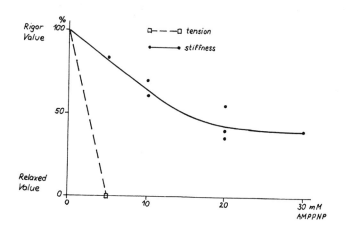

Figure 7. The behavior of rigor-induced tension with varying concentrations of β,γ-imino-ATP (APPNP). Note that the stiffness remains high at high analog concentration.

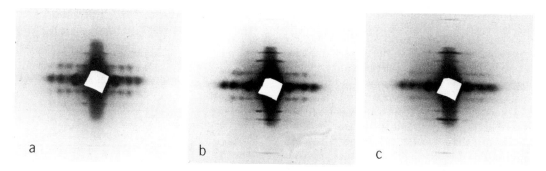

Figure 8. Low-angle X-ray diffraction photographs, conditions as in Fig. 3. (*a*) Muscle in rigor; (*b*) muscle in 10 mM AMPPNP; (*c*) muscle in relaxing solution. The appearance of (*c*) is typical (i.e., not a normal relaxed pattern) and is probably caused by the tight binding of AMPPNP.

is complete at 5 mM. Our X-ray findings using this analog are presented in Fig. 8. The pattern (Fig. 8b) obtained in the presence of 10 mM AMPPNP is remarkably crystalline. Notably the meridional 145 Å reflection is strong. However in contrast to what one finds in ATP relaxed muscle, it is still strongly sampled by a lattice. We infer that, in the presence of AMPPNP, the cross-bridges have swung away from their 45° position characteristic of rigor and have taken up a position most of the way to the 90° position characteristic of the presence of ATP (i.e., MADP·P$_i$). Furthermore the bridges have apparently done this without letting go of the actin filaments and without splitting the analog. The appearance of the diffraction patterns was similar in the presence of 5, 10, or 15 mM Mg AMPPNP.

Discussion

We would like to try to accommodate our observations within a unified hypothesis such as the cycle for cross-bridge activity shown in Fig. 9. This cycle is closely related to the model proposed by A. F. Huxley (1957) and has recently been published in the form given here by Lymn and Taylor (1971). On the basis of our experiments on APCPP we would suggest that cross-bridges in the 45° orientation exist either when myosin binds actin or, transiently, on binding ATP. However the stable form after ATP cleavage is the right-angled form. We hope to accommodate our observations

with the ATP-γS within this scheme. The maveric observations are those concerning AMPPNP. As a tentative explanation we propose that AMPPNP binds in a quite different way from ATP, APCPP, or ATP-γS and in some way mimics one of the states associated with MADPP$_i$ (i.e., state 2 in Fig. 9). Thus is brings about large changes in the cross bridge conformation (loss of rigor tension and strong meridional 145 Å reflection) without relaxing the muscle. If this explanation is correct then the analog may be a very useful tool for investigating the conformational states associated with the power stroke itself.

Our conclusions in their present form clearly need further experimental substantiation. However we expect that from such experiments a clearer picture of the correlation between the biochemical cycle and the structural cycle of a cross-bridge may be obtained.

References

GOODY, R. S. and F. ECKSTEIN. 1971. Thiophosphate analogues of nucleoside di- and tri-phosphates, *J. Amer. Chem. Soc.* **93**: 6252.

HUXLEY, A. F. 1957. Muscle structure and theories of contraction. *Prog. Biophys. Biophys. Chem.* **7**: 255.

LYMN, R. W. and E. W. TAYLOR. 1970. Transient state phosphate production in the hydrolysis of nucleoside triphosphates by myosin. *Biochemistry* **9**: 2975.

———. 1971. Mechanism of adenosine triphosphate hydrolysis by actomyosin. *Biochemistry* **10**: 4617.

MANNHERZ, H. G., J. BARRINGTON LEIGH, K. C. HOLMES, and G. ROSENBAUM. 1972. The identification of the transitory complex myosin-ATP by the use of α,β-methylene-ATP. *Nature*. In press.

MARSTON, S. B. and R. T. TREGEAR. 1972. Evidence for a complex between myosin and ADP in relaxed muscle fibres. *Nature New Biol.* **235**: 23.

REEDY, M. K., K. C. HOLMES, and R. T. TREGEAR. 1965. Induced changes in orientation of the cross bridges of glycerinated insect flight muscle. *Nature* **207**: 1276.

TAYLOR, E. W., R. W. LYMN, and G. MOLL. 1970. Myosin-product complex and its effect on the steady-state rate of nucleoside triphosphate hydrolysis. *Biochemistry* **9**: 2984.

TRENTHAM, D. R., R. G. BARDSLEY, J. F. ECCLESTONE, and A. G. WEEDS. 1972. Elementary processes of the magnesium ion-dependent adenosine triphosphate activity of heavy meromyosin. *Biochem. J.* **126**: 635.

YOUNT, R. G., D. OJALA, and D. BABCOCK. 1971. Interaction of P-N-P and P-C-P analogues of adenosine triphosphate with heavy meromyosin, myosin, and actomyosin. *Biochemistry* **10**: 2490.

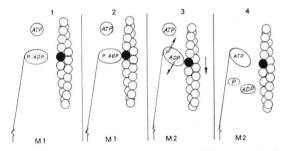

Figure 9. A schematic representation of the stages in the cyclic action of the cross-bridge. *M1* and *M2* refer to the right-angled and 45° positions, respectively, of the myosin cross-bridge (Mannherz et al., 1972). With some reservations our present X-ray observations can be accommodated within this scheme.

X-Ray Diagrams from Skeletal Muscle in the Presence of ATP Analogs

R. W. LYMN AND H. E. HUXLEY

MRC Laboratory of Molecular Biology, Cambridge CB2 2QH, England

Structural studies on skeletal muscle have revealed many of the events which must take place in a contraction. The outward movement of myosin cross-bridges to the actin filaments when a muscle goes into rigor (Huxley, 1968) is matched to some extent in a contracting, live muscle (Haselgrove, 1970). Similarly, the loss in rigor of the myosin layer lines so typical of live muscle is partially paralleled during contraction (Huxley and Brown, 1967; Haselgrove, 1970), indicating an increased disorder of the cross-bridges. These results have been interpreted in simple terms: The myosin cross-bridges in a resting live muscle are located on average very near the thick filament, their position determined by the ordered arrangement of the LMM (light meromyosin) parts of the

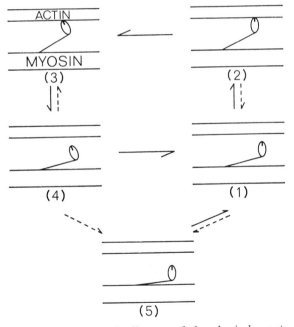

Figure 1. A schematic diagram of the physical events thought to take place during the contraction of a striated muscle. In step 1 the myosin cross-bridges are separated from the actin; step 2 is symbolic of attachment at a specific orientation, which then changes to that indicated by step 3, generating tension, and for a submaximal load, shortening. Step 4 represents the cross-bridge redissociated from actin, presumably by the binding of ATP. Step 5 represents the extremely well-ordered cross-bridge array that gives rise to the live, resting myosin X-ray pattern. This last may only be a more ordered form of step 1.

myosin which form the backbone; during a contraction, these cross-bridges move out, attach to actin, generate tension and movement, then detach again ready to start another cycle.

The scheme proposed recently (Lymn, 1970; Lymn and Taylor, 1971) to describe the mechanism whereby actomyosin hydrolyzes ATP agrees very well with this view. In the Lymn-Taylor model, ATP binds to an actomyosin link, causing separation to actin and myosin. The ATP bound to myosin is hydrolyzed. The resultant myosin-products complex recombines with actin, with subsequent loss of the products. The cycle is reinitiated by the binding of another ATP; the muscle is relaxed by inhibiting the recombination of actin and myosin; and, since loss of products from myosin alone is very slow, ATP splitting is switched off at the same time.

Although these schemes for chemical and physical events are so similar, details of the mechanism are elusive, particularly concerning two major steps in most sliding filament models of contraction: the movement of the myosin cross-bridge relative to actin when not attached (step 4 → 1 in Fig. 1), and the movement upon attachment necessary for contraction (step 2 → 3). About these steps both biochemistry and structural studies are still relatively uninformative. If, however, we can perturb an intact muscle in certain controlled ways, the highly enigmatic X-ray diagrams may yield more information. Thus we have chosen analogs of ATP, which have reasonably well-known effects on actomyosin in solution and which might arrest the cross-bridge cycle at some "chemically" defined position, and have added them to glycerinated muscle fibers to see what changes they might cause in the X-ray diagrams.

Methods and Results

As a model system for these experiments, we have used rabbit psoas muscle which had been glycerinated to render the membranes more permeable, thereby allowing free passage of the various reagents. The particular method of glycerination was that described by Rome (1972) except that we immersed muscles alternately in

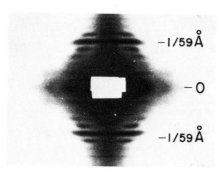

Figure 2. A rigor pattern from the glycerinated rabbit muscles used. The buffer was 0.1 M KCl, 0.005 M MgCl$_2$, 0.002 M EGTA, 0.01 M imidazole pH 7.2, 0°C. Fiber axis is vertical.

Figure 3. A muscle as in Fig. 2, but in the presence of 5 mM ATP, 2 mM EGTA. The myosin layer lines typical of a live, resting muscle are visible.

50% glycerol and the 0.5% Lubrol solution used by Julian (1971). The muscles were then stored in a glycerol solution at −15°C until use.

For individual experiments a small muscle bundle (about 0.5 mm diameter) is mounted in a Perspex cell and washed with our basic buffer, 0.1 M KCl, 0.01 M imidazole, 0.005 M MgCl$_2$, 0.002 M EGTA pH 7.2, 0°C. It is then in a normal rigor state and gives a typical rigor pattern, as shown in Fig. 2. This is dominated by the strong 59 Å actin reflection, which crosses the meridian, and the off-meridional reflections due to the rigor actin-myosin links of the cross-bridges.

Before one can accept results using ATP analogs with the model system of glycerinated muscle, it is necessary to show that a given preparation responds in an appropriate manner to the presence of ATP itself, i.e., by relaxing in the absence of Ca^{++} and contracting in its presence. Rome (1972) showed that glycerinated muscle which was relatively fresh (less than two months old) responded properly; its tension relaxed, it was readily extended; it gave X-ray diagrams typical of live muscle.

In our studies we have been able to obtain X-ray diagrams similar to those characteristic of live, relaxed muscle by adding 5 mM ATP to glycerinated muscle in the presence of 2 mM EGTA. One such diagram is shown in Fig. 3. The pattern of low-angle off-meridional reflections based on the actin lattice has disappeared; the myosin layer lines of a live, rest pattern have reappeared (though with only about one-third the intensity of a live muscle); and the 59 Å reflection has lost some of its near-meridional intensity. Diagrams like this can be obtained routinely.

We have primarily been looking at the effect of two analogs, PP$_i$ (inorganic pyrophosphate) and AMP-PNP (adenylylimido-diphosphate). Both of these cause a dissociation of actin and myosin in

solution, without apparent hydrolysis. Neither of these, at 5 mM concentration, has given any sign of a relaxed pattern; indeed neither has completely abolished the rigor cross-bridge reflections. They do, however, have quite visible effects on the axial pattern. This can be seen in Fig. 4, which is typical of both analogs at 5 mM, pH 7.2. The rigor pattern is still prominent. The 59 Å actin reflection has full rigor characteristics; namely, it crosses the meridian and has a high near-meridian intensity. But in comparison with Fig. 2, the pattern of attached cross-bridges is relatively weaker. This is seen most easily by looking at the near 72 Å off-meridional reflection and noting its intensity with respect to the 59 Å reflection. Using a microdensitometer we can measure the intensities of the different reflections. Table 1 gives values under different conditions. The relative intensity of the off-meridional 72 Å reflection has decreased by 30–40%. The 51 Å reflection has also become somewhat weaker, though not to the same extent. Since part of the 51 Å reflection is due to the actin filament and is seen in relaxed muscle, this is to be expected.

Another indication of change can be seen if we compare the relative intensities of the 365 Å off-meridional reflection (which arises from the "labeling" of the actin repeat by the myosin

Figure 4. As Fig. 2, but 5 mM PP$_i$. The rigor pattern is still visible, but the cross-bridge pattern, especially the 72 Å reflection, is visibly weaker.

Table 1. Relative Intensities of Off-Meridional Reflections

Buffer	I_{72}/I_{59}	I_{51}/I_{59}	I_{365}/I_{143}
Basic* (rigor)	0.47 ± 0.03	0.34 ± 0.05	
Basic + 4 mM AMP-PNP	0.33 ± 0.01	0.28 ± 0.02	
Basic* (rigor)	0.42 ± 0.02	0.33 ± 0.02	1.09 ± 0.15
Basic + 5 mM PP$_i$	0.25 ± 0.05	0.25 ± 0.01	0.70 ± 0.09

The intensities of the 72, 59, and 51 Å reflections were measured at $\pm 1/125$ Å from the meridian; the 365 Å reflection, at $\pm 1/220$ Å; the 143 Å reflection, on the meridian. All values show average deviation.

* Basic buffer is 0.1 M KCl, 0.01 M imidazole, 0.005 M MgCl$_2$, 0.002 M EGTA, pH 7.2 at 0°.

cross-bridges) and the 143 Å meridional reflection, which arises from the subunit repeat along the thick filaments. Measurements of these (Table 1) showed a sharp drop in the relative strength of the actin-based pattern.

This relative shift, however, could be at least partially due to an increase in the strength of the meridional myosin pattern. Careful examination of the X-ray diagrams shows a sharpening and rise in the intensity of the meridional 71.5 Å reflection. Certainly this reflection is more intense compared with the 59 Å meridional (Table 2).

In contrast to axial diagrams, equatorial diagrams show little or no change in the relative intensities of the two major reflections, the 10 and the 11. The ratios are given in Table 2. The addition of either 5 mM PP$_i$ or AMP-PNP gave no measurable change. These results represent a relatively small number of measurements and must be interpreted with caution. Earlier experiments, using old muscles and shorter sarcomere lengths, showed slight decreases in the 11/10 ratio, but the changes were within the experimental error. If a change occurs, it is small, less than the 20% uncertainty in measurement.

Since neither PP$_i$ nor AMP-PNP is hydrolyzed to supply a source of energy, possibly needed for a muscle to become relaxed (in which case a "relaxed' muscle is one at its highest potential energy), we tried using both cytidine and inosine triphosphates at 2.5 mM. Even though (or because) these two supported contraction, they did not give relaxed patterns, indicating that a source of energy is not in itself sufficient to relax muscle.

Table 2. Relative Intensities of Equatorial and Meridional Reflections

Buffer	Equatorial I_{11}/I_{10}	Meridional I_{72}/I_{59}
Basic* (rigor)	1.31 ± 0.09	0.85 ± 0.1
Basic + 4 mM AMP-PNP	1.34 ± 0.14	1.35 ± 0.13
Basic* (rigor)	1.31 ± 0.13	0.62 ± 0.2
Basic + 5 mM PP$_i$	1.24 ± 0.08	1.9 ± 0.7

* Basic buffer as for Table 1.

Discussion

In solution myosin and actin are in equilibrium:

$$A \cdot M \rightleftharpoons A + M \tag{1}$$

with the $A \cdot M$ complex the predominant species. The addition of certain myosin-binding ligands tends to dissociate $A \cdot M$ by adding a second line to the reaction:

$$A \cdot M + L \rightleftharpoons A \cdot M \cdot L \rightleftharpoons A + M \cdot L \tag{2}$$

If the ligand binds strongly to myosin, the second line dominates and, if there is sufficient ligand, most of the myosin will be complexed with it rather than the actin. In this case one can separate the actin and myosin by differential centrifugation or gel chromatography.

Since both PP$_i$ and AMP-PNP are known to cause such dissociation in 0.5 M salt, driving Eq. 2 to the right, an expected consequence is that they would cause some dissociation of actin-myosin links in whole muscle. Our measurements of the intensity of the off-meridional cross-bridge pattern show this to be so.

However, neither 5 mM PP$_i$ nor AMP-PNP causes a complete loss of the rigor cross-bridge pattern, whereas they cause almost complete dissociation in solution. This presumably means that in a whole muscle the effective free actin concentration is adequate to keep the reaction in Eq. 2 from proceeding all the way to the right. In fact the reaction seems to be far more toward the left. Had we sufficient analog, we could presumably drive the equilibrium further to the right, bringing about the complete loss of the rigor cross-bridge pattern. But in all cases we are only shifting an equilibrium, which means that each cross-bridge is going back and forth through all the steps in Eq. 2. The analogs we have used thus successfully perturb the muscle system, not by creating a new stable state, but by changing the relative frequency of transitions.

The intensity of the rigor cross-bridge pattern in axial diagrams decreases by 30–40%. This implies that the cross-bridges are, for a considerable proportion of their time, not ordered according to the helical repeat of the actin filament. The implication is that they spend this time either detached from actin or attached to actin in a disordered manner (e.g., tilted over a range of angles). The results from the equatorial diagrams (Table 2) argue that the cross-bridges are still spending most of their time near the actin filaments rather than the myosin filaments.

There is also a slight enhancement of the intensity of the meridional myosin pattern (Table 2). This can again be simply explained in two ways: first, some cross-bridges have detached completely from actin and have become ordered according to the myosin repeat in the thick filament; or second, some cross-bridges have detached from actin, moved through the cycle of Fig. 1, and recombined with actin in the perpendicular mode symbolized by step 2. Since there is no energy to support contraction, they will stay at this stage, alternating between the $A \cdot M$, $A \cdot M \cdot L$, and $A + M \cdot L$ complexes, contributing slightly to the myosin meridional reflections. At present we have no way of determining exactly the relative populations of the states just mentioned. However since the maximal shift in the equatorial intensities is 20% (the error in their measurement) and the off-meridional cross-bridge pattern intensity is down 30–40%, this may mean that there are some cross-bridges which are attached to the actin filament in such a disordered way that they do not contribute to the rigor axial cross-bridge pattern, but still contribute to the rigor-like equatorial pattern.

The enhancement of the meridional myosin pattern (in Fig. 4) raises another question: Why do we see no trace of the layer lines of the relaxed myosin pattern which is visible in Fig. 3? There are several immediate possibilities. First, the pattern, with so few cross-bridges (at most 40%) contributing, might simply be too weak. Second, it might be necessary for a very large proportion of the cross-bridges ($>90\%$) to be detached before they begin falling into the highly ordered state necessary to give rise to the pattern. Third, it might be necessary both to have the cross-bridges detached and ATP hydrolyzed for the ordered state to develop.

Figure 1 helps to illustrate. Step 3 is presumably the rigor configuration, with the myosin cross-bridge linked to actin at a specific angle, almost identical for each link, giving rise to the pattern in Fig. 2. Addition of analog causes the myosin to separate from the actin, changing to step 4. But the equatorials argue that it doesn't, on average, move very far away before returning to the actin.

Once rejoined to the actin, it does not stay at the single well-ordered orientation typified by step 3. If it were, the intensities in Fig. 4 would be higher than observed. Having some cross-bridges detached (step 4), which seems to occur under the conditions of Fig. 4, is not sufficient to give rise to the relaxed pattern. Something else, as argued above, is necessary to give the highly ordered array, symbolized by step 5, which gives the live, relaxed myosin pattern.

Turning to biochemical considerations, the data in Lymn (1970) and Lymn and Taylor (1970, 1971) indicate that during ATP hydrolysis by actomyosin in solution, the $A \cdot M \cdot Pr$ complex has an appreciable half-life. It is thus likely that the half-life of any other $A \cdot M \cdot L$ complex is also appreciable. There is no a priori reason why this complex should have the same physical characteristics as the $A \cdot M$ complex, considering that it is chemically excited and can go to either $A \cdot M$ or $A + M \cdot L$. Thus the complex could be physically excited, oscillating around the site of its attachment to actin. This physical movement would result in considerable disordering of the actin-myosin links and a concomitant lessening of the intensity of their pattern on the X-ray diagrams. On the other hand, the $A \cdot M$ complex would be unexcited, basically immobile, sitting in the rigor link of step 3, ordered neatly according to the actin helix and giving a strong rigor pattern.

This agrees well with our data. A wobbling $A \cdot M \cdot L$ complex would explain both the weaker off-meridional rigor pattern and the basically unchanged equatorial pattern. It is not necessary for each bridge to be continuously moving to explain these data; simply disordering the angle of cross-bridge attachment to actin would have the same result. But then the cause of disordering still needs to be explained.

Using current chemical data it is possible to calculate the half-life of the $A \cdot M$ complex in the presence of MgATP at physiological concentration. It has the shortest half-life of any of the steps in the simple Lymn-Taylor scheme, less than 1 msec. Thus the actin-myosin complex most common in a contracting muscle is probably $A \cdot M \cdot ADP \cdot P$. Thus, extending our interpretation, the $A \cdot M \cdot Pr$ complex is oscillating around some preferred attachment angle in a disordered way, becoming rigidly ordered only upon the loss of products; but this rigor-like state would last less than a millisecond before another ATP is bound. We have hypothesized that the excited state is, on the average, too disordered to give rise to a rigor-like off-meridional pattern. This would explain why no such pattern is seen in a contracting muscle, even though experiments (Haselgrove, 1970) indicate that as many as 50% of the cross-bridges in the region of overlap may be attached to myosin.

On the other hand, a substance such as ITP binds to actomyosin at less than one-tenth the rate of ATP (Finlayson et al., 1969), implying that in its presence at similar concentrations the rigor A · M state lasts at least ten times as long. Accordingly the rigor A · M links would be more common in a contracting muscle where ITP rather than ATP is the substrate, and one would expect to have better chance of seeing a rigor-like cross-bridge pattern. Using both ITP and CTP to support isometric contractions, we did indeed see such rigor patterns. The longer half-life of the A · M links in the presence of either substrate probably accounts both for the visibility of the rigor pattern in X-ray diagrams, and for the inefficiency of ITP in supporting contraction (Bendall, 1969), as they would hinder movement. If one lowered the ATP concentration tenfold, the results should be much the same.

With regard to proposed mechanisms of contraction, the possibility of an oscillating A · M · Pr complex is an attractive one, particularly if the oscillation is initiated at a site or angle different from the final one. This is illustrated by the steps in Fig. 1. In the presence of a nonhydrolyzed analog such as PP$_i$ or AMP-PNP, the initiation of recombination is presumably at the same point from which dissociation takes place, whereas the hydrolysis of a nucleoside triphosphate (ATP, ITP, CTP), which changes the nature of the M · L complex by increasing energy and redistributing charge, makes it possible (perhaps necessary) for the actin and myosin to recombine at a different site (step 2), from which it then oscillates towards step 3, where, upon loss of the products, it stays. The oscillations of the actin-myosin-products

complex could, of course, be those postulated in the recent model of Huxley and Simmons (1971) to explain transients of tension development.

Acknowledgments

R. W. L. gratefully acknowledges a British-American Fellowship of the American Heart Association and British Heart Foundation. We thank E. Rome for a prepublication copy of her recent paper. C. Bagshaw and D. Trentham graciously provided the AMP-PNP. J. C. Haselgrove generously printed the X-ray diagrams.

References

BENDALL, J. R. 1969. *Muscles, molecules, and movement*, p. 69. Heinemann, London.

FINLAYSON, B., R. W. LYMN, and E. W. TAYLOR. 1969. Studies on the kinetics of formation and dissociation of the actomyosin complex. *Biochemistry* **8**: 811.

HASELGROVE, J. C. 1970. *X-ray diffraction studies on muscle*. Ph.D. thesis, University of Cambridge.

HUXLEY, A. F. and R. M. SIMMONS. 1971. Proposed mechanism of force generation in striated muscle. *Nature* **233**: 533.

HUXLEY, H. E. 1968. Structural difference between resting and rigor muscle; evidence from intensity changes in the low-angle equatorial X-ray diagram. *J. Mol. Biol.* **37**: 507.

HUXLEY, H. E. and W. BROWN. 1967. The low-angle X-ray diagram of vertebrate striated muscle and its behaviour during contraction and rigor. *J. Mol. Biol.* **30**: 383.

JULIAN, F. J. 1971. The effect of calcium on the force-velocity relation of briefly glycerinated frog muscle fibres. *J. Physiol.* **218**: 117.

LYMN, R. W. 1970. *Kinetics of myosin and actomyosin action*. Ph.D. thesis, University of Chicago.

LYMN, R. W. and E. W. TAYLOR. 1970. Transient state phosphate production in the hydrolysis of nucleoside triphosphates by myosin. *Biochemistry* **9**: 2975.

———. 1971. Mechanism of adenosine triphosphate hydrolysis by actomyosin. *Biochemistry* **10**: 4617.

ROME, E. 1972. Relaxation of glycerinated muscle: Low-angle X-ray diffraction studies. *J. Mol. Biol.* **65**: 331.

The Biosynthesis of Sarcoplasmic Reticulum Membranes and the Mechanism of Calcium Transport

A. MARTONOSI, R. BOLAND, AND R. A. HALPIN

Department of Biochemistry, St. Louis University School of Medicine, St. Louis, Missouri 63104

The active transport of calcium by sarcoplasmic reticulum membranes involves a membrane bound ATPase enzyme of about 100,000 g molecular weight (Martonosi, 1969; Martonosi and Halpin, 1971; Martonosi et al., 1971a; MacLennan et al., 1971) which requires membrane phospholipids for activity (Martonosi, 1963, 1964, 1968; Martonosi et al., 1968, 1971b; Meissner and Fleischer, 1972). The physiological role of the Ca^{++} transport system is the regulation of the contractile state of skeletal and cardiac muscle (Weber, 1966; Ebashi and Endo, 1968; Martonosi, 1971).

The transport ATPase is activated by Mg^{++} and Ca^{++} (Hasselbach, 1964; Weber et al., 1966) and accepts a wide range of nucleoside triphosphates (Martonosi and Feretos, 1964; Makinose and The, 1965), acetylphosphate (De Meis, 1969) and carbamylphosphate (Pucell and Martonosi, 1971) as substrates.

The hydrolysis of ATP by sarcoplasmic reticulum membranes involves a phosphoprotein intermediate (Yamamoto and Tonomura, 1968; Makinose, 1969; Martonosi, 1969) which has some of the characteristics of an acylphosphate. The dependence of ATPase activity upon phospholipids is connected with the phospholipid requirement of the decomposition of phosphoprotein intermediate (Martonosi 1969; Martonosi et al., 1971a). Recent data suggest that phospholipids may position functional histidine residues at the active site of the enzyme which are involved in the hydrolytic decomposition of phosphoprotein intermediate.

The maximum concentration of protein-bound phosphate so far obtained is about 1 mole/10^5 g sarcoplasmic reticulum protein (Pucell and Martonosi, 1971), in good agreement with the molecular weight of the ATPase enzyme determined by polyacrylamide gel electrophoresis (Martonosi et al., 1971a; Martonosi and Halpin, 1971). As the transport ATPase consists of several subunits (Yu and Masoro, 1970; Martonosi and Halpin, 1971; Pucell and Martonosi, 1972), only one of the subunits is likely to serve as phosphate acceptor in the formation of phosphoprotein.

The protein and phospholipid components of the transport complex can be separated with reversible loss of ATPase activity and Ca^{++} transport; subsequent reconstitution of membrane proteins with lecithin, or lysolecithin, reactivates both functions (Martonosi et al., 1968; Meissner and Fleischer, 1972).

Self-assembly of membrane structures from microsomal components solubilized with deoxycholate is readily demonstrable (Martonosi, 1968; MacLennan et al., 1971). The morphological and functional characteristics of the reconstituted membranes are reminiscent of fragmented sarcoplasmic reticulum. This process of self-assembly may be the mechanism of the formation of sarcoplasmic reticulum membranes during development and could contribute to their turnover in adult muscles.

Interesting changes in sarcoplasmic reticulum functions occur in developing rabbit (Szabolcs et al., 1967; Holland and Perry, 1969) and chicken (Fanburg et al., 1968) muscles, which may suggest (Holland and Perry, 1969) that the formation of the transport ATPase precedes the appearance of Ca^{++} transport function. In an attempt to explain these findings we correlated developmental changes in the Ca^{++} transport and ATPase activity of sarcoplasmic reticulum membranes with their protein, phospholipid, and fatty acid composition and with the maximum concentration of phosphoprotein intermediate. The results suggest that the phospholipid-rich membranes of embryonic chick muscle are gradually converted during development into Ca^{++} transporting structures by stepwise insertion of the ATPase enzyme and possibly other components of the transport system. It appears that the functionally competent Ca^{++} transport apparatus requires protein and phospholipid constituents which are not present at early embryonic development. Their incorporation into the membrane at a later stage, results in the coupling of ATPase activity to Ca^{++} transport.

Experimental Procedures

White Leghorn eggs were incubated in a New Brunswick incubator shaker (Model G 27) at 38°C and a relative humidity of 64–70%. Microsomes were prepared from leg, superficial pectoralis, deep pectoralis, and heart muscles, obtained during 10 to 65 days of development beginning with the start of incubation. The muscles were excised, carefully cleaned of fat and connective tissue, and placed in an ice bath made of distilled water until sufficient amount was collected. The muscles were homogenized in a Virtis homogenizer or Waring Blendor using 4 volumes of 0.1 M KCl, 10 mM imidazol, and 0.3 M sucrose. The homogenates were centrifuged for 20 min at 8200 g in Lourdes GRA or VRA rotors. The sediment was discarded and the supernatant filtered through glasswool previously washed with the homogenizing medium. The filtrate was centrifuged for 30 min at 8200 g to remove the remaining mitochondria. The microsomes were sedimented from the supernatant at 50,000 g for 1 hr in a Spinco No. 30 rotor. The sediment was redispersed in a solution of 0.6 M KCl, 10 mM imidazol and 0.3 M sucrose, and after standing for 30 min at 2°C, it was centrifuged at 50,000 g for 1 hr. The sediment was dispersed in 0.3 M sucrose and centrifuged at 58,000 g for 30 min in a Spinco No. 30 rotor. The final sediment was dispersed in 0.3 M sucrose to a protein concentration of 5–8 mg/ml and used for assay of Ca^{++} uptake, ATPase activity, phosphoprotein intermediate, and protein composition. The remaining suspension was lyophilized and stored at about −20°C under nitrogen for analysis of phospholipids. In one series of experiments sucrose was omitted from the isolation media.

Rabbit skeletal muscle microsomes were prepared as described earlier (Pucell and Martonosi, 1971).

Sucrose density gradient centrifugation of microsomes was performed in a Spinco SW-39 swinging bucket rotor at 41, 319 g for 18–20 hr using a continuous sucrose gradient ranging from 0.66 M to 1.50 M sucrose concentration. The medium also contained 10 mM histidine (pH 7.3) as buffer. The microsome samples were applied at the top of the gradient in 0.2 ml solution containing 0.3 M sucrose.

ATPase activity was measured in three different media. Medium A contained 0.1 M KCl, 10 mM imidazol, 5 mM ATP, 5 mM MgCl$_2$, 0.5 mM EGTA, and 0.45 mM CaCl$_2$. Medium B was identical to A but with CaCl$_2$ omitted. Medium C contained in addition to KCl, imidazol and ATP, 5 mM CaCl$_2$. The reaction was started with the addition of microsomes (0.15 mg protein/ml) and after 5 min incubation at 25°C was stopped with trichloroacetic acid. The inorganic phosphate was determined according to Fiske and Subbarow (1925).

Ca^{++} uptake was measured on the day of microsome preparation at 25°C in a medium of 0.1 M KCl, 10 mM imidazol, 5 mM MgCl$_2$, 5 mM ATP, 5 mM oxalate, and 0.1 mM ^{45}CaCl$_2$ at several microsome concentrations ranging from 0.03–0.3 mg protein per ml. Samples were taken for Millipore filtration (Martonosi and Feretos, 1964) after 1, 2, 5, and 10 min of incubation. Uptake rates were evaluated usually on the basis of values obtained during the first minute.

The measurement of phosphoprotein intermediate was performed essentially as described earlier (Martonosi et al., 1971b). The assay system contained 0.05 M KCl, 5 mM imidazol, 0.5 mM [^{32}P]ATP, 0.3–0.6 mg microsomal protein per ml, and either 5 mM CaCl$_2$ or 5 mM MgCl$_2$ and 0.5 mM EGTA or 5 mM MgCl$_2$, 0.5 mM EGTA and 0.45 mM CaCl$_2$ as activators. Incubation was carried out in ice for 20 sec. The reaction was stopped and the samples processed as described earlier (Martonosi, 1969; Pucell and Martonosi, 1971).

Protein was measured by the Lowry (Lowry et al., 1951) or biuret (Gornall et al., 1949) methods. The method of Loftfield and Eigner (1960) was used for measurement of radioactivity.

Analytical polyacrylamide gel electrophoresis of microsomal proteins was carried out as described earlier (Martonosi and Halpin, 1971). The microsomal proteins were dissociated into subunits by ultrasonic treatment with a Branson Model W 185 sonifier (Heat Systems-Ultrasonics Inc., Long Island, New York) essentially as described by Pucell and Martonosi (1972). The gel electrophorograms were stained with Amido black and scanned using a Densicord (Photovolt) or a Chromoscan (Joyce-Loebl) densitometer.

For determination of phospholipids microsomes were extracted with chloroform-methanol according to Folch et al. (1957) and separated by thin layer chromatography on Silica gel G using chloroform:methanol:conc. NH$_4$OH (65:25:4) as solvent. The individual phospholipids were visualized and eluted from the chromatograms as described earlier (Martonosi and Halpin, 1972) and used for total phosphate determination according to Bartlett (1959).

For determination of fatty acid composition the total lipid extract was evaporated to dryness and subjected to transesterification in methanol:H$_2$SO$_4$ (95:5 v/v) under nitrogen for 90 min at 100°C (Ackman, 1969; Morrison and Smith, 1964). The resulting methyl esters of fatty acids were extracted and concentrated according to Keenan and Morre (1970) and applied for gas-liquid chromatography. Gas chromatography was performed on a Barber Colman Series 5000 instrument equipped with an electron capture detector, using a glass column (6 feet × 3 mm inner diameter) packed with 10% EGSS-X on 100 to 120 mesh Chromosorb Q, obtained from Applied Science Laboratory, State College, Pennsylvania. The flow rate of the carrier gas (argon, Chemetron Corporation, Chicago) was 40 ml per min. The carrier gas was dehydrated by passage through a molecular sieve column (Applied Science Labs). Column temperature: 180°C. Detector temperature: 270°C.

The methyl esters of fatty acids were identified by comparison of their relative retention times with those of known fatty acid methyl esters. The identification was confirmed by analysis of their mass spectra using an LKB model 9000 gas chromatograph-mass spectrometer equipped with a Becker-Ryhage molecule separator. A quantitative evaluation of fatty acid composition was performed by half-height analysis of the peaks of gas chromatograms.

Photooxidation was performed in a thermostated Plexiglas chamber containing the cylindrical reaction vessels with the oxygen electrodes of a YSI Model 53 oxygen monitor inserted into them. Microsome suspensions of 3–5 ml volume, containing 5×10^{-5} M methylene blue or 10^{-6} M rosebengal as catalysts, were illuminated with a 500 W projector bulb from a distance of about 25 cm, while the solution was stirred with magnetic stirrer. The uptake of oxygen was recorded continuously (Estabrook, 1967) and aliquots were taken at intervals for the determination of ATPase activity, amino acid composition (Spackman, 1967) sulfhydryl groups (Ellman, 1959), and tryptophan content (Spies and Chambers, 1948, 1949). Samples kept in dark under the above conditions showed no change in ATPase activity or phosphoprotein concentration during 30 min.

Turnover of microsomal proteins and phospholipids. White female rats of 100–120 g weight maintained on Purina Laboratory Chow No. 56 were injected intraperitoneally with 250 μC of [^{32}P]orthophosphate or 100 μCi L-4,5-[^{3}H]leucine dissolved in 0.9% NaCl at pH 7.0. At intervals following the injection groups of two animals were sacrificed by blow on the head, and the superficial leg muscles were homogenized in four volumes of 0.1 M KCl, 10 mM imidazol, pH 7.3. The myofibrillar fraction was sedimented at 3100 g for 30 min followed by the sedimentation of mitochondria (8,000 g, 30 min), the heavy microsomes (28,000 g, 1 hr), and light microsomes (54,000–100,000 g, 1 hr). The myofibrillar and mitochondrial fractions were washed with 0.1 M KCl, 10 mM histidine solution and the crude heavy and light microsome fractions with 0.6 M KCl, 10 mM imidazol.

For radioactivity measurements, small aliquots of the material were dispersed with 1 ml NCS reagent (Nuclear Chicago) and counted in a toluene-based scintillation fluid with 2,5-diphenyloxazole

(PPO) and 1,4-Bis-2-(4-methyl-5-phenyl-oxazolyl)-benzene (dimethyl POPOP) as scintillators.

For the determination of proteins on polyacrylamide gels, the gels were stained with 0.1% amidoblack in methanol:acetic acid:water (5:1:5 v/v), destained, and sectioned transversely into 25–26 slices of 3.2 mm average thickness. The protein-bound stain was eluted with 3 ml 1.0 N NaOH overnight, and the optical density of the eluate at 660 mμ was used as the index of protein content.

For measurement of radioactivity on polyacrylamide gels, the gels were sliced immediately after electrophoresis, the minced slices were placed in scintillation vials with 1 ml NCS, capped and after overnight standing at 65°C, counted in a toluene scintillation fluid containing PPO and dimethyl POPOP as scintillators.

Acetone-ether extraction of phospholipids was performed as follows: The particulate fraction was extracted repeatedly with 5% trichloroacetic acid at room temperature followed by washing with water. The phospholipids were isolated from the sedimented membrane material by extraction twice with 90% acetone, once with dry acetone, and twice with ether. The combined extracts were evaporated to dryness, dissolved in chloroform-methanol (2:1, v/v) and used for phosphate analysis or thin layer chromatography.

Materials

^{45}CaCl$_2$ and [^{32}P]ATP labeled in the terminal position were purchased from New England Nuclear Corp. Rosebengal and methylene blue were obtained from Fisher Scientific Co. All reagents were of analytical grade. Distilled-ion-exchanged water was used throughout.

Results

Changes in Ca^{++} transport and ATPase activity of sarcoplasmic reticulum. During early embryonic development (10–14 days) the Mg^{++} + Ca^{++}-activated ATPase activity and Ca^{++} transport of microsomes isolated from leg, superficial pectoralis, and deep pectoralis muscles are comparable to that of the heart (Fig. 1B). The ATPase activity rises rapidly in leg and superficial pectoralis muscle microsomes between 14–18 days of development reaching maximum values of 3–4 μmoles/mg protein/min at hatching, followed by a gradual decrease of ATPase activity to 1.5–2.0 μmoles/mg protein/min in 6-weeks-old chicks (Fig. 1B). During this period the rate of oxalate-potentiated Ca^{++} uptake steadily increased in leg, superficial pectoralis, and deep pectoralis muscles from 0.02–0.05 μmoles/mg protein/min at

14 days to 0.4–0.8 μmoles/mg protein/min at 65 days of development, with no indication of leveling off even at this time (Fig. 1A). Only relatively minor developmental changes were observed in the ATPase activity and oxalate-potentiated Ca^{++} transport of heart muscle microsomes (Fig. 1).

The Mg^{++}-activated (Fig. 2A) and Ca^{++}-activated (Fig. 2B) ATPase activities behaved during development in a qualitatively similar fashion to the Mg^{++} + Ca^{++}-activated ATPase (Fig. 1B).

The Ca^{++} activation of microsomal ATPase (Fig. 3) calculated as the difference between the Mg^{++} + Ca^{++}-activated (Fig. 1B) and Mg^{++}-activated (Fig. 2A) processes shows a reasonably close correlation with the rate of Ca^{++} transport (Fig. 1A).

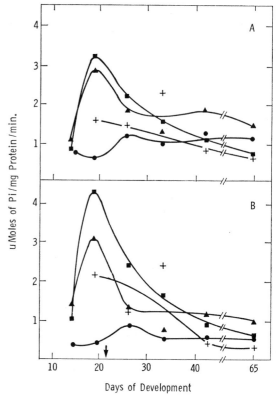

Figure 2. Changes in the Mg^{++} and Ca^{++}-activated microsomal ATPase activity during development. The assay system contained in addition to 0.1 M KCl, 10 mM imidazol and 5 mM ATP, 5 mM MgCl$_2$ and 0.5 mM EGTA (Fig. 2A) or 5 mM CaCl$_2$ (Fig. 2B) as activating ions. For symbols see legend to Fig. 1 and for other details the Experimental Procedures.

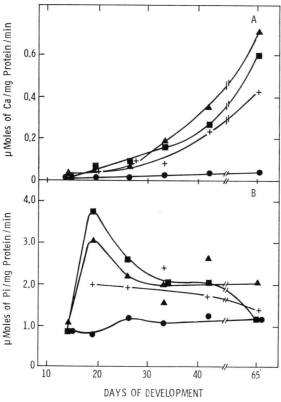

Figure 1. Changes in Mg^{++} + Ca^{++}-activated ATPase and Ca^{++} transport during development. *A*. For assay of Ca^{++} uptake the medium contained 0.1 M KCl, 10 mM imidazol, 5 mM oxalate, 5 mM MgCl$_2$, 5 mM ATP, and 0.1 mM ^{45}CaCl$_2$. Aliquots were taken for filtration through Millipore filters (Type HA) after 1, 2, 5, and 10 min incubation at 25°C, and the rate of Ca^{++} uptake was calculated from the linear portion of the curve. *B*. ATPase activity was measured in a medium of 0.1 M KCl, 10 mM imidazol, 5 mM ATP, 5 mM MgCl$_2$ 0.5 mM EGTA, and 0.45 mM CaCl$_2$ at 25°C as described under Experimental Procedures. Microsomes were isolated from(▲—▲) leg, (■—■) superficial pectoralis, (+—+) deep pectoralis, and (●—●) heart muscles as described under Experimental Procedures. The time of hatching is at 21 days.

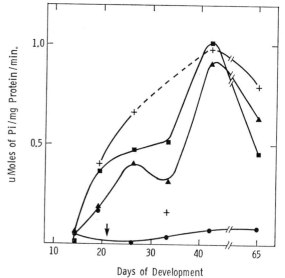

Figure 3. Ca^{++} activation of microsomal ATPase as a function of development. The difference between Mg^{++} + Ca^{++}-activated and Mg^{++}-activated ATPase activity was calculated from the data in Fig. 1B and Fig. 2A. Source of microsomes: (▲—▲) leg, (■—■) superficial pectoralis, (+—+) deep pectoralis, (●—●) heart.

It is noteworthy that the rise in ATPase activity between 14 and 19 days of development occurs much earlier than the increase in the rate of Ca^{++} transport, in agreement with the observations of Holland and Perry (1969).

In view of the structural complexity of the Ca^{++} transport system, the observed changes in ATPase activity during development may be related to one or more of the following factors: (1) changes in enzyme concentration; (2) stepwise assembly of the ATPase enzyme from its subunits with changes in specific activity; (3) delayed synthesis of a coupling factor which couples ATPase activity to Ca^{++} transport after hatching; (4) changes in the phospholipid composition of the membranes which may affect the specific activity of the enzyme and/or the Ca^{++} permeability of the sarcoplasmic reticulum, thus affecting ATPase activity. The following experiments were designed to distinguish between these alternatives.

Changes in phosphoprotein concentration during development.

In microsomes isolated from leg and superficial pectoralis muscles, the steady-state concentration of phosphoprotein measured with 5 mM MgCl$_2$–0.5 mM EGTA and 0.45 mM CaCl$_2$ (Fig. 4A) or with 5 mM CaCl$_2$

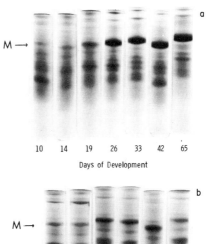

Figure 5. Protein composition of leg muscle microsomes during development. Polyacrylamide gel electrophoresis was carried out as described under Experimental Procedures. (a) Leg muscle microsomes; (b) heart muscle microsomes.

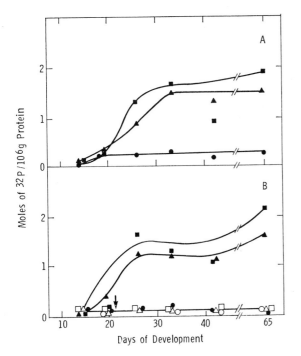

Figure 4. Concentration of phosphoprotein intermediate during development. The concentration of phosphoprotein was measured as described under Experimental Procedures. A, Activators were: 5 mM MgCl$_2$, 0.5 mM EGTA and 0.45 mM CaCl$_2$; B, 5 mM CaCl$_2$ (filled symbols) or 5 mM MgCl$_2$ and 0.5 mM EGTA (open symbols). (▲, △) Leg, (■, □) superficial pectoralis, (●, ○) heart.

(Fig. 4B) as activators rises during development, reaching maximum values of 1.5–1.9 moles per 10^6 g protein at 33 days of development (12-day-old chicks). The increase in phosphoprotein concentration during development (Fig. 4) roughly parallels the changes in "extra ATPase" activity (Fig. 3). Only marginal phosphoprotein concentrations were obtained with 5 mM MgCl$_2$ and 0.5 mM EGTA as activators (Fig. 4B).

The maximum steady state level of phosphoprotein in heart muscle was 6–10 times less than in skeletal muscle (Fig. 4AB) in keeping with its slow rate of Ca^{++} transport.

Protein composition of sarcoplasmic reticulum membranes during development.

Microsomes isolated from leg and heart muscles at various stages of development were subjected to polyacrylamide gel electrophoresis and the patterns shown in Fig. 5 were obtained. The M band, which is known to contain the transport ATPase enzyme (Martonosi, 1969; Martonosi et al., 1971a; Martonosi and Halpin, 1971), is only a minor component during 10 to 14 days of embryonic development when relatively large amounts of small molecular weight proteins are present (Figs. 5, 6). The increase in extra ATPase activity during subsequent

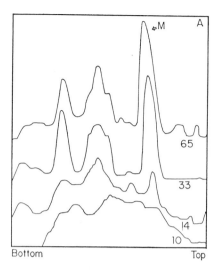

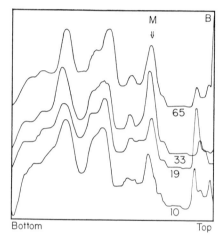

Figure 6. Changes in protein composition during development. Selected samples from Fig. 5 were scanned with Joyce-Loebl densitometer. *A.* Leg muscle microsomes. Traces from bottom to top represent samples obtained after 10, 14, 33, and 65 days of growth. *B.* Heart muscle microsomes. Traces from bottom to top represent samples obtained after 10, 19, 33, and 65 days of growth.

development of skeletal muscle is accompanied by an increase in the concentration of M protein which in 1 to 2-week-old chicks constitutes 60–70 percent of the microsomal proteins (Fig. 6). During this time there is some decrease in the relative amount of small molecular weight protein material (Fig. 5).

In heart microsomes the developmental changes in protein composition are much less pronounced (Figs. 5, 6) in accord with the nearly constant ATPase activity, Ca^{++} transport, and phosphoprotein concentration.

The protein composition of microsomes isolated from superficial and deep pectoralis muscles was similar to that of the leg.

Phospholipid composition of microsomes isolated from developing muscles. During the early embryonic period (10–14 days) the microsomal membranes isolated from leg muscles contain relatively large amounts of phospholipids (Fig. 7). The phospholipid:protein ratio declines during late embryonic development to 0.5–0.7 μmoles of lipid P/mg protein and remains at that level from the 26th to the 65th day. Part of this change in phospholipid:protein ratio is attributable to the appearance of the M protein. The increase in the protein content of microsomal membranes is reflected in their increased density as determined by sucrose density gradient centrifugation.

The principal phospholipid fractions were identified as phosphatidylcholine, phosphatidylethanolamine, phosphatidylserine, sphingomyelin, and phosphatidylinositides (Fig. 7). There is less phosphatidylcholine and more phosphatidylserine in chicken than in adult rat or rabbit skeletal muscle microsomes, and they both remain rather constant during development (Fig. 7). The concentration of phosphatidylethanolamine increases

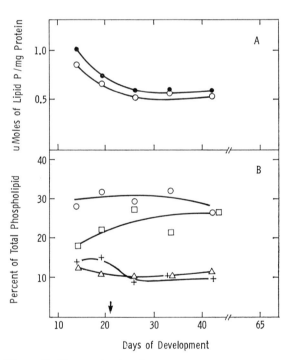

Figure 7. Changes in the phospholipid content of microsomes during development. *A.* The total phosphate content of microsomes (●—●) and of Folch extracts (○—○) was determined using the method of Bartlett (1959). *B.* For the analysis of individual phospholipids the lipid extracts were fractionated by thin layer chromatography on Silica Gel G using chloroform:methanol: conc. NH_2OH (65:25:4 v/v) as solvent and the individual phospholipids were determined as described under Experimental Procedures. (○) Lecithin, (□) phosphatidylethanolamine, (△) phosphatidylserine, (+) sphingomyelin + phosphatidylinositides.

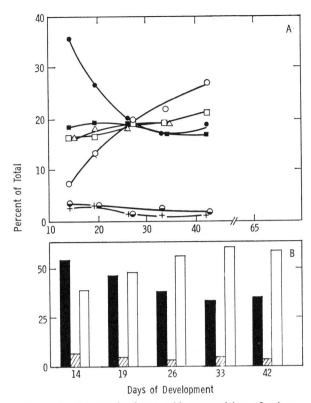

less than 5 % of the total fatty acids of microsomal lipids at 42 days of development (Fig. 8).

The net result of the developmental changes in fatty acid composition is that the early embryonic membranes containing primarily saturated fatty acids are converted by the 30th day of development into Ca^{++} transporting structures in which about 60 % of the fatty acids are unsaturated. The accumulation of unsaturated fatty acids at the expense of palmitate occurs with an increase in the average chain length, which may be of importance in maintaining the proper fluidity of the lipid phase of microsomal membranes at physiological temperature.

Turnover of sarcoplasmic reticulum membranes in adult rats. The turnover of phospholipids in microsomal membranes was determined from the rate of decline of radioactivity following a single intraperitoneal injection of [^{32}P]orthophosphate into adult animals (Fig. 9). Similar results were obtained with [^{14}C]acetate. The half-life of membrane phospholipids was estimated to be in the range of 9–12 days. A comparable turnover rate was obtained for the membrane proteins after injection of ^{14}C- or ^{3}H-labeled leucine (Fig. 10). Since most of the protein-bound radioactivity was connected with the M protein band both in heavy and light microsomes (Fig. 11), the turnover rate calculated for the total microsomal proteins also applies to

Figure 8. Changes in fatty acid composition of microsomes during development. For technical details see Experimental Procedures. *A.* (●—●) Palmitic, (■—■) stearic, (□—□) oleic, (○—○) linoleic, (△—△) arachidonic, (+—+) myristoleic, (◖—◖) palmitoleic. *B.* (■) Saturated fatty acids (stearate and palmitate); (▨) short chain (C-14 and C-16) unsaturated fatty acids (myristoleate and palmitoleate); (□) long chain (C-18 and C-20) unsaturated fatty acids (oleate, linoleate, arachidonate).

while the sum of sphingomyelin and phosphatidylinositides decreases during development; further work is required to establish the significance of these changes.

As the Ca^{++} permeability of artificial phospholipid membranes is strongly influenced by the unsaturation of fatty acids (Vanderkooi and Martonosi, 1971), the developmental changes in the fatty acid composition of sarcoplasmic reticulum membrane phospholipids were investigated. The principal fatty acids in microsomes isolated from 14-day-old chick embryos are two saturated fatty acids, palmitic acid and stearic acid (Fig. 8), representing about 54 % of the total esterified fatty acid content of microsomal lipids. During 14–33 days of development the concentration of palmitic and stearic acids decreases, while that of the long chain unsaturated fatty acids (oleic, linoleic, and arachidonic acids) increases (Fig. 8). The short-chain, unsaturated fatty acids (myristoleate and palmitoleate) are minor components representing

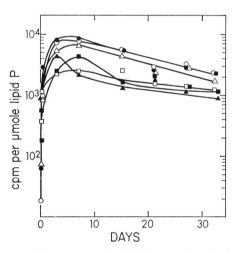

Figure 9. Turnover of ^{32}P-labeled microsomal phospholipids extracted with acetone-ether. Phospholipids were extracted with acetone-ether and separated by thin layer chromatography as described under Experimental Procedures. (○—○) Lipid extract of heavy microsomes, (●—●) lipid extract of light microsomes, (□—□) sphingomyelin, (■—■) phosphatidylserine, (△—△) phosphatidylcholine, (▲—▲) phosphatidylethanolamine. All purified phospholipids were obtained from heavy microsome preparation.

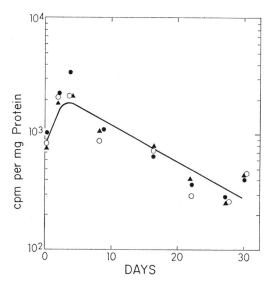

Figure 10. Turnover of [³H]leucine in particulate fractions of skeletal muscle. The radioactivity of particulate fractions was measured during a 30-day period following the intraperitoneal injection of [³H]leucine into rats. (▲—▲) Mitochondria, (○—○) heavy microsomes, (●—●) light microsomes.

the turnover of the transport ATPase enzyme. The similarity between the turnover of proteins and phospholipids makes it unlikely that turnover studies would yield decisive information about the mechanism of assembly of sarcoplasmic reticulum membranes.

Mode of involvement of phospholipids in ATPase activity.

The hydrolysis of ATP by sarcoplasmic reticulum membranes was resolved into the following elementary steps (Yamamoto and Tonomura, 1968; Makinose, 1969; Martonosi, 1969):

$$E + [^{32}P]ATP \leftrightarrows E \sim {}^{32}P + ADP \quad (1)$$

$$E \sim {}^{32}P + H_2O \leftrightarrows E + {}^{32}P_i \quad (2)$$

Although inhibition of either step would cause an inhibition of ATPase activity, selective inhibition of step 1 is expected to lower, while that of step 2 to increase, the steady-state concentration of phosphoprotein under all conditions when the phosphate acceptor sites are not saturated.

The inhibition of ATPase activity and Ca⁺⁺ transport caused by degradation of microsomal phospholipids with phospholipase A (*Crotalus terrificus terrificus*), phospholipase C (*Bac. cereus*), or phospholipase C (*Cl. welchii*) is accompanied by elevation of the steady state concentration of phosphoprotein with 5 mM MgCl₂ and 0.5 mM EGTA as activators (Table 1). The phosphoprotein concentration is only moderately affected by

phospholipase treatment under assay conditions (5 mM CaCl₂ or 5 mM MgCl₂ + 0.5 mM EGTA + 0.45 mM CaCl₂) where maximum steady state concentrations are permitted to develop (Table 1).

These observations form the basis of the conclusion (Martonosi, 1967, 1969; Martonosi et al., 1971b) that the primary requirement for phospholipids in ATP hydrolysis is related to the decomposition of phosphoprotein intermediate (step 2). The formation of phosphoprotein (step 1) proceeds even in the absence of phospholipids at sufficient rate to permit its accumulation. Since phospholipids are not involved as intermediates in ATP hydrolysis (Martonosi, 1964; Martonosi et al., 1968), their influence on ATPase activity is likely to be a conformational effect, perhaps the proper positioning of functional groups at the active site of the enzyme.

Possible participation of a histidine residue in hydrolysis of phosphoprotein intermediate.

Photooxidation of sarcoplasmic reticulum fragments with methylene blue or rosebengal as catalysts inhibits ATPase activity and Ca⁺⁺ transport (Yu et al., 1967). The inhibition of ATPase activity is accompanied by a rise in the concentration of phosphoprotein measured in the

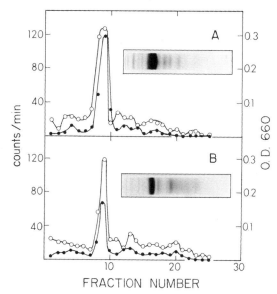

Figure 11. Polyacrylamide gel electrophoresis of heavy and light microsomal proteins obtained 25 hr after injection of radioactive leucine. Microsomes were solubilized with 0.5% sodium dodecylsulfate in 0.025 M sodium phosphate pH 6.0, and applied for electrophoresis in polyacrylamide gels at 5 mA per tube for 12 hr as described earlier under Experimental Procedures. The extruded gels were cut transversely into sections of 3.2 mm thickness for the determination of proteins and radioactivity. The fractions are numbered beginning at the top of the gel. *A*, Heavy microsomes; *B*, light microsomes. (●—●) Proteins, (○—○) radioactivity. Insets are photographs of gels prepared under identical conditions.

Table 1. Effect of Phospholipases on ATPase Activity, Ca[++] Transport, and Concentration of Phosphoprotein

Preparation	ATPase (μmole P_i/mg protein/min)	Ca transport (μmole Ca^{++}/mg protein)	Phosphorylated Intermediate (moles/10^6g protein)		
			5 mM $CaCl_2$	5 mM $MgCl_2$ 0.5 mM EGTA	5 mM $MgCl_2$ 0.5 mM EGTA 0.5 mM $CaCl_2$
Control microsomes	0.42	3.69	1.98	0.32	3.50
Microsomes digested with phospholipase C from *Cl. welchii*	0.06	0.09	2.69	1.98	2.75
Microsomes digested with phospholipase C from *Bac. cereus*	0.04	0.04	2.09	1.53	2.67
Microsomes digested with phospholipase A (*Crot. terr. terr.*)	0.018	0.01	2.72	2.62	2.77

For technical details see Experimental Procedures and the article by Martonosi et al., 1971b.

presence of 5 mM $MgCl_2$ and 0.5 mM EGTA, followed by a slow decline upon prolonged photo-oxidation (Fig. 12A). Only minor changes were found during the first 5 minutes of illumination with 5 mM $CaCl_2$ or 5 mM $MgCl_2$-0.5 mM EGTA and 0.45 mM $CaCl_2$ as activators, i.e., when the phosphoprotein concentration was near its maximal level.

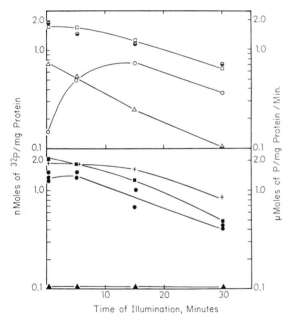

Figure 12. Effect of photooxidation upon ATPase activity and phosphoprotein formation in control and phospholipase C-treated microsomes. Photooxidation was carried out with 10^{-6} M rosebengal as catalyst at a protein concentration of 3.23 mg/ml for the control microsomes (*top*) and 3.70 mg/ml for the phospholipase C-treated microsomes (*bottom*), respectively. At intervals aliquots were taken for measurement of ATPase activity ($\triangle$, $\blacktriangle$) and phosphoprotein intermediate using 5 mM $MgCl_2$ and 5.0 mM EGTA ($\bigcirc$, $\bullet$); 5 mM $MgCl_2$, 0.5 mM EGTA and 0.45 mM $CaCl_2$ ($\square$, $\blacksquare$); or 5 mM $CaCl_2$ ($\ominus$, $+$) as activators. No change in ATPase activity was observed in dark controls during 30 min incubation under the above conditions. Phospholipase C treatment was performed as described earlier (Martonosi et al., 1971b).

The biphasic nature of these effects suggests that at least two classes of functional groups are involved in ATPase activity, which differ markedly in their sensitivity to photooxidation: (a) Modification of the more sensitive class causes the early inhibition of ATPase activity and an increase in the concentration of phosphoprotein when measured with 5 mM $MgCl_2$ and 0.5 mM EGTA as activators. (b) The second class of functional groups is modified only slowly, causing the progressive decline of phosphoprotein concentration under all assay conditions with further inhibition of ATPase activity.

Depletion of microsomal phospholipids by treatment with phospholipase C does not influence the sensitivity of various groups to photooxidation.

The inhibition of ATPase activity during photooxidation with rosebengal as catalyst was accompanied by uptake of oxygen (Fig. 13) and marked decrease in histidine content (Fig. 14, Table 2). Lesser changes were found in cystein (Fig. 13) and a marginal decline in methionine (Table 2). The puzzling decrease in valine and isoleucine content requires further investigation. Other amino acids (Table 2), including tryptophan (Fig. 13), remained unaltered.

SH groups are known to be required for the ATPase activity of sarcoplasmic reticulum (Hasselbach, 1966; Hasselbach and Seraydarian, 1966), raising the possibility that the observed effects of photooxidation may be due to the oxidation of SH groups. This seems unlikely since blocking of SH groups with salyrgan or *N*-ethylmaleimide, although inhibiting ATPase activity, does not produce the characteristic increase in phosphoprotein concentration which was observed after photooxidation (Fig. 15).

No significant oxidation of unsaturated fatty acids was detected during 30-minute photooxidation with rosebengal as catalyst.

Therefore we tentatively attribute the early effects of photooxidation to the modification of

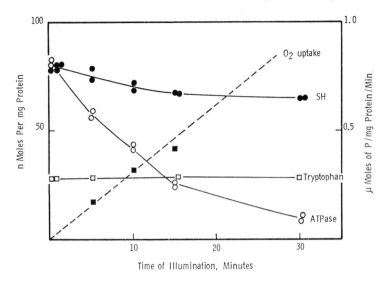

Figure 13. Correlation between oxygen uptake, ATPase activity, SH group and tryptophan content of microsomal membranes. Photooxidation was performed with 10^{-6} M rosebengal (as catalyst) at a protein concentration of 3.65 mg/ml. Aliquots were taken at intervals for the determination of ATPase activity, the various amino acids and SH groups as described under Experimental Procedures. The oxygen uptake was measured separately, using a Model 53 YSI oxygen analyzer. During measurement of oxygen uptake air was permitted to enter during dark periods in order to maintain the oxygen partial pressure close to equilibrium with atmospheric level.

Figure 14. Change in histidine content during photooxidation. Photooxidation was carried out at a protein concentration of 3.8 mg/ml, using 10^{-6} M rosebengal as catalyst. After times stated in the abscissa, aliquots were taken for the determination of amino acid composition using a Beckman Model 120 C amino acid analyzer according to Spackman (1967). The measurement of ATPase activity and phosphoprotein concentration were performed as described under Experimental Procedures. (□—□) Histidine content, (○—○) ATPase activity, (△—△) phosphoprotein concentration.

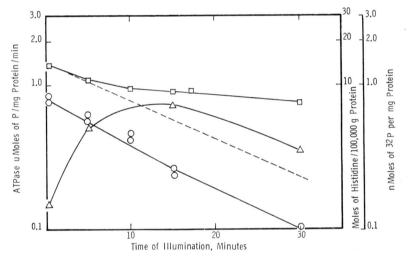

Table 2. Amino Acid Content of Photooxidized Microsomes

| | Controls | Photooxidized Microsomes (minutes) | | | |
		5	10	15	30
Lysine	53.02	53.68	49.64	54.41	54.14
Histidine	13.77	11.07	9.17	9.29	7.77
Arginine	43.57	47.45	41.76	45.74	44.70
Aspartic acid	81.64	86.36	87.63	85.62	90.09
Threonine	50.86	51.83	50.94	53.00	54.46
Serine	47.98	50.36	49.28	52.70	53.26
Glutamic acid	110.90	113.57	115.60	111.48	113.45
Proline	43.12	38.95	37.81	40.39	39.04
Glycine	67.96	65.79	67.40	65.53	66.90
Alanine	87.05	83.70	84.80	86.01	85.50
1/2 Cystine	9.09	10.20	10.87	11.15	12.26
Valine	70.30	65.59	66.81	60.48	61.37
Methionine	23.85	22.67	21.80	24.72	20.17
Isoleucine	52.30	49.12	46.55	45.71	46.32
Leucine	91.73	92.12	97.94	94.77	93.77
Tyrosine	17.10	20.83	20.59	21.46	21.38
Phenylalanine	35.92	32.29	33.94	34.19	33.92

Photooxidation was performed under conditions described in Fig. 12 and under Experimental Procedures. Average values of 2–4 determinations are presented.

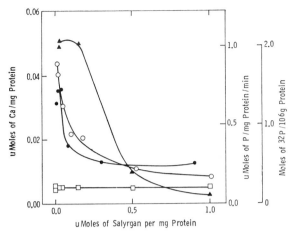

Figure 15. Effect of salyrgan on ATPase activity, Ca⁺⁺ release, and concentration of phosphoprotein. For measurement of Ca⁺⁺ release microsomes (0.17 mg/ml) were incubated in a medium of 0.1 M KCl, 10 mM histidine, 5 mM MgCl₂, 10⁻⁴ M ⁴⁵CaCl₂ and 3.3 × 10⁻⁴ M ATP for 3 min at room temperature. Salyrgan was added in concentrations stated on the abscissa, and incubation was continued for 5 min followed by Millipore filtration (○—○). Prior to measurement of ATPase activity microsomes were preincubated with salyrgan for 3 min at concentrations stated on the abscissa in a medium of 0.1 M KCl, 10 mM imidazol, 5 mM MgCl₂ and 5 × 10⁻⁵ M CaCl₂. The reaction was started with addition of ATP to a final concentration of 3 mM, and after incubation for 6 min the liberated inorganic phosphate was measured (●—●). For the measurement of the phosphoprotein intermediate the microsomes were preincubated with salyrgan for 5 min in a medium of 0.1 M KCl, 10 mM imidazol, and either 5 mM MgCl₂–0.5 mM EGTA (□—□) or 5 mM CaCl₂ (▲—▲). Reaction was started with [³²P]ATP and after 30 sec at 2°C the phosphoprotein was determined as described under Experimental Procedures.

catalytic histidine residues at the active site of the enzyme. Since the early inhibition of ATPase activity was accompanied by an increase in the concentration of phosphoprotein, we postulate that the sensitive histidine groups are selectively involved in the decomposition of phosphoprotein intermediate.

The inhibition of phosphoprotein formation after prolonged photooxidation may or may not be due to oxidation of histidine since blocking of SH groups causes similar effects.

The early effects of photooxidation are similar to the consequences of the depletion of microsomal phospholipids by treatment with phospholipase C. In both cases inhibition of ATPase activity occurs with unaltered or elevated phosphoprotein levels. A possible interpretation of these effects is presented in Table 3. It is assumed that the phosphoprotein intermediate is an acylphosphate involving the aspartyl or glutamyl carboxyl groups of the enzyme (Martonosi, 1969) and that both the formation and decomposition of acylphosphate requires distinct catalytic groups on the active site of the ATPase (Boyer, 1965).

We propose that the principal role of phospholipids in ATPase activity may be the positioning of a catalytic histidine residue (histidine S in Table 3) at the active site in order to promote its

Table 3. Hypothetical Mechanism of ATP Hydrolysis by Sarcoplasmic Reticulum Membranes

$$E\!-\!\overset{O}{\overset{\|}{C}}\!-\!O^- + ATP \rightleftharpoons E \cdots \overset{O}{\overset{\|}{C}} \cdots O \cdots PO_3^{2-} + ADP \quad (1)$$

with SH groups and Histidine R

$$E\!-\!\overset{O}{\overset{\|}{C}}\!-\!O\!-\!PO_3^{2-} \rightleftharpoons E\!-\!\!-\!\overset{O}{\overset{\|}{C}} + HPO_3^{2-} \quad (2)$$

with SH and S groups and Histidine S

$$E\!-\!\!-\!\overset{O}{\overset{\|}{C}} \rightleftharpoons E\!-\!\overset{O}{\overset{\|}{C}}\!-\!OH$$

with S and SH groups

Adapted from Boyer (1965).

attack on the phosphoprotein bond. Additional functional groups, possibly histidine residues which are relatively resistant to photooxidation (histidine R in Table 3), may be involved in the formation of phosphoprotein.

Discussion

Developmental changes in the composition and function of muscle microsomes provide a slow-motion picture of the assembly of functionally competent sarcoplasmic reticulum membranes endowed with Ca⁺⁺ transport activity. At 10–14 days of embryonic development the ATPase and Ca⁺⁺ transport activities of microsomes isolated from various muscles were rather low, and the phospholipid-rich embryonic membranes contained primarily low molecular weight proteins. The fatty acid content of phospholipids was characterized by relatively high concentration of palmitic and stearic acids. During subsequent development the ATPase and Ca⁺⁺ transport activity of the membranes sharply increased, with elevated levels of phosphoprotein intermediate. The increase in enzymic activity is accompanied by a massive increase in the concentration of ATPase enzyme and by an increase in the chain length and unsaturation of the fatty acids of microsomal phospholipids.

The assembly of the ATPase enzyme may involve interaction between preformed protein subunits in the microsomal membrane as the concentration of low molecular weight membrane proteins decreases when the functional ATPase appears.

There are many likely mechanisms by which this process of enzyme synthesis may be regulated.

1. The synthesis of the several subunits of ATPase enzyme (Martonosi et al., 1971a; Martonosi and Halpin, 1971) may begin at different times during development, the rate-limiting step for the formation of active ATPase being the synthesis of the last subunit. This hypothesis is being tested by comparing the amino acid composition, fingerprint, and partial sequence of the subunits of fully developed ATPase enzyme with that of the low molecular weight proteins present at early embryonic development.

Holland and Perry observed (1969) that the ATPase activity reaches a maximum rate long before the Ca^{++} uptake, suggesting that the mechanism which links the two processes together develops at a fairly late stage. The "coupling factors" involved in the energization of Ca^{++} transport may be some of the protein subunits present only in fully developed membranes. Its identity could be established by testing the effect of isolated membrane proteins obtained from fully developed sarcoplasmic reticulum upon the Ca^{++} transport activity of microsomes isolated at 19–21 days of development when the ATPase activity is at its maximum.

2. Alternatively the various subunits of the transport ATPase may be synthesized simultaneously, but their assembly into a functional enzyme complex could proceed only when the phospholipid and fatty acid composition of the membrane becomes optimal. In this case all subunits of the ATPase enzyme should be present at early embryonic development among the low molecular weight proteins of microsomal membranes, and assembly of functionally competent enzyme may be induced in vitro by lipid extracts obtained from adult microsomal membranes.

The coupling of ATPase activity to Ca^{++} transport may also require some unsaturated fatty acids, although restoration of ATPase activity was achieved in phospholipid-depleted microsomes with synthetic lecithin or lysolecithin preparations that contained primarily saturated fatty acids.

Phospholipids may facilitate energized Ca^{++} transport by activation of the transport complex or by reducing the permeability of sarcoplasmic reticulum membranes to Ca^{++}, thereby preventing the loss of accumulated Ca^{++} from the interior of the tubules (Martonosi et al., 1968). Preliminary experiments suggest that the permeability function may be less important in explaining developmental changes in the Ca^{++} transport of sarcoplasmic reticulum, as no significant difference was observed in the Ca^{++} permeability between microsomes of 19-day-old embryos and 3-week-old chicks.

Considering the evidence so far available, it appears that the developmental changes in ATPase activity and enzyme concentration are consistent with a stepwise assembly of the enzyme molecule from preformed subunits which are present in the membrane before significant ATPase activity appears. Although the role of phospholipids in ATPase activity and Ca^{++} transport is well established (Martonosi, 1971), their effect upon the interactions between the enyzme subunits which lead first to the synthesis of active ATPase and then to its coupling with the Ca^{++} transport machinery requires further investigation.

It is frequently hoped that information may be gained about the mechanism of assembly of biological membranes by the study of the turnover of their constituents in adult animals. Our measurements on the turnover of microsomal phospholipids and proteins of adult rat muscles using $[^{32}P]$ortho-phosphate, $[^{14}C]$ or $[^{3}H]$leucine, and $[^{14}C]$acetate indicate that the half-life of the various classes of membrane phospholipids and proteins falls in the relatively narrow range of 9–12 days. In view of the well-known limitations of the methods, these data are consistent with most conceivable mechanisms of membrane assembly, although they certainly suggest that the rate of synthesis of lipid and protein constituents of the membrane may be coordinated.

The mode of involvement of phospholipids in the activity of membrane-bound enzymes is usually described in general terms as being related to the effect of hydrophobic environment upon the "conformation" of enzyme protein.

Our purpose in the present work was to define this general conformational effect in terms of the positioning of specific catalytic groups at the active site of the enzyme. Photooxidation of a sensitive histidine residue led to similar changes in ATPase activity and phosphoprotein concentration as gentle removal of membrane phospholipids by treatment with phospholipase C. These observations raise the possibility that one contribution of phospholipids may be the positioning of a catalytic histidine residue at the active site of the enzyme. Further work is required to establish the direct relationship between the two sets of observations and to clarify the mode of involvement of the postulated histidine residue in the decomposition of phosphoprotein intermediate. The relationship of the sensitive histidine residue to the active center phosphopeptide isolated previously (Martonosi et al., 1971a; Martonosi and Halpin, 1971) is being investigated.

Histidine was suggested to serve as an intermediate phosphate acceptor in the formation and decomposition of acylphosphates in several systems

(Boyer, 1965). It may be of importance in this regard that we have observed a small amount of protein-bound phosphate with the characteristic acid lability of phosphohistidine under various conditions in skeletal muscle microsomes hydrolyzing [^{32}P]ATP. This work is being continued. Alternatively histidine may have a simple catalytic function similar to that observed in several proteolytic enzymes.

Acknowledgments

This work was supported by research grants NS 07749 from the National Institute of Neurological Diseases and Stroke USPHS, GB-33867X from the National Science Foundation, and a Grant in Aid from the American Heart Association, Inc.

References

ACKMAN, R. G. 1969. Gas-liquid chromatography of fatty acids and esters. In *Methods in enzymology*, ed. J. M. Lowenstein, vol. 14, p. 329. Academic Press, New York.

BARTLETT, G. R. 1959. Phosphorus assay in column chromatography. *J. Biol. Chem.* **234:** 466.

BOYER, P. D. 1965. Carboxyl activation as a possible common reaction in substrate level and oxydative phosphorylation and in muscle contraction. In *Oxidases and related redox systems*, ed. T. E. King et al., vol. 2, p. 994. Wiley, New York.

DE MEIS, L. 1969. Ca uptake and acetylphosphatase of skeletal muscle microsomes. *J. Biol. Chem.* **244:** 3733.

EBASHI, S. and M. ENDO. 1968. Calcium ion and muscle contraction. *Prog. Biophys. Mol. Biol.* **18:** 123.

ELLMAN, G. L. 1959. Tissue sulfhydryl groups. *Arch. Biochem. Biophys.* **82:** 70.

ESTABROOK, R. W. 1967. Mitochondrial respiratory control and the polarographic measurement of ADP:O ratios. In *Methods in enzymology*, ed. R. W. Estabrook and M. E. Pullman, vol. 10, p. 41. Academic Press, New York.

FANBURG, B. L., D. B. DRACHMAN, and D. MOLL. 1968. Calcium transport in isolated sarcoplasmic reticulum during muscle maturation. *Nature* **218:** 962.

FISKE, C. H. and Y. SUBBAROW. 1925. The colorimetric determination of phosphorus. *J. Biol. Chem.* **66:** 375.

FOLCH, J., M. LEES, and G. H. SLOANE STANLEY. 1957. A simple method for the isolation and purification of total lipids from animal tissues. *J. Biol. Chem.* **226:** 497.

GORNALL, A. G., C. J. BARDAWILL, and M. M. DAVID. 1949. Determination of serum proteins by means of the biuret reaction. *J. Biol. Chem.* **177:** 751.

HASSELBACH, W. 1964. Relaxing factor and the relaxation of muscle. *Prog. Biophys. Biophys. Chem.* **14:** 167.

———. 1966. Structural and enzymatic properties of the calcium transporting membranes of the sarcoplasmic reticulum. *Ann. N.Y. Acad. Sci.* **137:** 1041.

HASSELBACH, W. and K. SERAYDARIAN. 1966. The role of sulfhydryl groups in calcium transport through the sarcoplasmic reticulum membranes of skeletal muscle. *Biochem. Z.* **345:** 159.

HOLLAND, D. L. and S. V. PERRY. 1969. The adenosine triphosphatase and calcium ion-transporting activities of the sarcoplasmic reticulum of developing muscle. *Biochem. J.* **114:** 161.

KEENAN, T. W. and D. J. MORRE. 1970. Phospholipid class and fatty acid composition of Golgi apparatus isolated from rat liver and comparison with other cell fractions. *Biochemistry* **9:** 19.

LOFTFIELD, R. B. and E. A. EIGNER. 1960. Scintillation counting of paper chromatograms. *Biochem. Biophys. Res. Comm.* **3:** 72.

LOWRY, O. H., N. J. ROSEBROUGH, A. L. FARR, and R. J. RANDALL. 1951. Protein measurement with the Folin phenol reagent. *J. Biol. Chem.* **193:** 265.

MacLENNAN, D. H., P. SEEMAN, G. H. ILES, and C. C. YIP. 1971. Membrane formation by the adenosine triphosphatase of sarcoplasmic reticulum. *J. Biol. Chem.* **246:** 2702.

MAKINOSE, M. 1969. The phosphorylation of the membranal protein of the sarcoplasmic vesicles during active calcium transport. *Europ. J. Biochem.* **10:** 74.

MAKINOSE, M. and R. THE. 1965. Calcium-Akkumulation und nucleosidtriphosphat-Spaltung durch die Vesikel des sarkoplasmatischen Reticulums. *Biochem. Z.* **343:** 383.

MARTONOSI, A. 1963. The activating effect of phospholipids on the ATPase activity and Ca^{2+} transport of fragmented sarcoplasmic reticulum. *Biochem. Biophys. Res. Comm.* **13:** 273.

———. 1964. Role of phospholipids in the ATPase activity and Ca^{2+} transport of fragmented sarcoplasmic reticulum. *Fed. Proc.* **23:** 913.

———. 1967. The role of phospholipids in the ATPase activity of skeletal muscle microsomes. *Biochem. Biophys. Res. Comm.* **29:** 753.

———. 1968. Sarcoplasmic reticulum. IV. Solubilization of microsomal ATPase. *J. Biol. Chem.* **243:** 71.

———. 1969. The protein composition of sarcoplasmic reticulum membranes. *Biochem. Biophys. Res. Comm.* **36:** 1039.

———. 1969. Sarcoplasmic reticulum. VII. Properties of a phosphoprotein intermediate implicated in calcium transport. *J. Biol. Chem.* **244:** 613.

———. 1971. The structure and function of sarcoplasmic reticulum membranes. In *Biomembranes*, ed. L. A. Manson, vol. 1, p. 191. Plenum Press, New York.

MARTONOSI, A. and R. FERETOS. 1964. Sarcoplasmic reticulum. I. The uptake of calcium by sarcoplasmic reticulum fragments. *J. Biol. Chem.* **239:** 648.

MARTONOSI, A. and R. A. HALPIN. 1971. Sarcoplasmic reticulum. X. The protein composition of sarcoplasmic reticulum membranes. *Arch. Biochem. Biophys.* **144:** 66.

———. 1972. Sarcoplasmic reticulum. XVII. The turnover of proteins and phospholipids in sarcoplasmic reticulum membranes. *Arch. Biochem. Biophys.* In press.

MARTONOSI, A., J. R. DONLEY, and R. A. HALPIN. 1968. Sarcoplasmic reticulum. III. The role of phospholipids in the ATPase activity and Ca^{2+} transport. *J. Biol. Chem.* **243:** 61.

MARTONOSI, A., A. G. PUCELL, and R. A. HALPIN. 1971a. Recent observations on the mechanism of Ca^{2+} transport by fragmented sarcoplasmic reticulum membranes, p. 175. In *Cellular mechanisms for calcium transfer and homeostasis*, ed. G. Nichols, Jr. and R. H. Wasserman. Academic Press, New York.

MARTONOSI, A., J. R. DONLEY, A. G. PUCELL, and R. A. HALPIN. 1971b. Sarcoplasmic reticulum. XI. The mode of involvement of phospholipids in the hydrolysis of ATP by sarcoplasmic reticulum membranes. *Arch. Biochem. Biophys.* **144:** 529.

MEISSNER, G. and S. FLEISCHER. 1972. The role of phospholipid in Ca²⁺ stimulated ATPase activity of sarcoplasmic reticulum. *Biochim. Biophys. Acta* **255:** 19.

MORRISON, W. R. and L. M. SMITH. 1964. Preparation of fatty acid methyl esters and dimethyl acetals from lipids with boron fluoride-methanol. *J. Lipid Res.* **5:** 600.

PUCELL, A. G. and A. MARTONOSI. 1971. Sarcoplasmic reticulum. XIV. Acetyl phosphate and carbamyl phosphate as energy sources for Ca⁺⁺ transport. *J. Biol. Chem.* **246:** 3389.

———. 1972. Sarcoplasmic reticulum. XV. Dissociation of the membrane ATPase enzyme of sarcoplasmic reticulum into subunits by ultrasonic treatment. *Arch. Biochem. Biophys.* **151:** 558.

SPACKMAN, D. H. 1967. Accelerated amino acid analysis. In *Methods in enzymology*, ed. C. H. W. Hirs, vol. 11, p. 3. Academic Press, New York.

SPIES, J. R. and D. C. CHAMBERS. 1948. Chemical determination of tryptophan. *Anal. Chem.* **20:** 30.

———. 1949. Chemical determination of tryptophan in proteins. *Anal. Chem.* **21:** 1249.

SZABOLCS, M., A. KÖVER, and L. KOVACS. 1967. Studies on the postnatal changes in the sarcoplasmatic reticular

fraction of rabbit muscle. *Acta Biochim. Biophys. Acad. Sci. Hung.* **2:** 409.

VANDERKOOI, J. M. and A. MARTONOSI. 1971. Sarcoplasmic reticulum. XVI. The permeability of phosphatidyl choline vesicles for calcium. *Arch. Biochem. Biophys.* **147:** 632.

WEBER, A. 1966. Energized calcium transport and relaxing factors. In *Current topics in bioenergetics*, ed. D. R. Sanadi, vol. 1, p. 203. Academic Press, New York.

WEBER, A., R. HERZ, and I. REISS. 1966. Study of the kinetics of calcium transport by isolated fragmented sarcoplasmic reticulum. *Biochem. Z.* **345:** 329.

YAMAMOTO, T. and Y. TONOMURA. 1968. Reaction mechanism of the Ca²⁺ dependent ATPase of sarcoplasmic reticulum from skeletal muscle. II. Intermediate formation of phosphorylprotein. *J. Biochem.* (*Tokyo*) **64:** 137.

YU, B. P. and E. J. MASORO. 1970. Isolation and characterisation of the major protein component of sarcotubular membranes. *Biochemistry* **9:** 2909.

YU, B. P., E. J. MASORO, and F. D. DE MARTINIS. 1967. Imidazole and sequestration of calcium ion by sarcoplasmic reticulum. *Nature* **216:** 822.

Isolation of Sarcoplasmic Reticulum Proteins

David H. MacLennan and C. C. Yip

Banting and Best Department of Medical Research, Charles H. Best Institute, Toronto, Canada

G. H. Iles and P. Seeman

Department of Pharmacology, University of Toronto, Toronto, Canada

The sarcoplasmic reticulum is specialized for binding and release of Ca^{++} (Ebashi et al., 1969; Martonosi, 1971; Weber, 1966). Because of this specialization, the membrane contains relatively few proteins (Martonosi, 1969; Martonosi and Halpin, 1971; MacLennan, 1970; MacLennan and Wong, 1971; MacLennan et al., 1971), and, as would be expected, many of these proteins interact strongly with Ca^{++} ion. We are engaged in a program of resolution, characterization, and reconstitution of proteins involved in the Ca^{++} transport process of sarcoplasmic reticulum (MacLennan, 1970; MacLennan et al., 1971; MacLennan and Wong, 1971). Resolution and characterization have, of necessity, preceded reconstitution, and we have only recently turned our attention to reassembly of the Ca^{++} transport system from isolated components.

We believe that some seven proteins could be components of the Ca^{++} transport system. Four of these proteins have been purified to at least 80% homogeneity; three of the proteins still exist in an unresolved mixture. Our experimental evidence for a role of these proteins in the calcium transport system is preliminary, and our hypotheses concerning the possible functions of these proteins must accordingly be considered as working models for further experimentation.

Separation of Intrinsic and Extrinsic Fractions

Figure 1 is a dodecyl sulfate-polyacrylamide gel (SDS gel) electrophoretic profile of our preparation of rabbit skeletal muscle sarcoplasmic reticulum prepared by differential centrifugation as described by MacLennan (1970) and neither salt-extracted (Martonosi, 1968) nor separated on a sucrose density gradient (Yu et al., 1968). The proteins which we believe to be involved either structurally or functionally in the Ca^{++} transport process are the ATPase of mol wt 102,000, an acidic protein of 54,000 daltons, calsequestrin of 44,000 daltons, a set of three acidic proteins with mol wts ranging between 20,000 and 32,000 and a proteolipid of mol wt 6000–12,000. All of these proteins are identifiable in sarcoplasmic reticulum following extraction with 0.6 M KCl or following sucrose density gradient centrifugation. Ikemoto et al. (1971a) have reported that sarcoplasmic reticulum contains major proteins of 125,000, 100,000, and 62,000 daltons. We have isolated sarcoplasmic reticulum according to their procedure and have found that in our hands the protein profile of their preparation in SDS gels is identical to that shown in Fig. 1.

The proteins can be separated into two fractions by disruption of the membrane with deoxycholate in the presence of 1 M KCl (MacLennan, 1970). Deoxycholate at a concentration as low as 0.1 mg per mg protein dissolves considerable amounts of protein while leaving the ATPase activity in insoluble form (Fig. 2). This extraction procedure solubilizes calsequestrin, the 54,000 dalton protein,

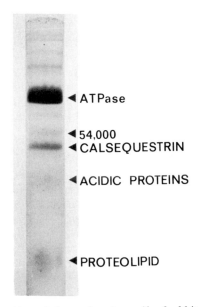

Figure 1. Disc gel electrophoretic profile of rabbit skeletal muscle sarcoplasmic reticulum after dissolution in sodium dodecyl sulfate (SDS). The polycrylamide gel was prepared in SDS according to the method of Weber and Osborn (1969) and stained with Coomassie blue.

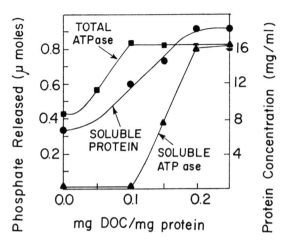

Figure 2. Effect of deoxycholate concentration on total ATPase activity, solubility of ATPase, and solubility of protein of sarcoplasmic reticulum. Sarcoplasmic reticulum was suspended at a protein concentration of 25 mg per ml in a solution containing 0.25 M sucrose, 1 M KCl, and 10 mM Tris-Cl, pH 8.0. Deoxycholate (10% solution, pH 8.0) was added in the ratio indicated and ATPase activity of 40 μg of protein was assayed. The samples were centrifuged at 165,000 g for 30 min and protein and ATPase activity in the soluble fraction were measured.

and the acidic proteins. Without disruption of the membrane none of these proteins can be isolated from sarcoplasmic reticulum membranes or from supernatants from which sarcoplasmic reticulum is isolated. All of these proteins remain monomeric and water soluble even upon removal of the detergent by dialysis and ion exchange chromatography. Green and Brucker (1972) have described such loosely bound membrane proteins as "extrinsic" membrane proteins, and they have described those proteins which are insoluble and interact with phospholipid as "intrinsic" membrane proteins. The ATPase is the major intrinsic membrane protein and it can be made soluble only by addition of more deoxycholate. Once dissolved, the ATPase can be purified by fractionation with ammonium acetate (MacLennan, 1970).

Intrinsic Membrane Proteins

ATPase. The ATPase, purified by fractionation with deoxycholate and ammonium acetate, has been obtained with ATPase activities as high as 42 μmoles of ATP hydrolyzed per min per mg protein, and this ATPase activity is completely dependent on both Mg^{++} and Ca^{++} (Fig. 3). The optimal Mg^{++} concentration is about 5 mM; the optimal Ca^{++} concentration in the presence of 0.1 mM EGTA is 0.1 mM. Thus the optimal free Ca^{++} concentration is as low as 10^{-6}–10^{-7} M (Portzehl et al., 1964). The absolute requirement of the ATPase for Ca^{++} and the high affinity of the enzyme for Ca^{++} are primary reasons for believing

that the ATPase is the Ca^{++} transport enzyme (Martonosi and Feretos, 1964).

In the presence of deoxycholate, the ATPase is a globular, dispersed preparation; upon removal of deoxycholate, the ATPase forms membranes in which its globular nature is still demonstrable by freeze-fracture (MacLennan et al., 1971). The symmetrical aspect of the two membrane faces is consistent with the possibility that the globular ATPase extends through the membrane.

Proteolipid. Membranes formed from the ATPase probably have the simplest composition of any membrane so far examined. It is therefore an interesting system in which to investigate the interaction between a globular protein and phospholipid. Upon disc gel electrophoresis of the ATPase, a band, staining with Coomassie blue and opalescent in oblique light (Fig. 4), migrates in the gels with an apparent mol wt of 6000 daltons. Studies with phospholipid extracted from the ATPase showed a similar blue-staining band and we initially concluded that the band was not protein but phospholipid. Subsequently we tested the preparation for proteolipid and discovered a low molecular weight proteolipid, which, by definition (Folch-Pi and Lees, 1951), is soluble in

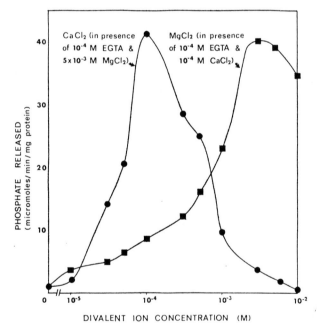

Figure 3. Divalent ion requirements for ATPase activity. The purified ATPase (10 μg) was incubated for 4 min at 37° in 1 ml of a solution containing 50 μmoles of Tris-Cl pH 7.5, 100 μmoles of KCl, 0.1 μmole of EGTA and 10 μmoles of ATP. CaCl₂ concentration was varied as indicated in the presence of 5 μmoles of MgCl₂; MgCl₂ concentration was varied as indicated in the presence of 0.1 μmole of CaCl₂. The reaction was stopped by addition of acid and phosphate release was measured by method D of Lindberg and Ernster (1956).

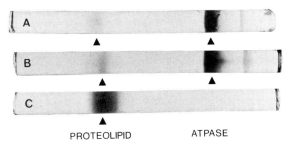

PROTEOLIPID   ATPASE

Figure 4. Dodecyl sulfate-polyacrylamide gel electrophoretic profiles of A, ATPase; B, ATPase plus proteolipid; C, purified proteolipid.

chloroform-methanol (2:1) but insoluble in the mixture upon addition of ethyl ether. We have isolated this proteolipid in two ways (MacLennan, 1972). It is dissolved in chloroform-methanol (2:1) together with the bulk of the phospholipid of sarcoplasmic reticulum. It is precipitated from this solution with 5 vol of ethyl ether, redissolved in chloroform-methanol (2:1) containing 10 mM HCl (acidified chloroform-methanol), and highly purified in this solvent. Alternatively the ATPase enzyme (or sarcoplasmic reticulum), suspended in sucrose solution, is treated with 9 vol of methanol. Phospholipid is extracted leaving the proteolipid behind. The residue is extracted sequentially with chloroform-methanol (2:1) to remove the remaining phospholipid and with acidified chloroform-methanol to dissolve the proteolipid. Proteolipid is precipitated with ethyl ether and subjected to thin layer chromatography on Silica Gel H with acidified chloroform-methanol as solvent. The proteolipid moves with an Rf of 0.3 to 0.5 and can be extracted from the plate into acidified chloroform-methanol. The protein, in purified form, is precipitated from solution with ethyl ether and dialyzed for several days against acidified chloroform-methanol to remove lipid.

In SDS gels (Fig. 4) the proteolipid appears as a single band with an Rf identical to that of insulin (mol wt 5700). We do not think that the molecular weight can be the same as insulin, however, since amino acid analysis (Table 1) shows that half residues of lysine and histidine are obtained, unless the molecular weight is estimated to be at least 12,000 daltons. We feel that the proteolipid binds SDS in a ratio greater than the expected 1.4 mg per mg protein. Increased negative charge, together with a relatively small size, probably causes the proteolipid to move faster in gels than would be expected for a molecule of 12,000 daltons. The amino acid analysis does not indicate any particularly low polarity (Capaldi and Vanderkooi, 1972) as was found for myelin proteolipids.

The protein interacts strongly with lipid; consequently lipid analysis is done after extensive

Table 1. Amino Acid Composition of the Proteolipid from Sarcoplasmic Reticulum

	Composition		
	g/100 g protein	Residues/mole Protein (a)	(b)
Lysine	1.25	0.55	1.10
Histidine	1.07	0.44	0.88
Arginine	13.90	5.00	10.00
Aspartic acid	5.25	2.56	5.12
Threonine	6.19	3.44	6.88
Serine	6.36	4.11	8.22
Glutamic acid	12.50	5.44	10.88
Proline	1.92	1.11	2.22
Glycine	1.46	1.44	2.88
Alanine	2.11	1.67	3.34
Cysteic acid	2.16	1.44	2.88
Valine	5.48	3.11	6.22
Methionine	2.33	1.00	2.00
Isoleucine	4.03	2.00	4.00
Leucine	17.24	8.56	17.12
Tyrosine	10.34	3.56	7.12
Phenylalanine	6.39	2.44	4.88
Total			95.84

(a) Based on a mol wt of 6000 daltons.
(b) Based on a mol wt of 12,000 daltons.

dialysis to remove loosely bound lipid. With a mol wt of 12,000 daltons we can assume 83 nmoles of proteolipid per mg of protein (Table 2). Phosphorus and glycerol analyses indicate that after dialysis only 5–10 nmoles of phospholipid are still bound per mg of protein or less than 1 mole per 10 moles of protein. Sphingosine and sugars are absent. Fatty acids, however, are consistently present at levels of 140–200 nmoles per mg protein or at levels of 1–2 moles per mole of proteolipid.

We have attempted to find a role for the protein in the ATPase. Since we have no means of estimating total proteolipid in a mixture, we cannot estimate its concentration in the ATPase preparation. Our purification scheme yields 2 mg from 800 mg of ATPase. Stain absorbed in the region of

Table 2. Composition of the Proteolipid of Sarcoplasmic Reticulum

	Content (nmoles/mg protein)
Protein[a]	83
Glucose[b]	0
Galactose[b]	0
Sphingosine[b]	0
Sialic acid[b]	0
Phosphorus	6–12
Glycerol[c]	5–8
Fatty acid[b]	143–198

[a] Based on a mol wt of 12,000 daltons.
[b] Determined by gas-liquid chromatography following digestion in methanolic-HCl.
[c] Determined enzymatically with glycerol kinase following digestion in methanolic-KOH (Garland and Randle, 1962).

proteolipid migration in SDS gels is only a few percent of that absorbed by the major ATPase protein. Some of this absorption may be due to phospholipid. Therefore, it is unlikely that the proteolipid exists in greater than a 1:1 ratio with the major ATPase protein, and it probably exists in less than 1:1 molar ratio. We have not been able to bring the proteolipid into water solution and consequently we have been unable to measure its Ca^{++}-binding activity directly. Its presence, however, does not affect the Ca^{++}-binding activity of phospholipid micelles. The proteolipid is not phosphorylated by $[^{32}P]ATP$: the 102,000 dalton protein carries ^{32}P upon separation of the phosphorylated ATPase by SDS gel electrophoresis at pH 6.0. There is no demonstrable dicyclohexyl-carbodi-imide (DCCD) or oligomycin sensitivity in the Ca^{++}-dependent ATPase. This is in contrast to the mitochondrial ATPase in which the activity is inhibited by DCCD and in which Cattell et al. (1971) were able to show that a proteolipid subunit of the mitochondrial ATPase was the site of DCCD binding. The rabbit protein was not antigenic in three guinea pigs, and we have not succeeded in obtaining antibodies with which to probe for function and localization of the proteolipid in the ATPase.

A current idea is that the proteolipid is involved in the interaction between the globular ATPase protein and phospholipid. The proteolipid could have hydrophobic and hydrophilic sites which could provide a nucleus for binding both protein and phospholipid (Tzagoloff, 1972). We have begun experiments to test this hypothesis. The ATPase is dissolved in 1% SDS and passed through Sephadex G-200 equilibrated with this solvent. The ATPase protein, eluted near the void volume, is depleted of phospholipid and is devoid of proteolipid (Fig. 5). It still exists as a 102,000 dalton subunit and it is still globular in appearance (Fig. 6). This experiment shows that the globular nature of the ATPase and the size of the molecule are unaffected by removal of proteolipid.

Reconstitution of the vesicular structure of the ATPase is being attempted through removal of

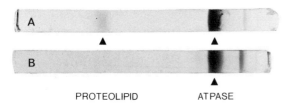

Figure 5. Dodecyl sulfate-polyacrylamide gel electrophoretic profiles of *A*, ATPase; *B*, ATPase after filtration through a 2.5 × 45 cm column of Sephadex G-200 equilibrated with a solution of 0.25 M sucrose, 0.01 M phosphate, pH 7.0, 1% SDS.

SDS followed by addition to the purified ATPase of phospholipid micelles (Fleischer and Fleischer, 1967) or of phospholipid micelles containing proteolipid in the ratio of 10:1 (W:W). We find that globular particles attach to the surface of lamellar lipid in both cases (Fig. 7A). Globular particles embedded in a lipid matrix have so far been seen only upon freeze-fracture of mixtures of the ATPase plus phospholipid plus proteolipid (Fig. 7B). Since our technique does not always permit formation of lipid bilayers, we do not feel that we have adequately tested the idea that proteolipid is required for vesicular structure, although our results to date are encouraging that it is.

Extrinsic Proteins

54,000 dalton protein. The protein fraction soluble in low levels of deoxycholate is a mixture of at least five acidic proteins. Calsequestrin (MacLennan and Wong, 1971) is the dominant protein in this fraction. When the solution is applied to DEAE cellulose and eluted with a KCl gradient, three major peaks are seen (Fig. 8). The first peak contains proteins which are not of sarcoplasmic reticulum origin. Phosphorylase is a major activity in this peak. The second peak contains the protein of mol wt 54,000. It is less acidic than calsequestrin but shares with it the property of being eluted upon extraction of sarcoplasmic reticulum with divalent ion chelators (Duggan and Martonosi, 1970). We have purified this protein to homogeneity (Fig. 9) by sequential passage through DEAE cellulose, Sephadex G-200, DEAE cellulose and hydroxylapatite (MacLennan, 1972).

Like calsequestrin the 54,000 dalton protein is acidic in amino acid composition (Table 3). Also, like calsequestrin, the protein binds Ca^{++}. Figure 10 shows that in the presence of high Ca^{++} (10^{-3} M), the 54,000 dalton protein binds about 470 nmoles Ca^{++} per mg or about one-half the maximal amount which calsequestrin can bind. The Ca^{++} concentration for 50% of maximal binding is about 100 μM for the 54,000 dalton protein as opposed to 40 μM for calsequestrin (MacLennan and Wong, 1971). The 54,000 dalton protein is present in sarcoplasmic reticulum in somewhat lower concentration than that of calsequestrin (cf. gel patterns in Duggan and Martonosi, 1970; MacLennan et al., 1971; and Fig. 1). The lower amount and the lower Ca^{++}-binding capacity suggest that if the 54,000 dalton protein played a role in calcium sequestration it could bind only a fraction of the amount of calcium bound by calsequestrin and could do so with a somewhat lower affinity for calcium.

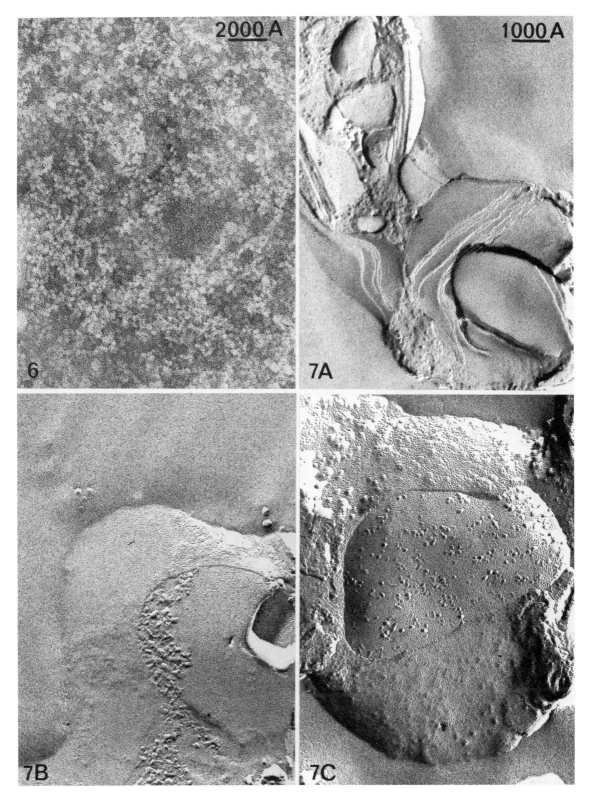

Figure 6. The ATPase enzyme after removal of proteolipid (see Fig. 5). The sample was negatively stained with phosphotungstate (×207,900).

Figure 7. Freeze-fracture profile of structures formed by interaction of the ATPase, freed of proteolipid, with *A*, a micelle of phospholipid prepared from sarcoplasmic reticulum and freed of proteolipid by ethyl ether fractionation. The fracture planes reveal either smooth "islands" of lipid or "islands" of globule regions, but little admixture of the lipid and globule-containing moieties. *B*. A micelle of phospholipid and proteolipid prepared by sonication of a mixture of phospholipid and proteolipid in a ratio of 10:1 (w/w). (See Fleischer and Fleischer, 1967, for details of phospholipid extraction and micelle formation.) The fracture plane of the membrane reveals a confluence of both globule and lipid (smooth surface) regions. For comparison 7*C* is a freeze-fracture profile of unfractionated sarcoplasmic reticulum (×103,500).

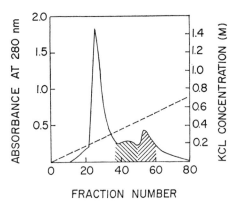

Figure 8. Fractionation of extrinsic proteins of sarco-plasmic reticulum on DEAE cellulose. Conditions of separation were as described by MacLennan (1972). ////////// 54,000 dalton protein fraction; ////////////// calse-questrin and acid protein fraction.

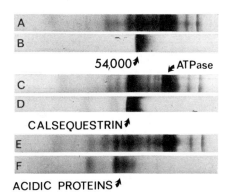

Figure 9. Dodecyl sulfate-polyacrylamide gel electro-phoretic profiles of: *A*, sarcoplasmic reticulum plus 54,000 dalton protein; *B*, 54,000 dalton protein; *C*, sarco-plasmic reticulum plus calsequestrin; *D*, calsequestrin; *E*, sarcoplasmic reticulum plus acidic proteins; *F*, acidic proteins.

Table 3. Amino Acid Composition of the 54,000 Dalton Protein from Sarcoplasmic Reticulum

	Composition	
	g/100 g protein	Residues/mole Protein[a]
Lysine	11.75	51
Histidine	2.47	10
Arginine	3.65	13
Aspartic acid	15.12	73
Threonine	3.27	18
Serine	5.16	33
Glutamic acid	21.12	91
Proline	5.93	34
Glycine	4.62	45
Alanine	2.14	28
Cysteic acid	0.61	4
Valine	4.45	25
Methionine	1.18	5
Isoleucine	4.07	20
Leucine	5.33	26
Tyrosine	4.11	14
Phenylalanine	5.03	19
Total		509

[a] Based on a mol wt of 54,000 daltons.

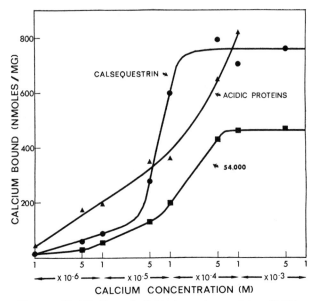

Figure 10. Binding of Ca^{++} by calsequestrin, 54,000 dalton protein, and acidic proteins as a function of the concentration of $CaCl_2$ in the presence of 5 mM Tris-Cl, pH 7.5. Conditions of assay were those described by MacLennan and Wong (1971).

The 54,000 dalton protein may be involved in calcium transport. We have prepared particles depleted in calsequestrin and the 54,000 dalton protein by extraction in the presence of 1 mM EGTA at pH 9.2 (Duggan and Martonosi, 1970). These particles are depleted in ability to transport Ca^{++}, but some restoration of the ability to transport and bind Ca^{++} can consistently be attained by addition of extracts and of the 54,000 dalton protein (Table 4). Neither calsequestrin nor the acidic proteins stimulate transport under the same conditions. The exact role of the 54,000 dalton protein in transport and binding awaits further study.

Acidic proteins. Upon elution from DEAE cellulose, the calsequestrin peak contains a group of acidic proteins which can be separated from calsequestrin upon elution from hydroxylapatite (Fig. 11). These proteins range in mol wt from 20,000–32,000 (Fig. 9). When filtered through Sephadex G-200 the proteins are eluted with an Rf indicative of monomeric forms. In fact some separation of the three proteins can be achieved by gel filtration (Fig. 12).

Two points concerning this group of proteins are of special interest. Ca^{++}-binding sites of high affinity are present in the crude mixture (Fig. 10). At Ca^{++} levels as low as 10^{-6} M, where neither calsequestrin nor the 54,000 dalton protein bind significant amounts of Ca^{++}, 35 nmoles of Ca^{++} are still bound per mg of the acidic proteins. We

Table 4. Effect of Extraction and Reconstitution on Ca^{++} Transport by Sarcoplasmic Reticulum

	Treatment	Binding[a]	Transport[b]
		(nmoles/mg protein)	
Expt. 1	None	119	2300
	EGTA	49	1760
	EGTA, pH 9.2 → 7.4	58	1920
	EGTA, pH 9.2	11	240
	EGTA, pH 9.2, supernatant, pH 7.4	13	342
Expt. 2	EGTA, pH 9.2	11	192
	EGTA, pH 9.2, 54,000 dalton protein	17	253

In experiment 1 particles were suspended at a protein concentration of 10 mg/ml in a solution of 0.25 M sucrose, 1 mM histidine and 10 mM Tris-Cl, pH 8.0 (buffer A). EGTA was added to 1 mM where indicated. The pH was adjusted to 9.2 with 1 N NH$_4$OH where indicated and after 15 min at 0°, the pH was readjusted to 7.4 with 1 N HCl where indicated. All samples were centrifuged at 105,000 g for 30 min and were resuspended either in buffer A or in the supernatant solution. Where suspension was in the supernatant solution, the pH was subsequently adjusted to 7.4. Particles were again centrifuged at 105,000 g for 15 min, washed in buffer A, suspended in buffer A at a protein concentration of 10 mg/ml, and assayed for transport as described by Sommer and Hasselbach (1967). [a] Binding was measured by omission of oxalate. [b] Transport was measured in the presence of oxalate.

In experiment 2 particles (4 mg) extracted with 1 mM EGTA at pH 9.2, washed, and suspended at 10 mg/ml in buffer A were incubated for 10 min at 24° in the presence of 0.5 mM CaCl$_2$ and 10 mM ATP in the presence and absence of 200 μg of the 54,000 dalton protein before 100 μg was taken for assay of ^{45}Ca binding.

estimate that there may be twelve nmoles of each of the three proteins in one mg (based on a 1:1:1 equivalence of proteins and on the approximate total of all three mol wts being about 80,000 daltons). The calcium-binding data would indicate that as much as three molecules of Ca^{++} are bound per molecule of each of the proteins at 10^{-6} M Ca^{++}. Such a Ca^{++}-binding affinity is of the same order of magnitude as the Ca^{++}-binding affinity of the Ca^{++}-dependent ATPase (Fig. 3).

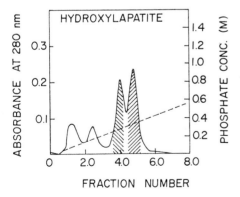

Figure 11. Fractionation of calsequestrin and acidic proteins on hydroxylapatite columns. Conditions of separation were as described by MacLennan (1972). ////// acidic protein fraction; \\\\\\ calsequestrin fraction.

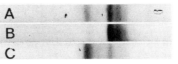

Figure 12. Dodecyl sulfate-polyacrylamide gel electrophoretic profiles of: *A*, acidic protein fraction; *B*, first peak upon filtration through Sephadex G-200; *C*, second peak upon filtration through Sephadex G-200. The acidic protein fraction (10 mg) obtained by elution from hydroxylapatite (see Fig. 11) was dissolved at 3 mg per ml in a solution of 0.15 M KCl, 0.01 M Tris-Cl, pH 8.0, 0.1 mM EGTA and filtered through a 2.5 × 90 cm column of Sephadex G-200 equilibrated with the same solution.

The second point of interest was gleaned from attempts to measure localization of sarcoplasmic reticulum proteins by iodination with lactoperoxidase as described by Philips and Morrison (1970). In this procedure, lactoperoxidase (mol wt 70,000) is incubated with vesicles in the presence of ^{125}I. Tyrosine groups of proteins on external surfaces are labeled in preference to tyrosine groups of proteins relatively inaccessible to lactoperoxidase. While the procedure works well on intact cells or intact mitochondria, labeling exterior proteins specifically, it is less useful with fragmented vesicles where lactoperoxidase does achieve some degree of penetration to the interior of the vesicle (M. Morrison, pers. commun.). Since sarcoplasmic reticulum is already fragmented, a relatively nonspecific labeling was anticipated. Figure 13 shows that most proteins of sarcoplasmic

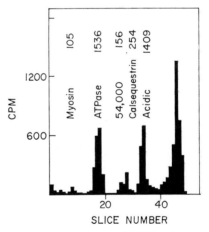

Figure 13. Incorporation of ^{125}I into proteins of sarcoplasmic reticulum. Sarcoplasmic reticulum (2.5 mg) was incubated for 1 hr at 24° in 1 ml of a solution containing 0.25 M sucrose, 0.01 M imidazole, pH 7.0, 10^{-7} M KI containing ^{125}I, and 10^{-7} M lactoperoxidase. At 15 min intervals H$_2$O$_2$ was added to 8 × 10^{-6} M. Samples were layered on top of gradients as described by Yu et al. (1968) and the fraction layering between 35% and 45% sucrose was isolated. The protein was diluted with water, centrifuged to a pellet, dissolved in a solution of 1% SDS, 10 mM PO$_4$, pH 7.0, and 1% 2-mercaptoethanol. One hundred μg of this sample were subjected to SDS gel electrophoresis for 2 hr at 8 mA per tube. Gels were stained with Coomassie blue to identify bands, and then 1 mM sections were cut and counted for ^{125}I content.

reticulum were, in fact, labeled with ^{125}I under the conditions described by Philips and Morrison (1970). However, acidic proteins were labeled to a much greater extent than would be calculated from their content in the membrane. This result suggests that the acidic proteins may be in a much more external position than any of the other proteins of sarcoplasmic reticulum and may well constitute the external projections of the sarcoplasmic reticulum membrane first described by Ikemoto et al. (1966).

Summary

The speculations advanced concerning the roles of the various proteins isolated from sarcoplasmic reticulum can be summarized in the tentative scheme of structure and function of the sarcoplasmic reticulum shown in Fig. 14. The ATPase, a phospholipid bilayer, and a proteolipid make up the membrane continuum. The ATPase is believed to be a globular protein extending through the sarcoplasmic reticulum membrane. The proteolipid may act as a bimodal molecule orienting the hydrophilic ATPase protein within the hydrophobic phospholipid bilayer. The three components ATPase, phospholipid, and proteolipid make up a Ca^{++} plus Mg^{++}-dependent ATPase; however, it is not as yet clear whether or not the proteolipid is functional in this complex. ATPase activity is dependent upon micromolar Ca^{++} in the presence of millimolar Mg^{++} and decimolar K^+ or Na^+: the affinity of the ATPase for Ca^{++} is high and specific. Martonosi and Feretos (1964) have shown a precise correlation between Ca^{++}-dependent ATPase activity and removal of Ca^{++} from solution by sarcoplasmic reticulum. This and earlier evidence (see Ebashi et al., 1969) have been taken as proof that the Ca^{++}-dependent ATPase is the Ca^{++} transport enzyme. Since the ATPase is activated by micromolar Ca^{++}, Ca^{++} concentrations in the sarcoplasm can be lowered by sarcoplasmic reticulum to the micromolar range, thereby leading to muscle relaxation (Ebashi et al., 1969).

The interior of sarcoplasmic reticulum is a separate compartment from the rest of the cell, and ion concentration within the reticulum need not be the same as ion concentration in the sarcoplasm. In fact, we and others (see MacLennan and Wong, 1971) have calculated that Ca^{++} concentrations in the interior of Ca^{++}-loaded sarcoplasmic reticulum may be as high as 20 mM. Calsequestrin and the 54,000 dalton protein, both of which are believed to be localized on the interior surface of the sarcoplasmic reticulum membrane and both of which bind vast quantities of Ca^{++}, are believed to play an important role in binding and sequestering the high concentration of Ca^{++} on the interior of the sarcoplasmic reticulum vesicle. Although these proteins bind large amounts of Ca^{++}, their affinities for Ca^{++} are not high and selectivity for Ca^{++} is not as great as has been reported for other Ca^{++}-binding proteins (MacLennan and Wong, 1971). This fact is no doubt troublesome to muscle biochemists used to thinking of Ca^{++} specificity of binding proteins in relation to the high Ca^{++}-binding affinity of troponin. However, the calcium-binding roles of troponin and calsequestrin are vastly different. Muscle contraction ensues when a few molecules of Ca^{++} are bound to a molecule of troponin in the presence of micromolar Ca^{++}, millimolar Mg^{++}, and decimolar K^+. The troponin Ca^{++}-binding sites must be of high affinity and specificity. Sarcoplasmic reticulum, on the other hand, must sequester large quantities of Ca^{++} in order to deplete the sarcoplasm of Ca^{++} and it must do so with a limited number of binding sites. Carvalho and Leo (1967) have shown that these sites can be occupied by a variety of cations but that in the presence of ATP they are selectively occupied by Ca^{++}. This observation indicates that a high innate selectivity for Ca^{++} is not a physiological feature of sarcoplasmic reticulum Ca^{++}-binding proteins. Quantity and not selectivity is essential in these binding proteins since the selectivity is provided by the Ca^{++} transport ATPase. As the ATPase pumps Ca^{++} inward, binding sites occupied by other ions are displaced because of increased internal Ca^{++} levels and because calsequestrin does have considerable selectivity for binding of Ca^{++} as compared with binding of Mg^{++} or K^+ (MacLennan and Wong, 1971). We postulate that calsequestrin and the 54,000 dalton protein, both of which have a large number of Ca^{++}-binding sites and intermediate affinities for Ca^{++}, fulfill a role in sequestering Ca^{++} on the interior of sarcoplasmic reticulum. We do not rule out the possibility that they also play a role in cross-membrane Ca^{++} transport.

The acidic protein fraction has a high Ca^{++} affinity and appears to exist on the exterior of the sarcoplasmic reticulum membrane. These proteins

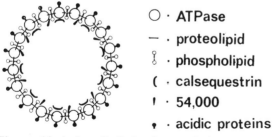

○ · **ATPase**

— · **proteolipid**

○ · **phospholipid**

(· **calsequestrin**

❙ · **54,000**

• · **acidic proteins**

Figure 14. A hypothetical scheme of the structural arrangement of proteins and lipid in vesicles of sarcoplasmic reticulum.

could bind Ca^{++} under the conditions of lowered sarcoplasmic Ca^{++} seen during relaxation. They may play an independent role in removal of Ca^{++} from the sarcoplasm or they may play the ancillary role of concentrating Ca^{++} at the active site of the transport ATPase. It is clear from the work of Ikemoto et al. (1971b) that the surface particles are not essential to the transport process or to ATPase activity, but it cannot be ruled out that in vivo the surface particles play a physiological role in regulation of free Ca^{++} concentrations in the sarcoplasm.

Acknowledgments

We thank Mr. Vijay Khanna, Miss Ming Chow, and Mr. Sambhu Nath Ghosh for expert technical assistance, Dr. S. J. Klebanoff for the gift of lactoperoxidase, Mr. G. Yogeeswaran for sialic acid and sphingosine determinations, and Dr. M. Halperin for glycerol determinations. This research was supported by grants MA-339, MT-2951, and ME-4617 from the Medical Research Council of Canada, and by Province of Ontario Department of Health grant P.R. 189. This is the fourth paper in a series "Resolution of Enzymes of Biological Transport."

References

CAPALDI, R. A. and G. VANDERKOOI. 1972. The low polarity of many membrane proteins. *Proc. Nat. Acad. Sci.* **69:** 930.

CARVALHO, A. P. and B. LEO. 1967. Effects of ATP on the interaction of Ca^{++}, Mg^{++}, and K^+ with fragmented sarcoplasmic reticulum isolated from rabbit skeletal muscle. *J. Gen. Physiol.* **50:** 1327.

CATTELL, K. J., C. R. LINDOP, I. G. KNIGHT, and R. B. BEECHEY. 1971. The identification of the site of action of NN^1-dicyclohexylcarbodi-imide as a proteolipid in mitochondrial membranes. *Biochem. J.* **125:** 169.

DUGGAN, P. F. and A. MARTONOSI. 1970. Sarcoplasmic reticulum. IX. The permeability of sarcoplasmic reticulum membranes. *J. Gen. Physiol.* **56:** 147.

EBASHI, S., M. ENDO, and I. OHTSUKI. 1969. Control of muscle contraction. *Quart. Rev. Biophys.* **2:** 4.

FLEISCHER, S. and B. FLEISCHER. 1967. Removal and rebinding of polar lipids in mitochondria and other membrane systems. *Methods Enzymol.* **10:** 406.

FOLCH-PI, J. and M. LEES. 1951. Proteolipides, a new type of tissue lipoproteins; their isolation from brain. *J. Biol. Chem.* **191:** 807.

GARLAND, P. B. and P. J. RANDLE. 1962. A rapid enzymatic assay for glycerol. *Nature* **196:** 987.

GREEN, D. E. and R. F. BRUCKER. 1972. The molecular principles of biological membrane construction and function. *Bioscience* **22:** 13.

IKEMOTO, N., F. A. SRETER, and J. GERGELY. 1966. Localization of calcium uptake and ATPase activity in fragments of sarcoplasmic reticulum. *Fed. Proc.* **25:** 465.

IKEMOTO, N., G. M. BHATNAGAR, and J. GERGELY. 1971a. Fractionation of solubilized sarcoplasmic reticulum. *Biochem. Biophys. Res. Comm.* **44:** 1510.

———. 1971b. Structural features of the vesicles of FSR. Lack of functional role in Ca^{++} uptake and ATPase activity. *Arch. Biochem. Biophys.* **147:** 571.

LINDBERG, O. and L. ERNSTER. 1956. Determination of organic phosphorus compounds by phosphate analysis. *Methods Biochem. Anal.* **3:** 1.

MACLENNAN, D. H. 1970. Purification and properties of an adenosine triphosphatase from sarcoplasmic reticulum. *J. Biol. Chem.* **245:** 4508.

———. 1972. Isolation of proteins of the sarcoplasmic reticulum. *Methods Enzymol.* **23:** In press.

MACLENNAN, D. H. and P. T. S. WONG. 1971. Isolation of a calcium-sequestering protein from sarcoplasmic reticulum. *Proc. Nat. Acad. Sci.* **68:** 1231.

MACLENNAN, D. H., P. SEEMAN, G. H. ILES, and C. C. YIP. 1971. Membrane formation by the adenosine triphosphatase of sarcoplasmic reticulum. *J. Biol. Chem.* **246:** 2702.

MARTONOSI, A. 1968. Sarcoplasmic reticulum. IV. Solubilization of microsomal adenosine triphosphatase. *J. Biol. Chem.* **243:** 71.

———. 1969. The protein composition of sarcoplasmic reticulum membranes. *Biochem. Biophys. Res. Comm.* **36:** 1039.

———. 1971. The structure and function of sarcoplasmic reticulum membranes. *Biomembranes* **1:** 191.

MARTONOSI, A. and R. FERETOS. 1964. Sarcoplasmic reticulum. II. Correlation between adenosine triphosphatase activity and Ca^{++} uptake. *J. Biol. Chem.* **239:** 659.

MARTONOSI, A. and R. A. HALPIN. 1971. Sarcoplasmic reticulum. X. The protein composition of sarcoplasmic reticulum membranes. *Arch. Biochem. Biophys.* **144:** 66.

PHILIPS, D. R. and M. MORRISON. 1970. The arrangement of proteins in the human erythrocyte membrane. *Biochem. Biophys. Res. Comm.* **40:** 284.

PORTZEHL, H., P. C. CALDWELL, and J. C. RUEGG. 1964. The dependence of contraction and relaxation of muscle fibres from the crab *Maia squinado* on the internal concentration of free calcium ions. *Biochim. Biophys. Acta* **79:** 581.

SOMMER, J. R. and W. HASSELBACH. 1967. The effect of glutaraldehyde and formaldehyde on the calcium pump of the sarcoplasmic reticulum. *J. Cell Biol.* **34:** 902.

TZAGOLOFF, A. 1972. A model of membrane biogenesis. *Bioenergetics* **3:** 39.

WEBER, A. 1966. Energized calcium transport and relaxing factors. *Current Topics Bioenergetics* **1:** 203.

WEBER, K. and M. OSBORN. 1969. The reliability of molecular weight determinations by dodecyl sulfate-polyacrylamide gel electrophoresis. *J. Biol. Chem.* **244:** 4406.

YU, B. P., F. D. DE MARTINIS, and E. J. MASORO. 1968. Isolation of Ca^{++} sequestering sarcotubular membranes from rat skeletal muscle. *Anal. Biochem.* **24:** 523.

Electrical Events in the T-System of Frog Skeletal Muscle

LEE D. PEACHEY

Department of Biology, University of Pennsylvania, Philadelphia, Pennsylvania 19104

It now is widely accepted that the transverse tubules (T-system) of skeletal muscle are the morphological path for excitation-contraction coupling between the surface and the interior of a muscle fiber. The evidence for this comes from a combination of comparative morphology and physiological experiments. Several reviews have been published during the development of this idea, and I will not retell old tales. Perhaps the best review up to the point where I will start is the first half of the Croonian Lecture delivered by A. F. Huxley in 1967 (Huxley, 1971).

My purpose here is to review the emergence of the idea that the T-system of the frog skeletal muscle fiber carries out its coupling function as a cable network propagating an action potential. This tubular action potential probably is based on active conductance changes much like those believed to function in the surface membranes of unmyelinated nerves, at the nodes of myelinated nerves, and at the surfaces of skeletal muscle cells. I will show that the formal treatment of Hodgkin and Huxley, developed for nerve cell action potentials, can geometrically be adapted to the T-system and can provide a description of electrical events that would be suitable for excitation-contraction coupling.

First I would like briefly to discuss ways of representing the complex geometry of the T-system in an electrical circuit, showing various current paths from the interior to the exterior of the muscle fiber. It is widely agreed that the walls of the T-tubules contribute a major part of the measured electrical capacity of the muscle fiber. A thorough discussion of this would be too lengthy, but I do think it worthwhile just to point out some of the possibilities.

Falk and Fatt (1964) considered both a lumped electrical model for the T-system, and a fully distributed one. In the former, the total capacitance of the T-system is lumped into a single capacitor, which in combination with a series resistor is connected across the surface capacitance and resistance (Fig. 1A). In the distributed model (Fig. 1B) the T-system capacitance is spread out along the T-system resistance. The latter seems to be more realistic in relation to the known morphology of the T-system, since some of the T-system

membrane is near the fiber surface and thus has little resistance in series with it, whereas that near the axis of the fiber would have a considerably larger series resistance. However Falk and Fatt (1964) could not find sufficient justification in their impedance data to choose the distributed model over the lumped one, though the lumped one did not perfectly explain their data. A contrary conclusion was reached by Schneider (1970), whose analysis of impedance measurements favored the distributed model.

The circuit shown in Fig. 1C is a hybrid between the lumped and fully distributed circuits. The T-system is considered to have its capacitance distributed along a resistor, but the combination is connected to the surface through an additional "access resistance." This extra resistance was introduced by Peachey and Adrian (1972) for reasons which will be discussed below. These authors discuss several possibilities for the morphological origin of the access resistance, but it must be admitted that the real source of this resistance, if it really exists, is not known.

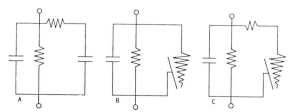

Figure 1. Three electrical representations of the surface membrane and T-system of skeletal muscle. In each, the capacitor and resistor on the left represent current paths from outside to inside the fiber through the surface membrane. The three circuits differ in their representation of the current path through the T-system, the limb on the right of each circuit. *A,* the T-system capacitance and resistance are lumped into a single resistor and capacitor in series. *B,* the T-system capacitance is fully distributed along the T-system resistance: the somewhat unorthodox electrical symbol is meant to suggest this diagrammatically. In this model some of the T-system capacitance (that near the fiber axis) is in series with essentially all of the tubular resistance, while some of it (near the top of the symbol) has essentially no resistance in series with it and represents T-tubules near the fiber surface. The circuit in *C* is a hybrid of *A* and *B.* The T-system capacitance is distributed along a resistance, but the combination is in series with an additional resistor which we have called an access resistance. The computations and arguments presented in the text favor the access resistance or partially distributed model for the T-system (circuit *C*) as the best representation of the muscle fiber.

479

Electron microscopy presents the T-system of skeletal muscle as a branching network of tubules extending from the surface of the fiber all the way to its axis or center. There is one such network at the Z-line level of each sarcomere in frog muscle (Fig. 2). Data on the caliber of the tubules, their volume relative to the fiber volume, and their surface/volume ratio have been presented for frog muscle (Peachey, 1965; Peachey and Schild, 1968). It should be mentioned that at present we assume that the T-system is uniform across the diameter of the fiber and that the tubules are randomly oriented with respect to the radial direction. If there is anisotropy or variation in properties at different places in the fiber, they are not immediately obvious, but to my knowledge these points have not specifically been checked in the electron microscope. An exception to the assumption of uniformity is the access resistance proposed to exist in the T-system near or at its junction with the fiber surface (Peachey and Adrian, 1972).

In 1969 Adrian and coworkers published two papers describing voltage-clamp experiments on frog muscle fibers bathed with solutions containing tetrodotoxin (10^{-6} g/ml). The purpose of the tetrodotoxin (TTX) was to eliminate propagated action potentials at the surface of the fiber, thereby preventing twitches of the whole fiber so that local contractions could be observed. As I remember it, there was one main argument behind thinking that the T-tubules were passive and that TTX would affect only the surface action potential and not the spread of potential changes in the T-tubules. This came from the local activation experiments of Huxley and Taylor (1958), which gave only graded, local contractions, not an all-or-none propagated response in a whole disc across the fiber. In effect it was assumed that there was no sodium conductance in the T-system, and since TTX is believed to affect only sodium channels, it would be without effect on the T-tubules. Therefore it seemed safe to use TTX to block surface activity in an experiment designed to study spread of electrical depolarization along the T-system. As we know now, this was wrong, but before going into that I would like to review these early experiments.

The main goal of the first set of experiments (Adrian et al., 1969a) was to find out what criteria an electrical stimulus at the surface of the fiber must satisfy if it is to give a contraction. The experiments employed a voltage clamp to control the fiber surface potential, and contractions were scored by observation with a binocular dissecting microscope at $100\times$ magnification. It seems reasonable to suppose that contractions were observed only when a considerable depth of the fiber was activated.

An important result of these experiments was the determination of a strength-duration relationship for critical depolarization for contraction using rectangular pulses. At about 4°C, rheobase was about -52 mV. For rectangular pulses to more than -10 mV, the criterion for contraction was that the area of the voltage-time curve above -30 mV be greater than 120 mV-msec. From earlier data these authors estimated the area above -30 mV for an action potential in frog muscle at 4°C to be about 250 mV-msec, or twice that needed for activation of contraction. I will use this criterion for contraction later in my discussion.

In the appendix of this first paper (Adrian et al., 1969a), a cable analysis of a passive T-system was presented. This analysis uses the morphological parameters derived from electron microscopy and makes the assumptions about uniformity mentioned earlier. The result is an equation (no. 17 in their paper) which describes the expected distribution of electrotonic potential across the walls of the T-tubules as a function of time and depth in the fiber when the surface potential is suddenly altered in a voltage step. To give a rough idea of the rapidity of changes expected from their model, they calculated that the potential 10 μ deep in a fiber of radius 40 μ would reach half its steady-state value in about 0.1 msec: at the center of the fiber this would occur at about 0.65 msec.

This analysis became the basis for a further set of experiments designed specifically to look at the inward spread of contraction activation along the T-system (Adrian et al., 1969b). These experiments employed a similar method for voltage-clamping the fiber surface. The microscopic system was improved to allow visualization of shortening of fibrils at various distances from the surface to the center of the fiber. Very small contractions involving only a few fibrils near the fiber surface could be seen. TTX was again included in the bathing solutions to inhibit action potentials.

With square steps of depolarization of long duration applied to the surface of the fiber, the depth of fiber activated was found to depend on the magnitude of the depolarization. Contraction of central fibrils required depolarization of the surface to potentials several millivolts more positive than required to activate myofibrils just under the fiber surface. This was as expected from the passive analysis of the earlier paper, since the central tubules would always be depolarized less than those near the surface. The central depolarization "sags" just as electrotonic potentials decay along any passive cable.

The extent of this "sag" could be estimated from the voltage-clamp experiments and used to

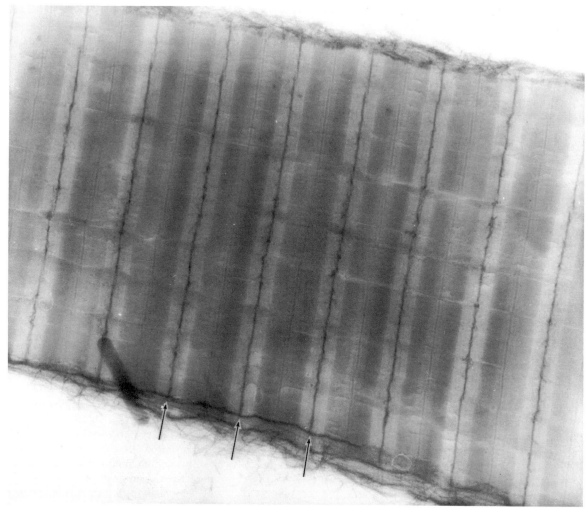

Figure 2. Frog (*Rana pipiens*) sartorius muscle fiber soaked in horseradish peroxidase and incubated according to the method of Karnovsky (1965). This method coats the surface of the fiber and fills the T-system with a dense precipitate, which appears dark in the electron micrograph. The section was 1 μ thick and the micrograph was made using an accelerating voltage of 800 kV. This section, which is about ten times thicker than used for electron microscopy at ordinary voltages, shows dense lines across the fiber at the levels of the Z-lines of the sarcomeres (arrows). Since the section thickness is about the same as the mesh size of the T-system, the images of multiple T-tubules superimpose to form an irregular but continuous dense line completely across the fiber image. $\times 8500$.

calculate a radial length constant, λ_T, for the T-tubule network. This was found to be about 60 μ for a 50 μ radius fiber and to vary in the same direction as fiber radius in smaller and larger fibers. Short pulses of a few milliseconds duration were found to be less effective than long pulses, again as expected from the analysis of the T-system as a passive cable network.

Overall, these experiments were in good agreement with what was expected for a network of passive tubules extending into the fiber from the surface membrane. There were, however, two disturbing results from these experiments. First, the total conductance of the T-system, as seen from the fiber surface, calculated from the values determined for λ_T, was approximately equal to the

input conductance for the fiber as a whole. Since the latter represents the conductance of the T-system parallel with the conductance of the surface membrane, this result would lead to the conclusion that the surface membrane of the fiber had essentially no conductance. Clearly this conclusion cannot be valid. In fact earlier estimates of the distribution of input conductance between T-system and surface (Eisenberg and Gage, 1969) suggested that the T-system should have only about one-half as much resting conductance as the fiber surface.

The second disturbing result was the observation that when the surface was voltage-clamped with the waveform of an action potential, an essentially full-sized action potential was required to achieve apparent activation of myofibrils on the fiber axis. This says that there is no safety factor in the action potential, which is an unlikely result from a biological point of view since most biological phenomena seem to have rather considerable safety factors.

A way out of the first difficulty, the excessively high T-system conductance calculated from the apparent λ_T, was presented in a later paper (Peachey and Adrian, 1972). Here it was proposed that there is an extra radial resistance in the T-system near the fiber surface. Possible structural sources of such a resistance were discussed in the paper just referred to. The important point for the present discussion is that this access resistance, as it was called, would increase the magnitude of the potential drop from the surface to the center of the fiber and thus decrease the apparent value of λ_T. Taking a value of access resistance of 100 ohm-cm^2 in recalculating potential profiles and applying the results to the experimental results of Adrian et al. (1969b), Peachey and Adrian (1972) obtained a λ_T of 85 μ for a fiber of 40 μ radius and an input conductance for the T-system which is about one-half that measured for the whole fiber. This is in much more satisfactory agreement with the results of Eisenberg and Gage (1969).

The most plausible explanation of the second difficulty, the marginal ability of an action potential waveform to cause contraction of axial myofibrils, is that normally there is a sodium conductance change in the T-tubule membranes when depolarized to near the contraction threshold. This proposed conductance change would need to have been eliminated by TTX in the voltage-clamp experiments of Adrian et al. (1969a,b) since it was in these experiments that no safety factor was observed. In some subsequent experiments (Costantin, 1970) strong support was obtained for the idea that the T-system membranes increase their conductance for sodium ions when depolarized and that the current carried by sodium ions into the fiber from the tubules aids in the spread of depolarization along the tubules. These experiments were similar to the earlier ones, except that TTX was not used. Action potentials were prevented from propagating along the fiber by lowering the external sodium concentration to about one-half its normal value. When this was done and the fiber surface was depolarized to the contraction threshold in a voltage clamp, a common observation was a spread of contraction all the way to the center of the fiber for a surface depolarization only one millivolt more positive than that needed to cause contraction of myofibrils near the fiber surface. In some cases the center myofibrils

were found to contract even when the surface ones did not. This was in striking contrast to the span of several millivolts found in the earlier experiments with TTX (Adrian et al., 1969b) and suggested an active conductance change in the tubular membranes that could be eliminated with TTX. The conclusion that this involved a sodium conductance change was supported by the observation that a further decrease in external sodium concentration eliminated this effect, presumably by removing sodium ions for carrying inward tubular current.

At this point in time, it seemed clear that the T-system of frog skeletal muscle was not a passive cable network, as assumed earlier and as found in experiments using TTX, but rather that it was an active membrane system capable of carrying its own propagated response based at least partly on a sodium activation mechanism.

An immediate problem became the construction of a theoretical model of an active cable network with the geometry of the T-system. The problem was that the cable equations not only had to mimic the T-system geometry, but they needed to have nonlinear coefficients representing changes in the sodium conductance in the tubular membranes. Adrian and I (Peachey and Adrian, 1972) attacked this problem using an approximation method for representing the cylindrical distribution of the T-system network. The radial distribution of T-tubules was represented as 16 elements in an equivalent circuit, each element scaled in relation to the content of tubules in each of 16 concentric shells of equal thickness from the surface to the center of the fiber (Fig. 3). Adjacent shells were linked together through resistances representing the radial resistance of the T-tubules. A numerical approximation method was used to calculate changes in potential with time in each shell.

After showing that the mathematical method gave sufficient spatial and temporal accuracy as used and that it accurately duplicated the analytic results for a passive T-system, we introduced a simple form of sodium activation and inactivation into the T-system of this model. The results were in good agreement with the experimental results of Costantin (1970). We could show how a small increase in surface potential above that needed to activate peripheral myofibrils could activate axial myofibrils; we also could show an inverted potential profile with the central T-tubules more depolarized than those near the fiber surface (in a voltage-clamp situation). Figures 4 and 5 show results of a computation for a voltage clamp with 3 msec rectangular pulses.

Rather than show more results of these early computations, I would like to show some results

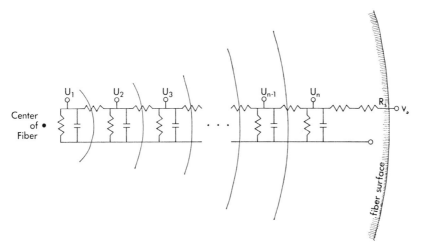

Figure 3. Electrical equivalent circuit used for computations of potential spread into a muscle fiber along the T-system. The T-tubules in each of 16 concentric shells of equal thickness are represented by a resistor and capacitor in parallel. Each such electrical unit is connected to adjacent units through a resistor representing the radial resistance of T-tubules from the center of one shell to the center of the next shell. The outermost shell is connected to the fiber exterior through an access resistance, R_s. The external potential is V_s. The potential in each shell is U_i, where i is the shell number. (From Peachey and Adrian, 1972.)

of a more detailed analysis of propagated action potentials that Richard Adrian and I have been working on for the last six months. The mathematical model used is an action potential propagating along the surface of a muscle fiber in the longitudinal direction and an active T-system attached to the fiber surface through an access resistance, usually taken to be 150 ohm-cm². Both the fiber surface and the T-system have conductance systems described by the Hodgkin-Huxley equations as applied to the muscle cell by Adrian et al. (1970). We have used the same values for surface membrane parameters and kinetic constants as used by Adrian et al. with three changes. First, we adjusted the six rate constants (α, β) to 20° using a Q_{10} of 2.5. Second, we increased $\bar{\alpha}_m$ by 15%, and third, we changed the potassium equilibrium potential from -70 mV to

Figure 4. Calculated potential profiles for imposed square waves at the surface of a fiber and for the resulting potential change across the T-system membranes at the center of the fiber. Calculations were done using the simple numerical method described briefly in the text and included sodium activation and inactivation in the T-system. Two surface square waves of 3 msec duration are shown, one 1 mV larger than the other. The smaller pulse is slightly sub-threshold, and the center of the fiber depolarizes only slightly. The larger surface pulse activates the T-system, which depolarizes to a level considerably above the surface depolarization. For further details, see Peachey and Adrian (1972), from which this figure is reproduced.

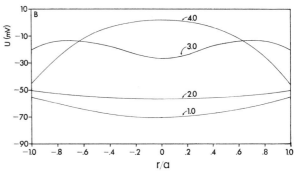

Figure 5. Potential profiles across the diameter of a muscle fiber for the larger pulse in Fig. 4. The fiber radius is a, and the distance from the center of the fiber is r. The numbers against the curves are times in msec after the beginning of the surface square pulse, which is 3 msec long. The inward spread of a wave of active depolarization is seen at 3 msec and an inverted potential profile is seen at 4 msec, with the center of the fiber more depolarized than the surface. Reproduced from Peachey and Adrian (1972).

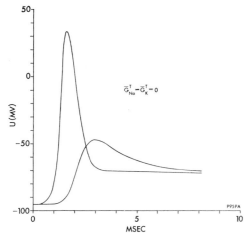

Figure 6. Calculation of a propagated surface action potential and the potential across the T-system at the center of the fiber for the case where the T-system membranes are passive. The surface action potential peaks earlier and much higher than does the central T-system. The surface action potential has a reasonable shape, comparing it to action potentials recorded from living fibers, but the central T-system depolarization is very small and probably would be inadequate for the activation of contraction of central fibrils. (Details of calculations shown in this and all subsequent figures will be published later.)

-85 mV. The same parameters were used for the T-system membranes in this model, except that we reduced the specific leak conductance and the saturating specific sodium and potassium conductances of the T-system membranes to obtain a potential change at the center of the fiber of a reasonable magnitude. Other parameters relating to the T-system were the same as used by Adrian et al. (1969a).

If the T-system sodium and potassium conductances are reduced to zero so that the T-system membranes have only a constant leak conductance, then the T-system acts as a passive cable. The results of a calculation using these conditions are shown in Fig. 6. This is similar to Fig. 2B in Adrian et al. (1969b), which was obtained from the solution to the differential equation for a passive T-system using slightly different electrical constants than used here. The two calculations agree in showing that a passive T-system carries only a very attenuated signal to the center of a fiber with an action potential on its surface. Clearly some activity is required in the T-system if the central depolarization is to activate contraction as well as does the surface action potential. Our next goal, then, was to find how much activity needed to be introduced into the T-system in the model to get a reasonable potential change at the center of the fiber. Our criterion for success was that the potential-time area above -30 mV at the center of the fiber be similar to that at the surface.

Activity in the T-system is introduced into our model by setting the specific saturating sodium and potassium conductances to values greater than zero. Figure 7 shows results of calculations using three values for specific saturating tubular sodium conductance. These represent scaling the surface value down by 16, 20, and 32 times for the T-system. In all three cases, the surface potassium and leak conductances are scaled down by 33 and 36 times respectively for the T-system. If we use the Adrian et al. (1969a) criterion of the area above -30 mV as important for contraction activation, then the intermediate value for tubular sodium conductance gives a central activation closest to that of the surface. The surface area is 39 or 40 mV-msec in each case, and the three different tubular sodium conductances give central areas of 16, 39, and 49 mV-msec. If we want the central value to be the same as the surface value, then the best choice for specific tubular saturating sodium conductance is 9 mmho/cm² tubule, a sodium channel density one-twentieth of that on the fiber surface. This result is in good agreement with the earlier results of Adrian et al. (1969a) in that 40 mV-msec at 20°C is equivalent to 120 mV-msec at 4°C for $Q_{10} = 2$.

We have included an access resistance of 150 ohm-cm² in these calculations. The need for an access resistance can be seen by the result in Fig. 8, which shows two computations using identical parameters except that in one $R_s = 0$, and in the other $R_s = 150$ ohm-cm². The latter is the same as one of the runs in Fig. 7. Reducing R_s to zero does

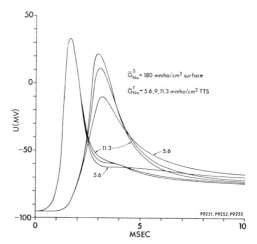

Figure 7. Calculated propagated action potentials as in Fig. 6, except that the T-system was made active by raising the saturating tubular potassium conductance to 1.27 mmho/cm² tubule, a value 33 times less than that on the surface, and by raising the saturating tubular sodium conductance, G_{Na}^T to three different values given in the figure.

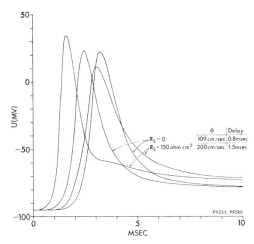

Figure 8. Two calculated propagated action potentials differing in the value of the access resistance, R_s, used. Without an access resistance, the surface action potential is slowed, as seen by its propagation velocity, θ. The delay between peak surface potential and the peak potential at the center of the fiber is reduced when $R_s = 0$.

not adversely affect propagation to the center of the fiber, and in fact it even reduces the delay. However, the surface action potential is adversely affected, having its velocity reduced from 200 cm/sec to only 109 cm/sec. Without R_s, the T-system apparently puts too much of a load on the surface action potential.

Next, I would like briefly to point out an effect of an active T-system on the after-potential at the surface. Figure 9 shows computations using our standard parameters for fiber radii of 50 μ and 60 μ. An interesting feature of the surface action potential of the larger fiber is a hump in the positive direction from about 3–5 msec after the

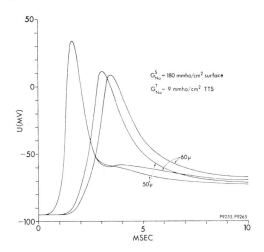

Figure 9. Two calculated propagated action potentials using the same specific conductances for both surface and T-system membranes, but for two different fiber radii, 50 and 60 μ.

start of the action potential. This hump was observed experimentally by Persson (1963), who presented evidence that it represented an increase in membrane conductance and suggested that a small increase in sodium permeability during the early after-potential could explain the hump he saw. It is also of interest that Gage and Eisenberg (1969), studying fibers treated with hypertonic glycerol, noted that the only part of the action potential appreciably changed by the treatment was the early after-potential. These two results suggest that the T-system contributes to the early after-potential and that sodium conductance changes are involved, and our model clearly shows how this comes about.

According to our model, the hump on the early after-potential, and indeed most of the entire early after-potential, is due to the delayed repolarization of the T-system after the surface action potential is largely complete. Positive current is flowing into the T-tubules from the external medium to supply current for the still open sodium channels in the T-system and to recharge the tubular capacity as the T-system repolarizes. This is happening at a time when the surface potassium channels, which are supplying the path for this current through the surface membrane, are closing. The result is a prolonged depolarization of the surface. When this happens after the surface potential has repolarized to a considerable extent, the result is a positive-going hump on the after-potential. When it happens earlier, it appears as a slowly falling repolarization phase of the surface action potential. Perhaps the most important aspect of this observation is that it gives us a method for obtaining with a microelectrode a record that reveals the activity of the T-system. This may be a valuable tool in future work on excitation-contraction coupling.

Finally, I would like to show you the overall time and radial dependence of potential in a muscle fiber as calculated in our current model. Figure 10 shows a three-dimensional plot of the distribution of potential across the diameter of a fiber from surface to surface as a function of time for a fiber with a passive T-system. The small amount of depolarization of the fiber center is quite apparent. Figure 11 shows a similar plot for a fiber with an active T-system. This is the same computation as the smaller fiber in Fig. 9, also appearing in Figs. 7 and 8. The activity of the fiber interior is clear from the large, delayed hump in T-system depolarization.

In summary, we now believe that the T-system propagates an active potential change into the interior of the muscle fiber. Inclusion of a Hodgkin-Huxley type conductance system into the membranes of the T-system leads to computed results

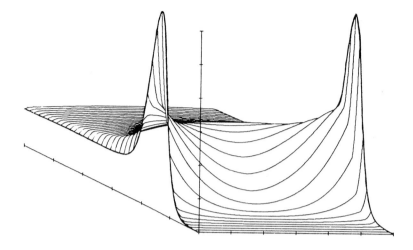

Figure 10. Three-dimensional plot of the potential changes from surface (*left*) to center (*center*) to surface (*right*) of a fiber with a passive T-system during a propagated action potential. Time extends from front to rear, for 8 msec. Internal potential is positive upward. This is the same computation as shown in Fig. 6.

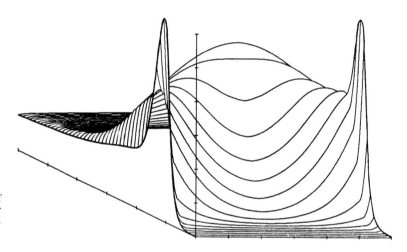

Figure 11. Three-dimensional plot of the computation for the smaller fiber in Fig. 9, also appearing in Figs. 7 and 8. Total time in this plot is 10 msec.

which describe a reasonable pattern of potential changes at the center of the fiber. To achieve these results it is necessary to scale down the maximum specific sodium and potassium conductances of the surface membrane before introducing them into the T-system. It also seems necessary to insulate partially the T-system from the surface membrane with an access resistance. If this is not done, the surface action potential propagation velocity is excessively slowed. Much of the early after-potential in muscle, including a frequently observed hump on the after-potential, is due to the T-system.

Acknowledgments

This work was supported by a grant from the National Science Foundation (GB-6975x). Computations were done at the University of Pennsylvania Medical School Computing Facility, supported by the National Institutes of Health (RR-15). High voltage electron micrographs were made at U.S. Steel Corporation under a contract from the N.I.H. (70-4136).

References

ADRIAN, R. H., W. K. CHANDLER, and A. L. HODGKIN. 1969a. The kinetics of mechanical activation in frog muscle. *J. Physiol.* **204**: 207.

———. 1970. Voltage clamp experiments in striated muscle fibres. *J. Physiol.* **208**: 607.

ADRIAN, R. H., L. L. COSTANTIN, and L. D. PEACHEY. 1969b. Radial spread of contraction in frog muscle fibres. *J. Physiol.* **204**: 231.

COSTANTIN, L. L. 1970. The role of sodium current in the radial spread of contraction in frog muscle fibers. *J. Gen. Physiol.* **55**: 703.

EISENBERG, R. S. and P. W. GAGE. 1969. Ionic conductances of the surface and transverse tubular membranes of frog sartorius fibers. *J. Gen. Physiol.* **53**: 279.

FALK, G. and P. FATT. 1964. Linear electrical properties of striated muscle fibres observed with intracellular electrodes. *Proc. Roy. Soc. (London)* B **160**: 69.

GAGE, P. W. and R. S. EISENBERG. 1969. Action potentials, afterpotentials, and excitation-contraction coupling in frog sartorius fibers without transverse tubules. *J. Gen. Physiol.* **53**: 298.

HUXLEY, A. F. 1971. The activation of striated muscle and its mechanical response. *Proc. Roy. Soc. (London)* B **178**: 1.

HUXLEY, A. F. and R. E. TAYLOR. 1958. Local activation of striated muscle fibres *J. Physiol.* **144:** 426.

KARNOVSKY, M. J. 1965. Vesicular transport of exogenous peroxidase across capillary endothelium into the T-system of muscle. *J. Cell Biol.* **27:** 49A.

PEACHEY, L. D. 1965. The sarcoplasmic reticulum and transverse tubules of the frog's sartorius. *J. Cell Biol.* **25:** 209.

PEACHEY, L. D. and R. H. ADRIAN. 1972. Electrical properties of the transverse tubular system, p. 1. In *Muscle structure and function*, ed. G. H. Bourne, vol. 3, 2nd ed. Academic Press, New York.

PEACHEY, L. D. and R. F. SCHILD. 1968. The distribution of the T-system along the sarcomeres of frog and toad sartorius muscles. *J. Physiol.* **194:** 249.

PERSSON, A. 1963. The negative after-potential of frog skeletal muscle fibres. *Acta Physiol. Scand.* **58:** suppl. 205.

SCHNEIDER, M. F. 1970. Linear electrical properties of the transverse tubules and surface membrane of skeletal muscle fibers. *J. Gen. Physiol.* **56:** 640.

Calcium and the Control of Contraction and Relaxation in a Molluscan Catch Muscle

Betty M. Twarog and Yojiro Muneoka

Department of Biology, Tufts University, Medford, Mass. 02155

The anterior byssal retractor muscle (ABRM) is a molluscan smooth muscle that when appropriately stimulated can be made to perform two distinct types of contraction: (1) a phasic contraction that relaxes upon cessation of stimulation, and (2) a contraction known as "catch" that continues long after stimulation has ended. The latter type is marked by an energy output much smaller than that during phasic contraction (Nauss and Davies, 1969; Baguet and Gillis, 1968); during catch there are neither signs of active state as measured by quick-release experiments (Jewell, 1959; Johnson and Twarog, 1960) nor electrical activity as measured by intracellular or extracellular experiments (Twarog, 1967a).

Catch can be relaxed by neural stimulation or by relaxing agents, 5-hydroxytryptamine (5-HT) or dopamine. Either or both of these are probably mediators released by the catch-relaxing nerves (Muneoka and Twarog, 1972). Relaxing nerves are in no sense inhibitory; when stimulated they selectively "shut off" catch tension and increase active tension. Figure 1 shows the increased phasic response to repetitive stimulation following stimulation of the relaxing nerves. Figure 2 shows that active tension, in response to ACh, is increased even though relaxation is speeded (York and Twarog, 1972).

The asterisk on the tension diagram of Fig. 3 designates the point at which catch begins and indicates the occurrence of an unknown process that controls the transition from active state to catch. The cross-bridges between the interacting proteins become fixed and they no longer turn over and use energy, yet tension levels remain high. The tension substrate has been a continuing source of disputation: Lowy et al. (1964) and Nauss and Davies (1966) argue that catch tension is maintained by interaction between actin and myosin, whereas Johnson (1962), Rüegg (1971), and Heumann and Zebe (1968) insist that the interaction is among paramyosin filaments and is independent of the actin-myosin contractile system. However the accumulation of evidence favoring the actin-myosin system (Szent-Györgyi et al., 1971) holds promise of a conclusive end to the argument in the near future.

In any case it is now fairly safe to assume that phasic contractions and perhaps catch are closely tied to increases in the level of intracellular free Ca^{++}. It has been speculated that in ABRM the level of intracellular free Ca^{++} is rapidly reduced in phasic contraction but remains high during catch (Twarog, 1966). As a muscle in catch is not in active state, prolonged contraction by virtue of continuing high intracellular free Ca^{++} levels requires the existence of an additional process that shuts off active state and stabilizes the bonds between the interacting proteins. In this regard it is worthy of note that the calcium binding regulatory protein of molluscan muscle is bound to the myosin (Kendrick-Jones et al., 1970). And in catch muscles, the myosin molecules are arranged on the surface of thick paramyosin filaments.

Alternatively it is conceivable that during catch ATP is sequestered, thereby producing a state of rigor. The fact that the total ATP level remains constant during catch (Rüegg and Strassner, 1963) could be explained by the ATP being in such a state that it is unavailable to the contractile mechanism.

In most muscles Ca^{++} is taken up into the sarcoplasmic reticulum (SR), thereby causing relaxation. In ABRM, whether or not catch is partially a function of continued high levels of intracellular free calcium, relaxation could be accomplished in several ways: (1) by unstabilizing the bonds among interacting proteins; (2) by

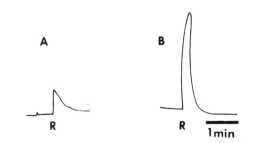

Figure 1. Increased tension and increased rate of relaxation of phasic contraction after stimulation of relaxing nerve endings within the muscle bundle, using field electrodes (see Methods). *A, B:* Tension response to 5 sec of repetitive stimuli at R (40 V, 1 msec, 10/sec). Between *A* and *B* repetitive stimulation for 1 min (80 V, 1 msec, 10/sec) caused accumulation of relaxing transmitter.

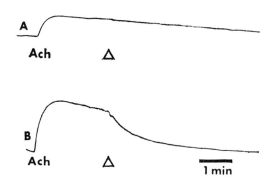

Figure 2. Increased tension and increased rate of relaxation of catch contraction after stimulation of relaxing nerves. *A, B:* Tension response to 10^{-5} M acetylcholine (ACh), washed at ($\triangle$). Between *A* and *B* repetitive stimulation for 1 min (80 V, 1 msec, 10/sec) caused accumulation of relaxing transmitter.

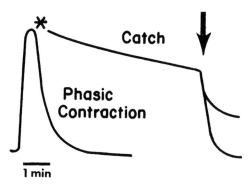

Figure 3. "Catch" vs. phasic contraction. Axes: tension, time. At * actin-myosin-paramyosin interaction modified, active state ceases, catch begins. At ↓ neural stimulation or relaxing agents cause partial or complete relaxation.

reducing the calcium affinity of regulatory proteins; or (3) by restoring sequestered ATP.

We have chosen as a working hypothesis that a high level of intracellular free Ca^{++} is at least in part responsible for maintaining catch and that a reduction in intracellular free Ca^{++} relaxes it; though conclusive evidence is not yet available, many observations support this. Pursuant to this, three aspects of current research on the molluscan catch muscle will be dealt with: (1) muscle structure; (2) activation of contraction by Ca^{++} (transition from active state to catch is not considered); and (3) recovery and relaxation.

Muscle Structure

The ABRM is made up of spindle-shaped fibers; each is approximately 5 μ in diameter and from 1–2 mm in length, has a single nucleus near its center, and may branch at one end. Formerly it was thought that the individual fibers ran continuously from one end of the muscle to the other (Fletcher, 1937).

Figure 4 is a longitudinal section showing the 400–600 Å thick filaments, with characteristic paramyosin periodicity, and thin filaments running into dense bodies. Vesicles that may correspond to elements of the SR are closely apposed to the membrane. There is connective tissue between the muscle fibers, and hemidesmosomes arranged on the peripheries of the muscle fibers are attached to connective tissue fibers (Twarog, 1967a; Heumann and Zebe, 1968).

When a muscle is exposed to a hypertonic fixing solution, the connective tissue and muscle fibers recede from each other except in the region of increased density, where the hemidesmosomes remain attached to the connective tissue fibers (Twarog, Dewey and Hidaka, 1973). Thus the muscle fibers neither run the length of the muscle, nor are they attached directly to each other; rather they are attached at the hemidesmosomes to connective tissue, which is in turn connected to other muscle fibers. Consideration of this arrangement would seem vital to any fruitful reassessment of conclusions based on previous studies of the mechanical properties of this muscle (Jewell, 1959; Rüegg, 1971).

Permanganate-fixed cross sections of ABRM show many regions where the muscle cells are in close apposition to one another. Figure 5 shows that at each of these regions there is a typical nexal junction similar to that shown to be responsible for the electrical connectivity among mammalian smooth muscle cells (Dewey and Barr, 1968). Since the space constant of ABRM (2.5 mm) is greater than the length of any individual cell (Twarog and Hidaka, 1971a), it is altogether likely that nexal junctions account for the electrical continuity of these cells too.

Activation of Contraction by Ca^{++}

Methods

We used three types of preparations in studying responses. (1) "Normally permeable;" that is, (a) whole muscle and attached nerve supply (Twarog, 1960) and (b) thin (300–700 μ diameter) muscle bundles (Twarog, 1967b; Muneoka and Mizonishi, 1969). In these preparations diffusion delays were minimal. Electrical stimuli were applied by field electrodes (Hidaka et al., 1967). On the whole, repetitive 2-msec pulses stimulated nerve branches; DC pulses excited the muscle directly. (2) "Highly permeable." A 300–700 μ bundle, depolarized in 0 Ca^{++} 0.54 M KCl plus 3 mM EGTA. This preparation is not electrically excitable but can be activated by Ca^{++} (pCa 3) or acetylcholine (ACh) (if the external Ca^{++} is equal to or above 1.0 mM);

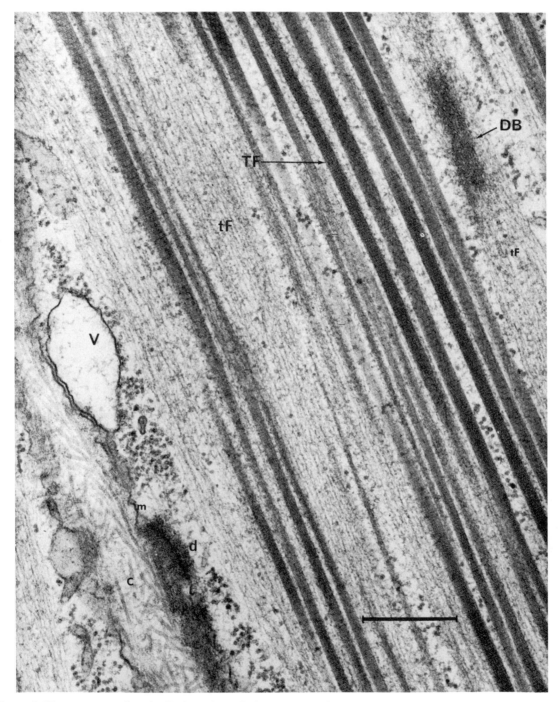

Figure 4. Ultrastructure: Longitudinal section of glutaraldehyde-fixed ABRM. *TF*, Thick (paramyosin) filament; *tF*, thin filaments which run into dense body, *DB; V*, vesicle (presumably SR) closely apposed to *m*, muscle membrane; *d*, hemidesmosome, connecting muscle membrane and *C*, connective tissue at region of increased electron density. Calibration: 0.5 *μ*.

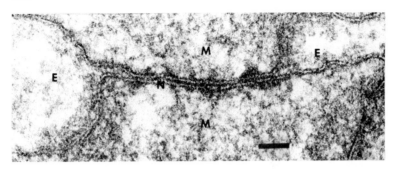

Figure 5. Nexal junction: Cross section of permanganate-fixed ABRM. *M*, Muscle fiber; *E*, extracellular space; *N*, nexus. Calibration: 530 Å. Electron micrograph by Prof. Maynard Dewey.

normal permeability can be restored by washing. (3) "Chemically skinned" (Winegrad, 1971; Julian, 1971). This preparation can be activated by Ca^{++} (pCa 7); the membrane is damaged, and normal permeability cannot be restored.

Activation

Table 1 shows that some, but not all, excitatory stimuli induce catch in normally permeable muscle bundles. In our experiments activation of either the highly permeable or chemically skinned muscle brought about active state; i.e., the muscle redeveloped tension when released during the course of an isometric contraction. Other authors have reported inducing a "catch-like" state with glycerinated ABRM (Johnson et al., 1959; Leenders 1966; Rüegg, 1971; Baguet and Sleewagen, 1971), but whether or not genuine catch was achieved in these preparations is still in doubt.

Figure 6 shows that the contractile elements are indeed activated by Ca^{++}. Mary Nichols, in our laboratory, found that full activation was brought about in the chemically skinned fiber bundle at pCa 6, whereas previous investigators using glycerinated ABRM achieved full activation only at pCa 5. In contrast activation in the highly permeable muscle is complete only at pCa 2.

Caffeine activation. As originally reported by Rüegg et al. (1963), contracture in ABRM can be activated by caffeine. Muneoka and Mizonishi (1969) described in detail the effect of caffeine on

ABRM and its dependence on Ca^{++}. Studies have shown that caffeine applied at 60 min intervals in artificial seawater (ASW) produces a very regular control response.

In our experiments using 0 Ca^{++} ASW, the initial test response to caffeine was as great as the control; this was so even when the bundle was perfused for several hours prior to testing. Further applications of caffeine, however, produced little or no response (Fig. 7); but after ASW containing 10 mM Ca^{++} was reintroduced, the caffeine response was fully restored. These experiments support our working hypothesis that caffeine releases Ca^{++} from a site that cannot be wholly depleted even by prolonged soaking in a calcium-free medium; and in order for the site to be replenished, Ca^{++} must be available in the external medium.

Figure 8 shows tension as percent of control response in 10 mM Ca^{++} ASW. The time axis indicates how long ABRM was soaked in high Ca^{++} solution before testing. In each instance the strength of the caffeine contracture increased with greater concentrations of Ca^{++}. Our evidence suggests that the effect of high Ca^{++} is a function

Table 1. Effects of Stimuli on "Normally Permeable" Muscle Bundles

	Catch	Stimulation of relaxing nerves
5 msec, repetitive pulses	No	Yes
DC pulses (10 sec or longer)		Yes
Acetylcholine		
in ASW	Yes	No
in 0 Na$^+$ ASW	No	No
High K$^+$ ASW	No	Yes
Caffeine	Yes	No

Production of catch as related to the possibility that relaxing nerves may be nonselectively stimulated. Note that depolarization by DC evokes catch whereas depolarization by high K$^+$ evokes only phasic contractions.

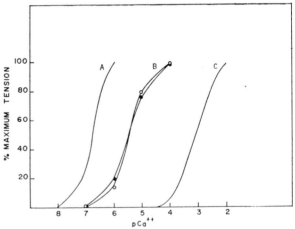

Figure 6. Activation by Ca^{++} in ABRM: *A*, chemically "skinned," treated with 120 mM KCl-EDTA, tested with Ca-EGTA (Nichols, unpublished); *B*, glycerinated, tested with Ca-EGTA; (●) 130 mM KCl (Leenders, 1966), (○) 50 mM KCl (Schadler, 1967); *C*, "highly permeable," treated with 540 mM KCl-EDTA, tested with 0.5 M KCl + CaCl$_2$ (Muneoka, unpublished).

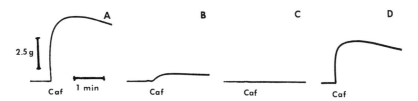

Figure 7. Caffeine contracture in 0 Ca^{++}. 10 mM caffeine applied at 60 min intervals at *Caf*. *A*, First response to caffeine after 60 min in 0 Ca^{++} ASW. *B*, Soaking in 0 Ca^{++} ASW continued, second caffeine response reduced. *C*, 0 Ca^{++} ASW continued, no response to third caffeine test. *D*, After 60 min in ASW (10 mM Ca^{++}), response restored.

of time: When Ca^{++} is increased at the same time that caffeine is applied, there is little or no increase in tension; but if Ca^{++} is increased 10–15 min before caffeine testing, the increase in response is almost maximal. This strongly implies that whatever the Ca^{++} concentration, the site at which caffeine releases Ca^{++} requires some critical length of time to be replenished; a further implication is that the site is an internal one, very probably the SR.

Figure 9 shows responses to caffeine applications given at 60 min intervals in ASW. During the final 60 min interval ABRM was exposed for 5 min to high K$^+$ ASW (containing 10 mM Ca^{++} as in the control ASW). After the period in high K$^+$, 10 mM Ca^{++}, caffeine response was greatly increased (Fig. 9D); by continuing brief exposure to K$^+$, 10 mM Ca^{++} during the interval, the increase was maintained. It was possible to reduce tension to the control level by excluding depolarization by K$^+$ from the interval in 10 mM Ca^{++}. This demonstrates that the replenishing of Ca^{++} that is available for release by caffeine is greatly speeded if the muscle is exposed to high K$^+$ ASW containing Ca^{++}.

Hagiwara and Nagai (1970) have shown that net influx of ^{45}Ca^{++} increases during exposure to high K$^+$. Since the site of Ca^{++} release can be replenished rapidly if net Ca^{++} influx is increased by exposure to high K$^+$, the evidence of these experiments lends weight to our hypothesis that Ca^{++} is stored internally and probably in the SR.

Figure 10 (Muneoka, 1969) shows that procaine, which is thought to block caffeine release of Ca^{++} from the SR in striated muscle (Feinstein, 1963; Weber and Herz, 1968), reduces contracture tension in the ABRM when applied simultaneously with caffeine. When the procaine is washed, the subsequent caffeine response is increased, an indication that procaine has reduced Ca^{++} release in ABRM much the same way as in striated muscle (Chiarandini et al., 1970). The figure further shows that once a caffeine contracture has been established, procaine does not relax it. This too suggests that once caffeine has released Ca^{++}, procaine has no further effect on tension.

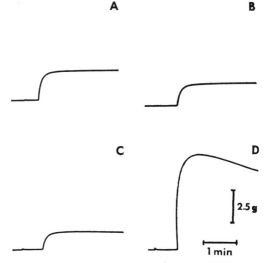

Figure 9. The effect of depolarization in the presence of Ca^{++} on subsequent caffeine contracture. 10 mM caffeine applied at 60 min intervals in ASW. Between *A*, *B*, and *C*, 60 min rest interval in 10 mM Ca^{++} ASW; between *C* and *D*, 60 min interval in 10 mM Ca^{++} ASW, including a 5 min exposure to high K$^+$, 10 mM Ca^{++} ASW, 30 min before caffeine test.

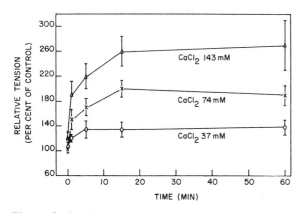

Figure 8. Caffeine contracture tension as a function of time of exposure to high Ca^{++} during rest interval. 10 mM caffeine applied after 60 min rest intervals. Control tension taken after first interval in ASW. At various times before the end of the rest interval, high Ca^{++} ASW was substituted for ASW. Total time of exposure to high Ca^{++} shown on abscissa. Each point: Mean of 10 experiments ±SD.

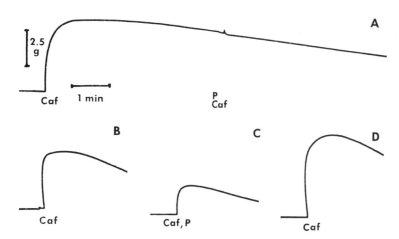

Figure 11 summarizes caffeine action on ABRM. Calcium flows across the membrane in both directions. At the upper right is a representation of a nexal junction. In the center is a vesicle of the SR. At bottom a paramyosin filament is diagramed with myosin cross-bridges and associated Ca^{++}-binding protein. The evidence thus far presented strongly suggests that Ca^{++} is slowly taken into

and released from an internal site, which, as stated, is here assumed to be the SR. Electron micrographs (see Fig. 4) reveal vesicles in close apposition to the membrane that could well be elements of the SR. Similar structures in oyster muscle have been identified as possible SR by Hanson and Lowy (1961). Stössel and Zebe (1968) isolated a fraction from ABRM which accumulated calcium with accompanying breakdown of ATP, and Heumann (1969) observed that calcium oxalate could be precipitated in vesicles of ABRM, solid evidence that the SR does indeed take up calcium. And Muneoka (1969 and unpublished) found that Co^{++}, Ni^{++}, Mn^{++}, and La^{3+} applied externally to a normally polarized muscle increase the effectiveness of caffeine release of Ca^{++}, another parallel between vertebrate striated muscle and ABRM.

Neural excitation: DC. Stimulation of ABRM by 500 msec current pulses produces depolarization, and at a critical level an action potential (spike) is brought about. The spike amplitude is proportional to the Ca^{++} concentration in the external medium (Twarog and Hidaka, 1971b). Apparently the current is carried by the influx of calcium ions (Fig. 12). The slope of the increase in spike amplitude approaches 29 mV for a tenfold change in Ca^{++} (cf. Fig. 24).

Figure 13 shows stimulation of ABRM in ASW, first by DC pulses and then by repeated 2-msec pulses. The DC stimulation results in catch, the

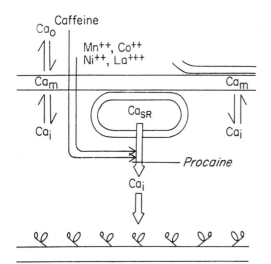

Figure 11. Activation by caffeine (see text): Ca_o External Ca^{++}; Ca_i, intracellular free Ca^{++}; Ca_m, Ca^{++} associated with muscle membrane; Ca_{SR}, Ca^{++} available for release by caffeine from an internal site, presumably SR. Caffeine release of Ca^{++} is potentiated by Mn^{++}, Co^{++}, Ni^{++}, La^{3+} and blocked by procaine.

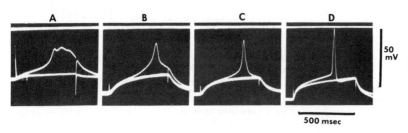

Figure 12. Spike amplitude as a function of Ca^{++} concentration. Upper trace: zero potential; lower traces: membrane potential. Electrotonic potentials and spike evoked by 500 msec current pulses in various concentrations of Ca^{++}. *A*, 2 mM Ca^{++}; *B*, 5 mM Ca^{++}; *C*, 10 mM Ca^{++}; *D*, 50 mM Ca^{++}. Note depolarization of resting membrane in 2 mM Ca^{++}, hyperpolarization in 50 mM Ca^{++}.

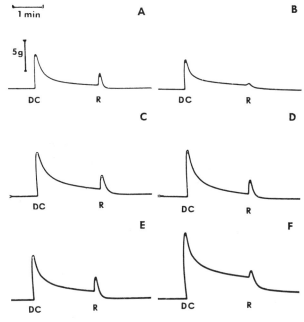

Figure 13. Potentiation of DC contraction by brief depolarization in high Ca⁺⁺. Stimulation in ASW by cathodal direct current (*DC*) (40 V, 5 sec) and repetitive pulses (*R*) (40 V, 2 msec, 10/sec for 5 sec). 20 min intervals between each test. Between *B* and *C*, *D* and *E*, exposed to 10 mM Ca⁺⁺ ASW; between *A* and *B*, 10 mM Ca⁺⁺ ASW and a 5 min exposure to high K⁺, 10 mM Ca⁺⁺ ASW; between *C* and *D*, 10 mM Ca⁺⁺ ASW and a 5 min exposure to 70 mM Ca⁺⁺ ASW; between *E* and *F*, 10 mM Ca⁺⁺ ASW and a 5 min exposure to high K⁺, 70 mM Ca⁺⁺ ASW.

repetitive stimulation in phasic contraction, and activation of the nerves that relax catch. The muscle was stimulated every 20 min in ASW, but during the 20 min interval between stimuli it was exposed to ASW containing varying Ca⁺⁺ concentrations. During some rest intervals the influx of Ca⁺⁺ was increased by depolarization by raising the

K⁺ concentration of the solution for 5 min. After a rest interval which included 5 min in 70 mM Ca⁺⁺ ASW, the response to DC was slightly increased. After exposure for 5 min to high K⁺ in 70 mM Ca⁺⁺ ASW, the response to DC was greatly increased. This suggests that the Ca⁺⁺ that enters the fibers plays a significant role in the response to DC stimulation; in other words, DC releases Ca⁺⁺ from an internal source, quite likely the SR.

In ASW responses to DC stimulation at 20 min intervals (Fig. 14) remain constant over extended periods of time, but producing a caffeine contracture during the 20 min interval somewhat diminishes the response. In 0 Ca⁺⁺ ASW the DC contraction will gradually lessen but is never blocked entirely. Following a caffeine contracture induced in 0 Ca⁺⁺ ASW, however, DC contraction is blocked. That the DC response in ASW persists even after Ca⁺⁺ stores have been depleted by caffeine leads naturally to the conclusion that Ca⁺⁺ from the external medium enters in quantities sufficient for activation. But in 0 Ca⁺⁺ ASW, caffeine liberation of Ca⁺⁺ precludes subsequent response to DC, implying that under such conditions DC response is dependent on internal stores of Ca⁺⁺.

Acetylcholine. Figure 15 summarizes ACh action. Acetylcholine is released by excitatory nerves and acts on receptors; the resultant increased conductance to Na⁺ and Ca⁺⁺ ions produces depolarization, which produces a further increase in Ca⁺⁺ influx. In an excitable muscle, spikes will be generated, and at the level of depolarization produced by the spike, bound Ca⁺⁺ is released, presumably from the SR. Strong depolarization by DC also releases bound Ca⁺⁺.

In 10 mM Ca⁺⁺ ASW, ACh triggers a rapid, strong contraction (Fig. 16A); the tension of this

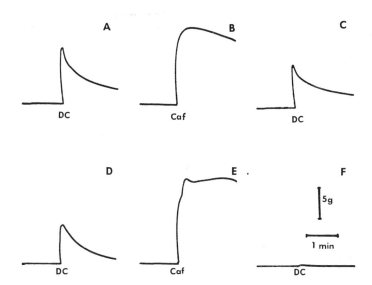

Figure 14. Block of DC contraction following caffeine contracture in 0 Ca⁺⁺. Stimulation by cathodal direct current (*DC*) (50 V, 5 sec) and by 10 mM caffeine (*Caf*). *A, B, C* in ASW; *D, E, F* in 0 Ca⁺⁺ ASW. 10 min interval between *A* and *B*, *D* and *E*; 20 min between *B* and *C*, *E* and *F*.

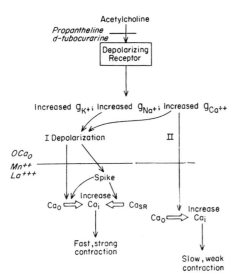

Figure 15. Activation by acetylcholine (see text): g_{K+}, g_{Na+}, g_{Ca++}: K⁺, Na⁺, Ca⁺⁺ conductance, respectively; Ca_o, Ca_a, Ca_{SR} as in Fig. 11. Propantheline and d-tubocurarine block the ACh receptor. The conductance increase to Na⁺ and Ca⁺⁺ evoked by ACh in ASW causes depolarization (I), which augments Ca⁺⁺ influx and may release Ca⁺⁺ from internal sites. In 0 Na⁺, ACh increases Ca⁺⁺ conductance (II) and stimulates Ca⁺⁺ influx without depolarization. Ca⁺⁺ influx is blocked by Ca⁺⁺-deficient ASW, Mn⁺⁺, and La³⁺.

contraction is slightly reduced in high Ca⁺⁺. Depolarization by ACh is blocked in 0 Na⁺ ASW; in this case ACh produces a slow, weak contraction at 10 mM Ca⁺⁺ (Fig. 16C), but the speed and strength of contraction increase as external Ca⁺⁺ is increased. When ACh triggers contraction in ASW

at concentrations of external Ca⁺⁺ of 10 mM or higher, the anesthetic effect of high Ca⁺⁺ on membrane excitability predominates (Shanes, 1958; Muneoka, 1966), and activation is seemingly independent of the concentration of external Ca⁺⁺. In contrast the tension developed by ACh in 0 Na⁺ is strongly influenced by the external concentration of Ca⁺⁺. This suggests that in the latter case all activating Ca⁺⁺ enters through the surface membrane. (Co⁺⁺, Ni⁺⁺, Mn⁺⁺, and La³⁺ reduce the response to ACh in 0 Na⁺ (Muneoka, unpublished), probably by inhibiting Ca⁺⁺ influx.)

Figure 17 (cf. Fig. 8) shows tension as percent of control response to ACh in 0 Na⁺, 10 mM Ca⁺⁺. The time axis indicates how long ABRM was soaked in 0 Na⁺ high Ca⁺⁺ solution before testing. In each instance the strength of the ACh contracture increases with greater concentrations of Ca⁺⁺. It is apparent that increasing the Ca⁺⁺ concentration increases tension with virtually no time lag. This supports the conclusion that Ca⁺⁺ entering through the surface membrane from the external medium activates ACh contraction in 0 Na⁺ (Muneoka, unpublished).

Experiments with La³⁺ have produced results consistent with our explanation of sources of Ca⁺⁺ in activation by high K⁺, by ACh, and by DC (Fig. 18). Exposure to 5 mM La³⁺ completely blocks excitation by high K⁺ and repetitive stimulation, almost completely blocks excitation by ACh in ASW, and reduces the response to DC pulses (DC is not blocked even after extended periods of soaking in La³⁺). In squid axon (van

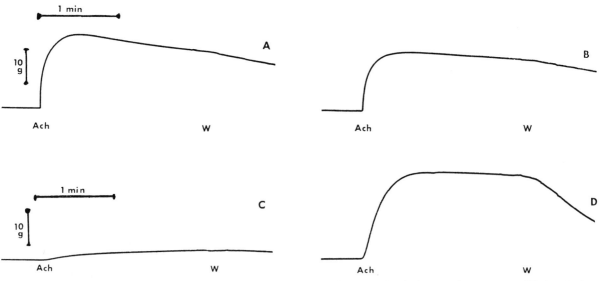

Figure 16. Effect of high Ca⁺⁺ on ACh contraction in ASW (in which ACh depolarizes) and in 0 Na⁺ ASW (in which ACh depolarization is blocked). ACh 10⁻⁴ M applied at 30 min intervals (*ACh*), ACh washed off (*W*) by ASW in *A* and *B*, by 0 Na⁺ ASW in *C, D*. *A*, 10 mM Ca⁺⁺ ASW; *B*, 50 mM Ca⁺⁺ ASW; *C*, 10 mM Ca⁺⁺ 0 Na⁺ ASW; *D*, 50 mM Ca⁺⁺ 0 Na⁺ ASW.

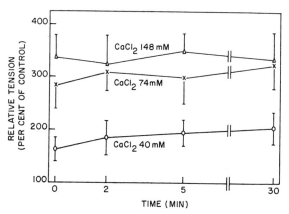

Figure 17. Acetylcholine contraction in 0 Na⁺ ASW as a function of duration of exposure to high Ca⁺⁺ during rest interval. 10^{-5} M ACh applied after 30 min rest interval. Control tension taken after first interval in ASW. At various times before the end of the rest interval, high Ca⁺⁺ ASW was substituted for ASW. Total time of exposure to high Ca⁺⁺ shown on abscissa. Each point shows mean of 10 experiments ±SD.

Breeman and De Weer, 1970) and also in mammalian smooth muscle (van Breeman, 1969), La³⁺ reduces both influx and efflux of Ca⁺⁺. The absence of any response to high K⁺ ASW in 5 mM La³⁺ is fully in accord with the conclusion that all acti-vating Ca⁺⁺ is, in this case, external in origin. Block of the response to repetitive stimulation is more complex and may be at nerve terminals. The persistent response to ACh and DC in 5 mM La³⁺ ASW indicates that a significant portion of the Ca⁺⁺ is released from an internal site. When the muscle is briefly depolarized in the presence of 5 mM La³⁺, the response to ACh is totally blocked (cf. Fig. 26B). Presumably La³⁺ enters the depolarized muscle. If La³⁺ concentration within the cell is sufficiently elevated, release of Ca⁺⁺ from the SR is apparently blocked.

Recovery and Relaxation

Recovery: net gain of Ca⁺⁺? All the foregoing experiments lead to the following scheme of recovery during rest, if it can be assumed that the response to caffeine can serve as a measure of internal Ca⁺⁺ (Fig. 19). Calcium, entering from the external medium, is taken up by the SR. Thus during recovery there is a slow net increase in internal Ca⁺⁺. Divalent cations apparently inhibit the influx of Ca⁺⁺ during recovery, just as they do during depolarization (Muneoka, 1969). Given enough time and enough external Ca⁺⁺, caffeine tension is restored to control level. Worthy of note

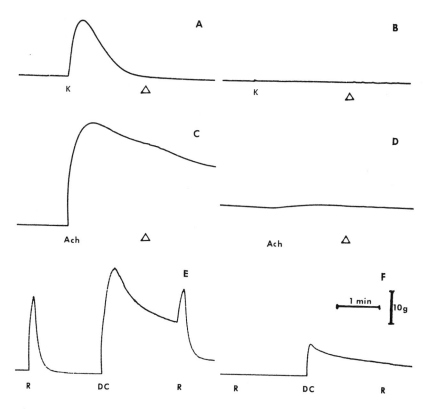

Figure 18. Effect of La³⁺ on tension induced by various stimuli. *A*, *C*, and *E*, control responses to high K⁺ ASW (*K*), 10^{-4} M ACh (*ACh*) and electrical stimulation, respectively. *B*, *D* and *F*, responses to the same stimuli after 25 min in 5 mM La³⁺. (△) Indicates wash in ASW. Repetitive pulses (*R*) (80 V, 2 msec, 10/sec for 5 sec); DC pulse (*DC*) (80 V, 5 sec).

here is that the time required for complete restoration of maximum contracture tension is quite long, more than 3 hr (Muneoka and Mizonishi, 1969), whereas in crustacean muscle, recovery time is around 20 min (Chiarandini et al., 1970). This slow replenishment of internal Ca^{++} stores in ABRM implies a time lag in the uptake of intracellular free calcium by the SR. The studies of Stössel and Zebe (1968) confirm the relatively weak Ca^{++}-accumulating ability of SR fractions from ABRM.

It would appear that ABRM has a unique system of relaxing nerves; and a muscle in which the SR operates so slowly could well be expected to be equipped with a mechanism for speeding the rate of removal of intracellular Ca^{++} ions, possibly by increased Ca^{++} uptake by the SR or possibly by active extrusion of Ca^{++}.

Relaxation: net loss of Ca^{++}? The cardinal focus of our studies continues to be the relaxing system; our hope is that an understanding of its operations will lend insight into the mechanism of control both of activation and of catch.

Figure 20 shows two highly effective catch-relaxing agents, 5-HT and dopamine. They are both found in *Mytilus* ganglia, and at least one of them, 5-HT, is found in the muscle (Muneoka and Twarog, 1972; York and Twarog, 1972). Both agents may be transmitters released by relaxing nerves. Structure-action studies indicate that the darkened parts of the molecules are necessary for relaxing action (Twarog and Cole, 1972).

In Fig. 21 it is seen that 5-HT, unlike procaine, relaxes caffeine contracture, and when applied simultaneously with caffeine it suppresses the contracture. After 5-HT was washed off, the subsequent response to caffeine was equal to the

Serotonin (5-Hydroxytryptamine)

Dopamine

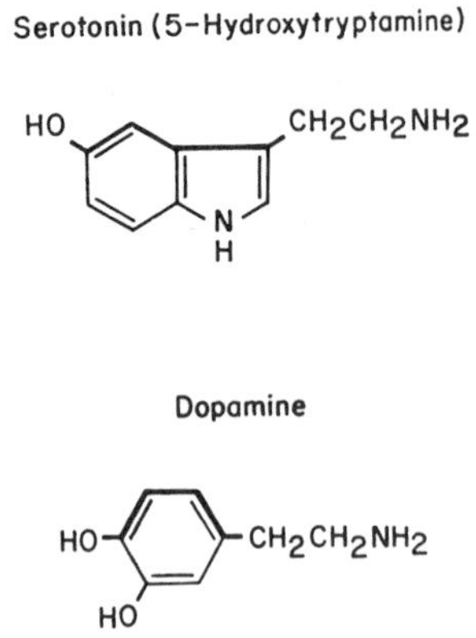

Figure 20. Catch-relaxing agents. The darkened portions of the molecules are required for relaxing action.

control or (in supraximal concentrations of 5-HT) reduced; this is in marked contrast to the increased response to caffeine observed after procaine was washed off. Thus it is unlikely that 5-HT blocks Ca^{++} release as does procaine. Certainly the suppression of Ca^{++} release cannot by itself explain why 5-HT, again unlike procaine, relaxes an already established caffeine contracture. Indeed it is more likely that 5-HT does not affect Ca^{++} release at all but rather reduces the level of intracellular free Ca^{++} by another mechanism, perhaps by speeding up Ca^{++} extrusion or uptake into SR.

The foregoing evidence suggests that 5-HT relaxes catch by reducing intracellular free Ca^{++}. Acceptance of this, however, evokes the further hypothesis that Ca^{++} controls catch tension by a yet unknown mechanism, since during catch the muscle is not in active state. This leads to the question of whether the removal of Ca^{++} from the external medium can relax catch. Figure 22 shows that in a normally polarized muscle, removal of Ca^{++} has no effect on relaxation, whereas 5-HT, applied in the 0 Ca^{++} solution, relaxes.[1] Sugi (1971), however, reported that a 0 Ca^{++}, 0 Mg^{++}, 0.5 mM EGTA solution relaxes catch, a finding supported by the figure. Recordings we made with microelectrodes indicate that in 0 Ca^{++}, 0 Mg^{++}, 3 mM EGTA the membrane is depolarized. After relaxation ASW caused contraction, indicating that the membrane was permeable to Ca^{++}. It may well be that in the depolarized and presumably permeable muscle, low concentrations of external Ca^{++} induce the efflux of Ca^{++}, thus relaxing catch. It is, however, alternatively possible that this solution depolarizes nerve endings and releases the relaxing transmitter;

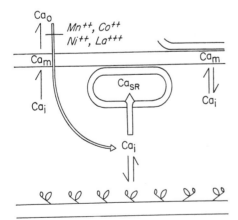

Figure 19. Rest and recovery (see text): Ca_o, Ca_i, and Ca_{SR} as in Fig. 11. Intracellular Ca^{++} is taken up at an internal site, presumably SR. There is a net influx of Ca^{++} from the outside, which is blocked by Mn^{++}, Co^{++}, Ni^{++}, or La^{3+}.

[1] Our observation that the relaxing action of 5-HT on these muscle bundles is not affected by 20–30 min exposures to 0 Ca^{++} (Twarog, 1967b) contradicts the report that in whole muscle preparations in 0 Ca^{++}, 5-HT action is blocked (Leenders, 1967).

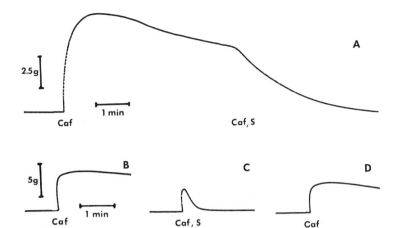

Figure 21. Effect of 5-HT on caffeine contracture. 10 mM caffeine (*Caf*) applied in ASW at 60 min intervals; 10 mM caffeine and 10^{-8} M 5-HT (*Caf, S*) applied as indicated. *A*, 5-HT relaxes caffeine contracture; *B*, control; *C*, 5-HT partially blocks caffeine contracture and increases rate of relaxa· tion; *D*, the first contracture after 5-HT block (after 60 min in ASW) is less than the control.

that is, the action is presynaptic. Since mersalyl (salyrgan), in concentrations sufficient to block the action of relaxing nerves, slightly reduces but does not block the relaxing action of a 0 Ca++, 0 Mg++, 3 mM EGTA solution, a presynaptic action seems quite improbable (Muneoka and Twarog, unpublished).

Some years ago a relationship between intracellular free Ca++ levels and relaxation was suggested, based on studies of the effects of 5-HT on excitability (Twarog, 1966). Detailed pharmacological investigations have shown that 5-HT and dopamine act on the same relaxing receptor (Twarog and Cole, 1972). Mersalyl selectively blocks the relaxing action of 5-HT, bromo-LSD blocks both 5-HT and dopamine, and both agents

block neurally induced relaxation (Muneoka and Twarog, 1972).

Figure 23 represents the dose-response curves for the two relaxing agents. Each causes an increase in excitability, that is, an increase in spike amplitude (Fig. 24), a decrease in threshold, and an increase in membrane conductance (Hidaka et al., 1967; Hidaka, 1969). The excitability increase in the case of each transmitter occurs at a dose level that produces almost total relaxation of catch. Increased excitability in barnacle muscle has been demonstrated to be related to the reduction of intracellular free Ca++ to a critical level (Hagiwara and Naka, 1964). And if relaxing agents reduce intracellular free Ca++ levels in direct proportion to their concentrations, this associated effect on

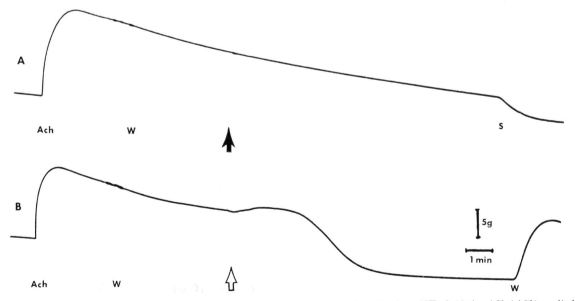

Figure 22. Effect of 0 Ca++ ASW and 0 Ca++, 0 Mg++ ASW on catch and on relaxation by 5-HT. *A*, 10^{-4} M ACh (*ACh*) applied in ASW, washed with ASW (*W*). 0 Ca++ 3 mM EGTA ASW applied at ▲ did not relax catch or block relaxation by 10^{-7} M 5-HT (*S*). *B*, 10^{-4} M ACh (*ACh*) applied in ASW, washed with ASW (*W*). 0 Ca++, 0 Mg++ 3 mM EGTA ASW applied at ⇧ caused contraction and then full relaxation of catch. Restoring 10 mM Ca++ ASW (*W*) produces a contraction.

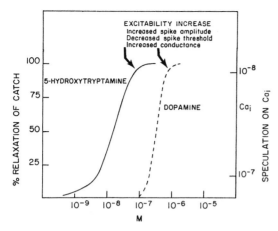

Figure 23. Dose-response curves for dopamine and 5-hydroxytryptamine relaxation of catch. Arrows indicate dose level at which effect on membrane appears with each agent. On the right, according to the speculation in the text, the values of intracellular Ca^{++} when these effects are seen.

excitability is to be expected when Ca^{++} is brought below a critical level.

There is evidence that relaxation involves an active energy consuming process that is blocked at very low temperatures ($-2°C$) (Muneoka and Twarog, unpublished). Applying 5-HT to ABRM Nauss and Davies (1966) noted an increase in arginine phosphate breakdown, and Baguet and Gillis (1964) found an increase in O_2 consumption. There is also some evidence that cyclic-AMP may mediate relaxation (Cole and Twarog, 1972).

Whereas La^{3+} blocks activation (cf. Fig. 18) the relaxing response to 5-HT or dopamine remains normal during long soaking in 5 mM La^{3+} (Fig. 25). On the other hand if ABRM is depolarized in the presence of La^{3+} for as little as 4 min, the relaxing

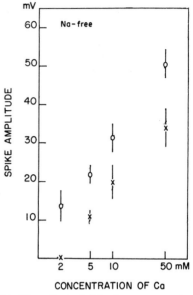

Figure 24. The effect of 5-hydroxytryptamine on spike amplitude. Spikes were evoked by 500 msec current pulses in various concentrations of Ca^{++} and recorded as in Fig. 12. At each Ca^{++} concentration, the amplitude of the spike was measured in 10^{-8} M 5-HT ($\times$) and in 10^{-7} M 5-HT ($\bigcirc$). Note constant increment of increase in spike amplitude in 10^{-7} M 5-HT, unchanged slope of curve. Each point is mean of 10 measurements $\pm$SD.

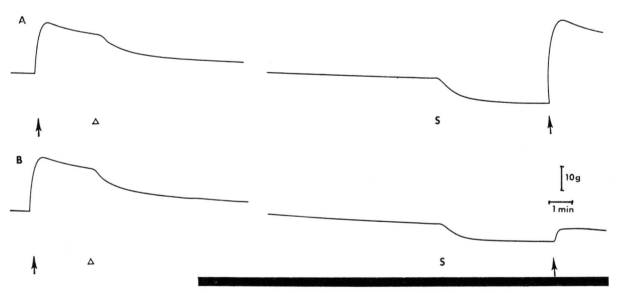

Figure 25. The effect of lanthanum ion (La^{3+}) applied in ASW on relaxation by 5-HT. *A*, (Control) ACh applied at ↑, washed at △; 10^{-7} M 5-HT (*S*) applied 25 min after washing; ACh applied 5 min later. *B*, As in *A* except that 5 min after washing, muscle was immersed in 5 mM La^{3+} ASW (solid line). Note that relaxation by 5-HT is unchanged, although ACh contraction is significantly reduced.

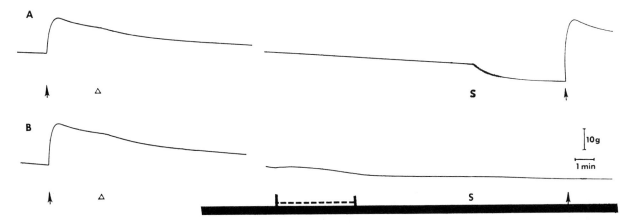

Figure 26. The effect of brief depolarization in the presence of La³⁺ on relaxation by 5-HT (*S*). *A* and *B* exactly as in Fig. 25 except that in *B*, during the 20 min exposure to 5 mM La³⁺ (solid line), a high K⁺, 5 mM La³⁺ solution was applied for 4 min (broken line). Note that relaxation by 5-HT is totally blocked and so also is the ACh contraction.

response is completely blocked (Fig. 26). From this it can be argued that only from within the cell can La³⁺ block either Ca⁺⁺ uptake into an internal reservoir or active extrusion of Ca⁺⁺ from the cell.

Experimental results are as yet inconclusive, but Fig. 27 summarizes some possible mechanisms of

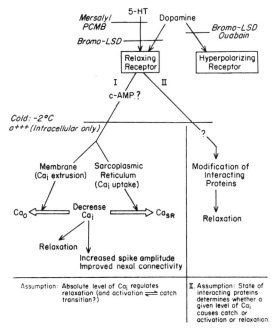

Figure 27. The action of relaxing agents and possible mechanisms of relaxation of catch. 5-Hydroxytryptamine and dopamine combine with a relaxing receptor. Relaxation by 5-HT is selectively blocked by mersalyl and PCMB; relaxation by both is blocked by bromo-LSD. (Dopamine also hyperpolarizes, but this effect can be blocked by ouabain without blocking relaxation.) Events triggered by the relaxing receptor are blocked by cold and by La³⁺ if it goes within the cell; c-AMP may be involved. The hypothesis that the relaxing mechanism decreases intracellular free Ca⁺⁺ is outlined in detail as well as one of the alternative hypotheses (see text).

relaxation. If the primary effect of 5-HT is to stimulate Ca⁺⁺ extrusion, the predicted net loss of Ca⁺⁺ should be measurable. Nauss and Davies (1966) measured increased efflux of ⁴⁵Ca⁺⁺ from muscles during relaxation by 5-HT. Bloomquist and Curtis (1972) measured increased efflux of ⁴⁵Ca⁺⁺ from resting muscles as a response to 5-HT. But since neither group measured influx, the question of whether there is a *net* loss or gain of Ca⁺⁺ is yet to be investigated.

For the time being our working hypothesis is that relaxing mediators decrease intracellular free Ca⁺⁺. This entails an increased action-potential amplitude, brought about perhaps by decreasing the ratio Ca⁺⁺ inside/Ca⁺⁺ outside or perhaps by increasing nexal connectivity (cf. Fig. 27). (An overshooting action potential has not yet been observed in ABRM. However it seems possible that the cell is damaged when impaled by the microelectrode and that some of the recorded voltage reflects activity in adjacent, uninjured cells that are connected by nexal junctions.) Loewenstein et al. (1967) have demonstrated in epithelia that reduced intracellular free Ca⁺⁺ increases junctional conduction. This is illuminating when seeking an explanation of the apparently paradoxical observation that 5-HT and dopamine, both of which relax catch, actually increase active tension in response to excitation (cf. Figs. 1, 2).

Increasing the amplitude of the action potential would account for the increased total influx of Ca⁺⁺; improving the electrical connections at nexal junctions would increase the total number of cells responding to each stimulus. And that relaxing mediators potentiate largely by recruitment of greater numbers of muscle fibers is supported by observations that only submaximal responses are markedly potentiated.

Of course available evidence permits entirely different conclusions: for instance, that the relaxing mechanism modifies the interacting contractile proteins, the Ca^{++} sensitizing system associated with them or the availability of ATP.

Our present hypothesis supposes that Ca^{++} in some way regulates the activation-catch transition. As testable hypotheses arise we shall pursue them.

Acknowledgment

This investigation was supported by U.S. National Institutes of Health Grant AM 11996 from the Division of Arthritis and Metabolic Diseases.

References

BAGUET, F. and J. M. GILLIS. 1964. Stimulation de la respiration des muscles de Lamellibranches par la 5-Hydroxytryptamine. *Arch. Int. Physiol. Biochim.* **72:** 351.

————. 1968. Energy cost of tonic contraction in a lamellibranch catch muscle. *J. Physiol.* **198:** 127.

BAGUET, F. and S. SLEEWAGEN. 1971. Ionic strength and pH effects on mechanical properties of a glycerol extracted catch muscle (ABRM). *Proc. Int. Cong. Physiol. Sci.* **9:** 84.

BLOOMQUIST, E. and B. A. CURTIS. 1972. The action of serotonin on calcium-45 efflux from the anterior byssal retractor muscle of *Mytilus edulis*. *J. Gen. Physiol.* **59:** 476.

CHIARANDINI, D. J., J. P. REUBEN, L. GIRARDIER, G. M. KATZ, and H. GRUNDFEST. 1970. Effects of caffeine on crayfish muscle fibers. II. Refracturiness and factors influencing recovery (repriming) of contractile responses. *J. Gen. Physiol.* **55:** 665.

COLE, R. A. and B. M. TWAROG. 1972. Relaxation of catch in a molluscan smooth muscle. I. Effects of drugs which act on the adenyl cyclase system. *Comp. Biochem. Physiol.* **43A:** 321.

DEWEY, M. M. and L. BARR. 1968. Structure of vertebrate intestinal smooth muscle. In *Handbook of physiology* (ed. C. F. CODE) sect. 6, vol. 4, p. 1629. American Physiological Society, Washington, D.C.

FEINSTEIN, M. B. 1963. Inhibition of caffeine rigor and radiocalcium movements by local anesthetics in frog sartorius muscle. *J. Gen. Physiol.* **47:** 151.

FLETCHER, C. M. 1937. Action potentials recorded from an unstriated muscle of simple structure. *J. Physiol.* **90:** 233.

HAGIWARA, E. and T. NAGAI. 1970. ^{45}Ca movements at rest and during potassium contracture in *Mytilus* ABRM. *Jap. J. Physiol.* **20:** 72.

HAGIWARA, S. and K. NAKA. 1964. The initiation of spike potential in barnacle muscle fibers under low intracellular Ca^{++}. *J. Gen. Physiol.* **48:** 141.

HANSON, J. and J. LOWY. 1961. *Proc. Roy. Soc. (London)* B **154:** 173.

HEUMANN, H. G. 1969. Calcium Akkumulierende Strukturen in einem glatten Wirbellosenmuskel. *Protoplasma* **67:** 111.

HEUMANN, H. G. and E. ZEBE. 1968. Uber die Funktionsweise glatter Muskelfasern. *Z. Zellforsch.* **85:** 534.

HIDAKA, T. 1969. Dopamine hyperpolarizes and relaxes *Mytilus* muscle. *Amer. Zool.* **9:** 251.

HIDAKA, T., T. OSA, and B. M. TWAROG. 1967. The action of 5-hydroxytryptamine on *Mytilus* smooth muscle. *J. Physiol.* **192:** 869.

JEWELL, B. R. 1959. The nature of the phasic and the tonic responses of the anterior byssal retractor muscle of *Mytilus*. *J. Physiol.* **149:** 154.

JOHNSON, W. H. 1962. Tonic mechanisms in smooth muscles. *Physiol. Rev.* 42 (Suppl) **5:** 113.

JOHNSON, W. H. and B. M. TWAROG. 1960. The basis for prolonged contractions in molluscan muscles. *J. Gen. Physiol.* **43:** 941.

JOHNSON, W. H., J. S. KAHN, and A. G. SZENT-GYÖRGYI. 1959. Paramyosin and contraction of "catch muscles." *Science* **130:** 160.

JULIAN, F. J. 1971. Ca effect on muscle force and velocity. *J. Physiol.* **218:** 117.

KENDRICK-JONES, J., W. LEHMAN, and A. G. SZENT-GYÖRGYI. 1970. Regulation in molluscan muscles. *J. Mol. Biol.* **54:** 313.

LEENDERS, H. J. 1966. Kontraktion und Spannungsrückstand an glycerinextrahierten Muskelfasern (ABRM) von *Mytilus edulis*. *Naturwiss.* **23:** 617.

————. Ca-coupling in the anterior byssal retractor muscle of *Mytilis L. J. Physiol.* **192:** 681.

LOEWENSTEIN, W. R., M. NAKAS, and S. J. SOCOLAR. 1967. Junctional membrane uncoupling: permeability transformations at a cell membrane junction. *J. Gen. Physiol.* **50:** 1865.

LOWY, J., B. M. MILLMAN, and J. HANSON. 1964. Structure and function in smooth tonic muscles of lamellibranch muscles. *Proc. Roy. Soc. (London)* B **160:** 525.

MUNEOKA, Y. 1966. Effects of $CaCl_2$ on the response of anterior byssal retractor muscle of *Mytilus edulis* to acetylcholine. *Biol. Bull. Hiroshima Univ.* **33:** 25.

————. 1969. Effects of some divalent cations, serotonin and procaine on caffeine contracture of anterior byssal retractor muscle of *Mytilus edulis*. *Zool. Mag.* **78:** 127.

MUNEOKA, Y. and T. MIZONISHI. 1969. Effects of changes in external calcium concentration on the caffeine contracture of anterior byssal retractor muscle of *Mytilus edulis*. *Zool. Mag.* **78:** 101.

MUNEOKA, Y. and B. M. TWAROG. 1972. Dopamine and serotonin as possible neurotransmitters in *Mytilus*. *Fed. Proc.* **31:** 333 (Abstr.).

NAUSS, J. M. and R. E. DAVIES. 1966. Changes in inorganic phosphate and arginine during the development, maintenance and loss of tension in the anterior byssus retractor muscle of *Mytilus edulis*. *Biochem. Z.* **345:** 173.

RÜEGG, J. C. 1971. Smooth muscle tone. *Physiol. Rev.* **51:** 201.

RÜEGG, J. C. and E. STRASSNER. 1963. Sperrtonus und Nucleosidtriphosphate. *Z. Naturforsch.* **18:** 133.

RÜEGG, J. C., R. W. STRAUB, and B. M. TWAROG, 1963. Inhibition of contraction in a molluscan smooth muscle by thiourea, an inhibitor of the actomyosin contractile mechanism. *Proc. Roy. Soc. (London)* B **158:** 156.

SCHÄDLER, M. 1967. Proportionale Aktivierung von ATPase-Aktivität und Kontraktionsspannung durch Calciumionen in isolierten contrakilen Strukturen verschiedener Muskelarten. *Pflügers Archiv.* **296:** 70.

SHANES, A. M. 1958. Electrochemical aspects of physiological and pharmacological action in excitable cells. Part II. The action potential and excitation. *Pharmacol. Rev.* **10:** 165.

STÖSSEL, W. and E. ZEBE. 1968. Zur intracellulären Regulation der Kontraktionsaktivität. *Pflügers Arch.* **302:** 38.

SUGI, H. 1971. Contracture of molluscan smooth muscle during calcium deprivation. *Proc. Jap. Acad.* **47:** 683.

SZENT-GYÖRGYI, A. G., C. COHEN, and J. KENDRICK-JONES. 1971. Paramyosin and the filaments of molluscan "catch" muscles. *J. Mol. Biol.* **56**: 239.

TWAROG, B. M. 1960. Innervation and activity of a molluscan smooth muscle. *J. Physiol.* **152**: 220.

————. 1966. Catch and the mechanism of action of 5-hydroxytryptamine on molluscan muscle: A speculation. *Life Sciences* **5**: 1201.

————. 1967a. The regulation of catch in molluscan muscle. *J. Gen. Physiol.* **50**: 157.

————. 1967b. Factors influencing contraction and catch in *Mytilus* smooth muscle. *J. Physiol.* **192**: 847.

TWAROG, B. M. and R. A. COLE. 1972. Relaxation of catch in a molluscan smooth muscle. II. Effects of serotonin, dopamine and related compounds. *Comp. Biochem. Physiol.* **43A**: 331.

TWAROG, B. M., M. M. DEWEY, and T. HIDAKA. 1973. The structure of *Mytilus* smooth muscle and electrical constants of the resting muscle membrane. *J. Gen. Physiol.* In press.

TWAROG, B. M. and T. HIDAKA. 1971a. Electrical constants and structure of molluscan smooth muscle. *Proc. Int. Cong. Physiol. Sci.* **9**: 572.

————. 1971b. The calcium spike in *Mytilus* muscle and the action of serotonin. *J. Gen. Physiol.* **57**: 252.

VAN BREEMEN, C. 1969. Blockade of membrane calcium fluxes by lanthanum in relation to vascular smooth muscle contractility. *Arch. Int. Physiol. Biochim.* **77**: 710.

VAN BREEMEN, C. and P. DE WEER. 1970. La^{+++} inhibition of ^{45}Ca fluxes across the axolema of the squid giant axon. *Nature* **226**: 760.

WEBER, A. and R. HERZ. 1968. The relationship between caffeine contracture of intact muscle and the effect of caffeine on reticulum. *J. Gen. Physiol.* **52**: 750.

WINEGRAD, S. 1971. Studies of cardiac muscle with a high permeability to calcium produced by treatment with ethylenediaminetetraacetic acid. *J. Physiol.* **58**: 71.

YORK, B. and B. M. TWAROG. 1972. Evidence for the release of serotonin by relaxing nerves in molluscan muscle. *Comp. Biochem. Physiol.* In press.

Length Dependence of Activation of Skinned Muscle Fibers by Calcium

M. ENDO*

Department of Biological Sciences, Purdue University, Lafayette, Indiana 47907

It is well known that the active tetanic tension of a living skeletal muscle fiber decreases linearly with increase of fiber length beyond its slack length (Ramsey and Street, 1940; Gordon et al., 1966). Skinned skeletal muscle fibers also behave similarly at high concentrations of calcium (Hellam and Podolsky, 1969). This has been successfully attributed to the decreased number of interacting sites between thick (myosin-containing) and thin (actin-containing) filaments on the basis of the sliding filament theory of contraction (Gordon et al., 1966).

On the other hand, the activation of skeletal muscle fiber seems to be increased by stretch, as indicated by the following facts. First, the active state following an action potential was shown to be longer in stretched fibers (Ritchie, 1954; Edman and Kiessling, 1966). Second, unlike tetanus, tension produced by a stimulus of a frequency lower than tetanus is greater in stretched fibers in spite of less overlap between thick and thin filaments (Rack and Westbury, 1969). Similarly, the isometric tension in a twitch sometimes increases by stretch (Close, 1972), although this does not always occur. Third, the mechanical threshold of potassium contracture of skeletal muscle fibers is decreased and the contracture tension in a certain range of potassium concentration is increased by stretch (Gonzalez-Serratos et al., 1971).

The increase in activation by stretch has generally been considered to be due to the increase in the amount of calcium released from the sarcoplasmic reticulum. However, it has been found and will be shown below that the activation of skinned skeletal muscle fibers by a given concentration of calcium increases with the increase of fiber length beyond its slack length. Thus the increase in activation of living muscle fibers by stretch may, at least partly, be the property of the contractile system itself.

Steady Isometric Tension

Single fibers were isolated from iliofibularis muscle of the African clawed toad, *Xenopus laevis*,

and skinned in a relaxing solution. The latter usually contained 21.5 mM K_2SO_4, 12 mM Na_2SO_4, 20 mM Tris-20 mM maleate-NaOH pH 6.8, 4 mM ATP, 4 mM $MgSO_4$, 4 mM EGTA, and 93 mM sucrose. To determine the cross-sectional area of skinned fibers before the experiments were run, the optical path difference across the fiber was measured by using a Zeiss interference microscope with transmitted light and was integrated along the transverse direction of the fiber (Endo, 1967, 1972).

The skinned fibers were then set up in a trough through which a solution could be perfused rapidly. Both ends of the fibers were fixed between pieces of Scotch tape and isometric tension was recorded with a mechanoelectronic transducer (RCA 5743). Sarcomere length was measured before the application of calcium by observing striation spacings under a microscope and taking photomicrographs of the striations.

Fibers were pretreated for 10 min with a solution containing 10^{-6} M free calcium in the presence of 12 mM $MgSO_4$, which was below the threshold for contraction (Ebashi and Endo, 1968), in order to load the sarcoplasmic reticulum with calcium to an appreciable extent (Endo et al., 1970). Without this pretreatment, the time course of tension development after a sudden increase in calcium ion concentration in the medium was slower, and under certain circumstances the steady tension reached was also smaller.

An activating solution containing various calcium ion concentrations was then applied to the fiber, and after tension reached its steady state value, a relaxing solution was applied, usually resulting in the complete disappearance of the tension developed by the activating solution. The activating solution was different from the relaxing solution only in that 10 mM EGTA-Ca^{++} buffer was used in place of 4 mM EGTA to give a desired concentration of free calcium, and the increase in the ionic strength by this replacement was compensated by reducing a suitable amount of Na_2SO_4. Free calcium ion concentration was calculated by using Ogawa's (1968) binding constant of Ca^{++}-EGTA.

For quantitative studies, I used primarily the first contraction of each fiber and the second only if the first was a weak contraction induced by

---

* Present address: Department of Pharmacology, University of Tokyo, Bunkyo-ku, Tokyo 113, Japan.

lower concentrations of calcium. This precaution was taken because initially uniform sarcomere spacings became nonuniform during and after a single contraction, especially if it was induced by a high concentration of calcium—this was probably due to longitudinally nonhomogeneous activation— and also because the tension-producing ability of skinned fibers at an average length of about 2.2 μm per sarcomere somehow deteriorated much faster than that at greater lengths. For qualitative studies, however, single skinned fibers could be used repeatedly.

Most of the experiments were conducted at about 0°C by perfusing a 1:4 mixture of methyl alcohol and water at −3°C below the glass base of the trough.

Figure 1 shows the length-tension relation of skinned muscle fibers in the range longer than slack length under various concentrations of calcium. It is clearly seen that stretch decreases isometric tension of the skinned fiber at higher concentrations of calcium, as already reported by Hellam and Podolsky (1969), although the linearity was not as good as Gordon et al.'s (1966) careful study on living fibers probably because of nonhomogeneity of sarcomere spacings throughout the skinned fiber. However it is also clearly seen that at lower concentrations of calcium (some fibers with 10^{-6} M calcium and all fibers so far tested with calcium

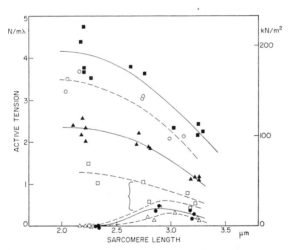

Figure 1. Length-tension relation of skinned muscle fibers of *Xenopus laevis* in the range longer than slack length under various concentrations of free calcium ion. A part of the same experiments described in Fig. 2 in Endo (1972). Free calcium ion concentration: (■) 3×10^{-5} M; (○) 3×10^{-6} M; (▲) 1.5×10^{-6} M; (□) 10^{-6} M; (●) 8.5×10^{-7} M; (△) 7×10^{-7} M. Each symbol indicates individual fibers. However, usually a long skinned fiber was prepared and cut into two or three parts and tested with the same free calcium at different lengths. Tension is referred to the cross-sectional area at sarcomere length of 2.1 μm, expressed by the integrated optical path difference ($m \cdot \lambda$) or corresponding cross-sectional area of the fiber in normal Ringer (m^2).

less than 10^{-6} M) stretch increased the isometric tension of *Xenopus* fiber in spite of the decrease in the overlap between thick and thin filaments. Essentially the same result was obtained with frog fibers.

This phenomenon is also shown in Fig. 2, in which experiments were repeated on one skinned fiber. If the fiber was stretched to an average length of 2.9 μm per sarcomere, 10^{-6} M calcium in the medium produced an appreciable tension (A1). However if the fiber was set at an average length of 2.3 μm per sarcomere, the same concentration of calcium induced no visible tension at all (A2). The effect of stretch was quite reversible, as shown in A3.

Although the sarcoplasmic reticulum could affect the free calcium ion concentration in the fiber space under certain circumstances, (Endo, 1967; Hellam and Podolsky, 1969; Ford and Podolsky, 1970; Endo et al., 1970), the increase in activation of skinned fibers by stretch described above cannot be attributed to the sarcoplasmic reticulum, since essentially the same phenomenon was obtained in skinned fibers after their sarcoplasmic reticulum was functionally destroyed by UV irradiation (Nagai et al., 1960; Endo, 1965) or by treatment with a detergent, Brij-58 (Orentlicher et al., 1972). Figure 2 (B series) shows an example of a Brij-treated fiber. The fiber was from the same single fiber as used in the experiments of Fig. 2A. It responded to 10^{-6} M calcium in exactly the same way as without Brij treatment, but response to caffeine applied under the same conditions as in Fig. 2A (except that a lower concentration of EGTA was used before and during caffeine treatment to enhance the response) was abolished.

Another possibility is that the calcium sensitivity is nonhomogeneous throughout the fiber, and at lower concentrations of calcium only a small fraction of sarcomeres was activated and shortened, while other nonactivated sarcomeres were stretched. This may result in no tension at slack length, but in appreciable tension at longer lengths. However this possibility was excluded since the change in sarcomere spacings during the contraction produced by a lower concentration of calcium was much too small to account for the greater tension production at longer lengths, as hypothesized above (Endo, 1972). It is, therefore, concluded that greater tension-producing ability at longer lengths with low calcium ion concentration must be a real property of the contractile system.

Time Course of Tension Development

The rate of tension development after a sudden increase in calcium ion concentration in the medium was very much dependent on the level

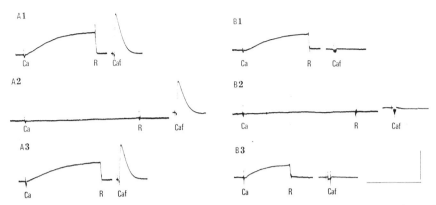

Figure 2. Response of skinned fibers to a low concentration of calcium at different lengths. *A*, Without Brij-treatment (fiber 20529A). *B*, After a treatment with 0.5% Brij-58 for 30 min (fiber 20529B). Both fibers were from one and the same fiber. Average sarcomere length: *A1* and *A3*, 2.9 μm; *A2*, 2.3 μm. *B1* and *B3*, 2.8 μm; *B2*, 2.2 μm. At *Ca*, the calcium ion concentration was raised from about 2×10^{-9} M to 10^{-6} M; at *R*, it was lowered again to about 2×10^{-9} M. Fibers were pretreated before each calcium treatment with 10^{-6} M free calcium in the presence of 12 mM total Mg^{++} (see text). At *Caf*, caffeine 25 mM was applied in the presence of 10 mM (*A1–A3*, and *B3*) or 2 mM (*B1* and *B2*) EGTA. Time: 5 min for calcium response, 1 min for caffeine response; tension: 1 mN. Note that while caffeine response was abolished in *B*, calcium response was essentially the same in *B* as in *A*.

of free calcium ion, as shown in Fig. 3A. Although this figure only serves to indicate the difference qualitatively, it can still be safely concluded that with 3×10^{-6} M calcium, half-time is about

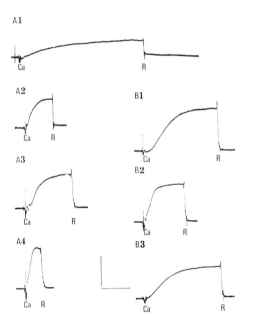

Figure 3. Time course of tension development of skinned fibers after a sudden increase in calcium ion concentration. Fibers were treated with 0.5% Brij-58 for 30 min before experiments. At *Ca*, the calcium ion concentration was raised from about 2×10^{-9} M (EGTA 10 mM without addition of calcium) to various levels (total EGTA 10 mM), and at *R* lowered again to about 2×10^{-9} M. *A*, the effect of calcium ion concentration (fiber 20528A). Free calcium ion: *A1*, 10^{-6} M; *A2*, 1.2×10^{-6} M; *A3*, 1.5×10^{-6} M; *A4*, 3×10^{-6} M. Experiment A1 was done first and A4 last. *B*, The effect of the fiber length (fiber 20528B). Calcium ion concentration: 1.5×10^{-6} M. Average sarcomere length: *B1* and *B3*, 2.2 μm; *B2*, 2.8 μm. Time: 2.5 min for *A1* and *A2*, 1 min for the rest. Tension: 1 mN.

several seconds, whereas with 10^{-6} M calcium it is of the order of several minutes. Similar results on glycerol-treated fibers have already been described by Julian (1971). Again, this is not due to the property of the sarcoplasmic reticulum, since the experiments in Fig. 3A and Julian's experiments were done on fibers treated with a detergent (Brij or Lubrol-WX, respectively). It should also be noted that, as is seen from the comparison of Fig. 2A with 2B, if the skinned fiber was properly pretreated with calcium and if free calcium ion was buffered with a high concentration of EGTA-Ca^{++} buffer, almost the same rate of tension development was obtained with normal fiber as that with the Brij-treated fiber in which the function of the sarcoplasmic reticulum was completely destroyed. Skinned fibers cannot start developing tension instantaneously after an activating solution is applied since the concentration of total calcium or Ca^{++}-EGTA complex in the fiber space must reach a certain value, for example about 2.3 mM in the experiment of Fig. 3A, to reach the threshold for contraction. This certainly makes the rate of tension development with lower Ca^{++} ion concentration slower, but this cannot explain such a big difference as shown in Fig. 3A. Thus the rate of tension development also appears to be dependent on the magnitude of activation.

When the Ca^{++} ion concentration was about $1.2–1.5 \times 10^{-6}$ M, the steady tension developed by a skinned fiber of *Xenopus* at slack length was greater than at longer lengths (Fig. 1; also Endo, 1972). However the rate of tension development was much slower at slack length than at longer lengths (Fig. 3B). The time required for Brij-treated fibers to develop half of steady tension in

Table 1. Comparison of Rate of Tension Development of Skinned Muscle Fibers at Different Sarcomere Lengths

A. Ca⁺⁺ 1.5 × 10⁻⁶ M

Fiber Reference (1)	Average Sarcomere Length (μm) (2)	Cross-sectional Area (× 10⁻⁴ mλ) (3)	Half-time (sec) (4)	(4)/(3) (5)	Average Sarcomere Length (μm) (6)	Cross-sectional Area (× 10⁻⁴ mλ) (7)	Half-time (sec) (8)	(8)/(7) (9)
20523B	2.19	7.9	24	3.0	2.79	6.2	10	1.6
20525A	2.20	4.5	32	7.1	2.82	3.5	12	3.4
20525B	2.20	5.6	33	5.9	2.78	4.4	14	3.2
20525C	2.22	2.1	22	10.5	2.82	1.7	6	3.5
20528B	2.2*	7.6	41	5.4	2.8*	6.0	16	2.7
20528C	2.23	3.7	35	9.5	2.86	2.9	15	5.2
Average ±SE			31.2 ±2.9	6.90 ±1.13			12.2 ±1.5	3.27 ±0.48

B. Ca⁺⁺ 3 × 10⁻⁶ M

Fiber Reference (1)	Average Sarcomere Length μm (2)	Cross-sectional Area (× 10⁻⁴ mλ) (3)	Time for 80% Saturation (sec) (4)	(4)/(3) (5)	Average Sarcomere Length μm (6)	Cross-sectional Area (× 10⁻⁴ mλ) (7)	Time for 80% Saturation (sec) (8)	(8)/(7) (9)
10818A-B	2.04	9.5	7.2	0.76	3.02	6.4	3.6	0.56
10825A-C	2.03	5.4	5.8	1.07	2.98	3.4	3.4	1.00
10826C-A	2.16	4.6	2.9	0.63	3.11	3.5	2.4	0.69
20531A	2.34	6.2	15.5	2.5	3.14	4.6	12	2.6
20531B	2.11	6.9	23	3.3	3.06	4.8	15	3.1

* Measured visually only.

A, Ca⁺⁺ 1.5 × 10⁻⁶ M. *B*, Ca⁺⁺ 3 × 10⁻⁶ M. Total EGTA 10 mM. All fibers except 108 series were pretreated with 0.5% Brij-58 for 30 min. For all fibers except 20523B and 108 series, the experiment at one length was repeated after that at the different length and the average values are shown. For 20523B the values are from only one experiment. For 108 series each value is of different fibers. However in each row the experiments were done on skinned fibers from the same single fiber.

the presence of 1.5 × 10⁻⁶ M Ca⁺⁺ was measured at different lengths and is given in Table 1A. All the fibers showed much longer half-times at about 2.2 μm than at about 2.8 μm.

The slower rate at slack length was not due to greater thickness at this length than at longer lengths, which would demand longer time for calcium buffer to diffuse into the fiber space. If this were the explanation, the half-times should be proportional to the square of the fiber thickness, and hence, as a first approximation, to the fiber cross-sectional area. In columns (5) and (9) in Table 1A, half-time was divided by the cross-sectional area, but all fibers still showed much larger value at slack length than at a length of about 2.8 μm per sarcomere. This is in contrast with the results obtained with 3 × 10⁻⁶ M calcium or higher. Table 1B shows that with this concentration of calcium, the time required for fibers to develop 80% of steady tension was not different at slack length from that at a longer length if the change in thickness of the fiber was taken into account. It should also be noted that while the time course of tension development with 3 × 10⁻⁶ M calcium or higher is not far from what would be expected for simple diffusion, the rate of tension

development with 1.5 × 10⁻⁶ M calcium or lower is altogether much slower.

The above results indicate that even with 1.5 × 10⁻⁶ M calcium, at which steady tension is greatest at slack length, the rate of development of activation is smaller at slack length than at longer lengths.

Shortening Velocity

The greater responsiveness of the contractile mechanism to calcium at longer sarcomere lengths may also be demonstrated by isotonic shortening experiments. One end of a skinned fiber of the frog, *Rana japonica*, which was fixed on a part of the trough in the isometric experiments, was attached to the lever of a moving-coil microammeter. By feeding a suitable magnitude of current to the microammeter, constant load condition could be obtained. At first a fiber was made to contract isometrically in an activating solution; after tension was fully developed, the load was quickly reduced to a very small value, and the time course of shortening of the fiber was recorded by detecting the movement of the lever of the microammeter with a photoelectric tube. Gordon et al. (1966) have shown that, unlike the isometric tension, the shortening velocity of intact muscle fiber under a very small load is virtually independent of the magnitude of overlap between thick and thin filaments above slack length. The same result was also obtained in glycerinated fibers under a high concentration of calcium (Julian, 1971). Similarly, skinned fibers at a length of about 3.0 μm per sarcomere activated by high concentrations of calcium shortened with almost constant velocity until it reached a length less than 2.0 μm per sacomere. Under a low concentration of calcium, however, the shortening velocity was not constant but started to decline much before the fiber reached the length of about 2.0 μm per sarcomere. This behavior of the fiber might reflect the same fact as described above, that the activation of skinned fibers by a given low concentration of calcium is greater at longer sarcomere length.

Discussion

While it still remains to be elucidated whether stretch of the skeletal muscle fiber also increases the amount of calcium released by the sarcoplasmic reticulum, the stretch-induced increase in activation of the contractile system by a given concentration of calcium described above appears to explain many of the features of the increase in activation by stretch in living fiber, especially if the effect of stretch on the time course of tension development is taken into account. For example, Close (1972) found that the stretch-induced increase in activation in a twitch almost disappeared in post-tetanically potentiated twitch. This is in accordance with the above result that with a rather small increase in calcium ion concentration from 1.5×10^{-6} to 3×10^{-6} M, the stretch effect disappeared.

The relation between active tension and length under low concentration of calcium was shown to have a positive slope up to about 2.9 μm per sarcomere (Fig. 1), as was the relation between total tension (resting plus active tension) and length over an entire range of length (Endo, 1972). This may be useful to the animal since a negative slope implies instability.

The dependence of the rate of tension development on calcium ion concentration, together with the already steep relation between free calcium and tension may contribute, at least to some extent, to the well-known sharp relation between depolarization and tension in living fibers.

The large dependence of the rate of tension development on free calcium concentration and on length of the sarcomere at lower concentrations of calcium might suggest that the characteristics of myosin-actin interaction at low concentrations of calcium is different from that of fully activated fiber. At least one additional condition seems to be required for the interaction to be effective at lower concentrations of calcium.

Nothing is known at present about the mechanism of stretch-induced increase in activation of the contractile system by calcium. It appears very unlikely that the amount of calcium bound to troponin increases by stretch. The smaller distance between thick and thin filaments at longer lengths might facilitate the myosin-actin interaction under a low concentration of calcium. However, it was reported that in frog's skinned fiber, unlike an intact fiber, the distance between thick and thin filament did not change much by stretching (Matsubara and Elliot, cited by Rome, 1972). My preliminary experiments also showed that, by adding a high concentration of large particles such as glycogen or albumin, the width of the skinned fiber decreased by approximately the same magnitude as it was stretched and hence probably the distance between myofilaments was reduced. However no increase in calcium sensitivity was found under these circumstances. Of course, both of these results are not yet conclusive, but while the interfilamental spacing appears to be a plausible explanation, there is no direct evidence so far to support it.

Acknowledgments

The author wishes to express his sincere thanks to Prof. S. Ebashi and Dr. S. Nakajima for their constant encouragement and helpful discussions, and to Prof. A. F. Huxley for helpful comments. This work was supported in part by PHS grant NS-08601 and HSAA No. 5-SO4-RR06013 from NIH, and grants from the Muscular Dystrophy Association of America and the Iatrochemical Foundation, Tokyo.

References

CLOSE, R. I. 1972. The relations between sarcomere length and characteristics of isometric twitch contractions of frog sartorius muscle. *J. Physiol.* **220:** 745.

EBASHI, S. and M. ENDO. 1968. Calcium and muscle contraction. *Prog. Biophys. Mol. Biol.* **18:** 123.

EDMAN, K. A. P. and A. KIESSLING. 1966. The time course of the active state of the frog skeletal muscle fibre in relation to the sarcomere length. *Acta Physiol. Scand.* **68,** (Suppl. 277): 41.

ENDO, M. 1965. The effect of ultraviolet light on single muscle fibres of the frog. *Abstract, 23 Int. Cong. Physiol. Sci.,* communication No. 810.

———. 1967. Regulation of contraction-relaxation cycle of muscle (in Japanese). *Proc. 17 Gen. Assn. Japan Med. Congr.* **1:** 193.

———. 1972. Stretch-induced increase in activation of skinned muscle fibres by calcium. *Nature,* in press.

ENDO, M., M. TANAKA, and Y. OGAWA. 1970. Calcium-induced release of calcium from the sarcoplasmic reticulum of skinned skeletal muscle fibres. *Nature* **228:** 34.

FORD, L. E. and R. J. PODOLSKY. 1970. Regenerative calcium release within muscle cells. *Science* **167:** 58.

GONZALEZ-SERRATOS, H., R. VALLE, and A. CILLERO. 1971. Effect of muscle stretching on tension and mechanical threshold during contractures. *Proc. Int. Union Physiol. Sci.* **9:** 211.

GORDON, A. M., A. F. HUXLEY, and F. J. JULIAN. 1966. The variation in isometric tension with sarcomere length in vertebrate muscle fibres. *J. Physiol.* **184:** 170.

HELLAM, D. C. and R. J. PODOLSKY. 1969. Force measurement in skinned muscle fibres. *J. Physiol.* **200:** 807.

JULIAN, F. J. 1971. The effect of calcium on the force-velocity relation of briefly glycerinated frog muscle fibres. *J. Physiol.* **218:** 117.

NAGAI, T , M. MAKINOSE, and W. HASSELBACH. 1960. Der physiologische Erschlaffungsfaktor und die Muskelgrana. *Biochim. Biophys. Acta* **43:** 223.

OGAWA, Y. 1968. The apparent binding constant of glycoletherdiamine-tetraacetic acid for calcium at neutral pH. *J. Biochem.* **64:** 255.

ORENTLICHER, M., J. P. REUBEN, and P. W. BRANDT. 1972. Morphology and physiology of detergent treated crayfish muscle fibers. *Biophys. Soc. Abstr.* FPM-I-14.

RACK, P. M. H. and D. R. WESTBURY. 1969. The effect of length and stimulus rate on tension in the isometric cat soleus muscle. *J. Physiol.* **204:** 443.

RAMSEY, R. W. and S. F. STREET. 1940. The isometric length-tension diagram of isolated skeletal muscle fibers of the frog. *J. Cell. Comp. Physiol.* **15:** 11.

RITCHIE, J. M. 1954. The effect of nitrate on the active state of muscle. *J. Physiol.* **126:** 155.

ROME, E. 1972. Relaxation of glycerinated muscle: Low angle X-ray diffraction studies. *J. Mol. Biol.* **65:** 331.

M and Z Band Components and the Assembly of Myofibrils

J. D. ETLINGER AND D. A. FISCHMAN

Departments of Biophysics, Biology, and Anatomy, University of Chicago, Chicago, Illinois 60637

Our inadequate understanding of myofibrillar assembly is in large part related to the paucity of information detailing the transverse bonds which hold myofilaments in the double hexagonal lattice. One approach to this problem is to follow the intracellular assembly of myofibrils in developing muscle, applying various perturbations to the cell and its environment, and noting the effects on fibrillogenesis. Another approach is to selectively extract materials from isolated, mature fibrils and then to examine possible alterations in the cohesive forces between myofilaments. Both of these approaches have been followed in this and other laboratories, and some of these observations will be discussed below.

Studies with Embryonic Muscle

The subject of myofibrillogenesis in embryonic muscle has been reviewed in detail (Fischman, 1970). In skeletal muscle nascent myofibrils can be detected first by immunofluorescent (Holtzer et al., 1957) and electron microscopic (Fischman, 1967) studies beneath the plasma membrane along the lateral surfaces of the myogenic cells (Fig. 1). Although the large polyribosomes implicated in myosin synthesis (see Morris et al., this volume) are sometimes located adjacent to the thick myofilaments, this is not a constant observation. In fact polysomes appear to be randomly distributed in the sarcoplasm, whereas myofibrillar assembly almost always occurs at the cell cortex. This positional separation of thick myofilaments and large polyribosomes suggests that the assembly of myosin filaments probably occurs after release of the nascent peptide subunits from the ribosomes. However almost nothing is known of the interaction in vivo of myosin, C-protein (see Offer, this volume) and M-band protein(s). In thin sections of embryonic striated muscle both single thick and thin filaments can be recognized before myofibrils have assembled (Fig. 2). Presumably the polymerization of the contractile proteins into thick and thin filaments occurs as a process independent of sarcomere construction. However the possibility has not been ruled out that interaction between thick and thin filaments plays some role, as yet unidentified, in the determination of myo-

filament length. Franzini-Armstrong (1970) in her studies of crab muscle has noted that in muscles with variable A band length there is a constant relationship between thin filament length and the length of the cross-bridge-bearing zone of adjacent thick filaments. Vernier models of thin filament polymerization (Huxley, 1963) may be inadequate to explain this puzzling observation.

A number of authors have noted (see review by Fischman, 1970) that formation of the double hexagonal lattice can occur before the appearance of dense M and Z bands. As noted by Page (1965) certain frog muscles contain myofilaments arranged in an hexagonal lattice, but lack M bands.

Based on the observation of Wollenberger (1964) that myofibrils are dispersed intracellularly during the trypsin-induced dissociation of embryonic myocardia into single cells, we have examined the requirements for protein synthesis during reassembly of the fibrillar lattice (Fischman and Zak, 1971). Using the method of Moscona (1961), the ventricles of 6- or 7-day embryonic chick hearts were dissociated with trypsin into a suspension of single cells. When such cells are incubated in Erlenmeyer flasks on a gyratory shaker, the cells aggregate, form spherical multicellular clusters, and resume pulsatile, contractile activity (Fischman and Moscona, 1971). During tissue dissociation, by a mechanism which is still obscure, the preexisting myofibrils are dispersed into randomly organized arrays of both thick (160 Å diameter) and thin (60 Å diameter) filaments (Figs. 3 and 4). Within 2–3 hr of incubation in the Erlenmeyer flasks, myofibrils can be recognized within the cells and beating has resumed. If this incubation, after tissue dissociation, takes place in the presence of 100 μg/ml of cycloheximide (a concentration which completely inhibits protein synthesis), myofibrils are still reassembled in 2–3 hr and the cells beat for at least 6 hr. In addition considerable reassembly of myofibrils occurs at 4°C. We conclude that positioning of myofilaments into the fibrillar lattice can occur in the absence of protein synthesis and may represent a case of self-assembly analogous to that described in simple spherical and cylindrical plant viruses (Caspar and Klug, 1962).

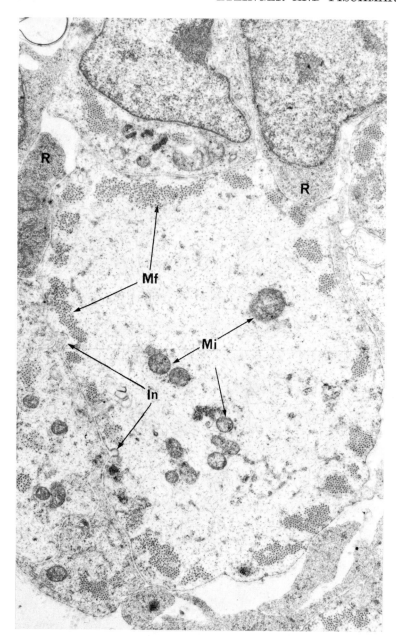

Figure 1. Cross section of embryonic leg muscle (12-day chick embryo) showing peripheral deposition of myofibrils within each myotube. Note the double hexagonal lattice of the myofilaments, even in those myofibrils containing few filaments. Glutaraldehyde-OSO₄ fixation; araldite embedding; uranyl acetate-lead citrate staining. *R*, Ribosomes; *Mf*, myofibrils; *Mi*, mitochondria; *In*, invaginations of sarcolemma. × 20,000.

Another example of intracellular myofibrillar disruption has been noted by Manasek et al. (1972) after treatment of hearts from chick embryos (11–13 somite stages) with cytochalasin B. Interestingly, hearts from later stage embryos were unaffected by the drug, suggesting increasing stability of the myofibrils with increasing embryonic age. A similar phenomenon has been noted in our own laboratory; although myofibrils are dispersed easily in 6 or 7-day embryonic chick hearts after trypsinization of the tissue, myofibrils in chick hearts of 10 days incubation or older are not disrupted by equivalent tissue dissociation procedures.

Studies with Adult Rabbit Muscle

With fresh isolated mature myofibrils we have made the following observations: (1) Myofibrils homogenized in 0.1 M KCl, 2 mM MgCl₂, 2 mM EGTA, 2 mM ATP, 0.01 M Tris-maleate buffer pH 6.8, 1 mM DTT (dithiothreitol), (i.e., a relaxing solution where cross-bridge interaction is minimized and where myofibrils do not contract) will only separate partially ($<10\%$) into a suspension of single myofilaments. The quantity of filaments released is, however, quite sufficient for electron microscopy. (2) Myofibrils extracted with DOC (sodium desoxycholate) of Z and M band dense materials and homogenized in a relaxing solution

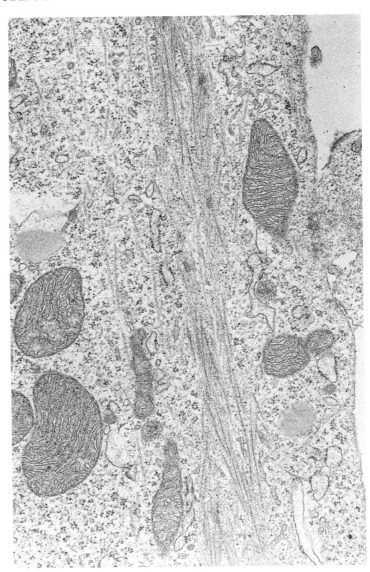

Figure 2. Longitudinal section of cultured, embryonic chick muscle illustrating the alignment of thick and thin myofilaments at a stage which we interpret as preceding fibrillar assembly. A diffuse, electron-dense material can be seen in association with some thin filament bundles; this is interpreted as nascent Z band. Fixation, embedding, and staining as in Fig. 1. × 22,400.

also do not fall apart into separated filaments. (3) Myofibrils digested with papain (0.1 M KCl pH 7, 0.03 mg/ml, 15 min) and extracted of S_1 with 0.1 M KCl, 2 mM Mg - pyrophosphate pH 7, will fall apart easily after gentle homogenization in a standard salt solution or relaxing solution. Z and M bands are also removed by this treatment. Clearly Z and M band dense materials are not sufficient to completely account for the interfilament cohesive forces, and although the papain experiment suggests that myosin cross-bridges play an important role in lattice stability, there may be additional proteolytic damage to the myofibril which cannot be evaluated at this time. It is conceivable that A-segments (Hanson et al., 1971) contain transverse linkages at the ends of the thick filaments which, in addition to M band material, prevent the fraying-out of these filaments.

We will present evidence that at least five high molecular weight protein subunits (greater than 150,000 daltons) are solubilized with myosin when A bands are extracted from uncontracted myofibrils. Some of these proteins are also extracted from whole myofibrils with DOC. One of these proteins may be identical to the M band protein (Masaki and Takaiti, 1972) and another may be the C-protein recently described by Starr and Offer (1971). A third source of these high molecular weight subunits may be the transverse element at the edges of the A-segments (Hanson et al., 1971).

Using DOC we have been able to remove Z band dense material yet still observe branching and overlap of thin filaments at the Z band. The major protein subunits solubilized from the Z band appear to be 95,000 and 85,000 daltons; the 95,000 dalton band is probably a subunit of 6S α-actinin

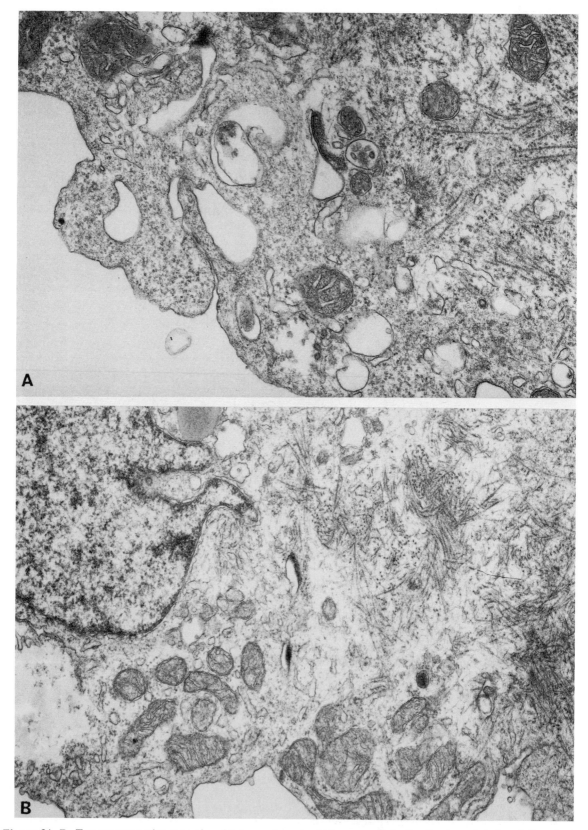

Figure 3A, B. Two representative examples of 7-day embryonic chick myocardial cells after trypsin-induced dissociation of the ventricles. Although well-formed myofibrils are present in cells of this stage heart prior to tissue dissociation, most myofibrils break down during the release of single cells, resulting in an intracellular pool of thick and thin myofilaments in the suspended cells. Adherent junctions are split during trypsin treatment and invaginated within phagosomes. *A,* ×23,000; *B,* ×20,000.

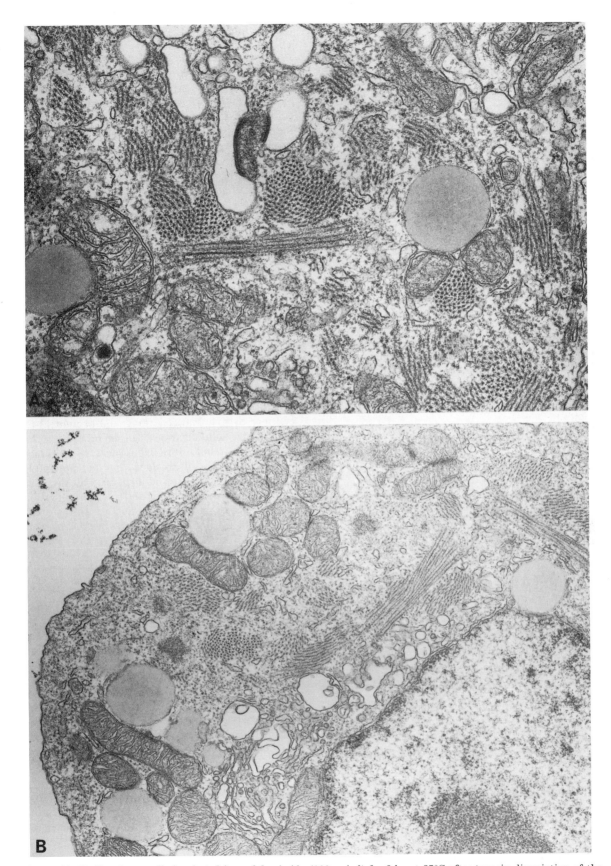

Figure 4A, B. Two heart cells incubated in cycloheximide (100 μg/ml) for 3 hr at 37°C after trypsin dissociation of the myocardia. Note the realignment of myofilaments with reformation of the myofibrillar, hexagonal lattice. These cells exhibit contractile activity for at least 6 hr in the absence of protein synthesis. *A*, ×34,000; *B*, ×10,000.

(Scopes and Penny, 1971). After extraction the Z band is seen to consist of thin filament branching which tends to maintain I band stability. The components removed with DOC will be compared with myofilament preparations, A + M band extracts, and whole fibrils. We have previously reported the use of DOC in the isolation of thin filaments (Etlinger and Fischman, 1971). In all experiments where extracts were prepared for electrophoresis, relaxed myofibrils were first isolated and washed extensively to minimize contamination with nonmyofibrillar proteins. In addition measures were taken to eliminate any proteolysis and aggregation both before and during electrophoresis.

Isolation and purification of myofibrils. Myofibrils from rabbit skeletal muscle were prepared in the following way (slightly modified from the method of Zak et al., 1972). The animals were killed by cervical dislocation, exsanguinated, and the back muscles were immediately excised and placed in relaxing buffer with pyrophosphate (RB + PP): 0.1 M KCl, 2 mM MgCl$_2$, 2 mM EGTA, 1 mM DTT, 2 mM Na$_4$P$_2$O$_7$, 0.01 M Tris-maleate pH 6.8 on ice. After trimming off the connective tissue and cutting into small pieces, the RB + PP solution was changed and the muscle was disrupted in a Waring blendor (top speed for 30 sec) and the solution filtered through cheese cloth. The myofibrils were then washed ten times by resuspension in RB (minus Na$_4$P$_2$O$_7$) and centrifuging at 800 g for 10 min. One wash solution contained 0.02% Triton-X-100 in RB, another was in 0.02% DOC in RB. Each wash was done with a 1:10 ratio of wet pellet weight to solution volume.

The use of a small amount of Na$_4$P$_2$O$_7$ in the initial soaking and homogenizing solution in addition to Mg^{++} and EGTA was found to be the most successful method for obtaining uncontracted myofibrils from fresh muscle. The presence of EGTA throughout the whole purification also minimized myofibrillar proteolysis (Busch et al., 1972). Myofibrils were homogenized in between washes until no bundles remained and were treated with low concentrations of detergents to remove sarcoplasmic reticulum and associated sarcoplasmic material. These fibrils showed excellent preservation of structure by electron microscopy and, except for some nuclei, exhibited very little cross contamination with other organelles. The fibrils can be purified of nuclei using sucrose gradients by the methods of Zak et al. (1972), but such purification does not eliminate any of the protein bands seen on SDS gels to be discussed below.

Extraction of Z and M band components with DOC. Although various methods exist in the literature for removal of M and Z band material, none were found suitable for both isolated fibrils and the subsequent electrophoresis of protein extracts and fibrillar residue. Trypsin and other proteases can remove Z and M band material but myosin is partially digested and cross-bridges released (Ashley et al., 1951; Harsányi and Garamvölgyi, 1969; Goll et al., 1969; Stromer et al., 1967). These methods are clearly unsuitable for the analysis of new proteins whose susceptibility to proteolysis is unknown. Extraction of Z bands with urea (Rash et al., 1968) is not sufficiently specific, for myosin is also solubilized. Although acetone has been shown to extract Z bands (Walcott and Ridgeway, 1967), it renders the A band insoluble for it denatures myosin. The use of low ionic strength and high pH extraction (Corsi and Perry, 1958; Stromer et al., 1969) is a rather slow procedure for Z band removal and results in severe myofibrillar swelling and distortion, rendering ultrastructural analysis difficult. We have found that the controlled use of sodium desoxycholate (DOC) can be used to specifically remove the densely staining materials present within the M and Z bands. The extractions are done in the presence of KCl, MgCl$_2$, EGTA, and at neutral pH to minimize solubilization of other proteins and to preserve residual fibrillar structure. Although DOC has a tendency to slowly gel and precipitate in the presence of salts at neutral pH, the short time (less than one minute) and low concentrations of DOC necessary for extraction present no technical problem. Figure 5 illustrates percent of total

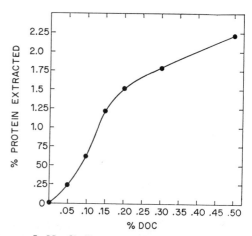

Figure 5. Myofibrillar protein extracted with DOC. Myofibrils were previously washed as described in the text including washes with 0.02% Triton-X-100 and 0.02% DOC. The extraction was performed by adding the desired amount of 10% DOC solution to 0.10 M KCl, 2 mM EGTA, 2 mM MgCl$_2$, 0.01 M Tris-maleate, 1 mM DTT pH 7. The weight ratio of wet pellet to extracting solution was 1:10. Protein was measured by the Biuret reaction (Gornall et al., 1949).

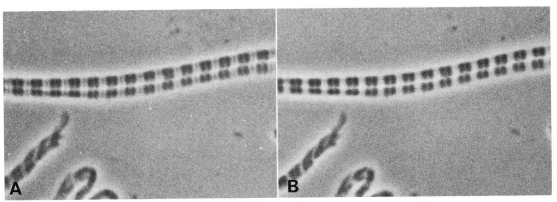

Figure 6. Phase contrast light micrographs of the same myofibrils before (*A*) and after (*B*) extraction with 0.15% DOC in RB for 1 min. Extraction was performed by irrigating the myofibrils beneath the coverslip. ×1800.

protein extracted from myofibrils by increasing concentrations of DOC. Between 0 and 0.15% DOC there is a rapid solubilization of 1.25% total myofibrillar protein, and additional proteins are extracted more slowly with higher concentrations. The extent of extraction is also dependent on the ratio of protein to DOC. Solution volume:pellet weight was 10:1 in these experiments. When followed in the phase contrast microscope, there is a rapid removal of Z band densities at 0.10–0.15% DOC; but upon further increase in DOC concentration, the density in the rest of the I band is reduced. That portion of the curve (Fig. 5) above 0.15% DOC partially reflects the slow solubilization of relaxing proteins associated with the thin filaments. This is supported by evidence from gel electrophoresis. M band density is partially extracted at 0.15% DOC and is increasingly

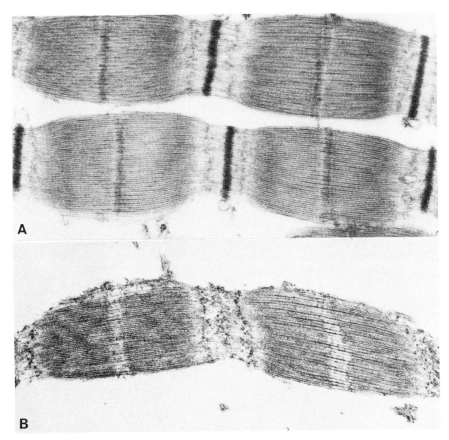

Figure 7. Electron micrographs of purified myofibrils before (*A*) and after (*B*) extraction with 0.25% DOC in RB for 10 min. Fixed in glutaraldehyde-OsO$_4$, embedded in araldite, stained with uranyl acetate and lead citrate. × 26,400.

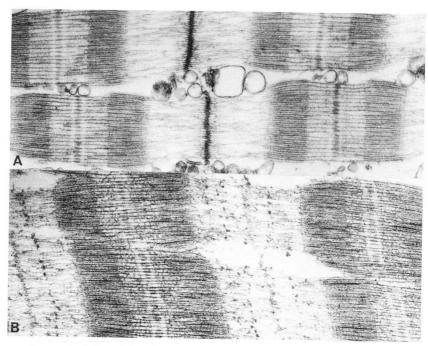

Figure 8. Electron micrographs before (A) and after (B) a 15 min DOC extraction of teased fiber bundles previously treated with 50% glycerol in RB for 48 hr. Fixed in glutaraldehyde-OsO₄, flat embedded in araldite, stained with uranyl acetate and lead citrate. × 22,400.

extracted with higher concentrations of the detergent. Thick filaments and thin filaments in the overlap region seem, however, highly resistant to DOC solubilization under these conditions, even up to 0.5% DOC. Figure 6 demonstrates the same fibril before and after 0.15% DOC extraction. Electron microscopy of isolated fibrils illustrates removal of M and Z band material (Fig. 7).

To obtain a clear understanding of fibrillar ultrastructure, particularly in the Z band region, it was necessary to perform extractions on oriented fiber bundles. In these experiments small muscle strips were isolated and soaked in the same relaxing solution described for myofibrils, tied to sticks, and then extracted for 48 hr with 50% glycerol in RB with two changes of solution. Small fiber bundles

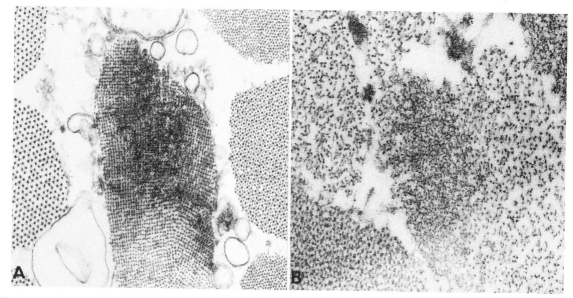

Figure 9. Z band cross sections of the same preparations used in Fig. 8. *A*, Cross section through unextracted fiber. ×41,000; *B*, cross section through 0.15% DOC-extracted fiber, ×53,000.

were teased out and extracted with RB containing 0.15% DOC. A branched network remains after removal of most of the dense Z band substance (Fig. 8). In cross section it appears (Fig. 9) that the 11-nm Z band lattice is partially disrupted after exposure to 0.15% DOC and there is a marked reduction in its electron density. In many areas, however, thin filaments appear to retain a branched structure connecting with other thin filaments entering the opposite side of the Z line. This is consistent with published models of the Z band (Knappies and Carlsen, 1972; Franzini-Armstrong and Porter, 1964; Rowe, 1971).

Fibrillar composition and protein localization. To elucidate fibrillar composition, the following SDS gel electrophoresis systems were used: 0.1% SDS acrylamide gels were made according to Weber and Osborn (1969), except that the acrylamide concentration was adjusted to 5%. A high pH buffer system, Tris-glycine pH 8.8, was used in the gels and in the electrode buffer (Paterson and Strohman, 1970). Specimen preparation and incubation were done by a modification of the method of Fairbanks et al. (1971) in which DTT and EDTA are included to maximize disulfide bond reduction and minimize proteolysis, respectively. In addition, all preparations were dissociated for 1 min at 100°C (Maizel, 1969). Finally the proteins were alkylated with an excess of iodoacetamide to prevent any reaggregation of proteins during electrophoresis (Shapiro et al., 1967). This electrophoresis system permits entry of all loaded protein into the gel and resolves more high molecular weight myofibrillar protein subunits than in previous reports (Scopes and Penny, 1971; Sender, 1971).

The preparations were run and characterized by calibrating the gels with independent protein markers and by coelectrophoresis of extracts with well-characterized myofibrillar proteins prepared by methods in the literature (Tsao, 1953; Arakawa et al., 1970; Dowben et al., 1965; Hartshorne and Mueller, 1969). Figure 10 illustrates the SDS gels obtained after several extraction procedures. The gel on the right labeled *MF* is the pattern obtained after loading on 100 μg of purified myofibrils before any extractions. By using varying amounts of protein and electrophoresing for different lengths of time, we have observed that other bands can be resolved as indicated in Fig. 10. The gel labeled *A + M band extract* is obtained by bathing uncontracted myofibrils for 10 min in 5 volumes of the following solution: 0.5 M KCl, 2 mM MgCl$_2$, 2 mM EGTA, 2 mM Na$_4$P$_2$O$_7$, 1 mM DTT, 0.1 M PO$_4$ pH 6.8. The fibrils appear to lose A and M band structure completely by this procedure and

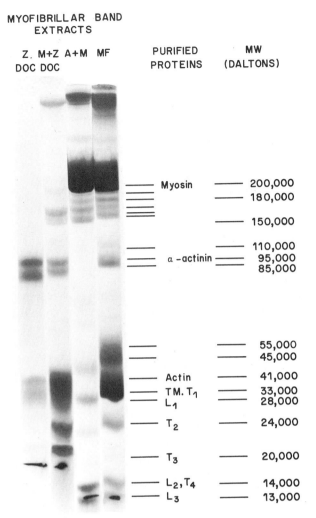

Figure 10. SDS-polyacrylamide gels prepared as described in the text. In addition, gels were stained with Coomassie blue in isopropanol-acetic acid and destained with a series of isopropanol-acetic acid solutions (Fairbanks et al., 1971). The mark near the bottom of the gels indicates the migration front of the tracking dye. The DOC extracts were seen to run in farther with respect to the tracking dye, although the mobilities of extracted proteins with respect to coelectrophoresed markers were not changed. *Tm*, Tropomyosin; *T*, troponin; *L*, light molecular weight components of myosin.

only I-Z-I segments remain, which often seem to be held together by a structure not resolvable with the light microscope. Extractions longer than 30 min, however, begin to extract Z and I band proteins.

If native myofibrils are extracted with 0.15% DOC in RB, as discussed above, the *M + Z DOC extract* is obtained. This extract is seen to include at least three bands which are also extracted with the whole A band discussed above. We presume these bands to be A band components, presumably some also in the M band. If the A + M band extraction is performed first (and repeated three times), followed by extraction with DOC, the *Z-DOC extract* is devoid of the bands above 150,000. This Z-DOC extract contains two bands at 95,000 and 85,000 daltons which were also present in the *M + Z band extract*. These latter proteins are presumably located in or near the Z band. Extraction of some proteins 13,000–45,000 daltons is also apparent above 0.10% DOC. Presumably these include relaxing proteins associated with the thin filament, based on analysis of molecular weight and the prior demonstration by Briggs and Fuchs (1963) that treatment of myofibrils with DOC can reduce EGTA sensitivity. The release of this material can be reduced with a brief 0.10% DOC extraction.

Also there are two relatively weak bands at 45,000 and 55,000 daltons which have a lower mobility than actin and which are sometimes enriched in either DOC extracts or high-salt A band extracts. The nature and localization of these proteins is not yet clear, although an M protein of subunit mol wt 43,000 has been reported recently (Morimoto and Harrington, 1972). The nature of the other bands above 200,000 daltons is unknown although these are always present in myofibrillar preparations. These bands are unlabeled in Fig. 10 since no suitable molecular weight markers exist for this region. A faint band at 110,000 daltons is sometimes seen in fibrils; however this band is substantially reduced during fibrillar purification and is not likely to be a component of the fibril.

Discussion

The major proteins released with DOC and with high ionic strength extraction are suggested to originate from the M band since M band material is removed in each case. M band protein has been isolated previously by ammonium sulfate fractionation of a crude Hasselbach-Schneider extract of muscle (Masaki and Takaiti, 1972). Although their original estimate of 110,000 mol wt for this subunit is inconsistent with our findings, it has been reported very recently that this preparation can be fractionated further into 155,000 and 90,000 dalton components. Only the heavy component, however, was found to give an immunological reaction with the M band. We have found that when a Hasselbach-Schneider type extraction is performed on either contracted myofibrils, impure fibrils, or for extended periods, additional bands appear on SDS gels. Since Masaki and Takaiti extracted muscle for two hours which had been homogenized in only buffered 0.1 M KCl, we suspect that their lower weight band may have been a nonmyofibrillar contaminant. The question of an aggregating factor (Masaki et al., 1968) which is claimed to promote side-to-side aggregation of LMM rods is of some

interest, although no firm evidence regarding the nature of this component has appeared. It is possible that this factor accounts for one of the high molecular weight bands in DOC or high ionic strength extracts noted in our gels. The 43,000 dalton subunit (Morimoto and Harrington, 1972) may also be an M band component and aggregating factor. It is reasonable to expect at least two protein subunits to appear in the extraction of the M band on ultrastructural grounds since, in the M band model proposed by Knappeis and Carlsen (1968), there are both M bridges which connect a thick filament to six surrounding thick filaments, and there are also M filaments which run parallel to the thick filaments and connect the M bridge networks. In addition, cross sections through parts of the M line often suggest the presence of an additional amorphous dense material in-between the M bridge structures analogous to the Z band ultrastructure. Therefore this would suggest that we might expect to remove at least three components in solubilizing M bands. It has not yet been ruled out that one of the 150–180,000 dalton DOC extractable proteins is also extracted from the rest of the A band.

The question of Z band composition is an important unsolved question. Estimates of dry mass using interference microscopy (Huxley and Hanson, 1957) have estimated that the Z band material, above that attributed to the rest of the I band, is 6% of the myofibrillar mass. It has been noted (Robson et al., 1970) that less than 1% of the myofibril consists of 6S α-actinin which they determined to be the active component of α-actinin. In addition α-actinin by itself gives only partial Z band reconstitution (Stromer et al., 1969). We have determined, on the basis of molecular weight and Z band removal, that one of the main proteins extracted by DOC is 6S α-actinin. We have also found that less than 1% of the total fibrillar protein can be accounted for by this protein. The nature of the second component removed by DOC, moving at 85,000 daltons, is uncertain. The possibility that one of these proteins is phosphorylase cannot yet be eliminated but is under investigation (Arakawa et al., 1970). We have found that with both DOC and extensive low ionic strength-high pH extractions, almost one-half of both the 95,000 and 85,000 dalton bands are retained by the myofibril. It is possible that some material extracted from the Z band binds to another part of the fibril. In any case, when Z band and even partial I band extraction is performed with high concentrations of DOC (0.5%), only 2% of the total fibrillar protein is extracted. The nature of the remaining backbone material is not clear. Antibodies to tropomyosin and actin do not bind to the Z band. When, however, we perform electrophoresis on fibrillar residues left after exhaustive A band and Z band extraction, only thin filament proteins remain. Perhaps antibody binding experiments performed on fibrils extracted of Z band dense material with DOC may show anti-actin and anti-tropomyosin binding to the Z band due to the unmasking of antigenic sites. Although 6S α-actinin and possibly another protein 85,000 daltons are removed as the dense material is removed, the possibility that DOC extracts contain a nonprotein component is now under investigation. Lipid and carbohydrate components of the Z band have been suggested in the past (Rash et al., 1968; Harsányi and Garamvölgyi, 1969; Walcott and Ridgway, 1967). The fact that a substantial amount of the 95,000 and 85,000 dalton proteins remain in the fibril after extraction, but that the dense material seen with phase contrast microscopy is removed with a rapid course of about 15 sec, suggests that another unidentified component is being released quickly during DOC extraction.

Finally the relationship of both Z and M band materials to myofibrillar assembly is uncertain. We have briefly discussed the cohesive nature of the myofibril and have suggested that at least certain M and Z components are not necessary for the formation of the myofilament lattice. This assembly information is most likely contained in the filaments themselves. A number of authors have noted the similarity of Z band density in the electron microscope to the electron-dense material associated with the inner surface of the plasma membrane at intercalated discs, desmosomes, and *zonalae adherentes.* (Franzini-Armstrong and Porter, 1964; D. Kelly, 1967; Rash et al., 1968). Some authors have suggested that Z bands are derived from the cell surface (Heuson-Stiennon, 1964; Auber, 1969; Hagopian and Spiro, 1970) and, as noted above, many investigators have commented upon the fact that fibrillar assembly is most frequently observed at the cell cortex (see review by Fischman, 1970). An important question, still unresolved, is whether the assembly of myofibrils, particularly initiation and longitudinal growth, is dependent upon a factor or factors (Z band material?) derived from the plasma membrane. Now that one can perform a partial biochemical dissection of the Z band, it is possible to utilize similar procedures in analyzing both the composition, structure, and formation of the dense materials at specializations of the plasma membrane and relate such to myofibrillar assembly.

Acknowledgments

This work was supported by USPHS grant #HE 13505-02, NSF grant #GB 24586 and the Harry Levine Memorial Foundation grant. J. E. acknowledges support from USPHS Biophysics training grant #GM 780. The excellent technical assistance of Miss R. Yambot and Mrs. L. Williams is gratefully acknowledged.

References

ARAKAWA, N., R. M. ROBSON, and D. E. GOLL. 1970. An improved method for the preparation of α-Actinin from rabbit striated muscle. *Biochim. Biophys. Acta* **200**: 284.

ASHLEY, C. A., K. R. PORTER, D. E. PHILPOTT, and G. M. HASS. 1951. Observations by electron microscopy on contraction of skeletal myofibrils induced with adenosinetriphosphate *J. Exp. Med.* **94**: 9.

AUBER, J. 1969. La myofibrillogenèse du muscle strié. II. Vertébrés. *J. Microscopie* (Paris) **8**: 367.

BRIGGS, F. N. and F. FUCHS. 1963. The nature of the muscle relaxing factor. *J. Gen. Physiol.* **46**: 883.

BUSCH, W. A., M. H. STROMER, D. E. GOLL, and A. SUZUKI. 1972. Ca^{++}-specific removal of Z lines from rabbit skeletal muscle. *J. Cell Biol.* **52**: 367.

CASPAR, D. L. D. and A. KLUG. 1962. Physical principles in the construction of regular viruses. *Cold Spring Harbor Symp. Quant. Biol.* **27**: 1.

CORSI, A. and S. V. PERRY. 1958. Some observations on the localization of myosin, actin and tropomyosin in the rabbit myofibril. *Biochem. J.* **68**: 12.

DOWBEN, R. M., W. M. LURRY, K. M. ANDERSON, and R. ZAK. 1965. Studies on Actin-Azomercurial complexes. *Biochemistry* **4**: 1264.

ETLINGER, J. D. and D. A. FISCHMAN. 1971. Preparation and characterization of thick and thin filaments from myofibrils. *Amer. Soc. Cell Biol. Abstr.*, p. 85.

FAIRBANKS, G., T. L. STECK, and D. F. H. WALLACH. 1971. Electrophoretic analysis of the major polypeptides of the human erythrocyte membrane. *Biochemistry* **10**: 2606.

FISCHMAN, D. A. 1967. An electron microscope study of myofibril formation in embryonic chick skeletal muscle. *J. Cell Biol.* **32**: 557.

———. 1970. The synthesis and assembly of myofibrils in embryonic muscle. In *Current topics in developmental biology*, ed. A. A. Moscona and A. Monroy, vol. 5, p. 235. Academic Press, New York.

FISCHMAN, D. A. and A. A. MOSCONA. 1971. Reconstruction of heart tissue from suspensions of embryonic myocardial cells: Ultrastructural studies on dispersed and reaggregated cells. In *Cardiac hypertrophy*, ed. N. R. Alpert. Academic Press, New York.

FISCHMAN, D. A. and R. ZAK. 1971. Evidence for the assembly of myofibrils in the absence of protein synthesis. *J. Gen. Physiol.* **57**: 245 (Abstr).

FRANZINI-ARMSTRONG, C. 1970. Natural variability in the length of thin and thick filaments in single fibers from a crab, *Portunus depurator*. *J. Cell Sci.* **6**: 559.

FRANZINI-ARMSTRONG, C. and K. R. PORTER. 1964. The Z disk of skeletal muscle fibrils. *Z. Zellforsch. Mikrosk. Anat.* **61**: 661.

GOLL, D. E., W. F. H. M. MOMMAERTS, M. K. REEDY, and K. SERAYDARIAN. 1969. Studies on α-actinin-like proteins liberated during trypsin digestion of α-actinin and of myofibrils. *Biochim. Biophys. Acta* **175**: 174.

GORNALL, A. G., C. J. BARDAWILL, and M. M. DAVID. 1949. Determination of serum proteins by means of the Biuret reaction. *J. Biol. Chem.* **177**: 751.

HAGOPIAN, M. and D. SPIRO. 1970. Derivation of the Z line in the embryonic chick heart. *J. Cell Biol.* **44**: 683.

HANSON, J., E. J. O'BRIEN, and P. M. BENNETT. 1971. Structure of the myosin-containing filament assembly (A-segment) separated from frog skeletal muscle. *J. Mol. Biol.* **58**: 865.

HARSÁNYI, V. and N. GARAMVÖLGYI. 1969. On the Z-substance of striated muscle. *Acta Biochim. Biophys. Acad. Sci. (Hung.)* **4**: 259.

HARTSHORNE, D. J. and H. MUELLER. 1969. The preparation of tropomyosin and troponin from natural actomyosin. *Biochim. Biophys. Acta* **175**: 301.

HEUSON-STIENNON, J. A. 1964. Morphogénese de la cellule musculaire striée au microscope électronique. I. Formation des structures fibrillaires. *J. Microscopie* (Paris) **4**: 657.

HOLTZER, H., J. M. MARSHALL, and H. FINCK. 1957. An analysis of myogenesis by the use of fluorescent antimyosin. *J. Biophys. Biochem. Cytol.* **3**: 705.

HUXLEY, H. E. 1963. Electron microscope studies of the structure of natural and synthetic protein filaments from striated muscle. *J. Mol. Biol.* **7**: 281.

HUXLEY, H. E. and J. HANSON. 1957. Quantitative studies on the structure of cross-striated myofibrils. *Biochim. Biophys. Acta* **23**: 229.

KELLY, D. E. 1967. Models of muscle Z-band fine structure based on a looping filament configuration. *J. Cell Biol.* **34**: 827.

KNAPPEIS, G. G. and F. CARLSEN. 1968. The ultrastructure of the M line in skeletal muscle. *J. Cell Biol.* **38**: 202.

MAIZEL, J. V. 1969. *Fundamental techniques of virology*, ed. K. Habel and N. P. Salzman, p. 334. Academic Press, New York.

MANASEK, F. J., B. BURNSIDE, and J. STROMAN. 1972. The sensitivity of developing cardiac myofibrils to cytochalasin-B. *Proc. Nat. Acad. Sci.* **69**: 308.

MASAKI, R., S. TAKAITI, and S. EBASHI. 1968. "M-substance," a new protein constituting the M-line of myofibrils. *J. Biochem.* (Tokyo) **64**: 909.

MASAKI, T. and O. TAKAITI. 1972. Purification of M-protein. *J. Biochem.* (Tokyo) **71**: 355.

MORIMOTO, K. and W. F. HARRINGTON. 1972. Isolation and physical chemical properties of an M-line protein from skeletal muscle. *J. Biol. Chem.* **247**: 3052.

MOSCONA, A. A. 1961. Rotation-mediated histogenetic aggregation of dissociated cells: a quantifiable approach to cell interactions *in vitro. Exp. Cell Res.* **22**: 45.

PAGE, S. 1965. A comparison of the fine structure of frog slow and twitch muscle fibers. *J. Cell Biol.* **26**: 477.

PATERSON, B. and R. C. STROHMAN. 1970. Myosin structure as revealed by simultaneous electrophoresis of heavy and light subunits. *Biochemistry* **9**: 4094.

RASH, J. E., J. W. SHAY, and J. J. BIESELE. 1968. Urea extraction of Z-bands, intercalated discs and desmosomes. *J. Ultrastructr. Res.* **24**: 181.

ROBSON, R. M., D. E. GOLL, N. ARAKAWA, and M. H. STROMER. 1970. Purification and properties of α-actinin from rabbit skeletal muscle. *Biochim. Biophys. Acta* **200**: 296.

ROWE, R. W. 1971. Ultrastructure of the Z line of skeletal muscle fibers. *J. Cell Biol.* **51**: 674.

SCOPES, R. K. and I. F. PENNY. 1971. Subunit sizes of muscle proteins as determined by sodium dodecyl sulphate gel electrophoresis. *Biochim. Biophys. Acta* **236**: 409.

SENDER, P. M. 1971. Muscle fibrils: Solubilization and gel electrophoresis. *FEBS Letters* **17**: 106.

SHAPIRO, A. L., E. VINUELA, and J. V. MAIZEL, JR. 1967. Molecular weight estimation of polypeptide chains by electrophoresis in SDS-polyacrylamide gels. *Biochem. Biophys. Res. Comm.* **28**: 815.

STARR, R. and G. OFFER. 1971. Polypeptide chains of intermediate molecular weight in myosin preparations. *FEBS Letters* **15**: 40.

STROMER, M. H., D. E. GOLL, and L. E. ROTH. 1967. Morphology of rigor-shortened bovine muscle and the effect of trypsin on pre- and post-rigor myofibrils. *J. Cell Biol.* **34**: 431.

STROMER, M. H., D. J. HARTSHORNE, H. MUELLER, and R. V. RICE. 1969. The effect of various protein fractions on Z- and M-line reconstitution. *J. Cell Biol.* **40**: 167.

TSAO, T. C. 1953. Fragmentation of the myosin molecule. *Biochim. Biophys. Acta* **11**: 368.

WALCOTT, B. and E. B. RIDGWAY. 1967. The ultrastructure of myosin-extracted striated muscle fibers. *Amer. Zool.* **7**: 499.

WEBER, K. and M. OSBORN. 1969. The reliability of molecular weight determination by dodecyl sulfate-polyacrylamide gel electrophoresis. *J. Biol. Chem.* **244**: 448.

WOLLENBERGER, A. 1964. Rhythmic and arrhythmic contractile activity of single myocardial cells cultured *in vitro. Circ. Res.* **14**: 184.

ZAK, R., J. ETLINGER, and D. A. FISCHMAN. 1972. Studies on the fractionation of skeletal and heart muscle, p. 163. In *Research in muscle development and the muscle spindle*, ed. B. Q. Banker et al. Ekcerpta Medica, Amsterdam.

Functions of Cytoplasmic Fibers in Non-Muscle Cell Motility

ROBERT D. GOLDMAN AND DAVID M. KNIPE*

Department of Biology, Case Western Reserve University, Cleveland, Ohio 44106

The finding of several types of fibers in the cytoplasm of a wide variety of cells has led to an increased interest in the general problem of non-muscle cell motility. The most extensively studied of these cytoplasmic fibers is the microtubule, which is thought to be directly involved in such diverse aspects of motility as chromosome movements during mitosis (Inoué and Sato, 1967), intracellular transport of organelles (Bickle et al., 1965; Freed and Lebowitz, 1970), and in the movement of cells during morphogenesis (Tilney and Gibbins, 1969; Byers and Porter, 1964). Microtubules are also considered to be the major component of a cytoskeletal system involved in the formation and maintenance of asymmetric cellular processes (Tilney and Gibbins, 1969; Tilney and Porter, 1965, 1967; Tilney et al., 1966; Goldman, 1971). Microfilaments, a second type of cytoplasmic fiber, are thought to function in cytoplasmic streaming, cytokinesis, membrane ruffling, cell locomotion, endocytosis and exocytosis (Schroeder, 1970; Wessells et al., 1971; Spooner et al., 1971; Orr et al., 1972; Goldman and Follett, 1969; Follett and Goldman, 1970; Goldman, 1972). A third type of cytoplasmic fiber, which we have called filament, has also been found in several types of cells (Goldman and Follett, 1969, 1970; Taylor, 1966). Filaments are intermediate in size between the large diameter microtubules and the small diameter microfilaments. They are similar in morphology to neurofilaments which are thought to be related to axoplasmic transport in nerves (Wuerker and Palay, 1969; Weiss and Mayr, 1971) and to the 100–120 Å filaments of unknown function found in smooth muscle cells (Cooke and Chase, 1971; Rice, this volume).

We have found all three types of fibers in a baby hamster kidney fibroblast cell line (BHK21). The microtubules of these cells are ∼250 Å in diameter, the filaments are ∼100 Å in diameter, and the microfilaments are ∼60 Å in diameter (Goldman, 1971, 1972; Goldman and Follett, 1969, 1970; Follett and Goldman, 1970). The fact that changes in the intracellular distribution of microtubules, microfilaments, and filaments have been observed

to be coincident with changes in the pattern of birefringence found in living BHK21 cells provides us with the rare opportunity to compare structural changes in the life cycle of single living cells with those changes observed with the electron microscope. In this paper we present evidence which supports the idea that these fibers play important roles in the spreading of cells on a substrate and in the attainment of cell shape. Attempts are currently being made to ascertain the specific roles of each of the fibers in these motile processes.

Cell Spreading

Observations with the light microscope. One of the simplest forms of motility which can be observed in cultures of BHK21 cells is the spreading of rounded cells into the characteristic shape of fibroblasts. Cells suspended in culture medium following their removal from petri dishes by trypsinization (Goldman and Follett, 1969; Follett and Goldman, 1970) are spherical in shape when observed with the light microscope. When placed on a solid substrate such as glass or plastic in medium maintained at 37°C, the cells attach and begin to spread over the substrate. During this spreading process there is active membrane ruffling (Figs. 1c, d). A large spherical region which excludes most large cytoplasmic particles is seen in a juxtanuclear position during the early stages of cell spreading. This spherical region is birefringent when observed with polarized light optics (Figs. 1a, b, d). As cells continue to spread, the birefringent sphere becomes diffuse and appears to give rise to some of the birefringent material seen along the long axes of fully spread cells (Fig. 2) (Goldman and Follett, 1970).

Observations with the electron microscope. Cells were fixed at various stages during spreading and were processed for electron microscopy (Goldman, 1971). Many cells observed within 15–30 minutes after contact with a substrate contain a spherical region which excludes most large cytoplasmic organelles, such as mitochondria (Fig. 3a). At higher magnification this spherical region, which is the birefringent sphere seen within living cells, is seen to contain 100 Å filaments (Fig. 3b). Microtubules are few in number in these cells and

* Present address: Department of Biology, Massachusetts Institute of Technology, Cambridge, Mass.

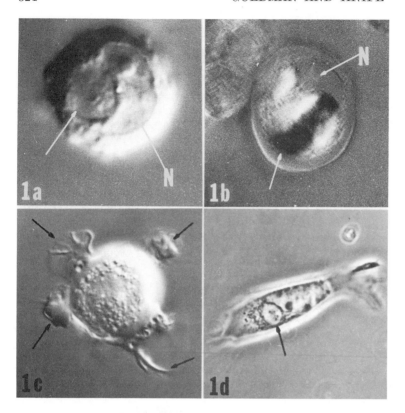

Figure 1. A series of light micrographs of living BHK21 cells during the early stages of spreading. *a.* A cell observed 15 min after being placed on a glass coverslip in medium maintained at 37°C. Note spherical shape of cell and the large spherical region (arrow) adjacent to the nucleus (*N*). Nomarski optics, ×2800. *b.* A similar cell observed with polarized light optics, showing birefringence (arrow) in a juxtanuclear (*N*) position. ×1440. *c.* Cell observed 30 min after being placed on a coverslip. Membrane ruffling is obvious in spreading regions (arrows). Nomarski optics, ×1600. *d.* Cell observed with phase contrast optics 30 min following attachment to glass. Note the juxtanuclear sphere (arrow). ×400.

when found are usually scattered randomly in the cytoplasm (Fig. 4). There are very few recognizable microfilaments. Those present are found in microvilli at the cell surface only (Fig. 5) (Follett and Goldman, 1970).

The electron microscopy of cells which are well spread on a substrate reveals the presence of large numbers of microtubules, filaments, and submembranous bundles of microfilaments. These three fibers are frequently arranged longitudinally along the long axis of the fibroblastic cell processes (Fig. 6). This arrangement indicates that they are at least in part responsible for the birefringence seen along the long axes of living BHK21 cells (Goldman, 1971; Goldman and Follett, 1969,

1970). Microtubules and filaments are frequently associated with each other in the fibroblastic processes. In cross sections of these processes the filaments are usually located outside a clear zone which surrounds microtubules (Fig. 7).

From these observations of the spreading process we may conclude that rounded up cells contain few morphologically visible microtubules and microfilaments but contain a large juxtanuclear accumulation of filaments (the birefringent sphere region). As cells spread on the substrate, large numbers of microfilaments and microtubules are formed and filaments are distributed throughout the spreading cell from their juxtanuclear position. These observations suggest that the assembly of microtubules

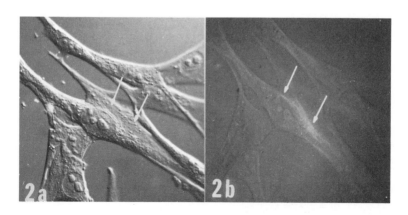

Figure 2a, b. A living, fully spread BHK21 cell in Nomarski (*a*) and polarized light (*b*) optics. Note birefringent region along long axis of cell (arrows). ×400.

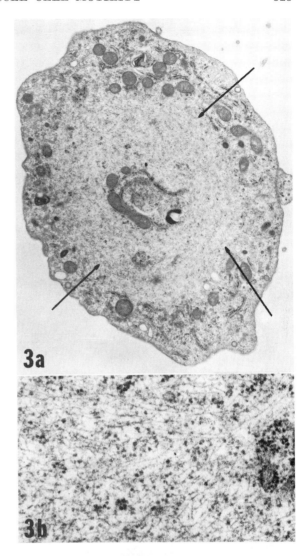

Figure 3a, b. Electron micrograph of a thin section through a cell fixed within 15 min after attachment to the substrate. A large spherical region is apparent due to the exclusion of most of the large cell organelles (arrows). At higher magnification this region is seen to contain filaments (*b*). *a*, ×8,000; *b*, ×41,900.

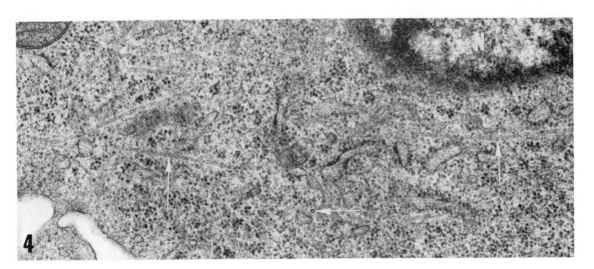

Figure 4. Electron micrograph of the nuclear (*N*) region of a cell 15 min following attachment. Note the randomly distributed microtubules in both longitudinal and cross section (arrows). ×33,000.

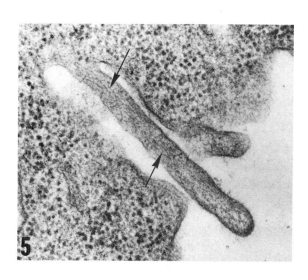

Figure 5. Electron micrograph of a microvillus found at the surface of a cell 15 min after being placed on a substrate. Microfilaments (arrows) are seen along its long axis. ×48,000.

Figure 6. An electron micrograph of a longitudinal section through the long axis of a fibroblastic cell process. The cell was flat-embedded on a plastic petri dish and thin sections were taken parallel to the substrate. Note bundles of microfilaments (*mf*), microtubules (*mt*), and filaments (*f*). ×48,000.

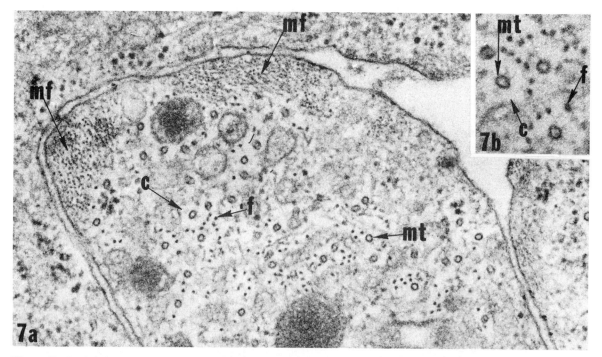

Figure 7a, b. An electron micrograph of a cross section through a fibroblastic cell process. Microtubules (*mt*) are usually surrounded by a clear zone (*c*) and filaments (*f*). Cross sections through the submembranous bundles of microfilaments (*mf*) are also present. Figure 7b is a blow up of the same electron micrograph to show more clearly the microtubule-filament relationship. Fine fibrous material is frequently seen in the clear zone (*c*). *a*, ×72,700; *b*, ×156,000.

and microfilaments, as well as the redistribution of filaments, play important roles in the normal cell spreading process.

Assembly of Fibers during Cell Spreading

Since spreading begins immediately after the attachment of cells to a substrate and since large numbers of oriented microfilament bundles and microtubules form very rapidly, it is of interest to know whether or not precursors exist in the cytoplasm from which microtubules and microfilaments are assembled during cell spreading. To answer this question, conditions were established where de novo synthesis of precursors to microfilaments and microtubules could not occur. The drug cycloheximide at a concentration of 20 μg/ml inhibits over 95% protein synthesis in BHK21 cells within less than 30 min after addition to a cell culture. The effects of this drug were assessed by observing the incorporation of [³H]leucine into cell protein utilizing radioautographic techniques.

Cells from a suspension were placed onto a glass or plastic substrate in medium containing 20 μg of cycloheximide/ml and were observed for periods up to 8 hr. The cells attach and spread normally. Electron microscopy reveals that these cells contain large numbers of microfilament bundles and microtubules. In addition the filaments disperse through the cytoplasm when cell spreading proceeds in the presence of cycloheximide. Cells washed free of cycloheximide after 8 hr in the drug grow normally, and within one or two days confluent monolayers of cells are formed. Cells left in cycloheximide do not grow. Thus under conditions in which over 95% of protein synthesis is inhibited, microtubules and microfilaments are formed apparently normally and filaments are distributed normally within the cytoplasm.

Based on these experiments with cycloheximide, it seems likely that pools of precursors do exist in the cytoplasm of round cells, and that these are involved in the assembly of at least two of the fibers, namely, the microtubules and the microfilaments. In contrast to the assembly of microfilaments and microtubules which occurs during cell spreading, the filaments appear to be stored as fully formed fibers in a juxtanuclear position in the rounded up (or suspended) cells.

Colchicine as a Probe for Microtubule Function

Colchicine has been used as a specific probe for microtubule function in a wide variety of cells. When cells are treated with colchicine, cytoplasmic microtubules disappear (Tilney and Gibbins, 1969; Goldman, 1971; Tilney, 1968). The drug is thought to act by binding to microtubule subunit protein and subsequently causing the breakdown or depolymerization of microtubules (Borisy and Taylor, 1967). As the effects of the drug are completely reversible, colchicine makes an ideal tool with which to probe microtubule function in living cells.

The general effects of colchicine on BHK21 cells have been described in detail in a previous publication (Goldman, 1971). In order to examine the role of microtubules in the cell spreading process BHK21 cells were taken from a suspension and placed on a substrate in medium containing 40 μg colchicine/ml. The cells still attach and spread, but the typical fibroblastic cell shape is not attained by the cells. Instead most of the cells appear more epithelial-like due to the lack of the very long cellular processes which are characteristic of fibroblasts (Fig. 8). Microtubules are not found in cells which have spread in colchicine, and the

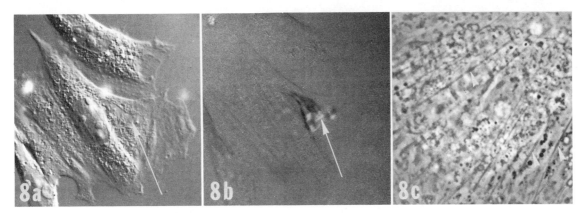

Figure 8a, b, c. BHK21 cell which has attached and spread in medium containing 40 μg colchicine/ml. The cell does not possess the very long fibroblastic processes seen in normal BHK21 cells. The juxtanuclear accumulation of filaments is seen as a relatively clear cap in Nomarski optics (arrow, a) and is birefringent when viewed with polarized light optics (arrow, b). Fine birefringent fibers are also seen running through the cytoplasm in (b), and similar fibers are seen with phase contrast in another cell (c) which has spread in colchicine. a and b, ×560. c, ×400.

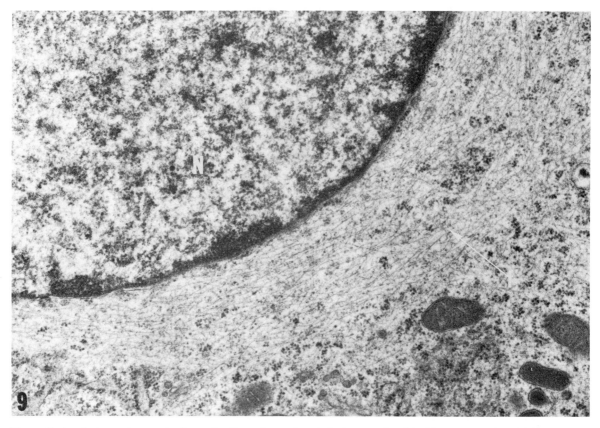

Figure 9. An electron micrograph of a section through a region of the juxtanuclear birefringent cap of a colchicine-treated cell. This cap contains masses of oriented arrays of filaments (*f*). ×30,250.

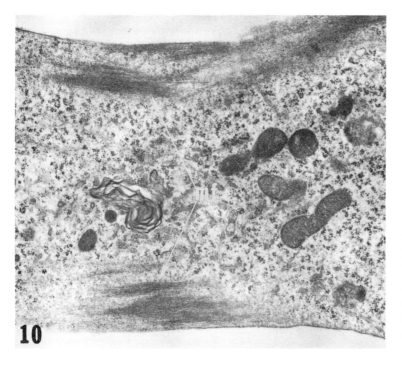

Figure 10. Electron micrograph of a thin section of a flat embedded cell which has spread in 40 μg colchicine/ml. The section was taken parallel to the substrate and just below the cell membrane. Several prominent bundles of microfilaments are seen (*mf*). Note the absence of microtubules and filaments, which would normally be seen in similar sections through untreated spread cells. ×16,500.

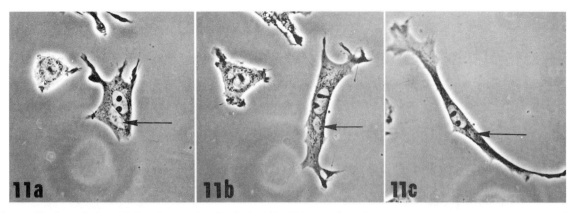

Figure 11a, b, c. Series of light micrographs of a single cell which was allowed to spread and remain in 40 μg colchicine/ml for 18 hr. The colchicine was then replaced with normal medium and the cell was observed as the colchicine effect was reversed. *a.* Cell in colchicine containing a prominent juxtanuclear cap (arrow). *b.* The same cell 50 min following the removal of colchicine. Note that the cap (arrow) is smaller and that the cell is forming fibroblastic processes. *c.* The same cell 100 min following the removal of colchicine. The cap (arrow) is more elongated and much less prominent. The cell has returned to its normal fibroblastic configuration. Phase contrast optics, ×400.

filaments remain arrested in a juxtanuclear position (Fig. 9) (Goldman, 1971). This accumulation of filaments may be seen as a juxtanuclear cap in living cells viewed with Nomarski and polarized light optics (Figs. 8a, b). Living cells which have spread in colchicine also contain oriented arrays of fine birefringent fibers distributed through the cytoplasm (Fig. 8b). These fibers are also apparent in phase contrast optics (Fig. 8c). The electron microscopy of similar cells reveals that the fine fibers seen in living cells appear to consist of submembranous bundles of microfilaments (Fig. 10). The colchicine effect is rapidly reversible, and once the colchicine is removed from the cells major fibroblastic processes are formed and the juxtanuclear cap of filaments is dispersed throughout the cytoplasm (Fig. 11).

These experiments with colchicine indicate that microtubules are an integral part of a cytoskeletal system necessary for the formation of normal fibroblastic cell shape. The effects of colchicine also demonstrate that the microtubule-filament association seen in normal cells (see Figs. 7a, b) may be of functional significance because in the absence of microtubules the filaments are not dispersed throughout the cytoplasm as cells spread. Another conclusion which may be drawn from these experiments is that normal-appearing microfilament bundles form as cells spread in colchicine. This finding indicates that microfilaments may be more directly involved in the spreading process than either microtubules or filaments.

HMM Binding and Actin-like Nature of Microfilaments

Ishikawa et al. (1969) found actin-like microfilaments in a wide variety of non-muscle cells using the technique of skeletal muscle HMM binding in glycerinated models of cells. Glycerinated models of spread BHK21 cells were prepared according to the techniques of Ishikawa et al. (1969). Rabbit skeletal muscle HMM (prepared according to the techniques of Szent-Györgyi [1951] and Pollard et al. [1970]) was added to the glycerinated BHK21 cells and then the cells were fixed and prepared for electron microscopy (Goldman, 1971). Figure 12a demonstrates that submembranous microfilaments are found which interact with HMM to form "decorated" actin-like arrays. Controls not subjected to HMM possess normal-appearing microfilament bundles (Fig. 12b). Microtubules and filaments do not bind HMM (Fig. 12c).

We had proposed in an earlier paper (Goldman and Follett, 1969) that the submembranous microfilament bundles were at least part of a contractile layer responsible for certain motile activities of the cell membrane, including membrane ruffling, pinocytosis, etc. Therefore based on the finding of submembranous microfilaments which interact with HMM, the microfilaments provide us with one good candidate for a contractile system beneath the membrane.

Effects of Cytochalasin B

Numerous recent papers have put forth the idea that cytochalasin B (CB) selectively disrupts microfilament function in a wide variety of non-muscle cells. This is based on the finding that morphological changes in microfilaments are associated with CB treatment in some cellular systems (Wessells et al., 1971). We have attempted to test the functions of microfilaments during cell spreading by using CB in concentrations ranging

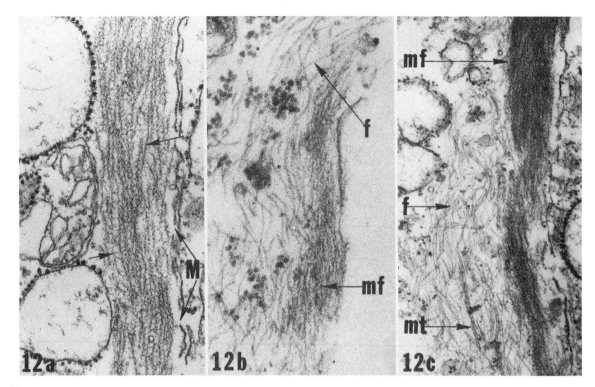

Figure 12. *a.* An electron micrograph of a cell which has been glycerinated and treated with HMM. A large bundle of "decorated" microfilaments is seen just below the cell membrane (*M*). These microfilaments are 130–150 Å thick and have many fine projections sticking out at angles from their long axes. Arrowhead configurations (arrows) are apparent in some regions, indicating that the HMM-microfilament interaction is very similar to the interaction of HMM and skeletal muscle actin. ×44,000. *b.* An electron micrograph of a cell which has been glycerinated but not treated with HMM. A submembranous bundle of ∼60 Å microfilaments is apparent (*mf*). Filaments (*f*) are also present. ×64,000. *c.* A glycerinated cell treated with HMM showing a thick bundle of decorated microfilaments (*mf*) and normal appearing filaments (*f*) and microtubules (*mt*). ×30,250.

from 0.1–10.0 µg/ml. When cells are suspended in medium containing 0.1 µg/ml CB and then placed immediately on a substrate, the cells attach, spread, and grow normally. At concentrations of 0.5–1.0 µg/ml, the cells attach and spread; however membrane ruffling is inhibited (Fig. 13a). Electron microscopy of cells which have spread in 1 µg/ml CB reveals a distribution of submembranous microfilament bundles, microtubules, and filaments

indistinguishable from normal spread cells. A large proportion of cells treated with 0.5–1.0 µg/ml CB for 18–24 hr are binucleate, indicating that cytokinesis is inhibited. Time lapse movies of similar cell populations demonstrate that cell locomotion is also inhibited at these lower drug concentrations.

BHK21 cells suspended in 5.0–10.0 µg CB/ml and then placed on a solid substrate do not spread

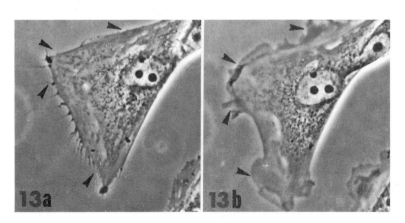

Figure 13. *a.* Cell which has attached and spread in 1.0 µg CB/ml and observed 12 hr later. This type of spread cell would normally show active ruffling at its free edges (arrows). *b.* The same cell observed 60 sec following the removal of CB. Note the immediate formation of expanding and ruffling membranes (compare arrows in both micrographs). ×400.

well on the substrate, and as a result the morphology of these cells is grossly altered (Goldman, 1972). The cells spread more slowly but eventually spread into spiky configurations (Fig. 14a). The portions of cells which have spread over the substrate (the spikes) always contain longitudinally oriented microtubules and filaments. Microfilament bundles, however, are extremely difficult to find in these cells, indicating a great reduction in the number of microfilament bundles formed (Fig. 14c). Bundles of microfilaments are found with greater ease in cells to which 10 μg CB/ml has been added following normal spreading (Goldman, 1972). In other words it would appear as if, once formed, microfilament bundles might be more resistant to change upon the addition of high concentrations of CB. The CB effects at all concentrations are rapidly reversible (Figs. 13a, b and 14a, b).

Attempts were made to determine the nature of action of CB in several other ways. Since rounded cells in a suspension contain very few recognizable microfilaments and since the cycloheximide experiments suggest that precursors for microfilaments exist in the cytoplasm of rounded cells, suspended cells should provide us with the ideal conditions for testing whether or not cytochalasin B acts by binding to microfilament subunits, thereby preventing their assembly into microfilaments. If the drug acts by binding to subunits of microfilaments to prevent their assembly, long term pretreatment of suspended cells with effective concentrations of CB should prevent the formation of microfilament bundles. Cells were suspended into either 1 or 10 μg CB/ml for 30–60 min. The cells were then placed on substrates. Cells still attached and spread into their "normal" CB configuration; that is, cells in 1 μg/ml become well spread (see Fig. 13a) and in 10 μg/ml become spiky (see Fig. 14a). Cells treated in this way were examined in the electron microscope and were found to contain bundles of microfilaments. It must be reemphasized, however, that cells pretreated with 10 μg/ml CB contained very few recognizable microfilament bundles.

If CB disrupts microfilament function, the formation of decorated microfilaments with HMM might also be altered. Cells treated for 18–24 hours with 1 or 10 μg CB/ml were glycerinated and treated with HMM in the presence of CB. These cells contain decorated microfilaments indistinguishable from decorated microfilaments of untreated cells (see Fig. 12a).

Based on these experiments with CB, we conclude that further studies are necessary before we can definitely determine whether CB can be used as a reliable and specific probe for microfilament function in BHK21 cells. Our reluctance regarding the use of CB stems mainly from the observation that cells treated with low doses of the drug which inhibit membrane ruffling, movement, and cytokinesis, contain an apparently normal distribution

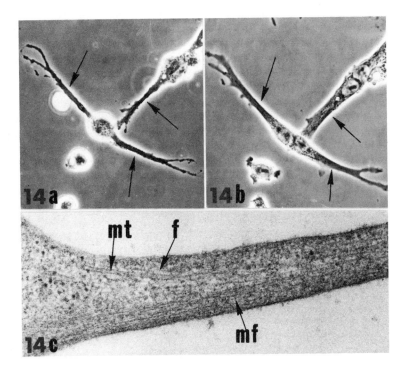

Figure 14. *a.* Cells which have attached and spread into spiky configurations in the presence of 10 μg CB/ml (arrows) and observed 12 hr later. *b.* The same cells observed 10 min following the removal of CB by washing with normal medium. Normal fibroblastic processes are apparent in both cells (arrows). Note that both of these cells possess two nuclei indicating that cytokinesis has been inhibited. Phase contrast optics, ×400. *c.* An electron micrograph through one of the small spikes formed in a cell which has spread in 10 μg CB/ml. Microfilaments (*mf*), microtubules (*mt*), and filaments (*f*) are present. Spikes such as this one which contain all three fibers are very difficult to locate. Most of the spikes contain filaments and microtubules, but there are few, if any, recognizable microfilaments. ×44,000.

of microfilaments. However at higher drug concentrations it is apparent that both the formation of microfilament bundles and cell spreading are inhibited, indicating that some microfilament precursors might interact with CB to prevent their assembly.

The possibility exists that low doses of CB act in a manner which does not result in obvious ultrastructural alterations in microfilaments. The fact that G actin can polymerize in the presence of CB (Spudich and Lin, 1972; Spudich, this volume) and that HMM is bound to F actin in vitro in the presence of very high concentrations of CB (Forer et al., 1972) agrees with our observations that microfilaments of BHK21 cells are assembled and interact with skeletal muscle HMM in the presence of CB. Spudich and Lin (1972) have also demonstrated that the intrinsic viscosity of skeletal muscle F actin drops significantly in the presence of low levels of CB (5 μg and below). Decreases in intrinsic viscosity do not necessarily indicate a significant breakdown in the fibrous structure of F actin or actin-like proteins, but may upset the physiological function of actin-like microfilaments in vivo. This type of finding may help to explain why there is no obvious effect on microfilaments at low concentrations of CB, even though cell motility is inhibited.

Considering the preliminary and uncertain nature of the evidence available regarding the molecular effects of CB, it is impossible to determine at the present time the exact site of drug action. Indeed there are many arguments in favor of a specific CB effect at the outer cell surface (glycocalyx) or at the plasma membrane which, secondarily, might result in changes in microfilament structure and function. These studies have demonstrated that CB inhibits the transport of glucose (Estensen and Plagemann, 1972), the synthesis of mucopolysaccharides (Sanger and Holtzer, 1972), and perhaps membrane interactions during cytokinesis (Bluemink, 1971; Estensen, 1971).

Discussion and Summary

As we have emphasized only the phenomenon of cell spreading in this paper, it is worthwhile noting that rounding up and respreading of BHK21 cells is a normal part of the growth cycle of these cells and of most normal cells in culture. Spread cells round up just prior to cell division and, following mitosis, each rounded daughter cell respreads into a fibroblastic configuration. Therefore the disassembly of microtubules and microfilaments and accompanying changes in the distribution of filaments are probably important factors in the rounding up of cells for division. It would also be expected that the controlled assembly of cytoplasmic fibers would be at least partially responsible for the respreading of daughter cells formed as a result of the mitotic process.

The submembranous bundles of microfilaments are most likely to be involved in motile activities of the cell surface, which include pinoctyosis, cytokinesis, and cell locomotion and in the general configuration and topography of the cell surface (Schroeder, 1970; Wessells et al., 1971; Goldman and Follett, 1969; Follett and Goldman, 1970). The fact that skeletal muscle HMM interacts with similarly located fibers seen in glycerinated BHK21 cells provides evidence that these microfilaments are structurally and functionally similar to actin. In addition biochemical studies have demonstrated that ten percent or more of the total extractable protein of a wide range of non-muscle cells is structurally very similar to actin (Fine and Bray, 1971; Bray, this volume; Anderson and Gesteland, 1972; Adelstein and Conti, this volume). Thus it also seems likely, on biochemical grounds, that the large number of submembranous microfilaments found in spread cells are actin-like.

Proteins similar to myosin have been isolated from a variety of non-muscle cells, including cultured fibroblasts (Adelstein and Conti, this volume; Adelman and Taylor, 1969; Pollard and Korn, this volume; Adelstein et al., 1971). However the localization of myosin-like protein is unknown. If this protein were present in small amounts relative to the actin-like microfilament protein and if it were organized within the microfilament bundles, it would be difficult to detect in our electron microscope preparations. If myosin-like proteins were localized in the submembranous layer to form an actomyosin-like contractile system, then this layer could account for many of the motile phenomena attributed to cell surfaces.

The microtubules of BHK21 cells, based on their localization and on the effects of colchicine, play an essential role in the formation of the long asymmetric processes characteristic of spread fibroblasts. Microtubules are also an important factor in the structural maintenance of the processes once they have formed. This has been demonstrated by treating fully spread BHK21 cells with colchicine. The cells rapidly lose their long processes as the microtubules break down. This phenomenon is completely reversible and may be repeated many times with the same cell (Goldman, 1971).

The filaments are related to motility primarily due to the changes in their localization during cell spreading. Since there are no known agents available which interact with filaments so as to disrupt their activities, very little is known about their in vivo function. Since filaments are parallel to the microtubules which lie along the long axis of the fibroblastic processes of spread cells, we have suggested that this association might be a functional one (Goldman and Follett, 1969). One possible function is that the microtubules and filaments form a complex involved in the intracellular transport of organelles such as mitochondria, which move in straight tracks along the length of the fibroblastic processes of living BHK21 cells. Similar associations between microtubules and filaments are seen in nerve cells where they might be related to the organelle movements seen in axons (Wuerker and Palay, 1969; Weiss and Mayr, 1971; Burdwood, 1965). Filaments have also been isolated from nerve and smooth muscle cells. Preliminary biochemical studies suggest that they contain protein which is not analogous to actin, myosin, or microtubule protein (Cooke and Chase, 1971; Rice, this volume; Huneeus and Davson, 1970).

In conclusion it appears obvious that one of the major difficulties involved in studying non-muscle cell motility is the lack of a relatively stable and highly organized structure such as is seen in striated muscle. This is due to the fact that the fibers in growing BHK21 cells are dynamic and that during active growth they are constantly being assembled and disassembled, which makes a study of their structure and function extremely difficult.

Acknowledgments

We thank Anne Bushnell and Germaine Berg for their excellent technical assistance, and Elizabeth Jones and Anne Goldman for helping with the manuscript. Lois Dickerman provided the autoradiographic results for the cycloheximide experiments. This work was supported by grants from the American Cancer Society (E-639), the Damon Runyon Memorial Fund for Cancer Research, Inc. (DRG1083A) and the National Science Foundation (GB23185). R.D.G. is the recipient of a National Institutes of Health Public Health Service Career Development Award (1-K4-GM-32,249-01).

References

ADELMAN, M. R. and E. W. TAYLOR. 1969. Further purification and characterization of slime mold myosin and slime mold actin. *Biochemistry* **8**: 4976.

ADELSTEIN, R. S., T. D. POLLARD, and W. M. KUEHL. 1971. Isolation and characterization of myosin and two myosin fragments from human blood platelets. *Proc. Nat. Acad. Sci.* **68**: 2703.

ANDERSON, C. W. and R. F. GESTELAND. 1972. Pattern of protein synthesis in monkey cells infected by Simian virus 40. *J. Virol.* **9**: 758.

BICKLE, D., L. TILNEY, and K. R. PORTER. 1965. Microtubules and pigment migration in the melanophores of *Fundulus heteroclitus* L. *Protoplasma* **61**: 322.

BLUEMINK, J. G. 1971. Cytokinesis and cytochalasin-induced furrow regression in the first-cleavage zygote of *Xenopus laevis*. *Z. Zellforsch.* **121**: 102.

BORISY, G. G. and E. W. TAYLOR. 1967. The mechanism of action of colchicine. Binding of colchicine-H³ to cellular protein. *J. Cell Biol.* **34**: 525.

BURDWOOD, W. O. 1965. Rapid bidirectional particle movements in neurons. *J. Cell Biol.* **27**: 115A.

BYERS, B. and K. R. PORTER. 1964. Oriented microtubules in elongating cells of the developing lens rudiment after induction. *Proc. Nat. Acad. Sci.* **52**: 1091.

COOKE, P. H. and R. H. CHASE. 1971. Potassium insoluble microfilaments in vertebrate smooth muscle cells. *Exp. Cell Res.* **66**: 417.

ESTENSEN, R. D. 1971. Cytochalasin B. I. Effect on cytokinesis of Novikoff hepatoma cells. *Proc. Soc. Exp. Biol. Med.* **136**: 1256.

ESTENSEN, R. D. and P. PLAGEMANN. 1972. Cytochalasin B: Inhibition of glucose and glucosamine transport. *Proc. Nat. Acad. Sci.* **69**: 1430.

FINE, R. E. and D. BRAY. 1971. Actin in growing nerve cells. *Nature New Biol.* **243**: 115.

FOLLETT, E. A. C. and R. D. GOLDMAN. 1970. The occurrence of microvilli during spreading and growth of BHK-21/C13 fibroblasts. *Exp. Cell Res.* **59**: 124.

FORER, A., J. EMMERSEN, and D. BEHNKE. 1972. Cytochalasin B: Does it affect actin-like filaments? *Science* **175**: 774.

FREED, J. J. and M. M. LEBOWITZ. 1970. The association of a class of saltatory movements with microtubules in cultured cells. *J. Cell Biol.* **45**: 334.

GOLDMAN, R. D. 1971. The role of three cytoplasmic fibers in BHK-21 cell motility. I. Microtubules and the effects of colchicine. *J. Cell Biol.* **51**: 752.

————. 1972. The effects of Cytochalasin B on the microfilaments of baby hamster kidney (BHK-21) cells. *J. Cell Biol.* **52**: 246.

GOLDMAN, R. D. and E. A. C. FOLLETT. 1969. The structure of the major cell processes of isolated BHK-21 fibroblasts. *Exp. Cell Res.* **57**: 263.

————. 1970. Birefringent filamentous organelle in BHK-21 cells and its possible role in cell spreading and motility. *Science* **169**: 286.

HUNEEUS, F. C. and P. F. DAVSON. 1970. Fibrillar proteins from squid axons. *J. Mol. Biol.* **52**: 415.

INOUÉ, S. and H. SATO. 1967. Cell motility by labile association of molecules. The nature of mitotic spindle fibers and their role in chromosome movement, p. 259. In *The contractile process*. N.Y. Heart Ass. Symp. Little, Brown, Boston.

ISHIKAWA, H., R. BISCHOFF, and H. HOLTZER. 1969. Formation of arrowhead complexes with heavy meromyosin in a variety of cell types. *J. Cell Biol.* **43**: 312.

ORR, T. S. C., D. E. HALL, and A. C. ALLISON. 1972. Role of contractile microfilaments in the release of histamine from mast cells. *Nature* **236**: 350.

POLLARD, T. D., E. SHELTON, R. WEIHING, and E. KORN 1970. Ultrastructural characterization of F-actin isolated from *Acanthamoeba castellanii* and identification of cytoplasmic filaments as F-actin by reaction with rabbit heavy meromyosin. *J. Mol. Biol.* **50**: 91.

SANGER, J. W. and H. HOLTZER. 1972. Cytochalasin B: Effects on cell morphology, cell adhesion and mucopolysaccharide synthesis. *Proc. Nat. Acad. Sci.* **69**: 253.

SCHROEDER, T. E. 1970. The contractile ring. I. Fine structure of dividing mammalian (HeLa) cells and the effects of Cytochalasin B. *Z. Zellforsch.* **109**: 43.

SPOONER, B. S., K. M. YAMADA, and N. K. WESSELLS. 1971. Microfilaments and cell locomotion. *J. Cell Biol.* **49**: 595.

SPUDICH, J. A. and S. LIN. 1972. Cytochalasin B, its interaction with actin and actomyosin from muscle. *Proc. Nat. Acad. Sci.* **69:** 442.

SZENT-GYÖRGYI, A. C. 1951. *The chemistry of muscular contraction*, 2nd Ed. Academic Press, N.Y.

TAYLOR, A. C. 1966. Microtubules in the microspikes and cortical cytoplasm of isolated cells. *J. Cell Biol.* **28:** 155.

TILNEY, L. G. 1968. Studies on the microtubules in Heliozoa. IV. The effect of colchicine on the formation and maintenance of the axopodia of *Actinosphaerium nucleofilum. J. Cell Sci.* **3:** 549.

TILNEY, L. G. and J. R. GIBBINS. 1969. Microtubules and filaments in the Filopodia of secondary mesenchyme cells of *Arbacia puntulata* and *Echinarachnius parma. J. Cell Sci.* **5:** 195.

TILNEY, L. G. and K. R. PORTER. 1965. Studies on the microtubules in Heliozoa. I. Fine structure of *Actinosphaerium* with particular reference to axial rod structure. *Protoplasma* **60:** 317.

TILNEY, L. G. and K. R. PORTER. 1967. Studies on the microtubules in Heliozoa. II. The effect of low temperature on these structures in the formation and maintenance of axopodia. *J. Cell Biol.* **34:** 327.

TILNEY, L. G., Y. HIRAMOTO, and D. MARSLAND. 1966. Studies on the microtubules in Heliozoa. III. A pressure analysis of the role of these structures in the formation and maintenance of the axopodia of *Actinosphaerium nucleofilum* (Barrett). *J. Cell Biol.* **29:** 77.

WEISS, P. A. and R. MAYR. 1971. Organelles in neuroplasmic ("Axonal") flow: Neurofilaments. *Proc. Nat. Acad. Sci.* **68:** 846.

WESSELLS, N. K., B. S. SPOONER, J. F. ASH, M. O. BRADLEY, M. A. LUDUENA, E. L. TAYLOR, J. T. WRENN, and K. M. YAMADA. 1971. Microfilaments in cellular and developmental processes. *Science* **171:** 135.

WUERKER, R. B. and S. L. PALAY. 1969. Neurofilaments and microtubules in anterior horn cells of the rat. *Tissue and Cell* **1:** 387.

Myosin Messenger RNA: Studies on Its Purification, Properties and Translation during Myogenesis in Culture

Glenn E. Morris,* Elizabeth A. Buzash, Arthur W. Rourke, Katherine Tepperman, William C. Thompson, and Stuart M. Heywood

Biological Sciences, Genetics and Cell Biology Section, University of Connecticut, Storrs, Connecticut 06268

Although a great deal is known concerning the chemistry and function of the contractile proteins, it has been only recently that progress has been made in attempts to gain a detailed understanding of the biosynthesis of these proteins. It is our ultimate goal to elucidate the control mechanisms involved in myofibrillar protein synthesis, to understand the role of these mechanisms in the development of muscle, and to establish the relationship between the synthesis of these proteins and their ultimate assembly into the contractile apparatus. Although we are currently working on the synthesis of many of the structural proteins of chick muscle, the relative ease by which polysomes synthesizing myosin heavy chain can be isolated on sucrose density gradients (Heywood and Rich, 1968) has led us to analyze myosin synthesis in some detail.

An RNA fraction, sedimenting at approximately 26S, can be extracted from the large myosin-synthesizing polysomes. This RNA fraction has been shown by a number of criteria to contain messenger RNA (mRNA) coding for the large molecular weight subunit (200,000) of myosin (Heywood and Nwagwu, 1969; Heywood, 1969; Rourke and Heywood, 1972). Peptide analysis by two-dimensional high voltage electrophoresis as well as by ion exchange chromatography has suggested that the myosin mRNA is translated with a high degree of fidelity in a cell-free amino acid incorporating system (Rourke and Heywood, 1972).

We report here some further studies on myosin synthesis in cell-free systems and some observations on the relationship between cell fusion and myosin biosynthesis in differentiating cell cultures.

We also describe a method for the isolation of polysomal myosin mRNA by removal of 28S ribosomal RNA (rRNA) and some preliminary studies on its properties and biosynthesis in chick muscle cell cultures. We hope that this approach will be useful in analyzing directly the role of mRNA synthesis during muscle differentiation.

* Present address: School of Biological Sciences, University of Sussex, Falmer, Brighton, Sussex, Great Britain.

Results and Discussion

It has been amply demonstrated that most, if not all, proteins incorporate methionine into the N-terminal position during the initiation of protein synthesis (Lucas-Lenard and Lipmann, 1971; Lengyel and Soll, 1969). In prokaryotes formylmethionine is utilized, whereas in eukaryotes methionine is incorporated into the N-terminal position. In a significant number of cases in both

Table 1. N-Terminal Amino Acids Obtained from Myosin-synthesizing Polysomes and Small Polysomes

Polysome Fraction	Edman Degradation Step	[^{35}S]Met (cpm)	[^{3}H]Ser (cpm)
Myosin-synthesizing polysomes	1	18	460
	2	7	1530
Small polysomes	1	72	400
	2	20	460

Embryonic chicks (14-day) were pulse-labeled with a mixture of [^{35}S]methionine and [^{3}H]serine for 5 min. Myosin-synthesizing polysomes and smaller polysomes were prepared by sucrose density gradient centrifugation (Heywood and Rich, 1969). Short nascent chains were recovered by phenol extraction (Jackson and Hunter, 1970) and subsequently submitted to Edman degradation (Gray, 1967).

prokaryotes and eukaryotes, the N-terminal methionine is cleaved off during polypeptide chain growth or soon after completion of the synthesis of the polypeptide (Lucas-Lenard and Lipmann, 1971). In order to determine if myosin heavy chain synthesis also initiates with methionine, 14-day embryos were pulse-labeled in vivo with a mixture of [^{35}S]methionine and [^{3}H]serine. Small nascent chains, selectively isolated by phenol extraction (Jackson and Hunter, 1970), were subsequently subjected to Edman degradation (Gray, 1967). As shown in Table 1, a fraction of the nascent peptides in both polysome fractions have an N-terminal methionine. The amount of serine present may imply that some of the longer chains have already been cleaved of their N-terminal amino acid. The fact that the Edman degradation is only 75% effective (Jackson and Hunter, 1970) accounts for the small amount of methionine observed in the

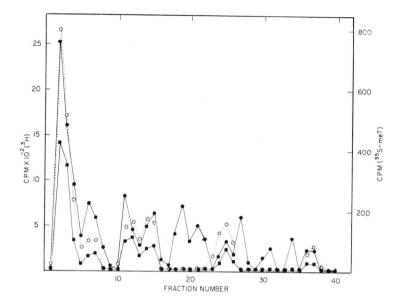

Figure 1. Analysis of the proteolytic digest of the 200,000 mol wt subunit of myosin by ion exchange chromatography on Aminex A-5 resin (Bio-Rad Laboratories). Cell-free synthesis of myosin, proteolytic digestion of the large myosin subunits, and elution of the peptides from the ion exchange column were as previously described (Rourke and Heywood, 1972). Peptides from myosin labeled in vivo with [3H]-amino acid mixture (●—●) were counted by liquid scintillation. A low background gas flow counter was used to count the [35S]methionine-labeled peptides from myosin labeled in vivo (○—○) and in the cell-free system (■—■).

second step of degradation. The large increase in serine in the nascent chains from myosin-synthesizing polysomes at the second cleavage suggests an N-terminal sequence of Met-Ser for growing myosin polypeptide chains. This agrees well with findings of Offer (personal communication) suggesting an N-terminal sequence for myosin heavy chain as *N*-acetyl-Ser-Ser-Asp-.

If a proteolytic digest of [35S] methionine-labeled myosin, synthesized in a cell-free amino acid incorporating system, is chromatographed on an ion exchange column and compared to in vivo labeled [3H]myosin (using a mixture of 3H-amino acids) and in vivo labeled [35S]methionine, a good correspondence of in vivo and in vitro labeled myosin peptides is obtained (Fig. 1). These results suggest that, within the limits of the column separation of the peptides, myosin is synthesized with a high degree of fidelity in the cell-free system and contains no additional methionine peptides. Therefore, unless there is an extra methionine peptide in the unresolved first peak, the N-terminal methioinine is cleaved off in the in vitro system as well as in the intact cell.

The myosin message is by far the largest eukaryotic mRNA that has been used to program a cell-free system and once its translation is initiated, we might expect a considerable lag, as nascent chains grow, before completed chains are released from the ribosomes. As shown in Fig. 2, there is a lag period of about 8 minutes before completed myosin chains are released, while endogenous globin synthesis is measurable at the earliest times tested.

Assuming that initiation occurs at the start of the incubation (addition of mRNA to the incubation mixture), myosin chain growth occurs at the rate

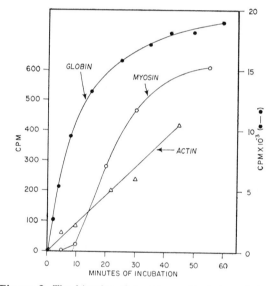

Figure 2. The kinetics of myosin synthesis in a heterogeneous cell-free amino acid incorporating system. The conditions for amino acid incorporation using chicken erythroblast ribosomes were as previously described (Rourke and Heywood, 1972). Three different incubation mixtures were used: (1) to measure globin system without added mRNA (●—●); (2) with added 26S myosin mRNA to measure myosin synthesis (○—○)—in this reaction mixture, globin synthesis was similar to the incubation mixture without added mRNA; and (3) with added 10–17S muscle mRNA to measure actin synthesis. Globin synthesis was taken as the total incorporation due to endogenous mRNA of the system. Myosin synthesis was determined as previously described by acrylamide gel electrophoresis (Rourke and Heywood, 1972), and the actin-like protein was measured from the radioactivity which was extracted from an acetone powder (0.2 mM ATP, 1 mM dithiothreitol, pH 7.2) of the incorporation system and migrating with carrier actin on SDS acrylamide gels. The radioisotope used was 30 μCi of a mixture of 14C-amino acids (New England Nuclear, NEC-445). Radioactivity was measured on a low background, gas flow counter.

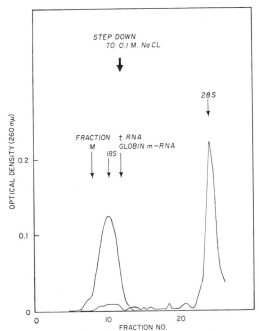

Figure 3. Separation of RNA species by a Sepharose 2B column. 18S and 28S RNA were prepared separately from chick embryonic leg muscle ribosomes by sucrose density gradient centrifugation (see Fig. 5). Chick erythroblast globin mRNA was prepared as described by Heywood (1970) and chick muscle tRNA as described by von Ehrenstein (1968). Fraction *M* refers to the radioactivity from polysomes which we identify as myosin mRNA. The column dimensions were 50 × 1 cm and 120 drop fractions (approx. 3 ml) were collected. The elution buffer was 500 mM NaCl, 20 mM Tris, 2.5 mM EDTA, 0.5% SDS, pH 7.5. To elute 28S rRNA, the salt concentration was reduced to 100 mM as indicated.

of 4–5 amino acids per second, which is comparable to the rate of 4–6 amino acids per second estimated for globin on the basis of observed translation times of the order of half a minute (Lamfrom and Knopf, 1965; Hunt et al., 1969).

The synthesis of an actin-like protein, responsive to the addition of 10–17S muscle mRNA, was also followed in the amino acid incorporating system (Fig. 2). Some radioactivity is observed to be incorporated into this protein in the absence of added mRNA, and the synthesis of the actin-like protein did not require the addition of muscle initiation factors as does the synthesis of myosin (Rourke and Heywood, 1972). Although preliminary, these results suggest that chicken erythroblasts may synthesize a small amount of an actin-like protein in addition to their main synthetic product hemoglobin.

The myosin mRNA fraction obtained by sucrose density gradient centrifugation contains a large contaminant of 28S rRNA. In order to investigate the details of myosin mRNA synthesis and appearance into polysomes during muscle differentiation, we have sought a method of purifying intact myosin mRNA.

We have taken advantage of the observation that 28S rRNA is bound to Sepharose 2B at high ionic strength (Petrovic et al., 1971), and have found that mRNA is not bound in this way. Figure 3 shows the behavior of several RNA species on a Sepharose 2B column. More than 98% of added 28S RNA is retained by the Sepharose at high ionic strength and released when the ionic strength is reduced. When a pulse-labeled polysomal RNA fraction expected to contain 28S RNA and myosin mRNA is subjected to Sepharose 2B chromatography, a considerable proportion (fraction M) is not retained by the Sepharose and elutes slightly ahead of an 18S marker. In the same way a preparation of 9S globin mRNA was not bound to the Sepharose, but because of its smaller size, was considerably retarded relative to 18S RNA.

Our procedure for isolating fraction M and partially characterizing it as myosin mRNA is shown in Table 2. A gradient of polysomes obtained from cultured embryonic chick thigh muscle cells

Table 2. Procedure for Isolation of Myosin mRNA

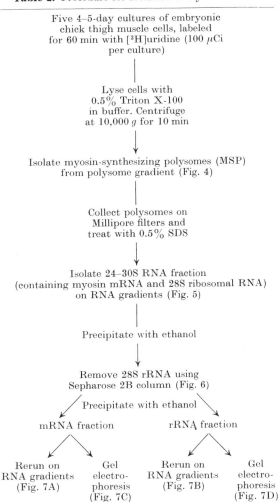

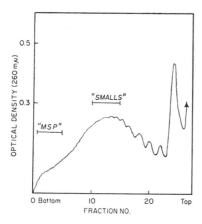

Figure 4. Sucrose density gradient of polysomes from cultured chick muscle cells. 0.2–0.3 ml of packed cells were lysed in 1 ml of TAM buffer (10 mM Tris, 250 mM NH$_4$Cl, 10 mM MgCl$_2$, pH 7.4) containing 0.5% Triton X-100. The 10,000 g supernatant was placed on a 15–40% sucrose gradient in TAM buffer and centrifuged at 25,000 rpm for 84 min at 2°C.

(Morse et al., 1971) is shown in Fig. 4. No changes in the pattern of the optical density profiles have been observed between one and five days in culture. Hosick and Strohman (1971) observed no change in the ratio of polysomes to monomers over this period. The myosin-synthesizing polysome fraction was collected on Millipore filters (Heywood and Nwagwu, 1969) and the RNA from it was analyzed by gradient centrifugation (Fig. 5). Fraction A (24–30S RNA) from the gradient was then subjected to Sepharose 2B chromatography (Fig. 6).

Both the mRNA and rRNA fractions were divided into two and analyzed on sucrose gradients and polyacrylamide gels as shown in Fig. 7. On sucrose gradients all the radioactivity in the mRNA fraction sediments as a symmetrical peak at about 26S (Fig. 7A), whereas the radioactivity from the rRNA fraction sediments with the carrier 28S rRNA (Fig. 7B). The absence of lower molecular weight material indicates that the RNA is not degraded during Sepharose column chromatography.

When the other half of the mRNA fraction is analyzed on polyacrylamide gels (Fig. 7C), it migrates as a symmetrical peak of very high apparent molecular weight. The second peak in Fig. 7C is not always present. The 28S RNA fraction on gels migrated as a well-defined peak (Fig. 7D). The aggregated material near the top of the gel does not correspond to the mRNA peak of Fig. 7C but may be related to aggregated material seen on the sucrose gradient (Fig. 7B). To emphasize that the mRNA peaks on gradients and gels are the same material, 628 cpm were recovered from the gels and 606 cpm from the gradients. The corresponding figures for the RNA fraction were 6600 and 7600 cpm, respectively.

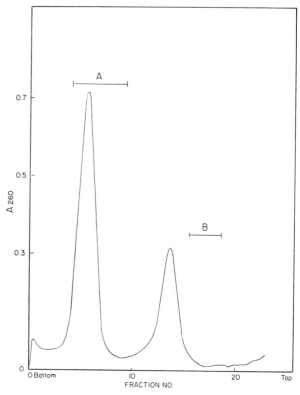

Figure 5. Sucrose density gradient of RNA from SDS-treated polysome fractions. 10–30% sucrose gradients in buffer (50 mM Tris-acetate, 5 mM EDTA, 0.5% SDS, pH 7.4) were centrifuged at 22,000 rpm for 16 hr at 2°C. Embryonic chick muscle ribosomes were added to the sample as carrier if necessary.

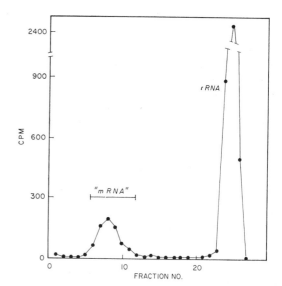

Figure 6. Sepharose 2B elution profile (Fig. 3) of fraction A RNA (24–30S; Fig. 5) from myosin-synthesizing polysomes (MSP; Fig. 4), showing the proposed separation of 26S myosin mRNA from 28S rRNA. Samples were taken up in Triton scintillation fluid (0.5% PPO and 30% Triton X-100 in toluene) and counted in a Nuclear Chicago liquid scintillation counter at 30% efficiency.

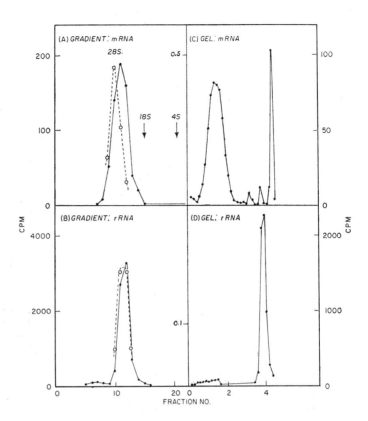

Figure 7. Analysis of RNA fractions from Sepharose 2B columns by sucrose density centrifugation and polyacrylamide gel electrophoresis. The two fractions from Sepharose 2B (mRNA and rRNA Fig. 6) were made 0.25 M in Tris-HCl pH 7.4 and precipitated with ethanol. Fifty μg of tRNA were added as carrier to the mRNA fraction. Half of each fraction was analyzed on gradients and the other half on gels. Analysis by gradient centrifugation was performed as in Fig. 5. One-ml fractions, after A$_{260}$ determination (○—○) were taken up in Aquasol (New England Nuclear) and radioactivity was determined by liquid scintillation. Gels were 10 cm, 2.5% acrylamide/0.125% bis-acrylamide in 40 mM Tris-acetate, 1 mM EDTA, 0.5% SDS, pH 7.6 and electrophoresis was at 4°C with 5 mA/tube. Only the first 4 cm of gel is shown for all the radioactivity is in this area of the gel. Gels were frozen and cut into 3 × 0.5 mm fractions, which were shaken for 6 hr in 1 ml of 0.6 M sodium acetate, 0.5% SDS, pH 6, at 37°C. Radioactivity was determined in Aquasol.

Myosin mRNA, coding for a protein of 200,000 daltons mol wt, should itself have a mol wt of the order of 2 × 10⁶ daltons compared with a value of 1.6 × 10⁶ estimated for chick 28S RNA (Loening, 1968). The relatively low apparent S value as determined by sucrose density gradients and the relatively low electrophoretic mobility on acrylamide gels suggest that myosin mRNA may have a more extended configuration than 28S rRNA under the conditions used. Additional studies, utilizing conditions assuring the complete unfolding of RNA molecules, will be required before a precise determination of the molecular weight of myosin mRNA can be obtained.

Further properties of the myosin mRNA fraction are shown in Table 3. The ratio of mRNA to rRNA

from the Sepharose column is somewhat variable between experiments, probably because of varying degrees of resolution of MSP on the polysome gradients. The proportion of mRNA falls off drastically with longer labeling periods, which may reflect a delay in the appearance of newly synthesized rRNA in the heavy polysomes. Myosin-synthesizing polysomes (MSP) contain a high proportion of the mRNA fraction relative to smaller polysomes, indicating that the RNA is not merely nonspecific RNA cosedimenting with the polysomes.

The absolute requirement for a proposed mRNA species is that it direct the synthesis of its corresponding protein in a heterologous cell-free system. Sepharose column RNA fractions were prepared as

Table 3. Analysis of Labeled Polysomal-bound mRNA from In Vivo and In Vitro Labeled Muscle

Source of RNA	Labeling Time (hr)	Polysome[a] Fraction	RNA Gradient Fraction[b]	Percent mRNA[c] (cpm)
³²P-labeled 14-day embryos	1.5	MSP	A	10
	5	MSP	A	2
	1.5	Smalls	B	90
[³H]uridine-labeled 3–4-day cultured muscle cells	1	MSP	A	48
	1	Smalls	A	10

[a] See Fig. 4. [b] See Fig. 5. [c] The radioactivity in the mRNA fraction of Fig. 6 is expressed as a percentage of the total radioactivity eluted from the Sepharose 2B column.

Table 4. Cell-free Synthesis of Myosin from mRNA Fractionated on Sepharose 2B Column

RNA	Experiment 1	2	3
—	4	0	1
Unfractionated	78	36	—
mRNA fraction	62	37	446
rRNA fraction	4	2	8

cpm incorporated into the 200,000 mol wt subunit of myosin.

The 24–30S RNA fraction from myosin-synthesizing polysomes was prepared from 20–25 14-day, chick embryo leg muscle as previously described (Heywood and Nwagwa, 1969). The cell-free synthesis and acrylamide gel analysis of myosin heavy chain synthesis was performed as previously described (Rourke and Heywood, 1972). In Experiment 1 rRNA (28S) was added to the incubation mixture to an equal concentration as that eluted from the Sepharose 2B column. Radioactivity was measured on a low backgound (2 cpm) gas flow counter.

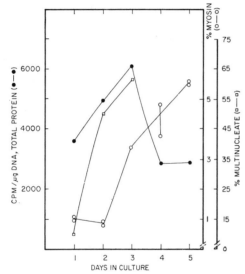

Figure 8. Relationship between myosin synthesis and the appearance of multinucleate muscle cells in culture. Muscle cells were prepared by a mechanical dissociation of leg muscle from 11-day chick embryos. Gels were plated at 5×10^6 cells per 15 cm plate. The cultures were assayed at 24 hr intervals. Four hours prior to analysis the cultures were fed and labeled with 5 μCi ^{3}H-amino acid mixture (New England Nuclear, NET-445) per plate. The number of plates per assay was 5 plates for 1 day, 5 on day 2, 5 on day 3, 3 on day 4, and 2 on day 5. Cells were harvested and lysed as previously described (Morse et al., 1970) except the buffer used was 0.47 M KCl, 0.01 M KH$_2$PO$_4$, 0.01 M K$_4$P$_2$O$_7$, pH 6.2. Myosin was extracted, purified, and analyzed by acrylamide gel electrophoresis as previously described (Rourke and Heywood, 1972). Radioactivity of myosin heavy chain was determined after electrophoresis by slicing the gel in 1-mm slices, solubilizing in NCS (Nuclear Chicago) and counting by liquid scintillation. Total readioactivity was determined from an aliquot of the whole lysate. DNA was determined by the method of Burton (1956). The percent multinucleated cells was determined from representative cultures at 24 hr intervals after fixing in methanol. Binucleate cells were not scored. (●—●) Total cpm incorporated per μg DNA; (□—·□) percent nuclei scored in multinucleated cells; (○—○) percent of total cpm incorporated into myosin heavy chain.

described in Table 2 using RNA prepared from 14-day embryonic leg muscle. The RNA fractions were tested in a cell-free system derived from chicken erythroblasts. As shown in Table 4, over 98 % of the messenger activity is found in the mRNA fraction and, when this activity is compared with that of an equivalent amount of unfractionated RNA, little or no mRNA activity appears to be lost during chromatography.

To summarize, our isolation procedure produces an apparently homogeneous RNA species with many of the expected properties of an active, undegraded myosin mRNA. We hope it will allow for an analysis of the synthesis of this specific transcription product during muscle cell differentiation. As a beginning in the attempt to understand the molecular events occurring during the terminal differentiation of muscle, an analysis of myosin synthesis occurring in differentiating cultured cells was undertaken preliminary to looking for the appearance of myosin mRNA during differentiation. As shown in Fig. 8, a low but significant amount of myosin synthesis is occurring in one-day muscle cell cultures. As cell fusion occurs between days 1 and 3 there is an apparent lag between fusion and increase in myosin synthesis (up to 6 % of total protein synthesis), suggesting that the increase in myosin synthesis is not closely associated with cell fusion. The rate of incorporation into total protein per plate slowly decreases after three days of culture (not shown). The relatively rapid drop in cpm per μg DNA is due to an increase in DNA per plate. These results suggest the presence of active myosin mRNA in mononucleated myogenic cells. In addition, analysis of radioactive polysomal mRNA on Sepharose 2B as described in this paper suggests the possibility that myosin mRNA synthesis is occurring in the one-day cell cultures.

Acknowledgments

This work has been supported by NIH Grant HDO3316-04. G.E.M. is a Beit Memorial Research Fellow. S.M.H. was the recipient of NIH Research Career Development Award, 5KO4 GM18904-02.

References

BURTON, K. 1956. Study of the conditions and mechanism of the diphenylamine reaction for the colorimetric estimation of deoxyribonucleic acid. *Biochemistry* **62:** 315.

GRAY, W. R. 1967. Sequential degradation plus dansylation *Methods Enzymol.* **11:** 469.

HEYWOOD, S. M. 1969. Synthesis of myosin on heterologous ribosomes. *Cold Spring Harbor Symp. Quant. Biol.* **34:** 799.

———. 1970. Specificity of mRNA binding factor in eukaryotes. *Proc. Nat. Acad. Sci.* **67:** 1782.

HEYWOOD, S. M. and M. NWAGWU. 1969. A partial characterization of presumptive myosin mRNA. *Biochemistry* **8:** 3839.

HEYWOOD, S. M. and A. RICH. 1968. In vitro synthesis of native myosin, actin, and tropomyosin from embryonic chick polysomes. *Proc. Nat. Acad. Sci.* **59**: 590.

HOSICK, H. L. and R. C. STROHMAN. 1971. Changes in ribosome-polyribosome balances in chick muscle cells during tissue dissociation, development in culture and exposure to simplified culture-medium. *J. Cell. Physiol.* **77**: 145.

HUNT, T., T. HUNTER, and A. MUNRO. 1969. Control of hemoglobin synthesis: Rate of translation of mRNA for the α- and β-chains. *J. Mol. Biol.* **43**: 123.

JACKSON, R. and T. HUNTER. 1970. Role of methionine in the initiation of haemoglobin synthesis. *Nature* **227**: 672.

KNOPF, P. M. and H. LAMFROM. 1965. Changes in the ribosome distribution during incubation of rabbit reticulocytes in vitro. *Biochim. Biophys. Acta* **95**: 398.

LENGYEL, P. and D. SÖLL. 1969. Mechanism of protein biosynthesis. *Bacteriol. Rev.* **33**: 264.

LOENING, U. E. 1968. Molecular weights of ribosomal RNA in relation to evolution. *J. Mol. Biol.* **38**: 355.

LUCAS-LENARD, J. and F. LIPMANN. 1971. Protein biosynthesis. *Annu. Rev. Biochem.* **40**: 759.

PETROVIC, S., A. NOVAKOVIC, and J. PETROVIC. 1971. Separation of ribosomal RNAs on agarose gels. *Biochim. Biophys. Acta* **254**: 493.

MORSE, R., H. HERRMANN, and S. M. HEYWOOD. 1971. Extraction with Triton X-100 of active polysomes from monolayer cultures of embryonic muscle cells. *Biochim. Biophys. Acta* **232**: 403.

ROURKE, A. and S. M. HEYWOOD, 1972. Myosin synthesis and specificity of eukaryote initiation factors. *Biochemistry* **11**: 2061.

VON EHRENSTEIN, G. 1968. Isolation of sRNA from intact *Escherichia coli* cells. *Methods Enzymol.* **12**: 588.

Gene Expression during Differentiation
of Contractile Muscle Fibers

DAVID YAFFE AND HAVIV DYM

Department of Cell Biology, The Weizmann Institute of Science, Rehovoth, Israel

Although a great deal of information exists regarding the chemistry and structure of the muscle proteins, little is known about the regulation of their syntheses and its relation to the developmental events in the cell. The ability to grow isolated muscle precursor cells and to a great extent control their differentiation in cell cultures makes this system very useful for studying aspects of the regulation of gene expression during differentiation.

Myoblasts from newborn rat skeletal muscle grow in a monolayer of mononucleated cells when first plated in cultures. During this initial period of development the cells proliferate, but very little fusion occurs. This phase lasts about 50 hr in standard culture conditions and is followed by a period in which the cells fuse rapidly and form multinucleated fibers. The transition from proliferation to the formation of multinucleated fibers is associated with distinct changes in synthetic activities, such as the cessation of DNA synthesis, the synthesis of the contractile proteins and large changes in the activity of many enzymes (Cooper and Konigsberg, 1961; Coleman and Coleman, 1968; O'Neill and Strohman, 1969; Shainberg et al., 1971; Stockdale and Holtzer, 1961).

Experiments to determine the nature of the lag period prior to fusion have indicated the existence of intrinsic covert developmental changes preceding cell fusion; only myoblasts which complete these changes can participate in the process of cell fusion. It was also found that the transition of the cells from the proliferative stage to fusion can be controlled to a great extent by the composition of the nutritional medium. Thus cells growing in a medium enriched with fetal calf serum and 8% embryo extract (designated FE medium) will continue to proliferate for an extended period without fusion. When the medium is changed from FE medium to standard medium (S medium, containing horse serum and only 1% EE), there is a lag period of about 18 hr before the onset of a phase of intense cell fusion and fiber formation (Yaffe, 1971).

This procedure enables one to control precisely the time of initiation of a specific period preceding

cell fusion during which myoblasts are directed from a course of proliferation to one of terminal differentiation. The fusion which follows this treatment of the cells is synchronized to a far greater extent compared to that which occurs in cultures allowed to fuse spontaneously and thus affords conditions that permit a detailed analysis of the sequence of changes in DNA, RNA, and protein synthesis which are associated with cell fusion (Lavie and Yaffe, unpublished; Yaffe, 1971; Yaffe et al., 1972).

We have previously shown the correlation between the morphological differentiation of rat muscle cultures and the increase in activity of creatine phosphokinase (CPK), myokinase (MK), and glycogen phosphorylase (GPh) (Shainberg et al., 1971). A comparison of the changes in activity of CPK during the differentiation of

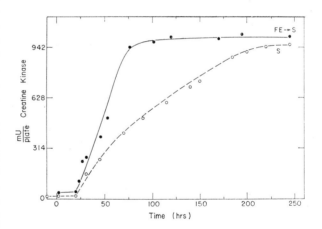

Figure 1. The effect of growth medium on the rate of increase in creatine kinase activity during differentiation of muscle cell cultures. Primary cultures were grown in FE medium and were switched to S medium 50 hr after plating (●—●). At various times after the change of medium, cultures were frozen and subsequently assayed for CPK activity (as described in Shainberg et al., 1971). Fusion started in these cultures 18 hr after change of medium. Each point represents the mean of 2 plates. The abscissa indicates the time after change from FE to S medium. The results obtained with the cultures continuously grown in S medium (○– – –○) were plotted on the same scale and aligned with the other group to match the time of the onset of fusion; cultures grown continuously in S medium started to fuse 52 hr after plating.

primary cultures that were allowed to fuse spontaneously and cultures in which the onset of fusion was controlled by FE medium is shown in Fig. 1. While the final levels of activities are similar under both conditions, the rate of increase is greater in the cultures which were grown in FE medium and then transferred to S medium. The relationship between the onset of cell fusion and the initiation of the rapid increase in the activity of the three enzymes is shown in Fig. 2. The curve representing CPK activity almost coincides with that indicating the progression of cell fusion. The initiation of the increase in activity of glycogen phosphorylase and MK lags several hours behind that found for CPK. Such a lag was consistently observed.

The increases in enzymic activities as well as the process of cell fusion are promptly stopped by addition of inhibitors of protein synthesis (e.g., cyclohexamide). However treatment with actinomycin D (AM), at doses which prevent more than 97% of RNA synthesis, did not interfere with cell fusion or with the increase in activity of CPK for at least 6 hr after addition of the drug (Figs. 3, 4). This temporary independence from RNA synthesis was also seen when AM was applied prior to the onset of fusion. These results indicate that the RNA

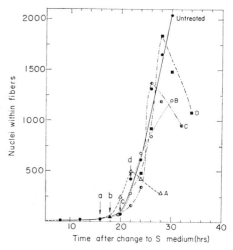

Figure 3. Cell fusion following application of actinomycin D. Cultures grown in FE medium were transferred 50 hr after plating to S medium. Actinomycin D (2 μg/ml) was added to groups of cultures at different times. *a, b, c, d* indicate the time of addition of the inhibitor to groups *A, B, C, D*, respectively. At various times aliquots of cultures were fixed and the number of nuclei within fibers was counted as described in Fig. 2. Cell fusion continued for several hours following application of actinomycin D.

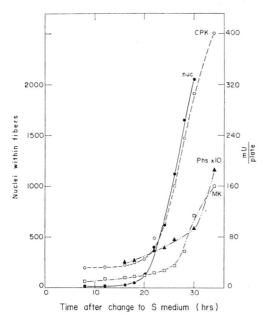

Figure 2. Temporal relation between cell fusion and the initiation of rapid increases in activity of creatine kinase (*CPK*), myokinase (*MK*), and glycogen phosphorylase (*PhS*). Cultures were grown in FE medium and changed to S medium 50 hr after plating. At subsequent time intervals 2 cultures were frozen and enzymic activity assayed. Two other cultures fixed at each time interval were stained and the number of nuclei within fibers in 5 randomly selected fields were counted. Each point represents the average of 2 plates.

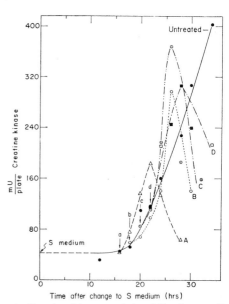

Figure 4. Creatine kinase activity following application of actinomycin D. Aliquots from the same cultures referred to in Fig. 3 were collected and assayed for CPK activity. CPK activity in the AM-treated groups was higher during the first 6–8 hr following application of the drug than the groups not treated with AM.

Figure 5. Acrylamide gel electrophoresis of a leucine ^{14}C-labeled myosin extract from cultures labeled at the onset of fusion. In group A cultures grown in FE medium were changed to S medium. At the onset of cell fusion 19 hr later the cultures were exposed to a 3 hr pulse of 2 μCi/plate [^{14}C]leucine given in low leucine S medium (leucine concentration reduced to 1/100). Group B received 2 μg/ml actinomycin together with the [^{14}C]leucine. At the end of this period the cultures were rinsed with cold phosphate-buffered saline and frozen at $-70°$C. The cells were then scraped and extracted for myosin (Baril et al., 1966; Dym and Yaffe, unpublished). The extract was treated overnight in a mixture of 1% SDS, 1% mercaptoethanol, and 8 M urea. The extract from 4 plates was put on one 10% polyacrylamide gel (Weber and Osborn, 1969) and run at 8 mA/gel for 6 hr. The gels were then frozen and sliced. The 1-mm slices were then solubilized in NCS (Amersham/Searle) and then put in toluene scintillation fluid and counted after sitting 48 hr in the dark. No myosin peak is detectable.

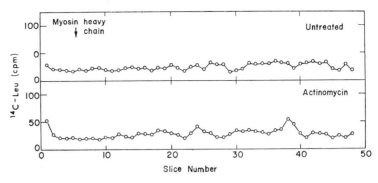

molecules which are required for the synthesis of proteins essential for fusion, as well as those which specify the increases in enzymic activities, were present in the mononucleated cells prior to their fusion (Dym and Yaffe, unpublished; Yaffe et al., 1972).

Synthesis of Myosin

In the experiments described above, changes in synthesis of enzymes were inferred from changes in their activities. The changes in synthesis of myosin during differentiation were studied by direct measurements of amino acid incorporation into the isolated polypeptide chains. Cultures in which development prior to fusion was controlled by the composition of the medium were pulsed at various stages of differentiation with radioactive amino acids. They were then extracted for myosin and the proteins were electrophoresed in 10% SDS acrylamide gels as described in Fig. 5. The gels were sliced and the radioactivity was measured. The migration of myosin chains was determined by simultaneously running gels containing purified rat skeletal muscle myosin.

The radioactive peaks identified as myosin heavy chains could not be detected in cultures exposed to [^{14}C]leucine prior to cell fusion or during the first three hours following the initiation of cell fusion (Fig. 5). Subsequently the peak identified as myosin became progressively more apparent. Figure 6 shows a typical gel pattern of a culture exposed to [^{14}C]leucine between 25 and 28 hr after the change to the standard medium (i.e., 6–9 hr after the onset of cell fusion) in which the heavy and light myosin chains are distinct. The temporal relationship between the onset of cell fusion and the incorporation of [^{14}C]leucine into myosin is shown in Fig. 7. The incorporation of labeled [^{14}C]leucine into the myosin lags about 4 hr subsequent to the onset of cell fusion. It was found that, although myosin synthesis is first detectable

at this time, application of AMD shortly before or at the onset of fusion did not prevent the appearance of the labeled myosin peak in the gels. Furthermore in most experiments myosin synthesis became detectable earlier in cultures treated with AMD than in the untreated cultures (Figs. 7, 8). These results show that the RNA which specified the synthesis of myosin is present in the mononucleated cells prior to fusion.

The accumulation of labeled myosin during the

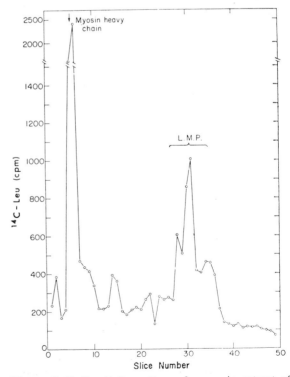

Figure 6. Radioactivity pattern of a myosin extract of cultures labeled at 25–28 hr after the change to S medium. Cultures were labeled with [^{14}C]leucine (2 μCi/plate) for 3 hr starting at 6 hr after the onset of cell fusion. Labeling and assay procedure were as described under Fig. 5. *LMP*, low molecular weight proteins associated with myosin.

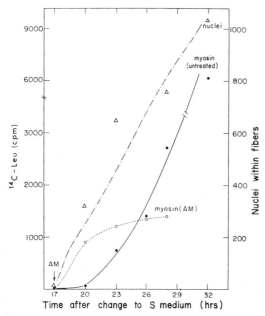

Figure 7. Time relation between the onset of cell fusion and myosin synthesis. Cultures grown in FE medium were changed to S medium. After 17 hr one group was given 2 μg/ml of actinomycin D. Four cultures of each group were exposed at time intervals to a 3 hr pulse of [¹⁴C]leucine then harvested and assayed for incorporation of label into the myosin peak as described under Fig. 5. Two plates not treated with actinomycin D were fixed at each corresponding time, stained, and the number of nuclei within fibers was counted. The ordinate indicates counts per minute in the myosin heavy chain peak electrophoresed on 10% polyacrylamide gels.

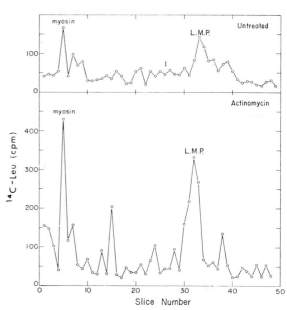

Figure 8. [¹⁴C]Leucine incorporation into myosin in cultures treated with actinomycin D. Cultures grown in FE medium were changed to S medium 50 hr after plating. After 19 hr when the cultures approached the phase of rapid cell fusion, actinomycin D (2 μg/ml) was given to one group. Both groups were exposed to a 3 hr pulse of [¹⁴C]leucine (2 μCi/plate) starting 3 hr after application of actinomycin D.

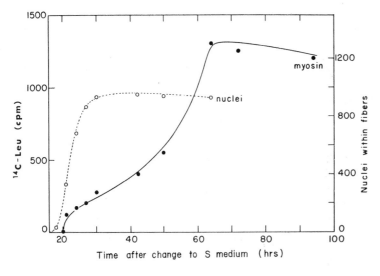

Figure 9. Accumulation of labeled myosin during advanced differentiation of the cultures. [¹⁴C]Leucine (2 μCi/plate) was given to cultures 18 hr following change from FE medium to S medium. Aliquots taken at time intervals were either fixed in methanol, stained, and the nuclei within fibers were counted or were extracted for myosin. The total [¹⁴C]leucine counts per minute incorporated into the myosin heavy chain peak from 10% polyacrylamide gels are given.

differentiation of the cultures is described in Fig. 9. Although initiation of a phase of intense myosin synthesis is closely correlated with cell fusion, the accumulation of labeled myosin continues long after the cessation of the phase of rapid cell fusion. A recent study of myosin synthesis in chick cultures indicates a similar conclusion (Paterson and Strohman, 1972).

Discussion

The experiments described above show that the synthesis of myosin and the increase in activity of several enzymes closely follow the fusion of mononucleated skeletal muscle myoblasts into multinucleated fibers. Although it seems as if in these cells the process of fusion triggers these changes, it is clear that this is not the initial event in the activation of the genes which specify the synthesis of these proteins. Rather, it was seen that when myoblasts enter the phase of fusion, the transcription of at least part of the information required for subsequent differentiation had already occurred. Thus terminal differentiation may proceed in the absence of further RNA synthesis.

Experiments in which actinomycin D was applied to cultures at different times prior to fusion indicate that the required RNA molecules are present in the cells at least 6 hr before fusion. Since actinomycin D may disturb the synthesis of vital protein unrelated to differentiation, it is difficult to determine with great certainty whether these RNA species are present even earlier than 6 hr before fusion. The time at which specific mRNAs involved in the expression of differentiation start to be transcribed can be found only by direct measurements of the synthesis of these RNA molecules. Investigation of DNA synthesis and the influence of inhibitors of DNA synthesis on fiber formation indicate that the myoblasts divide once during the 18 hr lag after the change of FE medium to S medium and cell fusion; furthermore, the cells become most responsive to the signal produced by this medium change during their DNA synthetic period (Lavie and Yaffe, unpublished). It is of great interest to test the relation of the time of initiation of the synthesis of mRNAs required for the changes in protein synthesis which follow fusion to this last period of DNA synthesis, and experiments to investigate this question are in progress.

In an earlier study actinomycin D was added to cultures containing differentiated muscle fibers, and it was observed that muscle fibers increased their rate of contractions (Yaffe and Feldman, 1964). In the experiments described here we observed that during the first hours following treatment with actinomycin D, myosin synthesis was enhanced relative to control values. Similar observations on enhancing effects of actinomycin D in different cell systems have been reported (Moscona et al., 1966; Tomkins et al., 1969). Such results cannot be attributed to effects of actinomycin D that are not directly related to synthesis of the specific proteins studied (e.g., changes in amino acid pool or decreased synthesis of other proteins, etc.) because three different independent parameters were applied: (a) rate of cell fusion, (b) changes in activity of enzymes, and (c) incorporation of labeled amino acids into proteins. Furthermore in these experiments the enhancing effect of actinomycin D was calculated as absolute increase per culture rather than as relative changes. One way to explain these data is that during the phase of development of the myoblasts preceding cell fusion, the mononucleated cells are actively synthesizing messenger RNA molecules which are then stored to provide sufficient amounts of templates to enable the intensive changes in the synthesis of proteins associated with the differentiation following cell fusion. The utilization of this stored information is inhibited by a system which appears to be sensitive to treatment with actinomycin.

Acknowledgments

The skillful technical assistance of Ruth Babliki and Pnina Fisher is gratefully acknowledged. Thanks are due to Drs. S. Kaufman and R. Singer for their valuable comments and suggestions during the preparation of the manuscript. The work was supported in part by a grant in aid from the Muscular Dystrophy Association of America, Inc. and by grant #DRG 10007 from the Damon Runyon Memorial Fund for Cancer Research.

References

BARIL, E. F., D. S. LOVE, and H. HERMANN. 1966. Investigation of myosin heterogeneity observed during chromatography on diethylaminoethyl cellulose. *J. Biol. Chem.* **241**: 822.

COLEMAN, J. R. and A. W. COLEMAN. 1968. Muscle differentiation and macromolecular synthesis. *J. Cell. Physiol.* **72**: (suppl. 1) 19.

COOPER, W. G. and I. R. KONIGSBERG. 1961. Succinic dehydrogenase activity of muscle cells grown *in vitro*. *Exp. Cell Res.* **23**: 576.

MOSCONA, A. A., M. MOSCONA, and N. SAENZ. 1966. Enzyme induction in embryonic retina: The role of translation and transcription. *Proc. Nat. Acad. Sci.* **61**: 160.

O'NEILL, M. and R. C. STROHMAN. 1969. Changes in DNA polymerase activity associated with cell fusion in cultures of embryonic muscle. *J. Cell. Physiol.* **73**: 61.

PATERSON, B. and R. C. STROHMAN. 1972. Myosin synthesis in culture and differentiating chicken embryo skeletal muscle. In press.

SHAINBERG, A., G. YAGIL, and D. YAFFE. 1971. Alterations of enzymatic activities during muscle differentiation *in vitro*. *Develop. Biol.* **25**: 1.

STOCKDALE, F. and H. HOLTZER. 1961. DNA synthesis and myogenesis. *Exp. Cell Res.* **24**: 508.

TOMKINS, G. M., T. D. GELEHRTER, D. GRANNER, D. MARTIN, H. H. SAMUELS, and E. B. TOMPSON. 1969. Control of specific gene expression in higher organisms. *Science* **166**: 1674.

WEBER, K. and M. OSBORN. 1969. The reliability of molecular weight determinations by dodecyl sulfate polyacrylamide gel electrophoresis. *J. Biol. Chem.* **244**: 4406.

YAFFE, D. 1971. Developmental changes preceding cell fusion during muscle differentiation *in vitro*. *Exp. Cell Res.* **66**: 33.

YAFFE, D. and M. FELDMAN. 1964. The effect of actinomycin D on heart and thigh muscle cells grown *in vitro*. *Develop. Biol.* **9**: 347.

YAFFE, D., A. SHAINBERG, and H. DYM. 1972. Studies on the prefusion stage during formation of multinucleated muscle fibers *in vitro*. In *Research concepts in muscle development and the muscle spindle*, ed. B. Banker et al. Excerpta Medica, Amsterdam.

Selected Topics in Skeletal Myogenesis

H. Holtzer, J. W. Sanger, H. Ishikawa, and K. Strahs

Department of Anatomy, School of Medicine, University of Pennsylvania, Philadelphia, Pennsylvania 19104

and

Tokyo University, Tokyo, Japan

This report covers several overlapping aspects of myogenesis: (1) replication-dependent events that shift cells from the presumptive myoblast compartment into the myoblast compartment; (2) the initiation of the coordinated synthesis of myosin, actin, and tropomyosin, and possibly of acetylcholine receptors and creatine phosphokinase; (3) factors affecting the orientation of individual myofibrils and the problem of in vivo self-assembly of thick and thin filaments; (4) the effects of colcimide and cytochalasin-B; and (5) considerations concerning whether actin or an actin-like molecule is a constitutive component of the cortex of many cell types.

Requirement for DNA Synthesis and Nuclear Division for Overt Myogenesis

Presumptive myoblasts are the penultimate generation of cells in the skeletal myogenic lineage. They replicate but do not bind fluorescein-labeled antibodies to myosin or tropomyosin (Okazaki and Holtzer, 1965, 1966). Though rich in microtubules and 100 Å filaments, presumptive myoblasts do not display thick myosin or thin actin filaments in the cytoplasm (Ishikawa et al., 1968, 1969; Fischman, 1970, 1972). Depending upon environmental cues, a presumptive myoblast may undergo a "proliferative" cell cycle, yielding two more replicating presumptive myoblasts, or alternatively undergo a "quantal" cell cycle, yielding one or two postmitotic myoblasts. Four hours after having emerged from a quantal mitosis, mononucleated myoblasts have synthesized and assembled into thick and thin filaments sufficient quantities of myosin, actin, and tropomyosin to be detected both with labeled antibodies and with the electron microscope (Holtzer, 1970a,b).

The following experiments demonstrate that: (1) there is an obligatory requirement for DNA synthesis if mother presumptive myoblasts are to yield daughter myoblasts; and (2) presumptive myoblasts held at the G1-S interface do not have the option of initiating the coordinated synthesis of myosin, actin, and tropomyosin for myofibrils. Brachial or thoracic segments of 3-day chick embryos were transected in the midline. Either the

right or left was incubated 10 hr in normal medium, whereas the contralateral half was placed in normal medium plus 10^{-6} M FdU (5-fluorodeoxyuridine) to block DNA synthesis. After 10 hr, or several hours more than required for one complete cell cycle (Bischoff and Holtzer, 1968), both halves were glycerinated and treated with fluorescein-labeled antimyosin or antitropomyosin. Squashes of the myotomes were prepared (Fig. 1) and the total number of mononucleated, postmitotic myoblasts with striated myofibrils counted (Holtzer et al., 1957; Holtzer and Sanger, 1972). The data is shown in Table 1. Clearly the transformation of presumptive myoblasts into myoblasts which synthesize myofibrillar proteins is a maturation step dependent on DNA synthesis. Table 1 also demonstrates that when blocked presumptive myoblasts are allowed to enter S and divide, they yield perfectly normal daughter myoblasts. A similar obligatory requirement for DNA synthesis and nuclear division has been demonstrated for passage of hematocytoblasts into the erythroblast compartment (Holtzer et al., 1972b). In other embryonic lineages "maturation" has also been attributed to a small number of quantal cell cycles (Holtzer, 1963, 1970b; Holtzer et al., 1972a).

The experiments with FdU do not distinguish between the requirement for DNA synthesis and the subsequent events of cytokinesis. To determine whether cytokinesis per se was required for the initiation of the genetic program of myoblasts, replicating presumptive myoblasts were grown in cytochalasin-B. Cytochalasin-B is an antibiotic that allows DNA synthesis and nuclear division but blocks cytokinesis (Carter, 1967). Cultures of myogenic cells grown in cytochalasin-B formed large numbers of binucleated myotubes. These binucleated myotubes exhibited all the terminal properties normally associated with myoblasts and myotubes in vitro: well-defined striated myofibrils, postmitotic nuclei, and spontaneous contractions (Sanger et al., 1971; Sanger and Holtzer, 1972).

These experiments with FdU and cytochalasin-B have been interpreted to mean that changes in the state of the DNA in the presumptive myoblast

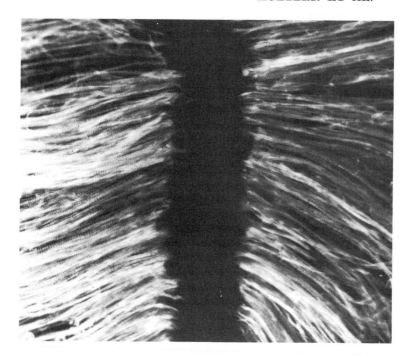

Figure 1. Low power, whole mount, photomicrograph of two consecutive segments of a 3-day chick embryo trunk treated with fluorescein-labeled antimyosin. In this microscopic field there are skin, dermal, somite, notochord, and nerve cells. Of these several hundreds of thousands of cells only the A-bands in the developing myofibrils of approximately 1.7×10^3 myoblasts bind the labeled antimyosin. These mononucleated, post-mitotic myoblasts achieve a length of over 300 microns and by the end of day 4 begin to fuse with neighboring myogenic cells to form the first binucleated myotubes.

must be induced during S, but the consequences of such a changed state in terms of transcription and translation require a G1 cytoplasm for expression. Normally this G1 cytoplasm is established as the result of cytokinesis, but the G1 cytoplasm of postmitotic binucleated myotubes serves equally well.

Experiments with 5-bromodeoxyuridine (BrdU) also demonstrate that changes in the DNA of mother presumptive myoblasts affects the capacity of daughter cells to fuse and to initiate the coordinated synthesis of at least three contractile proteins (Stockdale et al., 1964; Okazaki and Holtzer, 1965; Bischoff and Holtzer, 1970). One round of DNA synthesis in the presence of BrdU in relative high concentrations blocks the emergence

of daughter cells that fuse or synthesize myosin, actin, tropomyosin, or creatine phosphokinase (Holtzer et al., 1972a). By reducing the percent substitution of the analog in the DNA of replicating presumptive myoblasts, it is possible to uncouple the capacity to fuse from the capacity to synthesize the terminal luxury molecules of muscle. Low doses of BrdU blocked the emergence of surface components required for fusion, but these same postmitotic, mononucleated cells developed normal striated myofibrils. That the capacity to fuse, on the one hand, and the repression of DNA synthesis and the initiation of the coordinated synthesis of the contractile proteins, on the other, are not tightly coupled events has been emphasized before (Holtzer et al., 1957; Holtzer, 1970a; Holtzer and Sanger, 1972). These findings raise the issue of whether the cessation of DNA synthesis and the drop in DNA polymerase activity in maturing muscle precedes fusion and the synthesis of the contractile proteins or is the consequence of these events. O'Neill and Strohman (1969) favor the latter view, whereas Holtzer and Bischoff (1970) favor the former.

Thus far conditions have not been found that uncouple or differentially suppress the synthesis of myosin or actin or tropomyosin or creatine phosphokinase. The percent substitution of BrdU for thymidine that blocks the synthesis of one of these proteins blocks the synthesis of the others as well. However there must be genetic and physiological mechanisms that allow different muscles to accumulate greater or lesser amounts of myoglobin,

Table 1. Numbers of Striated Myoblasts/Posterior 3-Day Myotome

Control[a]	FdU-treated[a]	FdU-treated,[b] Not Reversed	FdU-treated,[b] Reversed
960	520	380	840
1220	830	650	1050
810	410	430	1010
1350	750		

[a] Three-day trunks were transected into right and left halves and reared as organ cultures for 10 hr in normal medium plus 10^{-6} M FdU for 10 hr. After treatment with fluorescein-labeled antimyosin, the myotomes were squashed and the individual striated mononucleated myoblasts counted.
[b] The trunks were transected into right and left halves and organ cultured in FdU for 10 hr. Either the right or left half was removed, washed, and then grown in normal medium with excess cold TdU for an additional 15 hr; the other half remained in FdU.

creatine phosphokinase, mitochondria, etc. Preliminary work with myoglobin, for example, suggests that the synthesis of this molecule is initiated relatively late in the postmitotic myotubes.

The striking phenomenon of fusion and the formation of multinucleated myotubes (Holtzer, 1958; Stockdale and Holtzer, 1961) is also tightly coupled to the cell cycle. The demonstration that myogenic cells in S, G2, or M do not fuse has been confirmed in many laboratories (Okazaki and Holtzer, 1965; Bischoff and Holtzer, 1968; Cox and Simpson, 1970; Yaffee et al., 1972). Nevertheless many provocative issues related to fusion remain unanswered. For example Bischoff and Holtzer (1969) reported that following a given mitosis, a period of 2 or 3 hr in G1 was required before myogenic cells acquired the cell surfaces and associated machinery which allowed melding of competent surfaces. O'Neill and Stockdale (1972a) claim that this period may be less than 0.5 hr. This reported difference of some 2 hr is most likely due to the fact that the two groups used different criteria for judging when cells have, in fact, fused. During the early stages of fusion it is impossible with the light microscope to determine when cells are truly fused in contrast with being merely intimately apposed.

A more intriguing and still more difficult issue to resolve was raised by Okazaki and Holtzer (1965) and by Ishikawa et al. (1968): Does the unique ability of myogenic cells to fuse "wax and wane" with their position in the cell cycle? Does a quantal mitosis, which yields a myoblast competent to fuse, yield a postmitotic myoblast able to synthesize contractile proteins? Is there a generation of myogenic cells which have the option to fuse and as a consequence of fusion, enter the postmitotic state, or failing to fuse, reenter the mitotic cycle? The technical difficulties involved in analyzing this issue are formidable. In addition to the considerable ambiguity of timing fusion accurately, there are the statistical problems associated with precise measurements of S, G2, and G1. For example, the cell cycle values of presumptive myoblasts of 4.5 hr for S and 1.5 hr for G2 that Bischoff and Holtzer (1968) reported for presumptive myoblasts were *average* values for one set of culture conditions; it is highly likely that a goodly number of cells will have an S and G2 less than 4.5 and 1.5 hr, respectively, particularly under other culture conditions. In addition there is a relatively high incidence of abnormal binucleated cells in frequently subcultured cells plated at low density. Thus, though the problem of "waxing and waning" has been the subject of several recent investigations (Simpson and Cox, 1972; Yaffe et al., 1972; Konigsberg, 1972; O'Neill and Stockdale, 1972b),

it is not in our opinion solved. If myogenic cells competent to fuse in G1 should be shown to have the option of entering S, it would be an elegant demonstration of the connection between DNA synthesis and change in the state of the cell surface (Holtzer, 1970a,b). It would also suggest that the quantal cell cycle that leads to the first daughters capable of fusion may be followed by variable numbers of proliferative cell cycles which perpetuate this capacity.

In a recent series of experiments protein synthesis was depressed to varying degrees and the effect on myogenesis observed (Weintraub and Holtzer, unpublished). The questions posed were: (1) Will the process of fusion be affected more or less than the synthesis of contractile proteins? (2) Will the synthesis of the different contractile proteins be equally suppressed or will some be more inhibited than others? Myogenic cultures were set up in concentrations of cyclohexamide that depressed protein synthesis between 40–60%. The cultures were maintained in cyclohexamide between 6 and 8 days. Fusion was relatively unimpaired, but the multinucleated myotubes did not lengthen properly, tending to form "myosacs" (Bischoff and Holtzer, 1968). The formation of striated myofibrils in cells maintained in the inhibitor was drastically curtailed, but those that did form were perfectly normal. This inhibitory effect was reversible; within 2 or 3 days after removing the cyclohexamide, the myotubes were indistinguishable from controls. Only when protein synthesis was inhibited by more than 40% was there a noticeable block to the overall amount of fusion.

In summary: A variety of experiments suggests that the genetic program of the presumptive myoblast is very different from that of the myoblast. One of the critical events initiating this metabolic generation gap between mother and daughter cells may occur during the S period of the presumptive myoblast, though its expression in terms of transcription and translation normally occurs in the G1 of the daughter cells. It is also suggested that this event involving DNA synthesis in the presumptive myoblast does not directly involve the structural genes for myosin, actin, or tropomyosin but involves a higher-order chromosomal unit, one regulating the whole program of "terminal myogenesis." Only after this higher-order genetic unit is activated is it possible for the myoblast to respond to cytoplasmic cues and to begin the coordinated synthesis of myosin, actin, and tropomyosin, as well as the synthesis of such molecules as creatine phosphokinase (Reporter et al., 1963; Shainberg et al., 1971) myoglobin and acetylcholine receptors (Fambrough and Rash, 1971).

Orientation of Myofibrils and Problem of in vivo Self-Assembly

In vivo multinucleated myotubes invariably are long, unbranched structures and invariably the striated myofibrils are assembled so as to run parallel to the long axis of the myotube. Figure 2 illustrates a well-developed myotube formed in tissue culture demonstrating the linear organization of the component myofibrils. Little is known of the biological factors ordering this linear orientation of the myofibrils or of how the forces ordering the sizing and stacking of alternating thick and thin filaments, as well as their alignment into tandem sarcomeres, interact. One approach to these problems is to take advantage of the fact that in vitro myotubes often bifurcate as they lengthen. In addition formed myotubes often fuse with each other at different angles. Figure 3 illustrates the parallel alignment of individual myofibrils in an unbranched region of myotubes growing in vitro. Figures 4–6 illustrate more complex arrangements taken from the same culture that could be due both to fusion of myotubes and to bifurcation of elongating myotubes. Observe how individual myofibrils maintain their integrity though they interweave with myofibrils coursing in many different directions. It is also worth stressing that these different myofibrils are assembled at different times and that it would be accurate to describe them as belonging to different generations. Often a single myofibril bends around the bifurcation site so that the same myofibril courses from one arm of the Y to the other. The major difference between the earliest striated myofibrils, which may consist of several hundred alternative thick and thin filaments, and the mature myofibril consisting of many millions is due to the alternate deposition of thick and thin filaments at the periphery of each myofibril; in brief the myofibrils grow in tree-ring fashion. The fact that the interweaving myofibrils of the kind illustrated in Fig. 5 increase manyfold

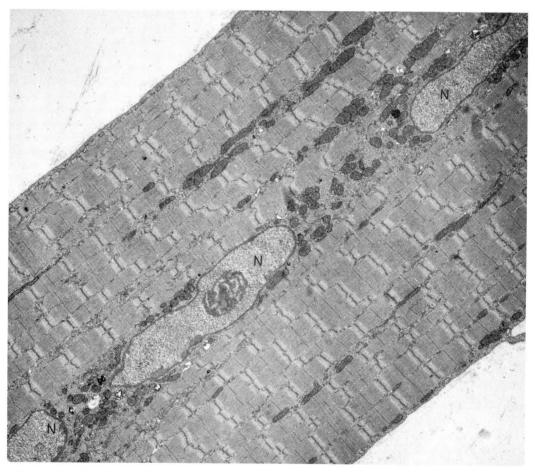

Figure 2. A well-developed myotube from a 16-day-old culture. Though the nuclei are still centrally located, in most other respects this is a terminally differentiated muscle fiber. Fibers of this kind contract continuously in the absence of innervation and show no signs of atrophy. 4,600 ×.

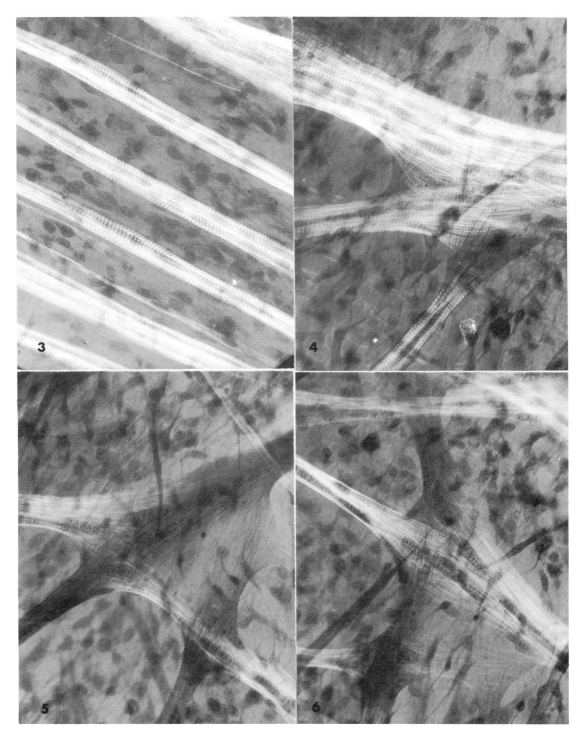

Figures 3–6. Photomicrographs taken with the polarizing microscope of living cultured breast skeletal muscle cells. Various alignments of myotubes: (*3*) parallel array of adjacent myofibrils and myotubes; (*4*) forking of myofibrils from large bundle into smaller bundles; (*5, 6*) turning of myofibrils in the area of bifurcation site so that the same myofibril courses from one branch to another branch. Observe the complex interweaving of individual myofibrils.

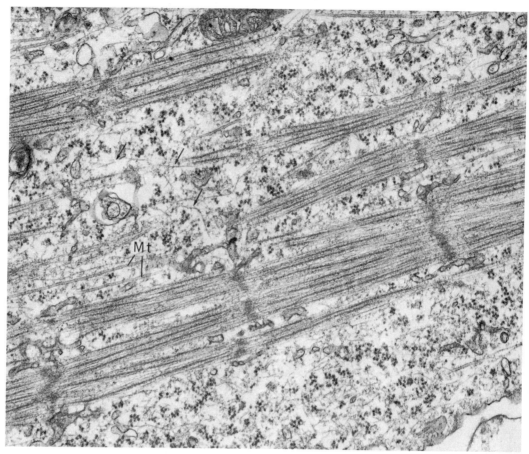

Figure 7. Early forming striated myofilaments in a 4-day-old myotube. Note there is no stage in development where thick and thin filaments are randomly oriented. The vast majority of thick and thin filaments at all stages in normal muscle are oriented parallel to one another, one thick filament surrounded by six thin filaments. Note the randomly oriented 100 Å filaments (arrows) and the longitudinally oriented microtubules (Mt). 33,000 ×.

in diameter while they maintain their integrity suggests that the myofibril is a crystalloid-like structure and that the myofibril is a basic unit of construction. These considerations raise interesting questions if it is assumed that: (1) monomers of myosin and actin are randomly distributed throughout the sarcoplasm; and (2) polymerization into thick and thin filaments is controlled by local pH and ionic composition (Sanger, 1971).

Experiments with colcimide support the view that those hexagonal arrays which are formed first serve as nucleation sites for the further deposition of thick and thin filaments and that the linear coursing of myofibrils is somehow oriented by the anisodiametric geometry of the multinucleated unit (Holtzer, 1970a). Young myotubes exposed to 10^{-6} M colcimide fragment within 6 hr into multinucleated ameboid structures termed "myosacs" (Bischoff and Holtzer, 1968). These irregularly shaped myosacs contain anywhere from two to over 100 nuclei. The effects of this

fragmentation into myosacs on the myofibrils may be judged by contrasting Figs. 7 and 8. Figure 7 is a photomicrograph illustrating the fine structure of the early sarcomeres in a normal 4-day myotube; the sarcomeres are reasonably well defined, though neither the H-zone nor the M-band has yet emerged. Cross sections through such a myotube reveal that considerably less than 10% of the cross-sectional area is occupied by myofibrils and that they tend to be located toward the cell surface. Figures 8 and 9 are photomicrographs of 4-day myotubes that had been treated with colcimide for 10 and 48 hr, respectively. Observe the dismantling and scrambling of the previously formed and stacked thick and thin filaments. It is not probable that the colcimide chemically dissociates the previously assembled myofibrils, but the scattering of filaments is likely to be a secondary consequence of the fragmentation of the myotube into myosac. The considerably greater number of filaments in the hexagonal arrays illustrated in

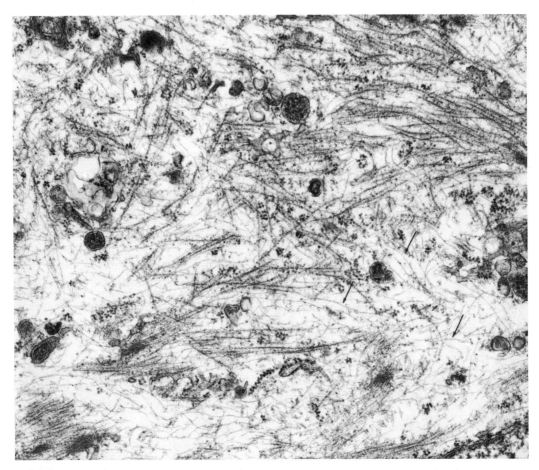

Figure 8. Irregularly oriented myofilaments in a 4-day-old myotube grown in 10^{-8} M colcimide for 10 hr. Thick and thin filaments have been dissociated. There are numerous 100 Å filaments (arrows) randomly distributed throughout the cytoplasm. 27,000 ×.

Fig. 9 is evidence that myosin, actin, and tropomyosin continue to be synthesized and assembled into thick and thin filaments though the cells were maintained in colcimide. Irrespective of the scrambled arrangement of the myofilaments, spontaneous contractions are observed as early and as frequently in myosacs as in normal myotubes. In brief colcimide disturbed the parallel alignment of the myofibrils, but it blocks neither the synthesis of contractile proteins nor their assembly into hexagonal arrays of thick and thin filaments.

Myosacs that have been in colcimide over a week may display four or five major groups of myofibrils with their axes oriented at different angles to one another. Often a single Z-band serves as anchor point for intersecting fibrils. Within the territory of a given sarcomere, however, the organization is remarkably normal—hexagonal arrays, H-zones, M-bands, and Z-bands—indicating local autonomy of those forces ordering the distribution of contractile proteins and associ-

ated molecules. This local autonomy must extend for hundreds of micra along the long axis, since a single myofibril is remarkably uniform in girth up and down its length. The fact that a given myofibril maintains a constant width in spite of interweaving among neighboring myofibrils may be explained by assuming that the polysomes synthesizing myosin, actin, and tropomyosin are aggregated evenly around the periphery of each myofibril.

The number of major bundles of myofibrils in the older myosacs is less than in the younger. Whether this is due to the elimination of many of the more randomly oriented small myofibrils or is due to smaller myofibrils secondarily amalgamating into larger units is still unclear.

That the overall alignment of myofibrils reflects the long axis of the myotube is shown by removing the colcimide from older myosacs. Under these conditions there is often a redistribution of the nuclei, and a long, unbranched typical myotube

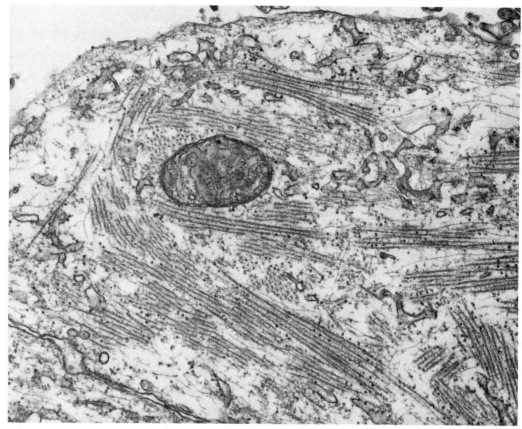

Figure 9. Irregular arrangement of myofibrils in a myosac taken from a 4-day-old culture treated with 10^{-8} M colcimide for 48 hr. In spite of the distorted coursing of myofibrils, the stacking of thick and thin filaments is reasonably normal. These ameboid-like, multinucleated structures contract spontaneously for as long as cultured. 33,000 ×.

structure grows out of the myosac. The orientation of the newly assembled myofibrils in the outgrowth is invariably parallel to the long axis of the elongating unit, whereas the formed myofibrils in the persisting myosac remain scrambled (Bischoff and Holtzer, 1968).

The elongation of a myotube, like the elongation of a nerve fiber, requires adhesion of the growth top to the substrate. This adhesion is nonspecific, and there is no compelling evidence indicating that it is uniquely dependent on the intervention of exogeneous molecules such as collagen or any hormone. However it is not easy to understand how the tension arising from adhesion and movement of the myotube growth tip is transduced into forces conditioning the alignment of myofibrils. Efforts to detect cytoplasmic streaming have thus far been negative. Alternatively the orienting mechanisms for the developing myofibrils may reside in the cell surface. The first myofibrils to appear form in close proximity to the cell surface (Holtzer et al., 1957; Fischman, 1967, 1972), and it may be that these early formed hexagonal arrays subsequently are displaced into the sarco-

plasm where they serve as nucleation sites for further growth. In the myosacs the detached early myofibrils would quickly lose their orientation as the ameboid structure changed its contours.

Induction of 100 Å Filaments and Dense Bodies by Colcimide

Myosacs may be induced by allowing mononucleated presumptive myoblasts to fuse in the presence of colcimide or by allowing the inhibitor to fragment already formed myotubes (Bischoff and Holtzer, 1968). In either case the resulting myosacs are densely packed with large numbers of 100 Å filaments (Fig. 10). The length of these 100 Å filaments is still unknown, though several over 2μ in length have been observed. Modest numbers of these 100 Å filaments are found in many types of normal and abnormal cell types (Ishikawa et al., 1968; Anderson et al., 1970), but in cells treated with colcimide their number is greatly increased. Large areas of myosacs consist exclusively of these 100 Å filaments, giving the impression that thick and thin filaments are excluded from such a domain. It would be interesting to learn if myosin

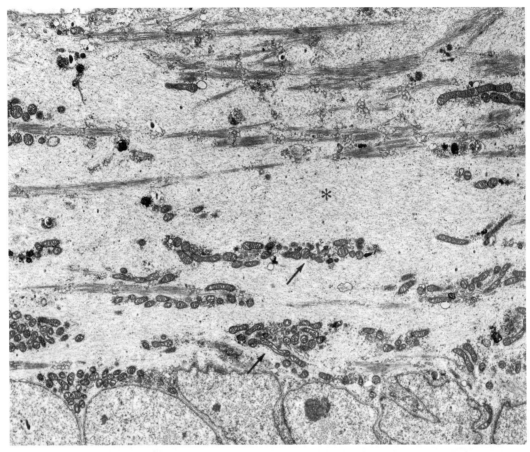

Figure 10. A 4-day-old myotube treated with 10^{-6} M colcimide for the last 24 hr. Massive aggregates of 100 Å (*) are prominent. Microtubules are not present. Given the modest number of microtubules in normal myotubes, it is difficult to understand how their breakdown could yield such large numbers of 100 Å filaments. The arrows indicate islands of mitochondria in the aggregates of 100 Å filaments. 5300 ×.

and actin are not synthesized, whether they are physically excluded, or whether they cannot polymerize in these areas. The induction of 100 Å filaments has been observed by Wisniewski et al. (1968) and Bunge and Bunge (1968) in nerve fibers and by Strahs, Sanger, and Holtzer (unpublished) in cardiac myoblasts and fibroblasts.

Long microtubules, present in myotubes (Fig. 7) are not present in myosacs, and this inverse relationship raises questions of interconversion between microtubules and 100 Å filaments. However given the relatively modest number of microtubules, it is difficult to understand how their disassembly or the failure of the normal numbers of monomers to polymerize could yield the large numbers of 100 Å filaments present in treated cells.

Another consequence of treating myotubes or cardiac myoblasts with colcimide is the induction of dense bodies. As shown in Figs. 11 and 12 these bodies may or may not be associated with fine filaments in the range of 50–70 Å. The dense bodies vary from irregularly shaped structures of some

200 Å to well over 2000 Å. Often they are found in tandem separated by varying distances. The core proper appears to consist of filaments approximately 120 Å in diameter. In places the thick core filaments and the thin radiating filaments appear to be continuous, and possibly the larger filaments result from additional material deposited on the thin filaments. Experiments are in progress to determine if the thin filaments radiating from the dense bodies form arrowhead complexes with HMM (Ishikawa et al., 1969). These amorphous dense bodies resemble the material around the Z-band in various normal (Hagopian and Spiro, 1970) and pathological states (Gonatas et al., 1966; Shafiq et al., 1969). Fine filaments and amorphous bodies reminiscent of the kind shown here have also been observed in the cortex of cultured chondroblasts and fibroblasts, and their relationship to myogenesis is unclear (Fig. 13).

With respect to the manner in which colcimide might induce these bodies, it is worth emphasizing that the migratory behavior and adhesive

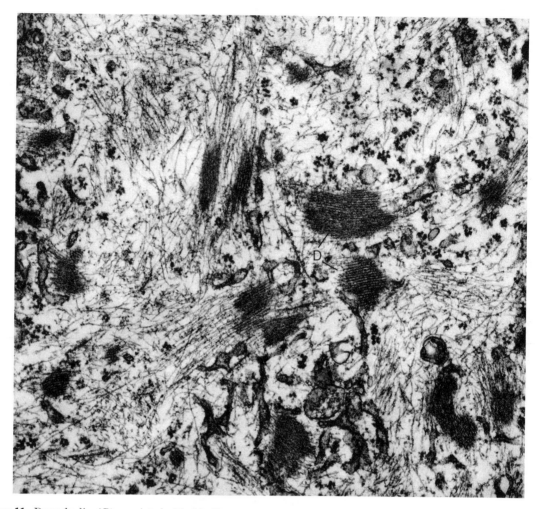

Figure 11. Dense bodies (*D*) associated with thin filaments are common in myosacs. Note the crystalloid structure of the dense bodies. Only a few thick filaments are associated with these thin filaments. This is a micrograph of a 6-day-old culture grown in 10^{-6} M colcimide for the last 24 hr. 37,000 ×.

properties of interphase cells in culture is altered by colcimide (Bischoff and Holtzer, 1968). Whether these changes are secondary to the dismantling of the microtubules or whether the colcimide acts directly on the cell surface is still unknown. If the latter, then other agents perturbing the cell surface of cells assembling myofibrils might also induce dense bodies (see next section on induction of dense bodies by cytochalasin-B).

In conclusion, in addition to microtubules, myoblasts and myotubes display modest numbers of 100 Å filaments as well as finer filaments associated with the cell cortex. Currently nothing is known about the composition or interrelationships between these seemingly discrete populations of filaments.

Effects of Cytochalasin-B on Cytokinesis and Myogenesis

Carter (1967) reported that cytochalasin-B inhibited ameboid movement and blocked cyto-

kinesis. These observations have been confirmed by many workers. However the observations and claims by Schroeder (1970) on the action of the drug on microfilaments of the contractile ring and the claims of Wessels and coworkers (Wessels et al., 1971; Spooner et al., 1971) that the drug blocks the contractile microfilament machinery of many cell types are less convincing.

Unequivocal evidence that cytochalasin-B does not block the initiation but only the very terminal stages of cell division is shown in Figs. 14–19. Replicating cells in the antibiotic enter anaphase and telophase, but the centripetally advancing furrow appears unable to cleave the midbody. If a contractile ring consisting of microfilaments is involved in cell division, then for the most part this contractile system functions quite normally in cytochalasin-B. It is only the *terminal* stage of cell division that is blocked by the drug. In any event the action of cytochalasin-B on the putative

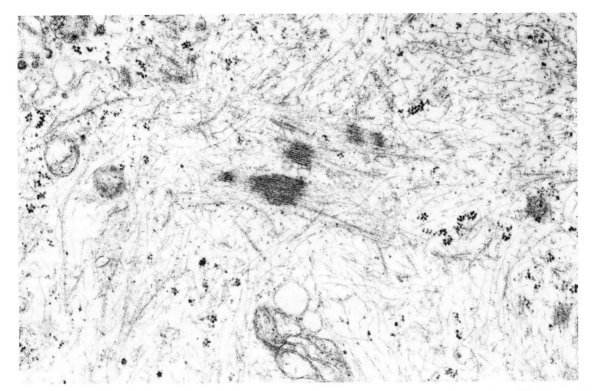

Figure 12. A Chick embryonic heart treated with 10^{-6} M colcimide. The heart was isolated from a 5-day-old embryo and cultured for three days with colcimide. Massive numbers of 100 Å filaments surround the dense bodies. 27,000 ×.

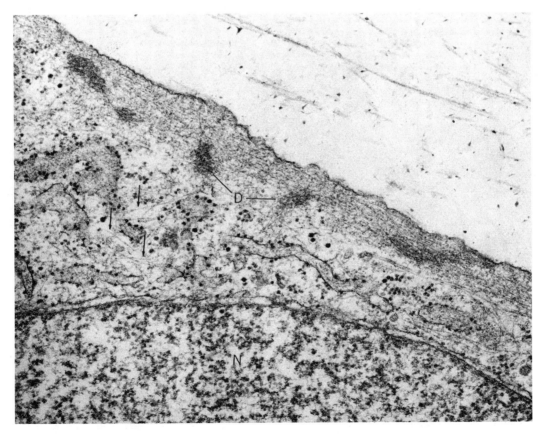

Figure 13. Thin filaments with dense bodies (*D*) are observed in chondroblasts in vitro. This chondroblast also contains numerous 100 Å filaments (arrows). Observe the skein in the cell cortex. After treatment with HMM this area would exhibit arrow head complexes. 39,000 ×.

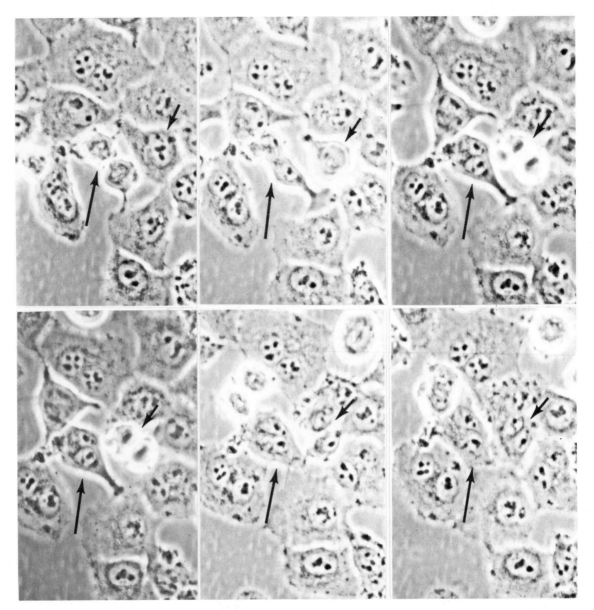

Figures 14–19. HeLa cells undergoing nuclear divisions in the presence of cytochalasin B (1 μg/ml). Two cells (indicated by arrows) show that during division furrowing proceeds normally until the mid-body is reached. In time the contracted furrows relax and a binucleated cell is formed. During the action of the cleavage furrow large bubbles are observed on the surface of the cells. Clearly the contractile ring functions in a reasonably normal fashion in cytochalasin-B. 850 ×.

primitive microfilaments is not analogous to the action of colcimide on microtubules.

It is worth stressing that the frequency of "bubbling" and the size of the "bubbles" observed in the cells temporarily blocked in telophase are greatly enhanced in cytochalasin-B. Often such bubbles approximate the size of one of the daughter cells (Figs. 16, 17). It may be the channeling of available cell surface into such bubbles that interferes with the completion of cytokinesis.

The claim that cytochalasin-B inhibited the contraction of embryonic cardiac and smooth muscle cells leads to the proposition that the putative microfilaments in a variety of cell types involved an actomyosin-like system. However Sanger et al. (1971) and Sanger and Holtzer (1972) observed that cytochalasin-B did not block the spontaneous contractions of isolated cardiac, skeletal, or smooth muscle cells. In another set of experiments (in collaboration with Drs. A. V. Somlyo and A. P. Somlyo) progesterone-primed uteri from mature rats were incubated in cytochalasin-B (50 μ/ml) for over 4 hr with no effect on the rate or amplitude of the contractions.

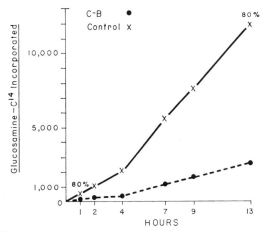

Figure 20. Incorporation of [¹⁴C]glucosamine into the mucopolysaccharide and glycoprotein fractions of amnion cultures. Equal numbers of cells were plated out in normal media. The cells were permitted to grow for 1 day. The medium was removed and fresh medium containing [¹⁴C]glucosamine with and without 0.5 µg/ml of cytochalasin-B added. At the indicated time, the cultures (medium and cells) were collected, processed, and counted.

These observations suggest that at moderate doses cytochalasin-B has little effect on the intracellular interaction of myosin and actin. Similar conclusions were reached by Forer et al. (1972).

The observation that tissues incubated in cytochalasin-B tended to dissociate into individual cells led to the testing of the effect of the antibiotic on sugar metabolism and the possibility that its effect on ameboid movement and cell adhesions might involve polysaccharide components of the cell surface. Figure 20 illustrates that the antibiotic rapidly blocks the uptake of [³H]glucosamine into cell-bound and secreted polysaccharides. Table 2 demonstrates that the glycogen content of treated cells is rapidly depleted. Experiments by Mizel and Wilson (1972), Hirsch and Zigmond (1972), and Kletzien et al. (1972) confirm and extend these observations. These investigators report that cytochalsin-B immediately inhibited the transport of [³H]glucose, [³H]glucosamine, and [³H]-2-deoxyglucose. Hexokinase activity or total ATPase was unaffected by the drug.

As shown in Figs. 21–24 cytochalasin-B has striking effects on the cytology of presumptive myoblasts and fibroblasts. Twenty-four hours in cytochalasin-B transforms these cells into the "arborized" state shown in Fig. 22. The maintenance of the arborized state requires the presence of the antibiotic, for within 30 min after its removal the arborized cell transforms into the "pancake" condition shown in Fig. 22. The shift from the arborized to pancake state does not require protein synthesis since it occurs in the presence of cyclohexamide. This transformation, however, requires energy, for it is reversibly blocked in the presence of NaCN. Cells in the arborized or pancake state do not exhibit ameboid movements, though the processes of the arborized cells are studded with microspikes.

At concentrations of cytochalasin-B (0.5 µ/ml) known to block the termination of cytokinesis and to inhibit ameboid movement, the drug has only modest effects on myogenesis (Sanger et al., 1971). Fusion is reduced in the antibiotic. Whether this is due to a direct action on the surface molecules required for fusion or is secondary to inhibiting the ameboid movement preceding fusion, is currently unclear. Recent data from Fambrough (unpublished), who has studied the resting potential, acetylcholine sensitivity, and the binding of ¹²⁵I-α-bungarotoxin of cells treated with cytochalasin-B, complement the cytological data. The arborized cells have a resting potential of −10 to −12 mV, display no ACh sensitivity, and do not bind ¹²⁵I-α-bungarotoxin. In contrast the control myotubes and the myotubes in cytochalasin-B have a resting potential around 55 mV and bind the labeled bungarotoxin. In brief, cells whose cytology changes in cytochalasin-B also show changes in their electrical physiology, whereas myotubes which show little cytological change in cytochalasin-B do not change their physiological properties.

Intact hearts from 4-day chick embryos cease contracting in cytochalasin-B in 10–15 hr, whereas trypsinized, isolated cardiac myoblasts beat and divide for as long as 3 days in the antibiotic (Sanger

Table 2. Effects of Cytochalasin-B on Incorporation of [³H]-D-Glucose

Time (hr)	mµ Moles Glucose Hydrolyzed from Glycogen per Dish (% control) ½C-B	Into Glycogen (cpm/dish) Control	½C-B	Into Glycoproteins and Mucopolysaccharides (cpm/dish) Control	½C-B
1	76	450	31	324	168
2	61	2395	39	248	216
4	40	4452	86	1312	468
8	41	4545	46	2000	896
12	30	5282	112	4128	1308

Amnion cells were prepared from 10-day chick embryos and cultured. [³H]-D-Glucose was added to cultures for various times to monitor the incorporation of [³H]-D-glucose into glycogen and secreted glycoproteins and mucopolysaccharides.

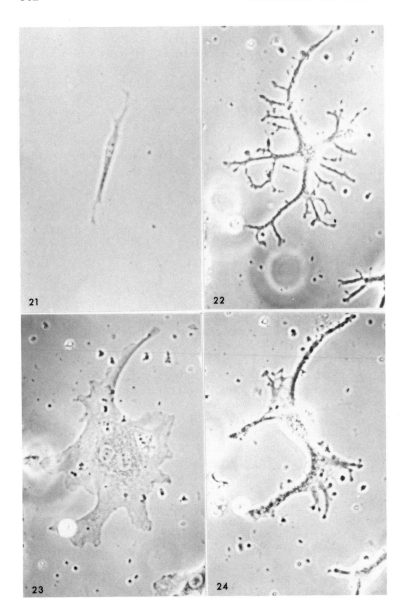

Figures 21–24. A cell isolated from 10-day-old breast muscle of a chick embryo. 600 ×. (*21*) 1 day in normal medium; (*22*) appearance of same cell in medium containing 5 μg/ml of cytochalasin-B after 2 days; (*23*) appearance of the cell 30 min after removal of the cytochalasin-B and normal medium added; (*24*) the normal medium was again removed and cytochalasin-B added. This picture was taken 30 min later.

and Holtzer, 1972). These findings, plus the observation that the antibiotic tends to dissociate cells, prompted an investigation of the effects of the drug on tight junctions, desmosomes, and other structures associated with the cell cortex of the intact heart. Thus far, if there are effects on such structures, they are not apparent in our preparations. However as illustrated in Fig. 25, cytochalasin-B is a potent inducer of dense bodies. The dense bodies induced by cytochalasin-B are indistinguishable from those induced by colcimide. In contrast to colcimide, cytochalasin-B has no obvious effect on microtubules nor does it induce 100 Å filaments. The fact that cytochalasin-B does not fragment myotubes into myosacs and does not dismantle microtubules is another instance of a positive correlation with cell elongation and microtubules.

If these dense bodies are related to "nemaline" bodies and are associated with Z-bands and the anchoring of myofibrils to the cell cortex, then their displacement into the sarcoplasm might explain the findings of Manasek et al. (1972). These investigators reported that the alignment of the very earliest myofibrils in 2-day chick heart was disrupted by the antibiotic, whereas developing myofibrils in slightly older heart cells were resistant. If cytochalasin-B interferes directly or indirectly with the formation or attachment of the Z-band complex to the cell surface, the small numbers of "loose" myofibrils in the cytoplasm of very early cardiac myoblasts would be more readily detected.

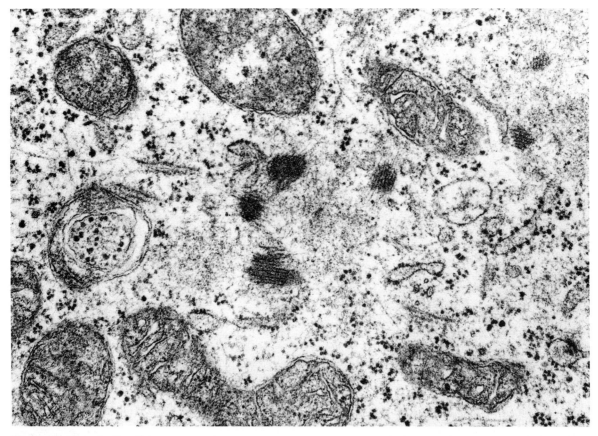

Figure 25. Dense bodies elicited by cytochalasin-B in an embryonic heart cell. Chick embryonic hearts were isolated from 5-day-old embryos and cultured intact for 2 days with cytochalasin-B (5 μg/ml). 60,000 ×.

Actin or Actin-like Molecules in Cell Cortex

Myosin-like and actin-like proteins have been isolated from many kinds of cells—nerve, sperm, fibroblasts, blood platelets, slime mold, and ameba. Efforts to localize myosin or tropomyosin by the use of fluorescein-labeled antibodies in cells other than skeletal or cardiac cells invariably have been negative. The modest antigenic differences between skeletal and cardiac myosins or even between the myosins of amphibians (Holtzer, 1958) and humans permit considerable cross-reactions; however antibody to skeletal myosin or tropomyosin neither precipitates nor is bound by smooth muscle myosin.

Results with antibody to actin have been ambiguous owing to difficulties of removing trace contaminants of tropomyosin, troponin, and other impurities for the preparation of antisera. Whether an antibody specific for actin has been prepared is still not clear (Holtzer, 1961, 1970a). Accordingly to determine whether actin in embryonic myoblasts and myotubes is similar to actin in older muscles and whether actin is present in other kinds of cells, Huxley's technique (1963) of decorating with

HMM was adapted for sectioned material (Ishikawa et al., 1969). The rationale of these experiments was that if free actin filaments were present in the cytoplasm of either presumptive myoblasts or myoblasts or other kinds of cells, then after treatment with HMM and staining with 1 % uranyl acetate their presence would be indicated by arrowhead complexes. Following exposure of glycerinated cells to HMM, arrowhead complexes formed all along the thin filaments of the most nascent as well as the most mature myofibrils both in mononucleated myoblasts and in multinucleated myotubes. However randomly oriented, arrowhead complexes did not form either in myoblasts or myotubes as might be expected if actin were stored in the cytoplasm prior to interacting with thick filaments. HMM did not complex to microtubules, 100 Å filaments, or to the filaments associated with desmosomes. However in addition to forming arrowhead complexes on preexisting actin filaments in myofibrils, considerable numbers of complexes emerged in association with the cell cortex of presumptive myoblasts, and myotubes. Furthermore similar arrowhead complexes were induced in the cortex of chondroblasts, fibroblasts, nerve cells, skin cells, and intestinal epithelial cells. If the binding of HMM to form arrowhead complexes is a unique protein-protein interaction and is diagnostic for actin and only actin, then actin must be a constitutive molecule present in the cell cortex, and possibly cytoplasm, of virtually all cells (Holtzer et al., 1972a).

The induction of arrowhead complexes by HMM in many cell types may have profound implications, not only for the physiology of contraction, but for understanding some aspects of cell differentiation. Accordingly the following observations should be considered: (1) There is no obligatory relationship between the existence of *preformed* 60 Å *filaments* and the appearance of arrowhead complexes after HMM. Prior to treatment with HMM, presumptive myoblasts, fibroblasts, and chondroblasts do not appear to contain significant numbers of 60 Å filaments either in their cortex or their cytoplasm. Yet after interacting with HMM, the area associated with the cell cortex is as dense with arrowhead complexes as is the cortex of myoblasts or myotubes. (2) The length of the preformed, 60 Å actin filament in the myofibril is 1 μ. The arrowhead complexes induced by HMM in the cortex of many kinds of cells is clearly more than 1 μ. This finding would imply that some force operates at the level of the individual sarcomeres to regulate the length of the actin filament. (3) A given arrowhead complex in the cell cortex exhibits the same polarity throughout its length. However no simple relationship (e.g., parallel or antiparallel) has been observed between the polarities of adjacent complexes.

The induction of arrowhead complexes by HMM in many kinds of cells raises the issue of whether there is but one kind of actin molecule or whether there is a family of actin-like molecules (Hatano et al., 1967; Marchesi and Steers, 1968; Mazai and Ruby, 1968). Experiments with BrdU demonstrated a tight coupling of the synthesis of myosin, tropomyosin, and actin destined for myofibrils. More recently we have found that arrowhead complexes formed in BrdU suppressed myogenic, chondrogenic, and fibrogenic cells, If there is a single structural gene for actin, then in BrdU-suppressed myogenic cells the actin molecules destined for the cell cortex are properly processed and distributed, whereas in the same cell the accumulation of actin molecules destined for myofilaments is blocked. In this scheme BrdU would block the synthesis of myosin, and because myosin failed to "trap" the actin monomers, no cytoplasmic filaments would be observed in the suppressed cells. An alternative possibility is that there are two or more structural genes, one for actin for myofilaments and one for the actin, or actin-like molecules, destined for the cell cortex.

The actin-like gene transcribed and translated in most kinds of cells is not repressed by BrdU, whereas the actin gene for transcribing the monomers for thin filaments is blocked by BrdU. This "two-gene" concept implies that the second gene for myofilament actin is derepressed only after the terminal quantal cell cycle and in conjunction with the genes for myosin and tropomyosin.

Clearly much has yet to be learned before we are confident of the number of structural genes for actin present and functioning in a given cell. Nevertheless the fact that actin or actin-like molecules are present in many cells, but that roughly equal numbers of myosin or myosin-like molecules are not present, invites speculation about filaments and contraction. The concept of sliding filaments is based on some sort of parity between actin and myosin. If, in fact, actin in non-muscle cells is greatly in excess of a myosin-like molecule, it may be that in these situations the function of actin is to regulate cytoplasmic gelation and solation and does not require interaction between either preformed actin or myosin *filaments*. The emergence of differentiated muscle cells in evolution and ontogeny would involve the deployment of actin monomers into preformed, reasonably stable, structural filaments.

In this connection it is worth stressing that non-myogenic cells do not bind antitropomyosin. Presumptive myoblasts do not bind antitropomyosin nor do BrdU-suppressed myogenic cells bind this labeled antibody. The failure to bind skeletal antitropomyosin in areas where arrowhead complexes can be induced could indicate immunological differences from the tropomyosin found in muscle. Alternatively however it could also indicate that tropomyosin was not present. The latter possibility also suggests differences in the composition and function of cortical filaments from their counterparts in myofibrils.

References

ANDERSON, H. C., S. CHARKO, J. ABBOTT, and H. HOLTZER. 1970. The loss of phenotypic traits by differentiated cells *in vitro*. VII. Effects of 5-bromodeoxyuridine and prolonged culturing on fine structure of chondrocytes. *Amer. J. Path.* **60:** 289.

BISCHOFF, R. and H. HOLTZER. 1968. The effect of mitotic inhibitors on myogenesis in vitro. *J. Cell Biol.* **36:** 111.

———. 1969. Mitosis and the processes of differentiation of myogenic cells in vitro. *J. Cell Biol.* **41:** 188.

———. 1970. Inhibition of myoblast fusion after one round of DNA synthesis in 5-bromodeoxyuridine. *J. Cell Biol.* **44:** 134.

BUNGE, R. and M. BUNGE. 1968. Electron microscopic observations on colchicine induced changes in neuronal cytoplasm. *Anat. Rec.* **160:** 323.

CARTER, S. B. 1967. Effects of Cytochalasins on mammalian cells. *Nature* **213:** 261.

COX, P. G. and S. B. SIMPSON. 1970. A microphotometric

study of myogenic lizard cells grown *in vitro*. *Develop. Biol.* **23**: 433.

FAMBROUGH, D. and E. RASH. 1971. Development of acetylcholine sensitivity during myogenesis. *Develop. Biol.* **26**: 55.

FISCHMAN, D. A. 1967. An electron microscope study of myofibril formation in embryonic chick skeletal muscle. *J. Cell Biol.* **32**: 557.

———. 1970. The synthesis and assembly of myofibrils in embryonic muscle. *Current Topics Develop. Biol.* **5**: 235.

———. 1972. The fine structure of muscle differentiation in monolayer culture, p. 88. In *Research in muscle development and the muscle spindle*, ed. B. Q. Banker et al. Excerpta Medica, Amsterdam.

FORER, A., J. EMMERSEN, and O. BEHNKE. 1972. Cytochalasin B: Does it effect actin-like filaments? *Science* **175**: 774.

GONATAS, N. K., G. M. SHY, and E. H. GODFREY. 1966. Nemaline myopathy: The origin of nemaline structures. *New Eng. J. Med.* **274**: 535.

HAGOPIAN, M. and D. SPIRO. 1970. Derivation of the Z line in the embryonic chick heart. *J. Cell Biol.* **44**: 683.

HATANO, S., T. TSUYOSHI, and F. OOSAWA. 1967. Polymerization of plasmodium actin. *Biochem. Biophys. Acta* **140**: 109.

HIRSCH, J. G. and S. ZIGMOND. 1972. Effects of cytochalasin B in uptake of 2-deoxyglucose uptake. *Science* In press.

HOLTZER, H. 1958. The development of mesodermal structures in regeneration and embryogenesis. In *Regeneration in vertebrates*, ed. C. Thornton. Univ. Chicago Press, Chicago.

———. 1961. Aspects of chondrogenesis and myogenesis, p. 35. In *Synthesis of molecular and cellular structures*, ed. D. Rudnick. Ronald Press, New York.

———. 1963. Comments on induction during cell differentiation. In *Induktion und morphogenise*, p. 127. 13th Mosbacher Colloq. Ges. physiol. Chemie. Springer-Verlag, Heidelberg.

———. 1970a. Myogenesis, p. 476. In *Cell differentiation*, ed. O. Schjeide and J. Vellis. Van Nostrand Reinhold New York.

———. 1970b. Proliferative and quantal cell cycles in the differentiation of muscle, cartilage, and red blood cells. *Symp. Int. Soc. Cell. Biol.* **9**: 69.

HOLTZER, H. and R. BISCHOFF. 1970. Mytosis and myogenesis. In *Physiology and biochemistry of muscle as a food*, ed. E. Buskey and K. Cassens. Vol. 2 p. 29. University of Wisconsin Press, Madison.

HOLTZER, H. and J. W. SANGER. 1972. Myogenesis: old views rethought, p. 122. In *Research in muscle development and the muscle spindle*, ed. B. Q. Banker et al. Excerpta Medica, Amsterdam.

HOLTZER, H., J. MARSHALL, and H. FINCK. 1957. An analysis of myogenesis by the use of fluorescent antimyosin. *J. Biophys. Biochem. Cytol.* **3**: 705.

HOLTZER, H., R. MAYNE, and H. WEINTRAUB. 1972a. Replication dependent events in cell differentiation. In *Biochemistry of gene expression in higher organisms*, ed. J. Lee. In press.

HOLTZER, H., H. WEINTRAUB, R. MAYNE, and B. MOCHAN. 1972b. The cell cycle, cell lineages and cell differentiation. *Current Topics Develop. Biol.* **7**: In press.

HUXLEY, H. E. 1963. Electron microscope studies on the structure of natural and synthetic filaments from striated muscle. *J. Mol. Biol.* **7**: 281.

ISHIKAWA, H., R. BISCHOFF, and H. HOLTZER. 1968. Mitosis and intermediate-sized filaments in developing skeletal muscle. *J. Cell Biol.* **38**: 538.

———. 1969. Formation of arrowhead complexes with heavy meromyosin in a variety of cell types. *J. Cell Biol.* **43**: 312.

KLETZIEN, R. F., J. PERDUE, and A. SPRINGER. 1972. Cytochalasin A and B. Inhibition of sugar uptake in cultured cells. *J. Biol. Chem.* **247**: 2964.

KONIGSBERG, I. R. 1972. Diffusion-mediated control of myoblast fusion. *Develop. Biol.* **26**: 133.

MANASEK, F. J., B. BURNSIDE, and J. STROMAN. 1972. The sensitivity of developing cardiac myofibrils to cytochalasin-B. *Proc. Nat. Acad. Sci.* **69**: 308.

MARCHESI, V. and S. STEERS. 1968. Selective solubilization of the protein component of the red cell membrane. *Science* **159**: 203.

MAZIA, D. and A. RUBY. 1968. Dissolution of erythrocyte membranes in water and comparison of the membrane protein with other structural proteins. *Proc. Nat. Acad. Sci.* **61**: 1005.

MIZEL, S. B. and L. WILSON. 1972. Inhibition of the transport of several hexoses in mammalian cells by cytochalasin B. *Science*. In press.

OKAZAKI, K. and H. HOLTZER. 1965. An analysis of myogenesis using fluorescein labelled antimyosin. *J. Histochem. Cytochem.* **13**: 726.

———. 1966. Myogenesis: fusion, myosin synthesis and the mitotic cycle. *Proc. Nat. Acad. Sci.* **56**: 1484.

O'NEILL, M. and F. E. STOCKDALE. 1972a. A kinetic analysis of myogenesis in vitro. *J. Cell Biol.* **52**: 52.

———. 1972b. Differentiation without cell division in cultured skeletal muscle. *Develop Biol.* In press.

O'NEILL, M. and R. C. STROHMAN. 1969. Changes in DNA polymerase activity associated with cell fusion in cultures of embryonic muscle. *J. Cell Physiol.* **73**: 61.

REPORTER, M. C., I. R. KONIGSBERG, and B. STREHLER. 1963. Kinetics of accumulation of creative phosphokinase activity in developing embryonic skeletal muscle *in vivo* and in monolayer culture. *Exp. Cell Res.* **30**: 410.

SANGER, J. W. 1971. Formation of synthetic myosin filaments: Influence of pH, ionic strength, cation substitution, dielectric constant and method of preparation. *Cytobiology* **4**: 450.

SANGER, J. W. and H. HOLTZER. 1972. Cytochalasin B.: Effects on cell morphology cell adhesion, and mucopolysaccharide synthesis. *Proc. Nat. Acad. Sci.* **69**: 253.

SANGER, J. W., S. HOLTZER, and H. HOLTZER. 1971. Effects of cytochalasin B on muscle cells in tissue culutre. *Nature New Biol.* **229**: 121.

SCHROEDER, T. E. 1970. The contractile ring. I. Fine structure of dividing mammalian (HeLa) cells and the effects of cytochalasin B. *Z. Zellforsch.* **109**: 431.

SHAFIQ, S. A., M. A. GORYCKI, S. A. ASIEDU, and A. T. MILHORAT. 1969. Tenotomy: Effect on the fine structure of the soleus of the rat. *Arch. Neurol.* **20**: 625.

SHAINBERG, A., G. YAGIL, and O. YAFFEE. 1971. Alterations of enzymatic activities during muscle differentiation *in vitro*. *Develop. Biol.* **25**: 1.

SIMPSON, B. S. and P. G. COX. 1972. Studies on lizard myogenesis, p. 72. In *Research in muscle spindle*, ed. B. Q. Banker et al. Excerpta Medica, Amsterdam.

SPOONER, B. S., K. M. YAMADA, and N. K. WESSELLS. 1971. Microfilaments and cell locomotion. *J. Cell Biol.* **49**: 595.

STOCKDALE, F. and H. HOLTZER. 1961. DNA synthesis and myogenesis. *Exp. Cell Res.* **24**: 508.

STOCKDALE, F., K. OKAZAKI, M. NAMEROFF, and H. HOLTZER. 1964. 5-Bromodeoxyuridine: effect on myogenesis *in vitro*. *Science* **146**: 533.

WESSELS, N. K., B. S. SPOONER, J. F. ASH, M. D. BRADLEY, M. A. LUDUENA, E. TAYLOR, J. T. WRENN, and K. M. YAMADA. 1971. Microfilaments in cellular and developmental processes. *Science* **171**: 135.

WISNIEWSKI, H., M. L. SHELANSKI, and R. D. TERRY. 1968. Effects of mitotic spindle inhibitors on neurotubules and neurofilaments in anterior horn cells. *J. Cell Biol.* **38**: 224.

YAFFE, D., A. SHAINBERG, and H. DYM. 1972. Studies on the prefusion stage during formation of multinucleated muscle fibers *in vitro*, p. 110. In *Research in muscle development and the muscle spindle*, ed. B. Q. Banker et al. Excerpta Medica, Amsterdam.

Cytoplasmic Actin: A Comparative Study

D. BRAY

Medical Research Council, Laboratory of Molecular Biology, Hills Road, Cambridge, England

There can be no doubt now that many types of cell that neither contract nor contain organized myofibrils nevertheless contain muscle-like proteins. A very long list could be made of the sources, ranging from vertebrate brain and spermatazoa to ameba and green algae, which have been shown to contain actin-like filaments or to yield actomyosin-like extracts. In the best studied of these, such as blood platelets (Bettex-Galland and Lüscher, 1965; Adelstein et al., 1971), the similarities to the muscle system are impressive and include proteins which purify together with muscle actin or myosin, have the same molecular weights, solubilities and enzymatic properties, and which cross-react with their muscle analogs.

Given these extensive similarities, it is possible to ask whether the proteins from noncontractile cells might be not only similar to those found in skeletal muscle, but identical. Of course their distribution within the cell and their overall function must be distinct, but these features could be due to the action of ancillary molecules. Just as troponin and actinin are thought to modify the actomyosin of skeletal muscle to provide for its

regulation or regular organization, so other proteins in other cells might harness the same molecular machinery in a variety of other ways.

In fact, it is likely that the myosins of different tissues are not identical. Structural analyses of the protein from nonmuscle sources have not yet been made, but those from various kinds of muscle have already shown differences in the sensitivity to proteases (Tada et al., 1969) and amino acid sequence (Huszar and Elzinga, 1972) of their heavy chains, as well as pronounced changes in their light chains (Lowey and Risby, 1971; Weeds and Pope, 1971). If smooth, cardiac, and skeletal muscle possess altered forms of myosin, then it seems likely that nonmuscle tissues will show even greater changes.

But the same is not true of actin, and, despite a large number of tests, no tissue-specific forms have been found. Actin-like proteins from blood platelets (Bettex-Galland and Lüscher, 1965; Adelstein et al., 1971) and vertebrate brain (Berl and Puszkin, 1970; Fine and Bray, 1971), in one or both cases, have been shown to be the same as muscle actin in molecular weight, amino acid composition,

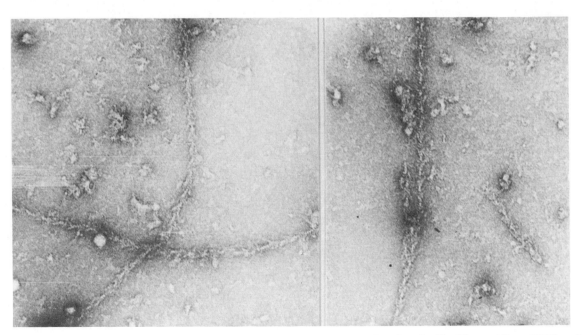

Figure 1. Decorated actin filaments from chick brain. The actin was prepared as described by Fine and Bray (1971) and decorated with the S1 fragment of chicken myosin as described by Nachmias et al. (1970).

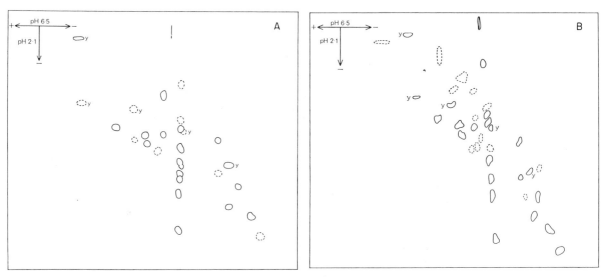

Figure 2. Comparison of the peptide maps of brain and muscle actin (from Fine and Bray, 1971). Composite pictures based on 5 pairs of fingerprints showing the position of strongly staining peptides (closed circles), weakly staining peptides (broken circles), and yellow staining peptides (marked with y). Only small amounts of the brain protein (A) were available, and some of the faint spots given by the muscle protein (B) did not appear. With these exceptions we could find no significant differences in the two maps.

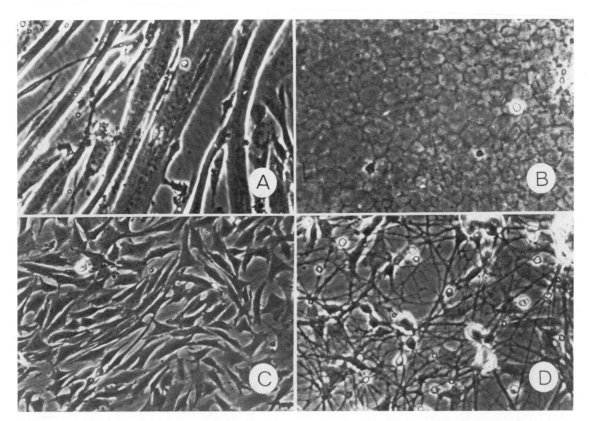

Figure 3. Four types of cell culture from 10–12 day old chick embryos. A, Skeleta muscle cells. Breast muscle was dissociated into single cells with trypsin and incubated for 4 hr in plain dishes to allow the fibroblasts to attach (C. R. Slater, personal communication). Myoblasts were then collected by gentle pipetting and allowed to grow in minimum essential medium (Flow labs) containing 10% embryo extract. Fusion occurred within 2–3 days and spontaneous contractions within a week of explantation. B, Lens epithelial cells. Small pieces of lens were grown on collagenized dishes containing minimal essential medium and 10% calf serum and were harvested after one week. C, Skin fibroblasts obtained by trypsinization of whole skin and grown for 8–10 generations in minimal essential medium containing 10% calf serum. D, Sympathetic neurons grown in medium containing nerve growth factor as previously described by Fine and Bray (1971).

3-methylhistidine content, (Crawford, personal communication), nucleotide content, ability to polymerize, ability to be decorated with muscle myosin fragments (Fig. 1), binding to muscle myosin with stimulation of ATPase activity and increase of viscosity and its release by ATP, and in the two dimensional fingerprints obtained by trypsin digestion (Fig. 2). The possibility that these are, in fact, muscle actin clearly exists; but in the absence of genetical tests, identity is hard to establish. Until the complete amino acid sequences of the proteins have been determined (and conceivably not even then), the biochemical evidence of identity must be made by default, by the accumulated absence of differences.

In the present report the comparison of the structure of actin-like proteins from a wide range of tissues and cell types and from several different species has been made. This has been carried out using a newly developed method which permits the peptide mapping of small quantities of protein from acrylamide gels and has shown that most kinds of vertebrate tissue contain a protein that is structurally similar to muscle actin. Although the peptide maps of muscle and non-muscle actins of the same species are indistinguishable, those from scallop were found to be slightly different from either chicken or dogfish. It is argued on the basis of this and other evidence that muscle and cytoplasmic actins are probably the same proteins.

Fingerprint Comparisons

To illustrate the methods that have been used, the comparison of actin-like protein from four cell types of a chick embryo will be described. Skin fibroblasts, skeletal muscle cells, sympathetic neurons, and lens epithelial cells were chosen because they can be grown as single cell types and because they vary widely in their morphology (Fig. 3). It seems likely that the variations in overall function and cytological location of the actin in these cells will be as great as any normally encountered.

Actin-rich precipitates were prepared from sonicated extracts of the cells by treatment with vinblastine sulfate as previously described (Fine and Bray, 1971). They were analyzed by detergent gel electrophoresis (Fig. 4), and all were found to contain prominent protein species which migrated in the position expected for muscle actin. Fingerprints of the proteins, together with samples of purified muscle actin and brain tubulin for comparison, were prepared by iodination and digestion with trypsin. The details of this method will be presented elsewhere, but in outline it involves the elution of the protein of a stained band into detergent solution, its reaction with [125I]sodium

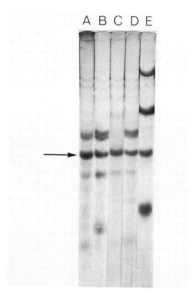

Figure 4. Actin-rich extracts of four types of culture fractionated on SDS-acrylamide gels. A, Muscle cells, B, epithelial cells, C, fibroblasts, D, neurons, E, mixture of β-galactosidase, serum albumin, muscle actin (arrow), and chymotrypsinogen A. The cultures described in the previous figure were harvested, washed and suspended in dilute buffer (10 mM sodium phosphate pH 6.5, 10 mM MgCl$_2$, 0.24 M sucrose) and disrupted by a brief sonication. Particulate material was removed by centrifugation at 100,000 g and the supernatant fraction treated with 1 mg/ml viblastine sulfate. The precipitates formed after 16 hr at 37°C were collected, dissolved in 1% SDS and 1% mercaptoethanol, and run on 7.5% gels prepared as described by Weber and Osborn (1969).

iodide and chloramine T, essentially as described by Hunter and Greenwood (1962), and the purification of the peptides released by trypsin digestion on a Sephadex G25 column. The method gives clear-cut and reproducible fingerprints from as little as 10 μg of protein, and the patterns obtained from the four tissue samples are shown in Fig. 5. Electrophoresis at both pH 6.5 and pH 3.5 gave a series of peptide bands which were indistinguishable from those of muscle actin.

Comparisons of this kind were carried out on presumptive actins from the following sources (the method used to prepare the extract is shown in parentheses as either (VP), the vinblastine precipitation step (Fine and Bray, 1971), or (AM), extraction into Weber-Edsall solution followed by precipitation at low ionic strength, neutral pH (Szent-Györgyi, 1951). From chick embryos (10–12 days): lens, lung, liver, kidney, stomach, muscle,* brain,* heart, skin, pancreas (VP), and cultures of fibroblasts* and muscle cells.* From adult mouse:

brain and pancreas* (VP). From dogfish: muscle,*
brain,* and liver (AM). From scallop: muscle,*
testes,* ovary, and gill* (AM). All of the samples
were compared by one-dimensional maps prepared,
as in Fig. 5, in two electrophoresis buffers, and
those marked with an asterisk* were in addition
prepared as two-dimensional maps. In the whole
of this series the actin fingerprints were strikingly
similar and are illustrated by the results in Figs. 5
and 6. In only three cases were reproducible changes
observed; in the peptide maps of the scallop actins
one peptide was absent and another shifted in
position (Fig. 6).

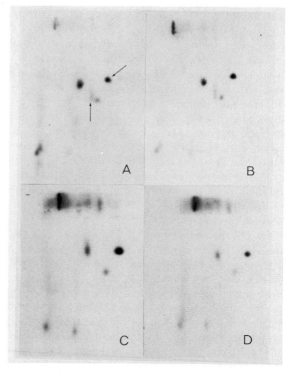

Figure 6. Two-dimensional fingerprints of iodinated
actins: (*A*) chicken actin, (*B*) chick fibroblasts, (*C*)
scallop muscle, (*D*) scallop gill. Samples of (*A*) pure actin
or (*B–D*) actomyosin extracts were run on 7.5% acrylamide gels and the actin bands extracted and fingerprinted
as described in the previous figure. Electrophoresis was at
pH 6.5 (left to right) and pH 3.5 (top to bottom). The two
peptides which differ between chick and scallop are
indicated by arrows.

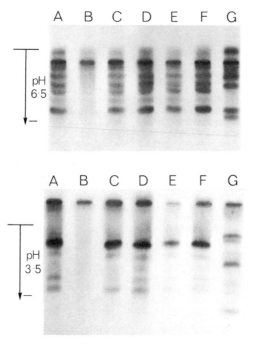

Figure 5. One-dimensional peptide maps prepared by the
iodination procedure: (*A*) chicken muscle actin (the gift
of Dr. A. Weeds), (*B*) the same but with omission of the
trypsin digestion, (*C*) presumptive actin from cultured
muscle cells, (*D*) from epithelial cells, (*E*) from fibroblasts,
(*F*) from neurons, (*G*) chick brain tubulin purified by the
method of Weisenberg et al., 1968. Stained bands were cut
from the gels, chopped into small pieces, and shaken for
16 hr at 37°C in 0.05 M sodium phosphate pH 7.5, 0.1%
SDS and 1 mM phenylmethyl sulfonylfluoride. The extracted protein was then precipitated with KCl (0.2 M),
washed in acetone, and suspended in 20 μl of 0.1% SDS,
0.05 M sodium phosphate pH 7.5. It was iodinated by the
addition of 1 μl (about 100 μCi) of [125I]sodium iodide and
5 μl of chloramine T solution (5 mg/ml). After 20 min at
room temperature, the reaction was stopped by the
addition of 50 μl of sodium metabisulfite solution (5 mg/
ml). The radioactive protein was collected by precipitation
with 10% trichloroacetic acid, washed in acetone, then
oxidized and digested with trypsin by normal procedures.
After the digestion the sample was mixed with dextran
blue and applied to a 1.2 × 10 cm G25 Sephadex column
and eluted with 0.02 M ammonium bicarbonate. The
void volume containing the dextran blue was discarded
and the next 4 ml of eluate collected, lyophilized, and
electrophoresed.

Discussion

The first point to make about these results is
that they clearly demonstrate the ubiquity of
actin-like proteins. Of the ten embryonic tissues
tested, each one gave a prominent protein species
which had the same migration rate on SDS-acrylamide gels (and hence molecular weight; Weber
and Osborn, 1969) and the same iodinated fingerprint. Also they were produced in large amounts by
procedures known to extract actin, and no large
discontinuities were observed between tissues which
were muscle-like and those which were not. The
possibility that these proteins had all originated
from small amounts of muscle contamination in
these tissues was ruled out by the similar findings
from the pure cultures of fibroblasts, epithelial
cells, and nerve cells. This ubiquity was less
obvious in adult tissues, and some sources, such as
mouse liver, failed to show prominent actin-like
band on acrylamide gels. It is possible that in adult
tissues the relative amounts of specialized proteins
may become so great as to hide the cytoplasmic
actin in the simple extractions performed here.

Interspecies comparisons between chicken, mouse, dogfish, and scallop showed the actins to be very similar structurally, as one would expect from the obviously close relationship of even primitive actins from slime molds (Nachmias et al., 1970; Adelman and Taylor, 1969) and ameba (Weihing and Korn, 1971) to those from vertebrates. A previous comparison of the ninhydrin-stained fingerprints of a range of muscle actins (Carsten and Katz, 1964) showed no differences between avian and mammalian actins and only minor ones in the case of fish, frog, and scallop. In the present case in which only the 10–12 tyrosine-containing peptides could be examined, there were no obvious differences in dogfish and chicken actins and only small ones in the case of scallop (Fig. 6).

Perhaps the most interesting result in the present context is that the same species differences were observed both in scallop muscle and in the two nonmuscle tissues, gill and testes. A similar finding was made by Katz and Carsten (1963) for the cardiac and skeletal muscle actins of dog compared to fish, and both results suggest that the actins of different tissues are more closely related than are the muscle actins of the different species. Since these changes are not correlated with function, in which respect tissue differences are far greater than species, it is likely that they represent neutral mutations and are a measure of the relative phylogenetic distance separating the various actins. Provided that the evolution of skeletal (or cardiac) muscle occurred only once, before the points of divergence of the two species, this is evidence against the presence of two forms of actin.

There is no theoretical reason why any part of the many different kinds of cell motility mechanism should be invariant. Constraints will exist, of course, since the various systems have all had to evolve from a common primitive actomyosin, and many of them must be able to function together in the same cell; but these are impossible to quantitate. A factor likely to be of major importance is the number of interactions that depend upon a given protein, since this will determine how easily it can change its structure while still fulfilling its physiological obligations. Skeletal muscle actin is involved in a large number of specific interactions—with itself, with myosin and tropomyosin, and perhaps with troponin and actinin. If the same is true of actin in other systems, then this may be sufficient reason why, as I believe to be the case, the same protein is used over and over again.

Acknowledgments

This work arose out of a collaboration with Richard Fine to whom I am indebted in many ways. I also thank Susan Brownlee for technical assistance and Francis Crick for criticism.

References

ADELMAN, M. R. and E. W. TAYLOR. 1969. Isolation of an actomyosin-like protein complex from slime mold plasmodium and the separation of the complex into actin- and myosin-like fractions. *Biochemistry* **8**: 4964.

ADELSTEIN, R. S., T. D. POLLARD, and W. M. KUEHL. 1971. Isolation and characterization of myosin and two myosin fragments from human blood platelets. *Proc. Nat. Acad. Sci.* **68**: 2703.

BERL, S. and S. PUSZKIN. 1970. Mg^{2+}-Ca^{2+}-activated adenosine triphosphatase system isolated from mammalian brain. *Biochemistry* **9**: 2058.

BETTEX-GALLAND, M. and E. F. LÜSCHER. 1956. Thrombosthenin, the contractile protein from blood platelets and its relation to other contractile proteins. *Advanc. Protein Chem.* **20**: 1.

CARSTEN, M. E. and A. M. KATZ. 1964. Actin: a comparative study. *Biochem. Biophys. Acta* **90**: 534.

FINE, R. E. and D. BRAY. 1971. Actin in growing nerve cells. *Nature* **234**: 115.

HUNTER, W. M. and F. C. GREENWOOD. 1962. Preparation of iodine 131 labelled human growth hormone of high specific activity. *Nature* **194**: 495.

HUSZAR, G. and M. ELZINGA. 1972. Homologous methylated and nonmethylated histidine peptides in skeletal and cardiac myosins. *J. Biol. Chem.* **247**: 745.

KATZ, A. M. and M. E. CARSTEN. 1963. Actin from heart muscle. Studies on amino acid composition. *Circulation Res.* **13**: 474.

LOWEY, S. and D. RISBY. 1971. Light chains from fast and slow muscle myosins. *Nature* **234**: 81.

NACHMIAS, V. T., H. E. HUXLEY, and D. KESSLER. 1970. Electron microscope observations on actomyosin and actin preparations from *Physarum polycephalum*, and on their interaction with heavy meromyosin subfragment I from muscle myosin. *J. Mol. Biol.* **50**: 83.

SZENT-GYÖGYI, A. 1951. *The chemistry of muscle contaction*, p. 151. Academic Press, New York.

TADA, M., G. BAILIN, K. BÁRÁNY, and M. BÁRÁNY. 1969. Proteolytic fragmentation of bovine heart meromyosin. *Biochemistry* **8**: 4842.

WEBER, K. and M. OSBORN. 1969. The reliability of molecular weight determinations by dodecyl sulfate-polyacrylamide gel electrophoresis. *J. Biol. Chem.* **244**: 4406.

WEEDS, A. G. and P. POPE. 1971. Chemical studies on light chains from cardiac and skeletal muscle myosins. *Nature* **234**: 85.

WEIHING, R. R. and E. D. KORN. 1971. *Acanthamoeba* actin. Isolation and properties. *Biochemistry* **10**: 590.

WEISENBERG, R. C., G. G. BORISY, and E. W. TAYLOR. 1968. The colchicine binding protein of mammalian brain and its relation to microtubules. *Biochemistry* **7**: 4466.

The "Contractile" Proteins of *Acanthamoeba castellanii*

Thomas D. Pollard* and Edward D. Korn

National Heart and Lung Institute, Laboratory of Biochemistry, Section on Cellular Biochemistry and Ultrastructure, Bethesda, Maryland 20014

The goal of our research is to understand the molecular and ultrastructural events responsible for the type of cellular movements characteristic of amebae and which occur in virtually all other living cells. Our approach has been to identify, purify, and characterize the presumptive "contractile" proteins of the small soil ameba, *Acanthamoeba castellanii*. We have found that the contractile proteins of this primitive cell share many important properties with muscle actin and myosin, but also that significant differences exist, especially in the myosin. These results, taken together with those of others on the slime mold *Physarum polycephalum* (Hatano and Oosawa, 1966; Adelman and Taylor, 1966a, b; Hatano and Ohnuma, 1970; Nachmias et al., 1970; Nachmias, 1972) and platelets (Adelstein et al., 1971; Behnke et al., 1971; Zucker-Franklin and Grusky, 1972), suggest that the actin-myosin based contractile mechanism of muscle is a highly specialized example of a widely distributed fundamental mechanism for generating force for movement in cells.

Certain features of the contractile proteins appear to have been highly conserved during evolution. In particular, *Acanthamoeba* actin closely resembles muscle actin in chemical composition, physical properties, ultrastructure, and ability to interact with muscle myosin and muscle troponin-tropomyosin complex. *Acanthamoeba* myosin has ATPase activity very similar to that of muscle myosin, binds reversibly to F-actin, and has its Mg^{++}-ATPase activity stimulated by F-actin. On the other hand, the differences between the *Acanthamoeba* and muscle contractile proteins are even more fascinating to consider. In spite of the overall similarities, the primary structure of *Acanthamoeba* actin has small but significant differences from muscle actin, and *Acanthamoeba* myosin differs from muscle myosin in being smaller (mol wt $\simeq$ 180,000 vs. 460,000), in being more soluble in dilute buffers, in requiring an additional "cofactor" protein for actin activation of its Mg^{++}-ATPase, and in interacting differently with the muscle control proteins—the troponin-tropomyosin complex.

Background

It has long been speculated that actin and myosin might exist in non-muscle cells and be responsible for causing their movement. In support of this idea, crude preparations of proteins were extracted from a number of non-muscle cells and found to share some properties with muscle actomyosin. Shortly before we began our investigation, Hatano and Oosawa (1966) provided the first rigorous proof for contractile proteins in non-muscle cells when they purified actin from the acellular slime mold *Physarum polycephalum*. Subsequently both Adelman and Taylor (1969a, b) and Hatano and Ohnuma (1970) described methods for purifying a myosin-like enzyme from *Physarum*. We turned to *Acanthamoeba castellanii* for our biochemical and ultrastructural studies because it exhibits active ameboid movements and phagocytosis and because it can be grown on soluble media in essentially unlimited quantities. In addition it was known (Bowers and Korn, 1968) to contain numerous 8-nm wide filaments which were thought to be *Acanthamoeba* actin.

Acanthamoeba Actin

Purification. The actin from *Acanthamoeba* can be isolated by precipitation with muscle myosin in a modification of Hatano's procedure (Hatano and Oosawa, 1966). After acetone treatment to denature the muscle myosin, the actin is further purified by isoelectric precipitation, sedimentation as F-actin, and gel filtration of depolymerized G-actin (Weihing and Korn, 1969, 1971). The resulting *Acanthamoeba* actin is at least 90% pure judging from polyacrylamide gel electrophoresis (Weihing and Korn, 1972) and short-column equilibrium analytical ultracentrifugation.

Physical properties and chemical composition. The globular *Acanthamoeba* actin monomer has a mol wt of 46,000 and a sedimentation coefficient of 3S. The overall amino acid composition closely resembles, but is not identical to, that of other actins. After cleaving the molecule with cyanogen bromide, Weihing and Korn (1972) isolated three large peptides from the *Acanthamoeba* actin (CB-10, CB-16, CB-17) which together

* Present address: Department of Anatomy, Harvard Medical School, Boston, Massachusetts 02115.

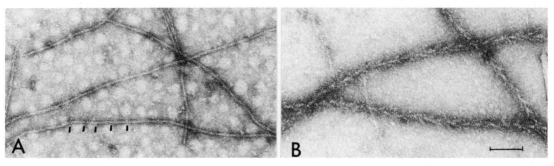

Figure 1. Electron micrographs of specimens negatively stained with 1% uranyl acetate. *A, Acanthamoeba* F-actin Note the periodic constrictions (marked by bars) at the cross-over points of the double-helical substructure. *B, Acanthamoeba* F-actin with muscle heavy meromyosin. The complex has the characteristic repeating arrowhead appearance. Magnification = 100,000. Bar = 0.1 μm.

account for about 25% of the molecule and which closely resemble in amino acid composition the three corresponding cyanogen bromide peptides from muscle actin. Like muscle actin peptide CB-10, *Acanthamoeba* actin CB-10 contains one mole of the rare amino acid 3-methylhistidine. *Acanthamoeba* actin CB-17 contains one mole of another unusual amino acid, N^ε-methyllysine, which is not found in other actins.

Functional properties. Like muscle actin the *Acanthamoeba* actin monomers polymerize at physiological salt concentrations to form double-helical 6-nm wide filaments (Fig. 1) which bind muscle heavy meromyosin to form distinctive polarized arrowhead-shaped complexes (Pollard et al., 1970). Muscle myosin is dissociated from the actin filaments by Mg^{++}-ATP, but some interaction

occurs even in the presence of ATP, because *Acanthamoeba* actin stimulates the Mg^{++}-ATPase activity of muscle heavy meromyosin (Weihing and Korn, 1971). In addition to forming a hybrid complex with muscle myosin, *Acanthamoeba* actin also interacts with the muscle troponin-tropomyosin complex to form a functioning unit whose interaction with muscle heavy meromyosin requires Ca^{++} (Eisenberg and Weihing, 1970).

Localization in the cell. It is not yet clear how much actin is present in the *Acanthamoeba*. Only 0.2% of the cell's protein is isolated as purified actin by the method described; if all of the cell's 3-methylhistidine were contained in actin (which is unlikely), it would constitute about 20% of the cell's protein. The true value probably lies between these extremes according to estimates from densitometric tracings of SDS polyacrylamide gel electrophoresis patterns. As shown in Fig. 2, the major polypeptide in the ameba, which accounts for about 10% of the stained protein, coelectrophoreses with purified actin.

The 8-nm wide filaments which are a prominent feature of the ameba's cytoplasm have been identified as actin (Pollard et al., 1970) using the procedure of Ishikawa et al. (1969) to bind heavy meromyosin to these filaments in glycerol-extracted amebae. Actin filaments are clearly the major component of the hyaline ectoplasm adjacent to the plasma membrane, of the distinctive small spiky pseudopods (acanthopods) which characterize this cell, and of the hyaline "locomotor" pseudopod at the anterior of the motile cell. Detailed studies of the distribution of filamentous actin within a motile ameba have not been made. Part of the *Acanthamoeba* actin may also occur in the monomeric form in the cell.

Recently, we (Pollard and Korn: 1972a; Korn and Wright, 1972) have found that a considerable

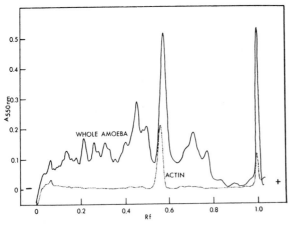

Figure 2. 7.5% polyacrylamide gel electrophoresis according to Neville (1971). Samples in 1.25% SDS. The solid line is the density pattern of polypeptides from the whole *Acanthamoeba*. The large peak with a relative mobility of 1.0 contains all of the proteins with molecular weights less than 25,000. The dotted line is purified *Acanthamoeba* actin, which coelectrophoreses with the major polypeptide in the whole ameba.

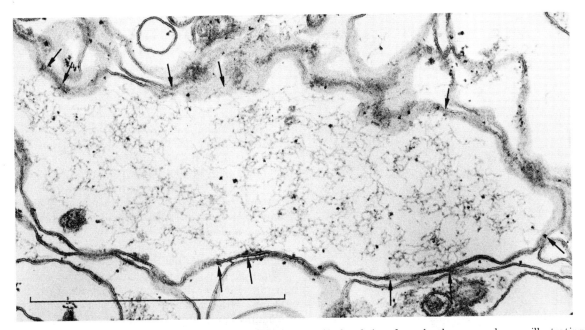

Figure 3. Electron micrograph of a thin section through a pellet of isolated *Acanthamoeba* plasma membranes, illustrating the associated F-actin filaments. Comparison of this micrograph with its stereo pair verified that the filaments make physical contact with the concave (cytoplasmic) surface of the membrane at the points marked by arrows. Magnification = 70,000. Bar = 1 μm.

number of actin filaments are associated with some preparations of purified *Acanthamoeba* plasma membranes (Fig. 3). These filaments were identified as actin by their dimensions, the fact that they depolymerize in dilute buffers, and because they bind heavy meromyosin in the absence, but not the presence, of Mg^{++} pyrophosphate. A major polypeptide band with mobility identical to actin was observed on SDS polyacrylamide gel electrophoresis of such preparations. These filaments are associated principally with the cytoplasmic surface of the plasma membrane, although a few filaments become adsorbed to the external surface during the isolation of the membrane. Electron micrographs clearly show that the filaments are attached to, but not an integral part of, the plasma membrane. Actin filaments are also observed close to the plasma membrane in electron micrographs of intact amebae. These observations suggest, but of course do not prove, that some of the cell's actin is attached to the plasma membrane in vivo.

To summarize, these studies establish that *Acanthamoeba* contains actin closely resembling muscle actin in physical, chemical, and functional properties. In the cell, the actin is present as 8-nm wide filaments, some of which appear to be associated with the plasma membrane. The remarkable similarities between *Acanthamoeba* actin and muscle actin suggest that the function of *Acanthamoeba* actin within the cell is to interact with myosin to generate force for movement.

Acanthamoeba Myosin[1]

Identification of *Acanthamoeba* myosin. The presence of actin implied that the *Acanthamoeba* should also contain a myosin-like enzyme, but from several observations it was apparent that the presumptive *Acanthamoeba* myosin must differ significantly from its muscle counterpart. (1) Electron micrographs of the intact ameba never show thick filaments resembling myosin filaments of striated or smooth muscle (or the thick myosin-like filaments of other amebae [Pollard and Ito, 1970] or platelets [Behnke et al., 1971]). (2) Although ameba homogenates contain actin, they exhibit no actomyosin-like viscosity changes on the addition of ATP, unless muscle myosin is first added (Weihing and Korn, 1971). (3) It is impossible to make typical actomyosin from *Acanthamoeba* using the procedures described for the isolation of muscle or *Physarum* actomyosin. (4) Gel filtration chromatography of *Acanthamoeba* homogenates reveals that the cell contains no ATPase with the same size as muscle myosin or heavy meromyosin (Fig. 4). Instead, extracts of the amebae have an ATPase with a partition coefficient corresponding to a globular protein with

[1] We suggest that the term "myosin" be used as a class name, preceded by the name of the cell of origin to identify it specifically. Minimal criteria to identify a protein as myosin might be (1) evidence of ATPase activity inhibited by Mg^{++} and activated by F-actin, and (2) ability to bind reversibly to F-actin.

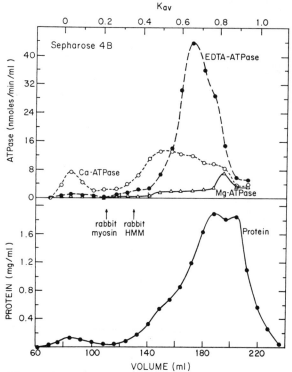

Figure 4. Gel filtration on a 2.5 × 45 cm column of Sepharose 4B. 7.5 ml of the 100,000 *g* supernatant from a 0.5 M KCl, 10 mM imidazole-HCl pH 7.0, 3 mM β-mercapto-ethanol extract of *Acanthamoeba* were applied to the column and eluted with the same buffer. ATPase activities were measured in 0.5 M KCl, 10 mM imidazole-HCl pH 7.0 with either 2 mM EDTA, 10 mM CaCl₂ or 10 mM MgCl₂. Arrows mark the partition coefficients of muscle myosin and heavy meromyosin.

a mol wt of about 180,000. This enzyme was identified as the *Acanthamoeba* myosin by the effects of ions on its activity and by the fact that it interacts with muscle F-actin. The enzyme binds to actin in the absence, but not in the presence, of ATP and its Mg⁺⁺-ATPase activity is activated by actin (Pollard, 1971).

Purification. The major problem in studying the *Acanthamoeba* myosin has been to devise a reliable procedure for its purification. We have been successful using a combination of ion exchange chromatography on DEAE-cellulose, ammonium sulfate fractionation, adsorption to agarose, and hydroxylapatite chromatography. The critical step in this procedure is the highly selective adsorption of both the enzyme and the cofactor protein (discussed below in detail) to the agarose column and their subsequent separation by elution of the column with a salt gradient (Fig. 5). Final purification on hydroxylapatite removes traces of cofactor and other contaminating proteins and yields a protein peak with constant ATPase specific activity of about 3.5 μmoles P_i/min/mg protein across the peak.

This procedure results in a 300- to 400-fold purification of the enzyme. The *Acanthamoeba* myosin accounts for about 0.25% of the cell's total protein (about 0.4% of the soluble proteins) so that it is likely that there is a large (> 10-fold) molar excess of actin in this ameba.

Subunit composition. Electrophoresis on polyacrylamide gels in the presence of sodium dodecyl sulfate (SDS) or urea shows that three polypeptide chains with mol wts of about 140,000, 16,000, and 14,000 copurify with the enzyme activity (Fig. 6). If the native mol wt of about 180,000 (estimated by gel filtration on Sephadex and Agarose at high ionic strength) is correct, the native molecule must consist of a single heavy polypeptide chain and at least one of each of the light polypeptide chains. Alternatively, one or both of the light polypeptide chains may be contaminants. However these small polypeptides chromatograph with the heavy chain on gel filtration columns, and in actin binding experiments the 140,000, 16,000, and 14,000 mol wt polypeptides all sediment together with F-actin.

Size of the native *Acanthamoeba* **myosin.** It is clear that the purified *Acanthamoeba* myosin is considerably smaller than other myosins (muscle mol wt = 460,000 [Lowey et al., 1969]; slime mold mol wt = 460,000 [Adelman and Taylor, 1969b]; platelet mol wt = 450,000 [Booyse et al., 1971]), causing us to question whether we have isolated the native *Acanthamoeba* myosin or perhaps only a proteolytic fragment of the molecule. Several control experiments are all consistent with this small myosin being the native molecule: Both brief and prolonged extraction of the homogenate give the same yield of enzyme with the same partition coefficient on gel filtration; gel filtration of the whole homogenate reveals no larger ATPase which might correspond to the hypothetical larger native myosin; there is no production of ninhydrin-reacting material during the extraction and initial centrifugation of the homogenate, showing that there is no measurable proteolysis occurring at that time; muscle myosin is highly sensitive to cleavage by a number of proteases, but, when incubated with the ameba homogenate, muscle myosin loses no ATPase activity and is not degraded into heavy meromyosin or subfragment-1 as judged by gel filtration; high concentrations of the protease inhibitor DFP do not affect the yield of enzyme or its size; and the purified enzyme consists of only one large and two small polypeptide chains, in contrast to the proteolytic digestion products of muscle myosin, heavy meromyosin, and subfragment-1, which are heterogeneous in denaturing solvents because the muscle myosin has suffered

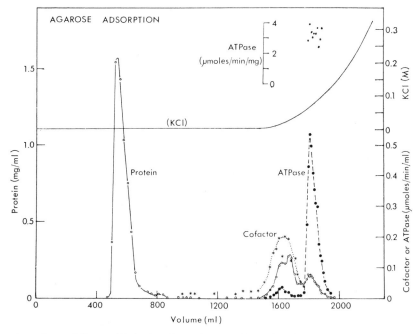

Figure 5. Adsorption of partially purified *Acanthamoeba* myosin and cofactor to a 5 × 75 cm column to Bio-Rad A1.5 m, 200–400 mesh (8% agarose). The proteins were first purified by chromatography on DEAE-cellulose and by ammonium sulfate fractionation. The proteins and the column were equilibrated with 2 mM Tris-HCl pH 7.6, 0.2 mM ATP, 1 mM DTT. After applying the protein (39.5 ml), the column was eluted with a concave KCl gradient in starting buffer. Myosin ATPase (●—●) was measured in 0.5 M KCl, 2 mM EDTA, 1 mM ATP, 10 mM imidazole-HCl pH 7.0. Cofactor activity (* · · · *) was assayed with 0.1 mg/ml muscle F-actin, 0.005 mg/ml *Acanthamoeba* myosin from this column, 1 mM Mg++-ATP, 10 mM KCl, 5 mM imidazole-HCl pH 7.0.

multiple cleavages. Although it is impossible, in practice, to rule out absolutely the possibility that we have isolated a fragment of the native molecule,

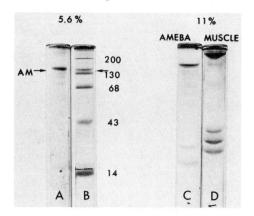

Figure 6. Polyacrylamide gel electrophoresis on 5.6% (*A* and *B*) and 11.0% (*C* and *D*) gels according to Fairbanks et al. (1971). Samples were dissolved in 1% SDS. *A*, Purified *Acanthamoeba* myosin. *B*, Purified *Acanthamoeba* myosin with several protein standards whose molecular weights are given in thousands: 200 = muscle myosin heavy chain; 130 = β-galactosidase; 68 = bovine serum albumin; 43 = ovalbumin; 14 = lysozyme. *C*, Purified *Acanthamoeba* myosin showing the low molecular weight polypeptides that always copurify with the 140,000 mol wt polypeptide. *D*, Muscle myosin showing the 200,000 mol wt heavy chain and the three light chains with mol wt of about 25,000, 18,000, and 16,000.

it is likely that the *Acanthamoeba* myosin is actually the smallest myosin yet discovered.

Solubility. Myosins from muscle and platelets aggregate at low ionic strength to form thick filaments, but *Acanthamoeba* myosin is soluble in dilute buffers and is monomeric in 0.1 M KCl pH 7. No thick filaments or other aggregates have been observed by electron microscopy.

Chemical compositon. The ultraviolet spectrum of purified *Acanthamoeba* myosin is typical of a protein and shows no evidence of nucleic acid contamination. A preliminary amino acid analysis is remarkable only in that no methylated histidines or lysines have been detected.

Enzyme activity. Most of our interest has been directed toward the enzymic activity of the molecule (Table 1). It hydrolyzes the terminal phosphate from ATP, its preferred substrate, but also slowly hydrolyzes other nucleotide triphosphates (CTP, GTP, and ITP). Adenosine diphosphate inhibits the ATPase activity. The enzyme does not hydrolyze ADP, AMP, pyrophosphate, or creatine phosphate. The ATPase activity of the enzyme is very similar to that of muscle myosin under a variety of conditions: activity is highest, about 3.5 to 4.0 μmoles of P_i/min/mg protein, in

Table 1. Enzymic Activity of *Acanthamoeba* Myosin

	%K+-EDTA Activity		
	EDTA (2 mM)	Ca++ (10 mM)	Mg++ (10 mM)
Standard conditions[a]	100	12	1
0.05 M KCl[b]	17	10	1
0.5 M NaCl[b]	<1	10	
NEM[c]	3	2	
ITP[d]	26		
GTP[d]	26		
CTP[d]	26		
ADP[d]	0		
ADP + ATP[e]	58		
AMP[d]	0		
Pyrophosphate[d]	0		
Creatine phosphate[d]	0		

[a] 0.5 M KCl, 10 mM imidazole-HCl pH 7.0, 2 mM ATP at 29°C. Under these conditions with 2 mM EDTA the purified *Acanthamoeba* myosin hydrolyzes 3.2–4.0 μmoles ATP/min/mg protein = 100%.

[b] These salts substituted for 0.5 M KCl.

[c] 0.45 mg/ml *Acanthamoeba* myosin incubated with 0.25 mM N-ethylmaleimide for 20 min.

[d] Alternate substrates at concentrations of 2 mM substituted for ATP.

[e] 2 mM ATP + 1 mM ADP.

the presence of EDTA and high concentrations of KCl. Under these conditions the energy of activation is 9 kcal/mole of ATP. Sodium chloride will not substitute for KCl in stimulating the enzyme.

Calcium inhibits the enzyme 80–90% and Mg++ inhibits the enzyme over 95%, independent of the concentration of monovalent cations. The sulfhydryl reagent N-ethylmaleimide rapidly destroys the K+-EDTA- and Ca++-ATPase activity of the enzyme.

Actin binding. We consider the ability of the enzyme to interact with actin its most important property because the actin-myosin complex presumably forms a functional unit in the living cell. We have found that the purified enzyme binds to actin filaments in the absence, but not in the presence, of ATP. This has been demonstrated in two ways: (1) Without ATP the myosin sediments with F-actin in the ultracentrifuge, but with Mg++-ATP present the myosin does not bind to the actin and remains in the supernatant. (2) Electron microscopy of mixtures of filamentous actin and purified *Acanthamoeba* myosin show that the myosin binds to the actin and laterally cross-links actin filaments to form doublets or higher orders of parallel arrays (Fig. 7). In a few cases the molecules of myosin have been seen to form arrowhead-shaped complexes along the actin filaments, but the tendency of the myosin to aggregate the actin generally obscures this pattern. When the actin

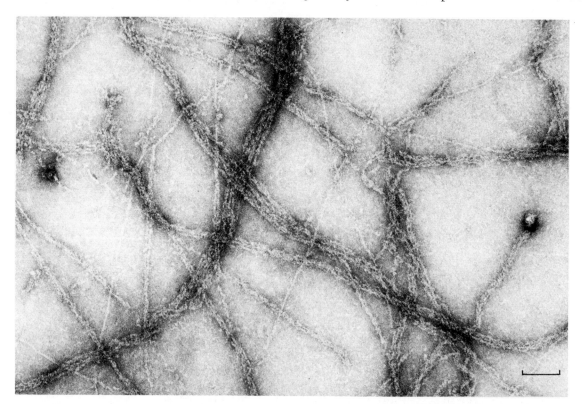

Figure 7. Electron micrograph of muscle F-actin and *Acanthamoeba* myosin negatively stained with 0.5% uranyl acetate. In the absence of ATP *Acanthamoeba* myosin binds to the actin filaments and cross-links neighboring filaments to form the parallel arrays seen here. When a sample like this is treated with Mg++-ATP prior to staining, only bare actin filaments lying separately on the grid are observed. Magnification = 100,000. Bar = 0.1 μm.

and myosin are prepared for electron microscopy in the presence of Mg^{++}-ATP, only bare actin filaments are observed.

Actin activation and discovery of a cofactor.

Especially important is the ability of actin to stimulate the Mg^{++}-ATPase of the *Acanthamoeba* myosin because the enzyme would otherwise be completely inactive at the high Mg^{++} concentrations found in cells. In contrast to muscle myosin which is activated up to 200-fold by actin alone, *Acanthamoeba* myosin requires a cofactor for actin to activate its Mg^{++}-ATPase (Table 2) (Pollard and Korn, 1972b). The cofactor is thought to be a protein because it is nondialyzable, is inactivated by boiling, elutes as a single peak of activity from DEAE-cellulose, has a partition coefficient on gel filtration corresponding to a mol wt of about 140,000, and can be highly purified by hydroxylapatite chromatography. By SDS polyacrylamide gel electrophoresis, the hydroxylapatite-purified cofactor is at least 80% pure and has a mol wt of about 95,000 (Fig. 8). The hydroxylapatite-purified cofactor lacks any independent ATPase activity.

The interaction of *Acanthamoeba* myosin, cofactor, and actin resulting in enzyme activity is rather complex and is not understood in any detail, but it is clear that the ATPase rate is dependent on the concentration of each of these components (Table 2). At low concentrations of cofactor there is a marked lag of 2–3 minutes before

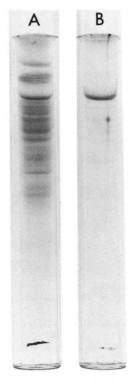

Figure 8. Polyacrylamide gel electrophoresis on 9% gels according to Fairbanks et al. (1971). *A*, Crude *Acanthamoeba* cofactor from an agarose column such as shown in Fig. 5. *B*, *Acanthamoeba* cofactor purified by chromatography of crude cofactor on hydroxylapatite. The major polypeptide has a mol wt of about 95,000.

Table 2. Interaction of *Acanthamoeba* Myosin, Crude Cofactor, and Actin

	Mg^{++}-ATPase Activity[a]
Complete System[b]	100%
Acanthamoeba myosin	
Crude cofactor[c]	
Muscle F-actin	
—F-actin	0
+0.5 mg/ml	80%
+1.0 mg/ml	100%
—Crude cofactor	3%
+0.026 mg/ml	65%
+0.065 mg/ml	100%
+0.065 mg/ml boiled	2%
—*Acanthamoeba* myosin	14%[c]
+0.002 mg/ml	46%
+0.005 mg/ml	100%
+0.2 mM EGTA	114%
+0.1 mM CaCl$_2$	90%

[a] 100% = 1.3 μmoles/min/mg *Acanthamoeba* myosin.
[b] The complete system contained 0.005 mg/ml *Acanthamoeba* myosin, 0.065 mg/ml *Acanthamoeba* crude cofactor, 1.0 mg/ml muscle F-actin, 10 mM KCl, 1 mM MgCl$_2$, 1 mM ATP, 5 mM imidazole-HCl pH 7.0. Incubation for 30 min at 29°C.
[c] This crude cofactor preparation was contaminated with a small amount of *Acanthamoeba* myosin.

the reaction rate increases to become constant for the remainder of the reaction. The lag is absent or very short at high concentrations of cofactor. The reason for the lag is unknown, but it suggests that the formation of the ternary complex is cooperative. The dependence of the Mg^{++}-ATPase rate on cofactor concentration follows a sigmoidal curve at a given actin and enzyme concentration. The dependence of the Mg^{++}-ATPase on the actin concentration is complex (Fig. 9), in contrast to muscle acto-HMM or acto-subfragment-1 where dependence of rate on actin concentration follows a simple hyperbolic curve (Eisenberg and Moos, 1968). At low concentrations of actin (0.1 mg/ml) very high rates (up to 1.8 μmoles P$_i$/min/mg *Acanthamoeba* myosin) are attained. Increasing the actin concentration to 0.3 mg/ml results in a decrease in activity. Concentrations of actin greater than 0.3 mg/ml give higher rates. The biphasic nature of this curve is accentuated at low cofactor concentrations where the rates at high actin concentrations are lower than those at 0.1 mg/ml actin. The ATPase rate of the actin-*Acanthamoeba* myosin-cofactor complex is strongly inhibited as the ionic strength is raised. It is interesting to note that the *Acanthamoeba* cofactor

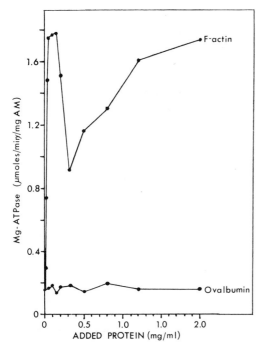

Figure 9. Dependence of *Acanthamoeba* myosin-cofactor complex Mg^{++}-ATPase on actin concentration. The *Acanthamoeba* proteins were obtained from an agarose column on which the cofactor and myosin were not separated from each other; the *Acanthamoeba* myosin K^{+}-EDTA ATPase had a specific activity of 3.0 μmoles/min/mg protein in this preparation. Assay conditions were 0.01 mg/ml *Acanthamoeba* myosin-cofactor complex, 2 mM MgCl$_2$, 1 mM ATP, 8 mM KCl, 10 mM imidazole-HCl pH 7.0 with the various concentrations of either muscle F-actin or ovalbumin shown on the graph. Similar results have been obtained using recombined purified *Acanthamoeba* myosin and purified *Acanthamoeba* cofactor.

also activates the muscle subfragment-1-actin-Mg^{++}-ATPase, but only by about 50%.

The mechanism of action of the cofactor is unknown, although it is clear from the actin binding experiments with the purified *Acanthamoeba* myosin that the cofactor is not required for *Acanthamoeba* myosin to bind to actin or to be dissociated from it by Mg^{++}-ATP. It is also clear that the cofactor does not bind tightly to the enzyme because the two may be at least partially separated under mild conditions on gel filtration columns. More complete separation is achieved during salt gradient elution after both have been adsorbed to agarose and by phosphate gradient elution from hydroxylapatite.

Although we do not know how the cofactor stimulates the actin activation of the *Acanthamoeba* myosin Mg^{++}-ATPase, the cofactor is of interest because it is in a position to regulate the energy-transducing function of actin and myosin in the cell and, presumably, the movement of the ameba.

Any additional factors modulating this actin-myosin-cofactor system remain unknown; Ca^{++} has no effect on the system.

Interaction with muscle troponin-tropomyosin complex. To explore other possible modes of control we examined the effect of the muscle troponin-tropomyosin complex on the actin-*Acanthamoeba* myosin-cofactor Mg^{++}-ATPase and found evidence for strong, but unusual, interaction. Like its effect on muscle myosin, the troponin-tropomyosin complex ("native tropomyosin") inhibited the actin activation of *Acanthamoeba* myosin-cofactor in the absence of Ca^{++}, but unlike its effect on muscle myosin, the troponin-tropomyosin complex also inhibited the ameba ATPase activity in the presence of Ca^{++}. This phenomenon has been investigated using purified tropomyosin and troponin components prepared by Drs. Eisenberg and Kielley (1972) (Table 3). Tropomyosin, troponin I, and troponin C function normally in the hybrid system with muscle F-actin and *Acanthamoeba* myosin and cofactor but troponin T does not. Tropomyosin has either no effect or inhibits Mg^{++}-ATPase activity slightly. Tropomyosin with troponin I completely blocks the Mg^{++}-ATPase in the presence or absence of Ca^{++}. Troponin C blocks the inhibitory effect of troponin I. Addition of troponin T to the other components blocks the effect of troponin C, allowing troponin I to inhibit the reaction. With muscle myosin, troponin T functions only in the absence of Ca^{++} so that, in the presence of Ca^{++}, troponin C can still block troponin I and actin and myosin are allowed to interact even when troponin T is present (Greaser and Gergely, 1971). With *Acanthamoeba* myosin and cofactor, troponin T blocks troponin C in both the presence and absence of Ca^{++}. This explains why the complete troponin-tropomyosin complex inhibits actin activation of *Acanthamoeba* myosin-cofactor in both the presence and absence of Ca^{++}. Apparently the *Acanthamoeba* myosin-cofactor complex lacks some previously unrecognized property of muscle myosin which is required for the normal function of the troponin-tropomyosin complex.

At the present time there is no evidence for troponin or tropomyosin in the *Acanthamoeba*. If they do exist in the ameba, they must differ from their muscle counterparts or else they would block the activity of the other ameba contractile proteins.

To summarize, *Acanthamoeba* contains a low molecular weight, soluble myosin which has been highly purified. The *Acanthamoeba* myosin binds reversibly to F-actin, and in electron micrographs appears to cross-link actin filaments. The enzyme is strongly inhibited by Mg^{++}, but in the presence

Table 3. Comparison of the Effect of Muscle Troponin-Tropomyosin on the Mg^{++}-ATPase of Actin-Activated *Acanthamoeba* Myosin-Cofactor Complex and Actin-Activated Muscle Subfragment-1

Additions[b]	ATPase Rates[a]			
	Acto-*Acanthamoeba* Myosin-Cofactor		Acto-Muscle Subfragment-1	
	EGTA	Ca^{++}	EGTA	Ca^{++}
None	100%	100%	100%	113%
Tropomyosin	96%		89%	
Tropomyosin + Troponin I	3%	3%	13%	18%
Tropomyosin + Troponin I + Troponin C	63%	42%	92%	189%[c]
Tropomyosin + Troponin I + Troponin C + Troponin T	4%	3%	23%	222%[c]

[a] ATPase rates are expressed as a percent of the rates without troponin-tropomyosin fractions: 100% was 7.5 nmoles/ min/mg of *Acanthamoeba* myosin-cofactor; 100% was 45 nmoles/min/mg of muscle subfragment-1. ATPase rates were measured at 29°C in 1.4 mM $MgCl_2$, 30 mM KCl, 8.3 mM imidazole-HCl pH 7.0 with either 1.0 mM EGTA or 0.5 mM $CaCl_2$. All samples contained 0.1 mg/ml muscle F-actin. The *Acanthamoeba* myosin-cofactor complex (0.8 mg/ml) was purified through the DEAE-cellulose and ammonium sulfate precipitation steps and had a K^+-EDTA ATPase activity of 200 nmoles/min/mg. (When assayed with 1.0 mg/ml muscle F-actin in 2 mM $MgCl_2$, 1 mM ATP, 10 mM imidazole-HCl pH 7.0, ATPase activity was 100 nmoles/min/mg; but with the conditions of this experiment, which are necessary to show the troponin-tropomyosin effect, the ATPase rate was lower due to the use of a lower concentration of actin and a higher concentration of KCl.) Subfragment-1 (0.015 mg/ml) was prepared from rabbit muscle myosin by papain digestion.

[b] Tropomyosin and troponin components (prepared by Drs. Eisenberg and Kielley) were used at these final concentrations: tropomyosin, 0.05 mg/ml; troponin I, 0.05 mg/ml; troponin C, 0.07 mg/ml; troponin T, 0.05 mg/ml.

[c] We found a variable amount of activation of the acto-subfragment-1 ATPase by these troponin-tropomyosin fractions in the presence of Ca^{++}.

of a newly discovered cofactor protein, actin activates the Mg^{++}-ATPase activity over 100-fold. These properties of the enzyme are consistent with the idea that *Acanthamoeba* myosin interacts with the cell's actin filaments to transduce the chemical energy of ATP into force for movement.

Discussion

The discovery of muscle-like "contractile" proteins in the *Acanthamoeba* establishes the biochemical basis for the cell's motility. Ultimately the movement of the ameba must be explained by the interaction of these proteins. Figure 10 is a tentative, hypothetical model illustrating one way in which the *Acanthamoeba* contractile proteins might interact to produce movement. This model is presented here to summarize the available facts and to point out where our knowledge is incomplete.

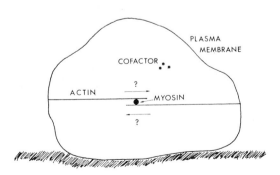

Figure 10. A hypothetical model showing possible relationships of *Acanthamoeba* "contractile" proteins within the cell.

At least part of the cell's actin is present as F-actin, some of which can be shown to be associated with isolated ameba plasma membranes. Presumably this is due to attachment of the actin filaments to the plasma membrane in vivo. The polarity of these actin filaments is unknown, but if their polarity were unique, the ameba plasma membrane would be analogous to the Z-line in striated muscle. In the absence of ATP *Acanthamoeba* myosin binds to F-actin and cross-links neighboring filaments. It is not known whether one or more myosin molecules participate in each cross-link, although the distance between adjacent actin filaments and the low tendency of *Acanthamoeba* myosin to aggregate are consistent with as few as one *Acanthamoeba* myosin forming a crosslink. Under physiological conditions the ATPase activity of *Acanthamoeba* myosin is inhibited but, in the presence of a third protein (the cofactor), actin strongly activates the myosin ATPase. The resulting hydrolysis of ATP makes the chemical energy of ATP available for work, but it is not known whether these ameba proteins are capable of transducing this energy into force for movement like their counterparts in muscle. If they are capable of this energy transduction, it is not unreasonable to postulate that the resulting force might result in motion of the cross-linked actin filaments relative to each other in a manner similar to the sliding interaction of actin and myosin in muscle. It must be emphasized that there is no direct evidence for the postulated energy transduction, force production, or movement of the actin filaments.

Because the *Acanthamoeba* and mammalian muscle contractile proteins share many properties in spite of their great phylogenetic separation, it seems possible to us that the common features might have been preserved during evolution because of their importance to the contractile mechanism. If this assumption is true, it should be possible to enumerate those properties that are essential and those that are nonessential to the contractile process by comparing the actins and myosins from muscle and primitive cells.

Muscle actin, *Acanthamoeba* actin, and several other non-muscle actins which are less completely characterized are remarkably similar (see Pollard, 1972, for review). Common properties include the presence of 3-methylhistidine and the ability to form polar double-helical filaments which bind both myosin and the troponin-tropomyosin complex and which stimulate the myosin ATPase. That the filament structure and ability to interact with myosin and tropomyosin has been preserved during evolution is not surprising because it seems obvious that these properties are important for contraction, but this comparative approach now suggests that even the presence of 3-methylhisti-dine may be essential to the contractile mechanism.

Myosin varies much more from species to species, but even here, several properties are common to all of the myosins that have been examined. These include the inhibition of ATPase activity by Mg^{++} and the ability of the enzymes to bind to and to be activated by actin. Other properties such as the presence of a tail region are variable. Myosins from muscle and non-muscle cells of higher organisms, such as platelets and fibroblasts (Adelstein and Conti, this volume), have extended tail regions which aggregate under physiological conditions to form thick filaments; *Physarum* myosin is a large molecule with a long tail and two heads (Hatano and Takahashi, 1971) although it is soluble under physiological conditions; *Acanthamoeba* myosin is small, appears to lack a tail region, and is soluble under physiological conditions. This suggests that the tail region of muscle myosin is not essential for contraction, although the tail clearly is important for the specialized arrangement of myosin in striated muscle cells.

Perhaps the most fascinating difference among the myosins is the requirement for a cofactor protein for actin to activate the *Acanthamoeba* myosin ATPase. The cofactor has not been described in other cells, although the very limited actin activation described for myosins from *Physarum* (Hatano and Ohnuma, 1970), platelets, and fibroblasts (Adelstein and Conti, this volume) might indicate that cofactor is present in these cells and has been purified away from the myosin.

Although the mechanism of action of the *Acanthamoeba* cofactor is unknown, it is possible that a similar function is incorporated directly into the muscle myosin molecule, where a third protein is not needed to activate the ATPase of the actomyosin. Perhaps the incorporation of this essential property into muscle myosin accounts, at least in part, for the higher molecular weight of muscle myosin.

References

ADELMAN, M. R. and E. W. TAYLOR. 1969a. Isolation of an actomyosin-like protein complex from slime mold plasmodium and the separation of the complex into actin- and myosin-like fractions. *Biochemistry* **8:** 4964.

———. 1969b. Further purification and characterization of slime mold myosin and slime mold actin. *Biochemistry* **8:** 4976.

ADELSTEIN, R. S., T. D. POLLARD, and W. M. KUEHL. 1971. Isolation and characterization of myosin and two myosin fragments from human blood platelets. *Proc. Nat. Acad. Sci.* **68:** 2703.

BEHNKE, O., B. I. KRISTENSEN, and L. E. NIELSEN. 1971. Electron microscopical observations on actinoid and myosinoid filaments in blood platelets. *J. Ultrastr. Res.* **37:** 351.

BOOYSE, F. M., T. P. HOVEKE, D. ZSCHOCKE, and M. E. RAFELSON. 1971. Human platelet myosin. Isolation and properties. *J. Biol. Chem.* **246:** 4291.

BOWERS, B. and E. D. KORN. 1968. The fine structure of *Acanthamoeba castellanii*. I. The trophozoite. *J. Cell Biol.* **39:** 95.

EISENBERG, E. E. and W. W. KIELLEY. 1972. Reconstitution of active troponin-tropomyosin complex in the absence of urea from its four column-purified components. *Fed. Proc., Fed. Amer. Soc. Exp. Biol.* **31:** 1630. (Abstr.).

EISENBERG, E. E. and C. MOOS. 1968. The adenosine triphosphatase activity of acto-heavy meromyosin. A kinetic analysis of actin activation. *Biochemistry* **7:** 1486.

EISENBERG, E. E. and R. R. WEIHING. 1970. Effect of skeletal muscle native tropomyosin on the interaction of amoeba actin with heavy meromyosin. *Nature* **228:** 1092.

FAIRBANKS, G., T. L. STECK, and D. F. H. WALLACH. 1971. Electrophoretic analysis of the major polypeptides of the human erythrocyte membrane. *Biochemistry* **10:** 13.

GREASER, M. L. and J. GERGELY. 1971. Reconstitution of troponin activity from three protein components. *J. Biol. Chem.* **246:** 4226.

HATANO, S. and J. OHNUMA. 1970. Purification and characterization of myosin A from the myxomycete plasmodium. *Biochim. Biophys. Acta* **205:** 110.

HATANO, S. and F. OOSAWA. 1966. Isolation and characterization of plasmodium actin. *Biochim. Biophys. Acta* **127:** 488.

HATANO, S. and K. TAKAHASHI. 1971. Structure of myosin A from the myxomycete plasmodium and its aggregation at low salt concentrations. *J. Mechanochem. Cell Motility* **1:** 7.

ISHIKAWA, H., R. BISCHOFF, and H. HOLTZER. 1969. Formation of arrowhead complexes with heavy meromyosin in a variety of cell types. *J. Cell Biol.* **43:** 312.

KORN E. D. and P. L. WRIGHT. 1972. *J. Biol. Chem.* In press.

LOWEY, S., H. S. SLAYTER, A. G. WEEDS, and H. BAKER. 1969. Substructure of the myosin molecule. I. Subfragments of myosin by enzymatic degradation. *J. Mol. Biol.* **42**: 1.

NACHMIAS, V. T. 1972. Electron microscopic observations on myosin from *Physarum polycephalum*. *J. Cell Biol.* **52**: 648.

NACHMIAS, V. T., D. KESSLER and H. E. HUXLEY. 1970. Electron microscope observations on actomyosin and actin preparations from *Physarum polycephalum* and on their interaction with heavy meromyosin subfragment I from muscle myosin. *J. Mol. Biol.* **50**: 83.

NEVILLE, D. M. 1971. Molecular weight determination of protein-dodecyl sulfate complexes by gel electrophoresis in a discontinuous buffer system. *J. Biol. Chem.* **246**: 6328.

POLLARD, T. D. 1971. An EDTA/Ca-ATPase from *Acanthamoeba*. *Fed. Proc., Fed. Amer. Soc. Exp. Biol.* **30**: 1309. (Abstr.)

———. 1972. Progress in understanding amoeboid movement at the molecular level. In *Biology of amoeba*, chap. 9, ed. K. Jeon. Academic Press, New York.

POLLARD, T. D. and S. ITO. 1970. Cytoplasmic filaments of *Amoeba proteus*. I. The role of filaments in consistency changes and movement. *J. Cell Biol.* **46**: 267.

POLLARD, T. D. and E. D. KORN. 1972a. *J. Biol. Chem.* In press.

POLLARD, T. D. and E. D. KORN. 1972b. A protein cofactor required for the actin activation of a myosin-like ATPase of *Acanthamoeba castellanii. Fed. Proc., Fed. Amer. Soc. Exp. Biol.* **31**: 502. (Abstr.)

POLLARD, T. D., E. SHELTON, R. R. WIHING, and E. D. KORN. 1970. Ultrastructural characterization of F-actin isolated from *Acanthamoeba castellanii* and identification of cytoplasmic filaments as F-actin by reaction with rabbit heavy meromyosin. *J. Mol. Biol.* **50**: 91.

WEIHING, R. and E. D. KORN. 1969. Ameba actin: The presence of 3-methylhistidine. *Biochem. Biophys. Res. Comm.* **35**: 906.

———. 1971. *Acanthamoeba* actin. Isolation and properties. *Biochemistry* **10**: 590.

———. 1972. *Acanthamoeba* actin: Composition of the peptide that contains 3-methylhistidine and a peptide that contains N^ε-methyllysine. *Biochemistry* **11**: 1538.

ZUCKER-FRANKLIN, D. and G. GRUSKY. 1972. The actin and myosin filaments of human and bovine blood platelets. *J. Clin. Invest.* **51**: 419.

Effects of Cytochalasin B on Actin Filaments

JAMES A. SPUDICH

Department of Biochemistry and Biophysics, University of California, San Francisco, California 94122

Cytochalasin B, an alkaloid from the fungus *Helminthosporium dematioideum* (Aldridge et al., 1967), inhibits a wide variety of cellular movements including cell division (Carter, 1967; Schroeder, 1970; Estensen, 1971), cytoplasmic streaming (Wessells et al., 1971), changes in cell shape during embryonic development (Spooner and Wessells, 1970; Wrenn and Wessells, 1970), elongation of nerve axons (Yamada et al., 1970), phagocytosis (Allison et al., 1971; Davis et al., 1971; Axline and Reaven, 1971), hormone secretion (Schofield, 1971), release of the neurotransmitter, norepinephrine (Thoa et al., 1972), blood clot retraction (Shepro et al., 1970), and under certain circumstances muscle contraction (Manasek et al., 1972). Microfilaments, which are thought to be involved in these movements (for reviews, see Jahn and Bovee, 1969; Wessells et al., 1971), have been observed to lose their filamentous form in the presence of cytochalasin B (for example, see Schroeder, 1970; Wessells et al., 1971).

There is evidence that microfilaments are very similar to actin filaments from muscle. First, they both have a diameter of 50 Å–70 Å. Furthermore actin-like protein has been isolated from many different types of cells where microfilaments are prevalent (Bettex-Galland and Lüscher, 1965; Weihing and Korn, 1971; Fine and Bray, 1971; review by Jahn and Bovee, 1969). Even the slime molds contain actin-like protein so similar to muscle actin that it interacts in vitro with muscle HMM to form an arrowheaded complex (Nachmias et al., 1970; Woolley, 1970) which is characteristic of muscle acto-HMM (Huxley, 1963). Also several investigators have reported that microfilaments in situ can be decorated with HMM to form an arrowheaded complex (Ishikawa et al., 1969; Shepro et al., 1969; Pollard et al., 1970; Perry et al., 1971; Tilney and Mooseker, 1971).

The above considerations suggest that cytochalasin B might inhibit movement by interacting directly with actin-like filaments and altering their morphology. We reported earlier (Spudich and Lin, 1972) the first indications that cytochalasin B does interact with purified muscle actin and reconstituted actomyosin. We have extended that investigation and have included studies with platelet actin, the purification of which we report

here. We have examined the effects of cytochalasin B on (1) actin activation of HMM ATPase activity, (2) polymerization of G-actin, and (3) the structure of actin filaments. The alkaloid does not affect significantly the actin activation of HMM ATPase or the G-actin polymerization. However electron microscopy and viscosity experiments suggest that the structure of purified filamentous actin is altered by incubation with the concentrations of cytochalasin B which are required to inhibit blood clot retraction. This structural change is inhibited by tropomyosin-troponin, which is known to form a complex with actin (Laki et al., 1962; Ebashi et al., 1969).

Materials and Methods

Assay for actin-activated HMM ATPase. Actin-activated HMM ATPase activity was determined in 0.5 ml of reaction mixture consisting of 25 mM Tris-Cl, 2.5 mM $MgCl_2$, 0.2 mM $CaCl_2$ (for assays in the absence of Ca^{++}, $CaCl_2$ was replaced with 0.5 mM EGTA), 10 mM KCl, HMM and actin (about 0.1 mg/ml each), 1.0 mM $[\gamma\text{-}^{32}P]$ATP (about 0.05 mCi per mmole), and dimethyl sulfoxide (Me_2SO) or cytochalasin B. The reaction mixture (pH 8) was incubated for 15–60 min at 25°C, and the reaction was stopped by addition of 0.5 ml of a mixture of 0.5 N perchloric acid and 1 mM potassium phosphate. Orthophosphate was selectively precipitated (Sugino and Miyoshi, 1964) by addition of 0.3 ml of a mixture of 27 mM ammonium molybdate and 33 mM triethylamine-HCl pH 5, and the precipitate was collected on a GF/C glass filter. The filter was washed three times with 2 ml of a mixture (4°C) of 0.2 N perchloric acid, 30 mM ammonium molybdate, 0.1 mM potassium phosphate and 10 mM triethylamine-HCl (the mixture was clarified by filtration before use). The filter was dried and counted on a Nuclear Chicago gas-flow counter. The actin-activated HMM ATPase activity was obtained by subtracting the ATPase activity of HMM alone from the total activity. One unit of actin-activated HMM ATPase activity is 1 μmole P_i liberated/hr/mg of HMM.

Muscle protein preparations. Actin was purified to electrophoretic homogeneity from an acetone powder of rabbit striated muscle according to the 0.6 M KCl purification procedure of Spudich and Watt (1971). Myosin was prepared from rabbit striated muscle as described by Tonomura et al., (1966). HMM was obtained by limited cleavage of myosin with trypsin (2 min incubation) followed by precipitation of light meromyosin and of the remaining myosin at low ionic strength (Lowey et al., 1969). TM·TP (the complex of tropomyosin and troponin) was prepared as described previously (Spudich and Watt, 1971); the preparation was shown to inhibit actin activation of HMM ATPase in the absence of Ca^{++} by more than 80% at a ratio of 0.4 mg of TM·TP per mg of actin.

Purification of platelet actin. Unless otherwise noted, all operations were carried out at 0–4°C. Porcine or bovine platelets were isolated from about 20 liters of fresh citrated blood (4.7 g sodium citrate per liter). The whole blood was centrifuged at 900 g for 15 min at 22°C. The supernatant fluid was centrifuged again at 900 g for 15 min, and the platelet-rich plasma was centrifuged at 3300 g for 15 min at 4°C to sediment the platelets which were then washed with 9 mg/ml NaCl, 3 mg/ml sodium citrate, 3 mM dithiothreitol and 1 mM EDTA by repeated centrifugations at 3300 g for 15 min. The platelets, which were essentially free from leucocyte and erythrocyte contamination, were stored at −20°C for two months. The platelets (15 g wet weight) were thawed and mixed vigorously with 100 ml of acetone at 22°C for 10 min and then were collected by filtration through Whatman 3 MM filter paper under a vacuum. The residue was resuspended with acetone and collected again by filtration. This wash was repeated two times, and the residue was dried overnight at 22°C and then stored desiccated at −20°C. The yield was about 2 g of acetone powder.

Platelet actin was purified from this acetone powder essentially according to the procedure described previously for purification of muscle actin (Spudich and Watt, 1971). The platelet acetone powder (1.3 g) was mixed with 26 ml of Buffer A (5 mM Tris-Cl, 0.2 mM ATP, 0.2 mM CaCl$_2$, 0.5 mM β-mercaptoethanol pH 8 at 22°C) and stirred for 1 hr at 22°C. After centrifugation at 12,000 g for 1 hr the clear supernatant fluid was mixed with KCl to 50 mM and MgCl$_2$ to 2 mM, left at 22°C for 3 hr and then at 4°C overnight (crude extract; 23 ml, 253 mg protein, 1150 units actin-activated HMM ATPase activity). The actin was sedimented by centrifugation at 100,000 g for 4 hr, resuspended by homogenization in 6 ml of Buffer A, mixed with KCl (to 0.6 M final concentration) and then stirred for 20 min. After an initial centrifugation at 12,000 g for 20 min to clarify the actin solution, actin was sedimented by centrifugation at 100,000 g for 4 hr, resuspended by homogenization in 2.5 ml of Buffer A, and dialyzed three times against a total of 1000 ml of Buffer A for 4 days in order to depolymerize the actin. The preparation was centrifuged at 100,000 g for 2 hr, and the supernatant fluid (2.5 ml, 13 mg protein, 400 units activity) was applied to a G-200 Sephadex column (4.9 cm^2 × 67 cm, equilibrated with Buffer A). Fractions (4 ml each) were collected over a two day period and assayed for protein and for the ability to activate muscle HMM ATPase activity. Two peaks of protein and activity were obtained. Peak I (28 ml, 3.4 mg protein, 112 units activity) was eluted in the void volume, whereas peak II (40 ml, 6 mg protein, 64 units activity) was eluted near the position expected for a protein with a mol wt of about 50,000. The specific activity (for activation of muscle HMM ATPase) of peaks I and II was not altered by removal of Ca^{++} from the ATPase reaction mixture.

Essentially the same results were obtained for purification of platelet actin starting with bovine platelets in place of porcine platelets. Two peaks were eluted from Sephadex G-200 (Fig. 1), both of which activated muscle HMM ATPase activity. Peak I may be a polymeric form of peak II which was unable to completely depolymerize. The mobility of these proteins on SDS-acrylamide gels (Fig. 2) is consistent with this possibility. However the possibility that the active component of peak I is a different protein from that of peak II has not been ruled out. As judged by SDS-acrylamide gel electrophoresis (Fig. 2), peak II appears to be homogeneous and has a molecular weight nearly identical to that of muscle actin (48,000 daltons). Peak II was further identified as platelet actin by electron microscopy. Polymerization of the protein in peak II by addition of KCl to 50 mM yielded filaments very similar in appearance to muscle F-actin, and addition of MgCl$_2$ to 25 mM yielded paracrystals (Fig. 3) like those obtained with muscle actin (Hanson, 1968).

The peak II Sephadex fractions were used as purified porcine and bovine platelet actin in the studies described below.

Polyacrylamide gel electrophoresis. Protein preparations were electrophoresed on 10% polyacrylamide gels

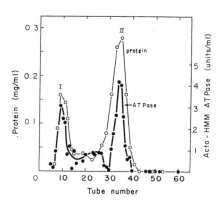

Figure 1. Sephadex filtration of platelet actin. A bovine platelet actin preparation (5 ml, 3.7 mg/ml protein), dialyzed as described for porcine platelet actin (see Materials and Methods), was filtered through a Sephadex G200 column (4.9 cm^2 × 67 cm) with Buffer A at a flow rate of about 10 ml/hr (4 ml/tube). Aliquots from various tubes were assayed for their ability to activate muscle HMM ATPase. Precalibration of the column with dextran blue and DNP-lysine indicated the void volume (about tube 10) and the bed volume (about tube 70), respectively.

containing 0.1% SDS, as described by Davies and Stark (1970).

Electron microscopy. Preparations were negatively stained with 1% uranyl acetate essentially according to the procedures described by Huxley and Zubay (1960) and Huxley (1963) and were examined in a Philips 300 electron microscope operated under conventional conditions at 60 kV.

Other methods and materials. Protein concentration was determined by the method of Lowry et al. (1951) after acid precipitation. Cytochalasin B (mol wt = 479) was purchased from Imperial Chemical Industries, Ltd.,

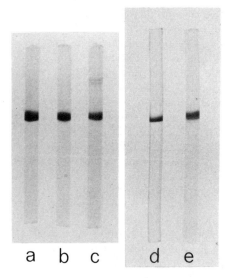

Figure 2. SDS-polyacrylamide gels of platelet actin. (a) 25 μg of bovine platelet actin (Sephadex peak II). (b) 20 μg of porcine platelet actin (Sephadex peak II). (c) 20 μg of porcine Sephadex peak I. (d) 5 μg of bovine platelet actin (Sephadex peak II) mixed with 5 μg of muscle actin. (e) 5 μg of porcine platelet actin (Sephadex peak II) mixed with 5 μg of muscle actin.

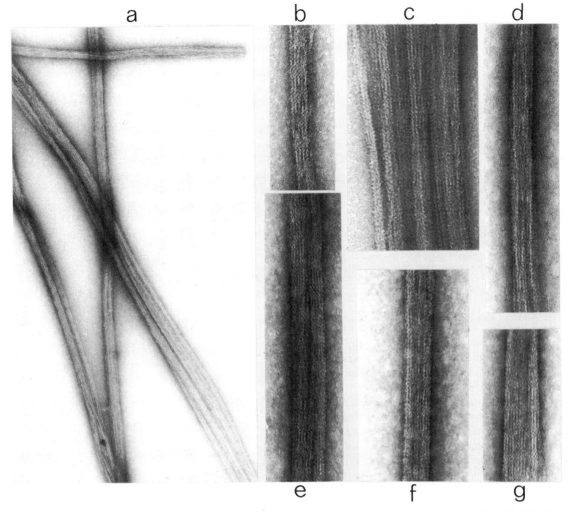

Figure 3. Electron micrographs of Mg^{++} paracrystals of platelet actin. (*a*) Bovine Sephadex peak II; ×73,000. (*b*) and (*c*) Bovine Sephadex peak II; ×165,000. (*d*) through (*g*) Porcine Sephadex peak II; ×165,000. The paracrystals were formed in Buffer A containing 50 mM KCl and 25 mM MgCl$_2$ and were stained with 1% uranyl acetate at 22°C. Protein concentration was about 0.2 mg/ml.

Cheshire, England, and was stored at 4°C as a 10 mg/ml stock solution in Me$_2$SO. [γ-^{32}P]ATP was prepared according to the method of Glynn and Chappell (1964).

Results

Effect of cytochalasin B on actin activation of HMM ATPase activity. We previously reported (Spudich and Lin, 1972) that 0.05 mM cytochalasin B (a concentration which will inhibit most types of in vivo movement) had essentially no effect on the ability of muscle actin to activate HMM ATPase activity, although high concentrations of cytochalasin B (0.3 mM) inhibited acto-HMM ATPase activity by as much as 60%.

Blood platelets, which are responsible for clot retraction, contain high levels of an actomyosin-like complex called thrombosthenin (Bettex-Galland and Lüscher, 1961) which can be separated into actin-like and myosin-like moieties (Bettex-Galland and Lüscher, 1965; Booyse et al., 1971; Adelstein et al., 1971). Since blood clot retraction is inhibited by low concentrations of cytochalasin B (Shepro et al., 1970), we purified platelet actin to homogeneity (see Materials and Methods) and examined its interaction with the alkaloid. We found that platelet actin does activate muscle HMM, but this activation was not inhibited by high concentrations (0.4 mM) of cytochalasin B (Table 1).

Although cytochalasin B conceivably could stop movement by inhibiting the ability of actin to activate myosin ATPase, the above results do not support this possibility. However these experiments do not rule out this mode of action, since platelet actin activation of muscle HMM ATPase activity might respond to cytochalasin B differently than

Table 1. Muscle Actin and Platelet Actin Activation of Muscle HMM ATPase in Presence and Absence of Cytochalasin B

Source	Specific Activity of Acto-HMM ATPase		Ratio +CB/−CB
	(units/mg actin) −CB	+CB	
Rabbit muscle	201	108	0.5
Porcine platelets, Seph. peak I	33	34	1.0
Seph. peak II	12	43	3.6
Bovine platelets, Seph. peak I	13	23	1.8
Seph. peak II	13	22	1.7

Muscle actin (0.10 mg), porcine platelet actin (0.030 mg of Sephadex peak I and 0.038 mg of Sephadex peak II), and bovine platelet actin (0.011 mg of Sephadex peak I and 0.025 mg of Sephadex peak II) were each mixed with 0.010 ml of 10 mg/ml CB or with 0.010 ml of Me_2SO. HMM (0.045 mg) and ATP were added, and actin activation of HMM ATPase was assayed in the absence of Ca^{++}, as described under Materials and Methods. Addition of $CaCl_2$ to the reaction mixture had very little effect on the ATPase specific activities.

would platelet actin activation of platelet myosin ATPase activity.

Effect of cytochalasin B on polymerization of G-actin.

The helical filamentous form of actin is assumed to be required for whatever role actin serves in cellular movement. The following experiments show that high levels of cytochalasin B do not inhibit polymerization of G-actin. In the presence of 0.35 mM cytochalasin B a solution of muscle G-actin (reduced viscosity $\simeq$ 0.05 ml/mg) became highly viscous (reduced viscosity = 0.5 ml/mg) upon addition of KCl to 0.1 M (see Fig. 3 of Spudich and Lin, 1972), indicating conversion of the actin to a high molecular weight form. Another method to follow polymerization of G-actin is to measure the concomitant hydrolysis of ATP (Straub and Feuer, 1950) according to the reaction $nG\text{-actin} + nATP \rightarrow (G\text{-actin-ADP})_n + nP_i$ (Mommaerts, 1952). As shown in Fig. 4 the ATP hydrolysis associated with G-actin polymerization was not inhibited by 0.33 mM cytochalasin B. It should be emphasized that these experiments do not rule out the possibility that in the presence of cytochalasin B G-actin polymerizes into a morphologically aberrant form or to an average length different from that obtained in the absence of the alkaloid.

Effect of cytochalasin B on the structure of actin filaments.

Cytochalasin B changes the morphology of actin filaments as judged by electron microscopy. We have consistently observed alteration of the structure of purified F-actin in the presence of cytochalasin B, although to varying degrees. In some experiments the filaments

appeared to collapse in the presence of the alkaloid, as shown in Fig. 5 using platelet actin and in Fig. 6 using muscle actin. In other experiments the actin appeared to remain more filamentous, but the filaments seemed to aggregate and often appeared shorter when incubated with cytochalasin B (see Fig. 6b).

Cytochalasin B decreases the intrinsic viscosity of purified muscle F-actin. In the presence of low levels of cytochalasin B, the intrinsic viscosity of muscle F-actin was significantly decreased (Fig. 7). The sensitivity of the actin viscosity (50% of the maximal decrease in viscosity occurred at about 0.025 mM cytochalasin B) compares favorably with the sensitivity of blood clot retraction (50% inhibition of clot retraction occurred at about 0.010 mM cytochalasin B; Fig. 7). This decrease in viscosity reached a limit of about 0.3 ml/mg, indicating that in the presence of an excess of cytochalasin B the actin was still in a high molecular weight form (the viscosity of monomeric actin is only about 0.05 ml/mg). These results are consistent with limited fragmentation or collapse of the actin filaments.

Tropomyosin-troponin inhibits the effect of cytochalasin B on the structure of actin filaments. TM·TP forms a complex with actin filaments by binding along their length, apparently just outside each of the two grooves of the F-actin helix (Spudich et al., 1972). We have consistently observed that filaments of highly purified actin appear less uniform than filaments of actin-TM·TP, as judged

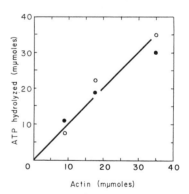

Figure 4. Extent of ATP hydrolysis associated with actin polymerization in the presence and absence of cytochalasin B. Reaction mixtures (0.6 ml each) consisted of (listed in order of addition) 0.5 µmole [γ-^{32}P]ATP, 0.01 ml of 10 mg/ml cytochalasin B (●) or 0.01 ml of Me_2SO (○), muscle G-actin, 12.5 µmoles Tris-Cl pH 8, 1.25 µmoles $MgCl_2$, 0.1 µmole $CaCl_2$ and 30 µmoles KCl. Incubation was at 25°C for 30 min to allow complete polymerization of the actin (i.e., the extent of reaction was measured, not the initial rate). The mµmoles of $^{32}P_i$ released was measured by filtration, as described under Materials and Methods. Each point represents an average of two to four independent determinations. The line is that expected for one mole of ATP hydrolyzed per mole of G-actin polymerized.

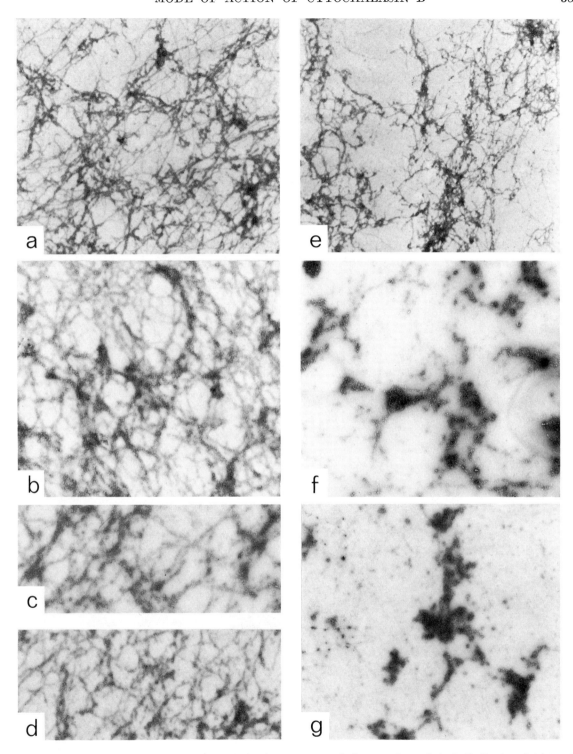

Figure 5. Electron micrographs of platelet actin in the presence and absence of cytochalasin B. Bovine platelet actin (Sephadex peak II) was concentrated by sedimentation out of 25 mM $MgCl_2$ and was resuspended in Buffer A containing 50 mM KCl. The actin (0.47 mg/ml or 0.01 mM) was then incubated at 22°C in Me_2SO (0.003 ml/ml of solution) (a) for 15 min, (b) for 35 min, (c) for 60 min, and (d) for 70 min in the absence of Me_2SO. Actin was simultaneously incubated at 22°C in 0.07 mM cytochalasin B (e) for 10 min, (f) for 25 min, (g) for 40 min. The preparations were stained with 1% uranyl acetate. × 14,000.

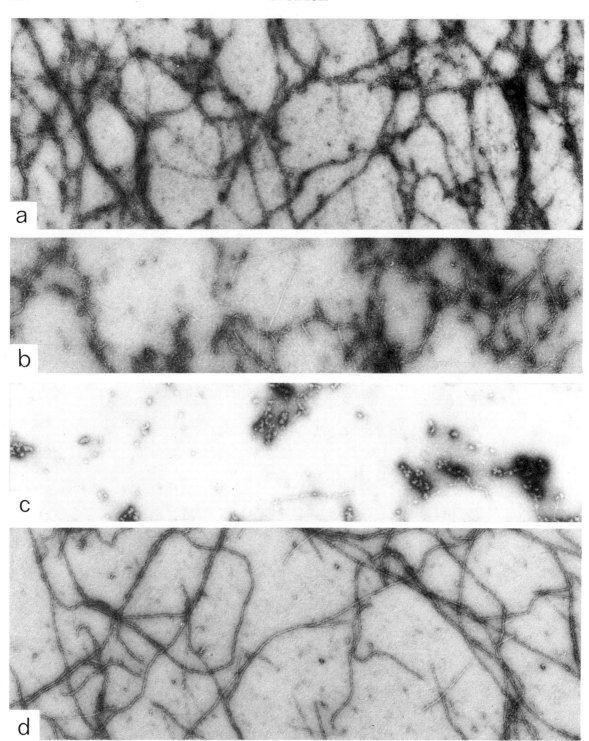

Figure 6. Electron micrographs of cytochalasin-treated muscle actin and of cytochalasin-treated actin-TM·TP. Muscle actin (0.006 mM) in Buffer A (diluted 1 to 10) containing 50 mM KCl was incubated (*a*) in Me₂SO (0.03 ml/ml of solution) for 45 min, (*b*) in 0.06 mM cytochalasin B for 45 min, (*c*) in 0.06 mM cytochalasin B for 3 hr at 22°C (the actin remained filamentous after 3 hr incubation in Me₂SO alone). For (*d*), actin (1.1 mg/ml or 0.02 mM) was incubated at 25°C with TM·TP (0.8 mg/ml) in Buffer A containing 70 mM KCl for 2 hr; cytochalasin B was then added to a final concentration of 0.35 mM. The mixture was incubated for 2 days at 22°C and for 3 days more at 4°C and then was mixed with 9 volumes of 50 mM KCl and immediately applied to electron microscope grids which were then stained with 1% uranyl acetate. All photographs are ×36,500.

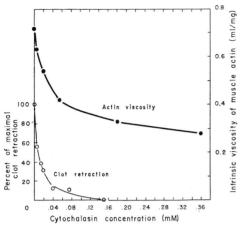

Figure 7. Decrease of intrinsic viscosity of F-actin caused by cytochalasin B. Reduced viscosities of muscle F-actin were measured at 25°C immediately after addition of various concentrations of cytochalasin B; concentrations of actin ranged from 0.007 to 0.028 mM. Extrapolation of the reduced viscosity [$\eta_{reduced}$ is $(\eta_{rel} - 1)/c$, where η_{rel} is the flow time for the protein solution divided by the flow time for the corresponding buffer, and c is the protein concentration in mg/ml] to zero actin concentration yielded the intrinsic viscosity at each concentration of cytochalasin B. These data are reproduced from Spudich and Lin (1972). For comparison, the sensitivity of blood clot retraction to cytochalasin B is shown. Aliquots (1 ml) of bovine platelet-rich plasma (see Materials and Methods; 4×10^8 platelets/ml) were mixed with the concentrations of cytochalasin B shown or with Me_2SO (0.02 ml); 0.4 ml of 0.1 M $CaCl_2$ was added to initiate clot formation, and the clots were incubated at 37°C for 1 hr. The serum expressed by contraction of the clot was measured by use of a syringe, taking care not to puncture the clot. Under these conditions, the clot in the absence of cytochalasin B (with or without Me_2SO) contracted to 44% of its original volume; the percent of maximal contraction in the presence of cytochalasin B was measured relative to this value.

by negative staining with uranyl acetate (compare Fig. 6a and 6d). Does TM·TP inhibit the structural changes in actin caused by cytochalasin B? As judged by viscosity measurements (Table 2) and

Table 2. Specific Viscosity of Actin and Actin-TM·TP in Presence and Absence of Cytochalasin B

		Specific Viscosity		Ratio +CB/−CB
		−CB	+CB	
Actin,	Exp. 1	0.57	0.37	0.65
	Exp. 2	0.52	0.32	0.62
Actin-TM·TP,	Exp. 1	0.68	0.63	0.93
	Exp. 2	0.60	0.58	0.97

The viscosity of F-actin (3.0 ml, 1.1 mg/ml in Buffer A containing 70 mM KCl), with and without TM·TP (0.8 mg/ml), was measured at 25.5°C with an Ostwald viscometer. Cytochalasin B was added to a final concentration of 0.35 mM, and the viscosity of each solution was measured over a period of 24 hr. After an initial rapid (within 5 min) decrease in viscosity, no further change was observed. F-actin was stored at 4°C at a concentration of 10 mg/ml in Buffer A with 50 mM KCl for 3 weeks (experiment 1) and for 2 months (experiment 2) before it was used in these experiments. The specific viscosity is η_{rel}-1, where η_{rel} is the flow time for the protein solution divided by the flow time for the corresponding buffer.

by electron microscopy (Fig. 6), 0.35 mM cytochalasin B did not significantly affect the structure of actin-TM·TP filaments. Since TM·TP is an integral part of the thin filaments in myofibrils, these results provide one explanation for the observations that muscle contraction is not inhibited by cytochalasin B (Sanger and Holtzer, 1972).

Discussion

Platelet actin (Sephadex peak II) is similar to muscle actin. It has a mol wt of about 48,000. The polymeric form appears to be a two-stranded rope capable of forming Mg^{++}-paracrystals, and it activates muscle HMM ATPase activity. Its specific activity for activation of muscle HMM, however, was found to be only 5–10% that of muscle actin. This may reflect the presence of inactive actin in the preparation or may be due to a real difference between muscle actin and platelet actin. Actin from different sources might be expected to have different properties, depending on the kind of movement in which the actin is involved. Even in the same cell, there may be several kinds of actin-like proteins which are involved in different types of movements. In this regard, the two peaks of actin-like activity shown in Fig. 1 might represent two different actin-like proteins which serve different functions in the platelet.

The primary objective of this investigation was to study the effects of cytochalasin B on platelet actin and on muscle actin. We observed that cytochalasin B can cause the collapse of filaments of both muscle actin and platelet actin, apparently to a form which maintains a high molecular weight. There are many reports of "disruption" of microfilaments (which are thought to be actin-like protein filaments) in vivo resulting from incubation of cells with cytochalasin B, and in many instances the microfilaments appear to be replaced by patches of amorphous material (Spooner and Wessells, 1970; Wrenn and Wessells, 1970; Yamada et al., 1970; Schroeder, 1970; Spooner et al., 1971; Orci et al., 1972; Manasek et al., 1972; Auersperg, 1972; Lash et al., 1970). It seems likely from our in vitro results that these patches observed in vivo are actin filaments which have collapsed as a result of direct interaction with cytochalasin B. It remains to be determined whether this effect of the alkaloid is responsible for its effect on cellular movements.

The structural studies presented here are in direct contradiction to the electron microscopy studies of Forer et al. (1972), who reported that the structure of F-actin from muscle was not affected by incubation with 1 mM cytochalasin B for 1 hr

We have consistently observed by electron microscopy a change in morphology of purified F-actin in the presence of cytochalasin B, albeit to varying degrees. Although we have not examined all the factors involved in the sensitivity of actin filaments to cytochalasin B, TM·TP, which often contaminates conventional actin preparations, inhibits the effect of the alkaloid on actin morphology. A detailed and thorough study of the effects of various conditions on the cytochalasin B sensitivity of the actin structure might lead to reconciliation of our electron microscope studies with those of Forer et al.

The observations reported here suggest that cytochalasin B might inhibit cellular movements by interacting with actin-like filaments and changing their morphology. We wish to emphasize, however, that further work is required to evaluate the physiological significance of the structural changes in F-actin which we have observed. There are other possible modes of action of cytochalasin B. For example the primary effect of the alkaloid might be to decrease the average length of actin-like filaments in vivo, which might result in release of the filaments from their presumed point of attachment to membranes. This change in length need not be accompanied by a change in the helical parameters of the actin filaments and would not necessarily be detected upon examination of microfilament morphology in sections of cells fixed for electron microscopy. Another possible mode of action of cytochalasin B might be to inhibit mitochondrial oxidative phosphorylation. The resultant decrease in energy charge may then inhibit the energy-utilizing contractile system. A third alternative is that cytochalasin B might alter the properties of the plasma membrane, perhaps by interacting directly with a membrane component (Estensen et al., 1971; Sanger and Holtzer, 1972; Estensen and Plagemann, 1972). In order to distinguish between these and other possible modes of action, we are currently investigating whether [^{3}H]cytochalasin B enters cells and to what it binds.

Acknowledgments

I wish to thank Kathy Lord for technical assistance, Ross Beirne for purification of tropomyosin-troponin, Anna Spudich for help in purification of platelet actin, and Shin Lin for isolation of blood platelets. This work was supported by a grant from the American Cancer Society (E-627).

References

ADELSTEIN, R. S., T. D. POLLARD, and M. KUEHL. 1971. Isolation and characterization of myosin and two myosin fragments from human blood platelets. Proc. Nat. Acad. Sci. 68: 2703.

ALDRIDGE, D. C., J. J. ARMSTRONG, R. N. SPEAKE, and W. B. TURNER. 1967. The structures of cytochalasins A and B. J. Chem. Soc. 17: 1667.

ALLISON, A. C., P. DAVIES, and S. DE PETRIS. 1971. Role of contractile microfilaments in macrophage movement and endocytosis. Nature New Biol. 232: 153.

AUERSPERG, N. 1972. Microfilaments in epithelial morphogenesis. J. Cell Biol. 52: 206.

AXLINE, S. G. and E. P. REAVEN. 1971. Inhibition of phagocytosis by cytochalasin B. Abstr. 11th Annu. Meet. Amer. Soc. Cell Biol., New Orleans, Louisiana. No. 24, p. 16.

BETTEX-GALLAND, M. and E. F. LÜSCHER. 1961. Thrombosthenin, a contractile protein from thrombocytes. Its extraction from human blood platelets and some of its properties. Biochim. Biophys. Acta 49: 536.

————. 1965. Thrombosthenin, the contractile protein from blood platelets and its relation to other contractile proteins. In Advances in protein chemistry, ed. C. B. Anfinsen, Jr. et al., vol. 20, p. 1. Academic Press, New York.

BOOYSE, F. M., T. P. HOVEKE, D. ZSCHOCKE, and M. E. RAFELSON, JR. 1971. Human platelet myosin. Isolation and properties. J. Biol. Chem. 246: 4291.

CARTER, S. B. 1967. Effects of cytochalasins on mammalian cells. Nature 213: 261.

DAVIES, G. E. and G. R. STARK. 1970. Use of dimethyl suberimidate, a cross-linking reagent, in studying the subunit structure of oligomeric proteins. Proc. Nat. Acad. Sci. 66: 651.

DAVIS, A. T., R. ESTENSEN, and P. G. QUIE. 1971. Cytochalasin B. III. Inhibition of human polymorphonuclear leucocyte phagocytosis. Proc. Soc. Exp. Biol. Med. 137: 161.

EBASHI, S., M. ENDO, and I. OHTSUKI. 1969. Control of muscle contraction. Quart. Rev. Biophys. 2: 351.

ESTENSEN, R. D. 1971. Cytochalasin B. I. Effect on cytokinesis of Novikoff hepatoma cells. Proc. Soc. Exp. Biol. Med. 136: 1256.

ESTENSEN, R. D. and P. G. W. PLAGEMANN. 1972. Cytochalasin B: Inhibition of glucose and glucosamine transport. Proc. Nat. Acad. Sci. 69: 1430.

ESTENSEN, R. D., M. ROSENBERG, and J. D. SHERIDAN. 1971. Cytochalasin B: Microfilaments and contractile processes. Science 173: 356.

FINE, R. E. and D. BRAY. 1971. Actin in growing nerve cells. Nature New Biol. 234: 115.

FORER, A., J. EMMERSEN, and O. BEHNKE. 1972. Cytochalasin B: Does it affect actin-like filaments? Science 175: 774.

GLYNN, I. M. and J. B. CHAPPELL. 1964. A simple method for the preparation of ^{32}P-labelled adenosine triphosphate of high specific activity. Biochem. J. 90: 147.

HANSON, J. 1968. In Budapest symposium on muscle, ed. E. Ernst and F. B. Straub, vol. 8, p. 99. Akademiai Kiado, Budapest.

HUXLEY, H. E. 1963. Electron microscope studies on the structure of natural and synthetic protein filaments from striated muscle. J. Mol. Biol. 7: 281.

HUXLEY, H. E. and G. ZUBAY. 1960. Electron microscope observations on the structure of microsomal particles from Escherichia coli. J. Mol. Biol. 2: 10.

ISHIKAWA, H., R. BISCHOFF, and H. HOLTZER. 1969. Formation of arrowhead complexes with heavy meromyosin in a variety of cell types. J. Cell. Biol. 43: 312.

JAHN, T. L. and E. C. BOVEE. 1969. Protoplasmic movements within cells. Physiol. Rev. 49: 793.

LAKI, K., K. MARUYAMA, and D. R. KOMINZ. 1962. Evidence for the interaction between tropomyosin and actin. Arch. Biochem. Biophys. 98: 323.

LASH, J., R. A. CLONEY, and R. R. MINOR. 1970. Tail resorption in Ascidians: Effects of cytochalasin B. *Biol. Bull.* **139**: 427.

LOWEY, S., H. S. SLAYTER, A. G. WEEDS, and H. BAKER. 1969. Substructure of the myosin molecule. I. Subfragments of myosin by enzymic degradation. *J. Mol. Biol.* **42**: 1.

LOWRY, O. H., N. J. ROSEBROUGH, A. L. FARR, and R. J. RANDALL. 1951. Protein measurement with the Folin phenol reagent. *J. Biol. Chem.* **193**: 265.

MANASEK, F. J., B. BURNSIDE, and J. STROMAN. 1972. The sensitivity of developing cardiac myofibrils to cytochalasin B. *Proc. Nat. Acad. Sci.* **69**: 308.

MOMMAERTS, W. F. H. M. 1952. The molecular transformations of actin. III. The participation of nucleotides. *J. Biol. Chem.* **198**: 469.

NACHMIAS, V. T., H. E. HUXLEY, and D. KESSLER. 1970. Electron microscope observations on actomyosin and actin preparations from *Physarum polycephalum*, and on their interaction with heavy meromyosin subfragment I from muscle myosin. *J. Mol. Biol.* **50**: 83.

ORCI, L., K. H. GABBAY, and W. J. MALAISSE. 1972. Pancreatic Beta-cell web: Its possible role in insulin secretion. *Science* **175**: 1128.

PERRY, M. M., H. A. JOHN, and N. S. T. THOMAS. 1971. Actin-like filaments in the cleavage furrow of newt egg. *Exp. Cell Res.* **65**: 249.

POLLARD, T. D., E. SHELTON, R. R. WEIHING, and E. D. KORN. 1970. Ultrastructural characterization of F-actin isolated from *Acanthamoeba castellanii* and identification of cytoplasmic filaments as F-actin by reaction with rabbit heavy meromyosin. *J. Mol. Biol.* **50**: 91.

SANGER, J. W. and H. HOLTZER. 1972. Cytochalasin B: Effects on cell morphology, cell adhesion, and mucopolysaccharide synthesis. *Proc. Nat. Acad. Sci.* **69**: 253.

SCHOFIELD, J. G. 1971. Cytochalasin B and release of growth hormone. *Nature New Biol.* **234**: 215.

SCHROEDER, T. E. 1970. The contractile ring. I. Fine structure of dividing mammalian (HeLa) cells and the effects of cytochalasin B. *Z. Zellforsch. Mikrosk. Anat.* **109**: 431.

SHEPRO, D., F. C. CHAO, and F. A. BELAMARICH. 1969. Heavy meromyosin coupling with thrombocyte filaments. *J. Cell Biol.* **43**: 129a.

SHEPRO, D., F. A. BELAMARICH, L. ROBBLEE, and F. C. CHAO. 1970. Antimotility effect of cytochalasin B observed in mammalian clot retraction. *J. Cell Biol.* **47**: 544.

SPOONER, B. S. and N. K. WESSELLS. 1970. Effects of cytochalasin B upon microfilaments involved in morphogenesis of salivary epithelium. *Proc. Nat. Acad. Sci.* **66**: 360.

SPOONER, B. S., K. M. YAMADA, and N. K. WESSELLS. 1971. Microfilaments and cell locomotion. *J. Cell. Biol.* **49**: 595.

SPUDICH, J. A. and S. LIN. 1972. Cytochalasin B, its interaction with actin and actomyosin from muscle. *Proc. Nat. Acad. Sci.* **69**: 442.

SPUDICH, J. A. and S. WATT. 1971. The regulation of rabbit skeletal muscle contraction. I. Biochemical studies of the interaction of the tropomyosin-troponin complex with actin and the proteolytic fragments of myosin. *J. Biol. Chem.* **246**: 4866.

SPUDICH, J. A., H. E. HUXLEY, and J. T. FINCH. 1972. The regulation of skeletal muscle contraction. II. Structural studies of the interaction of the tropomyosin-troponin complex with actin. *J. Mol. Biol.* In press.

STRAUB, F. B. and G. FEUER. 1950. Adenosine triphosphate, the functional group of actin. *Biochim. Biophys. Acta* **4**: 455.

SUGINO, Y. and Y. MIYOSHI. 1964. The specific precipitation of orthophosphate and some biochemical applications. *J. Biol. Chem.* **239**: 2360.

THOA, N. B., G. F. WOOTEN, J. AXELROD, and I. J. KOPIN. 1972. Inhibition of release of dopamine-β-hydroxylase and norepinephrine from sympathetic nerves by colchicine, vinblastine, or cytochalasin B. *Proc. Nat. Acad. Sci.* **69**: 520.

TILNEY, L. G. and M. MOOSEKER. 1971. Actin in the brushborder of epithelial cells of the chicken intestine. *Proc. Nat. Acad. Sci.* **68**: 2611.

TONOMURA, Y., P. APPEL, and M. MORALES. 1966. On the molecular weight of myosin. *Biochemistry* **5**: 515.

WEIHING, R. R. and E. D. KORN. 1971. *Acanthamoeba* actin. Isolation and properties. *Biochemistry* **10**: 590.

WESSELLS, N. K., B. S. SPOONER, J. F. ASH, M. O. BRADLEY, M. A. LUDUENA, E. L. TAYLOR, J. T. WRENN, and K. M. YAMADA. 1971. Microfilaments in cellular and developmental processes. *Science* **171**: 135.

WOOLLEY, D. E. 1970. An actin-like protein from amoebae of *Dictyostelium discoideum*. *Fed. Proc.* **29**: 667.

WRENN, J. T. and N. K. WESSELLS. 1970. Cytochalasin B: Effects upon microfilaments involved in morphogenesis of estrogen-induced glands of oviduct. *Proc. Nat. Acad. Sci.* **66**: 904.

YAMADA, K. M., B. S. SPOONER, and N. K. WESSELLS. 1970. Axon growth: Roles of microfilaments and microtubules. *Proc. Nat. Acad. Sci.* **66**: 1206.

Studies on the Contractile Proteins from Blood Platelets

Joel Abramowitz, Mazhar N. Malik, A. Stracher, and T. C. Detwiler

*Department of Biochemistry, State University of New York Downstate Medical Center,
Brooklyn, New York 11203*

Platelets are anucleate, disc-shaped blood cells with a diameter of about 3 microns in humans. They play critical roles in the related processes of hemeostasis and thrombosis. On stimulation by diverse agents, such as thrombin, ADP, epinephrine, and collagen, they undergo a sequence of profound changes. These changes include pseudopod formation, centralization of granules, secretion of specific constituents, aggregation, and finally contraction of the aggregated platelet. When the contracted platelet aggregate adheres to the exposed surfaces of a severed vessel, it is referred to as a platelet plug and it represents the primary defense against the loss of blood. When it forms on the diseased lining of an intact blood vessel, it is referred to as a thrombus and may completely block blood flow. For each of the above processes in platelet function, as well as for maintaining the normal disc shape, a role of contractile proteins has been suggested. In vitro, the most dramatic evidence for a contractile process linked to blood platelets is clot retraction, but the physiological role of clot retraction is uncertain. (For a thorough review of platelet contractility see Johnson, 1971.)

In 1959 Bettex-Galland and Luscher succeeded in extracting from human blood platelets a contractile protein, which they called thrombosthenin and which in its physical and biological properties resembled muscle actomyosin in many respects. Because of this latter similarity there has been wide interest in comparing the properties of these two actomyosins, in the hope that understanding platelet contractility may lead to a more general understanding of the contractile process, of which muscle represents a highly specific example.

Preparation of Platelet Actomyosin

Platelets from 1 liter of freshly drawn human or bovine blood were prepared by differential centrifugation and approximately 2–5 ml of a concentrated suspension was obtained, human blood giving the greater yield. The platelet suspension was then extracted with a mixture of 0.6 M KCl–0.05 M Tris pH 7.5. In order to facilitate lysis of the cells n-butanol (3%) was added to the extraction media. Extraction of the lysed cells was carried out overnight in the cold, the suspension was centri-fuged to remove membrane debris, and the pH of the supernatant was lowered to 6.3–6.4 at the same time the ionic strength was decreased to 0.05–0.10 M.

The optimal conditions for pH and ionic strength were determined in a separate study where it was found that, under the conditions specified, protein with the highest specific ATPase activity was obtained, although in somewhat lower yield.

The precipitate that formed was collected by centrifugation, redissolved in 0.5 M KCl, and the process repeated twice more. Approximately 20 mg of protein was obtained. In our hands thrombosthenin ATPase appears to be stable on storage in 0.5 M KCl–0.05 M Tris pH 7.5 at 4°C for periods of up to several weeks. Dithiothreitol does not appear to enhance the stability of the protein.

Enzyme assay. The ATPase activity was determined in a medium of 0.085 M KCl or 0.6 M KCl, 0.05 M Tris-HCl buffer, pH 7.2, 2.5 mM $MgCl_2$ or $CaCl_2$ and 1 mM ATP. After 10 or 15 min incubation at 37°C the reaction was stopped with cold 10% TCA. The tubes were centrifuged, and the inorganic phosphate in the supernatant fraction was determined by the method of Marsh (1959). Protein concentration (final) ranged from 0.15–0.40 mg/ml. Departures from these conditions are indicated in the legends.

Effect of Ca^{++} on ATPase activity. As shown in Fig. 1 platelet actomyosin is activated by Ca^{++} at both high and low ionic strength. The ATPase reaches a maximum value of about 100–140 nmoles P_i/mg/min at high ionic strength at approximately 10^{-2} M Ca^{++} and then remains constant. This is similar to the effect seen with smooth muscle actomyosin (Needham and Shoenberg, 1967). With most preparations our values of ATPase are around 100 nmoles P_i/mg/min at high ionic strength. This is about 30–40 times greater than that reported by Bettex-Galland and Luscher (1961) but not as high as for skeletal actomyosin (Needham and Shoenberg, 1967).

Effect of ionic strength on ATPase activity. For skeletal muscle actomyosin increasing ionic strength results in a substantial decrease in either

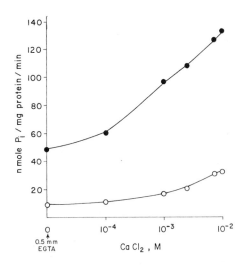

Figure 1. The effect of CaCl$_2$ on the ATPase activity of platelet actomyosin. The ATPase activity was determined as described in the text. (●—●) ATPase at 0.6 M KCl; (○—○) ATPase at 0.085 M KCl.

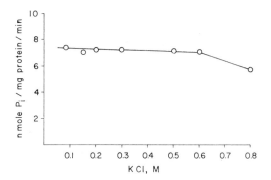

Figure 3. The effect of KCl on the Mg^{++}-moderated ATPase activity of platelet actomyosin. The ATPase activity was determined as described in the text.

Ca^{++}- or Mg^{++}-activated ATPase (Nihei et al., 1966). For smooth muscle actomyosin increasing ionic strength leads to an increase in Ca^{++}-activated ATPase and has little effect in the presence of Mg^{++} (Needham and Shoenberg, 1967). As can be seen in Figs. 2 and 3, platelet actomyosin more nearly resembles smooth muscle actomyosin in its ionic strength dependence of ATPase in the presence of both Ca^{++} and Mg^{++}. This is in contrast to Bettex-Galland and Luscher's earlier observations (1961).

Effect of Mg^{++} on ATPase. As can be seen from Fig. 4, increasing the Mg^{++} concentration from 10^{-6} to 10^{-2} M at low ionic strength results in a 50% decrease in the ATPase activity for platelet actomyosin. This is in contrast to skeletal muscle actomyosin which shows a significant increase in ATPase with increasing Mg^{++} concentration (Nihei et al., 1966).

At high ionic strength the ATPase is enhanced at very low Mg^{++} concentrations (10^{-6}–10^{-5} M) and then is gradually inhibited and approaches the value observed at low ionic strength as we further increase the Mg^{++} concentration.

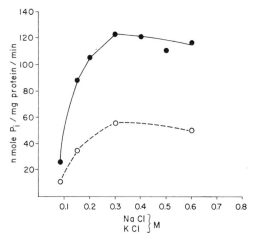

Figure 2. The effect of KCl or NaCl on the Ca^{++}-activated ATPase activity of platelet actomyosin. (●—●) KCl; (○—○) NaCl.

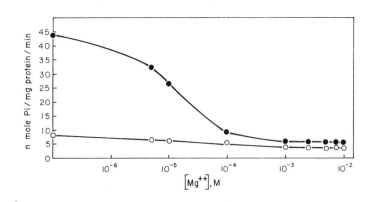

Figure 4. The effect of MgCl$_2$ on the ATPase activity of platelet acto-myosin. The ATPase activity was determined as described in the text. (●—●) 0.6 M KCl; (○—○) 0.085 M KCl.

Dissociation of Platelet Actomyosin

We next turned our attention to preparing platelet myosin from platelet actomyosin. Several investigators have already published reports on the isolation of platelet myosin (Adelstein et al., 1971; Booyse et al., 1971; Cohen et al., 1969). Our initial attempts were similar to those used by Rafelson and Booyse who added MgATP to their preparation and then spun their solution in the ultracentrifuge according to the procedure described for skeletal actomyosin by Weber (1956). In our hands this procedure was unsuccessful since the supernatant was still highly contaminated with actin as determined by disc-gel electrophoresis in SDS. Chromatography on DEAE-Sephadex led to very poor recoveries, as did gel diffusion on Sephadex G-200. Separation of F-actin and myosin by sucrose density gradient appeared to be a promising method, and it was this technique which was next used (Dow, 1971). Figure 5 shows that skeletal muscle actomyosin can easily be separated into its component F-actin and myosin parts in the presence of ATP and Mg^{++}. A sharp peak for myosin is obtained which by disc-gel electrophoresis in SDS can be shown to be essentially free of actin. F-actin forms a pellet at the bottom of the tube.

When platelet actomyosin is run by the same technique (Fig. 6) in 0.5 M KCl, the bulk of the

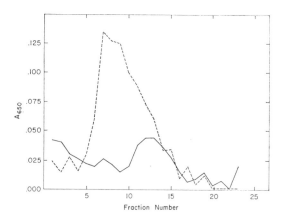

Figure 6. Sucrose density gradients of platelet actomyosin. Procedure same as in Fig. 5.

material forms a pellet at the bottom of the tube as would be expected for platelet actomyosin. However a slower moving peak is sometimes seen which, when run on SDS-gel, moves in the same position as G-actin. This is not seen in the skeletal muscle actomyosin sucrose gradients (Fig. 5). In the presence of MgATP the myosin peak does not appear to be as sharp as for skeletal actomyosin. SDS gels on the pellet and across the band show that the pellet is F-actin exclusively; the faster moving part of the main peak contains mainly myosin with significant actin contamination; the central portion, a mixture of both; and the trailing portion, mainly a band in the position of G-actin.

Discussion

The results of these studies on the enzymic properties of platelet actomyosin lead to several interesting conclusions with regards to this system when compared to muscle actomyosins. First, as seen in Figs. 1–4, in all cases the variation of ATPase with increasing concentrations of Ca^{++}, Mg^{++}, and ionic strength parallels closely the response of smooth muscle actomyosins (Needham and Shoenberg, 1967). Second, the specific activities are of interest to note. In the presence of Ca^{++} at high ionic strength the specific ATPase of platelet actomyosin is approximately 100 nmoles P$_i$/mg/ min. Under the same conditions, in the presence of 1 mM Mg^{++}, the specific ATPase of smooth muscle actomyosins is approximately 7 nmoles P$_i$/mg/min. At physiological ionic strengths (0.15 M) the values are about the same. Platelets are characterized by high concentrations of Ca^{++} intracellularly compared to Mg^{++} (approximately 10:1), whereas in muscle cells the reverse situation exists (Marcus and Zucker, 1967). At low ionic strengths increasing the Mg^{++} concentration from zero to

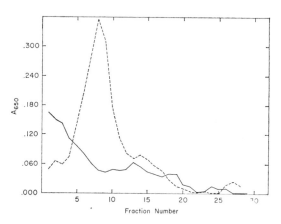

Figure 5. Sucrose density gradient of skeletal muscle actomyosin. 0.25 ml of a 3 × ppt skeletal muscle actomyosin solution (2.0 mg/ml) in 0.5 M KCl, 10 mM imidazole pH 7.6 or in 0.5 M KCl, 10 mM Mg-ATP, 10 mM imidazole pH 7.6 was layered on a 5 ml sucrose gradient 5–20% made up in the same buffer in which the sample was dissolved. The gradients were then centrifuged at 49,000 rpm for 14 hr at 3°C in a Beckman SW 50.1 rotor. At the end of the run, 8-drop fractions were collected, and an aliquot of each fraction was analyzed for protein by the Lowry method against serum albumin. (——) Sample and gradient contained 0.5 M KCl, 10 mM imidazole pH 7.6; (– – –) sample and gradient contained 0.5 M KCl, 10 mM Mg-ATP, 10 mM imidazole pH 7.6.

10^{-2} M actually results in a decrease in the specific ATPase from 7 to 4 nmoles P_i/mg/min (Fig. 4). With skeletal muscle actomyosin there is a marked activation of ATPase with increasing Mg^{++} concentration (Nihei et al., 1966). No comparable studies have been reported for smooth muscle actomyosin.

When the pH of the supernatant from the original precipitation of platelet actomyosin is further lowered to 5.3, a precipitate is obtained which appears to be exclusively a Mg-ATPase (Malik et al., unpublished observations). Thus the very low Mg-ATPase observed with platelet actomyosin may be the result of a slight contamination with this material and thus Mg-ATP may not be a substrate for platelet actomyosin. It would seem that if this protein were to be a Mg^{++}-activated ATPase, increasing Mg^{++} concentration would lead to some activation, as is the case for skeletal muscle actomyosin, rather than an inhibition as is observed here (Fig. 4). Further studies are in progress to clarify this point.

The experiments using sucrose density gradients show that, although actin-free myosin can be prepared from skeletal muscle actomyosin in the presence of Mg-ATP by this technique, it is more difficult to do with platelet actomyosin. Although a pellet containing actin is found at the bottom of the tube, SDS gels on the myosin peak still show the presence of a protein resembling actin that is associated with the myosin peak and that is not dissociated by Mg-ATP. Whether this protein is another form of actin or a protein resembling actin is not clear at the present time, and attempts to prepare myosin free of this protein are now in progress.

Acknowledgments

This work was supported by Grants from the New York Heart Association and the National Heart and Lung Institute (HL 10099 and HL 14020).

References

ADELSTEIN, R. S., T. D. POLLARD, and W. M. KUEHL. 1971. Isolation and characterization of myosin and two myosin fragments from human blood platelets. *Proc. Nat. Acad. Sci.* **68**: 2703.

BETTEX-GALLAND, M. and E. F. LUSCHER. 1961. Thrombosthenin—a contractile protein from thrombocytes. Its extraction from human blood platelets and some of its properties. *Biochim. Biophys. Acta* **49**: 536.

BOOYSE, F. M., T. P. HOVEKE, P. HITCHCOCKE, and M. E. RAFELSON, JR. 1971. Human platelet myosin. Isolation and properties. *J. Biol. Chem.* **246**: 4291.

COHEN, I., Z. BOHAK, A. DEVRIES, and E. KATCHALSKI. 1969. Thrombosthenin M. Purification and interaction with thrombin. *Europe. J. Biochem.* **10**: 388.

DOW, J. 1971. Structure of myosin in developing systems. Ph.D. dissertation, State University of New York, Downstate Medical Center.

JOHNSON, S. A. 1971. *The circulating platelet.* Academic Press, New York.

MARCUS, A. J. and M. B. ZUCKER. 1965. *The physiology of blood platelets. Recent biochemical, morphological and clinical research.* Grune and Stratton, New York.

MARSH, B. B. 1959. The estimation of inorganic phosphate in the presence of ATP. *Biochim. Biophys. Acta* **32**: 357.

NEEDHAM, D. M. and C. F. SHOENBERG. 1967. The biochemistry of the myometrium, p. 291. In *Cellular biology of the uterus*, ed. R. M. Wynn. Appleton-Century-Crofts, New York.

NIHEI, T., M. MORRIS, and A. L. JACOBSEN. 1966. Activation and inhibition of myosin B ATPase by Mg^{++} and Ca^{++} at low concentrations of KCl. *Arch. Biochem. Biophys.* **113**: 45.

WEBER, A. 1956. The ultracentrifugal separation of L-myosin and actin in an actomyosin sol under the influence of ATP. *Biochim. Biophys. Acta* **19**: 345.

The Characterization of Contractile Proteins from Platelets and Fibroblasts

ROBERT S. ADELSTEIN* AND MARY ANNE CONTI*

Section on Cellular Physiology, Laboratory of Biochemistry, National Heart and Lung Institute N.I.H., Bethesda, Maryland 20014

Cell motility, as well as more specialized cell functions such as clot retraction, appears to be mediated by contractile proteins. These proteins, once vaguely characterized as being actin-like and myosin-like, now appear to be quite similar on a molecular level to their well characterized muscle counterparts.

Thrombosthenin has long been known to consist of two fractions which have the properties of actin (thrombosthenin A) and myosin (thrombosthenin M) (Bettex-Galland and Lüscher, 1965; Lüscher and Bettex-Galland, 1971). During the fractionation of thrombosthenin, we identified intact myosin, actin, and two additional proteins corresponding to the rod and head portion of the myosin molecule. Characterization of these proteins has shown the close resemblance of the platelet contractile proteins to the corresponding muscle proteins (Adelstein et al., 1971). We have extended our observations on human platelet myosin and actin to show that: (a) the ATPase activity of the myosin head is activated by rabbit skeletal actin; (b) the intact myosin molecule contains at least two different light chains of mol wt 16–18,000; (c) platelet actin, purified by reversible polymerization, is capable of activating the ATPase activity of rabbit skeletal muscle heavy meromyosin (HMM); and (d) the activation of HMM by platelet actin is dependent on calcium in the presence of rabbit troponin-tropomyosin.

In contrast to human platelets where the contractile proteins constitute approximately 15% of the total cell proteins, mouse fibroblast cells grown in vitro contain substantially less actomyosin. Recent studies by Johnson et al. (1972) concerning the effects of cyclic AMP on fibroblast motility led us to study (in collaboration with Johnson and Pastan) the contractile proteins of these cells. In this paper we describe a method for isolating actomyosin from cloned mouse fibroblasts. After purification, the myosin was characterized by sodium dodecyl sulfate (SDS)-polyacrylamide gel

electrophoresis, ATPase assay with and without rabbit muscle actin, actin binding, and electron microscopy.

Methods

Preparation of Proteins

The purification of contractile proteins from both platelets and fibroblasts is based on (a) high ionic strength (0.6 M KCl) extraction of the lysed platelets and fibroblasts followed by low ionic strength (0.1 M KCl) precipitation at pH 6.4, repeated three times (for details see Adelstein et al., 1971); (b) centrifugation of the final high ionic strength solution at 100,000 g for 1 hr. In the preparation of platelet proteins this centrifugation is done in the presence of Mg-ATP and the ammonium sulfate fractionation is performed on this supernatant. In isolating the fibroblast contractile proteins it was advantageous, because of the small amounts of actomyosin present, to sediment the actomyosin complex first at 100,000 g in the absence of ATP. This precipitate was then dissolved in 0.6 M KCl-15 mM Tris, 5 mM dithiothreitol (DTT) in the presence of Mg-ATP, recentrifuged at 100,000 g for 1 hr and the supernatant used for ammonium sulfate fractionation. The final purification step in the case of both myosins (platelet and fibroblast) was chromatography of the appropriate ammonium sulfate fraction on Sepharose 4B.

Platelet actin was purified from the 0–25% ammonium sulfate fraction. Actin was polymerized in the presence of 0.6 M KCl and centrifuged at 100,000 g for 3 hr in 20 mM Mg-ATP pH 7.5. The resulting pellet was homogenized in 0.5 mM ATP, 0.1 mM CaCl$_2$, 2 mM Tris pH 7.7, 5 mM DTT and depolymerized by dialysis in the same buffer. Prior to use it was centrifuged for 3 hr at 100,000 g and the supernatant used in the experiments described below. For experiments that required F-actin the G-actin was dialyzed against 10 mM imidazole-HCl pH 7.5, 2 mM MgCl$_2$ and 5 mM DTT.

Characterization of Contractile Proteins

To characterize myosin isolated from platelets and fibroblasts and compare it to rabbit skeletal

---

* Present address: Section on Molecular Cardiology, Cardiology Branch, National Heart and Lung Institute, National Institutes of Health, Bethesda, Maryland 20014.

muscle myosin, the following parameters were investigated: (a) Migration in 0.1% SDS-polyacrylamide gels of various percent acrylamide, but usually 7.5%. (b) ATPase activity; myosin alone was measured in 0.6 M KCl in the presence of K-EDTA, Ca⁺⁺ or Mg⁺⁺. Actin activation was measured at low ionic strength (15 mM KCl) in the presence of Mg⁺⁺. (c) Binding of platelet and fibroblast myosin to rabbit skeletal actin in the presence and absence of ATP. (d) Appearance of 1% uranylacetate negatively stained preparations in the electron microscope (performed by Dr. Thomas Pollard).

Actin was characterized by its ability to show birefringence in high salt and to activate the ATPase activity of rabbit muscle HMM in the absence and presence of rabbit muscle troponin-tropomyosin.

Results

Platelet Contractile Proteins

Myosin. When the 35–60% ammonium sulfate fraction of the Mg-ATP 100,000 g supernatant is chromatographed on Sepharose 4B, two major peaks of ATPase activity result (Fig. 1). Peak 1 contains intact myosin (labeled M in Fig. 1) which has the following properties: (1) a heavy chain of 200,000 mol wt; (2) an ATPase activity stimulated by K-EDTA and Ca⁺⁺ and inhibited by Mg⁺⁺; (3) the ability to bind to rabbit muscle actin

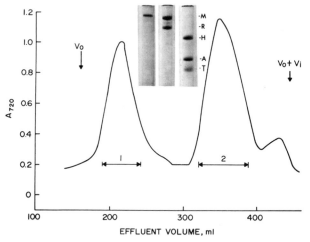

Figure 1. Profile of Sepharose 4B agarose filtration of the 35–60% ammonium sulfate fraction of platelet actomyosin. The equilibration buffer used was 0.6 M KCl-15 mM Tris pH 7.5; 3 mM DTT; 0.5 mM MgCl₂, and 0.5 mM ATP. An 11-mg sample was applied in 11 ml of 0.6 M KCl-15 mM Tris; 3 mM DTT; 5 mM ATP; 5 mM MgCl₂ to a 2.5 × 90 cm column. The column was eluted with the equilibration buffer. The small peak at the salt boundary is from the ATP in the applied sample. The photographs are of 0.1% SDS-5% acrylamide gels of the two pooled fractions and of rabbit skeletal muscle myosin (RM). (Adelstein et al., 1971).

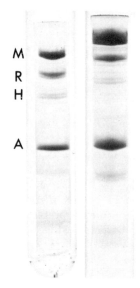

Figure 2. 0.1% SDS-7.5% acrylamide gels of platelet actomyosin following 16 hr (*left*) and ½ hr extraction. Migration is from top to bottom. M identifies myosin heavy chain; R, rod; H, head; A, actin.

in the absence, but not in the presence, of ATP; (4) the formation of short bipolar thick filaments in 0.1 M KCl as observed in the electron microscope. Peak 1 also contains the rod portion (R) of the myosin molecule which does not bind to actin and aggregates in 0.1 M KCl to form fibrous paracrystals lacking lateral projections.

The second peak contains the head portion (H) of the myosin molecule which, like the intact molecule, possesses enzymic activity and binds to rabbit skeletal muscle actin. In the electron microscope this fragment binds to F-actin to form the familiar "arrowhead" pattern with both platelet and rabbit skeletal muscle actin. The second peak also contains platelet actin (A) and platelet tropomyosin (T). The tropomyosin, identified and characterized by Cohen and Cohen (1972), is remarkable because its molecular weight is 5000 less than the 35,000 mol wt subunit of rabbit muscle tropomyosin.

Production of myosin fragments. To determine if proteolytic cleavage occurred during the initial 16 hr high-salt extraction, this extraction was shortened to one-half hour using freshly drawn platelets. Figure 2 shows SDS-polyacrylamide gels of the 100,000 g supernatant following a 16 hr and a one-half hour extraction. The presence of similar amounts of rod in each gel indicates that a limited proteolytic cleavage occurred either during or prior to, but not after, the one-half hour extraction.

The possibility that thrombin might be responsible for the production of the myosin fragments has been suggested by a number of authors (Cohen

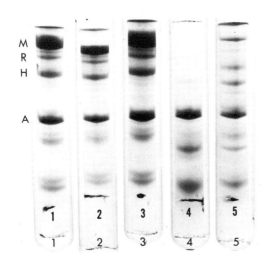

Figure 3. 0.1% SDS-7.5% acrylamide gels of platelet actomyosin. Gel 1, control showing myosin heavy chain (M), rod (R), head (H), and actin (A); gel 2, following thrombin incubation in presence of 9 mM ATP; gel 3, thrombin incubation without ATP; gel 4, trypsin incubation in presence of ATP; gel 5, trypsin without ATP. Actomyosin, 0.8–1.2 mg/ml; thrombin, 14–40 units/ml; trypsin, 0.005 mg/ml; 37°C pH 7.5, 1 hr.

et al., 1969; Baenziger et al., 1971) and reported by one (Booyse ét al., 1972). Indeed initial experiments in our laboratory using impure thrombin resulted in a decrease of intact myosin and an increase in the head and rod fragments. However when a thrombin digestion was carried out using plasmin-free thrombin (a gift of Dr. David Aaronson), no evidence was found for the digestion of the intact myosin molecule. The platelet actomyosin was incubated with thrombin or trypsin in the presence or absence of ATP. Although the gel pattern (Fig. 3) was unaltered after incubation with thrombin (1st three gels), the gel pattern following tryptic digestion changed markedly. In the presence of ATP intact myosin head and rod were extensively digested by trypsin. In the absence of ATP the rod portion of the molecule, which is not bound to F-actin, appears to undergo a more extensive tryptic digestion than head and intact myosin, which are bound.

Activation of platelet myosin by rabbit skeletal muscle actin. Rabbit skeletal muscle

Table 1. Rabbit Skeletal Muscle Actin Activation of Platelet Myosin Head

	Specific Activity
Platelet myosin	0.02 μm P$_i$/mg myosin/min
Platelet myosin + muscle actin	0.07 μm P$_i$/mg myosin/min

Conditions: 5 mM imidazole — HCl, 2.7 mM MgCl$_2$, 1.0 mM ATP, 1.0 mM EGTA, 16 mM KCl. Actin: 0.8 mg/ml. Myosin: 0.08 mg/ml.

Table 2. Effect of Rabbit Muscle Troponin-Tropomyosin on Rabbit Muscle Actin Activation of Platelet Myosin Head

	ATPase μm P$_i$/mg myosin/min	
	EGTA	Calcium
Platelet myosin	0.02	0.02
Platelet myosin + muscle actin	0.07	0.07
Platelet myosin + muscle actin + muscle troponin-tropomyosin	0.03	0.08

Conditions for assay are the same as Table 1; CaCl$_2$ 0.1 mM.

actin activates the low-salt Mg^{++}-stimulated ATPase activity of platelet myosin head approximately 3.5-fold. The conditions for this activation are outlined in Table 1. The addition of rabbit muscle troponin-tropomyosin to this mixture of rabbit skeletal actin and platelet myosin head conferred calcium sensitivity on the system as shown in Table 2. The presence of the troponin-tropomyosin complex led to inhibition of the ATPase activation of myosin in the absence of Ca^{++}.

Light chains of platelet myosin. Fast muscle myosin contains three light chains of 25,000, 18,000, and 16,000 mol wt (Sarker et al., 1971; Lowey and Risby, 1971; Weeds and Pope; 1971), whereas slow and cardiac muscle myosin contain two small chains of mol wt 27,000 and 20,000. The light chains of platelet myosin are shown in Fig. 4. The first gel on the left is of fast skeletal muscle myosin showing the three light chains. The next gel is platelet actomyosin prior to Sepharose 4B chromatography. It clearly shows at least two

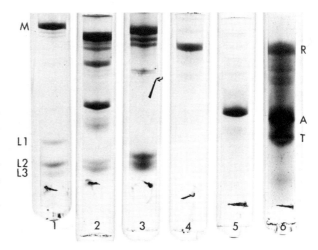

Figure 4. 0.1% SDS-7.5% acrylamide gels showing light chains in platelet myosin. Gel 1, rabbit skeletal muscle myosin; gel 2, platelet actomyosin prior to Sepharose chromatography; gel 3, intact myosin molecule after Sepharose chromatography; gel 4, platelet rod; gel 5, platelet actin; gel 6, rod (R), actin (A), and tropomyosin (T) isolated from ether-acetone powder of platelets. M indicates myosin heavy chain and L, light chains.

Table 3. Effect of Muscle Troponin-Tropomyosin on Platelet Actin Activation of Muscle HMM

	ATPase μm P_i/mg/min	
	EGTA	Ca^{++}
Muscle HMM	0.07	0.10
Muscle HMM + platelet actin	1.40	1.49
Muscle HMM + muscle troponin-tropomyosin	0.20	0.21
Muscle HMM + platelet actin + muscle troponin-tropomyosin	0.32	1.13

Conditions: 5 mM imidazole — HCl, 2.7 mM MgCl$_2$, 1.0 mM ATP, 13.0 mM KCl, 1.0 mM EGTA, 0.1 mM CaCl$_2$. Actin: 0.20 mg/ml. HMM: 0.03 mg/ml.

light chains in the molecular weight range of 16–18,000 but none at 20–27,000. The third gel, which is purified intact myosin following Sepharose chromatography, shows that the light chains are bound to intact platelet myosin. In contrast the last three gels, which are purified platelet rod, platelet actin, and a mixture of rod, actin, and platelet tropomyosin respectively, do not show the presence of light chains. These results indicate that platelet myosin has at least two light chains of mol wt 16–18,000 bound to the 200,000 mol wt heavy chains.

Actin. Platelet actin was purified from the 0–25% ammonium sulfate fraction. This actin is similar to rabbit skeletal muscle actin in both mol wt (46,000) and molar content of the unusual amino acid 3-methylhistidine. In the presence of 0.1 M KCl platelet actin shows flow birefringence. This birefringence (and concomitant increased viscosity) is markedly pH dependent and is reversibly lost at a pH greater than 8.3. Table 3 shows the activation of rabbit muscle HMM by platelet actin in the presence and absence of rabbit troponin-tropomyosin. This latter protein again conferred Ca^{++} sensitivity on the hybrid acto-myosin complex.

Fibroblast Myosin

Isolation and purification. Two different sources of cloned L-929 fibroblasts were used. Initially cells were grown in monolayer, but in order to obtain larger amounts of protein 4-liter spinner bottle cultures were used. The procedure for preparing the fibroblast actomyosin is outlined above. The cells were harvested by either scraping the bottles or centrifuging the suspended cultures.

In the initial experiment 12 bottles of monolayer cells yielded 58 mg of protein in the 0.6 M KCl-15 mM Tris extract supernatant. (See Table 4 for comparison with 22 liters of spinner bottle cells.) The 100,000 g supernatant, after solubilization of the protein in the presence of Mg-ATP, yielded 1.3 mg of protein with a high-salt K-EDTA stimulated ATPase specific activity of 0.06 μmoles of P_i/mg/min. A scan of the SDS-polyacrylamide gel at A_{600} of this fraction is shown in Fig. 5. The positions of the two peaks which coelectrophorese with muscle actin or myosin are indicated. The two large peaks near the bottom of the gel have been tentatively identified as histones in 8 M urea polyacrylamide gels run at acid pH.

This initial experiment using monolayer cultures indicated the presence of the two contractile proteins, actin and myosin, and suggested the use of cells from spinner bottle cultures in order to increase the yield of these proteins. Thus 20 g of fibroblasts were harvested from 22 liters of spinner bottles. (Sixteen g of these cells were washed and stored at −15°C for 1 week prior to use.) The purification procedure, yields of proteins and ATPase activity, as well as the specific activities of the various fractions are shown in Table 4. The 25–70% ammonium sulfate fraction (Fig. 6) was chromatographed on a 2.5 × 170 cm column of Sepharose 4B (Fig. 7) and the resulting fractions assayed for ATPase activity.

Table 4. Fibroblast Myosin: Purification and Enzyme Activity

	Protein (total mg)	Total EDTA-ATPase*	Specific Activity†		
			EDTA-ATPase	Ca-ATPase	Mg-ATPase
Extract supernatant	1260	5.0	0.004	—	—
100,000 g supernatant††	91	0.3	0.003	0.003	0.005
100,000 g supernatant + Mg-ATP	178	0.9	0.005	0.021	0.011
Ammonium sulfate					
0–25%	74.	0.07	0.001	0.004	0.003
25–50%	26	0.75	0.029	0.034	0.007
50–70%	20	0.34	0.017	0.010	<0.001
Sepharose 4B (25–70%) fraction	0.60	0.26	0.43	0.50	<0.01

* Total EDTA-ATPase is expressed in μmoles P_i released/min.

† Specific activity is expressed in μmoles of P_i released/mg protein per min.

†† This fraction is discarded.

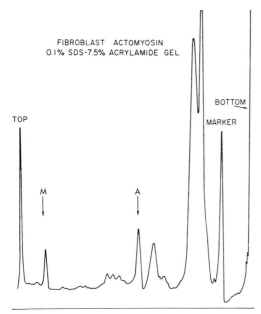

Figure 5. A_{600} scan of 0.1% SDS-7.5% acrylamide gel of fibroblast actomyosin. Electrophoresis is from left to right. M indicates peak coelectrophoresing with muscle myosin heavy chain and A with actin.

In contrast to platelet myosin only one peak with ATPase activity was found. This peak, which had the same elution properties as rabbit skeletal muscle myosin on Sepharose 4B, was pooled as indicated in Fig. 7. The fraction was concentrated with an Amicon Ultrafiltration cell, assayed for the ATPase specific activities shown in Table 4, and analyzed by SDS-polyacrylamide gel electrophoresis (Figs. 6 and 7). This column-purified material was used for the characterization of fibroblast myosin.

Characterization of fibroblast myosin. First the myosin was tested for its ability to bind to rabbit skeletal muscle actin. After centrifugation of fibroblast myosin with rabbit muscle F-actin in the presence and absence of ATP, the supernatants were analyzed by gel electrophoresis and assay of

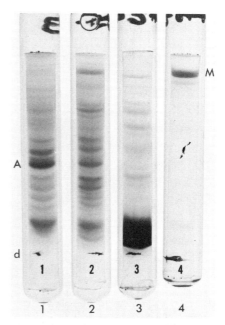

Figure 6. Purification of fibroblast myosin. 0.1% SDS-7.5% acrylamide gels. Gel 1, 0–25% $(NH_4)_2SO_4$ fraction; gel 2, 25–50% fraction; gel 3, 50–70% fraction; gel 4, fibroblast myosin following Sepharose 4B chromatography. M coelectrophoreses with muscle myosin heavy chain and A with muscle actin. The extent of dye migration is indicated by d.

ATPase activity to determine if the protein bound to actin and sedimented with it. When ATP was present in the initial mixture of fibroblast myosin and rabbit muscle actin, the supernatant retained both the myosin (Fig. 8) and the ATPase activity. When ATP was not included in the mixture, however, all the ATPase activity and the protein (as indicated by gel 2, Fig. 8) was lost from the supernatant (a small portion of rabbit muscle actin which did not sediment can be seen in both gels). This experiment indicates that fibroblast myosin binds to rabbit muscle actin and is dissociated by ATP.

Figure 7. Profile of Sepharose 4B agarose filtration of 25–70% ammonium sulfate fraction of fibroblast myosin. The equilibration buffer was 0.6 M KCl-15 mM Tris pH 7.5, 1 mM ATP, 1 mM $MgCl_2$ and 5 mM DTT. The two 2.5×85 cm columns connected in series were eluted with equilibration buffer. The ordinate indicates A_{720} of ATPase assay. The flow rate was 20 ml/hr. The photograph is of a 0.1% SDS-7.5% acrylamide gel from the indicated pooled fraction. Migration is from top to bottom. The extent of bromphenol blue dye migration is marked near the gel bottom (d).

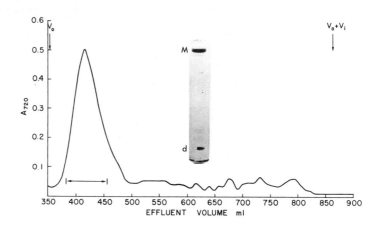

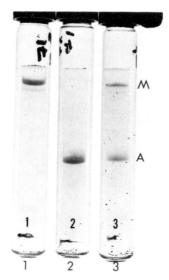

Figure 8. Supernatant fractions from the binding of rabbit skeletal muscle actin to fibroblast myosin. 0.1% SDS-7.5% acrylamide gels. Gel 1, fibroblast myosin (for comparison); gel 2, supernatant of incubation of fibroblast myosin with actin in the absence of ATP; gel 3, supernatant of incubation of fibroblast myosin with actin and 5 mm ATP. Some of the actin (A) remains in the supernatant of gels 2 and 3 and can be used as internal standard for size of the sample applied to gel. A small amount of Bio-Rad AG1-X2 used for destaining can be seen on the gels.

The activation of fibroblast myosin ATPase activity by rabbit muscle actin in the presence of Mg^{++} at low ionic strength was determined under conditions outlined in Table 5. A marked activation was found, but due to the relatively large amounts of fibroblast myosin required, this experiment was only performed once.

Table 5. Rabbit Skeletal Muscle Actin Activation of Fibroblast Myosin

	Specific Activity
Fibroblast myosin	$0.01~\mu m~P_i$/mg myosin/min
Fibroblast myosin + muscle actin	$0.09~\mu m~P_i$/mg myosin/min

Conditions: 3 mm imidazole — HCl, 2.7 mm $MgCl_2$, 1.0 mm ATP, 1.0 mm EGTA, 14.0 mm KCl. Actin: 0.30 mg/ml. Myosin: 0.03 mg/ml.

The remainder of the myosin was used for electron microscopy carried out by Dr. Thomas Pollard. In 0.1 m KCl the fibroblast myosin forms short (200-nm long), thin (10-nm wide) bipolar aggregates. These aggregates have a smooth central shaft and tufted ends similar in appearance to short muscle myosin filaments formed in vitro (Fig. 9).

Discussion

Platelet myosin is remarkably similar to muscle myosin in structure and function. Both have heavy chains of 200,000 mol wt, and both contain light chains tightly bound to the head portion of the molecule. Platelet myosin differs from muscle myosin in the size of light chains. It contains at least two chains of 16–18,000, but does not contain the 25,000 chain of fast muscle or the 27,000 chain of slow and cardiac muscle. The lower ATPase activation by actin of platelet myosin (as well as the different light chains) suggests a closer similarity to smooth muscle myosin. It is interesting to note that the ATPase activity of platelet myosin (assayed alone) also shows a salt dependence similar to smooth muscle myosin (Adelstein and Conti, unpublished) in the presence of 10 mm $CaCl_2$.

The appearance of the rod and head portion of

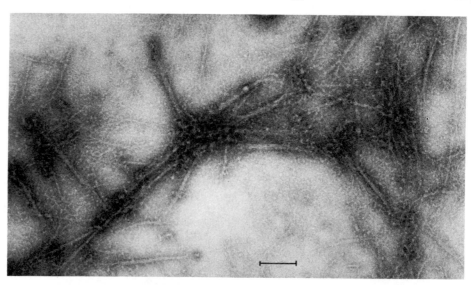

Figure 9. An electron micrograph of fibroblast myosin in 0.1 m KCl, 10 mm imidazole-HCl pH 7.0, negatively stained with 1.0% uranyl acetate. The fibroblast myosin forms bipolar aggregates about 200 nm long with a bare central shaft and tufted ends. Magnification 100,000. Bar = 0.1 μm. Micrograph by T. D. Pollard of the N.I.H.

the molecule in platelets suggests that a limited proteolytic digestion has taken place in vivo or during the early part of the isolation procedure. Thrombin, which has been proposed as the enzyme causing this cleavage, does not digest intact myosin in solution. Digestions that have been reported (Booyse et al., 1972) may have been due to contamination of the thrombin preparation with plasmin, an enzyme similar to trypsin in its substrate requirements.

Platelet actin has been isolated from the 0–25% ammonium sulfate fraction and shows both flow birefringence and increased viscosity in the presence of 0.1 M KCl. Platelet actin activated the ATPase activity of rabbit muscle HMM approximately 15-fold, and in the presence of rabbit troponin-tropomyosin required Ca^{++} to achieve activation. This observation, along with the results above indicating that the platelet myosin head alone does not possess EGTA sensitivity and the work of Cohen and Cohen (1972) on the EGTA sensitivity of platelet actomyosin, suggests that troponin-tropomyosin plays a role in the interaction of platelet actin and platelet myosin. However final elucidation of the mechanism for Ca^{++} activation in the platelet system is not yet established. To do this the troponins (or other regulatory molecules) must first be isolated and characterized.

The possibility that the effects of cyclic AMP on fibroblast motility were mediated by the contractile proteins led us to attempt the isolation of myosin from cloned mouse fibroblasts. The presence of actin in chick fibroblast cultures had been reported by Yang and Perdue (1972) and Ishekawa et al. (1969). The small content of contractile proteins made the task of isolation more formidable than isolation of the corresponding platelet proteins. However the same general approach was used to isolate fibroblast myosin.

Fibroblast myosin is similar to muscle myosin in that it contains a heavy chain of 200,000 daltons possesses an ATPase activity that is activated by muscle actin, is dissociated from muscle actin by ATP, and forms short bipolar aggregates in 0.1 M KCl.

Summary

Myosin has been isolated and characterized from two different kinds of non-muscle cells. Actin has been isolated and characterized from platelets. In both cases the proteins are remarkable for their resemblance to their muscle counterparts (see also Bray, this volume).

The crucial questions remain: (a) What is the role of these proteins in these cells? One specialized role in the platelets appears to be clot retraction but this does not preclude other roles, which are sug-

gested by the presence of EGTA sensitivity and the apparent excess of actin. (b) How do these proteins function? In clot retraction the proteins apparently function extracellularly. In fibroblasts the exact location of the myosin is not yet certain and it is therefore difficult to discern how the myosin and actin filaments interact. (c) How are these proteins regulated? Only indirect evidence at present indicates a role for troponin-tropomyosin, and the possibility of other modes of regulation (e.g., phosphorylation) should be considered.

Answers to these as well as other important questions concerning cellular actomyosin from non-muscle sources will, hopefully, be forthcoming in the next few years.

Acknowledgments

We thank Drs. W. W. Kielley, George S. Johnson, Ira H. Pastan, and Thomas D. Pollard for helpful discussions. We are indebted to Mrs. Elsie Yanchulis for her help in preparing the platelet concentrates and Mr. Charles Buckler for the fibroblasts.

References

ADELSTEIN, R. S., T. D. POLLARD, and W. M. KUEHL. 1971. Isolation and characterization of myosin and two myosin fragments from human blood platelets. Proc. Nat. Acad. Sci. 68: 2703.

BAENZIGER, N. L., G. N. BRODIE, and P. W. MAJERUS. 1971. A thrombin-sensitive protein of human platelet membranes. Proc. Nat. Acad. Sci. 68: 240.

BETTEX-GALLAND, M. and E. F. LÜSCHER. 1965. Thrombosthenin, the contractile protein from blood platelets and its relation to other contractile proteins. In Advances in protein chemistry, ed. Anfinsen et al., vol. 20, p. 1. Academic Press, New York.

BOOYSE, F. M., T. P. HOVEKE, D. KISIELESKI, and M. E. RAFELSEN, JR. 1972. Mechanism and control of platelet-platelet interactions. Microvasc. Res. 4: 199.

COHEN, I., Z. BOHAK, A. DE VRIES, and E. KATCHALSKI. 1969. Thrombosthenin M, purification and interaction with thrombin. Europe J. Biochem. 10: 388.

COHEN, I. and C. COHEN. 1972. A tropomyosin-like protein from human platelets J. Mol. Biol. 68: 383.

ISHIKAWA, H., R. BISCHOFF, and H. HOLTZER. 1969. Formation of arrowhead complexes with heavy meromyosin in a variety of cell types. J. Cell. Biol. 43: 312.

JOHNSON, G. S., W. D. MORGAN, and I. PASTAN. 1972. Regulation of cell motility by cyclic AMP. Nature 235: 54.

LOWEY, S. and D. RISBY. 1971. Light chains from fast and slow muscle myosin. Nature 234: 81.

LÜSCHER, E. F. and M. BETTEX-GALLAND. 1971. Thrombosthenin, p. 225. In The circulating platelet, ed. Shirley A. Johnson. Academic Press, New York.

SARKER, S., F. A. SRETER, and J. GERGELY. 1971. Light chains of myosins from white, red, and cardiac muscles. Proc. Nat. Acad. Sci. 68: 946.

WEEDS, A. G. and B. POPE. 1971. Chemical studies on light chains from cardiac and skeletal muscle myosin. Nature 234: 85.

YANG, Y. and J. F. PERDUE. 1972. The isolation and characterization of an actin-like protein from cultured chick fibroblasts. Fed. Proc. 31: 4038 (Abstr.).

Physarum Myosin: Two New Properties

V. T. NACHMIAS

Department of Biology, Haverford College, Haverford, Pennsylvania 19041

Strong but circumstantial evidence implicates actomyosins in certain forms of cell motility, especially ameboid types of movement and in cytoplasmic streaming. Pollard (1972) has recently critically reviewed the field of ameboid movement from the viewpoint of molecular biology. Without citing the extensive work that has been done on other types of cells, we may say briefly that evidence for *both* actin-like and myosin-like proteins is being accumulated. Fine-structure studies show that thin or microfilaments occur widely in motile eukaryotic cells. In a number of cases, this class of filaments has been shown to react with heavy meromyosin from muscle myosin to form polarized complexes (Ishikawa et al., 1969; Pollard et al., 1970; Alléra et al., 1971). Actin-like filaments isolated from actively streaming *Physarum, Acanthamoeba*, or *A. proteus* react similarly (Nachmias et al., 1970; Pollard et al., 1970; Pollard, 1971). Actin-like proteins as defined by stringent criteria (Pollard, 1972) have been isolated from motile cells (Hatano and Oosawa, 1966; Weihing and Korn, 1969; Adelman and Taylor, 1969a).

Thick filaments are less widely encountered. Perhaps myosin-like proteins have undergone more changes during evolution, especially in the aggregating, or tail, portion. However characteristic 15–25 nm, 0.5 μm long, solid, tapered filaments have been observed in sections of the cytoplasm of several species of amebae (Nachmias, 1964; Wolpert et al., 1964; Bhowmick, 1967; Pollard and Ito, 1970; Holberton and Preston, 1970). A number of reports have appeared on the isolation of actomyosins or myosin-like proteins from amebae and other motile cells: *Dictyostelium discoideum* (Wooley, 1970), leukocytes (Shibata et al., 1972), and platelets (Adelstein et al., 1971), to cite only some of the sources.

For further work on the characterization and recombination of the components of streaming systems, *Physarum polycephalum* is excellent material. At one stage of its life cycle, the organism exists as a large multinucleate plasmodium which is easily cultured. The fluid part of the cytoplasm, contained in gelled veins or cords, streams at rates up to 1350 μ per sec, changing direction every 1 to 2 min (Kamiya, 1959). Loewy (1952) first demonstrated that high-salt alkaline extracts of *Physarum* behaved viscometrically like actomyosin from rabbit striated muscles. Recently purification of actin and myosin from *Physarum* has been made possible by the pioneering work of Hatano and coworkers (Hatano and Oosawa, 1966; Hatano and Tazawa, 1968; Hatano and Ohnuma, 1970), and by the work of Adelman and Taylor (1969a,b), using a different approach to the problem. Although certain discrepancies between their results have not been clarified, the main results are not in dispute and, in particular, these investigators are in agreement, or complement one another, on several major points in regard to the *myosin* component of *Physarum*. These points are summarized as follows:

Properties of *Physarum* Myosin

1. *Physarum* myosin combines with muscle actin to form rapidly sedimenting complexes (Adelman and Taylor, 1969b). These complexes take the form of polar interactions (Nachmias, 1972a).

2. The myosin has an $S_{20,w}$ value close to 6 (Adelman and Taylor, 1969b; Hatano and Ohnuma, 1970; Hatano and Takahashi, 1971).

3. The shape and length of the molecule is similar to muscle myosin as determined by metal shadowing (Hatano and Takahashi, 1971) and (length) by negative staining of myosin-enriched actomyosin (Nachmias and Ingram, 1970). Hatano and Takahashi find a mean length of 1210 $\pm$ 340 Å.

4. The myosin elutes close to muscle myosin on Agarose 4B gel chromatography (Adelman and Taylor, 1969b; Nachmias, 1972b). It is estimated to have a molecular weight close to 458,000 (Adelman and Taylor, 1969b).

5. The myosin is a calcium-activated, EDTA- and magnesium-inhibited ATPase which is sensitive to sulfhydryl reagents (Adelman and Taylor, 1969b; Hatano and Ohnuma, 1970). Activation of the MgATPase by actin has been studied only by Adelman and Taylor (1969b). With their preparation they found no activation.

6. The myosin is soluble in 0.03–0.05 M KCl (physiological ionic strength) but forms under these conditions small aggregates of S values 8–15 (Adelman and Taylor, 1969b; Hatano and Takahashi, 1971) whose nature is not known.

We add here:

7. The purified myosin in the absence of actin forms 0.45–0.5 μm long bipolar filaments in 1–10 mM calcium salts at physiological ionic strengths. Magnesium salts are somewhat less effective.

8. The major subunit weight of the myosin appears to be 250,000 daltons ±10% by SDS gel electrophoresis.

New Findings

Thick filaments. The solubility of the myosin would at first appear to argue against its ability to cross-link actin. However, unlike the myosin, the *actomyosin* from *Physarum* is insoluble at low ionic strengths, at least when prepared by the method of Hatano and Tazawa (1968) and Nachmias (1972a). The soluble actomyosin obtained by Adelman and Taylor (1969a) may be explained by the persistence of pyrophosphate from the initial extract. Pyrophosphate solubilizes not only muscle myosin (Richards et al., 1967), but also *Physarum* actomyosin (unpublished observations in our laboratory). This interpretation seems likely from point (1) above. Furthermore examination of *Physarum* myosin-enriched actomyosin by negative staining shows the existence of cross-links between actin filaments (Fig. 1). These findings (cf. also Nachmias, 1972a) are similar to those seen in muscle actomyosins at low ionic strengths (Ikemoto et al., 1968). This suggests that the small aggregates formed by myosin alone in physiological salt solutions may well represent tail aggregations of either tail-to-tail or head-to-tail type. We have not been able to demonstrate these. However we have been able to show that highly purified myosin will aggregate to form 0.45 μm long bipolar thick filaments in the presence of 1–10 mM calcium salts. Magnesium salts are somewhat less effective (Nachmias, 1972b). Figure 2 shows a typical result of the final purification step after preparing partially purified myosin by the method of Hatano and Ohnuma (1970). The peak fractions in 0.5 M KCl, devoid of actin by gel electrophoresis and electron microscopy, had calcium ATPase activities of 1 μM P_i/min/mg (Adelman and Taylor, 1969a). Samples of these fractions from several columns were either diluted or dialyzed with water or divalent salt solutions to bring the KCl concentration to 0.05 M. Turbidity begins to develop immediately when 5–10 mM divalent salts are added. Figures 3–5 are of negatively stained samples. The method has been described (Nachmias et al., 1970).

Figure 3 provides an example of the appearance of the clear solutions in 0.05 M KCl or ammonium acetate in the absence of divalent ions. Figures 4 and 5 show the type of aggregates formed in the presence of magnesium or calcium chloride, respectively. Figure 5c, d are examples of filaments formed from muscle myosin by rapid dilution, for comparison. From the length of 0.45 μm and from the estimated length of the *Physarum* myosin molecule, it is evident that both tail-to-tail and head-to-tail interactions occur in these *Physarum* filaments. The nature of the filaments therefore can be said to provide more than enough evidence that *Physarum* myosin is capable of cross-linking actin filaments at low ionic strength when the

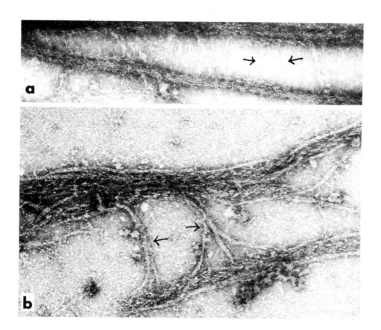

Figure 1. Myosin-enriched *Physarum* actomyosin dialyzed against: (*a*) 0.01 M ammonium acetate; ×80,000. (*b*) 0.01 M potassium chloride; ×100,000. In (*a*) note fine extensions between heavy cords (arrows). In (*b*) extensions are thicker and longer. No correlation between these particular types of association and the salt used has been shown, however. Figure 1b by Philip J. Krape.

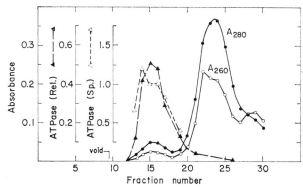

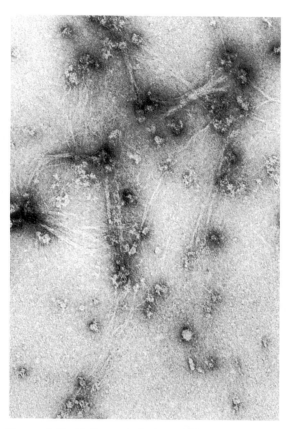

Figure 2. Purification of partially purified myosin by gel chromatography on Agarose 4B. Ten mg of partially purified *Physarum* myosin in 2 ml was applied to a Sepharose 4B column 1.8 × 30 cm preequilibrated and eluted with 0.5 M KCl, 0.05 M imidazole pH 7 and 0.1 mM dithiothreitol. Flow rate was 10 ml/hr. 2.2 ml fractions were collected and absorbance at 280 and 260 nm estimated in a Zeiss PMQ II spectrophotometer. Calcium-activated ATPase measurements were made using a semimicro version (Nachmias, 1972a) of the assay conditions of Adelman and Taylor (1969a). Reprinted by permission of the *Proceedings of the National Academy of Sciences*.

Figure 4. Purified myosin in 0.5 M KCl dialyzed against 9 volumes of 5 mM MgCl$_2$. 10 mM salt has a similar effect. Definite aggregation is present. ×80,000

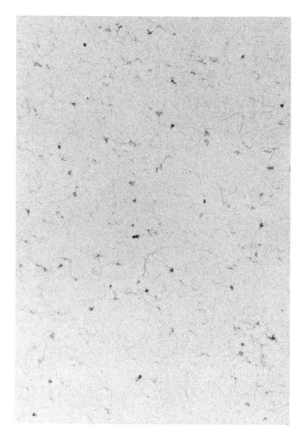

Figure 3. Purified myosin from a column as in Fig. 2 in 0.05 M ammonium acetate without addition of divalent salts. The appearance of myosin in 0.01–0.05 M KCl is indistinguishable from this. ×40,000.

myosin heads are attached to actin. The dimensions of these thick filaments are just those of the thick filaments found in fixed and sectioned ameboid cells (*vide supra*). Similar filaments have been found in *Physarum* in glycerinated microplasmodia undergoing spherule formation (Kessler, 1972). Their role remains to be determined.

The finding that divalent ions can produce aggregations of fibrous molecules is not new; the work of Cohen et al. (1970) on segments from the rod portion of myosin suggested their use here. Hinssen (1970) found thick filaments in *Physarum* actomyosin preparations treated for 18–24 hr with 8 mM ATP and 6 mM magnesium chloride. Possibly the magnesium salt was the essential component, although we found it less effective in inducing formation of thick filaments from purified myosin than calcium. Schoenberg reported (1969) that divalent ions or low pH were required for thick filaments to appear in diluted actomyosin extracted from smooth muscles. Low pH is not a sufficient condition for *Physarum* myosin to aggregate into thick filaments (Nachmias, 1972a).

Subunit weight. Myosins from striated muscles or from platelets when unfolded in sodium

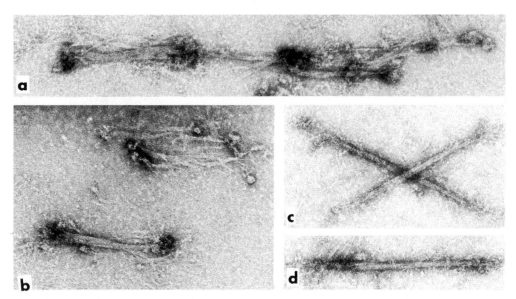

Figure 5. (*a, b*) Typical filaments made from purified *Physarum* myosin in 0.05 M KCl by dialysis against 9 volumes of 10 mM CaCl$_2$. The length of the completely formed filaments is 0.45 μm. ×80,000. (*c, d*) Filaments made from muscle myosin by dilution to 0.1 M KCl. ×80,000. Although these filaments were made by dilution and *a* and *b* were made by dialysis, these muscle filaments are longer than those made from *Physarum* myosin.

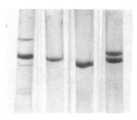

Figure 6. Photograph of SDS gel electrophoresis in 5% acrylamide. *Left to right:* column input, peak fraction from column, muscle myosin, and peak fraction and myosin run together.

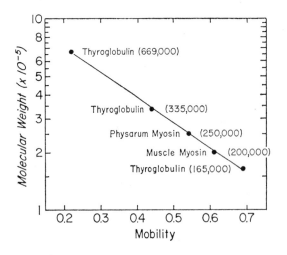

Figure 7. Estimate of the subunit weight of *Physarum* myosin on SDS gels.

dodecyl sulfate (SDS) or urea and a reducing agent and then electrophoresed in the presence of SDS (Reynolds and Tanford, 1970) have been found to yield polypeptides of mol wt 210,000 ± 5% (Weber and Osborne, 1969; Adelstein et al., 1971). We have found that the subunit weight of *Physarum* myosin under these conditions appears to be 250,000 ± 10%. We consistently observed band splitting on both 5% and 10% gels when reduced *Physarum* myosin was coelectrophoresed with reduced rabbit muscle myosin (Fig. 6). In different experiments β-mercaptoethanol or dithiothreitol were used. No double bands were observed when *Physarum* myosin was run alone. Estimates of the molecular weight were made using thyroglobulin as a marker and assuming a major subunit of weight 335,000, a minor dimer of 670,000 (Edelhoch and deCrumbugghe, 1966), and a subunit of 168,000 on sulfonation (Figure 7). Our estimate will have to be confirmed, preferably by unfolding in guanidine hydrochloride. However there is no obvious reason for artifact in our observations, which have been repeated on a number of preparations. It is evident that the estimated molecular weight of the native molecule can be accounted for by two heavy subunits without the contribution of any light chains. Nevertheless on 12 or 15% gels we obtain one persistent small molecular weight component in our myosin preparations. Its weight is estimated as 13,000–15,000 by comparison with cytochrome *c* or hemoglobin. We cannot rule out the possibility of a proteolytic fragment.

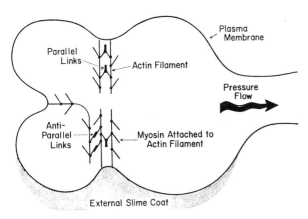

Figure 8. Schematic diagram of a possible mechanism for a contractile phase in the production of cytoplasmic streaming. Although myosin is depicted here as dimers, oligomers could be substituted. The scheme implies a) that the plasmalemma exerts an orientating effect on actin filaments and b) that only antiparallel myosin links between actins lead to relative movement of these filaments.

Summary

We have shown that the actomyosin from an organism that moves itself about and exhibits shuttle streaming, rather than moving other tissues as muscles do, is nevertheless a cross-linked structure in the sense that myosin molecules appear to attach both to other myosin molecules and to actin filaments. In keeping with this conclusion is the finding that purified *Physarum* myosin can form 0.45 μm long thick bipolar filaments in the presence of divalent salts. No longer filaments were observed. From SDS gel electrophoresis the myosin appears to be constituted of a double peptide chain of a molecular weight some 20% greater than that of muscle myosin subunit chains. There may be a small molecular weight component. On the basis of these findings, we offer with temerity one possible model for a means of producing cytoplasmic streaming in a pressure flow system (Kamiya, 1959) like *Physarum* (Fig. 8). Obviously it is grossly oversimplified and especially does not take account of any gel-sol interchange nor the intriguing effects of myosin on the flexibility of actin (Fujime and Ishiwata, 1971). Nevertheless it may be that a *contractile* step (Nagai and Kamiya, 1966) can be separated from a *flow* step in cytoplasmic streaming of this type.

Acknowledgments

This investigation was supported by Grant #AM-15132 from the National Institutes of Health. I am indebted to Pharmacia, Inc., for providing purified thyroglobulin. Thanks go to Mr. P. J. Krape and Mr. R. Ingram for assistance in several aspects of the work.

References

ADELMAN, M. R. and E. W. TAYLOR. 1969a. Isolation of an actomyosin-like protein complex from slime mold plasmodium and the separation of the complex into actin- and myosin-like fractions. *Biochemistry* **8**: 4964.

———. 1969b. Further purification and characterization of slime mold myosin and slime mold actin. *Biochemistry* **8**: 4976.

ADELSTEIN, R. S., T. D. POLLARD, and W. M. KUEHL. 1971. Isolation and characterization of myosin and two myosin fragments from human blood platelets. *Proc. Nat. Acad. Sci.* **68**: 2703.

ALLÉRA, A., R. BECK, and K. E. WOHLFARTH-BOTTERMANN. 1971. Identification of the plasma filaments in *Physarum polycephalum* as F-actin by *in situ* binding of heavy meromyosin. *Cytobiologie* **4**: 437.

BHOWMICK, D. K. 1967. Electron microscopy of *Trichamoeba villosa* and ameboid movement. *Expt. Cell Res.* **45**: 570.

COHEN, C., S. LOWEY, R. G. HARRISON, J. KENDRICK-JONES, and A. G. SZENT-GYÖRGY. 1970. Segments from myosin rods. *J. Mol. Biol.* **47**: 605.

EDELHOCH, H. and B. deCRUMBUGGHE. 1966. The properties of thyroglobulin. XIII. The structure of reduced alkylated thyroglobulin. *J. Biol. Chem.* **241**: 4357.

FUJIME, S. and S. ISHIWATA. 1971. Dynamic study of F-actin by quasielastic scattering of laser light. *J. Mol. Biol.* **62**: 251.

HATANO, S. and J. OHNUMA. 1970. Purification and characterization of myosin A from the myxomycete plasmodium. *Biochim. Biophys. Acta* **205**: 110.

HATANO, S. and F. OOSAWA. 1966. Extraction of an actin-like protein from the plasmodium of a myxomycete and its interaction with myosin A from rabbit striated muscle. *J. Cell. Physiol.* **68**: 197.

HATANO, S. and K. TAKAHASHI. 1971. Structure of myosin A from the myxomycete plasmodium and its aggregation at low salt concentrations. *J. Mechanochem. Cell Motil.* **1**: 7.

HATANO, S. and M. TAZAWA. 1968. Isolation, purification and characterization of myosin B from myxomycete plasmodium. *Biochim. Biophys. Acta* **154**: 507.

HINSSEN, J. 1970. Synthetische myosin filamente von Schleimpilz-Plasmodien. *Cytobiologie* **2**: 326.

HOLBERTON, D. V. and T. M. PRESTON. 1970. Arrays of thick filaments in ATP-activated *Amoeba* model cells. *Expt. Cell Res.* **62**: 473.

IKEMOTO, N., S. KITAGAWA, A. NAKAMURA, and J. GERGELY. 1968. Electron microscopic investigations of actomyosin as a function of ionic strength. *J. Cell Biol.* **39**: 620.

ISHIKAWA, H., R. BISCHOFF, and H. HOLTZER. 1969. Formation of arrowhead complexes with heavy meromyosin in a variety of cell types. *J. Cell Biol.* **43**: 312.

KAMIYA, N. 1959. Protoplasmic streaming. *Protoplasmatologia* **8**: 1.

KESSLER, D. 1972. On the location of myosin in the myxomycete *Physarum polycephalum* and its possible function in cytoplasmic streaming. *J. Mechanochem. Cell Motil.* In Press.

LOEWY, A. G. 1952. An actomyosin-like substance from the plasmodium of a myxomycete. *J. Cell. Comp. Physiol.* **40**: 127.

NACHMIAS, V. T. 1964. Fibrillar structures in the cytoplasm of *Chaos chaos*. *J. Cell Biol.* **23**: 183.

———. 1972a. Electron microscope observations on myosin from *Physarum polycephalum*. *J. Cell Biol.* **52**: 648.

NACHMIAS, V. T. 1972b. Filament formation by purified *Physarum* myosin. *Proc. Nat. Acad. Sci.* **69.** In press.

NACHMIAS, V. T., H. E. HUXLEY, and D. KESSLER. 1970. Electron microscope observations on actomyosin and actin preparations from *Physarum polycephalum* and on their interactions with heavy meromyosin subfragment I from muscle myosin. *J. Mol. Biol.* **50:** 83.

NACHMIAS, V. T. and W. C. INGRAM. 1970. Actomyosin from *Physarum polycephalum:* Electron microscopy of myosin-enriched preparations. *Science* **170:** 143.

NAGAI, R. and N. KAMIYA. 1966. Electron microscopic studies on fibrillar structures in the plasmodium. *Proc. Jap. Acad.* **42:** 934.

POLLARD, T. D. 1971. Filaments of *Amoeba proteus.* II. Binding of heavy meromyosin by thin filaments in motile cytoplasmic extracts. *J. Cell Biol.* **48:** 216.

———. 1972. Progress in understanding ameboid movement at the molecular level. In *Biology of the Amoeba,* ed. K. Jeon. In press.

POLLARD, T. D. and S. ITO. 1970. Cytoplasmic filaments of *Amoeba proteus.* I. The role of filaments in consistency changes and movement. *J. Cell Biol.* **46:** 267.

POLLARD, T. D., E. SHELTON, R. R. WEIHING, and E. D. KORN. 1970. Ultrastructural characterization of F-actin isolated from *Acanthamoeba castellanii* and identification of cytoplasmic filaments as F-actin by reaction with rabbit heavy meromyosin. *J. Mol. Biol.* **50:** 91.

REYNOLDS, J. A. and C. TANFORD. 1970. The gross conformation of protein-sodium dodecyl sulfate complexes. *J. Biol. Chem.* **245:** 5161.

RICHARDS, E. G., C.-S. CHUNG, D. B. MENZEL, and H. S. OLCOTT. 1967. Chromatography of myosin on diethyl amino ethyl-sephadex A-50. *Biochemistry* **6:** 528.

SHIBATA, N., N. TATSUMI, K. TANAKA, Y. OKAMURA, and N. SENDA. 1972. A contractile protein possessing Ca^{2+} sensitivity (natural actomyosin) from leucocytes. *Biochim. Biophys. Acta* **256:** 565.

SHOENBERG, C. F. 1969. An electron microscope study of the influence of divalent ions on myosin filament formation in chicken gizzard extracts and homogenates. *Tissue and Cell* **1:** 83.

WEBER, K. and M. OSBORN. 1969. The reliability of molecular weight determinations by dodecyl sulfate-polyacrylamide gel electrophoresis. *J. Biol. Chem.* **244:** 4406.

WEIHING, R. and E. D. KORN. 1969. Ameba actin: The presence of 3-methyl histidine. *Biochim. Biophys. Res. Comm.* **35:** 906.

WOLPERT, L., C. M. THOMPSON, and C. H. O'NEILL. 1964. Studies on the isolated membrane and cytoplasm of *Amoeba proteus* in relation to ameboid movement. In *Primitive motile systems in cell biology,* ed. R. D. Allen and N. Kamiya. Academic Press, New York.

WOOLLEY, D. E. 1970. Extraction of an actomyosin-like protein from amoebae of *Dictyostelium discoideum.* *J. Cell. Physiol.* **76:** 185.

Heat, Work, and Phosphocreatine Splitting during Muscular Contraction

CLAUDE GILBERT, K. M. KRETZSCHMAR, AND D. R. WILKIE

Department of Physiology, University College, London WC1E 6BT, England

Our previous work on the moment-to-moment chemical changes during an isometric tetanus has shown that, during the first couple of seconds (frog muscle, 0°C), a substantial amount of heat appears that cannot be accounted for by the expected chemical reactions—the hydrolysis of phosphocreatine (PCr) and ATP (Gilbert et al., 1971). During isometric contraction only a relatively small amount of the total energy (about 10 % in the present experiments) ever appears as mechanical work in stretching the compliance of the apparatus, the mechanical connections, and the muscle itself. The observed breakdown of phosphocreatine would be sufficient to account for this relatively small amount of work so the unknown energy-yielding process may merely produce heat. If this is the case, the unknown process may be relatively remote from the actual energy transformation in the cross-bridges. If, on the other hand, mechanical work is in fact not "paid for" by concurrent phosphocreatine splitting, so that work also comes from an unidentified chemical source, then this source must be sought close to the cross-bridges. The present experiments were therefore undertaken in order to find out whether or not performance of work was accompanied by concomitant splitting of phosphocreatine.

There is already a good deal of evidence from the work of Davies and his collaborators (see especially Kushmerick and Davies, 1969) to show that performance of work does lead to extra chemical breakdown. However these experiments differed from those in our laboratory in several important respects. Quite apart from the fact that they were on a different species of frog (*R. pipiens*) poisoned with fluorodinitrobenzene (FDNB) and nitrogen, whereas ours are on unpoisoned *R. temporaria*, Kushmerick et al. find such a large chemical breakdown during isometric contraction that there is some doubt whether or not their results are compatible with an "energy gap" such as is described above (but see p. 186 of Gilbert et al., 1971). Of course this question cannot be resolved satisfactorily without concomitant heat measurements, which are not available.

Later experiments by Gilbert and Kushmerick (1970) used conditions similar to those of Gilbert et al. (1971) save that the muscles were poisoned with iodoacetate. Experiments in which the stimulated muscle had shortened for 1 sec against an ergometer, thus performing a large amount of external mechanical work, showed significantly greater breakdown of PCr than did an isometric contraction of the same duration (1.1 sec). The main uncertainty in interpreting these experiments was that at the end of the contraction the total energy output of the working muscle was also much greater than that of the isometric control because of the production of shortening heat and of external work. Thus it was not clear whether the extra chemical change was linked specifically to work or whether it merely reflected the extra output of energy. A way of resolving this difficulty was suggested by R. C. Woledge. It involves comparing two patterns of contraction, one against the ergometer (Erg) and the other isometric (Im); the duration of the latter is to be adjusted so that the total output of energy, heat + work, is similar in the two cases. In the Erg contraction a relatively large fraction of this total would be work.

Experimental Method

This was similar to that described by Gilbert et al. (1971) save for the following modifications.

Myothermic. Heat production was measured on a thin thermopile with rapid response (constructed by K. M. K.). It is of Hill-Downing type with a maximum of 36 flattened chromel-constantan junctions, each of which is brought out to a separate external connection so that the active region of the pile can be selected according to the job in hand.

Calibration. In experiments where heat is to be compared with work and with chemical change, the question of absolute calibration of the heat measurements is obviously of paramount importance; defective procedures have in the past led to grave errors (Hill and Woledge, 1962). One fundamental difficulty is that a thermopile essentially measures rise of temperature; to convert this into heat produced one must multiply by the combined thermal capacity of the part of the thermopile on which the muscle is lying and of the muscle itself together with whatever Ringer's solution happens to be present. Accurate methods are available for determining the temperature calibration of the thermopile (see Hill, 1965); and though the determination of the thermal capacity of the pile is less accurate, this is of relatively little importance because with thin thermopiles it is a small quantity in comparison with the much larger thermal capacity of a pair of frog sartorii. However myothermic experiments are today (fortunately) not confined to frog sartorii; they have been performed, for example, on very small mammalian papillary muscles weighing only 3–10 mg (Gibbs et al., 1967). With such small muscles uncertainty about the thermal capacity of the thermopile becomes a serious matter.

There is in any case serious uncertainty in determining the thermal capacity of the muscle since the only method generally available is to remove it at the end of the experiment and weigh it, with the hope that the muscle, and the Ringer's fluid carried with it, has not altered since the heat production was recorded. This weight could then be multiplied by an assumed value for the specific heat to obtain the thermal capacity.

In the integrating thermopile (Wilkie, 1968) the thermal capacity of muscles + pile was determined directly during the course of each experiment by heating the central area of the pile at a known rate, W (watts), by passing current through an inbuilt resistance wire. From the resulting steady temperature rise, $(\Delta\theta)$, the thermal conductance away from the center of the pile, $W/\Delta\theta$ (watts · deg⁻¹), could easily be calculated. When the heating current is switched off, the muscles cool exponentially with time constant T(sec) which can be measured. Since $T = S/(W/\Delta\theta)$ the thermal capacity of muscles + pile, S joule · deg⁻¹, can be calculated; the value obtained applies to the actual conditions of the experiment.

Peltier method. Hill-Downing type thermopiles do not have a calibrated resistance wire, nor could they be equipped with one without losing their virtues of thinness and smoothness. However precisely the same effect can be achieved by passing current through the thermojunctions from an external battery. In this case a known rate of heating or cooling is produced at the junctions by the Peltier effect and can be made use of for calibration exactly as outlined above. In the range of powers (a few mW) encountered in myothermic work, the Peltier effect is relatively very large and completely swamps the Joule heating in the wires. When the relevant equations are written (Kretz-

schmar and Wilkie, 1972), an unexpected bonus is found. The thermoelectric property of the junction appears twice over: once as the Peltier coefficient Π volts in determining how much heat is moved per coulomb during calibration; and again as the Seebeck coefficient α volts · deg⁻¹ in determining how much current flows through the galvanometer per unit temperature difference during measurement. The thermojunction behaves like a reversible heat engine from which Kelvin deduced $\Pi = \alpha\theta$ where θ is the absolute temperature. As a consequence it turns out that absolute measurements can be made without knowing the thermoelectric properties of the junctions or their number. This could be especially useful for thermopiles made by the convenient "plating" process (Gibbs et al., 1967) for we have shown that the same equations apply to junctions whose thermoelectric power has been reduced by partial internal shunting.

The heat measurements in this report have all been based on the Peltier method; a comparison will ultimately be made with the weighing method of calibration to see if there are any systematic differences. Incidentally this new method of calibration renders the integrating thermopile (Wilkie, 1968) obsolete since any thermopile can be rapidly and temporarily converted into an integrating pile simply by attaching suitable strips of silver to it with grease.

The heat measurements were corrected for heat flow, heat loss, thermoelastic effect, and stimulus energy. All these corrections were small.

Mechanical measurements. These were made using our usual transducers (Jewell et al., 1967), the isotonic lever being interposed between the Levin-Wyman ergometer and the muscle so as to record the changes of length and tension. The muscles were set up under a force of 20 mN (2 g wt), which stretched them slightly beyond body-length but did not store any appreciable quantity of energy in the parallel elastic element. This was the starting length for the ergometer (Erg) experiments: the isometric (Im) experiments were made with the muscle 5 mm shorter. The shorter length corresponds approximately to $l_0 - 1$ mm. The mechanical arrangement differed slightly in the four types of experiment (thermopile Im and Erg, hammer Im and Erg; see below) so careful measurements were made of the compliance of the apparatus in each case. The compliance of the wet thread connection was measured separately by imposing tension fluctuations with a time-course similar to those produced by contracting muscle and measuring the corresponding changes in length. In this way errors due to hysteresis were minimized. The internal compliance of the muscle itself was taken from Jewell and Wilkie, 1958, Fig. 8. Curves E, F, and G are the most accurate and they almost superimpose if matched at the high-stress end. The standard curve thus obtained was matched to each individual muscle according to its length and weight.

The force-length curve of each compliant element was fitted by a suitable polynomial, which could readily be integrated to obtain the elastic energy. The appropriate elements corresponding to each type of experiment were then combined, taking due account of whether one muscle or two had been employed.

Chemical measurements. The only chemical measurements reported here were of phosphocreatine (PCr) and total creatine (Cr). This was estimated by our usual method (see Dydyńska and Wilkie, 1966) save that it had been automated using a Technicon Autoanalyser. All experiments were carried out on *R. temporaria* at 0°C in oxygen.

Experimental design. The first problem was to decide on a suitable pattern of contraction for the Isometric (Im) and Ergometer (Erg) contractions. The latter is fixed because muscles have an optimal speed of shortening of about 10 mm/sec; they work most effectively over a length range of about 10 mm and they take about 0.1 sec to achieve the appropriate tension (Fig. 1, middle curve, broken line). Accordingly they were stimulated at 15/sec, the ergometer was released at 0.1 sec, and the muscles were frozen by the hammer apparatus (Kretzschmar and Wilkie, 1969) at 1.1 sec. The corresponding heat production is shown by the upper interrupted curve in Fig. 1. At the instant of freezing, the Erg muscle has produced extra energy compared with the Im muscle in the form of shortening heat and of mechanical work. The former can be seen directly in the figure; the latter must be calculated from the middle and lower curves.

A previous attempt at this experiment (Oct. to Dec. 1971) had shown that after 1.7 sec the Im muscle had produced the same amount of (heat + work) as the Erg muscle, so this instant was chosen for freezing it.

A batch of about 60 frogs were kept alive in a refrigerator in small trays; none died from natural causes. Eleven frogs were used for measurements of heat and work, both sartorii being mounted together on the thermopile. At the end of each experiment the muscles were weighed and analyzed for creatine. Thirty-six other pairs of sartorii were used in the hammer apparatus for chemical determinations in experiments of three types: Im versus unstimulated control (C); Erg versus C; Im versus Erg (both muscles stimulated); 12 experiments of each type. The four types of

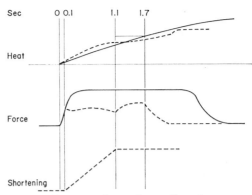

Figure 1. Direct tracings of recordings from a typical experiment. Ergometer (Erg) and isometric (Im) tracings have been superimposed. The Erg tracing is identified in each case by interruptions. The middle tracings show changes of tension. In the Erg experiments the tension rises isometrically for 0.1 sec until the start of the controlled release shown in the bottom tracing: the muscle is frozen as soon as shortening ends, at 1.1 sec. The upper tracings show the corresponding (uncorrected) heat records and illustrate the need to freeze the Im muscle later, at 1.7 sec, to compensate for the difference in the external work.

experiment were distributed systematically over the experimental period (March–May 1972) so as to avoid bias.

Statistical treatment. The aim was to derive from the chemical results the best estimates of I, the PCr breakdown (ΔPCr) in an isometric contraction, and most especially of E, the *extra* ΔPCr (if any) in the working contraction. Clearly E can be estimated in two independent ways, as the difference between the mean results for the Im vs. C and the Erg vs. C experiments, or as the mean of the Im vs. Erg experiments. The latter would presumably be more precise because some of the frog-to-frog variability is avoided. What was really needed was a single way of combining the results so as to obtain best estimates from all of them together. We could not find a standard statistical test that exactly fitted our experimental plan so we devised one from first principles. Its basis is that each of our measurements of ΔPCr consists of a true part (which we should like to find out) and an error. Thus if the true ΔPCr in an isometric contraction is I and the true *extra* breakdown in the working contraction is E, each individual IC result can be written $(I + e)$, each EC result is $(I + E + e)$ and each IE result is $(E + e)$, the errors, e, being different in each case. By suitable algebraic manipulation I and E can be found such that $\Sigma(e^2)$ is minimized. Using the notation Δ_{IC}, Δ_{EC}, Δ_{IE} for the individual results and N for the number of each type (12 in this case)

$$I = \frac{2\Sigma\Delta_{IC} + \Sigma\Delta_{EC} - \Sigma\Delta_{EI}}{3N}$$

and

$$E = \frac{2\Sigma\Delta_{EI} + \Sigma\Delta_{EC} - \Sigma\Delta_{IC}}{3N}$$

that is, the more direct estimate is given double the weight of the less direct one. Similarly the best estimate of

$$(I + E) = \frac{2\Sigma\Delta_{EC} + \Sigma\Delta_{IC} + \Sigma\Delta_{EI}}{3N}$$

Results and Discussion

Our results concerning PCr splitting are still being processed statistically and the further chemical analyses for inorganic phosphate and for ATP have not yet been performed. Nevertheless some firm conclusions can be drawn which are summarized in Fig. 2.

The first conclusion is that we did not do the experiment exactly right, for as comparison of the first two columns shows, the total output of energy was 10% greater in the Erg experiments; this difference is significant at 0.05 level. Evidently we have been victims once again of the batch-to-batch variations that dog these experiments and we should have used a somewhat longer isometric tetanus.

Fortunately the fault, though irritating, is not fatal. A much larger fraction of the total energy (35%) appears as mechanical work in the Erg experiments than in the Im ones (10%). This difference is highly significant and has a marked effect on the chemical changes, which are shown in the right-hand part of Fig. 2. The actual experimental results in the three types of experiment are shown by the middle columns and the "best estimates" obtained by our new statistical treatment are shown on the right. In the chemical experiments the weights of the muscles are not known, but for easy comparison with the energy measurements, all the chemical results have been scaled into energy units per gram by assuming (as in Gilbert et al., 1971) that the muscles contained 25 mM total creatine and that the hydrolysis of 1 mole of PCr results in an enthalpy change of −46 kJ (−11 kcal).

Isometric contractions. Comparison of the energy output with chemical change provides further confirmation for the conclusion of Gilbert et al. (1971). At this moment in a tetanus there is a large "energy gap" because the observed splitting of PCr can account for only about 35% of the energy output. If, following Woledge (this volume), a smaller value than −46 kJ/mole is adopted for the molar enthalpy change, the discrepancy becomes even more glaring.

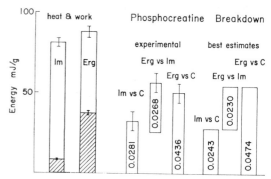

Figure 2. Heat and work and phosphocreatine splitting during isometric (Im) and ergometer (Erg) contractions. Work (shaded area) and heat are shown in mJ/g. The chemical measurements are actually of Δ(PCr/Cr) and the mean values of this number are given in the boxes. However for purposes of comparison they have been plotted in energy units by making the assumption that the total creatine content is 25 mM and that each mole of PCr split yields 46 kJ (11 kcal). The three right-hand blocks indicate the best estimates obtained from the pooled chemical results by the statistical technique outlined in the text. The error bars indicate ±1 SE in each case.

Working contractions. Splitting of PCr has increased by 95.5% compared with the isometric case, whereas the total energy output was greater by only 10%. It therefore appears that splitting of PCr *is* specifically linked to the performance of work and that the question posed in the introduction has been answered.

Beyond these main conclusions there are three points which upon further calculation, albeit speculative, seem justified. First, it appears that in the working contraction, as in the isometric one, there is a statistically significant gap between the physical energy observed and that available from the concurrent splitting of PCr. In the working contraction the gap appears smaller, 34 mJ/g instead of 54 mJ/g, but further information would be needed to establish whether this difference is genuine and if so, whether it arises merely from the fact that the working contraction is of shorter duration. Certainly it appears from Figs. 1 and 2 of our earlier paper (Gilbert et al., 1971) that the energy gap increases rapidly over the first couple of seconds of activity, so this may well be the correct explanation.

Second, it is interesting to explore the possibility that *all* the extra PCr split in the working contraction may be associated with the extra work performed. This is probably a conservative assumption because relatively less PCr would be used for other purposes since the working contractions were of shorter duration. The extra 22.3 mJ/g of work is associated with an extra 0.575 μmoles/g of PCr split. The ratio 39 kJ/mole (9.3 kcal/mole) is probably quite a large fraction of the likely free energy change for this reaction. This makes us wish even more strongly than usual that dependable and applicable physicochemical measurements of this quantity were available so that we could estimate the genuine thermodynamic efficiency (Wilkie, 1960).

The third speculation is an extension of the second to the situation during the isometric contraction. Assuming that the elastic work in this case (8.5 mJ/g) is also paid for at 39 kJ/mole, the work would account for 0.22 μmoles/g out of the total of 0.675 μmoles/g of PCr split—almost one-third, and very little heat production would be associated with this 0.22 μmoles/g.

Since we have evidence (Gilbert et al., 1971) that in the early stages of contraction about 0.1 μmole/g of ADP are rephosphorylated, presumably at the expense of PCr via the Lohmann reaction, so a further 0.1 μmoles/g of PCr split does not lead to heat production. Indeed the Lohmann reaction absorbs heat when it proceeds in this direction (see Woledge, this volume). Regardless of details the consequence must be that less than half the PCr split can be giving out its full heat of reaction, so the gap between heat produced and identified chemical sources for it yawns even wider.

Until this heat is accounted for it is hard to feel much confidence in theories of contraction that neglect its existence. The explanation may well turn out to be something remote from the contractile system, such as the unbinding of PCr from the sites where it is segregated, as suggested by Woledge; but equally well, it may not. In the meantime it would help considerably if more calorimetric determinations were made on reaction systems that might be involved. A start has been made on the calcium-pumping system (Gibbs and Seraydarian, 1972); it presents several difficulties but they do not seem to be insuperable.

Accepting that there are gaps in our knowledge, it seems to us that the results from several laboratories working on muscle energetics already impose serious constraints on the freedom to speculate about the mode of action of cross-bridges. The general assumption seems to be that one ATP molecule is hydrolyzed every time the bridge makes a complete cycle, with the possible exception of the first cycle after activation when the bridge may be already "charged up." If this is so, the net rate of splitting of ATP or PCr should be given by the rate of cycling multiplied by the fraction of bridges active. The energetic evidence then dictates that this product must be high only when the muscle is shortening at medium speed and performing a large amount of mechanical work. When shortening freely and when exerting isometric force, this product must be much lower.

Quite possibly in the former case a few bridges are cycling rapidly while in the latter many are cycling slowly. When the active muscle is being stretched, the turnover rate must be lower still, despite the fact that a large force is being exerted, as shown in the succeeding paper by Curtin and Davies.

Leaving all speculation aside, one conclusion that arises directly from our experimental results is that the rate of PCr splitting at any instant is generally *not* proportional to the rate of production of (heat + work) at that instant. The former can therefore not be inferred directly from the latter, which creates difficulties for those theories that had made an altogether reasonable assumption to the contrary. An incomplete list of theories that may need adjustment in consequence are those of A. F. Huxley (1957), Wilkie and Woledge (1967), objecting to Caplan (1966; Caplan's theory did not make this assumption), Bornhorst and Minardi (1970), and Chaplain and Frommelt (1971).

Acknowledgments

Our thanks are due to the Medical Research Council for continued material support and to the Grass Foundation for providing one of us with time to think and write.

References

Bornhorst, W. J. and J. E. Minardi. 1970. A phenomenological theory of muscular contraction. I. Rate equations at a given length based on irreversible thermodynamics. *Biophys. J.* **10**: 137.

Caplan, S. R. 1966. A characteristic of self-regulated linear energy converters. The Hill force-velocity relation for muscle. *J. Theoret. Biol.* **11**: 63.

Chaplain, R. A. and B. Frommelt. 1971. A mechanochemical model for muscular contraction. I. The rate of energy liberation at steady state velocities of shortening and lengthening. *J. Mechanochem. Cell Motil.* **1**: 41.

Dydyńska, M. and D. R. Wilkie. 1966. The chemical and energetic properties of muscles poisoned with FDNB. *J. Physiol.* **184**: 751.

Gibbs, C. L. and K. Seraydarian. 1972. Microcalorimetry of isolated skeletal muscle sarcoplasmic reticulum. *Proc. Aust. Physiol. Pharmacol. Soc.* **3**: 38.

Gibbs, C. L., N. V. Ricchiuti, and W. F. H. M. Mommaerts. 1967. Energetics of cardiac contractions. *J. Physiol.* **191**: 25.

Gilbert, C. and M. J. Kushmerick. 1970. Energy balance during working contractions of frog muscle. *J. Physiol.* **210**: 146.

Gilbert. C., K. M. Kretzschmar, D. R. Wilkie, and R. C. Woledge. 1971. Chemical change and energy output during muscular contraction. *J. Physiol.* **218**: 163.

Hill, A. V. 1965. *Trails and trials in physiology.* Arnold London.

HILL, A. V. and R. C. WOLEDGE. 1962. An examination of absolute values in myothermic measurements. *J. Physiol.* **162**: 311.

HUXLEY, A. F. 1957. Muscle structure and theories of contraction. *Prog. Biophys. Chem.* **7**: 255.

JEWELL, B. R. and D. R. WILKIE. 1950. An analysis of the mechanical components in frog's striated muscle. *J. Physiol.* **143**: 515.

JEWELL, B. R., K. M. KRETZSCHMAR, and R. C. WOLEDGE. 1967. Length and tension transducers. *J. Physiol.* **191**: 10.

KRETZSCHMAR, K. M. and D. R. WILKIE. 1969. A new approach to freezing tissues rapidly. *J. Physiol.* **202**: 66P.

KRETZSCHMAR, K. M. and D. R. WILKIE. 1972. A new method for absolute heat measurement, utilizing the Peltier effect. *J. Physiol.* **224**: 18P.

KUSHMERICK, M. J. and R. E. DAVIES. 1969. The chemical energetics of muscle contraction. II. The chemistry, efficiency and power of maximally working sartorius muscles. *Proc. Roy. Soc. (London)* B **174**: 315.

WILKIE, D. R. 1960. Thermodynamics and the interpretation of biological heat measurements. *Prog. Biophys.* **10**: 260.

————. 1968. Heat, work and phosphorylcreatine breakdown in muscle. *J. Physiol.* **195**: 157.

WILKIE, D. R. and R. C. WOLEDGE. 1967. The application of irreversible thermodynamics to muscular contraction. Comments on a recent theory by S. R. Caplan. *Proc. Roy. Soc. (London)* B **169**: 17.

Chemical and Mechanical Changes during Stretching of Activated Frog Skeletal Muscle

Nancy A. Curtin*

Graduate Group on Molecular Biology, University of Pennsylvania 19104

R. E. Davies

Department of Animal Biology, School of Veterinary Medicine and Graduate Group on Molecular Biology, University of Pennsylvania 19104

The type of contraction that is performed by an active muscle depends on the mechanical conditions imposed on it; it can develop tension if it is held at a constant length, which is the isometric condition, or it can shorten or be stretched. These conditions are all "physiological" in that all of them are involved in the movements of the body. Stretching of active muscle, which we have studied, is a prominent feature of movements such as climbing down stairs or lowering an object with the arms.

We were interested in the mechanical and chemical events that occurred during movement in the "stretching" direction. In some recent studies of the energetics of the activation process (Homsher et al., 1972; Smith, 1972) muscles were prestretched to the length at which the thick and thin filaments no longer overlap with each other. Unlike those experiments, ours were concerned with the processes that occur *during* stretching of a stimulated muscle. In our chemical experiments the muscles were never stretched beyond 100% l_0, which we measure as the distance from tendon to tendon of the resting muscle in situ. In the tension-velocity experiments the muscles were stretched to 110% l_0; but even at 110% l_0 the passive tension in resting muscle was negligible.

Several important studies of the properties of muscle during stretching have already been done. A. V. Hill and his colleagues (Abbott et al., 1951; Abbott and Aubert, 1951; Hill and Howarth, 1959) measured the heat released by isolated muscles as they were stretched. They found that the net energy output (that is, the heat output minus the absolute value of the work done on the muscle during stretching) was lower than the energy output during isometric contraction; it could be close to zero under some conditions of stretching. One possible interpretation of these results was that the energy from the work done on the muscle was used to reverse chemical reactions in the muscle.

This idea was tested by chemical measurements, but a net increase in the ATP or phosphorylcreatine content of stretched muscle was never found (Aubert and Maréchal, 1963; Infante et al., 1964c; Maréchal, 1964a,b; Maréchal and Beckers-Bleukx, 1965; Wilkie, 1968; Curtin et al., 1969, 1970; Butler et al., 1972). Recent experiments with ^{32}P-labeled phosphate (Gillis and Maréchal, 1969, 1971a,b; Mannherz, 1970; Ulbrich and Rüegg, 1971) and ^{18}O-labeled water (Maréchal et al., 1971) have shown that ATP exchange reactions do occur during stretching, but the reaction also occurred during all the other types of contractions tested—isometric, shortening, unloaded shortening—in which work was *not* being done on the muscle. So all of the chemical experiments show that there is no evidence that ATP is formed at the expense of energy received by the muscle as work done on it during stretching. A more likely explanation for the low energy output during stretching is that the work done on the muscle appears as heat and that stretching suppresses the heat-releasing chemical reactions (Abbott et al., 1951; Wilkie, 1960).

However, even with the question of the reversal of the ATP reaction settled, other interesting questions remained. For example, muscles are able to develop large tensions during stretching (Gasser and Hill, 1924; Katz, 1939; Abbott and Aubert, 1952; Aubert, 1956; Joyce et al., 1969; Joyce and Rack, 1969; Sugi, 1969, 1971; Chaplain, 1972) even though they seem to be in a state of low metabolic activity. In addition, one of the experiments by Aubert and Maréchal (1963) showed that more chemical breakdown, rather than less, occurred during stretching than during isometric contraction.

Our experiments were designed to determine the relationships between the tension produced and the amount of chemical change and the velocity of stretching.

* Present address: Department of Physiology, University College London, Gower Street, London WC1.

Results

For all of our experiments we used the sartorius muscle of the frog, *Rana pipiens*. The average l_0 was 3.00 cm; P_0, the maximum isometric tension at $100\% l_0$, was about 60 g or 600 g/g wet weight of muscle, or 2 kg/cm² cross-sectional area; the wet weight was about 0.100 g; v_{max}, the maximum velocity of shortening of an unloaded muscle, was about 6.0 cm/sec, that is, about 2.0 l_0/sec at 0°C. In some of the chemical experiments the muscles were treated with 2,4-dinitrofluorobenzene (DNFB) to inhibit the enzyme, ATP: creatine phosphotransferase (Cain and Davies, 1962).

Tension-Velocity Relationship

The tension-velocity relationship was determined for isovelocity (constant velocity) shortening and stretching of untreated muscle at 0°C. The movement of the muscle at velocities up to v_{max} was controlled by a Levin-Wyman ergometer, and the tension produced at $100\% l_0$ during the movement was measured. Figure 1 shows the results of two typical experiments. For shortening, the results fit the solid line calculated from the Hill equation (1938)

$$v/l_0 = b/l_0 \frac{(1 - P/P_0)}{(P/P_0 + a/P_0)}$$

where v/l_0 is the velocity in l_0/sec, P is the observed tension and P_0 is the maximum tension under isometric conditions. The values of the constants a/P_0 and b/l_0 were, respectively, 0.246 and 0.564. For stretching at low velocities, more tension was produced by the muscle than was predicted by the Hill equation (Katz, 1939; Aubert, 1956). The observed tension reached a peak of about $140\% P_0$ at a velocity of about 0.2 l_0/sec, and then as the velocity increased, the tension started to decrease.

This decrease in the tension at high velocities did not seem to be due to fatigue or damage of the muscles for two reasons: (1) The $P_{0 \text{ final}}$ was close to the $P_{0 \text{ initial}}$; the mean $P_{0 \text{ final}}$ was $95\% P_{0 \text{ initial}}$. (2) Muscles that had been stretched at high velocities (about 1 l_0/sec) and had developed tensions of about $100\% P_0$ could subsequently develop higher tension ($140\% P_0$) when they were stretched at a lower velocity.

In the experiment shown in Fig. 1 the tension was about $80\% P_0$ during stretching at 2 l_0/sec, which is almost equivalent to v_{max}. During shortening at this velocity the tension was only about $10\% P_0$. In fact, the tension produced during stretching was always greater than that produced during shortening at the same velocity.

Chemical Experiments

ATP breakdown. The first series of chemical experiments was done on untreated muscles at 0°C. The experimental muscle of each pair was stimulated at 25 pulses/sec and stretched a distance of 1.2 cm from $60\% l_0$ to $100\% l_0$ by a Levin-Wyman ergometer. The control muscle was not stimulated. The muscles were frozen by immersion in Freon at -180°C. The inorganic phosphate (P_i) content of the muscles was measured and the difference between the P_i content of the experimental and control muscles represented the total ATP breakdown by the contractile element ATPase and the Ca^{++} pump ATPase of the sarcoplasmic reticulum. The results are listed in Table 1.

In Fig. 2 the rates of ATP breakdown (on a per second basis) for isometric contraction at $100\% l_0$ and for stretching at 0.08 l_0/sec and 0.66 l_0/sec are given. Although it was not clearly shown by these results, other experiments (Fig. 4) do show that the rate of breakdown increased as the velocity of stretching increased.

Table 1. Chemical and Mechanical Changes in Untreated Frog Sartorius Muscles during Contraction under Isometric Conditions and during Stretching at 0°C

Velocity of Stretching N (l_0/sec)		ΔP_i			Negative Work		
		per stretch ±SE (μmole/g)	per cm stretched ±SE (μmole/g/cm)	per sec ±SE (μmole/g/sec)	per stretch ±SE (g·cm/g)	per cm stretched ±SE (g·cm/g/cm)	per sec stretching ±SE (g·cm/g/sec)
0	8	—	—	1.05 ± 0.08	—	—	—
0.08	16	1.78 ± 0.24	1.49 ± 0.20	0.37 ± 0.05	546 ± 42	455 ± 35	114 ± 8.8
0.66	10	0.40 ± 0.11	0.33 ± 0.10	0.56 ± 0.16	255 ± 31	212 ± 26	425 ± 52

The type of contraction performed by the experimental (E) muscles of each group is listed in the first column. They were either stimulated under isometric conditions at $100\% l_0$, or were stimulated as they were stretched 1.2 cm, from $60\% l_0$ to $100\% l_0$ at a velocity of either 0.08 l_0/sec or 0.66 l_0/sec as listed in the first column. The control (C) muscles were not stimulated. The mean of the difference between the inorganic phosphate content ($\Delta P_i = E - C$) and the mean of the amount of negative work done by the experimental muscles for each experiment are listed. The standard errors of the means ($\pm$SE) are also given. In the second column the number of muscle pairs (N) in each experiment is given. See Figs. 2, 5 and 6.

As mentioned before, the total ATP breakdown measured in these experiments included that used for pumping Ca^{++}. A second group of experiments was designed to make allowance for this ATP breakdown, so that the net ATP breakdown represented only that used by the ATPase of the contractile elements.

In these experiments DNFB-treated muscles were used. The experimental muscle of each pair was stimulated and stretched or was held under isometric conditions. The control muscle was held at 60 % l_0 and was stimulated for the same period of time as the experimental muscle. The control muscles did not develop any tension. Since the ATP breakdown under these conditions is not signifi-

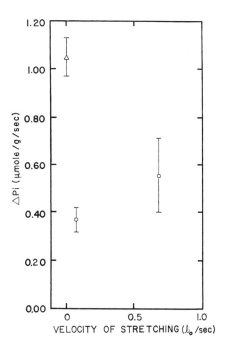

Figure 2. Variation in the rate of ATP breakdown with the velocity of stretching of untreated frog sartorius muscle at 0°C. Each point represents the mean ΔP_i/sec of stretching ($\bigcirc$) or of contraction under isometric conditions at 100% l_0 ($\triangle$). The bars represent $\pm$ one standard error of the mean. See Table 1.

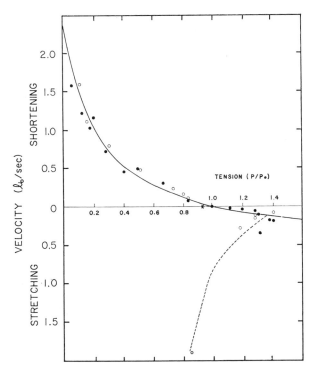

Figure 1. Tension-velocity relation for two untreated frog sartorius muscles at 0°C. Each muscle was stimulated and was kept under isometric conditions, or was allowed to shorten or was stretched at a constant velocity. The extent and velocity of movement was controlled by a Levin-Wyman ergometer. The tension produced at 100% l_0 during movement is expressed as P/P_0 initial, and the results are shown as ($\bullet$) for one muscle and as ($\bigcirc$) for the other muscle. The P_0 final, the isometric tension at 100% l_0 recorded at the end of the experiment for muscle ($\bigcirc$), is shown as ($\otimes$). Muscle ($\bullet$) shortened from 110% l_0 to 94% l_0 and was stretched from 94% l_0 to 110% l_0. Its l_0 was 2.70 cm. Muscle ($\bigcirc$) shortened from 110% l_0 to 90% l_0 and was stretched from 90% l_0 to 110% l_0. Its l_0 was 2.50 cm, its wet weight (m) was 0.06965 g, and P_0 P_0 initial l_0/m was 2.07 kg-cm/g. This value is roughly equivalent to the tension per cross-sectional area of the muscle in units of kg/cm². The solid line was calculated from Hill's equation (see text); the broken line is the observed relationship during stretching.

cantly different from the amount used when the muscle has been extended to the length at which the thick and thin filaments do not overlap (Infante et al., 1964a,b; Sandberg and Carlson, 1966; Kushmerick et al., 1969; Butler et al., 1972; Homsher et al., 1972, Smith, 1972), the ATP used in the control muscle was taken to be the amount for Ca^{++} pumping. The results of the experiments are listed in Table 2.

Figure 3 shows the rate of ATP breakdown (with allowance made for Ca^{++} pumping) for stretches at different velocities. The rate for contraction under isometric conditions is given also. The relationship was the same as was found for the untreated muscles. At low velocities of stretching the rate of breakdown was very much lower than during isometric contraction at 100 % l_0. The rate of breakdown reached a minimum of $0.07 \pm$ SE 0.02 μmole/g/sec at a velocity of 0.13 l_0/sec. Then as the velocity of stretching increased, the rate of breakdown increased. At about 2 l_0/sec, it was equal to that for isometric contraction. This strong velocity dependence of the rate of ATP breakdown may explain why the rate for stretching was greater than that for isometric contraction in the study by Aubert and Maréchal (1963), whereas in the others it was less than for the isometric case.

Table 2. Chemical and Mechanical Changes in DNFB-treated Frog Sartorius Muscles during Contraction under Different Conditions at 0°C

Velocity of Stretching (l_0/sec)	N	ΔP_i			Negative Work		
		per stretch $\pm$SE (μmole/g)	per cm stretched $\pm$SE (μmole/g/cm)	per sec $\pm$SE (μmole/g/sec)	per stretch $\pm$SE (g-cm/g)	per cm stretched $\pm$SE (g-cm/g/cm)	per sec stretching $\pm$SE (g-cm/g/sec)
0	10	—	—	0.64 ± 0.13	—	—	—
0.08	10	0.34 ± 0.13	0.56 ± 0.21	0.14 ± 0.05	168 ± 25	280 ± 42	70 ± 10
0.13	27	0.19 ± 0.05	0.18 ± 0.04	0.07 ± 0.02	232 ± 23	215 ± 21	86 ± 8
0.18	10	0.20 ± 0.07	0.16 ± 0.05	0.08 ± 0.03	293 ± 36	225 ± 30	122 ± 15
0.33	7	0.14 ± 0.17	0.12 ± 0.14	0.12 ± 0.14	333 ± 49	277 ± 41	277 ± 41
0.66	18	0.23 ± 0.10	0.19 ± 0.08	0.39 ± 0.16	261 ± 31	218 ± 26	435 ± 52
1.33	10	0.10 ± 0.05	0.09 ± 0.04	0.35 ± 0.17	451 ± 39	376 ± 33	1500 ± 130
2.00	18	0.15 ± 0.07	0.12 ± 0.06	0.74 ± 0.35	277 ± 21	231 ± 18	1380 ± 100

The type of contraction performed by the experimental (E) muscle of each group is listed in the first column. They were stimulated under isometric conditions at 100% l_0 for one second, or were stimulated as they were stretched at a particular velocity. For the velocity 0.08 l_0/sec, the muscles were stretched 0.6 cm, from 80% l_0 to 100% l_0; for 0.13 l_0/sec, some of the muscles were stretched 1.0 cm, from 66% l_0 to 100% l_0; the rest of the muscles stretched at 0.13 l_0/sec and at all the other velocities were stretched 1.2 cm, from 60% l_0 to 100% l_0. The control (C) muscle was held at 60% l_0 and was stimulated for the same period of time as the experimental muscle of the pair. The mean of the difference between the inorganic phosphate content ($\Delta P_i = E - C$) for each experiment and the mean of the amount of negative work done by the experimental muscles are listed. The standard errors of the means ($\pm$SE) are also given. In the second column the number of muscle pairs (N) in each experiment is given. See Figs. 3, 4, and 6.

In Fig. 4 the chemical results are shown as the ATP breakdown per cm distance stretched. The amount of chemical change was quite constant for all the velocities greater than 0.13 l_0/sec, even though the durations of the contractions were quite different. The fastest stretches only lasted 0.2 sec, whereas the slow ones were fifteen times longer, 3.0 sec.

This gives a more clear picture of what actually occurred during stretching than the plot of the per second rate of the ATP breakdown did. The per second rate increased at high velocities because the total amount of breakdown per stretch was the same, and during fast stretches this breakdown occurred during a much briefer time than it did during slow stretches.

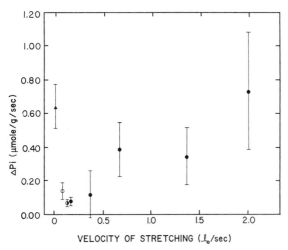

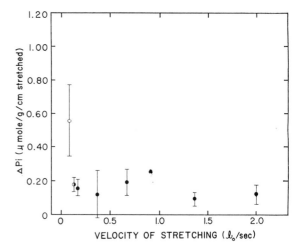

Figure 3. Variation in the rate of ATP breakdown (allowance has been made for Ca^{++} pumping) with the velocity of stretching of DNFB-treated muscles at 0°C. Each point ($\bigcirc$ $\circleddash$ $\bullet$) represents the mean ΔP_i/sec of stretching at a particular velocity. The muscles were stretched 0.6 cm from 80% l_0 to 100% l_0 at 0.08 l_0/sec, 1.0 cm from 66% l_0 to 100% l_0 for some of the stretches at 0.13 l_0/sec, and 1.2 cm from 60% l_0 to 100% l_0 for the rest of the stretches at 0.13 l_0/sec and at all other velocities. The value for isometric contraction at 100% l_0 ($\blacktriangle$) is included for comparison. See Table 2. The bars represent $\pm$ one standard error of the mean.

Figure 4. Variation in the ATP breakdown (allowance has been made for Ca^{++} pumping) with the velocity of stretching of DNFB-treated frog sartorius muscles at 0°C. Each point ($\bigcirc$ $\circleddash$ $\bullet$) represents the mean ΔP_i/cm distance stretched at a particular velocity. The chemical change has been normalized with respect to the distance stretched. The muscles were stretched 0.6 cm from 80% l_0 to 100% l_0 at 0.08 l_0/sec, 1.0 cm from 66% l_0 to 100% l_0 for some of the stretches at 0.13 l_0/sec, and 1.2 cm from 60% l_0 to 100% l_0 for the rest of the stretches at 0.13 l_0/sec and at all other velocities. See Table 2. The bars represent $\pm$ one standard error of the mean.

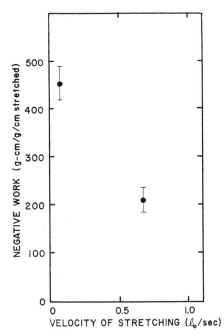

Figure 5. Variation in the amount of negative work done with the velocity of stretching of untreated frog sartorius muscles at 0°C. Each point (●) represents the mean negative work done per cm distance stretched at a particular velocity. The bars represent ± one standard error of the mean. Note that the negative work has been normalized with respect to the distance stretched, which was 1.2 cm (from 60% l_0 to 100% l_0) in these experiments. See Table 1.

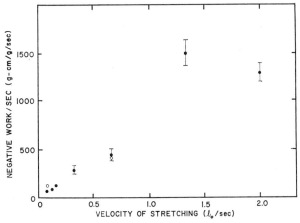

Figure 6. Variation in the negative work/second with the velocity of stretching of DNFB-treated and untreated muscles at 0°C. Each point represents the mean negative work/sec for a particular velocity of stretching. (●) DNFB-treated muscles; (○) untreated muscles, Where ± one standard error of the mean is greater than the point, it is indicated by a bar. See Tables 1 and 2.

Negative work. Figure 5 shows the amount of negative work done by the untreated muscles per cm distance stretched, and Fig. 6 shows the rate at which negative work was done by the muscles. As would be expected from the tension-velocity curve (Fig. 1) the rate of negative work increased as the velocity of stretching increased.

Discussion

Lack of Correlation between ΔATP and Negative Work, and ΔATP and Mechanical Impulse, $\int Pdt$

The amount of ATP breakdown during shortening has been shown to be correlated with the amount of positive work done by the muscle, even though the relationship is complex (Cain et al., 1962; Mommaerts et al., 1962; Cain and Davies, 1962; Carlson et al., 1963; Maréchal, 1964a,b; Infante et al., 1965; Davies et al., 1967; Kushmerick and Davies, 1969; Gilbert and Kushmerick, 1970); and for isometric contraction in a given muscle, the ATP breakdown has been correlated with the amount of tension produced and the duration of the contraction (Maréchal and Mommaerts, 1963; Infante et al., 1964a,b; Maréchal 1964a,b; Sandberg and Carlson, 1966; Carlson et al., 1967; Mommaerts 1969, 1970; Goldspink et al., 1970; Gilbert et al., 1971; Homsher et al., 1972). However, the ATP breakdown during stretching showed no significant correlation with either of these parameters. The product moment correlation coefficient (r) of ΔATP and negative work was −0.14 and its P value was greater than 0.10. For ΔATP and the mechanical impulse, $\int Pdt$, the value of r was +0.108 and its P value was greater than 0.20. The data used in these calculations were from the experiments which are summarized in Table 2. One hundred sets of numbers from individual muscle pairs were used in each calculation. In terms of models of muscle contraction, this lack of correlation may mean that the ATP breakdown step and the cross-bridge linkage which is responsible for the production of tension are not related to each other during stretching as they are during other types of contraction.

Is ATP Breakdown Required for Formation of Cross-Bridge Links during Stretching?

One question about cross-bridge operation that can be answered using the results given here is whether ATP must be broken down when each cross-bridge forms each link during stretching. If the process of linking did require the breakdown of ATP, the number of cross-bridge links would be equal to the number of ATP molecules broken down (other than those used for pumping Ca^{++}). Dividing the mechanical impulse, $\int Pdt$, which is produced by the linked cross-bridges, by the amount of ATP breakdown would be equivalent to dividing by the number of linked bridges. The resulting number would be a measure of the mechanical impulse produced per linked cross-bridge. This calculation was done for stretching, using the results of the experiments reported here, and for shortening,

Table 3. Comparison of Mechanical Impulse/ATP Breakdown during Isovelocity Stretching and Shortening

Velocity of Stretching (l_0/sec)	$\int Pdt/\Delta$ATP (g-sec/μmole)	Velocity of Shortening (l_0/sec)	$\int Pdt/\Delta$ATP (g-sec/μmole)	$\dfrac{\int Pdt/\Delta\text{ATP}_{\text{stretching}}}{\int Pdt/\Delta\text{ATP}_{\text{shortening}}}$
0.08	1980	0.10	659	3.00
0.13	3000			
0.18	2700	0.20	477	5.66
0.33	2380	0.33	222	10.7
0.66	570	0.66	155	3.68
1.33	1130	1.33	49.7	22.7
2.00	307	2.13	18.1	17.0

Allowance has been made for Ca^{++} pumping.

The values for the ΔATP for stretching are given in Table 2, and the values for shortening are taken from Kushmerick and Davies (1969, Table 6, p. 344). The values for the mechanical impulse, $\int Pdt$, were calculated from the equation $\int Pdt = [(\text{work in g-cm/g})/(\text{distance moved in cm})] \times [\text{duration of the movement in sec}]$. The data for the work, distance, and duration are given or can be calculated from the information in Table 2 for stretching or in Table 6 of the paper by Kushmerick and Davies (1969) for shortening.

using the results of Kushmerick and Davies (1969). See Table 3. The values for stretching were always greater than for shortening at the same velocity. For movement at 1.33 l_0/sec, the mechanical impulse per link was 23 times greater for stretching than for shortening at that velocity. So if the formation of each link during stretching did require ATP breakdown, then the links would have to produce a mechanical impulse which is 23 times greater during stretching than during shortening at 1.33 l_0/sec, or considered from another point of view, each link would exhibit only $\frac{1}{23}$ of its possible capability when the muscle shortens at 1.33 l_0/sec. Since this seemed very unlikely, we concluded that the postulate is not true. Thus during stretching, links can form and maintain a force over their range of movement *without* ATP breakdown.

Does All ATP Breakdown Occur at Beginning of Stretching?

When the values for the ATP breakdown per stretch, which are quite constant for all the velocities of stretching (see Table 2), were pooled, the mean was $0.19 \pm$ SE 0.03 μmole/g/stretch. This number is very interesting because it is equal to about 2 ATPs per cross-bridge or 1 ATP per heavy meromyosin (HMM) unit. It seemed possible that each HMM unit may be "charged" by the breakdown of a molecule of ATP early during contraction. This would result in the formation of about 0.16 μmole of inorganic phosphate per gram of muscle early in the contraction. If stretching prevented further breakdown of ATP, this would explain our finding that a constant amount of inorganic phosphate is formed during stretching, regardless of the velocity of stretching.

An experiment was done on DNFB-treated muscles to test this possibility. The experimental muscle of each pair was stretched from 60% l_0

to 100% l_0 at 0.66 l_0/sec. The control muscle was stretched from 60% l_0 to 80% l_0. So the difference between the experimental and the control muscle is the amount of ATP used during stretching of the experimental muscle from 80% l_0 to 100% l_0, or in other words, the second half of the stretch. If the contractile elements only broke down ATP at the beginning of the stretch, then the difference between the experimental and the control should be equal to the ATP used for Ca^{++} pumping. In fact, the observed ATP breakdown, experimental–control $= +0.23 \pm$ SE 0.08 μmole/g, was significantly greater ($P = 0.025$) than the amount that would be used for Ca^{++} pumping, $+0.03 \pm$ SE 0.008 μmole/g. This excluded the possibility that the entire ATP breakdown by the contractile elements occurred at the beginning of stretching.

Summary

We have shown that the breakdown of ATP is reduced in activated muscle during stretching. The amount per stretch is independent of the velocity of stretching even though the duration of stretching ranges from 0.2 sec up to 3.0 sec. The amount of ATP breakdown per stretch is equal to about 1 ATP per heavy meromyosin unit, but the significance of this finding is unknown.

Any theory of muscle contraction must be able to account for the ability of the muscle to develop tension even greater than P_0 and to maintain it over long distances of stretching with very little breakdown of ATP.

References

ABBOTT, B. C. and X. M. AUBERT. 1951. Changes of energy in a muscle during very slow stretches. *Proc. Roy. Soc.* B **139**: 104.

————. 1952. The force exerted by active striated muscle during and after change of length. *J. Physiol.* (*London*) **117**: 77.

ABBOTT, B. C., X. M. AUBERT, and A. V. HILL. 1951. The absorption of work by a muscle stretched during a single twitch or a short tetanus. *Proc. Roy. Soc. B* **139**: 86.

AUBERT, X. 1956. *Le couplage énergétique de la contraction musculaire.* Editions Arsica, Bruxelles.

AUBERT, X. and G. MARÉCHAL. 1963. Le bilan énergétique des contractions musculaires avec travail positif ou négatif, *J. Physiol. (Paris)* **55**: 186.

BUTLER, T. M., N. A. CURTIN, and R. E. DAVIES. 1972. Comparison of ATP usage in muscle during isometric tetanus and activated isovelocity stretch. *Fed. Proc.* **31**: 337.

CAIN, D. F. and R. E. DAVIES. 1962. Breakdown of adenosine triphosphate during a single contraction of working muscle. *Biochem. Biophys. Res. Comm.* **8**: 361.

CAIN, D. F., A. A. INFANTE, and R. E. DAVIES. 1962. Chemistry of muscle contraction. *Nature* **196**: 214.

CARLSON, F. D., D. HARDY, and D. R. WILKIE. 1963. Total energy production and phosphocreatine hydrolysis in the isotonic twitch. *J. Physiol. (London)* **46**: 851.

———. 1967. The relation between heat produced and phosphorylcreatine split during isometric contraction of frog's muscle. *J. Physiol. (London)* **189**: 209.

CHAPLAIN, R. A. 1972. The force-velocity relation of frog sartorius muscle at constant velocities of lengthening. *Experientia* **28**: 292.

CURTIN, N. A., D. D. DROBNIS, R. E. LARSON, and R. E. DAVIES. 1969. Very low ATP usage with very high tension in activated muscles during a slow stretch. *Fed. Proc.* **28**: 711.

CURTIN, N. A., S. M. M. SVENSSON, and R. E. DAVIES. 1970. Force development and the braking mechanism in stretching activated muscle is not limited by the energy from ATP splitting. *Fed. Proc.* **29**: 714.

DAVIES, R. E., M. J. KUSHMERICK, and R. E. LARSON. 1967. Professor A. V. Hill's "Further challenge to biochemists": ATP, activation, and the heat of shortening of muscle. *Nature* **214**: 148.

GASSER, H. S. and A. V. HILL. 1924. The dynamics of muscular contraction. *Proc. Roy. Soc. B* **96**: 398.

GILBERT, C. and M. J. KUSHMERICK. 1970. Energy balance during working contractions of frog muscle. *J. Physiol. (London)* **210**: 146P.

GILBERT, C., K. M. KRETZSCHMAR, D. R. WILKIE, and R. C. WOLEDGE. 1971. Chemical change and energy output during muscular contraction. *J. Physiol. (London)* **218**: 163.

GILLIS, J. M. and G. MARÉCHAL. 1969. Resynthesis of ATP in glycerinated fibres stretched during contraction. *3rd Int. Biophys. Congr.,* Cambridge, Mass. (Abstr.) p. 271.

———. 1971a. Influence of tension on the incorporation of ATP in glycerinated muscle fibres. *J. Physiol. (London)* **214**: 41P.

———. 1971b. Incorporation of inorganic phosphate into ATP, catalysed by tension, in glycerinated fibres. *25th Int. Congr. Physiol.* (Abstr.) **9**: 598.

GOLDSPINK, G., R. E. LARSON, and R. E. DAVIES. 1970. Thermodynamic efficiency and physiological characteristics of the chick anterior latissimus dorsi muscle. *Z. vergl. Physiologie.* **66**: 379.

HILL, A. V. 1938. The heat of shortening and the dynamic constants of muscle. *Proc. Roy. Soc. B* **126**: 136.

HILL, A. V. and J. V. HOWARTH. 1959. The reversal of chemical reactions in contracting muscle during an applied stretch. *Proc. Roy. Soc. B* **151**: 169.

HOMSHER, E., W. F. H. M. MOMMAERTS, N. V. RICCHIUTI,

and A. WALLNER. 1972. Activation heat, activation metabolism and tension-related heat in frog semitendinosus muscles. *J. Physiol. (London)* **220**: 601.

INFANTE, A. A., D. KLAUPIKS, and R. E. DAVIES. 1964a. Length, tension and metabolism during short isometric contractions of frog sartorius muscle. *Biochim. Biophys. Acta* **88**: 215.

———. 1964b. Relation between length of muscle and breakdown of phosphorylcreatine in isometric tetanic contractions. *Nature* **201**: 620.

———. 1964c. Adenosine triphosphate: Changes in muscle doing negative work. *Science* **144**: 1577.

———. 1965. Phosphorylcreatine consumption during single-working contractions of isolated muscle. *Biochim. Biophys. Acta* **94**: 504.

JOYCE, G. C. and P. M. H. RACK. 1969. Isotonic lengthening and shortening movements of cat soleus muscle. *J. Physiol. (London)* **204**: 475.

JOYCE, G. C., P. M. H. RACK, and D. R. WESTBURY. 1969. The mechanical properties of cat soleus muscle during controlled lengthening and shortening movements. *J. Physiol. (London)* **204**: 461.

KATZ, B. 1939. The relation between force and speed in muscular contraction. *J. Physiol. (London)* **96**: 45.

KUSHMERICK, M. J. and R. E. DAVIES. 1969. The chemical energetics of muscle contraction. II. The chemistry, efficiency and power of maximally working sartorius muscles. *Proc. Roy. Soc. B* **174**: 315.

KUSHMERICK, M. J., R. E. LARSON, and R. E. DAVIES. 1969. The chemical energetics of muscle contraction. I. Activation heat, heat of shortening and ATP ultilization for activation-relaxation processes. *Proc. Roy. Soc. B* **174**: 293.

MANNHERZ, H. G. 1970. On the reversibility of the biochemical reactions of muscular contraction during the absorption of negative work. *FEBS Letters* **10**: 233.

MARÉCHAL, G. 1964a. *Le métabolisme de la phosphorylcréatine et de l'adénosine triphosphate durant la contraction musculaire.* Editions Arscia, Bruxelles.

———. 1964b. Phosphorylcreatine and ATP changes during shortening and lengthening of stimulated muscle. *Arch. Int. Physiol. Biochim.* **72**: 306.

MARÉCHAL, G. and G. BECKERS-BLEUKX. 1965. La phosphorylcréatine et les nucléotides adényliques d'un muscle strié à la fin d'un étirement. *J. Physiol. (Paris)* **57**: 652.

MARÉCHAL, G. and W. F. H. M. MOMMAERTS. 1963. The metabolism of phosphocreatine during an isometric tetanus in frog sartorius muscle. *Biochim. Biophys. Acta* **70**: 53.

MARÉCHAL, G., W. F. H. M. MOMMAERTS, and K. SERAYDARIAN. 1971. Incorporation of inorganic phosphate into ATP and phosphorylcreatine of stretched muscle measured by ^{18}O tracer. *J. Physiol. (London)* **214**: 40P.

MOMMAERTS, W. F. H. M. 1969. Energetics of muscular contraction. *Physiol. Rev.* **49**: 427.

———. 1970. What is the Fenn-effect? *Naturwissenschaften* **57**: 326.

MOMMAERTS, W. F. H. M., K. SERAYDARIAN, and G. MARÉCHAL. 1962. Work and chemical change in isotonic muscular contractions, *Biochim. Biophys. Acta* **57**: 1.

SANDBERG, J. A. and R. D. CARLSON. 1966. The length dependence of phosphorylcreatine hydrolysis during an isometric tetanus. *Biochemische Z.* **345**: 212.

SMITH, I. C. H. 1972. Energetics of activation in frog and toad muscle. *J. Physiol. (London)* **220**: 583.

SUGI, H. 1969. The mode of tension development by stretch in active frog muscle fibres. *Proc. Japan. Acad.* **45**: 413.

———. 1971. The mode of tension changes by stretch in frog muscle fibres. *25th Int. Congr. Physiol.* (Abstr.) **9**: 543.

ULBRICH, M. and J. C. RÜEGG. 1971. Stretch induced formation of ATP-^{32}P in glycerinated fibres of insect flight muscle. *Experientia* **27**: 45.

WILKIE, D. R. 1960. Thermodynamics and the interpretation of biological heat measurements. *Prog. Biophys. Biophys. Chem.* **10**: 259.

———. 1968. Heat, work and phosphorylcreatine breakdown in muscle. *J. Physiol.* (*London*) **195**: 157.

Comments on Activation Heat and Its Relation to Activation

Allan Fraser

Department of Biophysics, Johns Hopkins University, Baltimore, Md. 21218

Frog's sartorius muscles were restimulated during relaxation from twitches and tetani. The increase in tension caused by restimulation was not accompanied by a burst of "activation heat" under some conditions. These findings eliminate some reactions as possible sources of activation heat and show that transient activation heat is not necessarily caused by activation. Recent papers by Homsher et al. (1972) and Smith (1972) should serve the reader as background for this note.

Method

An infrared radiometer (Fraser, 1971) was used to measure changes in muscle temperature. The few millisecond time resolution of this device makes it suitable for the analysis of the time course of twitch heat production in frog's muscle at room temperature. The studies involved monitoring muscle temperature changes at the time of restimulation of relaxing muscle. Since the twitch-tetanus tension ratio was near 0.5, large tension increases resulted from restimulations. Isometric bullfrog's (*Rana catesbeiana*) sartorius muscles were used at 15°C.

Results

Figure 1 shows the tension (lower trace) and temperature change (upper trace) record produced by a muscle that was restimulated at the peak of twitch tension. The temperature record shows no discontinuity in rate at the time of the second stimulation. There was a great increase in heat rate with the first stimulation, a burst of activation heat, but no comparable increase in heat rate with the second stimulation. This was true even though there was a considerable reactivation of the muscle, and the tension rate caused by the second stimulus was two-thirds of that caused by the first stimulus. Thirteen muscles were doubly twitched as in Fig. 1. Six of these produced records that had constant temperature rates through the second stimulation, and seven showed a slight increase after the second stimulation. On the average the heat rate measured in the 20 msec prior to the second stimulus was 0.44 (± 0.21 SD) of the maximum heat rate of the initial burst of activation heat, and the rate in the 20 msec after the second stimulus was 0.55 (± 0.17 SD) of the initial burst rate.

Many muscles were restimulated at times after the peak but before the end of twitch tension. Figure 2 shows the heat and tension records for one such stimulation pattern. It does show a substantial increase in heat rate after the second stimulation. Attempts were made to correlate with both time and tension the ability of the bullfrog's sartorius muscle to produce second bursts of heat on restimulation. Scatter in data collected from various muscles prevented analysis. With individual muscles, however, increasing the time interval

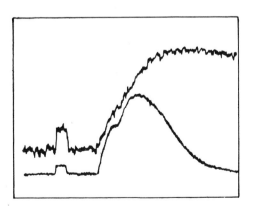

Figure 1. Fused double twitch. *Lower trace*, isometric tension with a rectangular 10 g calibration pulse at the left; *upper trace*, temperature change with a 1×10^{-3}°C calibration marker. The traces are 0.5 sec wide and are averages of 25 repetitions from one bullfrog's sartorius muscle with 2.45 μm sarcomere length at 15°C.

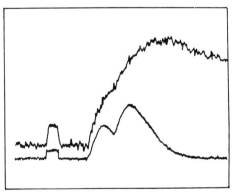

Figure 2. Unfused double twitch. *Lower trace*, isometric tension with a 10 g calibration pulse; *upper trace*, temperature change with a 1×10^{-3}°C calibration marker. Traces are 0.5 sec wide and are averages of 25 repetitions from a bullfrog's sartorius muscle at 2.48 μm sarcomere length.

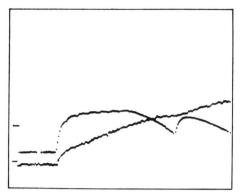

Figure 3. Twitch after tetanus. Unaveraged tension and temperature change record from a bullfrog's sartorius muscle at 106% of rest length. *Upper left*, tension calibration, 20 g; *lower left*, temperature calibration, $1 \times 10^{-3}°C$.

between stimuli always produced larger bursts of heat production after the second stimulus.

The burst heat rate caused by second stimuli compared to the burst heat rate evoked by first stimuli did not change with sarcomere length from 2.4 μm to 3.0 μm. The trend of little or no second burst at peak tension and burst rates increasing continually during relaxation was observed.

When muscles were restimulated after short tetani, traces like Fig. 3 resulted. Bursts of heat were never observed with stimuli delivered at about the time of half of full tension after a short tetanus. It is notable that the heat rate is very slight during the entire rise of tension of the last twitch on Fig. 3. This lack of a burst of activation heat is distinct from the absence of a labile heat (Aubert, 1956). Labile heat is a much slower component of muscle heat production.

Discussion

The activation of a muscle at any given time may be measured as the fraction of tetanic tension that the muscle generates if its sarcomeres are isometric, as the fraction of the maximum unloaded shortening velocity that the muscle achieves on release, or by similar means (Jewell and Wilkie, 1960). These measures of activation would show that activation was about half or less at the times of the final stimuli in the three kinds of experiments described above.

The burst of heat production immediately after isometric stimulation is thought to be related to the rise of activation and is called activation heat (Hill, 1949; Homsher et al., 1972; Smith, 1972). No other established source of heat has a significant rate at that time.

Figures 1 and 3 show muscles that produced normal bursts of activation heat with the first stimuli, but no bursts of activation heat with the last stimuli, even though the last stimuli caused reactivations of 0.5 or more. Therefore activation can occur without a burst of activation heat, and transient bursts of activation heat must be related to something other than simply activation of contractile activity. This does not imply that there is not a net heat for the last stimuli shown, reflecting energetic cost of initiating and stopping the last tensions generated.

A second stimulation after 50 msec delay in toad's muscle at 12°C produces about half as much free Ca^{++} as is released by a first stimulus (Jöbsis and O'Connor, 1966). The large increases in tension and tension rate caused by the restimulation as reported in this paper also imply that large incremental amounts of Ca^{++} are released by such restimulations. If no very unusual Ca^{++} binding behavior occurs and the Ca^{++} release produced by restimulations as shown in Figs. 1 and 3 is of the same order as for the first stimuli, then the release of Ca^{++} and its binding to tropomyosin must be of slight thermal importance. Thus Ca^{++} release cannot be the source of activation heat. The hypothesis that activation heat is the result of Ca^{++} being pumped back through the sarcoplasmic reticulum is not eliminated, however, if the pumping or binding reactions that produce heat were already operating at their maximum steady state rate when the additional Ca^{++} became available.

Acknowledgments

This work was made possible by Dr. F. D. Carlson and USPHS Grant GM 03723-18. Howard Carney and Robert Bonner improved the manuscript.

References

AUBERT, X. 1956. Le couplage energetique de la contraction musculaire. Editions Arsica, Brussels.

FRASER, A. 1971. Myothermic radiometry. *Rev. Sci. Instr.* **42**: 22.

HILL, A. V. 1949. The heat of activation and the heat of shortening in a muscle twitch. *Proc. Roy. Soc.* (London) B **136**: 195.

HOMSHER, E., W. F. H. M. MOMMAERTS, N. V. RICCHIUTTI, and A. WALLNER. 1972. Activation heat, activation metabolism, and tension related heat in frog semitendinosus muscles. *J. Physiol.* **220**: 601.

JEWELL, B. R. and D. WILKIE. 1960. The mechanical properties of relaxing muscle. *J. Physiol.* **152**: 30.

JÖBSIS, F. F. and M. V. O'CONNOR. 1966. Calcium release and reabsorption in the sartorius muscle of the toad. *Biochem. Biophys. Res. Comm.* **25**: 246.

SMITH, I. C. H. 1972. Energetics and activation in frog and toad muscle. *J. Physiol.* **220**: 583.

In Vitro Calorimetric Studies Relating to the Interpretation of Muscle Heat Experiments

R. C. Woledge

Department of Physiology, University College, London

The heat production of contracting muscle has been studied for a century. If the accumulated knowledge is to continue to play a part in ideas of contractility, interpretation in terms of specific chemical reactions is needed. This type of understanding of muscle heat requires three kinds of measurements: of the heat and work ($h + w$, Joules) produced by the living muscle; of the amounts (η, moles) of each of the n chemical reactions thought to be occurring in significant net quantity; and of the enthalpy change (ΔH, Joules per mole) for each of these. Armed with all these results we can then use the following statement of the first law of thermodynamics to test whether our list of reactions is complete.

$$(h + w) = \sum_0^n \eta_i \, \Delta H_i \tag{1}$$

Because accumulation of errors would prevent a decisive result except where n is small, attention is at present directed to situations where a whole, or an identifiable part, of $(h + w)$ is thought to derive from a single process. In such a case Eq. 1 can be written as

$$(h + w)/\eta = \Delta H \tag{2}$$

In many cases even the framing of a reasonable hypothesis is hampered by ignorance of ΔH values. An example is the recent discussion of the "activation" heat by Smith (1972) and by Homsher et al. (1972). Rather, more progress has been made in considering the $(h + w)$ derived from PCr splitting. Biochemical evidence suggests that in IAA-poisoned muscles in nitrogen at 0°C splitting of PCr might be the only reaction to have occurred, in significant quantity, a minute or so after contraction ends. Initially, energy balance studies seemed to support this. Wilkie (1968) showed that $(h + w)/\eta$ could remain constant in spite of wide variations of η and of w/h. The ratio obtained (-46 kJ/mole) was also equal to Meyerhof and Schulz (1935) estimate of ΔH, though higher than Carlson's (1963) more recent calculation (-37 kJ/mole). There is some danger that Wilkie's value of $(h + w)/\eta$ might be taken as an in vivo determination of ΔH and used via Eq. 2 to justify the belief

that PCr splitting alone is indeed the only net process occurring. This would be a circular argument. To break the circle it is necessary to determine the actual value of ΔH under the conditions inside the muscle cells. This paper is an interim report of an attempt to determine this value, which has led also to a study of the heat produced in the Lohmann reaction.

Although PCr splitting has been referred to as a single process, it is more convenient to write the chemical representation of what is thought to occur in muscle as a series of linked equilibria. The reactions I am considering are the following ones:

$$PCr^= \rightarrow Cr + HPO_4^= \tag{1}$$

$$HPO_4^= + H^+ \rightarrow H_2PO_4^- \tag{2}$$

$$Buffer\ H^+ \rightarrow Buffer + H^+ \tag{3}$$

$$HPO_4^= + Mg^{++} \rightarrow MgHPO_4 \tag{4}$$

$$PCr^= + Mg^{++} \rightarrow MgPCr \tag{5}$$

(The involvement of calcium ion is probably negligible because of the low free calcium concentration in muscle.) If the conditions inside the muscle were accurately known, this dissection of the overall reaction would be unnecessary; but because of uncertainty about these conditions we need not just a single value for ΔH for PCr splitting, but knowledge of how the value depends on pH and pMg. This is best found by obtaining values of ΔH for each of the above reactions and of the equilibrium constants of reactions 2, 4, and 5. Reaction 1 can be studied alone using magnesium-free alkaline conditions. The enthalpy change for the overall process (ΔH_{obs}) can then be calculated as a function of pH and pMg. It is prudent also to obtain values of ΔH_{obs} under several different conditions to confirm the calculated result.

The value of ΔH_3 requires special consideration. The intracellular buffering in the physiological range of pH is largely contributed by the imidazole group in the dipeptide carnosine and in the muscle proteins. The ionization heat of these groups in the proteins has not been determined and the possibility exists that it might differ considerably from the value which is the same for free histidine and carnosine. For the present this possibility has been discounted because of the experiments by Stella (1928) who determined ΔH_3 on living muscle by acidification with CO_2. His result, after correction (Hill and Woledge, 1962) for a calibration error (31 kJ/mole at 19°C), agrees with the ionization heat of histidine and carnosine (32 kJ/mole at 25°C, Table 1). Because of the importance of the buffer reaction in all determinations of ΔH, however, it would be desirable to repeat Stella's experiments with modern apparatus.

Methods

Calorimetric observations have been made mostly with an LKB Batch microcalorimeter, although in some early experiments a Calvet instrument was used. Four types of experiments have been made.

1. A small volume of acid is mixed with a larger volume of a solution containing about twice the equivalent quantity of base which had previously been one-quarter neutralized. Correction was made for heat of dilution of the acid by subtracting the heat observed in control experiments without base present. No account was taken of the heat of dilution of the base which should have been negligible.

2. Titration experiments in which successive additions of very small volumes (usually 10 or 20 μl) of 1 M $MgCl_2$ solution were made to about 3 ml of 30 mM Na_2HPO_4 or Na_2PCr. Correction was made for dilution of the $MgCl_2$ into a solution of the same ionic strength. In principle these experiments could yield values for the equilibrium constant (k) as well as for ΔH. In practice the titration curves were not sufficiently close to the theoretical shapes for k values to be obtained. These values were therefore determined from the shift in pK of the H^+ ion dissociation curves, produced by adding magnesium.

3. Enzymic splitting of phosphocreatine was investigated using alkaline phosphatase from *E. coli*. About 3 ml of a buffer solution containing about 1 mg of this enzyme preparation was mixed in the calorimeter with about 10 μM of Na_2PCr dissolved in 0.5 ml of water at pH 7.5–8. When magnesium was present, it was in equal total concentration with both enzyme and substrate. Heat production was complete in about one hour. It was confirmed by analysis that the PCr splitting had gone to completion and that there was no loss of creatine. In control experiments without enzyme present the heat produced was negligible. No correction was therefore made for dilution heat. Control experiments were made using a solution of Cr and Na_2HPO_4 in place of PCr.

4. Observation on the Lohmann reaction. About 0.5 mg of creatine phosphotransferase in 3 ml of buffer (containing 10 mM/liter of $MgCl_2$) were mixed in the calorimeter with 1 ml of the same buffer solution at the same pH containing 2.5 μM each of either ADP and PCr, or ATP and Cr. The observations were corrected for dilution heat and the disturbance of magnesium equilibrium on mixing by subtracting the amount of heat produced by the control observations in which no enzyme was present.

Mostly reactions have been studied only at 25°C but observations were also made of reaction 2 and 3 at 1°C. It is intended to study reactions 1, 4, and 5 at lower temperatures also; at present, assumptions have been made about the way ΔH might vary with temperature.

Calculation for Figs. 2 and 3 was made from the formulas given by Alberty (1968, 1969).

Heat of Splitting of Phosphocreatine

Details of the values obtained for reactions 1 through 5 are given in Table 1 and the calculated dependence of ΔH_{obs} on pH and pMg (at 0°C) is shown in Fig. 1. An experiment was made to check

Table 1. Data for Calculation of ΔH_{obs} for PCr Splitting

Reaction	ΔH at 25°C	pK at 25°C	ΔH at 0°C	pK at 0°C
1. $PCr^= \rightarrow Cr + HPO_4^=$	-40.0 ± 1.1		-40.0	
2. $HPO_4^= + H^+ \rightarrow H_2PO_4^-$	-7.3 ± 0.6	6.75	-11.2 ± 0.3	6.91
3. Buffer $H^+ \rightarrow H^+ +$ Buffer	$+32 \pm 0.6$		$+27 \pm 2$	
4. $HPO_4^= + Mg^{++} \rightarrow MgHPO_4$	$+10$	~ 2	$+3.7$	1.89
5. $PCr^= + Mg^{++} \rightarrow MgPCr$	$+10$	~ 1.4	$+3.7$	1.29
6. $PCr^= + H^+ \rightarrow HPCr^-$	-7	4.52	-11	4.21

Observations have been made of ΔH_1, ΔH_4 and ΔH_5 at 25°C only, of ΔH_2 and ΔH_3 at both 25°C and 0°C, and of pK_4 and pK_5 at 25°C. For reaction 3 carnosine was used as a representative buffer. The data for reactions 4 and 5 are only approximate because of difficulties experienced in interpreting the enthalpy titrations. pK_2 values are from Bates and Acree (1945), pK_6 at 25°C is from Alberty (1968). ΔH_6 has been taken as equal to ΔH_2; this is justifiable because reaction 6 has very little effect near physiological pH values. pK_4, pK_5, and pK_6 at 0°C have been calculated from the ΔH values and pK at 25°C. ΔH_1 has been assumed independent of temperature because the reaction involves no change in the number of charged particles. ΔH_4 and ΔH_5 have been assumed to decrease with temperature at 250 J/mole °C. These assumptions about temperature dependence of ΔH values were suggested by Alberty (personal communication).

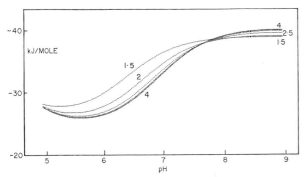

Figure 1. Calculation of the ΔH_{obs} for PCr splitting as a function of pH and pMg. Curves calculated according to the formulas given by Alberty (1968, 1969). Each line refers to the pMg value shown by the number on the line. The calculation uses the data given in Table 1 for 0°C.

the predicted variation of ΔH_{obs} with pH in the absence of magnesium. The result is shown in Fig. 2. It is clear that ΔH_{obs} does indeed decrease as the pH is lowered, as is predicted from the fact that ΔH_3 is greater than ΔH_2. The observed variation agrees well with the calculated curve. From the difference between these observations and control experiments in which PCr was replaced by a mixture of Cr and inorganic phosphates, ΔH_1 is obtained at different pH values; as expected there is no significant variation. Two single observations were also made with 10 mM/liter of magnesium chloride present, to check the prediction that the stronger binding of magnesium to $HPO_4^=$ than to $PCr^=$ will lead to lower values of ΔH_{obs} when magnesium is present. The value obtained

was $\Delta H_{obs} = -32 \pm 4$ kJ/mole (mean and estimated SE). This agrees reasonably with the predicted value of -36 kJ/mole.

Comparison with earlier values. Gellert and Sturtevant (1960) and Pin (1965) both report values of ΔH_{obs} under alkaline conditions in the presence of Mg^{++} (as they neglect Mg^{++} involvement they gave their results as values of ΔH_1). Their values are -38 ± 2 kJ/mole (pMg $\simeq 3$) and 35.6 ± 0.5 kJ/mole (pMg $\simeq 2$), respectively. The values calculated from my results for these conditions are -39 kJ/mole and -37 kJ/mole. There thus seems to be reasonable agreement. My result for reaction 2 at 1°C differs starkly from that of Bernhardt (1956) but agrees well with that of Bates and Acree (1945). This discrepancy is illustrated in Fig. 3. Scrutiny of Bernhardt's paper suggests that the pK values he reports are not sufficient in number or accuracy to support the calculation which he makes of ΔH via the Van't Hoff equation and that his results should not be accepted. Carlson's calculation of ΔH_{obs} at pH 7 (-40 kJ/mole) is larger than my result because he accepts Bernhardt's value for ΔH_2. It is hard to know what to make of the experiments of Meyerhof and Schulz (1935) from which they report a value of -46 kJ/mole for ΔH_{obs} obtained by calorimetric observation of the splitting of PCr by a "second water extract" of muscle. The predominant buffer in this extract would have been inorganic phosphate and so ΔH_3 would not differ much from ΔH_2. (This is confirmed by Meyerhof, 1930.) Under these conditions a value of slightly less than -40 kJ/mole would be expected. I can advance no explanation for his higher result. The discrepancy remains intriguing.

Conclusion. The best value now available for ΔH_{obs} for creatine phosphate splitting at 0°C, pH 7, pMg 2.5 is -34 kJ/mole. The principle

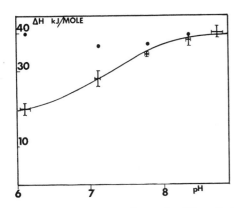

Figure 2. Effect of pH on ΔH_{obs} for PCr splitting. The crossbars each show the mean result (and SE in each direction) from four observations of the heat produced by enzymic splitting of phosphocreatine in histidine/ammonium buffer at various pH values. No magnesium was present. The experiments were made at 25°C. (——) Variation in ΔH_{obs} predicted from the dissociation heats of phosphate and of the buffer; (●) result of subtraction from these observations of the amount of heat produced in control experiments in which inorganic phosphate and creatine were added to the enzyme buffer solution. They are thus values for ΔH_1.

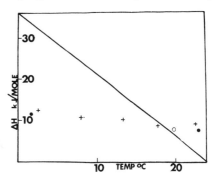

Figure 3. Determination of the heat of deionization of orthophosphate (reaction 2). (+) Calculated from the pK data observed by Bates and Acree (1945) in the presence of 0.1 M KCl; (——) calculated from pK data of Bernhardt (1956); (○) calorimetric measurement by Podolsky and Morales (1956); (●) this work, Table 1.

Table 2. Observations on the Lohmann Reaction at pH 7, pMg $\simeq$ 2

$$MgaPCr + Mg_bH_cADP + H^+ \rightarrow Mg_dH_eATP + Cr \quad 13$$
$$Buffer\ H^+ \rightarrow Buffer + H^+ \quad 3$$

Assuming a, b, c, d, and e to be independent of the nature of the buffer we calculated ΔH_{13} as follows.

	Phosphate Buffer	Carnosine Buffer
ΔH_3	$+8.23$	$+31.9$ kJ/mole
OBS^D in forward reaction	-9.33	$+12.55$ kJ/mole
OBS^D in backward reaction	$+0.63$	-1.24 kJ/mole
H_{obs} (= $\Delta H_3 + \Delta H_{13}$)	-9.96	$+13.79$ kJ/mole
ΔH_{13}	-18.19	-18.11 kJ/mole
pK_{13}	2.34	2.01

uncertainties about this value are the temperature dependence of ΔH_1, ΔH_4, and ΔH_5 and the possibility that magnesium ion buffering ought to be considered. It is extremely unlikely that these uncertainties can explain the difference between -34 kJ/mole and the value reported by Wilkie for $(h + w)/\eta$ (-46 kJ/mole) in IAA/N$_2$-poisoned muscle at 0°C. The discrepancy is only increased by the more recent observations of Wilkie and Kretzschmar (personal communication) that $(h + w)/\eta$ is in some groups of frogs as high as -60 kJ/mole. We must conclude that reactions 1 through 5 do not completely describe the processes occurring in muscle during and immediately after contraction. Calorimetry cannot directly tell us whether the missing reaction is an "interesting" or a "trivial" one. Since in muscle PCr splitting occurs by way of ATP splitting followed by the Lohmann reaction, more information can be gained by considering these processes separately.

Heat of the Lohmann Reaction

The Lohmann reaction can be represented by the following coupled equilibria:

$$H^+ + ADP^{3-} + PCr^= \rightarrow ATP^{4-} + Cr \quad (6)$$
$$PCr^= + Mg^{++} \rightarrow MgPCr \quad (5)$$
$$ATP^{4-} + H^+ \rightarrow HATP^{3-} \quad (7)$$
$$ADP^{3-} + H^+ \rightarrow HADP^= \quad (8)$$
$$Buffer\ H^+ \rightarrow Buffer + H^+ \quad (3)$$
$$ATP^{4-} + Mg^{++} \rightarrow ATPMg^= \quad (9)$$
$$ADP^{3-} + Mg^{++} \rightarrow ADPMg^- \quad (10)$$
$$HATP^{3-} + Mg^{++} \rightarrow MgHATP^- \quad (11)$$
$$HADP^{++} + Mg^{++} \rightarrow MgHADP \quad (12)$$

Complete calorimetric investigation of all these reactions is a formidable undertaking. Instead observations have just been made of ΔH_{obs} near pH 7 in 10 mMolar magnesium. As the Lohmann reaction does not go to completion, separate observations were made of the relatively small

amount of the "backwards" reaction, i.e., starting with ATP and Cr instead of ADP and CrP. As a check on the results two sets of experiments were performed, one in phosphate buffer and the other in carnosine buffer. The results of these experiments are shown in Table 2. To make the comparison between the two series of experiments, reaction 6 can be considered together with reaction 4, and 7 through 12 as reaction 13. The two series of experiments can then be seen to agree well in the value for those parts of the overall process not dependent on the buffer. The result obtained for ΔH_{obs} in carnosine buffer should (if we can justifiably neglect the temperature difference) give the value reasonably appropriate for muscle at 0°C, the pH and pMg having been chosen to represent the in vivo values. A calculation has also been made of the effects of variations in pH and pMg on ΔH_{obs} using data on reactions 7 through 10 given by Alberty (1968, 1969). The result shown in Fig. 4 is that these effects are not very large. For likely muscle conditions the Lohmann reaction in the direction in which it

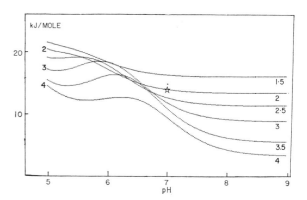

Figure 4. Calculation of ΔH_{obs} for the Lohmann reaction as a function of pH and pMg. Curves calculated according to the formulas given by Alberty (1968). Each line refers to the pMg value shown by the number on the line. The calculation refers to 25°C and uses the data given by Alberty (1968, 1969) for reactions 7 through 12 and the data in Table 1 for reactions 3 and 4. The value of ΔH_1 used was chosen to make ΔH_{obs} at pH 7, pMg 2, equal to the observed value of $+14$ kJ/mole in the carnosine buffer (Table 2). The star marks these experimental conditions.

proceeds during contractions should be endothermic to the extent of between 8 and 15 kJ/mole.

Comparison with in vivo observations. In principle the heat production due to the Lohmann reaction can be isolated in living muscle by the use of FDNB. As shown originally by Cain and Davies (1962) this inhibitor selectively blocks the Lohmann reaction. Comparison of the heat production before and after poisoning should thus reveal this contribution to the total heat production. The comparison has been made by Aubert (1964). He states that the steady rate of heat production during a tetanus is reduced 30% by the poisoning, although there is no reduction in the tension developed. Taken at face value this observation suggests that the overall process prevented by poisoning is exothermic by about 15 kJ/mole and is thus not just the Lohmann reaction as defined by the equations above. However Aubert reports a reduced rate of tension development and relaxation which suggests that the poisoning might have reduced the rate at which the muscle split ATP during contraction and hence decreased the heat production. We can only conclude that these experiments provide no support for the idea that the net process prevented by poisoning is endothermic. It seems very possible that the unknown exothermic process which accompanies PCr splitting in muscle is also prevented by the poisoning. The simplest explanation of this would be to suppose the unknown process to be linked to PCr splitting but not to ATP splitting.

Heat Produced by Splitting of ATP

I have made no observations of the heat produced by ATP splitting. The available information has been summarized by Alberty (1968, 1969) who, after taking account of Mg^{++} and H^+ ion involvement (but not the buffer reaction), suggests that the value appropriate to pH 7, pMg 2.5 at 25°C is −18 kJ/mole. Alberty has also made calculations for 0°C (personal communication) which give a result of −20 kJ/mole. Adding to this ($\eta_h \times -27$) kJ/mole for the buffer reaction, we obtain an estimate of −41 kJ/mole for ΔH_{obs} for ATP splitting at 0°C in vivo conditions. Another estimate can be obtained by subtracting my result for ΔH_{obs} for the Lohmann reaction (+12 kJ/mole) from that for PCr splitting (−34 kJ/mole). This gives −46 kJ/mole in reasonable proximity to the result obtained from Alberty.

Comparison with in vivo observations. Heat production and ATP splitting in FDNB muscles have been compared recently by Homsher et al. The value they obtain, −42 ± 4 kJ/mole, is certainly rather close to the estimated ΔH_{obs} values

for ATP splitting. There is no evidence in this comparison of the extra exothermic process apparently accompanying PCr splitting in muscle. The approximate agreement is all the more surprising in that the studies of Kushmerick and Davies (1969) and Dydynska and Wilkie (1966) have shown that ATP splitting in FDNB-poisoned muscle is accompanied by extra processes: myokinase activity, deamination of the AMP formed, and formation of hexose phosphates. All these processes are probably exothermic (Woledge, 1971). This preliminary result for ATP splitting is so different from that for PCr splitting that a closer study of the situation is called for.

Discussion

The results presented above suggest strongly that an unknown exothermic process accompanies PCr splitting in muscle but probably does not accompany ATP splitting. For this reason the Lohmann reaction appears exothermic in muscle, whereas it is endothermic in vitro. If this is correct the unknown process cannot be absolutely essential to contraction for in FDNB-poisoned muscle contraction, albeit slowed, occurs without it. There could be many hypotheses about the nature of this reaction. It is known not to be ATP splitting, formation of hexose phosphates, or formation of lactic acid. One hypothesis worth mentioning is that the missing reaction is an unbinding of PCr from muscle protein when it is split. That this occurs has already been suggested by Hill and Parkinson (1931) to explain the anomalously large change in osmotic pressure they observed during muscle contraction. The autoradiographic studies of D. K. Hill (1962) suggested that PCr was highly localized within the sarcomeres and therefore presumably bound. If it is indeed true that the majority of PCr is bound, and if the binding reaction should happen to absorb 15–20 kJ/mole, the calorimetric discrepancies would be largely resolved. Until the discrepancy about PCr splitting is resolved by calorimetric experiments on these or other possible reactions, it would be wise to suspend judgement as to the energetic balance in IAA/N_2-poisoned muscles. This means that measurements of the amount of PCr split during contraction cannot at present be confidently used to assess the energetic costs of contraction and that more complicated studies, such as those of Gilbert et al. of the relation of PCr splitting to contraction, are for the present extremely hard to interpret.

Acknowledgments

I should like to express my thanks to Dr. Paul Canfield for his help with the PCr splitting experiments, to Mr. Stuart Bruce for his help with the

Lohmann reaction experiments, to Mrs. Claude Gilbert for performing the PCr analyses, to Dr. R. D. Keynes for the loan of the Calvet calorimeter, and to the M.R.C. for the purchase of the LKB calorimeter.

References

ALBERTY, R. A. 1968. Effect of pH and metal ion concentration on the equilibrium hydrolysis of adenosine triphosphate to adenosine diphosphate. *J. Biol. Chem.* **243**: 1337.

———— 1969. Standard Gibbs free energy, enthalpy and entropy changes as a function of pH and pMg for several reactions involving adenosine phosphates. *J. Biol. Chem.* **244**: 3290.

AUBERT, X. 1964. Tension and heat production of frog muscle tetanized after intoxication with FDNB. *Pflügers Arch. ges Physiol.* **281**: 13.

BATES, R. G. and S. F. ACREE. 1945. pH of aqueous mixtures of potassium dihydrogen phosphate and disodium hydrogen phosphate at 0°C to 60°C. *J. Res. Nat. Bur. Std.* **34**: 373.

BERNHARDT, S. A. 1956. Ionisation constants and heat of tris and phosphate buffers. *J. Biol. Chem.* **218**: 961.

CAIN, D. F. and R. E. DAVIES. 1962. Breakdown of ATP during a single contraction of working muscle. *Biochem. Biophys. Res. Comm.* **8**: 361.

CARLSON, F. D. 1963. The mechanochemistry of muscular contraction, a critical revaluation of *in-vivo* studies. *Prog. Biophys.* **13**: 262.

DYDYNSKA, M. and D. R. WILKIE. 1966. The chemical and energetic properties of muscles poisoned with FDNB. *J. Physiol.* **184**: 751.

GELLERT, M. and J. M. STURTEVANT. 1960. The enthalpy change in the hydrolysis of creatine phosphate. *J. Amer. Chem. Soc.* **82**: 1497.

GILBERT, C., M. KRETZSCHMAR, D. R. WILKIE, and R. C.

WOLEDGE. 1971. Chemical change and energy output during muscular contraction. *J. Physiol.* **218**: 163.

HILL, A. V. and R. L. PARKINSON. 1931. Heat and osmotic change in muscular contraction without lactic acid formation. *Proc. Roy. Soc.* (*London*) B **108**: 148.

HILL, A. V. and R. C. WOLEDGE. 1962. An examination of absolute values in myothermic experiments. *J. Physiol.* **162**: 311.

HILL, D. K. 1962. The location of creatine phosphate in frog's striated muscle. *J. Physiol.* **164**: 31.

HOMSHER, E., W. F. H. M. MOMMAERTS, N. V. RICCHIUTI, and A. WALLNER. 1972. Activation heat, activation metabolism and tension-related heat in frog semitendinosus muscles. *J. Physiol.* **220**: 601.

KUSHMERICK, M. J. and R. E. DAVIES. 1969. The chemical energetics of muscle contraction. II. The chemistry, efficiency and power of maximally working sartorius muscles. *Proc. Roy. Soc.* (*London*) B **174**: 315.

MEYERHOF, O. 1930. *Die chemischen Vorgänge in Muskel.* Springer, Berlin.

MEYERHOF, O. and W. SCHULZ. 1935. Uber die enzymatische Synthese der Kreatinphosphosäure und die biologishe Reakionsform des Zuckers. *Biochem. Z.* **281**: 292.

PIN, P. 1965. Determination de l'enthalpie d'hydrolyse de phosphagenes. *J. Chim. Phys.* **62**: 591

PODOLSKY, R. J. and M. F. MORALES. 1956. The enthalpy change of adenosine triphosphate hydrolysis. *J. Biol. Chem.* **218**: 945.

SMITH, I. C. H. 1972. Energetics of activation in frog and toad muscle. *J. Physiol.* **220**: 583.

STELLA, G. 1928. The combination of carbon dioxide in muscle, its heat of neutralisation and its dissociation curve. *J. Physiol.* **68**: 49.

WILKIE, D. R. 1968. Heat, work and phosphoryl-creatine breakdown in muscle. *J. Physiol.* **195**: 157.

WOLEDGE, R. C. 1971. Heat production and chemical change in muscle. *Prog. Biophys. Mol. Biol.* **22**: 37.

Regulation of Force and Speed of Shortening in Muscle Contraction

F. J. JULIAN AND M. R. SOLLINS

Department of Muscle Research, Boston Biomedical Research Institute, Boston, Mass. 02114 and
Department of Neurology, Harvard Medical School, Boston, Mass. 02115

The work of Gordon et al. (1966a,b) provided strong evidence for believing that the fundamental event in striated muscle force generation is an interaction between the cross-bridges from the myosin-containing thick filaments and sites on the actin-containing thin filaments. (See A. F. Huxley, 1971, and H. E. Huxley, 1971, for details and present status of sliding filament theory for muscle contraction.) It is clear from the work of Gordon et al. that, in the tetanic steady state condition, the amount of force developed over most of the length-tension diagram can be simply related to the number of cross-bridges available for interaction with thin filament sites. They were also able to account very well for a plateau in the sarcomere length-tension relation by showing that it coincided with the region in the middle of the thick filaments where cross-bridges are absent (H. E. Huxley, 1963).

There is also much evidence for believing that the sole effect of electrical excitation of contraction is to release activating calcium from an internal store in the muscle (see Ebashi and Endo, 1968; Ebashi et al., 1969, for a complete review and discussion of the regulation of contraction by calcium). Then, if one further assumes that calcium activation influences only the number of cross-bridges interacting with thin filament sites, plots of shortening speed against load should produce curves all having the same extrapolated intercept on the velocity axis, i.e., the same V_{max}. The intercept on the force axis would produce an estimate of P_0, which would vary according to the level of activation. These conclusions would be true regardless of whether the points used to construct curves were obtained from a tetanus or at some time during a twitch. It will be shown that this simple description of how calcium might regulate contraction is not sufficient to explain the results presented here.

There is already evidence available indicating that the regulation of contraction by calcium is a complex process. The results shown in Figs. 1 and 2 summarize the main findings of a recent paper by Julian (1971). The contractions described in this paper were induced by adding varying amounts of

calcium to the medium bathing single muscle fibers whose sarcolemmas had been made very permeable by a brief exposure to a glycerine solution. The very steep S-shaped relation between isometric steady force level and pCa shown in Fig. 1 can be explained by supposing that the amount of force is proportional to the number of cross-bridges interacting with thin filament sites, and that the number of thin filament sites available for interaction varies in a positively cooperative way with the pCa. The results given in Fig. 2 show that this explanation is not enough because it is evident that raising the pCa not only decreased the P_0, but also

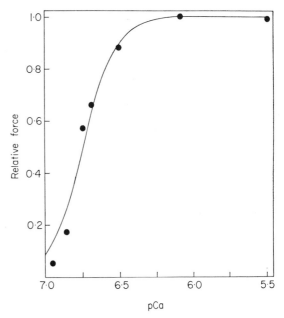

Figure 1. Steady state force as a function of pCa. The data points were obtained from a frog muscle fiber which had been briefly glycerinated and treated with detergent (for details, see Julian, 1971). Bathing medium contained (in mM): KCl, 100; $MgCl_2$, 1; ATP, 4; EGTA, 4; imidazole buffer, 10, pH 7.0. Contractions were induced by adding varying amounts of $CaCl_2$. Temperature 4°C. The curve was obtained using the Hill (1910) equation having a form relative force = $[Ca^{++}]^n/(K + [Ca^{++}]^n)$. The value used for n is 4, and $K = [Ca^{++}]_{0.5}$ where the subscript 0.5 indicates the $[Ca^{++}]$ for relative force = 0.5. From the data an estimate for $[Ca^{++}]_{0.5}$ was obtained: 1.8×10^{-7}. Finally the curve was drawn by transforming values for $[Ca^{++}]$ to pCa. (Modified from Julian, 1971.)

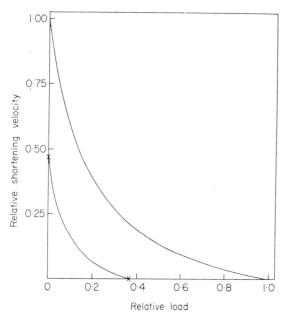

Figure 2. Force-velocity curves as a function of pCa. The data used to draw the curves were obtained from frog muscle fibers which had been briefly glycerinated and treated with a detergent and then placed in a bathing medium similar to the one described in the legend for Fig. 1 (see Julian, 1971, for details). Temperature 4°C. The upper curve describes the data obtained at pCa 5.5 and 6.1. In this case, $a/P_0 = 0.18$ where P_0 is the isometric force in low pCa solutions, and $b = 0.43$ ML/sec so that the extrapolated value for V_{max} is 2.39 ML/sec. The lower curve describes the data obtained at an average value for the pCa of 6.7. Here, $a/P_0 = 0.24$, where P_0 is the isometric force in high pCa solutions and $b = 0.27$ ML/sec, and the extrapolated value for V_{max} is 1.12 ML/sec. Note that in this graph all forces and velocities are expressed relative to the P_0 and V_{max} of full activation obtained in the low pCa solutions. In the high pCa solutions the steady isometric force developed decreased from the P_0 obtained in the low pCa solutions by about 0.6, and the extrapolated value for V_{max} decreased from that obtained in the low pCa solutions by only about 0.5. (Modified from Julian, 1971.)

decreased the extrapolated value for V_{max}. It should be noted, however, that there is a stronger effect on P_0 than on V_{max}. This means that calcium, in addition to regulating the number of thin filament sites available for cross-bridge interaction, must also influence the kinetics of the cross-bridge interaction cycle. The original publication (Julian, 1971) should be consulted for further evidence and discussion concerning regenerative release of calcium from an intact sarcoplasmic reticulum, presence of resting tension or a hidden internal load, and the conflict existing with other published work (Podolsky and Teichholz, 1970).

Work with Living Muscle Fibers

The foregoing indicates that activating calcium seems to have an effect on both force and speed of shortening in muscle contraction. Additional

evidence bearing on this important problem would be very useful. One possibility would be to compare the force-velocity properties obtained from living, electrically excited muscles in the tetanic, presumably fully activated, steady state with those obtained from the same muscles in states in which the level of activation is presumably decreased, e.g., at various times in the twitch. This has already been done for whole muscle by Jewell and Wilkie (1960), but their results do not seem to be completely decisive (Podolsky and Teichholz, 1970; Julian, 1971). The work to be described next was undertaken with the hope that it would lead to a more nearly certain conclusion regarding the effect of various levels of activation on the force-velocity properties of muscle.

Methods

Single twitch fibers together with pieces of tendon at either end were dissected from the anterior tibial muscles of frogs (*Rana pipiens*). The anterior tibial muscle was used because it is possible to obtain short fibers (L_0 about 5–10 mm) from this muscle (Blinks, 1965). The total V_{max} of these short fibers was smaller than for long fibers. After the dissection, small holes were cut in each tendon. The fiber was then transferred to the experimental chamber and stainless steel wire hooks from the force transducer and servo motor shaft were passed through the holes in the tendons. The compliance of the tendons was reduced by firmly knotting the tendons to the wires using 10-0 nylon suture material. The nylon loops forming the knots were placed about 0.15 mm distal to the insertions of the fiber into the tendons (see Fig. 3). The Ringer solution contained (in mM): NaCl, 115; KCl, 2.5; $CaCl_2$, 1.8; Na_2HPO_4, 2.15; NaH_2PO_4, 0.85; pH 7.2. The temperature of the solution in the experimental chamber was continuously monitored with a thermistor placed in the channel immediately behind the servo arm. The temperature was kept at 0°C for these experiments.

An isolated fiber was examined microscopically while in the experimental chamber before an experiment began. The muscle length was adjusted so that the average sarcomere length was 2.25 μ. An internal shortening of, at most, about 0.015 times L_0 probably occurred during the isometric contraction phase so that the average sarcomere length decreased to about 2.2 μ before shortening began in most cases. The total shortening allowed was prevented from exceeding about 0.07 times L_0, which means that the average sarcomere length did not decrease below 2.05 μ; i.e., force was generated and shortening allowed only in the plateau region of the sarcomere length-tension diagram. Also resting tension was negligible in the plateau

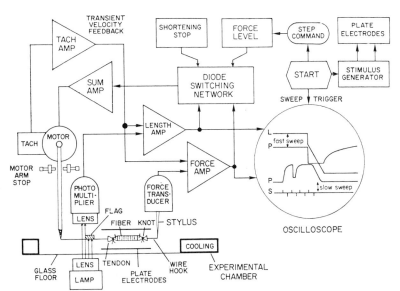

Figure 3. A block diagram of the controlling, stimulating, and recording system used to obtain force-velocity data from living, electrically excited single frog muscle fibers. *START* refers to a push button used to trigger in sequence the oscilloscope sweep, stimulus generator, and command signal for the servo system. *TACH* stands for tachometer, *AMP* for amplifier, *L* for length, *P* for force, and *S* for stimulus. Additional information is included in the text. See also Julian (1971) and Gordon et al. (1966a). (Modified from Julian, 1971.)

region. At the end of an experiment, the striation spacing was checked to see that no significant non-uniformities had developed.

The force transducer used in these experiments was of the capacitance change type similar to that described by Schilling (1960). It had a sensitivity of about 0.7 mV/mg and its drift was negligible. The resolution was high enough to measure relative loads at least as low as 0.05. The response to a very sudden displacement of the tip was complete in, at most, 2 msec. The length measuring system has already been described (Julian, 1971; see also Fig. 3). The sensitivity was set to 2 mV/μ. This made it possible to resolve relative length changes of less than 0.01. Electrical stimuli were delivered to the fiber through a pair of platinum plate electrodes, each of which was 1.5 cm long and 0.5 cm wide. The electrodes were spaced about 1 cm apart and arranged so that the fiber was in the center. The stimuli were capacitor coupled to the plate electrodes. The rising phase of the stimulus pulse was very rapid; the falling phase had a time constant of about 3 msec. In a tetanus, the stimulating pulses were alternately negative and positive going. The stimulus strength was set at 1.5 times the value obtained for the twitch threshold.

All of the experimental records (one series is shown in Fig. 4) have the same format. The lowest pair of traces are slow sweeps of the force and stimulus. It was possible to trigger the fast sweep shown in the upper parts of the records at any time after the slow sweep had begun. In this way it was possible to obtain expanded records of the length and force during rapid changes. The slow trace was used to measure forces and loads. A force baseline was obtained by passing a straight line through the initial and final segments of the slow force trace. The velocity of shortening was obtained from the fast length trace by calculating the slope of a straight line passed through the segment of length trace beginning about 5 msec after the force step was applied and ending just before shortening was brought to a halt by the shortening stop. The fast force trace was used to record the details of the force step.

The experiments were carried out by allowing a stimulated fiber to develop force under servo length control up to the predetermined time at which a force step was imposed and shortening under various clamped loads was allowed. It was necessary to include a transient velocity feedback loop, as shown in Fig. 1 and described by Gordon et al. (1966a), in order to prevent oscillations during the force step. The tachometer signal was passed through a small series condenser so that significant damping was present only during very rapid transients. A small amount of shortening was allowed and then the servo automatically switched back into length control. This sequence of events can be followed in the experimental records shown in Fig. 4.

Data were obtained from all fibers in an initial tetanus run, in a final tetanus run, and in a twitch series which was carried out in the interval between the two tetanus runs. The fiber was allowed to rest at least 5 min between stimulations. The following notation is used in this paper. P_0 is the steady maximum force developed during an isometric tetanic contraction. P_m is the maximum force developed in an isometric twitch. $P(t)$ is the force developed in a twitch just before the instant at which a force step is applied and shortening allowed. P_1 is the clamp force, or load, under which the fiber is allowed to shorten. In these experiments

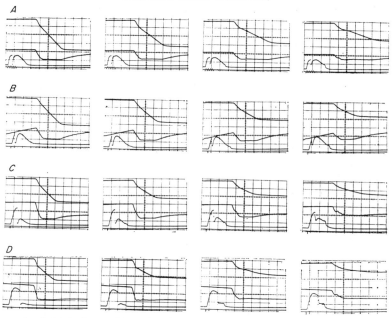

Figure 4. Typical records of shortening under various loads obtained from one of the six fibers of the series. Each photograph contains four traces: the lower pair were recorded at a slow-sweep speed, and the upper pair show the response at the time of release to the imposed load recorded at a fast sweep. In the fast sweep, the upper trace is the length and the lower is force; in the slow sweep, the upper trace is the force and the lower shows the times of occurrence of the stimulus. In each row loads are progressively heavier from left to right. A, initial tetanic series. Fast-sweep speed, 10 msec/div; slow-sweep speed, 500 msec/div. Length calibration, 0.1 mm/div; force calibration, 150 mg/div or 3.0 kg/cm²/div. B, series before the twitch peak. Fast-sweep speed, 10 msec/div; slow-sweep speed, 200 msec/div. Length calibration, 0.1 mm/div; force calibration, 75 mg/div or 1.5 kg/cm²/div. C, series at the twitch peak. D, series after the twitch peak. For C and D. the sweep speeds and length and force calibrations are the same as for B. Final control tetanic series not shown because of very close similarity to initial series. Single muscle fiber: initial length L_0, 5.9 mm; initial average sarcomere length, 2.25 μ. Temperature 0°C.

the average value for the ratio P_m/P_0 was 0.8. The distance between the points of insertion of the fiber into the tendons is L_0. The sudden length decrease required to drop the force to the desired load is ΔL. The average value for $\Delta L/L_0$ to drop P_1/P_0 to about 0.05 was found to be 0.015. The velocity is given in muscle lengths/sec (ML/sec). The velocity in μ/half-sarcomere/sec can be calculated from the values given for L_0 and sarcomere length.

In each fiber, values for a/P_0 and b were calculated using a linear form of the Hill (1938) equation, and a hyperbola was fitted to the data obtained from the initial and final tetanus runs. Data were also obtained in the intervening twitch series. The ratio between the twitch velocity and the velocity obtained from the steady state tetanic curve at the same load in the same fiber was calculated. The resulting ratios were used in Figs. 6 to 8 to locate the twitch velocity points properly with respect to the average tetanic hyperbolic relation given by the solid curve. That is, for a given load, the plotted value is $(V_t/V_T)\bar{V}_T$, where V_t and V_T are twitch and tetanus velocities, respectively, in the same fiber and $\bar{V}_T$ is the average over all fibers.

An initial series of experiments was done during the winter of 1970–1971. However stored winter

frogs tend to be listless, have little blood, and their muscles are pale and granular. A series of experiments was carried out in the spring and summer of 1971 in order to make certain that the results did not depend upon the condition of the frogs. The frogs used in the second series of experiments were freshly caught, very active and contained copious amounts of blood. Nevertheless the average value for P_m/P_0 was only 0.8. This could have been, in addition to a series elastic element effect, partly the result of a failure to activate the whole of a fiber cross section by a single stimulus at 0°C. A core of inactive myofibrils would not have seriously affected the validity of the results for two reasons. First, the fibers could be passively shortened and extended over most of the range of shortening allowed by applying only negligibly small forces. This indicates that a core of inactive myofibrils would not have produced a significant internal load. Second, in the final analysis, the twitch loads were expressed relative to the tetanic P_0 and a core of inactive myofibrils would only have contributed to an underestimation of the extrapolated P_0 appropriate for a given level of activation in the twitch. The main results obtained from the summer frogs did not differ from those obtained from the winter frogs. In this paper results are presented which were obtained from muscle fibers dissected from six different summer frogs.

Results

The aim of this work was to determine the relative force-velocity relation at various times in a twitch and compare it with that obtained from the tetanus. A typical experiment obtained from one of the six fibers forming the series is shown in Fig. 4. The experiment begins and ends with a sequence of loads applied in the tetanus (final tetanus not shown). In between, loads were applied at various times in the twitch as indicated in the figure. Note, in general, that the shortening records have segments of reasonably constant slope, i.e., constant velocity, in the interval following the initial transient and before shortening was brought to a halt by the shortening stop. This provides evidence for believing that the level of activation did not change significantly in this interval. Also measurements were never made at or after the time of appearance of a "shoulder" (A. F. Huxley and Simmons, 1970) in the falling phase of the twitch. These constraints of constant velocity of shortening and avoidance of the shoulder made it impossible to obtain force-velocity data late in the twitch.

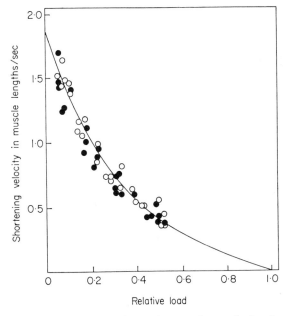

Figure 5. Pooled tetanic steady state force-velocity data obtained from six different single muscle fibers. (○) First tetanic series; (●) final control tetanic series. Relative load refers to P_1/P_0. In each of the experiments, values for a/P_0 and b were calculated and a hyperbola was fitted to the data using points from both the initial and final control series. The average value for a/P_0 is 0.39; the average for b is 0.72 ML/sec. The solid curve was drawn using these average values. Note that the experimental points are well fitted by the curve, and that the extrapolated value for V_{max} is almost 2 ML/sec at 0°C.

The results obtained from the analysis of the experimental records are shown in Figs. 5 to 8. The data from tetani are presented in Fig. 5. The hyperbola drawn was calculated using the average of the values obtained for a/P_0 and b and used to fit the data from tetani in each individual fiber. The extrapolated value for V_{max} is nearly 2 ML/ sec. This agrees well with Hill's recent results (1970) obtained by suddenly introducing a small amount of slack during a tetanus and measuring the time taken to redevelop force. The average values for a/P_0 and b/V_{max} are larger than those previously reported for frog sartorius muscle at the same temperature (Hill, 1938). However these values are very similar to those reported by Edman and Grieve (1966) who used single fibers obtained from frog semitendinosus muscles. It should be noted in Figs. 6 and 8 that the solid curve shown in each of these figures was obtained from Fig. 5. Also the twitch velocities were plotted using the technique given in the Methods section. This technique is used mainly because it is a convenient way of locating the twitch velocity points with respect to the average hyperbolic relation for tetani. Results are the same if the data obtained from each of the individual fibers are plotted separately.

The simple hypothesis being tested is that varying the number of regulating sites filled with calcium on the myofibrillar proteins, i.e., varying the level of activation, influences only the number of cross-bridges interacting with thin filament sites in a force generating process. P_0 for the tetanus and $P(t)$, as defined in this paper and by Podolsky and Teichholz (1970) for the twitch, are taken to be valid indicators of the number of cross-bridges interacting with thin filament sites just prior to the application of a load, P_1. If the loads are plotted relatively, either P_1/P_0 for the tetanus or $P_1/P(t)$ for the twitch, then the load per cross-bridge is the same and meaningful comparisons can be made of the shortening speeds observed. If calcium determined only the number of cross-bridges acting, this method of plotting should cause all of the points obtained from the twitch series to fall along the solid curve fitted to the data obtained from the tetani.

It is evident from an examination of Fig. 6 that in no case do most of the twitch points fall near the steady state tetanic curve. The points obtained before the twitch peak tend to fall above the tetanic curve. The points obtained at the twitch peak generally fall below the steady state curve. The points obtained after the twitch peak fall considerably below the steady state tetanic curve. In particular, after the twitch peak the shortening speed at very light load is definitely decreased

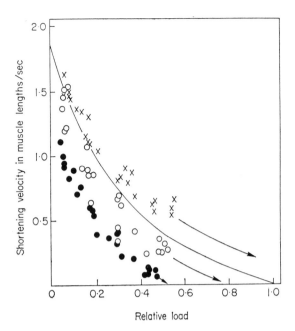

Figure 6. Pooled force-velocity data obtained from six single muscle fibers. (×) Before the twitch peak; (○) at the twitch peak; (●) after the twitch peak. The loads are expressed as $P_1/P(t)$ as described in the text. The solid curve is taken from Fig. 5 and shows the steady state tetanic relation. Note the generally poor fit of the points to the curve. The arrows indicate that in no case, except possibly at the twitch peak, would extrapolated intercepts on the relative force axis be near the value 1.0 applicable to the tetanic state.

below the value given by the tetanic curve, and it would appear that the shortening speed would approach zero for relative loads greater than about 0.5. These last two points are critical in deciding whether or not the hypothesis mentioned above concerning the role of calcium in the activation of contraction is correct.

The results shown in Fig. 6 indicate that $P(t)$ in a twitch is not a useful scale factor for expressing loads. In Figs. 7 and 8 the twitch data were replotted using the tetanic P_0 as a scale factor, i.e., P_1/P_0. This method of plotting gives the advantage that the extrapolated intercepts on the force and velocity axes of the curve fitted to the data points produce estimates of the P_0 and V_{max} capabilities for a certain level of activation at a particular time in the twitch. Obviously in contrast to the results shown in Fig. 6, no points can fall above the tetanic curve obtained in the fully activated state. The results given in Fig. 7 for before the twitch peak and at the twitch peak show that the points fall near the tetanic curve, particularly those associated with lighter loads, suggesting that the P_0 capability is more sensitive to changes in the level of activation. Presumably at these times in a twitch the level of activation is

still high and a clear separation from the steady state curve is not obtained.

The data given in Fig. 8 for after the twitch peak show an obvious difference from the tetanic fully activated curve. The small scatter in the data together with the availability of points for loads and velocities near zero means that the extrapolations of the computer-fitted curve give reliable estimates of the P_0 and V_{max} capabilities at this point in the falling phase of the twitch. The fact that the extrapolated value on the velocity axis is well below the tetanic V_{max} indicates that the level of activation also influences cross-bridge kinetics. It is important to note that the extrapolation on the relative load axis of the curve passing through the solid circles is decreased from the tetanic P_0 by about 0.6, whereas its extrapolation on the velocity axis is decreased by only 0.34 times the tetanic V_{max}. This is not in close agreement with the expectation given in the first part of this paper in which a decrease of about the same amount in the P_0 capability was associated with a decrease of 0.5 times the V_{max} of the fully active state. However the evidence given earlier was obtained under conditions in which the level

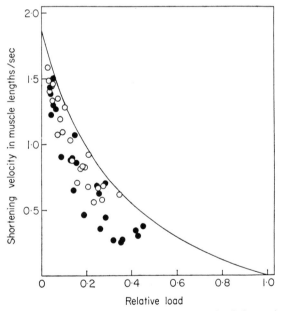

Figure 7. Pooled force-velocity data obtained from six single muscle fibers. (○) Before the twitch peak; (●) at the twitch peak. The loads are expressed as P_1/P_0 as described in the text. The solid curve is taken from Fig. 5 and shows the steady state tetanic relation. Note that all points fall below the curve, but that points associated with lighter loads fall closer to the curve. Before the twitch peak the ratio of the force developed just prior to the step to the twitch maximum ($P(t)/P_m$) was on average 0.68, at which point the time was on average 0.55 times that required to reach the twitch peak. At the twitch peak, $P(t)/P_m$ was equal to 1.

of activation was kept constant, whereas in the twitch work the level of activation was continuously decreasing in the falling phase.

The foregoing suggests an explanation for not finding in the twitch a decrease of at least 0.5 times the tetanic V_{max} when the estimated P_0 was decreased by about 0.6. There is in the living fiber a very pronounced transient in the length record following a sudden decrease in the force level before a constant shortening velocity develops. This transient becomes progressively longer as the load becomes heavier (see Fig. 4). This had the effect, particularly in the falling phase of the twitch, of progressively increasing the time required before a velocity reading could be obtained as the loads became heavier. Velocity readings for the heaviest loads were usually obtained 40–50 msec later than the readings for the lightest loads. The way in which this could influence the results is shown in Fig. 9. It can be seen that in effect each

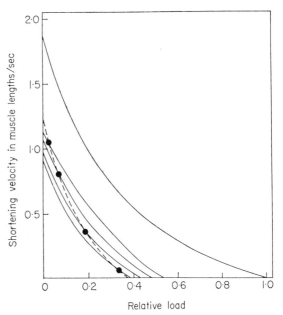

Figure 9. The effect of a decrease in the level of activation on the extrapolated values for P_i and V_i. The upper solid curve was obtained from Fig. 5 and gives the steady state tetanic relation. The solid circles belong to the dotted curve fitted to the P_1/P_0 points obtained after the twitch peak as shown in Fig. 8; they were chosen arbitrarily so that they fell along the curve at suitably spaced intervals. It is assumed that the level of activation decreases as the load increases because the transient is longer for the higher loads, and therefore the steady state velocity measurement had to be made at a later time. The lower family of solid curves was constructed in the following way: A V_i was estimated for the lightest load solid circle, a P_i was obtained from the equation $(P_0 - P_i)/P_0 = (6/5)[(V_{max} - V_i)/V_{max}]$, and a hyperbola was then passed through the three points. Similarly a P_i was estimated for the heaviest solid circle, a V_i was obtained from the equation $(V_{max} - V_i)/V_{max} = (5/6)[(P_0 - P_i)/P_0]$, and a hyperbola was passed through the three points. Hyperbolae were selected to pass through the remaining two solid circles and to satisfy the general criterion that $(V_{max} - V_i)/V_{max} = (5/6)[(P_0 - P_i)/P_0]$. A measure of the decrease in the level of activation is given by the difference between the P_i values for the solid hyperbolae passing through the lightest load and heaviest load solid circles, respectively; this is about 0.14. Notice that each solid circle belongs to a hyperbola whose intercepts give the best estimates for P_0 and V_{max} appropriate to a particular level of activation. The hyperbola described by the dashed line also passes through the four solid circles; its V_i value more nearly gives an estimate for the V_{max} appropriate to the higher level of activation, while its P_i is close to the estimate for the P_0 appropriate to the lower level of activation.

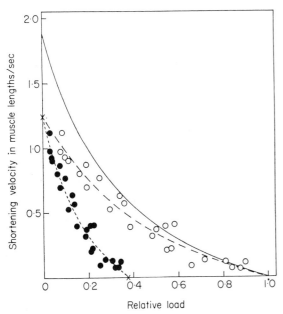

Figure 8. Pooled force-velocity data obtained from six single muscle fibers after the twitch peak. (●) Loads expressed as P_1/P_0 as described in the text; (○) loads expressed as P_1/P_i, where P_i is the intercept on the force axis of the curve passing through the solid circles. The solid curve is taken from Fig. 5 and shows the steady state tetanic relation. The dotted curve passing through the solid circles was fitted by a digital computer programmed to use a Newton-Raphson iteration to find values for a/P_0, b, and P_1/P_0 which produced the least square error between the fitted curve and the data points. The dashed curve passing through the open circles was drawn by shifting the force coordinates of the dotted curve to the right by a factor equal to P_0/P_i. Note that the dotted curve intercept on the force axis, P_i, is decreased from P_0 by 0.62, whereas the intercept on the velocity axis, V_i, is decreased by only 0.34 times V_{max}. The ratio of the force developed just prior to the step to the twitch maximum $(P(t)/P_m)$ was on average 0.84 at which point the time was on average 1.8 times that required to reach the twitch peak.

load belongs to the hyperbola describing the force-velocity relation at the time the velocity reading for the load was obtained. The intercept of the dashed line on the relative load axis gives an overestimate of the decline in P_0 capability with respect to the decrease in V_{max} capability approximately given by its intercept on the velocity axis.

Discussion

The results presented in Fig. 6 do not allow a simple explanation to be made of the relation between force and speed, and level of activation in a twitch. The following model work explores the effects of various schemes for activation, and this makes it possible to provide some plausible explanations for the data. A considerable amount of work has already been done with the model to be used here (Julian, 1969). The basic assumptions made, as proposed by Hill (1938), are that a force generator is connected to a series elastic element through which mechanical output is transmitted to the exterior. In addition various activation schemes are built into the model (see Julian, 1969) so that time-varying states, e.g., a twitch, can be studied. The force generator is the same as that given by A. F. Huxley (1957); it is based on a sliding filament mechanism in which force is produced by cross-bridges on the thick filaments interacting with sites on the thin, actin-containing filaments. The paper by Julian (1969) should be consulted for details concerning the model and for a description of the mathematical and numerical techniques used in the calculations. An appendix is included in this paper which describes the pertinent assumptions made concerning each of the activation schemes used. In particular it should be understood that "activation" as used here (and by Julian, 1969) refers very simply to the fraction of myofibrillar regulating sites assumed to be filled with calcium.

The first activation scheme, I, assumes that the level of activation determines only the number of actin sites available for interaction with cross-bridges. This would seem to be the most popular current view of the way in which calcium activates contraction. Activation scheme II includes scheme I. In addition it assumes that the level of activation also influences the value of g_2, A. F. Huxley's rate constant for breaking links which are opposing shortening. That is, g_2 is also multiplied by the activation function, which varies between zero and one. Justification for this kind of activation scheme can be obtained from the recent biochemical work of Lymn and Taylor (1971), in which it is shown that the dissociation of hydrolysis products is the rate-limiting step in that part of the mechanochemical cycle leading to the separation of actin from myosin by ATP.

One other comment concerning scheme II is worth making, since it will be shown that variation of g_2 produces a strong effect on V_{max}. According to Bárány (1967), mechanical V_{max} is proportional to the specific activity of the Mg^{++} and actin-activated ATPase of myosin, suggesting that this myosin ATPase is the rate-limiting step in a reaction which determines the maximum speed of shortening in a given muscle. It may be, however, in light of the work of Lymn and Taylor (1971) mentioned above, that the true rate-limiting

factor is the degree to which combination with actin promotes displacement of the products of hydrolysis: more rapid ejection of products leading to faster dissociation of a complex between cross-bridge and actin, opposing shortening, thereby raising V_{max}. Furthermore if the activation of contraction by calcium also influences the ejection of hydrolysis products and the dissociation of a cross-bridge and actin site interaction, then a variation of V_{max} with the level of activation could be explained.

It is assumed that in a twitch a single stimulus is sufficient to raise the level of activation from zero to one, or maximum, as shown in parts A and B of Fig. 10. Furthermore it is assumed that in a tetanus successive stimuli act to keep the level of activation at a maximum. If the level of activation is at a maximum in a steady state, then the two schemes are indistinguishable; e.g., they would produce the same force-velocity relation. In fact examination of part A and B of Fig. 10 shows that the twitches produced by the two different activation schemes postulated are quite similar as are the activation functions used in each case. Parts A and B of Fig. 10 also show the effect of the series elastic element on the rate of rise and peak value attained by the force in a twitch. The effect is substantial even though in the model with the elastic element present, it would take a sudden length decrease of only about 0.02 times L_0 to drop the steady tetanic force to near zero.

The time course of the generator stiffness, which is given by the number of cross-bridges linked to thin filament sites at any instant during a twitch, are also shown in parts A and B of Fig. 10. It is important to notice that generator stiffness and force output do not follow the same time course in a normal twitch. This is so in the kind of model being discussed because stiffness depends only on the number of cross-bridges linked to thin filament sites, while the force generated depends upon the distribution about the normal position of all the cross-bridges linked to thin filament sites. It is particularly noteworthy that, during the falling phase of the twitch, the stiffness tends to fall more rapidly than does the force output. This is because, even though there are progressively fewer cross-bridges linked to thin filament sites during the phase of declining tension in the twitch, the remaining linked cross-bridges bear a greater load as a result of the stretch produced by the relaxation of the series elastic element.

The model results presented in parts C, D, E, and F of Fig. 10 may help in interpreting the experimental data presented in the first part of this paper in Figs. 7 and 8. The model results were obtained by collecting force-velocity data during the model twitch in a way exactly the same as that

Figure 10. Computed model responses. Activation scheme I is shown in parts A, C, E; it assumes that the level of activation influences only the number of thin filament sites available for interaction with cross-bridges. Activation scheme II is shown in parts B, D, F; it assumes that the level of activation influences to a similar extent both the number of thin filament sites available for interaction with cross-bridges and the rate constant for breaking these interactions (see Appendix for details). A, B, isometric responses with normal (□) and very stiff series elastic element (△), and a normal generator stiffness (●). Note that the activation rises and falls faster than does normal tension, that there is a more rapid rise of tension to a higher level when the series elastic element is made very stiff, and that the generator stiffness decays more rapidly than does twitch tension. The value 1.0 on the ordinate refers to the maximum level of activation or to the maximum in tension and stiffness reached by the model in the tetanic steady state. C, D, force-velocity data obtained in the model twitch in the same way as that described for the experimental situation. In both graphs the solid curves describe the steady state, full activation force-velocity relations. Note that the two curves are identical indicating that the activation schemes are indistinguishable when activation is complete. For both curves, $a/P_0 = 0.22$ and $b = 0.40\ \mu$/half-sarcomere/sec. (○) Before the twitch peaks when for both schemes the force had risen to about one-half its maximum value and the activation function had the value 0.9 at the time of the release; (●) at the twitch peaks when the activation functions had the value 0.85 at the time of the release. Loads were plotted relative to the P_0 obtained in the tetanic, full activation state. Note the close similarity in the distribution of points in the two activation schemes, again indicating that the schemes are difficult to distinguish when activation is at or near a maximum. Note also the similarity to the

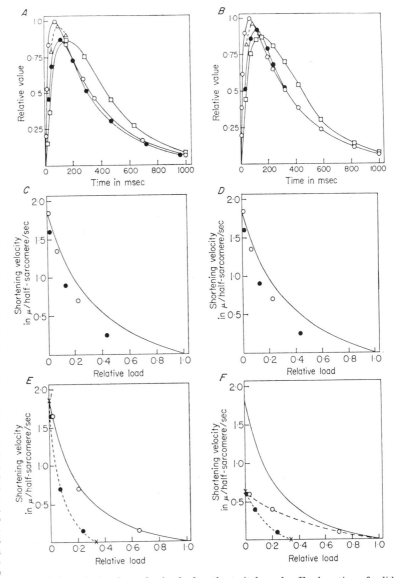

distribution found experimentally. E, F, model force-velocity data obtained after the twitch peaks. Explanation of solid curves as described in parts C and D. In both schemes the force had declined by about one-half its peak value and the activation function had the value 0.34 at the time of the release. (●) Loads plotted relative to the full activation P_0; (○) loads expressed relative to the intercept on the force axis of the dashed curves passing through the solid circles. Note that the resulting shift along the force axis causes the open circles to fall along the steady state curve in scheme I; in scheme II the failure of the shifted open circles to reach the steady state curve clearly brings out the dependence of V_{max} on the level of activation.

described in the Methods section for living fibers so that records were obtained similar to those shown in Fig. 4. Examination of parts C and D of Fig. 10 show that the points obtained using either activation scheme I or II fall near the steady state tetanic curve in a way similar to that obtained from living fibers (see Fig. 7). This is expected in the model because before the twitch peak and at the twitch peak the level of activation is high for both activation schemes so that a difference in force-velocity properties is difficult to detect.

This is not the case for later in the model twitch

when the level of activation is low. As shown in Parts E and F of Fig. 10, the difference between the two schemes becomes clearly apparent. For scheme I, which varied only the number of sites, the lightly loaded twitch point is very near the tetanic curve and the shifted points fall on the steady state curve. In scheme II, which also varied V_{max}, the lightly loaded twitch point is well below the tetanic curve, and the shift does not change the extrapolated value for V_{max}. This is very similar to the results obtained from living fibers shown in Fig. 8.

The model twitch loads were also plotted using $P(t)$ as a normalizing factor. The results were similar to those shown in Fig. 6 for living fibers. This may be a somewhat surprising result for activation scheme I in particular. However the presence of even a small series elastic element severely distorts the force output in a model twitch, and this makes $P(t)$ of little use for expressing loads even if the activation scheme varies only the number of sites available for interaction.

Conclusions

The results presented in Figs. 1, 2, and 8 indicate that the way in which calcium regulates contraction is complex (assuming in Fig. 8 that the sole effect of electrical stimulation is to release activating calcium). The model work using activation scheme II provides a plausible explanation for at least some of the effects observed in the activation of contraction by calcium. Perhaps the most important conclusion to be drawn is that it is not yet clear what factors determine V_{max}, which is in contrast to the relatively satisfactory current view that the number of cross-bridges interacting with thin filament sites determines P_0. It should be noted that V_{max} is determined in the model presented here by a balance of forces between cross-bridges aiding and opposing contraction. Even though the external net force is zero at V_{max}, there are still a considerable number, though fewer than at the isometric P_0, of cross-bridges attached to the filament sites. V_{max} is a consequence of the close coupling between the mechanical system of moving filaments and the biochemical steps of the reaction cycle between cross-bridges and thin filament sites, so that as the speed of shortening rises, increasing numbers of attached cross-bridges fail to break before they exert forces opposing contraction. It remains for future work to determine whether or not this view of how V_{max} arises is applicable to living muscle. It seems that such information must be available before much progress can be made toward understanding how calcium could affect both force and speed in muscle contraction.

It has also been proposed that V_{max} might be set by some kind of internal load, e.g., friction between the thick and thin filaments. However we have unpublished experiments in which very rapid length changes were applied to resting single muscle fibers whose sarcomere length was in the plateau region. This produced only negligible force changes, indicating that an internal load sufficient to set V_{max} does not exist in resting muscle. In addition we have other unpublished experiments indicating that a curve fitted to high load velocity points

obtained from a tetanus extrapolates very near to the isometric P_0 showing that an appreciable friction load is not present. The results of Gordon et al. (1966b), in which it was shown that the lightly loaded speed of shortening did not vary with sarcomere length, are not consistent with a view that V_{max} is set by a fixed internal load. Julian (1971) presented evidence against the possibility that a hidden internal load was responsible for his finding that V_{max} as well as P_0 depended on pCa. A. F. Huxley has shown (1971) that following very rapid force changes the velocity of shortening can be transiently much higher than the final steady state value (see also the model work of Julian et al., this volume). The very high transient speed of shortening implies that the steady state value is not determined by an internal load. These considerations argue against the view that internal load is an important determinant of V_{max}. One final comment concerns the model work using activation scheme II. In effect making the breaking rate constant g_2 vary with activation introduced an internal load in the force generator which was inversely dependent on the level of activation. This kind of internal load is different from those just discussed, which were either fixed or varied with activation, and its presence led, as described in the model work, to a variation of V_{max} with the level of activation.

Appendix

The activation schemes described in the Discussion were derived as follows (see A. F. Huxley, 1957 and Julian, 1969 for background).

n' = number of attached cross-bridges in a half sarcomere.

N = number of cross-bridges, or thin filament sites, whichever is assumed to be rate limiting, available for attachment in a half sarcomere.

$n'(x)$ = number of attached cross-bridges, in bridges/cm, having configuration x in a half sarcomere.

k = cross-bridge spring constant.

The force, F, generated in a half-sarcomere in dynes is:

$$F = \int_{-\infty}^{\infty} kn'(x)x\,dx$$

The generator stiffness, k_g, is given by:

$$k_g = \int_{-\infty}^{\infty} kn'(x)\,dx$$

Further: $n'(x)/N = n(x)$ = the fraction of cross-bridges having configuration x, in cm^{-1}.

Also: let $u = x/h$, where h is the positive maximum distance away from a normal configuration beyond which a cross-bridge cannot form an attachment to a thin filament site.

The following equation is identical to equation 1 given in the paper by Julian (1969):

$$\frac{\partial n'(x, t)}{\partial t} = f[N - n'(x, t)] - gn'(x, t) \qquad (1)$$

where f and g are rate constants defined by A.F. Huxley (1957).

Dividing through by N and h:

$$\frac{\partial n(u, t)}{\partial t} = f[1 - n(u, t)] - gn(u, t) \qquad (2)$$

Activation scheme I assumes a variation in N, which is taken to be the total number of thin filament sites in a half-sarcomere available for interaction with cross-bridges. In Eq. 1, N is replaced by $\gamma(t)N$ where $\gamma(t)$ is the postulated activation function varying between zero and one. Equation 2 then becomes

$$\frac{\partial n(u, t)}{\partial t} = f[\gamma(t) - n(u, t)] - gn(u, t) \qquad (3)$$

and the solution equivalent to equation 3 in the paper by Julian (1969) is

$$n(u, t) = \gamma(t)\frac{f}{f + g} - \frac{\gamma(t)f - (f + g)n_0}{(f + g)\exp[(f + g)t]} \qquad (4)$$

which holds only for the strictly isometric case.

Substituting into Eq. 4 the $f(u)$ and $g(u)$ functions given by A. F. Huxley (1957) yields the following solutions for the three regions of the u axis.

$$u < 0: n(u, t) = n_0(u) \exp(-g_2 t) \qquad (5)$$

$$0 \leq u \leq 1: n(u, t) = \gamma(t)\frac{f_1}{f_1 + g_1}$$
$$- \frac{\gamma(t)f_1 - (f_1 + g_1)n_0(u)}{(f_1 + g_1)\exp[(f_1 + g_1)ut]} \qquad (6)$$

$$u > 1: n(u, t) = n_0(u) \exp(-g_1 ut) \qquad (7)$$

In activation scheme II, g_2, as well as N, was multiplied by the activation factor $\gamma(t)$ so that in the full solution Eq. 5 becomes

$$n(u, t) = n_0(u) \exp[-\gamma(t)g_2 t] \qquad (8)$$

Eqs. 6 and 7 remaining unchanged.

The methods used in the computations carried out on a XDS Sigma 3 digital computer were very similar to those described by Julian (1969). However a very much more rapid method for computing $n_0(u)$, which is equal to $n(u + \Delta L)$, from $n(u)$ was devised. Distributions of $n(u)$ were represented in the computer by arrays of discrete points, and the transformation of $n(u)$ into $n(u + \Delta L)$ was done simply at each time step of the calculation by beginning at one end of the u axis (depending on the sign of ΔL) and finding the value for n at $(u + \Delta L)$ by using a linear interpolation. This value was then stored in the n_0 array in the location corresponding to u. All positions in the n_0 array were quickly filled in this way.

Acknowledgments

The main support for this work came from the USPHS Research Grant AM 09891 from the National Institute of Arthritis and Metabolic Diseases. Additional valuable support was provided by USPHS Career Development Award NS 06193 from the National Institute of Neurological Diseases and Stroke, Massachusetts Heart Association, Inc., grant number 1012 and American Heart Association, Inc., grant number 71-898. (All grants were awarded to F.J.J.)

References

BÁRÁNY, M. 1967. ATPase activity of myosin correlated with speed of muscle shortening. *J. Gen. Physiol.* **50:** 197.

BLINKS, J. R. 1965. Influence of osmotic strength on cross-section and volume of isolated single muscle fibers. *J. Physiol.* **177:** 42.

EBASHI, S. and M. ENDO. 1968. Calcium ion and muscle contraction. *Prog. Biophys. Mol. Biol.* **18:** 123.

EBASHI, S., M. ENDO, and I. OHTSUKI. 1969. Control of muscle contraction. *Quart. Rev. Biophys.* **2:** 351.

EDMAN, K. A. P. and D. W. GRIEVE. 1966. The mechanical parameters of the contraction of single muscle fibres of the frog. *J. Physiol.* **184:** 21.

GORDON, A. M., A. F. HUXLEY, and F. J. JULIAN. 1966a. Tension development in highly stretched vertebrate muscle fibers. *J. Physiol.* **184:** 143.

———. 1966b. The variation in isometric tension with sarcomere length in vertebrate muscle fibers. *J. Physiol.* **184:** 170.

HILL, A. V. 1910. The possible effects of the aggregation of the molecules of haemoglobin on its dissociation curve. *J. Physiol.* **40:** 4.

———. 1938. The heat of shortening and the dynamic constants of muscle. *Proc. Roy. Soc. (London)* B **126:** 136.

———. 1970. *First and last experiments in muscle mechanics*, Cambridge University Press.

HUXLEY, A. F. 1957. Muscle structure and theories of contraction. *Prog. Biophys. Biophys. Chem.* **7:** 255.

———. 1971. The activation of striated muscle and its mechanical response. *Proc. Roy. Soc. (London)* B **178:** 1.

HUXLEY, A. F. and R. M. SIMMONS. 1970. Rapid "give"

and the tension "shoulder" in the relaxation of frog muscle fibers. *J. Physiol.* **210**: 32.

HUXLEY, H. E. 1963. Electron microscope studies on the structure of natural and synthetic protein filaments from striated muscle. *J. Mol. Biol.* **7**: 281.

———. 1971. The structural basis of muscular contraction. *Proc. Roy. Soc. (London)* B **178**: 131.

JEWELL, B. R. and D. R. WILKIE. 1960. The mechanical properties of relaxing muscle. *J. Physiol.* **152**: 30.

JULIAN, F. J. 1969. Activation in a skeletal muscle contraction model with a modification for insect fibrillar muscle. *Biophys. J.* **9**: 547.

JULIAN, F. J. 1971. The effect of calcium on the force-velocity relation of briefly glycerinated frog muscle fibers. *J. Physiol.* **218**: 117.

LYMN, R. W. and E. W. TAYLOR. 1971. Mechanism of adenosine triphosphate hydrolysis by actomyosin. *Biochemistry* **10**: 4617.

PODOLSKY, R. J. and L. E. TEICHHOLZ. 1970. The relation between calcium and contraction kinetics in skinned fibers. *J. Physiol.* **211**: 19.

SCHILLING, M. O. 1960. Capacitance transducer for muscle research. *Rev. Sci. Instrumt.* **31**: 1215.

An Interpretation of the Effects of Fiber Length and Calcium on the Mechanical Properties of Insect Flight Muscle

Roger H. Abbott

Agricultural Research Council Unit of Insect Physiology, Department of Zoology, Oxford University, England

Insect fibrillar flight muscle has many properties in common with vertebrate skeletal muscle. However, it is not capable of the constant velocity shortening at constant load which is the mechanical characteristic of vertebrate striated muscles, but is adapted to produce rapid oscillatory contractions of small amplitude. Consequently, the classical type of mechanical study, involving the measurement of force-velocity diagrams and the fitting of a hyperbola to them, is not suitable for investigations of the properties of insect flight muscle. Instead, periodic length changes must be used. The result of forcing a sinusoidal length change of low amplitude on insect flight muscle is that, over a certain range of frequencies, the resulting sinusoidal tension changes lag behind the length changes (Machin and Pringle, 1960; Jewell and Rüegg, 1966). It is this delay which allows the muscle to deliver work continuously to a resonant load without neural control. One possible way of measuring the mechanical properties of insect fibrillar muscle is to measure the phase and amplitude of the sinusoidal tension changes at different forced oscillation frequencies. Alternatively, step changes of length may be imposed and the form of the resulting tension changes observed. In either case the major feature of the response is an exponential delay between length changes and tension change, such as might result from a first-order reaction. The most important parameters of the response to small forced length changes are the magnitude of this delayed tension and the rate constant of its development.

Using glycerol-extracted dorsal longitudinal muscle of the water bug *Lethocerus cordofanus*, the effects of mean fiber length and calcium ion concentration on these parameters have been measured. This paper is concerned with the interpretation of these results in terms of a model of cross-bridge action.

Methods

The basic measurements which are required are the size and rate constant of the delayed tension under different conditions. This is done by measuring and then analyzing the frequency response of the muscle in each of the conditions concerned. A preparation is required in which the chemical conditions can be readily changed, and glycerinated fibers were therefore used in the experiments. They were bathed in a solution which contained 10 mM ATP, 10 mM $MgCl_2$, 70 mM KCl, 4 mM EGTA, 20 mM phosphate buffer pH 7.0, and a variable amount of $CaCl_2$ to give various degrees of activation.

The frequency response measurements must be made as rapidly as possible, and to this end a computer-controlled mechanical apparatus has been constructed (Abbott, 1970). The sinusoidal driving signals were obtained from a Solartron JM1600 digital oscillator, and this instrument also measured directly the response of the muscle at each frequency. The amplitude of the sinusoidal length changes was 0.2% peak to peak, which is equivalent to about 20 Å per half-sarcomere. This is only one-fifth of the amplitude of motion usually ascribed to the cross-bridges (Huxley and Simmons, 1971; Podolsky and Nolan, 1971), so the muscle was working under nearly isometric conditions.

Frequency responses may be plotted either on a Bode plot, in which the phase and amplitude are plotted separately against the logarithm of the frequency, or on a Nyquist plot. Only one graph is needed when the information is plotted in Nyquist form; at each frequency a point is plotted whose distance from the origin represents the amplitude, and the angle between the line from the origin to the point and the horizontal axis is the phase angle. Figure 1 illustrates the relationship between the sinusoidal changes, the Bode plot, and the Nyquist plot. The points on the Nyquist plot may also be referred to by their Cartesian coordinates; these have been called the Elastic Modulus and the Viscous Modulus by Machin and Pringle (1960). In engineering circles they are called the in-phase and quadrature components. The plots obtained from insect flight muscle are approximately semicircular in shape, the same shape as is produced by a first-order delay. For an exactly exponential response, the magnitude of the delayed tension is the radius of the semicircle and the rate constant

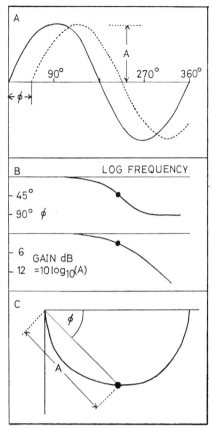

Figure 1. The graphical representation of frequency response measurements. A shows one cycle of the response at one particular frequency. The phase angle ϕ is 45°. In B, this one measurement is represented by the two points on the Bode diagram. The curves are similar to those produced by a simple first-order delay, in which the phase angle is 45° when the angular frequency is equal to the rate constant. In C, the point and curve are the Nyquist plot equivalents of those in B. If the point is specified by its cartesian coordinates, the horizontal axis is the in-phase axis and the vertical axis is the quadrature axis. The values of the coordinates are $A \cos \phi$ (in-phase) and $A \sin \phi$ (quadrature).

is the angular frequency at the lowest point. In the analysis of the force-velocity curves obtained from vertebrate muscle, an hyperbola with four parameters, a, b, P_0 and V_0, is drawn as close as possible to the experimental points. In an analogous way, a differential function whose major feature is an exponential delay was fitted to the frequency response data obtained from fibrillar muscle (Abbott, 1969; see also the appendix to this paper).

The experimental apparatus actually measures the response in Nyquist plot form, and this is the form in which the fitting process was carried out, so that each reading was given an equal weight. Trial values of the parameters (there are five, including the rate constant and magnitude of the delayed tension) were put into the differential function. At each frequency which was used in the

experiment, the position of the point on the Nyquist plot of the function was calculated. The distance between this point and the point actually measured was squared, and the squares for each frequency were then added. By a process of iteration, the values of the parameters were adjusted until this error was a minimum. Thus, each Nyquist plot yielded a least squares fit value of the magnitude and rate constant of the delayed tension.

Experimental Results

The experiment shown in Fig. 2 illustrates the effect on the frequency response of the muscle of the mean fiber length upon which the oscillations are superimposed. Each Nyquist plot is the average of 12 measurements. As the fibers are stretched the Nyquist plot grows in size, but does not change its shape, and the frequency points do not move around the curve. This means that the magnitude of the delayed tension for a given small superimposed length change is increased, but the rate constant of the first-order process causing the delay

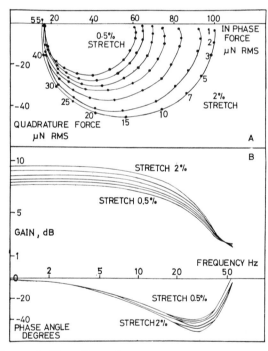

Figure 2. The effect of mean muscle length on the Nyquist and Bode plots of activated fibers. In A, the data is shown in Nyquist plot form. The smallest plot corresponds to the least fiber extension, and each successively larger plot is the result of stretching the fibers a further 0.25%. The amplitude of sinusoidal oscillation is the same in all cases. The figures around the largest curve indicate the frequencies at which the points were measured; the order is the same in all the plots. In B, the same data is replotted in Bode form. Note that there is only a small shift in the frequency at which the phase angle is minimum. This frequency is not directly related to the rate constant of the delay. The temperature was 25°C.

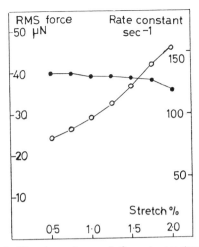

Figure 3. The dependence of the rate constant (●—●) and magnitude (○—○) of the delayed tension on the mean length of the fibers. The values were obtained from the Nyquist plots of Fig. 2. The small reduction of the rate constant at 2% stretch is not a consistent feature of the results and is probably due to a slight tendency of the fibers to go into the "high tension state" when they are highly activated (Jewell and Rüegg, 1966).

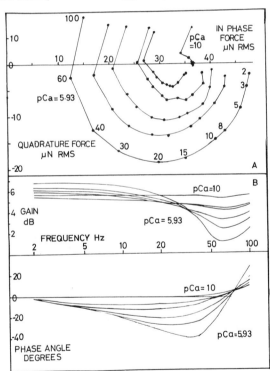

Figure 4. The effect of calcium ion concentration on the Nyquist and Bode plots of activated fibers, the same type of presentation as Fig. 2. In the virtual absence of calcium (pCa = 10) there is no delay and hence no oscillatory work at any frequency. Note that the gain and phase up to 40 Hz are almost constant in this relaxed condition. Each successively larger Nyquist plot (the average of 7) was measured at an increased calcium ion concentration; the values are shown in Fig. 5. In contrast to Fig. 2, the Bode plot (B) shows an increase in the frequency at which the phase angle is a minimum with an increase in calcium ion concentration.

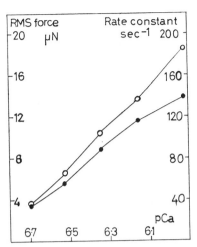

Figure 5. The dependence of the rate constant (●—●) and magnitude (○—○) of the delayed tension on the calcium ion concentration. Contrasting with the effect of stretch (Fig. 3), both are influenced by calcium.

is not affected. For comparison, the same data is reproduced in Fig. 2B in Bode plot form. The rate constants and magnitudes estimated from the results of Fig. 2 are shown in Fig. 3. The rate constant is hardly affected by stretch; the slight fall at the highest mean length is not a consistent feature of the experimental results. The magnitude of the delayed tension is increased markedly by stretching the muscle. In this example the increase is approximately linear, but at higher calcium ion concentrations the curve may be convex downwards.

All the results shown in Fig. 4 were measured at the same mean fiber length. The calcium ion concentration was varied by immersing the fibers in a series of solutions containing different amounts of $CaCl_2$. In the absence of added $CaCl_2$ there is no oscillatory work at any frequency. As the calcium ion concentration is increased, the amount of delayed tension, and hence oscillatory work, increases. At the same time there is a shift in the frequency at which the viscous modulus is minimum; as the calcium ion concentration is raised, this frequency increases. This means that calcium increases the rate constant of the process which limits the speed with which the active tension responds to a change in muscle length. Figure 5 shows the rate constant and magnitudes of the delayed tension changes extracted from the results of Fig. 4.

Interpretation of Results

One line of approach in attempting to understand the mechanism of muscle contraction is centered around the chemical reactions of the cross-bridge, or head of the myosin molecule. A scheme of reactions is drawn out, containing all the intermediate complexes which are thought to be necessary, and rate constants are ascribed to as many of the reactions as possible. In other models of the cross-bridge devised to explain mechanical results, the rate constants which have been considered are the rates of attachment and detachment (called f and g by Huxley, 1957) of the cross-links between the actin and myosin filaments (Podolsky and Nolan, 1971; Julian, 1971). Hopefully it will eventually be possible to identify f and g with two of the rate constants of the biochemical scheme. Insect flight muscle allows a rate constant to be determined directly, and the identification of this rate constant with the detachment rate of the cross-bridges was proposed by Thorson and White (1969). One thing is quite clear from the results presented here: Whichever process is involved, the mechanically measured rate constant is altered markedly by the calcium ion concentration. An alternative hypothesis, that calcium acts just as a switch controlling the number of bridges in action, was put forward by Podolsky and Teichholz (1970) because they found no effect of calcium on the maximum shortening velocity of skinned muscle fibers. However, there is no reason why the process whose rate is controlled by calcium should necessarily limit the maximum shortening speed. On the other hand, Julian (1971) believes that calcium does affect the maximum shortening speed, and so the evidence obtained from vertebrate muscle is so far inconclusive. Having established that calcium does affect a rate constant in fibrillar flight muscle, the next step is to try to identify it.

A very simple mechanical model of the cross-bridge is that there are two states, attached and tension producing, and detached and not tension producing. The transitions between the states are first-order processes; that is to say, the rate at which detached cross-bridges attach is proportional to the number of detached bridges, and the rate at which attached bridges detach is proportional to the number of attached bridges. The constants of proportionality are the rate constants of the two processes. In order to change the muscle tension it is necessary for the rate constants to be changed. We shall examine whether it is possible to explain the results described in the previous section on the basis of this model.

The following is a simplified version of the cross-bridge cycle of Thorson and White (1969). We assume in the first instance that the attachment rate is governed by the calcium ion concentration and muscle length

$$f = Qx \qquad (1)$$

where x is the muscle length and Q increases with calcium ion concentration. The tension, T, is assumed to be proportional to the fraction, n, of attached bridges

$$T = Fn. \qquad (2)$$

Finally, the rate at which the number of attached bridges is changing is equal to the rate at which they are attaching minus the rate at which they are detaching:

$$\frac{dn}{dt} = (1 - n)f - ng. \qquad (3)$$

Combining these we get

$$\frac{T}{x} = \frac{QF}{d/dt + (f + g)}. \qquad (4)$$

Thus, there is a first-order delay between length changes and tension changes, with a rate constant $f + g$. Thorson and White (1969) showed that if g is made to decrease with an increase of length, instead of assuming an increase in f, an exponential delay with the same rate constant is predicted. In the steady state, dn/dt is zero, and we see from (3) that

$$n = f/(f + g) \qquad (5)$$

The turnover rate, r, is proportional either to $(1 - n)f$ or ng, both of which give

$$r = fg/(f + g) \qquad (6)$$

Dividing (6) by (5), we see that the ratio of turnover frequency to tension is proportional to g. Making one further assumption, that during each cycle of attachment and detachment a fixed number of molecules of ATP is split, it follows that the ATPase activity required to maintain a given tension, the "tension cost" discussed by Pybus and Tregear (this volume), is proportional to the cross-bridge detachment rate.

What constraints can be put on this cycle by the results mentioned earlier? The rate constant of the delayed tension depends on $f + g$, but stretch does not affect it. Obviously, stretch must affect f or g, otherwise there would be no stretch activation of the ATPase. It must be that either one of the rate constants is much larger than the other and stretch affects the slow one only, so that the sum is hardly affected, or that f and g are affected equally in opposite directions. It is not possible for stretch to affect only the faster step, because it would not then influence the ATPase activity, which is limited by the slowest step. It is known that stretch does not alter the tension cost, since

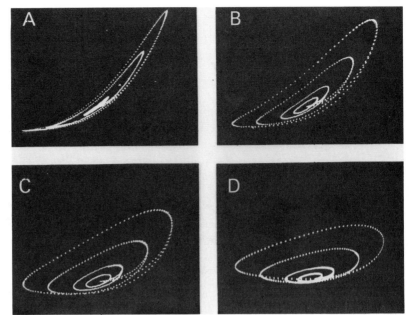

Figure 6. Computer solutions of the two-state model differential equation (3), when f is made proportional to the cube of muscle length. n, to which tension is assumed proportional, is plotted vertically, and length horizontally. The coefficient Q of the modified Eq. 1 is chosen so that at the highest amplitude of oscillation, the average value of f is 5. In all cases $g = 50$. The four diagrams differ in the frequency of oscillation: A, 1.5 Hz; B, 5 Hz; C, 10 Hz (this is very near the frequency of maximum work per cycle); D, 20 Hz.

ATPase activity and tension rise in proportion as the muscle is extended (Rüegg and Stümpf, 1969; Pybus and Tregear, this volume). Therefore, stretch cannot affect g, so it must affect f. It follows that f is much smaller than g, and therefore that only a small fraction of the cross-bridges is attached at any time. A typical turnover rate of the enzyme, which gives f, is 5 sec^{-1}, and a typical mechanically measured value for $f + g$ is 50–100 sec^{-1}. The fraction of attached bridges is therefore typically 0.1. This was shown by Thorson and White (1969) to be consistent also with the rate of change of ATPase activity with fiber extension. It also follows that the rate constant measured mechanically is g.

Now we can say which rates are affected by calcium. Since it increases the ATPase activity, it must increase f. Since it increases the mechanically measured rate constant, it must also affect g.

The experimental results show that as mean muscle length is increased, the amount of delayed tension obtained for a given small superimposed length change is increased. The straightforward interpretation of this result is that f is not proportional to x, as was assumed in Eq. 1, but is a nonlinear function of x, such as x^m, where m is greater than one. A linear increase of delayed tension with mean muscle length would require $m = 2$, but other results indicate that m must sometimes be higher. The test of this interpretation is to solve Eq. 4 with x replaced by x^m, and with large amplitudes of length change, and then compare the calculated tension response with measured ones. The equation is easily solved by numerical integration. Starting with the equilibrium value of n and $x = 0$, dn/dt is worked out from (1) (modified by putting x^m for x) and (3). Taking a small time interval, a new value of n can thus be worked out ($= n + (dn/dt) \Delta t$, where Δt is the time interval chosen). The new value of x can also be worked out, since the muscle is being driven at a fixed frequency. This is all the information needed to work out n after a further time interval Δt. Proceeding in this way, the complete response is built up. Figure 6 shows the results of this calculation at four different frequencies of length change, each at four different amplitudes of length change, with $m = 3$. For comparison, the high amplitude response of a fiber preparation is shown in Fig. 7. Further examples can be seen in papers by Rüegg (1968) and Steiger and Rüegg (1969). Qualitatively, several of the nonlinear features are reproduced by the modified model: (a) the curvature of the length tension diagram at low frequencies, (b) the increasing departure from an oval shape of the loops as the amplitude is increased, (c) the constancy of position of the lower limit of the loop as the amplitude is increased. (d) The actual shape of the loop, with the upper limb flattened and the lower one sagging, is also a feature of many experimental results.

One effect of this nonlinearity is to increase the efficiency of the muscle, because cross-bridge attachment is confined more to the times when the extension is greatest and the muscle is about to shorten, and therefore a greater proportion of the tension generation occurs when the muscle is shortening. Figure 8 shows length/tension loops

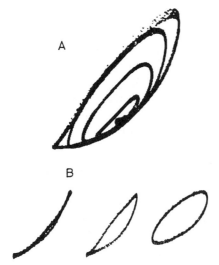

Figure 7. Actual high amplitude responses of muscle, taken from Pringle and Tregear (1969). *A* shows the effect of increasing the amplitude of oscillation at a fixed frequency. The result may be compared with *C* or *D* of Fig. 6. *B* shows the effect of a change in the frequency. The lowest frequency is at the left, the middle one is at twice this frequency, and the one on the right at four times the frequency. Note the curvature of the narrow loop at the lowest frequency, which may be compared with Fig. 6A, and the "sag" in the loop at the medium frequency, as in Fig. 6B.

calculated at the frequency of maximum work per cycle for $m = 1$, $m = 2$, and $m = 3$. The values of Q (Eq. 1) were adjusted so that the average value of f during the cycle was constant. This means that the ATPase activities corresponding to the three loops would be about the same, and this is confirmed by the fact that the calculated average tension is about the same (g is constant). There is an increase

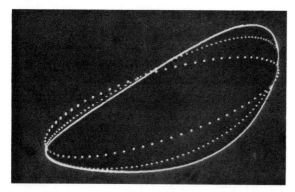

Figure 8. The increase in work output obtained from the two-state model when f is made a nonlinear function of stretch. The length tension loop with the widest spaced points is the result obtained when f is a linear function of the muscle length. The curve with the closer points is the result of making f depend on the square of the length, and the continuous curve results when f depends on the cube of the length. It is obvious that the area of the loop is increased by the nonlinearity. In each case the coefficient Q in the modified Eq. 1 was chosen so that the average value of f during the cycle was 5. g remained constant at 50. The frequency is 10 Hz.

in the area of the loops as m is increased, and hence an increase in efficiency.

It has been shown that on the basis of the simple two-state model, both f and g must be increased by the presence of calcium. It follows that as the calcium ion concentration is increased, the tension cost should rise, so that at higher degrees of activation a greater ATPase activity is needed to maintain a given tension. The only evidence available at the moment comes from experiments in which tension and ATPase have been simultaneously measured as the calcium ion concentration is raised (Schädler, 1967; Rüegg and Stümpf, 1969). On the basis of the model, the graph of ATPase against tension should curve upwards, but there is no sign of this in the results, and if anything the tension cost decreases with increasing calcium. A more sensitive test would be to obtain the tension cost at different levels of calcium by varying the muscle length as done by Pybus and Tregear (this volume). The data obtained by these authors was all at a high calcium ion concentration, and is therefore not relevant.

It is therefore not possible to account for all the results on the basis of the assumptions made earlier. One possible modification would be to define tension to be proportional to ATPase activity. This means writing, instead of (2)

$$T = F'ng$$

where F' is different from F.

The qualitative aspects of the mechanical properties predicted by the model are not altered by this substitution, because under conditions in which the length only is changing g is a constant. The effect of an increase in the calcium ion concentration will still be to increase both the magnitude and rate constant of the delayed tension. However, although this modification to the model serves to explain the results mentioned so far, it brings it into conflict with others. Several influences are known to alter the tension cost with a correlated change in the mechanically measured rate constant (Pybus and Tregear, this volume), which is a strong reason for believing that tension is related to the number of attached bridges. The problem is that calcium seems to increase the mechanical rate constant without increasing the tension cost, while other factors alter both.

The simple two-state model by itself cannot explain all the results, as there are not enough rate constants available to be affected differently by stretch, calcium, and the other factors such as temperature and ionic stretch used by Pybus and Tregear (this volume). The answer may be to split the attached state into two substates, one producing tension and the other not, so that tension cost could be varied separately from the overall detachment rate. This idea is very similar to the one proposed by White (this volume) to explain the tension transients observed when the phosphate buffer is replaced by histidine, and may be compared also with the proposal of Huxley and Simmons (1971). Alternatively, the two-state cross-bridge model may be able to account for the results when it is built into a distributed model of the sarcomere, as was done by Thorson and White (1969), with local stress or strain determining the rate constants instead of total muscle length as has been assumed here.

Appendix

In estimating the rate constant and magnitude of the delayed tension, allowance must be made for the fact that the resting elasticity of insect flight muscle contributes to the Nyquist plot of the active muscle. Up to frequency of about 30 Hz, the phase angle of the passive muscle is very small, and to a good approximation the muscle can be considered to be acting like a spring. Above this frequency, the phase angle becomes increasingly positive (see Fig. 4). To allow for the effect of the parallel elasticity, a purely elastic term is included in the differential function used to fit the results. There are two ways in which the interfering effects of the phase angle on the estimate of the rate constant may be reduced. The data may be restricted to frequencies at which the phase angle of the passive muscle is small, and this was done in fitting the experimental results obtained at different degrees of stretch. If the Nyquist plot of the passive muscle is available, it may be subtracted from the Nyquist plot of the active muscle before the fitting process is carried out. This was done for the experiments in which the calcium ion concentration was changed. These two procedures yield very similar values of the magnitude and rate constant of the delayed tension. In either case, the Nyquist plot is fitted to the function

$$\frac{T}{x} = A + B \frac{p_2}{(j\omega + p_2)} \cdot \frac{\left(1 + \left(-\frac{j\omega}{p_1}\right)^k\right)}{\left(1 + \left(\frac{j\omega}{p_1}\right)^k\right)}$$

where ω is the radian frequency; A is the pure elasticity in parallel with the active mechanism; B is the magnitude of the delayed tension, as the frequency-dependent part of the function is unity when ω is small; and p_1 is the rate constant of the dominating delay which causes oscillatory work. The term containing p_2, a further exponential delay, is introduced because the Nyquist plots are not confined to the region in which the quadrature

component is negative. The fifth parameter, k, is introduced to provide a distributed rate constant rather than a single valued one. j stands for $(-1)^{1/2}$.

Usually, the fitted value of k is nearly unity, and the value of p_2 is at least 10 times the value of p_1. If the data is restricted to the range in which the term containing p_2 is essentially unity, it is fitted almost as well by a simplified form of the function. In these circumstances, the term containing p_2 may be omitted, and k can be set to unity with little increase in the errors. The function then reduces to

$$\frac{T}{x} = A' + \frac{B'}{(j\omega + p_1)}$$

The value of p_1 is the same as before, but A' and B' are different from A and B. This simplified version of the function is identical in form with Eq. 4, which was derived from the assumptions of the two-state model of Thorson and White (1969).

The function which was actually fitted to the data is linear in A and B only, and the best values of the other three parameters must be found by trial and error. The calculation was performed by the same digital computer which was used to control the experiments, a PDP8-I, and the program was written in Fortran.

References

ABBOTT, R. H. 1969. The mechanism of the oscillatory contraction of insect fibrillar flight muscle. Ph.D. thesis, Oxford University.

———. 1970. Computer control of mechanical experiments on muscle. *Biochem. J.* **121:** 3P.

HUXLEY, A. F. 1957. Molecular structure and theories of contraction. *Prog. Biophys.* **7:** 255.

HUXLEY, A. F. and R. M. SIMMONS. 1971. Proposed mechanism of force generation in striated muscle. *Nature* **233:** 533.

JEWELL, B. R. and J. C. RÜEGG. 1966. Oscillatory contraction of insect fibrillar muscle after glycerol extraction. *Proc. Roy. Soc. (London)* B **164:** 428.

JULIAN, F. J. 1971. The effect of calcium on the force-velocity relation of briefly glycerinated frog muscle fibres. *J. Physiol.* **218:** 117.

MACHIN, K. E. and J. W. S. PRINGLE. 1960. The physiology of insect fibrillar flight muscles. III. The effects of sinusoidal changes of length on a beetle flight muscle. *Proc. Roy. Soc. (London)* B **152:** 311.

PODOLSKY, R. J. and A. C. NOLAN. 1971. Cross-bridge properties derived from physiological studies of frog muscle fibers, p. 247. In *Contractility of muscle cells and related processes*, ed. R. J. Podolsky. Prentice-Hall, Englewood Cliffs, New Jersey.

PODOLSKY, R. J. and L. E. TEICHHOLZ. 1970. The relation between calcium and contraction kinetics in skinned muscle fibres. *J. Physiol.* **211:** 19.

PRINGLE, J. W. S. and R. T. TREGEAR. 1969. Mechanical properties of insect fibrillar muscle at large amplitudes of oscillation. *Proc. Roy. Soc. (London)* B **174:** 33.

RÜEGG, J. C. 1968. Oscillatory mechanism in fibrillar insect flight muscle. *Experientia* **24:** 529.

Rüegg, J. C. and H. Stümpf. 1969. Activation of myofibrillar ATPase activity by extension of glycerol extracted insect fibrillar muscle. *Pflügers Arch. gesamte Physiol.* **305**: 36.

Schädler, M. 1967. Proportionale Aktivierung von ATPase-Aktivität und Kontraktionsspannung durch Calciumionen in isolierten contraktilen Strukturen verschiedener Muskelarten. *Pflügers Arch. gesamte Physiol.* **296**: 70.

Steiger, G. J. and J. C. Rüegg. 1969. Energetics and efficiency in the isolated contractile machinery of an insect fibrillar muscle at various frequencies of oscillation. *Pflügers Arch. gesamte Physiol.* **307**: 1.

Thorson, J. and D. C. S. White. 1969. Distributed representations for actin-myosin interactions in the oscillatory contraction of muscle. *Biophys. J.* **9**: 360.

Estimates of Force and Time of Actomyosin Interaction in an Active Muscle and of the Number Interacting at Any One Time

Judith Pybus and Richard Tregear

Agricultural Research Council Unit of Insect Physiology, Department of Zoology, Oxford University, England

It is now generally accepted that muscle generates force and does work because of the hydrolysis of ATP at the hydrolytic site on myosin. Moreover it is known that this hydrolytic event itself is extremely rapid and that in isolated myosin or relaxed, glycerol-extracted muscle the ADP-myosin complex so formed is relatively stable (Lymn and Taylor, 1971; Marston and Tregear, 1972). However the subsequent course of events remains unclear; there is not even sure evidence that myosin attaches to actin during muscular contraction, although various changes in the X-ray diffraction pattern indicate that it does (Huxley, 1967; Armitage et al., this volume) and it certainly does so in the absence of ATP (Reedy, 1968). In this work we have *assumed* that actomyosin interaction occurs during contraction and that this causes the force and work observed. On this basis we have sought to assess the characteristics of the interaction.

We have used glycerol-extracted muscle fibers for this purpose. Such preparations have the advantage that their chemical situation can be simplified and controlled while retaining the muscle's capacity to exert force and do work. Their disadvantage is that activation and relaxation can only be achieved by changing the bathing solution, the intrinsic system being destroyed. In one preparation this disadvantage can be overcome; fibrillar insect flight muscle is only slightly activated by the addition of Ca^{++} and is fully activated by a slight extension of the muscle (Rüegg and Tregear, 1965). This process, which has been termed "stretch-activation," is employed in the animal's flight; it is not unique to such muscles but is most fully developed in them (Rüegg et al., 1970). In the present instance we were not concerned with the mechanism of stretch-activation but only with its use as a tool; by this means one is able to vary the activation of the muscle without alteration of the bathing solution. Thus work can be continuously generated and compared to the amount of ATP hydrolyzed by the muscle. Several such studies have been undertaken by Rüegg and his collaborators (Rüegg and Tregear, 1965;

Steiger and Rüegg, 1969). The present study was intended primarily to disentangle the relations between force, work, and ATP hydrolysis. These measurements do not directly lead to the estimates mentioned in the paper's title. In order to obtain them one must postulate the nature of the mechanical event. In this paper we have assumed, on the basis of mechanical studies on frog muscle (Huxley and Simmons, 1971a), that the mechanical energy is stored in a spring-like structure within the individual actomyosin interaction. The interaction is assumed to detach by a first-order process.

Experiments

The basic experiment. One or two glycerol-extracted fibers from the dorsal longitudinal muscle of the water bug *Lethocerus cordofanus* were mounted on a mechanical apparatus, immersed in a small volume of Mg^{++}-ATP solution; the full composition of the solution is shown in the legend to Fig. 1. The fibers were either kept still or oscillated at such a frequency as to give out mechanical work. In this way a considerable proportion of the myosin molecules present were activated during each cycle of the oscillation (Pringle and Tregear, 1969). After 20–30 min the oscillation was stopped and the fibers transferred to another bath of the same medium. In any one bath the mean tension in the fibers was maintained constant; on changing to a fresh bath it was altered by slight extension of the fibers.

The ADP produced during each incubation was measured by quantitative oxidation of NADH (Bergmeyer, 1965). In control experiments it was established that the ADP and phosphate produced in the incubations were the same, despite the undoubted presence of adenylate kinase, and that the mitochondrial ATPase was fully inhibited by the added azide. Hence the ADP produced was taken as a true measure of the myosin's ATP hydrolysis.

From these results the ATP hydrolyzed was related to the tension in the muscle, whether at rest or under oscillation, and in the latter case to the power which the muscle produced.

Effects of extension. It is well established
that as a flight muscle is extended both ATP
hydrolysis rate (ATPase) and tension rise; this is
the stretch-activation phenomenon, which was
clearly visible in the present experiments. Since
both parameters rise with extension, it is no great
surprise to find that they show an approximately
linear relation to one another as extension is
gradually increased (Fig. 1). The power produced
on oscillation also rises as the mean extension is
increased up to a certain value, and then falls as the
muscle becomes overextended (Pringle and Tregear,
1969). Under these conditions the tension continues
to rise, and so does the ATPase (Fig. 1); thus the
linear relationship of tension to ATPase persists
even when that between power and ATPase (the
so-called Fenn effect) is lost.

Chemomechanical efficiency. As long as
the muscle is not overextended it is reasonable to
relate power and ATPase to one another. The results
from a series of similar experiments plotted together show, under these restrictions, an approximately linear relation between power and ATPase,
with a small positive intercept of the regression
line on the power axis (Fig. 2). It is difficult to
know whether to accept the slope of this line as the
best estimate of the chemomechanical efficiency, as
Rüegg has done (Steiger and Rüegg, 1969); this
presupposes that the ATPase when no power is

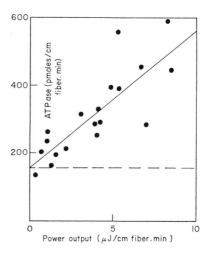

Figure 2. Relationship between ATPase and power output. Conditions as in Fig. 1, save that temperature is 16°C.

present—i.e., in an unstretched muscle—is still
present, in parallel, when it is doing work. Alternatively one can assume that this ATPase should
not be subtracted, in which case there is no unique
measure of efficiency, and the maximum value is
best estimated from the average ratio of power to
ATPase of the upper points of the graph. In the
present experiments the slopes of the lines obtained
as in Fig. 2, under various conditions, were commonly in the range 14–17 kJ/mole ATP hydrolyzed.
The optimum conditions (in the presence of
phosphate at 16°C) gave a slope of 24 kJ/mole, as
shown in Fig. 2; if the "baseline" ATPase is not
subtracted, this value falls to approximately
18 kJ/mole. The free energy available from ATP
hydrolysis is in the region of 42 kJ/mole (Kushmerick and Davies, 1970). It would thus appear
that glycerol-extracted fibers can convert chemical
to mechanical energy with an approximately 50%
efficiency.

Tension cost. Elementary observation reveals
that tension and ATPase are monotonically, and
approximately linearly, related (Fig. 1). We have
taken the regression of ATPase on tension as the
best estimate of this relation because most of the
error is in the ATPase measurements. As in the
relation with power, this gives a positive intercept
on the ATPase axis; there is some ATPase left at
zero tension (Fig. 3). Just as in the previous case,
this leaves two possibilities in interpretation,
dependent on whether the zero-tension ATPase is
legitimately subtractable. Our further calculations
assume that it is. On this basis the slope of the
regression line, the incremental ratio of tension to
ATPase, has been termed the tension cost. It is
clearly greater in an oscillating muscle fiber than

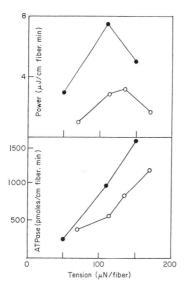

Figure 1. The effect of small degrees of extension on the
tension, power output, and ATPase of a glycerol-extracted
insect flight muscle fiber; the power and ATPase are
plotted against tension. Open and closed circles refer to two
separate experiments. Conditions: bathing medium 5 mM
EGTA + 4mM CaCl₂ pCa 5.5, 10 mM phosphate pH 6.95,
5 mM ATP + 15 mM MgCl₂, 10 mM NaN₃, 20°C. Static
extension 2–8% of rest length, oscillation amplitude 2%
peak/peak, oscillation frequency 7 Hz (○) or 8 Hz (●).

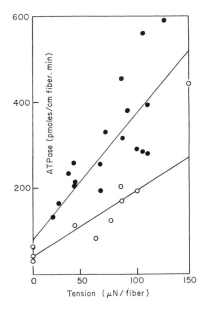

Figure 3. Relation between ATPase and tension in oscillatory (●) and static (○) muscle. Conditions as in Fig. 2.·

in a static one (Fig. 3). The tension cost is one measure of the rate constant (r) of detachment of the actomyosin interaction; $r = RP/T$, where T is the tension observed, P the tension per actomyosin interaction, and R the ATPase per half-sarcomere.

Optimum frequency. The above experiments were conducted with the muscle either still or sinusoidally oscillated at such a frequency as to give the greatest power output. Either during these experiments or during parallel experiments on muscle fibers from the same specimen in the same bathing solution, the muscle was oscillated for short periods at a variety of frequencies so that the frequency at which the maximum work per cycle was produced could be estimated (Fig. 4). This was termed the optimum frequency. The experiments were conducted at the amplitude used in the ATPase experiments, 2% of the muscle's length. They were also repeated at a low amplitude, 0.2%, since the mechanical behavior at low amplitudes of oscillation has an established interpretation (Abbott, this volume). The optimum frequency was higher at low amplitude (Fig. 4); the mean ratio between the optimum frequencies at high and low amplitude was 0.68.

Again on a simple theoretical basis the optimum frequency can be related to the rate constant of actomyosin detachment; this has been done exactly for the low-amplitude case (Thorson and White, 1969). Thus the tension cost and the optimum frequency allegedly provide measures of the same

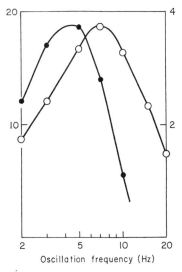

Figure 4. Relation between work output (nJ/cm fiber) per cycle and oscillation frequency in muscle oscillated at 2% (●) or 0.2% (○) amplitude. Conditions as in Fig. 2.

parameter and should be proportional to one another.

Relation between tension cost and frequency. In order to test the proportionality of these two parameters, the experiments were repeated under different conditions: the temperature was varied between 12 and 30°C, and various anions were added which reduced the optimum frequency of the muscle. The combined data are shown in Fig. 5, in which the static and oscillatory tension costs

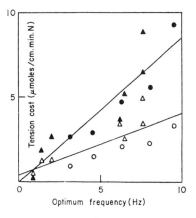

Figure 5. The tension cost in oscillatory (solid symbols) or static muscle (open symbols) related to the optimum working frequency in a 2% amplitude oscillation. (●, ○) Temperature varied from 12 to 30°C, chemical conditions as in Fig. 1; (▲, △) temperature 20°C, chemical composition varied by the substitution of histidine for PO_4 buffer and/or the addition of 1.5 mM pyrophosphate or 10 mM SO_4; pCa and pH invariant throughout. Points taken in order from left to right: histidine + sulphate, histidine + pyrophosphate, histidine, phosphate (ionic strength varied), phosphate, phosphate + pyrophosphate.

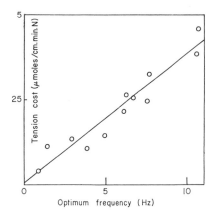

Figure 6. Test of linearity between tension cost and optimum frequency. Oscillatory and static tension costs combined and optimum frequencies at 0.2% and 2% amplitude combined as described in the text. The line shown is the regression of tension cost on optimum frequency.

are compared to the optimum frequency at high amplitude. The data are somewhat scattered and do not provide a good test of linear proportionality of the two parameters. However, the calculated regression lines show very small intercepts on the axes, as expected on the hypothesis. The dynamic tension cost is greater than the static cost; regression of dynamic tension cost against static gave an average ratio between the two of 2:1.

In order to obtain a better test of the alleged proportionality between the parameters, the data were combined, on the assumption that the observed scatter was largely due to experimental error. For each condition the static and dynamic tension costs were combined by dividing the dynamic value by 2.1 and taking the average of the two values; similarly the optimum frequencies at high and low amplitudes were combined by multiplying the latter by 0.68 and averaging them. The regularity of the relationship between the parameters was considerably increased by this process (Fig. 6). The relationship appears to be close to linearity and the intercepts on the axes are close to zero.

Discussion

The simplest explanation of the observed proportionality between tension cost and optimum working frequency is that they both measure the same thing. If tension is generated by an actomyosin interaction, they should do so because both are determined by the rate constant of detachment of the interaction. The higher this is, the higher will be the tension cost since one ATP molecule is hydrolyzed in the interval between the actomyosin interactions (Lymn and Taylor, 1971), and the higher will be the optimum frequency since the

detachment rate constant determines the mechanical response time of the muscle (Abbott, this volume). Thus the proportionality supports the existence of an actomyosin interaction.

There are two possible explanations of the greater tension cost observed when the muscle oscillates; either the rate constant of detachment might be increased, or the force exerted by the interaction might be reduced by the muscle's movement. In the first case one would expect the optimum working frequency to rise with amplitude of oscillation. Actually it falls, so that this explanation is unlikely. The second effect is expected if the actomyosin interactions store energy in a spring-like fashion, as suggested by Huxley and Simmons (1971a) and shown diagrammatically in Fig. 7. For work to be performed in an oscillation, attachment must occur preferentially when the muscle is elongated (Thorson and White, 1969) so that the extended springs may shorten and give out their stored energy. Since the tension in a spring is proportional to its extension, allowing it to shorten

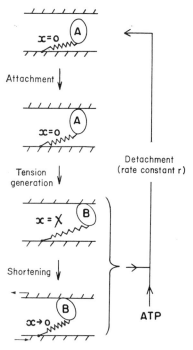

Figure 7. Model of actomyosin interaction used in the numerical deductions. When the head of the myosin molecule changes from state A to state B, the spring is stretched by χ, generating tension. If the filaments move the stored energy is released. The myosin detaches from the actin with a first-order rate constant r. Consequent ATP hydrolysis returns the myosin to the state A. This model is based on the work of Huxley and Simmons(1971a); the spring has been visualized as within subfragment 2, but its structural origin is really unknown. The model differs from that of Huxley and Simmons (this volume) in that energy is here considered to be stored in a single event rather than a series of steps.

reduces its average tension during its life time. Hence oscillation will decrease the average tension produced by each interaction and raise the tension cost. In order to get all the stored energy out of the interaction the spring must shorten to its unstrained length and so deliver, on average, half the tension that it would if the muscle did not move. The tension cost is in practice approximately doubled by an oscillation which gives the best chemomechanical coupling. Thus the observations are consistent with the existence of a spring-like energy store.

The rate constant of detachment of these spring-like interactions (r in Fig. 7) may be deduced directly from the optimum working frequency of the muscle during low-amplitude oscillations; Thorson and White (1969) showed that it is simply the angular frequency at optimum work per cycle. Under the various conditions used in these experiments it ranged between 8 and 100 sec^{-1}, equivalent to interaction lifetimes of 10–125 msec. At high amplitudes the response to oscillation is nonlinear and no mathematical model of the response has been worked out, so no deduction is possible.

The extension of the spring when energy is stored within it (χ in Fig. 7) cannot be deduced from these experiments, although it is probably less than the amplitude of the chemomechanically efficient oscillations (24 nm/half-sarcomere). In frog semitendinosus muscle Huxley and Simmons showed by rapid release of an activated muscle that χ was in the range 8–12 nm. No similar observations have been made on *Lethocerus* flight muscle, but White (1970) showed that such muscle in rigor had to be released by 8–16 nm/half-sarcomere to lose its tension. We have taken $\chi = 10$ nm for further calculation.

Using this value the force in the extended spring can be calculated. The energy stored during its extension (step A:B in Fig. 7) is greater than or equal to the greatest mechanical energy released per ATP hydrolysis, and less than or equal to the free energy of that hydrolysis. The former, taken from our observations, is 18 kJ/mole or 3×10^{-20} J/molecule, and the latter, 42 kJ/mole or 7×10^{-20} J/molecule (Kushmerick and Davies, 1970) ATP 5 mM, phosphate 10 mM, ADP in the fibers 1 mM. The force in the 10-nm extended spring is therefore between 6 and 14 pN ($P = 2\,E/\chi$). Since on any simple model some of the springs detach before they have given out all their energy, it is probable that the true value of P lies well above the lower end of this range and we have taken 10 pN for further calculation. All this logic is based on the present model, in which the energy is transferred to the spring in a single event (Fig. 7). If energy is delivered in the series of steps suggested by

Huxley and Simmons (1971b), then the calculated force is approximately half as great, 5 pN.

The observed tension cost can now be converted into a second estimate of the rate constant of detachment. The tension cost at a high-amplitude optimum frequency of 10 Hz (equivalent to a low-amplitude value of 17 Hz; $r = 110$ sec^{-1}) is 3.8 μmoles/cm-min-N (Fig. 6). On conversion this becomes 4.6 molecules ATP/half-sarcomere-s-pN (sarcomere length = 2.4 μm). Assuming all interactions within one half-sarcomere to act in parallel and each to exert a force of 10 pN, then the second estimate of r is 46 sec^{-1}. Since the first estimate was 110 sec^{-1}, there is a clear discrepancy. If the force is only 5 pN, the discrepancy is greater. We do not know whether this discrepancy is real, indicating that the muscle's speed of contraction is limited by a shorter process than its tension generation, or is simply a product of cumulative error in the estimates made in calculation. If the former is correct, it throws doubt on the whole concept of the spring-like interaction; if the latter, it indicates that either the tension estimate used is too low, by a factor or two, or the mechanically derived rate constant is too high by a similar factor.

The tension in the spring also allows calculation of the fraction of myosin molecules which interact with actin at any one moment during contraction. On current assumptions there are some 2×10^{11} myosin molecules in each half-sarcomere/mm^2 (head-bearing half-filaments 1100 nm long, with 6 molecules/14.5 nm on a 53-nm hexagonal lattice; Squire, 1971). If each of these produced 10 pN, the total force would be 2 N/mm^2, whereas if both subfragments in each interacted separately, it would be 4 N/mm^2. The force observed in an oscillating muscle is up to 0.2 N/mm^2, equivalent to 0.4 N/mm^2 of *static* force on the spring assumption (see earlier discussion). Thus the apparent proportion attached is 10 or 20%; if the interaction force were only 5 pN, these values would be doubled. The range of values deduced fits with the deductions either from X-ray diffraction (Armitage et al., this volume) or from mechanical analysis (Abbott, this volume); few of the myosin molecules are attached to actin at any one time.

We have considered our results in isolation, but there are similar results from many different muscles. Schadler (1967) demonstrated a proportionality between ATPase and tension in glycerol-extracted rabbit psoas, dog heart, guinea pig taenia coli, bovine carotid artery and mussel byssus retractor muscle. Infante et al. (1964) showed a parallelism between ATPase and tension in intact but poisoned frog muscle, and Peterson et al. (1972) showed a close proportionality between oxygen consumption in intact, unpoisoned bovine mesenteric vein and the tension produced. The concept of tension cost is therefore generally applicable.

The proportionality between tension cost and detachment rate constant has not been demonstrated before, but similar measurements were made by Bárány (1967) and Bárány and Close (1971). They showed that the unloaded contraction velocity of different muscles, or of the same muscle under different innervation, was proportional to the ATPase of myosin maximally activated by actin. Again, both parameters should, on a simple theory, be measures of the detachment rate constant.

For these reasons we believe that our results have a general application. Myosin probably does interact with actin giving rise to a spring-like energy store from which mechanical energy can be abstracted. The force probably is in the range 5–10 pN and the spring extension 10 nm, and only a small proportion of the molecules present exert tension at any one time.

References

BÁRÁNY, M. 1967. ATPase activity of myosin correlated with speed of shortening. *J. Gen. Physiol.* **50**: (suppl.) 197.

BÁRÁNY, M. and R. I. CLOSE. 1971. The transformation of myosin in cross-innervated muscles. *J. Physiol.* **213**: 455.

BERGMEYER, H. U. 1965. *Methods of enzymatic analysis.* Academic Press, New York.

HUXLEY, A. F. and R. M. SIMMONS. 1971a. Mechanical properties of the cross-bridges of frog striated muscle. *J. Physiol.* **218**: 59P.

———. 1971b. Proposed mechanism of force generation in striated muscle. *Nature* **233**: 533.

HUXLEY, H. E. 1967. Recent X-ray diffraction and electron microscope studies of striated muscle. *J. Gen. Physiol.* **50**, Suppl.: 71.

INFANTE, A., D. KLAUPIKS, and R. E. DAVIES. 1964. Length, tension and metabolism during short isometric contractions of frog sartorius muscle. *Biochim. Biophys. Acta* **88**: 215.

KUSHMERICK, M. and R. E. DAVIES. 1970. Chemical energetics of muscle contractions. *Proc. Roy. Soc.* B **174**: 293.

LYMN, R. W. and E. W. TAYLOR. 1971. Mechanism of ATP hydrolysis by actomyosin. *Biochemistry* **10**: 4617.

MARSTON, S. and R. T. TREGEAR. 1972. Evidence for a complex between myosin and ADP in relaxed muscle fibers. *Nature New Biol.* **235**: 23.

PETERSON, J. W., R. J. PAUL, and S. R. CAPLAN. 1972. Oxygen consumption as an index of contractile energetics in vascular smooth muscle. *Biophys. J.* **12**: 83a.

PRINGLE, J. W. S. and R. T. TREGEAR. 1969. Mechanical properties of insect fibrillar muscle at large amplitudes of oscillation. *Proc. Roy. Soc.* B **174**: 33.

REEDY, M. K. 1968. Rigor crossbridges in glycerinated insect flight muscle. *J. Mol. Biol.* **31**: 155.

RÜEGG, J. C. and R. T. TREGEAR. 1965. Mechanical factors affecting the ATPase activity of glycerol-extracted insect fibrillar muscle. *Proc. Roy. Soc.* B **165**: 497.

RÜEGG, J. C., G. J. STEIGER, and M. SCHADLER. 1970. Mechanical activation of the contractile system in skeletal muscle. *Pflügers Arch. ges. Physiol.* **319**: 139.

SCHADLER, M. 1967. Proportional activation of ATPase and tension by calcium ions in isolated contractile structures of various muscles. *Pflügers Arch. ges. Physiol.* **296**: 70.

STEIGER, G. J. and J. C. RÜEGG. 1969. Energetics and efficiency in the isolated contractile machinery of an insect flight muscle. *Pflügers Arch. ges. Physiol.* **307**: 1.

SQUIRE, J. M. 1971. General model for the structure of all myosin-containing filaments. *Nature* **233**: 457.

THORSON, J. and D. C. S. WHITE. 1969. Distributed representations for actin-myosin interaction in the oscillatory contraction of muscle. *Biophys. J.* **9**: 360.

WHITE, D. C. S. 1970. Rigor contraction in glycerinated insect flight and vertebrate muscle. *J. Physiol.* **208**: 583.

Muscle Contraction Transients, Cross-Bridge Kinetics, and the Fenn Effect

R. J. Podolsky and A. C. Nolan

Laboratory of Physical Biology, National Institute of Arthritis and Metabolic Diseases, Bethesda, Maryland 20014

The mechanical activity of muscle fibers must reflect, in some way, the kinetic properties of the cross-bridges, and a number of attempts have recently been made to discover the rules for this transform. We have recently found a set of rules that can account for the results of a number of experiments in considerable detail. These rules are testable in the sense that they suggest new experiments and make definite predictions about the way these experiments ought to come out.

The Huxley Model

The first attempt to quantitatively relate the physiological properties of muscle fibers to the kinetics of cross-bridge turnover was made by A. F. Huxley in 1957. Huxley assumed that projections from the thick filaments react with sites on the thin filament to form cross-bridges and that these bridges act independently to produce force and motion. He evaluated the rate functions for making and breaking cross-bridges by fitting the steady state force-velocity and force-energy relations of the model to the empirical relations found by Hill (1938) for frog muscle. To complete the analysis it was necessary to know the relation between cross-bridge turnover and energy production, and Huxley assumed that one ATP molecule was split each time a cross-bridge turned over. The rate function for making cross-bridges then had to be relatively small, so that as the contraction speed increased, a greater fraction of the thin filament sites slipped past the thick filament projections without forming cross-bridges, which gave rise to the Fenn effect (Fenn, 1923). Another consequence of the small rate function for making cross-bridges was that the instantaneous number of cross-bridges present at any time decreased as the speed of contraction increased.

The Podolsky-Nolan Model

We made use of the equations put forward by Huxley but changed the boundary conditions on the system (Podolsky et al., 1969; Podolsky and Nolan, 1971). Instead of ad usting the cross-bridge parameters to fit the force-energy relation, we tried

to fit the nonsteady motions that precede steady shortening after step changes in load (Podolsky, 1960; Civan and Podolsky, 1966), which were discovered after Huxley worked out his model. These velocity transients are shown in Fig. 1. The upper traces are displacement and the lower traces are force. Step decreases in force from P_0, the steady isometric force, to $P < P_0$ are followed by transients resembling damped sine waves that in some cases last as long as 40 msec. For the largest force steps ($0.75 P_0$ and $0.87 P_0$) noise obscures the first few milliseconds of the displacement trace, but for the others it is evident that the fiber contracts initially at a rate faster than the steady speed. The arrows mark the points at which the actual motion intersects the back extrapolation of the steady motion; the times corresponding to these points are called null times. The first null time becomes shorter as the force step increases. For the larger force steps, a second null time can be seen, which also becomes shorter as the force step increases. By sighting along the displacement traces, third null times can often be detected.

The first question we asked was whether the nonsteady motions produced by the entire range of force steps could be accounted for by a single set of rate functions and a single force function. Figure 2 shows the fit that was finally obtained (Podolsky and Nolan, 1971). It was not at all obvious at the outset that a good fit was necessarily obtainable. For example, if the motion of the fiber were due to phenomena other than cross-bridge turnover, or if the rate functions depended on velocity, a good fit might not have been possible. The goodness of fit suggests—but does not prove—that the nonsteady motion is due primarily to cross-bridge turnover and that the rate functions used to calculate the motion probably have some basis in fact.

The rate and force functions used in these calculations are shown in Fig. 3. The upper line shows an actin site moving past a myosin projection. The lower curve shows the force function of a cross-bridge; the origin of the coordinate system is the point at which the force is zero. The

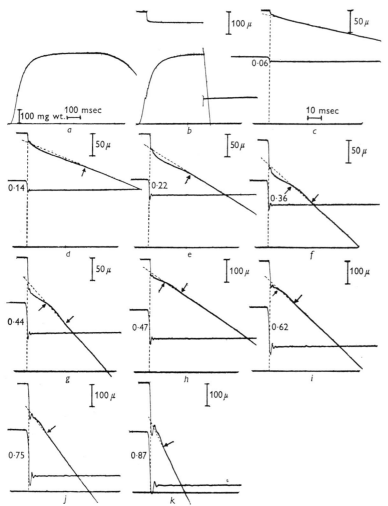

Figure 1. Response of frog muscle fibers to step changes in load. Upper trace is displacement, lower trace is force, base line is zero force; temperature 3°C. Force step magnitude as fraction of P_0 is given alongside the force trace. Arrows mark the null points. (From Civan and Podolsky, 1966.)

force function k is asymmetric, so that bridges exerting negative force are stiffer than those on the other side of the origin.

f is the rate function for making a bridge; it is positive over a range of 120 Å. (This distance can be increased to 180 Å without affecting the overall response very much.) The rate function for breaking a bridge, g, has two very different values. The part that overlaps the f function, as well as the first 63 Å on the left of the origin, is small. Beyond 63 Å, the g function is very large. The sharp discontinuity in the value of g suggests that two different processes are involved in breaking cross-bridges, one acting on the right of -63 Å, and another acting on the left. This point will come up again in connection with the energetics of contraction.

When the muscle is isometric and the myofilaments do not move relative to each other, the steady state distribution of cross-bridges given by these rate functions is shown at the top of Fig. 4. In this case cross-bridges are present only in the

region where f differs from zero. However, when the load is less than P_0 (lower three panels of Fig. 4), the relative motion of the myofilaments carries cross-bridges formed on the right of the origin towards the left, where they are ultimately broken by the g function. The solid lines show the steady state distribution of cross-bridges for various loads. An obvious property of this set of distribution functions is that the area increases as the load is decreased and the steady speed increases, which means that the instantaneous number of cross-bridges increases with contraction speed.

The shaded areas show the distribution just after the load on an isometrically contracting muscle has been reduced from P_0 to the indicated value. These distributions represent the isometric distribution displaced by various amounts. The non-steady isotonic motion comes about because these distributions differ from those of the steady state. This point of view assumes that cross-bridges move quickly (on a millisecond time scale) in response to

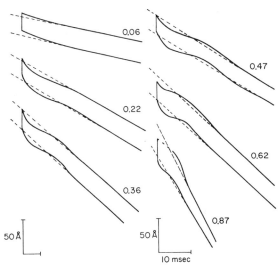

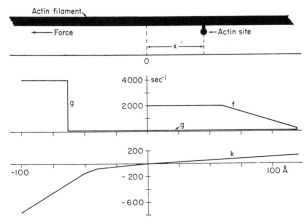

Figure 2. Contraction following step changes in load. For each pair of displacement traces, the upper is the motion of the living fiber shown in Fig. 1 for the indicated force step; the lower trace is the computed motion for the same force step. The interrupted straight line is the back extrapolation of the steady motion. For force step 0.87, the initial 5 msec of the experimental record is uncertain (Podolsky and Nolan, 1971).

changes in force, so that cross-bridge movement makes a significant contribution to the "series elastic element" observed in quick release experiments.

The distribution functions are all close to 1 at the origin, regardless of how fast the myofilaments move. This means that when an actin site moves past a myosin projection, the interaction prob-

Figure 3. *Top panel.* Diagram of the coordinate system: x is the distance between a site on the actin filament and the position at which an attached cross-bridge exerts zero force. The Z line to which the actin filament is connected is off the diagram to the right. *Middle panel.* Dependence of f, the probability of making a cross-bridge, and g, the probability of breaking a cross-bridge, on x. g_2 is the value of g for -63 Å $< x < 0$; g_3 is the value of g for $x < -63$ Å. *Lower panel.* Dependence of the force function, k, on x. (From Podolsky and Nolan, 1971.)

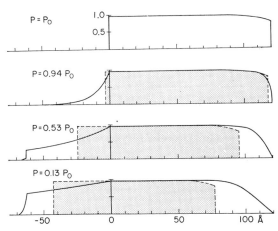

Figure 4. Change in the distribution of cross-bridge lengths during isotonic contraction for various force steps. The distribution function at the start of isotonic contraction is shown by the shaded area; the steady state distribution is given by the solid line. Note that in the steady state the area of the distribution function is greater than that at the start of isotonic contraction (Podolsky and Nolan, 1971).

ability is so high that a cross-bridge will almost always be made. Therefore for an extensive contraction the number of cross-bridges that turn over will be essentially the same for all contraction speeds. (This assumes that the influence of velocity on the turnover rate of cross-bridges on the right of the origin is small relative to that on the left of the origin. Calculation shows that this is the case for the rate functions shown in Fig. 3. It is also true for similar cases in which g $(0 \leq x \leq 120) < 60$ sec^{-1}.)

If one ATP molecule were hydrolyzed each time a cross-bridge turned over, the energetics of the system would not conform with the Fenn effect, according to which the release of chemical energy is associated with work production rather than movement per se. However, if when cross-bridges are broken on the left of the origin an ATP molecule were split only in the region of the small g (g_2), ATP hydrolysis for an extensive contraction would be greater when the muscle is heavily loaded and doing work, which is the kind of behavior implied by the Fenn effect. Breakage by the large g (g_3) would then be attributed to a mechanical rather than a chemical process. The fact that the stress in the cross-bridge at -63 Å (which is the boundary between the small and the large values of g) is more than twice the maximum value on the right of the origin is consistent with this idea, since physiological studies suggest that cross-bridges "give" when subjected to forces of about $2P_0$ (Katz, 1939).

We see that in this model ATP turnover is not tightly coupled with cross-bridge turnover. This uncoupling is needed because to generate the correct contraction kinetics, particularly the high velocity transient just after the force step, the rate constant for making cross-bridges must be relatively large, which tends to make cross-bridge turnover independent of contraction speed. This is a fundamental difference between our model and that of Huxley (1957).

Chemistry of Contraction

Recent biochemical studies (Eisenberg et al., 1968; Tonomura et al., 1969; Lymn and Taylor, 1971; Trentham et al., 1972) suggest that the rate constants for the model can be associated with the following cycle of cross-bridge activity:

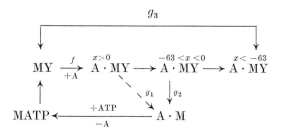

MY is an activated form of the myosin molecule. This species can react with an actin site A to produce a cross-bridge, $A \cdot MY$. The rate constant for this process, f, is finite only when the reaction of MY with A results in a force generating cross-bridge, i.e., only when $x > 0$. The motion of the myofilament then carries the bridge into the region of negative force. In the first 63 Å of this region, the cross-bridge converts with rate constant g_2 to the species AM, which is quickly dissociated by ATP into $M \cdot ATP$ and A. If this process does not occur by the time the bridge has been strained to -63 Å, another process occurs in which the bridge is broken without the intervention of ATP (rate constant g_3). The driving force for this mechanical process is provided by other cross-bridges attached to the myofilament.

We also allow for a process (rate constant g_1) that leads to the dissociation by ATP of cross-bridges that exert positive force. This reaction is presumably reflected in the fall of force at the end of an isometric contraction. For convenience, we have taken $g_1 = g_2$ in the present calculations, which probably overestimates g_1 (Huxley and Simmons, 1970). However, as long as g_1 is much less than f, as would be so in either case, its value has very little effect on the contraction kinetics.

Length Step Experiments

Huxley and Simmons (1971) recently reported experiments with frog muscle fibers in which the

time course of force redevelopment was recorded after a series of length steps. They found that when the fiber was suddenly shortened so that force dropped to a fraction of P_0, redevelopment of isometric force seemed to have two components, one requiring a few milliseconds for completion and the other a much longer time. They attributed the faster process to relaxation of the cross-bridges and the slower one to cross-bridge turnover.

A fast phase of force redevelopment in an isometric experiment is equivalent to a high velocity shortening transient in an isotonic experiment. Therefore, the effects seen by Huxley and Simmons are manifestations of the same processes that give rise to the initial part of the isotonic velocity transients in force step experiments. With their data available, it became of interest to see how well the rate functions we derived from the entire time course of isotonic velocity transients account for the first few milliseconds of force redevelopment in length step experiments.

The calculations were made according to the steps shown in Fig. 5, which is patterned after the method used by Hill (1938) to describe isometric contraction. The right side of the figure shows, reading from top to bottom: (i) the myofilament configuration for steady isometric force; (ii) the state immediately after the length step, which is partitioned between movement of the myofilaments and relaxation of the passive elasticity, ε, in series with the myofilaments; and (iii) the state after ε has been restretched to its original length by myofilament movement.

The left side of Fig. 5 shows the distribution of cross-bridge lengths at various phases of this

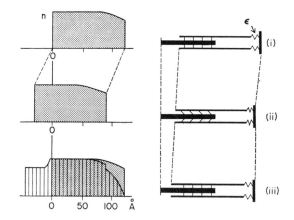

Figure 5. Method of calculating redevelopment of force following step changes in length. Distribution functions (*left*, dotted areas) and myofilament configurations (*right*) at the initial length (i), just after the force step (ii), and after force has redeveloped (iii). ε is a passive elastic element. The lined area in (iii) is a representative distribution function for the cross-bridges during the redevelopment of force.

procedure. Initially the distribution is that corresponding to steady isometric force (top). The length step results in a displacement of the initial distribution (middle), which reduces the force of the cross-bridges. Turnover of cross-bridges changes the distribution so that the contractile force tends to exceed the stress in ε; mechanical balance is continuously restored by increasing the strain in ε by relative movement of the myofilaments. In the bottom panel the lined area shows a transitional value of the distribution function, and the shaded area is the final value, corresponding to the redevelopment of full isometric force.

Since the length step is taken up by the passive elasticity as well as the cross-bridges, the force at the end of the step depends on the magnitude of this compliance. The rate of force redevelopment also depends on the magnitude of this compliance. This is brought out in Fig. 6, which shows the force traces calculated from our model using two

different values for the series compliance. An exponential compliance was used for the lower traces. When this compliance was added to the movement of the cross-bridges during the step, the force-displacement relation recorded by Huxley and Simmons (1971) ($T_1(y)$ in their Fig. 3) was obtained. We also made the calculation for the case of no compliance other than that of the cross-bridges; these are the upper curves. With this assumption the calculated "instantaneous" force-displacement relation does not fit the experimental results of Huxley and Simmons, but it is a useful limiting case, as it maximizes the rate of force redevelopment for any given set of rate constants for cross-bridge turnover.

The dotted lines in Fig. 6 are the experimental data reported by Huxley and Simmons (1971). These traces are very nearly bracketed by those calculated from our model. Two quantitative features of the data are evident in the traces calculated with the exponential elastic element (lower curves): (1) the force rises much more rapidly during the first few msec than it does during the next 15 msec, so that there is a phase of "quick recovery," and (2) the difference between the force just after the step and the force approached at the end of the quick recovery ($T_2(y)$ in the nomenclature of Huxley and Simmons) passes through a maximum at an intermediate value of the length step. The two sets of traces differ in that at the end of the quick recovery phase the recorded force is greater than that of the model. It is apparent from Fig. 6, however, that the amplitude (but not the shape) of the calculated force trace is very sensitive to the stiffness of the passive series elasticity, and that a closer fit to the experimental data would have been obtained if the true "instantaneous elasticity" of the sarcomere were stiffer than the value recorded by Huxley and Simmons. It is of interest in this connection that the authors suggested that this might actually have been the case in their experiments.

In spite of the differences, it seems fair to say that the quick redevelopment of force after a length step can be accounted for in considerable detail by the rapid turnover of cross-bridges. This is particularly noteworthy for the two largest length steps, since in the corresponding force-step experiments the first few milliseconds of the motion were obscured by noise, and the degree to which our rate functions could account for the motion was not known. While this does not rule out the cross-bridge relaxation process postulated by Huxley and Simmons, it does show that their interpretation of the phenomena underlying the response in the first few milliseconds following a length step is not unique.

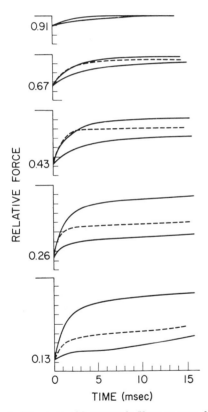

Figure 6. Response of frog muscle fibers to step changes in length. Interrupted traces are the experimental data of Huxley and Simmons (1971); the numbers alongside the ordinate give the force at the end of the length step. The solid traces are forces computed according to the method diagrammed in Fig. 5, using the rate functions in Fig. 3. For the lower solid traces the passive elastic element was $P = P_0 \exp(-\Delta\zeta \ln 2/29.5)$, where $\Delta\zeta$ (in Å/$\frac{1}{2}$-sarcomere) is the difference in strain at P_0 and at P; for the upper solid traces the elastic element was completely rigid, so that $\Delta\zeta = 0$.

Double Force Step Experiments

If the basis of the mechanical transients is a shift in cross-bridge distribution from one configuration to another, the nonsteady motion at a given force should be affected by the history of the fiber. In particular, the motion at a given load should depend on whether the fiber was previously contracting isometrically or isotonically. For example, the bottom panel in Fig. 4 shows the initial and final distributions following a force step from P_0 to $0.13P_0$. If instead of going directly to the final load, the load was first reduced to $0.53P_0$, say, long enough to set up the steady state and then reduced further to $0.13P_0$, the initial distribution at $0.13P_0$ would be the steady state distribution for $0.53P_0$ displaced to the left by about 10 Å rather than the shaded distribution.

To see exactly how the two motions would be expected to differ, we made some calculations using the rate functions for our model. These are shown in Fig. 7. The number above each trace segment is the relative load for that part of the motion. The traces ending on the right side of the figure show the motion at relative load 0.68 with two different initial conditions: (i) steady shortening at relative load 0.79, and (ii) steady isometric contraction. The amplitude of the transient after the double

step is about half of that after the single step. This is probably due to the fact that the difference between the steady and the nonsteady velocities depends on the difference in the initial and final distribution functions, which is greater when the force steps are greater. An unexpected result, however, is that the null times are almost the same in both cases, which indicates that, at least for these rate functions, the null time depends more on the final load than on the initial conditions.

The other set of traces shows motion at relative load 0.28 with various initial conditions. Here, too, the null times at the final load are close to each other, and the amplitude of the transient is decreased where the initial condition is isotonic rather than isometric.

The response of the muscle to the same force steps, shown in Fig. 8, is very much like that of the model both with respect to null time (Pitts et al., 1971) and transient amplitude. This is additional evidence that the transient response of the muscle fiber is due primarily to cross-bridge turnover and that the rate functions shown in Fig. 3 are reasonably accurate.

Concluding Remarks

I would like to close by retracing the evolution of quantitative cross-bridge models. They generally

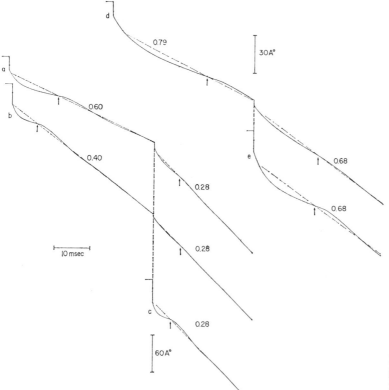

Figure 7. Computed motion following single and double force steps. Relative loads are indicated by numbers alongside the traces. Arrows locate the first null times.

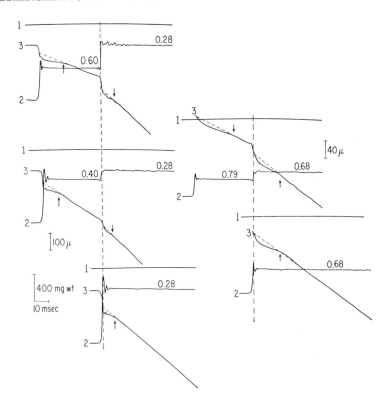

Figure 8. Response of frog muscle fibers to single and double force steps. Bundle of about 12 fibers; fiber length excluding tendon, 15 mm; 2 mm tendon; sarcomere length, 2.2 μ; temperature, 4°C. In each panel, trace 3 corresponds to the calculated motion in Fig. 7. Trace 1, zero of force; trace 2, force; trace 4, delayed displacement triggered just before the second release; trace 5, position of lever before preparation is stimulated (W. R. Pitts, Jr., A. C. Nolan, and R. J. Podolsky, unpublished experiments).

suppose that myosin and actin sites act independently; this assumption is supported by several strong lines of evidence. In the Huxley (1957) model, cross-bridge properties were deduced by taking as boundary conditions for the kinetic equations the steady state force-velocity and force-energy relations of Hill. It was assumed that (i) cross-bridges are in rapid mechanical equilibrium with applied forces, and (ii) one ATP molecule is hydrolyzed each time a cross-bridge turns over. A shortcoming of this model is that it does not give the correct nonsteady motion after step changes in force.

We took the equations used above and substituted the isotonic velocity transients for the force-energy relation as a primary boundary condition. We kept the assumption that cross-bridges are in rapid mechanical equilibrium. The rate functions so derived account reasonably well for the responses of muscle fibers to both length-step and double force-step experiments. The force-energy relation was obtained by loosening the linkage between cross-bridge turnover and ATP hydrolysis.

Huxley and Simmons (1971) modified the Huxley (1957) model so that it could produce the rapid component of the transients that follow step changes in length by assuming that the cross-bridges require a few milliseconds to equilibrate mechanically. It remains to be seen, however, how well this hybrid model accounts for the transient behavior of muscle fibers beyond the first few milliseconds, which is important to know since some force steps produce transients that last ten times longer than that.

A testable difference between the Huxley (1957) and the Huxley and Simmons (1971) models on the one hand, and ours on the other, is the influence of velocity on the instantaneous number of cross-bridges, which as noted above is given by the area of the distribution function. In the Huxley models this number decreases when the velocity is increased, while in our model it increases.

A possible technique for measuring the influence of velocity on cross-bridge number is to make use of the fact that the relative intensity of the 1, 0 and 1, 1 equatorial reflections in the X-ray diffraction pattern of muscle fibers depends on the distribution of mass around the actin and the myosin filaments (Huxley, 1968). One would have to compare the X-ray diffraction pattern of isometrically contracting muscle fibers with that of actively shortening fibers. An advantage of this method is that it is not necessary to perturb the motion of the fiber to obtain the result.

Another way of distinguishing between the

models is to find out how velocity affects the instantaneous elasticity of the fiber, since this parameter is a measure of cross-bridge number. Instantaneous elasticity can be measured by quickly changing the length of the fiber and measuring the associated change in force. In principle, this can be done in either quick release or a quick stretch. A release appears to be preferable, however, since in this maneuver the motion of the cross-bridges is similar to that of shortening and the force function of the bridge is presumably known. Stretches, on the other hand, force some cross-bridges into configurations that are never seen during shortening. Thus during stretch cross-bridges are carried into the region beyond the right hand limit of the f function ($x > 120$ Å in Fig. 3) where the force function of the cross-bridge is not known. It is possible that the bridge is weakened in this region, in which case it would contribute less to the force than one might have expected. It seems prudent, therefore, to give more weight to data obtained by releases and to interpret stretch data with caution until we know more about the properties of overstrained cross-bridges.

References

CIVAN, M. M. and R. J. PODOLSKY. 1966. Contraction kinetics of striated muscle fibres following quick changes in load. *J. Physiol.* **184**: 511.

EISENBERG, E., C. R. ZOBEL, and C. MOOS. 1968. Subfragment 1 of myosin: Adenosine triphosphatase activation by actin. *Biochemistry* **7**: 3186.

FENN, W. O. 1923. A quantitative comparison between the energy liberated and the work performed by the isolated sartorius muscle of the frog. *J. Physiol.* **58**: 175.

HILL, A. V. 1938. The heat of shortening and the dynamic constants of muscle. *Proc. Roy. Soc. London* B **126**: 136.

HUXLEY, A. F. 1957. Muscle structure and theories of contraction. *Prog. Biophys. Biophys. Chem.* **7**: 255.

HUXLEY, A. F. and R. M. SIMMONS. 1970. Rapid "give" and the tension "shoulder" in the relaxation of frog muscle fibers. *J. Physiol.* **210**: 32P.

———. 1971. Proposed mechanism of force generation in striated muscle. *Nature* **233**: 533.

HUXLEY, H. E. 1968. Structural difference between resting and rigor muscle; evidence from intensity changes in the low-angle equatorial X-ray diagram. *J. Mol. Biol.* **37**: 507.

KATZ, B. 1939. The relation between force and speed in muscular contraction. *J. Physiol.* **96**: 45.

LYMN, R. W. and E. W. TAYLOR. 1971. Mechanism of adenosine triphosphate hydrolysis by actomyosin. *Biochemistry* **10**: 4617.

PITTS, W. R., JR., A. C. NOLAN, and R. J. PODOLSKY. 1971. Isotonic velocity transients in frog muscle fibers following double force steps. *Biophys. J.* **11**: 236a.

PODOLSKY, R. J. 1960. The approach to the steady state. *Nature* **188**: 666.

PODOLSKY, R. J. and A. C. NOLAN. 1971. Cross-bridge properties derived from physiological studies of frog muscle fibers, p. 247. In *Contractility of muscle cells*, ed. R. J. Podolsky. Prentice-Hall, Englewood Cliffs, New Jersey.

PODOLSKY, R. J., A. C. NOLAN, and S. A. ZAVELER. 1969. Cross-bridge properties derived from muscle isotonic velocity transients. *Proc. Nat. Acad. Sci.* **64**: 504.

TONOMURA, Y., H. NAKAMURA, N. KINOSHITA, H. ONISHI, and M. SKIGEKAWA. 1969. The pre-steady state of the myosin-adenosine triphosphate system. X. The reaction mechanism of the myosin-ATP system and a molecular mechanism of muscle contraction. *J. Biochem.* (*Tokyo*) **66**: 599.

TRENTHAM, D. R., R. G. BARDSLEY, J. F. ECCLESTON, and A. G. WEEDS. 1972. Elementary processes of the magnesium ion-dependent adenosine triphosphatase activity of heavy meromyosin. *Biochem. J.* **126**: 635.

Mechanical Transients and the Origin of Muscular Force

A. F. HUXLEY AND R. M. SIMMONS

Department of Physiology, University College, London WC1E 6BT

The invitation to contribute to a symposium of this kind gives an opportunity for writing in a more discursive, and a more speculative, way than is accepted in papers for regular scientific journals. An invitation in 1955 to contribute to *Progress in Biophysics* gave one of us a similar opportunity (A. F. Huxley, 1957) to present some ideas on the details of the contraction mechanism that had arisen through thinking in terms of the sliding-filament theory. Our experimental work since then has consisted largely in following up some of the consequences of those ideas, and it is therefore appropriate to combine a discussion of our present ideas on the origin of force in a contracting muscle fiber with a review of the scheme put forward in 1957.

The 1957 Scheme

The scheme contained three main propositions which are separable in the sense that one may be correct while another is not. These are:

a. Each zone of overlap between thick and thin filaments contains uniformly distributed independent sites, each of which generates a relative force (which may fluctuate) between the filaments, and the total force on a thin filament is the sum of the forces contributed by the sites between that filament and the neighboring thick filaments.

b. Each site consists of a movable side-piece elastically connected to the thick filament, which undergoes cycles of (i) attachment to a thin filament site with a moderate rate constant, (ii) exerting a force on the thin filament, and (iii) detachment, the rate constant for the latter becoming large when the sliding motion of the filaments has brought the side-piece to a position where the force it exerts is near to zero.

c. The force arises because the location of an enzyme required for attachment is such that attachment occurs only when a cross-piece is displaced by Brownian movement in the direction which causes the restoring force due to its elastic element to be in the pulling direction.

The following paragraphs will outline our present attitude toward these propositions.

Independent force generators in overlap zone. This idea was proposed by A. F. Huxley

and Niedergerke (1954) on the grounds (i) that Ramsey and Street (1940) had found that isometric tetanus tension (T_0) declines linearly with increase of length beyond the optimum (i.e., where overlap declines with stretch), and (ii) that Jasper and Pezard (1934), comparing muscles of arthropods with different sarcomere lengths, had found speed of shortening to be correlated with shortness of sarcomeres (i.e., large number of overlap zones in series per unit length of fiber). The relation between T_0 and overlap was reinvestigated by Gordon et al. (1966), who found a very satisfactory proportionality between T_0 and overlap, the latter being taken specifically as overlap of each thin filament with that part of the thick filament which carries cross-bridges (H. E. Huxley, 1957), suggesting strongly that the independent force generators not only exist but are to be identified with the cross-bridges seen in electron micrographs.

Gordon et al. also found that the speed of unloaded shortening was independent of the amount of overlap, a result which is to be expected if the sites act independently and the resistance to be overcome is negligible.

Regarding the characteristics of muscle fibers with different sarcomere lengths, we do not know of any recent systematic measurements, but there are a few observations, all of which are in the expected direction. Thus, Atwood et al. (1965) have confirmed Jasper and Pezard's observation that speed of shortening is inversely correlated with sarcomere length. T_0 would be expected to vary directly with A-band length (proportional to length of overlap zones); a correlation of this kind in crab fibers was found in unpublished work by C. M. Armstrong when working at University College, and Zachar and Zacharová (1966) found that both T_0 (per unit cross-sectional area) and thick filament length are rather more than twice as great in a crayfish fiber as in frog twitch fibers.

We regard these results as justifying acceptance, at least provisionally, of the idea of independent force generators in the overlap zone. The idea appears at present to be widely, but by no means universally, accepted; for example, the theories of contraction proposed by Ullrick (1967), by Elliott et al. (1970) and by Yu et al. (1970) are not of this kind at all.

Cyclic attachment and detachment of side-pieces. The case worked out by A. F. Huxley (1957) showed that a set of simple and plausible values for the rate constants for attachment and detachment led to relations between load, speed of shortening, and rate of energy liberation that agreed, within experimental error, with the results of Hill (1938). An even simpler, though less plausible, set of assumptions within the same general framework, proposed by Deshcherevskii (1968), leads exactly to the formulae used by Hill (1938) to describe his results. We do not know of any other type of theory which leads in a natural way to these relationships, so we regard the idea of cyclic attachment and detachment as strongly (though not conclusively) supported.

A real difficulty for these formulations has recently been raised by Chaplain and Frommelt (1971). They pointed out that although on Hill's (1938) formulae the total rate of energy liberation increases progressively as the load is decreased toward zero, this is not the case when the revised formula for shortening heat given by Hill (1964a) is used. In this formula the shortening heat coefficient α decreases as the load is reduced, causing the total rate of energy liberation to pass through a maximum and then to decline as the load approaches zero. There does not seem to be any simple way of accommodating this result without additional hypotheses, though it could be explained by a reduction of the degree of activation as the tension is lowered, a possibility that has been suggested in other connections (Jewell and Wilkie, 1960; Hill, 1964b; Pringle, 1967, p. 49; and see below).

Another respect in which this aspect of the 1957 theory needs revision is the value taken for the rate at which the attached side-pieces turn over during isometric contraction. This was derived by assuming that the rate constant for detachment during the isometric state was equal to the rate constant of decline of tension at the end of an isometric contraction; i.e., relaxation was assumed to consist simply in stopping attachments from being formed, with detachment continuing with the same rate constant as during isometric contraction. The basis for making this assumption has, however, been destroyed by our observation (Huxley and Simmons, 1970b) that the exponential phase of relaxation begins at the moment when a part of the length of the fiber—almost invariably one end—begins to lengthen, so that the fiber is no longer in the isometric condition at the stage when the rate constant is measured. The converse of this observation is illustrated in Fig. 1, which shows that when the "spot-follower" device is used to hold the

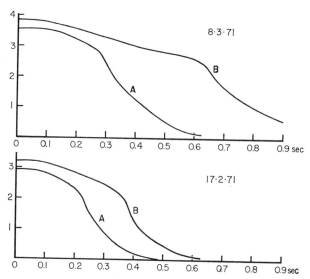

Figure 1. Prolongation of relaxation phase of an isometric tetanus when the spot-follower is used; two experiments on single frog fibers, 4°C. Ordinate: tension, arbitrary scale; abscissa: time measured from last shock of tetanus of about 1 sec duration. *A*, tendons held stationary; *B*, servo system and spot-follower in operation so that the middle 50% or so of the fiber is kept at constant length in spite of elongation of one end of the fiber which begins at about the time of the "shoulder" in the corresponding *A* record. Presumably some part of the fiber within the controlled segment begins to elongate when the "shoulder" occurs in the *B* record.

middle part of the fiber at constant length, despite the elongation of one or both ends, the onset of the rapid exponential phase of tension decline is greatly delayed in comparison with a contraction when the tendon ends are held stationary. It does not follow necessarily that the rate of turnover during isometric contraction is as slow as the low rate of tension decline before the "shoulder" might suggest, since it is possible that this time course represents the decay of "active state" when no shortening occurs. But there is no longer a basis for taking the exponential rate constant of the late part of relaxation as representing anything that occurs in a truly isometric contraction.

Force already present at time of attachment. The assumption that the force was already present (as a result of thermal agitation) in the side-piece at the moment when it became attached to the thin filament was made because it is the simplest available within the general framework: Only a single mode of attachment is required, and no mysterious and unobserved shape change in either protein has to be assumed.

A mechanism of this kind invites the question whether the side-piece will be displaced thermally, through the distance required, often enough to account for the rates of shortening and of tension

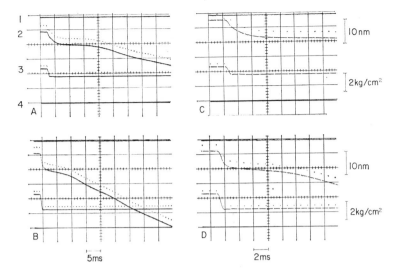

Figure 2. Transient responses to a stepwise reduction of load. Isolated frog fiber, 2.5°C. Each record begins 0.4 sec after the start of a tetanus. Trace 1, baseline for length, corresponding to striation spacing of 2.14 μm. Trace 2, length, obtained with spot-follower; bar on right indicates 10 nm length change per half-sarcomere. Trace 3, tension. Trace 4, baseline for tension. C and D are similar to A and B except that they are recorded on a faster time base. Dots above the traces show 1 msec intervals. From Huxley (1971) by permission of the Royal Society.

rise that are actually found. Some rough calculations have suggested that the speed of frog muscle at low temperature could be explained on this basis provided it is assumed that a side-piece attaches every time it is displaced to the requisite distance, but this is an unlikely assumption since it is difficult to see how the side-piece could be guided in precisely the direction needed to reach the site on the thin filament whenever the absolute amount of its deflection is large enough. Also even if this assumption is made, it is likely that the speeds of more rapid muscles would be difficult to explain in this way.

It will be pointed out later in this article that the 1957 theory does not account for the transient responses that are seen when the length or the load is suddenly altered during a contraction. The type of explanation for these phenomena that we favor is one in which the production of tension (or at least a large part of it) occurs as a step distinct from, and subsequent to, the attachment itself. If this is correct, it removes the difficulty about the absolute values of the rates of attachment. Podolsky and his colleagues, however, favor an explanation based on a modification of the 1957 theory in which the rate constants for attachment need to be even higher; this will give rise to an even more acute difficulty in explaining the real speeds of muscles, unless the actual production of tension is put into a step subsequent to the attachment but so very rapid that it is not detectable in experiments that have so far been carried out.

Responses to Step Change of Length or Load

This discussion shows that, even if we provisionally accept that force is generated by independent sites in the overlap zone, operating by cycles of

attachment, pulling, and detachment, there is serious doubt about the simplifying assumption in the 1957 theory that force in the side-piece is created by thermal agitation and is exerted on the thin filament as soon as attachment occurs. Examples of the responses of isolated frog muscle fibers to step changes of tension or length, which are not adequately explained on these assumptions, are shown in Figs. 2, 3, and 4. We originally used tension steps (Fig. 2) on the grounds that their interpretation is simpler if the fiber contains a true "series elastic component," since the length of this component will remain constant from the moment when the tension has reached its new value. We switched to length steps largely because it is extremely difficult to get the tension to change in a way which is a satisfactory approximation to a

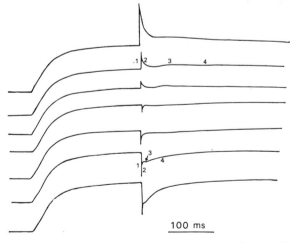

Figure 3. Tension changes in response to step changes of length. Isolated frog fiber, 4°C. Tracings from a family of records with length steps ranging from stretch of 5.4 nm/half-sarcomere to release of 8.8 nm/half-sarcomere. The numbers refer to the stages of the response listed in Table 1.

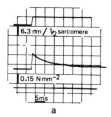
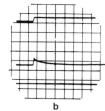
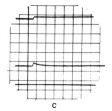

a   b   c

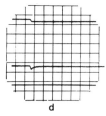

d   e   f

Figure 4. Tension changes in response to step changes of length. Isolated frog fiber, 4°C. Family of records showing the early part of the response on a fast time-base. Upper trace: length. Lower traces: tension and baseline for tension. Stretches: a, 5.4; b, 2.7; c, 1.4 nm/half-sarcomere. Releases: d, 1.4; e, 2.7; f, 8.7 nm/half-sarcomere.

step; the records in Fig. 2 are exceptionally good and the problem becomes progressively more difficult with larger steps. The difficulty arises from the fact that we use a servo system for controlling the tension, and the complicated and nonlinear characteristics of the fiber itself are within the servo loop. Nonlinear elements have to be incorporated in the control circuits, and its parameters have to be adjusted afresh, for each size of step, by successive approximation over a series of contractions in order to obtain records like those in Fig. 2. It is no solution to use direct mechanical control of tension instead of a servo system, since it is impracticable to make the natural frequency of the moving parts high enough to get records comparable in quality with those shown in Fig. 2. When it is the length that is controlled, however, the characteristics of the fiber become unimportant; if it is the total length between tendon attachments that is controlled, fiber characteristics do not affect the system at all, and when the "spot-follower" (Gordon et al., 1966) is used so that the quantity controlled is the length of a middle segment of the

fiber defined by markers that are stuck on to it, the fiber characteristics come in only as a second-order factor insofar as the alteration of tension influences the length of the extensible structure of the tendons.

The responses to tension steps (Fig. 2) show the same features that have been described by Podolsky (1960) and Civan and Podolsky (1966). Four stages in the response can be recognized and corresponding stages are seen in the tension responses to a step change of length, as is shown in Table 1. The two types of experiment are clearly showing two manifestations of the same underlying behavior in the fiber, and the choice of which type to use is primarily a matter of experimental convenience and ease of interpretation, although it is clearly important that theories based on one type of experiment should be checked against results of the other type.

Effect of altering the amount of overlap.
At first (Huxley and Simmons, 1970a) we thought it likely that stage 1—the tension drop which occurs simultaneously with the length step itself—represented shortening of the filaments themselves, without appreciable sliding, and that stage 2, the early rapid recovery, represented sliding due to action of the cross-bridges accompanied by some reextension of the filaments. The other extreme possibility would be that the filaments are effectively rigid, and that the instantaneous elasticity which shows itself in stage 1 is in the cross-bridges themselves. To resolve this question, we recorded responses to length steps carried out with the fiber at different initial lengths, so as to vary the amount of overlap between the filaments and thus the number of sites by which they are connected. The result shown in Fig. 5 was surprisingly simple (Huxley and Simmons, 1971a): For a given absolute size of length step, the time courses of the tension

Table 1. Stages of Responses to Stepwise Reductions of Load or Length in Frog Muscle Fibers

	Tension Step	Length Step
1. During step change	Simultaneous shortening	Simultaneous drop of tension
2. Next 1–2 msec	Rapid early shortening	Rapid early tension recovery
3. Next 5–20 msec	Extreme reduction or even reversal of shortening speed	Extreme reduction or even reversal of rate of tension recovery
4. Remainder of the response	Shortening at steady speed, sometimes with superposed damped oscillation	Gradual recovery of tension, approaching T_0 asymptotically

Times refer to experiments at 4°C.

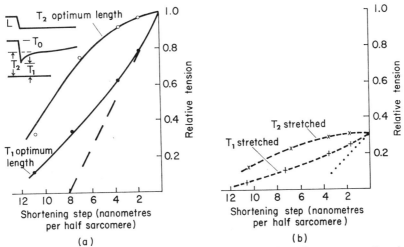

Figure 5. Effect of varying the amount of overlap on the early stages of tension recovery after shortening steps. Isolated frog fiber, 4°C. *Inset:* method of measuring T_1, the extreme tension reached during the step, and T_2, the tension approached in early rapid recovery. (*a*) Sarcomere length 2.2 μm. T_1 and T_2 plotted against step size for a series of releases. (*b*) Sarcomere length 3.2 μm. *Crosses:* T_1 and T_2 in series of releases at 3.2 μm expressed as fractions of T_0 at 2.2 μm. *Broken lines:* T_1 and T_2 curves from (*a*) scaled down in proportion to isometric tension at 3.2 μm. *Dots:* Curve expected for T_1 if all the instantaneous elasticity was in a series structure; this is the lower part of the T_1 curve from (*a*) shifted horizontally. (From Huxley and Simmons, 1971a.)

changes were similar, and their magnitude varied in direct proportion to the amount of overlap and thus to the isometric tension. The simple interpretation of this result is that the instantaneous elasticity resides in the connections between thick and thin filaments in the overlap zone. Compliance in the filaments themselves and in the Z line cannot, of course, be actually zero; but we do not think that it is likely to lead to serious errors of interpretation if as a first approximation these series compliances are assumed negligible and the tension changes are regarded as directly showing the responses of the side-pieces when the filaments are stationary after a brief sliding motion during the actual step change of fiber length.

Insofar as this neglect of filament compliance is justifiable, it is now the length step rather than the tension step that is the simpler to interpret, since it shows directly the changes in mean tension in the cross-bridges when their ends attached to the filaments are stationary. In the tension step the filaments are undergoing relative motion with a complex time course at the same time as the unknown changes in the cross-bridges are occurring, and it is not even justifiable to suppose that the tension in any one side-piece remains constant, since the control of tension ensures only that the sum of the contributions from many cross-bridges is constant.

Interpretation of tension response to stepwise shortening.

Our present ideas about the events within the cross-bridges, corresponding to the stages of the response set out in Table 1, are as follows. The stages are taken in the order 1, 4, 3, 2.

a. *Stage 1.* This is a change of length in an elastic component of each cross-bridge. Neither detachment nor attachment occurs to an appreciable extent. Ideally there is also no length change in parts of the cross-bridge in series with its elastic element, but these changes (which we regard as the basis of stage 2, see below) become very rapid in large shortening steps, and we are not yet able to make our steps fast enough to avoid having these changes begin before the length change is complete.

b. *Stage 4.* Cross-bridges detach and reattach further along the filament until the distribution of force among them is the same as it was before the step.

So far, these interpretations are the same as would be given by the 1957 theory.

c. *Stage 3.* So far as we are aware, this stage is not satisfactorily explained by any of the current theories. We have in mind three possible types of explanation for it: Each of these could give rise to the delayed fall of tension which is sometimes seen after a shortening step (Armstrong et al., 1966) and which is not to be expected on the basis of either the 1957 theory or the modification of it by Podolsky et al. (1969), which does allow a drop of shortening speed below its steady value in a tension-step experiment.

(i) If the f (attachment) and g (detachment) functions (Huxley, 1957) are assumed to have the form shown in Fig. 6A, the distribution of the variable x immediately after the shortening step will be as shown by the full line in Fig. 6B. The rate of detachment is therefore larger than in the isometric state, while the rate of attachment is the

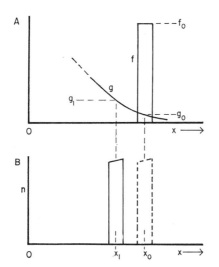

Figure 6. *Above:* an example of functions for attachment (f) and detachment (g) which could lead to a delayed fall of tension after a shortening step. Abscissa: x as defined in A. F. Huxley (1957), i.e., displacement of side-piece from its equilibrium position. *Below:* distributions of attached side-pieces corresponding to the f and g functions above. (– – –) in isometric state; (———) immediately after a shortening step. g_0 and g_1 are mean values of g for the two distributions; the product gx is more than twice as large after the step as it is during isometric contraction.

same, being f_0 multiplied by the proportion of bridges free, which has not had time to change. The rate of loss of tension due to detachment is proportional to kx_1g_1, and the rate of increase due to attachment to kx_0g_0, so that provided $g_1x_1 > g_0x_0$, there will be a net decline of tension. This will slow down and reverse as the number of bridges in positions near x_1 declines and the number of free bridges available for attachment near x_0 rises.

(ii) The effect described under (i) would be much exaggerated if one or more chemical reactions had to take place after detachment before a cross-bridge could reattach or at least before it could produce tension again. This would delay the rise in frequency of formation of new attachments.

(iii) The initial tension drop may cause some "de-activation" such as is generally supposed to be the basis of the oscillatory behavior of asynchronous muscles of insects (Pringle, 1967, p. 49). Other reasons for thinking that some such process may exist in frog muscle were mentioned above, and a point which makes this kind of explanation attractive is that this stage of recovery is much more conspicuous in some preparations than others, so that it is unlikely to be closely related to essential events in the force-generating process itself.

d. *Stage 2.* Although an early phase of fairly rapid recovery after a shortening step is expected on the 1957 theory (due to rapid detachment of cross-bridges that have been displaced to a

position where they exert negative tension), this is far less clearly separated from the later recovery than what is found experimentally. A modification to that theory proposed by Podolsky et al. (1969) to explain the responses to step changes of load gives a very much better approximation, and the early phase of tension recovery after a single shortening step can probably be adequately represented in this way by making very specific assumptions about the functions f and g which govern the rates of attachment and detachment of the cross-bridges.

Certain features of the responses to step changes of length did, however, suggest to us a different kind of explanation. This was published in *Nature* last autumn (Huxley and Simmons, 1971b), and an account of it will be given in the next section. It assumes that this rapid phase of recovery takes place without detachment or attachment of cross-bridges.

Mechanism Proposed to Account for Early Quick Phase of Recovery

Two striking kinds of nonlinearity can be seen in the early quick phase of recovery in the records in Figs. 3 and 4. These are shown separately in Figs. 7 and 8. In Fig. 7, tension is plotted against the size of the length step. Curve T_1 shows the extreme tension reached during the release or stretch; it varies with size of step in much the way that is found with most biological materials; i.e., it has a linear range but becomes apparently less stiff as

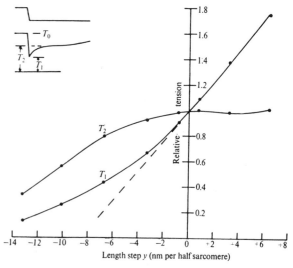

Figure 7. Nonlinearity of the amount of the quick tension recovery. T_1 and T_2, measured as in Fig. 5 and plotted against size of length step (stretches positive, releases negative). (– – –) Extrapolation of the part of the T_1 curve which refers to small stretches and releases. In more recent experiments in which the spot-follower was used, the T_1 curve is closer to the extrapolated line. (From Huxley and Simmons, 1971b, by permission of *Nature*.)

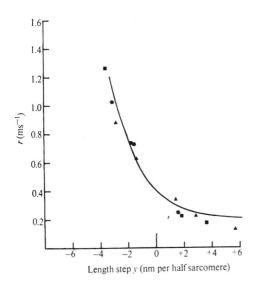

Figure 8. Variation of rate constant of quick phase of tension recovery after length steps of various sizes, from three experiments using the spot-follower; temperature 4°C. The curve is $r = 0.2(1 + \exp - 0.5y)$. (From Huxley and Simmons, 1971b, by permission of *Nature*.)

the tension becomes less. It is not clear how much of the latter tendency is real, since the initial peak is certainly being reduced by the start of the early quick recovery, and this effect is greatest in the large releases where the recovery is fastest. Curve T_2 shows the tension level which is approached during the early quick recovery. This curve has the unexpected and unusual features of being nearly flat in the region of small steps and concave downwards for releases of moderate size.

Figure 8 illustrates the other aspect of the nonlinearity of the early recovery, namely, the fact that its speed becomes greater the larger the size of a release, and slower the larger the size of a stretch, there being no discontinuity between the two sides of the curve.

The assumptions that we showed able to account for this type of behavior are: (i) The movement by which a cross-bridge performs work during the period while it is attached takes place in a small number of steps from one to the next of a series of stable positions with progressively lower potential energy; and (ii) there is a virtually instantaneous elasticity within each cross-bridge allowing it to shift from one of these stable positions to the next without a simultaneous displacement of the whole thick and thin filaments relative to one another.

Representation as a Voigt element. These two elements in the cross-bridge must be thought of as being in series, in the sense that any change in the separation between the two attached ends of the cross-bridges due to sliding of the filaments is the

sum of the movement (i) and the length change in the elastic element (ii). The element (i) generates force because of its tendency to spend more time in positions of lower potential energy; it can therefore be represented by the spring K_1 in Fig. 9B, which is drawn with a dashpot D in parallel to indicate that shifts of this element from one of its stable positions to another take place with a finite rate constant. The spring K_2 in series represents the elastic element (ii). The whole is thus formally equivalent to a "Voigt element," whose response to a step change of length is well known to be similar to stages 1 and 2 of the response of a muscle fiber, i.e., a relatively large initial tension change followed by partial recovery with a finite rate constant.

An ideal Voigt element consisting of springs which obey Hooke's law and a dashpot with Newtonian viscosity characteristics would, of course, give a linear response, i.e., when the step size is altered, the tension changes will alter in direct proportion and the time constant of recovery will be unchanged. It was shown, however, in our *Nature* article (Huxley and Simmons, 1971b), by a simplified mathematical treatment that will not be repeated here, that assumptions (i) and (ii) lead to responses which are nonlinear in the same kind of way as the responses of real muscle fibers. The elements K_1 and D in Fig. 9B must therefore be thought of as having stiffness and viscous resistance which vary with the total length L of the two elements in series.

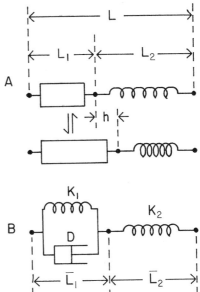

Figure 9. *A.* The theory proposes that each cross-bridge contains an element (i), with length L_1, which undergoes reversible stepwise changes, in series with an element (ii) which shows instantaneous elasticity. *B.* When the time average ($\overline{L_1}$) of L_1 is considered, the behavior of element (i) resembles that of a spring (K_1) whose length changes are damped by the dashpot D.

*Qualitative deduction of nonlinear charac-
teristics.* Figure 10A shows a potential energy
diagram representing the stable positions that the
cross-bridge is able to adopt, according to assumption (i). The abscissa corresponds to the length L_1
in Fig. 9A. If the filaments are held stationary (L
constant), thermal agitation causes the cross-
bridge to jump from one to another of its stable
positions, the proportion of time spent in each
depending on the tension in the elastic element K_2
and therefore on L. Plotting the time-average of
the potential energy against $\overline{L_1}$, the time-average
of L_1, will round off the sharp features of 10A to
give a curve like 10B. The slope of this is 10C
which therefore represents the effective force-

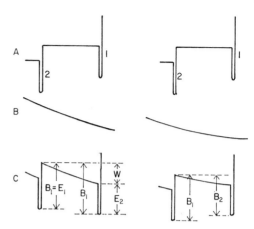

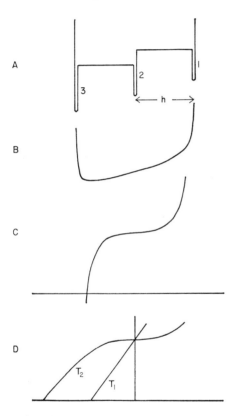

Figure 10. Relation between potential energy diagram of
cross-bridge states and the extent of the quick tension
recovery. *A*, Schematic diagram of potential energy of the
element (i), assumed to have 3 stable positions. Abscissa
is L_1 (Fig. 9). *B*, Time average of potential energy shown
in *A*, plotted against $\overline{L_1}$ (Fig. 9), the time average of L_1.
C, Slope of B_1 which gives the tension due to the tendency
for the element to spend more time in states of lower
potential energy. *D*, T_1: Length-tension relation of the
elastic element (ii); since it is assumed linear, this line
applies both for the instantaneous tension as a function
of L_2 and for the time-average tension as a function of $\overline{L_2}$.
T_2: Length-tension relation obtained by combining in
series the T_1 line and the curve *C*, i.e., their abscissae are
added for a given tension. This curve gives the expected
curve for the two elements in series when equilibrium has
been reached, i.e., at the end of the quick recovery phase;
it is seen to resemble the experimental T_2 curves in Figs.
5 and 7.

Figure 11. Explanation of the dependence of the rate
constant of early recovery on size of step. *A*, Part of the
potential energy diagram of Fig. 10A, relating to states
1 and 2. Abscissa is L_1 (Fig.9). *B*, Potential energy diagram
of the elastic component, plotted against $-L_2$ so that
points vertically above one another in *A* and *B* correspond
to the same value of L ($= L_1 + L_2$). *C*, Sum of *A* and *B*,
giving total potential energy of the cross-bridge as a function of L_1, at constant L (i.e., in absence of sliding between
the filaments). The left-hand set of curves are for a value
of L which makes the cross-bridge spend more time in
state 1 than state 2 (bottom of right-hand well lower than
left-hand). The right-hand curves refer to a smaller value
of L, i.e., curve *B* is shifted to the left. The result is a
reduction of W, reducing the activation energy B_2 for the
transition from state 1 to state 2.

extension curve of the element (i). This element is in
series with the elastic element K_2 whose force-
extension curve (assumed Hookean) is shown as T_1
in Fig. 10D; the combined force-extension curve
is obtained by taking a value of tension and
adding the extensions of the two elements that
correspond to this tension; the result is the curve
T_2 in Fig. 10D, which is seen to resemble the real
T_2 curve in Fig. 7.

As regards the dependence of the rate constant
on the size of step, consider the equilibrium between
two successive states of the element (i), say,
states 1 and 2 (Fig. 10A). The relevant part of its
potential energy diagram is reproduced in Fig.
11A; the total length L is thought of as constant
(no sliding), and the possibility of the system going
to other states (e.g., state 3) is disregarded because
the force in K_2 will be assumed sufficient to make
this transfer improbable. An increase in L_1 implies
a decrease in L_2, so the total potential energy of
the cross-bridge will include also a contribution
representing the energy stored in the elastic
element; the latter is shown in Fig. 11B and is
added to 11A to give the total potential energy in
11C. The quantities B_1 and B_2 in the figure are the
activation energies for the transfers $2 \rightarrow 1$ and
$1 \rightarrow 2$ respectively. B_1 is simply the depth of the
potential well in Figs. 10A and 11A and has no
appreciable contribution from the elastic element
because the well is extremely narrow, so that the

elastic element would only change in length by a very small amount (and therefore do very little work) when the system goes from the bottom of well 2 to the maximum of the potential energy diagram. Hence the rate constant for transfer from state 2 to state 1 is unaffected by a change in the force in the elastic element. B_2 however contains the quantity W (Fig. 11C), which varies with the force in the elastic element; it is equal to $h\bar{F}$ where h is the change in L_1 when the system goes from state 1 to state 2, and $\bar{F}$ is the force in the elastic element when the system is midway between the two states. Hence the rate constant for the transfer from 1 to 2 is proportional to $\exp(-h\bar{F}/kT)$, increasing as the force is reduced by a shortening step and decreasing as it is increased in a stretching step. The rate constant of approach to the new equilibrium distribution after the step is the sum of the rate constants for the transfers in the two directions and will therefore have a form similar to the curve in Fig. 8, which is seen to fit the observations well.

This descriptive account of how the proposed system would work has led to the same conclusions as were reached in the formal treatment in our article in *Nature*.

Structural basis of the proposed mechanism.

In our article in *Nature* we discussed our mechanism in terms of a particular identification of its functional elements with structures that are believed to be present on quite independent grounds. Figure 12 shows the model that was discussed there; it is the result of taking the diagram of H. E. Huxley (1969) and assuming (i) that the limited number of stable positions required in assumption (i) of our theory correspond to different orientations of the myosin head relative to the actin molecule to which it is attached, the stable positions occurring when any two consecutive members of the series of sites M_1, M_2, etc., are attached to the corresponding sites A_1, A_2, etc., on the actin molecule; and (ii) that the elastic

element K_2 of Fig. 9 is the part of the myosin between the hinge and the attachment of the head (subfragment 2). It was indicated in our article, and it is important to emphasize again, that this is only one of numerous ways in which our kinetic results could be related to actual structures. Even within the possibilities suggested by H. E. Huxley's diagram, there are several alternatives, in addition to the one discussed in our article in *Nature*. Sliding of the filaments, with displacement of one end of the myosin molecule (attached to thin filament) relative to the other (embedded in thick filament) can be taken up by change in: (A) length of subfragment 2 (between hinge and head), (B) tilting of myosin head about its point of attachment to the thin filament, or (C) a shape change within the myosin head, which may be thought of as equivalent to bending at a second (unknown) hinge within the head. The model discussed in our *Nature* article put the stepwise change (i) in (B) and the elastic structure (ii) in (A), but it is also possible to imagine the following alternative possibilities:

a. (i) is in (A) and (ii) in (B)

b. (i) is in (C) and (ii) in (A)

c. (i) is in (C) and (ii) in (B)

d. (i) is in (B) and (ii) in (C)

e. (i) is in (A) and (ii) in (C)

It is also conceivable that both the stepwise and the elastic component of length change in the cross-bridge take place within the myosin head, perhaps at two different bending sites. This gives a total of seven possibilities, even if we restrict ourselves to models that can be accommodated within H. E. Huxley's (1969) diagram of a cross-bridge. Measurements of overall tension and length changes are unlikely to distinguish between these (and other) possibilities; evidence will have to be looked for from chemical kinetic studies, X-ray diffraction, light scattering and suchlike approaches.

Relation to Podolsky's theory.

The theory set out in the preceding sections has more than one point of close resemblance to that of Podolsky et al. (1969), so it is not surprising that both lead to similar-looking transient responses; we have little doubt that both theories can be refined to the point where they give an adequate account of the time course of the response to a single step. These resemblances are (i) a small shortening step allows some of the cross-bridges to transfer from a situation in which they exert little or no force to one in which they exert a substantial force, (ii) the rate constant for this transfer increases with the size of the shortening step, and (iii) rapid detachment does not begin immediately after the start

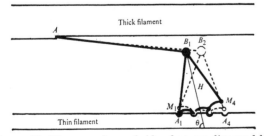

Figure 12. Diagram of cross-bridge features discussed by Huxley and Simmons (1971b) as a structural basis for their theory. The myosin head H is connected to the thick filament by a link AB containing the undamped elasticity. Solid line shows head in position where M_1A_1 and M_2A_2 attachments are made; broken lines show position where M_2A_2 and M_3A_3 attachments are made. (From Huxley and Simmons, 1971b, by permission of *Nature*.)

of shortening (this is not explicitly stated in our *Nature* article, but it is a necessary consequence of the assumption of finite rate constants for transfer from one of the stable states to another).

The principal difference between the theories is that on Podolsky's theory the transition referred to in (i) above is made by attachment of cross-bridges which were free in the isometric state, whereas on our theory the cross-bridges are attached throughout and the transfer is from one attached state to a different one in which more force is exerted. A direct consequence of this difference is that Podolsky's theory, unlike ours, predicts an increased number of attached cross-bridges, and therefore an increased stiffness, at the end of the early rapid recovery; Podolsky et al. (1969) also show that the number of attached cross-bridges should be greater during steady shortening than in the isometric condition. Clearly, one or other of the theories (or perhaps both) will be disproved when good measurements of the stiffness changes during the transient responses, and in steady shortening, are made.

A number of records which might in principle show stiffness changes have been made in our laboratory, and we have seen no suggestion of an increase in stiffness such as is expected on Podolsky's theory. However, we do not regard our experiments as good enough to be decisive, and we are at present reconstructing our apparatus so as to get the required precision.

One phenomenon, however, that we have regularly seen, and which is not subject to uncertainty due to imperfections in the apparatus, does suggest that the number of attached cross-bridges is less during steady shortening than in the isometric state. This is the "pull-out" seen when, for example, a load equal to the isometric tension is suddenly reapplied at the end of a period of rapid shortening. An example of this phenomenon is shown in Fig. 13. The pull-out resembles the rapid

extension that occurs when the load on an isometrically contracting muscle is suddenly increased by a factor of 1.8 or more (Katz, 1939), suggesting that in a case like that shown in Fig. 13 the number of cross-bridges attached is not more than 1/1.8 of the number attached during isometric contraction.

Interpretation of Other Phenomena

We have suggested (i) that much of the rapid elasticity of muscle resides in the cross-bridges and (ii) that each cross-bridge contains, besides an elastic element, a structure which undergoes step-wise changes of length. These propositions have implications in connection with phenomena that have previously been supposed to occur in structures running continuously through the sarcomere. Some of these implications will be discussed in the remaining sections of this article.

"Series elasticity" and rise of isometric tension. In our experiments using the spot-follower, compliance outside the fiber (in tendons, attachments, and measuring devices) is altogether eliminated, and our interpretation of the instantaneous component of the response was that it corresponded to a length change in the cross-bridges themselves; that is to say, it was within the essential contractile apparatus rather than being in series with it in any useful sense of the words.

On the standard interpretation of the time course of tension rise in an isometric tetanus (Hill, 1938), the contractile element is shortening and stretching the series-elastic component, the former having a definite force-velocity relation and the latter a definite force-extension relation. On this basis, the response to step shortening during isometric tension rise should be similar to the response to an equal-sized step imposed during steady shortening at the same tension. We found that this is not the case (Fig. 14), but another very simple result is found: The response to a step during isometric tension rise is closely similar to that in the plateau of an isometric tetanus except that it is scaled down in proportion to the tension which exists at the moment when the step is applied. This result is similar to the effect of reducing the amount of overlap (Fig. 5), and the natural interpretation of it is that the rise of tension during the onset of an isometric tetanus corresponds directly to the increasing number of cross-bridges attached.

This implies that the rapid elasticity is not a separable quantity in series; indeed, Fig. 14a shows directly that its characteristics alter as the tension rises: The curves can be fitted by scaling up, and not by shifting horizontally as would be the case for a series elastic component with properties that do not change.

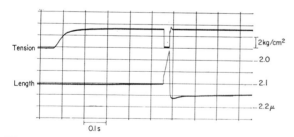

Figure 13. Records showing the pull-out phenomenon when a load equal to the isometric developed tension is suddenly reapplied at the end of a period of rapid shortening. Isolated frog fiber, 3.5°C, spot-follower not being used. Upper trace, tension; lower trace, length change (shortening upwards; figures on right give scale of sarcomere length). The "pull-out" is the very rapid fall of the lower trace (lengthening) which coincides with the reapplication of tension. (From A. F. Huxley, 1971, by permission of the Royal Society.)

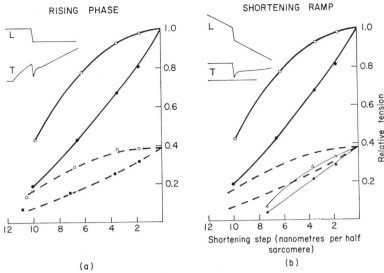

Figure 14. Responses to length steps (a) during isometric tension development and (b) during shortening. T_1 and T_2 defined as in Figs. 5 and 7. Isolated frog fibers, 4°C: sarcomere length about 2.2 μm. (a) Shortening steps during tension development. (■, □) T_1 and T_2 levels in shortening steps at 40 msec after start of a tetanus, when tension was 0.39 T_0. Circles, solid lines: T_1 and T_2 levels in shortening steps during the plateau, at T_0. Broken lines: T_1 and T_2 curves for the plateau scaled down by factor 0.39, i.e., in proportion to tension before the steps. (b) Shortening steps superposed on steady shortening (see inset); same experiment as (a). Triangle, lower solid lines: T_1 and T_2 in shortening steps during steady shortening when tension was 0.38 T_0. Circles, upper solid and dashed lines copied from (a). The squares in (a) fit well on the broken line but the triangles in (b) clearly do not. The transient responses are thus different although the tension levels are practically the same.

Thermoelastic phenomena in a sliding-filament system.

Hill (1953) and Woledge (1961) have shown that there is a small temperature rise in a contracting muscle when the load on it is suddenly reduced; this is distinct from the effect of shortening heat which is liberated continuously as long as the load remains below isometric and the muscle is shortening. Hill interpreted it as a thermoelastic phenomenon in the contractile material. On thermodynamic grounds, thermoelastic heat should be liberated in an amount ΔQ equal to $-\alpha Tl \Delta P$, where α is the coefficient of thermal expansion, T the absolute temperature, l the length of the sample, and ΔP the change of load. The coefficient $\Delta Q/l \Delta P$ was found by Hill to be about -0.018, roughly the value expected from the coefficient of expansion of a material such as ebonite; he pointed out that the effect was in the wrong direction for a material with rubber-like elasticity.

These calculations assume that the tension is carried by continuous structures within the muscle fiber (Hill's paper was published before the appearance of the sliding-filament theory). They are still at least approximately valid for thermoelastic heat produced in the filaments themselves, but the position is very different for thermoelastic heat produced within cross-bridges. If there are n cross-bridges between a thick and a thin filament in an overlap zone, the load on each is P/n, where P is the load on each filament. If each cross-bridge has a length l and is composed of material with

coefficient of expansion α, the thermoelastic heat liberated in the whole overlap zone is $Q = -n \cdot \alpha Tl \cdot \Delta P/n$, or $-\alpha Tl \Delta P$. Notice that the length l is that of a single cross-bridge; if both filaments were made of similar material, the heat liberation would be $-\alpha T \cdot s/2 \cdot \Delta P$, which is $s/2l$—perhaps 20—times larger. Alternatively, the expression $\alpha' Ts/2 \cdot \Delta P$ could be used for the thermoelastic effect due to cross-bridges, provided the value taken for α' is the fractional increase in length of the half-sarcomere due to expansion of the cross-bridges, i.e., $2\alpha l/s$.

Thermoelastic effects in the two components of a cross-bridge.

If, in spite of this geometrical factor, thermoelastic effects within the cross-bridge are appreciable, the theory we have proposed would introduce another complication, namely, that the effect might be related either to the step-wise length change in the element (i) in Fig. 9a, or to the length change in the elastic component (ii). The latter would be directly related to the tension change while the former would more likely be related to the enthalpy change associated with each step. This provides a basis on which a distinction might be made experimentally: If a series of shortening steps of increasing size are made at a speed low enough that tension follows along the T_2 curve (Fig. 7), thermoelastic heat in the elastic element should follow the tension change $(T_0 - T_2)$, whereas heat related to the stepping process will vary with the amount of length change associated with this process which is the horizontal distance between the T_2 and T_1 curves, or the abscissa of the curve in Fig. 10C. The former only begins to increase rapidly as the step size becomes larger than 5 nm per half-sarcomere or so; the latter is large at first and flattens off at about this size of step.

If some or all of the observed heat liberation on reduction of load is related to the stepping process (e.g., going from position 1 to position 2, Fig. 10), and if the heat still appears when the shortening is performed slowly enough so that the distribution between these positions remains in equilibrium all the time, then the entropy of the contractile material must decrease from position 1 to position 2, and a rise of temperature would displace the equilibrium in favor of position 1. A suggestion of an effect in this direction is that some preliminary measurements of the temperature coefficients of the rate constants of the quick recovery phase show a higher value for stretches $(2 \rightarrow 1)$ than for releases $(1 \rightarrow 2)$.

Thermoelastic effect during isometric tension rise. Insofar as thermoelastic heat is produced in a structure which is stretched passively by the tension in a muscle fiber, the effect during isometric tension rise should be the reverse of the effect when load is reduced. If, however, the effect is due to a process by which tension or shortening is generated, such as stepping from position 1 to position 2 in our theory, the effect will be in the same direction in the two cases: Stepping in this direction is required both for the production of tension after a cross-bridge has attached during isometric tension rise and for the quick phase of tension recovery after a shortening step. Woledge (1961) interpreted his experiments as indicating a thermoelastic absorption of heat during isometric rise, and he has pointed out to us that Fig. 5 of his later paper (1968) shows a marked reduction of heat liberation, below the maintenance heat rate, during isometric tension recovery after a period of shortening. This certainly suggests that the structure principally concerned is one which is stretched during isometric tension rise, but the matter deserves further attention.

References

ARMSTRONG, C. M., A. F. HUXLEY, and F. J. JULIAN. 1966. Oscillatory responses in frog muscle fibres. *J. Physiol.* **186**: 26P.

ATWOOD, H. L., G. HOYLE, and T. SMYTH, JR. 1965. Mechanical and electrical responses of single innervated crab-muscle fibres. *J. Physiol.* **180**: 449.

CHAPLAIN, R. A. and B. FROMMELT. 1971. A mechano-chemical model for muscular contraction. I. *J. Mechanochem. Cell Motil.* **1**: 41.

CIVAN, M. M. and R. J. PODOLSKY. 1966. Contraction kinetics of striated muscle fibres following quick change of load. *J. Physiol.* **184**: 511.

DESHCHEREVSKII, V. J. 1968. Two models of muscular contraction. *Biophysics* **13**: 1093.

ELLIOTT, G. F., E. M. ROME, and M. SPENCER. 1970. A type of contraction hypothesis applicable to all muscles. *Nature* **226**: 417.

GORDON, A. M., A. F. HUXLEY, and F. J. JULIAN. 1966. The variation in isometric tension with sarcomere length in vertebrate muscle fibres. *J. Physiol.* **184**: 170.

HILL, A. V. 1938. The heat of shortening and the dynamic constants of muscle. *Proc. Roy. Soc. (London)* B **126**: 136.

――――. 1953. The "instantaneous" elasticity of active muscle. *Proc. Roy. Soc. (London)* B **141**: 161.

――――. 1964a. The effect of load on the heat of shortening of muscle. *Proc. Roy. Soc. (London)* B **159**: 297.

――――. 1964b. The effect of tension in prolonging the active state in a twitch. *Proc. Roy. Soc. (London)* B **159**: 583.

HUXLEY, A. F. 1957. Muscle structure and theories of contraction. *Prog. Biophys.* **7**: 255.

――――. 1971. The activation of muscle and its mechanical response. *Proc. Roy. Soc. (London)* B **178**: 1.

HUXLEY, A. F. and R. NIEDERGERKE. 1954. Interference microscopy of living muscle fibres. *Nature* **173**: 971.

HUXLEY, A. F. and R. M. SIMMONS. 1970a. A quick phase in the series-elastic component of striated muscle, demonstrated in isolated fibres from the frog. *J. Physiol.* **208**: 52P.

――――. 1970b. Rapid "give" and the tension "shoulder" in the relaxation of frog muscle fibres. *J. Physiol.* **210**: 32P.

――――. 1971a. Mechanical properties of cross-bridges of frog striated muscle. *J. Physiol.* **218**: 59P.

――――. 1971b. Proposed mechanism of force generation in striated muscle. *Nature* **233**: 533.

HUXLEY, H. E. 1957. The double array of filaments in cross-striated muscle. *J. Biophys. Biochem. Cytol.* **3**: 631.

――――. 1969. The mechanism of muscular contraction. *Science* **164**: 1356.

JASPER, H. H. and A. PEZARD. 1934. Relation entre la rapidité d'un muscle strié et sa structure histologique. *Compt. Rend. Acad. Sci.* **198**: 499.

JEWELL, B. R. and D. R. WILKIE. 1960. The mechanical properties of relaxing muscle. *J. Physiol.* **152**: 30.

KATZ, B. 1939. The relation between force and speed in muscular contraction. *J. Physiol.* **96**: 45.

PODOLSKY, R. J. 1960. Kinetics of muscular contraction: the approach to the steady state. *Nature* **188**: 666.

PODOLSKY, R. J., A. C. NOLAN, and S. A. ZAVELER. 1969. Cross-bridge properties derived from muscle isotonic velocity transients. *Proc. Nat. Acad. Sci.* **64**: 504.

PRINGLE, J. W. S. 1967. The contractile mechanism of insect fibrillar muscle. *Prog. Biophys. Mol. Biol.* **17**: 1.

RAMSEY, R. W. and S. F. STREET. 1940. The isometric length-tension diagram of isolated skeletal muscle fibers of the frog. *J. Cell. Comp. Physiol.* **15**: 11.

ULLRICK, W. C. 1967. A theory of contraction for striated muscle. *J. Theoret. Biol.* **15**: 53.

WOLEDGE, R. C. 1961. The thermoelastic effect of change of tension in active muscle. *J. Physiol.* **155**: 187.

――――. 1968. The energetics of tortoise muscle. *J. Physiol.* **197**: 685.

YU, L. C., R. M. DOWBEN, and K. KORNACKER. 1970. The molecular mechanism of force generation in striated muscle. *Proc. Nat. Acad. Sci.* **66**: 1199.

ZACHAR, J. and D. ZACHAROVÁ. 1966. Potassium contractures in single muscle fibres of the crayfish. *J. Physiol.* **186**: 596.

The Conversion of Osmotic into Chemical Energy Coupled with Calcium Translocation across the Sarcoplasmic Membrane

M. MAKINOSE

Max-Planck-Institut für medizinische Forschung, Abteilung Physiologie, Heidelberg, Germany

In the last year it was demonstrated that the sarcoplasmic calcium pump can run backwards (Barlogie et al., 1971; Makinose, 1971; Makinose and Hasselbach, 1971). Typical experimental conditions for the backward reaction are shown in Fig. 1. The vesicles were loaded with calcium in a solution containing 2 mM acetyl phosphate, 7 mM MgCl$_2$, 20 mM inorganic phosphate, and 0.2 mM calcium. An addition of 1 mM EGTA after the completion of the calcium uptake induces only a very slow calcium release. When 2 mM ADP is added to this system, one observes a rapid calcium efflux and a simultaneous synthesis of ATP from ADP and inorganic phosphate present in the solution. For every two calcium ions released from the vesicles, one molecule of ATP is synthesized. This result and further analysis show that, for the rapid calcium efflux coupled with ATP synthesis, the following conditions must be fulfilled: ADP, orthophosphate, and magnesium must be present, and the vesicles must be loaded with calcium.

Hence the reaction scheme and the overall reaction for the sarcoplasmic calcium pump can be

formulated as

$$2Ca_0 + ATP + E \xrightleftharpoons{Mg^{++}} Ca_2E \sim P + ADP \quad (1)$$

$$Ca_2E \sim P \xrightleftharpoons{Mg^{++}} 2Ca_i + PO_4 + E \quad (2)$$

$$2Ca_0 + ATP \xrightleftharpoons{Mg^{++}} 2Ca_i + ADP + PO_4 \text{ (overall)}$$

where Ca_0 and Ca_i are calcium ions outside and inside of the vesicles, respectively, and E and $E \sim P$ free and phosphorylated enzyme, respectively. This means that all the reaction products of the calcium accumulation are required as reactants for the reverse reaction. Magnesium is needed apparently as the activator of the reaction in both directions.

The reaction scheme implies a common phosphorylated intermediate which connects the first and the second reaction steps. In fact if the calcium accumulation is driven by acetyl phosphate, the vesicle protein incorporates the orthophosphate present in the medium (Fig. 1). Immediately after

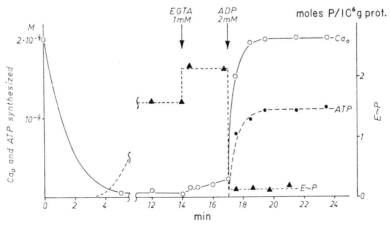

Figure 1. Reversal of the sarcoplasmic calcium pump and [^{32}P]orthophosphate incorporation of the membrane protein. Reaction mixtures contain 2 mM acetyl phosphate, 7 mM MgCl$_2$, 20 mM orthophosphate, 100 mM glucose, 0.2 mM CaCl$_2$, 0.02 mg/ml hexokinase and 0.5 mg/ml vesicles protein, pH 7. The mixture for the measurement of the calcium translocation was labeled with [^{45}Ca]CaCl$_2$ and those for the measurement of ATP synthesis and phosphoprotein formation with [^{32}P]orthophosphate.

the addition of EGTA, the level of incorporated phosphate usually increases up to 20 %. Addition of ADP results in a sudden drop of the phosphorylation level, and simultaneously the system begins to synthesize ATP. This phosphoprotein complex formed (E ∼ P) is stable at acidic pH and is easily decomposed in the presence of hydroxylamine at pH 5.3. In other words it is not possible to distinguish this phosphoprotein from that which is formed during the active calcium accumulation by the transfer of the γ-phosphate of nucleoside triphosphates (Makinose, 1966, 1969; Martonosi, 1967; Yamamoto and Tonomura, 1967, 1968). Apparently the vesicle protein forms an energy-rich phosphate bond with inorganic phosphate in the medium, presumably an acyl phosphate bond. Equation 1 describes its dephosphorylation by ADP resulting in an ATP synthesis.

According to this scheme the phosphoprotein can be formed independently of any energy-rich phosphate complex such as ATP or acetyl phosphate. The experiment illustrated in Fig. 2 shows that this is actually the case. The vesicles are incubated in a solution containing 5 mM orthophosphate, 7 mM MgCl₂ and 3 mM CaCl₂, but no energy-rich phosphate. During the incubation time of 60 min the calcium concentration inside the vesicles increases presumably to the concentration of calcium outside. As shown in Fig. 2 addition of excess EGTA causes formation of phosphoprotein in the vesicle membrane. This phosphoprotein vanishes immediately after the ADP addition, and at the same time a certain amount of ATP is

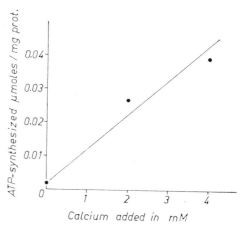

Figure 3. ATP synthesis by sarcoplasmic membrane in varied calcium concentrations. Reaction mixtures consist of the same components as described in Fig. 2 except the concentration of calcium added. After 120 min incubation together with 2 mM ADP, 6 mM EGTA is added to the mixtures containing 0 and 2 mM calcium, and 12 mM EGTA to the mixture containing 4 mM calcium. After 5 min the reaction is stopped by the addition of perchloric acid, and the amount of synthesized ATP is determined by measuring the glucose-6-phosphate formed in the system.

synthesized from ADP and orthophosphate in the system. Figure 3 shows that the higher the calcium concentration in the incubation solution, the more ATP is synthesized.

The phosphoprotein obtained in this way shows the same properties as the other phosphoproteins described (Makinose, 1966, 1969; Martonosi, 1967; Yamamoto and Tonomura, 1967, 1968; Fig. 1). Clearly the transfer of the orthophosphate to the protein does not result from the high calcium concentration in the vesicles itself, but from the high concentration gradient of the calcium ions across the membrane. Its magnitude provides enough energy to form an energy-rich phosphate bond such as an acyl phosphate.

This simple system shows, as a matter of fact, that the sarcoplasmic membrane has a perfect mechanism which converts the osmotic energy directly into chemical energy.

According to a personal communication from Tonomura, he and his coworkers reached the same conclusion based on similar experiments.

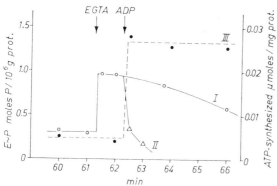

Figure 2. Phosphoprotein formation in sarcoplasmic membrane in the absence of energy-rich phosphoryl substrate. The sarcoplasmic vesicles (0.5 mg/ml) are incubated for an hour in a medium containing 20 mM histidine, 5 mM [³²P]orthophosphate, 100 mM KCl, 3 mM CaCl₂, 100 mM glucose, 7 mM MgCl₂, and 0.02 mg/ml hexokinase, pH 7. At 61 min 15 sec, 10 mM EGTA was added. The EGTA solution used is alkalized previously to neutralize the proton liberated from EGTA at neutral pH. Immediately after the addition of EGTA phosphoprotein is formed in a significant amount and disappears slowly (I). Subsequent addition of 2 mM ADP causes a rapid drop of the phosphoprotein level (II) leading to ATP synthesis (III).

References

BARLOGIE, W., W. HASSELBACH, and M. MAKINOSE. 1971. Activation of calcium efflux by ADP and inorganic phosphate. *FEBS Letters* **12**: 267.

MAKINOSE, M. 1966. Die phosphorylierung der membran des sarkoplasmatischen retikulum unter den bedingungen des aktiven Ca-transportes. *2nd Int. Biophys. Cong.* (IOPAB), Vienna.

———. 1969. The phosphorylation of the membranal protein of the sarcoplasmic vesicles during active calcium transport. *Europe. J. Biochem.* **10**: 74.

———. 1971. Calcium efflux-dependent formation of ATP

from ADP and orthophosphate by the membranes of the sarcoplasmic vesicles. *FEBS Letters* **12**: 269.

MAKINOSE, M. and W. HASSELBACH. 1971. ATP synthesis by the reverse of the sarcoplasmic calcium pump. *FEBS Letters* **12**: 271.

MARTONOSI. A. 1967. The role of phospholipids in the ATPase activity of skeletal muscle microsomes. *Biochem. Biophys. Res. Comm.* **29**: 753.

YAMAMOTO, T. and Y. TONOMURA. 1967. Reaction mechanism of the Ca++-dependent ATPase of sarcoplasmic reticulum from skeletal muscle. I. Kinetic studies. *J. Biochem.* **62**: 558.

———. 1968. Reaction mechanism of the Ca++-dependent ATPase of sarcoplasmic reticulum from skeletal muscle. II. Intermediate formation of phosphoryl protein. *J. Biochem.* **64**: 137.

A Model for Muscle Contraction in Which Cross-Bridge Attachment and Force Generation Are Distinct

F. J. Julian, K. R. Sollins, and M. R. Sollins

Department of Muscle Research, Boston Biomedical Research Institute, Boston, Mass. 02114 and
Department of Neurology, Harvard Medical School, Boston, Mass. 02115

This model is based on that recently presented by A. F. Huxley and R. M. Simmons (1971). Its most basic assumption is that the attachment and force-generating steps are completely separate; i.e., there is no force present when a cross-bridge attaches to the thin filament; the force is generated, following attachment, by a flip of the cross-bridge head into a different conformational state. In order to model other known muscle phenomena besides those studied by Huxley and Simmons, the model was extended to include attachment and detachment of cross-bridges and continuous motion of the filaments. The model can be related to current ideas of the biochemistry of the contraction cycle. A possible chemical scheme is shown in Fig. 1A.

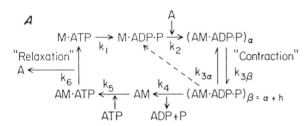

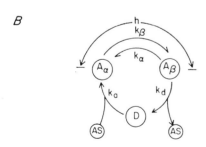

Figure 1. *A,* Possible biochemical scheme for the contraction cycle. In the figure, *A* represents actin; *M,* myosin; and the dot, a complex. The reversible reaction between states α and β of the complex $AM \cdot ADP \cdot P$ corresponds to the force-generating step. The extension of the elastic element in the cross-bridge changes by *h* between states α and β. The dashed line indicates the simplification used in the present form of the model. *B,* Simplified scheme used in the model. A_α and A_β give the number of cross-bridges attached in either the α or β state, and *D* gives the number in the detached state. The *k*'s are rate constants for state transitions. *AS* stands for actin sites on the thin filaments. As in *A, h* gives the change in elastic element extension between states α and β.

Figure 1B is a diagram of the simplified state of the model as it was actually programmed for a digital computer, an XDS Sigma 3. The simplification, which is equivalent to assuming that the rate constants k_4, k_5, k_6 and k_1 can be lumped, produced a substantial decrease in programming complexity and cost of computer time.

The model is taken to represent a volume of muscle one half-sarcomere in length and one cm² in cross section. Length changes are confined to the plateau region of the sarcomere length-tension relation. Steady state, maximally activated conditions are assumed. Each cross-bridge consists of a head (assumed to correspond to the S_1 subfragment of HMM) attached to the myosin filament by an elastic element (assumed to correspond to the S_2 subfragment). The head is capable of attachment, governed by the rate constant k_a, to the sites on the thin filament in one (α) of two stable states. Attachment is made to depend on the extension of the cross-bridge (represented by the variable *u*) in such a way that it only occurs when there is no extension in the elastic element or spring. Once attached, a cross-bridge can reversibly change state (flip) to the second (β) stable state. The flip causes an extension of the spring by a distance *h*, resulting in generation of tension in the cross-bridge. Detachment, which can only occur from the β state, is governed by the rate constant k_d, a function of *u* (see Fig. 2D).

As described by Huxley and Simmons (1971), the rate constants for flipping, k_α and k_β, depend on the depth of potential energy wells associated with the stable α and β states and the energy changes in the spring. In order to derive expressions for k_α and k_β, equations must be written for the work done on the spring in a flip of distance *h* as a function of *u*, the initial extension of the spring. The resulting equations, given in the legend for Fig. 2A, are linear functions of *u*. Note that if the filaments are moving with respect to each other, the *u* value of a cross-bridge in state α is not necessarily zero, and that of a cross-bridge in state β is not necessarily *h*. In order to visualize the way in which the depth of the potential energy wells and work term combine to form the activation

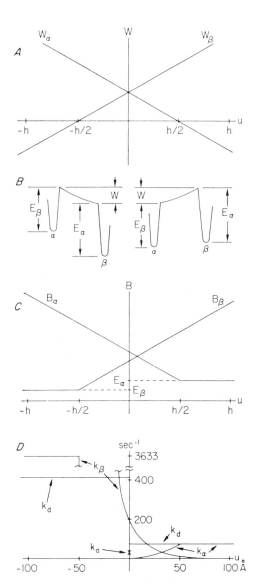

energy B, two cases are shown in Fig. 2B. The left hand part illustrates a case in which a cross-bridge in state α has a u value more negative than $-h/2$. In this case W_β is negative, and the activation energy is given simply by the well depth E_β. Alternatively, if the cross-bridge is in state β and has a u value less than $h/2$, W is positive and the activation energy includes the work term as well as the well depth. It is important to realize that in Fig. 2B the energies are plotted against some variable measuring the extent to which a flip has progressed (e.g., rotation of the head) as opposed to parts A, C, and D, in which u indicates the initial extension of a cross-bridge. The right-hand part of Fig. 2B illustrates another situation, where the u value for the α state is less negative than $h/2$ and that for β is greater than $h/2$, which can be analyzed in a similar way. Figure 2C shows the actual activation energies for a flip to state α, B_α, or to state β, B_β, the plateaus reflecting the fact that a negative work term does not influence the activation energy. The flipping rate constants plotted in Fig. 2D are inversely proportional to the exponential function of the activation energy, i.e.,

$$k_\alpha = c_\alpha \exp\left(-B_\alpha/kT\right) =$$
$$\begin{cases} c_\alpha' & u > h/2 \\ c_\alpha' \exp\left(-W_\alpha/kT\right) & u \le h/2, \end{cases}$$

where $c_\alpha' = c_\alpha \exp\left(-E_\alpha/kT\right)$,
and

$$k_\beta = c_\beta \exp\left(-B_\beta/kT\right) =$$
$$\begin{cases} c_\beta' & u < -h/2 \\ c_\beta' \exp\left(-W_\beta/kT\right) & u \ge -h/2, \end{cases}$$

where $c_\beta' = c_\beta \exp\left(-E_\beta/kT\right)$.

The way in which the rate constants depend on u and their values is given in Fig. 2D. This, together with the state diagram in Fig. 1B, allows differential equations to be written for A_α and A_β, the number of cross-bridges attached in states α and β. A_α and A_β are functions of time and u, the extension of the elastic element; i.e., A_α (or A_β) is the number of cross-bridges in state α (or β), at a

Figure 2. Derivation of rate constants. In all cases the subscript α or β indicates the state to which a cross-bridge is flipping. For example, E_β (B and C) contributes to the activation energy for a flip to state β, even though E_β is the depth of the well associated with the α state. A, Work required to stretch spring. W_α and W_β give the work done on the spring in a change of state to α and to β as a function of the extension of the spring just prior to the transition. When W is negative, energy is given off by the spring. The equations are:

$$W_\alpha = (K/2)(-2uh + h^2),$$
$$W_\beta = (K/2)(2uh + h^2),$$

where K is the stiffness of the cross-bridge spring. B, Energy considerations for flips between states α and β. The well associated with each state is labeled. Two examples are presented to illustrate the way in which activation energy for a flip depends on well depth and the work required by the spring, as discussed in the text. C, Activation energies for state changes. B_α and B_β represent the activation energy required for a change of state to the α and β states as a function of the cross-bridge's position before the flip. D, Rate constants in

sec^{-1} for flipping, attachment, and detachment of cross-bridges.

$$k_a = 29 \qquad u = 0$$

$$k_d = \begin{cases} 412.5 & u < 0 \\ 75 & u \ge 0 \end{cases}$$

$$k_\alpha = \begin{cases} 75 & u > h/2 \\ 75 \exp\left[(-K/2)(-2uh + h^2)/kT\right] & u \le h/2 \end{cases}$$

$$k_\beta = \begin{cases} 3633 & u < -h/2 \\ 3633 \exp\left[(-K/2)(2uh + h^2)/kT\right] & u \ge -h/2 \end{cases}$$

where K was chosen to be 2.2×10^{-9} dyne/Å and h, 100 Å or 10 nm.

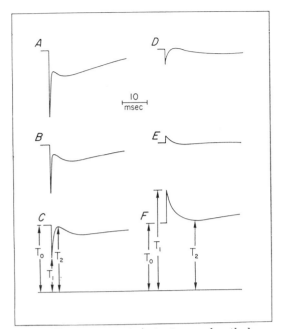

Figure 3. Force responses to instantaneous length changes. T_0 is the steady isometric force. T_1, the extreme force attained, is concurrent with the length step, and T_2 is the force at the end of the quick phase of the recovery. In all cases the force returned to T_0 in a time consistent with that observed in real muscle. The size of the length changes (stretch taken as positive) are in Å: A, -56; B, -43; C, -29; D, -14; E, 6; F, 29.

specific time, having extension u. Therefore the cross-bridges in each state can be considered as distributed over u. For example, in the isometric steady state the distribution of A_α is a spike at $u = 0$, and that of A_β, a spike at $u = h$. The reason for this is that cross-bridges attach only into state α with $u = 0$ and flip to state β at $u = h$. Detachment occurs only from state β. On the other hand, in the case of steady shortening under a constant load, both distributions spread out in the direction of more negative u.

The model's behavior was compared to that of real muscle by observing the force responses to sudden length changes and, conversely, the length responses to imposed loads. In extending Huxley and Simmons' (1971) model, it was necessary to maintain the transient force responses to sudden length changes which their model was designed to fit. The results are presented in Fig. 3. Beginning in the isometric steady state, an instantaneous length change is imposed on the model which has the effect of instantly shifting the A_α and A_β distributions by the amount of the length change. This produces an instantaneous change in the force. The force redevelopment at the new length has two distinct phases: first, a rapid change toward the original level, caused by reequilibration of the shifted A_α and A_β; then, a slow change back to the steady isometric level, T_0, caused by

detachment, attachment, and flipping. A plot of T_1 and T_2 (as defined in Fig. 3; see also Huxley and Simmons, 1971) versus size of length step produces results similar to those of Huxley and Simmons. It should be noted that the time for recovery to T_2 is shortest for the largest release and progresses to longest for the largest stretch.

Conversely, the length response to a sudden maintained change in force level or load can be determined. Following the sudden change in force, the model is constrained to hold the load constant by shortening at an appropriate speed. The complete responses given in Fig. 4 show three effects reported for real muscles by A. F. Huxley (1971): first, following the passive instantaneous elastic effect, a quick phase in which the shortening velocity is much greater than the steady state velocity; second, a plateau phase in which the velocity is near zero; third, an approach to the steady state velocity, sometimes with a damped oscillation. The first phase is caused by reequilibration of the shifted A_α and A_β. The plateaus are related to the relative loads in that the heavier

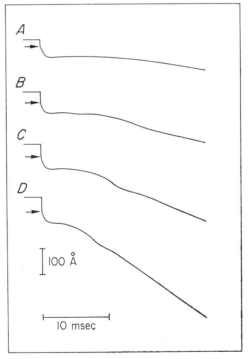

Figure 4. Length responses to imposed loads. The arrows indicate the size of the initial instantaneous length step required to drop the force to the desired load. Thereafter three features of the length responses are to be noted: first, a short period of very rapid shortening; second, a plateau phase in which practically no shortening occurs; finally, the attainment of a constant velocity of shortening. The loads relative to T_0 and steady state velocities relative to V_{max}, are, respectively: A, 0.50 and 0.18; B, 0.30 and 0.34; C, 0.15 and 0.57; D, 0.01 and 0.95. The extrapolated value for V_{max} is 1.93 μ/half-sarcomere/sec. The length calibration, 100 Å, applies per half-sarcomere.

loads are associated with longer plateaus. These effects can be seen to be the duals of those in the length traces shown in Fig. 3. The relation between steady state shortening velocity and relative load is very near that reported for living, electrically excited frog muscle fibers by Julian and Sollins (this volume). A very important characteristic of the model is that the stiffness (proportional to the number of cross-bridges attached) decreases from the isometric level as the velocity approaches V_{max}.

The version of the model presented here, together with the values given for the parameters and rate constants, is not necessarily final. Further tests of the model's usefulness are needed; e.g., it is important that it be able to account for the observed relation between power input and heat and power output. Also, it would be interesting to test the model's generality by determining the extent of the modifications required to obtain a satisfactory fit to insect fibrillar muscle behavior.

Acknowledgments

The main support for this work came from the United States Public Health Service Research Grant #AM 09891 from the National Institute of Arthritis and Metabolic Diseases. Additional valuable support was provided by USPHS Career Development Award NS 06193 from the National Institute of Neurological Diseases and Stroke, by the Massachusetts Heart Association, Inc. (#1012), and by the American Heart Association, Inc. (#71-898). (All grants were awarded to F. J. J.).

References

HUXLEY, A. F. 197'. The activation of striated muscle and its mechanical response. *Proc. Roy. Soc.* (London) (B) **178**: 1.

HUXLEY, A. F. and R. M. SIMMONS. 1971. Proposed mechanism of force generation in striated muscle. *Nature* **233**: 533.

Muscle 1972: Progress and Problems

H. E. HUXLEY

MRC Laboratory of Molecular Biology, Cambridge, England

I think we would all agree that we have had an exceptionally good meeting and that we are very greatly indebted to Jim Watson and the staff of Cold Spring Harbor Laboratory for making it possible. I would especially like to express our thanks to Barbara Bankey and her assistants who have done such a marvellous job in looking after our needs, with efficiency and cheerfulness, and so really helped to keep us in a good frame of mind and able to concentrate all our attention in the right direction.

We have been exposed to a wealth of very relevant information about all aspects of muscular contraction, and I think the striking feature of this meeting is the way in which these different lines of work seem to be coming together at last so that one can begin to relate other peoples' results, obtained with different approaches, in a rational and sometimes even numerical way with one's own work. I think that we now have the feeling that we are part of one community. As in other types of communities, we want to be free to do our own thing, but at the same time we need to feel that what we are doing plays a useful and recognized part in furthering the overall aims of the group.

Thus we have to look after three things. We have to see what overall problems we are trying to solve. We have to get as far as we can in our own particular area and try to establish within it common standards of evidence. And we have to make every effort to explain and discuss our results with people in related areas.

Given that our objective is to understand the mechanism of muscular contraction, one might first divide the problem into three main sections:

(1) How is the splitting of ATP converted into mechanical force?
(2) How is this conversion switched on and off?
(3) How are the necessary structures assembled and maintained?

These questions were originally restricted to muscle fibers, but in the last few years another and perhaps even more vital question has arisen. It looks as though many other forms of directed biological movement are mediated by protein structures having, in part at least, remarkably close resemblances to elements of the muscle system. So one must ask, can we use our experience with muscle to help us understand these broader problems of cellular movement?

What I would like to do very briefly here is to summarize my own impression of progress towards these various objectives as it has emerged at this meeting. Obviously this can only be a personal, biased, and incomplete account, but it may help to provide at least one type of pathway through some of the very fertile fields that have been described.

Perhaps the most general characteristic of current work is the way in which, within a given area, results are now becoming sufficiently reliable, quantitative, and comparable for rational argument —rather than distant shouts or avoidance of issues —to take place between people whose results differ and sometimes for convincing agreement to be reached. I think the main reason for this, apart simply from the accumulation of facts, is that well-defined and controllable systems are becoming increasingly available in which one can study not only the overall phenomena of, say, ATP-splitting or force-velocity relationships, but also all the important details which it is necessary to know if one is to properly test one's ideas. Thus, there has been a tremendous advance due to the fact that people can now compare their protein preparations according to the criteria of the chain weight of the various components, so that, even though this criterion of homogeneity or identity is still a relatively crude one, they at least know when totally different components are present. It is fortunate that this development in technique took place at a time when other studies on muscle protein systems were becoming sufficiently refined to show up the awkward but illuminating complications of behavior introduced by the additional protein components. The physiologists, too, seem to be increasingly successful at producing skinned and unskinned fiber preparations of various sorts in which highly relevant details of cross-bridge behavior can be made accessible to measurement and manipulation.

I think the next general characteristic that strikes one—and it is of course a rather gratifying aspect of the whole meeting—is that virtually all the work reported takes for granted that muscles contract by a mechanism in which actin and myosin filaments slide past each other as a consequence of longitudinal forces developed by cyclically operating cross-bridges. This model seems to provide a satisfactory framework or background for a good deal of work, whether it is directly concerned with various aspects of the model or not. To spell this out in a bit more detail, I think nearly everyone—perhaps everyone?—agrees that the system is of the following kind: myosin molecules are assembled into thick filaments and arranged in them with opposite structural polarity on each side of the midpoint of the filament. The heads of the myosin molecules, the S_1 subunits, project out sideways and in contraction can attach cyclically to sites on the thin actin-containing filaments. These attached bridges exert a longitudinal force for a certain distance during each cycle of their action, in which probably one molecule of ATP is split, and draw the actin filaments along towards the center of the A-band. The actin molecules are arranged in the thin filaments with appropriate polarity to interact with the nearby myosin heads, and the thin filaments are attached with opposite polarity on either side of the Z-lines. Thus the Z-lines are drawn together when myosin-actin interaction takes place. Interaction between actin and myosin can be regulated in vertebrate striated muscle by the tropomyosin-troponin system located in the thin filaments. This system is sensitive to the level of free calcium ions in the muscle, and the muscle is switched on by release of calcium from the sarcoplasmic reticulum (following the depolarization of the outside membrane of the fiber) and switched off again when calcium is rebound by the reticulum at the end of stimulation.

Given that this is the basic mechanism, then our first problem, that of the conversion of chemical energy into mechanical force, can be divided according to the following questions:

1. What properties does the cross-bridge mechanism need to have in order to explain the observed mechanical behavior of contracting muscle?

2. To what extent can these properties be accounted for in terms of the physical and·biochemical properties of the protein molecules and molecular assemblies involved?

3. How are those properties in turn to be accounted for in terms of detailed molecular structure?

Cross-Bridge Properties

Earlier work had demonstrated the physical existence of cross-bridges and had shown that the observed relationships between tension, speed of shortening, and heat production could be accounted for reasonably well by a model involving cyclically acting cross-bridges with appropriately chosen rate constants for attachment and detachment which vary according to the relative position of the bridge and the attachment site on the actin. However, in its original form, this model appears to

have been rather less satisfactory as a basis for attempts to account for the responses of contracting muscles to sudden changes in length or load, i.e., to account for the transient response of the system. Two different attempts have now been made (Huxley and Simmons; Podolsky and Nolan)* to account for these transients. In both models, the overall distance of movement of the cross-bridge is of the order of 80–120 Å. In one model, it is assumed that the tension generated by an attached cross-bridge can vary in a series of steps depending on the effective angle of attachment of the S_1 portion to actin and the degree of extension of the S_2 linkage (which connects S_1 to the LMM backbone of the thick filament). Transitions between these steps, if they take place with a rapid but finite rate, will influence the rapid transient response of the muscle and can apparently account for the detailed nature of these responses rather satisfactorily. Attachment and detachment of bridges are relatively slow and are therefore not involved in these short term disturbances (a few milliseconds). In this model, in fact, attachment of bridges is essentially the rate-limiting step that controls the speed of shortening at moderate loads. Thus the number of attached cross-bridges at any particular moment decreases as the velocity of shortening increases.

In the second model, the rate of attachment is much higher so that changes in the *number* of attached cross-bridges can account for the rapid transient response without invoking an internal change of appropriate time-constant in a cross-bridge itself. In this model, nearly all cross-bridges attach each time they return to the beginning of their cycle; but tension and ATP-splitting are made to vary in the appropriate way with shortening velocity by arranging the rate constants so that at high shortening velocities, many bridges are carried into the region where they exert negative tension, and where it is postulated they can detach without completing the cycle of ATP-splitting. The present writer prefers the first model, but the second is certainly an ingenious and interesting one, and it appears that further rather sophisticated mechanical experiments will be necessary before one can hope to eliminate one of them. Of course, it may be possible to make a decision on other grounds—for instance, by measuring by some other technique how the number of attached cross-bridges varies as a function of speed of shortening. Unfortunately, this is not at present a parameter which emerges with any clarity from the X-ray data, for the relative intensity of the equatorial reflections is a measure of the proportion of cross-bridges *near* to the actin filaments but not necessarily of those

---

* All references cited are to articles in this volume.

attached. One should also bear in mind that one of the assumptions of these kinds of models hitherto has been that the cross-bridges are independent tension generators and that the various rate constants which define their interactions with actin are independent of the number of neighboring attached bridges. However, as we will see later, there are strong grounds for believing that this is not the case at least in some in vitro systems, so perhaps both the steady state and the transient analyses will have a lot of further ramifications.

A good deal of somewhat similar activity is concerned with accounting for the special properties of insect oscillatory muscle in terms of cross-bridge properties and correlating these with X-ray observations of cross-bridge behavior (White; Abbott; Pybus and Tregear; Armitage et al.).

In all these cases, one would obviously like to relate the rate constants needed in the muscle model to the rate constants for the various steps in the actin-myosin-ATP cycle as measured biochemically. This is possible in principle for some steps, as we will see, though at the moment lack of biochemical data on frog actomyosin prevents a direct comparison. However, several of the steps in the cycle, especially those involving attachment and detachment, will have rate constants which may be very sensitive to the physical constraints present in muscle; one cannot say, for example, whether the attachment rate will be that appropriate for the corresponding concentrations of actin and myosin in solution or whether it will be lower or higher.

Experiments on cross-bridge-actin interaction from a biochemical point of view fall mainly into two groups, those dealing with enzyme kinetics of actomyosin ATPase and those dealing with the mode of operation of the troponin-tropomyosin regulatory system.

Enzyme Kinetics. It is generally recognized that basically the Taylor-Lymn scheme gives a rather satisfying and attractive picture of why myosin ATPase is slow and shows the "early burst" phenomenon (representing release of products, when the reaction is stopped by acid, from a long-lived enzyme-intermediate complex) and why actin can accelerate the ATPase (by speeding up the breakdown of the intermediate complex when the myosin combined with actin). It had been settled that this intermediate complex was not simply a myosin-products complex, a conclusion also borne out by electron spin resonance studies (Seidel and Gergely) which can distinguish the two states. It is appealing to suppose that this intermediate complex corresponds to a cross-bridge in a "strained" activated state, ready to attach to actin, switch back to its original conformation, and

pull the actin along. However, there are still a number of uncertainties which have to be clarified. According to one type of solution experiment (Eisenberg and Kielly), in an actin-HMM system in which the actin concentration is sufficiently high to give 85% of maximum activation of the HMM ATPase, only 40% of the HMM is found to be combined with actin from its sedimentation behavior, and there was no reason to think that a large amount of HMM was denatured, since it combined with actin in absence of ATP. This experiment would seem to mean that there is a relatively slow step in the cycle, which takes place after ATP has bound to actomyosin and dissociation has occurred, and which has to take place *before* the myosin-intermediate complex can attach to actin again. If the rate-limiting step does occur at this point, it avoids certain other difficulties that arise if the rate-limiting step at high actin concentration is some process occurring while the cross-bridge is attached. In that case, even the maximum rates of ATP-splitting observed would correspond to a relatively long time constant for that step. But at high velocities of shortening, cross-bridges can be attached only for a relatively short time.

Another rather basic question here is whether the energy produced in a whole muscle can be accounted for, as regards both its total amount and its time course, by the breakdown of ATP and CP (Gilbert et al.; Curtin and Davies; Woledge). It appears that the in vivo heat of hydrolysis of CP (the overall process in a muscle in which the ATP level is maintained constant by the kinase system) is too high to be readily accounted for by estimates and measurements in vitro of the various known processes involved. Moreover, during a contraction considerably more heat is evolved early in the process than can be accounted for even on the basis of the in vivo heat of hydrolysis of CP averaged over a longer series of contractions. Thus some other exothermic process, which is not closely coupled in time with ATP- or CP-splitting, must accompany contraction; whether or not this is calcium movement by the regulatory system remains to be seen. However, it does appear that when work is done by the contractile system, that work must be paid for at the time by splitting of ATP or CP.

Regulatory Mechanisms

Turning now to the question of regulation, there has been a great deal of success in elucidating the nature and function of all the components involved in the troponin-tropomyosin system (Ebashi et al.; Hartshorne and Dreizen; Greaser et al.; Perry et al.) and there is general agreement on the main features of the system. There is some divergence about nomenclature, and although I hesitate to intrude into this field, I think everyone might agree that there are advantages to be gained from a common nomenclature that may outweigh the disadvantages of losing some of the symbols to which one has become attached personally. I would therefore like to propose that the following scheme be generally adopted: that the Ca-binding component of troponin (mol wt ~18,000) be called Tp C; that the inhibitory component (mol wt ~23,000) be called Tp I and that the tropomyosin-binding component (mol wt ~37,000) be called Tp T.

There is general agreement that only Tp C binds significant amounts of calcium over the range of Ca^{++} concentrations in which regulation occurs (10^{-7}–10^{-5} M), and that Tp C on its own has no effect on the ATPase of actin plus myosin plus tropomyosin. Tp I acts as an inhibitor of actomyosin ATPase whether calcium is present or not, and the presence of Tp T is not required for this inhibition. The inhibition can also be observed in the absence of tropomyosin, though larger amounts of Tp I are then required. Tp C neutralizes the inhibition produced by Tp I whether tropomyosin is present or not, but the two components alone do not give normal Ca-sensitivity in an AMTM system. The presence of Tp T is needed for the normal functions of the original troponin complex to be fully reconstituted, but there are still some residual divergences about the precise role of this component. However, there is general agreement that the troponin complex contains one unit of each of the three components mentioned, and that in the presence of one mole of tropomyosin, such a complex can regulate the activity of about seven actin monomers. Very interesting cooperative effects are exhibited by the system (Bremel et al.), from which it emerges that the binding of myosin to actin in the "rigor" configuration (i.e., the type of binding which occurs when no nucleotide is present on the myosin and which probably corresponds to the situation at the end of the working stroke by the cross-bridge, before the next ATP molecule has come along to dissociate the bridge) alters the Ca-binding by troponin in the thin filaments and switches on neighboring actins even in the absence of calcium. Moreover, such rigor links (and possibly active ones too) can increase the ATPase-activating activity of regulated actin which has already been switched on by calcium. Besides building up a fascinating picture of a complex allosteric system, these findings have important repercussions in all aspects of cross-bridge behavior.

The mechanism of action of this regulatory system appears to be to block attachment, in the absence of calcium, of myosin cross-bridges to actin when they are loaded with ATP (i.e., when

myosin-nucleotide is in the form of the intermediate complex). There are a number of different models for the process by which this might be achieved. All models would assume that the binding of calcium causes a change in the troponin complex, but some would then suppose that this change directly alters the position of the tropomyosin strands (which run in the long pitch grooves of the actin filament), thereby exposing the active sites on actin which previously had been blocked; others would suppose that the change acts more directly on the configuration of the active site on the actin itself, and that the tropomyosin acts to cause this change to spread from one actin monomer to its neighbors. At present it seems that the available evidence is insufficient for us to make a decisive judgement between models. There is another problem which it is perhaps churlish to raise when the regulatory system has been reconstituted with such great success, i.e., up to 90%, or sometimes more, inhibition of actomyosin ATPase. Nevertheless, the regulatory system in a muscle is a lot better than 99.9% effective, and one would like to know the reason for the difference; is it because the present components are not fully "native" or is it because some additional mechanism, or additional part of the mechanism, has not yet been identified and preserved?

The regulatory mechanism can also be studied rather informatively by X-ray diffraction and electron microscope techniques (Cohen et al.; Hanson et al.; Haselgrove; Lowy and Vibert; Huxley), and again there is general agreement on a number of major issues although several of them need further confirmation. Tropomyosin molecules run along the two long-pitch grooves of the actin filaments, with a periodic repeat of approximately 385 Å which corresponds very closely to the distance occupied by seven actin monomers in either of the long-pitch actin helices. The tropomyosin repeat defines the intervals (again 385 Å) at which the troponin complex is attached to the filaments. The position of tropomyosin in the groove is affected by changes in the calcium level, and it moves nearer to the center of the groove when calcium becomes bound by troponin. Computation from model structures indicate that a movement by about 20 Å would account for the effects seen in the X-ray diagrams. This is certainly compatible with a steric blocking model but cannot exclude changes in actin too.

Whilst the troponin system acting on the thin filaments is, as far as can be seen at present, universal among vertebrates, a different regulatory system exists in molluscs, in which calcium acts not on the thin filaments but on the myosin (Lehman et al.). Apart from the evolutionary insights provided by species comparisons, I think the most remarkable aspect of this whole story is that the actin filaments from these myosin-regulated muscles, which, as it were, have never seen troponin (though they contain tropomyosin) can participate perfectly normally in a troponin-regulated synthetic system. This seems to mean that early in evolution actin possessed these attributes even before they were used, and that the actin structure has since then changed very little—as indeed other present evidence suggests.

Protein Structures

The task of accounting for the remarkable properties of the protein molecules involved in contraction in terms of their internal structure has made some progress, though of course the lack of crystalline preparations for detailed X-ray studies is still the main obstacle that needs to be removed. Sequence studies are at last making real headway (Elzinga and Collins; Hodges et al.). Perhaps the most directly interpretable of these is the large section of the tropomyosin sequence, which is of enormous interest both as the first example of the sequences compatible with a coiled-coil structure and also of the pattern of sequences with which actin and troponin have to interact in the regulating mechanism. The chemistry of the light chains of myosin and their involvement in the functioning of the intact molecule is being steadily disentangled (Weeds and Frank; Lowey and Holt; Dreizen and Richards; Kendrick-Jones et al.) and perhaps soon someone will be brave enough to start on the main chain sequence of S_1. However, there is a very large gap in our present knowledge, unfortunately right at the heart of the whole problem. It concerns the nature of the structural changes in the myosin head accompanying the different stages of ATP-splitting which are thought to produce the changes in the effective angle of attachment to actin during the operation of the cross-bridge. Spectroscopic and possibly X-ray evidence exists for the occurrence of such structural changes in myosin (e.g., Seidel and Gergely; Botts et al.; Leigh et al.; Lymn and Huxley; Bagshaw et al.) but not for their nature (e.g., how much of the structure is involved in the change?).

A very gratifying new clue to the problem of the assembly of myosin molecules into filaments of defined length is provided by the identification of a new protein, present in combination with myosin in the thick filaments, and giving rise to the 430–440 Å periodicity seen in certain regions of the A-band (Pepe; Offer; Rome). No one knows, of course, how the length-determining mechanism works, nor indeed whether this protein is involved in it, but it is certainly easier to invent possible mechanisms if one has more than one protein species to juggle

with! This is particularly apparent for the actin-containing filaments, which in some muscles (e.g., *Pecten* striated) appear to have a very simple protein composition (actin and tropomyosin only) and yet have a sharply defined length; this raises again the question whether or not the subunit repeats of actin and tropomyosin have an exactly integral relationship to each other.

Contraction in Non-Muscle Systems

Finally, there is the question of the occurrence of mechanisms, analogous to those used in muscle, in non-muscular motile systems. In the last few years, it had already been established decisively that such analogies do exist, and what perhaps is remarkable now is in how many different situations they occur, and how close many aspects of the mechanisms are to the actin-cross-bridge system in muscle (Bray; Pollard and Korn; Abramowitz et al.; Adelstein and Conti; Goldman and Knipe). The similarities in the actin component have now been shown to extend to close correspondence of several parts of their amino acid sequences and to allow regulation by the troponin-tropomyosin system of muscle, though whether actin-filament regulation is used in the non-muscle systems is not yet known; and the example of molluscan muscle shows that ability of the actin filaments present to be regulated by troponin does not necessarily mean that regulation is actually effected this way in vivo. There is evidence now for variations in the myosin-like components so that while they share the common feature of having an ATPase which is activated by actin—which presumably reflects the ability of the molecule to change shape and exert force while attached to actin—the size, shape, and mode of association of the molecules appears to vary. Again this is an extremely active and exciting area at present,

* * *

These are just a few of the aspects of the meeting that I found I could relate most readily to my own images of the problem, and no doubt there are many other pathways that are just as valid.

We have all found many new colleagues whose results and ideas are very important to what we are doing ourselves, and I think there is a good case for having more frequent workshop-type meetings on both sides of the Atlantic.

Finally, I think it is clear that this subject is really coming into its stride, having most of the necessary techniques and ideas to get solid data that can answer direct and detailed questions about the molecular mechanism of contraction in a satisfying way. So I think we are fortunate to be working in this area at present.

Subject Index

A

A band, fluorescent staining of, 99–103
ABRM (anterior byssal retractor muscle)
  activation of, 490–497
  relaxation mechanism, 497–502
  structure of, 490
  X-ray patterns of, 353–358
Acetylcholine, 495–497
Actin
  amino acid sequence of, 1–6
  crystallization of, 6
  effect on myosin ESR spectrum 190–191
  in *Acanthamoeba*, 573–575, 582
  isolation from platelets, 585–586, 599, 602
F-actin
  binding to HMM, 280–285
  effect of cytochalasin B on, 588–591
  flexibility of, 277–285
  helical structure of, 333–335, 342–345, 379–382
  interaction with myosin,
    effect of C-protein, 92–93
    inhibition by troponin system, 215–222
  structural change in, during contraction, 277–285
G-actin, 4, 6, 195–197, 317, 588, 597
Actin filament
  effect of cytochalasin B on, 585–592
  in vertebrate smooth muscle, 429
  structural change in, during contraction, 341–351, 356–358
  regulated, cooperative behavior in, 267–273
  X-ray diffraction patterns of, 341–351, 356–358
Actin-like proteins, 523–533, 563–571
  in fibroblasts, 599–605
  in platelets, 585–592, 602
Actin-myosin interaction in intact muscle, 157–166; *see also* Cross-bridge cycle
α-Actinin, 215, 222, 520–521
β-Actinin, 215, 222
Actinomycin D, effect on differentiating cells, 543–547

Action potentials, propagation of, 483–486
Activation, length dependence of, 505–509
Activation heat, 627–628
Acto-HMM ATPase, *see* HMM ATPase, actin-activated
Actomyosin
  ATPase activity of,
    Ca^{++} sensitivity, 182, 191; *see also* Troponin, subunits of
    effect of NEM on, 153–156
    effect of troponin-tropomyosin, 232–233, 236, 254–258
    transient kinetics, 175–185
    steady state kinetics, 443–444, 452
  ATPase site of, 121–122
  from platelets, 595–598
Acto-S-1 ATPase, *see* Subfragment-1 ATPase, actin-activated
ADP
  content of muscle, 161–162
  kinetics of binding to S-1, 129–130
Alkali light chains, *see* Myosin, light chains of
ATP analogs, 113–119, 127–135, 138–139, 189, 197–198, 444–453
ATP-binding peptide, 121–125
ATP hydrolysis
  and conformational changes, 188–190
  by sarcoplasmic reticulum, 455–459, 462–465
  in FDNB-poisoned muscle, 633
  in intact muscle, 620–624

B

Bode plot of frequency response, 647–649
Buffers, ionization heat of, 629–630

C

Caffeine contracture of ABRM, 492–494
Calcium activation
  of ABRM, 490–497
  of skinned fibers, 505–509
  and force-velocity relation, 635–636

Calcium-binding protein, *see also* Troponin, subunits of
  amount in myofibril, 260–261
  interaction with inhibitory protein, 257–260
  of sarcoplasmic reticulum, 472–476
Calcium, effect on
  delayed tension, 647–653
  F-actin flexibility, 279–281
  platelet actomyosin ATPase, 595, 605
  rigidity of myosin, 62
  tension development, 506–508
Calcium, interaction with light chains, 37–39
Calcium transport
  and ATPase of sarcoplasmic reticulum, 457–459, 465–466, 470
  and energy conversion, 681–682
Cardiac muscle
  myosin in
    light chains of, 11–13
    proteolytic fragments of, 56–57
  troponin in, 243
Catch muscle, *see* ABRM
Cathepsins in troponin preparations, 225, 230, 246, 253, 257
Cell fusion, 543–546, 550–551
Cell spreading, 523–533
Chemomechanical efficiency, 651–652, 656
Colchicine as probe of microtubule function, 527–529
Colcimide
  effect on myofibril assembly, 554–556
  induction of dense bodies, 556–558
Contraction, regulation of
  and force-velocity relation, 635–644, 678
  in invertebrates, 47–52, 319–327, 358
  theories of, 150–151, 164–165, 199, 317
C-protein
  antibody to, 89–91, 98–100, 335–338
  binding to myosin, 93–95, 335
  function of, 92–93
  interaction with myosin and LMM, 93–95
  isolation of, 87–88

Name Index